Lecture Notes in Computer Science   1988
Edited by G. Goos, J. Hartmanis and J. van Leeuwen

Springer
*Berlin
Heidelberg
New York
Barcelona
Hong Kong
London
Milan
Paris
Singapore
Tokyo*

Lubin Vulkov Jerzy Waśniewski
Plamen Yalamov (Eds.)

Numerical Analysis and Its Applications

Second International Conference, NAA 2000
Rousse, Bulgaria, June 11-15, 2000
Revised Papers

Series Editors

Gerhard Goos, Karlsruhe University, Germany
Juris Hartmanis, Cornell University, NY, USA
Jan van Leeuwen, Utrecht University, The Netherlands

Volume Editors

Lubin Vulkov
Plamen Yalamov
University of Rousse
Department of Computer Science
7000 Rousse, Bulgaria
E-mail:{vulkov/yalamov}@ami.ru.acad.bg

Jerzy Waśniewski
UNI-C, Danish Computing Center for Research and Education
DTU, Building 304
2800 Lyngby, Denmark
E-mail: unijw@uni-c.dk

Cataloging-in-Publication Data applied for

Die Deutsche Bibliothek - CIP-Einheitsaufnahme

Numerical analysis and its applications : second international
conference ; revised papers / NAA 2000, Rousse, Bulgaria, June 11 -
15, 2000. Lubin Vulkov ... (ed.). - Berlin ; Heidelberg ; New York ;
Barcelona ; Hong Kong ; London ; Milan ; Paris ; Singapore ; Tokyo :
Springer, 2001
  (Lecture notes in computer science ; Vol. 1988)
  ISBN 3-540-41814-8

CR Subject Classification (1998): G.1, F.2.1, G.4, I.6

ISSN 0302-9743
ISBN 3-540-41814-8 Springer-Verlag Berlin Heidelberg New York

This work is subject to copyright. All rights are reserved, whether the whole or part of the material is
concerned, specifically the rights of translation, reprinting, re-use of illustrations, recitation, broadcasting,
reproduction on microfilms or in any other way, and storage in data banks. Duplication of this publication
or parts thereof is permitted only under the provisions of the German Copyright Law of September 9, 1965,
in its current version, and permission for use must always be obtained from Springer-Verlag. Violations are
liable for prosecution under the German Copyright Law.

Springer-Verlag Berlin Heidelberg New York
a member of BertelsmannSpringer Science+Business Media GmbH

http://www.springer.de

© Springer-Verlag Berlin Heidelberg 2001
Printed in Germany

Typesetting: Camera-ready by author, data conversion by DaTeX Gerd Blumenstein
Printed on acid-free paper   SPIN 10782010   06/3142   5 4 3 2 1 0

Preface

This volume of the Lecture Notes in Computer Science series contains the proceedings of the Second Conference on Numerical Analysis and Applications, which was held at the University of Rousse, Bulgaria, June 11-15, 2000. The conference was organized by the Department of Numerical Analysis and Statistics at the University of Rousse with support from the University of Southern Mississippi, Hattiesburg. The conference was co-sponsored by SIAM (Society for Industrial and Applied Mathematics) and ILAS (International Linear Algebra Society). The official sponsors of the conference were Fujitsu America, Inc., Hewlett-Packard GmbH, and Sun Microsystems. We would like to give our sincere thanks to all sponsors and co-sponsors for the timely support.

The second conference continued the tradition of the first one (1996 in Rousse) as a forum, where scientists from leading research groups from the "East" and "West" are provided with the opportunity to meet and exchange ideas and establish research cooperation. More than 120 scientists from 31 countries participated in the conference.

A wide range of problems concerning recent achievements in numerical analysis and its applications in physics, chemistry, engineering, and economics were discussed. An extensive exchange of ideas between scientists who develop and study numerical methods and researchers who use them for solving real-life problems took place during the conference.

We are indebted to our colleagues who helped us in the organization of this conference. We thank the organizers of the mini-symposia for attracting active and highly qualified researchers.

October 2000

Lubin Vulkov
Jerzy Waśniewski
Plamen Yalamov

Table of Contents

Numerical Analysis and Its Applications

Sensitivity Analysis of the Expected Accumulated Reward
Using Uniformization and IRK3 Methods 1
H. Abdallah and M. Hamza

Spectral Properties of Circulant Band Matrices Arising in ODE Methods ... 10
P. Amodio

A Parameter Robust Method for a Problem with
a Symmetry Boundary Layer 18
A. R. Ansari, A. F. Hegarty and G. I. Shishkin

An Algorithm Based on Orthogonal Polynomial Vectors
for Toeplitz Least Squares Problems 27
M. Van Barel, G. Heinig and P. Kravanja

From Sensitivity Analysis to Random Floating Point Arithmetics –
Application to Sylvester Equations 35
A. Barraud, S. Lesecq and N. Christov

Construction of Seminumerical Schemes:
Application to the Artificial Satellite Problem 42
R. Barrio

Stability Analysis of Parallel Evaluation of Finite Series
of Orthogonal Polynomials 51
R. Barrio and P. Yalamov

On Solving Large-Scale Weighted Least Squares Problems 59
V. Baryamureeba

A Hybrid Newton–GMRES Method for Solving Nonlinear Equations 68
S. Bellavia, M. Macconi and B. Morini

Comparative Analysis of Marching Algorithms
for Separable Elliptic Problems 76
G. Bencheva

Inexact Newton Methods and Mixed Nonlinear Complementary Problems 84
L. Bergamaschi and G. Zilli

Skew-Circulant Preconditioners for Systems of LMF-Based ODE Codes 93
D. Bertaccini and M. K. Ng

New Families of Symplectic Runge–Kutta–Nyström Integration Methods 102
S. Blanes, F. Casas and J. Ros

Convergence of Finite Difference Method for Parabolic Problem
with Variable Operator 110
D. Bojović

Finite Volume Difference Scheme for
a Stiff Elliptic Reaction-Diffusion Problem with a Line Interface 117
I. A. Braianov

Nested-Dissection Orderings for Sparse LU with Partial Pivoting 125
I. Brainman and S. Toledo

Fractional Step Runge–Kutta Methods for the Resolution of
Two Dimensional Time Dependent Coefficient
Convection–Diffusion Problems 133
B. Bujanda and J. C. Jorge

Variable Stepsizes in Symmetric Linear Multistep Methods 144
B. Cano

Preliminary Remarks on Multigrid Methods for Circulant Matrices 152
S. S. Capizzano and C. T. Possio

Computing the Inverse Matrix Hyperbolic Sine 160
J. R. Cardoso and F. S. Leite

Robust Preconditioning of Dense Problems from Electromagnetics 170
B. Carpentieri, I. S. Duff and L. Giraud

A Mathematical Model for the Limbic System 179
L. Cervantes and A. F. Collar

Understanding Krylov Methods in Finite Precision 187
F. Chaitin-Chatelin, E. Traviesas and L. Plantié

A Rational Interpolation Approach to Least Squares Estimation
for Band-TARs 198
J. Coakley, A.-M. Fuertes and M.-T. Pérez

Uniqueness of Solution of the Inverse Electroencephalographic Problem ... 207
A. F. Collar, J. J. O. Oliveros and A. I. Grebénnikov

Exploiting Nonlinear Structures of
Computational General Equilibrium Models 214
Ch. Condevaux-Lanloy, O. Epelly and E. Fragnière

Constitutive Equations and Numerical Modelling of
Time Effects in Soft Porous Rocks 222
M. Datcheva, R. Charlier and F. Collin

Solvers for Systems of Nonlinear Algebraic Equations –
Their Sensitivity to Starting Vectors 230
D. Dent, M. Paprzycki and A. Kucaba-Pietal

The Min-Max Portfolio Optimization Strategy:
An Empirical Study on Balanced Portfolios 238
C. Diderich and W. Marty

Convergence Rate for a Convection Parameter Identified
Using Tikhonov Regularization 246
G. Dimitriu

Local Refinement in Non-overlapping Domain Decomposition 253
V. Dobrev and P. Vassilevski

Singularly Perturbed Parabolic Problems on Non-rectangular Domains 265
R. K. Dunne, E. O'Riordan and G. I. Shishkin

Special Types of Badly Conditioned Operator Problems
in Energy Spaces and Numerical Methods for Them 273
E. G. D'yakonov

Proper Weak Regular Splitting for M-Matrices 285
I. Faragó

Parameter-Uniform Numerical Methods for a Class
of Singularly Perturbed Problems with a Neumann Boundary Condition .. 292
*P. A. Farrell, A. F. Hegarty, J. J. H. Miller, E. O'Riordan
and G. I. Shishkin*

Reynolds–Uniform Numerical Method for Prandtl's Problem
with Suction–Blowing Based on Blasius' Approach 304
B. Gahan, J. J. H. Miller and G. I. Shishkin

Multigrid Methods and Finite Difference Schemes
for 2D Singularly Perturbed Problems 316
F. Gaspar, F. Lisbona and C. Clavero

Recursive Version of LU Decomposition 325
K. Georgiev and J. Waśniewski

Inversion of Symmetric Matrices in a New Block Packed Storage 333
G. Georgieva, F. Gustavson and P. Yalamov

The Stability Boundary of Certain Two-Layer and
Three-Layer Difference Schemes 341
A. V. Goolin

High Order ε-Uniform Methods for
Singularly Perturbed Reaction-Diffusion Problems 350
J. L. Gracia, F. Lisbona and C. Clavero

A Grid Free Monte Carlo Algorithm for
Solving Elliptic Boundary Value Problems 359
T. Gurov, P. Whitlock and I. Dimov

Newton's Method under Different Lipschitz Conditions 368
J. M. Gutiérrez and M. A. Hernández

Positive Definite Solutions of the Equation $X + A^*X^{-n}A = I$ 377
V. Hassanov and I. Ivanov

Fast and Superfast Algorithms for Hankel-Like Matrices Related to
Orthogonal Polynomials 385
G. Heinig

Acceleration by Parallel Computations of Solving High-Order
Time-Accurate Difference Schemes for Singularly Perturbed
Convection-Diffusion Problems 393
P. W. Hemker, G. I. Shishkin and L. P. Shishkina

Experience with the Solution of
a Finite Difference Discretization on Sparse Grids 402
P. W. Hemker and F. Sprengel

Topology Optimization of Conductive Media Described
by Maxwell's Equations 414
R. H. W. Hoppe, S. I. Petrova and V. H. Schulz

Finite Element Simulation of Residual Stresses
in Thermo-coupled Wire Drawing Process 423
R. Iankov, A. Van Bael and P. Van Houtte

Construction and Convergence of Difference Schemes for
a Modell Elliptic Equation with Dirac-delta Function Coefficient 431
B. S. Jovanovic, J. D. Kandilarov and L. G. Vulkov

Operator's Approach to the Problems with Concentrated Factors 439
B. S. Jovanović and L. G. Vulkov

A Method of Lines Approach to the Numerical Solution
of Singularly Perturbed Elliptic Problems 451
J. D. Kandilarov, L. G. Vulkov and A. I. Zadorin

Sobolev Space Preconditioning of
Strongly Nonlinear 4th Order Elliptic Problems 459
J. Karátson

Numerical Techniques for the Recovery of
an Unknown Dirichlet Data Function in Semilinear Parabolic Problems
with Nonstandard Boundary Conditions 467
R. Van Keer and M. Slodička

A Generalized GMRES Iterative Method 475
D. R. Kincaid, J.-Y. Chen and D. M. Young

AMLI Preconditioning of Pure Displacement
Non-conforming Elasticity FEM Systems 482
T. Kolev and S. Margenov

Computationally Efficient Methods for Solving SURE Models 490
E. J. Kontoghiorghes and P. Foschi

Application of Boundary Collocation Method in
Fluid Mechanics to Stokes Flow Problems 498
A. Kucaba-Pietal

Strang-Type Preconditioners for Differential-Algebraic Equations 505
S.-L. Lei and X.-Q. Jin

Solvability of Runge-Kutta and Block-BVMs Systems Applied
to Scalar ODEs 513
G. Di Lena and F. Iavernaro

On the Local Sensitivity of the Lyapunov Equations 521
S. Lesecq, A. Barraud and N. Christov

A Level Set-Boundary Element Method for Simulation
of Dynamic Powder Consolidation of Metals 527
Z. Li and W. Cai

Parallel Performance of a 3D Elliptic Solver 535
I. Lirkov, S. Margenov and M. Paprzycki

Schwarz Methods for Convection-Diffusion Problems 544
H. MacMullen, E. O'Riordan and G. I. Shishkin

Matrix Computations Using Quasirandom Sequences 552
M. Mascagni and A. Karaivanova

On the Stability of the Generalized Schur Algorithm 560
N. Mastronardi, P. Van Dooren and S. Van Huffel

Stability of Finite Difference Schemes on
Non-uniform Spatial-Time-Grids 568
P. P. Matus, V. I. Mazhukin and I. E. Mozolevsky

Matrix Equations and Structures:
Efficient Solution of Special Discrete Algebraic Riccati Equations 578
B. Meini

A Numerical Comparison between Multi-revolution Algorithms for
First-Order and Second-Order ODE Systems 586
M. Begoña Melendo

A Robust Layer-Resolving Numerical Method for
Plane Stagnation Point Flow 594
J. J. H. Miller, A. P. Musgrave and G. I. Shishkin

On the Complete Pivoting Conjecture for
Hadamard Matrices of Order 16 602
M. Mitrouli

Regularization Method by Rank Revealing QR Factorization
and Its Optimization 608
S. Nakata, T. Kitagawa and Y. Hosoda

A Fast Algorithm for High-Resolution Color Image Reconstruction
with Multisensors 615
M. K. Ng, W. C. Kwan and R. H. Chan

A Performance Study on a Single Processing Node
of the HITACHI SR8000 628
S. Nishimura, D. Takahashi, T. Shigehara, H. Mizoguchi and T. Mishima

Estimation of the Wheat Losses Caused by
the Tropospheric Ozone in Bulgaria and Denmark 636
T. Ostromsky, I. Dimov, I. Tzvetanov and Z. Zlatev

A Homotopic Residual Correction Process 644
V. Y. Pan

Parallel Monte Carlo Methods for Derivative Security Pricing 650
G. Pauletto

Stability of a Parallel Partitioning Algorithm for
Special Classes of Banded Linear Systems 658
V. Pavlov

Numerical Solution of ODEs with Distributed Maple 666
D. Petcu

The Boundary Layer Problem of Triple Deck Type 675
L. Plantié

Cellular Neural Network Model for Nonlinear Waves in Medium
with Exponential Memory 684
P. Popivanov and A. Slavova

Numerical Analysis of the Nonlinear Instability of
One-Dimensional Compound Capillary Jet 692
St. Radev, M. Kaschiev, M. Koleva, L. Tadrist and F. Onofri

Modelling of Equiaxed Microstructure Formation in
Solidifying Two–Component Alloys 702
N. Sczygiol

A Posteriori and *a Priori* Techniques of Local Grid Refinement for
Parabolic Problems with Boundary and Transition Layers 710
G. I. Shishkin

On a Necessary Requirement for *Re*-Uniform Numerical Methods to
Solve Boundary Layer Equations for Flow along a Flat Plate 723
G. I. Shishkin, P. A. Farrell, A. F. Hegarty, J. J. H. Miller
and E. O'Riordan

A Godunov-Ryabenkii Instability for a Quickest Scheme 732
E. Sousa

Modelling Torsional Properties of Human Bones by
Multipoint Padé Approximants 741
J. J. Telega, S. Tokarzewski and A. Gałka

Numerical Algorithm for Studying Hydrodynamics in
a Chemical Reactor with a Mixer 749
I. Zheleva and A. Lecheva

A Domain Decomposition Finite Difference Method for
Singularly Perturbed Elliptic Equations in Composed Domains 756
I. V. Tselishcheva and G. I. Shishkin

Numerical Analysis of Solid and Shell Models of Human Pelvic Bone 764
A. John

FEM in Numerical Analysis of Stress and
Displacement Distributions in Planetary Wheel of Cycloidal Gear 772
M. Chmurawa and A. John

Author Index 781

Sensitivity Analysis of the Expected Accumulated Reward Using Uniformization and IRK3 Methods

Haïscam Abdallah and Moulaye Hamza

IRISA
Campus de Beaulieu, 35042 Rennes cedex, France
{abdallah.mhamza}@irisa.fr

Abstract. This paper deals with the sensitivity computation of the expected accumulated reward of stiff Markov Models. Generally, we are faced with the problem of computation time, especially when the Markov process is stiff. We consider the standard uniformization method for which we propose a new error bound. Because the time complexity of this method becomes large when the stiffness increases, we then suggest an ordinary differential equations method, the third order implicit Runge-Kutta method. After providing a new way of writing the system of equations to be solved, we apply this method with a stepsize choice different from the classical one in order to accelerate the algorithm execution. Finally, we compare the time complexity of both of the methods on a numerical example.

1 Introduction

As the use of computing systems increases, the requirement of analyzing both their performance and reliability have become more important. Reward Markov models are common tools for modelling such systems behaviour. Doing so, a Continuous-Time Markov Chain (CTMC) is used to represent changes in the system's structure, usually caused by faults and repairs of its components, and reward rates are assigned to the states of the model. Each reward represents the state performance of the system in a particular configuration. For these models, it may be of interest to evaluate not only some instantaneous transient measures, but also some cumulative ones such as the Expected Accumulated Reward (EAR) over a given interval $[0, t]$, t being the system's mission time. As the input parameters used to define the Markov models (fault rates, repair rates, etc.) are most of the time estimated from few experimental observations, the transient solutions are subject to uncertainties. Therefore, it becomes necessary to introduce parametric sensitivity analysis, the computation of derivatives of system measures with respect to input parameters. Generally, we are faced with the problem of computation time, especially when the Markov model is stiff, i.e., when the failure rates are much smaller than the repair rates. In this paper, we focus on the computation of the sensitivity of the EAR of stiff Markov models.

We consider two numerical methods: the Standard Uniformization (SU) method and the third order Implicit Runge-Kutta (IRK3) method.

The SU method consists in expressing the EAR [1] and its sensitivity in the form of an infinite sum. The main advantage of this method is that for a given tolerance, the truncation of the previous infinite sum allows to bound the global error. We propose to derive a new error bound. Unfortunately, when the models are stiff and the mission time is large, the computation time becomes prohibitive. We then suggest the L-stable Ordinary Differential Equations (ODE) IRK3 method. This method has been used to compute the instantaneous state probability vector [2] and its sensitivity [3]. In order to compute the sensitivity of the EAR by this method, first we provide a new way of writing a non homogeneous ODE in a system of the form $y' = \lambda y$, where λ is a constant. Next, we choose a new stepsize to accelerate the execution of the IRK3 algorithm.

The paper is organized as follows: the following section sets the problem. In Section 3, the SU technique is presented and a new bound is provided. Section 4 is devoted to the IRK3 method and the new stepsize choice. A concrete example and a comparison of both of the methods from a time complexity point of view are given in Section 5.

2 Problem Formulation

Consider a computing system modelled by a CTMC, say $\mathbf{X} = \{X_t,\ t \geq 0\}$, defined over a finite state space $\mathbb{E} = \{1, 2, ..., M\}$. Let $R = (r_i)$ be the reward rate vector; r_i denotes the reward rate assigned to state i of $\mathbf{X}$. We suppose the transition rates depend on a parameter θ (failure rate, repair rate, etc.). The infinitesimal generator (or transition rate matrix) of the CTMC $\mathbf{X}$ is denoted by $Q(\theta) = (q_{ij}(\theta))$. Let $\Pi(\theta, t)$ be the instantaneous state probability vector. The EAR over the interval $[0, t]$ is defined by

$$E[Y(\theta,t)] = \sum_{i=1}^{M} r_i L_i(\theta,t) = L(\theta,t)R \text{ where } L(\theta,t) = \int_0^t \Pi(\theta,s)ds. \quad (1)$$

The sensitivity of $E[Y(\theta,t)]$ is its partial derivative relatively to θ. From (1), we get

$$\frac{\partial}{\partial \theta}E[Y(\theta,t)] = \frac{\partial}{\partial \theta}[L(\theta,t)R] = \left[\frac{\partial}{\partial \theta}L(\theta,t)\right]R, \quad (2)$$

given that reward rates are supposed to be constant.

It is known that the vector $\Pi(\theta, t)$ is the solution of the Chapman-Kolmogorov first order linear differential equations :

$$\frac{\partial}{\partial t}\Pi(\theta,t) = \Pi(\theta,t)Q(\theta);\ \ \Pi(\theta,0) = \Pi(0) \text{ is given.} \quad (3)$$

Then we have

$$\Pi(\theta,t) = \Pi(0)P(\theta,t), \quad (4)$$

where
$$P(\theta,t) = e^{Q(\theta)t} = \sum_{n=0}^{\infty} Q(\theta)^n \frac{t^n}{n!}.$$

The computation of the sensitivity of $L(\theta,t)$ may be done in two ways. The first one consists in computing $\Pi(\theta,t)$, integrating it over $[0,t]$, and deriving that expression relatively to θ. This is the case for the SU method. The other one integrates system (3). A new system of equations, whose solution is $L(\theta,t)$, is obtained. That new system is then derived with respect to θ and the solution of the final system of equations is $\frac{\partial}{\partial \theta}L(\theta,t)$. In that case, we use the IRK3 method.

3 The SU Method

The SU technique [1], [4] transforms $Q(\theta)$ into the stochastic matrix $\tilde{P}(\theta) = I + Q(\theta)/q$ where I is the identity matrix and q is a constant such that $q > max_i \mid q_{ii}(\theta) \mid$. It follows that $Q(\theta) = q(\tilde{P}(\theta) - I)$ and

$$P(\theta,t) = e^{qt\tilde{P}(\theta)-qtI} = e^{-qt}Ie^{qt\tilde{P}(\theta)} = e^{-qt}e^{qt\tilde{P}(\theta)}. \tag{5}$$

The matrix $P(\theta,t)$ may then be writte:

$$P(\theta,t) = \sum_{n=0}^{\infty} p(n,qt)\tilde{P}(\theta)^n \text{ where } p(n,qt) = e^{-qt}\frac{(qt)^n}{n!}. \tag{6}$$

From relation (4), we get

$$\Pi(\theta,t) = \sum_{n=0}^{\infty} p(n,qt)\Pi(0)\tilde{P}(\theta)^n. \tag{7}$$

Defining the vector $\tilde{\Pi}^{(n)}(\theta)$ by $\tilde{\Pi}^{(n)}(\theta) = \Pi(0)\tilde{P}(\theta)^n$, we have recursively:

$$\tilde{\Pi}^{(n)}(\theta) = \tilde{\Pi}^{(n-1)}(\theta)\tilde{P}(\theta), n \geq 1; \quad \tilde{\Pi}^{(0)}(\theta) = \Pi(0).$$

The expression of the cumulative distribution $L(\theta,t)$ is obtained by integrating relation (7):

$$L(\theta,t) = t\sum_{n=0}^{\infty} p(n,qt)\frac{1}{n+1}\sum_{k=0}^{n} \tilde{\Pi}^{(k)}(\theta). \tag{8}$$

The derivation of (8) with respect to θ gives the sensitivity of $L(\theta,t)$, denoted by $S_L(\theta,t)$, as follows:

$$S_L(\theta,t) = t\sum_{n=0}^{\infty} p(n,qt)\frac{1}{n+1}\sum_{k=0}^{n} \frac{\partial}{\partial \theta}\tilde{\Pi}^{(k)}(\theta). \tag{9}$$

The vectors $\frac{\partial}{\partial \theta}\tilde{\Pi}^{(k)}(\theta)$, $k \geq 1$, are such that

$$\frac{\partial}{\partial \theta}\tilde{\Pi}^{(k)}(\theta) = \left[\frac{\partial}{\partial \theta}\tilde{\Pi}^{(k-1)}(\theta)\right]\tilde{P}(\theta) + \tilde{\Pi}^{(k-1)}(\theta)\left[\frac{\partial}{\partial \theta}\tilde{P}(\theta)\right], \frac{\partial}{\partial \theta}\tilde{\Pi}^{(0)}(\theta) = 0. \tag{10}$$

In practical implementations, the previous infinite series (9) is truncated at a step N_S. Let F_L be the error vector on $S_L(\theta, t)$. We have

$$F_L = t \sum_{n=N_S+1}^{\infty} p(n, qt) \frac{1}{n+1} \sum_{k=0}^{n} \frac{\partial}{\partial \theta} \tilde{\Pi}^{(k)}(\theta).$$

The infinite norm of the vector F_L verifies:

$$\|F_L\|_\infty \leq t \sum_{n=N_S+1}^{\infty} p(n, qt) \frac{1}{n+1} \sum_{k=0}^{n} \left\| \frac{\partial}{\partial \theta} \tilde{\Pi}^{(k)}(\theta) \right\|_\infty.$$

From relation (10), it may be established by recurrence that for all $k \in \mathbb{N}$, $\left\| \frac{\partial}{\partial \theta} \tilde{\Pi}^{(k)}(\theta) \right\|_\infty \leq k \left\| \frac{\partial}{\partial \theta} \tilde{P}(\theta) \right\|_\infty$. It follows that

$$\|F_L\|_\infty \leq t \sum_{n=N_S+1}^{\infty} p(n, qt) \frac{1}{n+1} \sum_{k=0}^{n} k \left\| \frac{\partial}{\partial \theta} \tilde{P}(\theta) \right\|_\infty$$

$$= t \sum_{n=N_S+1}^{\infty} p(n, qt) \frac{1}{n+1} \frac{n(n+1)}{2} \left\| \frac{\partial}{\partial \theta} \tilde{P}(\theta) \right\|_\infty$$

$$= \frac{t}{2} \left\| \frac{\partial}{\partial \theta} \tilde{P}(\theta) \right\|_\infty \sum_{n=N_S+1}^{\infty} n p(n, qt)$$

$$= \frac{t}{2} \left\| \frac{\partial}{\partial \theta} \tilde{P}(\theta) \right\|_\infty qt \sum_{n=N_S}^{\infty} p(n, qt)$$

$$= \frac{qt^2}{2} \left\| \frac{\partial}{\partial \theta} \tilde{P}(\theta) \right\|_\infty \sum_{n=N_S}^{\infty} p(n, qt)$$

$$= \frac{t^2}{2} \left\| \frac{\partial}{\partial \theta} Q(\theta) \right\|_\infty \sum_{n=N_S}^{\infty} p(n, qt).$$

Taking into account this bound and relation (2), the error on $\frac{\partial}{\partial \theta} E[Y(\theta, t)]$, denoted by F_Y, is such that

$$F_Y \leq \frac{t^2}{2} \|R\|_\infty \left\| \frac{\partial}{\partial \theta} Q(\theta) \right\|_\infty \sum_{n=N_S}^{\infty} p(n, qt). \tag{11}$$

If E_L is the error vector on $L(\theta, t)$ (relation (8)) and N_L is the infinite truncation step, we can easily show that:

$$\|E_L\|_\infty \leq t \left[1 - \sum_{n=0}^{N_L} p(n, qt) \right].$$

Thus we can bound the error on $E[Y(\theta, t)]$, denoted by E_Y, as follows:

$$E_Y \leq t \left[1 - \sum_{n=0}^{N_L} p(n, qt) \right] \|R\|_\infty. \tag{12}$$

The infinite truncation error and the time complexity of the SU method for computing $\frac{\partial}{\partial\theta}E[Y(\theta,t)]$ (and $E[Y(\theta,t)]$) depend on the truncation strategy. Remember that, for a given tolerance error ε, the infinite sum (7) will be truncated after term N_T such that

$$\varepsilon \geq 1 - \sum_{n=0}^{N_T} p(n,qt). \tag{13}$$

A first strategy consists in truncating the infinite sums $\Pi(\theta,t)$, $L(\theta,t)$, and $S_L(\theta,t)$ at the same point, that is to say $N_T = N_L$. From relations (12) and (11), we have:

$$E_Y \leq t\varepsilon \|R\|_\infty \tag{14}$$

and

$$F_Y \leq \frac{t^2}{2} \|R\|_\infty \left\|\frac{\partial}{\partial\theta}Q(\theta)\right\|_\infty [\varepsilon + p(N_T, qt)]. \tag{15}$$

Another strategy allows the computation of $E[Y(\theta,t)]$ with an absolute maximal error ε_Y. This is equivalent to set

$$\varepsilon = \frac{\varepsilon_Y}{t\|R\|_\infty}.$$

Note that, it is also possible to compute $\frac{\partial}{\partial\theta}E[Y(\theta,t)]$ after truncation in N_S i.e., by bounding relation (11) by a given tolerance error ε_S. It is clear that $\varepsilon < \varepsilon_Y < \varepsilon_S$ and that the deviations become important when the mission time t increases. The values of N_L and N_S may then be much greater than qt and the time complexity may considerably raise. To avoid that, we consider the first strategy.

The computation of $\frac{\partial}{\partial\theta}E[Y(\theta,t)]$ requires essentially 3 vector-matrix products per iteration. The time complexity of the SU method is then $O\left(3N_T M^2\right)$. Using a compact storage of $Q(\theta)$ and its derivative $\frac{\partial}{\partial\theta}Q(\theta)$, that time complexity may be reduced to $O\left(N_T(2\eta+\eta_s)\right)$ where η and η_s denote the number of non-null elements in $Q(\theta)$ and $\frac{\partial}{\partial\theta}Q(\theta)$. When the stiffness (thus qt) and the state space cardinality M increase, the computation time becomes prohibitive because from (13), $N_T > qt$. The following method we propose deals efficiently with that class of problems.

4 The IRK3 Method

Generally, ODE methods apply to systems of equations of the form $y'(t) = f(t,y(t))$. They consist in dividing the solution interval $[0,t]$ into $\{0 = t_0, ..., t_n = t\}$ and computing an approximated solution of the unknown function $y(t)$ at each point t_i, $i \geq 1$. Let $y(t_i)$ (resp. y_i) be the exact (resp. approximated) solution of the differential equation at t_i. The stepsize is defined as $h_i = t_{i+1} - t_i$.

In order to deal with the stiffness, we consider an ODE L-stable (or stiffly stable) method such that IRK3. More details on ODE methods and L-stability property may be found in [5]. The integration of system (3) gives

$$\frac{\partial}{\partial t} L(\theta, t) = L(\theta, t) Q(\theta) + \Pi(0); \quad L(\theta, 0) = L(0) = 0. \tag{16}$$

Deriving this relation with respect to θ, we obtain

$$\frac{\partial}{\partial t} S_L(\theta, t) = S_L(\theta, t) Q(\theta) + L(\theta, t) \frac{\partial}{\partial \theta} Q(\theta); \quad S_L(\theta, 0) = S_L(0) = 0. \tag{17}$$

Putting altogether (16) and (17), we get the following ODE system of the form $y' = \lambda y$

$$\frac{\partial}{\partial t} V(\theta, t) = V(\theta, t) B(\theta) \tag{18}$$

where $V(\theta, t) = (S_L(\theta, t) \ L(\theta, t) \ 1)$ and

$$B(\theta) = \begin{pmatrix} Q(\theta) & 0 & 0 \\ \frac{\partial}{\partial \theta} Q(\theta) & Q(\theta) & 0 \\ 0 & \Pi(0) & 0 \end{pmatrix}$$

The initial condition is such that $V(\theta, 0) = (0\ 0\ 1)$.

Applied to equation (18), the IRK3 method gives V_{i+1} as solution of the linear system equations

$$V_{i+1} \left[I - \frac{2}{3} h_i B(\theta) + \frac{1}{6} h_i^2 B(\theta)^2 \right] = V_i \left[I + \frac{1}{3} h_i B(\theta) \right]. \tag{19}$$

At time $t + h_i$, the local error vector is $\varepsilon(h_i) = \frac{h_i^4}{72} V(t) B(\theta)^4$. The Local Truncation Error (LTE_i) is its norm (the infinite norm for example). At each step i, LTE_i must be less than a specified tolerance τ. The stepsize h_i must satisfy $h_{min} \leq h_i \leq h_{max}$; the bounds h_{min} and h_{max} are fixed to avoid too many steps (if h_i if too small) and a bad precision (if h_i is too large). For example, in [6], $h_{min} = 10^{-7}$ and $h_{max} = 10$. Moreover, h_i is chosen such that the solution at $t + h_i$ meet the specified tolerance. A commonly used technique is

$$h_i \leq h_{i-1} \left(\frac{\tau}{LTE_i} \right)^{\frac{1}{r+1}} \tag{20}$$

where r is the order of the method ($r = 3$).

Usually, the LTE_i computation is done after solving the linear system (19) and must satisfy $LTE_i \leq \tau$. It is important to note that when a step is rejected, the computation time for solving the system is then useless. When the stepsize is small (e.g. $h_0 = 10^{-7}$), it is accepted for the following steps increasing the execution time. To avoid these drawbacks, we considered the expression of the error vector $\varepsilon(h_i)$. We observed that, at any step, it only depends upon some known variables of the previous step. At each step, LTE_i may be calculated first and the optimal stepsize chosen by using the formula (20). When doing so, *we automatically have the biggest h_i for which the $LTE_i \leq \tau$*. The stepsize h_i is rejected if it is bigger than h_{max}, and instead we take it to be h_{max}, or if it is less than h_{min} and it will be set to h_{min}. With this technique, we got stepsizes varying from 0.1 to 23, with an initial stepsize $h_0 = 10^{-7}$, a mission time $t = 10^5$ and state space size $M = 100$; the average stepsize was 1.5 (see following section). Let us note the average stepsize decreases when M increases. Thus, setting $t = 10^5$ and $M = 400$, the average stepsize becomes 0.99.

The IRK3 method requires essentially the resolution of the linear system of equations (19). The square matrix $B(\theta)$ is of order $2M+1$. Very often, it is stored with a compact scheme and the system is solved using an iterative method like Gauss-Seidel. The time complexity depends on the number of steps, denoted by p, and the number of iterations per step of average $\bar{I}$. Let η' be the number of non-null elements in $B(\theta)^2$, the time complexity of the IRK3 method is then $O\left(\bar{I}p\eta'\right)$.

5 Numerical Results

We consider a fault-tolerant multiprocessor system including n processors and b buffer stages. The system is modelled as an $M/M/n/n+b$ queuing system. Jobs arrive at rate Λ and are lost when the buffer is full. The job service rate is Θ. Processors (resp. buffer stages) fail independently at rate λ (resp. γ) and are repaired singly with rate μ (resp. τ). Processor failure causes a graceful degradation of the system (the number of processors is decreased by one). The system is in a failed state when all processors have failed or any buffer stage has failed. No additional processor failures are assumed to occur when the system is in a failed state. The model is represented by a $CTMC$ with the state-transition diagram shown in figure 1. At any given time the state of the system is (i,j) where $0 \leq i \leq n$ is the number of nonfailed processors, and j is zero if any of the buffer stage is failed, otherwise it is one. An appropriate reward rate in a given state is the steady-state throughput of the system with a given number of nonfailed processors [7]. The reward rate is zero in any system failure state. In

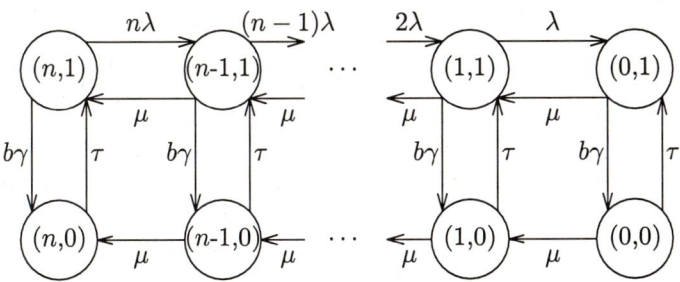

Fig. 1. State-transition diagram for an n-processors system

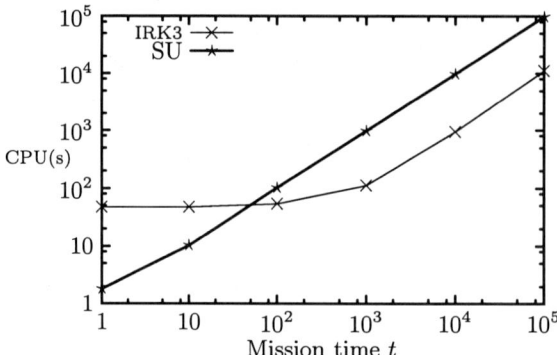

Fig. 2. CPU time vs mission time for $M = 100$

this experiment, the number of states is $M = 2(n+1)$. We shall choose $\lambda = \gamma = 10^{-6}$ per hour and $\mu = \tau = 100$ per hour, in order to produce an extremely stiff Markov process. The numerical results are obtained by executing the algorithms on a SUN Ultra-30, 295 MHZ station in numerical double precision. For the SU method, the tolerance ε is 10^{-10} for for all the values of t. The local tolerance τ for The IRK3 method is also set to 10^{-10}. These values give an acceptable precision for both of the presented methods [8].

First of all, the EAR sensitivity was computed for a moderate state space cardinality, $M = 10$. The mission time t was varied from 1 to 10^5 hours. We concluded that, in this case, the SU method performs very well. When we increased the number of state to $M = 100$, the SU method was better only for t less than 100 (figure 2). Beyond that limit, the IRK3 method was faster than the SU technique. To show how far IRK3 method resists to great values of M and t, we executed it for $M = 400$ and t still varying from 1 to 10^5 hours. The results are plotted in figure 3. We realized that the computation of the EAR sensitivity

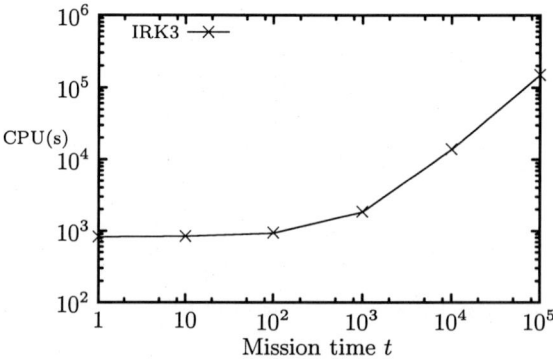

Fig. 3. CPU time vs mission time for $M = 400$

for $M = 400$ and $t = 10^5$ took about 27 hours CPU time when using IRK3 method while it was practically infeasible by the SU method. We conclude that even if it allows the global error control, the SU method remains usable only for moderate values the mission time. When stiffness and mission time increase, the IRK3 method may be recommended.

References

1. Reibman, A., Trivedi, K. S.: Transient Analysis of Cumulative Measures of Markov Model Behavior. Communication Statistics-Stochastic Models **5**, num. 4 (1989) 683–710
2. Malhotra, M. : A Computationally Efficient Technique for Transient Analysis of Repairable Markovian Systems. Performance Evaluation **24**, num. 4, (1996) 311–331
3. Abdallah, H., Hamza, M.: Sensitivity analysis of instantaneous transient measures of highly reliable systems. 11th European Simulation Symposium (ESS'99). Erlangen-Nuremberg, Germany, october 26-28, (1999) 652–656
4. Trivedi, K. S., Riebman, A. L.: Numerical Transient Analysis of Markov Models. Computer and Operations Research **15**, num. 1, (1988) 19–36
5. Hairer, E., Norsett, S. P., Wanner, G.: Solving ordinary differential equations. Springer, New York, 1993
6. Lindemann, C., Malhotra, M., Trivedi, K. S.: Numerical Methods for Reliability Evaluation of Markov Closed Fault-Tolerant Systems. IEEE Transactions on Reliability **44**, num. 4, (1995) 694–704
7. Trivedi, K. S.: Probability and Statistics with Reliability, Queueing and Computer Science Applications. Prentice-Hall, Englewood Cliffs, N. J, 1982
8. Abdallah, H., Hamza, M.: Sensibilité de mesures transitoires instantanées des systèmes informatiques hautement fiables. Publication Interne **1232**, Février, 1999 IRISA, Campus de Beaulieu, Rennes, France

Spectral Properties of Circulant Band Matrices Arising in ODE Methods*

Pierluigi Amodio

Dipartimento di Matematica, Università di Bari,
Via E. Orabona 4, I-70125 Bari, Italy
amodio@dm.uniba.it

Abstract. We investigate interesting spectral properties of circulant matrices with a band structure by analyzing the roots of an associated polynomial. We also derive practical conditions about the curve containing the eigenvalues of the matrix which can be used to study the stability domain of some numerical methods for the solution of ODEs.

Keywords: circulant matrices, ordinary differential equations, linear multistep methods

AMS(MOS) subject classification: 65F10, 65L05, 65L20, 65F15

1 Introduction

Circulant matrices are a quite useful tool of linear algebra to emphasize periodical behaviour of several phenomena. This means that circulant matrices often arise in problems of physics, probability and statistics, geometry, and numerical analysis.

The number of known properties of this class of matrices is enormous [5]. The main one is that any operation (sum, product, inverse, transpose) involving circulant matrices still gives a circulant matrix. For this reason, when working with Toeplitz matrices which constitute a larger class of matrices, one often resorts to circulant matrices [4,7].

Circulant matrices are a subclass of normal matrices which are diagonalized by the Fourier matrix. The eigenvalues of such matrices are on a curve which can be obtained by means of an explicit formula involving its elements.

This last property allows us to relate circulant band matrices with some numerical methods for the solution of ordinary differential equations. Let us consider a k-step linear multistep method

$$\sum_{j=-\nu}^{k-\nu} \alpha_j y_{n+j} = h \sum_{j=-\nu}^{k-\nu} \beta_j f(t_{n+j}, y_{n+j}) \qquad (1)$$

with ν initial and $k-\nu$ final conditions. The idea of selecting a number of initial conditions different from $\nu = k$ has been used to define Boundary Value Methods

---

* Work supported by MURST.

(BVMs) which constitute an important class of methods for the solution of initial value problems [3,6]. By considering the functions

$$\rho(z) = \sum_{j=-\nu}^{k-\nu} \alpha_j z^j, \quad \sigma(z) = \sum_{j=-\nu}^{k-\nu} \beta_j z^j, \qquad (2)$$

the boundary locus (the boundary of the linear stability domain) of (1) is given by the curve of the complex plane $\rho(e^{i\theta})/\sigma(e^{i\theta})$ for $\theta \in [0, 2\pi]$ (i is the imaginary unit).

The functions $\rho(e^{i\theta})$ and $\sigma(e^{i\theta})$ also represent the curves containing the eigenvalues of the circulant band matrices

$$A = \begin{pmatrix}
\alpha_0 & \alpha_1 & \cdots & \alpha_{k-\nu} & 0 & \cdots & 0 & \alpha_{-\nu} & \cdots & \alpha_{-1} \\
\alpha_{-1} & \alpha_0 & \ddots & & & & & & \ddots & \vdots \\
\vdots & \ddots & \ddots & & & \ddots & 0 & \ddots & & \alpha_{-\nu} \\
\alpha_{-\nu} & & & & & & & \ddots & & 0 \\
0 & \ddots & & & & & & & \ddots & \vdots \\
\vdots & \ddots & & & & & & & \ddots & 0 \\
0 & & & & & & & & & \alpha_{k-\nu} \\
\alpha_{k-\nu} & \ddots & 0 & & & & & \ddots & & \vdots \\
\vdots & \ddots & & & & & \ddots & & \alpha_0 & \alpha_1 \\
\alpha_1 & \cdots & \alpha_{k-\nu} & 0 & \cdots & 0 & \alpha_{-\nu} & \cdots & \alpha_{-1} & \alpha_0
\end{pmatrix} \qquad (3)$$

and B defined analogously. The matrix (3) is banded since we suppose that k is much smaller than the dimension of the matrix itself. The above consideration implies that the boundary locus coincides with the curve containing the eigenvalues of $B^{-1}A$.

Example 1. The explicit Euler method $y_n = y_{n-1} + hf(t_{n-1}, y_{n-1})$ has a stability domain given by the circle with center (-1,0) and radius 1. The same curve may be obtained from the matrices

$$A = \begin{pmatrix} 1 & & & -1 \\ -1 & 1 & & \\ & \ddots & \ddots & \\ & & -1 & 1 \end{pmatrix}, \quad B = \begin{pmatrix} 0 & & & 1 \\ 1 & 0 & & \\ & \ddots & \ddots & \\ & & 1 & 0 \end{pmatrix},$$

by considering the spectrum of the family of matrices

$$B^{-1}A = \begin{pmatrix} -1 & 1 & & & \\ & -1 & \ddots & & \\ & & \ddots & 1 & \\ 1 & & & -1 \end{pmatrix}.$$

As a further example the trapezoidal rule has an unbounded stability domain because the matrix B is singular.

In analogy with what has been done with the linear multistep method (1), in this paper we analyze some properties of circulant band matrices (3) by using the information given by the roots of their associated polynomials

$$\hat{\rho}(z) = z^\nu \rho(z), \qquad \hat{\sigma}(z) = z^\nu \sigma(z). \tag{4}$$

Our aim is also to derive in a simple way some important properties about the boundary locus of linear multistep methods.

2   Conditioning of Circulant Band Matrices

From the study of the conditioning of Toeplitz band matrices [1], it has been derived that the family of matrices

$$\begin{pmatrix} \alpha_0 & \alpha_1 & \cdots & \alpha_{k-\nu} & & & \\ \alpha_{-1} & \alpha_0 & \ddots & & \ddots & & \\ \vdots & \ddots & \ddots & \ddots & & \alpha_{k-\nu} & \\ \alpha_{-\nu} & & \ddots & \ddots & \ddots & & \vdots \\ & \ddots & & \ddots & & \alpha_0 & \alpha_1 \\ & & & \alpha_{-\nu} & \cdots & \alpha_{-1} & \alpha_0 \end{pmatrix}_{n \times n} \tag{5}$$

is well conditioned (the condition numbers are uniformly bounded with respect to n) if the associated polynomial (4) has ν roots of modulus smaller than 1 and the remaining of modulus larger than 1. On the other hand, it is weakly well conditioned (the condition numbers grow as a small power of n) if (4) has exactly either ν roots of modulus smaller than 1 or $k-\nu$ of modulus larger than 1, i. e. possible roots of unit modulus are all among the first ν or the remaining $k-\nu$.

The same properties cannot be generalized to a family of nonsingular circulant band matrices since the condition number of any matrix in this class is independent of the size of the matrix. Anyway, by considering the matrix (3) which is generated by the same elements of the corresponding Toeplitz matrix (5), a number of interesting properties can be derived. Let us start from the following basic results, whose proof follows by straightforward calculation:

Theorem 1. *Let A be the circulant matrix (3) and $\hat{\rho}(z)$ its associated polynomial as defined in (4). If $z_1, z_2, \ldots, z_k$ are the roots of $\hat{\rho}$, then A may be decomposed in the form*

$$A = \alpha_{k-\nu} \prod_{j=1}^{\nu} C_j \prod_{j=\nu+1}^{k} E_j,$$

where C_j and E_j are the following elementary matrices

$$C_j = \begin{pmatrix} 1 & & & -z_j \\ -z_j & 1 & & \\ & \ddots & \ddots & \\ & & -z_j & 1 \end{pmatrix}, \quad E_j = \begin{pmatrix} -z_j & 1 & & \\ & -z_j & \ddots & \\ & & \ddots & 1 \\ 1 & & & -z_j \end{pmatrix}. \quad (6)$$

The eigenvalues of an elementary matrix are on the circle centered at the diagonal element of the matrix and radius equals to the modulus of the off-diagonal element. Indeed, the eigenvalues of the matrix C_j and E_j are, respectively, $\lambda_l^{(j)} = 1 - z_j w^{-l}$ for $j = 1, \ldots, \nu$, and $\lambda_l^{(j)} = -z_j + w^l$ for $j = \nu+1, \ldots, k$, where w is the nth root of unity, $w = e^{i(2\pi/n)}$. For what concerns the eigenvalues of the matrix A, the following result holds:

Theorem 2. *The eigenvalues of the matrix A in (3) are given by*

$$\lambda_l = \alpha_{k-\nu} \prod_{j=1}^{k} \lambda_l^{(j)}, \quad l = 0, \ldots, n-1$$

where $\lambda_l^{(j)}$ are the eigenvalues of the elementary matrices (6).

Proof. The thesis follows from the fact that the eigenvector corresponding to the eigenvalue $\lambda_l^{(j)}$ is the lth column of the Fourier matrix. Therefore, any eigenvalue of the product of circulant matrices C_j and E_j is given by the product of the corresponding eigenvalues. □

We are now in a position to easily derive the following

Corollary 1. *The family of circulant matrices (3) is nonsingular if the associated polynomial (4) has no roots of unit modulus. The condition number of (3) depends on the distance of the roots from the unit circumference.*

We observe that if one root is equal to 1, then the circulant matrix is always singular since the corresponding matrix C_j or E_j is singular.

A complementary result to that of Corollary 1 should be the calculation of the minimum eigenvalue which corresponds to the 2-norm of the inverse of A. A practical criterion can be derived by analyzing the function

$$f(\theta) = \sum_{j=-\nu}^{k-\nu} \alpha_j \cos(j\theta). \quad (7)$$

If f is strictly monotone in $(0, \pi)$ and, in addition, $f(0)$ and $f(\pi)$ have the same sign, the following two properties can be deduced:

- the curve containing the eigenvalues entirely lies in the real positive or in the real negative half plane;
- $|\lambda_{min}| = \min(|f(0)|, |f(\pi)|)$.

The first property is quite useful to check whether the boundary locus of a linear multistep method (1) is entirely in the real positive half plane and, therefore, the method is $A_{\nu,k-\nu}$-stable (a property that corresponds to A-stability for linear multistep methods (1) with $\nu = k$ [3]).

To obtain practical conditions ensuring the monotonicity of $f(\theta)$ it is convenient to use just the coefficients α_j rather than the roots of $\hat{\rho}$. By considering the variable transformation $\cos \theta = t$, function (7) can be recast as the polynomial, of degree $k - \nu$, $\hat{f}(t)$. Since the function $f(\theta)$ is strictly monotone if and only if $f'(\theta) \neq 0$ in $(0, \pi)$, that is $\hat{f}'(t) \neq 0$ for $t \in (-1, 1)$, we need to check that all the roots of $\hat{f}'(t)$ are greater than 1 in modulus.

From the usual substitutions $\cos 0 = 1$ and $\cos(n+1)\theta \equiv T_{n+1}(t) = 2tT_n(t) - T_{n-1}(t)$ we obtain the following expression for $\hat{f}'(t)$ associated to the matrix with bandwidth $\max(\nu, k - \nu) \leq 5$:

$$\hat{f}'_{10}(t) = 80(\alpha_5 + \alpha_{-5})t^4 + 32(\alpha_4 + \alpha_{-4})t^3 + 12(\alpha_3 + \alpha_{-3} - 5(\alpha_5 + \alpha_{-5}))t^2 \\ + 4(\alpha_2 + \alpha_{-2} - 4(\alpha_4 + \alpha_{-4}))t + \alpha_1 + \alpha_{-1} - 3(\alpha_3 + \alpha_{-3}) + 5(\alpha_5 + \alpha_{-5}).$$

As an example, almost circulant tridiagonal matrices satisfy $\hat{f}'(t) \neq 0$ when (by considering $\alpha_2 = \ldots = \alpha_5 = \alpha_{-2} = \ldots = \alpha_{-5} = 0$ in the above expression for $\hat{f}'(t)$) $\alpha_1 \neq \alpha_{-1}$, while the coefficients of almost pentadiagonal circulant matrices need to satisfy

$$|\alpha_1 + \alpha_{-1}| \geq 4 |\alpha_2 + \alpha_{-2}|.$$

For matrices with a bandwidth larger then 2, the above condition should be checked numerically.

3 Stability Domain of ODE Methods

In this section we analyze the boundary locus of some known linear multistep methods used as BVMs by using the properties of their associated circulant band matrix given in the previous section. The obtained results are in general not new, but are here re-derived in a quite simple way.

Since each linear multistep method (1) satisfies $\rho(1) = 0$, then the corresponding matrix A as defined in (3) is always singular and the curve containing the eigenvalues crosses the origin of the complex plane.

By recalling that the boundary locus of a linear multistep method (1) is equivalent to the curve representing the spectrum of $B^{-1}A$, where B is associated to the function $\sigma(z)$, for Corollary 1 it is sufficient that one root of $\hat{\sigma}(z)$ is equal to 1 in modulus in order to obtain an unbounded boundary locus. This is the case, for example, of the Extended Trapezoidal Rules of the second kind [3]

$$\sum_{j=-\nu}^{\nu-1} \alpha_j y_{n+j} = \frac{h}{2}(f(t_{n+1}, y_{n+1}) + f(t_n, y_n)),$$

where the coefficients α_j are chosen in order to obtain the maximum attainable order.

On the other hand, the obtained results are not useful to state that these methods are perfectly $A_{\nu,k-\nu}$-stable (the boundary locus coincides with the imaginary axis).

The previous methods can be generalized by using any value of ν and still have an unbounded boundary locus. In general they are used as initial and final methods to obtain a BVM (see [3]).

A different family of methods that is easy to analyze is that of GBDFs (Generalized BDFs, see [2]) defined as ($l = 1, 2$)

$$\sum_{j=-\nu}^{\nu-l} \alpha_j y_{n+j} = hf(t_n, y_n). \tag{8}$$

This family of methods has the matrix B equal to the identity matrix and hence the associated boundary loci are bounded curves. Here it is more convenient to use the variable change $\cos\theta = t + 1$ since the obtained polynomial, expressed for $\nu = 5$ and $l = 1$ by means of the formula ($k = 2\nu - 1$)

$$\hat{f}_k(t) = 16\alpha_{-5}t^5 + 8(\alpha_4 + \alpha_{-4} + 10\alpha_{-5})t^4 + 4(\alpha_3 + \alpha_{-3} + 8(\alpha_4 + \alpha_{-4})$$
$$+35\alpha_{-5})t^3 + 2(\alpha_2 + \alpha_{-2} + 6(\alpha_3 + \alpha_{-3}) + 20(\alpha_4 + \alpha_{-4}) + 50\alpha_{-5})t^2$$
$$+ \sum_{i=-5}^{4} i^2 \alpha_i t + \sum_{i=-5}^{4} \alpha_i = \sum_{i=0}^{5} a_i t^i,$$

has in general all the coefficients $a_0 = \ldots = a_{\nu-1} = 0$. In fact, formula (8) has order $2\nu - l$ and hence, among the others, the conditions

$$\sum_{j=-\nu}^{\nu-l} \alpha_j \, j^{2s} = 0, \qquad s = 0, \ldots, \nu - 1. \tag{9}$$

must be satisfied. Conditions (9) are expressed in matrix form by the following homogeneous linear system

$$\begin{pmatrix} 1 & 1 & 1 & \ldots & 1 \\ 0 & 1 & 4 & \ldots & \nu^2 \\ 0 & 1 & 16 & \ldots & \nu^4 \\ \vdots & \vdots & \vdots & & \vdots \\ 0 & 1 & \nu^{2(\nu-1)} & \ldots & \nu^{2(\nu-1)} \end{pmatrix} \begin{pmatrix} \alpha_0 \\ \alpha_{-1} + \alpha_1 \\ \alpha_{-2} + \alpha_2 \\ \vdots \\ \alpha_{-\nu} + \alpha_\nu \end{pmatrix} = \begin{pmatrix} 0 \\ 0 \\ 0 \\ \vdots \\ 0 \end{pmatrix}$$

where $\alpha_\nu = 0$ and, if $l = 2$, also $\alpha_{\nu-1} = 0$. By applying Gaussian elimination to the above system one has (the coefficient matrix is $\nu \times (\nu+1)$)

$$\begin{pmatrix} 1 & 1 & 1 & \cdots & 1 & 1 \\ 1 & 4 & \cdots & & (\nu-1)^2 & \nu^2 \\ & 12 & \cdots & \nu(\nu-1)^2(\nu-2) & \nu^2(\nu^2-1) \\ & \ddots & & \vdots & \vdots \\ & & & \gamma_\nu & 2\nu\gamma_\nu \end{pmatrix} \begin{pmatrix} \alpha_0 \\ \alpha_{-1}+\alpha_1 \\ \alpha_{-2}+\alpha_2 \\ \vdots \\ \alpha_{-\nu}+\alpha_\nu \end{pmatrix} = \begin{pmatrix} 0 \\ 0 \\ 0 \\ \vdots \\ 0 \end{pmatrix}$$

where γ_ν is a constant which depends on the size of the matrix, and, by considering a suitable row scaling, one then obtains ν linear combinations of the coefficients α_i corresponding to the previously stated $a_i = 0$, for $i = 0, \ldots, \nu-1$ (this was proved by direct computation up to $\nu = 15$). Therefore, $\hat{f}_k(t) = 2^{\nu-1}\alpha_{-\nu}t^\nu$, that is $f_k(\theta) = 2^{\nu-1}\alpha_{-\nu}(\cos\theta - 1)^\nu$.

The value $\theta = 0$ is the only root of f_k and this means that f_k is strictly monotone in $(0, \pi)$. Moreover, since $f_k(\pi) > 0$, all the methods are $A_{\nu,k-\nu}$-stable. We observe that the higher the multiplicity of θ as root of f_k, the more the boundary locus of the GBDF is flattened on the imaginary axis (see Fig. 1).

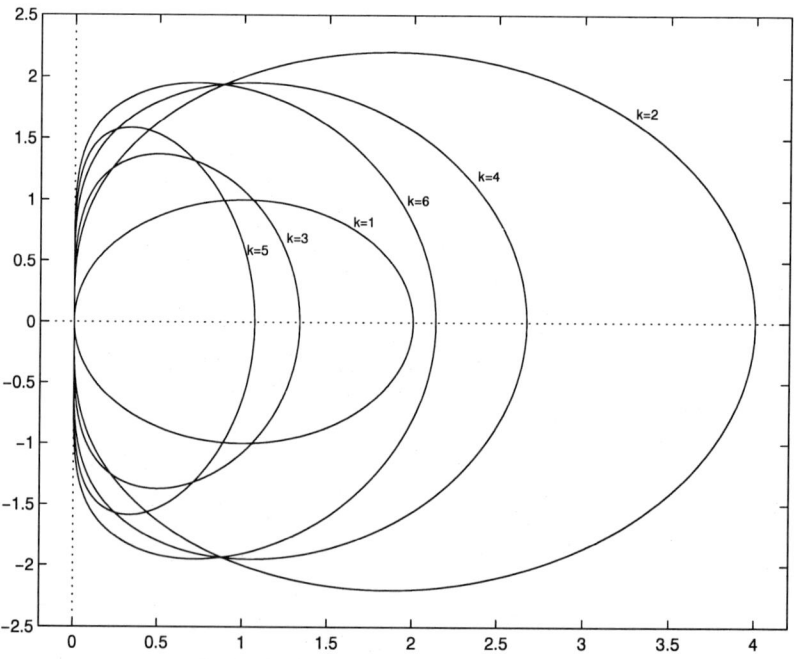

Fig. 1. Boundary locus of GBDFs for $k = 1, \ldots, 6$

References

1. P. Amodio, L. Brugnano, *The conditioning of Toeplitz band matrices*, Math. Comput. Modelling **23 (10)** (1996), 29–42.
2. L. Brugnano, D. Trigiante, *Convergence and stability of Boundary Value Methods*, J. Comput. Appl. Math. **66** (1996), 97–109.
3. L. Brugnano, D. Trigiante, *Solving ODEs by Linear Multistep Initial and Boundary Value Methods*, Gordon & Breach, Amsterdam, (1998).
4. T. F. Chan, *An optimal circulant preconditioner for Toeplitz systems*, SIAM J. Sci. Stat. Comput. **9** (1988), 766–771.
5. P. J. Davis, *Circulant matrices*, John Wiley & Sons, New York, (1979).
6. F. Iavernaro, F. Mazzia, *Block-Boundary Value Methods for the solution of Ordinary Differential Equations*, Siam J. Sci. Comput. **21** (1999), 323–339.
7. V. V. Strela and E. E. Tyrtyshnikov, *Which circulant preconditioner is better?*, Math. Comput. **65 (213)** (1996), 137–150.

A Parameter Robust Method for a Problem with a Symmetry Boundary Layer*

Ali R. Ansari[1], Alan F. Hegarty[1], and Grigorii I. Shishkin[2]

[1] Department of Mathematics & Statistics, University of Limerick
Limerick, Ireland,
ali.ansari@ul.ie, alan.hegarty@ul.ie
[2] Institute of Mathematics and Mechanics, Russian Academy of Sciences
Ekaterinburg, Russia
grigorii@shishkin.ural.ru

Abstract. We consider the classical problem of a two-dimensional laminar jet of incompressible fluid flowing into a stationary medium of the same fluid [2]. The equations of motion are the same as the boundary layer equations for flow over an infinite flat plate, but with different boundary conditions. Numerical experiments show that, using an appropriate piecewise uniform mesh, numerical solutions are obtained which are parameter robust with respect to both the number of mesh nodes and the number of iterations required for convergence.

1 Introduction

Numerical methods for the solution of various linear singular perturbation problems, which are uniformly convergent with respect to the perturbation parameter, were developed in, *inter alia*, [5,6,7]. The key idea in these methods is the use of piecewise uniform meshes, which are appropriately condensed in the boundary layer regions. It is of interest to determine whether these ideas can be used for nonlinear problems, in particular flow problems. We thus apply the technique to simple model problems, the exact solutions of which are available . In [7] it was shown that, for the flat plate problem of Blasius [1,2], the method is uniformly convergent with respect to the perturbation parameter.

Here we examine analogously the classical two-dimensional laminar jet problem [2]. A two-dimensional jet of fluid emerges from a narrow slit in a wall into static medium of the same fluid. If the jet is thin, such that u the horizontal component of velocity varies much less rapidly along the jet *i.e.*, the x-axis, than across it, we have a boundary layer at $y = 0$, *i.e.*, the axis of the jet [3,4]. The pressure gradient is zero in the jet since it is zero in the surrounding fluid. The equations of motion are therefore the same as the Prandtl boundary layer equations [2], *i.e.*,

$$-\nu u_{yy} + uu_x + vu_y = 0$$
$$u_x + v_y = 0$$
(1)

---
* This work was supported in part by the Enterprise Ireland grants SC-97-612 and SC-98-612 and by the Russian Foundation for Basic Research (grant No. 98-01-00362).

but with the different boundary conditions

$$u_y(x,0) = v(x,0) = = 0 \ \forall \, x \geq 0$$

$$\lim_{y \to \pm\infty} u(x,y) = 0 \ \forall \, x \in \mathbb{R} \tag{2}$$

The primary equation of motion, involving the second derivative of u and the viscosity ν, is clearly a singularly perturbed differential equation with ν as the perturbation parameter. Our objective here is to obtain numerical solutions to this problem that are robust with respect to ν. The sensitivity of classical numerical methods to the perturbation parameter is reflected in the maximum pointwise errors becoming unacceptably large for small ν. This has been shown for linear problems, *e.g.*, in [5] where it is also seen that inappropriate condensing of the mesh in the boundary layer region also fails to resolve the difficulty.

The approach adopted here will involve a piecewise uniform mesh [6], which, when used in conjunction with an upwind finite difference method, leads to *parameter robust* solutions, *i.e.*, numerical solutions where the maximum pointwise error tends to zero independently of the perturbation parameter, while the work required to obtain the solutions is also independent of ν. As analytical solutions of this particular problem are achievable we will use them to compute the discretisation errors in the L^∞ norm.

It should be noted that Prandtl's boundary layer equations are valid approximations to the Navier-Stokes equations only for a small range of values of ν. As there is no known *parameter robust* method for solving the Navier-Stokes equations, even for this simple geometry, it is worthwhile considering the solution of the simpler model, even for values of ν where it is not physically valid. Numerical results will verify that the numerical method is indeed parameter robust.

2 The Analytical Solution

As mentioned in the previous section it is possible to obtain analytical solutions to the jet problem under consideration here [2,3,4] . The solutions for u and v are given here without derivation

$$u = 6\nu\varphi^2 \frac{x}{y^2} \operatorname{sech}^2 \varphi \tag{3}$$

$$v = 2\nu\varphi \left[2\varphi \operatorname{sech}^2 \varphi - \tanh \varphi \right] \tag{4}$$

where $\varphi = \frac{1}{2} \left(\frac{1}{6}\right)^{1/3} \left(\frac{J_0}{\rho\nu^2}\right)^{1/3} \frac{y}{x^{2/3}}$, ν is the viscosity, ρ is the density and J_0 is defined as

$$\int_{-\infty}^{\infty} \rho u^2 \, dy = J_0 = \text{constant}.$$

Furthermore, some simple analysis [1,2] shows that the thickness of the boundary layer ξ is

$$\xi \sim \left(\frac{\rho\nu^2}{J_0}\right)^{1/3} x^{2/3}. \tag{5}$$

Both ρ and J_0 are constants and we set $\rho = 1 = J_0 = 1$ here.

3 The Numerical Solution

To begin with we must decide on a domain of solution. We confine consideration to a finite rectangle $\Omega = (a, A) \times (0, B)$, where the constants a, A and B are fixed and independent of the perturbation parameter ν. We fix $a > 0$ as the equations are singular at $x = 0$ (this is apparent from (3) & (4)). The size of the near-wall subdomain where the equations are not appropriate increases with $1/\nu$ and thus allowing a to increase as $\nu \to 0$ would make the problem easier. However, we require the method to work well on a fixed domain and thus fix a. We denote the boundary of Ω by $\Gamma = \Gamma_L \cup \Gamma_R \cup \Gamma_T \cup \Gamma_B$ where $\Gamma_L, \Gamma_R, \Gamma_T$ and Γ_B denote the left, right, top and bottom edges of Ω respectively.

We are now in a position to define the computational mesh for this problem. On the rectangular domain Ω we place the piecewise uniform rectangular mesh $\Omega_\nu^{\mathbf{N}}$ which is defined as the tensor product $\Omega_\nu^{\mathbf{N}} = \Omega^{N_x} \times \Omega_\nu^{N_y}$ where $\mathbf{N} = (N_x, N_y)$. Here Ω^{N_x} is a uniform mesh over the interval $[a, A]$ with N_x mesh intervals, while $\Omega_\nu^{N_y}$ is a piecewise uniform fitted mesh with N_y mesh intervals on the interval $[0, B]$. The interval $[0, B]$ is divided into two subintervals $[0, \sigma]$ and $[\sigma, B]$, and $\frac{1}{2} N_y$ uniform mesh intervals are assigned to each subinterval. Note that in this paper we set $N_x = N_y = N$.

The transition point σ is of significance as, by reducing σ as ν decreases, the mesh in the neighbourhood of the x-axis will be condensed. σ is chosen, following the principles set out in [6] and [7] as

$$\sigma = \min \left\{ \frac{1}{2} B, 2\nu^{2/3} \ln N \right\}.$$

The choice of $\nu^{2/3}$ is motivated from (5), while the particular choice of the constant 2 is based on experimental work, which seems to suggest this as a near optimal value giving reasonable convergence rates for the iterative process.

Note that though (5) shows that the jet spreads out as x increases, the choice of σ ignores this. The reason for this is that the errors dominate near $x = a$; when the jet spreads beyond $y = \sigma$, the velocity and errors are much reduced. This reiterates the simplicity of the solution technique.

We linearise the first equation by adapting the continuation algorithm set out in [7] for the problem of flow past a flat plate. In the case of the jet problem, we encounter stability difficulties and thus we need to generalise the algorithm from [5], as elaborated below.

After linearisation and discretisation of (1) and the associated boundary conditions (2) we have the sequence of discrete linear problems for $m = 0, 1, \ldots$:

$$-\nu \delta_y^2 U_\nu^m(x_i, y_j) + \bar{U}_\nu^{m-1} D_x^- U_\nu^m(x_i, y_j)^m + \bar{V}_\nu^{m-1} D_y^\pm U_\nu^m(x_i, y_j) = 0, \qquad (6)$$
$$D_x^- U_\nu^m(x_i, y_j) + D_y^- V_\nu^m(x_i, y_j) = 0,$$

with boundary conditions

$$D^0 U_\nu^m(x_i, y_0) = 0, \quad V_\nu^m(x_i, y_0) = 0, \quad U_\nu^m(x_i, y_N) = 0 \qquad (7)$$

where
$$D_y^+ U_\nu^m(x_i, y_j) \equiv \frac{U_\nu^m(x_i, y_{j+1}) - U_\nu^m(x_i, y_j)}{y_{j+1} - y_j},$$
$$D_y^- U_\nu^m(x_i, y_j) \equiv \frac{U_\nu^m(x_i, y_j) - U_\nu^m(x_i, y_{j-1})}{y_j - y_{j-1}},$$
with analogous definition of $D_x^- U_\nu^m(x_i, y_j)$ and $D_y^- V_\nu^m(x_i, y_j)$,
$$D_y^0 U_\nu^m(x_i, y_j) \equiv \frac{U_\nu^m(x_i, y_{j+1}) - U_\nu^m(x_i, y_{j-1})}{y_{j+1} - y_{j-1}},$$
$$\delta_y^2 U_\nu^m(x_i, y_j) \equiv \frac{D_y^+ U_\nu^m(x_i, y_j) - D_y^- U_\nu^m(x_i, y_j)}{(y_{j+1} - y_{j-1})/2},$$
and where
$$D_y^\pm U_\nu^m(x_i, y_j) \equiv \begin{cases} D_y^- U_\nu^m(x_i, y_j) & \text{for } V_\nu^m(x_i, y_j) > 0, \\ D_y^+ U_\nu^m(x_i, y_j) & \text{for } V_\nu^m(x_i, y_j) < 0 \end{cases}$$

In addition,
$$\bar{U}_\nu^{m-1}(x_i, y_j) = \theta_1 U_\nu^{m-1}(x_i, y_j) + (1 - \theta_1) U_\nu^{m-2}(x_i, y_j)$$
$$\bar{V}_\nu^{m-1}(x_i, y_j) = \theta_2 V_\nu^{m-1}(x_i, y_j) + (1 - \theta_2) V_\nu^{m-2}(x_i, y_j)$$
where the parameters $0 \leq \theta_1, \theta_2 \leq 1$ are selected to stabilise the iterative process as ν becomes smaller. For large ν we set $\theta_1 = \theta_2 = 1$. Experimentally, it has been noted that when $\nu < 2^{-12}$ the number of iterations starts to increase but this problem is easily overcome by appropriate choice of θ_1, θ_2.

4 Numerical Results

The analytical solution has a singularity at $x = 0$. This means that the choice for constants that define the x-range of the domain i.e. $x \in [a, A]$ needs to be restricted to $a > 0$ to avoid the singularity. Here we (arbitrarily) set $a = 0.1$, $A = 1.1$ and $B = 1$. The piecewise uniform mesh for this problem, $\Omega_\nu^N \equiv \{(x_i, y_j)\}$, with the above constants is
$$x_i = x_{i-1} + h$$
$$y_j = \begin{cases} 2i\sigma/N & i = 0, 1, 2, \ldots N/2 \\ \sigma + 2(i - N/2)(1 - \sigma)/N & i = N/2, \ldots, N \end{cases}$$
where
$$\sigma = \min\left\{\frac{1}{2}, 2\nu^{\frac{2}{3}} \ln N\right\}.$$

At this point we summarise the problem as

$$P_\nu^N \begin{cases} \text{Find } (U_\nu, V_\nu) \text{ such that } \forall (x_i, y_j) \in \Omega_\nu^N \\ -\nu \delta_y^2 U_\nu^m(x_i, y_j) + \bar{U}_\nu^{m-1} D_x^- U_\nu^m(x_i, y_j) + \bar{V}_\nu^{m-1} D_y^\pm U_\nu^m(x_i, y_j) = 0 \\ D_x^- U_\nu^m(x_i, y_j) + D_y^- V_\nu^m(x_i, y_j) = 0 \\ D_y^o U_\nu^m(x_i, y_0) = 0 \text{ and } V_\nu^m(x_i, y_0) = 0 \text{ on } \Gamma_B \\ U_\nu^m = u \text{ on } \Gamma_L \cup \Gamma_T \end{cases}$$

The algorithm for solving P_ν^N sweeps across the domain Ω from Γ_L to Γ_R. At the ith stage of the sweep, we compute the values of (U_ν, V_ν) on $X_i = \{(x_i, y_j), 0 \le j \le N\}$, where (U_ν, V_ν) are known on X_{i-1}. This is achieved by solving the first linearised equation for U_ν, followed by a solution of the second linear equation for V_ν.

In order to solve the first equation on X_i we need values of U_ν on X_{i-1}, boundary values for U_ν^m on $\Gamma_B \cup \Gamma_T$ and an initial guess at U_ν^0 on X_i. On each X_i, the 2 point boundary value problem for $U_\nu^m(x_i, y_j)$ is solved for $0 \le j \le N-1$. Since $U_\nu^m(x_i, y_0)$ is thus an unknown the term $D_y^\pm U_\nu^m(x_i, y_j)$ can and does introduce the value $U_\nu^m(x_i, y_{-1})$, which is eliminated by implementing the central difference approximation of the Neumann condition, so that all instances of $U_\nu^m(x_i, y_{-1})$ are replaced by $U_\nu^m(x_i, y_1)$. The initial guess to start the algorithm i.e. U_ν^0 on X_1 is taken from the prescribed boundary condition for U_ν (the analytical solution) on Γ_L. For each X_i, V_ν^0 is set to be zero.

Once the solution to the tridiagonal system of equations for U_ν^m is obtained we then solve the linear system

$$D_x^- U_\nu^m(x_i, y_j) + D_y^- V_\nu^m(x_i, y_j) = 0, \quad 1 \le j \le N,$$

for V_ν. The process here is trivial as U_ν is known from the previous step and V_ν is initialised using the boundary condition i.e. $V_\nu = 0$ on Γ_B.

This process is continued until a stopping criterion is achieved. This involves setting the tolerance tol, for the difference between two successive iterates i.e.

$$\max(|U_\nu^m - U_\nu^{m-1}|_{\Omega_\nu^N}, |V_\nu^m - V_\nu^{m-1}|_{\Omega_\nu^N}) \le tol$$

where we take tol to be 10^{-6}. We let $m = M$ for all instances where the stopping criterion is met. Once this happens we set $U_\nu = U_\nu^M$ and $V_\nu = V_\nu^M$ on X_i proceed to the next step X_{i+1} using $U_\nu(x_i, y_j)$ as the initial guess for $U_\nu^m(x_{i+1}, y_j)$.

Graphs of the solution (U_ν, V_ν) using this direct method with $N = 32$ for $\nu = 1$ and $\nu = 2^{-30}$ are shown in Figs. 1 and 2. Graphs of the errors in the numerical solutions for $\nu = 2^{-30}$ are shown in Fig. 3. Additionally, we approximate the scaled partial derivatives $\nu^{\frac{1}{3}}\frac{\partial u}{\partial x}$, $\nu\frac{\partial u}{\partial y}$ and $\frac{\partial v}{\partial x}$ by the corresponding scaled discrete derivatives $D_x^- U_\nu, D_y^- U_\nu$ and $D_x^- V_\nu$. Note that $\frac{\partial u}{\partial x} = -\frac{\partial v}{\partial y}$ and correspondingly $D_x^- U_\nu = -D_y^- V_\nu$.

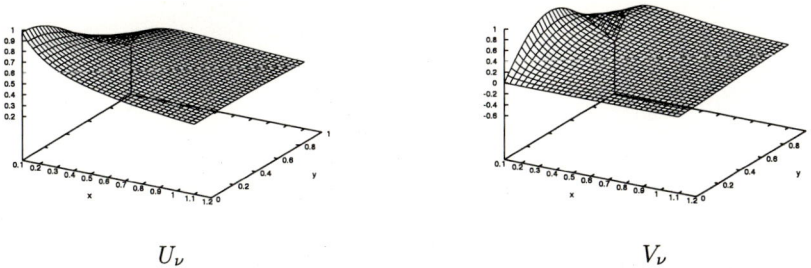

Fig. 1. Surface plot of numerical solutions on $\Omega_\nu^\mathbf{N}$; $\nu = 1$, $N = 32$

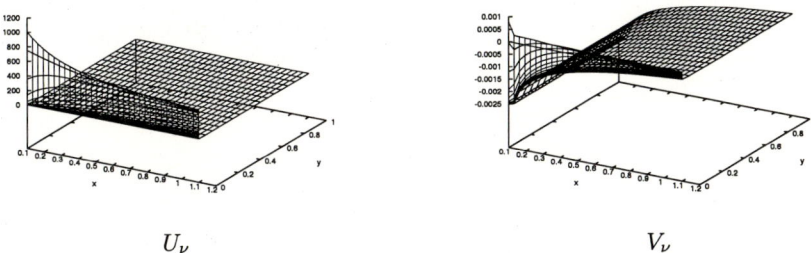

Fig. 2. Surface plots of numerical solutions on $\Omega_\nu^\mathbf{N}$; $\nu = 2^{-30}$, $N = 32$

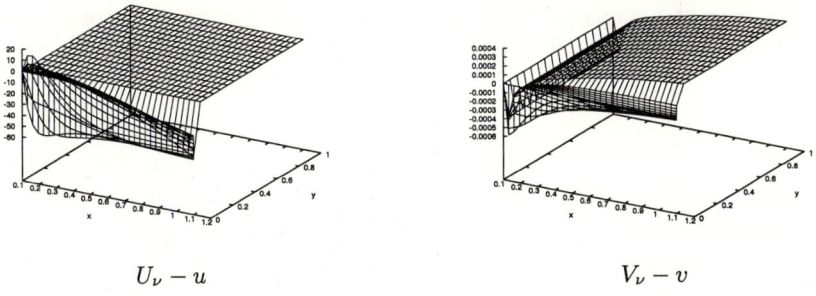

Fig. 3. Surface plots of errors on $\Omega_\nu^\mathbf{N}$; $\nu = 2^{-30}$, $N = 32$

Table 1 lists the maximum errors and corresponding ν-uniform convergence rates for the velocity components (u, v) and their scaled derivatives on $\Omega_\nu^\mathbf{N}$. It is evident that all the results are robust apart from the scaled approximation to v_x, which is robust only for a subdomain of $\Omega_\nu^\mathbf{N}$ which excludes a neighbourhood of $x = a$, for example $\tilde{\Omega}_\nu^\mathbf{N} = \Omega_\nu^\mathbf{N} \cap (0.2, 1.1] \times [0, 1]$ as in the last 2 rows of Table 1.

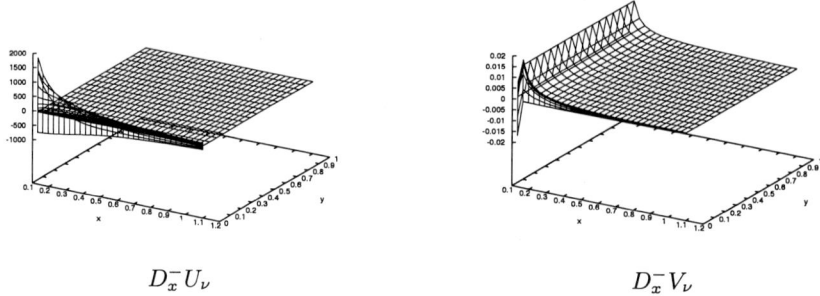

Fig. 4. Surface plots of approximations to the derivatives on Ω_ν^N; $\nu = 2^{-30}$, $N = 32$

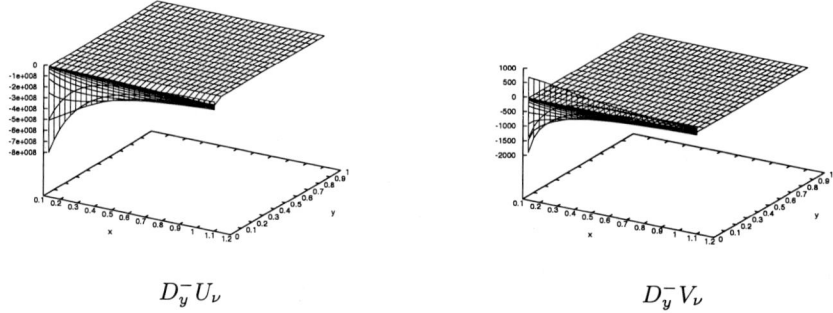

Fig. 5. Surface plot of approximations to the derivatives on Ω_ν^N; $\nu = 2^{-30}$, $N = 32$

Table 2 shows that with a simple choice of $\theta_1 = \theta_2 = 0.75$ for all N and ν the number of iterations per 'time-like' step X_i (*i.e.*, total number of iterations divided by N) increases only very slightly with $1/\nu$.

5 Summary

We have demonstrated through experimental results that the numerical method and associated algorithm gives solutions for the velocity terms and their scaled discrete derivatives which appear to be uniformly convergent with respect to the viscosity ν. The number of iterations of the algorithm depends weakly on ν but it is believed that this can also be rectified. However, the method is not claimed to be optimal, and future work will involve the investigation of alternative methods of solution of the nonlinear system of equations. Other matters for further investigation include the dependence of the numerical solutions on the distance

Table 1. Maximum pointwise errors and associated ν-uniform rates

	32	64	128	256	512
$\nu^{1/3}\|U_\nu - u\|_{\Omega_\nu^N}$	0.53(−01)	0.31(−01)	0.18(−01)	0.10(−01)	0.57(−02)
		0.78	0.80	0.81	0.83
$\dfrac{1}{\|V_\nu\|}\|V_\nu - v\|_{\Omega_\nu^N}$	0.23(+00)	0.14(+00)	0.11(+00)	0.81(−01)	0.53(−01)
		0.73	0.33	0.47	0.62
$\nu^{1/3}\|D_x^- U_\nu - u_x\|_{\Omega_\nu^N}$	0.45(+00)	0.44(+00)	0.32(+00)	0.21(+00)	0.13(+00)
		0.02	0.46	0.62	0.74
$\nu\|D_y^- U_\nu - u_y\|_{\Omega_\nu^N}$	0.37(+00)	0.33(+00)	0.22(+00)	0.14(+00)	0.77(−01)
		0.15	0.57	0.73	0.81
$\nu\|D_x^- V_\nu - v_x\|_{\Omega_\nu^N}$	0.17(+02)	0.27(+02)	0.34(+02)	0.38(+02)	0.40(+02)
		−0.67	−0.30	−0.18	−0.08
$\nu^{1/3}\|D_y^- V_\nu - v_y\|_{\Omega_\nu^N}$	0.45(+00)	0.44(+00)	0.32(+00)	0.21(+00)	0.13(+00)
		0.02	0.46	0.62	0.74
$\nu\|D_x^- V_\nu - v_x\|_{\hat\Omega_\nu^N}$	0.12(+01)	0.82(+00)	0.53(+00)	0.30(+00)	0.17(+00)
		0.54	0.63	0.82	0.84

Table 2. Number of one-dimensional linear solves to attain a solution (scaled by factor $1/N$) with $\theta_1 = \theta_2 = 0.75$

ν	32	64	128	256	512
1	6	6	6	5	5
2^{-4}	13	13	13	12	12
2^{-8}	15	15	15	16	17
2^{-12}	14	14	14	14	16
2^{-16}	14	14	14	13	14
2^{-20}	15	15	15	14	14
2^{-24}	17	17	16	16	16
2^{-28}	19	18	18	17	18

from the wall a and a comparison of the value of J_0 at Γ_R with the imposed value at Γ_L.[1]

References

1. Prandtl, L., Tietjens, O. G., *Applied Hydro- and Aeromechanics*, Dover Publications, New York (1957)
2. Schlichting, H., *Boundary-layer theory*, 7th ed. McGraw Hill, New York (1979)
3. Acheson, D. J., *Elementary Fluid Dynamics*, Oxford University Press, Oxford (1990)
4. Warsi, Z. U. A., *Fluid dynamics: theoretical and computational approaches*, CRC Press, Boca Raton (1993)

---

[1] The authors would like to thank the referee for some valuable comments on this paper.

5. Hegarty, A. F., Miller J. J. H., O'Riordan E., Shishkin, G. I., *Special meshes for finite difference approximations to an advection-diffusion equation with parabolic layers*, Journal of Computational Physics, Vol. 117, (1995) 47–54
6. Miller, J. J. H., O'Riordan, E., Shishkin, G. I., *Fitted numerical methods for singular perturbation problems*, World Scientific, London (1996)
7. Farrell, P. A., Hegarty, A. F., Miller, J. J. H., O'Riordan, E., Shishkin, G. I., *Robust Computational Techniques for Boundary Layers*, Chapman & Hall/CRC Press, Boca Raton (2000)

An Algorithm Based on Orthogonal Polynomial Vectors for Toeplitz Least Squares Problems

Marc Van Barel[1]*, Georg Heinig[2], and Peter Kravanja[1]

[1] Department of Computer Science, Katholieke Universiteit Leuven
Celestijnenlaan 200A, B-3001 Heverlee, Belgium
Marc.VanBarel@cs.kuleuven.ac.be
Peter.Kravanja@na-net.ornl.gov
[2] Department of Mathematics, Kuwait University,
POB 5969, Safat 13060, Kuwait
georg@mcs.sci.kuniv.edu.kw

Abstract. We develop a new algorithm for solving Toeplitz linear least squares problems. The Toeplitz matrix is first embedded into a circulant matrix. The linear least squares problem is then transformed into a discrete least squares approximation problem for polynomial vectors. Our implementation shows that the normwise backward stability is independent of the condition number of the Toeplitz matrix.

1 Toeplitz Linear Least Squares Problems

Let $m \geq n \geq 1$, $t_{-n+1}, \ldots, t_{m-1} \in \mathbb{C}$ and

$$T := [t_{j-k}]_{j=0,\ldots,m-1}^{k=0,\ldots,n-1}$$

a $m \times n$ Toeplitz matrix that has full column-rank. Let $b \in \mathbb{C}^m$. We want to solve the corresponding Toeplitz linear least squares problem (LS-problem), i.e., we want to determine the (unique) vector $x \in \mathbb{C}^n$ such that

$$\|Tx - b\| \text{ is minimal} \tag{1}$$

where $\|\cdot\|$ denotes the Euclidean norm.

Standard algorithms for least squares problems require $\mathcal{O}(mn^2)$ floating point operations (flops) for solving (1). The arithmetic complexity can be reduced by taking into account the Toeplitz structure of T. Several algorithms that require only $\mathcal{O}(mn)$ flops have been developed. Such algorithms are called *fast*. One of the first fast algorithms was introduced by Sweet in his PhD thesis [10]. This method is not numerically stable, though. Other approaches include those by Bojanczyk, Brent and de Hoog [1], Chun, Kailath and Lev-Ari [3], Qiao [9],

---

* The work of the first and the third author is supported by the Belgian Programme on Interuniversity Poles of Attraction, initiated by the Belgian State, Prime Minister's Office for Science, Technology and Culture. The scientific responsibility rests with the authors.

Cybenko [4,5], Sweet [11] and many more. None of these algorithms has yet been shown to be numerically stable and for several approaches there exist examples indicating that the method is actually unstable.

Recently, Ming Gu [7] has developed fast algorithms for solving Toeplitz and Toeplitz-plus-Hankel linear least squares problems. In his approach, the matrix is first transformed into a Cauchy-like matrix by using the Fast Fourier Transform or trigonometric transformations. Then the corresponding Cauchy-like linear least squares problem is solved. Numerical experiments show that this approach is not only efficient but also numerically stable, even if the coefficient matrix is very ill-conditioned.

In this paper we will also develop a numerically stable method that works for ill-conditioned problems—in other words, for problems that cannot be solved via the normal equations approach. We proceed as follows. The original LS-problem is first embedded into a larger LS-problem. The coefficient matrix of the latter problem has additional structure: it is a circulant block matrix. This LS-problem is then (unitarily) transformed into a LS-problem whose coefficient matrix is a coupled Vandermonde matrix. The latter LS-problem is then solved by using the framework of orthogonal polynomial vectors.

2 Embedding of the Original LS-Problem

We embed the original LS-problem (1) in the following way. Let A and B be matrices and let a and y be vectors. The extended LS-problem is formulated as follows: determine the vectors x and y such that the norm of the vector

$$r := \begin{bmatrix} A & B \\ T & 0 \end{bmatrix} \begin{bmatrix} x \\ y \end{bmatrix} - \begin{bmatrix} a \\ b \end{bmatrix}$$

is minimal. (We assume, of course, that A, B, a and y have appropriate sizes.) If the matrix B is nonsingular, then the first 'component' x of the solution $\begin{bmatrix} x \\ y \end{bmatrix}$ of the extended LS-problem coincides with the solution x of the original LS-problem for any choice of A, B and a. We can always choose A and B such that the two block columns

$$C_1 := \begin{bmatrix} A \\ T \end{bmatrix} \quad \text{and} \quad C_2 := \begin{bmatrix} B \\ 0 \end{bmatrix}$$

are circulant matrices. For example, we can choose B equal to the identity matrix of order $n-1$ and we can choose A as the $(n-1) \times n$ Toeplitz matrix

$$A := [t_{-n+1+j-k}]_{j=0,\ldots,n-2}^{k=0,\ldots,n-1}$$

with $t_{-n-k} = t_{m-k-1}$ for $k = 0, 1, \ldots, n-1$. We take a to be the zero vector. However, we can also choose the size of B larger to obtain a number of rows M for the two circulant matrices C_1 and C_2 such that the discrete Fourier transform of size M can be computed efficiently. For example, we could choose M as the smallest power of two larger than or equal to $m+n-1$. The matrices A and B are now chosen to have sizes $(M-m) \times n$ and $(M-m) \times (M-m)$, respectively. Note that B is square and assumed to be nonsingular.

3 Transformation of the Extended LS-Problem

Define C_3 as the vector
$$C_3 := -\begin{bmatrix} a \\ b \end{bmatrix} \in \mathbb{C}^M.$$

The vector C_3 can be interpreted as the first column of a circulant matrix. The extended LS-problem can therefore be formulated as follows: determine the vectors x and y such that the norm of the vector

$$r = \begin{bmatrix} C_1 & C_2 & C_3 \end{bmatrix} \begin{bmatrix} x \\ y \\ 1 \end{bmatrix} \in \mathbb{C}^M$$

is minimal. Note that the matrix $\begin{bmatrix} C_1 & C_2 & C_3 \end{bmatrix}$ is of size $M \times (n+M-m+1)$. It is well-known that a $p \times p$ circulant matrix C can be factorized as

$$C = \mathcal{F}_p^H \Lambda \mathcal{F}_p$$

where Λ is a $p \times p$ diagonal matrix containing the eigenvalues of C and $\mathcal{F}_p$ denotes the $p \times p$ Discrete Fourier Transform matrix (DFT-matrix)

$$\mathcal{F}_p := \left[\omega_p^{jk} \right]_{j,k=0,\ldots,p-1}$$

where $\omega_p := e^{-2\pi i/p}$ and $i = \sqrt{-1}$. Similarly, if C is of size $p \times q$, where $p \geq q$, then C can be factorized as

$$C = \mathcal{F}_p^H \Lambda \mathcal{F}_{p,q}$$

where Λ is again a $p \times p$ diagonal matrix and where $\mathcal{F}_{p,q}$ denotes the $p \times q$ submatrix of $\mathcal{F}_p$ that contains the first q columns of $\mathcal{F}_p$.

By applying the Discrete Fourier Transform to r, the norm of r remains unchanged: $\|r\| = \|\mathcal{F}_M r\|$. The following holds:

$$\mathcal{F}_M r = \mathcal{F}_M \begin{bmatrix} C_1 & C_2 & C_3 \end{bmatrix} \begin{bmatrix} x \\ y \\ 1 \end{bmatrix} \tag{2}$$

$$= \begin{bmatrix} \Lambda_1 \mathcal{F}_{M,n} & \Lambda_2 \mathcal{F}_{M,s} & \Lambda_3 \mathcal{F}_{M,1} \end{bmatrix} \begin{bmatrix} x \\ y \\ 1 \end{bmatrix} \tag{3}$$

where $s := M - m$ and where $\Lambda_j =: \operatorname{diag}(\lambda_{j,k})_{k=1}^M$ is a $M \times M$ diagonal matrix for $j = 1, 2, 3$.

We will now translate the extended LS-problem into polynomial language. Define $x(z)$ and $y(z)$ as

$$x(z) := \sum_{k=0}^{n-1} x_k z^k \quad \text{and} \quad y(z) := \sum_{k=0}^{s-1} y_k z^k.$$

Here x_k and y_k denote the components of the vectors x and y. The DFT-matrix $\mathcal{F}_M$ can be interpreted as a Vandermonde matrix based on the nodes $z_k = \omega_M^k$, $k = 0, 1, \ldots, M-1$. Equation (3) now implies that the extended LS-problem can be formulated in the following way: determine the polynomials $x(z)$ and $y(z)$, where $\deg x(z) \leq n-1$ and $\deg y(z) \leq s-1$, such that

$$\sum_{k=0}^{M-1} |\lambda_{1,k} x(z_k) + \lambda_{2,k} y(z_k) + \lambda_{3,k} 1|^2 \tag{4}$$

is minimal.

4 Orthogonal Polynomial Vectors

The minimisation problem (4) can be solved within the framework of orthogonal polynomial vectors developed by Van Barel and Bultheel [2,12,13,14]. The following notation will be used: to indicate that the degree of the first component of a polynomial vector $P \in \mathbb{C}[z]^{3\times 1}$ is less than or equal to α, that the degree of the second component of P is less than 0 (hence, this second component is equal to the zero polynomial), and that the degree of the third component is equal to β, we write

$$\deg P = \begin{bmatrix} \alpha \\ -1 \\ \beta \end{bmatrix}.$$

We consider the following inner product and norm.

Definition 1 (inner product, norm). *Consider the subspace $\mathcal{P} \subset \mathbb{C}[z]^{3\times 1}$ of polynomial vectors P of degree*

$$\deg P = \begin{bmatrix} n \\ s \\ 0 \end{bmatrix}.$$

Given the points $z_k \in \mathbb{C}$ and the weight vectors

$$F_k = \begin{bmatrix} \lambda_{1,k} & \lambda_{2,k} & \lambda_{3,k} \end{bmatrix} \in \mathbb{C}^{1\times 3}, \quad k = 1, 2, \ldots, M,$$

we define the discrete inner product $\langle P, Q \rangle$ for two polynomial vectors $P, Q \in \mathcal{P}$ as follows:

$$\langle P, Q \rangle := \sum_{k=1}^{M} P^H(z_k) F_k^H F_k Q(z_k). \tag{5}$$

The norm $\|P\|$ of a polynomial vector $P \in \mathcal{P}$ is defined as:

$$\|P\| := \sqrt{\langle P, P \rangle}.$$

A necessary and sufficient condition for (5) to be an inner product in $\mathcal{P}$, is that $\mathcal{P}$ is a subspace of polynomial vectors such that a nonzero polynomial vector $P \in \mathcal{P}$ for which $\langle P, P \rangle = 0$ (or equivalently: $F_k P(z_k) = 0$, $k = 1, 2, \ldots, M$) does not exist. Our original LS-problem can be now stated as the following discrete least squares approximation problem: determine the polynomial vector $P^\star \in \mathcal{P}'$ such that $\|P^\star\| = \min_{P \in \mathcal{P}'} \|P\|$ where $\mathcal{P}'$ denotes all vectors belonging to $\mathcal{P}$ and having their third component equal to the constant polynomial 1.

In [14], Van Barel and Bultheel formulated a fast algorithm for computing an orthonormal basis for $\mathcal{P}$. The degree sequence of the basis vectors B_j, $j = 1, 2, \ldots, \delta$, is as follows:

$$\begin{bmatrix} 0 & 1 & \cdots & n-s & n-s & n-s+1 & n-s+1 & \cdots & n & n & n \\ -1 & -1 & \cdots & -1 & 0 & 0 & 1 & \cdots & s-1 & s & s \\ -1 & -1 & \cdots & -1 & -1 & -1 & -1 & \cdots & -1 & -1 & 0 \end{bmatrix}.$$

Every polynomial vector $P \in \mathcal{P}'$ can be written (in a unique way) as:

$$P = \sum_{j=1}^{\delta} a_j B_j$$

where $a_1, \ldots, a_\delta \in \mathbb{C}$. The coordinate a_δ is determined by the fact that the third component polynomial of P has to be monic and of degree 0. The following holds:

$$\|P\|^2 = \langle P, P \rangle$$
$$= \left\langle \sum_{j=1}^{\delta} a_j B_j, \sum_{j=1}^{\delta} a_j B_j \right\rangle$$
$$= \sum_{j=1}^{\delta} |a_j|^2 \quad \text{(since } \langle B_i, B_j \rangle = \delta_{ij}\text{)}.$$

It follows that $\|P\|$ is minimized by setting $a_1, \ldots, a_{\delta-1}$ equal to zero. In other words,

$$P^\star = a_\delta B_\delta \quad \text{and} \quad \|P^\star\| = |a_\delta|.$$

The discrete least squares approximation problem can therefore be solved by computing the orthonormal polynomial vector B_δ. We obtain $P^\star$ by scaling B_δ to make its third component monic.

5 Numerical Experiments

We have implemented our approach in Matlab (MATLAB Version 5.3.0.10183 (R11) on LNX86). The numerical experiments that we will present in this section are similar to those done by Ming Gu in [7]. The computations have been done in double precision arithmetic with unit roundoff $u \approx 1.11 \times 10^{-16}$. We have considered two approaches:

- QR: the QR method as implemented in Matlab. This is a classical approach for solving general dense linear least squares problems;
- NEW: the approach that we have described in the previous sections.

We have compared the two approaches QR and NEW for two types of Toeplitz matrices:

- Type 1: the entries t_k are taken uniformly random in the interval $(0, 1)$;
- Type 2: $t_0 := 2\omega$ and $t_k := \frac{sin(2\pi \omega k)}{\pi k}$ for $k \neq 0$ where $\omega := 0.25$. This matrix is called the Prolate matrix and is very ill-conditioned [6,15].

The right-hand side vector b has been chosen in two ways:

- Its entries are generated uniformly random in $(0, 1)$. This generally leads to large residuals.
- The entries of b are computed such that $b = Tx$ where the entries of x are taken uniformly random in $(0, 1)$. In this case, we obtain small residuals.

To measure the normwise backward error, we have used the following result of Waldén, Karlson and Sun [16]. See also [8, section 19.7].

Theorem 1. *Let $A \in \mathbb{R}^{m \times n}$, $b \in \mathbb{R}^m$, $0 \neq x \in \mathbb{R}^n$, and $r := b - Ax$. Let $\theta \in \mathbb{R}$. The normwise backward error*

$$\eta_F(x) = \min\{ \| [\Delta A, \theta \Delta b] \|_F \ : \ \|(A + \Delta A)x - (b + \Delta b)\|_2 = \min \}$$

is given by

$$\eta_F(x) = \min\{ \eta_1, \sigma_{\min}([A \ \ \eta_1 C]) \}$$

where

$$\eta_1 := \frac{\|r\|_2}{\|x\|_2}\sqrt{\mu}, \quad C := I - \frac{rr^T}{r^T r} \quad \text{and} \quad \mu = \frac{\theta^2 \|x\|_2^2}{1 + \theta^2 \|x\|_2^2}.$$

We have computed $\eta_F(x)$ with $\theta := 1$.

The numerical results are shown in Tables 1 and 2 for the two possible choices of the right-hand side vector b.

6 Conclusions

The numerical experiments show that the current implementation is still not accurate enough to be comparable with QR or with the algorithms developed by Ming Gu. However, the results show that the normwise backward error does not depend on the condition number of the Toeplitz matrix. We are currently working on improving the accuracy as well as the speed of the implementation to obtain a viable alternative for the algorithms of Ming Gu where the Toeplitz matrix can range from well-conditioned to very ill-conditioned.

Table 1. Normwise backward error (small residuals)

Matrix type	Order		$\kappa(T)$	$\eta_F(x)/u$	
	m	n		QR	NEW
1	160	150	5.4×10^2	1.9×10^2	1.7×10^4
	320	300	3.4×10^2	7.5×10^2	9.1×10^4
	640	600	7.7×10^2	5.9×10^2	3.3×10^5
2	160	150	2.1×10^{16}	3.9×10^1	2.7×10^2
	320	300	1.5×10^{16}	2.5×10^0	5.5×10^2
	640	600	1.3×10^{16}	2.8×10^0	1.5×10^3

Table 2. Normwise backward error (large residuals)

Matrix type	Order		$\kappa(T)$	$\eta_F(x)/u$	
	m	n		QR	NEW
1	160	150	5.4×10^2	4.1×10^1	3.0×10^3
	320	300	3.4×10^2	1.3×10^2	2.5×10^4
	640	600	7.7×10^2	1.1×10^2	1.4×10^5
2	160	150	2.1×10^{16}	1.3×10^2	3.9×10^0
	320	300	1.5×10^{16}	1.5×10^0	8.2×10^0
	640	600	1.3×10^{16}	2.7×10^0	2.3×10^1

References

1. A. BOJANCZYK, R. BRENT, AND F. DE HOOG, *QR factorization of Toeplitz matrices*, Numer. Math., 49 (1986), pp. 81–94.
2. A. BULTHEEL AND M. VAN BAREL, *Vector orthogonal polynomials and least squares approximation*, SIAM J. Matrix Anal. Appl., 16 (1995), pp. 863–885.
3. J. CHUN, T. KAILATH, AND H. LEV-ARI, *Fast parallel algorithms for QR and triangular factorization*, SIAM J. Sci. Statist. Comput., 8 (1987), pp. 899–913.
4. G. CYBENKO, *A general orthogonalization technique with applications to time series analysis and signal processing*, Math. Comp., 40 (1983), pp. 323–336.
5. ———, *Fast Toeplitz orthogonalization using inner products*, SIAM J. Sci. Statist. Comput., 8 (1987), pp. 734–740.
6. I. GOHBERG, T. KAILATH, AND V. OLSHEVSKY, *Fast Gaussian elimination with partial pivoting for matrices with displacement structure*, Math. Comp., 64 (1995), pp. 1557–1576.
7. M. GU, *Stable and efficient algorithms for structured systems of linear equations*, SIAM J. Matrix Anal. Appl., 19 (1998), pp. 279–306.
8. N. HIGHAM, *Accuracy and Stability of Numerical Algorithms*, SIAM, 1996.
9. S. QIAO, *Hybrid algorithm for fast Toeplitz orthogonalization*, Numer. Math., 53 (1988), pp. 351–366.
10. D. SWEET, *Numerical Methods for Toeplitz matrices*, PhD thesis, University of Adelaide, Adelaide, Australia, 1982.
11. ———, *Fast Toeplitz orthogonalization*, Numer. Math., 43 (1984), pp. 1–21.

12. M. VAN BAREL AND A. BULTHEEL, *A parallel algorithm for discrete least squares rational approximation*, Numer. Math., 63 (1992), pp. 99–121.
13. ———, *Discrete linearized least squares approximation on the unit circle*, J. Comput. Appl. Math., 50 (1994), pp. 545–563.
14. ———, *Orthonormal polynomial vectors and least squares approximation for a discrete inner product*, Electron. Trans. Numer. Anal., 3 (1995), pp. 1–23.
15. J. VARAH, *The Prolate matrix*, Linear Algebra Appl., 187 (1993), pp. 269–278.
16. B. WALDÉN, R. KARLSON, AND J.-G. SUN, *Optimal backward perturbation bounds for the linear least squares problem*, Numerical Linear Algebra with Applications, 2 (1995), pp. 271–286.

From Sensitivity Analysis to Random Floating Point Arithmetics – Application to Sylvester Equations

Alain Barraud[1], Suzanne Lesecq[1], and Nicolai Christov[2]

[1] Laboratoire d'Automatique de Grenoble
BP46, 38402 Saint Martin d'Hères, France
[2] Department of Automatics, Technical University of Sofia
1756 Sofia, Bulgaria

Abstract. Classical accuracy estimation in problem solving is basically based upon sensitivity analysis and conditionning computation. Such an approach is frequently much more difficult than solving the problem itself. Here a generic alternative through the concept of random arithmetic is presented. These two alternatives are developped around the well know Sylvester equations. Matlab implentation as a new object class is discussed and numerically illustrated.

1 Introduction

The Sylvester matrix equations (SME) are among some fundamental problems in the theory of linear systems. That is why, the question of their reliable solution, including evaluation of their precision, is of great practical interest. The conditioning of SME is well studied and different types of condition numbers are derived [1]. Unfortunately, perturbation bounds, based on condition numbers, may eventually produce pessimistic results, although better bounds based upon local non linear analysis are now available [2]. In any case, only global results are given but not component wise analysis. Lastly, this approach is usually much more difficult from a numerical computation point of view than the problem itself. Basically, their memory cost is $O(n^4)$, and their flops count is $O(n^6)$, where n is the problem size (assuming for simplicity square matrix unknown).

Random arithmetic is considered, here, as an alternative approach to compute simultaneously the solution of a given problem, and its accuracy. This technique is fundamentally component wise. Furthermore, its cost is basically unchanged compared with the use of standard floating point, except that the new unit is not a flop but a Random Flop which is designed here by "Rflop". In our Matlab implementation, one Rflop is a small multiple of one flop, and some overhead computations. So, this generic technique is, a priori, very competitive compared with more classical accuracy scheme.

The following notations are used later on: $\mathcal{R}^{m \times n}$ – the space of real $m \times n$ matrices; I_n – the unit $n \times n$ matrix; $A^\top = [a_{ji}]$ – the transpose of the matrix $A = [a_{ij}]$; $\text{vec}(A) \in \mathcal{R}^{mn}$ – the column-wise vector representation of the matrix

$A \in \mathcal{R}^{m \times n}$; $A \otimes B = [a_{ij}B]$ – matrices A and B Kronecker product; $\|\cdot\|_2$ – the spectral (or 2-) norm in $\mathcal{R}^{m \times n}$; $\|.\|_F$ – the Frobenius (or F-) norm in $\mathcal{R}^{m \times n}$.

2 Problem Statement and Notations

Consider the standard Sylvester equation :

$$AX + XB + C = 0 \qquad (1)$$

where $A \in \mathcal{R}^{n \times n}$, $B \in \mathcal{R}^{m \times m}$, and $X, C \in \mathcal{R}^{n \times m}$. We suppose that $0 \notin \{\lambda_i(A) + \lambda_k(B) : i \in \overline{1,n},\ k \in \overline{1,m}\}$ where $\lambda_i(M)$ are the eigenvalues of the matrix M. Under this assumption, the equation (1) has a unique solution. Let the matrices A, B and C be perturbed as $A \mapsto A + \Delta A\ B \mapsto B + \Delta B\ C \mapsto C + \Delta C$ and let the perturbed Sylvester be defined by :

$$(A + \Delta A)Y + Y(B + \Delta B) + (C + \Delta C) = 0 \qquad (2)$$

The perturbed equation (2) has an unique solution $Y = X + \Delta X$, in the neighborhood of X if the perturbations $(\Delta A, \Delta B, \Delta C)$ are sufficiently small. Denote by :

$$\Delta := [\Delta_A, \Delta_B, \Delta_B]^T \in \mathcal{R}_+^3 \qquad (3)$$

the vector of absolute norm perturbations $\Delta_A := \|\Delta A\|_F$, $\Delta B := \|\Delta B\|_F$ and $\Delta_C := \|\Delta C\|_F$ in the data matrices $A\ B,\ C$; and $a = \|A\|_F, b = \|A\|_F, c = \|C\|_F, x = \|X\|_F$ the Frobenius norms of the data and solution matrices. Lastly, it is usefull to define the relative perturbation vector $\widetilde{\Delta} := [\Delta_A/a, \Delta_B/b, \Delta_B/c]^T = \left[\widetilde{\Delta_A}, \widetilde{\Delta_B}, \widetilde{\Delta_B}\right]^T \in \mathcal{R}_+^3$.

3 Sensitivity Analysis

Here, we consider local bounds for the perturbation $\Delta_X := \|\Delta X\|_F$ in the solution of (1). These are bounds of the type

$$\Delta_X \leq f(\Delta) + O(\|\Delta\|^2),\ \Delta \to 0 \qquad (4)$$
$$\Delta_X/x \leq f(\Delta)/c + O(\|\Delta\|^2),\ \Delta \to 0 \qquad (5)$$

where f is a continuous function, non-decreasing in each of its arguments and satisfying $f(0) = 0$. Particular cases of (4) and (5) are the well known linear perturbation bounds [1]. Denote by M_X, M_A, M_B and M_C the following operators $M_X = I_m \otimes A + B^T \otimes I_n$, $M_A = X^T \otimes I_n$, $M_B = I_m \otimes X$, $M_C = I_{nm}$. Then absolute condition numbers are given by :

$$\begin{cases} K_A = \|M_X^{-1}M_A\|_2,\ K_B = \|M_X^{-1}M_B\|_2,\ K_C = \|M_X^{-1}M_C\| \\ \qquad K_S = \|M_X^{-1}[M_A, M_B, M_C]\|_2 \end{cases} \qquad (6)$$

and the corresponding linear perturbation estimations are :

$$\Delta_X \leq K_A \Delta_A + K_B \Delta_B + K_C \Delta_C \quad \text{and} \quad \Delta_X \leq K_S \|\Delta\|_2 \tag{7}$$

In the same way relative condition numbers and estimation will be :

$$\begin{cases} \widetilde{K_A} = \left\|M_X^{-1}\widetilde{M_A}\right\|_2/x, \; \widetilde{K_B} = \left\|M_X^{-1}\widetilde{M_B}\right\|_2/x, \; \widetilde{K_C} = \left\|M_X^{-1}\widetilde{M_C}\right\|_2/x. \\ \widetilde{K_S}\widetilde{K_S} = \|M_X^{-1}\left[\widetilde{M_A}, \widetilde{M_B}, \widetilde{M_C}\right]\|_2/x \end{cases} \tag{8}$$

where $\widetilde{M_A} = aM_A$, $\widetilde{M_B} = bM_B$, $\widetilde{M_C} = cM_C$, and lastly :

$$\Delta_X/x \leq \widetilde{K_A}\Delta_A + \widetilde{K_B}\Delta_B + \widetilde{K_C}\Delta_C \quad \text{and} \quad \Delta_X \leq \widetilde{K_S}\left\|\widetilde{\Delta}\right\|_2 \tag{9}$$

4 A Random Arithmetic Approach

Each floating point operation produces a round off error, hence there are potentially two results, one by lack, the other by excess. They both legitimately represent the exact result. Consequently, if a given algorithm contains k arithmetic operations there are 2^k results r_i, which are all equally representing the theoretical result r. Let us define $\bar{r}$ the mean of the r_i. Then, the basic idea is that the accuracy of the numerical result given by the considered algorithm can be deduced from the dispersion of the r_i, i.e. from its standard deviation σ. From a practical point of view, some questions must be considered. Firstly how to obtain the so called r_i, secondly how many r_i must be computed, and lastly how to compute a confidence interval [4]. It is currently admitted that rounding errors are uniformly distributed on $[-1/2, +1/2]$ ulp, for rounded floating point arithmetic as IEEE standard ($[0, +1]$ ulp for chopped arithmetic), where ulp means *Unit in the Last Place*. Now, for the simplicity sake, it is supposed that rounded arithmetic is used. Consider the kth elementary floating point operation: $z = fl(x \diamond y)$. Then a particular r_i can be obtained by perturbing this result as :

$$\hat{z} = rfl(x \diamond y) = z + e \tag{10}$$

where rfl (random floating operation) is an alternative notation for fl. The random perturbation e consists in adding 1, or substracting 1 to the last bit of z with a probability 1/4, and leaving z unchanged with a probabilty 1/2. Practically it is sufficient to generate 3 to 5 realisations of $\hat{z}$. Let us define N this number. Consequently each standard floating point variable of an algorithm is substituted by a set of N values and computed as follows :

$$\hat{z}_i = rfl(x_i \diamond y_i), \; i = 1, ..., N \tag{11}$$

Now, let us introduce the following notations :

$$z = E(\hat{z}) = \left[\sum_{i=1}^{N} z_i\right]/N \quad \text{and} \quad \sigma^2 = \left[\sum_{i=1}^{N}(z_i - z)^2\right]/(N-1)$$

Then the **estimated number of "significant" bits** is :

$$nb(z) = \min\left[\left(\max(\log_2 \frac{|z|}{\sigma \tau_p \sqrt{N}}, 0), t\right)\right] \quad (12)$$

where τ_p is the value of the Student's law for a $p\%$ confidence interval of. Clearly the number of decimal digits is obtained with log_{10}. The numerical result $\hat{z}$ of an algorithm can be defined as follows, to the first order in β^{-t}, where z_{th} is the theoretical result, β the arithmetic base, t the number of base β digits, $u_i(d)$ are constants depending only on the data and the considered algorithm, α_i are the values lost at the rounding step (standard floating point arithmetic effect), e_i the applied perturbations (random floating point effect). The fundamental point is that the following result must be valid :

$$\hat{z} = z_{th} + \sum_{i=1}^{n} u_i(d)\beta^{-t}[\alpha_i - e_i] + O(\beta^{-2t}) \quad (13)$$

The theoretical justifications can be found for example, in [3]. Consequently $E(z_i) = z_{th}$. In practice the following hypothesis must be verified : the exponent and the sign of each floating point result do not depend on the random perturbation, the number of operations rfl must be much larger than the number of data on which the algorithm operates, the mantissa of the data must be (sufficiently) randomly distributed. These hypothesis are usually true for real life industrial problems. On the contrary, computing the mean of n equal terms does not agree with some of these conditions. However, the validity of the first order approximation (13) may decrease when the computations accuracy decreases, so it can be observed that $E(z_i) \neq z_{th}$. This situation can be dynamically checked with $nb(z)$. As a consequence, the algorithm must be stopped, for example, when a divide by a non significant value is attempted (not necessarily 0), or several operands or data are non significant. At each computation step, the number of "significant" bits is now available. What happens when some of the operands of a rfl operation (10) have no significant bits ? The concept of **"numerical zero"** ($\bar{0}$) offers an easy to implement response, according to the definition :

$$\hat{z} = \bar{0} \Leftrightarrow z = 0 \text{ or } nb(z) = 0 \quad (14)$$

This fundamental notion induces some other **basic properties** which are the foundations of the random floating point arithmetic. Some of them are the logical tests : $\neq, \leq, <$, specified by :

$$\begin{cases} \hat{a} \neq \hat{b} \Leftrightarrow a \neq b \text{ and } nb(b-a) > 0 \; ; \; \hat{a} \leq \hat{b} \Leftrightarrow a \leq b \text{ or } nb(b-a) = 0 \\ \hat{a} = \hat{b} \Leftrightarrow b-a = 0 \text{ or } nb(b-a) = 0 \; ; \; \hat{a} < \hat{b} \Leftrightarrow a < b \text{ and } nb(b-a) > 0 \end{cases}$$

Further, $nb(a) = nb(b) = 0$ must be considered as a fatal error. Another consequence is that, for example, a test like "$if \; det(A) = 0, break$" is now well defined in random floating point arithmetic. Computing the determinant of the

Hilbert matrices and the Hilbert inverses gives $det(A) = 0$ "true" for dimensions greater than 13 in double precision IEEE arithmetic, although their values are very small and respectively very large, but with no significant bits. It is more important that the rfpa objects does not agree whith the mathematical rules. This explains why floating point cannot be view as the numerical counterpart of the set $\mathbb{R}$. It has to be noticed for **random arithmetic** that :

" $>$ " is the negation of " $\leq$ " and " $<$ " is transitive
" $\leq$ " is **not transitive** and " $=$ " is **not transitive**

These fundamental properties explain why this approach is always successful until the "practical" hypothesis are fulfilled. Furthermore, it can be verified that if $\Pr\left(z_{th} \in \left[z - \frac{t_p \sigma}{\sqrt{N}}, z + \frac{t_p \sigma}{\sqrt{N}}\right]\right) = 0.95$ then :

$$\Pr|z_{th} - z| \geq 10\frac{t_p \sigma}{\sqrt{N}} = 0.00054 \quad \text{and} \quad \Pr|z_{th} - z| \leq 0.1\frac{t_p \sigma}{\sqrt{N}} = 0.61$$

This means that $nb(z)$, the number of the estimated significant bits (12), has a probability 0.39 to be pessimistic by more than one decimal place (underestimation), and a probability ~ 1 to be never optimistic by more than one decimal place (overestimation).

5 A Matlab Implementation

In order to numerically exhibit how random floating point arithmetic works, a Matlab (Mathworks product) implementation has been developped as a new object class called "rfpa" for Random Floating Point Arithmetic. All the basic operators working on the default class "double" have been overloaded in order to be able to execute standard m files. It has been chosen to apply the definition (11) to more complex operators thanthe elementary operations. This idea has been applied to built in functions such as trigonometrics, basic linear algebra operators (det, eig, schur, \,...). So, our rfpa implementation mixed true random arithmetic and more global ones. There is practically no differences until the considered algorithms are (approximatively) backward stable. In the last case, perturbations are applied to the data before each of the N executions are run. Default random parameter values are $N = 3$, and $p = 95\%$. However these values can be changed dynamically.

6 Solving Sylvester Equations

Here are reported some numerical examples to illustrate the previously discussed two appoaches. Our Sylvester test equation is defined by the Matlab expression $A = invhilb(n); Z = zeros(n,n); J = ones(n,n); A = [A, Z; J, A]; nn = length(A); B = invhilb(m); Z = zeros(m,m); J = ones(m,m); B = [B, Z; J, B]; mm = length(B); X = ones(nn, mm); C = -(A * X + X * B)$. The size of the final A and B matrices are $nn = 2n$ and respectively $mm = 2m$, where m, n are parameters controlling the global difficulty to numerically solve these equations, because their condition number increases very quickly with m and n. It must be noticed that A, B and C are exact floating point numbers, so no perturbation is introduced to solve this problem. A first Matlab output (on a PC with Matlab 5.3) is obtained with $n = 2$ and $m = 3$: no optimistic estimation of significant digit; maximum pessimistic estimation 0.8 decimal place, structured condition number : 7.92e+002 ; mean number of significant digits : 12.8 from K_S (6), 13.7 from random arithmetic, and 14.1 truly. These three estimations are respectively called nK, nRf, and nTr. The first one is defined by $nK = -log_{10}(\varepsilon K_S)$, where ϵ is the machine precision. Now, the following table is a synthesis of some 8 other runs of increasing ill conditionned problems.

n m	K_S	nK	nRf	nTr	n m	K_S	nK	nRf	nTr
3 2	7.9152e+002	12.8	13.2	14.5	4 4	1.0746e+005	10.6	12	12.5
4 1	1.5247e+004	11.5	12.6	13.1	5 4	2.8724e+006	9.2	11.7	12.2
5 1	5.3501e+005	9.9	11.6	12.5	5 6	1.4135e+008	7.5	9.6	10
3 4	5.1993e+004	10.9	12.1	12.8	6 8	2.3443e+011	4.3	6	6.4

Clearly the definition of nK implies (implicitly) that the Bartels - Stewart algorithm is backward stable, which is not allways true. It is well known that there are pathological cases where εK_S must be replaced by something like $N\varepsilon K_S$ with N>>1. In any case, our comparison argument remains true a fortiori.

7 Conclusion

A new **generic** approach has been presented to estimate accuracy in computed problem solution. This technique offers a componant wise analysis and is basically the less pessimistic estimate and "never" optimistic more than one decimal place. For comparison purpose, only the global result (mean number of significant digits) has been reported here, although individual number of significant digits is obtained for each solution component X_{ij}. Evaluating precision via condition number computation has usually a complexity greater than the problem solving itself. Consequently, random arithmetic is basically cheaper and much less difficult than an approach via any sensitivity technique. Artificial perturbation, an old concept [4] must be considered as an alternative in most of control theory problems.

References

1. N. J. Higham, Perturbation theory and backward error for AX-XB=Cn BIT,33,124-136,1993.
2. M. Konstantinov, M. Stanislavova and P. PetKov, Perturbation bounds and characterisation of the solution of the associated algebraic Riccati equation, Lin. Alg. and Appl., vol 285, pp 7-31, 1998.

3. M. Pichat, J. Vignes, Ingénierie du contrôle de la précision des calculs sur ordinateurs, Technip,1993.
4. J. Vignes, R. Alt, An efficient stochastic method for roundoff arror analysis, In Accurate Scientific Computations, W. L. Miranker, R. A. Toupin, Eds, Springer Verlag, 1985.

Construction of Seminumerical Schemes: Application to the Artificial Satellite Problem

Roberto Barrio

GME, Depto. Matemática Aplicada, Edificio de Matemáticas
University of Zaragoza, E–50009 Zaragoza, Spain
rabarrio@posta.unizar.es

Abstract. In this paper we study the combination of averaging theories and the numerical integration of the averaged equations by means of Chebyshev series methods, that permits to obtain the numerical solution as a short Chebyshev series. The proposed scheme is applied to the artificial satellite problem.

1 Introduction

In the study of long term evolution of celestial bodies in Celestial Mechanics (like in very long time integration of the Solar System [10]) different averaging techniques are usually employed. Most of them are special algebraic and analytical techniques developed to facilitate the computation of averaged systems.

In this paper we present the construction of seminumerical schemes in the numerical integration of systems of differential equations by mixing averaging theories and a series method for the numerical integration of the averaged system. The approach that we follow employs the modified perturbation method proposed in [5], that uses the Lie series formalism in a way that permits to split the differential system in two parts: one that follows a Hamiltonian structure and another one that is non-Hamiltonian.

Afterwards, in the numerical integration of the averaged equations, we consider a family of symmetric integrators. In particular, we use Runge-Kutta collocation methods based on Chebyshev polynomials, that give a dense output in the form of a Chebyshev series, situation required if we are interested in obtaining an "analytical" expression of the solution.

In the last section, the method is applied to the important problem of the orbital analysis of Earth's artificial satellites subject to Hamiltonian (Earth potential) and non-Hamiltonian perturbations (the air-drag).

2 Application of Lie Transforms in Averaging Systems of Ordinary Differential Equations

The typical problem in averaging theory consist of solving the differential system

$$\dot{\boldsymbol{x}}(t) = \boldsymbol{f}(t, \boldsymbol{x}, \varepsilon) \equiv \varepsilon\, \boldsymbol{f}_1^{(0)}(t, \boldsymbol{x}) + \ldots + \varepsilon^k\, \boldsymbol{f}_k^{(0)}(t, \boldsymbol{x}) + \varepsilon^{k+1}\, \widehat{\boldsymbol{f}}(t, \boldsymbol{x}, \varepsilon), \quad \boldsymbol{x}(0, \varepsilon) = \boldsymbol{x}_0.$$

with f periodic in t. Let

$$\dot{y}(t) = f^*(y,\varepsilon) \equiv \varepsilon\, f_0^{(1)}(y) + \ldots + \varepsilon^k\, f_0^{(k)}(y)$$

be its truncated averaged system calculated by any perturbation method. For this system, there is given a general theorem about the validity of the averaging method [11,12] that establishes that, under several conditions (among them, $f(t,x,\varepsilon)$ smooth and periodic in t), there exist constants c, ε_0, T such that $\|x(t,\varepsilon) - y(t,\varepsilon)\| \leq c\varepsilon^k$ for $0 \leq \varepsilon \leq \varepsilon_0$, and $0 \leq t \leq T/\varepsilon$.

The final attempt of any averaging method is to find the near-identity transformation that gives us the averaged system, but most of the proposed methods (see [11,12]) only give explicitly the direct transformation, and not the direct and the inverse one. So, these theories do not give good initial averaged conditions.

A perturbation method that gives both, the direct and inverse transformations and the averaged system is the Lie-Deprit method [7] for Hamiltonian systems and its adaptations to general differential systems [9]. Here we use a modification given in [5] where the Hamiltonian and vectorial treatment of Lie transforms theory are combined for a differential equation system, for which part of the perturbing terms have Hamiltonian nature.

Theorem 1. [5] *Given a differential system*

$$\frac{dx}{dt} = f(x,\varepsilon) = \sum_{i \geq 0} \frac{\varepsilon^i}{i!} f_i^{(0)}(x) \tag{1}$$

such that $x = (q,p)$ *represents a set of canonical variables of coordinates* q *and momenta* p. *Moreover the functions* $f_i^{(0)}$, $i \geq 0$ *are decomposed in two different parts* $f_i^{(0)}(x) = f_{Hi}^{(0)}(x) + f_{NHi}^{(0)}(x)$, *such that* $f_{Hi}^{(0)}$ *come from a Hamiltonian* $\mathcal{H}_i^{(0)}$, *then Eq. (1) is transformed, through the generating function*

$$W(x,\varepsilon) = \sum_{i \geq 1} \frac{\varepsilon^i}{i!} \left(W_i^H(x) + W_i^{NH}(x) \right) \tag{2}$$

into another differential equation

$$\frac{dy}{dt} = f^*(y,\varepsilon) = \sum_{i \geq 0} \frac{\varepsilon^i}{i!} \left(f_{H0}^{(i)}(y) + f_{NH0}^{(i)}(y) \right), \tag{3}$$

such that $y = (q',p')$ *is also a set of canonical variables with coordinates* q *and momenta* p. *Now, the terms* $f_{H0}^{(i)}$, $i \geq 1$ *are obtained by calculating* $f_{Hj}^{(i)} = \mathcal{J} \cdot \mathrm{grad}_x \mathcal{H}_j^{(i)}$, *with* $\mathcal{J}$ *the symplectic matrix* $\mathcal{J} = \begin{pmatrix} 0 & \mathbb{I} \\ -\mathbb{I} & 0 \end{pmatrix}$, *where* $\mathbb{I}$ *is the identity matrix, and* $\mathcal{H}_j^{(i)}$ *by using the algorithm of Lie transforms for Hamiltonians*

$$\mathcal{H}_j^{(i)} = \mathcal{H}_{j+1}^{(i-1)} + \sum_{0 \leq k \leq j} \binom{j}{k} \{\mathcal{H}_{j-k}^{(i-1)}; V_{k+1}\}, \tag{4}$$

for $j \geq 0$, $i \geq 1$; $\{\cdot\,;\,\cdot\}$ the Poisson bracket and $\mathcal{V}(\boldsymbol{q},\boldsymbol{p},\varepsilon)=\sum_{i\geq 1}\frac{\varepsilon^i}{i!}\mathcal{V}_i$ ($\mathcal{V}$ a scalar generating function). Finally, the terms $\boldsymbol{f_{NH}}_0^{(i)}$, $i \geq 1$ are calculated with

$$\boldsymbol{f_{NH}}_j^{(i)} = \boldsymbol{f_{NH}}_{j+1}^{(i-1)} + \sum_{0 \leq k \leq j} \binom{j}{k}\left((\mathcal{L}_{k+1}^H + \mathcal{L}_{k+1}^{NH})\boldsymbol{f_{NH}}_{j-k}^{(i-1)} + \mathcal{L}_{k+1}^{NH}\boldsymbol{f_H}_{j-k}^{(i-1)}\right),$$

also for $j \geq 0$, $i \geq 1$. The Lie operators $\mathcal{L}^H$ and $\mathcal{L}^{NH}$ are defined by

$$\mathcal{L}_j^\circ \boldsymbol{s} = \frac{\partial \boldsymbol{s}}{\partial \boldsymbol{x}} \cdot \boldsymbol{W}_j^\circ - \frac{\partial \boldsymbol{W}_j^\circ}{\partial \boldsymbol{x}} \cdot \boldsymbol{s}, \qquad \text{with} \quad \circ = H \text{ or } NH. \tag{5}$$

Besides, now $\boldsymbol{W}^H = \mathcal{J} \cdot \mathrm{grad}_{\boldsymbol{x}} \mathcal{V}$ is the generating function built from $\mathcal{V}$.

It is important to remark that, in general, the generating function is unknown and it has to be obtained order by order by solving a linear system of first order partial differential equations, the "homological equation" [13]:

$$\left[\boldsymbol{f}_0^{(0)}\,;\,\boldsymbol{W}_i\right] = \boldsymbol{f}_0^{(i)} - \boldsymbol{F}\left(\boldsymbol{f}_0^{(0)},\ldots,\boldsymbol{f}_i^{(0)},\boldsymbol{W}_1,\ldots,\boldsymbol{W}_{i-1}\right),$$

where $\boldsymbol{F}$ is a function of the previous orders, $\boldsymbol{f}_0^{(i)}$ is taken as the average of $\boldsymbol{F}$ and $[\,\cdot\,;\,\cdot\,]$ stands for the Poisson bracket or the Lie operators, depending on the nature of the terms. The solvability of the homological equation can be assured if we assume that $\boldsymbol{f}_0^{(0)}$ possesses several properties [13]. In our case, a suitable election of the canonical set of variables permits to reduce the solution of the homological equation to quadratures.

Once we have found the generating function $\boldsymbol{W}$, it is possible to obtain the direct and inverse transformations of the variables:

Proposition 1. [9] *The direct transformation* $\boldsymbol{x} = \sum \frac{\varepsilon^i}{i!} \boldsymbol{y}_0^{(i)}(\boldsymbol{y})$ *is given by*

$$\boldsymbol{y}_j^{(i)} = \boldsymbol{y}_{j+1}^{(i-1)} + \sum_{0 \leq k \leq j}\binom{j}{k}\mathcal{L}_{k+1}\boldsymbol{y}_{j-k}^{(i-1)}(\boldsymbol{y}), \quad \text{where} \quad \mathcal{L}_j \boldsymbol{s} = \frac{\partial \boldsymbol{s}}{\partial \boldsymbol{y}} \cdot \boldsymbol{W}_j$$

with $\boldsymbol{y}_0^{(0)} = \boldsymbol{y}$, $\boldsymbol{y}_j^{(0)} = 0$ $(j > 0)$, *and the inverse transformation* $\boldsymbol{y} = \sum \frac{\varepsilon^j}{j!}\boldsymbol{x}_j^{(0)}(\boldsymbol{x})$

$$\boldsymbol{x}_j^{(i)} = \boldsymbol{x}_{j-1}^{(i+1)} - \sum_{0 \leq k \leq j-1}\binom{j-1}{k}\mathcal{L}_{k+1}\boldsymbol{x}_{j-k-1}^{(i)}(\boldsymbol{x}), \quad \text{where} \quad \mathcal{L}_j \boldsymbol{s} = \frac{\partial \boldsymbol{s}}{\partial \boldsymbol{x}} \cdot \boldsymbol{W}_j$$

with $\boldsymbol{x}_0^{(0)} = \boldsymbol{x}$, $\boldsymbol{x}_0^{(i)} = 0$ $(i > 0)$.

An important property of the Lie-Deprit method is that applied to Hamiltonian systems it generates a canonical transformation [7]. Therefore, as a consequence, we have that the modified method (Theorem 1) also generates a canonical transformation applied only to the Hamiltonian part, and thus, the composition of the perturbation theory and a symplectic numerical integration scheme will generate a symplectic seminumerical theory. The problem is that usually the systems suitable for averaging (periodic standard form or angular standard form) are not separable Hamiltonians and, then, the symplectic integrators are implicit. Besides, we are interested in the integration of differential systems with Hamiltonian and non-Hamiltonian perturbations. Thus, in this paper we will not consider symplectic integrators; instead, we use a particular family of symmetric methods that also have interesting qualitative properties.

3   Collocation Method

In this section we formulate a collocation method for the solution of the averaged system

$$\dot{\boldsymbol{y}}(t) = \boldsymbol{f}^*(\boldsymbol{y}(t)), \qquad \boldsymbol{y}(t_0) = \boldsymbol{y}_0 \equiv \sum_{j\geq 0} \frac{\varepsilon^j}{j!} \boldsymbol{x}_j^{(0)}(\boldsymbol{x}_0). \qquad (6)$$

The formulation here presented consists of calculating, on each integration step, an approximation of the solution by means of the interpolation polynomial at the extrema of a Chebyshev polynomial of the first kind. Thus, the solution is given by means of the coefficients of this collocation polynomial.

This formulation [4] follows the idea, used by Clenshaw, of approximating the second member of the differential equation on each integration step $[t_k, t_{k+1}]$ at the initial conditions $\boldsymbol{y}_k$ by means of a finite series of Chebyshev polynomials of the first kind $\{T_i(u)\}$, that is to say,

$$\boldsymbol{f}^*(\boldsymbol{y}(t)) = \sum_{k=0}^{n-1} {}' \boldsymbol{c}_k \, T_k(u), \qquad \text{with} \quad -1 \leq u \leq 1, \qquad (7)$$

where u is given by the map $u = ((t - t_k) - (t_{k+1} - t))/(t_{k+1} - t_k)$, in order to use the standard interval $[-1, 1]$. The prime in the sum symbol means that the first term in the series must be halved.

An approximation of the Fourier–Chebyshev coefficients c_i are obtained by means of numerical calculation of the quadratures. In our case the coefficients are computed with the Gauss–Lobatto formula

$$\boldsymbol{c}_k = \frac{\zeta_k}{n-1} \sum_{i=0}^{n-1} {}'' T_k(\eta_i^{(n-1)}) \, \boldsymbol{f}^*(\boldsymbol{y}(\eta_i^{(n-1)})), \qquad \zeta_k = \begin{cases} 1, & k = n-1 \\ 2, & 0 \leq k \leq n-2 \end{cases} \qquad (8)$$

where $\eta_i^{(n-1)} = \cos(i\pi/(n-1))$ are the extrema of $T_{n-1}(u)$ and the double prime means that the first and last terms must be halved.

Once the second member of the differential system is approximated, we integrate the series to obtain an approximation of the solution

$$\boldsymbol{y}_0 + \frac{t_{i+1} - t_i}{2} \int_{-1}^{x} \sum_{k=0}^{n-1} {}' \boldsymbol{c}_k \, T_k(v) \, dv = \sum_{k=0}^{n} {}' \boldsymbol{a}_k \, T_k(u), \qquad (9)$$

where the coefficients a_k are obtained by using the recursive formulas for the integration of the Chebyshev polynomials

$$a_n = \frac{t_{i+1} - t_i}{2} \frac{c_{n-1}}{2n}, \qquad a_{n-1} = \frac{t_{i+1} - t_i}{2} \frac{c_{n-2}}{2(n-1)},$$

$$a_r = \frac{t_{i+1} - t_i}{2} \frac{1}{2r} (c_{r-1} - c_{r+1}), \qquad \text{for } 1 \leq r \leq n-2. \tag{10}$$

The first coefficient a_0 is calculated by using the initial conditions y_0 of the problem on the integration step $[t_k, t_{k+1}]$ through

$$\tfrac{1}{2} a_0 = a_1 - a_2 + a_3 - \ldots + (-1)^{n-1} a_n + y_0. \tag{11}$$

Note that in Eq. (8), the values of the solution $y(t)$ are required. However, the function $y(t)$ is unknown, hence, the method is implicit. Therefore, an iterative method is needed, as well as a good initial estimation of the solution to begin with. Besides, since the collocation methods are based on approximations of the right hand member of Eq. (6), the lesser variations of it, the better would be the convergence. In our problem, as the differential system is the averaged one, the variations of the second member of the differential system are very small and, therefore, a very low number of iterations (1 or 2) is needed. Besides, we can take very big stepsizes with a low number of terms in the series.

These methods (ChRK) have several properties, among them, it is interesting to remark that they are Runge-Kutta collocation methods, are A-stable [3], generate P-stable indirect Runge-Kutta-Nyström collocation methods for special second-order initial-value problems and exhibit linear growth in time of the global error for time-reversible systems due to their symmetric structure. Other interesting features are that they can be easily formulated using variable stepsizes and in a matrix form suitable for parallel implementation.

4 Seminumerical Integration Scheme

The combination of the analytical theories (to obtain the averaged system and the averaged initial conditions and to recover the osculating elements from the averaged ones) and the RK collocation method with a Chebyshev series as output (ChRK), will give us a seminumerical method that computes in a fast way (for low precision) an "analytical" solution of the differential system.

SEMINUMERICAL INTEGRATION SCHEME

Step 1: Determination of the averaged system: $f \longrightarrow f^*$ (Theorem 1).
Step 2: Numerical integration
    2-i: Averaged initial conditions: $x_0 \longrightarrow y_0$ (Proposition 1)
    2-ii: Numerical integration (ChRK): $y(t) \simeq \sum_{i=0}^{n}{'} a_i\, T_i(u(t))$.
Step 3: Recovering of osculating elements: $y(t) \longrightarrow x(t)$ (Proposition 1).

5 Application to the Artificial Satellite Problem

The scheme presented in this paper has been applied to the artificial satellite problem, modeled as the two body equations perturbed by the J_2 term of the Earth potential and by the air-drag (only for low satellites). The analytical transformations are based on Theorem 1 and they have been taken from [2].

One of the first things to do is the selection of the canonical set of variables. Here, as in [6], we use the Delaunay variables $\{\ell, g, h, L, G, H\}$. As a consequence, the homological equation (for the vector components) that must be solved by computing some quadratures with respect to the mean anomaly ℓ, is

$$\begin{cases} n\dfrac{\partial (\boldsymbol{W}_j)_i}{\partial \ell} + \left(\boldsymbol{f}_0^{(j)}\right)_i = \left(\widetilde{\boldsymbol{f}}_0^{(j)}\right)_i & \text{for } 2 \leq i \leq 6, \\ 3a^{-2}(\boldsymbol{W}_j)_4 + n\dfrac{\partial (\boldsymbol{W}_j)_1}{\partial \ell} + \left(\boldsymbol{f}_0^{(j)}\right)_1 = \left(\widetilde{\boldsymbol{f}}_0^{(j)}\right)_1, \end{cases} \quad (12)$$

where $n = \sqrt{\mu/a^3}$, a is the semimajor axis of the orbit, $\widetilde{\boldsymbol{f}}_0^{(j)}$ is obtained in the precedent steps and $\boldsymbol{f}_0^{(j)}$ is chosen according to the simplification (the averaging), that is, removing the mean anomaly (fast angle variable):

$$\boldsymbol{f}_0^{(j)} = \left\langle \widetilde{\boldsymbol{f}}_0^{(j)} \right\rangle = \frac{1}{2\pi} \int_0^{2\pi} \widetilde{\boldsymbol{f}}_0^{(j)}(\ell, g, h, L, G, H)\, d\ell.$$

5.1 Numerical Tests

In the numerical tests we have applied the seminumerical scheme to a low altitude satellite that we call Low and a geostationary type satellite that we call Geo. The analytical theory used in the simulations has 145 terms in the averaged equations up to second order in the small parameter $\varepsilon \simeq J_2$, and the generator 936 terms. The direct and inverse transformations have 1123 terms. Let us note

Table 1. Integration time (T), number of revolutions (NR), number of steps (NS), number of function evaluations (NF) and CPU time in seconds using the ChRK and DOP853 [8] in the seminumerical integration of the Low and Geo orbits and in the numerical integration of the non-averaged equations (DOP853*)

	T	NR	DOP853 NS/NF	DOP853 CPU	ChRK NS/NF	ChRK CPU	DOP853* NS/NF	DOP853* CPU
Low	30 days	467	12/182	0.78	5/112	0.02	6,698/94,188	7.69
Geo	1 year	365	12/182	1.60	5/98	0.02	1,510/21,028	2.24
Geo	100 years	36,500	17/257	1.74	8/140	0.09	152,691/2,118,395	106.98

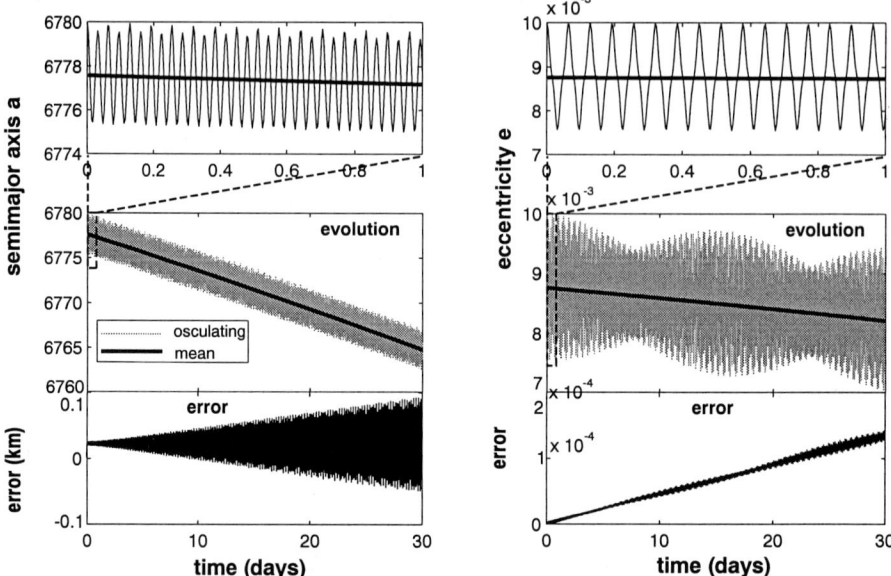

Fig. 1. Figures on the top: evolution of the osculating and mean values for the semimajor axis a (in Km.) and the eccentricity e for 30 days. Figures on the bottom: error in the semimajor axis a (in Km.) and in the eccentricity e for 30 days in a seminumerical integration of the Low orbit

that for a complete analytical theory of an artificial satellite is usual to have thousands or, even, millions of terms [1].

In Figure 1 we show the evolution, in the osculating and averaged elements, of the semimajor axis and the eccentricity of the Low orbit calculated with the seminumerical scheme. In the Figure 2 we show the error depending on the initial conditions, that is, if we use as the initial conditions for the averaged system the osculating ones (as several averaging methods do) or the transformed mean initial conditions. From the figures it is clear the necessity of the transformation.

Finally, in Table 1 we show the number of steps and the CPU time on a PC PII-333Mhz using the ChRK and a standard RK (DOP853 [8]). For the ChRK we have taken $n = 7$ and in all steps it was needed, in average, only two iterations to reach the tolerance level. Also, for comparison, we present the CPU time for the DOP853 with the non-averaged equations. All the tests have been done for a relative error of 10^{-5} in the variable stepsize implementations (we remark that we have not used reversible stepsize strategies, but due to the symmetric nature of the ChRK it will be desirable). From the table it is clear the difference between using averaged and non-averaged equations. Besides, we remark the good performance of the implicit ChRK compared with the explicit DOP853, due to the smoothness of the averaged equations and because for the

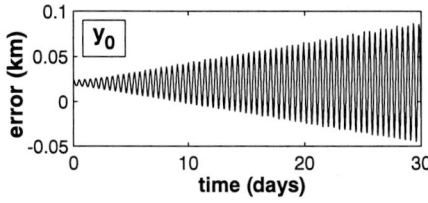 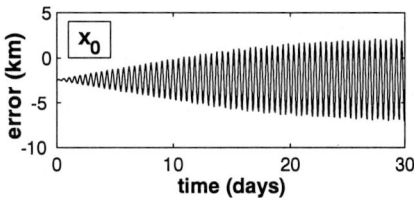

Fig. 2. Errors in the semimajor axis (Km.) in a seminumerical integration scheme by using osculating (x_0) and mean elements (y_0) as initial conditions for the Low orbit

ChRK the **Step 3** (recovering of osculating elements) it is done after integration and only at the required points by means of the dense polynomial solution.

As conclusions, we remark that the combined use of analytical theories and collocation integrators permits to obtain very fast integrators (seminumerical integration schemes) that also give us an "analytical" solution as a polynomial.

Acknowledgements

The author is supported partially by the Spanish Ministry of Education and Science (Project #ESP99-1074-CO2-01) and by the Centre National d'Études Spatiales at Toulouse (France).

References

1. Abad, A., Elipe, A, Palacián, J. and San Juán, J. F.: ATESAT: A symbolic processor for artificial satellite theory, Math. Comp. Simul., **45** (1998), 497–510.
2. Barrio, R.: Lie Transforms for the Noncanonical Case: Application to an Artificial Satellite Subject to Air Drag and Earth Potential, Center National d'Etudes Spatiales, Toulouse, CT/TI/MS/MN/93–154, (1993).
3. Barrio, R.: On the A-stability of RK collocation methods based on ultraspherical polynomials, SIAM J. Numer. Anal., **36** (1999), 1291–1303.
4. Barrio, R, Elipe, A and Palacios, M.: Chebyshev collocation methods for fast orbit determination, Applied Mathematics and Computation, **99** (1999), 195–207.
5. Barrio, R. and Palacián, J.: Lie Transforms for Ordinary Differential Equations: Taking Advantage of the Hamiltonian Form of Terms of the Perturbation, Int. J. Numer. Meth. Engng., **40** (1997), 2289–2300.
6. Barrio, R. and Palacián, J.: Semianalytical methods for high eccentric orbits: zonal harmonics and air drag terms, Adv. Astr. Scien., **95** (1997), 331–339.
7. Deprit, A.: Canonical Transformations Depending on a Small Parameter, Celes. Mech., **1** (1969), 12–30.
8. Hairer, E., Nørsett, S. P. and Wanner, G.: Solving Ordinary Differential Equations I, Ed. Springer–Verlag (2nd. edition), New York (1993).
9. Kamel, A. A. A.: Perturbation Method in the Theory of Nonlinear Oscillations, Celes. Mech., **3** (1970), 90–106.

10. Laskar, J.: Secular evolution of the Solar System over 10 million years, Astron. Astrophys., **198** (1988), 341–362.
11. Murdock, J. A.: Perturbations, Theory and Methods, Classics in Applied Mathematics 27, Ed. SIAM (2nd. edition), Philadelphia (1999).
12. Perko, L. M.: Higher Order Averaging and Related Methods for Perturbed Periodic and Quasi-Periodic Systems, SIAM J. Appl. Math., **17** (1968), 698–724.
13. Reich, S.: Dynamical Systems, Numerical Integration, and Exponentially Small Estimates, Habilitationsschrift, Freie Universitat Berlin (1998).

Stability Analysis of Parallel Evaluation of Finite Series of Orthogonal Polynomials[*]

Roberto Barrio[1] and Plamen Yalamov[2]

[1] GME, Depto. Matemática Aplicada, Edificio de Matemáticas
University of Zaragoza, E–50009 Zaragoza, Spain
rabarrio@posta.unizar.es

[2] Center of Applied Mathematics and Informatics
University of Rousse, 7017 Rousse, Bulgaria
yalamov@ami.ru.acad.bg

Abstract. In this paper we study the rounding errors in the parallel evaluation of a finite series of a general family of orthogonal polynomials. Both, the theoretical bounds and the numerical tests present an almost similar behavior between the sequential and the parallel algorithms.

1 Introduction

The evaluation of polynomials is one of the most common problems in scientific computing and, with the development of parallel computers, it is interesting to design parallel algorithms to evaluate polynomials. Recently general algorithms for the parallel evaluation of polynomials written as a finite series of general orthogonal polynomials have been proposed in [1,3].

Usually the parallel algorithms are more unstable than the sequential algorithms for the same problem, but, for particular triangular systems the parallel algorithms possess stability properties similar to those of the Gaussian elimination [7]. Thus, an important task is the stability analysis of the new parallel algorithms.

In this paper we analyse the stability of the parallel algorithms for the evaluation of polynomials given in [1,3]. The analysis shows that the parallel algorithms are almost as stable as their sequential counterparts for practical applications. Extensive numerical experiments applied to Jacobi polynomials series confirm the theoretical conclusions.

The algorithms that we study evaluate finite series $p_n(x) = \sum_{i=0}^{n} c_i \, \Phi_i(x)$ of a family of orthogonal polynomials $\{\Phi_n(x)\}$ which satisfy the triple recurrence relation

$$\begin{aligned}&\Phi_0(x) = 1, \qquad \Phi_1(x) = \alpha_1(x),\\ &\Phi_k(x) - \alpha_k(x)\,\Phi_{k-1}(x) - \beta_k\,\Phi_{k-2}(x) = 0, \quad k \geq 2,\end{aligned} \qquad (1)$$

with $\alpha_k(x)$ a linear polynomial of x.

---

[*] The first author is supported partially by the Spanish Ministry of Education and Science (Project #ESP99-1074-CO2-01) and by the Centre National d'Études Spatiales at Toulouse (France). The second author is supported partially by Grants MM-707/97 and I-702/97 from the Bulgarian Ministry of Education and Science.

Let us assume the standard model of roundoff arithmetic with a guard digit [5]:
$$\mathrm{fl}(x * y) = x * y(1 + \sigma), \quad |\sigma| \le \rho_0, \quad * \in \{+, -, \times, /\},$$
where ρ_0 is the machine precision. Also we denote $\gamma_n := n\rho_0/(1 - n\rho_0) = n\rho_0 + \mathcal{O}(\rho_0^2)$. By tilde we denote computed results in the following.

2 Parallel Algorithms

In [1] there were presented four parallel algorithms to evaluate a series of a general family of orthogonal polynomials. Two algorithms, PC and PF, are based on parallel methods applied to the matrix formulation of the sequential algorithms of Clenshaw and Forsythe. The other two algorithms, MPC and MPF, are based on a matrix product formulation of the sequential recurrences of the sequential algorithms.

Let us briefly describe the PC algorithm in a form suitable for the stability analysis [4]. It is based on the sequential Clenshaw algorithm that can be written as the solution of a tridiagonal triangular linear system $Sq = c$ where S is given below and c is the vector of coefficients of the polynomial series. In the following we suppose that $n = kp$, with p the number of processors.

For the purpose of the stability analysis of the parallel algorithms the entries of S are rearranged in order to make the analysis easier. Before the permutation matrix S is as follows:

$$S = \begin{pmatrix} A_1 & B_1 & & & & \\ & C_1 & D_1 & & & \\ & & A_2 & B_2 & & \\ & & & C_2 & D_2 & \\ & & & & \ddots & \ddots \\ & & & & & A_p & B_p \\ & & & & & & C_p \end{pmatrix},$$

where $A_i \in \mathbb{R}^{2 \times 2}$ and $B_i \in \mathbb{R}^{2 \times (k-2)}$ have the following structure,

$$A_i = \begin{pmatrix} 1 & -\alpha_{(i-1)k+2} \\ 0 & 1 \end{pmatrix}, \quad B_i = \begin{pmatrix} -\beta_{(i-1)k+3} & 0 & 0 \ldots 0 \\ -\alpha_{(i-1)k+3} & -\beta_{(i-1)k+4} & 0 \ldots 0 \end{pmatrix},$$

$C_i \in \mathbb{R}^{(k-2) \times (k-2)}$ is upper triangular tridiagonal as the original matrix S, and $D_i \in \mathbb{R}^{(k-2) \times 2}$ looks as follows:

$$C_i = \begin{pmatrix} 1 & -\alpha_{(i-1)k+3} & -\beta_{(i-1)k+4} & & \\ & 1 & -\alpha_{(i-1)k+4} & \ddots & \\ & & \ddots & \ddots & -\beta_{ik} \\ & & & 1 & -\alpha_{ik} \\ & & & & 1 \end{pmatrix}, \quad D_i = \begin{pmatrix} 0 & 0 \\ \vdots & \vdots \\ 0 & 0 \\ -\beta_{ik+1} & 0 \\ -\alpha_{ik+1} & -\beta_{ik+2} \end{pmatrix}.$$

The coefficients $\{\alpha_i\}$ and $\{\beta_i\}$ are the coefficients of the triple recurrence relation (1) that defines the particular family of orthogonal polynomials that we use.

In the parallel algorithm we permute the rows and columns of S in such a way that the permuted matrix is as follows:

$$S = \begin{pmatrix} C & D \\ B & A \end{pmatrix},$$

where $C = \mathrm{diag}\{C_1, \ldots, C_p\}$, $A = \mathrm{diag}\{A_1, \ldots, A_p\}$, $B = \mathrm{diag}\{B_1, \ldots, B_p\}$, and

$$D = \begin{pmatrix} 0 & D_1 & & \\ & \ddots & \ddots & \\ & & \ddots & D_{p-1} \\ & & & 0 \end{pmatrix}.$$

By using the introduced block structure the PC algorithm can be given in the following way:

Step 1: Compute in parallel: $S = LU$, where

$$L = \begin{pmatrix} C & 0 \\ B & \mathbb{I}_{2p} \end{pmatrix}, \quad U = \begin{pmatrix} \mathbb{I}_{n-2p} & C^{-1}D \\ 0 & A - BC^{-1}D \end{pmatrix}.$$

Step 2: Solve $Ly = c$.
Step 3: Solve $Uq = y$.
Step 4: $\sum_{r=0}^{n} c_r \Phi_r(x) = \beta_2 \, q_{p(k-2)+2} + q_{p(k-2)+1} \, \Phi_1(x) + c_0$.

In [3] two other parallel algorithms are proposed (ChPC and ChPF) which are suitable for the parallel evaluation of Chebyshev series. These algorithms are much more efficient than the general parallel algorithms proposed above because they are especially designed for the evaluation of a polynomial series of a particular family of orthogonal polynomials, i.e., Chebyshev polynomials. As above the ChPC algorithm can be formulated by using a block matrix notation. Let us have $T \in \mathbb{R}^{(n+1)\times(n+1)}$

$$T = \mathrm{diag}\{\underbrace{T_p, T_p, \ldots, T_p}_{p \text{ times}}\}, \qquad (2)$$

with $T_p \in \mathbb{R}^{k \times k}$

$$T_p = \begin{pmatrix} 1 & -2T_p(x) & 1 & & & \\ & \ddots & \ddots & \ddots & & \\ & & 1 & -2T_p(x) & 1 & \\ & & & 1 & -2T_p(x) \\ & & & & 1 \end{pmatrix}, \qquad (3)$$

Also we define the vectors $e_{p+1}, \Phi_{0:p} \in \mathbb{R}^{p+1}$ and $q, c \in \mathbb{R}^{n+1}$ given by $e_{p+1} = (0, \ldots, 0, 1)^\top$, $T_{0:p} = (T_0(x), \ldots, T_p(x))^\top$ (the values of the Chebyshev polynomials) and $c = (c_k^0, c_k^1, \ldots, c_k^{p-1})^\top$ with $c_k^i = (c_i, c_{i+p}, c_{i+2p}, \ldots, c_{i+(k-1)p})$ (the vector of polynomial coefficients). Besides, for the initialization process we need the matrix $T_1 \in \mathbb{R}^{(p+1) \times (p+1)}$

$$T_1 = \begin{pmatrix} 1 & -2x & 1 & & & \\ & \ddots & \ddots & \ddots & & \\ & & 1 & -2x & 1 & \\ & & & 1 & -x & \\ & & & & 1 & \end{pmatrix}, \qquad (4)$$

Thus the ChPC algorithm can be rewritten as:

Step 1: Solve $T_1 \Phi_{0:p} = e_{p+1}$.
Step 2: Solve in parallel $T q = c$.
Step 3: $\sum_{r=0}^{n} c_r T_r(x) = \sum_{i=0}^{p-1} (q_{ik+1} T_i(x) - q_{ik+2} T_{p-i}(x))$.

---

3 Rounding Error Bounds

In [4] the accumulation of rounding error for the PC and PF algorithms is studied, and in [2] the ChPC and ChPF algorithms are analyzed in a similar way. Below we present some results that state relative forward error bounds for some of these algorithms. It is interesting to remark that in the following results we use the Skeel's componentwise condition number: $\text{cond}(M) = \| |M^{-1}| |M| \|_\infty$ instead of $\mu(M) = \|M^{-1}\|_\infty \|M\|_\infty$.

Theorem 1. [4] *The relative normwise forward error in the solution q of system $Sq = c$ obtained by the PC algorithm is bounded as follows:*

$$\frac{\|\delta q\|_\infty}{\|\tilde{q}\|_\infty} \leq \gamma_{18} \, \text{cond}(S) \, \frac{\|C^{-1} D\|_\infty}{1 - \gamma_2 \, \text{cond}(C)}.$$

This theorem is applied to the particular family of Gegenbauer polynomials [6]. It is found in [4] that the PC algorithm is almost insensible to the number of processor inside the interval $(-1, 1)$, while the rounding error of the parallel PC (near $x \pm 1$) and PF ($\forall x \in [-1, 1]$) algorithms decreases when p grows.

Theorem 2. [2] *The relative normwise forward error in the solution q of system $Tq = c$ obtained by the ChPC algorithm is bounded as follows:*

$$\frac{\|\delta q\|_\infty}{\|\tilde{q}\|_\infty} \leq \rho_0 \cdot \min\left\{ 2 + 4(p+1)(p+2), \, 2 + \frac{8(p+1)}{\sqrt{1-x^2}} \right\} \text{cond}(T_p) + \mathcal{O}(\rho_0^2).$$

A detailed analysis [2] of this result tells us that the rounding errors are almost similar in sequential and in parallel but for a special set of points $\{\cos(i\,\pi/p) \mid i = 1, \ldots, p-1\}$ the rounding error grows in the parallel algorithm.

3.1 Numerical Tests

In the theoretical analysis the MPC and MPF algorithms are not studied. Besides, the PC and PF algorithms are only studied in detail [4] for the Gegenbauer polynomials. Therefore, the goals of the present paper is to compare the behaviour between a Gegenbauer family and a Jacobi family of orthogonal polynomials and to study the MPC and MPF algorithms.

We have tested the PC, PF, MPC and MPF algorithms in order to analyze the effects of rounding errors. In the simulations we have studied the algorithms with Jacobi polynomial series ($p_n(x) = \sum_{i=0}^{n} c_i P_i^{(\alpha,\beta)}$) of degree $n = 4096$. For each type of series we have used two sets of coefficients: set **S1** of monotonically decreasing coefficients ($c_i = 1/(i+1)^2$) and set **S2** of random coefficients normally distributed with mean 0 and variance 1. All the tests are done on a workstation SUN ULTRASPARC 1 and the programs have been written in FORTRAN 77 in double precision with unit roundoff $\rho_0 \simeq 2.2 \times 10^{-16}$.

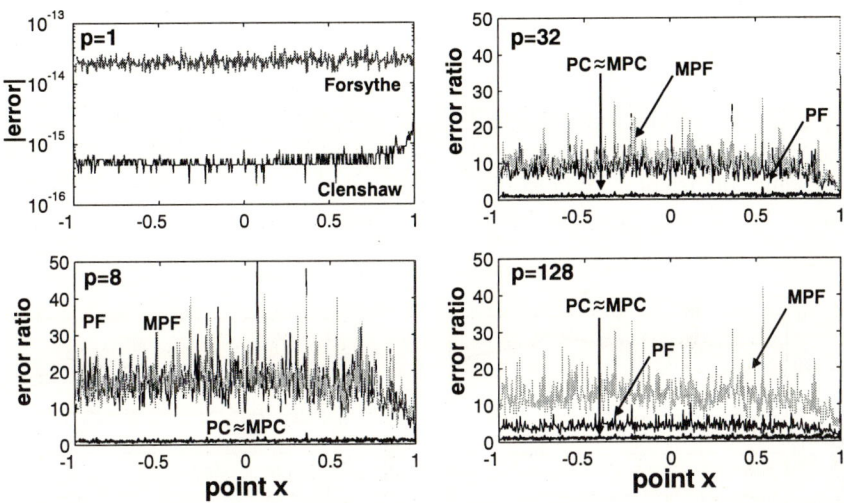

Fig. 1. Error in the evaluation of Legendre series (set **S1**) on one processor and error ratio between the rounding error on p processors and on one processor

In Figure 1 we show the rounding error ratio between parallel and sequential algorithms in the evaluation of a Legendre series. We can see that when the number of processors grows the performance of all the algorithms is similar. Only for the MPF algorithm the rounding errors decrease much slower than the PF one. In Figures 2, 3, 4 and 5 we compare the performance for two Jacobi series. In the examples with set **S1** the behavior is essentially the same as for Legendre series (that are members of the Gegenbauer family), while for the set **S2** the

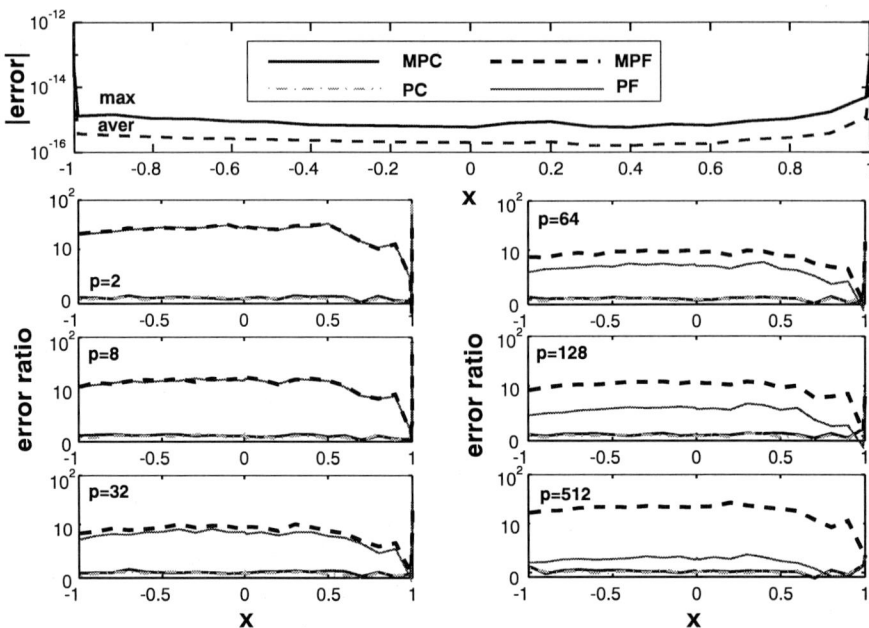

Fig. 2. Figure on the top: relative average rounding error in the evaluation of a series of Jacobi polynomials ($\alpha = \sqrt{2}/10+1$, $\beta = \sqrt{2}/10+1$) with monotonically decreasing coefficients. The rest of the figures show the ratio between the average rounding error on p processors and on one processor

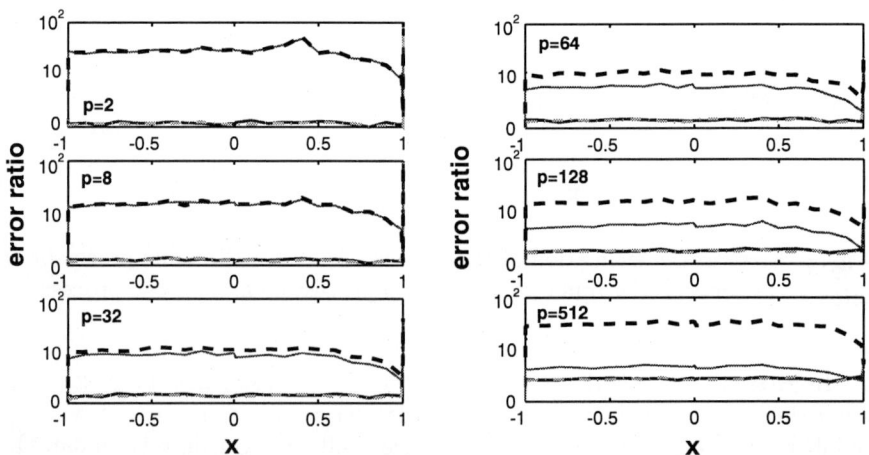

Fig. 3. Error ratio between the average rounding error on p processors and on one processor in the evaluation of a series of Jacobi polynomials ($\alpha = \sqrt{2}/10$, $\beta = \sqrt{2}/10$) with monotonically decreasing coefficients

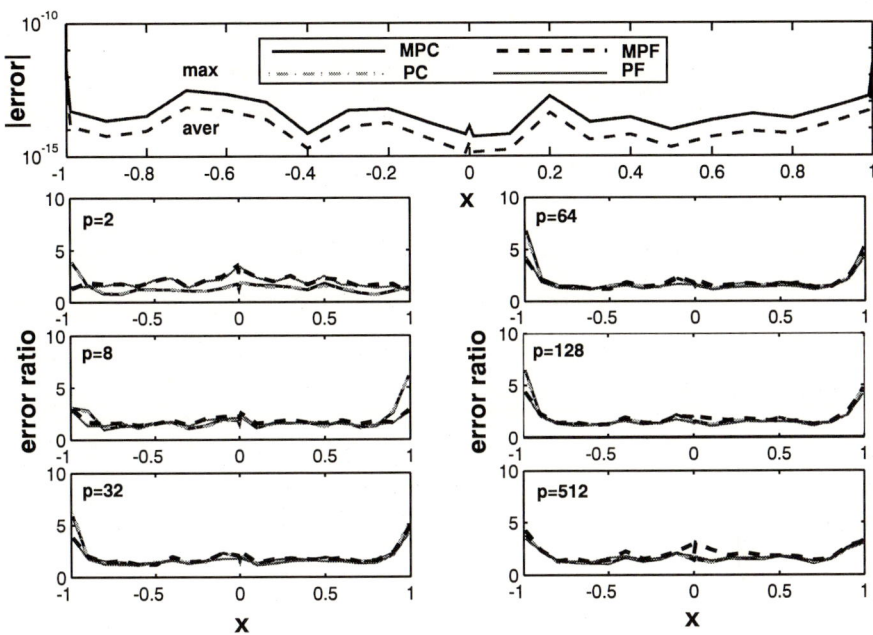

Fig. 4. Figure on the top: relative average rounding error in the evaluation of a series of Jacobi polynomials ($\alpha = \sqrt{2}/10 + 1$, $\beta = \sqrt{2}/10 + 1$) with random coefficients. The rest of the figures show the ratio between the average rounding error on p processors and on one processor

performance of any algorithm is similar. Besides, the growth of the parameters (α, β) that define the family of Jacobi polynomials seems not to influence the performance.

Finally, in Figure 6 we show the evolution of the rounding error ratio depending on the number of processors p in the evaluation of a Legendre series.

From the figures we conclude that from the numerical tests the behaviour detected for the Gegenbauer polynomials in [4] can be extended to the Jacobi polynomial series and that the behaviour of the PC and MPC algorithms is similar. A more detail analysis is needed for the MPF algorithm.

References

1. Barrio, R.: Parallel algorithms to evaluate orthogonal polynomial series, to appear on SIAM J. Sci. Comput.
2. Barrio, R.: Stability of parallel algorithms to evaluate Chebyshev series, submitted to Comput. Math. Appl.
3. Barrio, R. and Sabadell, J.: Parallel evaluation of Chebyshev and Trigonometric series, Comput. Math. Appl., **38** (1999), 99–106.

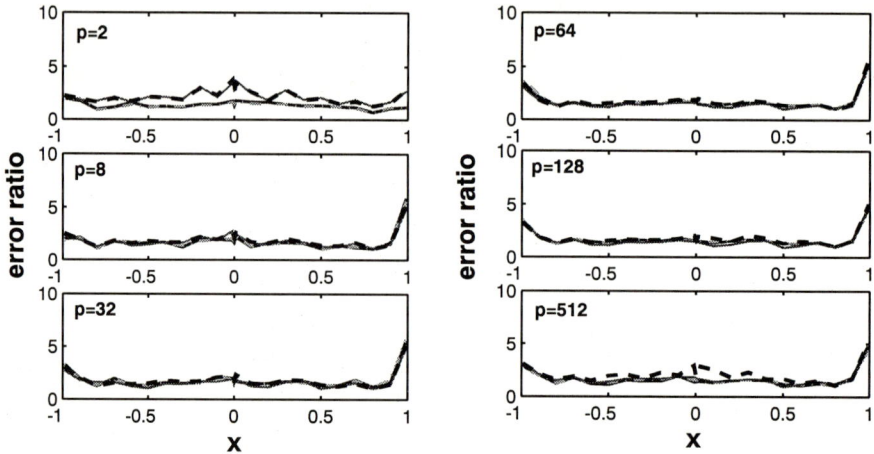

Fig. 5. Error ratio between the average rounding error on p processors and on one processor in the evaluation of a series of Jacobi polynomials ($\alpha = \sqrt{2}/10$, $\beta = \sqrt{2}/10$) with random coefficients

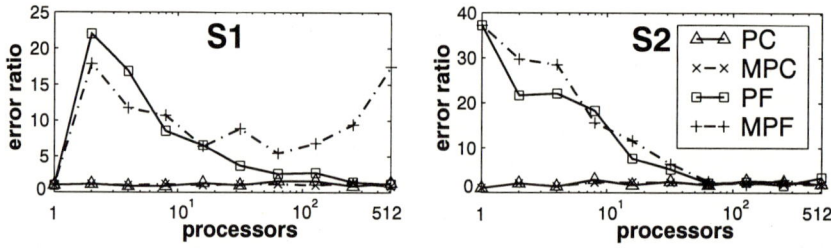

Fig. 6. Error ratio between the rounding error in parallel and in sequential in the evaluation of Legendre series at the point $x = 0$. with the sets **S1** and **S2**

4. Barrio, R. and Yalamov, P. Y.: Stability of parallel algorithms for polynomial evaluation, submitted to SIAM J. Sci. Comput.
5. Higham, N. J.: Accuracy and stability of numerical algorithms, SIAM, Philadelphia, (1996).
6. Magnus, W., Oberhettinger, F. and Soni, R. P.: Formulas and Theorems for the Special Functions of Mathematical Physics, Springer-Verlag, (1966).
7. Yalamov, P. Y.: Stability of a Partitioning Algorithm for Bidiagonal Systems, Parallel Computing., **23** (1997), 333–348.

On Solving Large-Scale Weighted Least Squares Problems

Venansius Baryamureeba

Department of Informatics, University of Bergen
5020 Bergen, Norway

Abstract. A sequence of least squares problems of the form $\min_y \|G^{1/2}(A^T y - h)\|_2$ where G is an $n \times n$ positive definite diagonal weight matrix, and A an $m \times n$ ($m < n$) sparse matrix with some dense columns; has many applications in linear programming, electrical networks, elliptic boundary value problems, and structural analysis. We discuss a technique for forming low-rank correction preconditioners for such problems. Finally we give numerical results to illustrate this technique.

Keywords: Least squares, Conjugate gradients, Preconditioner, Dense columns.

1 Introduction

Consider a sequence of weighted least squares problems of the form

$$\min_y \|A^T y - h\|_G, \qquad (1)$$

where $y \in \Re^m, h \in \Re^n, G \in \Re^{n \times n}$ is a positive definite diagonal weight matrix, and $A \in \Re^{m \times n} (m < n)$ is a sparse matrix with some dense columns. The weight matrix G and the vector h vary from one computation step (in interior point algorithms, a computation step is equivalent to interior point iteration) to another while A is kept constant. Throughout this paper, we assume A to have full rank m. Define $\|(.)\|_G = \|G^{1/2}(.)\|_2$. Then the solution of (1) is given by the normal equations

$$AGA^T y = v, \qquad (2)$$

where $v = AGh$. Let $A = [A_1, A_2]$ be the matrix whose columns have been reordered into two block matrices A_1 and A_2. After reordering of A, let $\mathcal{J}_1 = \{1, 2, \ldots, r\}$ and $\mathcal{J}_2 = \{r+1, r+2, \ldots, n\}$ be column indices of A corresponding to A_1 and A_2 respectively. Let G be partitioned such that G_1 and G_2 are block (square) submatrices corresponding to $\mathcal{J}_1$ and $\mathcal{J}_2$ respectively. Then

$$AGA^T = A_1 G_1 A_1^T + A_2 G_2 A_2^T. \qquad (3a)$$

In this notation, (2) becomes

$$(A_1 G_1 A_1^T + A_2 G_2 A_2^T) y = v. \qquad (3b)$$

Let $H \in \Re^{n \times n}$ be positive definite and diagonal. Likewise, we partition H such that

$$AHA^T = A_1 H_1 A_1^T + A_2 H_2 A_2^T. \tag{3c}$$

The matrices A_1 and A_2 consist of the sparse columns in A and the dense columns in A respectively. The matrix $A_1 H_1 A_1^T$ is the sparse part of the coefficient matrix of the normal equations (or the sparse part of the preconditioner) from the previous computation step with a known factorization.

Let $A_1 \in \Re^{m \times r}(r \leq n)$ have full rank m. The main issue we want to address in this paper is given (3a) and (3c) how should we form an efficient low-rank correction preconditioner?

1.1 Organization and Notation

In Section 2, we construct preconditioners based on low-rank corrections. Section 3 is on numerical experiments and Section 4 on concluding remarks.

Throughout this paper we use the following notation. The symbol $\min_i$ or $\max_i$ is for all i for which the argument is defined. For any matrix A, A_{ij} is the element in the i-th row and j-th column, and A_j is the j-th column. The symbol 0 is used to denote the number zero, the zero vector, and the zero matrix. For any square matrix X with real eigenvalues, $\lambda_i(X)$ are the eigenvalues of X arranged in nondecreasing order, $\lambda_{\min}(X)$ and $\lambda_{\max}(X)$ denote the smallest and largest eigenvalues of X respectively; i.e.,

$$\lambda_{\min}(X) \equiv \lambda_1(X) \leq \lambda_2(X) \leq \cdots \leq \lambda_m(X) \equiv \lambda_{\max}(X).$$

We denote the spectral condition number of X by $\kappa(X)$ where by definition

$$\kappa(X) \equiv \lambda_{\max}(X)/\lambda_{\min}(X).$$

The letters L and D represent unit lower Cholesky factor and positive definite diagonal matrices respectively.

2 The Preconditioner

To attain rapid convergence for conjugate gradient type methods we require that either the spectral condition number of the preconditioned matrix be close to one in order for the error bound based on the Chebyshev polynomial to be small, or the preconditioned matrix have great clustering of eigenvalues. From the computational point of view, we require that the linear systems with the preconditioner as coefficient matrix be easier to solve, and the construction cost of the preconditioner be low.

2.1 The Class of Preconditioners

For a given index set

$$Q_1 \subseteq \{j : j \in \mathcal{J}_1 \text{ and } G_{jj} \neq H_{jj}\},$$

let the $r \times r$ diagonal matrices D_1 and K_1 be given by

$$D_{1_{jj}} = \begin{cases} G_{jj} - H_{jj} & \text{if } j \in Q_1 \\ 0 & \text{if } j \in \mathcal{J}_1 \setminus Q_1 \end{cases}$$

and

$$K_1 = H_1 + D_1. \tag{4}$$

Let $\bar{A}_1 \in \Re^{m \times q_1}$, where $q_1 = |Q_1|$, consist of all columns A_j such that $j \in Q_1$ and let $\bar{D}_1 \in \Re^{q_1 \times q_1}$ be the diagonal matrix corresponding to the nonzero diagonal elements of D_1. In this notation,

$$A_1 K_1 A_1^T = A_1(H_1 + D_1)A_1^T = A_1 H_1 A_1^T + \bar{A}_1 \bar{D}_1 \bar{A}_1^T,$$

namely, $A_1 K_1 A_1^T$ is a rank q_1-correction of $A_1 H_1 A_1^T$.

Given the index set

$$Q_2 \subseteq \{j : j \in \mathcal{J}_2\},$$

let the $(n-r) \times (n-r)$ diagonal matrix K_2 and the $n \times n$ diagonal matrix K be given by

$$K_{2_{ss}} = \begin{cases} G_{jj} & \text{if } j \in Q_2 \\ 0 & \text{if } j \in \mathcal{J}_2 \setminus Q_2 \end{cases}$$

and

$$K = \begin{bmatrix} K_1 & 0 \\ 0 & K_2 \end{bmatrix}, \tag{5}$$

where K_1 is defined in (4) and $s = j - r$. Let $\bar{A}_2 \in \Re^{m \times q_2}$, where $q_2 = |Q_2|$, consist of all columns A_j such that $j \in Q_2$ and let $\bar{G}_2 \in \Re^{q_2 \times q_2}$ be the diagonal matrix corresponding to the nonzero diagonal elements of K_2. In this notation,

$$A_2 K_2 A_2^T = \bar{A}_2 \bar{G}_2 \bar{A}_2^T.$$

Thus the general class of preconditioners is given by

$$AKA^T = \left(A_1 H_1 A_1^T + + \bar{A}_1 \bar{D}_1 \bar{A}_1^T\right) + \bar{A}_2 \bar{G}_2 \bar{A}_2^T. \tag{6}$$

Let $Q = Q_1 \bigcup Q_2$ and $q = q_1 + q_2$. The elements in the class of preconditioners (6) are determined by the choice of the index set Q.

In interior point methods for linear programming, low-rank correction preconditioners have been suggested (or discussed) [2,3,5,6]. However, these papers do not discuss the case when A contains some dense columns.

In applications where the problem matrix A contains some dense columns, an effective low-rank correction preconditioner is of the form (6). In this paper we establish bounds on the spectral condition number of the preconditioned matrix $(AKA^T)^{-1}AGA^T$, where AKA^T and AGA^T are given by (6) and (3a) respectively, and suggest how to form K.

Let $LDL^T = A_1 H_1 A_1^T$ be the given. Then the linear system with the preconditioner as coefficient matrix is of the form

$$\left(LDL^T + \bar{A}_1 \bar{D}_1 \bar{A}_1^T + \bar{A}_2 \bar{G}_2 \bar{A}_2^T\right) z = v. \tag{7}$$

For the techniques on how to solve (7) efficiently, see for example [1].

2.2 Bounds on Eigenvalues of the Preconditioned Matrix

Our interest is in bounding the spectral condition number of the preconditioned matrix. By the definition of the spectral condition number, this is equivalent to establishing a lower bound for the smallest eigenvalue and an upper bound for the largest eigenvalue of the preconditioned matrix.

Theorem 1. :[1] Let $G, H \in \Re^{n \times n}$ be positive definite diagonal matrices. Let AGA^T and AHA^T be partitioned as in (3a) and (3c) respectively. Let K_1 and K be defined in (4) and (5) respectively. Then

(i) $\min\left\{1, \min_{j \in \mathcal{J}_1 \backslash \mathcal{Q}_1} \left\{\frac{G_{jj}}{H_{jj}}\right\}\right\} \leq$

$$\lambda_i((A_1 K_1 A_1^T)^{-1} A_1 G_1 A_1^T) \leq \max\left\{1, \max_{j \in \mathcal{J}_1 \backslash \mathcal{Q}_1} \left\{\frac{G_{jj}}{H_{jj}}\right\}\right\}.$$

and

(ii) $\min\left\{1, \min_{j \in \mathcal{J}_1 \backslash \mathcal{Q}_1} \left\{\frac{G_{jj}}{H_{jj}}\right\}\right\} \leq$

$$\lambda_i((AKA^T)^{-1} AGA^T) \leq \max\left\{1, \max_{j \in \mathcal{J}_1 \backslash \mathcal{Q}_1} \left\{\frac{G_{jj}}{H_{jj}}\right\}\right\} + \psi,$$

where $\psi = \sum_{j \in \mathcal{J}_2 \backslash \mathcal{Q}_2} G_{jj} \|A_j\|_2^2 / \lambda_{\min}(AKA^T)$. □

2.3 Choosing the Index Set $\mathcal{Q}$

The idea is to choose the index set $\mathcal{Q}$ such that the upper bound on the spectral condition number of the preconditioned matrix $(AKA^T)^{-1}AGA^T$ is minimized. In particular, we choose $\mathcal{Q}_1$ such that the upper bound on $\kappa((A_1 K_1 A_1^T)^{-1} A_1 G_1 A_1^T)$ is minimized and $\mathcal{Q}_2$ such that ψ is minimized. This implies that $\mathcal{Q}_1$ consists of indices $j \in \mathcal{J}_1$ corresponding to the largest $G_{jj}/H_{jj} > 1$ and or the smallest $G_{jj}/H_{jj} < 1$ such that $\kappa(K_1^{-1} G_1)$ is minimized. Similarly, $\mathcal{Q}_2$ consists of indices $j \in \mathcal{J}_2$ corresponding to the largest $Gjj\|A_j\|_2^2$. Thus the required index set $\mathcal{Q} = \mathcal{Q}_1 \bigcup \mathcal{Q}_2$.

Baryamureeba, Steihaug, and Zhang [3] suggest to choose $\mathcal{Q}$ such that $\kappa(K^{-1}G)$ (for K positive definite) is minimized. Wang and O'Leary[5,6] suggest to choose $\mathcal{Q}$ to consist of indices j corresponding to largest values of $|G_{jj} - H_{jj}|$.

Table 1. Generated results for diagonal matrices G and H

Indices j	no. of indices	G_{jj}	H_{jj}	G_{jj}/H_{jj}	$|G_{jj} - H_{jj}|$
$1,\ldots,5$	5	10^7	10^2	10^5	1.00×10^7
$6,\ldots,10$	5	10^5	10^2	10^3	9.99×10^5
$11,\ldots,20$	10	10^3	10^1	10^2	9.90×10^2
$21,\ldots,30$	10	10^{-4}	10^{-3}	10^{-1}	1.00×10^{-3}
$31,\ldots,40$	10	10^1	10^1	1	0
$41,\ldots,44$	4	10^{-1}	10^{-5}	10^4	1.00×10^{-1}
$45,\ldots,47$	3	10^5	10^4	10^1	9.00×10^4
$48,\ldots,51$	4	10^{-3}	10^2	10^{-5}	1.00×10^2

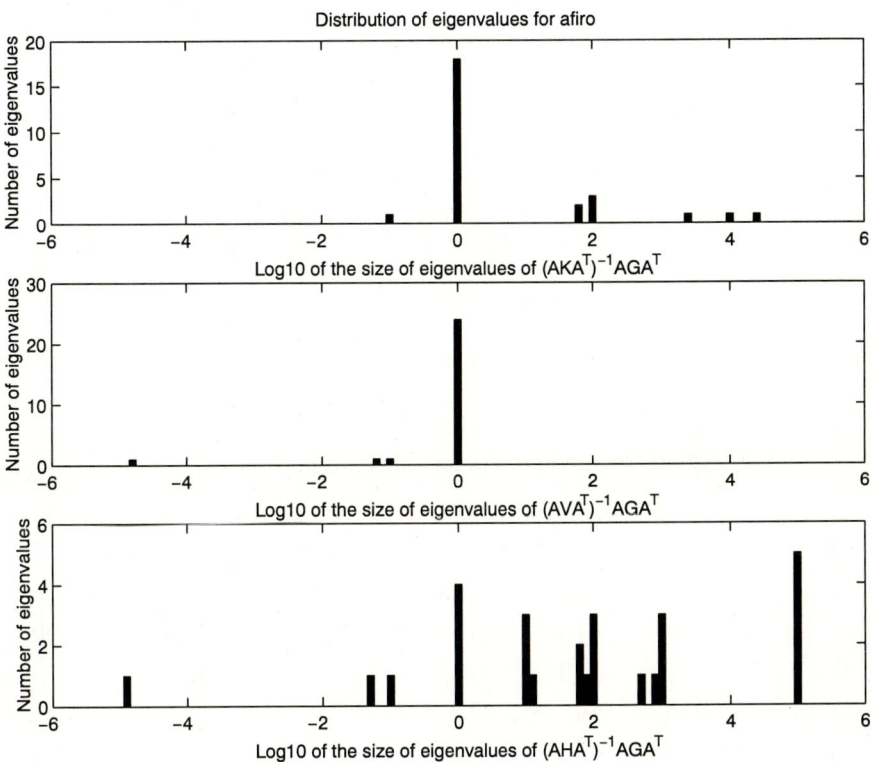

Fig. 1. For K we choose $\mathcal{Q}(q = 20)$ such that $\kappa(K^{-1}G)$ is minimized. Then we set $K_{jj} = 0$ for all $j \in \mathcal{J}_2 \setminus \mathcal{Q}_2$. For V we choose $\mathcal{Q}$ to consist of indices corresponding to 20 largest $|G_{jj} - H_{jj}|$. Then we set $V_{jj} = 0$ for all $j \in \mathcal{J}_2 \setminus \mathcal{Q}_2$

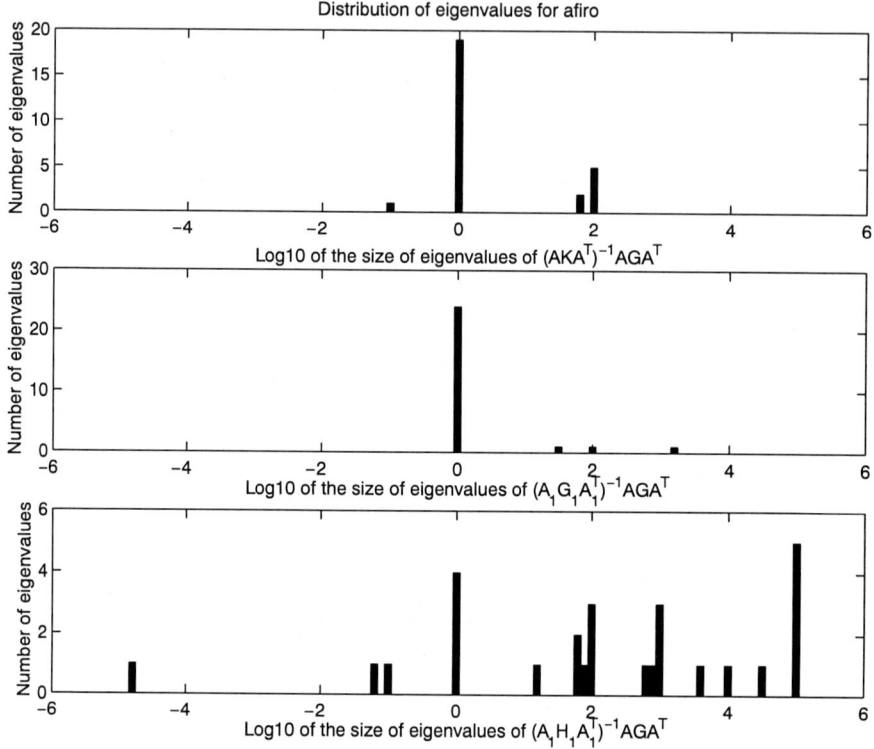

Fig. 2. We choose $\mathcal{Q}_1(q_1 = 17)$ such that $\kappa(K_1^{-1}G_1)$ is minimized, $\mathcal{Q}_2 = [45, 46, 47](q_2 = 3)$

Table 2. Generated results for diagonal matrices G and H

Indices j	no. of indices	G_{jj}	H_{jj}	G_{jj}/H_{jj}	$|G_{jj} - H_{jj}|$
$1,\ldots,10$	10	10^7	2×10^6	5	5.00×10^6
$11,\ldots,20$	10	10^3	10^1	10^2	9.90×10^2
$21,\ldots,30$	10	10^{-4}	10^{-3}	10^{-1}	1.00×10^{-3}
$31,\ldots,40$	10	10^1	10^1	1	0
$41,\ldots,44$	4	10^{-1}	10^{-5}	10^4	1.00×10^{-1}
$45,\ldots,47$	3	10^5	10^4	10^1	9.00×10^4
$48,\ldots,51$	4	10^{-3}	10^2	10^{-5}	1.00×10^2

3 Numerical Testing

We extract the matrix A from the netlib set [4] of linear programming problems. We use afiro ($m = 27, n = 51$) test problem in our numerical experiments.

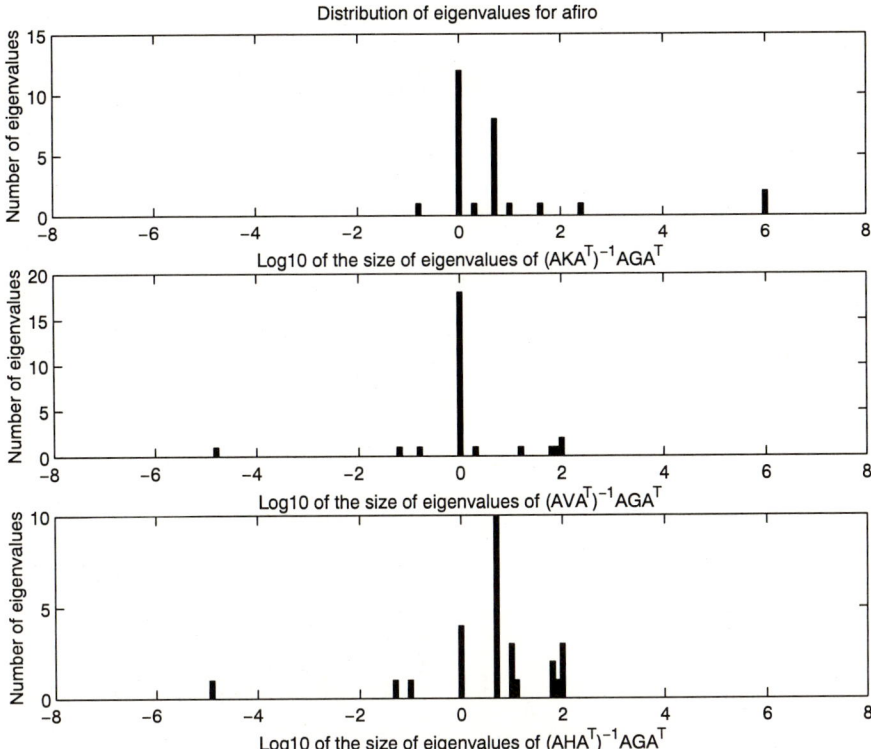

Fig. 3. For K we choose $\mathcal{Q}(q = 18)$ such that $\kappa(K^{-1}G)$ is minimized. Then we set $K_{jj} = 0$ for all $j \in \mathcal{J}_2 \setminus \mathcal{Q}_2$. For V we choose $\mathcal{Q}$ to consist of indices corresponding to 18 largest $|G_{jj} - H_{jj}|$. Then we set $V_{jj} = 0$ for all $j \in \mathcal{J}_2 \setminus \mathcal{Q}_2$

In Figure 1 and 2 we assume that the columns corresponding to indices $j = 45, 46, 47$, and 51 are dense and G, H are given by Table 1.

In Figure 3 and 4 we assume that the columns corresponding to indices $j = 2, 7, 21, 24, 33, 45, 47$, and 51 are dense and G, H are given by Table 2.

The results in Figure 1, 2, 3, and 4 strongly support the technique by Baryamureeba [1] (Theorem 1) for choosing the index set $\mathcal{Q}$ when A has some dense columns. Firstly, these results suggest that $\mathcal{Q}_1$ should be chosen based on the magnitude of $G_{1_{jj}}/H_{1_{jj}}$ (so that $\kappa(K_1^{-1}G_1)$ is minimized) instead of largest $|G_{1_{jj}} - H_{1_{jj}}|$. Secondly, the results show that it is not necessary to include in $\mathcal{Q}_2$ indices $j \in \mathcal{J}_2$ corresponding to $G_{jj} \ll 1$. Instead $\mathcal{Q}_2$ should consist of indices $j \in \mathcal{J}_2$ corresponding to the largest $G_{jj}\|A_j\|_2^2$ values. Furthermore, the results in Figure 2 and 4 show that the sparse part $A_1 G_1 A_1^T$ is not necessarily a good preconditioner for AGA^T.

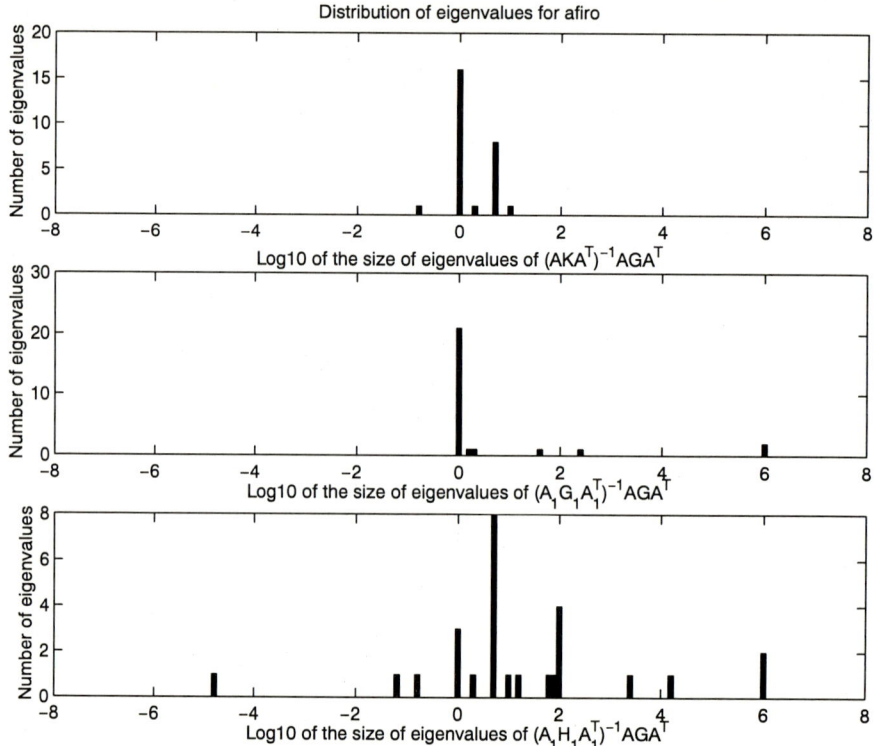

Fig. 4. We choose $Q_1(q_1 = 14)$ such that $\kappa(K_1^{-1}G_1)$ is minimized, $Q_2 = [2, 7, 45, 47](q_2 = 4)$

4 Concluding Remarks

We have given numerical results to show that the derived theoretical bounds on the eigenvalues of the preconditioned matrix (Theorem 1) are actually good bounds. Lastly, we believe that these results strongly support the technique by Baryamureeba [1] (Theorem 1) for solving large-scale linear programs, and the technique merits further study.

References

1. Baryamureeba, V.: Solution of large-scale weighted least squares problems. Technical Report No. 186, March 22, 2000 (Revised March 31, 2000) Department of Informatics, University of Bergen, 5020 Bergen, Norway. Submitted to Numerical Linear Algebra with Applications.
2. Baryamureeba, V., Steihaug, T.: Computational issues for a new class of preconditioners. Edited by Michael Griebel, Svetozar Margenov, and Plamen Yalamov, Large-Scale Scientific Computations of Engineering and Environmental Problems II, In Series Notes on Numerical Fluid Mechanics, VIEWEG. (2000) **73** 128–135.

3. V. Baryamureeba, T. Steihaug, and Y. Zhang, Properties of a class of preconditioners for weighted least squares problems. Technical Report No. 170, Department of Informatics, University of Bergen, 5020 Bergen, Norway, April 30, 1999 (Revised July 6, 1999). Submitted to Mathematical Programming.
4. Gay, D. M.: Electronic mail distribution of linear programming test problems. Mathematical Programming Society COAL Newsletter (1985) 10-12.
5. Wang, W., O'Leary, D. P.: Adaptive use of iterative methods in interior point methods for linear programming. Computer Science Department Report CS-TR-3560, Institute for Advanced Computer Studies Report UMIACS-95-111, University of Maryland, 1995.
6. Wang, W., O'Leary, D. P.: Adaptive use of iterative methods in predictor-corrector interior point methods for linear programming. Technical Report No. CS-TR-4011, Computer Science Department, University of Maryland, April 1999.

A Hybrid Newton-GMRES Method for Solving Nonlinear Equations[*]

Stefania Bellavia, Maria Macconi, and Benedetta Morini

Dipartimento di Energetica "S. Stecco"
via C. Lombroso 6/17, 50134 Firenze, Italia

Abstract. A subspace linesearch strategy for the globalization of Newton-GMRES method is proposed. The main feature of our proposal is the simple and inexpensive way we determine descent directions in the low-dimensional subspaces generated by GMRES. Global and local quadratic convergence is established under standard assumptions.

1 Introduction

We consider the problem of solving large-scale systems of equations

$$F(x) = 0, \qquad (1)$$

where F is a nonlinear function from $\mathbb{R}^n$ to $\mathbb{R}^n$ and propose a new iterative process in the context of the inexact methods [4]. Specifically, we consider the relevant framework of the Newton-Krylov methods. Well known convergence properties of these methods motivated the recent works to create robust and locally fast algorithms and to develop reliable and efficient software ([2,8,9]). The basic idea of a Newton-Krylov method is to construct a sequence of iterates $\{x_k\}$ such that, at each iteration k, the correction $\bar{s}_k = x_{k+1} - x_k$ is taken from a subspace of small dimension and satisfies

$$F'(x_k)\bar{s}_k = -F(x_k) + r_k, \qquad \|r_k\| \le \bar{\eta}_k \|F(x_k)\|, \qquad (2)$$

where F' is the system Jacobian, $\bar{\eta}_k$ is a suitable scalar in $[0, 1)$ called forcing term and r_k is commonly referred to as the residual vector.

Globally convergent modifications of Newton-Krylov methods are commonly called hybrid methods [2]. They are obtained using globally convergent strategies for the unconstrained minimization problem

$$\min_{x \in \mathbb{R}^n} f(x), \qquad (3)$$

where f is an appropriately chosen merit function whose global minimum is a zero of F.

---
[*] This work was partially supported by Murst Cofin98 "Metodologie numeriche avanzate per il calcolo scientifico", CNR "Progetto coordinato sistemi di calcolo di grandi dimensioni e calcolo parallelo", Rome, Italy

Hybrid methods based on a linesearch backtracking strategy are often used. At the k-th step of these methods, a direction p in $\mathbb{R}^n$ is chosen and the new iterate x_{k+1} has the form $x_{k+1} = x_k + \theta p$ where $\theta \in (0,1]$ is such that $f(x_{k+1}) < f(x_k)$. The existence of such a θ is ensured if p is a descent direction for f at x_k, i.e. $\nabla f(x_k)^T p < 0$ ([5]). Classical choices of f are $f = f_1$ or $f = f_2$ where $f_1 = \|F\|_2$ and $f_2 = \|F\|_2^2/2$. In both cases, a vector p satisfying $\|F'(x_k)p + F(x_k)\|_2 < \|F(x_k)\|_2$ is ensured to be a descent direction for f at x_k [3]. Using this result, several authors proposed globally convergent modifications of the basic Newton-Krylov method where backtracking linesearch procedures along the inexact step $\bar{s}_k$ are performed (see e.g. [2,3,6,8,9]).

Here we are concerned with global strategies that search for a decrease of the merit function in a low-dimensional subspace generated by the Krylov method. This strategy can be stated as follows: letting S be a subspace which contains $\bar{s}_k$ and is generated by the Krylov method, find a vector $\Delta_k \in S$ such that

$$\nabla f(x_k)^T \Delta_k < 0, \quad \text{and} \quad f(x_k + \Delta_k) < f(x_k).$$

Namely, we formulate a global strategy using the subspace approach that revealed to be a promising way for solving large-scale nonlinear systems ([1,2,3,7]). The particular Krylov method we consider is the well known iterative process GMRES [10].

In a Newton-GMRES context we propose an iterative method where a linesearch procedure along the inexact Newton step $\bar{s}_k$ is combined with an alternative linesearch strategy in a low-dimensional subspace S. The given approach is related to a curvilinear linesearch globalization procedure recently proposed in [1]. Specifically, first we use the direction of the inexact Newton step $\bar{s}_k$. If it does not work well, i.e. relatively few steps do not suffice to decrease the value of f, we fall back on a step obtained by a slower method. The direction used in the latter method is a descent direction which is selected among the coordinate vectors and the steepest descent direction of the merit function in S. A key feature of our proposal is the simple and inexpensive way to determine such alternative direction. In fact, we use only information that are built in the progress of GMRES iterations.

The given theoretical analysis shows that our method is globally convergent and close to the solution it reduces to Newton-GMRES method. Therefore, it retains fast local convergence rate of the Inexact Newton methods. Moreover, the proposed strategy is consistent with preconditioning techniques, too.

Through the paper, for any vector (matrix) $\|\cdot\|$ denotes the 2-norm of the specified vector (matrix) and e_j the j-th unit coordinate vector, with its dimension inferred from the context. The symbol $(x)_i$ represents the i-th component of a vector x. The condition number of a real matrix A is denoted by $K_2(A)$. Further, the gradient vector of a given smooth real function $f: \mathbb{R}^n \mapsto \mathbb{R}$, is denoted by $\nabla f(x)$. The closed ball with center y and radius δ is indicated by $N_\delta(y)$.

2 Descent Directions in a Subspace

We consider an iterative process where, at the k-th iteration, the linear system

$$F'(x_k)s = -F(x_k), \tag{4}$$

is approximately solved by GMRES as specified in (2).

Let $s_k^0 \in \mathbb{R}^n$ be the initial guess for the true solution of (4) and $r_k^0 = -F'(x_k)s_k^0 - F(x_k)$ the initial residual. GMRES generates a sequence $\{s_k^m\}$ of vectors such that each s_k^m is the solution to

$$\min_{s \in s_k^0 + K_m} \|F'(x_k)s + F(x_k)\|, \tag{5}$$

where K_m is the Krylov subspace

$$K_m = span\{r_k^0, F'(x_k)r_k^0, (F'(x_k))^2 r_k^0, \ldots, (F'(x_k))^{m-1} r_k^0\}.$$

In order to solve (5), Arnoldi process is used to construct an orthonormal basis $v_1, v_2, \ldots, v_m$ of K_m where $v_1 = r_k^0 / \|r_k^0\|$. Using this process, the matrices $V_m \equiv [v_1, v_2, \ldots, v_m] \in \mathbb{R}^{n \times m}$ and $V_{m+1} \equiv [v_1, v_2, \ldots, v_{m+1}] \in \mathbb{R}^{n \times (m+1)}$ verify the relevant relation

$$F'(x_k) V_m = V_{m+1} H_m, \tag{6}$$

where $H_m \in \mathbb{R}^{(m+1) \times m}$ is an upper Hessenberg matrix. Further, (5) reduces to

$$\min_{y \in \mathbb{R}^m} \|\rho_k e_1 - H_m y\|, \tag{7}$$

where $\rho_k = \|r_k^0\|$. In theory GMRES iterates are computed until the current vector s_k^m satisfies (2). Then, $\bar{s}_k$ is set equal to s_k^m and the new iterate $x_{k+1} = x_k + \bar{s}_k$ is formed. However, for large problems and large values of m, storage requirements for the basis of K_m may become prohibitive. This problem is overcome by a *restarted* GMRES, i.e. a process that uses GMRES iteratively and restarts it after a fixed number of iterations using the last computed iterate [10].

Now we turn our attention to the way one may select descent directions which are alternative to $\bar{s}_k$ and belong to a low-dimensional subspace generated by GMRES. We consider the case $f = f_2$, but we point out that, if $f = f_1$, analogous conclusions can be drawn.

Following [2], we restrict our search direction from x_k to be in the subspace

$$S_G = span\{v_1, v_2, \ldots, v_m, s_k^0\}.$$

Clearly, $\bar{s}_k \in S_G$ and if $s_k^0 \in K_m$ then S_G coincides with K_m.

Let $W = [w_1, w_2, \ldots, w_{m+1}]$ be the orthonormal basis of S_G such that $w_i = v_i$ for $i = 1, 2, \ldots, m$ and w_{m+1} is computed by Gram-Schmidt method. Thus, the global strategy assumes the form of the low-dimensional minimization problem

$$\min_{\tilde{y} \in \mathbb{R}^{m+1}} g(\tilde{y}) = f_2(x_k + W\tilde{y}),$$

with $g : \mathbb{R}^{m+1} \to \mathbb{R}$. Since $F'(x_k)W = [V_{m+1}H_m, F'(x_k)w_{m+1}]$, the vector $\nabla g(0)$ is given by

$$\nabla g(0) = (F'(x_k)W)^T F(x_k) = \begin{pmatrix} H_m^T V_{m+1}^T F(x_k) \\ w_{m+1}^T F'(x_k)^T F(x_k) \end{pmatrix}. \tag{8}$$

Now, we show that it is possible to exploit information gathered by GMRES in order to search descent directions alternative to $\bar{s}_k$ in S_G. In fact, we have the steepest descent direction $d_m = -\nabla g(0)$ for $g(y)$ at $y = 0$. Further, given the coordinate vectors $e_j \in \mathbb{R}^{m+1}$, $j = 1, \ldots, m+1$, we have $w_j = We_j$ and

$$\nabla f_2(x_k)^T w_j = \nabla g(0)^T e_j = (\nabla g(0))_j. \tag{9}$$

Hence, we can conclude that if $(\nabla g(0))_j < 0$, w_j is a descent direction for f_2 at x_k. An interesting additional observation is that if $s_k^0 = 0$, S_G coincides with K_m. Then, $W = V_m$, and

$$v_1 = r_k^0 / \|r_k^0\| = -F(x_k)/\rho_k,$$
$$\nabla g(0) = (F'(x_k)V_m)^T F(x_k) = (V_{m+1}H_m)^T F(x_k) = -\rho_k H_m^T e_1.$$

Therefore, at no additional cost, the elements $h_{1,j}$, $j = 1, \ldots, m$, of the first row of H_m, give us the directional derivatives of f_2 at x_k along v_j.

We remark that Krylov methods have the virtue of requiring no matrix evaluation. In fact, the action of the Jacobian F' on a vector v can be approximated by means of finite differences ([3,8]). An attractive feature of our global strategy restricted to S_G is that (8) does not require the Jacobian matrix explicitly and need only to compute the product $F'(x_k)w_{m+1}$.

Moreover, following [1], it can be shown that, if a right preconditioner P_r is used, the global strategy is performed in the subspace

$$S_G^p = span\{\tilde{v}_1, \tilde{v}_2, \ldots, \tilde{v}_m, s_k^0\},$$

where $\tilde{v}_1, \tilde{v}_2, \ldots, \tilde{v}_m$ is an orthonormal basis of the Krylov subspace

$$K_m^p = \{r_k^0, (F'(x_k)P_r)r_k^0, (F'(x_k)P_r)^2 r_k^0, \ldots, (F'(x_k)P_r)^{m-1} r_k^0\}.$$

3 The New Method

In this section we present a new globally convergent hybrid method which combines two linesearch backtracking strategies.

At each iteration the Inexact Linesearch Backtracking (ILB) strategy given by Eisenstat and Walker in [6] is tried first. If within a maximum number N_b of backtracks no progress is found in the merit function f_1, we leave the direction $\bar{s}_k$ and apply a new globalization process, called CD (Coordinate Directions) strategy. This alternative global method searches for a reduction in the merit function f_2 along a properly selected vector of the subspace S_G and uses the following quadratic model for $f_2(x_k + W\tilde{y})$ at x_k:

$$\hat{g}(\tilde{y}) = \|F'(x_k)W\tilde{y} + F(x_k)\|^2/2.$$

with $\hat{g} : \mathbb{R}^{m+1} \to \mathbb{R}$.

The resulting k-th iteration can be sketched as follows.

Algorithm

Given x_k, $\eta_{max} \in (0,1)$, $\bar{\eta}_k \in [0, \eta_{max}]$, $\alpha, \beta \in (0,1)$, $0 < \theta_m < \theta_M < 1$, $N_b > 0$.

1. Apply GMRES to compute $\bar{s}_k$ such that $\|F(x_k) + F'(x_k)\bar{s}_k\| \leq \bar{\eta}_k \|F(x_k)\|$.
2. Perform the ILB-strategy:
   2.1 Set $s_k = \bar{s}_k$, $\eta_k = \bar{\eta}_k$, $nb = 0$.
   2.2 While $f_1(x_k + s_k) > (1 - \alpha(1-\eta_k)) f_1(x_k)$ & $nb < N_b$ do:
      Choose $\theta \in [\theta_m, \theta_M]$
      Update $s_k = \theta s_k$, $\eta_k = 1 - \theta(1-\eta_k)$ and $nb = nb + 1$.
3. If
$$f_1(x_k + s_k) \leq (1 - \alpha(1-\eta_k)) f_1(x_k), \quad (10)$$
   Set $\Delta_k = s_k$. Go to step 5.
4. Perform the CD-strategy:
   4.1 Compute $\nabla g(0)$. Set $d_m = -\nabla g(0)$.
   4.2 Let j^* be such that $|(\nabla g(0))_{j^*}| = \max_{1 \leq j \leq m+1} |(\nabla g(0))_j|$.
   4.3 If $(\nabla g(0))_{j^*} > 0$
      Set $u = -w_{j^*}, \quad e = -e_{j^*}$
   Else
      Set $u = w_{j^*}, \quad e = e_{j^*}$.
   4.4 Compute $\alpha_e = \operatorname{argmin} \hat{g}(\alpha e), \quad \alpha_d = \operatorname{argmin} \hat{g}(\alpha d_m)$.
   4.5 If $f_2(x_k + \alpha_d W d_m) < f_2(x_k + \alpha_e W e)$
      Set $u = \alpha_d W d_m, \quad \nabla f_2(x_k)^T u = -\alpha_d \|d_m\|^2$.
   Else
      Set $u = \alpha_e W e, \quad \nabla f_2(x_k)^T u = \alpha_e e^T \nabla g(0)$.
   4.6 If
$$f_2(x_k + u) \leq f_2(x_k) + \alpha \nabla f_2(x_k)^T u, \quad (11)$$
$$\nabla f_2(x_k + u)^T u \geq \beta \nabla f_2(x_k)^T u \quad (12)$$
   Set $\Delta_k = u$. Go to step 5.
   4.7 Choose $\theta \in [\theta_m, \theta_M]$. Update $u = \theta u$. Go to step 4.6.
5. Set $x_{k+1} = x_k + \Delta_k$.

We remark that in the ILB-process we move along the direction $\bar{s}_k$ and we select successively shorter steps s_k of the form $s_k = \frac{1-\eta_k}{1-\bar{\eta}_k} \bar{s}_k$. In the CD-strategy, we select among $\pm w_1, \ldots, \pm w_{m+1}$ the vector w_{j^*} that produces the maximum local decrease of f_2. Then, we form $\alpha_d d_m$ and $\alpha_e e$, i.e. the minimizer of the quadratic model $\hat{g}(\tilde{y})$ along d_m and e respectively. The new direction u is chosen between $W d_m$ and $W e$ on the base of the minimum between the two values $f_2(x_k + \alpha_d W d_m)$ and $f_2(x_k + \alpha_e W e)$. Finally, a backtracking linesearch along u is performed until the Goldstein-Armijo conditions (11) and (12) are met. Since u is a descent direction for f_2 at x_k, there exists a point x_{k+1} such that (11) and (12) hold (see [5, Theorem 6.3.2]).

4 Convergence Results

Now we will address the convergence behaviour of the method described in the previous section. We will make the following assumptions:

- $F : \mathbb{R}^n \to \mathbb{R}^n$ is continuously differentiable;
- F' is Lipschitz continuous in $L = \{x \in \mathbb{R}^n : f_2(x) \le f_2(x_0)\}$.
- for each k, $F'(x_k)$ is nonsingular and $\|F(x_k) + F'(x_k)\bar{s}_k\| \le \eta_{max}\|F(x_k)\|$ holds.

Note that from the first two assumptions it follows that ∇f_2 is Lipschitz continuous in L. Furthermore, the last assumption avoids that the method breaks down. In fact, the $(k+1)$-th iterate can not be determined either if x_k is such that $F'(x_k)$ is singular or if there are no descent directions in S_G. On the other hand, if the step $\bar{s}_k$ provided by GMRES satisfies $\|F(x_k) + F'(x_k)\bar{s}_k\| \le \eta_{max}\|F(x_k)\|$, then the existence of descent directions in S_G is guaranteed. Note that this is not a serious restriction when the null starting vector for GMRES is used, since η_{max} can be taken arbitrarily near one.

We will show that, if there exists a limit point x^* of $\{x_k\}$ such that $F'(x^*)$ is invertible, then $F(x^*) = 0$ and $x_k \to x^*$. Further, for k sufficiently large, x_{k+1} has the form $x_{k+1} = x_k + \bar{s}_k$; then the ultimate rate of convergence depends on the choice of the forcing terms $\bar{\eta}_k$, as shown in [4].

In our analysis we will use the following two results that show the convergence behaviour of methods obeying (10) and (11)-(12), respectively.

Theorem 1 ([6]). *If $\{x_k\}$ is a sequence generated applying ILB-strategy, i.e. for each k (10) is satisfied, and x^* is a limit point such that $F'(x^*)$ is invertible, then $\|F(x)\| \to 0$. Further, let $\Gamma = 2\|F'(x^*)^{-1}\|(1+\eta_{max})/(1-\eta_{max})$ and $\delta > 0$ sufficiently small that, $\|F'(x)^{-1}\| \le 2\|F'(x^*)^{-1}\|$ whenever $x \in N_\delta(x^*)$, and also*

$$\|F(y) - F(x) - F'(x)(y-x)\| \le \frac{1-\alpha}{\Gamma}\|y-x\|, \tag{13}$$

if $x, y \in N_{2\delta}(x^*)$. Then, if $x_k \in N_\delta(x^*)$, x_{k+1} has the form $x_{k+1} = x_k + s_k(\eta_k)$ with $s_k(\eta_k) = (1-\eta_k)\bar{s}_k/(1-\bar{\eta}_k)$ and

$$1 - \eta_k \ge \min\{1 - \bar{\eta}_k, \frac{\theta_m \delta}{\Gamma\|F(x_k)\|}\}. \tag{14}$$

Theorem 2 ([5],Th. 6.3.3). *Let $x_0 \in \mathbb{R}^n$ be given and $\{x_k\}$ be a sequence such that for each $k > 0$, $\Delta_k = x_{k+1} - x_k$ satisfies (11) and (12) and $\nabla f(x_k)^T \Delta_k < 0$. Then, either $\nabla f(x_k) = 0$ for some k or $\lim_{k \to \infty} \frac{\nabla f(x_k)^T \Delta_k}{\|\Delta_k\|} = 0$.*

Now, main convergence results for the proposed hybrid method can be stated.

Theorem 3. *Assume that there exists a limit point x^* such that $F'(x^*)$ is invertible. Then, the sequence $\{\|F(x_k)\|\}$ converges to zero.*

Proof. Note that the sequence $\{\|F(x_k)\|\}$ is strictly decreasing and bounded from below by zero. Hence, it is convergent. Let $\{\tilde{x}_k\}$ be the subsequence such that $\{\tilde{x}_k\} \to x^*$.

If there exists an index $\bar{k}$ such that, for $k > \bar{k}$, $\tilde{x}_k$ is computed by ILB-strategy, from Theorem 1 we get that $\|F(\tilde{x}_k)\| \to 0$ and therefore $F(x^*) = 0$.

Otherwise, let $\tilde{K}$ be the set of indices such that, for $k \in \tilde{K}$, $\tilde{x}_k$ is computed by the CD-strategy and let $\{\tilde{x}_{l_k}\}$ be the subsequence of $\{\tilde{x}_k\}$ such that $l_k \in \tilde{K}$. A direct adaptation of Theorem 2 yields

$$\frac{\nabla f_2(\tilde{x}_{l_k})^T \Delta_{l_k}}{\|\Delta_{l_k}\|} \to 0.$$

Further, by construction, at step l_k-th CD-strategy gives

$$\frac{|\nabla f_2(\tilde{x}_{l_k})^T \Delta_{l_k}|}{\|\Delta_{l_k}\|} = |(\nabla g(0))_{j^*}| = \|W^T \nabla f_2(\tilde{x}_{l_k})\|_\infty,$$

if the direction w_{j^*} is selected, and

$$\frac{|\nabla f_2(\tilde{x}_{l_k})^T \Delta_{l_k}|}{\|\Delta_{l_k}\|} = \|d_m\| = \|W^T \nabla f_2(\tilde{x}_{l_k})\|,$$

if the steepest descent direction in S_G is chosen. Then, since from [3, Corollary 3.5] we have

$$\|W^T \nabla f_2(\tilde{x}_{l_k})\| \geq \frac{1 - \eta_{max}}{(1 + \eta_{max})k_2(F'(x_k))} \|\nabla f_2(\tilde{x}_{l_k})\|,$$

it follows

$$\frac{|\nabla f_2(\tilde{x}_{l_k})^T \Delta_{l_k}|}{\|\Delta l_k\|} \geq \frac{1}{\sqrt{n}} \frac{1 - \eta_{max}}{(1 + \eta_{max})k_2(F'(x_k))} \|\nabla f_2(\tilde{x}_{l_k})\|.$$

Due to the invertibility of $F'(x^*)$, for k sufficiently large $k_2(F'(x_k))$ can be bounded from zero, and as a consequence, $\|F(\tilde{x}_{l_k})\| \to 0$ and $\|F(x_k)\| \to 0$.

Theorem 4. *Assume that there exists a limit point x^* such that $F'(x^*)$ is invertible. Then, there exists a sufficiently large $\bar{k} > 0$ such that, for $k > \bar{k}$, $x_{k+1} = x_k + \bar{s}_k$. Further $x_k \to x^*$.*

Proof. ¿From Theorem 3 it follows $\|F(x_k)\| \to 0$. Hence $F(x^*) = 0$. Assume $\theta_m < 1/2$. Set $K = \|F'(x^*)^{-1}\|$, $\Gamma = 2K(1 + \eta_{max})/(1 - \eta_{max})$, and let δ sufficiently small that, if $x \in N_\delta(x^*)$, $\|F'(x)^{-1}\| \leq 2K$ and (13) holds whenever $x, y \in N_{2\delta}(x^*)$.

Since x^* is a limit point of $\{x_k\}$ and $F(x^*) = 0$ there exists a k sufficiently large that

$$x_k \in N_\epsilon \equiv \{y | \|y - x^*\| \leq \frac{\delta}{2}, \|F(y)\| \leq \frac{\theta_m \epsilon}{\Gamma}\},$$

where $\epsilon < \delta$ is such that $2K\theta_m \epsilon/\Gamma < \delta/2$. Clearly, $\theta_m \delta/(\Gamma \|F(x_k)\|) > 1$, then from (14) it follows that the k-iteration is successfully performed with $\eta_k = \bar{\eta}_k$, i.e. $x_{k+1} = x_k + \bar{s}_k$. To complete the proof, we show that $x_{k+1} \in N_\epsilon$. To this end, first note that from [6, Th. 6.1] it follows

$$\|\bar{s}_k\| \leq \Gamma(1 - \bar{\eta}_k)\|F(x_k)\|. \tag{15}$$

Further, since

$$\|\bar{s}_k\| \leq \Gamma(1 - \bar{\eta}_k)\|F(x_k)\| < \Gamma\|F(x_k)\| < \theta_m \delta < \frac{\delta}{2},$$

we have

$$\|x_{k+1} - x^*\| \leq \|x_k - x^*\| + \|\bar{s}_k\| < \delta,$$

which implies $x_{k+1} \in N_\delta(x^*)$. Finally, from (13) the following relation can be derived

$$\|x_k - x^*\| \leq 2K\|F(x_k)\|.$$

It yields

$$\|x_{k+1} - x^*\| \leq 2K\|F(x_{k+1})\| < 2K\|F(x_k)\| \leq \frac{2K\theta_m\epsilon}{\Gamma} < \frac{\delta}{2}.$$

Thus, $x_{k+1} \in N_\epsilon$ and we can conclude that there exists a $\bar{k} > 0$ such that, for $k > \bar{k}$, no backtracking is performed along $\bar{s}_k$, $x_k \in N_\epsilon$ and $x_k \to x^*$.

References

1. Bellavia S., Morini B.: A globally convergent Newton-GMRES subspace method for systems of nonlinear equations. Submitted for publication.
2. Brown P. N., Saad Y.: Hybrid Krylov Methods for nonlinear systems of equations. SIAM J. Sci. Stat. Comput. **11** (1990) 450–481.
3. Brown P. N., Saad Y.: Convergence Theory of Nonlinear Newton-Krylov algorithms. SIAM J. Optim. **4** (1994) 297–330.
4. Dembo R. S., Eisenstat S. C., Steihaug T.: Inexact Newton Methods. SIAM J. Numer. Anal. **19** (1982) 400–408.
5. Dennis J. E., Schnabel R. B.: Numerical Methods for Unconstrained Optimization and Nonlinear Equations. Prentice Hall, Englewood Cliffs, NJ, 1983.
6. Eisenstat S. C., Walker H. F.: Globally Convergent Inexact Newton Methods. SIAM J. Optim. **4** (1994) 393–422.
7. Feng D., Pulliam T. H.: Tensor-GMRES method for large systems of nonlinear equations. SIAM J. Optim. **7** (1997) 757–779.
8. Kelley C. T.: Iterative Methods for Linear and Nonlinear Equations. SIAM, Philadelphia, 1995
9. Pernice M., Walker H. F.: NITSOL: a new iterative solver for nonlinear systems. SIAM J. Sci Comput. **19** (1998) 302–318.
10. Saad Y., Schultz M. H.: GMRES: a generalized minimal residual method for solving nonsymmetric linear systems. SIAM J. Sci. Stat. Comput. **6** (1985) 856–869.

Comparative Analysis of Marching Algorithms for Separable Elliptic Problems

Gergana Bencheva

Central Laboratory of Parallel Processing, Bulgarian Academy of Sciences
Acad. G. Bontchev Str., Bl.25A, 1113 Sofia, Bulgaria
gery@cantor.bas.bg

Abstract. Standard marching algorithms (MA) and generalized marching algorithms (GMA) for 2D separable second order elliptic problems on rectangular $n \times m$ grids are described. Their numerical stability and computational complexity are theoretically and experimentally compared. Results of numerical experiments performed to demonstrate the stability of GMA versus the instability of MA are presented.

Keywords: fast elliptic solvers, marching algorithms, computational complexity of algorithms

AMS Subject Classification: 65F05, 65F30, 65N22

1 Introduction

After discretizing separable elliptic boundary value problems on rectangular domains, linear algebraic systems with special block banded structure are obtained. The *fast elliptic solvers* are highly efficient algorithms for their direct solution, and the so called *marching algorithms* is one class of them.

The goal of this study is to review the theoretical and experimentally compare the numerical stability and computational complexity of the standard marching algorithm (MA) and the generalized marching algorithm (GMA) for 2D separable elliptic problems discretized on a rectangular $n \times m$ grid. These two algorithms are first proposed in [2,3] and later reformulated in [5] by using the incomplete solution technique, which slightly reduces the asymptotical operation count of the GMA. The standard marching algorithm is optimal in the sense that its computational cost depends linearly on the dimension of the system, namely, the number of arithmetic operations for its implementation is of order $\mathcal{N}_{MA} = \mathcal{O}(nm)$. Unfortunately MA is unstable and hence is of practical interest for sufficiently small-sized problems, or more generally for $m \ll n$. The GMA is a stabilized version of the MA obtained (in [5]) by limiting the size of the marching steps and using the incomplete solution technique for problems with sparse right-hand sides in the second part of the algorithm. The total cost of the resulting algorithm, in the case when $m = n$, $n + 1 = p(k+1)$, $p, k \in \mathbf{Z}$, is of order $\mathcal{N}_{GMA} = \mathcal{O}(n^2 \log p + n^2)$.

The remainder of the paper is organized as follows. At the beginning of Section 2 the considered problem is formulated and the technique for incomplete solution of problems with sparse right-hand sides is briefly outlined. Next, the algorithms MA and GMA are described in the same section. Results of numerical experiments that confirm the theoretical estimates are given in Section 3.

At the end of the paper some concluding remarks about the applications of the presented algorithms as representatives of fast elliptic solvers are formulated.

2 Description of the Algorithms

In this section we present the standard marching algorithm (MA) and a stabilized version of it, the generalized marching algorithm (GMA), obtained by limiting the size of the marching steps combined with the so-called technique of incomplete solution of problems with sparse right-hand sides. The exposition in this section is based on the survey [1].

2.1 Formulation of the Problem and Preliminaries

A separable second order elliptic equation of the form:

$$-\sum_{s=1}^{2} \frac{\partial}{\partial x_s} \left(a_s(x_s) \frac{\partial u}{\partial x_s} \right) = f(x), \ x = (x_1, x_2) \in \Omega = (0,1)^2 ,$$
$$u = 0, \quad \text{on } \partial\Omega \qquad (1)$$

is discretized by finite differences or by piecewise linear finite elements on right-angled triangles. The following block banded system with tensor product matrix is obtained:

$$A\mathbf{x} = \mathbf{f}, \qquad (2)$$

where

$$A = B \otimes I_n + I_m \otimes T$$
$$\equiv \begin{pmatrix} T + b_{1,1} I_n & b_{1,2} I_n & & 0 \\ b_{2,1} I_n & T + b_{2,2} I_n & & 0 \\ & \ddots & \ddots & & \vdots \\ 0 & \cdots & & b_{m,m-1} I_n \ T + b_{m,m} I_n \end{pmatrix},$$

$$\mathbf{x} = (\mathbf{x}_1, \mathbf{x}_2, \ldots, \mathbf{x}_m)^T, \ \mathbf{f} = (\mathbf{f}_1, \mathbf{f}_2, \ldots, \mathbf{f}_m)^T,$$
$$\mathbf{x}_j = (x_{1,j}, x_{2,j}, \ldots, x_{n,j})^T, \ \mathbf{f}_j = (f_{1,j}, f_{2,j}, \ldots, f_{n,j})^T,$$
$$\mathbf{x}_j, \mathbf{f}_j \in \mathbf{R}^n, \ j = 1, \ldots, m.$$

Here, I_n is the identity $n \times n$ matrix and $\otimes$ is the Kronecker (or tensor) product of the matrices C and D, defined by $C_{m_1 \times n_1} \otimes D_{m_2 \times n_2} = (c_{i,j} D)_{i=1 \ j=1}^{m_1 \ n_1}$, where $C = (c_{i,j})_{i=1 \ j=1}^{m_1 \ n_1}, D = (d_{k,l})_{k=1 \ l=1}^{m_2 \ n_2}$.

The matrices $T = (t_{i,j})_{i,j=1}^{n}$ and $B = (b_{i,j})_{i,j=1}^{m}$ are tridiagonal, symmetric and positive definite, corresponding to a finite difference approximation of the one–dimensional operators $(-\partial/\partial x_s)(a_s(x_s)(\partial/\partial x_s)(\cdot))$, for $s = 1, 2$, respectively.

Here, we briefly describe the so-called incomplete solution technique applied to systems of the form (2) with a sparse right-hand side. That technique has independently been proposed by Banegas, Proskurowski and Kuznetsov. More details may be found in [1,4].

It is assumed that the right-hand side $\mathbf{f}$ has only d ($d \ll m$) nonzero block components and for some reason only r ($r \ll m$) block components of the solution are needed. Let for definiteness $\mathbf{f}_j = 0$ for $j \neq j_1, j_2, \ldots, j_d$. To find the needed components $\mathbf{x}_{j'_1}, \mathbf{x}_{j'_2}, \ldots, \mathbf{x}_{j'_r}$ of the solution, the well-known algorithm for separation of variables is applied taking advantage of the right–hand side sparsity:

ALGORITHM SRHS

Step 0. determine all the eigenvalues and d components of all the eigenvectors of the tridiagonal matrix B.

Step 1. compute the Fourier coefficients $\beta_{i,k}$ of $\mathbf{f}'_i$ from equations:
$$\beta_{i,k} = \mathbf{q}_k^T \cdot \mathbf{f}'_i = \sum_{s=1}^{d} q_{j_s,k} f_{i,j_s}, \; i = 1, \ldots, n, \; k = 1, \ldots, m.$$

Step 2. solve m $n \times n$ tridiagonal systems of linear equations:
$$(\lambda_k I_n + T) \eta_k = \beta_k, \; k = 1, \ldots, m.$$

Step 3. recover r components of solution per lines using
$$\mathbf{x}_j = \sum_{k=1}^{m} q_{j,k} \eta_k \text{ for } j = j'_1, j'_2, \ldots, j'_r.$$

Remark 1. Here, $\{\mathbf{q}_k, \lambda_k\}_{k=1}^{m}$ denote the eigenpairs of the tridiagonal matrix $B_{m \times m}$, i.e., $B\mathbf{q}_k = \lambda_k \mathbf{q}_k$, $k = 1, \ldots, m$.

The computational complexity of ALGORITHM SRHS is given in:

Proposition 21 *The* ALGORITHM SRHS *requires* $m[2(r+d)n+(5n-4)]$ *arithmetic operations in the solution part,* $m[n \text{devisions} + 3(n-1) \text{other operations}]$ *to factor the tridiagonal matrices* $\lambda_k I_n + T$, $k = 1, \ldots, m$ *in* $LD^{-1}U$ *form, and* $\mathcal{O}(dm^2) + 9m^2$ *arithmetic operations for computing all the eigenvalues and* d *components of all the eigenvectors of the matrix* B.

2.2 Marching Algorithm (MA)

We now describe the standard marching algorithm. The first block equation of the system (2) is placed at the bottom and the reordered system is rewritten in the following two-by-two block form:

$$\begin{pmatrix} U & G \\ C & 0 \end{pmatrix} \begin{pmatrix} \mathbf{x}' \\ \mathbf{x}_m \end{pmatrix} = \begin{pmatrix} \mathbf{f}' \\ \mathbf{f}_1 \end{pmatrix}. \tag{3}$$

Here, U is upper triangular matrix and admits the following form:

$$U = \begin{pmatrix} b_{2,1}I_n\, T + b_{2,2}I_n & b_{2,3}I_n & \cdots & 0 \\ 0 & b_{3,2}I_n\, T + b_{3,3}I_n & \cdots & 0 \\ \vdots & \ddots & \ddots & \vdots \\ 0 & \cdots & \cdots & 0\, b_{m,m-1}I_n \end{pmatrix}.$$

The remaining block matrices and vectors read as follows:

$$G = (0,\ldots,0, b_{m-1,m}I_n, T + b_{m,m}I_n)^T,\ C = (T + b_{1,1}I_n, b_{1,2}I_n, 0, \ldots, 0),$$
$$\mathbf{x}' = (\mathbf{x}_1, \mathbf{x}_2, \ldots, \mathbf{x}_{m-1})^T,\ \mathbf{f}' = (\mathbf{f}_2, \mathbf{f}_3, \ldots, \mathbf{f}_m)^T.$$

In order to find the solution of (3), one makes use of the following block factored form of the matrix,

$$\begin{pmatrix} U & 0 \\ C & -CU^{-1}G \end{pmatrix} \begin{pmatrix} I & U^{-1}G \\ 0 & I \end{pmatrix} \begin{pmatrix} \mathbf{x}' \\ \mathbf{x}_m \end{pmatrix} = \begin{pmatrix} \mathbf{f}' \\ \mathbf{f}_1 \end{pmatrix}. \tag{4}$$

This system is solved by successive solution of the following systems:

$$\begin{vmatrix} U\mathbf{y}_1 = \mathbf{f}' \\ -CU^{-1}G\mathbf{x}_m = \widetilde{\mathbf{f}_1} = \mathbf{f}_1 - C\mathbf{y}_1 \\ U\mathbf{x}' = -G\mathbf{x}_m + \mathbf{f}' \end{vmatrix}. \tag{5}$$

The standard backward recurrence is employed to solve systems with U. I.e., consider $U\underline{\xi} = \mathbf{g}$, where $\underline{\xi} = (\underline{\xi}_1, \underline{\xi}_2, \ldots, \underline{\xi}_{m-1})^T$, $\mathbf{g} = (\mathbf{g}_1, \mathbf{g}_2, \ldots, \mathbf{g}_{m-1})^T$. Then,

$$\underline{\xi}_{m-1} = \tfrac{1}{b_{m,m-1}} \mathbf{g}_{m-1};$$
for $i = m - 1$ down to 2
$$\underline{\xi}_{i-1} = \tfrac{1}{b_{i,i-1}} \left(\mathbf{g}_{i-1} - b_{i,i}\underline{\xi}_i - T\underline{\xi}_i - b_{i,i+1}\underline{\xi}_{i+1} \right)$$
end.

As readily seen, the computational cost is $\mathcal{O}(nm)$ operations.

The system with the Schur complement $-CU^{-1}G$ is equivalent to the incomplete solution of a system with the original matrix A and with a sparse right-hand side with only one non-zero block component; namely,

$$A\widetilde{\mathbf{x}} = \begin{pmatrix} \widetilde{\mathbf{f}_1} \\ 0 \\ \vdots \\ 0 \end{pmatrix}. \tag{6}$$

The only block component which is needed is $\widetilde{\mathbf{x}}_m = \mathbf{x}_m$. This is seen by the following argument; one may apply the same reordering and block factorization to (6) in the same manner as to the original system (2). Since now $\mathbf{f}' = 0$ the resulting system (5) will equal $-CU^{-1}G\mathbf{x}_m = \mathbf{f}_1$.

To solve the last system (6) ALGORITHM SRHS is used. Here $d = 1, j_1 = 1$ and $r = 1, j'_1 = m$. According to Proposition 21 this step of MA requires $\mathcal{O}(nm)$ operations.

For the computational complexity of MA we summarize:

Theorem 1. *The marching algorithm in combination with the separation of variables technique for the incomplete solution of the reduced system requires an optimal cost of operations for solving problems with separable variables (2); namely, the cost is $\mathcal{O}(nm)$ operations.*

2.3 The Generalized Marching Algorithm (GMA)

In [2,3] it is demonstrated that the recurrence used for solving systems with the upper triangular matrix U is unstable and hence the marching algorithm is unstable for large m. This makes the marching algorithm of practical interest only if the length of this recurrence is small, i.e., for $m \ll n$.

When $m = n$ or of same order, to solve the problem (2) one may use the generalized marching algorithm (GMA) which we now describe.

For ease of presentation, let $m = n$, and $n + 1 = p(k + 1)$ for some integers p and k. Consider now the following reordering of the unknown vector $\mathbf{x}$: all rows of $\mathbf{x}$ of multiplicity $k + 1$ form its second block component $\mathbf{x}^{(2)}$.

This reordering of $\mathbf{x}$ induces the following symmetric block odd–even reordering of the original matrix A; namely, one gets a reordered form $\widetilde{A}$ of A, where $\widetilde{A} = (\widetilde{A}_{r,s})_{r,s=1}^2$. Note, that the first block $\widetilde{A}_{1,1}$ is block–diagonal, since the block $\mathbf{x}^{(2)}$ is a separator; it partitions the $n \times n$ grid into p strips with k grid lines. More specifically, the following block form of $\widetilde{A}$ is obtained:

$$\widetilde{A} = \begin{pmatrix} A_1^{(k)} & & 0 & \vdots & \\ & \ddots & & \vdots & \widetilde{A}_{1,2} \\ 0 & & A_p^{(k)} & \vdots & \\ \cdots & \cdots & \cdots & & \\ & \widetilde{A}_{2,1} & & & \widetilde{A}_{2,2} \end{pmatrix} = \begin{pmatrix} \widetilde{A}_{1,1} & \widetilde{A}_{1,2} \\ \widetilde{A}_{2,1} & \widetilde{A}_{2,2} \end{pmatrix},$$

where corresponding blocks are defined by:

$\widetilde{A}_{1,1} = blockdiag(A_s^{(k)})_{s=1}^p,$
$A_s^{(k)} = I_k \otimes T + B_s^{(k)} \otimes I_n,$
$B_s^{(k)} = tridiag(b_{k_s+i,k_s+i-1}, b_{k_s+i,k_s+i}, b_{k_s+i,k_s+i+1})_{i=1}^k,$
$\widetilde{A}_{2,2} = blockdiag(T + b_{k_s+1,k_s+1} I_n), \ k_s = (s-1)(k+1), s = 1, \ldots, p.$

The components of the solution vector $\mathbf{x}$ and the right-hand side vector $\mathbf{f}$ are grouped as follows:

$$\mathbf{x} = \begin{pmatrix} \mathbf{x}^{(1)} \\ \mathbf{x}^{(2)} \end{pmatrix}, \ \mathbf{x}^{(1)} = \begin{pmatrix} \mathbf{x}_1^{(1)} \\ \vdots \\ \mathbf{x}_p^{(1)} \end{pmatrix}, \ \mathbf{x}_s^{(1)} = \begin{pmatrix} x_{(s-1)(k+1)+1} \\ \vdots \\ x_{(s-1)(k+1)+k} \end{pmatrix}, s = 1, \ldots, p,$$

$$\mathbf{f} = \begin{pmatrix} \mathbf{f}^{(1)} \\ \mathbf{f}^{(2)} \end{pmatrix}, \ \mathbf{f}^{(1)} = \begin{pmatrix} \mathbf{f}_1^{(1)} \\ \vdots \\ \mathbf{f}_p^{(1)} \end{pmatrix}, \ \mathbf{f}_s^{(1)} = \begin{pmatrix} f_{(s-1)(k+1)+1} \\ \vdots \\ f_{(s-1)(k+1)+k} \end{pmatrix}, s = 1, \ldots, p,$$

$$\mathbf{x}^{(2)} = \left(\mathbf{x}_{k+1}, \ldots, \mathbf{x}_{s(k+1)}, \ldots, \mathbf{x}_{(p-1)(k+1)}\right)^T,$$
$$\mathbf{f}^{(2)} = \left(\mathbf{f}_{k+1}, \ldots, \mathbf{f}_{s(k+1)}, \ldots, \mathbf{f}_{(p-1)(k+1)}\right)^T.$$

The reordered matrix $\widetilde{A}$ allows block factorization and the problem now has the form:
$$\begin{pmatrix} \widetilde{A}_{1,1} & 0 \\ \widetilde{A}_{2,1} & S \end{pmatrix} \begin{pmatrix} I & \widetilde{A}_{1,1}^{-1}\widetilde{A}_{1,2} \\ 0 & I \end{pmatrix} \begin{pmatrix} \mathbf{x}^{(1)} \\ \mathbf{x}^{(2)} \end{pmatrix} = \begin{pmatrix} \mathbf{f}^{(1)} \\ \mathbf{f}^{(2)} \end{pmatrix},$$

where $S = \widetilde{A}_{2,2} - \widetilde{A}_{2,1}\widetilde{A}_{1,1}^{-1}\widetilde{A}_{1,2}$ is the Schur complement of $\widetilde{A}$. Two systems with the block-diagonal matrix $\widetilde{A}_{1,1}$ and one with the Schur complement

$$\begin{vmatrix} \widetilde{A}_{1,1}\mathbf{y}^{(1)} = \mathbf{f}^{(1)} \\ (\widetilde{A}_{2,2} - \widetilde{A}_{2,1}\widetilde{A}_{1,1}^{-1}\widetilde{A}_{1,2})\mathbf{x}^{(2)} = \widetilde{\mathbf{f}}^{(2)} \equiv \mathbf{f}^{(2)} - \widetilde{A}_{2,1}\widetilde{A}_{1,1}^{-1}\mathbf{f}^{(1)} \\ \widetilde{A}_{1,1}\mathbf{x}^{(1)} = \mathbf{f}^{(1)} - \widetilde{A}_{1,2}\mathbf{x}^{(2)} \end{vmatrix} \quad (7)$$

have to be solved to compute the solution of the original system.

The systems with $\widetilde{A}_{1,1}$ are solved by applying p times the standard marching algorithm for the subproblems with $A_s^{(k)}$. This procedure requires $\mathcal{O}(npk)$ arithmetic operations.

The length of the recurrence needed for solving systems with the upper triangular blocks is $k - 1 = \frac{n+1}{p} - 2$, and can be controlled by choosing sufficiently large p.

The system with the Schur complement is equivalent to incomplete solution of a system with the original matrix and with a sparse right-hand side; namely the system

$$A\widehat{\mathbf{x}} = \widehat{\mathbf{f}}, \quad \text{where } \widehat{\mathbf{f}}_i = \begin{cases} \widetilde{\mathbf{f}}_s^{(2)}, & i = s(k+1) \\ 0, & i \neq s(k+1) \end{cases}, \quad (8)$$

have to be solved incompletely seeking only $\widehat{\mathbf{x}}_{s(k+1)} = \mathbf{x}_{s(k+1)}$, $s = 1, \ldots, p-1$. The last problem is handled by the fast direct solver called FASV and proposed in [6] (detailed description of FASV may be found in [4]).

Let for definiteness, $p = 2^l$. Since we have to perform l steps of the algorithm FASV, the cost for the incomplete solution of (8) by algorithm FASV is given by the following proposition:

Theorem 2. *The second step of the block-Gaussian elimination based on the incomplete solution of problem (8) using algorithm FASV requires $24n^2 l - 9n^2$ operations.*

Remark 2. The generalized marching algorithm in the form presented in Bank [2] requires in the second step of the block-Gaussian elimination $28n^2 l$ operations. This shows that the algorithm proposed in [5] and presented here has asymptotically a slightly smaller operation count.

Summarizing, one may see that the implementation of the generalized marching algorithm requires a total cost of

$$\mathcal{O}(npk) + 24n^2 l - 9n^2 = \mathcal{O}(n^2) + 24n^2 \log_2(p) - 9n^2$$

arithmetic operations.

3 Numerical Experiments

To experimentally compare the numerical stability of the described algorithms, some numerical tests have been performed (using the HP9000/C110 computer) for the following test problem.

Example: Application to the case $a_1(x_1) = a_2(x_2) \equiv 1$. I.e., the two dimensional Poisson equation with Dirichlet boundary conditions is considered:

$$\left| \begin{array}{l} -\Delta u(x_1, x_2) = f(x), \ x \in \Omega = (0,1) \times (0,1) \\ u = 0, \quad \text{on } \partial\Omega \end{array} \right. .$$

The solution is $u(x_1, x_2) = \sin(\pi x_1)\sin(\pi x_2)$, which implies that the right-hand side is $f(x_1, x_2) = 2\pi^2 \sin(\pi x_1)\sin(\pi x_2)$.

This problem is discretized using the five point finite difference scheme on uniform $n \times n$ ($m = n$, $h_1 = h_2 = h$) mesh with mesh parameter $h = 1/(n+1)$.

The discrete problem is solved using MA and GMA and the results are collected in separate tables (Table 1 and Table 2) for each algorithm, respectively. The first column of both tables shows the mesh–size. In the remaining two

Table 1. Results for algorithm MA

n	l_2-error	Max.error
3	2.651e-02	5.303e-02
7	6.475e-03	1.295e-02
15	1.609e-03	3.219e-03
31	6.086e+08	5.611e+09

columns of Table 1, the l_2– and C–norms of the vector of the pointwise error of the solution computed by MA are given. It is clearly seen from this table, that for this example MA is stable if $m \leq 15$.

The columns of Table 2 are in groups of 3 with similar data as in Table 1. Here we vary the step–size $k - 1$ of the recursion in the GMA. First group contains

Table 2. Results for algorithm GMA ($n = p(k+1) - 1$)

		$k = 3$			$k = 15$	
n	p	l_2-error	Max.error	p	l_2-error	Max.error
7	2	6.475e-03	1.295e-02			
15	4	1.609e-03	3.219e-03			
31	8	4.018e-04	8.036e-04	2	6.950e-04	5.653e-03
63	16	1.004e-04	2.008e-04	4	4.601e-04	6.329e-03
127	32	2.510e-05	5.020e-05	8	1.064e-04	1.900e-03
255	64	6.275e-06	1.255e-05	16	2.558e-05	5.469e-04
511	128	1.569e-06	3.137e-06	32	6.397e-06	1.369e-04
1023	256	3.922e-07	7.844e-07	64	1.798e-06	4.371e-05

the results obtained by GMA for $k = 3$. At any refinement step both l_2- and C-norms of the error in this case decrease 4 times, which means that the algorithm is stable. The results were similar for the case $k = 7$ (not shown in the table). The second group of columns in Table 2 shows that in the case $k = 15$ the stability of GMA is affected, but still the results are acceptable. That is, in practice depending upon the specific floating point arithmetic one has to choose the value of the parameter k very carefully, i.e., it is machine dependent.

4 Concluding Remarks

As demonstrated by the complexity estimates and numerical experiments, the advantage of GMA is clearly seen. In practice, apart from solving separable elliptic problems in a single rectangular domain, the development of fast elliptic solvers is strongly motivated by their potential application to the construction of efficient preconditioners for iterative solution of more general problems on more general domains and meshes, such as: problems with slowly varying coefficients and/or jumping coefficients corresponding to the case of multi-layer media; composite domains, e.g., L-shaped or T-shaped domains; as well as block-preconditioning of coupled systems of partial differential equations including elasticity and Stokes problems. In particular, a further implementation of here considered marching algorithms together with algorithms analyzed in [4] into the framework of domain decomposition and domain embedding methods is of interest in order to handle more general elliptic problems on more realistic general domains and meshes.

Acknowledgments

This research has been supported in part by the Bulgarian NFS Grant MY-I-901/99.

References

1. Axelsson O., Vassilevski, P. S.: Solution of Equations R^n (Part II): Iterative Solution of Linear Systems. In: Ciarlet, P., Lions, J. (eds.): Handbook on Numerical Methods, North Holland, to appear.
2. Bank, R.: Marching Algorithms for Elliptic Boundary Value Problems. II: The Variable Coefficient Case. SIAM J. Numer. Anal. **14** (1977) 950-970
3. Bank, R., Rose, D.: Marching Algorithms for Elliptic Boundary Value Problems. I: The Constant Coefficient Case. SIAM J. Numer. Anal. **14** (1977) 792-829
4. Bencheva, G.: Comparative Performance Analysis of 2D Separable Elliptic Solvers. Proceedings of XII Conference "Software and Algorithms of Numerical Mathematics", Nectiny Castle, Czech Republic (1999), to appear.
5. Vassilevski, P. S.: An Optimal Stabilization of Marching Algorithm. Compt. rend. de l'Acad. bulg. Sci. **41** (1988) No 7, 29-32
6. Vassilevski, P. S.: Fast Algorithm for Solving a Linear Algebraic Problem with Separable Variables. Compt. rend. de l'Acad. bulg. Sci. **37** (1984) No 3, 305-308

Inexact Newton Methods and Mixed Nonlinear Complementary Problems

L. Bergamaschi and G. Zilli

Dipartimento di Metodi e Modelli Matematici per le Scienze Applicate
Università di Padova, via Belzoni 7, 35131 Padova, Italy

Abstract. In this paper we present the results obtained in the solution of sparse and large systems of nonlinear equations by Inexact Newton-like methods [6]. The linearized systems are solved with two preconditioners particularly suited for parallel computation. We report the results for the solution of some nonlinear problems on the CRAY T3E under the MPI environment. Our methods may be used to solve more general problems. Due to the presence of a logarithmic penalty, the interior point solution [10] of a nonlinear mixed complementary problem [7] can indeed be viewed as a variant of an Inexact Newton method applied to a particular system of nonlinear equations. We have applied this inexact interior point algorithm for the solution of some nonlinear complementary problems. We provide numerical results in both sequential and parallel implementations.

1 The Inexact Newton-Cimmino Method

Consider the system on nonlinear equations

$$G(x) = 0 \qquad G = (g_1, ..., g_n)^T \qquad (1)$$

where $G : R^n \to R^n$ is a nonlinear C^1 function, and its Jacobian matrix $J(x)$. For solving (1) we use an iterative procedure which combines a Newton and a Quasi-Newton method with a row-projection (or row-action) linear solver of Cimmino type [11], particularly suited for parallel computation. Here below, referring to block Cimmino method, we give the general lines of this procedure.

Let $As = b$ be the linearized system to be solved. Let us partition A into p row-blocks: $A_i, i = 1, \ldots, p$, i.e. $A^T = [A_1, A_2, \ldots, A_p]$ and partition the vector b conformally. Then the original system is premultiplied (*preconditioning*) by

$$H_p = [A_1^+, \ldots, A_i^+, \ldots, A_p^+] \qquad (2)$$

where $A_i^+ = A_i^T (A_i A_i^T)^{-1}$ is the Moore-Penrose pseudo inverse of A_i.

We obtain the equivalent system $H_p A s = H_p b$,

$$(P_1 + \cdots + P_p)s = \sum_{i=1}^{p} A_i^+ A_i s = \sum_{i=1}^{p} A_i^+ b_i = H_p b, \qquad (3)$$

where for each $i = 1,\ldots,p$, $P_i = A_i^+ A_i$ is the orthogonal projection onto range(A_i^T). As A is non singular, the matrix $H_p A \sum_{i=1}^{p} A_i^+ A_i$ is symmetric and positive definite. Then the solution of (3) is approximated by the Conjugate Gradient(CG) method. The q (underdetermined) linear least squares subproblems in the pseudoresidual unknowns $\delta_{i,k}$

$$A_i \delta_{i,k} = (b_i - A_i s_k), \quad 1 \leq i \leq p \tag{4}$$

must be solved at *each* conjugate gradient iteration ($k = 1, 2\ldots$).

Combining the classic Newton method and the block Cimmino method we obtain the block Inexact Newton-Cimmino algorithm [11], in which at a major outer iteration the linear system $J(x_k)s = -G(x_k)$, where $J(x^*)$ is the Jacobian matrix, is solved in *parallel* by the block Cimmino method.

In [12,4] a simple p-block partitioning of the Jacobian matrix A was used for solving *in parallel* a set of nonlinear test problems with sizes ranging from 1024 to 131072 on a CRAY T3E under the MPI environment. The least squares subproblems (4) were solved *concurrently* with the iterative Lanczos algorithm LSQR.

In this paper (see in section 4) we adopt a suitable block row partitioning of the matrix A in such a way that $A_i A_i^T = I$, $i = 1,\ldots p$, and consequently, $A_i^+ = A_i^T$. This simplify the solution of the subproblems (4).

Due to the costly communication routines needed in this approach we have also implemented in parallel the preconditioned BiCGstab for the solution of the linearized system. As the preconditioner we choose AINV [2] which is based on the approximate sparse computation of the inverse of the coefficient matrix.

2 Inexact Newton Method for Nonlinear Complementary Problems

The methods of section 1 may be used to solve more general problems as the nonlinear mixed complementary problems [7] (including linear and nonlinear programming problems, variational inequalities, control problems, etc.).

Let us consider the following system of constrained equations:

$$F(v,s,z) = \begin{pmatrix} G(v,s,z) \\ SZe \end{pmatrix} = 0 \quad (s,z) \geq 0 \tag{5}$$

where $G : \mathbb{R}^{n+2m} \to \mathbb{R}^n$ is a nonlinear function of v, $S = \text{diag}(s_1,\ldots,s_m)$, $Z = \text{diag}(z_1,\ldots,z_m)$, $e = (1,\ldots 1)^T$. The interior point methods [10] for the solution of (5) require the solution of the nonlinear system $F(x) = 0$. Using the Newton method we have to solve at every iterations a linear system of the form

$$F'(x_k)\Delta x = -F(x_k) + \sigma_k \mu_k e_0 \tag{6}$$

where $\mu_k = (s_k^T z_k)/m$, $\sigma_k \in]0,1[$, that is an Inexact Newton method [6].

An interior point method in which the linearized system is solved approximately (by means of an iterative method) will be called *inexact (truncated) interior point method*. In this framework system (6) becomes

$$F'(x_k)\Delta x = -F(x_k) + \sigma_k \mu_k e_0 + r_k \qquad (7)$$

where r_k is the residual of the iterative method applied to the linear system satisfying $\|r_k\| \leq \eta_k \mu_k$, and η_k is, for every k, the *forcing term* of the inexact Newton method [6]. Global convergence is assured by means of backtracking [1].

3 Numerical Results I (Sequential)

We have applied the *inexact interior point methods* for the solution of two nonlinear complementarity problems: the (sparse) obstacle Bratu problem [5,8], and the (dense) Lubrication problem [9,5].

3.1 The Obstacle Bratu Problem

This problem can be formulated as a nonlinear system of equations:

$$f(v) = z_1 - z_2, \quad Z_1 S_1 e = 0, \quad Z_2 S_2 e = 0, \quad s_1 = v - v_l, \quad s_2 = v_u - v, \qquad (8)$$

with the constraint $s_i, z_i \geq 0$, $i = 1, 2$. The nonlinear function $f(v)$ is defined as

$$f(v) = Av - \lambda h^2 E(v) e, \qquad E(v) = \mathrm{diag}(\exp(v_1), \ldots, \exp(v_n)),$$

where A is the matrix arising from FD discretization of the Laplacian on the unitary square with homogeneous Dirichlet boundary conditions, v_l, v_u are the obstacles and h is the grid spacing.

The system (7) at step k can be written as

$$\begin{bmatrix} B & 0 & 0 & -I & I \\ 0 & Z_1 & 0 & S_1 & 0 \\ 0 & 0 & Z_2 & 0 & S_2 \\ -I & I & 0 & 0 & 0 \\ I & 0 & I & 0 & 0 \end{bmatrix} \begin{bmatrix} \Delta v \\ \Delta s_1 \\ \Delta s_2 \\ \Delta z_1 \\ \Delta z_2 \end{bmatrix} = \begin{bmatrix} -f + z_1 - z_2 \\ -Z_1 S_1 e + \sigma_k \mu_k e \\ -Z_2 S_2 e + \sigma_k \mu_k e \\ -s_1 + v - v_l \\ -s_2 + v_u - v \end{bmatrix} \equiv \begin{bmatrix} b_1 \\ b_2 \\ b_3 \\ b_4 \\ b_5 \end{bmatrix}$$

where $B = f'(v)$. Taking into account the simple structure of some of the block matrices, we can use a Schur complement approach to reduce the original system ($5n \times 5n$) to a system with n rows and n columns. In this way we obtain a system in the Δv unknown only: $C\Delta v = r$, where

$$C = B + S_1^{-1} Z_1 + S_2^{-1} Z_2, \quad r = b_1 + S_1^{-1}(b_2 - Z_1 b_4) - S_2^{-1}(b_3 - Z_2 b_5).$$

Once this nonsymmetric system has been solved (we used the BiCGstab solver), we can compute $\Delta z_1, \Delta z_2$ and $\Delta s_1, \Delta s_2$ by:

$$\Delta z_1 = S_1^{-1} Z_1 (b_2 - b_4 - \Delta v), \qquad \Delta s_1 = b_4 + \Delta v$$

Table 1. Results obtained with the *inexact interior point Newton method* for the obstacle Bratu problem with three different mesh sizes and four values of λ (nl= nonlinear, it= iterations, s=seconds on a 600 Mhz Alpha workstation)

λ	n	nl it.	tot lin it.	CPU (s)	$\|H\|$
1	1024	13	397	0.24	0.12371E-09
4	1024	11	335	0.23	0.21711E-08
6	1024	12	374	0.25	0.13904E-12
1	4096	14	873	2.61	0.12668E-08
4	4096	13	828	2.56	0.13289E-10
6	4096	12	840	2.57	0.44356E-11
1	16384	15	1779	36.21	0.34762E-08
4	16384	14	1700	34.46	0.30286E-09
6	16384	14	2021	40.68	0.79480E-09

$$\Delta z_2 = S_2^{-1} Z_2 (b_3 - b_5 + \Delta v), \qquad \Delta s_2 = b_5 - \Delta v.$$

We may note that matrix C is is obtained by adding to B the two nonnegative diagonal matrices $D_l^{-1} Z_1$ and $D_u^{-1} Z_2$ thus enhancing its diagonal dominance. The algorithm has been tested for different grids $h = 1/32, 1/64, 1/128$ with values of $n = 1024, 4096, 16384$, respectively, for different values of $\lambda = 1, 4, 6, 10$. The initial vectors for the experiments are $v^{(0)} = s^{(0)} = z^{(0)} = [1, \ldots, 1]^T$ with the obstacles $v_l = [0, \ldots, 0]^T$, $v_u = [4, \ldots, 4]^T$. For the last λ-value we reported a failure since a number of backtracking larger than the allowed maximum (=5) have been recorded. Actually, for $\lambda > 6.8$ the algorithm did not achieve convergence (this result is well documented in the literature, see [8]). The sequential results for the cases $\lambda = 1, 4, 6$ are reported in Table 1. The CPU times refer to the computation on a 600 Mhz Alpha workstation with 512 Mb RAM.

3.2 The Lubrication Problem

A very difficult problem from the point of view of nonlinearity is represented by the Elastohydrodynamic Lubrication Problem [5] which consists of two integral equations coupled with a differential equation – the Reynold's equation. Given the parameters α, λ and an inlet point x_a, find the pressure $p(x)$, the thickness $h(x)$, the free boundary x_b and the variable k satisfying:

$$h(x) = x^2 + k - \frac{2}{\pi} \int_{x_a}^{x_b} \ln|x-s|\, ds \quad \text{in } [x_a, \infty)$$

$$\frac{d}{dx}\left(\frac{h^3(x)}{e^{\alpha p}} \frac{dp}{dx}\right) = \lambda \frac{dh}{dx} \quad \text{in } [x_a, x_b], \qquad \frac{2}{\pi}\int_{x_a}^{x_b} p(s)\, ds = 1$$

with the free boundary conditions $p(x_a) = 0$ and $p(x_b) = \dfrac{dp}{dx}(x_b) = 0$. The discretization of this problems yields a highly nonlinear and dense system of equations. We solve the linearized system with a direct method (Lapack routines). In Table 2 we show the results obtained with the *inexact interior point*

Table 2. Results obtained for the Lubrication Problem with the *inexact interior point Newton method* with $\alpha = 2.832$, $\lambda = 6.057$

n	nl it.	CPU (s)	Jacobian	LU factor	LU solver	$\|H\|$
200	14	1.30	0.61	0.28	0.01	0.18840E-06
1000	19	69.41	20.75	35.59	0.68	0.13107E-06

Newton method with $\alpha = 2.832$, $\lambda = 6.057$ using $n = 200$ and $n = 1000$ points of the discretization of the interval $[-3, 2]$. The initial vector is chosen as

$$x_i^{(0)} = \frac{3}{4}\left(1 - \frac{|x_a + hi|}{2}\right), \quad h = \frac{5}{n}.$$

4 Numerical Results II (Parallel)

Parallel results for nonlinear problems. Here we show the results obtained in the solution of the nonlinear system (1) applying the Newton-Cimmino method. As we mentioned above at the end of section 1, to overcome the problem of the costly solution of the least square subproblems (4), we adopt a suitable *block row partitioning* of the matrix A in such a way that $A_i A_i^T = I$, $i = 1, \ldots q$, and consequently, $A_i^+ = A_i^T$. This partitioning [11] is always possible for every sparse matrix and produces a number q of blocks A_i whose rows are mutually orthogonal. The numerical results of Table 3 were obtained on a CRAY T3E under the MPI environment for two sparse problems (also solved in [12,4] adopting a simple block partitioning, using the iterative algorithm LSQR), which arise from Finite Difference discretization in the unit square Ω of the following PDEs:

1. Poisson problem $\quad -\Delta u - \dfrac{u^3}{1 + x^2 + y^2} = 0 \quad \text{in } \Omega, \quad + \text{b.c.} \quad (9)$

2. Bratu problem [8] $\quad -\Delta u - \lambda e^u = 0 \quad \text{in } \Omega, \quad \lambda \in \mathbb{R}, \quad + \text{b.c.} \quad (10)$

The linear system is solved using a tolerance $\varepsilon_2 = 10^{-5}$ while the Newton iteration stops whenever the relative residual norm is less than $\varepsilon_2 = 10^{-4}$. From Table 3 we can see that the speedups are not completely satisfactory, reaching the maximum value of 1.4 for $p = 4$ processors. This fact is mainly due to the cost of the communication routine MPI_ALLREDUCE which performs the communication of the local pseudoresiduals and their sums on every processor. This operation is costly, and its cost increases with the number of processors.

A most effective parallel solution was obtained with the use of a standard Krylov method with AINV as a preconditioner [2]. AINV is based on the incomplete construction of a set of biconjugate vectors. This process produces two triangular factors Z and W and a diagonal matrix D so that: $A^{-1} \approx ZD^{-1}W^T$. Therefore, application of the preconditioner consists in two matrix-vector products and a diagonal scaling. These matrix-vector products have been parallelized exploiting data locality as in [3], minimizing in this way the communication among processors. The incompleteness of the process is driven by a tolerance parameter ε. Previous (sequential) experimental results show that a choice of $\varepsilon \in [0.05, 0.1]$ leads to a good convergence of the Krylov subspace methods, very similar to that obtained using the ILU preconditioner. In our test cases we choose $\varepsilon = 0.05$.

In Table 4 we show the results when BiCGstab is employed as the linear solver using both AINV and the diagonal scaling (Jacobi) as the preconditioners. The CPU time on p processors (T_p) is measured in seconds on a CRAY T3E. From the results we note that for the small problem ($n = 4096$), as expected, the speedups S_p are not very high. However, for the $n = 65\,536$ problem they reach a value of 19 (AINV) and 21 (Jacobi) on 32 processors. Note that in all the

From the table we note that the major part of the computation is represented by the construction and factorization of the Jacobian matrix. This suggests that a Quasi-Newton approach may drastically reduce the CPU time of a single iteration. In Figure 1 the nonlinear convergence profile is provided, showing the superlinear rate of the convergence of the Inexact Newton method. Figures 2 and 3 display the plots of the film thickness and the pressure, respectively. They compare well with the results of the literature [9].

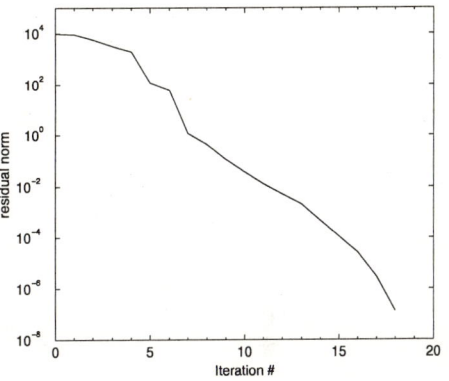

Fig. 1. Convergence profile

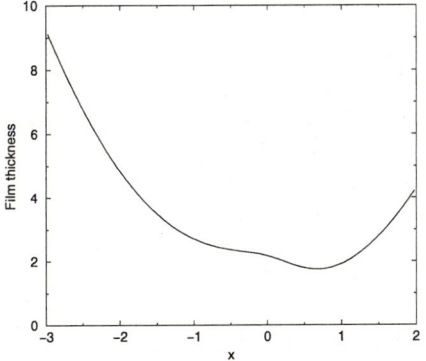

Fig. 2. Film thickness

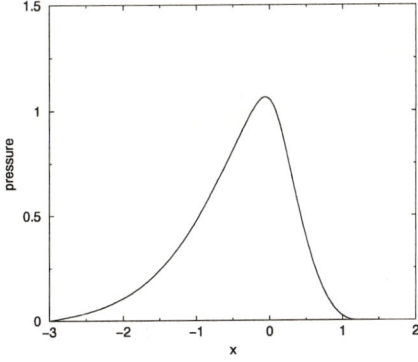

Fig. 3. Pressure

Table 3. Time (in seconds), speedups $S_p = T_1/T_p$, number of outer and inner iterations k_{NEWT}, k_{CG}, obtained for solving the two test problems, using the row-orthogonal partitioning on a CRAY T3E under the MPI environment

Poisson problem					
$n = 4096$	$p = 1$	$p = 2$	$p = 4$	$p = 8$	$p = 16$
Time (speedup)	7.22	5.74 (1.3)	5.79 (1.3)	6.05 (1.2)	6.83 (1.1)
k_{NEWT}	2	2	2	2	2
k_{CG}	1123	1123	1123	1123	1123

Bratu problem					
$n = 4096$	$p = 1$	$p = 2$	$p = 4$	$p = 8$	$p = 16$
Time (speedup)	6.70	5.11 (1.3)	4.96 (1.4)	5.33 (1.3)	8.01 (0.8)
k_{NEWT}	4	4	4	4	4
k_{CG}	630	630	630	630	630

runs the CPU time needed by AINV is less than the one required by Jacobi. Moreover, the AINV preconditioner shows a degree of parallelism comparable with that of the diagonal scaling.

Parallel results for nonlinear complementary problems. As in section 4, we also solved in parallel the obstacle Bratu problem (8) via the *inexact interior point method*, using the BiCGstab method as linear solver with AINV and Jacobi as preconditioners. In Table 5 we show the results obtained. The same considerations of section 4 hold, even with larger speedup values.

5 Conclusions and Future Topics

In this paper we experimented that the Inexact Newton method performs well in solving nonlinear problems and mixed nonlinear complementary problems both in sequential and in parallel computations. We adopted two different parallel preconditioners in the iterative solution of sparse problems: the row-action Cimmino method [11] and the incomplete inverse AINV [2]. While the latter obtains good results (speedup values up to 23 with 32 processor), the former heavily suffers for the overhead due to the MPI communication routines. Future work will address the parallel implementation of an Inexact Quasi-Newton interior point method applied to the solution of mixed nonlinear complementary problems.

Table 4. Results obtained on the CRAY T3E for the Bratu problem employing the AINV and Jacobi preconditioners

n	p	AINV(0.05) T_p	nl	lin	S_p	Jacobi T_p	nl	lin	S_p
4096	1	0.74	3	57	–	0.92	4	134	–
	2	0.47	3	57	1.58	0.57	4	133	1.61
	4	0.33	3	57	2.24	0.42	4	135	2.19
	8	0.25	3	57	2.96	0.33	4	134	2.78
16 384	1	5.06	3	107	–	5.46	4	243	–
	2	2.82	3	107	1.79	3.32	4	268	1.64
	4	1.53	3	107	3.31	1.82	4	270	3.00
	8	0.90	3	107	5.62	1.17	4	266	4.67
	16	0.61	3	107	8.29	0.74	4	257	7.38
65 536	1	40.75	4	224	–	41.56	4	497	–
	2	21.02	4	223	1.93	20.42	4	479	2.03
	4	11.00	4	225	3.70	13.31	5	582	3.12
	8	5.79	4	224	7.03	5.76	4	462	7.21
	16	3.48	4	224	11.70	3.24	4	466	12.82
	32	2.15	4	219	18.95	2.05	4	493	20.27

Table 5. Results obtained on a CRAY T3E for the obstacle Bratu problem employing the AINV and Jacobi preconditioners

n	p	AINV(0.05) T_p	nl	lin	S_p	Jacobi T_p	nl	lin	S_p
4096	1	5.30	14	402	–	5.82	14	946	–
	2	2.96	14	405	1.73	3.23	14	938	1.76
	4	1.69	14	400	2.75	1.99	14	941	2.60
	8	1.05	14	400	3.89	1.23	14	948	3.69
	16	0.73	14	403	4.83	0.92	14	943	4.72
16 384	1	41.66	15	823	–	46.05	15	1959	–
	2	21.56	15	820	1.93	24.11	15	1943	1.91
	4	11.10	15	795	3.75	13.12	15	1967	3.51
	8	6.21	15	819	6.71	7.13	15	1968	6.46
	16	3.75	15	828	11.11	4.43	15	1970	10.40
	32	2.56	15	821	16.27	2.88	15	1950	15.98
65 536	1	321.32	16	1688	–	346.82	16	4026	–
	2	162.25	16	1651	1.93	180.30	16	4022	1.92
	4	84.64	16	1676	3.79	94.47	16	4022	3.67
	8	50.22	16	1672	6.39	51.39	16	4046	6.74
	16	24.31	16	1706	13.21	28.28	16	4043	12.26
	32	14.04	16	1675	22.88	15.05	16	4027	23.04

References

1. S. Bellavia An Inexact Interior Point method *Journal of Optimization Theory and Applications*, vol. 96, 1 (1998).
2. M. Benzi and M. Tuma, A sparse approximate inverse preconditioner for nonsymmetric linear systems, *SIAM J. Sci. Comput.*, 19 (1998), pp. 968–994.
3. L. Bergamaschi and M. Putti. Efficient parallelization of preconditioned conjugate gradient schemes for matrices arising from discretizations of diffusion equations. In *Proceedings of the Ninth SIAM Conference on Parallel Processing for Scientific Computing*, March, 1999. (CD–ROM).
4. L. Bergamaschi, I. Moret, and G. Zilli, Inexact block Quasi-Newton methods for sparse systems of nonlinear equations, *J. Fut. Generat. Comput. Sys.* (2000) (in print).
5. I. Bongartz, I., A. R. Conn, N. I. M. Gould and P. L. Toint, CUTE:*Constrained and unconstrained testing environment*, Research Report, IBM T. J. Watson Research Center, Yorktown Heights, NY, 1993.
6. R. S. Dembo, S C. Eisenstat, and T. Steihaug, Inexact Newton methods, *SIAM. J. Numer. Anal.* 19, 400-408, 1982.
7. S. P. Dirske, and M. C. Ferris *MCLIB: A collection of nonlinear mixed complementary problems*, Tech. Rep., CS Depth., University of Winsconsin, Madison, WS, 1994.
8. D. R. Fokkema, G. L. G. Slejipen and H. A. Van der Vorst, Accelerated Inexact Newton schemes for large systems of nolinear equations, *SIAM J. Sci. Comput.*, 19 (2), 657-674, 1997.
9. M. M. Kostreva, Elasto-hydroninamic lubrication: A non-linear complementary problem. *International Journal for Numerical Methods in Fluids*, 4:377-397, 1984.
10. S. J. Wright, *Primal-Dual Interior-Point Methods*, Siam, Philadelphia, 1997.
11. G. Zilli, Parallel method for sparse non-symmetric linear and non-linear systems of equations on a transputer network, *Supercomputer*, 66–XII-4, 4-15, 1996.
12. G. Zilli and L. Bergamaschi, Parallel Newton methods for sparse systems of non-linear equations, *Rendiconti del Circolo Matematico di Palermo* II-58, 247-257, 1999.

Skew-Circulant Preconditioners for Systems of LMF-Based ODE Codes

Daniele Bertaccini[1]* and Michael K. Ng[2]**

[1] Dipartimento di Matematica, University of Firenze
viale Morgagni, 67/a, 50134 Firenze, Italy
bertaccini@na-net.ornl.gov
[2] Department of Mathematics, The University of Hong Kong
Pokfulam Road, Hong Kong
mng@maths.hku.hk

Abstract. We consider the solution of ordinary differential equations (ODEs) using implicit linear multistep formulae (LMF). More precisely, here we consider Boundary Value Methods. These methods require the solution of one or more unsymmetric, large and sparse linear systems. In [6], Chan et al. proposed using Strang block-circulant preconditioners for solving these linear systems. However, as observed in [1], Strang preconditioners can be often ill-conditioned or singular even when the given system is well-conditioned. In this paper, we propose a nonsingular skew-circulant preconditioner for systems of LMF-based ODE codes. Numerical results are given to illustrate the effectiveness of our method.

1 Introduction

In this paper, we consider the solution of ordinary differential equations (ODEs) by using implicit Linear Multistep Formulae (LMF). By applying the above formulae, the solution to a given ODE is given by the solution of a linear system

$$M\mathbf{y} = \mathbf{b}, \tag{1}$$

where M depends on the LMF used.

Here, we concentrate on the linear initial value problem

$$\begin{cases} \dfrac{d\mathbf{y}(t)}{dt} = J_m \mathbf{y}(t) + \mathbf{g}(t), \; t \in (t_0, T], \\ \mathbf{y}(t_0) = \mathbf{z}, \end{cases} \tag{2}$$

where $\mathbf{y}(t), \mathbf{g}(t) : \mathbb{R} \to \mathbb{R}^m$, $\mathbf{z} \in \mathbb{R}^m$, and $J_m \in \mathbb{R}^{m \times m}$ integrated using Boundary Value Methods (BVMs), a class of numerical methods based on the linear

---
* Research supported in part by Italian Ministry of Scientific Research.
** Research supported in part by Hong Kong Research Grants Council Grant No. HKU 7147/99P and UK/HK Joint Research Scheme Grant No. 20009819.

multistep formulae (LMF) (see [4] and references therein). A BVM approximates its solution by means of a discrete boundary value problem. By using a μ-step LMF over a uniform mesh $t_j = t_0 + jh$, for $0 \leq j \leq s$, with $h = (T - t_0)/s$, we have

$$\sum_{i=-\nu}^{\mu-\nu} \alpha_{i+\nu} y_{n+i} = h \sum_{i=-\nu}^{\mu-\nu} \beta_{i+\nu} f_{n+i}, \quad n = \nu, \ldots, s - \mu + \nu. \tag{3}$$

Here, $\mathbf{y}_n$ is the discrete approximation to $\mathbf{y}(t_n)$, $\mathbf{f}_n = J_m \mathbf{y}_n + \mathbf{g}_n$ and $\mathbf{g}_n = \mathbf{g}(t_n)$.

The BVM in (3) must be used with ν initial conditions and $\mu - \nu$ final conditions. That is, we need the values $\mathbf{y}_0, \cdots, \mathbf{y}_{\nu-1}$ at $t = t_0$ and the values $\mathbf{y}_{n+\mu-\nu-1}, \cdots, \mathbf{y}_n$ at $t = T$. The initial condition in (2) only provides us with one value. In order to obtain the other initial and final values, we have to provide additional $(\mu - 1)$ equations. The coefficients $\alpha_i^{(j)}$ and $\beta_i^{(j)}$ of these equations should be chosen such that the truncation errors for these initial and final conditions are of the same order as that in (3), see [4, p,132]. By combining (3) with the additional methods, we obtain a linear system as in (1).

The discrete problem (1) generated by the above process is given by

$$M\mathbf{y} \equiv (A \otimes I_m - hB \otimes J_m)\mathbf{y} = \mathbf{e}_1 \otimes \mathbf{z} + h(B \otimes J_m)\mathbf{g}, \tag{4}$$

where $\mathbf{e}_1 = (1, 0, \cdots, 0)^t \in \mathbb{R}^{(s+1)}$, $\mathbf{y} = (\mathbf{y}_0, \cdots, \mathbf{y}_s)^t \in \mathbb{R}^{(s+1)m}$, $\mathbf{g} = (\mathbf{g}_0, \cdots, \mathbf{g}_s)^t \in \mathbb{R}^{(s+1)m}$, and A and B are $(s+1)$-by-$(s+1)$ matrices given by:

$$A = \begin{pmatrix} 1 & \cdots & 0 & & & & & \\ \alpha_0^{(1)} & \cdots & \alpha_\mu^{(1)} & & & & & \\ \vdots & \vdots & \vdots & & & 0 & & \\ \alpha_0^{(\nu-1)} & \cdots & \alpha_\mu^{(\nu-1)} & & & & & \\ \alpha_0 & \cdots & \alpha_\mu & & & & & \\ & \alpha_0 & \cdots & \alpha_\mu & & & & \\ & & \ddots & \ddots & \ddots & & & \\ & & & \ddots & \ddots & & \ddots & \\ & & & & \alpha_0 & \cdots & \alpha_\mu & \\ & 0 & & & \alpha_0^{(s-\mu+\nu+1)} & \cdots & \alpha_\mu^{(s-\mu+\nu+1)} & \\ & & & & \vdots & & \vdots & \\ & & & & \alpha_0^{(s)} & \cdots & \alpha_\mu^{(s)} & \end{pmatrix}$$

and B can be defined similarly. The size of the matrix M is very large when h is small and/or m is large. If a direct method is used to solve the system (4), e.g., in the case of a d-level structure arising in d-dimensional partial differential equations, the operation count can be much higher for practical applications (see the numerical comparisons with a band solver in [2]).

In [1,2], Bertaccini proposed to use Krylov subspace methods such as the Saad and Schultz's GMRES method to solve (1). In order to speed up the convergence rate of Krylov subspace methods, he proposed circulant matrices as preconditioners. The first preconditioner proposed in [1,2] for the matrix M in (4) is the well-known T. Chan circulant preconditioner, see [5]. The second one proposed in [1,2] is a new preconditioner that he called the P-circulant preconditioner. Moreover, Bertaccini [2] and Chan et al. [6] proposed the generalized Strang preconditioner for (4). They showed theoretically and numerically that both the P-circulant and generalized Strang preconditioned systems converge very quickly. However, when J_m is singular (for instance in some ODEs, see [1]), the matrix S is singular. The main aim of this paper is to propose a nonsingular block skew-circulant preconditioner for M.

We stress that, in the current literature, there exist some algorithms for banded Toeplitz linear systems whose theoretical computational cost is lower (see e.g. [7] and references in [5]). They are very effective in the symmetric positive definite case. However, the linear system (4) is usually unsymmetric and can have a high condition number, even if slowly growing with s (at most linearly, see [3,4]). Thus, if the normal equations approach is used to solve (4), care should be used in order to avoid possible severe numerical instability and/or very slow convergence if an iterative solver is used, even for simple problems, as observed in [3]. Moreover, notice that the diagonalization of the Jacobian matrix J_m in (4), is usually very expensive (if possible) when m is large and can be an ill-conditioned problem. Thus, the use of a solver that involves explicitly the above decomposition can be not appropriate (see also implementation details in [2]).

The paper is organized as follows. In §2, we introduce the new block skew-circulant preconditioner in and give the convergence analysis of our method. Finally, numerical examples are given in §3.

2 Construction of Skew-Circulant Preconditioners

In [1], Bertaccini proposed to use Krylov subspace methods with block-circulant preconditioners for solving (4). Two preconditioners were considered. The first one is the T. Chan block-circulant preconditioner T. It is defined as

$$T = c(A) \otimes I_m - hc(B) \otimes J_m \qquad (5)$$

where $c(A)$ is the minimizer of $\|A - C\|_F$ over all $(s+1)$-by-$(s+1)$ circulant matrices C under the Frobenius norm $\|\cdot\|_F$, see [5], and $c(B)$ is defined similarly. More precisely, the diagonals $\hat{\alpha}_j$ and $\hat{\beta}_j$ of $c(A)$ and $c(B)$ are given by

$$\hat{\alpha}_j = \left(1 - \frac{j}{s+1}\right)\alpha_{j+\nu} + \frac{j}{s+1}\alpha_{j+\nu-(s+1)}, \quad j = 0, 1, \cdots, s,$$

and $\hat{\beta}_j$ similarly but with $\beta_{j+\nu}$ instead of $\alpha_{j+\nu}$, respectively. The second preconditioners proposed in [1,2] is called the P-circulant preconditioner. It is defined as

$$P = \tilde{A} \otimes I_m - h\tilde{B} \otimes J_m \qquad (6)$$

where the diagonals $\tilde{\alpha}_j$ and $\tilde{\beta}_j$ of $\tilde{A}$ and $\tilde{B}$ are given by

$$\tilde{\alpha}_j = \left(1 + \frac{j}{s+1}\right)\alpha_{j+\nu} + \frac{j}{s+1}\alpha_{j+\nu-(s+1)}, \quad j = 0, 1, \cdots, s,$$

and $\hat{\beta}_j$ similarly but with $\beta_{j+\nu}$ instead of $\alpha_{j+\nu}$, respectively. Bertaccini [1,2] and Chan et al. [6] considered using the following generalized Strang preconditioner for (4):

$$S = s(A) \otimes I_m - h s(B) \otimes J_m, \tag{7}$$

where $s(A)$ is given by

$$s(A) = \begin{pmatrix} \alpha_\nu & \cdots & \alpha_\mu & & & \alpha_0 & \cdots & \alpha_{\nu-1} \\ \vdots & \ddots & & \ddots & & & \ddots & \vdots \\ \alpha_0 & & \ddots & & \ddots & & & \alpha_0 \\ & \ddots & & \ddots & & \ddots & 0 & \\ & & \ddots & & \ddots & & \ddots & \\ & & & 0 & & \ddots & & \\ \alpha_\mu & & & & \ddots & & & \alpha_\mu \\ \vdots & \ddots & & & & \ddots & & \vdots \\ \alpha_{\nu+1} & \cdots & \alpha_\mu & & & \alpha_0 & \cdots & \alpha_\nu \end{pmatrix},$$

and $s(B)$ can be defined similarly. Due to consistency condition on coefficients of LMF: $\sum_{j=0}^{\mu} \alpha_j = 0$, $s(A)$ is always singular. If, for simplicity, J_m is diagonalizable, the eigenvalues of (7) are $\phi_j - h\psi_j\mu_r$, ϕ_j, ψ_j, μ_r eigenvalues of $s(A)$, $s(B)$, J_m, respectively. Then, we have the following result:

Lemma 1. *If some eigenvalues of J_m are zero, then the preconditioner S is singular.*

In this paper, we propose the following preconditioner for (4):

$$C = \tilde{s}(A) \otimes I_m - h\tilde{s}(B) \otimes J_m, \tag{8}$$

where $\tilde{s}(A)$ is given by

$$\tilde{s}(A) = \begin{pmatrix} \alpha_\nu & \cdots & \alpha_\mu & & & & -\alpha_0 & \cdots & -\alpha_{\nu-1} \\ \vdots & \ddots & & \ddots & & & & \ddots & \vdots \\ \alpha_0 & & & & \ddots & & & & -\alpha_0 \\ & \ddots & & & & \ddots & & & 0 \\ & & \ddots & & & & \ddots & & \\ & & & 0 & & & & \ddots & \\ -\alpha_\mu & & & & & & & & \alpha_\mu \\ \vdots & \ddots & & & & & & \ddots & \vdots \\ -\alpha_{\nu+1} & \cdots & -\alpha_\mu & & & & \alpha_0 & \cdots & \alpha_\nu \end{pmatrix},$$

and $\tilde{s}(B)$ can be defined similarly. We note that $\tilde{s}(A)$ and $\tilde{s}(B)$ are the Strang-type skew-circulant preconditioners of A and B respectively, see [5].

Now we are going to prove that C is invertible provided that the given BVM is $0_{\nu,\mu-\nu}$-stable. The stability of a BVM is closely related to two characteristic polynomials defined as follows:

$$\rho(z) = z^\nu \sum_{j=-\nu}^{\mu-\nu} \alpha_{j+\nu} z^j \quad \text{and} \quad \sigma(z) = z^\nu \sum_{j=-\nu}^{\mu-\nu} \beta_{j+\nu} z^j. \tag{9}$$

Note that they are μ-degree polynomials. A polynomial $p(z)$ of degree μ is an $N_{\nu,\mu-\nu}$-polynomial if

$$|z_1| \leq |z_2| \leq \cdots \leq |z_\nu| \leq 1 < |z_{\nu+1}| \leq \cdots \leq |z_\mu|,$$

being simple roots of unit modulus.

Definition 1. *[4, p.97] Consider a BVM with the characteristic polynomials $\rho(z)$ given by (9). The BVM is said to be $0_{\nu,\mu-\nu}$-stable if $\rho(z)$ is an $N_{\nu,\mu-\nu}$-polynomial.*

Definition 2. *[4, p.101] Consider a BVM with the characteristic polynomials $\rho(z)$ and $\sigma(z)$ given by (9). The region*

$$\mathcal{D}_{\nu,\mu-\nu} = \{q \in \mathbf{C} : \rho(z) - q\sigma(z) \text{ has } \nu \text{ zeros inside } |z| = 1$$
$$\text{and } \mu - \nu \text{ zeros outside } |z| = 1\}$$

is called the region of $A_{\nu,\mu-\nu}$-stability of the given BVM. Moreover, the BVM is said to be $A_{\nu,\mu-\nu}$-stable if

$$\mathbf{C}^- \equiv \{q \in \mathbf{C} : Re(q) < 0\} \subseteq \mathcal{D}_{\nu,\mu-\nu}.$$

Theorem 1. *If some eigenvalues of J_m are zero while the others are in $\mathbf{C}^-$ and the BVM for (2) is $0_{\nu,\mu-\nu}$-stable, then the preconditioner C is nonsingular.*

However, if all the eigenvalues of J_m are not equal to zero, then we can apply an $A_{\nu,\mu-\nu}$-stable BVM method and we have the following result similar to the Strang circulant preconditioner S (see [6,2]).

Theorem 2. *If the BVM for (2) is $A_{\nu,\mu-\nu}$-stable and $h\lambda_k(J_m) \in \mathcal{D}_{\nu,\mu-\nu}$, then the preconditioner C is nonsingular.*

Next we show that the spectrum of the preconditioned system is clustered around 1 and hence Krylov subspace methods will converge fast if applied to solving the preconditioned system, see [6].

Theorem 3. *All the eigenvalues of the preconditioned matrix $C^{-1}M$ are 1 except for at most $2m\mu$ outliers. When Krylov subspace methods are applied to solving the preconditioned system $C^{-1}M\mathbf{y} = \mathbf{b}$, the method will converge in at most $2m\mu + 1$ iterations in exact arithmetic.*

Regarding the cost per iteration, the main work in each iteration for Krylov subspace methods is the matrix-vector multiplication

$$C^{-1}M\mathbf{z} = (\tilde{s}(A) \otimes I_m - h\tilde{s}(B) \otimes J_m)^{-1}(A \otimes I_m - hB \otimes J_m)\mathbf{z}$$

Since A, B are banded matrices and J_m is assumed to be sparse, the matrix-vector multiplication $(A \otimes I_m - hB \otimes J_m)\mathbf{z}$ can be done very fast. To compute $C^{-1}(M\mathbf{z})$, since $\tilde{s}(A)$ and $\tilde{s}(B)$ are circulant matrices, we have the following decompositions $\tilde{s}(A) = DF\Lambda_A F^*D^*$ and $\tilde{s}(B) = DF\Lambda_B F^*D^*$, where

$$D = \mathrm{diag}(1, e^{-i\pi/(s+1)}, e^{-2i\pi/(s+1)}, \ldots, e^{-si\pi/(s+1)}),$$

Λ_A and Λ_B are diagonal matrices containing the eigenvalues of $\tilde{s}(A)$ and $\tilde{s}(B)$ respectively and F is the Fourier matrix, see [5]. It follows that

$$C^{-1}(M\mathbf{z}) = (D^*F^* \otimes I_m)(\Lambda_A \otimes I_m - h\Lambda_B \otimes J_m)^{-1}(DF \otimes I_m)(M\mathbf{z}).$$

This product can be obtained by using Fast Fourier Transforms and solving s linear systems of order m. Since J_m is sparse, the matrix $\Lambda_A \otimes I_m - h\Lambda_B \otimes J_m$ will also be sparse. Thus $C^{-1}(M\mathbf{z})$ can be obtained by solving s sparse m-by-m linear systems.

3 Numerical Tests

To compare the effectiveness of our preconditioner with various circulant approximations we have considered two test problems. We have omitted comparisons with the preconditioner based on the T. Chan circulant approximation because the T. Chan preconditioner can be very ill-conditioned, see [2].

We will compare the number of iterations needed to converge for the GMRES method. More numerical tests can be found in [1,2,3,6]. Some implementation details can be found in [1]. The initial guess for those iterative solvers is the zero vector. The stopping criterion is $\|r_j\|_2 < 10^{-6}\|b\|_2$, r_j true residual after j iterations. All experiments are performed in MATLAB.

Moreover, we list the condition numbers of the matrix of the underlying linear system and of the different block preconditioners by LINPACK estimated 1-norm procedure. We will see that the condition numbers of the original system, P-circulant and skew-circulant preconditioners are often about the same, differently to what happens for the Strang preconditioner.

Example 1: We consider the advection equation of first order with periodic boundary conditions

$$\begin{cases} \dfrac{\partial u}{\partial t} - \dfrac{\partial u}{\partial x} = 0, \\ u(x,0) = x(\pi - x), & x \in [0, \pi] \\ u(\pi, t) = u(0, t), & t \in [0, 2\pi] \end{cases}$$

We discretize the partial derivative $\partial/\partial x$ with the central differences and step size $\delta x = \pi/m$. We obtain a family of systems of ODEs with a $m \times m$ skew-symmetric Jacobian matrix.

The generalized Adam Method with $k = 3$ (order 4, see [4] for the coefficients), suitable for ODE problems whose Jacobian matrix has eigenvalues on the imaginary axis, is used to solve the above differential equation. The number of matrix-vector products required to solve the related linear system are given in Table 1. It can be observed that the skew-circulant-based block preconditioned iterations converge usually fast, while the Strang-based one cannot be used for odd m because the Jacobian matrix has an eigenvalue equal to zero.

Table 1. (Example 1) Number of matrix-vector multiplications required for convergence of GMRES, where * denotes that the preconditioner cannot be used and its condition number is undefined

m	s	No Precond. Iter.	No Precond. Cond.	Strang Iter.	Strang Cond.	P-circulant Iter.	P-circulant Cond.	Skew-circulant Iter.	Skew-circulant Cond.
25	8	157	170	*	*	23	130	30	150
	16	136	280	*	*	22	200	28	1700
	32	98	480	*	*	21	340	21	840
50	8	299	330	23	5700	20	230	36	570
	16	328	530	28	4000	23	340	30	790
	32	234	770	34	70000	28	580	24	2500
75	8	>500	450	*	*	20	330	38	9100
	16	>500	660	*	*	25	500	31	590
	32	430	1200	*	*	26	780	43	3400

Example 2: Let us consider the heat equation with a variable diffusion coefficient

$$\begin{cases} \dfrac{\partial u}{\partial t} - \dfrac{\partial}{\partial x}\left(a(x)\dfrac{\partial u}{\partial x}\right) = 0, & \\ u(0,t) = u(x_{max},t) = 0, & t \in [0, 2\pi] \\ u(x,0) = x, & x \in [0, \pi] \end{cases}$$

If we discretize the operator $\partial/\partial x$ with centered differences and stepsize $\delta x = \pi/(m+1)$. We obtain a system of m ODEs whose $m \times m$ Jacobian matrix is tridiagonal (Toeplitz if and only if $a(x)$ is constant). We note that the Jacobian matrix has real and strictly negative eigenvalues. Here, $a(x) = \exp(-x^\beta)$.

Table 2. (Example 2) Number of matrix-vector multiplications required for convergence of GMRES

		No Precond.		Strang		P-circulant		Skew-circulant	
m	s	Iter.	Cond.	Iter.	Cond.	Iter.	Cond.	Iter.	Cond.
20	8	75	1.9×10^3	14	3.0×10^{11}	13	1.3×10^3	9	7.7×10^2
	16	114	1.9×10^3	14	3.0×10^{11}	13	1.3×10^3	9	7.7×10^2
	32	159	1.9×10^3	14	3.0×10^{11}	14	1.3×10^3	9	7.7×10^2
50	8	193	1.2×10^4	44	1.0×10^{11}	14	8.0×10^3	10	4.8×10^3
	16	308	1.2×10^4	34	1.0×10^{11}	15	8.1×10^3	10	4.8×10^3
	32	453	1.2×10^4	55	1.0×10^{11}	15	8.1×10^3	10	4.8×10^3
100	8	>500	4.5×10^4	40	6.0×10^{13}	14	3.0×10^4	10	2.0×10^4
	16	>500	4.5×10^4	49	6.0×10^{13}	15	3.0×10^4	10	2.0×10^4
	32	>500	4.5×10^4	70	6.0×10^{13}	15	3.0×10^4	10	2.0×10^4

The generalized Adams Method with $k = 4$ (order 5, see [4] for the coefficients), suitable for stiff problems, is used to solve the differential problem. The number of matrix-vector products needed to solve the related linear system, when $\beta = 3$, are given in table 2. It can be observed that the Strang preconditioner can be used if β is between 0 and 1 and m is not too large. For instance, when $\beta = 3$, the number of iterations of using the Strang preconditioner increases significantly when m increases.

Moreover, we find that the ill-conditioning of the Strang circulant approximation gives polluted numerical results already when m is of the order of a hundred. For $\beta > 3$, the Strang block preconditioner cannot be used at all because it is severely ill-conditioned even if the double precision is in use. However, the new skew-circulant preconditioner performs very well.

Acknowledgments

The first author would like to thank The University of Hong Kong for the support and hospitality offered in January-February 2000, when part of this work was completed.

References

1. D. Bertaccini, *P-Circulant Preconditioners and the Systems of the ODE Codes*, Iterative Methods in Scientific Computation IV, D. Kincaid et al., Eds, IMACS Series in Computational Mathematics and Applied Mathematics, pp. 179–193, 1999.
2. D. Bertaccini, *A circulant preconditioner for the systems of LMF-based ODE codes*, to appear on SIAM J. Sci. Comput.
3. D. Bertaccini and M. Ng, *The Convergence Rate of Block Preconditioned Systems Arising From LMF-Based ODE Codes*, submitted.
4. L. Brugnano and D. Trigiante, *Solving Differential Problems by Multistep Initial and Boundary Value Methods*, Gordon and Berach Science Publishers, Amsterdam, 1998.
5. R. Chan and M. Ng, *Conjugate Gradient Methods for Toeplitz Systems*, SIAM Review, Vol. 38 (1996), 427–482.
6. R. Chan, M. Ng and X. Jin, *Circulant Preconditioners for Solving Ordinary Differential Equations*, to appear in Structured Matrices, D. Bini et al., eds. Nova Science Pub.
7. D. Bini, B. Meini, *Effective Methods for Solving Banded Toeplitz Systems*, SIAM J. Matr. Anal. Appl., vol. 20 (1999), 700–719

New Families of Symplectic Runge–Kutta–Nyström Integration Methods

S. Blanes[1], F. Casas[2], and J. Ros[3]

[1] DAMTP, University of Cambridge
Silver Street, Cambridge CB3 9EW, England
[2] Departament de Matemàtiques, Universitat Jaume I
12071-Castellón, Spain
[3] Departament de Física Teòrica and IFIC, Universitat de València
46100-Burjassot, Valencia, Spain

Abstract. We present new 6-th and 8-th order explicit symplectic Runge–Kutta–Nyström methods for Hamiltonian systems which are more efficient than other previously known algorithms. The methods use the processing technique and non-trivial flows associated with different elements of the Lie algebra involved in the problem. Both the processor and the kernel are compositions of explicitly computable maps.

1 Introduction

In Hamiltonian dynamics, a frequent special case occurs when the Hamiltonian function reads

$$H(\mathbf{q},\mathbf{p}) = \frac{1}{2}\mathbf{p}^T M^{-1}\mathbf{p} + V(\mathbf{q}) \ , \qquad (1)$$

with M a constant, symmetric, invertible matrix. In this situation the equations of motion are

$$\dot{\mathbf{q}} = M^{-1}\mathbf{p} \ , \qquad \dot{\mathbf{p}} = -\nabla_\mathbf{q} V(\mathbf{q}) \qquad (2)$$

or, after elimination of $\mathbf{p}$,

$$\ddot{\mathbf{q}} = -M^{-1}\nabla_\mathbf{q} V(\mathbf{q}) \ . \qquad (3)$$

It is therefore natural to consider Runge–Kutta–Nyström (RKN) methods when the second order system (3) has to be solved numerically. These methods can be rendered symplectic, thus preserving qualitative features of the phase space of the original Hamiltonian dynamical system. In fact, a number of symplectic RKN schemes of order ≤ 4 have been designed during the last decade which outperform standard non-symplectic methods (see [9] for a review), and the recent literature has devoted much attention to the integration of (1) by means of efficient high-order symplectic algorithms [5,6,8,11]. The usual approach is to compose a number of times the exact flows corresponding to the kinetic and potential energy in (1) with appropriately chosen weights to achieve the desired order. More specifically, if A and B denote the Lie operators

$$A = M^{-1}\mathbf{p}\nabla_\mathbf{q} \ , \qquad B = -(\nabla_\mathbf{q} V)\nabla_\mathbf{p} \qquad (4)$$

associated with $\frac{1}{2}\mathbf{p}^T M^{-1}\mathbf{p}$ and $V(\mathbf{q})$, respectively [1], then the exact solution of (2) can be written as

$$\mathbf{z}(t) = e^{t(A+B)}\mathbf{z}(0) \equiv e^{t(A+B)}\mathbf{z}_0 ,$$

where $\mathbf{z} = (\mathbf{q}, \mathbf{p})^T$, and the evolution operator $e^{t(A+B)}$ for one time step $h = t/N$ is approximated by

$$e^{h(A+B)} \simeq e^{hH_{\mathbf{a}}} \equiv \prod_{i=1}^{s} e^{ha_i A} e^{hb_i B} \tag{5}$$

with

$$e^{ha A}\mathbf{z}_0 = (\mathbf{q}_0 + ha M^{-1}\mathbf{p}_0, \mathbf{p}_0)^T \tag{6}$$
$$e^{hb B}\mathbf{z}_0 = (\mathbf{q}_0, \mathbf{p}_0 - hb\nabla_{\mathbf{q}} V(\mathbf{q}_0))^T .$$

Observe that the approximate solution $\mathbf{z}_a(t) = e^{tH_{\mathbf{a}}}\mathbf{z}_0$ evolves in the Lie group whose Lie algebra $L(A,B)$ is generated by A and B with the usual Lie bracket of vector fields [1].

The coefficients a_i, b_i in (5) are determined by imposing that

$$H_{\mathbf{a}} = A + B + \mathcal{O}(h^n) \tag{7}$$

to obtain an n-th order symplectic integration method. This makes necessary to solve a system of polynomial equations, which can be extraordinarily involved even for moderate values of n, so that various symmetries are usually imposed in (5) to reduce the number of determining equations. For instance, if the composition is left-right symmetric then H_a does not contain odd powers of h, but then the number of flows to be composed increases. Although additional simplifications also take place due to the vanishing of the Lie bracket $[B,[B,[B,A]]]$ for the Hamiltonian (1), the question of the existence of high-order RKN symplectic integrators more efficient than standard schemes is still open.

Recently, the use of the processing technique has allowed to develop extremely efficient methods of orders 4 and 6 [2]. The idea is to consider the composition

$$e^{h\mathcal{H}(h)} = e^P e^{hK} e^{-P} \tag{8}$$

in order to reduce the number of evaluations: after N steps we have $e^{t(A+B)} \simeq e^{t\mathcal{H}(h)} = e^P (e^{hK})^N e^{-P}$. At first e^P (the processor) is applied, then e^{hK} (the kernel) acts once per step, and finally e^{-P} is evaluated only when output is needed. Both the kernel and the processor are taken as composition of flows corresponding to A and B, in a similar way to (5).

In this paper, by combining the processing technique with the use of non-trivial flows associated with different elements of $L(A,B)$ we obtain optimal 6-th order RKN methods more efficient than others previously known and some 8-th order symplectic schemes with less function evaluations per step. The analysis can also be easily extended to a more general class of second order differential equations.

2 Analysis and New Methods

In addition to A and B there are other elements in $L(A, B)$ whose flow is explicitly and exactly computable. In particular, the flow corresponding to the operators

$$V_{3,1} \equiv [B, [A, B]] \qquad V_{5,1} \equiv [B, [B, [A, [A, B]]]] \qquad (9)$$
$$V_{7,1} \equiv [B, [A, [B, [B, [A, [A, B]]]]]] \quad V_{7,2} \equiv [B, [B, [B, [A, [A, [A, B]]]]]]$$

has an expression similar to the second equation of (6) by replacing $\nabla_{\mathbf{q}} V$ with an appropriate function $\mathbf{g}(\mathbf{q})$ [3]. Therefore it is possible to evaluate exactly $\exp(hC_{b,c,d,e,f})$, with

$$C_{b,c,d,e,f} = bB + h^2 c\, V_{3,1} + h^4 d\, V_{5,1} + h^6 (e V_{7,1} + f V_{7,2}) \;, \qquad (10)$$

b, c, d, e, and f being free parameters. We can then substitute some of the $e^{hb_i B}$ factors by the more general ones $e^{hC_{b_i, c_i, d_i, e_i, f_i}}$ both in the kernel and the processor in order to reduce the number of evaluations and thus improve the overall efficiency. The operator $C_{b,c,d,e,f}$ will be referred in the sequel as modified potential, and we simply write $C_{b,c}$ when $d = e = f = 0$.

By repeated application of the Baker-Campbell-Hausdorff formula [10] the kernel and processor generators K and P can be written as

$$K = A + B + \sum_{i=2}^{\infty} \left\{ h^{i-1} \sum_{j=1}^{d(i)} k_{i,j} E_{i,j} \right\}, \quad P = \sum_{i=1}^{\infty} \left\{ h^i \sum_{j=1}^{d(i)} p_{i,j} E_{i,j} \right\}, \qquad (11)$$

where $d(m)$ denote the dimension of the space spanned by brackets of order m of A and B (its first 8 values being 2,1,2,2,4,5,10,15) and $\{E_{m,j}\}_{j=1}^{d(m)}$ is a basis of this space. Therefore

$$\mathcal{H}(h) = e^P K e^{-P} = A + B + \sum_{i=2}^{\infty} \left\{ h^{i-1} \sum_{j=1}^{d(i)} f_{i,j} E_{i,j} \right\}, \qquad (12)$$

where the $f_{i,j}$ coefficients are given in terms of polynomials involving $k_{i,j}$ and $p_{i,j}$ [2]. Specific n-th order integration methods require that $f_{i,j} = 0$ up to $i = n$, and these equations impose restrictions to the kernel: it must satisfy $k(n) = d(n) - 1$ independent conditions ($n \geq 2$) [2], and $k(2n) = k(2n - 1)$ if it is a symmetric composition. The explicit form of these conditions and the coefficients $p_{i,j}$ of the processor P in terms of $k_{i,j}$ up to order 8 have been obtained in [3]. It has also been shown that the kernel completely determines the optimal method we can obtain by processing [2]. Here optimal means that the main term of the local truncation error attains a minimum.

As stated above, we take as processor of a RKN method the explicitly computable composition

$$e^P = \prod_{i=1}^{r} e^{h z_i A} e^{h y_i B} \;, \qquad (13)$$

where the replacement $\exp(hy_i B) \longmapsto \exp(hC_{y_i,v_i,\ldots})$ can be done when necessary, and the number r of B (or C) evaluations is chosen to guarantee that the $\sum_{i=1}^{n-1} d(i)$ equations $p_{i,j} = p_{i,j}(z_k, y_k)$ have real solutions.

As far as the kernel is concerned, due to the different character of the operators A and B, two types of symmetric compositions have been analyzed:

(i) Type ABA: $\left(\sum_{i=1}^{s+1} a_i = \sum_{i=1}^{s} b_i = 1\right)$

$$e^{hK} = e^{ha_1 A} e^{hb_1 B} e^{ha_2 A} \cdots e^{ha_s A} e^{hb_s B} e^{ha_{s+1} A} \qquad (14)$$

with $a_{s+2-i} = a_i$ and $b_{s+1-i} = b_i$.

(ii) Type BAB: $\left(\sum_{i=1}^{s} a_i = \sum_{i=1}^{s+1} b_i = 1\right)$

$$e^{hK} = e^{hb_1 B} e^{ha_1 A} e^{hb_2 B} \cdots e^{hb_s B} e^{ha_s A} e^{hb_{s+1} B} \qquad (15)$$

with $a_{s+1-i} = a_i$ and $b_{s+2-i} = b_i$.

A systematic analysis of the 6-th order case has been afforded in [3], where a number of optimal processed methods with modified potentials and $s = 2, 3$ were obtained. There also some methods involving only A and B evaluations with $s = 4, 5, 6$ were also reported, with their corresponding truncation error. Here we have generalized the study to seven stages ($s = 7$). Now the three free parameters allow to find an extremely efficient 6-th order processed method: it has error coefficients which are approximately 50 times smaller than the corresponding to the most efficient 6-th order symplectic non-processed RKN method with $s = 7$ given in [8]. In Table 1 we collect the coefficients of this new processed method and also of the most efficient 6-th order algorithm we have found involving the modified potential $C_{b,c,0,e,f}$ in the kernel and $C_{y,v}$ in the processor.

A similar study can be carried out, in principle, for the 8-th order case, although now the number of possibilities (and solutions) increases appreciably with respect to $n = 6$, so that the analysis becomes extaordinarily intricate. Here we have considered kernels with $s = 4, 5$ involving modified potentials and $s = 9, 10, 11$ when only A and B evaluations are incorporated. Taking into account the well known fact that methods with small coefficients have been shown to be very efficient [6], we apply this strategy for locating possible kernels. The coefficients of two of them are given in Table 1, although many others are available.

On the other hand, the coefficients z_k, y_k in the processor (13) have to satisfy 26 equations, but this number can be reduced by taking different types of compositions. For instance, if the coefficients in $e^Q = \prod_i e^{hz_i A} e^{hy_i B}$ are determined in such a way that $Q(h) = \frac{1}{2} P(h) + O(h^7)$, then $e^{Q(h)} e^{Q(-h)} = e^{P(h)} + O(h^8)$ because $P(h)$ is an even function of h up to order h^8. Then, only 16 equations are involved. Here also the criterium we follow is to choose the smallest coefficients z_k, y_k of $e^{Q(h)}$.

Table 1. Coefficients of the new symplectic RKN integrators with processing

Order 6; Type BAB; $s=7$; $r=8$		
$b_1 = 0.115899400930169$	$b_2 = -1.21532440212000$	$b_3 = 1.45706208067905$
$a_1 = 0.244868573793901$	$a_2 = -0.00214552789272415$	$a_3 = 0.301340867944477$
$z_1 = -0.350316247513416$	$z_2 = 0.0744434640156453$	$z_3 = -0.0369370026731913$
$z_4 = -0.0597184197245884$	$z_5 = 0.404915108936223$	$z_6 = -0.180941427380936$
$z_7 = -0.0346188279494959$	$z_8 = -\sum_{i=1}^{7} z_i$	
$y_1 = 0.218575120792731$	$y_2 = -0.370670464937763$	$y_3 = 0.342037685653768$
$y_4 = -0.225359207496863$	$y_5 = 0.0878524557495559$	$y_6 = 0.195239165175742$
$y_7 = -0.155222704734044$	$y_8 = -\sum_{i=1}^{7} y_i$	

Order 6; Type ABA; $s=3$; $r=6$; Modified potential		
$a_1 = -0.0682610383918630$	$b_1 = 0.2621129352517028$	$c_1 = d_1 = e_1 = f_1 = 0$
	$c_2 = 0.0164011128160783$	$d_2 = 0$
	$e_2 = 1.86194612413481 \cdot 10^{-5}$	$f_2 = -6.3155794861591 \cdot 10^{-6}$
$z_1 = 0.1604630501234888$	$y_1 = -0.012334538446142270$	$v_1 = 0.013816178183636998$
$z_2 = -0.1222126706298830$	$y_2 = -0.6610294848488182$	$v_2 = -0.050288359617427786$
$z_3 = 0.1916801124727711$	$y_3 = -0.023112349678219939$	$v_3 = -0.013462400168471472$
$z_4 = 0.5630722377955035$	$y_4 = 1.81521815949959 \cdot 10^{-4}$	$v_4 = 6.03819193361427 \cdot 10^{-4}$
$z_5 = -0.7612758792358986$	$y_5 = 2.3768244683666757$	$v_5 = -0.01$
$z_6 = -\sum_{i=1}^{5} z_i$	$y_6 = -\sum_{i=1}^{5} y_i$	$v_6 = 0.01$

Order 8; Type BAB; $s=11$; $r=8$		
$b_1 = 0.03906544126305366$	$b_2 = 0.216015988434324$	$b_3 = -0.126717696299036$
$b_4 = -0.04128542496526060$	$b_5 = 0.04458478096712717$	
$a_1 = 0.142940453575212$	$a_2 = 0.309791505162032$	$a_3 = 0.301210185530089$
$a_4 = -0.005822573683400349$	$a_5 = -0.344741324170165$	
$z_1 = -0.0295940574778285$	$z_2 = 0.0102454583206065$	$z_3 = 0.168519324003820$
$z_4 = -0.577391651425342$	$z_5 = 0.0991834279391326$	$z_6 = 0.0203810695211463$
$z_7 = -0.106234446989598$	$z_8 = -\sum_{i=1}^{7} z_i$	
$y_1 = 0.175492972679660$	$y_2 = -0.372698829093994$	$y_3 = -0.00224032125918971$
$y_4 = 0.0926169248899539$	$y_5 = -0.201446308655374$	$y_6 = 0.216983390044259$
$y_7 = -0.0918456713646654$	$y_8 = -\sum_{i=1}^{7} y_i$	

Order 8; Type BAB; $s=5$; $r=7$; Modified potential		
$d_1 = 0.0001219127419188233$		
$e_1 = 5.741889879702246 \cdot 10^{-6}$	$f_1 = -2.271708973531348 \cdot 10^{-6}$	
$b_2 = -0.1945897221635392$	$c_2 = 5.222572249380952 \cdot 10^{-4}$	
$a_1 = 0.6954511641703808$	$a_2 = -0.05$	
$z_1 = 0$	$y_1 = 0.3644761259072299$	$v_1 = 0.016298916362212911$
$z_2 = -0.004624860718237988$	$y_2 = -0.2849544383272169$	$v_2 = -0.019769812343547362$
$z_3 = 0.3423219445639433$	$y_3 = 0.2023898776842639$	$v_3 = 0.004608026684270971$
$z_4 = 0.1760176996772205$	$y_4 = -0.2743578195701579$	$v_4 = 0$
$z_5 = 0.3625045293826689$	$y_5 = -4.75975395524748 \cdot 10^{-3}$	$v_5 = 0$
$z_6 = -0.2729727321466362$	$y_6 = 0.1455974775779454$	$v_6 = 0$
$z_7 = -\sum_{i=1}^{6} z_i$	$y_7 = -\sum_{i=1}^{6} y_i$	$v_7 = 0$

3 A Numerical Example

To test in practice the efficiency of these new symplectic methods, we compare them with other schemes of similar consistency on a specific example. For order 6, these are the most efficient seven stage method designed by Okunbor and Skeel, OS6 [8], and the non-symplectic variable step RKN method, DP6, obtained in [4]. Concerning the 8-th order, we compare with the symplectic integrator due to Yoshida [11] (Yos8, 15 function evaluations per step), the method obtained by

McLachlan [6] (McL8, 17 stages) and the optimized symmetric scheme designed by Calvo and Sanz-Serna [5] (CSS8, 24 evaluations).

The example we consider is the perturbed Kepler Hamiltonian

$$H = \frac{1}{2}(p_x^2 + p_y^2) - \frac{1}{r} - \frac{\varepsilon}{2r^3}\left(1 - \frac{3x^2}{r^2}\right) \tag{16}$$

with $r = \sqrt{x^2 + y^2}$. This Hamiltonian describes in first approximation the dynamics of a satellite moving into the gravitational field produced by a slightly oblate planet. The motion takes place in a plane containing the symmetry axis of the planet [7].

We take $\varepsilon = 0.001$, which approximately corresponds to a satellite moving under the influence of the Earth, and initial conditions $x = 1 - e$, $y = 0$, $p_x = 0$, $p_y = \sqrt{(1+e)/(1-e)}$, with $e = 0.5$. We integrate the trajectory up to the final time $t_f = 1000\pi$ and then compute the error in energy, which is represented (in a log-log scale) as a function of the number of B evaluations.

Obviously, the computational cost of evaluating the modified potential must be estimated. This has been done by running the same program repeatedly with different types of modified potential and only with the evaluation of B. We observe that, for this problem, an algorithm using $C_{b,c,d,e,f}$ is twice as expensive as the same algorithm involving B evaluations, and only a 20% more computationally costly when $C_{b,c}$ are involved. This is so due to the reuse of certain calculations in the modified potentials.

With this estimate, we present in Fig. 1(a) the results obtained with the 6-th order processed methods of Table 1, in comparison with DP6 and OS6, whereas the relative performance of the 8-th order symplectic schemes is shown in Fig. 1(b). Solid lines denoted by pm$k$ and pk, $k = 6, 8$, are obtained by the new methods with and without modified potentials, respectively.

It is worth noticing the great performance of the symplectic processed schemes of Table 1 with respect to other standard symplectic and non-symplectic algorithms. This is particularly notorious in the case of the 6-th order integrators, due to the fact that a full optimization strategy has been carried out in the construction process. In the case of order 8, the new methods are also more efficient than other previously known symplectic schemes, although only a partial optimization has been applied. In this sense, there is still room for further improvement.

Finally, we should mention that the results achieved by p8 are up to two orders of magnitude better than those provided by McL8 for other examples we have tested. These include the simple pendulum, the Gaussian and the Hénon-Heiles potentials.

4 Final Comments

Although in the preceding treatment we have been concerned only with Hamiltonian systems, it is clear that essentially similar considerations apply to second

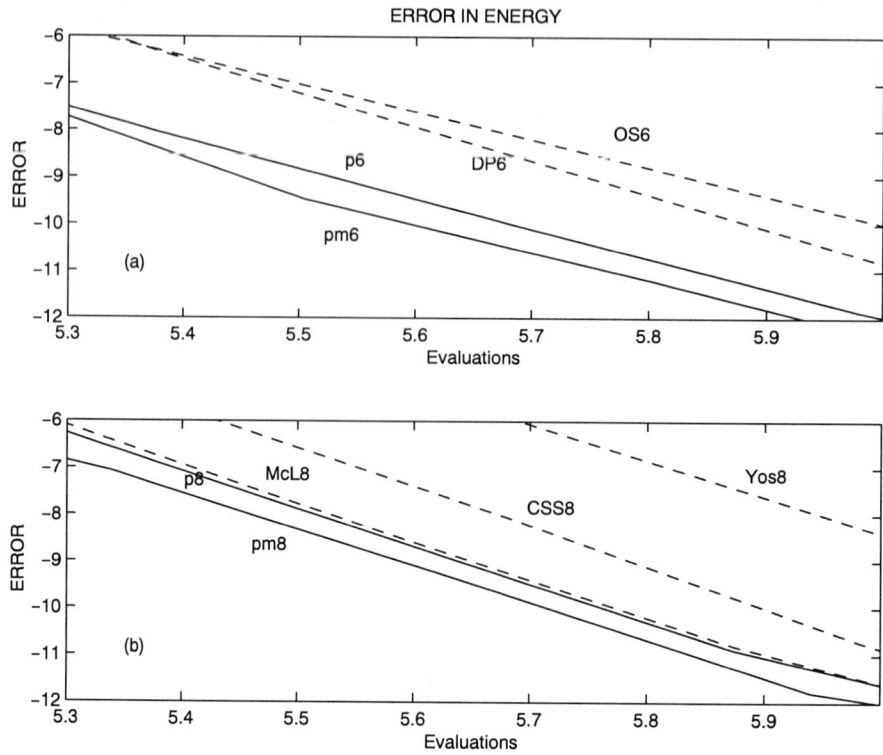

Fig. 1. Average errors in energy vs. number of evaluations for the sixth (a) and eighth (b) order processed symplectic RKN methods

order systems of ODE of the form

$$\ddot{\mathbf{x}} = \mathbf{f}(\mathbf{x}), \qquad \mathbf{x} \in \mathbb{R}^l, \qquad \mathbf{f}: \mathbb{R}^l \longrightarrow \mathbb{R}^l \qquad (17)$$

when it is required that some qualitative or geometric property of (17) be preserved in the numerical discretization. In fact, introducing the new variables $\mathbf{z} = (\mathbf{x}, \mathbf{v})^T$, with $\mathbf{v} = \dot{\mathbf{x}}$, and the functions $\mathbf{f}_A = (\mathbf{v}, \mathbf{0})$, $\mathbf{f}_B = (\mathbf{0}, \mathbf{f}(\mathbf{x})) \in \mathbb{R}^{2l}$, we have

$$\dot{\mathbf{z}} = \mathbf{f}_A + \mathbf{f}_B, \qquad (18)$$

with the systems $\dot{\mathbf{z}} = \mathbf{f}_A$ and $\dot{\mathbf{z}} = \mathbf{f}_B$ explicitly integrable in closed form. In this case the Lie operators A, B are given by

$$A = \mathbf{v} \cdot \nabla_{\mathbf{x}}, \qquad B = \mathbf{f}(\mathbf{x}) \cdot \nabla_{\mathbf{v}} \qquad (19)$$

and the methods of Table 1 can be directly applied for carrying out the numerical integration. This is so even for the physically relevant class of time-dependent non-linear oscillators of the form

$$\ddot{\mathbf{x}} + \delta \dot{\mathbf{x}} + \mathbf{f}_1(\mathbf{x}) = \mathbf{f}_2(t) \ . \qquad (20)$$

Acknowledgements

SB acknowledges the Ministerio de Educación y Cultura (Spain) for a post-doctoral fellowship and the University of Cambridge for its hospitatlity. FC is supported by the Collaboration Program UJI–Fundació Caixa Castelló 1999 under project 0I039.01/1. JR is supported by DGICyT, Spain (Grant no. PB97/1139).

References

1. V. I. Arnold, *Mathematical Methods of Classical Mechanics*, 2nd Ed., Springer, New York, 1989.
2. S. Blanes, F. Casas, and J. Ros, *Symplectic integrators with processing: a general study*, SIAM J. Sci. Comput., 21 (1999), pp. 711-727.
3. S. Blanes, F. Casas, and J. Ros, *High-order Runge-Kutta-Nyström geometric methods with processing*, to appear.
4. R. W. Brankin, I. Gladwell, J. R. Dormand, P. J. Prince, and W. L. Seward, *Algorithm 670. A Runge-Kutta-Nyström code*, ACM Trans. Math. Softw., 15 (1989), pp. 31-40.
5. M. P. Calvo and J. M. Sanz-Serna, *High-order symplectic Runge-Kutta-Nyström methods*, SIAM J. Sci. Comput., 14 (1993), pp. 1237-1252.
6. R. I. McLachlan, *On the numerical integration of ordinary differential equations by symmetric composition methods*, SIAM J. Sci. Comput., 16 (1995), pp. 151-168.
7. L. Meirovich, *Methods of Analytical Dynamics*, McGraw-Hill, New York, 1988.
8. D. I. Okunbor and R. D. Skeel, *Canonical Runge-Kutta-Nyström methods of orders five and six*, J. Comp. Appl. Math., 51 (1994), pp. 375-382.
9. J. M. Sanz-Serna and M. P. Calvo, *Numerical Hamiltonian Problems*, Chapman & Hall, London, 1994.
10. V. S. Varadarajan, *Lie Groups, Lie Algebras, and Their Representations*, Springer-Verlag, New York, 1984.
11. H. Yoshida, *Construction of higher order symplectic integrators*, Phys. Lett. A, 150 (1990), pp. 262-269.

Convergence of Finite Difference Method for Parabolic Problem with Variable Operator*

Dejan Bojović

University of Kragujevac, Faculty of Science
Radoja Domanovića 12, 34000 Kragujevac, Yugoslavia

Abstract. In this paper we consider the first initial-boundary value problem for the heat equation with variable coeficients in the domain $(0,1)^2 \times (0,T]$. We assume that the solution of the problem and the coefficients of equation belong to the corresponding anisotropic Sobolev spaces. Convergence rate estimates consistent with the smoothness of the data are obtained.

1 Introduction

For the class of finite difference schemes approximating parabolic initial-boundary value problems convergence rate estimates consistent with the smoothness of data, i.e.

$$\|u - v\|_{W_2^{r,r/2}(Q_{h\tau})} \leq C(h + \sqrt{\tau})^{s-r} \|u\|_{W_2^{s,s/2}(Q)}, \quad s \geq r. \tag{1}$$

are of the major interest. Here $u = u(x,t)$ denotes the solution of the original initial–boundary value problem, v denotes the solution of corresponding finite difference scheme, h and τ are discretisation parameters, $W_2^{s,s/2}(Q)$ denotes anisotropic Sobolev space, $W_2^{s,s/2}(Q_{h\tau})$ denotes discrete anisotropic Sobolev space, and C is a positive generic constant, independent of h, τ and u. For problems with variable coefficients constant C depends on the norms of coefficients.

Estimates of this type have been obtained for parabolic problems with coficient wich depends only from variable x [1]. In this paper we are deriving estimates for the parabolic problem with coeficients depending from variables x and t. Bramble–Hilbert lemma [2] is used in ours proof.

2 Initial–Boundary Value Problem and Its Aproximation

Let us define anisotropic Sobolev spaces $W_2^{s,s/2}(Q)$, $Q = \Omega \times I$, $I = (0,T)$, as follows [5]:

$$W_2^{s,s/2}(Q) = L_2(I, W_2^s(\Omega)) \cap W_2^{s/2}(I, L_2(\Omega)),$$

---

* Supported by MST of Republic of Serbia, grant number 04M03/C

with the norm

$$\|f\|_{W_2^{s,s/2}(Q)} = \left(\int_0^T \|f(t)\|^2_{W_2^s(\Omega)} dt + \|f\|^2_{W_2^{s/2}(I,L_2(\Omega))} \right)^{\frac{1}{2}}.$$

We consider the first initial–boundary value problem for parabolic equation with variable coefficients in the domain $Q = \Omega \times (0,T] = (0,1)^2 \times (0,T]$

$$\frac{\partial u}{\partial t} - \sum_{i=1}^{2} \frac{\partial}{\partial x_i} \left(a_i(x,t) \frac{\partial u}{\partial x_i} \right) = f, \quad (x,t) \in Q, \qquad (2)$$
$$u = 0, \ (x,t) \in \partial\Omega \times [0,T], \quad u(x,0) = u_0(x), \ x \in \Omega.$$

We assume that the generalized solution of the problem (2) belongs to the anisotropic Sobolev space $W_2^{s,s/2}(Q)$, $2 < s \leq 4$ (see [4]), with the right-hand side $f(x,t)$ which belongs to $W_2^{s-2,s/2-1}(Q)$. Consequently, coefficients $a_i = a_i(x,t)$ belong to the space of multipliers $M\left(W_2^{s-1,(s-1)/2}(Q)\right)$ [6], i.e. it is sufficient that [3]

$$a_i \in W_{4/(s-1)}^{s-1+\varepsilon,(s-1+\varepsilon)/2}(Q), \varepsilon > 0, \quad \text{for } 2 < s \leq 3,$$
$$a_i \in W_2^{s-1,(s-1)/2}(Q), \quad \text{for } 3 < s \leq 4.$$

We also assume that the coefficients $a_i(x,t)$ are decrasing functions in variable t, and $a_i(x,t) \geq c_0 > 0$.

Let $\bar\omega$ be the uniform mesh on $\bar\Omega = [0,1]^2$ with the step size h, $\omega = \bar\omega \cap \Omega$, $\gamma = \bar\omega \cap \partial\Omega$. Let θ_τ be the uniform mesh in $(0,T)$ with the step size τ, $\theta_\tau^+ = \theta_\tau \cup \{T\}$, $\bar\theta_\tau = \theta_\tau \cup \{0,T\}$. We define uniform mesh in Q: $Q_{h\tau} = \omega \times \theta_\tau$, $Q_{h\tau}^+ = \omega \times \theta_\tau^+$ i $\bar Q_{h\tau} = \bar\omega \times \bar\theta_\tau$.

It will be assumed that

$$c_1 h^2 \leq \tau \leq c_2 h^2, \quad c_1, c_2 = const > 0.$$

We define finite differences in the usual manner [7]:

$$v_{x_i} = \frac{v^{+i} - v}{h} = v_{\bar x_i}^{+i}, \quad v_t(x,t) = \frac{v(x,t+\tau) - v(x,t)}{\tau} = v_{\bar t}(x,t+\tau).$$

where $v^{\pm i}(x,t) = v(x \pm h r_i, t)$ and r_i is the unit vector along x_i axis. We also define the Steklov smoothing operators:

$$T_i^+ f(x,t) = \int_0^1 f(x + hx' r_i, t) \, dx' = T_i^- f(x + hr_i, t),$$
$$T_i^2 f(x,t) = T_i^+ T_i^- f(x,t) = \int_{-1}^1 (1 - |x'|) f(x + hx' r_i, t) \, dx',$$
$$T_t^+ f(x,t) = \int_0^1 f(x, t + \tau t') \, dt' = T_t^- f(x, t+\tau).$$

The initial-boundary value problem (2) will be approximated on $\overline{Q}_{h\tau}$ by the finite difference scheme

$$v_{\bar{t}} + L_h v = T_1^2 T_2^2 T_t^- f, \text{ in } Q_{h\tau}^+,$$
$$v = 0, \text{ on } \gamma \times \bar{\theta}_\tau, \quad v = u_0, \text{ on } \omega \times \{0\},$$
(3)

where

$$L_h v = -\frac{1}{2} \sum_{i=1}^{2} ((a_i v_{x_i})_{\bar{x}_i} + (a_i v_{\bar{x}_i})_{x_i}).$$

The finite–difference scheme (3) is the the standard symmetric scheme with the averaged right–hand side. Note that for $s \leq 4$ the right–hand side may be discontinuous function, so scheme without averaging is not well defined.

3 Convergence of the Finite–Difference Scheme

Let u be the solution of initial-boundary value problem (2) and v - the solution of finite difference scheme (3). The error $z = u - v$ satisfies the conditions

$$z_{\bar{t}} + L_h z = \sum_{i=1}^{2} \eta_i + \varphi, \text{ in } Q_{h\tau}^+,$$
$$z = 0, \text{ on } \omega \times \{0\}, \quad z = 0, \text{ on } \gamma \times \bar{\theta}_\tau,$$
(4)

where

$$\eta_i = T_1^2 T_2^2 T_t^- (D_i(a_i D_i u)) - \tfrac{1}{2}((a_i u_{x_i})_{\bar{x}_i} + (a_i u_{\bar{x}_i})_{x_i}), \text{ and}$$
$$\varphi = u_{\bar{t}} - T_1^2 T_2^2 u_{\bar{t}}.$$

We define discrete inner products

$$(v, w)_\omega = (v, w)_{L_2(\omega)} = h^2 \sum_{x \in \omega} v(x, t) w(x, t),$$
$$(v, w)_{Q_{h\tau}} = (v, w)_{L_2(Q_{h\tau})} = h^2 \tau \sum_{x \in \omega} \sum_{t \in \theta_\tau^+} v(x, t) w(x, t) = \tau \sum_{t \in \theta_\tau^+} (v, w)_\omega,$$

and discrete Sobolev norms:

$$\|v\|_\omega^2 = (v, v)_\omega, \quad \|v\|_{Q_{h\tau}}^2 = (v, v)_{Q_{h\tau}}, \quad \|v\|_{L_h}^2 = (L_h v, v)_\omega,$$

$$\|v\|_{W_2^{2,1}(Q_{h\tau})}^2 = \|v\|_{Q_{h\tau}}^2 + \sum_{i=1}^{2} \|v_{x_i}\|_{Q_{h\tau}}^2 + \sum_{i=1}^{2} \|v_{x_i \bar{x}_i}\|_{Q_{h\tau}}^2 + \|v_{\bar{t}}\|_{Q_{h\tau}}^2.$$

The following assertion holds true:

Lemma 1. *Finite–difference scheme (4) satisfies a priori estimate*

$$\|z\|_{W_2^{2,1}(Q_{h\tau})} \leq C \|\psi\|_{Q_{h\tau}},$$
(5)

where $\psi = \sum_{i=1}^{2} \eta_i + \varphi$.

Proof. Multiplying (4) by $L_h z = \frac{1}{2} L_h(z + \check{z}) + \frac{\tau}{2} L_h z_{\bar{t}}$, where $\check{z} = z(x, t - \tau)$ and summing through the nodes of ω we obtain

$$\frac{1}{2\tau}\left(\|z\|_{L_h}^2 - \|\check{z}\|_{L_h}^2\right) + \frac{\tau}{2}\|z_{\bar{t}}\|_{L_h}^2 + \|L_h z\|_\omega^2 = (\psi, L_h z)_\omega \leq \frac{1}{2}\|\psi\|_\omega^2 + \frac{1}{2}\|L_h z\|_\omega^2,$$

$$\|z\|_{L_h}^2 - \|\check{z}\|_{L_h}^2 + \tau^2 \|z_{\bar{t}}\|_{L_h}^2 + \tau \|L_h z\|_\omega^2 \leq \tau \|\psi\|_\omega^2,$$

$$\|z\|_{L_h}^2 - \|\check{z}\|_{\check{L}_h}^2 + \|\check{z}\|_{\check{L}_h}^2 - \|\check{z}\|_{L_h}^2 + \tau^2 \|z_{\bar{t}}\|_{L_h}^2 + \tau \|L_h z\|_\omega^2 \leq \tau \|\psi\|_\omega^2,$$

where $\check{L}_h(t) = L_h(t - \tau)$. Recalling the condition that $a_i(x, t)$ is decrasing function in variable t we simply deduce that $\|\check{z}\|_{\check{L}_h}^2 - \|\check{z}\|_{L_h}^2 \geq 0$. We thus obtain

$$\|z\|_{L_h}^2 - \|\check{z}\|_{\check{L}_h}^2 + \tau \|L_h z\|_\omega^2 \leq \tau \|\psi\|_\omega^2.$$

Summing through the nodes of θ_τ^+ we obtain

$$\|z(T)\|_{L_h(T)}^2 - \|z(0)\|_{L_h(0)}^2 + \tau \sum_\tau^T \|L_h z\|_\omega^2 \leq \tau \sum_\tau^T \|\psi\|_\omega^2.$$

Using the relations $\|z(T)\|_{L_h(T)}^2 \geq 0$ and $\|z(0)\|_{L_h(0)}^2 = 0$ we have

$$\tau \sum_\tau^T \|L_h z\|_\omega^2 \leq \tau \sum_\tau^T \|\psi\|_\omega^2. \tag{6}$$

Using the relation $\|z_{\bar{t}}\| \leq \|\psi\| + \|L_h z\|$ we have

$$\tau \sum_\tau^T \|z_{\bar{t}}\|_\omega^2 \leq 4\tau \sum_\tau^T \|\psi\|_\omega^2. \tag{7}$$

Finally, recalling wellknown relations

$$\|L_h z\|_\omega \geq C \|z_{x_i \bar{x}_i}\|_\omega, \quad \|z\|_\omega \leq C \|z_{x_i}\|_\omega \quad \text{and} \quad \|z_{x_i}\|_\omega \leq C \|z_{x_i \bar{x}_i}\|_\omega,$$

and relations (6) and (7) we simply obtain

$$\|z\|_{W_2^{2,1}(Q_{h\tau})} \leq C \|\psi\|_{Q_{h\tau}} \leq C \left(\sum_{i=1}^2 \|\eta_i\|_{Q_{h\tau}} + \|\varphi\|_{Q_{h\tau}} \right). \tag{8}$$

In a such a way, the problem of deriving the convergence rate estimate for finite–difference scheme (3) is now reduced to estimating the right-hand side terms in (8).

First of all, we decompose term η_i in the following way [3]: $\eta_i = \sum_{k=1}^{7} \eta_{ik}$, where

$$\eta_{i1} = T_1^2 T_2^2 T_t^-(a_i D_i^2 u) - (T_1^2 T_2^2 T_t^- a_i)(T_1^2 T_2^2 T_t^- D_i^2 u),$$

$$\eta_{i2} = (T_1^2 T_2^2 T_t^- a_i - a_i)(T_1^2 T_2^2 T_t^- D_i^2 u),$$

$$\eta_{i3} = a_i(T_1^2 T_2^2 T_t^- D_i^2 u - u_{x_i \bar{x}_i}),$$

$$\eta_{i4} = T_1^2 T_2^2 T_t^-(D_i a_i D_i u) - (T_1^2 T_2^2 T_t^- D_i a_i)(T_1^2 T_2^2 T_t^- D_i u),$$

$$\eta_{i5} = (T_1^2 T_2^2 T_t^- D_i a_i - 0.5(a_{i,x_i} + a_{i,\bar{x}_i}))(T_1^2 T_2^2 T_t^- D_i u),$$

$$\eta_{i6} = 0.5(a_{i,x_i} + a_{i,\bar{x}_i})(T_1^2 T_2^2 T_t^- D_i u - 0.5(u_{\bar{x}_i} + u_{x_i})),$$

$$\eta_{i7} = 0.25(a_{i,x_i} - a_{i,\bar{x}_i})(u_{\bar{x}_i} - u_{x_i}).$$

Let us introduce the elementary rectangles $e = e(x,t) = \{(\xi_1, \xi_2, \nu) : \xi_i \in (x_i - h, x_i + h), i = 1,2, \nu \in (t - \tau, t)\}$. The linear transformation $\xi_i = x_i + h x_i^*$, $i = 1, 2$, $\nu = t + \tau t^*$, defines a bijective mapping of the canonical rectangles $E = \{(x_1^*, x_2^*, t^*) : |x_i^*| < 1, i = 1, 2, -1 < t^* < 0\}$ onto e. We define $u^*(x^*, t^*) \equiv u^*(x_1^*, x_2^*, t^*) = u(x_1 + h x_1^*, x_2 + h x_2^*, t + \tau t^*)$, .

The value of η_{i1} at a mesh point $(x,t) \in Q_{h\tau}^+$ can be expressed as

$$\eta_{i1}(x,t) = \frac{1}{h^2}\left\{\iint_E k(x_1^*)k(x_2^*)a_i^*(x^*,t^*)D_i^2 u^*(x^*,t^*)\,dt^*dx^* \right.$$
$$\left. - \iint_E k(x_1^*)k(x_2^*)a_i^*(x^*,t^*)\,dt^*dx^* \times \iint_E k(x_1^*)k(x_2^*)D_i^2 u^*(x^*,t^*)\,dt^*dx^*\right\}$$

where $k(x_i^*) = 1 - |x_i^*|$.

Thence we deduce that η_{i1} is a bounded bilinear functional of the argument $(a_i^*, u^*) \in W_q^{\lambda, \lambda/2}(E) \times W_{2q/(q-2)}^{\mu, \mu/2}(E)$, where $\lambda \geq 0$, $\mu \geq 2$ and $q > 2$. Furthermore, $\eta_{i1} = 0$ whenever a_i^* is a constant function or u^* is a polynomial of degree two in x^* and degree one in t^*. Applying the bilinear version of the Bramble–Hilbert lemma [2], [8] we deduce that

$$|\eta_{i1}(x,t)| \leq \frac{C}{h^2}|a_i^*|_{W_q^{\lambda,\lambda/2}(E)}|u^*|_{W_{2q/(q-2)}^{\mu,\mu/2}(E)}, \quad 0 \leq \lambda \leq 1,\ 2 \leq \mu \leq 3,\ q > 2.$$

Returning from the canonical variables to the original variables we obtain

$$|a_i^*|_{W_q^{\lambda,\lambda/2}(E)} \leq Ch^{\lambda - \frac{4}{q}}|a_i|_{W_q^{\lambda,\lambda/2}(e)} \quad \text{and}$$

$$|u^*|_{W_{2q/(q-2)}^{\mu,\mu/2}(E)} \leq Ch^{\mu - \frac{2(q-2)}{q}}|u|_{W_{2q/(q-2)}^{\mu,\mu/2}(e)}.$$

Therefore,

$$|\eta_{i1}(x,t)| \leq Ch^{\lambda + \mu - 4}|a_i|_{W_q^{\lambda,\lambda/2}(e)}|u|_{W_{2q/(q-2)}^{\mu,\mu/2}(e)}, \quad 0 \leq \lambda \leq 1,\ 2 \leq \mu \leq 3,\ q > 2.$$

Summing over the mesh $Q_{h\tau}^+$ we obtain

$$\|\eta_{i1}\|_{Q_{h\tau}} \leq Ch^{\lambda + \mu - 2}\|a_i\|_{W_q^{\lambda,\lambda/2}(Q)}\|u\|_{W_{2q/(q-2)}^{\mu,\mu/2}(Q)}, \quad 0 \leq \lambda \leq 1,\ 2 \leq \mu \leq 3. \quad (9)$$

Now suppose that $3 < s \le 4$. Then the following Sobolev imbeddings hold:
$$W_2^{\lambda+\mu,(\lambda+\mu)/2}(Q) \subset W_{2q/(q-2)}^{\mu,\mu/2}(Q), \quad \text{for } \lambda \ge 4/q \text{ and}$$
$$W_2^{\lambda+\mu-1,(\lambda+\mu-1)/2}(Q) \subset W_q^{\lambda,\lambda/2}(Q), \quad \text{for } \mu \ge 3 - 4/q.$$

Setting $q = 4$, $\lambda = 1$, $\mu = s-1$ in (9), using previous imbeddings, we obtain:
$$\|\eta_{i1}\|_{Q_{h\tau}} \le Ch^{s-2} \|a_i\|_{W_2^{s-1,(s-1)/2}(Q)} \|u\|_{W_2^{s,s/2}(Q)}, \quad 3 < s \le 4. \tag{10}$$

In the case $2 < s \le 3$, setting $q = 4/(s-2)$, $\lambda = s-2$, $\mu = 2$ in (9) and using imbeddings
$$W_2^{\lambda+\mu,(\lambda+\mu)/2}(Q) \subset W_{2q/(q-2)}^{\mu,\mu/2}(Q), \quad \text{for } \lambda \ge 4/q \text{ and}$$
$$W_{4/(\lambda+\mu-1)}^{\lambda+\mu-1+\varepsilon,(\lambda+\mu-1+\varepsilon)/2}(Q) \subset W_q^{\lambda,\lambda/2}(Q), \quad \text{for } \lambda \le 4/q + \varepsilon,$$

we have
$$\|\eta_{i1}\|_{Q_{h\tau}} \le Ch^{s-2} \|a_i\|_{W_{4/(s-1)}^{s-1+\varepsilon,(s-1+\varepsilon)/2}(Q)} \|u\|_{W_2^{s,s/2}(Q)}, \quad 2 < s \le 3. \tag{11}$$

The term η_{i3} is a bounded bilinear functional of the argument $(a_i^*, x^*) \in C(\overline{E}) \times W_2^{s,s/2}(E)$, and $\eta_{i3} = 0$ whenever u^* is a polynomial of degree three in x_1^* and x_2^* and degree one in t^*. Recalling the Bramble-Hilbert lemma and imbeddings
$$W_{4/(s-1)}^{s-1+\varepsilon,(s-1+\varepsilon)/2}(Q) \subset C(\overline{Q}), \quad \text{for } 2 \le s \le 3 \text{ and}$$
$$W_2^{s-1,(s-1)/2}(Q) \subset C(\overline{Q}), \quad \text{for } 3 < s \le 4,$$

we obtain estimates of the form (10) and (11) for η_{i3}.

Using the same technique as before we obtain estimates of the form (10) and (11) for other terms η_{ik}. In a such a way we have estimates:
$$\|\eta_i\|_{Q_{h\tau}} \le Ch^{s-2} \|a_i\|_{W_{4/(s-1)}^{s-1+\varepsilon,(s-1+\varepsilon)/2}(Q)} \|u\|_{W_2^{s,s/2}(Q)}, \quad 2 < s \le 3, \tag{12}$$
$$\|\eta_i\|_{Q_{h\tau}} \le Ch^{s-2} \|a_i\|_{W_2^{s-1,(s-1)/2}(Q)} \|u\|_{W_2^{s,s/2}(Q)}, \quad 3 < s \le 4. \tag{13}$$

Applying the linear version of the Bramble–Hilbert lemma we simply obtain estimate of the term φ:
$$\|\varphi\|_{Q_{h\tau}} \le Ch^{s-2} \|u\|_{W_2^{s,s/2}(Q)}, \quad 2 < s \le 4. \tag{14}$$

Combining (8) with (12)-(14) we obtain the final result:

Theorem 1. *The difference scheme* (3) *converges in the* $W_2^{2,1}(Q_{h\tau})$ *norm, provided* $c_1 h^2 \le \tau \le c_2 h^2$. *Furthermore,*

$$\|u - v\|_{W_2^{2,1}(Q_{h\tau})} \le Ch^{s-2} \max_i \|a_i\|_{W_{4/(s-1)}^{s-1+\varepsilon,(s-1+\varepsilon)/2}(Q)} \|u\|_{W_2^{s,s/2}(Q)}, 2 < s \le 3,$$
$$\|u - v\|_{W_2^{2,1}(Q_{h\tau})} \le Ch^{s-2} \max_i \|a_i\|_{W_2^{s-1,(s-1)/2}(Q)} \|u\|_{W_2^{s,s/2}(Q)}, \quad 3 < s \le 4.$$

These estimates are consistent with the smoothness of the data.

References

1. Bojović, D., Jovanović, B. S.: Application of interpolation theory to the analysis of the convergence rate for finite difference schemes of parabolic type. Mat. vesnik **49** (1997) 99-107.
2. Bramble, J. H., Hilbert, S. R.: Bounds for a class of linear functionals with application to Hermite interpolation. Numer. Math. **16** (1971) 362–369.
3. Jovanović, B. S.: The finite–difference method for boundary–value problems with weak solutions. Posebna izdan. Mat. Inst. **16**, Belgrade 1993.
4. Ladyzhenskaya, O. A., Solonnikov, V. A., Ural'ceva, N. N.: Linear and Quasilinear Parabolic Equations. Nauka, Moskow 1967 (Russian)
5. Lions, J. L., Magenes, E.: Problèmes aux limites non homogènes et applications. Dunod, Paris 1968
6. Maz'ya, V. G., Shaposhnikova, T. O.: Theory of multipliers in spaces of differentiable functions. Monographs and Studies in Mathematics **23**. Pitman, Boston, Mass. 1985.
7. Samarski, A. A.: Theory of difference schemes. Nauka, Moscow 1983 (Russian).
8. Triebel, H.: Interpolation theory, function spaces, differential operators. Deutscher Verlag der Wissenschaften, Berlin 1978.

Finite Volume Difference Scheme for a Stiff Elliptic Reaction-Diffusion Problem with a Line Interface

Ilia A. Braianov

Center of Applied Mathematics and Informatics
University of Rousse, 7017 Rousse, Bulgaria
braianov@ami.ru.acad.bg

Abstract. We consider a singularly perturbed elliptic problem in two dimensions with stiff discontinuous coefficients of order $\mathcal{O}(1)$ and $\mathcal{O}(\varepsilon)$ on the left and on the right of interface, respectively. The solution of this problem exhibits boundary and corner layers and is difficult to solve numerically. The FVM is implemented on condensed (Shishkin's) mesh that resolves boundary and corners layers, and we prove that it yelds an accurate approximation of the solution both inside and outside these layers. We give error estimates in discrete energetic norm that hold true uniformly in the perturbation parameter ε. Numerical experiments confirm these theoretical results.

1 Introduction

Let consider the elliptic problem

$$L^-u \equiv -\Delta u(x,y) + q(x,y)u(x,y) = f(x,y),\ (x,y) \in \Omega^-, \tag{1}$$
$$L^+u \equiv -\varepsilon^2 \Delta u(x,y) + q(x,y)u(x,y) = f(x,y),\ (x,y) \in \Omega^+, \tag{2}$$
$$[u(x,y)]_\Gamma = 0,\ L^\Gamma u \equiv -\varepsilon^2 \frac{\partial u(+0,y)}{\partial x} + \frac{\partial u(-0,y)}{\partial x} = K(y),\ y \in (0,1), \tag{3}$$
$$u(x,0) = g_s(x),\ u(x,1) = g_n(x),\ u(-1,y) = g_w(y),\ u(1,y) = g_e(y), \tag{4}$$
$$\Omega^- = (-1,0) \times (0,1),\ \Omega^+ = (0,1) \times (0,1),\ \Gamma = 0 \times (0,1),$$

where

$$0 < q_0 \le q(x,y) \le q^0. \tag{5}$$

We suppose that all data in the problem are sufficiently smooth with possible discontinuity at the interface line Γ.

Problems of type (1)-(5) often are called "stiff", see [3]. It is well known the solution u of (1)-(5) has singularities at the corners of the square Ω, [7], [2]. Since Γ is supposed to be regular, the solution can also have corner singularities only at the intersection points of the boundary $\partial \Omega$ and the interface Γ. In order the solution to be sufficiently smooth some compatibility conditions should be fulfilled at this corners. Essential difficulties arise from the anisotropy of the coefficients. Since the diffusion coefficients are small in Ω^+ boundary and corner layer appears around the boundary of Ω^+. The interface conditions (3) on the right cause also weak corner singularities around two corners of Ω^- laying on the interface Γ.

Let β is a positive constant and $\beta \leq \sqrt{q_0/2}$. We assume that the solution of the problem (1)-(5) satisfies the following assumptions

Assumption 1 *Let the coefficients of problem (1)-(5) are sufficiently smooth and satisfy all necessary compatibility conditions. Then the solution u can be decomposed into regular part u^r that satisfies for all l, m-integer, $l \leq m$, $m = 0, \ldots, 3$ the estimates*

$$\left|D_x^{m-l} D_y^l u^r(x,y)\right| \leq C_m, \ (x,y) \in \bar{\Omega}^- \cup \bar{\Omega}^+, \tag{6}$$

and singular part u_s

$$u^s(x,y) = \begin{cases} E^{xy-}(x,y) & (x,y) \in \Omega^-, \\ E^x(x,y) + E^y(x,y) + E^{xy+}(x,y), & (x,y) \in \Omega^+, \end{cases} \tag{7}$$

that satisfies for all $m = 0, \ldots, 4$, the estimates

$$\left|D_x^{m-l} D_y^l E^{xy-}(x,y)\right| \leq C\varepsilon^{2-m} \left(\exp(-\beta(x+y)/\varepsilon) + \exp(-\beta(1-y+x)/\varepsilon)\right),$$
$$\left|D_x^{m-l} D_y^l E^x(x,y)\right| \leq C\varepsilon^{-(m-l)} \left(\exp(-\beta x/\varepsilon) + \exp(-\beta(1-x)/\varepsilon)\right),$$
$$\left|D_x^{m-l} D_y^l E^y(x,y)\right| \leq C\varepsilon^{-l} \left(\exp(-\beta y/\varepsilon) + \exp(-\beta(1-y)/\varepsilon)\right),$$
$$\left|D_x^{m-l} D_y^l E^{xy+}(x,y)\right| \leq C\varepsilon^{-m} \left(\exp(-\beta(x+y)/\varepsilon) + \exp(-\beta(1-x+y)/\varepsilon) \right.$$
$$\left. + \exp(-\beta(1+x-y)/\varepsilon) + \exp(-\beta(2-x-y)/\varepsilon)\right) \tag{8}$$

where C is independent of ε constant.

2 Numerical Solution

2.1 Grid and Grid Functions

It is well known that in singularly perturbed problems can not be achieved an uniform convergence on uniform mesh. In order to obtain ε-uniformly convergent difference scheme we construct a partially uniform mesh $\bar{w}$ condensed closely to the boundary of $\bar{\Omega}^+$, see Fig. 1. Denote by $\bar{w}^+$ the mesh in $\bar{\Omega}^+$ and by $\bar{w}^-$ the mesh in $\bar{\Omega}^-$. In $\bar{\Omega}^+$ we construct a condensed (Shishkin's) mesh similar as this one introduced in [4].

$$\bar{w}^+ = \{(x_i, y_j), \ x_i = x_{i-1} + h_i^x, \ y_j = y_{j-1} + h_j^y, \ j = 1, \ldots, N_2,$$
$$i = N_1 + 1, \ldots, N_1 + N_2 = N, \ x_{N_1} = 0, \ x_N = 1, \ y_0 = 0, \ y_{N_2} = 1\},$$
$$h_j^y = h_{i+N_1}^x = h_1 = 4\delta/N_2, \ i,j = 1, \ldots, N_2/4 \cup 1 + 3N_2/4, \ldots, N_2,$$
$$h_j^y = h_{i+N_1}^x = h_2 = (1 - 2\delta)/N_2, \ i,j = 1 + N_2/4, \ldots, 3N_2/4,$$
$$\delta = \min\{2\varepsilon \ln N_2/\beta, 1/4\}.$$

Since in the left part the problem is not singularly perturbed we construct a coarse mesh in $\bar{\Omega}^-$. Setting the conditions the mesh in $\bar{\Omega}^+$ to overlap this one in $\overline{\Omega^-}$, we chose $N_2 = 2^l$, $l \geq 2$, l-integer. Let $s_0 = \max s$, s-integer, that satisfies $h_2 \leq \delta/2^s$. Setting $m = 2^{s_0}$, we chose, $M_1 = 2m + N_2/2$ and $h_3 = \delta/m$. We also take $N_1 = \max n$, n-integer, that satisfies $1/n \geq h_2$, and $h_4 = 1/N_1$. Then the mesh in $\bar{\Omega}^-$ is

$$\bar{w}^- = \{(x_i, y_j),\ x_i = x_{i-1} + h_i^x,\ y_j = y_{j-1} + h_j^y,\ i = 1, ..., N_1,\ j = 1, ..., M_1,$$
$$x_0 = -1,\ x_{N_1} = 0,\ y_0 = 0,\ y_{M_1} = 1\},$$
$$h_i^x = h_4,\ i = 1, ..., N_1,\ h_j^y = h_2,\ j = 1 + m, ..., M_1 - m,$$
$$h_j^y = h_3,\ j = 1, ..., m \cup j = M_1 - m + 1, ..., M_1.$$

For each point (x_i, y_j) of w we consider the rectangle $e_{ij} = e(x_i, y_j)$, see Fig. 1,2. There are three types of grid points, boundary, regular and irregular, see Fig. 1.

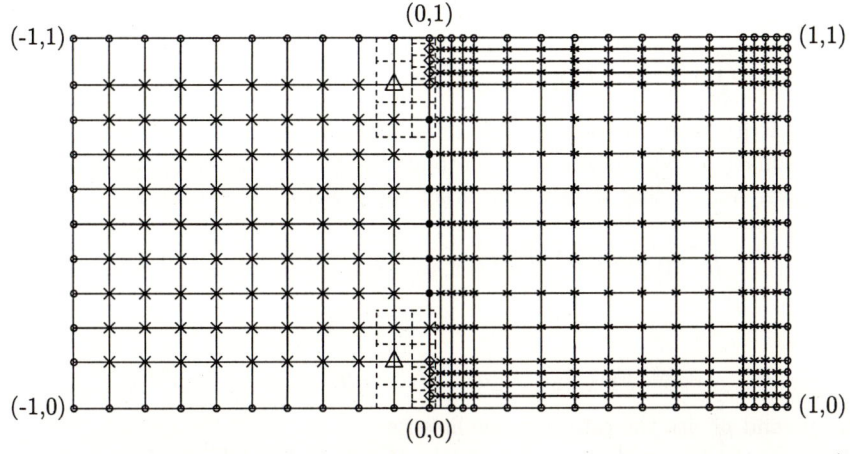

o-boundary points, ×-regular points in w^-, •-regular interface points, △-irregular points in w^- $-$,∗-regular points in w^+, $-$,◇-irregular interface points.

Fig. 1. Grid with local refinement on boundary and interior layers

Let u, v, g are given grid functions of a discrete arguments $(x_i, y_j) \in \bar{w}$. Denote $g_{ij} = g(x_i, y_j)$, $g_{i\mp 0,j} = g(x_i \mp 0, y_j)$. Further we shall use the standard notations

$$\bar{h}_i^x = \frac{h_i^x + h_{i+1}^x}{2},\quad \bar{h}_j^y = \frac{h_j^y + h_{j+1}^y}{2},\quad \bar{g}_{ij} = \frac{h_{i+1}^x g_{i+0,j} + h_i^x g_{i-0,j}}{2\bar{h}_i^x},$$

$$v_{\bar{x},ij} = \frac{v_{ij} - v_{i-1j}}{h_i^x},\quad v_{\hat{x},ij} = \frac{v_{i+1j} - v_{ij}}{\bar{h}_i^x},\quad v_{x,ij} = v_{\bar{x},i+1j}.$$

We shall also use the following discrete scalar product

$$(u,v)_{0,w} = \sum_{i=1}^{N_1-1}\sum_{j=1}^{M_1-1} h_i^x \bar{h}_j^y u_{ij} v_{ij} + \sum_{i=N_1}^{N-1}\sum_{j=1}^{M_2-1} \bar{h}_i^x \bar{h}_j^y u_{ij} v_{ij},$$

$$(u,v]_{x,\varepsilon} = \sum_{i=1}^{N_1-1}\sum_{j=1}^{M_1-1} h_i^x \bar{h}_j^y u_{ij} v_{ij} + \sum_{j=1}^{M_2-1} h_i^x \bar{h}_j^y u_{N_1,j} v_{N_1,j} +$$

$$+ \sum_{i=N_1+1}^{N}\sum_{j=1}^{M_2-1} \varepsilon^2 h_i^x \bar{h}_j^y u_{ij} v_{ij},$$

$$(u,v]_{y,\varepsilon} = \sum_{i=1}^{N_1-1}\sum_{j=1}^{M_1} \bar{h}_i^x h_j^y u_{ij} v_{ij} + \sum_{j=1}^{M_2} \frac{h_{N_1}^x + \varepsilon^2 h_{N_1+1}^x}{2} h_j^y u_{N_1,j} v_{N_1,j} +$$

$$+ \sum_{i=N_1+1}^{N-1}\sum_{j=1}^{M_2} \varepsilon^2 \bar{h}_i^x h_j^y u_{ij} v_{ij}, \tag{9}$$

and corresponding norms

$$\|u\|_{0,w} = \sqrt{(u,u)_{0,w}}, \quad \|u\|_{x,\varepsilon} = \sqrt{(u,u]_{x,\varepsilon}}, \quad \|u\|_{y,\varepsilon} = \sqrt{(u,u]_{y,\varepsilon}}. \tag{10}$$

2.2 Finite Difference Approximation

Balance Equation Further in the numerical approximation we will use the balance equation corresponding to problem (1)-(5). Integrating the equations (1), (2) over sell e_{ij} that does not interact the interface Γ we obtain

$$\int_{\partial e_{ij}} W_\nu ds = \int\int_{e_{ij}} (f(x,y) - q(x,y)u(x,y))\, dxdy, \tag{11}$$

where

$$W_\nu = -(p^x, p^y)(D_x u, D_y u), \quad W_1 = -p^x D_x, \quad W_2 = -p^y D_y.$$

where p^x and p^y are the diffusion coefficients.

Let now the rectangle e_{ij} interacts the interface Γ, and e_{ij}^-, e_{ij}^+ are left and right part of e_{ij} respectively. Denote by $S_{\Gamma,ij}$ the intersection of e_{ij} and Γ. Using the interface conditions (3), we obtain

$$\int_{\partial e_{ij}^-/S_{\Gamma,ij}} W_\nu ds + \int_{\partial e_{ij}^+/S_{\Gamma,ij}} W_\nu ds =$$
$$\int\int_{e_{ij}^-} (f - qu) dxdy + \int\int_{e_{ij}^+} (f - qu) dxdy + \int_{S_{\Gamma,ij}} K(y) dy. \tag{12}$$

Approximation at the Regular Points At the regular points in w that does not lay on the interface we will use the standard approximations, see [6]. Using (11) at the regular points of w^- and w^+ we obtain

$$-p^x U_{\bar{x}\hat{x},ij} - p^y U_{\bar{y}\hat{y},ij} + q_{ij} U_{ij} = f_{ij}. \tag{13}$$

At the regular points on the interface using (12) we get

$$-\frac{\varepsilon^2 U_{x,N_1 j} - U_{\bar{x},N_1 j}}{\bar{h}_{N_1}^x} - \frac{h_{N_1}^x + \varepsilon^2 h_{N_1+1}^x}{2\bar{h}_{N_1}^x} U_{\bar{y}\hat{y},N_1 j} + \bar{q}_{N_1 j} U_{N_1 j} = \bar{f}_{N_1 j} + \frac{K_j}{\bar{h}_{N_1}^x}. \tag{14}$$

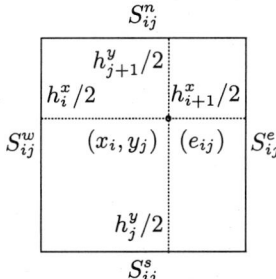
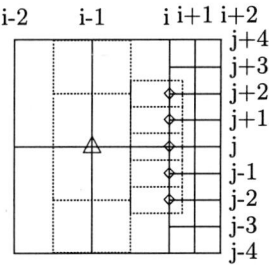

Fig. 2a. typical cell   Fig. 2b. irregular grid points

Fig. 2.

Approximation at the Irregular Points At the irregular interface points (x_i, y_j), can happen that the point (x_{i-1}, y_j) is not a grid point (see Fig. 1, Fig. 2b). Then we can not use the value of U_{i-1j} in the numerical approximation. The needed values could be obtained from the values at the coarse grid points on the left by piecewise polynomial interpolation. Below we shall use a piecewise constant interpolation. For ease of exposition consider the particular situation of Fig. 2b. From the figure we see that there are possible two cases.

1. The cell e_{ij+l} coincides with only one cell ($l = -1, 0, 1$). We shall suppose that the grid functions g_{ij} is extended over neighboring cell e_{i-1j} as a constant and we approximate U_{i-1j+l} by U_{i-1j}.
2. The cell e_{ij+l}, ($l = -2, +2$) coincides with two cells. Consider the case $l = 2$. We shall suppose that the grid functions g_{ij} is extended over neighboring cells e_{i-1j} and e_{i-1j+4} as a constant. We use the approximation $U_{i-1j+2} = (U_{i-1j} + U_{i-1j+4})/2$.

Denote
$$\nabla_{\bar{x}} U_{N_1 j} = \sum_{k=1}^{m_j} \frac{H^1_{j_k}(U_{N_1 j} - U_{N_1-1 j_k})}{\bar{h}^y_j h^x_{N_1}}$$

where $m_j = 1, 2$ is the number of neighboring cells (on the left) of cell $e_{N_1 j}$ and $H^1_{j_k}$ is the length of $S^w_{N_1 j}$ laying on the side $S^e_{N_1-1 j_k}$. Then at the irregular interface points we obtain the approximation (14) with $\nabla_{\bar{x}} U_{N_1 j}$ instead of $U_{\bar{x}, Nj}$.

In order to obtain the approximation at the irregular points in w^- we set the requirements that the finite difference scheme conserves the mass. For example, in the particular situation of Fig. 2b we have

$$\int_{S^e_{i-1j}} W_1 dy = \int_{y_{j-2}}^{y_{j-3/2}} W_1 dy + \sum_{k=-1}^{1} \int_{S^w_{ij+k}} W_1 dy + \int_{y_{j+3/2}}^{y_{j+2}} W_1 dy. \quad (15)$$

Denote
$$\nabla_x U_{N_1 j} = \sum_{k=1}^{n_j} \frac{H^2_{j_k}(U_{N_1 j_k} - U_{N_1-1 j})}{\bar{h}^y_j h^x_{N_1}},$$

where $n_j = 1, 2$ is the number of neighboring cells (on the right) of the cell e_{N_1-1j} and $H_{j_k}^2$ is the length of $S_{N_1 j_k}^w$ laying on the side $S_{N_1-1j}^e$. Then at the irregular points (x_{N_1-1}, y_j) we obtain the approximation (13) with $\nabla_x U_{N_1 j}$ instead of $U_{x,Nj}$.

Formulation of the Discrete Problem Setting the boundary conditions

$$U_{i0} = g_{s,i},\ U_{iM_1} = g_{n,i},\ U_{iM_2} = g_{n,i},\ U_{0j} = g_{w,j},\ U_{Nj} = g_{e,j}, \qquad (16)$$

we obtain the finite difference problem (FDP) (13), (14), (16). The FDP can be written as a system of linear algebraic equations

$$Au = F,\ (x_i, y_j) \in w, \qquad (17)$$

where in the right hand side F we have taken boundary conditions (16) into account. The following lemma shows that the matrix A is symmetric and positive definite and therefore invertible and problem (17) has unique solution.

Lemma 1. *The matrix A in (17) is symmetric and positive definite in the scalar product $(.,.)_{0,w}$ and for arbitrary discrete functions U, V on $\bar{w}$ satisfying zero boundary conditions (16) holds*

$$(U, V)_A \equiv (AU, U)_{0,w} = (\nabla_{\bar{x}} U, \nabla_{\bar{x}} V]_{x,\varepsilon} + (U_{\bar{y}}, V_{\bar{y}}]_{y,\varepsilon} + (QU, V)_{0,w}. \qquad (18)$$

Here Q is a diagonal matrix corresponding to q.

Since the matrix A is symmetric and positive definite, then it defines a norm called the energy norm.

$$\|u\|_A = (Au, u)_{0,w}^{\frac{1}{2}}. \qquad (19)$$

The matrix A is badly scaled, see Table 2 below. The condition number ρ_A of A tends to ∞ when $\varepsilon \to 0$. But simple diagonal preconditioning improves the situation significantly. Denote

$$D = diag\{d_{ii}\},\quad d_{ii} = A_{ii}^{-1}.$$

Then the condition number of the matrix DA is independent of ε, see [5].

Uniform Convergence Next theorem presents the main result in the paper.

Theorem 1. *Let $u \in C^3(\bar{\Omega}^-) \cup C^3(\bar{\Omega}^+) \cup C(\bar{\Omega}^+)$ is a solution of the differential problem (1)-(5) and satisfies Assumption 1. Let U is a solution of the discrete problem (17). Then the following ε-uniformly estimates holds*

$$\|U - u\|_A \leq C\left(N^{-\frac{1}{2}}\right), \qquad (20)$$

and if $\varepsilon = \mathcal{O}(N^{-1} \ln N)$

$$\|U - u\|_A \leq C\left(N^{-1} \ln N\right), \qquad (21)$$

Let in addition $u \in C^4(\bar{\Omega}^-) \cup C^4(\bar{\Omega}^+)$ and satisfies the Assumption 1. Then, if $\varepsilon = \mathcal{O}(N^{-3/2})$ the following estimates hold

$$\|U - u\|_A \leq C\left(N^{-\frac{3}{2}} \ln N\right), \quad \|U - u\|_{\infty,\bar{w}} \leq C\left(N^{-\frac{1}{2}} \ln N\right), \qquad (22)$$

for some positive constant C independent of the mesh and ε.

3 Numerical Results

Consider the problem (1)-(5) with coefficients

$$q(x,y) = 2, \ (x,y) \in \Omega^-, \quad q(x,y) = 1, \ (x,y) \in \Omega^+.$$

We took the right had side and the boundary conditions so that the exact solution to be

$$u(x,y) = \begin{cases} 2 + x - 2xy, & (x,y) \in \bar{\Omega}^-, \\ 1 + \exp(-y/\varepsilon) + \exp(-x/\varepsilon) + \\ \quad - \exp(-(x+y)/\varepsilon) - \exp(-(1+x-y)/\varepsilon), & (x,y) \in \bar{\Omega}^+. \end{cases}$$

Table 1. Error on Shishkin mesh

$N_2 \backslash \varepsilon$	$\varepsilon = 1$	$\varepsilon = 10^{-1}$	$\varepsilon = 10^{-2}$	$\varepsilon = 10^{-3}$	$\varepsilon = 10^{-4}$	$\varepsilon = 10^{-5}$	$\varepsilon = 10^{-6}$
$N_2 = 4, \ \|.\|_\infty$	3.879e-4	6.746e-2	8.929e-2	1.143e-1	1.169e-1	1.172e-1	1.172e-1
$N_2 = 8, \ \|.\|_\infty$	1.054e-4	2.437e-2	5.353e-2	5.246e-2	5.234e-2	5.233e-2	5.233e-2
$N_2 = 16, \ \|.\|_\infty$	2.683e-5	6.765e-3	3.044e-2	3.043e-2	3.042e-2	3.042e-2	3.042e-2
$N_2 = 32, \ \|.\|_\infty$	6.736e-6	1.737e-3	1.286e-2	1.286e-2	1.286e-2	1.286e-2	1.286e-2
$N_2 = 64, \ \|.\|_\infty$	1.686e-6	4.375e-4	4.843e-3	4.844e-3	4.844e-3	4.844e-3	4.844e-3
$N_2 = 4, \ \|.\|_A$	8.606e-4	4.633e-2	4.651e-2	5.913e-2	6.043e-2	6.056e-2	6.058e-2
$N_2 = 8, \ \|.\|_A$	2.400e-4	1.912e-2	1.149e-2	1.027e-2	1.029e-2	1.029e-2	1.029e-2
$N_2 = 16, \ \|.\|_A$	6.206e-5	5.628e-3	6.988e-3	2.734e-3	1.879e-3	1.772e-3	1.761e-3
$N_2 = 32, \ \|.\|_A$	1.570e-5	1.471e-3	3.416e-3	1.087e-3	4.494e-4	3.225e-4	3.069e-4
$N_2 = 64, \ \|.\|_A$	3.932e-6	3.768e-4	1.308e-3	4.163e-4	1.411e-4	6.750e-5	5.498e-5

First we investigate the convergence rate. Table 1 shows the maximum and energetic norm of the error on Shishkin mesh. The results in the table confirm the theoretical ones. They show that there is a convergence in maximum norm too. Approximate solution and error in the case $\varepsilon = 0.01$ and $N = 16$ are shown on Fig. 3. Table 2 gives the condition number of matrixes A and DA where D is a diagonal preconditioning matrix defined in Section 2. We can see from the table that this simple diagonal preconditioning improves the condition number significantly and it becomes independent of ε.

References

1. Ewing, R. E., Lazarov, R. D., Vassilevski, P. S.: Local refinement techniques for elliptic problems on cell centered grids. I. Error estimates. Math. Comp. **56** N 194 (1991) 437–461

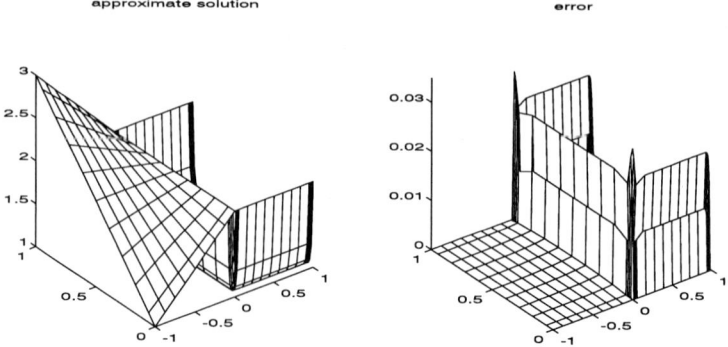

Fig. 3. Approximate solution and error

Table 2. Condition number

$N_2\backslash\varepsilon$	$\varepsilon = 1$	$\varepsilon = 10^{-1}$	$\varepsilon = 10^{-2}$	$\varepsilon = 10^{-3}$	$\varepsilon = 10^{-4}$	$\varepsilon = 10^{-5}$	$\varepsilon = 10^{-6}$
$N_2 = 4, \rho_A$	1.280e+1	1.228e+2	1.751e+2	1.473e+3	1.446e+4	1.443e+5	1.443e+6
$N_2 = 8, \rho_A$	5.168e+1	4.972e+2	5.872e+3	6.220e+5	6.259e+7	6.263e+9	6.264e+11
$N_2 = 16, \rho_A$	2.069e+2	1.995e+3	3.018e+4	3.936e+6	4.045e+8	4.056e+10	4.057e+12
$N_2 = 32, \rho_A$	8.282e+2	7.988e+3	9.553e+4	1.683e+4	1.782e+9	1.793e+11	1.794e+13
$N_2 = 4, \rho_{DA}$	1.280e+1	1.610e+1	2.749e+0	2.358e+0	2.319e+0	2.315e+0	2.315e+0
$N_2 = 8, \rho_{DA}$	5.158e+1	6.790e+1	1.904e+1	1.698e+1	1.678e+1	1.676e+1	1.676e+1
$N_2 = 16, \rho_{DA}$	2.068e+2	2.782e+2	8.975e+1	9.312e+1	9.695e+1	9.736e+1	9.740e+1
$N_2 = 32, \rho_{DA}$	8.281e+2	1.130e+3	3.894e+2	4.398e+2	4.844e+2	4.897e+2	4.902e+2

2. Han, H., Kellogg, R. B.: Differentiability properties of solutions of the equations $-\varepsilon^2 \Delta u + ru = f(x,y)$ in a square. SIAM J. Math. Anal., **21** (1990) 394–408.
3. Lions, J. L.: Perturbations singuliéres dans les probléms aux limite et en contrôle optimal. Springer-Verlag, Berlin, 1973.
4. Miller, J. J. H., O'Riordan, E., Shishkin, G. I.: Fitted numerical methods for singular perturbation problems. World scientific, Singapore, (1996)
5. Roos, H.-G.: note on the conditioning of upwind schemes on Shishkin meshes. IMA J. of Num. Anal., **16** (1996) 529–538.
6. Samarskii, A. A.: Theory of difference schemes. Nauka, Moscow, (1977) (in Russian)
7. Volkov, E. A.: Differentiability properties of solutions of boundary value problems for the Laplase and Poisson equations on a rectangle. Proc. Steklov Inst. Math., **77** (1965) 101–126

Nested-Dissection Orderings for Sparse LU with Partial Pivoting

Igor Brainman and Sivan Toledo

School of Mathematical Sciences, Tel-Aviv University
Tel-Aviv 69978, Israel
sivan@math.tau.ac.il
http://www.math.tau.ac.il/~sivan

Abstract. We describe the implementation and performance of a novel fill-minimization ordering technique for sparse LU factorization with partial pivoting. The technique was proposed by Gilbert and Schreiber in 1980 but never implemented and tested. Like other techniques for ordering sparse matrices for LU with partial pivoting, our new method preorders the columns of the matrix (the row permutation is chosen by the pivoting sequence during the numerical factorization). Also like other methods, the column permutation Q that we select is a permutation that minimizes the fill in the Cholesky factor of $Q^T A^T A Q$. Unlike existing column-ordering techniques, which all rely on minimum-degree heuristics, our new method is based on a nested-dissection ordering of $A^T A$. Our algorithm, however, never computes a representation of $A^T A$, which can be expensive. We only work with a representation of A itself. Our experiments demonstrate that the method is efficient and that it can reduce fill significantly relative to the best existing methods. The method reduces the LU running time on some very large matrices (tens of millions of nonzeros in the factors) by more than a factor of 2.

1 Introduction

Reordering the columns of sparse nonsymmetric matrices can significantly reduce fill in sparse LU factorizations with partial pivoting. Reducing fill in a factorization reduces the amount of memory required to store the factors, the amount of work in the factorization, and the amount of work in subsequent triangular solves. Symmetric positive definite matrices, which can be factored without pivoting, are normally reordered to reduce fill by applying the same permutation to both the rows and columns of the matrix. When partial pivoting is required for maintaining numerical stability, however, pre-permuting the rows is meaningless, since the rows are exchanged again during the factorization. Therefore, we normally preorder the columns and let numerical consideration dictate the row ordering. Since columns are reordered before the row permutation is known, we need to order the columns such that fill is minimized no matter how rows are exchanged. (Some nonsymmetric factorization codes that employ pivoting, such as UMFPACK/MA38 [2,3], determine the column permutation during the numerical factorization; such codes do not preorder columns so the technique in this paper does not apply to them.)

A result by George and Ng [6] suggests one effective way to preorder the columns to reduce fill. They have shown that the fill of the LU factors of PA is essentially contained in the fill of the Cholesky factor of $A^T A$ for every row permutation P. (P is a permutation matrix that permutes the rows of A and represents the actions of partial pivoting.) Gilbert [8] later showed that this upper bound on the fill of the LU factors is not too loose, in the sense that for a large class of matrices, for every fill element in the Cholesky factor of $A^T A$ there is a pivoting sequence P that causes the element to fill in the LU factors of A. Thus, nonsymmetric direct sparse solvers often preorder the columns of A using a permutation Q that minimizes fill in the Cholesky factor of $Q^T A^T A Q$.

The main challenge in column-ordering algorithms is to find a fill-minimizing permutation without computing $A^T A$ or even its nonzero structure. While computing the nonzero structure of $A^T A$ allows us to use existing symmetric ordering algorithms and codes, it may be grossly inefficient. For example, when an n-by-n matrix A has nonzeros only in the first row and along the main diagonal, computing $A^T A$ takes $\Omega(n^2)$ work, but factoring it takes only $O(n)$ work.

This challenge has been met for the class of reordering algorithms based on the minimum-degree heuristic. Modern implementations of minimum-degree heuristics use a *clique-cover* to represent the graph G_A of the matrix[1] A (see [5]). A clique cover represents the edges of the graph (the nonzeros in the matrix) as a union of cliques, or complete subgraphs. The clique-cover representation allows us to simulate the elimination process with a data structure that only shrinks and never grows. There are two ways to initialize the clique-cover representation of $G_{A^T A}$ directly from the structure of A. Both ways create a data structure whose size is proportional to the number of nonzeros in A, not the number of nonzeros in $A^T A$. From then on, the data structure only shrinks, so it remains small even if $A^T A$ is relatively dense. In other words, finding a minimum-degree column ordering for A requires about the same amount of work and memory as finding a symmetric ordering for $A^T + A$, the symmetric completion of A.

Nested-dissection ordering methods were proposed in the early 1970's and have been known since then to be theoretically superior to minimum-degree methods for important classes of sparse symmetric definite matrices. Only in the last few years, however, have nested-dissection methods been shown experimentally to be more effective than minimum-degree methods.

In 1980 Gilbert and Schreiber proposed a method for ordering $G_{A^T A}$ using nested-dissection heuristics, without ever forming $A^T A$ [7,9]. Their method uses *wide separators*, a term that they coined. They have never implemented or tested their proposed method.

The main contribution of this paper is an implementation and an experimental evaluation of the wide-separator ordering method, along with a new presentation of the theory of wide separators.

---

[1] The graph $G_A = (V, E)$ of an n-by-n matrix A has a vertex set $v = \{1, 2, \ldots, n\}$ and an edge set $E = \{(i,j) | a_{ij} \neq 0\}$. We ignore numerical cancellations in this paper.

Modern symmetric ordering methods generally work as follows:

1. The methods find a small vertex separator that separates the graph G into two subgraphs with roughly the same size.
2. Each subgraph is dissected recursively, until each subgraph is fairly small (typically several hundred vertices).
3. The separators are used to impose a coarse ordering. The vertices in the top-level separator are ordered last, the vertices in the second-to-top level come before them, and so on. The vertices in the small subgraphs that are not dissected any further appear first in the ordering. The ordering within each separator and the ordering within each subgraph has not yet been determined.
4. A minimum-degree algorithm computes the final ordering, subject to the coarse ordering constraints.

While there are many variants, most codes use this overall framework.

Our methods apply the same framework to the graph of $A^T A$, but without computing it. We find separators in $A^T A$ by finding *wide separators* in $A^T + A$. We find a wide separator by finding a conventional vertex separator and widening it by adding to it all the vertices that are adjacent to the separator in one of the subgraphs. Such a wide separator corresponds to a vertex separator in $A^T A$. Just like symmetric methods, our methods recursively dissect the graph, but using wide separators. When the remaining subgraphs are sufficiently small, we compute the final ordering using a constrained column-minimum-degree algorithm. We use existing techniques to produce a minimum-degree ordering of $A^T A$ without computing $G_{A^T A}$ (either the row-clique method or the augmented-matrix method).

Experimental results show that our method can reduce the work in the LU factorization by up to a factor of 3 compared to state-of-the-art column-ordering codes. The running times of our method are higher than the running-times of strict minimum-degree codes, such as COLAMD [10], but they are low enough to easily justify using the new method. On many matrices, including large ones, our method significanly reduces the work compared to all the existing column ordering methods. On some matrices, however, constraining the ordering using wide-separators increase fill rather than reduce it.

The rest of the paper is organized as follows. Section 2 presents the theory of wide separators and algorithms for finding them. Our experimental results are presented in Section 3. We discuss our conclusions from this research in Section 4.

2 Wide Separators: Theory and Algorithms

Our column-ordering methods find separators in $G_{A^T A}$ by finding a so-called *wide separator* in $G_{A^T + A}$. We work with the graph of $A^T + A$ and not with G_A for two reasons. First, this simplifies the definitions and proofs. Second, to the best of our knowledge all existing vertex-separator codes work with undirected graphs, so there is no point in developping the theory for the directed graph G_A.

A vertex subset $S \subseteq V$ of an undirected graph $G = (V, E)$ is a *separator* if the removal of S and its incident edges breaks the graph into two components $G_1 = (V_1, E_1)$ and $G_2 = (V_2, E_2)$, such that any path between $i \in V_1$ and $j \in V_2$ passes through at least one vertex in S. A vertex set is a *wide separator* if every path between $i \in V_1$ and $j \in V_2$ passes through a sequence of two vertices in S (one after the other along the path).

Our first task is to show that every wide separator in $G_{A^T + A}$ is a separator in $G_{A^T A}$. (proofs are omitted from this abstract due to lack of space)

Theorem 1. A wide separator in $G_{A^T + A}$ is a separator in $G_{A^T A}$.

The converse is not always true. There are matrices with separators in $G_{A^T A}$ that do not correspond to wide separators in $A^T + A$. The converse of the theorem is true, however, when there are no zeros on the main diagonal of A:

Theorem 2. If there are no zeros on the diagonal of A, then a separator in $G_{A^T A}$ is a wide separator in $G_{A^T + A}$.

Given a code that finds conventional separators in an undirected graph, finding wide separators is easy. The separator and its neighbors in either G_1 or G_2 form a wide separator:

Lemma 1. Let S be a separator in an undirected graph G. The sets $S_1 = S \cup \{i | i \in V_1, (i, j) \in E \text{ for some } j \in S\}$ and $S_2 = S \cup \{i | i \in V_2, (i, j) \in E \text{ for some } j \in S\}$ are wide separators in G.

The proof of the theorem is trivial. The sizes of S_1 and S_2 are bounded by $d|S|$, where d is the maximum degree of vertices in S. Given S, it is easy to enumerate S_1 and S_2 in time $O(d|S|)$. This running time is typically insignificant compared to the time it takes to find S.

Which one of the two candidate wide separators should we choose? A wide separator that is small and that dissects the graph evenly reduces fill in the Cholesky factor of $A^T A$, and hence in the LU factors of A. The two criteria are usually contradictory. Over the years it has been determined the the best strategy is to choose a separator that is as small as possible, as long as the ratio of the number of vertices in G_1 and G_2 does not exceed 2 or so.

The following method, therefore, is a reasonable way to find a wide separator: Select the smallest of S_1 and S_2, unless the smaller wide separator unbalances the separated subgraphs (so that one is more than twice as large as the other) but the larger does not. Our code, however, is currently more naive and always choose the smaller wide separator.

3 Experimental Results

3.1 Experimental Setup

The experiments that this section describe test the effectiveness and performance of several column-ordering codes. We have tested our new codes, which implement nested-dissection-based orderings, as well as several existing ordering codes.

Our codes build a hierarchy of wide separators and then use the separators to constrain a minimum-degree algorithm. We obtain the wide separators by widening separators in G_{A^T+A} that SPOOLES [1] finds. SPOOLES is a new library of sparse ordering and factorization codes that is being developed by Cleve Ashcraft and others. Our codes then invoke a column-mininum-degree code to produce the final ordering. One minimum-degree code that we use is SPOOLES's multi-stage-minimum-degree (MSMD) algorithm, which we run on the augmented matrix. The other minimum-degree code that we used is a version of COLAMD [10] that we modified to respect the constraints imposed by the separators.

The existing minimum-degree codes that we have tested include COLAMD, SPOOLES's MSMD (operating on the augmented matrix with no separator constraints), and COLMMD, a column minimum-degree code, originally written by Joseph W.-H. Liu and distributed with SuperLU.

We use the following acronims to refer to the ordering methods: MSMD refers to SPOOLES' minimum-degree code operating on the augmented matrix without constraints, WS+MSMD refers to the same minimum-degree code but constrained to respect wide separators, and similarly for COLAMD and WS+COLAMD.

In one set of experiments we first reduced the matrices to *block triangular form* (see [12]) applied the ordering and factorization to the diagonal blocks in the reduced form.

We always factor the reordered matrix using SuperLU [4,11], a state-of-the-art sparse-LU-with-partial-pivoting code. SuperLU uses the BLAS; we used the standard Fortran BLAS for the experiments. We plan to use a higher-performance implementation of the BLAS for the final version of the paper.

We conducted the experiments on a 500MHz dual Pentium III computer with 1 GByte of main memory running Linux. This machine has two processors, but our code only uses one processor.

We tested the ordering methods on a set of nonsymmetric sparse matrices from Tim Davis's sparse matrix collection[2]. We used all the nonsymmetric matrices in Davis's collection that were not too small (less than 0.1 second factorization time with one of the ordering methods) and that did not require more than 1Gbytes to factor. The matrices are listed in Table 1. For further details about the matrices, see Davis's web site (the final version of this paper will include a table listing the order and number of nonzeros for each matrix; the table is omitted from this abstract due to lack of space).

3.2 Results and Analysis

Table 1 summarizes the results of our experiments. The table shows experiments without reduction to block triangular form.

---

[2] http://www.cise.ufl.edu/~davis/sparse/

Columns 2–9 in the table show that wide-separator ordering techniques are effective. Wide separator (WS) orderings are the most effective ordering methods, in terms of work in the factorization, on 23 out of the 41 test matrices. WS orderings are the most effective on 9 out of the 10 largest matrices (largest in terms of work in the factorization). On the single matrix out of the 10 largest where a WS ordering was not the best, it required only 7% more flops to factor.

The reduction in work due to wide separators is often significant. On the larget matrix in our test suite, li, wide separators reduce factorization work by almost a factor of 2. The reduction compared to the unconstrained MD methods is also highly significant on raefsky3, epb3, and graham1.

When WS orderings do poorly compared to MD methods, however, they sometimes do significantly poorer. On ex40, for example, using wide separators requires 2.66 times the number of flops that COLAMD alone requires. The slowdowns on some of the lhr and bayer matrices are even more dramatic, but reduction to block triangular form often resolves these problems.

On lhr14c, for example, reduction to block triangular form prior to the ordering and factorization reduced the ordering time by more than a factor of 10 and reduced the number of nonzeros in MSMD+WS from 2.1e9 to 8.2e7 (and to 4.5e7 for MSMD alone). These experiments are not reported here in detail because we conducted them too late. The complete results will appear in the final version of the paper.

As columns 7–9 in the table show, reducing flop counts generally translates into reducing the running time of the factorization algorithm and reducing the size of the LU factors. The detailed comparisons between ordering methods other than COLAMD and WS+COLAMD are similar and are omitted from the table. Hence, our remarks concerning the flop counts above also apply to the running time of the factorization code and the amount of memory required to carry out the factorization and to store the factors.

Wide-separator orderings are more expensive to compute than strict minimum-degree orderings, but the extra cost is typically small compared to the subsequent factorization time. Column 10 in the table shows the cost of ordering relative to the cost of the factorization. The table shows that a few matrices take longer (sometimes much longer) to order than to factor. This happens to matrices that arise in chemical engineering (the bayer matrices and the lhr matrices). We hope to resolve this issue using reduction to block tridiagonal form. Another point that emerges from the table is that on small matrices, wide-separator orderings are expensive to compute relative to the cost of the factorization.

4 Conclusions and Future Work

Our main conclusion from this research is that hybrid wide-separator/minimum-degree column orderings are effective and inexpensive to compute. They often reduce substantially the amount of time and storage required to factor a sparse matrix with partial pivoting, compared to minimum-degree orderings such as COLAMD and COLMMD. They are more expensive to compute than minimum-

Table 1. A comparison of wide-separator and mimimum-degree column orderings. Columns 2–6 show the number of floating-point operations (flops) required to factor the test matrices using 5 different ordering methods. The flop counts for the most efficient method (or methods) are printed in bold. Columns 7–9 show the effectiveness of WS+COLAMD relative to that of COLAMD: $\%_T$ compares factorization running times (< 100 means that WS+COAMD is better), $\%_F$ compares flops, and $\%_Z$ compares number of nonzeros in the factors. The last column, denoted $\%_O$, show the time to find wide-separators as a percentage of the WS+COLAMD factorization time

Name	MSMD	WS+MSMD	COLMMD	COLAMD	WS+COLAMD	$\%_T$	$\%_F$	$\%_Z$	$\%_O$
bwm2000	**2.75E+04**	7.86E+04	**2.75E+04**	2.86E+04	2.83E+04	200	98	98	100
cavity04	9.57E+05	9.57E+05	1.30E+06	**6.37E+05**	**6.37E+05**	100	100	100	0
poli_large	**1.45E+05**	**1.45E+05**	1.65E+05	1.70E+05	1.70E+05	100	100		37
bayer10	1.04E+07	3.01E+07	1.24E+07	**1.01E+07**	1.45E+07	125	143		2040
lhr04c	**1.47E+07**	6.73E+07	1.68E+07	1.77E+07	3.25E+07	164	183	128	150
bayer02	1.09E+07	1.09E+07	9.72E+06	**9.28E+06**	**9.28E+06**	117	100		362
rw5151	3.12E+07	3.16E+07	3.29E+07	3.29E+07	**2.92E+07**	92	88	92	37
lhr07c	**2.78E+07**	2.22E+08	3.16E+07	3.06E+07	6.67E+07	196	217	135	132
bayer04	2.79E+07	2.79E+07	**2.41E+07**	2.51E+07	2.51E+07	100	100		447
lhr10c	3.72E+07	**3.32E+08**	3.98E+07	3.92E+07	1.31E+08	197	334	152	533
lhr11c	**4.77E+07**	**4.77E+07**	5.18E+07	5.22E+07	5.22E+07	116	100	100	343
memplus	**3.95E+07**	**3.95E+07**	4.01E+07	5.60E+09	5.60E+09	94	100	100	0
ex19	9.45E+07	1.12E+08	7.08E+07	**4.07E+07**	1.09E+08	230	267	151	83
lhr14c	8.68E+07	2.10E+09	**8.46E+07**	8.51E+07	2.60E+08	191	305	149	284
bayer01	6.12E+07	4.82E+07	6.47E+07	**4.76E+07**	1.11E+08	121	233		8857
ex35	1.03E+08	1.33E+08	9.25E+07	**5.65E+07**	1.38E+08	207	244	136	34
cavity26	1.77E+08	**1.39E+08**	1.71E+08	2.04E+08	1.48E+08	75	72	85	19
epb1	1.47E+08	1.22E+08	**1.02E+08**	1.43E+08	1.25E+08	116	87	95	27
goodwin	6.42E+08	5.77E+08	**5.06E+08**	1.91E+09	6.44E+08	34	33	57	15
epb2	7.14E+08	5.17E+08	7.14E+08	6.46E+08	**5.64E+08**	107	87	97	15
garon2	1.18E+09	1.20E+09	1.28E+09	**1.06E+09**	1.98E+09	184	186	119	5
shyy161	1.07E+09	9.00E+08	1.04E+09	1.03E+09	**7.56E+08**	77	73	92	34
graham1	1.69E+09	**9.24E+08**	1.42E+09	1.33E+09	9.54E+08	72	71	82	11
epb3	2.22E+09	**8.09E+08**	1.79E+09	2.06E+09	1.18E+09	77	57	83	27
olafu	3.16E+09	2.71E+09	2.96E+09	2.84E+09	**2.58E+09**	73	90	89	21
rim	2.89E+09	2.01E+09	2.12E+09	5.55E+09	**1.77E+09**	31	31	54	28
venkat50	**4.30E+09**	4.36E+09	5.84E+09	4.51E+09	4.91E+09	93	108	85	13
venkat25	**4.30E+09**	4.36E+09	5.84E+09	4.51E+09	4.91E+09	94	108	85	14
venkat01	**4.30E+09**	4.36E+09	5.79E+09	4.46E+09	4.87E+09	93	109	85	14
ex40	3.69E+09	3.39E+09	2.29E+09	**1.08E+09**	2.87E+09	268	265	146	8
af23560	5.33E+09	7.01E+09	4.95E+09	**4.52E+09**	9.50E+09	181	210	133	1
raefsky3	1.05E+10	**5.24E+09**	7.75E+09	1.04E+10	5.47E+09	45	52	64	11
ex11	1.55E+10	1.19E+10	1.43E+10	1.19E+10	**1.12E+10**	76	94	92	6
raefsky4	1.56E+10	**7.80E+09**	1.07E+10	1.10E+10	8.56E+09	62	77	80	9
psmigr_1	**1.48E+10**	**1.48E+10**	1.66E+10	1.68E+10	1.68E+10	94	100	100	0
psmigr_3	**1.58E+10**	**1.58E+10**	1.72E+10	1.74E+10	1.74E+10	95	100	100	0
psmigr_2	**1.56E+10**	**1.56E+10**	1.74E+10	1.76E+10	1.76E+10	94	100	100	0
wang3	3.12E+10	**1.55E+10**	3.47E+10	2.78E+10	2.45E+10	84	88	90	0
wang4	3.70E+10	**2.45E+10**	3.52E+10	3.37E+10	2.72E+10	81	80	89	0
bbmat	5.97E+10	4.77E+10	**4.46E+10**	**4.46E+10**	5.82E+10	109	130	112	4
li	1.59E+11	**8.10E+10**	2.17E+11	1.63E+11	8.15E+10	44	50	72	4

degree orderings, but the cost is typically small relative to the cost of the subsequent factorization.

The use of the block triangular decomposition of the matrices and ordering seems to resolve the problems with some of the chemical engineering problems, but we are still investigating this issue.

Acknowledgments

Thanks to John Gilbert for telling us about wide-separator orderings. Thanks to John Gilbert and Bruce Hendrickson for helpful comments on an ealy draft of the paper. Thanks to Cleve Ashcraft for his encouragement, for numerous discussions concerning this research, and for his prompt response to our questions concerning SPOOLES.

References

1. Cleve Ashcraft and Roger Grimes. SPOOLES: An object-oriented sparse matrix library. In *Proceedings of the 9th SIAM Conference on Parallel Processing for Scientific Computing*, San-Antonio, Texas, 1999. 10 pages on CD-ROM.
2. T. A. Davis and I. S. Duff. An unsymmetric-pattern multifrontal method for sparse lu factorization. *SIAM Journal on Matrix Analysis and Applications*, 19:140–158, 1997.
3. T. A. Davis and I. S. Duff. A combined unifrontal/multifrontal method for unsymmetric sparse matrices. *ACM Transactions on Mathematical Software*, 25:1–19, 1999.
4. James W. Demmel, Stanley C. Eisenstat, John R. Gilbert, Xiaoye S. Li, and Joseph W. H. Liu. A supernodal approach to sparse partial pivoting. *SIAM Journal on Matrix Analysis and Applications*, 20:720–755, 1999.
5. A. George and J. W. H. Liu. The evolution of the minimum-degree ordering algorithm. *SIAM Review*, 31:1–19, 1989.
6. Alan George and Esmond Ng. On the complexity of sparse QR and LU factorization on finite-element matrices. *SIAM Journal on Scientific and Statistical Computation*, 9:849–861, 1988.
7. John R. Gilbert. *Graph Separator Theorems and Sparse Gaussian Elimination*. PhD thesis, Stanford University, 1980.
8. John R. Gilbert. Predicting structure in sparse matrix computations. *SIAM Journal on Matrix Analysis and Applications*, 15:62–79, 1994.
9. John R. Gilbert and Robert Schreiber. Nested dissection with partial pivoting. In *Sparse Matrix Symposium 1982: Program and Abstracts*, page 61, Fairfield Glade, Tennessee, October 1982.
10. S. I. Larimore. An approximate minimum degree column ordering algorithm. Master's thesis, Department of Computer and Information Science and Engineering, University of Florida, Gainesville, Florida, 1998. Also available as CISE Tech Report TR-98-016 at ftp://ftp.cise.ufl.edu/cis/tech-reports/tr98/tr98-016.ps.
11. Xiaoye S. Li. *Sparse Gaussian Elimination on High Performance Computers*. PhD thesis, Department of Computer Science, UC Berkeley, 1996.
12. Alex Pothen and Chin-Ju Fan. Computing the block triangular form of a sparse matrix. *ACM Transactions on Mathematical Software*, 16(4):303–324, December 1990.

Fractional Step Runge–Kutta Methods for the Resolution of Two Dimensional Time Dependent Coefficient Convection–Diffusion Problems

B. Bujanda[1] and J. C. Jorge[2]

[1] Dpto. de Matemáticas y Computación, Univ. de La Rioja
Luis de Ulloa s/n 26.004, Logroño (La Rioja), Spain
bbujanda@dmc.unirioja.es
[2] Departamento de Matemática e Informática, Universidad Pública de Navarra,
Pamplona (Navarra), Spain
jcjorge@unavarra.es

Abstract. In this paper we obtain a unconditional convergence result for discretization methods of type Fractional Steps Runge-Kutta, which are highly efficient in the numerical resolution of parabolic problems whose coefficients depend on time. These methods combined with standard spatial discretizations will provide totally discrete algorithms with low computational cost and high order of accuracy in time. We will show the efficiency of such methods, in combination with upwind difference schemes on special meshes, to integrate numerically singularly perturbed evolutionary convection–diffusion problems.

1 Introduction

Let $u(\bar{x}, t)$ be the solution of a two-dimensional space evolution problem ($\bar{x} \in \mathbb{R}^2$) which admits an operational formulation in the form

$$\begin{cases} \frac{du(t)}{dt} + A(t)\, u(t) = g(t), & t \in [0, T], \\ u(0) = u_0, \end{cases} \tag{1}$$

where $A(t) : \mathcal{D} \subseteq H \longrightarrow H$, $t \in [0, T]$, are unbounded linear operators, in a Hilbert space H, with scalar product $((\cdot, \cdot))$ and associated norm $\|\cdot\|$, of functions defined in a domain $\Omega \subseteq \mathbb{R}^2$. Let us also suppose that $A(t)$ admits a natural decomposition in two simpler addends, $A_1(t)$, $A_2(t)$, such that $A_i(t) : \mathcal{D}_i \subseteq H \longrightarrow H$, for $i = 1, 2$, where $\mathcal{D}_1 \cap \mathcal{D}_2 = \mathcal{D}$ and $A_i(t)$ for $i = 1, 2$ are maximal and coercive operators for all $t \in [0, T]$, i.e.

$$\begin{cases} \forall f \in H,\ \exists v \in \mathcal{D}_i,\ \text{such that}\ v + A_i(t)v = f\ \text{and} \\ \exists \alpha_i > 0\ \text{such that}\ ((A_i(t)v, v)) \geq \alpha_i \|v\|^2, & \forall v \in \mathcal{D}_i. \end{cases}$$

In convection-diffusion problems, we will consider the case

$A(t) = A_1(t) + A_2(t)$ with

$$A_i(t) = -d_i(x_1, x_2, t)\frac{\partial^2}{\partial x_i^2} + v_i(x_1, x_2, t)\frac{\partial}{\partial x_i} + k_i(x_1, x_2, t),$$

with $d_i(x_1, x_2, t) \geq d_0 > 0$ and $k_i(x_1, x_2, t) \geq 0$, $i = 1, 2$.

In this paper we will show the advantages of Fractional Step Runge Kutta methods (abbreviately FSRK) to discretize the time variable joined to a standard spatial discretization via Finite Difference or Finite Element schemes. For simplicity we shall choose a simple upwind scheme on rectangular meshes to discretize the spatial variables of (1) (2) obtaining a totally discrete scheme of type:

$$
\begin{cases}
U_h^0 = [u_0]_h, \\
U_h^{m,1} = U_h^m + \Delta t\, a_{11}^1 \left(-A_{1h}(t_{m,1}) U_h^{m,1} + g_{1h}(t_{m,1}) \right), \\
U_h^{m,2} = U_h^m + \Delta t \sum_{i=1}^{2} a_{2i}^{n(i)} \left(-A_{n(i)h}(t_{m,i}) U_h^{m,i} + g_{n(i)h}(t_{m,i}) \right), \\
\vdots \\
U_h^{m,s} = U_h^m + \Delta t \sum_{i=1}^{s} a_{si}^{n(i)} \left(-A_{n(i)h}(t_{m,i}) U_h^{m,i} + g_{n(i)h}(t_{m,i}) \right), \\
U_h^{m+1} = U_h^m + \Delta t \sum_{i=1}^{s} b_i^{n(i)} \left(-A_{n(i)h}(t_{m,i}) U_h^{m,i} + g_{n(i)h}(t_{m,i}) \right), \\
\text{with } \begin{cases} n(i) = 1, & \text{if } i \text{ is even,} \\ n(i) = 2, & \text{if } i \text{ is odd,} \end{cases}
\end{cases}
\qquad (2)
$$

here, we are denoting with h the mesh size and $[\cdot]_h$ is the restriction of a function defined in Ω, to a rectangular mesh Ω_h that covers Ω. Using these algorithms we obtain approximations U_h^m to $[u(\bar{x}, t_m)]_h$ with $t_m = m\,\Delta t$. $A_{1h}(t)$ and $A_{2h}(t)$ are the difference operators that appear by discretizing, with the upwind scheme, the operators $A_1(t)$ and $A_2(t)$ given in (2). The intermediate approximations $U_h^{m,i}$ for $i = 1, \ldots, s$ are called stages values of the FSRK method, and we can consider them as approximations to $[u(\bar{x}, t_{m,i})]_h$ in $t_{m,i} = t_m + c_i\,\Delta t$. Finally we take $g_{n(i)h}(t_{m,i}) = [g_{n(i)}(t_{m,i})]_h$ with $g_1(t) + g_2(t) = g(t)$.

So, a FSRK method is determined by the choise of the coefficients c_j, $a_{ji}^{n(i)}$, $b_i^{n(i)}$. We will refer often to a FSRK method by means of its coefficients, sorted in a Butcher's table like follows:

$$
\begin{array}{c|c|c}
\mathcal{C}\,e & \mathcal{A}^1 & \mathcal{A}^2 \\ \hline
 & (b^1)^T & (b^2)^T
\end{array}
\qquad \text{where } e = \begin{pmatrix} 1 \\ \vdots \\ 1 \end{pmatrix}, \quad \mathcal{C} \equiv \begin{pmatrix} c_1 & & \\ & \ddots & \\ & & c_s \end{pmatrix},
$$

$\mathcal{A}^k \equiv (a_{ji}^k)_{j,i=1}^{s}$, $\qquad b^k \equiv (b_i^k)_{i=1}^{s}\ k = 1, 2$,

$$
\text{verifying } \begin{cases}
a_{ji}^k = 0, & \text{if } i > j, \quad k \in \{1, 2\}, \\
a_{ii}^{n(i)} > 0, & a_{ii}^{3-n(i)} = 0, \quad \forall i \in \{1, \ldots, s\}, \\
a_{ji}^{3-n(i)} = 0, & \forall i, j \in \{1, \ldots, s\}, \\
b_i^{3-n(i)} = 0, & \forall i \in \{1, \ldots, s\}.
\end{cases}
\qquad (3)
$$

For solving (2) we must attack a family of linear systems of the form $\left(I + \Delta t\, a\, A_{n(i)h}(t_{m,i})\right) U_h^{m,i} = f_{ih}$, where the second term, f_{ih}, is computed explicitly from previous stages and some evaluations of $g_{ih}(t)$, and where the matrices $\left(I + \Delta t\, a\, A_{n(i)h}(t_{m,i})\right)$ are tridiagonal. Therefore the computational cost resulting of the numerical integration with these methods is, in every time step, linearly dependent on the number of mesh points in Ω_h. This fact represents an important advantage with respect to classical implicit methods because it is possible to obtain unconditional convergence (i.e. without limitations between Δt and h) and the order of complexity of the algorithm is the same of an explicit method, while the classical implicit methods have a higher computational cost due to they have to resolve block tridiagonal linear systems.

It is well known that the time variation of the coefficients d_i, v_i and k_i hampers the analysis of the unconditional convergence when we use a numerical integrator in a parabolic problem (see [3], [4]). In fact, the use of a space semidiscretization in a parabolic problem results in a Stiff problem with the form

$$\begin{cases} u'(t) = J(t)u(t) + f(t), \\ u(0) = u_0; \end{cases} \qquad (4)$$

classically, methods that verify the AN-stability property have been used in the time integration of (4) in order to preserve the contractivity of their exact solutions independently of Δt and h. This property is preserved by simple integrators of low order (implicit Euler) or by high order Runge-Kutta methods if they are totally implicit (like Gauss methods), i.e., to obtain high orders in time with classical methods preserving AN-stability increases the computational cost because we must use non semiexplicit methods (see [5]).

Nevertheless, recent papers (see [4]) show that if we consider time variation of the form

$$\|(A(t) - A(s))u\| \leq L|t-s|^\alpha (\|u\| + \|A(s)u\|),\ u \in \mathcal{D},\ \alpha \in (0,1],\ \text{and}\ L > 0,$$

the AN-stability can be weakened to A-stability obtaining unconditional convergence for standard time integrators, like Runge-Kutta methods.

Since the FSRK methods which we propose are semiexplicit, the AN-stability will be preserved only by the simplest methods of low order (see [7]).

For the discretizations obtained via FSRK methods we give a stability result by imposing on the operators $A_i(t)$ the next condition:

$$\|A_i(t')\, u - A_i(t)\, u\| \leq |t-t'|\, M_i\, \|A_i(t)\, u\|, \quad \forall i = 1,2, \quad \forall t, t' \in [0,T], \qquad (5)$$

which is related to a Lipschitz variation in the coefficients of $A_i(t)$. In this case, the A-stability is a sufficient condition to guarantee the unconditional stability of scheme (2), at least in finite time intervals.

2 Convergence

To realize the study of the convergence in a simple form we decompose the analysis of global error in two components: on one hand, the contribution of the time semidiscretization process and on the other one, the contribution of the space semidiscretization stage.

2.1 Time Semidiscretization

Let $U^m \equiv U^m(\bar{x})$ be the solution, that approaches $u(\bar{x}, t_m)$, obtained with the scheme

$$\begin{cases} U^0 = u_0 \in \mathcal{D}, \\ U^{m,1} = U^m + \Delta t \, a_{11}^1 \left(- A_1(t_{m,1}) \, U^{m,1} + g_1(t_{m,1}) \right), \\ \quad \vdots \\ U^{m,s} = U^m + \Delta t \sum_{i=1}^{s} a_{si}^{n(i)} \left(- A_{n(i)}(t_{m,i}) \, U^{m,i} + g_{n(i)}(t_{m,i}) \right), \\ U^{m+1} = U^m + \Delta t \sum_{i=1}^{s} b_i^{n(i)} \left(- A_{n(i)}(t_{m,i}) \, U^{m,i} + g_{n(i)}(t_{m,i}) \right), \end{cases} \quad (6)$$

where the coefficients c_j, $a_{ji}^{n(i)}$, $b_i^{n(i)}$ verify the restrictions (3).

In order to study the scheme (6) we introduce the next tensorial notation: given $M \equiv (m_{ij}) \in \mathbb{R}^{s \times s}$, we define $\bar{M} \equiv (m_{ij} I_H) \in H^{s \times s}$, given $v \equiv (v_i) \in \mathbb{R}^s$ we define analogously $\bar{v} \equiv (v_i I_H) \in H^s$;

$$\hat{A}_k^m = \begin{pmatrix} A_k(t_{m,1}) & 0 & \cdots & 0 \\ 0 & A_k(t_{m,2}) & \cdots & 0 \\ \vdots & \vdots & \ddots & \vdots \\ 0 & 0 & \cdots & A_k(t_{m,s}) \end{pmatrix} \text{ and } G_k^m = \begin{pmatrix} g_k(t_{m,1}) \\ g_k(t_{m,2}) \\ \vdots \\ g_k(t_{m,s}) \end{pmatrix}, \quad k = 1, 2, \quad (7)$$

where I_H is the identity in H. In [1] the following three results are proved:

Theorem 1. *Let $\{A_i(t)\}_{i=1}^2$ be maximal and coercive operators, then the scheme (6) admits unique solution, bounded independently of Δt, which can be expressed as*

$$U^{m+1} = \tilde{R}(-\Delta t \hat{A}_1^m, -\Delta t \hat{A}_2^m) \, U^m + \Delta t \, \tilde{S}(\Delta t \hat{A}_1^m, \Delta t \hat{A}_2^m, \Delta t G_1^m, \Delta t G_2^m), \quad (8)$$

with $\tilde{R}(-\Delta t \hat{A}_1^m, -\Delta t \hat{A}_2^m) = \bar{I} - \sum_{i=1}^{2} \overline{(b^i)^T} \Delta t \hat{A}_i^m \left(\bar{I} + \sum_{j=1}^{2} \overline{A^j} \Delta t \hat{A}_j^m \right)^{-1} \bar{e}$ *and*

$$\tilde{S}(\Delta t \hat{A}_1^m, \Delta t \hat{A}_2^m, \Delta t G_1^m, \Delta t G_2^m) =$$

$$-\Delta t \sum_{i=1}^{2} \overline{(b^i)^T} \hat{A}_i^m \left(\bar{I} + \Delta t \sum_{j=1}^{2} \overline{A^j} \hat{A}_j^m \right)^{-1} \left(\sum_{k=1}^{2} \overline{A^k} G_k^m \right) + \sum_{i=1}^{2} \overline{(b^i)^T} G_i^m. \quad (9)$$

In (8) we have separated the solution of scheme (6) in two terms: the contribution of the solution in the previous instant, U^m, operated by

$\widetilde{R}(-\Delta t \hat{A}_1^m, -\Delta t \hat{A}_2^m)$, and the contribution of the source terms $g_i(t)$, that we have grouped in $\widetilde{S}(\Delta t \hat{A}_1^m, \Delta t \hat{A}_2^m, \Delta t G_1^m, \Delta t G_2^m)$. This decomposition permits us to deduce immediately that the contractivity of FSRK (i.e., $\|U^{m+1} - V^{m+1}\| \leq \|U^m - V^m\|$ where U^m and V^m are solutions obtained from different initial conditions U^0 and V^0) is equivalent to $\|\widetilde{R}(-\Delta t \hat{A}_1^m, -\Delta t \hat{A}_2^m)\| \leq 1$. A weaker stability condition can be introduced asfollows: a method of type Runge Kutta is said A-stable if for any two solutions U^m and V^m, obtained from the initial conditions U^0 and V^0, with non homogeneous terms $g_i(t)$ and $g_i(t) + \varepsilon_i(t)$ respectively it holds that

$$\|U^m - V^m\| \leq C(\|U^0 - V^0\| + \max_{t \in [0,T]} \sum_{i=1}^n \|\varepsilon_i(t)\|). \tag{10}$$

Using again (8), it is easy to check that

$$\|\widetilde{R}(-\Delta t \hat{A}_1^m, -\Delta t \hat{A}_2^m)\| \leq e^{\beta \Delta t}, \tag{11}$$

and $\|\widetilde{S}(\Delta t \hat{A}_1^m, \Delta t \hat{A}_2^m, \Delta t G_1^m, \Delta t G_2^m)\| \leq C \Delta t (\|G_1^m\| + \|G_2^m\|)$ are sufficient conditions to verify the stability property (10) for $m \leq \frac{T}{\Delta t}$.

Theorem 2. *Let* $(\mathcal{C}, \mathcal{A}^1, (b^1)^T, \mathcal{A}^2, (b^2)^T)$ *be a FSRK method given by (3), whose coefficients verify*

$$|R(z_1, z_2)| = \left|1 - \sum_{i=1}^2 (b^i)^T z_i (I + \sum_{j=1}^2 \mathcal{A}^j z_j)^{-1} e\right| \leq 1,$$

$$\forall z_1, z_2 \in \mathbb{C}, \text{ with } Re(z_i) \leq 0, \text{ for } i = 1, 2, \tag{12}$$

(this property is called A-stability of the FSRK) and let $A_1(t), A_2(t)$ *be maximal, coercive and commuting operators verifying (5). Then there exists a constant* β *independent of* Δt *such that (11) is verified.*

Main idea of proof

To bound $\widetilde{R}(-\Delta t \hat{A}_1^m, -\Delta t \hat{A}_2^m)$ we decompose this operator in the form $\widetilde{R}(-\Delta t \hat{A}_1^m, -\Delta t \hat{A}_2^m) = R(-\Delta t A_1(t_m), -\Delta t A_2(t_m)) + \Delta t P$ and we use that, under the hypotheses of this Theorem, it is verified that $\|R(-\Delta t A_1(t_m), -\Delta t A_2(t_m))\| \leq 1$ (see [7]), and also $\|P\| \leq C M$ (see [1]) to deduce that

$$\|\widetilde{R}(-\Delta t \hat{A}_1^m, -\Delta t \hat{A}_2^m)\| \leq 1 + C M \Delta t \leq e^{\beta \Delta t}$$

with $\beta = C M$ being $M = \max_{\substack{i=1,\ldots,s \\ k=1,2}} \{|c_i| M_k\}$

and C a constant that depend on the size of the coefficients of the FSRK. ◇

In [8] it is also proved that if the FSRK method is strongly A-stable, i.e., if it verifies (12) and there exists $c < 1$ and K, sufficiently large, such that

$|R(z_1, z_2)| < c$, if $Re(z_i) \leq 0$, for $i \in \{1,2\}$ and $|z_1| + |z_2| \geq K$, then the contractivity result $\|R(-\Delta t A_1(t_m), -\Delta t A_2(t_m))\| \leq 1$ can be improved to $\|R(-\Delta t A_1(t_m), -\Delta t A_2(t_m))\| \leq e^{-\beta' \Delta t}$ with $\beta' > 0$ and independent of $\Delta t \in (0, \Delta t_0]$. Because of this, in some cases the stability result (11) can also be improved to contractivity by using a strongly A-stable FSRK method. To be more precise, for $\Delta t \in (0, \Delta t_0]$ and M_i small enough[1] negative values of β can be considered in (11). Combining this contractivity property with the consistency property, that we will introduce next, the study of the convergence in infinite length intervals can be realized.

To study the consistency of scheme (6) we define the local error as

$$e^{m+1} = u(t_{m+1}) - \widetilde{R}(-\Delta t \hat{A}_1^m, -\Delta t \hat{A}_2^m) u(t_m) - \Delta t \, \widetilde{S}(\Delta t \hat{A}_1^m, \Delta t \hat{A}_2^m, \Delta t G_1^m, \Delta t G_2^m)$$

and we say that a FSRK method is consistent of order p if for sufficiently smooth data $u(t_m)$, G_1^m and G_2^m, it is verified that $\|e^{m+1}\| \leq C(\Delta t^{p+1})$, $\forall m \geq 0$, $\Delta t \to 0$, where C is a constant independent of Δt.

Theorem 3. *Let us consider a FSRK method satisfying the order conditions*

$$\begin{cases} (b^{i_1})^T (C)^{\rho_1} \mathcal{A}^{i_2} (C)^{\rho_2} \ldots \mathcal{A}^{i_r} (C)^{\rho_r} e = \prod_{j=1}^{r} \frac{1}{(r-j+1) + \sum_{k=j}^{r} \rho_k}, \\ \forall r = 1, \ldots, p, \forall (\rho_1, \ldots, \rho_r) \in \{0, \ldots p-1\}^r \text{ verifying } 1 \leq r + \sum_{k=1}^{r} \rho_k \leq p \\ \text{and } \forall (i_1, \ldots, i_r) \in \{1,2\}^r, \end{cases}$$

joined to the reductions $(C)^k e - k\mathcal{A}^i (C)^{k-1} e = 0$, $i \in \{1,2\}$, $k = 1, \ldots, k_0$, *and let us apply it to a problem of type (1), whose solution verifies the following smoothness requirements*

$$\begin{cases} \|A_{i_1}^{(\rho_1)}(t) \ldots A_{i_{l+1}}^{(\rho_{l+1})}(t) u_i^{(k)}(t)\| \leq C, & i_\bullet \in \{1,2\}, \\ l \in \{1, \ldots, p-k+1\}, \quad i \in \{1,2\}, \quad k_0 \leq k \leq p, \\ \text{and } \rho_1 + \ldots + \rho_l \leq p - k - l + 1; \\ \|u_i^{(p+1)}(t)\| \leq C, \text{ with } u_i'(t) = -(A_i(t)u(t) - g_i(t)). \end{cases}$$

Then

$$\|e^{m+1}\| \leq C(\Delta t)^{p+1}, \tag{13}$$

where C is a constant independent of Δt.

To realize the study of the convergence of the semidiscrete scheme we define the global error associated to the time semidiscretization as $E^{\Delta t} \equiv \sup_{m \leq \frac{T}{\Delta t}} (u(t_m) - U^m)$ and we say that the scheme (6) is convergent of order p, if $\|E^{\Delta t}\| \leq C(\Delta t)^p$ where C is a constant independent of Δt. It is immediate to check that, if the scheme (6) verifies (11) and (13), then it is convergent of order p.

---
[1] This property is related to a small time variation of the coefficients $d_i(x_1, x_2, t)$, $v_i(x_1, x_2, t)$ and $k_i(x_1, x_2, t)$, at least from sufficiently large values of t.

2.2 Total Discretization

The totally discrete scheme (2), that we propose, is obtained by discretizing in space (6) by using the simple upwind schemes defined on rectangular meshes.

To study the convergence of scheme (2), we define the global error associated to the total discretization (2) in the instant t_m as $E_h^m = \| [u(t_m)]_h - U_h^m \|_h$ [2], and we say that the discretization is convergent, of order p in time and of order q in space, if

$$E_h^m \leq C(h^q + \Delta t^p), \tag{14}$$

where C is a constant independent of Δt and of h.

To analyze the convergence of the total discretization we separate, in certain way, the contribution to the global error E_h^m of the time and of the space discretization stages; the contribution of the space discretization to the global error will be studied by means of the term, called local error of the spatial discretization, that we define as $\hat{e}_h^m = \| [\hat{u}^m]_h - \hat{U}_h^m \|_h$, where $\hat{u}^m$ is obtained giving a step with the semidiscrete scheme (6) by taking as initial point $U^{m-1} = u(t_{m-1})$, and $\hat{U}_h^m$ is obtained giving a step with the totally discrete scheme (2) and taking as initial point $U_h^{m-1} = [u(t_{m-1})]_h$.

Analogously to the study of the convergence realized in [2] we can prove the following

Theorem 4. *Let $u(t)$ be the unique solution of problem (1), with Ω sufficiently smooth, in such a way that $\{A_i(t)u(t)\}_{i=1}^2$ and $\{g_i(t)\}_{i=1}^2$ are $C^2(\Omega)$, and let us consider a FSRK method and the simple upwind discretizations $\{A_{ih}(t)\}_{i=1}^2$ of $\{A_i(t)\}_{i=1}^2$ in a uniform rectangular mesh Ω_h. Then*

$$\hat{e}_h^m \leq C \Delta t \, h, \tag{15}$$

where h is the mesh size of Ω_h.

To obtain the convergence of the totally discrete scheme (2), we bound the global error in the form

$$E_h^m \leq \| [u(t_m)]_h - [\hat{u}^m]_h \|_h + \| [\hat{u}^m]_h - \hat{U}_h^m \|_h + \| \hat{U}_h^m - U_h^m \|_h; \tag{16}$$

using the formula (8) for a step in the time integration, it is immediate that

$$\hat{U}_h^m - U_h^m = \widetilde{R}(-\Delta t \hat{A}_1^m, -\Delta t \hat{A}_2^m)([u(t_{m-1})]_h - U_h^{m-1}),$$

and applying this equality in (16), we obtain a recurrence law for the global errors, that under the necessary hypotheses for fulfilling (13), (15) and

$$\| \widetilde{R}(-\Delta t \hat{A}_{1h}^m, -\Delta t \hat{A}_{2h}^m) \|_h \leq e^{\beta \Delta t}, \tag{17}$$

with β independent of h, permits us to prove (14).

Observe that the bound (17) is a particular case of (11) if the operators $A_{1h}(t), A_{2h}(t)$ preserve the monotonicity and commutativity properties of

---
[2] $\| \cdot \|_h$ is a suitable norm for the space of discrete functions defined in Ω_h.

the operators $A_1(t), A_2(t)$. Such properties are easily checkable in some cases, like for example, simple Finite Difference schemes on rectangular meshes in problems of type (1) whose coefficients $d_i(x_1, x_2, t)$, $v_i(x_1, x_2, t)$ and $k_i(x_1, x_2, t)$ not depend on the spatial variable x_j with $i \neq j$. In other cases, for example, for arbitrary spatial variations in the coefficients of (1), we do not know theoretical results which permit us the obtaining of (11), nevertheless, in the numerical experiments realized in some non commuting operator cases, the obtained numerical solutions present also the same stable behaviour.

3 Numerical Results

3.1 A Parabolic Problem with Time–Dependent Coefficients

To integrate the following convection–diffusion problem:

$$\begin{cases} \frac{\partial u}{\partial t} - d_1 \frac{\partial^2 u}{\partial x^2} - d_2 \frac{\partial^2 u}{\partial y^2} + v_1 \frac{\partial u}{\partial x} + v_2 \frac{\partial u}{\partial y} + (k_1 + k_2)u = g(x,y,t), & (x,y,t) \in \Omega \times [0,5], \\ u(0,y,t) = u(1,y,t) = u(x,0,t) = u(x,1,t) = 0, & x \in [0,1], \quad y \in [0,1], \quad t \in [0,5], \\ u(x,y,0) = x^3(1-x)^3 y^3 (1-y)^3, & x,y \in [0,1], \\ g(x,y,t) = e^{-t} \sin(x(1-x)) \sin(y(1-y)), \end{cases}$$

with $d_1 = d_2 = (1+e^{-t})$, $v_1 = (1+x)(2+e^{-t})$, $v_2 = (1+y)(2-e^{-t})$, $k_1 = 1+x^2$ and $k_2 = 1+\sin(\pi y)$, we combine the third order FSRK method given by (18)[3].

---

[3] The details of the construction of this method can be seen in [1]; there it is proven that at least five stages are necessary to obtain third order and six stages are convenient. Note also that each RK that compose the FSRK can be reduced to a third order SDIRK method with three stages (see also [6])

$$\left(\frac{\mathcal{A}^1}{(b^1)^T}\right) = \begin{pmatrix} 0.435866521508459 & & & & & & \\ 0.435866521508459 & 0 & & & & & \\ 0.264133478491540 & 0 & 0.435866521508459 & & & & \\ 0.524203567293128 & 0 & -0.224203567293127 & 0 & & & \\ 0.054134244066592 & 0 & 0.0741327129164892 & 0 & 0.435866521508459 & & \\ 2.005981609913539 & 0 & 1.336337252930893 & 0 & -2.59231886284469 & 0 & \\ 2.838287230686191 & 0 & 2.207497360663944 & 0 & -4.04578459135012 & 0 & \end{pmatrix}$$

$$\left(\frac{\mathcal{A}^2}{(b^2)^T}\right) = \begin{pmatrix} 0 & & & & & & \\ 0 & 0.435866521508459 & & & & & \\ 0 & 0.170931386851894 & 0 & & & & \\ 0 & -0.13586652150846 & 0 & 0.435866521508459 & & & \\ 0 & 0.062944816984284 & 0 & -0.09326511998115 & 0 & & \\ 0 & -0.543014480247272 & 0 & 0.8571479587388134 & 0 & 0.435866521508459 & \\ 0 & -0.781001854745764 & 0 & 1.100752954088072 & 0 & 0.680248900657693 & \end{pmatrix} \quad (18)$$

$$\mathcal{C}\,e = \left(0.435866521508459, 0.435866521508459, 0.7, 0.3, 0.56413347849154, 0.75\right)^T.$$

Using this scheme we have computed the numerical errors

$$E_{N,\Delta t} = \max_{\substack{(x_i,y_j)\in \Omega_{\frac{1}{N}} \\ t_m = m\Delta t,\ m=1,2,\ldots,\frac{5}{\Delta t}}} |U^{N,\Delta t}(x_i, y_j, t_m) - U^{2N, \frac{\Delta t}{2}}(x_i, y_j, t_m)|,$$

where $U^{N,K}(x_i, y_j, t_m)$ is the numerical solution obtained in the spatial point (x_i, y_j) and in the time point t_m, using a uniform rectangular mesh $\Omega_{\frac{1}{N}}$ (with $N \times N$ points) and constant time step K. To obtain these numerical errors we have taken $\Delta t = \frac{0.1}{\sqrt[3]{N}}$ in order to the contributions, in the global error, of time and space discretization stages are of the same order.

Table 1. errors $(E_{N,\Delta t})$

N	8	16	32	64	128	256	512
$E_{N,\Delta t}$	$4.0206E-4$	$2.3801E-4$	$1.2921E-4$	$6.7694E-5$	$3.491E-5$	$1.782E-5$	$8.992E-6$

3.2 A Singular Perturbation Case

In the following singular perturbation problem

$$\begin{cases} \frac{\partial u}{\partial t} - d_1 \frac{\partial^2 u}{\partial x^2} - d_2 \frac{\partial^2 u}{\partial y^2} + v_1 \frac{\partial u}{\partial x} + v_2 \frac{\partial u}{\partial y} + (k_1 + k_2)u = g(x, y, t), & (x, y, t) \in \Omega \times [0, 5], \\ u(0, y, t) = u(1, y, t) = 0, & y \in [0, 1], \quad t \in [0, 5], \\ u(x, 0, t) = u(x, 1, t) = 0, & x \in [0, 1], \quad t \in [0, 5], \\ u(x, y, 0) = h(x)\, h(y), & x, y \in \Omega, \\ g(x, y, t) = e^{-t} h(x) + e^{-t} h(y), & (x, y, t) \in \Omega \times [0, 5], \end{cases}$$

with $d_1 = \varepsilon(2-e^{-t})(1+xy)$, $d_2 = \varepsilon(2-e^{-t})(2-y)$, $v_1 = (2+\sin(\pi t)\,e^{-t})(1+x^2)$, $v_2 = (2-\sin(\pi t)\,e^{-t})(2+\sin(\pi y))$, $k_1 = 1+x^2$, $k_2 = 1+\sin(\pi y)$ and $h(\zeta) = (e^{-\frac{1-\zeta}{\varepsilon}} - e^{-\frac{1}{\varepsilon}} - (1-e^{\frac{-1}{\varepsilon}})\zeta)$ we have used the FSRK method given by (18) for the time integration. For the spatial discretization we have used the simple upwind scheme and the rectangular Shishkin meshes given in [2].

In table 2 we show, for each ε, the numerical errors

$$E_{\varepsilon,N,\Delta t} = \max_{\substack{(x_i,y_j)\in\Omega_{\frac{1}{N}} \\ t_m=m\Delta t,\, m=1,2,\dots,\frac{5}{\Delta t}}} |U^{N,\Delta t}(x_i,y_j,t_m) - U^{2N,\frac{\Delta t}{2}}(x_i,y_j,t_m)|,$$

obtained taking $\Delta t = \frac{0.1}{\sqrt[3]{N}}$. We have evaluated these errors from $t = 0.1$ until $T = 5$ since an order reduction occurs only in the first step because of data in $t=0$ do not verify (13) for $p=3$.

Remark 1. When the meshes are non uniform, we have used bilinear interpolation in the spatial variables x and y to evaluate $U^{2N,\frac{\Delta t}{2}}(x_i,y_j,t_m)$ in the points (x_i,y_j) of the mesh $\Omega_{\frac{1}{N}}$. Note that the order is less than one (concretely $N^{-1}\log(N) + \Delta t^3$) due to the singular perturbation nature of the problem. More numerical tests with arbitrary time and space dependence on the coefficients can be seen in [1].

Table 2. errors ($E_{\varepsilon,N,\Delta t}$)

ε	$N=8$	$N=16$	$N=32$	$N=64$	$N=128$	$N=256$
1	$2.7102E-4$	$1.5276E-4$	$8.1128E-5$	$4.2080E-5$	$2.1447E-5$	$1.0873E-5$
10^{-1}	$7.9712E-3$	$1.1429E-2$	$7.6423E-3$	$4.5348E-3$	$2.5052E-3$	$1.3207E-3$
10^{-2}	$2.2611E-2$	$1.4857E-2$	$1.0849E-2$	$7.8887E-3$	$5.4609E-3$	$3.4323E-3$
10^{-3}	$2.2424E-2$	$1.6301E-2$	$1.1793E-2$	$8.4844E-3$	$5.8843E-3$	$3.7733E-3$
10^{-4}	$2.2604E-2$	$1.6466E-2$	$1.1915E-2$	$8.5733E-3$	$5.9517E-3$	$3.8365E-3$
10^{-5}	$2.2621E-2$	$1.6482E-2$	$1.1928E-2$	$8.5831E-3$	$5.9596E-3$	$3.8443E-3$
10^{-6}	$2.2623E-2$	$1.6484E-2$	$1.1930E-2$	$8.5841E-3$	$5.9603E-3$	$3.8451E-3$
10^{-7}	$2.2623E-2$	$1.6484E-2$	$1.1930E-2$	$8.5843E-3$	$5.9604E-3$	$3.8444E-3$
10^{-8}	$2.2623E-2$	$1.6484E-2$	$1.1930E-2$	$8.5843E-3$	$5.9604E-3$	$3.8444E-3$
10^{-9}	$2.2623E-2$	$1.6484E-2$	$1.1930E-2$	$8.5843E-3$	$5.9604E-3$	$3.8444E-3$
$E_{N,\Delta t}^{max}$	$2.2623E-2$	$1.6484E-2$	$1.1930E-2$	$8.5843E-3$	$5.9604E-3$	$3.8451E-3$

References

1. B. Bujanda, "Métodos Runge-Kutta de Pasos Fraccionarios de orden alto para la resolución de problemas evolutivos de convección-difusión-reacción", Tesis, Universidad Pública de Navarra, 1999.
2. C. Clavero, J. C. Jorge, F. Lisbona & G. I. Shishkin, "A fractional step method on a special mesh for the resolution of multidimensional evolutionary convection-difusion problems", App. Numer. Math., **27** (1998) 211–231.

3. M. Crouzeix, "Sur l'aproximation des èquations differentielles opèrationelles linèaires par des mèthodes de Runge-Kutta", These d'Etat, Univ. de Paris VI, 1975.
4. C. González & C. Palencia "Stability of Runge-Kutta methods for abstract time-dependent parabolic problems: the Hölder case", Departamento de Matemática Aplicada y Computación, Facultad de Ciencias, Universidad de Valladolid, 1996.
5. E. Hairer, S. P. Nørsett & G. Wanner, "Solving ordinary differential equations", Vol II, Springer-Verlag, 1987.
6. E. Hairer & G. Wanner, " Solving ordinary differential equations", Vol II, Springer-Verlag, 1987.
7. J. C. Jorge, "Los métodos de pasos fraccionarios para la integración de problemas parabólicos lineales: formulación general, análisis de la convergencia y diseño de nuevos métodos", Tesis, Universidad de Zaragoza, 1992.
8. J. C. Jorge & F. Lisbona, "Contractivity results for alternating direction schemes in Hilbert spaces" App. Numer. Math. **15**, (1994) 65–75.

Variable Stepsizes in Symmetric Linear Multistep Methods

B. Cano

University of Valladolid, Facultad de Ciencias
Prado de la Magdalena, s/n, 47005-Valladolid, Spain
bego@mac.cie.uva.es
http://www.mac.cie.uva.es

Abstract. It is well known the great deal of advantages of integrating reversible systems with symmetric methods. The correct qualitative behaviour is imitated, which leads also to quantitative advantageous properties with respect to the errors and their growth with time. More particularly, fixed stepsize symmetric linear multistep methods especially designed for second order differential equations can integrate very efficiently periodic or quasiperiodic orbits till long times. A study will be given on what happens when variable stepsizes are considered so as to deal with highly eccentric orbits.

1 Introduction

In the last years, an effort has been made to construct and analyse methods especially designed to integrate eccentric orbits with variable stepsizes. In particular, symmetric variable-stepsize one-step methods have been shown to lead to slow growth of error with time when integrating periodic orbits of reversible systems [3], and some efficient numerical techniques have been designed [8] [9] which take profit of this advantageous property when the system to integrate is also Hamiltonian. See also [1] for numerical comparisons among the different integrators and more particular analysis of error growth with time for them.

On the other hand, some symmetric fixed-stepsize linear multistep methods for second order differential equations of the type (2) have also been proved to lead to slow error growth with time for periodic orbits of reversible systems. (The key property these methods must satisfy for this is that the first characteristic polynomial has only 1 as a double root and all the others are single [4]). High-order explicit methods of this type have been suggested in [10] and they are very efficient when integrating not too much eccentric orbits as they just need one function evaluation per step in contrast with many more needed by one-step methods of the same order. It was also proved in [4] that symmetric linear multistep methods for first order differential equations lead to unstable numerical solutions, except some very particular cases described in [6]. That's why we will concentrate on this paper in symmetric linear multistep methods for second order differential equations, which we will also denote by **symmetric LMM2's**.

The aim of this paper is to begin a study on the construction and numerical behaviour of symmetric LMM2's when variable stepsizes are considered in order to deal with highly eccentric orbits. The techniques used in [8] to generalize numerical integrators to variable stepsizes by considering a suitable change of variables in time and modifications of Hamiltonians are not applyable here as the resulting initial value problem would not be of the form (2) and therefore, we would not be able to integrate it with a LMM2. Therefore, in the following paper, we consider a natural generalization to variable stepsizes of LMM2's, which is described in Section 2. Also in this section, necessary and sufficient conditions on the coefficients of the methods are given for symmetry. In Section 3, an explicit second-order variable-stepsize symmetric LMM2 is constructed, It does not mean to be an optimal method, but a first example of a procedure of constructing integrators of this type. In order to prove convergence for this particular integrator, consistency and stability is required. The former is given by the way the method is constructed. The latter is proved in Section 4 under mild conditions. Finally, in Section 5, some numerical results are described in order to see whether the advantageous error growth also applies for this method when variable stepsizes are considered.

2 Symmetry Conditions for Variable-Stepsize LMM2's

A fixed-stepsize linear kth-step method for second order differential equations is defined by a difference equation like this

$$\alpha_k y_{n+k} + \cdots \alpha_0 y_n = h^2[\beta_k f_{n+k} + \cdots + \beta_0 f_n], \qquad n \geq 0, \qquad (1)$$

and k starting values $y_0, y_1, \ldots, y_{k-1}$. In formula (1), h is the stepsize of the method, $\{y_m\}_{m \in \{0,1,2,\ldots\}}$ are the numerical approximations to the solution of an initial value problem of the form

$$\begin{aligned} \ddot{y}(t) &= f(y(t)), \\ y(t_0) &= u_0, \\ \dot{y}(t_0) &= v_0, \end{aligned} \qquad (2)$$

in times $\{t_m = t_0 + mh\}_{m \in \{0,1,2,\ldots\}}$, $\{f_m\}$ denotes $\{f(y_m)\}$ for the function f in (2), and $\{\alpha_l\}_{l=0}^k$ and $\{\beta_l\}_{l=0}^k$ are the constant coefficients (not depending on the problem (2) considered neither on the stepsize h) which determine the method.

A natural generalization of methods of this type to variable stepsizes is to consider the following change in (1):

$$\alpha_k(h_n, \ldots, h_{n+k-1}) y_{n+k} + \cdots + \alpha_0(h_n, \cdots, h_{n+k-1}) y_n$$
$$= h_{n+k-1}^2 [\beta_k(h_n, \cdots, h_{n+k-1}) f_{n+k} + \cdots + \beta_0(h_n, \cdots, h_{n+k-1}) f_n]. \qquad (3)$$

Now $\{y_m\}_{m \in \{0,1,2,\ldots\}}$ denotes the approximations to the solutions of (2) in times $\{t_m = t_0 + h_0 + \cdots + h_{m-1}\}_{m \in \{0,1,2,\ldots\}}$. So, obviously, h_m is the stepsize considered to go from y_m to y_{m+1}, and the coefficients of the method can't now be constants but functions which depend on the stepsizes given by the method in each particular case.

For fixed-stepsize methods, it is well known that the following conditions are sufficient and necessary to get a symmetric stable LMM2 [11]

$$\alpha_j = \alpha_{k-j}, \qquad \beta_j = \beta_{k-j}, \qquad j = 0, \ldots, k.$$

For their variable-stepsize counterparts, the following are sufficient conditions for symmetry

$$\alpha_j(h_0, \ldots, h_{k-1}) = \alpha_{k-j}(h_{k-1}, \ldots, h_0)$$
$$\beta_j(h_0, \ldots, h_{k-1}) = \frac{h_0^2}{h_{k-1}^2}\beta_{k-j}(h_{k-1}, \ldots, h_0), \qquad j = 0, \ldots, k. \qquad (4)$$

This is easily verified. If the method takes $y_n, \ldots, y_{n+k-1}$ to y_{n+k} by (3), then, by using (4), the following difference equation is satisfied

$$\alpha_0(h_{n+k-1}, \ldots, h_n)y_{n+k} + \cdots + \alpha_k(h_{n+k-1}, \cdots, h_n)y_n$$
$$= h_n^2[\beta_0(h_{n+k-1}, \cdots, h_n)f_{n+k} + \cdots + \beta_k(h_{n+k-1}, \cdots, h_n)f_n], \qquad (5)$$

which says that the method would take $y_{n+k}, \ldots, y_{n+1}$ to y_n if the same stepsizes in the reversed order had been considered. To assure that the method would have taken the same stepsizes when integrating backwards, it is also necessary to ask that the stepsize going forward from a numerical approximation y_n to y_{n+1} is the same as the one going backwards from y_{n+1} to y_n. (The same happens for one-step methods [3]). In other words, if the stepsize taken is just a function of the point of departure and the point of arrival $H(y_n, y_{n+1})$, it must be verified that

$$H(y_n, y_{n+1}) = H(y_{n+1}, y_n).$$

This is verified, for example, if the following arithmetic media is considered

$$h_n = \frac{\epsilon}{2}[s(y_n) + s(y_{n+1})], \qquad (6)$$

for any function s. (This function will be chosen so that the stepsize h_n is suitable for the integration of the problem.)

3 Construction of an Explicit Second-Order Variable-Stepsize Symmetric LMM2

Looking for a second-order explicit method which verifies the conditions in Section 1, we have seen that $k = 2$ leads to fixed-stepsize method and $k = 3$ to nearly the same conclusion, as only two different stepsizes would be possible in this case. Therefore, we have to look in 4th-step methods to get a variable stepsize-adaptive method.

In such a case, we have eight unknowns $\alpha_0, \ldots, \alpha_4, \beta_1, \beta_2, \beta_3$ for determined stepsizes and the conditions of order would be 4. However, as the symmetry conditions are difficult to treat by themselves because it is not established the kind of dependence on the stepsizes, we have forced symmetry by writing the order conditions in forward and backward forms and assuming then the conditions (4). More explicitly, we have made formula (3) exact for the polynomials $y(t) = t^m$ ($m = 0, 1, 2, 3$), when going from $t_0 = 0$ to $t_4 = h_0 + h_1 + h_2 + h_3$, and when going from t_4 to t_0. In such a way, a linear system of seven equations and eight unknowns turn up. By using $MATHEMATICA^{tm}$, we have solved it and found that the system says that

$$\alpha_4 = -\alpha_1 \frac{h_0}{h_0 + h_1 + h_2 + h_3} - \alpha_2 \frac{h_0 + h_1}{h_0 + h_1 + h_2 + h_3} - \alpha_3 \frac{h_0 + h_1 + h_2}{h_0 + h_1 + h_2 + h_3}$$
$$\alpha_0 = -\alpha_3 \frac{h_3}{h_0 + h_1 + h_2 + h_3} - \alpha_2 \frac{h_2 + h_3}{h_0 + h_1 + h_2 + h_3} - \alpha_1 \frac{h_1 + h_2 + h_3}{h_0 + h_1 + h_2 + h_3}$$
(7)

$$\beta_1 = \beta_3 \frac{h_2}{h_1} - \alpha_3 g_3(h_0, h_1, h_2, h_3) - \alpha_2 g_2(h_0, h_1, h_2, h_3) - \alpha_1 g_1(h_0, h_1, h_2, h_3)$$
$$\beta_2 = -\beta_3 \frac{h_1 + h_2}{h_1} - \alpha_3 p_3(h_0, h_1, h_2, h_3) - \alpha_2 p_2(h_0, h_1, h_2, h_3)$$
$$\quad - \alpha_1 p_1(h_0, h_1, h_2, h_3),$$
(8)

where all the coefficients correspond to the method going from $t = 0$ to $t = h_0 + h_1 + h_2 + h_3$ in this order, and where $\{g_i\}_{i=1,2,3}$ and $\{p_i\}_{i=1,2,3}$ are rational functions of the stepsizes.

By taking $\alpha_2 = 2$, $\alpha_1 = \alpha_3 = -2$, the first characteristic polynomial for fixed stepsize would be

$$\rho(x) = x^4 - 2x^3 + 2x^2 - 2x + 1.$$

This polynomial has single roots $\pm i$ and double root 1. Therefore, every LMM2 which has this as its first characteristic polynomial and which is implemented with fixed stepsize would lead to linear error growth with time for a great deal of problems including Kepler's, in contrast with more general methods which would lead to quadratic error growth [4].

Substituting these values of $\{\alpha_i\}_{i=1}^3$ in the formulas (8) for $\{\beta_i\}_{i=1}^2$, we have found that a possible symmetric choice for $\{\beta_i\}_{i=1}^3$ is

$$\beta_1 = \frac{-2h_3^2 h_2 - 2h_3^2 h_0 - 4h_3 h_2^2 + 2h_1 h_2 h_3 - 2h_0 h_2 h_3 + 2h_1 h_2^2}{6h_1 h_3^2},$$

$$\beta_2 = \frac{n(\beta_2)}{6h_1 h_2 h_3^2}, \quad \text{where}$$
$$n(\beta_2) = 2h_0^2 h_1^2 + 2h_0^2 h_1 h_3 + 4h_0 h_1^3 + 4h_0 h_1^2 h_2 + 2h_0 h_1^2 h_3 - 2h_1^3 h_2 + 6h_0 h_1 h_2 h_3$$
$$\quad - 6h_1^2 h_2^2 + 2h_0 h_2^2 h_3 + 4h_1 h_2^2 h_3 + 4h_2^3 h_3 + 2h_0 h_2 h_3^2 + 2h_2^2 h_3^2 - 2h_1 h_2^3,$$
$$\beta_3 = \frac{-2h_0^2 h_1 - 2h_0^2 h_3 - 4h_0 h_1^2 + 2h_2 h_1 h_0 - 2h_0 h_1 h_3 + 2h_2 h_1^2}{6h_2 h_0^2}.$$

We have found these coefficients by assuming that $6h_2 h_0^2 \beta_3$ has a homogeneous third-degree polynomial expression on the stepsizes and selecting a choice from all the possible ones which make (8) possible apart from the symmetry condition between β_1 and β_3 (4).

4 Analysis of Stability of the Previous Method

Stability of variable-stepsize LMM is not such an easy issue as stability of variable-stepsize one-step methods. There are some results, however, corresponding to some particular cases such as Adams or Störmer methods [5][2]. There are no results indeed for our particular method. Therefore, we proceed to study its stability.

The first step is to express the method as a one-step integrator by considering a wider phase-space. We introduce the variables $\{v_n\}_{n\in\{0,1,...\}}$ defined by

$$y_{n+1} = y_n + h_n v_n. \tag{9}$$

In such a way, it can be seen that the following difference equation is satisfied by $\{v_n\}_{n\in\{0,1,...\}}$

$$v_{n+3} = \frac{h_{n+1} + h_{n+3}}{h_n + h_{n+2}} \frac{h_{n+2}}{h_{n+3}} v_{n+2} - \frac{h_{n+1}}{h_{n+3}} v_{n+1} + \frac{h_{n+1} + h_{n+3}}{h_n + h_{n+2}} \frac{h_n}{h_{n+3}} v_n$$
$$+ h_{n+3} \frac{h_n + h_{n+1} + h_{n+2} + h_{n+3}}{2(h_n + h_{n+2})} (\beta_1 f_{n+1} + \beta_2 f_{n+2} + \beta_3 f_{n+3}). \tag{10}$$

In fact, (9) and (10) are the equations used to implement the method in practice, as this process (also called stabilization [7]) leads to much lower errors when the stepsize is small (roundoff errors are diminished significantly).

Considering then the augmented vector $Y_n = [y_{n+3}, v_{n+2}, v_{n+1}, v_n]^T$, the following recurrence relation is satisfied

$$Y_{n+1} = A_n Y_n + h_{n+3} F_n(Y_n), \tag{11}$$

where

$$A_n = \begin{pmatrix} 1 & \dfrac{h_{n+1} + h_{n+3}}{h_n + h_{n+2}} h_{n+2} & -h_{n+1} & \dfrac{h_{n+1} + h_{n+3}}{h_n + h_{n+2}} h_n \\ 0 & \dfrac{h_{n+1} + h_{n+3}}{h_n + h_{n+2}} \dfrac{h_{n+2}}{h_{n+3}} & -\dfrac{h_{n+1}}{h_{n+3}} & \dfrac{h_{n+1} + h_{n+3}}{h_n + h_{n+2}} \dfrac{h_n}{h_{n+3}} \\ 0 & 1 & 0 & 0 \\ 0 & 0 & 1 & 0 \end{pmatrix} \otimes I,$$

$$F_n(Y_n) = \begin{pmatrix} h_{n+3}(\beta_1 f_{n+1} + \beta_2 f_{n+2} + \beta_3 f_{n+3}) \\ \dfrac{h_n + h_{n+1} + h_{n+2} + h_{n+3}}{2(h_n + h_{n+2})} (\beta_1 f_{n+1} + \beta_2 f_{n+2} + \beta_3 f_{n+3}) \\ 0 \\ 0 \end{pmatrix} \otimes I.$$

By considering succesive powers of the matrices A_l $(A_n A_{n-1} \cdots A_{j+1})$, it can be proved that the infinity norm of this power is bounded independently of the number of factors $n-j$, whenever every stepsize h_l satisfies

$$h_l \geq ch_{l-2m}, \quad m = 1, 2, \ldots, \lfloor \tfrac{l}{2} \rfloor, \quad h_l \leq h, \quad l = 1, 2, \ldots$$

for some constants c and h. (These conditions are easily satisfied for c small and h large enough.)

Stability is given if, whenever (11) is slightly modified as well as the initial value Y_0, the vectors $\{Y_n\}$ obtained are also only modified accordingly. This happens due to the mentioned previous bound for the powers of A_l. A modification of (11) would be

$$\bar{Y}_{n+1} = A_n \bar{Y}_n + h_{n+3}(\bar{F}_n + \nu_n).$$

So, if E_n denotes $\bar{Y}_n - Y_n$, we have that

$$E_{n+1} = A_n E_n + h_{n+3}(\bar{F}_n - F_n) + h_{n+3}\nu_n,$$

and therefore

$$E_{n+1} = A_n \ldots A_0 E_0 + \sum_{j=0}^{n-1} A_n \cdots A_{j+1} h_{j+3}[(\bar{F}_j - F_j) + \nu_j] + h_{n+3}\nu_n,$$

Now if F_n is Lipschitz with respect to Y_n (as it happens in fact if f is and $h_n \leq Ch_{n-1}$ for some constant C), the following bound is verified

$$\|E_{n+1}\|_\infty \leq K\left(\|E_0\|_\infty + (T_f - t_0)\max_j \|\nu_j\|_\infty\right) + KL\sum_{j=0}^{n-1} h_{j+3}\|E_j\|_\infty,$$

where K is the bound for the powers $A_n \cdots A_{j+1}$, T_f is the final time till we integrate and L is the Lipschitz constant for F_n. A discrete Gronwall argument then says that

$$\|E_{n+1}\|_\infty \leq K e^{KL(T_f - t_0)}\left(\|E_0\|_\infty + (T_f - t_0)\max_j \|\nu_j\|_\infty\right),$$

which proves stability under the given assumptions.

5 Numerical Experiments and Error Growth with Time

We have implemented the method in Section 3 using stabilization [7] and compensated summation, in order to reduce roundoff error as much as possible. We have integrated Kepler's problem

$$\ddot{x}_i = -\frac{x_i}{(x_1^2 + x_2^2)}, \quad i = 1, 2,$$

with initial positions and velocities given by $x_1(0) = 1-\text{ecc}$, $x_2(0) = 0$, $\dot{x}_1(0) = 0$, $\dot{x}_2(0) = \sqrt{\frac{1+\text{ecc}}{1-\text{ecc}}}$. In such a way, the solution describes an orbit of a satellite around a planet in the form of an ellipse of eccentricity ecc. Here we have taken ecc = 0.9, as these variable-stepsize methods are constructed so as to integrate problems where a great variability in the solution makes it suitable to treat with more care some parts than others.

The stepsize function has been chosen according to (6), solving iteratively this equation to a given relative tolerance 10^{-3}. The function s has been chosen as

$$s(x_1, x_2) = \frac{\pi}{2\sqrt{2}}\sqrt{x_1^2 + x_2^2}.$$

This choice was suggested in [3]. (It means the time of a free fall into the centre from the current configuration.) It leads to smaller stepsizes in the pericentre and larger in the apocentre, as it is reasonable because of the velocity of the satellite in each case. Some other possible choices of s led to bigger errors, so we decided to take this for our experiments.

As it was proved in [4], the method considered in this paper, when implemented with fixed stepsizes lead to linear error growth with time for this problem. The main objective of this paper is to see whether the same phenomena happens for variable stepsizes. Our numerical experiments allow us to say that they do. Figure 1 shows how error grows with time when we measure the error at final times $10T, 30T, \ldots, 21870T$, $T = 2\pi$ being the period of the problem. Each line corresponds to a different tolerance ϵ in (6). More explicitely, $\epsilon = 2\pi 10^{-6}$ and $\epsilon = \pi 10^{-6}$. You can see that, when the final time is multiplied by 3, the errors are multiplied by the same number. Order 2 is also manifest in this Figure, as you can see that the errors corresponding to a same final time divide by 4 approximately when the stepsize is halved.

6 Conclusions

For the variable-stepsize symmetric LMM2 constructed, advantageous error growth is observed in the same way as its fixed-stepsize counterpart. A question arises on whether this happens for every generalization of the same kind.

The great advantage this type of methods can have over Runge-Kutta ones is that just one function evaluation per step is needed when they are explicit. Although a symmetric selection of the stepsizes makes the variable-stepsize mode implicit, no more function evaluations are needed for LMM's while they are needed for explicit Runge-Kutta methods because the stepsize is required in the evaluation of the stages. This fact can make variable-stepsize symmetric LMM2's interesting when problems of very costly function evaluations are considered. In this case, this part of the computation could be much more expensive than the calculus of the coefficients of the LMM.

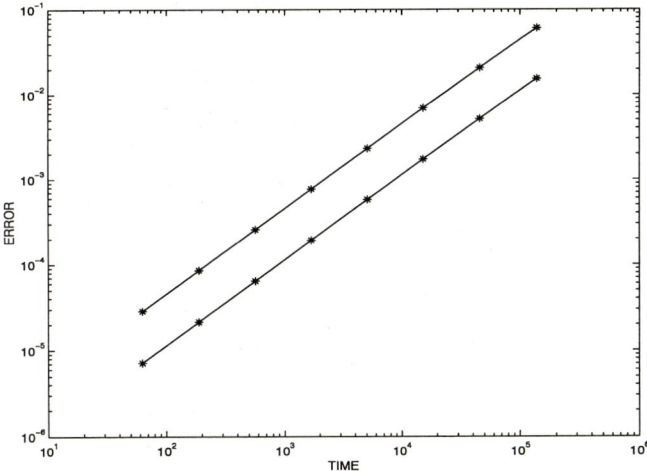

Fig. 1. Error growth with time for Kepler's problem with eccentricity 0.9

References

1. Calvo, M. P., López-Marcos M. A., Sanz-Serna, J. M.: Variable step implementation of geometric integrators. Appl. Num. Math. **28** (1998) 1–16
2. Cano, B., García-Archilla, B.: A generalization to variable stepsizes of Stömer methods for second-order differential equations. Appl. Num. Math. **19** (1996) 401–417
3. Cano, B., Sanz-Serna, J. M.: Error growth in the numerical integration of periodic orbits, with application to Hamiltonian and reversible systems. SIAM J. Num. Anal. **34** (1997) 1391–1417
4. Cano, B., Sanz-Serna, J. M.: Error growth in the numerical integration of periodic orbits by multistep methods, with application to reversible systems. IMA J. Num. Anal. **18** (1998) 57–75
5. Crouzeix, M., Lisbona, F. J.: The convergence of variable-stepsize, variable formula, multistep methods. SIAM J. Num. Anal. **21** (1984) 512–534
6. Evans, N. W., Tremaine, S.: Linear multistep methods for integrating reversible differential equations. Oxford Astrophysics OUTP-99-06 A (1999), submitted to Astron. J.
7. Hairer, E. Nörsett, S. P., Wanner, W. G.: Solving Ordinary Differential Equations I. Nonstiff Problems 2nd edn. Berlin: Springer (1993)
8. Hairer, E.: Variable time step integration with symplectic methods, Appl. Numer. Math. **25** (1997) 219–227
9. Leimkuhler, B.: Reversible adaptive regularization I: Perturbed Kepler motion and classical atomic trajectories. R. Soc. Lond. Philos, Trans. Ser. A Math. Phys. Eng. Sci. **357** (1999) 1101–1133
10. Quinlan, G. D., Tremaine, S. D.: Symmetric multistep methods for the numerical integration of planetary orbits. Astron. J. **100** (1990) 1694–1700
11. Stoffer, D.: On reversible and canonical integration methods. Res. Rep. 88–05, Applied Mathematics, Eidgenössische Technische Hochschule (ETH), Zürich (1988)

Preliminary Remarks on Multigrid Methods for Circulant Matrices

Stefano Serra Capizzano[1] and Cristina Tablino Possio[2]

[1] Dipartimento di Energetica "S. Stecco", Università di Firenze
Via Lombroso 6/17, 50134 Firenze, Italy
Serra@mail.dm.unipi.it

[2] Dipartimento di Scienza dei Materiali, Università di Milano Bicocca
Via Cozzi 53, 20125 Milano, Italy
cristina.tablino.possio@mater.unimib.it

Abstract. In this note we propose a multigrid approach to the solution of (multilevel) banded circulant linear system. In particular, we discuss how to define a "coarse-grid projector" such that the projected matrix at lower dimension preserves the circulant structure. This approach naturally leads to an optimal algorithm having linear cost as the size N of the system and so improving the the classical one based on Fast Fourier Transforms (FFTs) that costs $O(N \log N)$ arithmetic operations (ops). It's worth mentioning that these banded circulants are used as preconditioners for elliptic and parabolic PDEs (with Dirichlet or periodic boundary conditions) and for some $2D$ image restoration problems where the point spread function (PSF) is numerically banded. Therefore the use of the proposed multigrid technique reduces the overall cost from $O(k(\varepsilon,n)N \log N)$ to $O(k(\varepsilon,n)N)$, where $k(\varepsilon,n)$ is the number of Preconditioned Conjugate Gradient (PCG) iterations to reach the solution within a given accuracy of ε. The full analysis of convergence and the related numerical experiments are reported in a forthcoming paper [18].

Keywords: Circulant matrices, two-grid and multigrid iterations.

AMS(MOS) Subject Classification: 65F10, 65F15.

1 Prelude

Let f be a d-variate trigonometric polynomial defined over the hypercube Q^d, with $Q = (0, 2\pi)$ and $d \geq 1$ and having degree $c = (c_1, c_2, \ldots, c_d)$, $c_j \geq 0$ with regard to the variables $s = (s_1, s_2, \ldots, s_d)$. From the Fourier coefficients of f

$$a_j = \frac{1}{(2\pi)^d} \int_{Q^d} f(s) e^{-\mathbf{i}(j,s)} \, ds, \quad \mathbf{i}^2 = -1, \quad j = (j_1, \ldots, j_d) \in \mathbf{Z}^d \quad (1)$$

with $(j,s) = \sum_{k=1}^{d} j_k s_k$, $n = (n_1, \ldots, n_d)$ and $N(n) = \prod_{k=1}^{d} n_k$, one can build the sequence of Toeplitz matrices $\{T_n(f)\}$, where $T_n(f) = \{a_{j-i}\}_{i,j=e}^{n}$

$\in \mathbf{C}^{N(n) \times N(n)}$, $e = (1, \ldots, 1)^T \in \mathbf{N}^d$ is said to be the Toeplitz matrix of order n generated by f (see [20]).

It is clear that $a_j = 0$ if there exists an index i such that the absolute value of j_i exceeds c_i (i.e. if the condition $|j| \leq c$ is violated).

Accordingly, the d-level circulant matrix of order $N(n)$ generated by the same polynomial f (see e.g. [20]) is defined as

$$S_n(f) = \sum_{|j| \leq c} a_j Z_n^j = \sum_{|j_1| \leq c_1} \cdots \sum_{|j_d| \leq c_d} a_{(j_1, \ldots, j_d)} Z_{n_1}^{j_1} \otimes \cdots \otimes Z_{n_d}^{j_d} \quad (2)$$

where the matrix Z_m is the cyclic permutation Toeplitz matrix that generates the unilevel circulants, i.e. $(Z_m)_{s,t} = (t-s) \mod m$. If F_m denotes the m-th unilevel Fourier matrix whose (s,t) entry is given by the quantity $(e^{i\frac{st2\pi}{m}})/\sqrt{m}$, then it is well known that $S_n(f) = F_n D_f^{[n]} F_n^H$ where $F_n = F_{n_1} \otimes \cdots \otimes F_{n_d}$ is the d-level Fourier matrix and $D_f^{[n]} = \text{Diag}_{0 \leq j \leq n-e} f(2\pi j/n)$. Here the relation $0 \leq j \leq n - e^T$ and the expression $2\pi j/n = 2\pi(j_1/n_1, \ldots, j_d/n_d)$ are intended componentwisely.

Under the assumption that $c_i \leq \lfloor (n_i - 1)/2 \rfloor$, the matrix $S_n(f)$ is the Strang or natural circulant preconditioner of the corresponding Toeplitz matrix $T_n(f)$ [5]. We observe that the above mentioned assumption is fulfilled at least definitely since each c_i is a fixed constant and n_i is the size at level i, which is natural to think large when considering a discretization process.

Now let $c_n(f)$ be the n-th Cesaro sum of f given by

$$c_n(f) = \sum_{j=(0,\ldots,0)}^{n-e^T} \frac{F_j(f)}{N(n)}$$

with $(F_j(f))(s) = \sum_{|k| \leq j} a_k e^{i(k,s)}$ being the j-th Fourier expansion of f. Then $\text{degree}(c_n(f)) = \text{degree}(f)$ and $C_n(f) = S_n(c_n(f))$ is the T. Chan [6] optimal preconditioner of $T_n(f)$.

Besides Toeplitz linear systems (see for instance [4]), these banded circulant preconditioners have been used in the context of preconditioning of discretizations by Finite Differences of PDEs over hyperrectangular domains [2,11,12,14,15]. In this case, by the consistency condition, it is known that f is a nonnegative trigonometric polynomial which vanishes at $x = 0$ and can be chosen positive elsewhere (see e.g. [16]). Therefore $S_n(f)$ is singular, so that it is usually replaced by

$$\tilde{S}_n(f) = S_n(f) + \left(\min_{\|j\|_\infty = 1} f\left(\frac{2\pi j}{n}\right) \right) \frac{ee^T}{N(n)} \quad (3)$$

which is positive definite and can be used as preconditioner.

On the other hand, $C_n(f)$ is always positive definite since $c_n(f)$, with f being a nonnegative polynomial, can vanish if and only if f is identically zero (see [17]).

However, the clustering properties related to the modified Strang preconditioner are better than those of the optimal one in the case of nonnegative generating functions with zeros (for a rigorous analysis of this phenomenon refer to [19,7]).

Now if we consider PCG-like methods for the solution of a linear system $A_n \mathbf{x} = \mathbf{b}$, then the cost per iteration is given by

a. solution of $P_n \mathbf{y} = \mathbf{c}$ with the preconditioner P_n,
b. a constant number of matrix-vector products with matrix A_n,
c. computations of lower order complexity (vector-vector operations etc.).

In the case where $A_n = T_n(f)$, the overall cost of **b.** and **c.** is of $O(N(n))$ arithmetic operations (ops) due to the bandedness of $T_n(f)$ while the cost of **a.** is of $O(N(n) \log N(n))$ ops due the use of FFTs.

The method of multigrid type that we propose in this paper reduces the cost of **a.** to $O(N(n))$ ops when P_n is circulant and banded (for a proof of this claim see [18]).

Indeed the technique can be also extended [18] to the case where $\tilde{P}_n = P_n + \sum_{k=1}^{p} \vartheta_k f_{i_k} f_{i_k}^H$, P_n being circulant and banded, $\sum_{k=1}^{p} \vartheta_k f_{i_k} f_{i_k}^H$ being a special p rank corrections with f_q denoting the q-th Fourier column of F_n. Of course, this extension is of interest since it allows to treat the case of the modified Strang preconditioner given in (3).

The paper is organized as follows. In Section 2 we recall definitions and basic results concerning two-grid and multigrid iterations. Then, in Section 3, we define our multigrid technique for unilevel and multilevel circulants and we analyze in detail the properties of the projected "coarse-grid operators". A short Section 4 of conclusions ends the paper.

2 Premises

Consider the iterative method

$$x^{(j+1)} = \mathcal{V}_n x^{(j)} + b_1 := \mathcal{V}_n(x^{(j)}, b_1) \qquad (4)$$

for the solution of the linear system $A_n x = b$ where $A_n, M_n, \mathcal{V}_n := I_n - M_n^{-1} A_n \in \mathbf{C}^{n \times n}$, and $b, b_1 := M^{-1} b \in \mathbf{C}^n$. Given a full-rank matrix $p_n^k \in \mathbf{C}^{n \times k}$, with $k < n$, a Two-Grid Method (TGM) is defined by the following algorithm [10]

$$\underline{TGM(\mathcal{V}_n, p_n^k, \nu)(x^{(j)})}$$

1. $d_n = A_n x^{(j)} - b$
2. $d_k = (p_n^k)^H d_n$
3. $A_k = (p_n^k)^H A_n p_n^k$
4. Solve $A_k y = d_k$
5. $\hat{x}^{(j)} = x^{(j)} - p_n^k y$
6. $x^{(j+1)} = \mathcal{V}_n^\nu(\hat{x}^{(j)}, b_1)$

Step 6. consists in applying the "smoothing iteration" (4) ν times, while steps 1. $\to$ 5. define the "coarse grid correction", that depends on the projection operator p_n^k. The global iteration matrix of $TGM := TGM_n^k$ is then given by

$$TGM(V_n, p_n^k, \nu) = V_n^\nu \left[I_n - p_n^k \left((p_n^k)^H A_n p_n^k \right)^{-1} (p_n^k)^H A_n \right].$$

In [8,9], by using specific analytical properties of the generating function f, we defined a fast TGM for Toeplitz and τ problems (the τ class is the algebra associated to the most known sine transform [1] and is generated by the Toeplitz matrix $T_n(\cos(s))$). Here we propose the multigrid idea for multilevel banded circulant matrices and, in particular, we define the operator p_n^k and we analyze the projected matrix $A_k = (p_n^k)^H A_n p_n^k$.

3 Multigrid Method for Unilevel Circulant Matrices

For $A_n = S_n(f)$, with f being univariate trigonometric polynomial with a unique zero $x^0 \in (0, 2\pi]$, we consider the *smoothing* iteration (4), where the matrix V_n is defined by $V_n = I_n - A_n/\|f\|_\infty$ so that $V_n = S_n(1 - f/\|f\|_\infty)$. Then (4) takes the form of the relaxed method $x^{(j+1)} = x^{(j)} - [\|f\|_\infty]^{-1} \left(A_n x^{(j)} - c \right)$.

In order to provide a general method to obtain projectors from an arbitrary banded circulant matrix P_n, for some bandwidth d independent of n, we define the operator $T_n^k \in \mathbf{R}^{n \times k}$, $n = 2k$, such that

$$(T_n^k)_{i,j} = \begin{cases} 1 & \text{for } i = 2j-1, \ j = 1, \ldots, k, \\ 0 & \text{otherwise}. \end{cases} \quad (5)$$

Given any matrix P_n we obtain a projector $p_n^k \in \mathbf{R}^{n \times k}$ as $p_n^k = P_n \cdot T_n^k$.

For P_n too, we define the eigenvalue function $p(x)$, which sets the weights of the frequencies in the projector; in other words, the spectral behaviour of P_n selects the subspace in which the original problem is projected and solved. In this way we set $P_n = S_n(p)$.

If x^0 is a zero of f, then set $\hat{x} = (\pi + x^0) \bmod 2\pi$ and take the trigonometric polynomial

$$p(x) = (2 - 2\cos(x - \hat{x}))^{\lceil \beta/2 \rceil} \sim |x - \hat{x}|^{2\lceil \beta/2 \rceil} \quad \text{over } (0, 2\pi] \quad (6)$$

where

$$\beta = \operatorname{argmin} \left\{ \lim_{x \to x^0} \frac{(x - x^0)^{2i}}{f(x)} < +\infty \right\}, \quad (7)$$

$$0 < p^2(x) + p^2(\pi + x). \quad (8)$$

If f has more than one zero in $(0, 2\pi]$, then the corresponding polynomial p will be the product of the basic polynomials satisfying (6), (7) and (8) for any single zero. Of crucial importance is the following set of simple observations.

Remark 31 *Relations (7) and (8) impose some restrictions on the zeros of f. First, the zeros of f should be of finite order (by (7)). Secondly, if x^0 is a zero* of f, then $f(\pi + x^0) > 0$; otherwise relationship (8) cannot be satisfied with any polynomial p. However the second restriction depends on the fact that we half the dimension so that if f has some zeros in $(0, 2\pi]$ located with period π, then we have to change the "form" of the projection that is its smaller dimension. Compare for instance [8] and [3] concerning the case of symmetric Toeplitz structures: indeed in [3] for the generating function $f(x) = x^2(\pi^2 - x^2)$, the authors consider a "block form" of the projector proposed in [8]. This new choice works much finer and overcomes a problem due to the position of the zeros of $f(x) = x^2(\pi^2 - x^2)$. Finally we recall that a more general solution to the problem of the position of the zeros can be found in [13] where the author proposes to change the proportionality factor between the matrix sizes of the "finer" and of the "coarser" levels.

Remark 32 If $\mathrm{degree}(f) = c$, i.e., $f(z) = \sum_{j=-c}^{c} a_j z^j$ for some coefficients a_j, then f can have a zero of order at most $2c$, so that $\beta \leq c$ and therefore $\mathrm{degree}(p^2) \leq 2\lceil c/2 \rceil$.

3.1 Properties of T_n^k

First of all, let us consider a spectral decomposition of T_n^k. In analogy with the τ case proposed and analyzed in [8,9], the operator represents a spectral link between the space of the frequencies of size n and the corresponding space of frequencies of size k. Indeed, by observing that

$$[T_n^k]^T f_\mu^{[n]} = \frac{1}{\sqrt{2}} f_\mu^{[k]}, \quad \mu = 0, \ldots, k-1$$

and

$$[T_n^k]^T f_{\mu'}^{[n]} = \frac{1}{\sqrt{2}} f_\mu^{[k]}, \quad \mu' = k, \ldots, n-1, \ \mu' = \mu + k,$$

it directly follows that

$$[T_n^k]^T F_n = \frac{1}{\sqrt{2}} [1, 1] \otimes F_k \tag{9}$$

where F_m is the unilevel Fourier matrix of size m and T_n^k is the operator defined in (5).

In order to apply a full multigrid method, it is important to preserve the "structure" at the lower levels. Therefore, if we apply the MGM to $A_n := S_n(f)$ with f nonnegative, we require that the matrix at the lower level belongs to the circulants of different size k with nonnegative eigenvalue function. These and other properties are established in the following proposition. We remark that very similar statements have been proved [13,8,9] for other structures (matrices discretizing elliptic PDEs, τ matrices etc.).

Proposition 1. Let $n = 2k$, $p_n^k = S_n(p) T_n^k$, and let f be nonnegative.

1. The matrix $2(p_n^k)^H S_n(f) p_n^k$ coincides with $S_k(\hat{f})$ with $\hat{f}(x) = f(x/2)\, p^2(x/2)$ $+ f(\pi + x/2) p^2(\pi + x/2)$ for $x \in (0, 2\pi]$. If f is a polynomial then $\hat{f}$ is a polynomial having at most the same degree as f.
2. If x^0 is a zero of f then $\hat{f}$ has a corresponding zero y^0 where $y^0 = 2x^0 \bmod 2\pi$.
3. The order of the zero y^0 of $\hat{f}$ is exactly the same as the one of the zero x^0 of f, so that at the lower level the new projector is easily defined in the same way.

Proof.
The projected matrix $(p_n^k)^H S_n(f) p_n^k$ can be spectrally decomposed by taking into account relation (9). Indeed we have

$$(p_n^k)^H S_n(f) p_n^k = [T_n^k]^T S_n(p) S_n(f) S_n(p) T_n^k$$
$$= [T_n^k]^T S_n(p^2 f) T_n^k$$
$$= [T_n^k]^T F_n D_{p^2 f}^{[n]} F_n^H T_n^k$$
$$= \frac{1}{2} ([1,1] \otimes F_k) D_{p^2 f}^{[n]} ([1,1]^T \otimes F_k^H)$$
$$= \frac{1}{2} F_k \left(D_{1,p^2 f}^{[n]} + D_{2,p^2 f}^{[n]} \right) F_k^H$$

where

$$D_{1,p^2 f}^{[n]} = \underset{0 \le j \le k-1}{\mathrm{Diag}} \left((p^2 f)(x_j^{[n]}) \right)$$

and

$$D_{2,p^2 f}^{[n]} = \underset{k \le j \le n-1}{\mathrm{Diag}} \left((p^2 f)(x_j^{[n]}) \right).$$

Since $x_j^{[n]} = x_j^{[k]}/2$ for $j = 0, \ldots, k-1$ and $x_{j'}^{[n]} = x_j^{[k]}/2 + \pi$ for $j = 0, \ldots, k-1$ with $j' = j + k$, it follows that the matrix $2(p_n^k)^H S_n(f) p_n^k$ can be seen as $S_k(\hat{f})$ where $\hat{f}(x) = f(x/2)\, p^2(x/2) + f(\pi + x/2) p^2(\pi + x/2)$ for $x \in (0, 2\pi]$.

From the expression of $\hat{f}$ and since $p(\pi + x^0) = 0$ by (6), it directly follows that $y^0 = 2x^0 \bmod 2\pi$ is a zero of $\hat{f}$ (i.e. item 2. is proved).

Moreover, by (8), we deduce that $p^2(x^0) > 0$ since $p^2(x^0 + \pi) = 0$ and the order of the zero y^0 of $(p^2 f)(x/2)$ is the same as the order of $f(x)$ at x^0. But by (7) we can see that $p^2(x/2 + \pi)$ has at y^0 a zero of order at least equal to the one of $f(x)$ at x^0. Since both the contributions in $\hat{f}(x)$ are nonnegative the thesis of item 3. follows.

Finally we have to demonstrate the last part of item 1. Suppose that f is a nonnegative trigonometric polynomial (and then real-valued) of degree c. Consequently, by looking at f and p^2 as Laurent polynomials on the unit circle, we have $f(z) = \sum_{j=-c}^{c} a_j z^j$, $p^2(z) = \sum_{j=-l}^{l} b_j z^j$, with $z = e^{ix}$, $\overline{a_j} = a_{-j}$, $\overline{b_j} = b_{-j}$ and $x \in (0, 2\pi]$. By a straightforward calculation we deduce the following representations

$$(p^2 f)(x/2) = (p^2 f)(z^{1/2}) = \sum_{j=-(c+l)}^{c+l} g_j z^{j/2},$$

$$(p^2 f)(x/2 + \pi) = (p^2 f)(-z^{1/2}) = \sum_{j=-(c+l)}^{c+l} g_j (-1)^j z^{j/2},$$

with $\overline{g_j} = g_{-j}$ so that $\hat{f}$ is a polynomial since

$$\hat{f}(x) = \sum_{j=-\lfloor (c+l)/2 \rfloor}^{\lfloor (c+l)/2 \rfloor} 2g_{2j} z^j.$$

Now by Remark 32 we recall that l is at most equal to $2 \lceil c/2 \rceil$ and consequently

$$\lfloor (c+l)/2 \rfloor \leq \lfloor (c + 2\lceil c/2 \rceil)/2 \rfloor = c$$

so that the second part of item 1. is proved. □

4 Concluding Remarks

In the multilevel case the projector is simply defined as $K_n P_n$ where P_n is a multilevel circulant matrix generated by a d-variate polynomial p satisfying a d-variate version of conditions (6), (7) and (8), $n = (n_1, \ldots, n_d)$ and $K_n = T_{n_1} \otimes \cdots \otimes T_{n_d}$. Under these assumptions a d-variate rewriting of Proposition 1 holds true (see [18] for further details).

Concerning the cost per iteration, we observe that steps 1.−3., 5. and 6. in the procedure TGM costs $O(N(n))$ ops due to the bandedness of the involved multilevel matrices. Then the cost of TGM at dimension n is $c(n)$ with $c(n) \leq c(n/2) + qN(n)$, with $q > 0$ constant independent of n. The above relation trivially implies $c(n) \leq 2qN(n)$ and then the linear cost of the proposed technique, since the convergence rate is independent of the multiindex n as reported in [18]. Finally we point out that Proposition 1 and its multilevel generalization are crucial in order to define a full multigrid method since they allow a recursive application at the lower levels of the procedure TGM.

References

1. D. Bini and M. Capovani, "Spectral and computational properties of band symmetric Toeplitz matrices", *Linear Algebra Appl.*, **52/53** (1983), pp. 99–125.
2. R. H. Chan and T. F. Chan, "Circulant preconditioners for elliptic problems", *J. Numer. Linear Algebra Appl.*, **1** (1992), pp. 77–101.
3. R. H. Chan, Q. Chang and H. Sun, "Multigrid method for ill-conditioned symmetric Toeplitz systems", *SIAM J. Sci. Comp.*, **19-2** (1998), pp. 516–529.

4. R. H. Chan and M. Ng, "Conjugate gradient methods for Toeplitz systems", *SIAM Rev.*, **38** (1996), pp. 427–482.
5. R. H. Chan and G. Strang, "Toeplitz equations by conjugate gradients with circulant preconditioner", *SIAM J. Sci. Stat. Comp.*, **10** (1989), pp. 104–119.
6. T. F. Chan, "An optimal circulant preconditioner for Toeplitz systems", *SIAM J. Sci. Stat. Comp.*, **9** (1988), pp. 766–771.
7. F. Di Benedetto and S. Serra Capizzano, "A unifying approach to abstract matrix algebra preconditioning", *Numer. Math.*, **82-1** (1999), pp. 117–142.
8. G. Fiorentino and S. Serra, "Multigrid methods for Toeplitz matrices", *Calcolo*, **28** (1991), pp. 283–305.
9. G. Fiorentino and S. Serra, "Multigrid methods for symmetric positive definite block Toeplitz matrices with nonnegative generating functions", *SIAM J. Sci. Comp.*, **17-4** (1996), pp. 1068–1081.
10. W. Hackbusch, *Multigrid Methods and Applications*. Springer Verlag, Berlin, 1985.
11. S. Holmgren and K. Otto, "Iterative solution methods and preconditioners for block tridiagonal systems of equations", *SIAM J. Matrix Anal. Appl.*, **13** (1992), pp. 863–886.
12. S. Holmgren and K. Otto, "Semicirculant preconditioners for first order partial differential equations", *SIAM J. Sci. Comput.*, **15** (1994), pp. 385–407.
13. T. Huckle, "Multigrid preconditioning and Toeplitz matrices", *private communication*.
14. X. Q. Jin and R. Chan, "Circulant preconditioners for second order hyperbolic equations", *BIT*, **32** (1992), pp. 650–664.
15. I. Lirkov, S. Margenov and P. Vassilevsky, "Circulant block factorization for elliptic problems", *Computing*, **53** (1994), pp. 59–74.
16. S. Serra Capizzano and C. Tablino Possio, "Spectral and structural analysis of high precision Finite Difference matrices for Elliptic Operators", *Linear Algebra Appl.*, **293** (1999), pp. 85–131.
17. S. Serra, "A Korovkin - type Theory for finite Toeplitz operators via matrix algebras", *Numer. Math.*, **82-1** (1999), pp. 117–142.
18. S. Serra Capizzano and C. Tablino Possio, "Multigrid methods for multilevel circulant matrices", *manuscript*, (2000).
19. E. Tyrtyshnikov, "Circulant preconditioners with unbounded inverses", *Linear Algebra Appl.*, **216** (1995), pp. 1–23.
20. E. Tyrtyshnikov, "A unifying approach to some old and new theorems on distribution and clustering", *Linear Algebra Appl.*, **232** (1996), pp. 1–43.

Computing the Inverse Matrix Hyperbolic Sine*

J. R. Cardoso[1] and F. Silva Leite[2]

[1] Instituto Superior de Engenharia de Coimbra, Quinta da Nora
3030 Coimbra, Portugal
`jocar@sun.isec.pt`
[2] Departamento de Matemática, Universidade de Coimbra
3000 Coimbra, Portugal
`fleite@mat.uc.pt`

Abstract. We give necessary and sufficient conditions for solvability of the matrix equation $\sinh X = A$ in the complex and real cases and present some algorithms for computing one of these solutions. The numerical features of the algorithms are analysed along with some numerical tests.

Keywords: primary matrix function, inverse matrix hyperbolic sine, matrix exponentials, logarithms and square roots, Padé approximants

1 Introduction

The matrix hyperbolic sine of a real or complex square matrix X is defined by $\sinh X := (e^X - e^{-X})/2$. Inversely, if A is given, we call *inverse matrix hyperbolic sine* of A to any solution of the matrix equation $\sinh X = A$, which is denoted by $\sinh^{-1} A$. Since the matrix hyperbolic sine is based on matrix exponentials, it is a primary matrix function ([9], ch. 6). Properties of such functions are the key for obtaining conditions under which the matrix equation $\sinh X = A$ has solutions, in a way which is similar to the matrix equation $e^X = A$, whose solutions are logarithms. The Jordan canonical form also plays an important role for analysing theoretical aspects of the inverse matrix hyperbolic sine, as will be shown later.

The problem of computing solutions of $\sinh X = A$, when A is real, is analysed through two algorithms which are a result of careful manipulations of algorithms for computing matrix logarithms. For the special case when the matrix A is P-symmetric, one of them is structure preserving. The general algorithms work under a restriction on the spectrum of A. Since skew-symmetric matrices may not fit this assumption, they require different treatment.

The problem of computing matrix logarithms has received particular attention recently ([2,3,4,10,11] and [12]). However, as far as we know, the inverse matrix hyperbolic (or trigonometric) functions have not deserved the same interest.

---
* Work supported in part by ISR and research network contract ERB FMRXCT-970137.

This paper deals only with inverse matrix hyperbolic sines but a similar study may be done for other inverse matrix functions such as $\cosh^{-1}$, $\sin^{-1}$ and $\cos^{-1}$. Our interest on the inverse matrix hyperbolic sine was motivated by the work of Crouch and Bloch [1], where the matrix equation $XQ^T - QX^T = M$ appears associated with the generalized rigid body equations. In this case Q is orthogonal and M is skew-symmetric. It turns out that $X = (e^{\sinh^{-1}\frac{M}{2}})Q$ is a solution of that matrix equation.

This paper is organized as follows. In section 2, we present some hyperbolic and trigonometric primary matrix functions along with some of their properties. In section 3 we give necessary and sufficient conditions for a complex (resp. real) matrix to have a complex (resp. real) inverse matrix hyperbolic sine. We also make some considerations about the principal inverse matrix hyperbolic sine and present some algorithms in section 4. Finally, comments and examples on the implementations of the algorithms are given in section 5.

2 Some Hyperbolic and Trigononetric Matrix Functions

Using the corresponding scalar expressions as in the hyperbolic sine case, we may also define another primary matrix functions such as $\cosh X := (e^X + e^{-X})/2$, $\sin X := (e^{iX} - e^{-iX})/2i$ and $\cos X := (e^{iX} + e^{-iX})/2$, where X is any real or complex square matrix. Some of the identities holding in the scalar case extend to the matrix case under the assumption on that the matrices commute. For example, if $XY = YX$, it is easy to prove that

$\sinh(X \pm Y) = \sinh X \cosh Y \pm \cosh X \sinh Y;$
$\cosh(X \pm Y) = \cosh X \cosh Y \pm \sinh X \sinh Y;$
$\sin(X \pm Y) = \sin X \cos Y \pm \cos X \sin Y;$
$\cos(X \pm Y) = \cos X \cos Y \mp \sin X \sin Y.$

Setting $X = Y$ in the identities above, it is straightforward that

$\cosh^2 X - \sinh^2 X = I;$
$\sin^2 X + \cos^2 X = I;$
$\quad \sinh(2X) = 2 \sinh X \cosh X;$
$\quad \sin(2X) = 2 \sin X \cos X.$

When X is a *P- symmetric* or a *P-skew-symmetric* matrix (i.e., $X^T P = PX$ or $X^T P = -PX$, with X, P real and P nonsingular, respectively), we are particulary interested in studying the structure of the image of these matrices by the matrix functions defined above. We note that for particular case when $P = I$, we get the symmetric and the skew-symmetric matrices, respectively.

Using the definitions, it is easy to prove that the image of a P-symmetric matrix by any primary matrix function is still P-symmetric and that the image of a P-skew-symmetric matrix by the hyperbolic or trigonometric sine is also *P*-skew-symmetric. However, the image of a *P*-skew-symmetric matrix by the hyperbolic or trigonometric cossines is *P*-symmetric.

3 The Equation $\sinh X = A$

We start with some remarks about the scalar equation $\sinh x = a$, where $a, x \in \mathbb{C}$.
Given any $a \in \mathbb{C}$, this equation has always an infinity of solutions. In fact,

$$x = \log[a + (a^2 + 1)^{1/2}] \quad \text{and} \quad x + 2k\pi i \quad (k \in \mathbb{Z})$$

satisfy the equation $\sinh x = a$. If

$$a \notin \mathcal{E} = \{\alpha i : \alpha \in \mathbb{R}, |\alpha| \geq 1\},$$

then $\sinh x = a$ has a unique solution lying on the horizontal open strip in the complex plane defined by

$$\mathcal{D} = \{x \in \mathbb{C} : -\frac{\pi}{2} < \text{Im}(x) < \frac{\pi}{2}\}.$$

The key idea to show this fact is to observe that the real part of $a + (a^2+1)^{1/2}$, where $x^{1/2}$ denotes the complex square root that lies on the open right half plane, is positive and the real part of $a - (a^2 + 1)^{1/2}$ is negative. The remain of the proof is immediate.

As a consequence of this fact, we have the following result.

Lemma 1. *If A has no eigenvalues in $\mathcal{E}$ then there exists an unique inverse matrix hyperbolic sine of A with eigenvalues in $\mathcal{D}$.*

This inverse matrix hyperbolic sine is called the *principal* and is denoted by $\text{Sinh}^{-1} A$ (with capital case).

Contrary to the scalar case, the matrix equation $\sinh X = A$ may not have solutions. The following theorem gives a result to decide what conditions A must satisfy in order that this equation has a solution.

Theorem 1. *If A is a complex square matrix then $\sinh X = A$ has some solution in $\mathbb{C}$ if and only if the Jordan blocks of A with size ≥ 2 associated to the eigenvalues i and $-i$ occur in pairs of the form*

$$J_k(\lambda), J_k(\lambda)$$

or

$$J_k(\lambda), J_{k-1}(\lambda),$$

where $\lambda \in \{-i, i\}$ and $J_p(\lambda)$ denotes the Jordan block of A associated to the eigenvalue λ with size p.

If i or $-i$ are eigenvalues of A, then no inverse matrix hyperbolic sine is a primary matrix function of A.

Proof. First we prove the necessary condition. If there exists X such that $\sinh X = A$, then the eigenvalues of A are hyperbolic sines of eigenvalues of X. If $\lambda \in \{-i, i\}$ is an eigenvalue of A, then the corresponding eigenvalues in X are of the form
$$\mu = (2p \pm \frac{1}{2})\pi i, \ p \in \mathbb{Z}.$$

Let $J_l(\mu)$ be a Jordan block of X associated to μ. Applying a result from [9], p. 424, we may conclude that the Jordan decomposition of $\sinh[J_l(\mu)]$ gives rise to a pair of blocks of the form
$$J_{l/2}(\lambda), J_{l/2}(\lambda), \ \text{if } l \text{ is even,}$$
or
$$J_{\frac{l+1}{2}}(\lambda), J_{\frac{l-1}{2}}(\lambda), \ \text{if } l \text{ is odd.}$$

Now we prove the sufficient condition. Decomposing A in its Jordan canonical form, we may write $A = S \operatorname{diag}(A_1, A_2) \, S^{-1}$, where S is nonsingular, A_1 is a direct sum of Jordan blocks with eigenvalues not lying on $\{-i, i\}$ and A_2 is a direct sum (with an even number) of Jordan blocks associated to i or $-i$ which have the form described above. To get $\sinh^{-1} A$, it is enough to find $\sinh^{-1} A_1$ and $\sinh^{-1} A_2$. Using some results about primary matrix functions ([9], ch.6), there exists at least an inverse matrix hyperbolic sine $\sinh^{-1} A_1$ which is a primary matrix function of A_1. To show that there exists $\sinh^{-1} A_2$ we may suppose, without loss of generality, that A_2 is a pair of blocks of the form $J_k(i) \oplus J_k(i)$. Decomposing $\sinh(J_{2k}[(2p + \frac{1}{2})\pi i])$ in its Jordan canonical form, we have
$$\sinh(J_{2k}[(2p + \frac{1}{2})\pi i]) = T[J_k(i) \oplus J_k(i)]T^{-1},$$
for some nonsingular T, which implies that
$$\sinh^{-1}[J_k(i) \oplus J_k(i)] = T^{-1} J_{2k}[(2p + \frac{1}{2})\pi i] \, T.$$

This proves the sufficient condition. If A has eigenvalues i or $-i$ then no X such that $\sinh X = A$ can be a primary matrix function of A. In fact, if there exists a such function f satisfying
$$X = f(A),$$
then f can not transform two Jordan blocks into one.

The inverse matrix hyperbolic sine of A is not always a primary matrix function of A. However, if the spectrum of A, $\sigma(A)$, does not contain $\pm i$, there are some inverse matrix hyperbolic sines which are primary matrix functions of A, as guaranteed by the theorem. This type of functions enjoys some specials properties since they can be written as a polynomial in A. In particular, they commute with A.

The next corollary concerns the analysis of the matrix equation $\sinh X = A$ in the real case. Given a real matrix A, we want to know conditions under which this equation has a real solution. Before stating the result, we note that the nonreal eigenvalues of a real matrix occur in conjugate pairs. The number and the size of the Jordan blocks associated to a nonreal eigenvalue and to its conjugate are the same.

Corollary 1. *If A is a real square matrix then $\sinh X = A$ has some real solution if and only if the Jordan blocks of A with size ≥ 2 associated to the eigenvalue i occur in pairs of the form*

$$J_k(i), J_k(i)$$

or

$$J_k(i), J_{k-1}(i).$$

If i is an eigenvalue of A ($-i$ is also an eigenvalue) then no inverse matrix hyperbolic sine of A is a primary matrix function of A.

Proof. The necessary condition is a consequence of the previous theorem. To prove the sufficient condition, we consider the scalar complex function

$$f(x) = \begin{cases} \text{Log}(x + \sqrt{x^2 + 1}), & x \in \bigcap_{p=1}^{l} B_p \\ \text{Log}(x - \sqrt{x^2 + 1}), & x \in \bigcap_{q=1}^{m} C_q \\ \text{Log}(x + (x^2 + 1)^{1/2}), & \text{otherwise} \end{cases},$$

where

- $\text{Log}\, w$ denotes the principal logarithm;
- $\sqrt{w}$ denotes the square root of w which lies on the open left half plane;
- $w^{1/2}$ denotes the principal square root of w;
- $\lambda_1, \cdots, \lambda_l$ are the eigenvalues of A of the form αi, for some $\alpha > 1$;
- $\mu_1, \cdots, \mu_m$ are the eigenvalues of A of the form $-\alpha i$, for some $\alpha > 1$;
- $r = \min_{\lambda \neq \mu} |\lambda - \mu|$, λ and μ are eigenvalues of A;
- $B_p := \{x \in \mathbb{C} : |x - \lambda_p| < r/2\}$, $p = 1, \cdots, l$;
- $C_q := \{x \in \mathbb{C} : |x - \mu_q| < r/2\}$, $q = 1, \cdots, m$.

If i is not an eigenvalue of A, f is defined and has derivatives of any order for each $x \in \sigma(A)$. Thus $f(A)$ is a primary matrix function and it is not hard to prove that it is real. If i is an eigenvalue and the associated Jordan blocks occur in pairs as described above, then the equation $\sinh X = A$ has a real solution. This follows from the proof of the previous theorem. To conclude the proof, it is enough to use a similar argument as that in the proof of the theorem to conclude that there is not any inverse matrix hyperbolic sine which is a primary matrix function of A.

Since for any $a, x \in \mathbb{C}$, we have $\sinh x = a \Leftrightarrow e^x = a \pm (a^2 + 1)^{1/2}$, we may define an infinity of inverse hyperbolic sine functions. It depends on the branch we take for logarithm and for square root. Using the notations above, we may define the principal inverse hyperbolic sine as

$$\mathrm{Sinh}^{-1} a = \mathrm{Log}[a + (a^2 + 1)^{1/2}], \quad a \in \mathbb{C}.$$

Since this function is differentiable if $a^2 + 1 \notin \mathbb{R}_0^-$, that is, $a \notin \mathcal{E}$ (see the beginning of this section), we may define the corresponding primary matrix function as

$$\mathrm{Sinh}^{-1} A = \mathrm{Log}[A + (A^2 + 1)^{1/2}],$$

where A is a complex matrix such that $\sigma(A) \cap \mathcal{E} = \phi$, with $\sigma(A)$ denoting the spectrum of A. The eigenvalues of $\mathrm{Sinh}^{-1} A$ lie on the strip $\mathcal{D}$ and if A is real, then $\mathrm{Sinh}^{-1} A$ is also real. Moreover, if $\sigma(A) \cap \mathcal{E} = \phi$ then $\sinh(\mathrm{Sinh}^{-1} A) = A$ and if $\sigma(A) \subset \mathcal{D}$ then $\mathrm{Sinh}^{-1}(\sinh A) = A$.

We saw in section 2 that the hyperbolic sine of a P-skew-symmetric matrix is also P-skew-symmetric. In the following theorem we show that the opposite is true for the principal inverse matrix hyperbolic sine.

Theorem 2. *If A is a P-skew-symmetric matrix and $\sigma(A) \cap \mathcal{E} = \phi$, then $\mathrm{Sinh}^{-1} A$ is also a P-skew-symmetric matrix.*

Proof. It is enough to show that $B = A + (A^2 + I)^{1/2}$ is P-orthogonal (i.e., $B^T P B = P$) and use the fact that the principal matrix logarithm of a P-orthogonal matrix is P-skew-symmetric [2].

4 Algorithms

The main algorithm to be presented here (algorithm 1) for computing the principal inverse matrix hyperbolic sine involves the computation of matrix logarithms and matrix square roots. It is well known that one of the most suitable method for computing principal matrix square roots involves the Schur decomposition ([6,8]) and has been implemented in Matlab (version 5.2). An alternative method for the same purpose is the Denmam & Beavers iterative method in [7]. For the matrix logarithm there are some methods proposed in the literature but there has not been agreement in choosing the most suitable. See [3] for a comparaison among the methods and [12] for a new method. Here we use the so called Briggs-Padé method which combines an inverse squaring and scaling procedure with Padé approximants. The usual form of this method involves diagonal Padé approximants of the function $\log(1-x)$. However, in [2], we presented an improved algorithm for this method which instead uses diagonal Padé approximants of the function

$$\log(\frac{1+x}{1-x}) = 2\tanh^{-1} x.$$

There, we showed that these approximants are well conditioned with respect to matrix inversion and its use reduces the number of matrix square roots needed in the inverse squaring and scaling procedure. This reduction is important since it increases the accuracy in the resulting approximation.

Algorithm 1
This algorithm computes $\text{Sinh}^{-1} A$, when A is real and $\sigma(A) \cap \mathcal{E} = \phi$.
ε is a given tolerance.

1. Find the real Schur decomposition of A,
$$A = QRQ^T,$$
where Q is orthogonal and R is block upper triangular;
2. Set $T := R + (R^2 + I)^{1/2}$, where the matrix square roots may be computed by the function *sqrtm* of Matlab;
3. Set $B_j := (T^{\frac{1}{2^j}} - I)(T^{\frac{1}{2^j}} + I)^{-1}$ and $u_j := 2\|B_j\| \, [H(\|B_j^2\|) - t_{m-1,m}(\|B_j^2\|)]$, $j \in \mathbb{N}$, where
$$H(x) := \frac{1}{2x^{1/2}} \text{Log}\left(\frac{1 + x^{1/2}}{1 - x^{1/2}}\right) = \frac{1}{x^{1/2}} \tanh^{-1} x^{1/2},$$
$$t_{m-1,m}(x) := \frac{1}{2x^{1/2}} S_{2m,2m}(x^{1/2}),$$
and $S_{2m,2m}(x)$ is the $(2m, 2m)$ diagonal Padé approximant of $2\tanh^{-1} x$;
4. Compute k successive square roots of T until $\|B_k^2\| < 1$ and $u_k < \varepsilon$;
5. Compute $S_{2m,2m}(B_k)$;
6. Approximate $\text{Log}\, T$ using the relations
$$\text{Log}(T) = 2^k \text{Log}(T^{\frac{1}{2^k}}) \approx 2^k S_{2m,2m}(B_k).$$
7. Set $\text{Sinh}^{-1} A = Q(\text{Log}\, T) Q^T$.

The most expensive step in the algorithm is the first. The cost of computing the real Schur decomposition is about $25n^3$ flops ([5], 7.5.6). After this step all the matrices involved are block triangular and, in this case, the cost of taking one matrix square root is about $\frac{n^3}{3}$ flops [8]. To guarantee full precision in Matlab (with relative machine epsilon $\varepsilon \approx 2.2 \times 10^{-16}$), a good compromise between taking many square roots and increase the order of Padé approximants is to use the S_{88} Padé approximant.

When $\|B_k^2\| < 1$, the nonegative real number u_k in step 4 measures the approximation computed for the logarithm by the diagonal Padé approximant $S_{2m,2m}(B_k)$, since
$$\|\text{Log}[(I + B)(I - B)^{-1}] - S_{2m,2m}(B)\| \leq u_k.$$

The computation of $S_{2m,2m}(B_k) = P_{2m}(B_k)[Q_{2m}(B_k)]^{-1}$, where P_{2m} and Q_{2m} are polynomials of degree at most $2m$, needs about $(2r + s - 2)\frac{n^3}{3}$

flops if $sr = 2m$, with $s = \lceil\sqrt{2m}\rceil$, $r = \lfloor 2m/s \rfloor$ and $(2r+s)\frac{n^3}{3}$ flops otherwise [3].

We note that in the last step $\text{Sinh}^{-1}A = Q\,\text{Log}\underbrace{\left[R + (R^2 + I)^{1/2}\right]}_{T} Q^T$.

If in algorithm 1 we omit step 1 and compute the matrix square root using the Denman & Beavers method, we obtain a new algorithm, say **algorithm 2**, which also computes $\text{Sinh}^{-1}A$ whenever $\sigma(A) \cap \mathcal{E} = \phi$. One of the advantages of using this new algorithm, instead of the first one, is that it is structure preserving (in exact arithmetic) for all P-symmetric matrices that satisfy the spectral assumption. This is due to the fact that the iterative method of Denman & Beavers involves only inverses and sums, which preserve P-symmetry, and the diagonal Padé approximants used in the logarithm also preserve this kind of structure [2].

Since the spectrum of skew-symmetric matrices is purely imaginary, they may not satisfy the condition $\sigma(A) \cap \mathcal{E} = \phi$. For these matrices we propose a different algorithm to compute $\text{Sinh}^{-1}A$.

Algorithm 3
A is any real skew-symmetric matrix.

1. Find the real Schur form of A

$$A = QDQ^T,$$

where Q is orthogonal, $D = \text{diag}(0, \cdots, 0, A_1, \cdots, A_l)$, and $A_k = \begin{bmatrix} 0 & \alpha_k \\ -\alpha_k & 0 \end{bmatrix}$, $\alpha_k > 0$, $k = 1, \cdots, l$.

2. Set $\text{Sinh}^{-1}A = Q\,\text{diag}(0, \cdots, 0, X_1, \cdots, X_l)\,Q^T$, where, for all $k = 1, \cdots, l$,

$$X_k = \begin{cases} \begin{bmatrix} \ln[\alpha_k + (\alpha_k^2 - 1)^{1/2}] & \pi/2 \\ -\pi/2 & \ln[\alpha_k + (\alpha_k^2 - 1)^{1/2}] \end{bmatrix}, & \text{if } \alpha_k \geq 1 \\ \begin{bmatrix} 0 & \cos^{-1}(1 - \alpha_k^2)^{1/2} \\ -\cos^{-1}(1 - \alpha_k^2)^{1/2} & 0 \end{bmatrix}, & \text{if } 0 < \alpha_k < 1 \end{cases}.$$

Remark. When A is skew-symmetric and does not satisfy $\sigma(A) \cap \mathcal{E} = \phi$, we have no guarantee that the inverse matrix hyperbolic sine of A is skew-symmetric. A necessary and sufficient condition for $\sinh X = A$ to have a skew-symmetric solution is that the eigenvalues of A are of the form αi, with $|\alpha| \leq 1$. To prove the necessary condition, we suppose that X is a skew-symmetric matrix such that $\sinh X = A$. Then the eigenvalues of X are of the form $\pm \beta i$, $\beta \in \mathbb{R}$, and

the eigenvalues of A are hyperbolic sines of eigenvalues of X, that is, they are of the form
$$\sinh(\pm \beta i) = \pm i \sin \beta,$$
where $|\alpha| = |\sin \beta| \leq 1$. The sufficient condition is an immediate consequence of facts discussed previously.

5 Numerical Experiments

We have implemented the algorithms 1 and 2 in Matlab (with relative machine epsilon $\epsilon \approx 2.2 \times 10^{-16}$) on a Pentium II. We used the Frobenius norm, $(8,8)$ Padé approximants and a tolerance of $\varepsilon = \|A\| \times 10^{-16}$. The expressions for $S_{88}(x)$ and u_j are:

$$S_{88}(x) = \frac{-2x(15159x^6 - 147455x^4 + 345345x^2 - 225225)}{35(35x^8 - 1260x^6 + 6930x^4 - 12012x^2 + 6435)},$$

$$u_j = 2\|B_j\| \left[H(\|B_j^2\|) - t_{34}(\|B_j^2\|) \right],$$

where

$$H(x) = \frac{1}{2x^{1/2}} \log\left(\frac{1+x^{1/2}}{1-x^{1/2}}\right) \text{ and } t_{34}(x) = \frac{1}{2x^{1/2}} S_{88}(x^{1/2}).$$

In order to measure the relative error of the computed inverse hyperbolic sine $\bar{X}$, we used the quantity

$$error = \frac{\|\sinh \bar{X} - A\|}{\|A\|},$$

where the hyperbolic sine was computed using the function *expm* of Matlab.

We tested several Hilbert matrices of orders $n \leq 15$, which have a large condition number. In both algorithms the relative error varied between 10^{-15} and 10^{-14}. We also tested the matrix

$$A = \begin{bmatrix} 0.0001 & 0.9999 & 10^4 & 0.0001 \\ -0.9999 & 0.0001 & -10^5 & 10^6 \\ 0 & 0 & 0.0001 & 1.0001 \\ 0 & 0 & -1.0001 & 0.0001 \end{bmatrix},$$

which has a large condition number, $\text{cond}(A) = 1.0101 \times 10^{12}$, and eigenvalues close to i and $-i$. In this case, we noticed a loss of accuracy of about 8 significant digits. The relative error was about 5.6052×10^{-8} in both algorithms. This result was somewhat expected since the computation of $(R^2 + I)^{1/2}$ in step 2 has lost about six significant digits of accuracy.

To study the behaviour of both algorithms in which concerns to structure preserving of P-symmetric matrices, we considered several particular examples with $P = \begin{bmatrix} 0 & I_k \\ -I_k & 0 \end{bmatrix}$, with $2k = n$, and $P = \begin{bmatrix} I_p & 0 \\ 0 & -I_q \end{bmatrix}$, with $p + q = n$, of orders 6, 7 and 8. In both algorithms we observed that for matrices with nonlarge condition number the original structure was preserved, althoug the inverse was computed by a method that does not preserve such structure. For matrices with large condition number, our tests showed that algorithm 2 is slightly better for the first choice of P and that both algorithms rarely preserved the structure for the second choice of P.

Based on examples tested, we observe that a reduction in accuracy may occur when A has a large condition number.

References

1. A. Bloch and P. Crouch, Optimal control and geodesic flows, *Systems & Control Letters*, **28**, N. 3 (1996), 65-72.
2. J. R. Cardoso and F. Silva Leite, Theoretical and numerical considerations about logarithms of matrices. Submitted in 1999.
3. L. Dieci, B. Morini and A. Papini, Computational techniques for real logarithms of matrices, *SIAM Journal on Matrix Analysis and Applications*, **17**, N. 3, (1996), 570-593.
4. L. Dieci, B. Morini, A. Papini and A. Pasquali, On real logarithms of nearby matrices and structured matrix interpolation, *Appl. Numer. Math.*, **29** (1999), 145-165.
5. G. Golub and C. Van Loan, *Matrix Computations*. Johns Hopkins Univ. Press, 3rd ed., Baltimore, MD, USA, 1996.
6. N. J. Higham, Computing real square roots of a real matrix, *Linear Algebra and its Applications*, **88/89**, (1987), 405-430.
7. N. J. Higham, Stable iterations for the matrix square root, *Numerical Algorithms*, **15**, (1997), 227-242.
8. N. J. Higham, A new sqrtm for Matlab, *Numerical Analysis Report*, **336**, (1999), University of Manchester.
9. R. A. Horn and C. R. Johnson, *Topics in Matrix Analysis*. Cambridge University Press, 1994.
10. C. Kenney and A. J. Laub, Padé error estimates for the logarithm of a matrix, *International Journal of Control*, **50**, N. 3, (1989), 707-730.
11. C. Kenney and A. J. Laub, Condition estimates for matrix functions, *SIAM Journal on Matrix Analysis and Applications*, **10**, (1989), 191-209.
12. C. Kenney and A. J. Laub, A Schur-Frechet algorithm for computing the logarithm and exponential of a matrix, *SIAM Journal on Matrix Analysis and Applications*, **19**, N. 3, (1998), 640-663.

Robust Preconditioning of Dense Problems from Electromagnetics

B. Carpentieri[1], I.S. Duff[1,2], and L. Giraud[1]

[1] CERFACS
Toulouse, France,
`carpenti,duff,giraud@cerfacs.fr`
[2] RAL, Oxfordshire, UK

Abstract. We consider different preconditioning techniques of both implicit and explicit form in connection with Krylov methods for the solution of large dense complex symmetric non-Hermitian systems of equations arising in computational electromagnetics. We emphasize in particular sparse approximate inverse techniques that use a static nonzero pattern selection. By exploiting geometric information from the underlying meshes, a very sparse but effective preconditioner can be computed. In particular our strategies are applicable when fast multipole methods are used for the matrix-vector products on parallel distributed memory computers.

Keywords: Preconditioning techniques, sparse approximate inverses, nonzero pattern selection strategies, electromagnetic scattering applications.

AMS subject classification: 65F10, 65F50, 65N38, 65R20, 78A45, 78A50, 78-08

1 Introduction

A considerable amount of work has been recently spent on the simulation of electromagnetic wave propagation phenomena, addressing various topics ranging from radar cross section to electromagnetic compatibility, absorbing materials, and antenna design. The physical issue is to compute the diffraction pattern of the scattered wave, given an incident field and a scattering obstacle. The *Boundary Element Method (BEM)* is a reliable alternative to more classical discretization schemes like Finite Element Methods and Finite Difference Methods for the numerical solution of this class of problems. The idea of *BEM* is to shift the focus from solving a partial differential equation defined on a closed or unbounded domain to solving a boundary integral equation only over the finite part of the boundary. This approach leads to the solution of linear systems of equations of the form

$$Ax = b, \qquad (1)$$

where the coefficient matrix $A = [a_{ij}]$ is a large, dense, complex matrix of order n arising from the discretization. The coefficient matrix can be symmetric non-Hermitian in the EFIE (Electric Field Integral Equation) formulation, or unsymmetric in the CFIE (Combined Field Integral Equation) formulation. Direct dense methods based on Gaussian elimination are often the method of choice for solving such systems, because they are reliable and predictable both in terms of accuracy and cost. However, for large-scale problems they become impractical even on large parallel platforms because they require storage of n^2 double precision complex entries of the coefficient matrix and $\mathcal{O}(n^3)$ floating-point operations to compute the factorization. Iterative Krylov subspace based solvers can be a promising alternative provided we have fast matrix-vector multiplications and robust preconditioners. Here we focus on the design of robust preconditioning techniques. The paper is organized as follows: in Section 1 we introduce the problem and we discuss some issues addressed by the design of the preconditioner for this class of problems; in Section 2 we report on the results of our numerical investigations and finally, in Section 3, we propose a few tentative conclusions arising from the work.

1.1 The Design of the Preconditioner

A preconditioner M is any matrix that can accelerate the convergence of iterative solvers. The original system (1) is replaced with a new system of the form $M^{-1}Ax = M^{-1}b$ when preconditioning from the left, and $AM^{-1}y = b$, with $x = M^{-1}y$, when preconditioning from the right. A good preconditioner, to be effective, has to be a close approximation of A, easy to construct and cheap to store and to apply. For dense matrices, some additional constraints have to be considered. The choice of the *best* preconditioning family can require more effort than in the sparse case, because for dense systems there are far less results. When the coefficient matrix of the linear system is dense, the construction of even a very sparse preconditioner may become too expensive in execution time as the problem size increases. In some context like in the multipole setting all the entries of the coefficient matrix are not directly available and the preconditioner has to be constructed from a sparse approximation of A, possibly computed by accessing only local information. Thus a suitable pattern is required to select a representative set of the entries of A to build M.

The parallel issue suggests to consider also preconditioning techniques of explicit form that compute an approximation to the inverse of A, because then the application of the preconditioner reduces to perform at each step a M-V product, which is a highly parallelizable kernel on both shared and distributed memory machines. Some of these techniques require to prescribe a sparse pattern in advance for the approximate inverse, able to capture most of the large entries of A^{-1}. Thus in that case an effective pattern is required also for the preconditioner.

Algebraic Strategy. In the *BEM* context the matrices arising from the discretization of the problem exhibit regular structure: the largest entries are located on the main diagonal, and only a few adjacent bands have entries of high magnitude. Most of remaining entries have much smaller modulus. In Figure 1(a), we plot for a cylindric geometry the matrix obtained by scaling $A = [a_{ij}]$ so that $\max_{i,j} |a_{ij}| = 1$, and discarding from A all entries less than $\varepsilon = 0.05$ in modulus. This matrix has 16 non-zeros per row on the average and its size is 1080. Several heuristics based on algebraic information can be used to extract a sparsity pattern from A that retains the main contributions to the singular integrals [2].

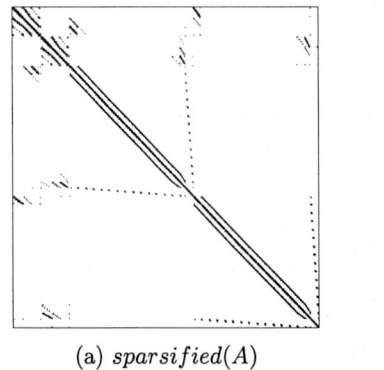 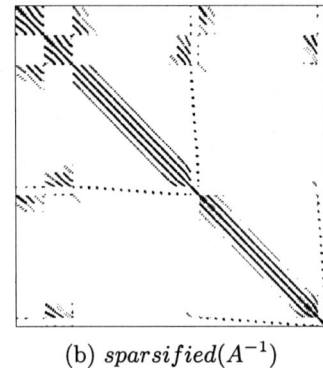

(a) $sparsified(A)$ (b) $sparsified(A^{-1})$

Fig. 1. Nonzero pattern for A (left) and A^{-1} (right) when the smallest entries are discarded. The test problem is a cylinder

On smooth geometries, due to the decay of the Green's function, the regular structure of A is generally maintained also for its inverse. Figure 1(b) shows the pattern of $sparsified(A^{-1})$, where A^{-1} has been computed using LAPACK library routines, and then sparsified, after scaling, with the same value of the threshold as the one used to produce Figure 1(a). This pattern selection strategy, referred to as the *algebraic strategy*, can be effective to construct preconditioners of both implicit and explicit form, but requires to access all the entries of the coefficient matrix and for large problems this can become too expensive or even not possible, like in a multipole framework.

Relevant information for the construction of the preconditioner can be extracted from the meshes of the underlying physical problem. In particular, two types of information are directly available:

the connectivity graph, describing the topological neighbourhood amongst the edges, and
the coordinates of the nodes in the mesh, describing geometric neighbourhoods amongst the edges.

Topological Strategy. In the integral equation context that we consider here, the object surface is discretized by a triangular mesh using the so-called flux finite elements or Rao-Wilton-Glisson elements [7]. Each degree of freedom (DOF), corresponding to an unknown in the linear system, represents the vectorial flux across each edge in the triangular network. Topological neighbourhoods can be defined according to the concept of level k neighbours, as introduced in [6]. Level 1 neighbours of a DOF are the DOF plus the four DOFs belonging to the two triangles that share the edge corresponding to the DOF itself. Level 2 neighbours are all the level 1 neighbours plus the DOFs in the triangles that are neighbours of the two triangles considered at level 1, and so forth. In Figure 2 we plot, for each DOF of the mesh for the same cylindric geometry considered before, the magnitude of the associated entry in A (the graph on the left) and in A^{-1} (the graph on the right) with respect to the level of its neighbours. In both cases the large entries derive from the interaction of a very localized set of edges in the mesh so that by retaining a few levels of neighbours for each DOF an effective pattern to approximate both A and A^{-1} is likely to be constructed. A pattern selection strategy based on topological information is referred to as *topological strategy*.

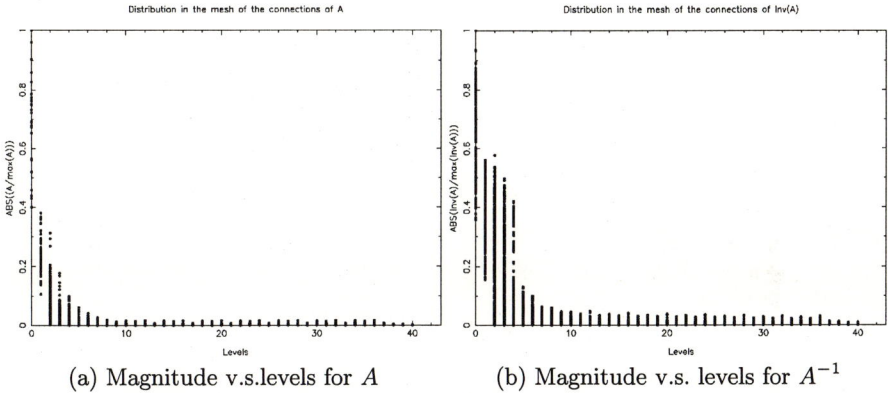

(a) Magnitude v.s.levels for A (b) Magnitude v.s. levels for A^{-1}

Fig. 2. Topological localization in the mesh for the large entries of A (left) and A^{-1} (right). The test problem is a cylinder and is representative of the general behaviour

Geometric Strategy. For the same scattering problem previously considered, we plot in Figure 3, for each pairs of edges in the mesh, the magnitude of their associated entries in A and A^{-1} with respect to their distance in terms of the wavelength of an incident electromagnetic radiation. The wavelength is a physical parameter affecting the complexity of the problem to be solved. For an accurate representation of the oscillating solution of Maxwell's equations, in fact, around ten points per wavelength need to be used for the discretization.

The largest entries of A and A^{-1} are strongly localized in a similar fashion. The pattern for constructing an approximation of A or A^{-1} can be computed by selecting for each edge all those edges within a sufficiently large sphere that defines our geometric neighbourhood. In the case of preconditioning techniques of explicit form, by using a suitable size for this sphere we hope to include the most relevant contributions to the inverse and consequently to obtain an effective sparse approximate inverse. When the surface of the object is very non-smooth, these large entries may come from the interaction of far-away or non-connected edges in a topological sense, which are neighbours in a geometric sense. Thus this approach is more promising to handle complex geometries where parts of the surface are not connected. This selection strategy will be referred to as the *geometric strategy*.

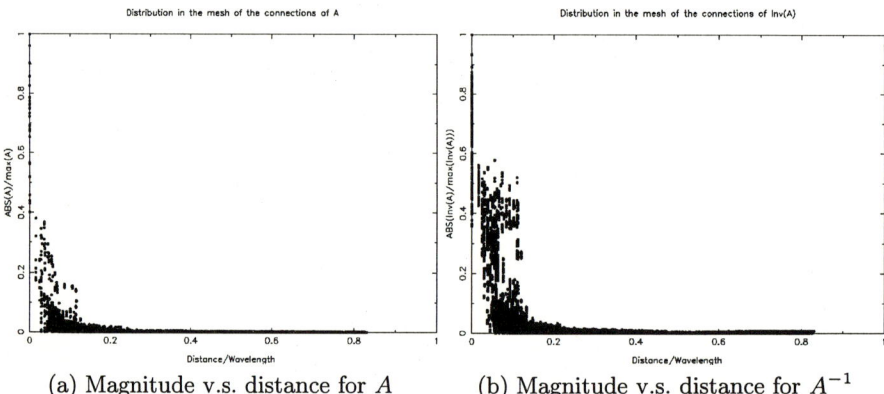

(a) Magnitude v.s. distance for A (b) Magnitude v.s. distance for A^{-1}

Fig. 3. Geometric localization in the mesh for the large entries of A (left) and A^{-1} (right). The test problem is a cylinder. This is representative of the general behaviour

2 Numerical Experiments

In this section we compare the performance of different preconditioning techniques in connection with Krylov solvers on a selected set of test problems. Amongst the test cases considered in [2], we select the three following examples, corresponding to bodies with different geometries:

Example 1: Cylinder with a hollow inside, a matrix of order $n = 1080$;
Example 2: Cylinder with a break on the surface, a matrix of order $n = 1299$;
Example 3: Sphere, a matrix of order $n = 1080$,

where, for physical consistency, we set the frequency of the wave so that there are about ten discretization points per wavelength.

We use, amongst Krylov methods, restarted GMRES [8], Bi-CGSTAB [9], symmetric and nonsymmetric QMR [4], TFQMR [3]. We consider the following preconditioning techniques, all computed by replacing A with its sparse approximation referred to as *sparsified(A)*, and implemented as right preconditioners:

- *SSOR*;
- $ILU(0)$, the incomplete LU factorization with zero level of fill-in, applied to *sparsified(A)*;
- *FROB*, a Frobenius norm minimization technique, with the pattern of *sparsified(A)* prescribed in advance for the approximate inverse;
- *SPAI*, introduced in [5], with the adaptive strategy implemented in the MI12 routine from HSL;
- *SLU*, a complete LU factorization of *sparsified(A)*, used as implicit preconditioner.

For comparison purpose, we also report on results for the unpreconditioned case and using a simple diagonal scaling. The stopping criteria in all cases consists in reducing the original residual by 10^{-5}. The symbol "-" means that convergence is not obtained after 500 iterations. In each case, we take as the initial guess $x_0 = 0$, and the right-hand side is such that the exact solution of the system is formed by all ones. All the numerical experiments refer to runs in double precision complex arithmetic on a SUN workstation.

The pattern to construct *sparsified(A)* and all the preconditioners are computed by using the geometric strategy, retaining all those entries within a sphere of radius 0.12 times the wavelength. We try to have the same number of nonzeros in the different preconditioners resulting from the various methods: in the incomplete LU factorization, no additional level of fill-in is allowed in the factors; in the Frobenius-norm minimization technique, the same sparsity pattern prescribed on A (and then exactly the same number of nonzero entries) is imposed on the preconditioner; with *SPAI* we choose *a priori*, for each column of M, the same fixed maximum number of nonzeros as in the computation of *sparsified(A)*; and finally for the *SLU* method, *sparsified(A)* is factorized using ME47, a sparse direct solver from HSL, and those exact factors are used as the preconditioner. The efficient implementation of the ME47 solver guarantees a minimal fill-in in the factors. We do not report on results with the *AINV* preconditioner [1] because they are discouraging.

Amongst different techniques Frobenius norm minimization methods are the most promising; they are highly parallelizable and numerically effective. The L-S solutions require some computational effort, but the patterns computed by the geometric strategy are generally very sparse, and the resulting least squares problems are small and can be effectively computed via a dense QR factorization. As it can be seen in Table 1, ILU preconditioners are not effective for such systems. In our tests, modifications of the coefficient matrix do not help to improve their robustness. Better performance can be obtained by allowing more fill-in in the factors but at the cost of increased computational cost and storage requirement. The *SLU* preconditioner represents in this sense an extreme case

Table 1. Number of iterations required by different preconditioned Krylov solvers to reduce the residual by 10^{-5}

Precond.	Example 1 - Density of $M = 5.03\%$					Bi - CGStab	UQMR	TFQMR
	GMRES(m)							
	m=10	m=30	m=50	m=80	m=110			
Unprec	-	-	-	251	202	293	258	170
M_j	-	-	465	222	174	239	210	169
SSOR	-	417	199	137	101	116	154	126
ILU(0)	-	-	-	-	-	-	-	-
FROB	134	83	49	49	49	53	57	47
SPAI	-	-	-	-	-	-	340	465
SLU	-	-	377	223	178	236	244	265
	Example 2 - Density of $M = 1.59\%$							
Precond.	GMRES(m)					Bi - CGStab	UQMR	TFQMR
	m=10	m=30	m=50	m=80	m=110			
Unprec	-	-	-	398	289	321	405	251
M_j	-	-	473	330	243	257	354	228
SSOR	-	363	236	157	126	153	246	136
ILU(0)	-	-	-	160	97	-	273	437
FROB	114	88	68	57	57	45	85	46
SPAI	-	-	-	-	-	-	-	-
SLU	-	-	-	318	206	412	499	-
	Example 3 - Density of $M = 1.50\%$							
Precond.	GMRES(m)					Bi - CGStab	UQMR	TFQMR
	m=10	m=30	m=50	m=80	m=110			
Unprec	202	62	61	57	57	75	69	40
M_j	175	71	67	59	59	80	71	46
SSOR	176	87	77	63	63	82	80	55
ILU(0)	-	-	-	470	330	-	284	217
FROB	15	14	14	14	14	10	19	10
SPAI	-	-	-	-	-	-	-	-
SLU	385	143	107	74	74	73	95	68

with respect to $ILU(0)$ since a complete fill-in is allowed in the factors. This approach, although not easily parallelizable, is generally effective on this class of applications for dense enough sparse approximations of A. But if the pattern is very sparse, approximate inverse techniques prove to be more robust. SSOR, compared to FROB, is generally slower in term of iterations, but is very cheap to compute. However it is not easily parallelizable, and the extra-cost for computing an approximate inverse can be overcome by the time saved in the iterations when solving the same linear system with many right-hand sides. This is often the case in electromagnetic applications, when illuminating an object with various waves corresponding to different angles of incidence.

3 Conclusions

Iterative methods can present an attractive alternative to direct methods even for the solution of this class of problems, especially when great accuracy for the solution is not demanded, as is often the case for physical problems. The behaviour of these techniques is strongly dependent on the choice of the preconditioner. Frobenius norm minimization methods are the most promising candidates to precondition effectively these problems; they deliver a good rate of convergence, and are inherently parallel. The numerical experiments have shown that, using additional geometric information from the underlying mesh, we can compute a very sparse but effective preconditioner. This pattern selection strategy does not require access to all the entries of the matrix A, so that it is promising for an implementation in a fast multipole setting where A is not directly available but where only the near field entries are computed.

Acknowledgments

The work of the first author was supported by I.N.D.A.M. (Rome, Italy) under a grant (Borsa di Studio per l'estero, Provvedimento del Presidente del 30 Aprile 1998).

References

1. M. Benzi, C. D. Meyer, and M. Tůma. A sparse approximate inverse preconditioner for the conjugate gradient method. *SIAM J. Scientific Computing*, 17:1135–1149, 1996.
2. B. Carpentieri, I. S. Duff, and L. Giraud. Sparse pattern selection strategies for robust frobenius norm minimization preconditioners in electromagnetism. Technical Report TR/PA/00/05, CERFACS, Toulouse, France, 1999. To Appear in Numerical Linear Algebra with Applications.
3. R. W. Freund. A transpose-free quasi-minimal residual algorithm for non-hermitian linear systems. *SIAM J. Scientific Computing*, 14(2):470–482, 1993.
4. R. W. Freund and N. M. Nachtigal. QMR: a quasi-minimal residual method for non-hermitian linear systems. *Numerische Mathematik*, 60(3):315–339, 1991.
5. M. Grote and T. Huckle. Parallel preconditionings with sparse approximate inverses. *SIAM J. Scientific Computing*, 18:838–853, 1997.
6. J. Rahola. Experiments on iterative methods and the fast multipole method in electromagnetic scattering calculations. Technical Report TR/PA/98/49, CERFACS, Toulouse, France, 1998.
7. S. M. Rao, D. R. Wilton, and A. W. Glisson. Electromagnetic scattering by surfaces of arbitrary shape. *IEEE Trans. Antennas Propagat.*, AP-30:409–418, 1982.

8. Y. Saad and M. H. Schultz. GMRES: A generalized minimal residual algorithm for solving nonsymmetric linear systems. *SIAM J. Scientific and Statistical Computing*, 7:856–869, 1986.
9. H. A. van der Vorst. Bi-CGSTAB: a fast and smoothly converging variant of Bi-CG for the solution of nonsymmetric linear systems. *SIAM J. Scientific and Statistical Computing*, 13:631–644, 1992.

A Mathematical Model for the Limbic System

Lucía Cervantes and Andrés Fraguela Collar

Facultad de Ciencias Físico Matemáticas
Benemérita Universidad Autónoma de Puebla. Puebla, Puebla. México
{lcervant,fraguela}@fcfm.buap.mx

Abstract. The limbic circuit, involving the prefrontal cortex, hippocampus and certain subcortical structures plays a determinant role in the emotional activity and for understand psychopathologies like schizophrenia and sensitization to certain psychostimulants.
In this work, we constructed a non-linear network representing the interaction between seven important nuclei of the limbic system.
This model is a first approach that allows to simulate different activities of the circuit, associated with the dopamine sensitization and the neurodevelopmental hypothesis for the neuropathology of schizophrenia

1 Introduction

Pathophysiological processes that underlie the profound neuropsychiatric disturbances in Schizophrenia are poorly understood[1]. However, considerable evidence from clinical, neuropsychological brain image and postmortem anatomical studies strongly implicates the prefrontal-temporo-limbic cortical dysfunctions in schizophrenia[1,5].

Physiologically speaking, the dopamine and the neurodevelopmental hypotheses of schizophrenia postulate that at least some forms of schizophrenia could have their origin in an early neurodevelopmental defect (damage in the ventral hippocampus and/or the prefrontal cortex) that may result in the prefrontal - temporo-limbic cortical dysfunctions (recovered lesion) in early adulthood (overactivity in neurotransmission from DA cell bodies,located in the ventral tegmental area (VTA) of the midbrain, hypofunction of the prefrontal cortex)[2,4,5].

We model a network composed by seven nuclei from the limbic circuit which is a prefrontal-temporolimbic cortical circuit involved in the pathology of schizophrenia[2,4,5]. The nuclei (fig. 1) are interconnected in an excitatory and inhibitory way, via glutamatergic, dopaminergic(DA) and gabaergic neurotransmitters. Using the physiological information obtained about the Limbic circuit, we constructed a model which describes the dynamics of the interaction between the nuclei involved.

The model is conformed by a system of non linear differential equations of first order , obtained from a general balance equation for the density of excitatory and inhibitory synaptic activity in each nucleus of the circuit (see 3).

The final system of differential equations obtained is the result of an asymptotic analysis with respect to a small parameter which is involved with the relative distribution of connections between the different nuclei of the circuit.

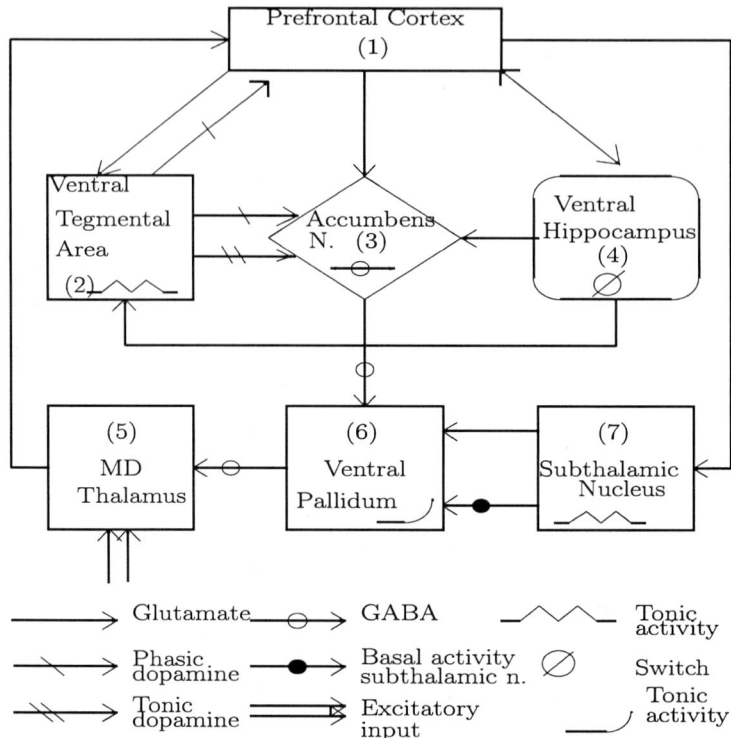

Fig. 1. The Limbic System with their nuclei and neurotransmitters involved

In the differential equations that compose the system, there are certain functions which play a fundamental role, they correspond to the non linearity of the system and they are associated with the activation ability of each nucleus, this activation ability is on its time related with intrinsic characteristics of each nucleus and the kind of neurotransmitter that is liberated by its neurons.

Different forms of the model allow us to simulate the situations corresponding to a prenatal lesion, healthy individual and recovered lesion and the chance of comparing the nuclei activity in each situation.

2 Unknown Functions and Model's Parameters

We start defining the function whose dynamic will describe the interaction between two given nuclei through the synaptic activity which exerts one nucleus over other.

Given two nuclei Ω_i, Ω_j we define : $h_{ij}^{\pm}(x,t)$ for $x \in \Omega_j$, like the excitatory (+) or inhibitory(−)density of synaptic activity in the location x of the nucleus Ω_j at the instant t, produced by the connections from the nucleus Ω_i.

We also consider the following functions:

$R_{ij}^{\pm}(x',x,v) :=$ The density of synaptic connections between the locations $x\prime \in \Omega_i$ and $x \in \Omega_j$ from the circuit of neurons whose bodies are in Ω_i and through such connections travel action potentials with a specific density of propagation velocities v and which produce over the neurons from an excitatory(+)or an inhibitory (−) effect.

The functions $R_{ij}^{\pm}$ keep macroscopic anatomical information from the circuit.

$\tilde{g}_i(x',t) :=$ The fraction of the whole of neurons from the nucleus Ω_i which shoot action potentials at the instant t.

The functions $\tilde{g}_i$ describe the activation ability of each nucleus, so in consequence they keep functional information of the nuclei.

We consider that the decreasing order of the connections (density of axons) between each pair of nuclei is of exponential type and that there exists only a velocity of propagation of the action potentials:V.In this case we can represent:

$$R_{ij}^{\pm}(x',x,v) = \alpha_{ij}^{\pm}(x')\, e^{-\lambda_{ij}^{\pm}|x-x'|}\, S_j(x)\, \delta(v-V) \qquad (1)$$

Where $S_j(x)$ represents the density of synaptic contacts in the location $x \in \Omega_j$, $\alpha_{ij}^{\pm}(x)$ is a normalization constant which we will define after and δ is the Dirac's delta function.

To the pairs (i,j) such that the nucleus Ω_i produces synaptic activity over the nucleus Ω_j, we will call an acceptable pair. We also write $(+)(i,j)$ when the synaptic activity is excitatory and $(-)(i,j)$ when it is inhibitory.

In this case we have the following acceptable pairs:

(+)(2,1)	(+)(1,2)	(+)(1,3)	(+)(1,4)	(−)(6,5)	(−)(3,6)	(+)(1,7)
(+)(4,1)	(+)(4,2)	(+)(2,3)			(+)(7,6)	
(+)(5,1)		(−)(3,3)				
		(+)(4,3)				

Note that in this list we did not included:
- The tonic activity that Ω_2 exerts over Ω_3
- The basal activity that Ω_7 exerts over Ω_6
- The outer excitatory or inhibitory inputs over Ω_5

we also consider lateral inhibition inside the nucleus Ω_3, it is the function h_{33}^{-}

3 Model of Interaction between Two Nuclei

Considering that $h_{ij}^{\pm}(x,t)$ is expressed in $(cm^3)^{-1}$, $R_{ij}^{\pm}(x',x,v)$ in $sec(cm^3)^{-7}$ and $\tilde{g}_i(x',t)$ is dimensionless, then it is clear that

$$R_{ij}^{\pm}(x',x,v)\, \tilde{g}_i\!\left(x',t - \frac{|x-x'|}{v}\right) dx'\, dv \qquad (2)$$

represents the density of active synapsis in the location $x \in \Omega_j$ produced by the neurons located between the points x' and x'+dx' from Ω_i and by whose fibers travel action potentials with velocities between v and $v + dv$ which arrive from the location x at the instant t, so we obtain :

$$h_{ij}^{\pm}(x,t) = \int_0^\infty dv \int_{\Omega_i} R_{ij}^{\pm}(x',x,v)\,\tilde{g}_i(x', t - \frac{|x-x'|}{V})dx' \qquad (3)$$

If we substitute the expression (1) in (3) we obtain,

$$h_{ij}^{\pm}(x,t) = S_j(x) \int_{\Omega_i} \alpha_{ij}^{\pm}(x')\, e^{\lambda_{ij}^{\pm}|x-x'|}\, \tilde{g}_i(x', t - \frac{|x-x'|}{V})dx' \qquad (4)$$

for every acceptable pair $(+)(i,j)$ or $(-)(i,j)$.

It is clear that for every j fixed and $x \in \Omega_j$ it is accomplished that:

$$\sum_{(+)(i,j)} \int_0^\infty dv \int_{\Omega_i} R_{ij}^+(x',x,v)dx' + \sum_{(-)(i,j)} \int_0^\infty dv \int_{\Omega_i} R_{ij}^-(x',x,v)dx' = S_j(x) \qquad (5)$$

substituting the expression (1) in (5) and assuming $\alpha_{ij}^{\pm}$ like a constant, we arrive to the following "condition of normalization" for such coefficients,

$$\alpha_{ij}^{\pm} = \frac{\lambda_{ij}^{\pm}}{2n_j} \qquad (6)$$

where n_j is the whole number of nuclei which are interacting with the nucleus j, $n_1 = 3, n_2 = n_6 = 2, n_3 = 4, n_4 = n_5 = n_7 = 1$.

Assuming that the density of glutamatergic and gabaergic fibers do not depend from the nuclei connected by such fibers, we can define

$$\lambda_1 := \lambda_{51}^+ = \lambda_{41}^+ = \lambda_{12}^+ = \lambda_{42}^+ = \lambda_{13}^+ = \lambda_{43}^+ = \lambda_{14}^+ = \lambda_{76}^+ = \lambda_{17}^+; \qquad (7)$$
$$\lambda_2 := \lambda_{65}^- = \lambda_{36}^- = \lambda_{33}^- \qquad (8)$$

However, for the dopaminergic fibers we consider that $\lambda_{21}^+ \neq \lambda_{23}^+$ and define:

$$\lambda_3 := \lambda_{23}^+ \qquad (9)$$
$$\lambda_4 := \lambda_{21}^+ \qquad (10)$$

Introducing the new unknown functions

$$H_{ij}^{\pm}(x,t) := \frac{h_{ij}^{\pm}(x,t)}{S_j(x)} \qquad (11)$$

and considering that such functions do not variate spatially inside each nucleus, so, after defining,

$$\lambda_1 \approx \lambda_2 \approx \lambda_3 = \lambda, \quad \epsilon = \frac{\lambda}{\lambda_4} \qquad (12)$$

and making the variable change
$$\lambda v\, t = \tau \tag{13}$$
we finally obtain the simplified equations
$$\frac{d^2 H_{ij}^\pm}{d\tau^2} + 2\frac{dH_{ij}^\pm}{d\tau} + H_{ij}^\pm = \frac{1}{n_j}\left[g_i^0 + \frac{dg_i^0}{d\tau}(\tau)\right] \tag{14}$$
for all the pairs (i,j) which appear in the equations (7), (8) and (9).
$$\epsilon^2 \frac{d^2 H_{ij}^\pm}{d\tau^2} + 2\epsilon \frac{dH_{ij}^\pm}{d\tau} + H_{ij}^\pm = \frac{1}{n_j}\left[g_i^0(\tau) + \epsilon \frac{dg_i^0}{d\tau}(\tau)\right] \tag{15}$$
for the pair $(2,1)$ from the expression (10).

4 Model for the Limbic System

According to the physiological information about the activation ability of each nucleus, we can write,

$$g_1^0(\tau) = G_1(u_1) = \frac{G_1^1(u_1^1) + (1-u_3^4)G_1^2(u_1^2)}{2} \tag{16}$$

$$g_2^0(\tau) = G_2(u_2) = G_2^1(u_2^1) \tag{17}$$

$$g_3^0(\tau) = G_3(u_3) = \frac{G_3^1(u_3^1) + G_3^2(u_3^2) + G_3^3(u_3^3) + (1-u_1^2)G_3^4(u_3^4)}{4} \tag{18}$$

$$g_4^0(\tau) = G_4(u_4) = G_4^1(u_4^1) \tag{19}$$

$$g_5^0(\tau) = G_5(u_5) = \frac{G_5^1(u_5^1) + G_5^2(u_5^2)}{2} \tag{20}$$

$$g_6^0(\tau) = G_6(u_6) = \frac{G_6^1(u_6^1) + G_6^2(u_6^2)}{2} \tag{21}$$

$$g_7^0(\tau) = G_7(u_7) = G_7^1(u_7^1) \tag{22}$$

where the functions G_j^k describe the percentage of neurons from the nucleus Ω_j which shoot action potentials like a result of the partial activation u_j^k of the nucleus defined by,

$$u_1^1 = \frac{H_{51}^+ + H_{21}^+}{2}, \qquad u_1^2 = H_{41}^+$$

$$u_2^1 = \frac{H_{12}^+ + H_{42}^+}{2}, \qquad u_3^1 = \frac{H_{13}^+ - H_{33}^-}{2}$$

$$u_3^2 = H_{23}^+, \qquad u_3^3 = D\cos^2(\omega\tau + \theta)$$

$$u_3^4 = H_{43}^+, \qquad u_4^1 = H_{14}^+$$

$$u_5^1 = -H_{65}^-, \qquad u_5^2 = \frac{E_+(\tau) - E_-(\tau)}{2}$$

$$u_6^1 = \frac{H_{76}^+ - H_{36}^-}{2}, \qquad u_6^2 = F\cos^2 2(\alpha\tau + \phi)$$

$$u_7^1 = H_{17}^+,$$

The functions u_3^3 and u_6^2, normalized with $0 \leq D, F \leq 1$, represents the spontaneous activity from the nucleus Ω_2 over Ω_3 and from Ω_3 over Ω_7 respectively. The expressions E_+, E_- in u_5^2 represents the excitatory an the inhibitory inputs to the circuit through the nucleus Ω_5. Also, we can take for a healthy individual,

$$m(u) = G_5^1(u) = G_6^1(u) = \frac{1}{1 + e^{3(\frac{1}{2}-u)}}$$

$$n(u) = G_2^1(u) = G_4^1(u) = G_7^1(u) = G_3^1(u) = G_3^4(u) =$$
$$= \frac{1}{1 + e^{2(\frac{3}{4}-u)}}$$

$$p(u) = G_3^2(u) = \frac{1}{1 + e^{3.5(\frac{1}{4}-u)}}$$

$$q(u) = G_1^1(u) = G_1^2(u) = \frac{1}{1 + e^{1.5(\frac{1}{8}-u)}}$$

$$G_3^3(u_3^3) = D\left(1 - \cos^2(\omega\tau + \theta)\right) = D \operatorname{sen}^2(\omega\tau + \theta) \tag{23}$$

$$G_5^2(u_5^2) = \left[\frac{E^+(\tau) - E^-(\tau)}{2}\right]^2 \tag{24}$$

$$G_6^2(u_6^2) = F\left(\frac{1 + \cos^2 2(\alpha\tau + \varphi)}{2}\right) = F\cos^2(\alpha\tau + \varphi) \tag{25}$$

If we define,

$$H_{ij}^{\pm}(0) = H_{ij}^{\pm 0} \quad , \quad \frac{dH_{ij}^{\pm}}{d\tau}(0) = H_{ij}^{\pm 1} \tag{26}$$

$$k_{ij}^{\pm} = H_{ij}^{\pm 1} + H_{ij}^{\pm 0} - \frac{1}{n_j} g_i^0(0) \tag{27}$$

$$M_{ij}^{\epsilon} = \epsilon H_{ij}^{+1} + H_{ij}^{+0} - \frac{1}{n_j} g_i^0(0) \tag{28}$$

also consider the system,

$$\frac{dH_{ij}^{\pm}}{d\tau} + H_{ij}^{\pm} - \frac{1}{n_j} g_i^0(\tau) = k_{ij}^{\pm} e^{-\tau} \tag{29}$$

$$\epsilon \frac{dH_{ij}^+}{d\tau} + H_{ij}^+ - \frac{1}{n_j} g_i^0(\tau) = M_{ij}^{\epsilon} e^{-\frac{\tau}{\epsilon}} \tag{30}$$

then we obtain the following

Theorem 1. *The model of the Limbic circuit for a healthy individual is reduced to the system(42),(43). This system is equivalent to the original one in the sense that each solution of the original system (14),(15) satisfying the initial conditions $H_{ij}^{\pm 0}, H_{ij}^{\pm 1}$ is equal to the solution of the system (42),(43) satisfying the same initial condition $H_{ij}^{\pm 0}$.*

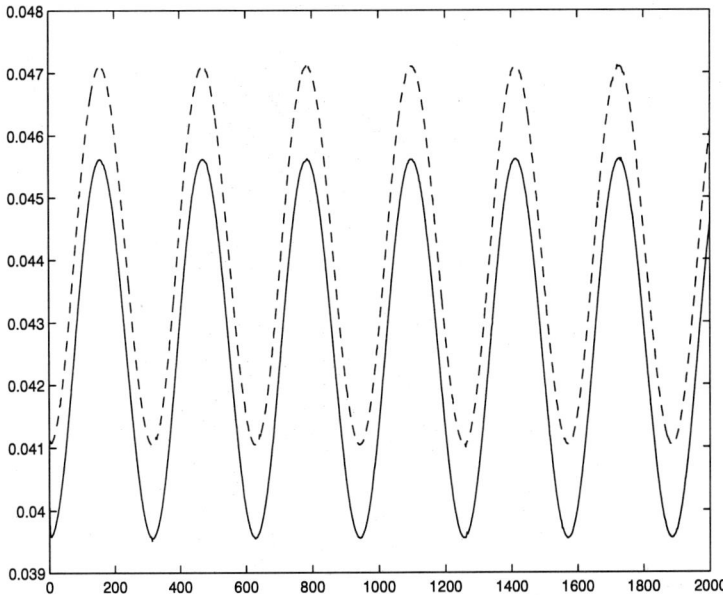

Fig. 2. The Accumbens Nucleus activity for a normal individual (low) and for a pathological one (recovered lesion)(high)

Making variations over certain coefficients in some functions G_j^k we obtain, from (42),(43) the systems which represents the cases of prenatal and recovered lesion.

The system (42),(43) conformed by 14 equations can be reduced through changes of variables and an asymptotic analysis to an equivalent system of seven equations for small ε and big τ, from whose solutions we can write the activation variables of each nucleus.

A numerical comparative analysis of the activation function of the different nuclei shows us changes comparative in the dynamics, specially an hyperactivity of the accumbens nucleus and an hypoactivity of the prefrontal cortex, according with the physiological hypothesis of the dopamine sensitization and the neurodevelopmental hypothesis for schizophrenia.(figs. 2,3)

Figs. 2,3. In the horizontal scale, 2000 is equivalent to 100 msec.

References

1. Duncan, Scheitman, Lieberman (1999) An integrated view of pathophysiological models of Schizophrenia. Brain research Reviews, 29:250-264.
2. Flores G., Barbeau D., Quirion R., Srivastava L (1996a) Decreased Binding of Dopamine D3 Receptors in Limbic Subregions after Neonatal Bilateral Lesion of Rat Hippocampus. The Journal of Neuroscience,16(6):2020-2026.

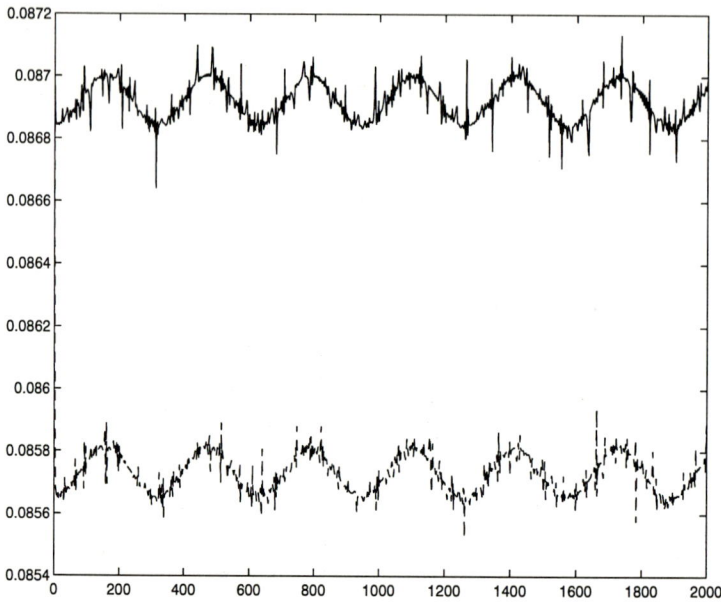

Fig. 3. The Prefrontal Cortex activity for a normal individual (high) and for a pathological one (low)

3. Fraguela A., Escamilla J. A. A mathematical model for study the activation process in the cerebral cortex. Numerical Analysis; Theory, applications and programs. Edited by M. V. Lomonosov State University. Moscow 1999.pp 47-55.
4. Lipska BK, Jaskiw GE, Weinberger DR.(1993a) Postpubertal emergence of hiperresponsiveness to stress and to amphetamine after neonatal excitotoxic hippocampal damage: a potential animal model of schizophrenia. Neuropsychopharmacology 9:67-75.
5. Weinberger DR, Lipska BK(1995) Cortical maldevelopment, anti-psychotic drugs, and schizophrenia: a search for common ground. Schizophrenia Research 16: 87-110.

Understanding Krylov Methods in Finite Precision[*]

Françoise Chaitin-Chatelin[1,2], Elisabeth Traviesas[1], and Laurent Plantié[1]

[1] CERFACS
42 Avenue Gaspard Coriolis, 31057 Toulouse cedex 01, France
{chatelin,travies,plantie}@cerfacs.fr
http://www.cerfacs.fr/algor
[2] Université Toulouse I, Toulouse, France

Abstract. Krylov methods are, since their introduction in the 1980s, the most heavily used methods to solve the two problems

$$Ax = b \quad \text{and} \quad Ax = \lambda x,\ x \neq 0$$

where the matrix A is very large.
However, the understanding of their numerical behaviour is far from satisfactory. We propose a radically new viewpoint for this longstanding enigma, which shows mathematically that the Krylov-type method works best when it is most ill-conditioned.

1 Introduction

Krylov methods have been, since their introduction in the 80s by Y. Saad, widely used worldwide to solve large scale problems such as

$$Ax = b \quad \text{and} \quad Ax = \lambda x,\ x \neq 0$$

which are the two basic problems associated with a large (often sparse) matrix A. Despite this widespread use, the understanding of their finite precision behaviour is far from satisfactory. The consequence is that even the best codes include heuristics and their convergence is not guaranteed.

Such a state of affairs is intellectually frustrating and, until now, the Krylov methods continue to challenge the best experts. This paper presents the programme undertaken at Cerfacs in the Qualitative Computing Group since 1996, which looks at the Krylov methods in a completely new and original way [2].

2 The First Step: The Basic Arnoldi Algorithm for the Hessenberg Decomposition

2.1 Irreducibility of H ?

Classically, the Hessenberg decomposition

$$A = VHV^*,\ H \text{ Hessenberg},\ VV^* = V^*V = I \qquad (1)$$

is considered under the assumption that H is **irreducible**.

---
[*] CERFACS Technical Report TR/PA/00/40

The main reason is that, from a mathematical standpoint, if H is not irreducible, it can be partitionned by means of two or more irreducible Hessenberg submatrices which are on the diagonal. The spectrum of A is the union of the corresponding spectra. The reasoning is, of course, impeccable in exact arithmetic. But, as we shall see, it may be misleading for finite precision computation, where no 0 is exact on the subdiagonal of H, hence no exact partitionning can be done in practice.

The explicit assumption that

$$H \text{ is irreducible} \qquad (2)$$

leads to very strong assumptions on H, hence on A:

a) if A is diagonalizable, then it should have *simple* eigenvalues, and
b) if A is defective, it should be *non derogatory* (that is, it should not have more than 1 eigenvector per distinct eigenvalue).

The assumption (2) is unrealistic because it artificially excludes matrices with multiple eigenvalues which are either diagonalizable or defective and derogatory. Moreover, when one is given a matrix which is exactly reducible, a simple backward analysis shows that $A + \Delta A$ will almost always fulfill (2).

Imposing the condition (2) on H (or A) seems therefore unreasonable. It led to the widespread belief that the Arnoldi (or Lanczos) algorithm cannot compute multiple eigenvalues in finite precision. However, anyone can easily experience the contrary on a workstation.

2.2 Happy Breakdown of the Algorithm

We define $H = (h_{ij})$, H_k being the $k \times k$ upper left submatrix, and $V_k = [v_1, ..., v_k]$ is an orthonormal basis for the Krylov subspace span$\{v_1, Av_1, ..., A^{k-1}v_1\}$ for $k = 1,...n$.

Another seemingly pressing reason to assume irreducibility for H a priori is that, if $h_{k+1\,k} = 0$ at step $k < n$ (in exact arithmetic), the mathematical algorithm stops: the vector Av_k and the previous orthogonal vectors $\{v_1, ..., v_k\}$ are *linearly dependant*. The basis for the Krylov subspace defined by A and v_1 has dimension k and cannot be expanded. Mathematically, the vectors $v_{k+1},..., v_n$ are not defined. It is well known that the Arnoldi algorithm realizes recursively a QR factorization of the sequence of rectangular matrices of size $n \times (k+1)$

$$B_{k+1} = [v_1, AV_k] = V_{k+1} \begin{pmatrix} 1 & \\ & \begin{array}{c} \\ H_k \\ h_{k+1\,k} \end{array} \end{pmatrix} = V_{k+1} R_{k+1} \qquad (3)$$

for $k = 1, ..., n-1$, where the triangular factor R_{k+1} is a triangle of size $k+1$.

For $k = n$, one has the particular formula

$$B_{n+1} = [v_1, AV_n] = V_n[e_1, H_n] \tag{4}$$

The QR factorization can be implemented via *i)* a Gram-Schmidt type algorithm or *ii)* the Householder algorithm.

When $h_{k+1\,k} = 0$, what happens algorithmically in exact arithmetic? The answer depends on the orthogonalization strategy:

i) stop because of division by 0 (for classical or modified GS as well)
ii) continue: division by 0 is avoided and a vector v_{k+1} orthogonal to $v_1,..., v_k$ is computed which initiates a new Krylov subspace.

In finite precision, however, the computed value $\tilde{h}_{k+1\,k}$ is $\neq 0$ and all three above implementations involve a division by a small quantity, i.e. the computed value of $h_{k+1\,k} = \|Av_k - \sum_{i=1}^{k} h_{ik}v_i\|$ ($= 0$ in exact arithmetic).

The event $h_{k+1\,k} = 0$ is a *singularity* for the algorithm: it signals a rank deficiency in the Krylov basis for A initiated with v_1 (dimension $k < n$). The algorithmic computation can be ill-conditioned in the neighborhood of the singularity because of the possible division by a small quantity. It is interesting to remark that Householder copes with the singularity only in exact arithmetic. The interested reader is referred to [4] where the sensitivity of the Arnoldi algorithm to the starting vector v_1 is studied by means of condition numbers for $h_{k+1\,k}$ and v_{k+1}.

The singularity $h_{k+1\,k} = 0$ has another feature: in exact arithmetic, its occurence allows to obtain at step k an *exact* solution for $Ax = b$ and a subset of k *exact* eigenelements for $Ax = \lambda x$, $x \neq 0$. This is why the event $h_{k+1\,k} = 0$ has been dubbed "happy breakdown" by software developpers [7] [8].

It is clear that the singular event $h_{k+1\,k} = 0$ occurs for $k \leq n$. The usual assumption (2) forces the event to occur *as late as possible*, that is for $k = n$. However, in software practice, as we shall see in the next section, one wants the event to occur *as soon as possible*, for k as small as possible. This fact explains why the (often implicit) assumption (2) is most unfortunate, since it forbids the occurence of the most wanted event $h_{k+1\,k} = 0$ for k very small with respect to n. Therefore, it fails to provide the appropriate conceptual framework to study the "convergence" of practical Krylov methods, which is the topic of the next section.

Remark 1. The particular factorization of $B_{n+1} = [v_1, AV_n]$, given in (4), shows that the final step $k = n$ of the Arnoldi algorithm can be interpreted as the *singular event* $h_{n+1\,n} = 0$. This is a consequence of the fact that A is a matrix of *finite* order n. If, more generally, we think of A as an operator in a functional space, then the above algorithmic process would continue endlessly, as long as $h_{k+1\,k} \neq 0$ [1].

Consequently, in the matrix case, there is always, in exact arithmetic, at least one singular event for $k = n$, and maybe one or several additional ones for $k < n$. The main difference between these two kinds of singularities is that the first is known to occur exactly for $k = n$, whereas for the second, one does not know in advance whether and when it may occur.

This is this lack of information which accounts for the fact that the complete Arnoldi-Householder algorithm can be ill-conditioned if a singular event occurs before $k = n$ (that is, if an *early* happy breakdown occurs).

3 Iterative or Restarted Version of the Incomplete Arnoldi Algorithm

For very large matrices, the Arnoldi algorithm is not allowed to run until completion, that is until $k = n$, mainly for practical reasons of cost. A maximal size $m \ll n$ for the Krylov subspace is imposed, either fixed a priori or determined dynamically. The exact information for $k = n$ can therefore never become available.

To compensate for that limitation, one uses the incomplete algorithm *iteratively*: it is restarted for another set of m steps with a *new* starting vector v_1, which is carefully computed from the information available from H_m at the previous iteration. The way the new $v_1^{(i)}$ is computed is with an "early happy breakdown" in mind.

The idea is to enrich the starting vector $v_1^{(i)}$ at each new iteration with information about the desired solution (solve $Ax = b$ or $Ax = \lambda x$) which has been computed during the $(i-1)^{\text{th}}$ incomplete Arnoldi iteration, $i = 1, 2, ...$

Example 1. Suppose that the r (simple) eigenvalues μ_j of A which lie in a given region of $\mathbb{C}$ are wanted. Suppose that $\hat{v}_1$ belongs to the invariant subspace of A associated with the μ_j, then the Krylov subspace generated by A and $\hat{v}_1$ has dimension $r \ll n$. Starting from any such $\hat{v}_1$, there is a happy breakdown for $k = r$ at most. In practice, one does not know such a $\hat{v}_1$. One aims at computing a sequence of starting vectors $v_1^{(i)}$, $i = 1, 2, ...$, which will progressively converge towards such a $\hat{v}_1$ which contains exactly the information which is sought for.

This examplifies the rationale behind restarted versions of the Arnoldi algorithm which are known under the generic name of Krylov methods. The convergence of $v_1^{(i)}$ towards $\hat{v}_1$ as i increases is monitored by the backward error on A associated with approximate solutions computed from the current Hessenberg matrix.

The previous example makes it clear that such backward errors can be small (with respect to machine precision) only if the $v_1^{(i)}$ is close enough to a vector $\hat{v}_1$. Therefore:

> Convergence of Krylov methods are best understood in the light of an early happy breakdown.

One sees fully now why the assumption (2) of irreducibility of H in the decomposition (1) goes against the appropriate mathematical framework to analyse the convergence of Krylov methods.

4 Detection of the Singular Event $h_{k+1\,k} = 0$, $k < n$, in Finite Precision

As already indicated, the singular event $h_{k+1\,k} = 0$ cannot easily be detected in finite precision because the computed value $\tilde{h}_{k+1\,k}$ is non zero and may be too large (due to ill-conditioning) to be considered as zero. This is a serious difficulty which is one of the major keys to unlock the analysis of convergence of Krylov methods.

We start our study by going back to the mathematical meaning of a singular event.

4.1 A Krylov Basis of Dimension $k < n$

The event $h_{k+1\,k} = 0$ means that the vectors $v_1, ..., v_k$ and Av_k are linearly dependant. Therefore, the matrix R_{k+1} in the factorization (3) is singular: indeed $h_{k+1\,k}$ is its $(k+1)^{\text{th}}$ diagonal element. In order to quantify the distance to singularity of any computed R_{j+1}, $j = 2, ..., m$, one can compute

$$\frac{1}{\text{cond}_2(R_{j+1})} = \frac{\sigma_{\min}(R_{j+1})}{\sigma_{\max}(R_{j+1})} = \text{dist}(R_{j+1}, \text{singularity})$$

in the 2-norm.

4.2 The Method Error in the Incomplete Arnoldi Algorithm

At step $k = 1, ..., n-1$, the following identity holds:

$$AV_k = V_k H_k + h_{k+1\,k} v_{k+1} e_k^T, \qquad (5)$$

or equivalently:

$$(A - h_{k+1\,k} v_{k+1} v_k^*) V_k = V_k H_k. \qquad (6)$$

V_k is an orthonormal basis for the perturbed matrix $A_k{}' = A - h_{k+1\,k} E_k$, where the structure of the deviation E_k is of rank one: $E_k = v_{k+1} v_k^*$. The k eigenvalues of H_k are a subset of the spectrum of $A_k{}'$. The form of the identity (6) calls for an interpretation in terms of **homotopic perturbations** of the type $t E_k$, $t \in \mathbb{C}$ (see [6] [9]). It is easily shown that $|h_{k+1\,k}|$, which is the (*absolute*) *homotopic backward error*, can be interpreted in terms of the (*absolute*) *method error* incurred when one wishes to represent A by its projected matrix H_k. As a by-product of this analysis, *all k eigenvalues of H_k have the same relative homotopic backward error* $|h_{k+1\,k}|/\|A\|_2$.

Remark 2. If one uses the Householder orthogonalization, the scalar $h_{k+1\,k}$ is not guaranteed to be real nonnegative.

4.3 The Stopping Criterion in Software Practice

The mathematical analysis has provided two possible quantities to detect a singular event:

i) $\alpha_k = 1/\mathrm{cond}_2(R_{j+1})$ and
ii) $\beta_k = |h_{k+1\,k}|/\|A\|_2$.

Software developpers, on the other hand, control the convergence by means of a third quantity based on the Arnoldi residual, that we write here in the case of an eigenproblem $Ax = \lambda x$:

iii) $\gamma_k = |h_{k+1\,k}||y_k|/\|A\|_2\|y\|_2$,

where y can be any eigenvector of H_k and $y_k = e_k^T y$, e_k being the kth canonical vector in $\mathbb{R}^k$.

It is clear that $\gamma_k \leq \beta_k$ since

$$\gamma_k = \beta_k \frac{|y_k|}{\|y\|_2} \quad \text{and} \quad |y_k|^2 \leq \sum_{i=1}^{k} |y_i|^2.$$

To γ_k, which represents the relative norwise backward error associated with the pair $(\mu, z = V_k y)$, where $Hy = \mu y$ in $\mathbb{R}^k$, we propose to add the relative norwise backward error associated with the scalar μ only, that is

iv) $\delta_k = 1/\|A\|_2\|(A - \mu I)^{-1}\|_2$,

which express the relative distance of $A - \mu I$ to singularity. Note that $\delta_k \leq \gamma_k$.

We discuss the numerical behaviour of these four indicators in the next section. We shall see that it is useful to consider the variant $\gamma_k' = \|Az - \mu z\|_2/\|A\|_2\|z\|_2$ of γ_k which is equal to γ_k in exact arithmetic. However, once convergence has been reached, a numerical artefact takes place if the algorithm is let running: γ_k can spuriously decrease, whereas γ_k' remains of the order of machine precision, as it should [3].

5 A Numerical Illustration

We consider, as a numerical example, the matrix Rose [3] which is the companion matrix A of the polynomial

$$p(x) = (x-1)^3(x-2)^3(x-3)^3(x-4).$$

The matrix is defective non derogatory with 3 multiple defective eigenvalues of multiplicity 3 equal to their index.

The Jordan form of $A = XJX^{-1}$ (with eigenvalues in the order 1, 2, 3, 4) is known and the starting vector v_1 is chosen such that a singular even occurs for $k = 3$. The starting vector v_1 is of the form:

$$v_1 = \frac{u}{\|u\|_2}, \quad u = Xc, \quad c = (10^p, 1, 1, 0, 0, 0, 0, 0, 0, 0)^T.$$

It is clear that u belongs to the invariant subspace associated to 1. Three values for p are selected: $p = 0$, 3 and 5, yielding 3 different starting vectors such that $k = 3$. Three orthogonalization strategies are chosen to implement the Arnoldi algorithm:

1) classical Gram-Schmidt (CGS) (marked below with ○)
2) modified Gram-Schmidt (MGS) (marked below with +)
3) Householder (H) (marked below with ×)

In order to fully compare the respective behaviours of Arnoldi in exact arithmetic and in finite precision in the presence of a singular event, the algorithm is run until completion, that is $k = 10$.

We plot the curves $k \to \alpha_k, \beta_k, \gamma_k, \gamma_k{'}$ and δ_k for the three implementations. See Figures 1 to 5.

To analyse the accuracy obtained for the three computed eigenvalues μ_i which are close to 1, we look at the errors $\epsilon_k = \max_{i=1,2,3} |\mu_{ik} - 1|$.

The accuracy history is summarized by the plot $k \to \epsilon_k$. See Figure 6.

The following conclusions can be drawn from this numerical example.

1) Detection of the early happy breakdown

α_3 is not sensitive to p whereas β_3 is (10^{-13}, 10^{-10}, 10^{-9}). See Figures 1 and 2. γ_k continues to decrease after machine precision has been reached (for $k > 5$), which is not the case for $\gamma_k{'}$. This latter indicator $\gamma_k{'}$ is therefore more reliable. See Figures 3 and 4.

2) Accuracy of computation.

The backward error δ_k on μ reaches 10^{-18} at least for $k > 5$ (see Figure 5). The direct error ϵ_k (on Figure 6) is of the order of 10^{-4} which is in alignement with the holderian condition number of $\lambda = 1$ as triple eigenvalue which is equal to 374 [5]

$$374 \times (10^{-18})^{1/3} \sim 4 \times 10^{-4}.$$

The variation of the results with p indicates that the value $k = 3$ for the singular event in exact arithmetic can be seen as $k = 1$ in finite precision when p is large enough: this reflects the fact that v_1 is then close to the eigenvector of 1. This fact is clearly seen on the Figures 7 and 8 which represent the plots $p \to |h_{2,1}|/\|A\|_2$ and $p \to |h_{4,3}|/\|A\|_2$ for p ranging from -15 to 15.

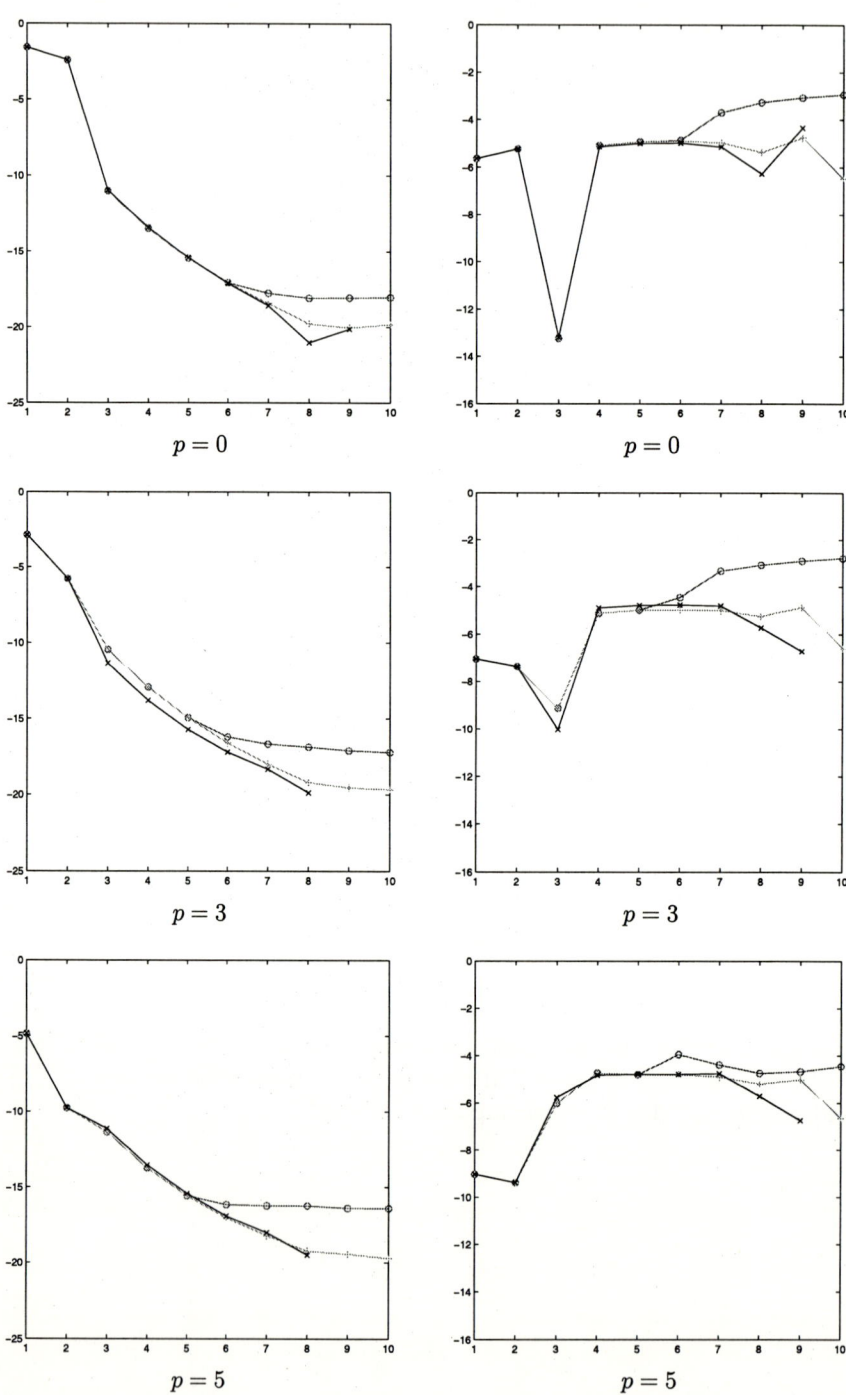

Fig. 1: α_k Fig. 2: β_k

Understanding Krylov Methods in Finite Precision

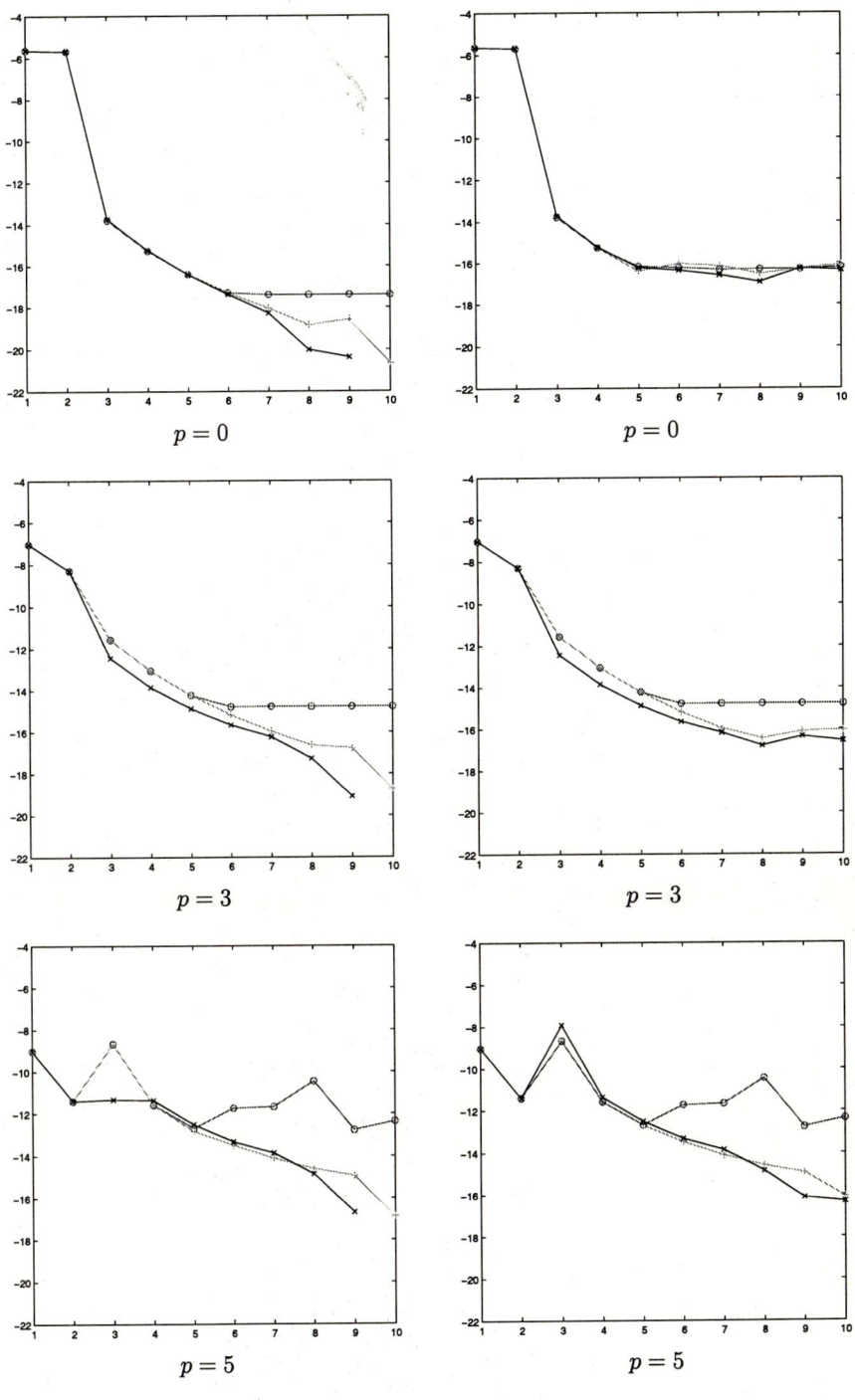

Fig. 3: γ_k **Fig. 4:** $\gamma_k{}'$

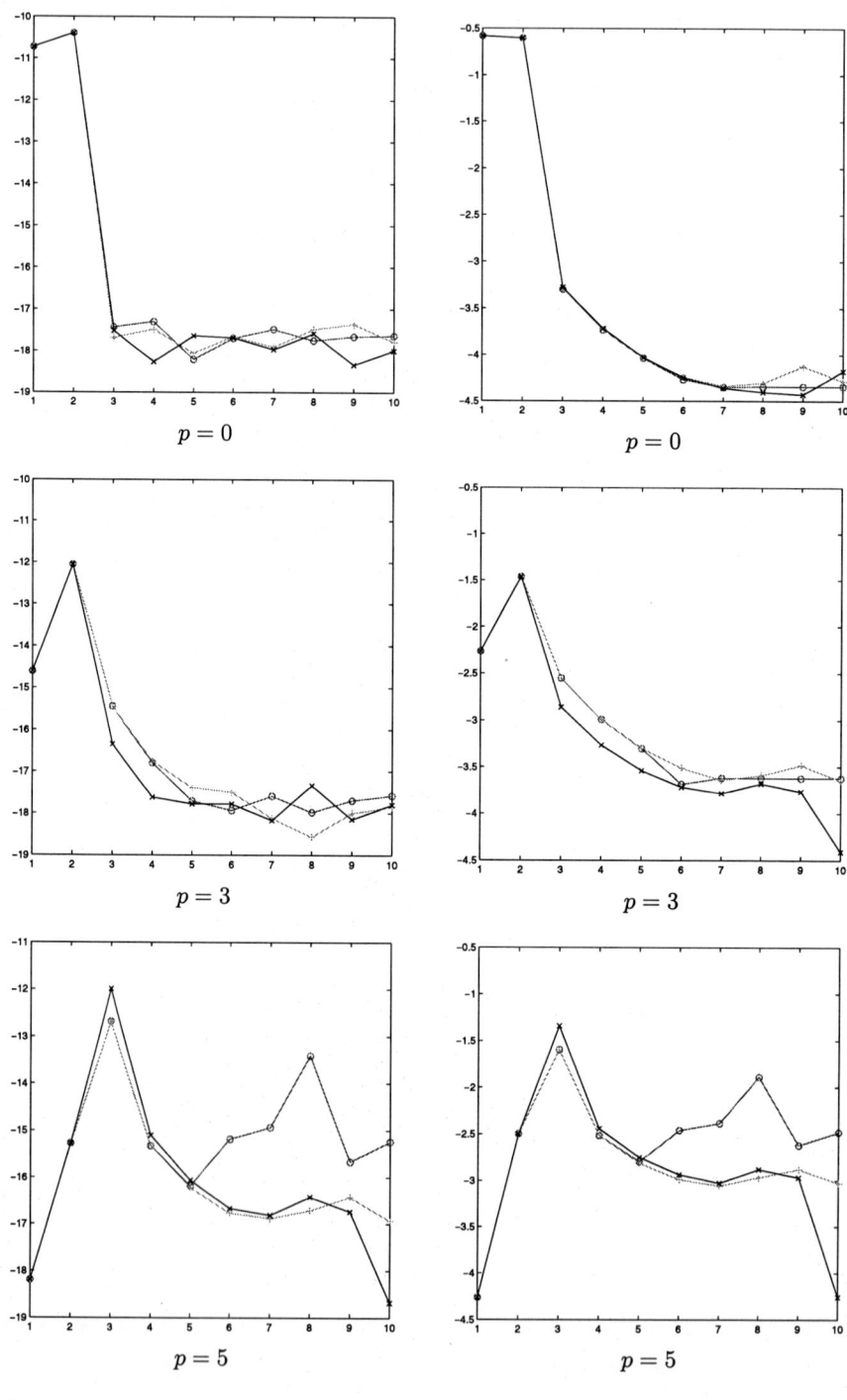

Fig. 5: δ_k **Fig. 6:** ϵ_k

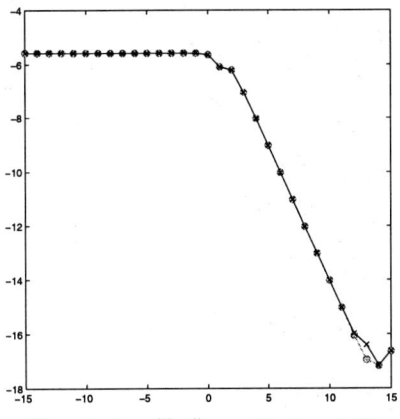
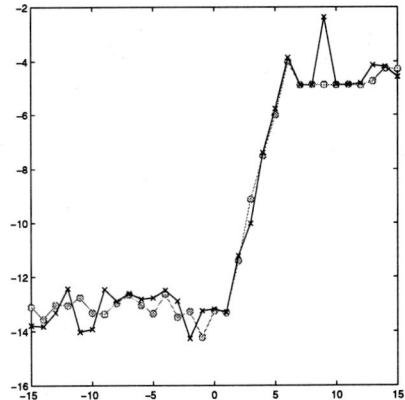

Fig. 7: $h_{2,1}/\|A\|_2$, $-15 \leq p \leq 15$
Fig. 8: $h_{4,3}/\|A\|_2$, $-15 \leq p \leq 15$

References

1. Chatelin, F.: Spectral approximation of linear operators. Academic Press (1983).
2. Chaitin-Chatelin, F.: Comprendre les méthodes de Krylov en précision finie : le programme du Groupe Qualitative Computing au CERFACS. CERFACS report TR/PA/00/11 (2000).
3. Chaitin-Chatelin, F., Frayssé, V.: Lectures on finite precision computations. SIAM (1996).
4. Chaitin-Chatelin, F., Gratton, S., Traviesas, E.: Sensibilité des méthodes de Krylov au vecteur de départ. Work in progress.
5. Chaitin-Chatelin, F., Harrabi, A., Ilahi, A.: About Hölder condition number and the stratification diagram for defective eigenvalues. To appear in IMACS J. on Math. of Comp. CERFACS report TR/PA/99/19 (1999).
6. Chaitin-Chatelin, F., Toumazou, V., Traviesas, E.: Accuracy assessment for eigencomputations: variety of backward errors and pseudospectra. Lin. Alg Appl. **309** (2000) 73–83.
7. Saad, Y.: Numerical methods for large eigenvalue problems. Algorithms and architectures for advanced scientific computing. Manchester University Press (1992).
8. Saad, Y.: Iterative methods for sparse linear systems. PWS, Minnesota (1995).
9. Traviesas, E.: Sur le déploiement du champ spectral d'une matrice, Ph.D. thesis, University Toulouse I and CERFACS (2000).

A Rational Interpolation Approach to Least Squares Estimation for Band-TARs[*]

Jerry Coakley[1], Ana-María Fuertes[2], and María-Teresa Pérez[3]

[1] Birkbeck College, Dept. of Economics
7-5 Gresse Street, London W1P 2LL, UK, jcoakley@econ.bbk.ac.uk
[2] London Guildhall University, Dept. of Economics
84 Moorgate, London EC2M 6SQ, UK, afuertes@lgu.ac.uk
[3] Universidad de Valladolid, Dept. de Matemática Aplicada a la Ingeniería
Paseo del Cauce s/n, 47011 Valladolid, Spain
terper@wmatem.eis.uva.es

Abstract. This paper shows that the residual sum of squares of Band-TAR models is a rational function of degree (4,2) of the threshold parameter. Building on this result a novel fitting approach is proposed which permits a continuous threshold space and employs QR factorizations and Givens updating. Its efficiency gains over a standard grid search are illustrated by Monte Carlo analysis.

1 Introduction

Threshold autoregressive (TAR) models (Tong 1983) have been widely applied in recent years to capture the nonlinear behavior of economic and financial variables. An m-regime TAR can be written as

$$z_t = \sum_{j=1}^{m} (\phi_0^j + \phi_1^j z_{t-1} + \ldots + \phi_{p_j}^j z_{t-p_j}) I_t(\theta^{j-1} \leq v_{t-d} < \theta^j) + \varepsilon_t, \quad (1)$$

for $t = 1, \ldots, N$, where $\varepsilon_t \sim \text{iid}(0, \sigma^2)$, $I_t(\cdot)$ is the indicator function, $-\infty = \theta^0 < \theta^1 < \ldots < \theta^m = \infty$ are threshold parameters, and p_j and d are autoregressive (AR) and threshold lag order, respectively. This is a nonlinear model in time but piecewise linear in the threshold space. It partitions the one-dimensional Euclidean space into m regimes, each of which is defined by an AR model, depending on the values taken by the threshold or switching variable, v_{t-d}.

A particular case of (1) is the following Band-TAR model

$$\Delta z_t = A(t,\theta)^- I_t(v_{t-d} < -\theta) + B(t) I_t(|v_{t-d}| \leq \theta) + A(t,\theta)^+ I_t(v_{t-d} > \theta) + \varepsilon_t, \quad (2)$$

---

[*] We are grateful to Erricos Kontoghiorghes and an anonymous referee for helpful comments.

with

$$A(t,\theta)^- = \alpha_1(z_{t-1} + \theta) + \alpha_2(z_{t-2} + \theta) + ... + \alpha_p(z_{t-p} + \theta),$$
$$A(t,\theta)^+ = \alpha_1(z_{t-1} - \theta) + \alpha_2(z_{t-2} - \theta) + ... + \alpha_p(z_{t-p} - \theta),$$
$$B(t) = \beta_0 + \beta_1 z_{t-1} + \beta_2 z_{t-2} + ... + \beta_q z_{t-q},$$

where $\theta > 0$ is an identifying restriction. Under particular stationarity conditions, (2) characterizes a process z_t that converges to the boundaries of the inner band which act as attractors. More specifically, the process mean-reverts to θ when $v_{t-d} > \theta$ and to $-\theta$ for $v_{t-d} < -\theta$. It generalizes the Band-TAR with $p = q = 1$ introduced by Balke and Fomby (1997) which is self-exciting ($v_{t-d} = z_{t-d}$) and assumes a random-walk inner band ($\beta_0 = \beta_1 = 0$).

Band-TARs have been applied to capture asymmetries, limit cycles and jump phenomena in the behavior of financial and economic variables (Coakley and Fuertes 1997; Obstfeld and Taylor 1997). The fitting approach proposed in this paper can be easily adapted to a number of related specifications such as the continuous (C-) TAR model of Chan and Tsay (1998):

$$\Delta z_t = \phi_1^1(z_{t-1} - \theta)I_t + \phi_1^2(z_{t-1} - \theta)(1 - I_t) + \sum_{j=1}^{p}\gamma_j \Delta z_{t-j} + \varepsilon_t,$$
$$I_t = \begin{cases} 1 \text{ if } v_{t-1} \geq 0 \\ 0 \text{ otherwise} \end{cases} \quad (3)$$

where $v_{t-1} = z_{t-1} - \theta$. Under the stationarity condition $-2 < (\phi_1^1, \phi_1^2) < 0$, the latter represents a process with differential adjustment towards the attractor θ depending on the sign of the past deviation.

Our goal is to fit model (2) to the observed time series $\{z_t\}_{t=1}^{N}$ and $\{v_t\}_{t=1}^{N}$. Ordinary least squares (LS) or, equivalently, conditional maximum likelihood (ML) under Gaussian innovations, lead to the minimization of the following residual sum of squares (RSS) function

$$\text{RSS}(\phi) = \sum_{t}^{n}(\Delta z_t - A(t,\theta)^-)^2 I_t(v_{t-d} < -\theta) + \sum_{t}^{n}(\Delta z_t - B(t))^2 I_t(|v_{t-d}| \leq \theta)$$
$$+ \sum_{t}^{n}(\Delta z_t - A(t,\theta)^+)^2 I_t(v_{t-d} > \theta)$$

with respect to $\phi = (\theta, \alpha', \beta', d, p, q)'$, where $\alpha = (\alpha_1, ..., \alpha_p)'$ and $\beta = (\beta_0, ..., \beta_q)'$ are the outer- and inner-band AR parameters, respectively, and $n = N - \max(d, p, q)$ the effective sample size.

Let us assume initially that (d, p, q) are known. Our goal is to estimate $(\theta, \alpha', \beta')$. The above RSS function is discontinuous in θ implying that standard gradient-based algorithms cannot be applied. If the threshold space Θ is small, a simple grid search (GS) can be effectively used to find the value $\hat{\theta} \in \Theta$ that minimizes the RSS (or some LS-related criterion) or maximizes the log-likelihood function. The estimates of the parameters α and β can be computed by standard LS conditional on $\hat{\theta}$.

The threshold space is the continuous region $\Theta \subseteq \mathbb{R}^+$. However, in practice the GS is applied to a feasible (discrete) range in Θ by fixing a number of threshold candidates which are usually the sample percentiles (or order statistics) of v_{t-d}, that is $\xi_{(t)} = \{v_{(\tau_0)} \leq \ldots \leq v_{(i)} \leq \ldots \leq v_{(\tau_1)}\} \subset \Theta$ where $v_{(\tau_0)}$ and $v_{(\tau_1)}$ are some bounds required to guarantee that each regime contains a minimum number of observations for the submodels to be estimable. However, since in principle any point in the continuous threshold space could maximize the log-likelihood, a full or detailed GS using $\xi_\lambda = \cup_{i=\tau_0}^{\tau_1-1} \{v_{(i)} < \theta_i^j < v_{(i+1)} : \theta_i^{j+1} = \theta_i^j + \lambda, j = 1, 2, \ldots\} \cup \xi_{(t)} \subset \Theta$ where λ is a step size, is preferable to a GS restricted to $\xi_{(t)}$. While a potential pitfall of the latter is that it may yield rather imprecise parameter estimates for small N, a practical problem with ξ_λ is that it may prove computationally very expensive for small step size λ when the data are widely dispersed. This calls for an estimation method capable of handling a continuous threshold range while keeping costs within tractable limits. The numerical algorithm proposed in this paper is in this spirit.

The organization of the paper is as follows. In §2 the Band-TAR is stated in arranged form to facilitate efficient estimation. The main results are given in §3. The proposed fitting approach is outlined in §4 and its efficiency gains are explored via small Monte Carlo simulation. A final section concludes.

2 Arranged Autoregression

For $p = q = L$, where L is a fixed upper bound, the observed data $\{z_t\}_{t=1}^N$ can be represented in AR form as $\boldsymbol{y} = f(\boldsymbol{X}) + \boldsymbol{\varepsilon}$, $\boldsymbol{X} = (\boldsymbol{x}_1, \boldsymbol{x}_2, \ldots, \boldsymbol{x}_L)'$, where $\boldsymbol{y}$ and $\boldsymbol{x}_j, j = 1, \ldots, L$ are n-vectors containing the sample observations of the variables Δz_t and z_{t-j}, respectively, $\boldsymbol{\varepsilon}$ is a disturbance n-vector and $n = N - \max(d, L)$ is the effective sample size. This formulation can easily be transformed into a change-point (or Band-TAR) problem by rearranging its cases (rows) according to v_{t-d}, yielding $\boldsymbol{y}^v = f(\boldsymbol{X}^v) + \boldsymbol{\varepsilon}^v$.

Let $\theta = \theta_k$ ($\theta_k > 0$) be a plausible threshold value such that two indices, k_1 and k_2 ($k_1 < k_2$), are associated with it satisfying $v_{(i)} < -\theta_k$ for $i = 1, 2, \ldots, k_1$, $v_{(i)} \geq \theta_k$ for $i = k_2, k_2 + 1, \ldots, n$, and $-\theta_k \leq v_{(i)} < \theta_k$ for $i = k_1 + 1, \ldots, k_2 - 1$. Using the above ordered-form notation the $s = k_2 - k_1 - 1$ cases classified into the *inner* regime of (2) can be written as $\Delta \boldsymbol{z}_s = Z_s^\beta \beta + \boldsymbol{\varepsilon}_s$ where:

$$Z_s^\beta = \begin{pmatrix} 1 & x_{k_1+1,1}^v & x_{k_1+1,2}^v & \cdots & x_{k_1+1,L}^v \\ 1 & x_{k_1+2,1}^v & x_{k_1+2,2}^v & \cdots & x_{k_1+2,L}^v \\ \vdots & \vdots & \vdots & & \vdots \\ 1 & x_{k_2-1,1}^v & x_{k_2-1,2}^v & \cdots & x_{k_2-1,L}^v \end{pmatrix}, \qquad (4)$$

$\Delta \boldsymbol{z}_s = (y_{k_1+1}^v, y_{k_1+2}^v, \ldots, y_{k_2-1}^v)'$ and $\boldsymbol{\varepsilon}_s = (\varepsilon_{k_1+1}^v, \varepsilon_{k_1+2}^v, \ldots, \varepsilon_{k_2-1}^v)'$. Likewise the $r = n - (k_2 - k_1 - 1)$ cases in the *outer* regime can be written as $\Delta \boldsymbol{z}_r =$

$Z_r^\alpha(\theta_k)\alpha + \varepsilon_r$ where:

$$Z_r^\alpha(\theta_k) = \begin{pmatrix} x_{11}^v + \theta_k & x_{12}^v + \theta_k & \cdots & x_{1L}^v + \theta_k \\ \vdots & \vdots & & \vdots \\ x_{k_11}^v + \theta_k & x_{k_12}^v + \theta_k & \cdots & x_{k_1L}^v + \theta_k \\ x_{k_21}^v - \theta_k & x_{k_22}^v - \theta_k & \cdots & x_{k_2L}^v - \theta_k \\ \vdots & \vdots & & \vdots \\ x_{n1}^v - \theta_k & x_{n2}^v - \theta_k & \cdots & x_{nL}^v - \theta_k \end{pmatrix}, \quad (5)$$

$\Delta z_r = (y_1^v, \ldots, y_{k_1}^v, y_{k_2}^v, \ldots, y_n^v)'$ and $\varepsilon_r = (\varepsilon_1^v, \ldots, \varepsilon_{k_1}^v, \varepsilon_{k_2}^v, \ldots, \varepsilon_n^v)'$. Note that the upper $k_1 \times L$ and lower $(n-k_2+1) \times L$ partition matrices of $Z_r^\alpha(\theta_k)$ correspond to the $A(t,\theta_k)^+$ and $A(t,\theta_k)^-$ outer AR schemes of (2), respectively.

It follows that any new threshold, $\theta \neq \theta_k$, changes the entries of the outer-regime regressor matrix, $Z_r^\alpha(\theta_k)$. In addition, some specific thresholds change the size (number of rows) of $Z_r^\alpha(\theta_k)$ and also of the inner-regime regressor matrix Z_s^β via the addition/deletion of cases. These specific threshold values are the order statistics of v_{t-d}, that is $\theta \in \xi_{(t)}$, which determine a countable number of continuous nonoverlapping intervals $[\theta_i, \theta_{i+1})$, where θ_i and θ_{i+1} denote consecutive order statistics. The latter define the threshold space, $\Theta = \{\cup_{i=\tau_0}^{\tau_1-1}[\theta_i, \theta_{i+1})\} \subset \mathbb{R}^+$. For $\theta \in [\theta_i, \theta_{i+1})$ matrix (5) can be rewritten as $Z_r^\alpha(\theta) = Z0_r + U_r^\theta$ where:

$$Z0_r = \begin{pmatrix} x_{11}^v & x_{12}^v & \cdots & x_{1L}^v \\ \vdots & \vdots & & \vdots \\ x_{k_11}^v & x_{k_12}^v & \cdots & x_{k_1L}^v \\ x_{k_21}^v & x_{k_22}^v & \cdots & x_{k_2L}^v \\ \vdots & \vdots & & \vdots \\ x_{n1}^v & x_{n2}^v & \cdots & x_{nL}^v \end{pmatrix}, \quad (6)$$

and $U_r^\theta = u_r u_\theta' = (1, \ldots, 1, -1, \ldots, -1)'(\theta, \ldots, \theta)$ is a rank-one matrix with u_r an r-vector whose first k_1 components are all 1 and the remaining $(n - k_2 + 1)$ components are all -1, and u_θ is an L-vector. Thus estimation of α entails a regressor matrix which depends on an unknown threshold parameter.

3 Parameter Dependent Least Squares Problem

Consider the linear regression model $y = X(\theta)\gamma + \epsilon$ and associated LS problem

$$\min_\gamma \| X(\theta)\gamma - y \|, \quad (7)$$

where y and γ are the $n \times 1$ and $m \times 1$ regressand and parameter vector, respectively, and $X(\theta)$ is a full-column rank $n \times m$ ($n \geq m$) regressor matrix which depends explicitly on a parameter θ. The solution of (7) can be written in terms of the Moore-Penrose inverse of X (Björck 1996) as

$$\hat\gamma^\theta = X(\theta)^\perp y. \quad (8)$$

Analogously, the RSS can be expressed as

$$\| \hat{e}^{\theta} \|_2^2 = y'(I - X(\theta)X(\theta)^{\perp})y, \tag{9}$$

where I is the $n \times n$ identity matrix. Since $X(\theta)$ is full-column rank, then $X(\theta)'X(\theta)$ is nonsingular and (9) can be calculated by

$$\| \hat{e}^{\theta} \|_2^2 = y'(I - X(\theta)(X(\theta)'X(\theta))^{-1}X(\theta)')y. \tag{10}$$

Suppose that $X(\theta)$ is a polynomial matrix, that is, its entries are polynomials in θ. We are interested in the case $X(\theta) = X_0 + X_1\theta$, where X_0 and X_1 are constant matrices (independent of θ) and rank$(X_1) = 1$, to which (5) belongs.

The following theorem applies to polynomial matrices.

Theorem 1. *Given an $n \times n$ polynomial matrix of degree r, $A(\theta) = A_0 + A_1\theta \ldots + A_r\theta^r$, where A_i, $i = 1, \ldots, r$ are rank-one matrices, then $\det A(\theta)$ is a polynomial of degree $r(r+1)/2$ if $n \geq r$ or $nr - n(n-1)/2$ if $n < r$.*

Following the proof of Theorem 1 in Coakley et al.([2000]), analogous results can be stated when some of the matrices A_i have rank different from one. We are interested in the case where $r = 2$ and the matrix A_1 is obtained as $\tilde{A}_1 + \tilde{A}_1'$ with $\tilde{A}_1$ a rank-one matrix. Then $\det A(\theta)$ is a degree-four polynomial and the next formula follows directly from the proof of Theorem 1

$$\det A(\theta) = \det A_0 + \theta(\sum_{i=1}^{n} \det A^i_{0(1)}) + \theta^2(\sum_{i=1}^{n} \det A^i_{0(2)} + \sum_{i,j=1}^{n} \det A^{i,j}_{0(1,1)})$$

$$+ \theta^3(\sum_{i,j=1}^{n} \det A^{i,j}_{0(1,2)}) + \theta^4(\sum_{i,j,k=1}^{n} \det A^{i,j,k}_{0(1,1,2)}), \tag{11}$$

where indexes i, j and k in the same sum are always different. Letting a_i^s denote the transpose of the s row of A_i then

$$A^i_{0(1)} = (a_1^0, \ldots, a_{i-1}^0, a_i^1, a_{i+1}^0, \ldots, a_n^0)',$$
$$A^i_{0(2)} = (a_1^0, \ldots, a_{i-1}^0, a_i^2, a_{i+1}^0, \ldots, a_n^0)',$$
$$A^{i,j}_{0(1,2)} = (a_1^0, \ldots, a_{i-1}^0, a_i^1, a_{i+1}^0, \ldots, a_{j-1}^0, a_j^2, a_{j+1}^0, \ldots, a_n^0)',$$

The matrices $A^{i,j}_{0(1,1)}$ and $A^{i,j,k}_{0(1,1,2)}$ are analogously defined. The following corollary particularizes equation (11) for the Band-TAR estimation problem.

Corollary 1. *If an $n \times n$ polynomial matrix of degree 2 is obtained as $A(\theta) = (B + C\theta)'(B + C\theta)$, where C is a rank-one matrix whose rows are vectors of the form $(1, \ldots, 1)$ or $(-1, \ldots, -1)$ then $\det A(\theta)$ is a polynomial of degree 2.*

Proof. For this particular polynomial matrix $A(\theta)$ the matrices A_1 and A_2 are

$$A_1 = \begin{pmatrix} d_1 + d_1 & d_1 + d_2 & \ldots & d_1 + d_n \\ d_2 + d_1 & d_2 + d_2 & \ldots & d_2 + d_n \\ \vdots & \vdots & & \vdots \\ d_n + d_1 & d_n + d_2 & \ldots & d_n + d_n \end{pmatrix}, \quad A_2 = \begin{pmatrix} n & \ldots & n \\ \vdots & & \vdots \\ n & \ldots & n \end{pmatrix}, \tag{12}$$

where $d_i = \sum_{j=1}^n (\delta_j b_{ji})$ and δ_j is $+1$ if the jth row of C is $(1,..,1)$ and -1 otherwise. The coefficient of θ^3 in (11) vanishes since for the above matrices

$$\det A^{i,j}_{0(1,2)} = -\det A^{j,i}_{0(1,2)}. \tag{13}$$

To show this, it suffices to notice that

$$\det A^{i,j}_{0(1,2)} = \det(a^0_1, \ldots, d_i \mathbf{1} + \mathbf{d}, \ldots, n\mathbf{1}, \ldots, a^0_n)'$$
$$= n d_i \det(a^0_1, \ldots, \mathbf{1}, \ldots, \mathbf{1}, \ldots, a^0_n)' + n \det(a^0_1, \ldots, \mathbf{d}, \ldots, \mathbf{1}, \ldots, a^0_n)',$$

where $\mathbf{d}$ and $\mathbf{1}$ denote the vectors $(d_1, \ldots, d_n)'$ and $(1, \ldots, 1)'$, respectively. Analogously, the coefficient of θ^4 is proven to be zero. □

We now state the main result of the paper which is a direct consequence of (10) and Corollary 1.

Theorem 2. *If the $n \times m$ ($n \geq m$) regressor matrix $X(\theta)$ in (7) is a polynomial matrix of degree 1, $X(\theta) = X_0 + X_1 \theta$, with X_1 of rank one and whose rows are $(1, \ldots, 1)$ or $(-1, \ldots, -1)$, then the residual sum of squares $\| \hat{e}^\theta \|_2^2$ is a rational function of degree (4,2) provided $X(\theta)$ is a full-column rank matrix.*

Next section outlines an estimation approach for Band-TARs which builds on these results. A more detailed description can be found in Coakley et al. (2000).

4 The Fitting Algorithm and Simulation Analysis

For each plausible threshold lag $d \in \{1, 2, ..., D\}$ the algorithm iterates as follows. For the *outer* regime:

1. Calculate the QR factorization of $(Z0^\alpha_r | \Delta z_r)$ for the initial threshold interval $[\theta_{\tau_1-1}, \theta_{\tau_1}) \subset \Theta$, where θ_{τ_1-1} and θ_{τ_1} represent (extreme) consecutive order statistics of the observed variable.
2. For each threshold interval $[\theta_i, \theta_{i+1})$ repeat for different $p \in \{1, 2, ..., L\}$:
   (a) Generate seven (for instance, equally spaced) values for the threshold in the current interval, θ_i^j $j = 1, \ldots, 7$.
   (b) Calculate the R factor of the matrix $(Z^\alpha_r(\theta_i^j) | \Delta z_r)$ for $j = 1, \ldots, 7$, by means of a rank-one correction update of the decomposition of $(Z0^\alpha_r | \Delta z_r)$. Use these R_j factors to calculate the $RSS_r(\theta_i^j)$.
   (c) Via rational interpolation on the points $(\theta_i^j, RSS_r(\theta_i^j))$ identify the $RSS(\theta)$ function associated with the current interval.
   (d) Minimize $RSS(\theta)$ over the current interval to obtain a (locally) optimal threshold, θ_i^*, compute the Akaike Information Criterion value at θ_i^*, AIC_α^*, and move to the next interval.

Table 1. Monte Carlo simulation results

				FGS approach				
DGP	σ_ε^2	$\sigma_{\hat\beta}^2$	B_μ	t_μ(mins.)	$RMSE$	B_τ	t_τ(mins.)	MAD
I	.2	.00049	.00077	**.1893**	.02210	-.00021	**.1873**	.01080
I	.4	.00108	.00109	**.4703**	.03278	-.00105	**.4624**	.01520
I	.9	.28120	.25780	**.6059**	.58919	.05593	**.6008**	.08144
II	.2	.03915	-.12880	**.3106**	.23592	-.04238	**.3088**	.05681
II	.4	.03030	-.03651	**.3685**	.19472	-.00064	**.3688**	.04549
II	.9	.04603	-.00773	**.4787**	.21448	-.01143	**.4767**	.05722
III	.2	.05104	.10290	**.5210**	.24805	.06048	**.5208**	.06048
III	.4	.30790	.31850	**.6430**	.63937	.13290	**.6326**	.13290
III	.9	1.1917	.98270	**.8470**	1.4680	.31150	**.8465**	.31530
				RF approach				
I	.2	.00038	-.00028	**.2425**	.01949	.00013	**.2335**	.01023
I	.4	.00108	.00060	**.2332**	.02940	-.00105	**.2341**	.01440
I	.9	.26110	.24211	**.2319**	.56500	.04996	**.2325**	.07308
II	.2	.03862	-.12870	**.2351**	.23480	-.04672	**.2365**	.06494
II	.4	.03635	-.04014	**.2357**	.17770	-.00126	**.2368**	.04246
II	.9	.04400	.00578	**.2359**	.20891	.00623	**.2370**	.05006
III	.2	.03246	.09917	**.2658**	.20280	.06753	**.2666**	.06753
III	.4	.21313	.30622	**.2659**	.54164	.13130	**.2667**	.13130
III	.9	1.1013	.91536	**.2664**	1.3917	.28592	**.2670**	.28601

The *inner* regime iterations are simpler since (4) does not depend on θ. The best fit Band-TAR parameters are those which minimize an overall Akaike, $AIC_\alpha^* + AIC_\beta$, over all threshold intervals in Θ.

The above algorithm and subsequent Monte Carlo experiments are programmed in GAUSS 3.26, a high-level matrix programming language with built-in statistical and econometric functions, and run in a 500MHz Pentium III. Three data generating processes (DGPs) are used to create N_0+100 observations, where the initial $N_0 = 200$ observations are discarded to minimize the effects of initialization (set at zero). The DGPs used are the following particularizations of the self-exciting ($v_{t-d} = z_{t-d}$) Band-TAR model (2):

I) $q = 2, p = 2, d = 1, \theta = 0.35, \beta' = \{0.5, -0.55, -0.75\}$ and $\alpha' = \{-0.8, -0.75\}$
II) $q = 1, p = 3, d = 2, \theta = 0.92, \beta' = \{0.4, -1.0\}$ and $\alpha' = \{-0.5, -0.73, -0.35\}$
III) $q = 3, p = 5, d = 1, \theta = 0.18, \beta' = \{-0.95, -1.65, 0.8, 0.45\}$ and $\alpha' = \{-1.8, 0.35, 0.4, -0.6, -0.75\}$

Three different error terms $\epsilon_t \sim$ iid $N(0, \sigma_\varepsilon^2)$, $\sigma_\varepsilon^2 = \{0.2, 0.4, 0.9\}$ are employed which combined with the above DGPs imply 9 different specifications. To focus on the *ceteris paribus* effect of the continuous RSS rational function component of our fitting approach (RF hereafter) we take as benchmark a fast GS method (FGS) with $\lambda = .001$, which also uses QR factorizations and Givens updates. The summary statistics used in the analysis are the sample variance

($\sigma_{\hat{\theta}}^2$), mean bias (B_μ), root mean squared error (RMSE), median of bias (B_τ), mean absolute deviation (MAD), mean computation time (t_μ) and median of computation time (t_τ). Table 1 reports the results based on $M = 500$ replications. A comparison of bias measures across methods reveals that, despite the small λ used, FGS generally yields threshold estimates more biased than those from RF. For instance, the $RMSE$ and MAD from RF are smaller than those from FGS in 9 and 7 cases, respectively, out of the 9 specifications explored.

The discontinuity imposed by the Heaviside function requires solving a number of LS problems sequentially to identify and estimate the Band-TAR model. While computation costs may not be an issue in *ad hoc* TAR fitting to a single time series, these are germane in inference analysis using simulation techniques. The growing evidence of nonlinear behaviour and in particular of regime-switching dynamics in economic time series has fostered the development of new tests — which can be viewed as extensions of existing linear tests — in a TAR framework. Exploring the small sample properties of these tests by Monte Carlo or bootstrap methods and/or estimating response surfaces with a sensible number of replications can quickly become intractable if the computation costs of TAR fitting are disregarded.

The latter underlines the importance of using efficient numerical tools in TAR fitting such as the QR approach and Givens rotations. Table 1 provides *prima facie* evidence of how these tools speed up Band-TAR fitting (Coakley *et al.* 2000). More interestingly perhaps, while the computation costs of the FGS method — as measured by t_μ and t_τ — increase with the innovation volatility (noise), these are invariant to the latter in the RF method and depend only on the sample size. This difference is likely to be relevant when fitting Band-TARs to highly volatile data such as those associated with financial variables.

5 Conclusions

This paper shows that the RSS of Band-TAR models is a continuous rational function of the threshold. Using this result we propose a novel fitting approach, which allows for a continuous range for the threshold while keeping computation costs within tractable limits. It uses standard minimization techniques and employs QR factorizations and Givens updates. Its efficiency gains over a fast grid search are illustrated via Monte Carlo experiments. As computation time is highly dependent on the rational interpolation algorithm used, we leave improvement of the latter for future research.

References

1997. Balke, N. S., Fomby, T. B.: Threshold Cointegration. International Economic Review, **38** (1997) 627-45.
1996. Björck, A.: Numerical Methods for Least Squares Problems. SIAM, Philadelphia (1996).

1998. Chan, K. S., Tsay, R. S.: Limiting properties of the Least Squares Estimator of a Continuous Threshold Autoregressive Model. Biometrika, **85** (1998) 413-26.
1997. Coakley, J., Fuertes, A. M.: Border Costs and Real Exchange Rate Dynamics in Europe. Journal of Policy Modelling, Forthcoming.
2000. Coakley, J., Fuertes, A. M., Pérez, M. T.: An Investigation of Numerical Issues in Threshold Autoregressive Modelling for Time Series. Birkbeck College Discussion Paper, University of London, Forthcoming.
1997. Obstfeld, M., Taylor, A. M. Taylor: Nonlinear Aspects of Goods-market Arbitrage and Adjustment. J. of the Japanese and International Economies, **11** (1997) 441-79.
1983. Tong, H.: Threshold Models in Non-linear Time Series Analysis, Springer-Verlag, Berlin (1983).

Uniqueness of Solution of the Inverse Electroencephalographic Problem*

Andrés Fraguela Collar**, José J. Oliveros Oliveros, and
Alexandre Ivánovich Grebénnikov

Facultad de Ciencias Físico Matemáticas
Av. San Claudio y 18 sur, 72570 Puebla, Pue., México
{fraguela,oliveros,agrebe}@fcfm.buap.mx

Abstract. A model with the brain and the other shell of the head like conductors medium with differents conductivities has been used for to estudy the inverse electroencephalographic problem. Technics of the potential theory has been used for to transform the model in a operational problem which under some conditions gives the uniqueness of recuperation of cortical neurons aggregate (sources) in the cerebral cortex from measurement of the potential in the scalp.

1 Introducción

The electroencephalography method is the more famous between the nondestructive methods of investigation of the brain and is based in the record of its electric activity. The scalp EEG is a valuable clinical tool. Furthermore, the evoked potential measured on the scalp shows promise in the diagnosis and treatment of central nervous system diseases ([10] pp. 5). The potential field produced by this electric activity open great posibilities of investigation ([10],[13]) that induce statement of inverse electroencephalographic problems (IEP) ([9],[10],[11]). Different statement of IEP can be consult in ([7],[10],[13]).

So the IEP consist, in outline, in to determinate, from measurement of potencial on the scalp of sources in the cerebral cortex. The IEP, lies in the collection of problems called **ill posed**.

2 Model of Conducting Medium for the Electroencephalographic Activity

We suppose, that the human head, considered as conductor medium, is divided in five disjoint zones as shown the figure 1, namely:

1. Ω_1 – Brain
2. Ω_2 – Muscles
5. Ω_5 – Scalp
3. Ω_3 – Intracraneal liquid
4. Ω_4 – Skull

---

* Parcially supported by CONACYT, proyect 28451A.
** Sabbatical year IIMAS-UNAM.

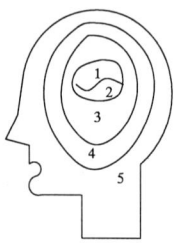

Fig. 1. Head is divided by shells with different conductivities

In the following, we will suppose that we have a conductor medium $\Omega = \bigcup_{i=1}^{5} \Omega_i$ as shown in the figure 1, where each component Ω_i has a constant conductivity σ_i, besides $\sigma_i \neq \sigma_j$ for $i \neq j$.

We have denoted by S_i the surfaces which compose the boundary of the Ω_i regions: $\partial\Omega_1 = S_0 \cup S_1$; $\partial\Omega_2 = S_0 \cup S_2$; $\partial\Omega_3 = S_1 \cup S_2 \cup S_3$; $\partial\Omega_4 = S_3 \cup S_4$; $\partial\Omega_5 = S_4 \cup S_5$ and by: $\Omega_6 = \overline{\Omega}^c = R^3 \backslash \overline{\Omega}$, $\widetilde{\Omega} = \overline{\Omega} \backslash S_5$.

We suppose that the current in the Ω region are produced only for the electric activity of the brain. Such current are: Ohmics and impressed.

The moving of charged ions through of the extracellular fluid and the diffusion current through of the neuronals membrane produced the Ohmics current and the impressed current, respectively.

We will denote by J, the volumetric density of impressed current in Ω_1 and by j, the superficial density of impressed current in S_1 (cerebral cortex).

So, the volumetric density current in the region Ω_1 ocupate for the brain is ([9], pp. 88): $J_T^1 = J + \sigma_1 E_1$ where E is the electric field generated, and for the Ohm's law, $\sigma_1 E$ denote the density of Ohmics currents. In the others regions we will consider ohmics current only.

It is possible to neglect the term $\dfrac{\partial \rho_j}{\partial t}$ in the continuity equation ([6]): $\nabla \cdot J_T^i + \dfrac{\partial \rho_i}{\partial t} = 0$ where J_T^i and ρ_i denote, the density of current and charge, in every region Ω_i $i = 1,..,5$, respectively.

In this way we obtain:

$$\nabla \cdot (J + \sigma_1 E_1) = 0 \quad en \quad \Omega_1 \qquad (1)$$

$$\nabla \cdot (\sigma_j E) = 0 \quad en \quad \Omega_j \quad j = 2,..,5. \qquad (2)$$

We can consider that the magnetic field B generated by the electric activity of the brain satisface that $\frac{\partial B}{\partial t} = 0$ ([13], pp. 206). Therefore, exist a electrostatic potential u such that $E = \nabla u$. The potential u satisface the following equation:

$$\Delta u = -\dfrac{1}{\sigma_1} \nabla \cdot J \quad (\Omega_1) \qquad (3)$$

$$\Delta u = 0 \quad (\Omega_i); \quad i = 2,...,5. \qquad (4)$$

We introduce the following notation: $u_i = u|_{\Omega_i}$, $i = 1,...,5$, n_0 is normal unitary vector outside to Ω_1 in S_0; n_i is normal unitary vector outside to Ω_i en S_i; $i = 1,...,5$, $f(x) = -\frac{1}{\sigma_1}\nabla \cdot J(x)$, $x \in \Omega_1$; $\varphi(x) = -(j \cdot n_1)(x)$, $x \in S_1$.

The boundary conditions are the continuity of the potentials and the continuity of the normal component of the current on the surfaces S_j $(j = 0,...,5)$ which separate the Ω_i regions ([14]). Such conditions take the form:

$$\left.\begin{array}{l} \sigma_1\dfrac{\partial u_1}{\partial n_0} = \sigma_2\dfrac{\partial u_2}{\partial n_0} \quad (S_0) \\[4pt] u_1 = u_2 \quad (S_0) \quad \sigma_1\dfrac{\partial u_1}{\partial n_1} = \sigma_3\dfrac{\partial u_3}{\partial n_1} + \varphi \quad (S_1) \\[4pt] u_1 = u_3 \quad (S_1) \quad \sigma_2\dfrac{\partial u_2}{\partial n_2} = \sigma_3\dfrac{\partial u_3}{\partial n_2} \quad (S_2) \\[4pt] u_2 = u_3 \quad (S_2); \\[4pt] u_3 = u_4 \quad (S_3) \quad \sigma_3\dfrac{\partial u_3}{\partial n_3} = \sigma_4\dfrac{\partial u_4}{\partial n_3} \quad (S_3) \\[4pt] u_4 = u_5 \quad (S_4) \quad \sigma_4\dfrac{\partial u_4}{\partial n_4} = \sigma_5\dfrac{\partial u_5}{\partial n_4} \quad (S_4) \\[4pt] \dfrac{\partial u_5}{\partial n_5} = 0 \quad (S_5) \end{array}\right\} \quad (5)$$

respectively, where $\dfrac{\partial u_i}{\partial n_j}$ denote the normal derivative of u_i in S_j with respect to n_j.

In the following we will study the boundary problem (3)-(5). We will call to this problem **Electroencephalographic Boundary Problem** (EBP).

3 Application of the Potential Theory Methods for to Obtain a Weak Solution of the EBP with $f \equiv 0$

For to resolve the problem of to find a harmonic function u in $\widetilde{\Omega}$ with boundary conditions of Dirichlet g, or Neumann h, we used techniques of the potential theory. The solution is search like a potential of doble or single layer, respectivily. This conditions g and h correspond to measurement of potential or current on the boundary S of $\widetilde{\Omega}$, respectivily.

If f and g are continuous functions on the boundary of $\widetilde{\Omega}$, that is to say, $f, g \in C(S)$, the problems are transform it to find a density of charge ρ for the boundary conditions of Neumann and a density of dipolars moments μ for the Dirichlet'sconditions which satisface operational equations of Fredholm of second kind ([8]).

Henceforth, we suppose that we have not volumetrics sources, since in this case the inverse problem has not a unique solution.

We will not considered, in general, that the EEG measure on scalp arise of a potential which is distributed continuously on itself because if this case occurs, the density of currents that produce such measurement will be distributed in a uniform way. But this fact in general no occurs, because the current is concentrated in the "active zone", which can to be distributed in a irregular way. For this reason, we will suppose that the boundary conditions in the EBP belong to $L_2(S_1)$. For to applied methods of the potential theory for boundary conditions in L_2 we need Sojovtsky's formulas in L_2, analogue to the Sojovtsky's formulas for the continuous case ([8], pp. 88).

We considered the single layer potential $V(x) = \frac{1}{4\pi} \int_S \frac{\rho(y)}{|x-y|} ds_y$. Of the next result ([12], Cap 1, §4): *if G if a measurable and bounded subset of $\mathbf{R}^m$ and K is a integral weakly singular operator on G with kernel $\frac{A(x,y)}{|x-y|^\lambda}$, $0 \leq \lambda < m$ where $\lambda p' > m$, $\frac{1}{p} + \frac{1}{p'} = 1$, then the operator $K : L_p(G) \to L_q(G)$ is compact for each q such that $1 \leq q < q_0^* = \frac{mp}{m-(m-\lambda)p}$,* we obtain that if S is a Liapunov's surface of clase $C^{1,\alpha}$ then the principal values V_0 y $\left(\frac{dV}{dn}\right)_0$ of the potencial of single layer and its normal derivative can be extended from $C(S)$ to compacts operators in $L_2(S)$. For the last result, the Sojovtsky's formulaes for the normal derivatives of a single layer, for boundary conditions in L_2 take the form:

$$\frac{dV}{dn_i} = \left(\frac{dV}{dn}\right)_0 - \frac{1}{2}I \qquad (6)$$

$$\frac{dV}{dn_e} = \left(\frac{dV}{dn}\right)_0 + \frac{1}{2}I \qquad (7)$$

where n represent the outside normal to the surface S and, $\frac{dV}{dn_i}, \frac{dV}{dn_e}$ represent the limits values of the normal derivatives interior and outside of the potential of single layer, respectively.

Definición 1 *(Solubility of the EBP with $f \equiv 0$).* Given a vector $\varphi \in L_2(S_1)$ such that $\int_{S_1} \varphi ds_1 = 0$, we say that EBP with $f \equiv 0$ is soluble, if exist a sucesion of classics solutions : $v_n \in \bigcap_{i=1}^{5} \{C^2(\Omega_i) \cap C^1(\overline{\Omega}_i)\} \cap C(\overline{\Omega})$, $n \in N$ of the problem $\Delta v_n(x) = 0$, $x \in \Omega$ where v_n satisface the boundary conditions (5) with $\varphi_n \in C(S_1)$ instead of φ y $\int_{S_1} \varphi_n ds_1 = 0$, $\varphi_n \to \varphi_i$ in $L_2(S_1)$. Under this conditions, if exist the limit in $L_2(\widetilde{\Omega})$ of the sucesion v_n, is called weak solution of EBP with $f \equiv 0$.

We begin with the suposition that the boundary condition of the EBP with $f \equiv 0$ is a continuous function on S_1 and we will search its clasic solution like a sum of potential of single layer with respect to continuous densities ρ_i on S_i, $i = 0, \ldots, 5$, that is to say,

$$R(\rho)(x) = u(x) = \frac{1}{4\pi} \sum_{i=1}^{5} \int_{\widetilde{S}_i} \frac{\widetilde{\rho}_i(\widetilde{y}_i)}{|x - \widetilde{y}_i|} d\widetilde{s}_i \qquad (8)$$

where $\widetilde{S_1} = S_0 \cup S_1$, $\widetilde{S_2} = S_0 \cup S_2$, $\widetilde{S_j} = S_j$ $j = 3, 4, 5$, $\widetilde{\rho}_j(x) = \rho_j(x)$; $x \in S_j$; $j = 3, 4, 5$.

$$\widetilde{\rho}_1 = \begin{cases} \rho_1(x)\,; x \in S_1 \\ \rho_0(x)\,; x \in S_0 \end{cases} ; \widetilde{\rho}_2 = \begin{cases} \rho_2(x)\,; x \in S_2 \\ \rho_0(x)\,; x \in S_0 \end{cases}$$

and $\boldsymbol{\rho} = (\rho_0(x_0), \rho_1(x_1), \rho_2(x_2), \rho_3(x_3), \rho_4(x_4), \rho_5(x_5))^T$.

For any choice of densities ρ_i the function $u(x)$ defined in (8) satisface the Laplace equation: $\Delta u(x) = 0$, $x \in \Omega$. If we seek that $u(x)$ comply with the boundary conditions for the continuity of the normal component of the current gives in (5), of the classics equation of Sojovtsky, we obtain a systems of integral Fredholm equations of second kind for to determinate the densities ρ_i ([3] y [4]), which can be writed in the matricial way:

$$(K + I)\boldsymbol{\rho} = \boldsymbol{J} \qquad (9)$$

where $\boldsymbol{J} = (0, \dfrac{2}{\sigma_1 + \sigma_3}\varphi, 0, 0, 0, 0)$, and $K = (K_{ij})$ is the matrix which component are the integral operators:

$K_{00} = \dfrac{2(\sigma_1-\sigma_2)}{\sigma_1+\sigma_2}\left(\dfrac{dV_1}{dn_0}\right)_{0,0}$ $K_{01} = \dfrac{(\sigma_1-\sigma_2)}{\sigma_1+\sigma_2}\left(\dfrac{dV_1}{dn_0}\right)_{0,1}$ $K_{02} = \dfrac{(\sigma_1-\sigma_2)}{\sigma_1+\sigma_2}\left(\dfrac{dV_2}{dn_0}\right)_{0,2}$

$K_{03} = \dfrac{(\sigma_1-\sigma_2)}{\sigma_1+\sigma_2}\left(\dfrac{dV_3}{dn_0}\right)_0$ $K_{04} = \dfrac{(\sigma_1-\sigma_2)}{\sigma_1+\sigma_2}\left(\dfrac{dV_4}{dn_0}\right)_0$ $K_{05} = \dfrac{(\sigma_1-\sigma_2)}{\sigma_1+\sigma_2}\left(\dfrac{dV_5}{dn_0}\right)_0$

$K_{10} = \dfrac{4(\sigma_1-\sigma_3)}{\sigma_1+\sigma_3}\left(\dfrac{dV_1}{dn_1}\right)_{0,0}$ $K_{11} = \dfrac{2(\sigma_1-\sigma_3)}{\sigma_1+\sigma_3}\left(\dfrac{dV_1}{dn_1}\right)_{0,1}$ $K_{12} = \dfrac{2(\sigma_1-\sigma_3)}{\sigma_1+\sigma_3}\left(\dfrac{dV_2}{dn_1}\right)_{0,2}$

$K_{13} = \dfrac{2(\sigma_1-\sigma_3)}{\sigma_1+\sigma_3}\left(\dfrac{dV_3}{dn_1}\right)_0$ $K_{14} = \dfrac{2(\sigma_1-\sigma_3)}{\sigma_1+\sigma_3}\left(\dfrac{dV_4}{dn_1}\right)_0$ $K_{15} = \dfrac{2(\sigma_1-\sigma_3)}{\sigma_1+\sigma_3}\left(\dfrac{dV_5}{dn_1}\right)_0$

$K_{20} = \dfrac{4(\sigma_2-\sigma_3)}{\sigma_2+\sigma_3}\left(\dfrac{dV_1}{dn_2}\right)_{0,0}$ $K_{21} = \dfrac{2(\sigma_2-\sigma_3)}{\sigma_2+\sigma_3}\left(\dfrac{dV_1}{dn_2}\right)_{0,1}$ $K_{22} = \dfrac{2(\sigma_2-\sigma_3)}{\sigma_2+\sigma_3}\left(\dfrac{dV_2}{dn_2}\right)_{0,2}$

$K_{23} = \dfrac{2(\sigma_2-\sigma_3)}{\sigma_2+\sigma_3}\left(\dfrac{dV_3}{dn_2}\right)_0$ $K_{24} = \dfrac{2(\sigma_2-\sigma_3)}{\sigma_2+\sigma_3}\left(\dfrac{dV_4}{dn_2}\right)_0$ $K_{25} = \dfrac{2(\sigma_2-\sigma_3)}{\sigma_2+\sigma_3}\left(\dfrac{dV_5}{dn_2}\right)_0$

$K_{30} = \dfrac{4(\sigma_3-\sigma_4)}{\sigma_3+\sigma_4}\left(\dfrac{dV_1}{dn_3}\right)_{0,0}$ $K_{31} = \dfrac{2(\sigma_3-\sigma_4)}{\sigma_3+\sigma_4}\left(\dfrac{dV_1}{dn_3}\right)_{0,1}$ $K_{32} = \dfrac{2(\sigma_3-\sigma_4)}{\sigma_3+\sigma_4}\left(\dfrac{dV_2}{dn_3}\right)_{0,2}$

$K_{33} = \dfrac{2(\sigma_3-\sigma_4)}{\sigma_3+\sigma_4}\left(\dfrac{dV_3}{dn_3}\right)_0$ $K_{34} = \dfrac{2(\sigma_3-\sigma_4)}{\sigma_3+\sigma_4}\left(\dfrac{dV_4}{dn_3}\right)_0$ $K_{35} = \dfrac{2(\sigma_3-\sigma_4)}{\sigma_3+\sigma_4}\left(\dfrac{dV_5}{dn_3}\right)_0$

$K_{40} = \dfrac{4(\sigma_4-\sigma_5)}{\sigma_4+\sigma_5}\left(\dfrac{dV_1}{dn_4}\right)_{0,0}$ $K_{41} = \dfrac{2(\sigma_4-\sigma_5)}{\sigma_4+\sigma_5}\left(\dfrac{dV_1}{dn_4}\right)_{0,1}$ $K_{42} = \dfrac{2(\sigma_4-\sigma_5)}{\sigma_4+\sigma_5}\left(\dfrac{dV_2}{dn_4}\right)_{0,2}$

$K_{43} = \dfrac{2(\sigma_4-\sigma_5)}{\sigma_4+\sigma_5}\left(\dfrac{dV_3}{dn_4}\right)_0$ $K_{44} = \dfrac{2(\sigma_4-\sigma_5)}{\sigma_4+\sigma_5}\left(\dfrac{dV_4}{dn_4}\right)_0$ $K_{45} = \dfrac{2(\sigma_4-\sigma_5)}{\sigma_4+\sigma_5}\left(\dfrac{dV_5}{dn_4}\right)_0$

$K_{50} = 4\left(\dfrac{dV_1}{dn_5}\right)_{0,0}$ $K_{51} = 2\left(\dfrac{dV_1}{dn_5}\right)_{0,1}$ $K_{52} = 2\left(\dfrac{dV_2}{dn_5}\right)_{0,2}$

$K_{53} = 2\left(\dfrac{dV_3}{dn_5}\right)_0$ $K_{54} = 2\left(\dfrac{dV_4}{dn_5}\right)_0$ $K_{55} = 2\left(\dfrac{dV_5}{dn_5}\right)_0$

where we have introduced the notation:

$$\left(\frac{dV_k}{dn_i}\right)_0(x_i) = \begin{cases} \frac{1}{4\pi}\left[v.p\int_{S_0}\frac{\partial}{\partial n_0^x}\left(\frac{1}{|x_0-y_0|}\right)\rho_0(y_0)ds_0+ \right. \\ \left. \int_{S_k}\frac{\partial}{\partial n_0^x}\left(\frac{1}{|x_0-y_k|}\right)\rho_k(y_k)ds_0\right] \quad i=0 \\ \frac{1}{4\pi}\left[v.p\int_{S_k}\frac{\partial}{\partial n_k^x}\left(\frac{1}{|x_k-y_k|}\right)\rho_k(y_k)ds_k+ \right. \\ \left. \int_{S_0}\frac{\partial}{\partial n_k^x}\left(\frac{1}{|x_k-y_0|}\right)\rho_0(y_0)ds_0\right] \quad i=k \end{cases}$$

with

$$\left(\frac{dV_k}{dn_0}\right)_0 = \left(\frac{dV_k}{dn_0}\right)_{0,0} + \left(\frac{dV_k}{dn_0}\right)_{0,k}; \quad k=1,2.$$

$$\left(\frac{dV_k}{dn_k}\right)_0 = \left(\frac{dV_k}{dn_k}\right)_{0,k} + \left(\frac{dV_k}{dn_k}\right)_{0,0}; \quad k=1,2.$$

$$\left(\frac{dV_k}{dn_i}\right)_0 = \left(\frac{dV_k}{dn_i}\right)_{0,k} + \left(\frac{dV_k}{dn_i}\right)_{0,0}; \quad k\neq i,$$

The matricial operator K is compact in $L_2(S_0) \times \cdots \times L_2(S_5)$ because the operators $K_{ij}: L_2(S_j) \to L_2(S_i)$ are compacts.

The proof of the following theorem can to be consulted in [3] and [4].

Teorema 1 *The weak solution of the EBP with $f \equiv 0$ exist and is unique for any boundary condition $\varphi \in L_2(S_1)$ such that $\int_{S_1} \varphi ds_1 = 0$. Furthermore, this solution not depend of the choice of the sucesion $\varphi_n \in C(S_1)$ if satisface the conditions $\int_{S_1} \varphi_n ds_1 = 0$ y $\|\varphi_n - \varphi\|_{L_2(S_1)} \to 0, n \to \infty$.*

Observaciones: Any eigenvector ρ od K for the eigenvalue $\lambda = -1$, comply that $V(\rho)(x)$ es constant in all space $\mathbf{R}^3$ since in this case $V(\rho)(x)$ is solution of the homogeneous EBP. For this reason, if we define the singular operator $R: L_2(S_0) \times \cdots \times L_2(S_5) \to L_2(S_5)$ by $R(\rho)(x_5) = V(\rho)|_{S_5}$, is easy to see that exist a unique eigenvector $\rho^0 = (\rho_0^0(x_0), \rho_1^0(x_1), \cdots, \rho_5^0(x_5))$ of K, for the eigenvalue $\lambda = -1$ such that

$$R(\rho^0)(x_5) \equiv 1; \quad x_5 \in S_5. \tag{10}$$

If we resolve the equation (10), for to obtain the eigenvector ρ^0, the weak solution of the IEP is reduced to study the system:

$$(K+I)\rho = J \tag{11}$$

$$R\rho = V \tag{12}$$

where V is a function given of $L_2(S_5)$ which correspond to the EEG generate by the electric activity in the cerebral cortex and measured on the scalp (S_5).

The next theorem garantize the uniqueness of recuperation of φ in the cerebral cortex of the EEG measured in the scalp V.

Teorema 2 *Given a measurement V on the scalp, exist a unique φ on the cerebral cortex which produce such measurement. Consequently, the injectivity of the operator R is proved and therefore the uniqueness of recuperation of φ from V.*

Proof: Let φ such that $V \equiv 0$ in S_5. $R(\rho)$ is a sum of potentials of single layer and, therefore, is harmonic in Ω_6. Furthermore like $V \equiv 0$ en S_5, by the uniqueness of the exterior Dirichlet problem, we have that $R(\rho) \equiv 0$ in $R^3 \backslash \overline{\Omega}$. Because $(K+I)\rho = J$ we have that $R(\rho) \equiv 0$ en Ω_5. The junction of this results say us that $R(\rho) \equiv 0$ en $R^3 \backslash \overline{\cup_1^4 \Omega_i}$. Appling the formulas (6) and (7) we deduce that $\rho_5 \equiv 0$. With a analogous analysis for the others densities we find that $\rho_k \equiv 0$. So, we conclude that $J \equiv 0$ and therefore that, $\varphi \equiv 0$. △ However, the inverse operator of R not is continuous and therefore, is required to apply algorithms of regularization for to obtain in a stable way the normal component of the current in the cerebral cortex.

References

1. Amir A.,1994. Uniqueness of the generators of brain evoked potential maps. IEEE transactions on Biomedical Engineering vol. 41, pp. 1-11.
2. Fraguela A., Morín M., Oliveros J., 1999. Statement of the inverse problem of localization of the parameters a source of neuronal current with a dipolo way. Mathematical Contribution, Serie Comunications 25, pp. 41-55 of the Mexican Mathematical Society (in Spanish).
3. Fraguela A., Oliveros J., 1998. Operational statement of the inverse electroencephalographic problem. Mathematical Contribution, Serie Comunications, 22, pp. 39-54, of the Mexican Mathematical Society (in Spanish).
4. Fraguela A., Oliveros J., Grebbennikov A., 2000. Operational statement and analysis of the inverse electroencephalographic problem. Mexican Review of Physic. (Unpublished).
5. Grebennikov A. I., Fraguela A. Statement and numerical analysis of some inverse problems of electroencephalography. Numerical Analysis Theory. applications, programms. Moscow, MSU,1999, pp. 28-46.
6. Heller L., 1990. Return Current in encephalography. Variational Principles. Biophysical Journal Volumen 57, pp. 601-607.
7. Koptelov, Yu. M., Zakharov, E. V., Inverse Problems in electroencephalography and their numerical solving. Ill-posed problems in natural sciences (moscow, 1991), 543-552, VSP, Utrecht, 1992. 92C55.
8. Kress R., 1989. Linear integral Equations. Springer Verlag.
9. Livanov M. N., 1972. Espacial organization of the brain's process. Moscú: Ed. Nauka (in Russian).
10. Nunez P. L., 1981. Electric Field of the brain. N. Y., Oxford Univ. Press.
11. Nunez P. L.,1995. Neocortical Dynamics and Human EEG Rhytmics. Oxford University Press, Inc.
12. Mikhlin S. G. 1977. Linear Partial Differential equations. Editorial Vischaya Schkola.
13. Plonsey R., Fleming D. G., 1969. Biolectric phenomena. N. Y. Mc Graw-Hill.
14. Sarvas J. 1987. Basic mathematical and electromagnetic concepts of the biomagnetic inverse problem. Phys. Med. Biol., Vol. 32, no. 1, pp. 11-22.

Exploiting Nonlinear Structures of Computational General Equilibrium Models

Christian Condevaux-Lanloy[1], Olivier Epelly[1], and Emmanuel Fragnière[2]

[1] University of Geneva, HEC, Unimail
102 Boulevard Carl-Vogt, 1204 Genève 11, Switzerland
{Christian.Condevaux-lanloy,Olivier.Epelly}@hec.unige.ch
http://ecolu-info.unige.ch/logilab
[2] University of Lausanne, HEC, BSFH1,
1015 Dorigny-Lausanne, Switzerland
Emmanuel.Fragniere@hec.unil.ch

Abstract. SETNL is a set of subroutines written in C++ that enables to manipulate nonlinear programming problems in different ways. Solution procedures which are usually implemented at the level of the optimization modeling languages can thus be moved into the algorithm. An example is presented where two embedded solution methods, one based one the economic theory (Negishi) and the other one on the operations research theory (Dantzig-Wolfe decomposition), are used to solve a dynamic general equilibrium model.

1 Introduction

Economic models are often formulated in Algebraic Modeling Languages (AMLs) because within this framework models can be described in a high level language which is very close to the mathematical standard notation. As a result, formulations in AMLs are brief and readable thanks to the use of indexes involved in parameters, variables and constraints definition. Once formulated in AMLs, models can be processed by different solvers without having to be modified. In this sense, an AML is a black-box, users being freed of implementation considerations as far as the solution process is concerned. However, large-scale or complex nonlinear programming models may need customized solution techniques, such as decomposition procedures, which cannot be easily implemented within this framework. To this end, we have developed a library of C++ routines, called SETNL (see http://ecolu-info.unige.ch/logilab/setnl), which can extract particular block structures of nonlinear programs and manipulate them in different ways during the solution process. SETNL enhances AMLs standard capability of processing nonlinear programs (NLP). This is performed via a flexible object-oriented approach which allows to implement a variety of solution techniques such as nested decomposition algorithms. The key point is that problem formulations have not to be changed, contrary to current practice which relies on the use of AMLs procedural statements such as `if-then-else` or `loop` integrated into models. SETNL allows to move such procedural statements contained in models into the solution algorithm itself. This removal of procedural statements makes models more readable and is expected to improve the solution process efficiency.

We present the SETNL capabilities through a model having a structure which can be exploited by specialized algorithms. The reference case is a demonstration version of MERGE, a dynamic General Equilibrium Model (GEM) developed by Manne and Richels [5]. It is a model for evaluating the regional and global effects of GreenHouse Gases (GHG) reduction policies. It quantifies alternative ways of thinking about climate change. The model may explore views on a wide range of contentious issues: costs of abatement, damages of climate change, valuation and discounting. This GEM is formulated in GAMS as a welfare-optimization problem, following Negishi [6]. Negishi weights are unknown when solution starts and are iteratively determined. Without using SETNL, this is performed by a loop defined in the GAMS model itself. However, the use of SETNL first enables to implement this loop outside the modeling environment, second to exploit the special structure involved in the model optimized at each iteration.

Indeed, each NLP that is solved inside each Negishi iteration displays a primal block angular structure which makes it a good candidate for the Dantzig-Wolfe decomposition [3]. The principle of decomposition is by now a well known idea in mathematical programming. In the early 1960's, decomposition techniques were proposed as a promising approach in addressing the limitations of computers in solving large-scale mathematical models. However, research and experiments were limited to linear and integer programming. By the 1990's, more attention was focused on nonlinear programs as computational power was making large-scale nonlinear programming possible. Recent results have shown that NLP decomposition methods can help not only to solve untractably large problems, but to drastically improve solution times and in some cases, remove numerical instabilities [1]. With the added potential benefits of parallel implementation, decomposition methods can provide an important tool for the economic modeler. The general idea of decomposition is to break down the original mathematical programming problem into smaller, more manageable problems. These subproblems can then be solved and iteratively adjusted so that their solutions can be patched together to derive the solution to the original problem.

As a result, SETNL is able to compute general equilibria via a Dantzig-Wolfe decomposition embedded within a Negishi loop, without expecting modelers complex reformulations of the original welfare optimum problem. Such a solution technique is designed to generally improve computational efficiency and even to enable their processing in case of very large models. This application also shows that two solution procedures, one based on the economic theory (Negishi) and the other one based on operations research theory (Dantzig-Wolfe) can be both handled at the algorithmic level.

The paper is organized as follows. Section 2 briefly presents how AMLs process mathematical programs and early experiments that precluded the current development. The concept SETNL is explained in section 3. The effectiveness of this approach is illustrated by an example of routines which are useful in a decomposition framework. The way SETNL is included in-between the optimization modeling language and the nonlinear solver is also detailed. In Section 4 we describe a case where a general equilibrium model is solved through SETNL. The algorithm developed here involves two solution procedures: Negishi and Decomposition. In Section 5, we give some concluding remarks.

2 Early Experiments with the GAMS Modeling Language

One of the most important aspects of AMLs is their capability of dealing with nonlinear programs (NLP). In the case of linear programs, all the data of a given problem is transferred in one move from an AML to a solver before the solution process starts. Nonlinear models involve a more complicated process. In this case, any solution algorithm generates, iterates, and needs to get update on information such as values, Jacobian and possibly Hessian of nonlinear functions. This role is assigned to the built-in nonlinear interpretor of AMLs since solvers have no algebraic knowledge of the model. Figure 1 describes the exchange of information occuring between an AML and a solver. In the case

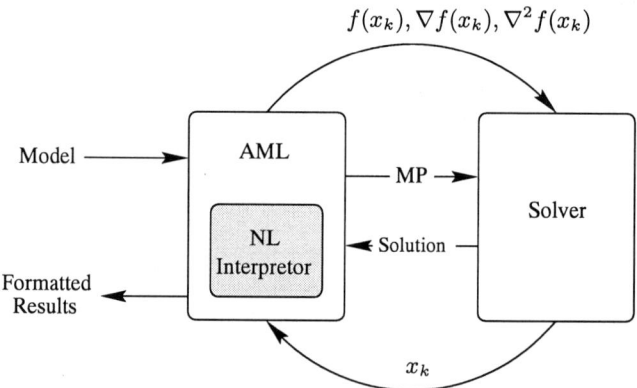

Fig. 1. Links between an AML and a solver

of large scale or complex non-linear models which necessitate customized solutions procedures, this transparent scheme can be a drawback. For example, an automatic decomposition is difficult due to the fact AMLs do not produce partial information in a structured way. Early experiments to retrieve the structure from nonlinear programming problems were realized by Chang and Fragniere [2]. A prototype, under the form of procedures called SPLITDAT and DECOMP, although very simple, enabled to solve the Ramsey model through a Benders decomposition (same idea as Dantzig-Wolfe however adapted to the dual block angular structure, see section 4). In such a situation some functions of the mathematical program need to be split into several separable pieces. Using an AML, it is necessary to call the nonlinear interpretor to solely handle parts of the entire problem. This means some customization of the nonlinear interpretor to avoid generating the full nonlinear information associated with the entire problem. Modeling languages such as GAMS or AMPL provides an I/O library. It is a set of functions allowing one to extract due course information needed to solve a given optimization problem formulated with AMLs: values, gradients, and Hessians of objective and constraint functions. The aim of the additional I/O subroutines SPLIDAT and DECOMP was to provide the GAMS user with the possibility of using Benders decomposition algorithms within the GAMS modeling language framework. SPLITDAT takes the original model and splits the data into a master problem and one or more subproblems. DECOMP uses the decomposition algorithm to determine the optimal solution of the original problem.

Experiments had been performed with an intertemporal aggregate growth model stemming from work done by Ramsey [7]. This model involves three decision variables in each time period t: Consumption, Investment, and Capital Stock (C_t, I_t, K_t). Consider an economy with a single agent acting as producer, consumer, investor, and saver. Given initial levels of our decision variables (C_0, I_0, K_0), and a Cobb-Douglas production function that is a function of capital and labor $(f(K_t, L_t) = A \times K_t^\beta L_t^{(1-\beta)})$, and given exogenous labor supplies, L_t, we want to find an optimal level of Consumption, Investment, and Capital Stock. To find this optimal level, we maximize a discounted logarithmic utility function under a capital stock constraint and a production constraint. With a fixed growth rate, g, and a utility discount factor, udf, the model is written as:

$$\begin{aligned}
\max & \sum_t (udf)^t \times log(C_t) \quad t \in T \quad \text{time periods} \\
\text{s.t.} \ & K_{t+1} = K_t + I_t \\
& a_t \times K_t^\beta = C_t + I_t \quad \text{where } a_t = A \times (L^t)^{1-\beta} \\
& \qquad \qquad \qquad \qquad \text{since labor}, L_t, \text{has a fixed growth rate} \\
& \qquad \qquad \qquad \qquad (L^t = L_0 \times (1+g)^t) \\
& C_t, K_t, I_t \geq 0
\end{aligned} \qquad (1)$$

In this particular case SPLITDAT store the problem according to the time dimension and then splits it into a master problem and one subproblem. This is done through the use of the I/O dictionary file. The GAMS I/O dictionary file is a character file containing the names of the sets, variables, and equations of the model. For the variables and constraints, each name is formed by the original name of the variable or constraint plus the corresponding set indices. SPLITDAT then recognizes that the capital stock equation is the only transition equation that carries over variables from one time period to the next. Once the master problem and subproblem have been fully formulated, DECOMP takes the information and iterates the Benders decomposition algorithm until an optimal solution is found. In the 20 period model (see Figure 2), Benders decomposition was able to solve this problem after just 5 iterations with a relative error tolerance of $eps = 10^{-4}$ (the gap between the upper and lower bound divided by the magnitude of the objective function).

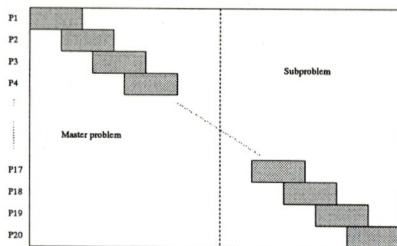

Fig. 2. Staircase structure solved with Benders decomposition

3 SETNL with AMPL and CONOPT

SETNL is a high-level C++ library allowing on the one hand to access standard information from the modeling system and on the other to provide the modeler with advanced features such as the automated exploitation of the structure and the access to partial nonlinear information. SETNL is made up of two parts. The first part is called SET and is described in [4]. SET enables one to retrieve block structures directly from the algebraic formulation of the original problem. The second part which is described in the present paper is about the manipulation of NLP problems.

Standard practice is that AMLs yield nonlinear information for solvers without allowing to distinguish subsets of variables or subsets of constraints. This information is however necessary in any decomposition procedure. SETNL allows to break into pieces nonlinear information such as functions values, Jacobian and possibly Hessian matrices of nonlinear equations.

Depending on the type of models, SETNL is able to extract a wide range of block structures such as splitting a given structured problem into several subproblems. These subproblems are sent to nonlinear solvers in the appropriate input format ready to start the optimization process. Every time a nonlinear solver needs updates on partial information, SETNL access the nonlinear interpretor to get them. We present below a sample of routines of different complexity which are useful for applying the decomposition scheme on the experiments presented in Section 2:

long get_nlnz();
// returns the number of NL non zero of the Jacobian matrix

double get_objvalue(real∗ X, expr∗ setcol);
// returns the current value of the objective value for a subset of columns "setcol" and a given solution X

void get_objgrad(real∗ X, expr∗ setcol);
// evaluates the gradient of a subset "setcol" of the objective function

int partition()
// determines the splitting of the problem

The previous example though simple is illustrative since as you can notice in Figure 1 once the rows and columns are ordered regarding the time dimension they are automatically associated with a subproblem of the staircase structure. There is solely the objective function that needs to be broken apart, in this particular case, in two pieces (i.e. the master problem and the subproblem).

Figure 3 shows how SETNL is integrated in the modeling system. The model formed by the mathematical formulation and the data are processed in the usual way (see Figure 1 for comparison). SETNL is currently linked with The I/O AMPL library and CONOPT (an NLP solver). Two I/O AMPL routines have been modified to enter as new parameters a given set of indexes that lists the variables associated to a given subproblem.

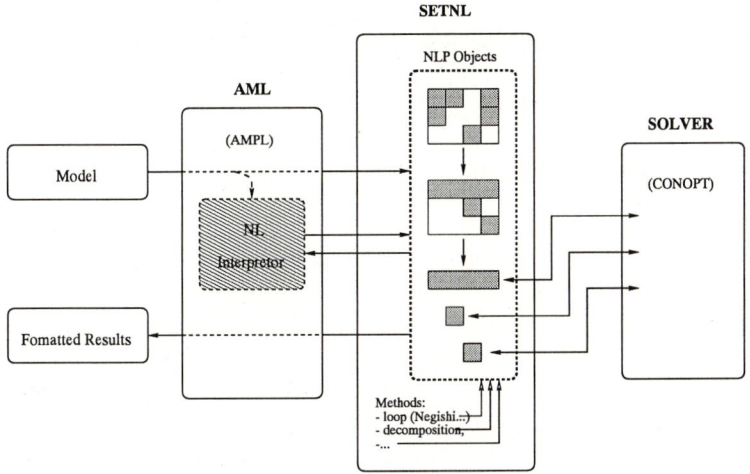

Fig. 3. SETNL

For instance in Figure 3 we see that the problem may display a primal block angular structure (see formulation 2). Once SETNL knows about the structure the problem is split into a master problem and several subproblems in the appropriate format in order to be red by the CONOPT nonlinear solver. SETNL implements in this case the Dantzig-Wolfe algorithm (see Section 4 for a brief explanation). Each time CONOPT asks for information updates (e.g. gradients, function values) SETNL communicates with the nonlinear interpretor to get uniquely the partial non linear information needed. The decompostion proceeds until the desired level of precision is attained.

4 Solving Equilibrium Problems as Welfare Optimization Problems

Our intentions here is to show that thanks to SETNL a model arising from the field of computational economics can be solved through embedded solution procedures arising from different academic disciplines. In the case of the MERGE model we must take into account two procedural statements: a first loop corresponding to The Negishi algorithm and a second one associated with the Decomposition algorithm which is embedded in the Negishi algorithm. Although formally coded with SETNL C++ routines, both loops here are explained in an intuitive manner, referring to computational economics concepts. This allows us to highlight difficulties encountered by the economic modeler when coding such an algorithm.

The Negishi algorithm is presented here in its GAMS original version as written by [5]. This section appears at the end of the model. The loop is defined for a certain number of iterations which is determined by the modeler (Negishi is indeed known to be a tatonnement approach). Within each iteration the model called here NWEL is optimized by the nonlinear solver. From the primal solutions (all the variables finishing with ".L") and the dual solutions (all the variables finishing with ".M") new Negishi weights, NW(), are computed for each region (RG). These Negishi weights are included in the new objective function of the nonlinear programming problem and the process continues until the chosen number of iterations is reached.

```
LOOP(ITER$(ORD(ITER) NE CARD(ITER)),
    SOLVE JM MAXIMIZING NWEL USING NLP;
    DISPLAY TRDBAL.M;
    PVPI(TRD,PP) =  TRDBAL.M(TRD,PP)/TRDBAL.M("NUM","2000");
    NW(RG)     =  SUM(PP, PVPI("NUM",PP)*C.L(RG,PP))
                + SUM((PP,TRD), PVPI(TRD,PP)*X.L(RG,TRD,PP));
    NW(RG) = NW(RG) / SUM(R, NW(R));
    NWTITR(ITER,RG) = NW(RG);
);
```

Then the nonlinear programming problem solved within each Negishi iteration corresponds to the solution of a large scale block-angular convex nonlinear programming problem of the following form

$$\begin{aligned} \text{maximize} \quad & \sum_{i=1}^{p} f_i(x_i) \\ \text{subject to} \quad & \sum_{i=1}^{p} g_i(x_i) \leq 0 \\ & h_i(x_i) \leq 0, \, i = 1, 2, \ldots, p, \end{aligned} \quad (2)$$

where $x_i \in S_i \subset \mathcal{R}^{n_i}, S_i, f_i, g_i$ and h_i are convex, with $f_i : \mathcal{R}^{n_i} \longmapsto \mathcal{R}$, $g_i : \mathcal{R}^{n_i} \longmapsto \mathcal{R}^{m_0}$, $h_i : \mathcal{R}^{n_i} \longmapsto \mathcal{R}^{m_i}$. We further assume that the interior of the feasible set X, defined by the above constraints, is not empty. The problem has $n = \sum_{i=1}^{p} n_i$ variables and $m = m_0 + \sum_{i=1}^{p} m_i$ constraints.

The Dantzig-Wolfe decomposition algorithm can be interpreted in an economic way (i.e. price-directed decomposition). Essentially, the subproblems pass extreme point proposals among its feasible set, and the master problem tries to find the right convex combinations of the feasible points to arrive at the optimal allocation of the shared resources among the subproblems. Specifically, the subproblems pass the maximum and resource usage in a given iteration. The master then returns a price vector telling the subproblems whether they are using too much or too little of the common resources. This procedure can iterate until the prices returned to the subproblems do not change, implying an optimal solution has been reached. This scheme fits with economic model containing multiple regions.

5 Conclusion

We showed in this paper that the SETNL library enables to code solution procedures for nonlinear models arising from the economic theory in an efficient algorithmic environment. Indeed the difficulties to handle nonlinearities force economic modelers to program those solution techniques with the help of modeling language syntaxes which is not the primary role of these languages. To illustrate the capabilities of SETNL, we developped an embedded Dantzig-Wolfe/Negishi algorithm to solve a computational general equilibrium model. The benefits of these developments are that the economic modeler can focus on the modeling process instead of the solution process.

References

1. D. CHANG, *Solving dynamic equilibrium models with multistage benders decomposition*. Stanford University, Department of Engineering Economic Systems and Operations Research, working paper, October 1997.
2. D. CHANG AND E. FRAGNIÈRE, *Splitdat and decomp: Two new GAMS I/O subroutines to handle mathematical programming problems with an automated decomposition procedure*. Stanford University, Department of Operations Research, working paper, August 1996.
3. G. B. DANTZIG AND P. WOLFE, *The decomposition algorithm for linear programming*, Econometrica, 29 (1961), pp. 767–778.
4. E. FRAGNIÈRE, J. GONDZIO, R. SARKISSIAN, AND J.-P. VIAL, *Structure exploiting tool in algebraic modeling languages*, tech. rep., Section of Management Studies, University of Geneva, June 1999. to appear in Management Science.
5. A. S. MANNE, R. MENDELSON, AND R. G. RICHELS, *MERGE - a model for evaluating regional and global effects of ghg reduction policies*, Energy Policy, 23 (1995), pp. 17–34.
6. T. NEGISHI, *Welfare economics and existence of an equilibrium for a competitive economy*, Metroeconomica, (1960), pp. 92–97.
7. F. P. RAMSEY, *A mathematical theory of saving*, Economics Journal, (1928).

Constitutive Equations and Numerical Modelling of Time Effects in Soft Porous Rocks

M. Datcheva[1], R. Charlier[2], and F. Collin[2]

[1] Institute of Mechanics, Bulgarian Academy of Sciences, Sofia
[2] University of Liege, Liege, Belgium

Abstract. A constitutive model is developed within the framework of Perzyna's viscoplasticity for predicting the stress-strain-time behaviour of soft porous rocks. The model is based on the hyperelasticity and multisurface viscoplasticity with hardening. A time-stepping algorithm is presented for integrating the creep sensitive law. An example of application to one-dimensional consolidation is presented. The objectives are to: 1. present a soft rock model which is capable of taking into account the rate sensitivity, time effects and creep rupture; 2. to discuss the use of an incremental procedure for time stepping using large time increments and 3. to extend the finite element code Lagamine (MSM-ULg) for viscoplastic problems in geomechanics.

1 Introduction

For solving geomechanical problems, such as well-bore stability, subsidence, hydraulic fracturing and ect., the most important is to deal with a proper constitutive model for the complex soft porous rock mechanical behaviour. Various models have been developed for the time independent behaviour of chalk as a typical soft porous rock. However the failure of an underground structure in this rock may occur due to creep deformation and therefore the use of conventional time-independent procedure for the interpretation of laboratory results and the analysis of geotechnical boundary-valued problems may result in solutions which do not properly capture the actual in situ response. The model proposed in this study is a time-dependent inelastic model for rocks and soils based on the Perzyna's elasto-viscoplastic theory, [6]. Motivations for adopting Perzyna's elastic–viscoplastic theory are:

1. The formulation is well accepted and well used;
2. The generality of the time–rate flow rule offers the capability of simulating time–dependent material behaviour over a wide range of loading;
3. The incorporation of the inviscid multisurface cap–failure–tension model, developed in DIG–ULg in the frame of PASACHALK project, was of interest ([1] and [2]); and
4. The formulation is readily adaptable to a numerical algorithm suitable for finite element procedure and particularly for implementation in Lagamine (MSM–ULg) finite element code.

2 Mechanical Model. Perzyna's Viscoplasticity

Perzyna's theory, [6] is a modification of classical plasticity wherein viscous–like behaviour is introduced by a time–rate flow rule employing a plasticity yield function. The strain rate tensor $\dot{\varepsilon}_{ij}$ is composed of elastic $\dot{\varepsilon}_{ij}^e$ and viscoplastic $\dot{\varepsilon}_{ij}^{vp}$ strains, or by the definition:

$$\dot{\varepsilon}_{ij} = \dot{\varepsilon}_{ij}^e + \dot{\varepsilon}_{ij}^{vp}. \tag{1}$$

The stress rate tensor $\dot{\sigma}_{ij}$ is related to the elastic strain rate via a linear elastic or hyperelastic constitutive tensor C_{ijkl}. Therefore, taking into account the relation (1) the elastoplastic constitutive relation between stress and strain rates reads:

$$\tilde{\sigma}_{ij} = C_{ijkl}\left(\dot{\varepsilon}_{kl} - \dot{\varepsilon}_{ij}^{vp}\right), \tag{2}$$

The Jaumann type objective time derivative of stress tensor is defined by $\tilde{\sigma}_{ij} = \dot{\sigma}_{ij} + w_{ik}\,\sigma_{kj} + \sigma_{ik}\,w_{kj}^T$ where $\boldsymbol{w}$ is the anti-symmetric part of the velocity gradient. The viscoplastic flow rule is expressed as:

$$\dot{\varepsilon}_{ij}^{vp} = \gamma\langle\Phi(f)\rangle\frac{\partial g}{\partial \sigma_{ij}}, \qquad \langle\Phi(f)\rangle = \begin{cases} \Phi(f), & f > 0 \\ 0, & f \leq 0 \end{cases} \tag{3}$$

in which γ is a fluidity parameter; Φ - viscous flow function; $g = g(\boldsymbol{\sigma},\varpi)$ - creep potential and $f = f(\boldsymbol{\sigma},\varpi)$ is any valid plasticity function, playing the role of *loading* surface. The parameter ϖ stays for some hardening function of the viscoplastic strain history, i.e., $\varpi = \varpi(\bar{\varepsilon}^{vp})$, where $\bar{\varepsilon}^{vp}$ is an equivalent viscoplastic strain representing the magnitude of the viscoplastic deformation. For a given value of ϖ, all states of stress that satisfy $f = 0$ form the current "static" yield surface. The "static" yield surface forms a boundary between elastic ($f \leq 0$) and viscoplastic ($f > 0$) domains. When a constant stress state is imposed such that $f > 0$, viscoplastic flow will occur. If f is a nonhardening yield function the flow will continue to occur at a constant rate. If f is a hardening function, viscoplastic flow occurs at a decreasing rate because as viscoplastic strain accumulates, $\varpi(\bar{\varepsilon}^{vp})$ changes in value such that $f(\boldsymbol{\sigma},\varpi) \to 0$ and thus $\dot{\varepsilon}_{ij}^{vp} \to 0$. In this way the static yield surface is moving out on a real time to eventually form a new static yield surface containing the imposed stress state. Once the new static yield surface has stabilized, the steady state solution $\dot{\varepsilon}_{ij}^{vp} = 0$ is achieved. The resulting strains accumulated during this loading would be identical to the corresponding time-independent plastic solution.

2.1 Application to the Soft Porous Rocks

The concept of two inelastic deformation mechanisms – collapse (volumetric or cap) and shear failure or deviatoric – has been applied to the viscoplastic analysis. Such a concept is based on the experimental observation for high porous rocks and for chalk especially (see [8]). Therefore:

$$\dot{\varepsilon}_{ij}^{vp} = \dot{\varepsilon}_{ij}^c + \dot{\varepsilon}_{ij}^d, \tag{4}$$

Indexes c and d are for collapse (cap or volumetric) and deviatoric or shear failure viscoplastic strain components and model parameters respectively. In the present work the loading surface and creep potential are functions of the first, second and third stress invariants: $I_\sigma = \sigma_{ij}\delta_{ij}$, $II_s = \sqrt{\frac{1}{2}s_{ij}s_{ij}}$ and $\beta = -\frac{1}{3}\sin^{-1}\left(\frac{3\sqrt{3}}{2}\frac{III_s}{II_s^3}\right)$, where $III_s = \frac{1}{3}s_{ik}s_{kj}s_{ji}$ and $s_{ij} = \sigma_{ij} - I_\sigma/3$ is the stress deviator. The static yield surface $f = 0$ and the creep potential function $g = 0$ are divided into two regions along the first stress invariant axis I_σ: the cap surface region ($I_\sigma < L = \frac{1}{2}\left(\frac{3c}{\tan\phi_C} - 3p_0\right)$) with f_c and g_c and the *failure* surface region ($I_\sigma \geq L$) with f_d and g_d. Here c is the cohesion, ϕ_C is the friction angle in the compression path and p_0 is the preconsolidation pressure. Such an approach overcomes difficulties, mentioned in [7] to extend the Perzyna's type viscoplasticity in the case of multisurfase inelasticity.

Cap loading surface f_c is a hardening surface defined by

$$f_c = II_s^2 + m^2\left(I_\sigma - \frac{3c}{\tan\phi_C}\right)(I_\sigma + 3p_0) = 0, \tag{5}$$

where

$$m = a(1 + b\sin 3\beta)^n, \tag{6}$$

$$b = \frac{[\sin\phi_C(3 + \sin\phi_E)]^{\frac{1}{n}} - [\sin\phi_E(3 - \sin\phi_C)]^{\frac{1}{n}}}{[\sin\phi_C(3 + \sin\phi_E)]^{\frac{1}{n}} + [\sin\phi_E(3 - \sin\phi_C)]^{\frac{1}{n}}}, \tag{7}$$

$$a = \frac{1}{\sqrt{3}}\frac{2\sin\phi_C}{3 - \sin\phi_C}(1 + b)^{-n}, \tag{8}$$

n is a model parameter and ϕ_E is the friction angle in the extension stress path. We assume that:

Hypothesis 1 *Hardening for the cap surface is due to the volumetric inelastic strain - ε_v^c and therefore $f_c = f_c(\boldsymbol{\sigma}, \varepsilon_v^c)$. The only hardening variable for the cap surface, p_0, depends only on ε_v^c.*

The hardening law is given as:

$$\dot{p}_0 = \frac{1+e}{\lambda - \kappa}p_0\dot{\varepsilon}_v^c \tag{9}$$

with e - the void ratio, λ - the slope of the virgin consolidation line and κ - the slope of swell/recompression line in $e - \ln(I_\sigma/3)$ space. For the cap deformation mechanism $f_c \equiv g_c$ and referring to the eq. (3), the associated viscoplastic law is:

$$\dot{\varepsilon}_{ij}^c = \gamma_c\langle\Phi_c(f_c)\rangle\frac{\partial f_c}{\partial \sigma_{ij}}, \tag{10}$$

where

$$\Phi_c = \left(\frac{f_c}{p_0}\right)^{\alpha_c} \quad \text{and} \quad \gamma_c = \omega\left(\frac{|I_\sigma|}{3p_a}\right)^\iota,$$

with ω, ι and p_a - material constants.

Failure loading surface is a hardening, Van Eekelen [4] yield function:

$$f_d = II_s + m\left(I_\sigma - \frac{3c}{\tan\phi_C}\right) = 0. \tag{11}$$

Hypothesis 2 *For the failure surface there is no hardening due to the cap type deformation. The equivalent deviatoric inelastic strain*

$$\bar{e}_d = \int_0^t \sqrt{\left(\dot{\varepsilon}_{ij}^d - \frac{1}{3}\dot{\varepsilon}_{kl}^d \delta_{kl}\delta_{ij}\right)\left(\dot{\varepsilon}_{ij}^d - \frac{1}{3}\dot{\varepsilon}_{kl}^d \delta_{kl}\delta_{ij}\right)}\, d\tau$$

is the only hardening parameter for the failure surface, so $f_d = f_d(\boldsymbol{\sigma}, \bar{e}_d)$ *and thus the internal state variables for the failure deformation mechanism – the friction angles and the cohesion - are functions only of* $\bar{e}_d$.

For the internal state variables the concrete expressions, explored in the present work are given like in [3]:

$$\phi_C = \phi_{C0} + (\phi_{Cf} - \phi_{C0})\frac{\bar{e}_d}{B_p + \bar{e}_d}, \tag{12}$$

$$\phi_E = \phi_{E0} + (\phi_{Ef} - \phi_{E0})\frac{\bar{e}_d}{B_p + \bar{e}_d}, \tag{13}$$

$$c = c_0 + (c_f - c_0)\frac{\bar{e}_d}{B_c + \bar{e}_d}, \tag{14}$$

where ϕ_{C0}, ϕ_{E0}, c_0 and ϕ_{Ef}, ϕ_{Ef}, c_f are initial and final friction angles and cohesion in compression (C) and extension (E) stress paths. Coefficients B_p and B_c have the values of the equivalent deviatoric inelastic strain for which half of the hardening on friction angles and cohesion is achieved, see [3].

The viscoplastic flow law is non-associated and taking into account (3) it is given by:

$$\dot{\varepsilon}_{ij}^d = \gamma_d \langle \Phi_d(f_d) \rangle \frac{\partial g_d}{\partial \sigma_{ij}}, \tag{15}$$

with $\Phi_d = \left(\frac{f_d}{p_0}\right)^{\alpha_d}$, $\gamma_d = \gamma_c\, a_2$, where α_d and a_2 are material constants.

The potential function g_d depends on the dilatancy angles ψ_C and ψ_E in compression and extension paths respectively, and is given as:

$$g_d = II_s + m'\left(I_\sigma - \frac{3c}{\tan\phi_C}\right), \tag{16}$$

where

$$m' = a'(1 + b'\sin 3\beta)^n, \tag{17}$$

$$b' = \frac{[\sin\psi_C(3 + \sin\psi_E)]^{\frac{1}{n}} - [\sin\psi_E(3 - \sin\psi_C)]^{\frac{1}{n}}}{[\sin\psi_C(3 + \sin\psi_E)]^{\frac{1}{n}} + [\sin\psi_E(3 - \sin\psi_C)]^{\frac{1}{n}}}, \tag{18}$$

$$a' = \frac{1}{\sqrt{3}}\frac{2\sin\psi_C}{3 - \sin\psi_C}(1 + b')^{-n} \tag{19}$$

Here the well known Taylor rule: $\phi_C - \psi_C = \phi_E - \psi_E = $ const is used, which is based on experimental evidences.

3 Numerical Algorithm

This section concerns a way for implementing the Perzyna's type viscoplasticity in the finite element code for large deformation inelastisic analysis Lagamine, [5]. It is presumed that the strain history is specified and the objectives is to determine the corresponding stress history. Using a step-by-step time integration scheme a numerical solution algorithm is developed at the constitutive level. The time increments realized are large and the nonlinearity is much higher than for the classical elastoplastic laws. The errors that are introduced by the integration scheme can be significantly reduced by sub-incrementation. In the code Lagamine the time increment $\Delta t = t^B - t^A$, where B indicates the end of the time step and A its beginning, is divided into a constant number N of sub-intervals with a length $\Delta_N t$. For each sub-interval we have to integrate the eq. (3), which an incremental form is:

$$\Delta\sigma = C(\sigma)(\Delta\varepsilon - \Delta\varepsilon^{vp}) \tag{20}$$

The right side of (20) depends on σ and the hardening function ϖ. The problem posed is therefore to know which stress state and value of ϖ to introduce in the right side of (20). A generalised mid-point algorithm is adopted here, where the viscoplastic strain increment is approximated by one-parameter time integration scheme as:

$$\Delta\varepsilon^{vp} = \Delta_N t \left[(1-\theta)\,\dot{\varepsilon}^{vp}|_A + \theta\dot{\varepsilon}^{vp}|_B\right], \tag{21}$$

$$0 \leq \theta \leq 1.$$

In each Gauss integration point the following operations would be carried:

1. Use the stress rate from the previous sub-interval $i-1$ and the stress σ^A at the beginning of the step Δt to evaluate a mid-point stress σ^θ:

$$\sigma^\theta = \sigma^A + \tilde{\sigma}_{i-1}\,\theta\,\Delta_N t \tag{22}$$

Hardening parameters, ε_v^c for the *cap* and $\bar{e}_d$ for the *failure* regions are evaluated in the same manner. Using the unified notation ϖ it reads:

$$\varpi^\theta = \varpi^A + \dot{\varpi}_{i-1}\,\theta\,\Delta_N t \tag{23}$$

For the first sub-interval, the stress rate and the hardening parameter rate are initialised through the explicit or forward-Euler scheme.

2. Call the constitutive law to calculate approximated values of the stress and hardening parameter rates:

$$\tilde{\sigma}_i = C(\sigma^\theta)\left(\dot{\varepsilon} - \dot{\varepsilon}^{vp}(\sigma^\theta, \varpi^\theta)\right), \tag{24}$$

$$\dot{\tilde{\varpi}}_i = \dot{\varpi}\left(\dot{\varepsilon}^{vp}(\sigma^\theta, \varpi^\theta)\right). \tag{25}$$

3. Calculate new stress and hardening parameter mid-point approximations:
$$\sigma^\theta = \sigma^A + \tilde{\sigma}_i \theta \Delta_N t, \quad (26)$$
$$\varpi^\theta = \varpi^A + \dot{\varpi}_i \theta \Delta_N t \quad (27)$$

4. Repeat 2. and 3. untill it converges. It is supposed that the convergence is obtained after two iterations.
5. Calculate the stress state at the end of the sub-interval:
$$\sigma^B = \sigma^A + \tilde{\sigma}_i \theta \Delta_N t, \quad (28)$$

and update the hardening function:
$$\varpi^B = \varpi^A + \dot{\varpi}_i \theta \Delta_N t \quad (29)$$

6. Take into account an Jaumann correction based on the mid-point stress value:
$$\sigma^B = \sigma^B + \omega \sigma^\theta + \sigma^\theta \omega^T \quad (30)$$

The above described updating procedure depends on the current value of the stress invariant I_σ, which dictates the type of the activated deformation mechanism. If $I_\sigma < L$ then *cap* constitutive equations (5)-(10) are employed in 2. and for $I_\sigma \geq L$ equations (11)-(19) are used.

4 Numerical Example

Finite element simulation of one–dimensional consolidation has been performed. Plain strain state has been considered. The material property data is: mass density of solid skeleton $\rho = 2.647 \ kNs^2/m^4$ and the initial porosity $n = 0.332$. Elastic properties are caracterised by constants: $E = 3.6 \times 10^5 \ kPa$, $\nu = 0.3$. For the viscoplastic response material constants are: $\phi_{Cf} = \phi_{C0} = 32°$, $\phi_{Ef} = \phi_{E0} = 52°$, $c_f = c_0 = 10 \ kPa$, $B_p = 0.0001$, $B_c = 0.0002$, $n = -0.229$, $\alpha_c = 0.9$, $\alpha_d = 0.1$, $\omega = 2.0 \times 10^{-5}$, $a_2 = 1.3$, $\iota = 0.52$ and the reference stress $p_a = 1.0 \times 10^8 \ kPa$. The sizes of the sample are 3.00x3.00 meters. Finite element mesh and boundary conditions are shown on Fig.1, a. For the numerical simulation a multistage rapid loading path followed by creep has been applied such that for $0 \leq t \leq 4 \ s$ there is loading up to 1.2 MPa, for $4 \ s < t \leq 236 \ s$ the load is kept constant, for $236 \ s < t \leq 240 \ s$ loading up to 2.4 MPa and for $t > 240 \ s$ there is a creep with a constant load of 2.4 MPa. Fig.1, b. illustrates a typical variation of the vertical displacement with the loading history at nodal points 7 and 9 .

5 Conclusions

The viscoplastic formulation and the numerical algorithm presented provide a general format for extending inviscid models to Perzyna-type viscoplasticic constitutive relationships suitable for finite element applications. The problem of

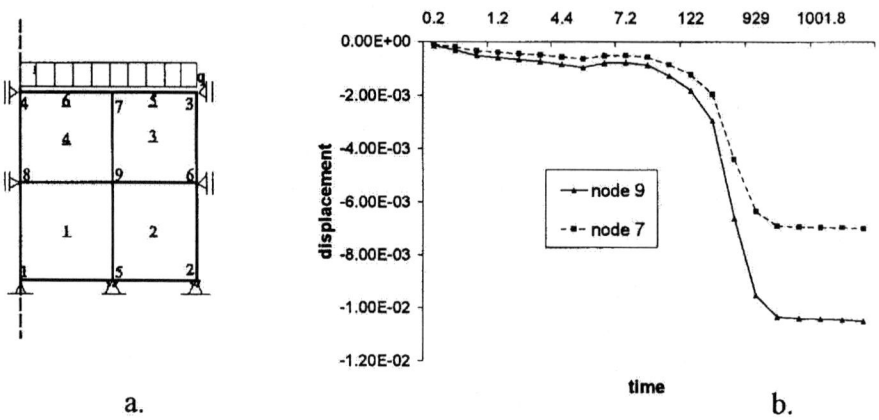

Fig. 1. a. Finite element mesh and boundary conditions for the numerical example. b. Displacement versus time in nodal points 7 and 9 - a representative result

Perzyna's viscoplasticity extension to multisurface inelasticity is solved by dividing the stress space on subspaces depending on the value of the first stress invariant and by defining for each subspace loading and potential functions properly modelling the deformatiom mechanism activated by stress states belonging to the given subspace. As an application of this concept, a viscoplastic model for high porous rocks capable of describing experimentally observed shear failure and collapse deformation mechanisms is presented. The model is implemented into the Lagamine finite element code. A numerical test example for solving one-dimentional nonlinear consolidation shows a reasonable prediction, qualitatively representing the experimental observations during oedometer creep tests on chalk, [8]. The experience with the Lagamine FE code shows that the selection of the time step length is very important for the accuracy of the solution. The variable time stepping scheme, realized by Lagamine automatic strategy is more advantageouse to achieve the solution accuracy. Further work is needed for evaluating proper identification techniques and experimental verification of both elastic and inelastic behaviour. The experimental data has to be suffitiant for performing a more precise least - square nonlinear estimation procedure. Thus it will be possible to compare not only qualitatively but also quantitavely the experimental and numerical results.

Acknowledgements

The NATO Science Committee is gratefully acknowledged for supporting the first author by Type B (Advanced) Fellowship to work in DIG, University of Liege, Belgium. Partially this work has been sponsored by the National Science Fund, Bulgaria within the Research project MM 903/99.

References

1. Collin, F., Radu, J-P., Charlier, R.: First development of the mechanical model. PASACHALK – 1st periodic twelve-monthly progress report . (1998)
2. Collin, F., Delage, P., Schroeder, C., Charlier, R.: Geomechanical constitutive modelling of a chalk partly saturated by oil and water. EUROCK 2000 Symposium. Aachen, Germany, 27-31 March. (2000)
3. Barnichon J.-D.: Finite Element Modelling in Structural and Petroleum Geology. PhD thesis, Université de Liège, (1998)
4. Van Eekelen H. A. M.: Isotropic yield surfaces in three dimensions for use in soil mechanics. International Journal for Numerical and Analytical Methods in Geomechanics. No. 4. pp. 98–101 (1980)
5. Charlier R.: Approche unifiée de quelques problèmes non linéaires de mécanique des milieux continus par la méthode des éléments finis. Université de Liège. Thèse de Doctorat. (1987)
6. Perzyna P.: Fundamental Problems in Viscoplasticity. Advances in Applied Mechanics, vol. 9, 244–368 (1966)
7. Simo J. C. and T. J. R. Hughes: Computational Inelasticity. Springer Verlag, Berlin (1998)
8. Shao J. F. and Henry J. P.: Development of elastoplastic model for porose rock. International Journal of Plasticity. vol. 7, 1–13, (1991)

Solvers for Systems of Nonlinear Algebraic Equations – Their Sensitivity to Starting Vectors

Deborah Dent[1], Marcin Paprzycki[1], and Anna Kucaba-Pietal[2]

[1] School of Mathematical Sciences, University of Southern Mississippi
Hattiesburg, MS 39406-5106
[2] Department of Fluid Mechanics and Aerodynamics,
Rzeszow University of Technology
Rzeszow, W.Pola 2, Poland

Abstract. In this note we compare the sensitivity of six advanced solvers for systems of nonlinear algebraic equations to the choice of starting vectors. We will report on results of our experiments in which, for each test problem, the calculated solution was used as the center from which we have moved away in various directions and observed the behavior of each solver attempting to find the solution. We are particularly interested in determining the best global starting vectors. Experimental results are presented and discussed.

1 Introduction

Recently we can observe a growing interest in engineering problems resulting in large systems of nonlinear algebraic equations. For instance, in a real-world problem originating from avionics [11,12] a realistic model would require solution of 500+ equations, but due to the lack of convergence, the programs and methods used by the authors were unable to solve systems of more than 64 equations.

The mathematical theory and computational practice are well established when a system of linear algebraic equations or a single nonlinear equation is to be solved [18]. This is clearly not the case for systems of nonlinear algebraic equations. Our current research has shown both a lack of libraries of solvers and standard sets of test problems (different researchers use different test problems with only a small overlap). In this context we have to remember that, until recently, in the engineering practice, only systems with relatively few equations have been solved. This explains one of the problems of existing "popular" test cases. Most of them have a very small number of equations (2-10) and only very few are defined so that they can reach 100 equations. In our earlier work [6,7,8,9] we have reported on our efforts to collect most of the existing solvers and apply them to up to 22 of standard test problems with the number of equations ranging from 2 to 200. We were able to locate solvers based on *Newton*'s method and its modifications, *Brown's*, *bisection*, *continuation*, *hybrid* algorithms, and the *homotopy*, and *tensor* methods and applied each solver to the test problems collected from the literature and the Internet. We were able to conclude that we can exclude the simple algorithms and in-house implementations from further testing, that methods like *homotopy* and *continuation* cannot be used as a blackbox approach without more work and the *tensor* method seemed to be the most robust.

When the test problems were considered, we were able to find that five of them are easily solvable by all approaches and thus they are useless for testing purposes. The results of the remaining test problems allowed us to observe that proper choice of the starting vector has a strong effects on the solution process (bad selection of the starting vector can result in lack of convergence). Because the likelihood of convergence depends on the solution method and the problem to be solved, we decided to perform a behavior comparison of six advanced solvers for the test problems identified earlier.

In this note, we will report on results of our experiments in which, for each test problem, the perturbed solution, and initial starting vectors of all ones, zeros and random numbers were used to observe the behavior of each solver attempting to find the solution. Based on these experiments we will try to establish which solvers can handle global convergence.

The paper will be organized as follows. Section 2 briefly describes the solvers that used in our work. In section 3 we introduce the test problems. Section 4 will summarize the results of our numerical experiments followed by a concluding remarks and description of future work.

2 Solvers and Algorithms for Systems of Nonlinear Algebraic Equations

As mentioned above, in our earlier work, we have found that only more sophisticated algorithms are capable of solving test systems of nonlinear algebraic equations (outside of the group of five easy ones). We are now focusing on non-commercial versions of codes based on a *hybrid* algorithm and the *Brown's*, *homotopy*, *continuation*, and *tensor* methods. These algorithms are all documented in ACM TOMS and briefly reviewed in [15]. Their implementations were obtained from the NETLIB repository [17]. It is appropriate to use [15] as the reference where brief descriptions of the code are given. Further, in the subsections of §2 the original works where the methods were proposed have to be referred. We have thus modified (to handle up to 200 equations) the following software packages: 1) HYBRD, 2) SOS, 3) CONTIN, 4) HOMPACK, and 5) TENSOLVE. Recently we have also discovered and added to this list the LANCELOT package, which is a part of the NEOS environment [16].

We will now briefly summarize these algorithms and the solvers (in all cases the references cited and [18] should be consulted for the details). We assume that a system of n nonlinear algebraic equations $f(\mathbf{x}) = \mathbf{0}$ is to be solved where $\mathbf{x}$ is n-dimensional vector and $\mathbf{0}$ is the zero vector.

2.1 HYBRD

HYBRD is part of the MINPACK-1 suite of codes [13,14]. HYBRD's design is based on a combination of a modified *Newton* method and the *trust region* method. Termination occurs when the estimated relative error less than or equal the defined by the user tolerance (we used the suggested default value of the square root of the machine precision).

2.2 SOS

SOS is a part of the SLATEC suites of codes [10]. SOS solves a system of N simultaneous nonlinear equations in N unknowns. It solves the problem $f(\mathbf{x}) = \mathbf{0}$ where x is a vector with components $x(1),...,x(N)$ and f is a vector of nonlinear functions. This code is based on an iterative method called the *Brown's* method [2] which is a variation of Newton's method using Gaussian elimination in a manner similar to the Gauss-Seidel process. All partial derivatives required by the algorithm are approximated by first difference quotients.. The convergence behavior of this code is affected by the ordering of the equations, and it is advantageous to place linear and mildly nonlinear equations first in the ordering. Convergence is roughly quadratic. This method requires a good choice for the starting vector $\mathbf{x}_0$.

2.3 CONTIN

CONTIN, also know as PITCON [19] implements a continuation algorithm with an adaptive choice of a local coordinate. A *continuation* method is designed to be able to target more complicated problems and is the subject of various research efforts [1,20]. This method is expected to be slower than *linesearch* and the *trust region* methods, but it is to be useful on difficult problems for which a good starting point is difficult to establish. The method defines an easy problem for which the solution is known along with a path between the easy problem and the hard problem that is to be solved. The solution of the easy problem is gradually transformed to the solution of the hard problem by tracing this path. The path may be defined as by introducing an addition scalar parameter λ into the problem and defining a function

$$h(\mathbf{x}, \lambda) = f(\mathbf{x}) - (1 - \lambda)f(\mathbf{x}_0) \qquad (1)$$

where $x_0 \in \mathbf{R}^n$. The problem $h(\mathbf{x}, \lambda) = \mathbf{0}$ is then solved for values of λ between 0 and 1. When $\lambda=0$, the solution is clearly $\mathbf{x} = \mathbf{x}_0$. When $\lambda=1$, we have that $h(\mathbf{x},1)=f(\mathbf{x})$, and the solution of $h(\mathbf{x},\lambda)$ coincides with the solution of the original problem $f(\mathbf{x})=\mathbf{0}$. The algorithm for constructing the path is given in [19]. The convergence rate of the *continuation* methods varies, but according to documentation, the method does not require a good choice of the initial vector $\mathbf{x}_0$.

2.4 HOMPACK

HOMPACK [21] is a suite of subroutines for solving nonlinear systems of equations by *homotopy* methods [4]. The *homotopy* and *continuation* methods are closely related. In the *homotopy* method, a given problem $f(\mathbf{x})=\mathbf{0}$ is embedded in a one-parameter family of problems using a parameter λ assuming values in the range $[0, \ldots, 1]$. Like the *continuation* method, the solution of an easy problem is gradually transformed to the solution of the hard problem by tracing a path. There are three basic path-tracking algorithms for this method: ordinary differential equation based (code FIXPDF), normal flow (code FIXPNF), and quasi *Newton* augmented Jacobian matrix (code FIXPQF). The code is available in both Fortran 77 and Fortran 90 [21]. The Fortran 77 version was used in our test. We tested all three approaches and since the results were very close, we will report FIXPDF results only.

2.5 TENSOLVE

TENSOLVE [3] is a modular software package for solving systems of nonlinear equations and nonlinear least-square problems using the *tensor* method. It is intended for small to medium-sized problems (up to 100 equations and unknowns) in cases where it is reasonable to calculate the Jacobian matrix or its approximations. This solver provides two different strategies for global convergence; a line search approach (default) and a two-dimensional trust region approach. The stopping criteria is meet when the relative size of $\mathbf{x}_{k+1} - \mathbf{x}_k$ is less than the $macheps^{\frac{2}{3}}$, or $\|f(x_{k+1})\|\infty$ is less than $macheps^{\frac{2}{3}}$, or the relative size of $f'(\mathbf{x}_{k+1})^T f(\mathbf{x}_{k+1})$ is less than $macheps^{\frac{1}{3}}$ and unsuccessfully if the iteration limit is exceeded.

2.6 LANCELOT

LANCELOT is one of the solvers available on the NEOS Web-based environment [5,16]. The NEOS environment is a high speed, socket-based interface for UNIX workstations that provide easy access to all the optimization solvers available on the NEOS Server. This tool allows users to submit problems to the NEOS Server directly from their local networks. Results are displayed on the screen. LANCELOT is a standard Fortran 77 package for solving large-scale nonlinearly constrained optimization problems. The areas covered by Release A of the package are: unconstrained optimization problems, constrained optimization problems, the solution of systems of nonlinear equations, and nonlinear least-squares problems.

The software combines a trust region approach adapted to handle the bound constraints, projected gradient techniques, and special data structures to exploit the (group partially separable) structure of the underlying problem. It additionally provides direct and iterative linear-solvers (for *Newton* equations), a variety of preconditioning and scaling algorithms for more difficult problems, quasi-*Newton* and *Newton* methods, provision for analytical and finite-difference gradients.

3 Test Cases for Systems of Nonlinear Algebraic Equations

In previous studies we were able to classify several of the test problems as easily solvable by all methods. These included the Rosenbrock's, Discrete Boundary Value, Broyden Tridiagonal, Broyden Banded and the Freudenstein-Roth functions [17]. Since the fact that a solver is capable of solving them introduces no new information we have decided to remove them from further considerations. In our search for test problems we have come across problems of least squares type as well as constrained and unconstrained optimization. We have decided to concentrate out attention strictly on systems of nonlinear algebraic equations and Table 1 contains the list of test problems used in our work.

Table 1. Test problems

1. Powell singular function [17]	10. Variably dimensioned function[17]
2. Powell badly scaled function [17]	11. Exponential/Sine Function[22]
3. Wood function [17]	12. Semiconductor Boundary Condition[22]
4. Helical valley function[17]	13. Gulf Research and Development[17]
5. Watson function[17]	14. Extended Powell Singular[17]
6. Chebyquad function[17]	15. Extended Rosenbrock[17]
7. Brown almost-linear function[17]	16. Dennis, Gay and VU[17]
8. Discrete integral equation [17]	17. Matrix Square Root [17]
9. Trigonometric function[17]	

All codes are implemented in Fortran 77 and were run in double precision on a PC with a Pentium Pro 200 MHz processor. When applying the five solvers we have kept the default settings of all parameters as suggested in the implementation (which matches our assumption of the solver being treated like black-box software).

3.1 Simple Test Case

In this study we examined the 17 test problems summarized in Table 1 by studying the sensitivity of the starting vectors. For each problem we used the default vector, all ones, all zeros and random numbers as our initial starting vectors. We used the default number of equations for each problem, which ranged from 2 to 10 equations.

We were able to observe behavior patterns from the problems the were able to converge which helped us determine which solvers are more adapt for global convergence. The results for each problem were typical to that of problem 8, the Brown Almost-linear function. Figure 1 shows the number of iterations required for convergence for each of the various testing methods used on this problem. This problem shows that it is easy to converge with any solver as long as the initial starting vector is in a certain range of the solution vector but once outside of that range, convergence did not occur for HYBRD, SOS, CONTIN, HOMPACK, and LANCELOT. We applied the same test to the other problems and found the pattern set by these problems to be consistent - solvability of the test problems depends on the solver and the starting vector except for TENSOLVE. It appears that TENSOLVE seems to be more robust and was able to achieve convergence regardless of the starting vectors.

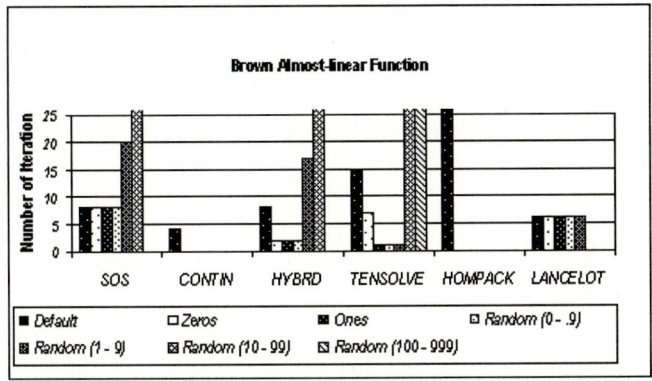

Fig. 1. Various Initial Starting Vectors for Problem 8

3.2 More Difficult Test Cases

We ran a set of test on all 17 problems. The starting vector were defined by adding percentages to a known solution in increments of 10%. For example, the solution set for Problem 8 is $[1.0, \ldots, 1.0]$ for $n=10$. We ran the problem 10 times using the initial values sets of $([1.1, \ldots, 1.1], [1.2, \ldots, 1.2], \ldots, [2.0, \ldots, 2.0])$. Next we repeated the process subtracting percentages from a known solution in increments of 10%. We then recorded the point when there was non-convergence in the positive direction (adding percentages) and the negative direction (subtracting percentages). If there was always a convergence, we recorded the results as 100%, otherwise, we record the exact percentage away from the exact solution that non-convergence occurred. We then noted that the behavior was similar for all the test problems and the convergence rate above and below the exact solution for each problem per solver was less than 20% as shown in Figure 2.

The results are rather interesting as they show that, in the experimental setup used in our experiments; the *tensor* method outperforms the other solvers (with the *combination trust region/projected gradient* technique coming second and
hybrid method third) when looking at global convergence. The TENSOLVE, LANCELOT, HYBRD and SOS codes appear to have been better designed to handle various initial starting vectors.

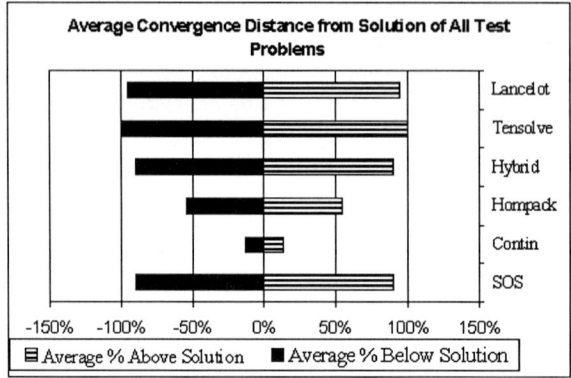

Fig. 2. Average Percentage Convergence in Dependence of Starting Vector of the Methods

4 Conclusions and Future Work

In this note we have briefly reported on our experiments analyzing the sensitivity of the initial starting vectors for standard test systems of up to 10 nonlinear algebraic equations solved by five advanced solvers. We have established a pattern with previous works and this set of experiments and has found that:

- solvability of the test problems depends on the solver and the starting vector,
- problems which are not solvable using one method may be solvable by another method, and
- of the solvers tested, the *tensor* method based solver appeared to be most robust.

Similar results were achieved for large number of equations in a previous publication [6].

Our future work will concentrate on expanding the *tensor* method based solver (as the most promising one) to handle very large systems. We will also continue our search for solvers that can handle medium to large systems of nonlinear algebraic equations as well as new interesting test problems that can be recommended to study the robustness of the nonlinear solvers. We will apply these solvers to the original avionics problem and observe their performance.

References

1. Allgowerr, E. and George, K.: Numerical Continuation Methods: An Introduction, Springer-Verlag, Berlin (1990) 365
2. Brown, K. M.: A quadratically convergent Newton-like method based upon Gaussian elimination, SIAM Journal on Numerical Analysis, 6 (1969) 560-569
3. Bouaricha, A., Schnabel, R.: Algorithm 768: TENSOLVE: A Software Package For Solving Systems Of Nonlinear Equations And Nonlinear Least-Squares Problems Using Tensor Methods. ACM Trans. Math. Software, **23**, 2 (1997), 174-195

4. Burden, R. L., Faries, J. D.: Numerical Analysis. PWS-Kent Publishing Company, Boston, (1993) 575-576
5. Conn, A. R., Gould, N. I. M and Toint, Ph. L.: LANCELOT: A Fortran package for large-scale nonlinear optimization (Release A), Springer Series in Computiational Mathematics 17, Springer-Vergag, (1992)
6. Dent, D., Paprzycki, M., Kucaba-Pietal, A.: Comparing Solvers for Large Systems of Nonlinear Algebraic Equations. Proceedings of the 16th IMACS World Congress, to be published
7. Dent, D., Paprzycki, M., Kucaba-Pietal, A,: Performance of Solvers for Systems of Nonlinear Algebraic Equations. Proceedings of 15th Annual Conf. on Applied Math (1999) 67-77
8. Dent, D., Paprzycki, M., Kucaba-Pietal, A,: Studying the Numerical Properties of Solvers for Systems of Nonlinear Equations. Proceedings Of The Ninth International Colloquium On Differential Equations (1999), 113-118
9. Dent, D., Paprzycki, M., Kucaba-Pietal, A,: Testing Convergence of Nonlinear System Solvers. Proceedings of the First Southern Symposium on Computing, (1998)
10. Fong, K. W., Jefferson, T. H., Suyehiro, T.,Walton, L.: Guide to the SLATEC Common Mathematical Library, Argonne, Ill., (1990)
11. Kucaba-Pietal, A., Laudanski, L.:Modeling Stationary Gaussian Loads. Scientific Papers of Silesian Technical University, Mechanics, **121** (1995) 173-181
12. Laudanski, L.: Designing Random Vibration Tests. Int. J. Non-Linear Mechanics, **31,** 5 (1996) 563-572
13. More, J. J., Garbow, B. S., Hillstrom, K. E.: User Guide for MINPACK-1, Argonne National Laboratory Report ANL-80-74, Argonne, Ill., (1980)
14. More, J. J., Sorensen, D. C., Hillstrom, K. E., Garbow, B. S.: The MINPACK Project, in Sources and Development of Mathematical Software, W. J. Cowell, ed., Prentice-Hall (1984)
15. NEOS Guide (1996) http://www-fp.mcs.anl.gov/otc/Guide/
16. NEOS Solvers (2000) http://www-neos.mcs.anl.gov/neos/server-solvers.html
17. Netlib Repository (1999) http://www.netlib.org/liblist.html
18. Rheinboldt, W. C.: Methods for Solving System of Nonlinear Equations. SIAM, Philadelphia (1998)
19. Rheinboldt, W. C., Burkardt, J.: Algorithm 596: A Program For A Locally Parameterized Continuation Process. ACM Trans. Math. Software, **9** (1983) 236-241
20. Stoer, J., Bulirsh, R.: Introduction to Numerical Analysis. Springer, New York, (1993) 521
21. Watson, L. T., Sosonkina, M., Melville, R. C., Morgan, A. P., Walker, H. F.: Algorithm 777:HOMPACK 90: Suite Of Fortran 90 Codes For Globally Convergent Homotopy Algorithms. ACM Trans. Math. Software **23**, 4 (1997), 514 – 549
22. Weimann, U. N.: A Family of Newton Codes for Systems of Highly Nonlinear Equations. ZIB Technical Report TR-91-10, ZIB, Berlin, Germany (1991)

The Min-Max Portfolio Optimization Strategy: An Empirical Study on Balanced Portfolios*

Claude Diderich and Wolfgang Marty

Credit Suisse Asset Management
CH-8070 Zurich, Switzerland

Abstract. Modern investment processes often use quantitative models based on Markowitz's mean-variance approach for determining optimal portfolio holdings. A major drawback of using such techniques is that the optimality of the portfolio structure only holds with respect to a single set of expected returns. Becker, Marty, and Rustem introduced the robust min-max portfolio optimization strategy to overcome this drawback. It computes portfolio holdings that guarantee a worst case risk/return tradeoff whichever of the specified scenarios occurs. In this paper we extend the approach to include transaction costs. We illustrate the advantages of the min-max strategy on balanced portfolios. The importance of considering transaction costs when rebalancing portfolios is shown. The experimental results illustrate how a portfolio can be insured against a possible loss without sacrificing too much upside potential.

1 Introduction

One of the most widely used models for determining optimal portfolio holdings with respect to a tradeoff between expected return and risk is the mean-variance approach introduced by Markowitz [3]. The basic argument behind the model is that investors hold portfolios with the highest expected return for a given level of variance. Although the mean-variance portfolio optimization model allows to compute optimal portfolio holdings, the optimality only holds with respect to a single set of expected returns. Furthermore, slight changes in the expected returns have a big implact on the optimal portfolio holdings.

In [5] Becker, Marty, and Rustem introduced the robust min-max portfolio optimization strategy (min-max strategy) to overcome this problem. Their framework considers a finite set of possible expected returns, called scenarios. It computes portfolio holdings that guarantee a worst case risk/return tradeoff whichever of the specified scenarios occurs. In contrast with Markowitz's approach, the min-max strategy only gives a lower bound on the expected return for a given variance or risk.

Using a framework like the min-max strategy proves very useful for determining short term portfolio structure changes, for example, for balanced portfolios[1].

---

* The views expressed in this paper are those of the authors and do not necessarily reflect the opinion of Credit Suisse Asset Management.
[1] A balanced portfolio is a portfolio consisting of equities, bonds, and cash.

It allows to over- and underweight certain assets or asset classes, based on different return forecasts. As any portfolio modification induces costs, we extended the min-max strategy to take into account transaction costs.

Other approaches currently under investigation for computing efficient portfolio holdings, with respect to a given utility function, are stochastic multi-period optimization models [2], models based on continuous time methods using partial differential equations [4], or factor models [1], to name just the most important ones.

2 The Min-Max Strategy

We review the min-max portfolio strategy introduced by Becker, Marty, and Rustem [5]. Consider a set of individual assets or asset classes. Consider a set S of return scenarios, each scenario representing a specific view of the market outcome. For example, one scenario could be based on interest rate raise expectations, or a slowdown of a specific region's economic growth. Let r_s be the expected returns of the asset classes considered, with respect to scenario $s \in S$. We assume that the investor is interested in performance relative to a given benchmark portfolio, rather than absolute performance[2]. Let p be the investor's initial portfolio holdings, b the considered benchmark portfolio, t the transaction costs in percents, and A, c representing general constraints. Let Q be the covariance matrix associated with the asset classes. It may be estimated using historical data or using volatility and/or correlation forecast models. For the sake of simplicity, only a single covariance matrix is used. In the most general setting, the min-max strategy allows for multiple covariance matrices, one for each forecasted return scenario.

The robust min-max portfolio optimization strategy can be formulated as

$$\begin{aligned}
&\min_{\omega,\omega^+,\omega^-} (\omega - b)'Q(\omega - b) \\
&\text{subj. to } \forall s \in S \colon r'_s (\omega - b) - t'(\omega^+ + \omega^-) \geq R \\
&\quad 1'\omega = 1 \\
&\quad \omega = p + \omega^+ - \omega^- \\
&\quad \omega \geq 0, \omega^+ \geq 0, \omega^- \geq 0 \\
&\quad A\omega \leq c.
\end{aligned} \qquad (1)$$

The unknown weights ω represent the portfolio holdings and ω^+ and ω^- the buy and sell decisions. R represents the lower bound of the expected return defining the risk/return tradeoff. It is called the *min-max return*. The portfolio holdings ω^*, solution of the problem (1) for a given R, are called *min-max optimal portfolio holdings*. We assume that the investor is fully invested and that no short positions are allowed. Instead of fixing R, most investors require $e(\omega) = \sqrt{(\omega - b)'Q(\omega - b)}$, the *portfolio tracking error*, to be bound by some constant. In this case, we iterativly solve problem (1) for different values of R by using a

---

[2] Most institutional investors evaluate their performance against the performance of the market, represented by an index or benchmark like, for example, the S&P 500.

Table 1. Considered indices modeling the five asset classes used, benchmark structure, initial portfolio holdings, as well as three return scenarios

Index	b	p	sce. 1	sce. 2	sce. 3
Salomon SFr. 3 month money market	5%	10%	2%	2%	2%
Salomon SFr. Gov. Bond 1+	40%	30%	3%	5%	5%
Salomon World Gov. Bond	15%	10%	3%	3%	5%
FT Switzerland	25%	30%	7%	7%	8%
FT World ex. Switzerland	15%	20%	7%	5%	7%

bi-sectioning algorithm. Such an approach can be used because $R = f(e(\omega^*))$ is a monotone increasing function for min-max optimal portfolio holdings. The graphical representation of f is called the *min-max frontier*.

It can be shown that adding any return scenario $r = \sum_{s \in S} \lambda_s r_s$, where $\sum_{s \in S} \lambda_s = 1$ and $\lambda_s \geq 0$, to S does not change the optimal solution of problem (1).

Although problem (1) does not have a complex structure, solving it is not an easy task. This is especially due to the fact that the Hessian matrix of the quadratic program, although positive semi-definite, is singular. Furthermore, for small values of R, the objective function takes values close to zero. For the experiments described in this paper, we relied on the CPLEX barrier quadratic programming algorithm.

3 Experimental Results

To illustrate the min-max strategy, we consider a portfolio based on the asset classes represented by the indices in Table 1. These asset classes are common for Swiss pension fund portfolios. The initial portfolio p as well as the considered benchmark b are also shown in Table 1. For illustrative purposes, we use the return scenarios shown in Table 1. All computations are done using a one month horizon. Input data as well as all the results, unless otherwiese stated, are annualized for reading convenience. The correlations between asset classes are estimated using ten years of historical monthly data (1988–97). All data is provided by Datastream, converted to Swiss francs where necessary.

3.1 The Min-Max Frontier

In Fig. 1 we illustrate the min-max frontier, which, for a given risk, represents the worst return to expect for a min-max optimal portfolio. Indeed, whichever of the three specified scenarios occurs, the relative return obtained is at least as large as the min-max return. Furthermore, the scenario giving a return equal to the min-max return is not always the same. Indeed, for small tacking error values, scenario one gives the smallest relative return, whereas for large tracking error values scenario two gives the smallest relative return. The min-max strategy maximizes the worst case expected relative return.

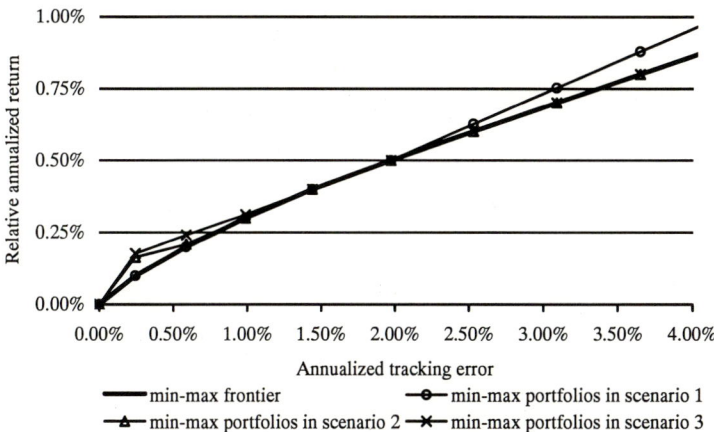

Fig. 1. A clipping of the min-max frontier as well as the relative return of the min-max optimal portfolio holdings evaluated using the three return scenarios

On the other hand, as illustrated in Fig. 2, consider the mean-variance efficient frontier using scenario one. Choosing a portfolio on this frontier but a different scenario occurring, for example the scenario two, will give a considerably worse return than the min-max optimal portfolio having the same risk. The min-max approach computes portfolios that guarantee a minimal relative return with respect to all the given scenarios. For this insurance agains a potential loss a certain premium has to be paid. In general the premium is less than the potential loss, but mathematically there exists no non-trivial relation between them.

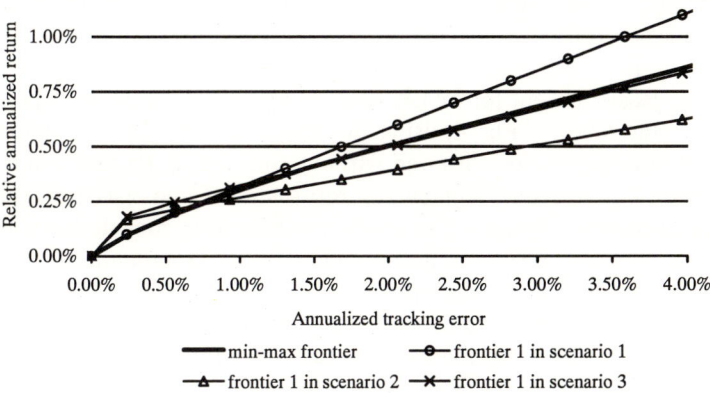

Fig. 2. Relative returns obtained from min-max optimal portfolio holdings compared to relative returns obtained from portfolio holdings selected on the mean-variance efficient frontier computed using scenario one

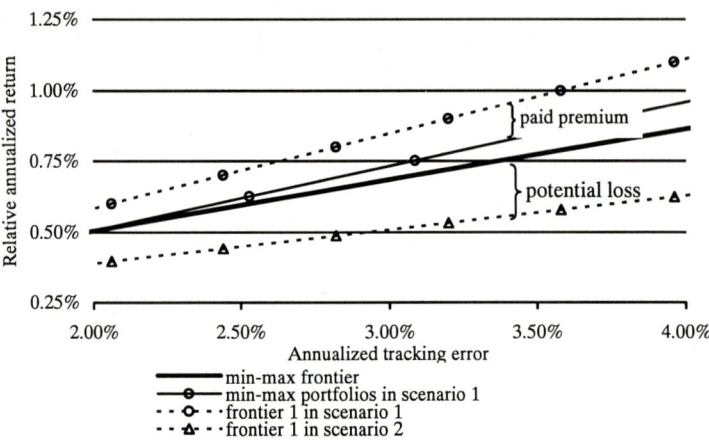

Fig. 3. Comparison of the premium paid using the min-max strategy to insure against potential loss

In Fig. 3 we illustrate the premium to be paid for the potential loss insurance. If scenario one occurs, the premium paid by choosing a min-max portfolio instead of a portfolio on the efficient frontier with respect to scenario one is about 12 basis points[3] at an annualized tracking error level of 3.4%. On the other hand, comparing the two same portfolios, but scenario two occurring, the min-max portfolio insures against a potential loss of 20 basis points annualized at the same tracking error level of 3.4%.

3.2 The Effect of Transaction Costs

Up to now we did not consider transaction costs. But, if the min-max strategy is used for tactical asset allocation[4], transaction costs must not be neglected. To illustrate this situation, we consider a portfolio which we wanted to rebalance to a min-max optimal portfolio. We set the transaction costs to be 50 basis points for all asset classes, except for money market where we use 20 basis points. The situation is illustrated in Fig. 4, using the scenarios from Table 1. We compute min-max optimal portfolios with and without considering transaction costs. We then compare the expected annualized returns with and without transaction cost, transaction cost adjusted. The additional gain from considering transaction costs for the example in Fig. 4 is around 50 basis points annualized, for tracking errors between 0.2% and 1%, the value increasing even further for larger tracking errors. The effect of transaction costs is sensible when the transaction costs are of the same order of magnitude as the expected relative returns. In any case, considering transaction costs does never deteriorate the solution.

---

[3] One basis point is an alternative notation for 0.01%.

[4] A tactical asset allocation decision is a decision to change the portfolio structure such as to take advantage of expected short term movements in the markets. Usually the horizon for tactical asset allocation decisions does not exceed one month.

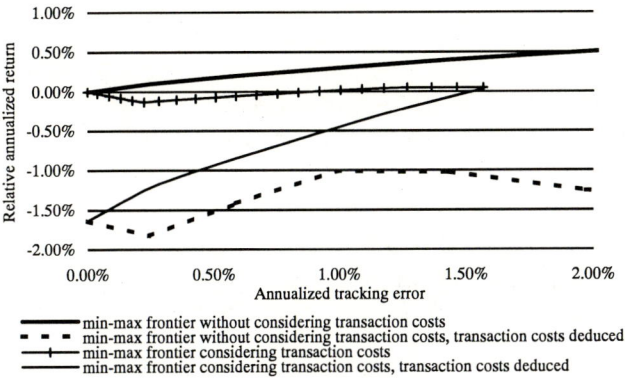

Fig. 4. The effect of transaction costs on the expected relative min-max return

3.3 Backtesting the Min-Max Strategy

To illustrate the validity of the min-max strategy in a dynamic context, we simulate its application over a two year horizon. The chosen timeperiod, from January 1998 up to May 2000, includes the Russian crises in fall 1998. The portfolio is rebalanced monthly, when necessary. For this experiment we assume the benchmark as well as the initial portfolio structure shown in Table 1. The covariance matrix is estimated as described previously and kept constant over the simulation horizon. We use three scenarios, which we calculate as 1) one month monthly historical returns, 2) three month monthly historical returns, and 3) one year monthly historical returns at the date of rebalancing. The portfolio selected each month is such that its annualized tracking error does not exceed 1%.

In Fig. 5 we show the evolution of the value of an investment of 100 Swiss francs starting in January 1998. Realized returns are used for the computation. Furthermore we present the returns obtained when choosing the portfolio on the efficient frontier associated with each single scenario.

Fig. 6 illustrates the structure of the min-max portfolio over the two year simulation horizon. From a portfolio manager's perspective, these changes are reasonable and implementable. During the optimization, no explicit restrictions on the maximal turnover, except for transaction costs of 50 basis points for all asset classes, except for money market (20 basis points), were used.

4 Conclusion

In this paper we have represented the min-max strategy for computing optimal portfolio holdings. The computed portfolio holdings guarantee a certain expected return with respect to a given set of scenarios. They represent portfolios insured against downside risk without sacrifycing too much upside potential. We have illustrated how to take into account transaction costs during the computation of efficient portfolios.

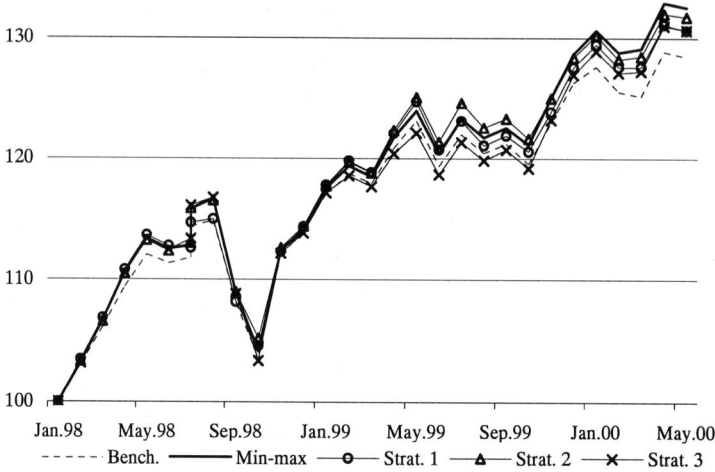

Fig. 5. Evaluation of the min-max strategy over a two year period and comparison with mean-variance efficient frontier strategies

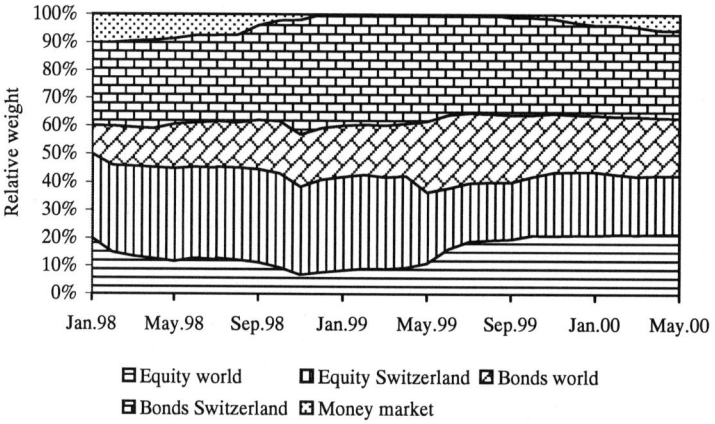

Fig. 6. Asset allocation structure of the min-max portfolios over the two year simulation period

References

1. J. Y. Campbell, A. W. Lo, and A. C. MacKinlay. *The Econometrics of Financial Markets*. Princeton University Press, Princeton, NJ, 1997.
2. G. Consigli and M. A. H. Dempster. The CALM stochastic programming model for dynamic asset-liability management. In *Worldwide Asset and Liability Modeling*, chapter 19, pages 464–500. Cambridge University Press, Cambridge, United Kingdom, 1988.

3. H. Markowitz. *Portfolio Selection: Efficient Diversification of Investments*. John Wiley, New York, NY, 1959.
4. R. C. Merton. *Continuous-Time Finance*. Blackwell, Malden, MA, 1992.
5. B. Rustem, R. Becker, W. Marty. Robust min-max portfolio strategies for rival forecast and risk scenarios. *Journal of Economic and Dynamic Control*, to appear.

Convergence Rate for a Convection Parameter Identified Using Tikhonov Regularization

Gabriel Dimitriu

University of Medicine and Pharmacy, Faculty of Pharmacy,
Department of Mathematics and Informatics, 6600 Iasi, Romania
dimitriu@umfiasi.ro

Abstract. In this paper we establish a convergence rate result for a parameter identification problem. We show that the convergence rate of a convection parameter in an elliptic equation with Dirichlet boundary conditions is $\mathcal{O}(\sqrt{\delta})$, where δ is a norm bound for the noise in the data.

1 Introduction

In this study we present a convergence rate result for a parameter identification problem. To be precise, we show that the convergence rate of the convection parameter b in the elliptic equation

$$-(au_x)_x + bu_x + cu = f \quad \text{in } (0,1), \tag{1}$$

with Dirichlet boundary conditions $u(0) = u(1) = 0$ is $\mathcal{O}(\sqrt{\delta})$, where δ is a norm bound for the noise in the data f. This parameter represents the solution of the identification problem associated with (1) and regularised by Tikhonov method. We take $f \in L^2(0,1)$, $(a,b,c) \in Q \subset \tilde{Q} = W^{1,2}(0,1) \times W^{1,2}(0,1) \times W^{1,2}(0,1)$ with $\tilde{Q}$ endowed with the Hilbert-space product topology and $Q = \{(a,b,c) \in \tilde{Q} : 0 < \underline{a} \le a(x), |a|_{W^{1,2}(0,1)} \le \mu, |b|_{W^{1,2}(0,1)} \le \mu, c(x) \ge \underline{c} > 0 \text{ a.e. in } (0,1)\}$.

2 Functional Framework

Following the functional framework described in [4] we consider the nonlinear ill-posed problem

$$F(q) = f_0. \tag{2}$$

By ill-posedness, we always mean that the solutions do not depend continuously on the data. Here $F : \text{Dom}(F) \subset X \to Y$ is a nonlinear operator between Hilbert spaces X and Y. We assume that the operator F satisfies the following conditions:

(i) F is continuous and
(ii) F is weakly (sequentially) closed, i.e. for any sequence $\{q_n\} \subset \text{Dom}(F)$, weak convergence of q_n to q in X and weak convergence of $F(q_n)$ to f in Y imply $q \in \text{Dom}(F)$ and $F(q) = f$.

We use the concept of an q^*-minimum-norm solution q_0 for the problem (2):

Definition 1. Let $q^* \in X$ be fixed. We say that q_0 is q^*-minimum-norm solution (q^*-MNS) for (2) if

$$\|F(q_0) - f_0\| = \min\{\|F(q) - f_0\| \; : \; q \in \text{Dom}(F)\} \tag{3}$$

and

$$\|q_0 - q^*\| = \min\{\|q - q^*\| \; : \; \|F(q) - f_0\| = \|F(q_0) - f_0\|\}. \tag{4}$$

A solution of (3) and (4) need not exist and, even if it does, it need not be unique, because of the nonlinearity of F. Also we note that q^* plays an important role in obtaining the solutions defined by (3) and (4). Thus, the choice of q^* can influence which (least-squares) solution we want to approximate. In the situation of multiple least-squares solutions, q^* plays the role of a selection rule. In what follows we assume existence of an q^*-minimum-norm least-squares solution for the unperturbed data $f_0 \in Y$.

To cope with the ill-posedness of the problem (2) we shall use the well known Tikhonov regularisation. By this method a solution for (2) is approximated by a solution of the nonlinear regularised optimization problem

$$\min_{q \in \text{Dom}(F)} \{\|F(q) - f_\delta\|^2 + \alpha \|q - q^*\|^2\}, \tag{5}$$

where $\alpha > 0$ is a small parameter, $f_\delta \in Y$ is an approximation to the exact right-hand side f_0.

From computational reasons, since problem (5) can only be solved approximately we slightly generalise it by considering the problem of finding an element $q_\alpha^{\delta,\eta} \in \text{Dom}(F)$ such that

$$\|F(q_\alpha^{\delta,\eta}) - f_\delta\|^2 + \alpha \|q_\alpha^{\delta,\eta} - q^*\|^2 \leq \|F(q) - f_\delta\|^2 + \alpha \|q - q^*\|^2 + \eta, \tag{6}$$

for all $q \in \text{Dom}(F)$, where $\eta \geq 0$ is a small parameter. Obviously, for $\eta = 0$, the problem (6) is equivalent to (5).

Aspects of stability, convergence and convergence rates (as $\alpha \to 0$) has been extensively studied in the literature, both in the linear and nonlinear case, e.g. in [1], [2], [3], [4], [6], [8]. Under the given assumptions on operator F and using compactness-type arguments it was proved in [4] that problem (6) admits a stable solution in the sense of continuous dependence of the solutions on the data f_δ and that the solutions of (6) converge towards a solution of (2) as $\alpha \to 0$ and $f_\delta \to f_0$.

3 Convergence Rate Result

We now focus on the convergence rate analysis. The theorem below gives sufficient conditions for a rate $\|q_\alpha^{\delta,\eta} - q_0\| = \mathcal{O}(\sqrt{\delta})$ for the regularised solutions.

Theorem 1. ([4]) *Let* $\mathrm{Dom}(F)$ *be convex, let* $f_\delta \in Y$ *with* $\|f_\delta - f_0\| \leq \delta$ *and let* q_0 *be an* q^*-*MNS. Moreover, let the following conditions hold:*

(i) F *is Fréchet differentiable,*

(ii) *there exists* $L > 0$ *such that*

$$\|F'(q_0) - F'(\bar{q})\| \leq L\|q_0 - \bar{q}\|, \quad \text{for all } \bar{q} \in \mathrm{Dom}(F),$$

(iii) *there exists* $w \in Y$ *satisfying*

$$q_0 - q^* = F'(q_0)^* w,$$

(iv) $L\|w\| < 1$.

Then for the choices $\alpha \sim \delta$ *and* $\eta = \mathcal{O}(\delta^2)$, *we obtain:*

$$\|q_\alpha^{\delta,\eta} - q_0\| = \mathcal{O}(\sqrt{\delta}).$$

If F is twice Fréchet differentiable, condition (ii) and (iv) may be replaced by the weaker condition

$$(ii)' \quad 2\left(w, \int_0^1 (1-t) F''[q_0 + t(q_\alpha^{\delta,\eta} - q_0)](q_\alpha^{\delta,\eta} - q_0)^2 \, dt\right) \leq \rho \|q_\alpha^{\delta,\eta} - q_0\|^2,$$

with $\rho < 1$.

To see this, note that the left-hand side of $(ii)'$ equals $2(w, r_\alpha^{\delta,\eta})$ with $r_\alpha^{\delta,\eta}$ as in relation

$$F(q_\alpha^{\delta,\eta}) = F(q_0) + F'(q_0)(q_\alpha^{\delta,\eta} - q_0) + r_\alpha^{\delta,\eta}.$$

Taking into account the conditions (i) and (ii) of Theorem 1 we have

$$\|r_\alpha^{\delta,\eta}\| \leq \frac{1}{2} L \|q - q_0\|^2.$$

In the specific setting given by problem (2) the parameter b plays the role of q and the operator F is given by the mapping *parameter* $\mapsto$ *solution*, that is $F(q) := u(b)$. For $b \in \mathrm{Dom}(F)$, let $A(b) : H^2(0,1) \cap H_0^1(0,1) \to L^2(0,1)$ be defined by

$$A(b)\varphi = -(a\varphi_x)_x + b\varphi_x + c\varphi.$$

To apply the convergence rate result given in Theorem 1, we next calculate the first and second order Fréchet derivatives of the function $b \to u(b)$.

Lemma 1. *The mapping* $b \to u(b)$ *from* $W^{1,2}(0,1)$ *into* $W^{2,2}(0,1)$ *is Fréchet differentiable with the Fréchet differential with increment* h *denoted by* $\delta_b u(b) h := \eta(h)$, *with* $\eta(h)$ *the unique solution of*

$$-(a\eta(h)_x)_x + b\eta_x(h) + c\eta(h) = -h u_x(b), \quad \text{in } (0,1), \tag{7}$$

with boundary conditions

$$\eta(h)(0) = \eta(h)(1) = 0.$$

Proof. The verification is quite standard but we include it for the purpose of completness (see [7]). We define the sets $B := \{b \in W^{1,2}(0,1) : |b|_{W^{1,2}} \leq \mu\}$ and $B_\delta := \{b \in W^{1,2}(0,1) : |b|_{W^{1,2}} \leq \mu + \delta\}$. Let $h \in W^{1,2}(0,1)$ and $b \in B$ and note that there exists $\varepsilon(h) > 0$ such that for any $\varepsilon \in (0, \varepsilon(h))$ the element $b + \varepsilon h \in B_\delta$ and $u(b + \varepsilon h)$ exists. Set $u^\varepsilon = \varepsilon^{-1}(u(b + \varepsilon h) - u(b))$ and observe that u^ε must satisfy

$$-(au^\varepsilon_x)_x + bu^\varepsilon_x + cu^\varepsilon = -hu_x(b + \varepsilon h), \tag{8}$$

$$u^\varepsilon(0) = u^\varepsilon(1) = 0.$$

It follows that $u(b + \varepsilon h) \to u(b)$ in $W^{1,2}(0,1)$ as $\varepsilon \to 0$ and that u^ε converges weakly in $W^{2,2}(0,1)$ and thus strongly in $W^{1,2}(0,1)$. We denote this limit by $\eta(h)$. As a consequence of u^ε satisfying (8) and the limit behavior of u^ε, we obtain that $u^\varepsilon \to \eta(b)$ as $\varepsilon \to 0$ strongly in $W^{2,2}(0,1)$ as well. Thus, the limit $\eta(h)$ is the Gâteaux derivative of the mapping $b \to u(b)$ with increment h, i.e. $\eta(h) = \delta u(b)h$, and it satisfies

$$-(a\eta(h)_x)_x + b\eta_x(h) + c\eta(h) = -hu_x(b), \quad \text{in } (0,1), \tag{9}$$

with boundary conditions

$$\eta(h)(0) = \eta(h)(1) = 0.$$

The application $h \to \eta(h)$ is a bounded linear operator from $W^{1,2}(0,1)$ to $W^{2,2}(0,1)$. That $\eta \to \eta(h)$ is the Fréchet differential of u at b with increment h can be verified as follows: let $h \in W^{1,2}(0,1)$ with $|h|_{W^{1,2}(0,1)} \leq \delta$, and set

$$\Delta(h) := |h|^{-1}_{W^{1,2}(0,1)}(u(b+h) - u(b) - \eta(h)).$$

We note that $\Delta(h)$ satisfies

$$-(a\Delta_x(h))_x + b\Delta_x(h) + c\Delta(h) = \frac{h}{|h|_{W^{1,2}(0,1)}}(u_x(b+h) - u_x(b)),$$

$$\Delta(h)(0) = \Delta(h)(1) = 0.$$

From $|u(q)|_{W^{2,2}(0,1)} \leq C|f|_{L^2(0,1)}$ we have

$$|\Delta(h)|_{W^{2,2}(0,1)} \leq C|u_x(b+h) - u_x(b)|_{L^2(0,1)}.$$

which implies that $|\Delta(h)|_{W^{2,2}(0,1)} \to 0$, whenever $|h|_{W^{1,2}(0,1)} \to 0$. Thus, the lemma is established. □

In a similar way one can prove that the second order Fréchet derivative of $b \to u(b)$ is the bilinear mapping denoted by $\xi(b)(h,h) := \delta^2_b u(b)(h,h)$, which satisfies equation

$$-(a\xi_x(h,h))_x + b\xi_x(h,h) + c\xi(h,h) = -2h\eta(h), \tag{10}$$

with boundary conditions

$$\xi(h,h)(0) = \xi(h,h)(1) = 0.$$

Then $F(b) = u(b) = A(b)^{-1}f$ and we have:

$$F'(b)h = -A(b)^{-1}(hu_x(b)), \tag{11}$$

$$F''(b)(h,h) = 2A(b)^{-1}[hA(b)^{-1}(hu_x(b))]. \tag{12}$$

Moreover, the adjoint $F'(b)^*$ is given by

$$F'(b)^*h = -u_x(b)A(b)^{-1}h.$$

Therefore, condition (iii) of Theorem 1 takes the form

$$b_0 - b^* = F'(b_0)^*w = -u_x(b_0)A(b_0)^{-1}w, \quad \text{for some } w \in L^2(0,1).$$

We note that such an element w exists if

$$\frac{b^* - b_0}{u_x(b_0)} \in H^2(0,1) \cap H_0^1(0,1) \tag{13}$$

and is given by

$$w = A(b_0)\frac{b^* - b_0}{u_x(b_0)}. \tag{14}$$

Turning to condition $(ii)'$ we shall show that a certain bound on

$$v := \frac{b^* - b_0}{u_x(b_0)}$$

will imply that for $\alpha \sim \delta$, $\eta = O(\delta^2)$ there exists $\rho < 1$ such that

$$2\left(w, \int_0^1 F''(b^t)(h_\alpha^{\delta,\eta}, h_\alpha^{\delta,\eta})(1-t)\,dt\right)_{L^2(0,1)} \leq \rho \|h_\alpha^{\delta,\eta}\|_{L^2(0,1)}^2, \tag{15}$$

for all $\delta > 0$ sufficiently small. We make the notations:

$$h_\alpha^{\delta,\eta} := b_\alpha^{\delta,\eta} - b_0 \quad \text{and} \quad b^t := b_0 + th_\alpha^{\delta,\eta}.$$

Since b plays the role of q, $b_\alpha^{\delta,\eta}$ is of course defined as $q_\alpha^{\delta,\eta}$ in the previous section.

The left-hand side of (15) will be denoted by $E_\alpha^{\delta,\eta}$. For the estimation of $E_\alpha^{\delta,\eta}$ we shall use the following facts. By Theorem 1, $b_\alpha^{\delta,\eta}$, and hence from $h_\alpha^{\delta,\eta}$, are uniformly bounded in $L^2(0,1)$ for $\delta > 0$ sufficiently small. Since (see [1]), $A(b)^{-1} : L^2(0,1) \to H^2(0,1) \cap H_0^1(0,1)$ is uniformly bounded for b in bounded sets of $L^2(0,1)$, this implies that there exists $\tilde{K}$ such that

$$\|u(b^t)\|_{H^2(0,1)} \leq \|A(b^t)^{-1}\|_{L^2(0,1), H^2(0,1) \cap H_0^1(0,1)} \|f\|_{L^2(0,1)} \leq \tilde{K}\|f\|_{L^2(0,1)},$$

for all $t \in [0,1]$.

Let $K := \|A(b_0)^{-1}\|_{L^2(0,1),H^2(0,1)\cap H_0^1(0,1)}$. In the following we use the estimates:

$$\|fg\|_{L^2(0,1)} \leq \|f\|_{L^2(0,1)}\|g\|_{L^\infty(0,1)} \text{ and } \|g\|_{L^\infty(0,1)} \leq \frac{1}{4\sqrt{3}}\|g\|_{H^2(0,1)}, \quad (16)$$

for all $f \in L^2(0,1)$ and $g \in H^2(0,1) \cap H_0^1(0,1)$. From the expression of the first order Fréchet derivative given by (11) we obtain that

$$A(b^t)^{-1}g = A(b_0)^{-1}g - t\int_0^1 A(b^{ts})^{-1}(h_\alpha^{\delta,\eta}(A(b^{ts})^{-1}g)_x)\,ds, \quad (17)$$

for any $g \in L^2(0,1)$. By Theorem 1, $b_\alpha^{\delta,\eta} \to b_0$, for $\alpha \sim \delta$ and $\eta = O(\delta^2)$, which we assume from now on. With

$$\overline{K} = \|A(b_0)\|_{H^2(0,1)\cap H_0^1(0,1),L^2(0,1)}$$

and using (16), (17) together with $u(b^t) = A(b^t)^{-1}f$, we obtain

$$E_\alpha^{\delta,\eta} \leq 4\|v\|_{L^2(0,1)} \sup_{t\in[0,1]} \|A(b_0)A(b^t)^{-1}[h_\alpha^{\delta,\eta}A(b^t)^{-1}(h_\alpha^{\delta,\eta}u_x(b^t))]\|_{L^2(0,1)}$$

$$\leq 4\|v\|_{L^2(0,1)} \left(\sup_{t\in[0,1]} \|h_\alpha^{\delta,\eta}A(b^t)^{-1}(h_\alpha^{\delta,\eta}u_x(b^t))\|_{L^2(0,1)} \right.$$

$$\left. + \overline{K} \sup_{t,s\in[0,1]} \|tA(b^{ts})^{-1}[h_\alpha^{\delta,\eta}A(b^{ts})^{-1}\left(h_\alpha^{\delta,\eta}A(b^t)^{-1}(h_\alpha^{\delta,\eta}u_x(b^t))\right)_x]\|_{H^2(0,1)} \right)$$

$$\leq 4\|v\|_{L^2(0,1)} \left(\|h_\alpha^{\delta,\eta}\|_{L^2(0,1)} \frac{1}{4\sqrt{3}} \sup_{t\in[0,1]} \|A(b^t)^{-1}(h_\alpha^{\delta,\eta}u_x(b^t))\|_{H^2(0,1)} \right.$$

$$\left. + \overline{K}\left(\frac{\tilde{K}\|h_\alpha^{\delta,\eta}\|_{L^2(0,1)}}{4\sqrt{3}}\right)^3 \tilde{K}\|f\|_{L^2(0,1)} \right)$$

$$\leq 4\|v\|_{L^2(0,1)} \left\{ \overline{K}\tilde{K}^4 \frac{\|f\|_{L^2(0,1)}}{192\sqrt{3}} \|h_\alpha^{\delta,\eta}\|_{L^2(0,1)}^3 \right.$$

$$+ \frac{\|h_\alpha^{\delta,\eta}\|_{L^2(0,1)}}{4\sqrt{3}} \left[K\|h_\alpha^{\delta,\eta}\|_{L^2(0,1)} \sup_{t\in[0,1]} \|u_x(b^t)\|_{L^2(0,1)} \right.$$

$$\left. \left. + \left(\frac{\|h_\alpha^{\delta,\eta}\|_{L^2(0,1)}}{4\sqrt{3}}\right)^2 \tilde{K}\|f\|_{L^2(0,1)} \right] \right\}$$

$$\leq 4\|v\|_{L^2(0,1)} \left[\left(\overline{K}\tilde{K}^4 \frac{\|f\|_{L^2(0,1)}}{192\sqrt{3}} + \tilde{K}\frac{\|f\|_{L^2(0,1)}}{192\sqrt{3}}\right) \|h_\alpha^{\delta,\eta}\|_{L^2(0,1)}^3 \right.$$

$$+ \frac{K}{4\sqrt{3}} \|h_\alpha^{\delta,\eta}\|_{L^2(0,1)}^2 \times \left(\|u_x(b_0)\|_{L^\infty(0,1)} + \frac{1}{4\sqrt{3}}\right).$$

$$\left. \cdot \sup_{t,s\in[0,1]} \|t\left(A(b^{ts})^{-1}(h_\alpha^{\delta,\eta}(A(b^{ts})^{-1}f)_x)\right)_x\|_{H^2(0,1)} \right]$$

$$\leq \frac{K}{\sqrt{3}}\|v\|_{L^2(0,1)}\|u_x(b_0)\|_{L^\infty(0,1)}\|h_\alpha^{\delta,\eta}\|^2_{L^2(0,1)}$$
$$+\frac{\|f\|_{L^2(0,1)}\|v\|_{L^2(0,1)}}{96\sqrt{3}}(\overline{K}\tilde{K}^4 + \tilde{K} + K\tilde{K}^2)\|h_\alpha^{\delta,\eta}\|^3_{L^2(0,1)}.$$

Therefore,

$$E_\alpha^{\delta,\eta} \leq \frac{K}{\sqrt{3}}\|v\|_{L^2(0,1)}\|u_x(b_0)\|_{L^\infty(0,1)}\|h_\alpha^{\delta,\eta}\|^2_{L^2(0,1)} + \mathcal{O}\left(\|h_\alpha^{\delta,\eta}\|^3_{L^2(0,1)}\right).$$

so that condition $(ii)'$ is satisfied for $\delta > 0$ sufficiently small, provided that

$$\frac{K}{\sqrt{3}}\left\|\frac{b^* - b_0}{u_x(b_0)}\right\|_{L^2(0,1)}\|u_x(b_0)\|_{L^\infty(0,1)} < 1,$$

or equivalently

$$\left\|\frac{b^* - b_0}{u_x(b_0)}\right\|_{L^2(0,1)} < \frac{\sqrt{3}}{K\|u_x(b_0)\|_{L^\infty(0,1)}} \tag{18}$$

The condition (18) can be interpreted in the following manner. The difference between b^* and b_0 has to be sufficiently small not only *globally*, by the complete estimate but also *locally*, in the sense that the estimate q^* has to be better where the expression $|u_x(b_0)|$ is small.

Remark 1. The general result concerning the convergence rate of the estimate parameter q in the operator equation $F(q) = f$, remains also valid in the case when F is a monotone and hemicontinuous operator (see [5]).

References

1. Colonius, F., Kunisch, K.: Stability for parameter estimation in two point boundary value problems. J. Reine Angewandte Math. **370** (1986) 1–29
2. Engl, H. W.: Discrepancy principles for Tikhonov regularization of ill-posed problems leading to optimal convergence rates. J. Opt. Theor. Appl. **52** (1987) 209–215
3. Engl, H. W., Hanke, M., Neubauer, A.: Regularization of Inverse Problems. Kluwer Academic Publishers (1996)
4. Engl, H. W., Kunisch, K., Neubauer, A.: Convergence rates for Tikhonov regularisation of nonlinear ill-posed problems. Inverse Problems **5** (1989) 523–540
5. Hou, Z., Yang, H.: Convergence rates of regularized solutions of nonlinear ill-posed operator equations involving monotone operators. Science in China (Series A) **41** No. 3 (1998) 252–259
6. Kravaris, C. Seinfeld, J. H.: Identification of parameters in distributed parameter systems by regularization. SIAM J. Control Opt. **23** (1985) 217–241
7. Kunisch, K, White, L. W.: Parameter estimations, regularity and the penalty method for a class of two point boundary value problems. SIAM J. Control and Optimization **25** No. 1 (1987) 100–120
8. Neubauer, A.: Tikhonov regularisation for non-linear ill-posed problems: optimal convergence rates and finite-dimensional approximation. Inverse Problems **5** (1989) 541–667

Local Refinement in Non-overlapping Domain Decomposition

Veselin Dobrev[1] and Panayot Vassilevski[2]

[1] Central Laboratory for Parallel Processing, Bulgarian Academy of Sciences,
Acad. G. Bonchev St., bl. 25 A, Sofia, 1113, Bulgaria
veso@cantor.bas.bg
[2] Center for Applied Scientific Computing, Lawrence Livermore National Laboratory,
Mail Stop L-560, 7000 East Avenue, Livermore, CA 94550, USA
panayot@llnl.gov

Abstract. Finite element spaces are constructed that allow for different levels of refinement in different subdomains. In each subdomain the mesh is obtained by several steps of uniform refinement from an initial global coarse mesh. The approximation properties of the resulting discrete space are studied.
Computationally feasible, bounded extension operators, from the interface into the subdomains, are constructed and used in the numerical experiments. These operators provide stable splitting of the composite (global) finite element space into local subdomain spaces (vanishing at the interior interfaces) and the "extended" interface finite element space. They also provide natural domain decomposition type preconditioners involving appropriate subdomain and interface preconditioners.
Numerical experiments for 3-d elasticity illustrating the properties of the proposed discretization spaces and the algorithm for the solution of the respective linear system are also presented.

1 Discretization

Let $\Omega \subset \mathcal{R}^3$ be a polyhedral domain and assume that it is subdivided into disjoint tetrahedra forming an initial coarse triangulation $\mathcal{T}_0$. Applying successively some refinement procedure to $\mathcal{T}_0$ we obtain a sequence of nested quasiuniform triangulations $\mathcal{T}_0$, $\mathcal{T}_1$, $\mathcal{T}_2$, ... which have geometrically decreasing mesh parameters $h_0 > h_1 > \ldots$.

Next, let $\{\Omega_i\}_{i=1}^s$ be a non-overlapping decomposition of Ω:

$$\overline{\Omega} = \bigcup_{i=1}^s \overline{\Omega}_i, \quad \Omega_i \cap \Omega_j = \emptyset \text{ for } i \neq j.$$

We will assume that each subdomain Ω_i is a coarse mesh domain, i.e., it is completely covered by elements from $\mathcal{T}_0$.

We use Lagrangian finite elements of a given polynomial degree $m \geq 1$ over the triangulations $\mathcal{T}_j$ to define the approximation spaces V_j.

For each subdomain Ω_i we choose a number of refinement levels l_i, for $i = 1, \ldots, s$.

Let $\{\hat{l}_1, \ldots, \hat{l}_{\hat{s}}\}$ be the list of level numbers $\{l_i\}_{i=1}^s$ sorted in ascending order. We define the following auxiliary domains:

$$\overline{\mathcal{O}}_i = \bigcup_{l_j \leq \hat{l}_i} \overline{\Omega}_j, \quad \overline{\mathcal{O}}_i^f = \bigcup_{l_j > \hat{l}_i} \overline{\Omega}_j, \text{ and } \quad \overline{\mathcal{O}}_i^e = \bigcup_{l_j = \hat{l}_i} \overline{\Omega}_j. \quad \text{for } i = 1, \ldots, \hat{s}.$$

Note that the original domain Ω can be divided into the following disjoint subsets $\overline{\Omega} = \overline{\mathcal{O}}_i \cup \overline{\mathcal{O}}_i^f = \overline{\mathcal{O}}_{i-1} \cup \overline{\mathcal{O}}_i^e \cup \overline{\mathcal{O}}_i^f$.

We now introduce the spaces

$$\widehat{V}_i = \left\{ v \in V_{\hat{l}_i}, \text{ such that } v|_{\mathcal{O}_{i-1}} \equiv 0 \right\}, \text{ for } i = 1, \ldots, \hat{s},$$

and then define the approximation space of our main interest by the sum

$$V_h = \widehat{V}_1 + \widehat{V}_2 + \cdots + \widehat{V}_{\hat{s}}.$$

This space consists of continuous functions and it is a subspace of $H^1(\Omega)$. Inside each subdomain Ω_i, $V_h|_{\Omega_i}$ consists of all the functions in $V_{l_i}|_{\Omega_i}$ whose trace on $\partial \Omega_i$ belongs to a coarser space depending on the levels of the neighboring subdomains. In particular $V_h|_{\partial \Omega_i} \subseteq V_{l_i}|_{\partial \Omega_i}$.

Theorem 1. *Let $u \in H^{m+1}(\Omega)$ where m is the degree of Lagrangian elements we used to define the spaces V_j. Denote by h_i the mesh parameter (diameter of tetrahedra) of the triangulation $\mathcal{T}_{\hat{l}_i}$ (which is the triangulation for all subdomains Ω_j such that $l_j = \hat{l}_i$), for $i = 1, \ldots, \hat{s}$. This implies that $h_{i+1} \leq q h_i$ for some fixed $q \in (0,1)$. Define also the boundaries $G_i = \overline{\mathcal{O}}_i \cap \overline{\mathcal{O}}_i^f$, $i = 1, \ldots, \hat{s} - 1$.*

The following estimate for the best approximation of u with functions from V_h holds:

$$\inf_{u_h \in V_h} \|u - u_h\|_{1,\Omega} \leq \sum_{i=1}^{\hat{s}} C h_i^m |u|_{m+1, \mathcal{O}_i^e} +$$

$$+ \sum_{i=1}^{\hat{s}-1} C(1 + q^m) h_i^m \inf_{\substack{w_i \in H^{m+1}(\mathcal{O}_i^f) \\ w_i|_{G_i} = u|_{G_i}}} |w_i|_{m+1, \mathcal{O}_i^f}.$$

The proof follows from standard arguments utilizing the approximation properties of the local subspaces.

2 Linear Elasticity

In this section we use the discrete space defined in the previous section to discretize a linear elasticity problem. The problem is posed as follows: let $\lambda, \mu \in L_\infty(\Omega)$ be uniformly positive in Ω functions which are called Lamé coefficients;

let also $\mathbf{f} \in (L_2(\Omega))^3$ be some given body force and $\mathbf{g} \in (L_2(\Gamma_N))^3$ be some given surface force on a part of the boundary $\Gamma_N \subset \partial\Omega$; the rest of the boundary $\Gamma_D = \partial\Omega \setminus \Gamma_N$ is assumed to have positive surface measure. The problem then reads:

Find the displacement $\mathbf{u} \in (H^1(\Omega))^3$ which satisfies:

$$a(\mathbf{u}, \mathbf{v}) = \Phi(\mathbf{v}), \quad \forall \mathbf{v} \in (H^1(\Omega))^3 : \mathbf{v}|_{\Gamma_D} = 0$$
$$\mathbf{u}|_{\Gamma_D} = 0,$$

where

$$a(\mathbf{u}, \mathbf{v}) = \int_\Omega 2\lambda\, \varepsilon(\mathbf{u}) : \varepsilon(\mathbf{v}) + \mu\, \mathrm{div}\, \mathbf{u}\, \mathrm{div}\, \mathbf{v},$$

$$\Phi(\mathbf{v}) = \int_\Omega \mathbf{f} \cdot \mathbf{v} + \int_{\Gamma_N} \mathbf{g} \cdot \mathbf{v}.$$

Here $\varepsilon(\mathbf{u}) = \{\varepsilon_{ij}(\mathbf{u})\}_{i,j=1}^3$ is the linearized strain tensor which is defined by the equality:

$$\varepsilon_{ij}(\mathbf{v}) = \frac{1}{2}(\partial_j v_i + \partial_i v_j), \quad \mathbf{v} = (v_1, v_2, v_3).$$

It is well known that $a(\cdot,\cdot)^{\frac{1}{2}}$ defines a norm on the space $\mathbf{V} = \{\mathbf{v} \in (H^1(\Omega))^3 : \mathbf{v}|_{\Gamma_D} = 0\}$ which is equivalent to the $(H^1(\Omega))^3$–norm; that is,

$$a(\mathbf{v}, \mathbf{v}) \simeq \|\mathbf{v}\|^2_{(H^1(\Omega))^3} \equiv \|v_1\|^2_{1,\Omega} + \|v_2\|^2_{1,\Omega} + \|v_3\|^2_{1,\Omega}, \quad \forall \mathbf{v} \in \mathbf{V}. \quad (1)$$

We discretize the problem by replacing the space $\mathbf{V}$ with its finite dimensional subspace $\mathbf{V}_h \cap \mathbf{V}$, where $\mathbf{V}_h = (V_h)^3$. The discrete problem reads:

Find $\mathbf{u}_h \in \mathbf{V}_h \cap \mathbf{V}$ such that:

$$a(\mathbf{u}_h, \mathbf{v}_h) = \Phi(\mathbf{v}_h), \quad \forall \mathbf{v}_h \in \mathbf{V}_h \cap \mathbf{V}. \quad (2)$$

Using the norm equivalence (1) it is easy to obtain an estimate for the error $\|\mathbf{u} - \mathbf{u}_h\|_{1,\Omega}$ similar to the error estimate for the scalar case given in the previous section.

3 Extension Mappings

Our aim is to define an efficient parallel algorithm for solving the system of linear equations (2), which will be based on the given non-overlapping domain decomposition $\{\Omega_i\}_{i=1}^s$. To handle the case of inexact subdomain solves (or preconditioners) we use the technique studied in [6], [4], which exploits computable extension mappings.

The union of all boundaries of the subdomains $\{\Omega_i\}_{i=1}^s$ we call *interface* and denote by Γ:

$$\Gamma = \bigcup_{i=1}^s \partial\Omega_i.$$

Let $\mathbf{E}_h : \mathbf{V}_h|_\Gamma \to \mathbf{V}_h$ be an extension operator, that is:

$$\left(\mathbf{E}_h \mathbf{v}_h^b\right)\big|_\Gamma = \mathbf{v}_h^b, \qquad \forall \mathbf{v}_h^b \in \mathbf{V}_h|_\Gamma.$$

Using $\mathbf{E}_h$, we represent $\mathbf{V}_h$ as a direct sum:

$$\mathbf{V}_h = \mathbf{E}_h \left(\mathbf{V}_h|_\Gamma\right) \oplus \mathbf{V}_h^0,$$

where
$$\mathbf{V}_h^0 = \{\mathbf{v}_h \in \mathbf{V}_h : \mathbf{v}_h|_\Gamma = 0\}.$$

The space $\mathbf{V}_h^0$ can also be represented as a direct sum of the following spaces:

$$\mathbf{V}_i^0 = \left\{\mathbf{v}_h \in \mathbf{V}_h : \mathbf{v}_h|_{\overline{\Omega}\setminus\Omega_i} = 0\right\}, \qquad i = 1, \ldots, s.$$

In this way, $\mathbf{V}_h$ is decomposed into the direct sum:

$$\mathbf{V}_h = \mathbf{V}_1^0 \oplus \mathbf{V}_2^0 \oplus \cdots \oplus \mathbf{V}_s^0 \oplus \mathbf{E}_h\left(\mathbf{V}_h|_\Gamma\right).$$

It is obvious that $\mathbf{V}_i^0$ and $\mathbf{V}_j^0$ are orthogonal with respect to the inner product $a(\cdot,\cdot)$ when $i \neq j$. In general, this is not true for the spaces $\mathbf{V}_h^0$ and $\mathbf{E}_h\left(\mathbf{V}_h|_\Gamma\right)$. That is why, we impose the following boundedness condition on $\mathbf{E}_h$:

$$a\left(\mathbf{E}_h\left(\mathbf{v}_h|_\Gamma\right), \mathbf{E}_h\left(\mathbf{v}_h|_\Gamma\right)\right) \leq \eta\, a\left(\mathbf{v}_h, \mathbf{v}_h\right), \qquad \forall \mathbf{v}_h \in \mathbf{V}_h \cap \mathbf{V}, \qquad (3)$$

with constant $\eta \geq 1$ independent of the discretization parameters $\{l_i\}_{i=1}^s$. Note that (3) is simply boundedness of $\mathbf{E}_h$ in energy norm. This condition is equivalent to the following strengthened Cauchy–Schwarz inequality:

$$a\left(\mathbf{E}_h\mathbf{v}_h^b, \mathbf{v}_h^0\right) \leq \left(1 - \frac{1}{\eta}\right)^{\frac{1}{2}} a\left(\mathbf{E}_h\mathbf{v}_h^b, \mathbf{E}_h\mathbf{v}_h^b\right)^{\frac{1}{2}} a\left(\mathbf{v}_h^0, \mathbf{v}_h^0\right)^{\frac{1}{2}}, \qquad (4)$$

$$\forall \mathbf{v}_h^b \in \left(\mathbf{V}_h \cap \mathbf{V}\right)|_\Gamma,\ \forall \mathbf{v}_h^0 \in \mathbf{V}_h^0.$$

We will consider vector extension mappings in which each of the scalar components in extended separately, that is $\mathbf{E}_h$ has the form:

$$\mathbf{E}_h \mathbf{v}_h^b = \left(E_h v_{h,1}^b, E_h v_{h,2}^b, E_h v_{h,3}^b\right), \qquad \mathbf{v}_h^b = \left(v_{h,1}^b, v_{h,2}^b, v_{h,3}^b\right),$$

where $E_h : V_h|_\Gamma \to V_h$ is a bounded scalar extension mapping:

$$\|E_h v_h^b\|_{1,\Omega} \leq C\, \|v_h^b\|_{\frac{1}{2},\Gamma} \equiv C \inf_{\substack{v \in H^1(\Omega) \\ v|_\Gamma = v_h^b}} \|v\|_{1,\Omega}, \qquad \forall v_h^b \in V_h|_\Gamma.$$

Using the last inequality and the norm equivalence (1), it is easy to prove that (3) holds with constant η independent of $\{l_i\}_{i=1}^s$. The extension mappings E_h are naturally defined subdomain by subdomain. We start with bounded extension mappings

$$E_h^i : V_{l_i}|_{\partial\Omega_i} \to V_{l_i}|_{\Omega_i},$$

which are defined on the uniformly refined space $V_{l_i}|_{\partial\Omega_i}$ and their image is also contained in an uniformly refined space – $V_{l_i}|_{\Omega_i}$. Such operators are easily constructed (as we will see later), and this is generally a well-established technique.

To define the global extension operator E_h we need local extension operators from the space $V_h|_{\partial\Omega_i}$ acting into $V_h|_{\Omega_i}$. The definition of V_h implies that:

$$V_h|_{\partial\Omega_i} \subseteq V_{l_i}|_{\partial\Omega_i} \quad \text{and} \quad V_h|_{\Omega_i} = \{v_h^i \in V_{l_i}|_{\Omega_i} : v_h^i|_{\partial\Omega_i} \in V_h|_{\partial\Omega_i}\}$$

and therefore

$$E_h^i(V_h|_{\partial\Omega_i}) \subset V_h|_{\Omega_i}.$$

This fact allows us to define E_h in the following way: if $v_h^b \in V_h|_\Gamma$ then

$$\left(E_h v_h^b\right)\big|_{\Omega_i} = E_h^i\left(v_h^b|_{\partial\Omega_i}\right), \qquad i = 1, \ldots, s.$$

One can estimate the norm of E_h, in a straightforward manner, in terms of the norm of the individual components E_h^i.

4 Multilevel Extension Mappings

In this section we briefly consider the definition of two types of *multilevel* extension operators (cf., [2], [3], and [5]).

For simplicity of notation, we will define an extension operator E_h from $\partial\Omega$ into the whole domain Ω at some arbitrary refinement level l: $E_h : V_l|_{\partial\Omega} \to V_l$.

In this section we will use the notation $V_k^b = V_k|_{\partial\Omega}$. A general multilevel extension operator is defined as follows: let $r_k : V_l^b \to V_k^b$ be linear operators (with $r_l = I$ and $r_{-1} = 0$) and $E_k^0 : V_k^b \to V_k$ be the trivial extension with zeros in the nodes of $\mathcal{T}_k$ inside Ω. The multilevel extension mapping $E_h : V_l|_{\partial\Omega} \to V_l$ based on the decomposition operators r_k is defined by the sum:

$$E_h = \sum_{k=0}^{l} E_k^0(r_k - r_{k-1}).$$

It is known that if $\{r_k\}$ satisfy the norm equivalence (where h_k stands for the mesh size of $\mathcal{T}_k$)

$$\|v_l^b\|_{\frac{1}{2},\partial\Omega}^2 \simeq \sum_{k=0}^{l} h_k^{-1}\|(r_k - r_{k-1})v_l^b\|_{0,\partial\Omega}^2, \qquad \forall v_l^b \in V_l^b \qquad (5)$$

then the corresponding multilevel extension operator E_h is uniformly bounded.

We next define the two computationally feasible decomposition operators that we used in the numerical experiments:

– let Φ_k^b be the set of the nodal basis functions of the space V_k^b and define the mappings $\tilde{q}_k : L_2(\partial\Omega) \to V_k^b$, $k = 0, 1, \ldots$ by the equality

$$\tilde{q}_k v = \sum_{\phi \in \Phi_k^b} \frac{(v, \phi)_{0,\partial\Omega}}{(1, \phi)_{0,\partial\Omega}} \phi, \qquad \forall v \in L_2(\partial\Omega).$$

If we take $r_k = \tilde{q}_k$ then (5) holds and the corresponding extension operator is bounded.

- to define the second example we introduce the discontinuous spaces

$$V_{k,+}^b = \{v \in L_2(\partial\Omega) : v|_T \in \mathcal{P}_m(T), \forall T \in \mathcal{T}_k^b\},$$

where $\mathcal{P}_m(T)$ stands for the set of all polynomials of degree $\leq m$ over the triangle T and $\mathcal{T}_k^b$ is the set of all triangles of the restricted to $\partial\Omega$ triangulation $\mathcal{T}_k$. Note that V_k^b is a proper subset of $V_{k,+}^b$. We define the projections $p_k : V_{k,+}^b \to V_k^b$ by averaging about the nodes x,

$$\left(p_k v_{k,+}^b\right)(x) = \sum_{\substack{T \in \mathcal{T}_k^b \\ \bar{T} \ni x}} |T| \lim_{\substack{y \to x \\ y \in T}} v_{k,+}^b(y) \bigg/ \sum_{\substack{T \in \mathcal{T}_k^b \\ \bar{T} \ni x}} |T|,$$

where $|T|$ is the measure (the area) of T. If we denote by $q_{k,+}$ the $L_2(\partial\Omega)$-orthogonal projection on $V_{k,+}^b$ then we take $r_k = p_k q_{k,+}$. It can be proven that $\{r_k\}$ are uniformly bounded (in $\|\cdot\|_{0,\partial\Omega}$ norm) projection operators and, as a corollary, that the norm equivalence (5) holds.

5 Preconditioning

In order to solve the discrete problem (2) we have to reformulate it into matrix-vector form by choosing a basis in the space $\mathbf{V}_h \cap \mathbf{V}$. Let $\Phi_1, \Phi_2, \ldots, \Phi_s$, and Φ^b be bases respectively in the spaces

$$\mathbf{V}_1^0, \quad \mathbf{V}_2^0, \quad \cdots \quad \mathbf{V}_s^0, \quad \text{and} \quad (\mathbf{V}_h \cap \mathbf{V})|_\Gamma,$$

then the set

$$\Phi = \Phi_1 \cup \Phi_2 \cup \cdots \cup \Phi_s \cup \mathbf{E}_h \Phi^b$$

is basis in the space $\mathbf{V}_h \cap \mathbf{V}$. In this basis the stiffness matrix has the following 2×2 block structure:

$$A = \begin{pmatrix} A_0 & A_{0b} \\ A_{b0} & A_b \end{pmatrix} \begin{matrix} \} \ \Phi_1 \cup \Phi_2 \cup \cdots \cup \Phi_s \\ \} \ \mathbf{E}_h \Phi^b \end{matrix}$$

The inequality (4) is equivalent to the strengthened Cauchy inequality for A:

$$v_0^T A_{0b} v_b \leq \left(1 - \frac{1}{\eta}\right)^{\frac{1}{2}} \left(v_b^T A_{bb} v_b\right)^{\frac{1}{2}} \left(v_0^T A_{00} v_0\right)^{\frac{1}{2}} \quad \forall v_b, v_0,$$

and therefore A is spectrally equivalent to its block diagonal part. Moreover, if M_0 and M_b are spectrally equivalent to A_0 and A_b respectively then the block additive and block multiplicative preconditioners

$$M_A = \begin{pmatrix} M_0 & 0 \\ 0 & M_b \end{pmatrix} \quad M_M = \begin{pmatrix} M_0 & 0 \\ A_{b0} & M_b \end{pmatrix} \begin{pmatrix} I & M_0^{-1} A_{0b} \\ 0 & I \end{pmatrix}$$

are also spectrally equivalent to A.

The block A_0 is easily preconditioned because it is block diagonal with blocks corresponding to the spaces $\mathbf{V}_i^0$ ($i = 1, \ldots, s$) which have multilevel structure

$$\mathbf{V}_{i,0}^0 \subset \mathbf{V}_{i,1}^0 \subset \cdots \subset \mathbf{V}_{i,l_i}^0 = \mathbf{V}_i^0,$$

where

$$\mathbf{V}_{i,k}^0 = \left\{\mathbf{v} \in (V_k)^3 : \mathbf{v}|_{\overline{\Omega} \setminus \Omega_i} = 0\right\}.$$

Therefore multilevel and multigrid methods can be used for the preconditioning of the blocks of A_0. In the numerical experiments we used $\mathcal{V}$-cycle multigrid with one pre- and one post-smoothing iteration per level.

The preconditioning of the block A_b is a more complicated task. Without going into details we will give just an idea of the algorithm we used in the numerical experiments. Namely, we apply the idea of multigrid preconditioning of locally refined spaces considered in [1], but here we apply it to the interface space $\mathbf{V}_h|_\Gamma$. In the multigrid algorithm the following sequence of nested spaces is used:

$$\mathbf{V}_{h,0}|_\Gamma \subset \mathbf{V}_{h,1}|_\Gamma \subset \cdots \subset \mathbf{V}_{h,l}|_\Gamma = \mathbf{V}_h|_\Gamma,$$

where the spaces $\mathbf{V}_{h,k} = (V_{h,k})^3$ are defined exactly as the space $\mathbf{V}_h$ with the only difference that the levels $\{l_i\}_{i=1}^s$ in the subdomains $\{\Omega_i\}_{i=1}^s$ are replaced with the coarser levels $\{l_{i,k} = \min(l_i, k)\}_{i=1}^s$; the last level l is chosen to be the smallest number for which $\mathbf{V}_{h,l}|_\Gamma = \mathbf{V}_h|_\Gamma$, that is for which

$$\Gamma = \bigcup_{l_i \leq l} \partial \Omega_i.$$

In the spaces $\mathbf{V}_{h,k}|_\Gamma$ the following varying (non–inherited) symmetric, positive definite forms are used to define the multigrid algorithm:

$$A_k(\mathbf{u}_{h,k}^b, \mathbf{v}_{h,k}^b) = a(\mathbf{E}_{h,k}\mathbf{u}_{h,k}^b, \mathbf{E}_{h,k}\mathbf{v}_{h,k}^b), \qquad \forall \mathbf{u}_{h,k}^b, \mathbf{v}_{h,k}^b \in \mathbf{V}_{h,k}|_\Gamma,$$

where the extension mappings $\mathbf{E}_{h,k} : \mathbf{V}_{h,k}|_\Gamma \to \mathbf{V}_{h,k}$ are defined in a way similar to the way $\mathbf{E}_h$ was defined.

In the space $\mathbf{V}_{h,k}|_\Gamma$ we smooth only in the region of Γ where $\mathbf{V}_{h,k}|_\Gamma$ is finer than $\mathbf{V}_{h,k-1}|_\Gamma$. This region is the non-empty set $\Gamma \setminus \bigcup_{l_i < k} \partial \Omega_i$.

We finish this section with the remark that both the multiplication of A with a vector and the solution of a system with M_A (or M_M) can be carried out in parallel. Each subdomain corresponds to a processor that calculates the local actions (of A or M_A^{-1}). In addition communications between neighbor subdomains are required for the assembling of the global actions.

6 Numerical Experiments

We present numerical results for two linear elasticity problems in the unit cube $\Omega = (0,1)^3$. The kth level triangulation $\mathcal{T}_k$ is obtained in the following way: first

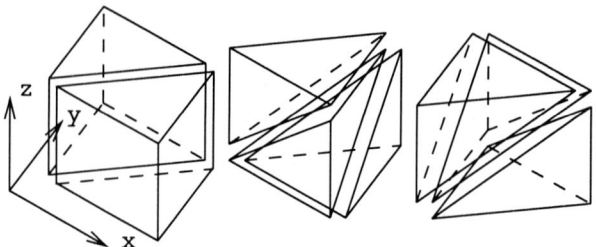

Fig. 1. Cube partitioning into six tetrahedra

we divide Ω into $2^k \times 2^k \times 2^k$ equal cubes and then each cube is partitioned into six tetrahedra as illustrated in Figure 1. With this triangulations we use *quadratic* Lagrangian finite elements to define the spaces V_k, i. e. $m = 2$.

Note that when quadratic FE are used in the 2-dimensional spaces $\partial \Omega_i$ some of the nodal basis functions ϕ have vanishing integral, i. e. $(1, \phi)_{0,\partial\Omega_i} = 0$. Therefore the operators $\tilde{q}_k$ can not be defined in this case. Instead, we used $\tilde{q}_{k+1}$ defined for linear FE with the triangulation $\mathcal{T}_{k+1}$, that is we used the decomposition operators $r_k = I_k^{21} \, \tilde{q}_{k+1} \, I_k^{12}$ where I_k^{21} and I_k^{12} are the operators defining the natural bijection between the space of linear FE over $\mathcal{T}_{k+1}$ and the space of quadratic FE over $\mathcal{T}_k$. Namely, these two spaces have the same set of nodes and this bijection simply replaces the two bases functions – the linear and the quadratic. This is illustrated in Figure 2.

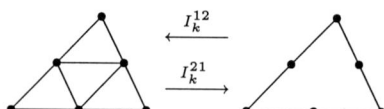

Fig. 2. Replacing piecewise linear function with quadratic function and vice versa

The second decomposition operator we defined $p_k q_{k,+}$ can be defined for both linear and quadratic FE. In Table 1 we give the three different extension mappings used in the numerical experiments. Comparing the results for E_2 and E_3 we can see the effect of the replacement of quadratic FE with linear.

Table 1. Extensions used

	E_1	E_2	E_3
r_k	$I_k^{21} \, \tilde{q}_{k+1} \, I_k^{12}$	$p_k^2 q_{k,+}^2$	$I_k^{21} \, p_{k+1}^1 q_{k+1,+}^1 \, I_k^{12}$

To solve the linear systems we used the preconditioned conjugate gradient (PCG) algorithm. The stopping criterion was

$$r^T M^{-1} r \leq 10^{-18} r_0^T M^{-1} r_0$$

where r is the current residual, r_0 is the initial one, and M is the preconditioner used (M_A or M_M).

The *Massage Passing Interface* (MPI) was used for the parallel implementation of the algorithm.

Test problem 1. We take the following geometry and Lamé coefficients:

$$\Omega = (0,1)^3, \quad \Gamma_N = \{0 < x, y < 1, z = 1\}, \quad \Gamma_D = \partial\Omega \setminus \Gamma_N, \quad \lambda = \frac{3}{4}, \quad \mu = \frac{3}{8}$$

and the following components for the displacement:

$$u_1(x,y,z) = 0$$
$$u_2(x,y,z) = \sin(\pi x)\sin(\pi y)\sin(\pi z)$$
$$u_3(x,y,z) = (1-x)x(1-y)y(1-z)z.$$

We divide the domain into $s = 2 = 1 \times 1 \times 2$, $s = 4 = 2 \times 2 \times 1$, $s = 8 = 2 \times 2 \times 2$, and $s = 16 = 2 \times 2 \times 4$ subdomains. In each subdomain we take equal number of refinement levels $l_i = l$, $i = 1, \ldots, s$. Thus the mesh is uniform in the whole domain. In Table 2 we give the number of iterations made by the PCG algorithm when the three different extension mappings were used. One can see

Table 2. Iterations with E_1, E_2, and E_3 and additive preconditioner

s	\multicolumn{4}{c}{h}	s	\multicolumn{4}{c}{h}	s	\multicolumn{4}{c}{h}
	1/4	1/8	1/16	1/32		1/4	1/8	1/16	1/32		1/4	1/8	1/16	1/32
2	32	35	37	38	2	44	52	56	60	2	40	47	50	52
4	29	34	37	38	4	43	55	62	66	4	42	52	57	59
8	28	33	36	37	8	44	53	61	65	8	41	51	56	59
16	–	43	43	44	16	–	68	70	71	16	–	55	58	61

that the number of iterations increases when the mesh parameter h decreases and when s increases, but there is a tendency for stabilizing. Notice the slight jump of iterations when s is increased from 8 to 16. This is due to the change of the initial level — when $s = 16$, T_0 has $6 \times 4 \times 4 \times 4$ tetrahedra, while for $s = 2, 4$, and 8, T_0 has $6 \times 2 \times 2 \times 2$ tetrahedra. When we compare the extensions, we see that the one based on $\tilde{q}_k$ (E_1) is better than the other two. The comparison of E_2 and E_3 shows that the transition from quadratic FE to linear improves the number of iterations slightly. In Table 3, the number of iterations with E_1 and multiplicative preconditioner (M_M) are given. Comparing these numbers with those from the additive (M_A) version, we see that M_M is almost two times better

Table 3. Iterations with E_1 and multiplicative preconditioner

s	\multicolumn{4}{c}{h}			
	1/4	1/8	1/16	1/32
2	18	18	19	20
4	17	18	19	20
8	16	18	19	20
16	–	26	24	24

than M_A, but M_M requires the solution of two systems with M_0 (preconditioners inside the subdomains) while with M_A requires just one.

Test problem 2. For this test we choose

$$\Omega = (0,1)^3 \qquad \Gamma_D = \partial\Omega \qquad \lambda = \frac{2}{5} \qquad \mu = \frac{2}{5}.$$

We take the exact solution as a sum of two functions – one smooth and one rough which has support in $(0, \frac{1}{2})^3$ (see Figure 3):

$$\begin{aligned} u_1(x,y,z) &= \Phi(x,y,z) \\ u_2(x,y,z) &= \Phi(x,y,z) \\ u_3(x,y,z) &= \Phi(x,y,z) + (1-x)x(1-y)yz \end{aligned} \qquad \begin{aligned} \Phi(x,y,z) &= 5.10^5 \, \phi(x)\phi(y)\phi(z) \\ \phi(a) &= \begin{cases} a^2(\frac{1}{2}-a)^2 & a \in (0,\frac{1}{2}), \\ 0 & a \notin (0,\frac{1}{2}). \end{cases} \end{aligned}$$

The first decomposition we consider with this test problem has $s = 8 = 2 \times 2 \times 2$

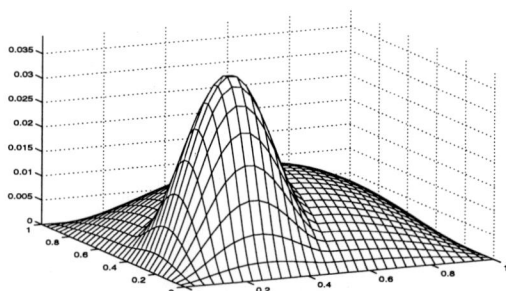

Fig. 3. Graphic of $u_3(x, y, \frac{1}{4})$

subdomains. In this way the rough component of the solution is contained in $\Omega_1 = (0, \frac{1}{2})^3$. We take two different levels of refinement $l_i = l$, for $i = 2, \ldots, 8$ and a finer level l_1 for Ω_1. In Table 4 are given the discrete energy norms of the error and the respective number of iterations for some spaces with refinement in Ω_1. We see that for fixed h (the mesh size outside Ω_1) when the mesh inside Ω_1 is refined the error decreases. At some level the error inside Ω_1 and the error outside

Table 4. Discrete energy norm of the error ($\times 10^{-3}$) and number of iterations with E_1 and multiplicative preconditioner

	h_1			
h	1/4	1/8	1/16	1/32
1/4	8.1067	2.6031	0.9491	0.8320
1/8		2.6151	0.4737	0.1554
1/16			0.4885	0.0698
1/32				0.0747

	h_1			
h	1/4	1/8	1/16	1/32
1/4	12	13	13	13
1/8		13	13	13
1/16			14	14
1/32				14

it are balanced (for example $h = 1/4$ and $h_1 = 1/16$) and more refinement in Ω_1 does not improve the approximation (compare $h_1 = 1/16$ and $h_1 = 1/32$ for $h = 1/4$). This behavior is in agreement with the error estimate presented above. The number of the iterations made by the PCG algorithm is again independent of the mesh sizes, which is natural because we used multigrid algorithms for the preconditioning.

One disadvantage of the discretizations with refinement in Ω_1 (i. e. when $l_1 > l$ or $h_1 < h$) is that the number of the unknowns in Ω_1 is approximately 8^{l_1-l} times larger than those in the other subdomains. Therefore the processor corresponding to Ω_1 has to do much more computations than the rest because the number of computations increases linearly with the number of the unknowns. To avoid this unbalanced discretization we divide Ω_1 into $2 \times 2 \times 2$ equal subdomains. The remaining 7 subdomains remain the same. Thus we obtain a balanced discretization for the case when the mesh size inside $(0, \frac{1}{2})^3$ is two times smaller than that outside of it. Note that the discrete space V_h does not change. In Table 5 the number of the iterations with two balanced discretizations are given. For the

Table 5. Iterations with balanced discretizations (with M_A and E_1)

	h_1			
h	1/4	1/8	1/16	1/32
1/4	21	35	–	–
1/8		24	29	–
1/16			26	30
1/32				27

case $h_1 = h$ the discretization is unchanged ($s = 8$) and for the case $h_1 = \frac{1}{2}h$ we subdivide $(0, \frac{1}{2})^3$ into $2 \times 2 \times 2$ subdomains (i. e., $s = 15$). We see that the balancing procedure we applied does not deteriorate the convergence rate of the PCG algorithm.

References

1. Bramble, J.: Multigrid methods. Pitman Research Notes in Mathematics v. 294, Longman Scientific & Technical (1993).
2. J. H. Bramble, J. E. Pasciak and P. S. Vassilevski, "*Computational scales of Sobolev norms with application to preconditioning*", Math. Comp. 69 (2000), 463-480.
3. V. Dobrev and P. S. Vassilevski, "*Non–mortar finite elements for elliptic problems*", Proceedings of the Fourth Intern. Conference on Numerical Methods and Applications (NMA'98), "Recent Advances in Numerical Methods and Applications" (O. Iliev, M. Kaschiev, S. Margenov, Bl. Sendov and P. S. Vassilevski, eds.), World Scientific, Singapore, 1999, pp. 756–765.
4. G. Haase, U. Langer, A. Meyer, and S. V. Nepomnyaschikh, Hierarchical extension operators and local multigrid methods in domain decomposition preconditioners, *East–West J. Numer. Math.* 2(1994), 173–193.
5. S. V. Nepomnyaschikh, Optimal multilevel extension operators, *Report SPC 95–3, Jan, 1995*, Technische Universität Chemnitz–Zwickau, Germany.
6. P. S. Vassilevski and O. Axelsson, "*A two–level stabilizing framework for interface domain decomposition preconditioners*", in: Proceedings of the Third International Conference $O(h^3)$, Sofia, Bulgaria, August 21–August 26, Sofia, Bulgaria, "**Advances in Numerical Methods and Applications**", (I. T. Dimov, Bl. Sendov and P. S. Vassilevski, eds.), World Scientific, Singapore, New Jersey, London, Hong Kong, 1994, pp. 196–202.

Singularly Perturbed Parabolic Problems on Non-rectangular Domains*

R. K. Dunne[1], E. O'Riordan[1], and G. I. Shishkin[2]

[1] School of Mathematical Sciences, Dublin City University
Dublin, Ireland
[2] Institute for Mathematics & Mechanics, Russian Academy of Sciences
Ekaterinburg, Russia

Abstract. A singularly perturbed time-dependent convection-diffusion problem is examined on non-rectangular domains. The nature of the boundary and interior layers that arise depends on the geometry of the domains. For problems with different types of layers, various numerical methods are constructed to resolve the layers in the solutions and the numerical solutions are shown to converge independently of the singular perturbation parameter.

1 Introduction

We consider the following class of singularly perturbed parabolic problems

$$(P_\varepsilon) \qquad L_\varepsilon u(x,t) \equiv (\varepsilon u_{xx} + a u_x - b u_t - d u)(x,t) = f(x,t) \quad \text{on} \quad D, \qquad (1a)$$

$$u(x,t) = g(x,t) \quad \text{on} \quad \overline{D} \setminus D, \qquad (1b)$$

$$a \geq \alpha, \qquad b \geq \beta > 0, \qquad d \geq \delta \geq 0 \qquad (1c)$$

where $D = (\phi_1(t), \phi_2(t)) \times (0, T]$ is a non-rectangular domain bounded by the curves $x = \phi_1(t)$, $x = \phi_2(t)$ such that

$$\phi_1(0) = 0, \quad \phi_2(0) = 1, \quad \phi_1(t) < \phi_2(t), \quad \forall t,$$

and $0 < \varepsilon \leq 1$ is the perturbation parameter. We also assume that the data a, b, d, f, g and ϕ_1, ϕ_2 are sufficiently smooth, and f, g satisfy sufficient compatibility conditions at the corners of the domain.

In order to generate numerical approximations to the solutions of problems in P_ε, the problem is transformed to one on a rectangular domain. This is achieved by introducing the new co-ordinate system $(\hat{x}, \hat{t})$ and the change of variables

$$\hat{x} = \hat{x}(x,t) \equiv \frac{x - \phi_1(t)}{\phi_2(t) - \phi_1(t)}, \qquad \hat{t} = t. \qquad (2)$$

---

* This research was supported in part by the National Centre for Plasma Science and Technology Ireland, by the Enterprise Ireland grant SC-98-612 and by the Russian Foundation for Basic Research under grant No. 98-01-00362.

The transformed class of problems is then

$$(\hat{P}_\varepsilon) \quad \hat{L}_\varepsilon \hat{u}(\hat{x},\hat{t}) \equiv (\varepsilon \hat{u}_{\hat{x}\hat{x}} + \hat{a}\hat{u}_{\hat{x}} - \hat{b}\hat{u}_{\hat{t}} - \hat{d}\hat{u})(\hat{x},\hat{t}) = \hat{f}(\hat{x},\hat{t}) \quad \text{on } \hat{D}, \quad (3a)$$

$$\hat{u}(\hat{x},\hat{t}) = \hat{g}(\hat{x},\hat{t}) \quad \text{on } \overline{\hat{D}} \setminus \hat{D} \quad (3b)$$

where

$\hat{D} = \Omega \times (0,T]$, $\Omega = (0,1)$, $\hat{u}(\hat{x},t) = u(x(\hat{x},t),t)$, $\hat{g}(\hat{x},t) = g(x(\hat{x},t),t)$,
$\hat{a} = a(x,t)(\phi_2 - \phi_1) - b(x,t)(\phi_1'(x - \phi_2) - \phi_2'(x - \phi_1))$,
$\hat{b} = b(x,t)(\phi_2 - \phi_1)^2$, $\hat{d} = d(x,t)(\phi_2 - \phi_1)^2$, $\hat{f} = f(x,t)(\phi_2 - \phi_1)^2$,
$\phi_i = \phi_i(t)$, $\phi_i' = \phi_i'(t)$, $x = x(\hat{x},t) \equiv \hat{x}(\phi_2 - \phi_1) + \phi_1$.

Notice that irrespective of ϕ_1 and ϕ_2, $\hat{b} > 0$ and $\hat{d} \geq 0$. However, in general, the sign of $\hat{a}$ may differ from a at certain points of the domain. Thus the sign of $\hat{a}$, which is crucial in selecting a suitable numerical method for $\hat{P}_\varepsilon$, depends on the shape of the original domain and the original coefficient functions a and b.

2 Straight Line Walls

As the expression for $\hat{a}$ is quite complicated in the general case, we assume that the functions ϕ_1 and ϕ_2 are linear. That is, assume that

$$\phi_1(t) = -m_1 t, \quad \phi_2(t) = 1 - m_2 t. \quad (4)$$

The resulting problem class, $\hat{P}_\varepsilon^1$, is thus

$$\hat{P}_\varepsilon^1 \subset \hat{P}_\varepsilon \quad (5)$$

where

$\hat{a} = (1 - (m_2 - m_1)t)(a(x,t) - b(x,t)(\hat{x}(m_2 - m_1) + m_1))$,
$\hat{b} = b(x,t)(1 - (m_2 - m_1)t)^2$, $\hat{d} = d(x,t)(1 - (m_2 - m_1)t)^2$,
$\hat{f} = f(x,t)(1 - (m_2 - m_1)t)^2$, $\hat{g} = g(x,t)$,
$x = x(\hat{x},t) \equiv \hat{x}(1 - (m_2 - m_1)t) - m_1 t$.

We now deal with two special cases of the above problem class.

2.1 Parallel Straight Line Walls

The first special case we consider is when the side walls are parallel, i.e $m_1 = m_2 = m$ and that the coefficient functions a and b are constant. That is

$$a(x,t) = \alpha, \quad b(x,t) = \beta, \quad (x,t) \in \overline{D}. \quad (6)$$

The problem class, $\hat{P}_\varepsilon^P$, is thus

$$\hat{P}_\varepsilon^P \subset \hat{P}_\varepsilon^1 \qquad (7)$$

where
$$\hat{a} = \alpha - m\beta, \quad \hat{b} = \beta, \quad \hat{d} = d(x,t), \quad \hat{f} = f(x,t), \quad \hat{g} = g(x,t), \quad x = \hat{x} - mt.$$

Depending on the values of α and β, $\hat{P}_\varepsilon^P$ falls naturally into one of three distinct problem classes:

$$\hat{P}_\varepsilon^P \subset \hat{P}_\varepsilon^+ \cup \hat{P}_\varepsilon^0 \cup \hat{P}_\varepsilon^- \qquad (8)$$

where

$$\hat{P}_\varepsilon^+ = \{\hat{P}_\varepsilon \mid \hat{a}(\hat{x},\hat{t}) > 0, \quad \forall (\hat{x},\hat{t}) \in \overline{\hat{D}}\}, \qquad (9a)$$

$$\hat{P}_\varepsilon^0 = \{\hat{P}_\varepsilon \mid \hat{a}(\hat{x},\hat{t}) = 0, \quad \forall (\hat{x},\hat{t}) \in \overline{\hat{D}}\}, \qquad (9b)$$

$$\hat{P}_\varepsilon^- = \{\hat{P}_\varepsilon \mid \hat{a}(\hat{x},\hat{t}) < 0, \quad \forall (\hat{x},\hat{t}) \in \overline{\hat{D}}\}. \qquad (9c)$$

For problems from the first and third classes, $\hat{P}_\varepsilon^+$ and $\hat{P}_\varepsilon^-$, the solution possesses a regular boundary layer, in a neighbourhood of $\hat{x} = 0$ in the former and in a neighbourhood of $\hat{x} = 1$ in the latter. In the second case, $\hat{P}_\varepsilon^0$, the solution has parabolic boundary layers in a neighbourhood of both $\hat{x} = 0$ and $\hat{x} = 1$.

Clearly

$$\hat{P}_\varepsilon^P \subset \hat{P}_\varepsilon^+ \quad \text{if} \quad \alpha > m\beta,$$
$$\hat{P}_\varepsilon^P \subset \hat{P}_\varepsilon^0 \quad \text{if} \quad \alpha = m\beta,$$
$$\hat{P}_\varepsilon^P \subset \hat{P}_\varepsilon^- \quad \text{if} \quad \alpha < m\beta.$$

2.2 Non-parallel Straight Line Walls

The next special case we consider is when both ϕ_1 and ϕ_2 are still straight lines, but are now no longer parallel. The former is sloped as before but the latter will be positioned vertically, i.e., $m_1 = m, m_2 = 0$ and we also assume that $m > 0$. As before we assume that a and b are constant.

The problem class, $\hat{P}_\varepsilon^V$, is thus

$$\hat{P}_\varepsilon^V \subset \hat{P}_\varepsilon^1 \qquad (10)$$

where
$$\hat{a} = (1 + m\hat{t})(\alpha + \beta m(\hat{x} - 1)), \quad \hat{b} = \beta(1 + m\hat{t})^2, \quad \hat{d} = d(x,t)(1 + m\hat{t})^2,$$
$$\hat{f} = f(x,t)(1 + m\hat{t})^2, \quad \hat{g} = g(x,t), \quad x = \hat{x}(1 + mt) - mt.$$

Again depending on the values of α and β, $\hat{P}_\varepsilon^V$ will fall into a particular problem class. We can identify three types of problem subclasses

$$\hat{P}_\varepsilon^{l+} \cup \hat{P}_\varepsilon^{l0} \cup \hat{P}_\varepsilon^{l-} \subset \hat{P}_\varepsilon^1 \qquad (11)$$

where
$$\hat{P}_\varepsilon^{l+} = \{\hat{P}_\varepsilon^V \mid \alpha > m\beta\}, \tag{12a}$$
$$\hat{P}_\varepsilon^{l0} = \{\hat{P}_\varepsilon^V \mid \alpha = m\beta\}, \tag{12b}$$
$$\hat{P}_\varepsilon^{l-} = \{\hat{P}_\varepsilon^V \mid \alpha < m\beta,\ \alpha > 0\}. \tag{12c}$$

For problems from the class $\hat{P}_\varepsilon^{l-}$, the solution exhibits no boundary layer (due to the compatibility conditions), while for problems from the two classes $\hat{P}_\varepsilon^{l0}$ and $\hat{P}_\varepsilon^{l+}$ we have a boundary layer in a neighbourhood of $\hat{x} = 0$ (more precisely, a parabolic layer in the former).

Note that we have
$$\hat{P}_\varepsilon^V \subset \hat{P}_\varepsilon^+ \cup \hat{P}_\varepsilon^l \cup \hat{P}_\varepsilon^i \cup \hat{P}_\varepsilon^r \cup \hat{P}_\varepsilon^-$$
where, for $\zeta \in (0,1)$ and $\gamma > 0$, we define
$$\hat{P}_\varepsilon^l = \{\hat{P}_\varepsilon \mid \hat{a}(\hat{x},\hat{t}) \geq \gamma\hat{x},\ \forall (\hat{x},\hat{t}) \in \overline{\hat{D}}\},$$
$$\hat{P}_\varepsilon^r = \{\hat{P}_\varepsilon \mid \hat{a}(\hat{x},\hat{t}) \leq -\gamma(1-\hat{x}),\ \forall (\hat{x},\hat{t}) \in \overline{\hat{D}}\}$$
$$\hat{P}_\varepsilon^i = \left\{\hat{P}_\varepsilon \mid \hat{a}(\hat{x},\hat{t}) \begin{cases} < 0 & \hat{x} < \zeta,\ \forall \hat{t} \in [0,T] \\ = 0 & \hat{x} = \zeta,\ \forall \hat{t} \in [0,T] \\ > 0 & \hat{x} > \zeta,\ \forall \hat{t} \in [0,T] \end{cases} \right\}.$$

Note also that $\hat{P}_\varepsilon^{l+} \subset \hat{P}_\varepsilon^+$, $\hat{P}_\varepsilon^{l0} \subset \hat{P}_\varepsilon^l$, $\hat{P}_\varepsilon^{l-} \subset \hat{P}_\varepsilon^i$.

In the next section we construct numerical methods that resolve the layers that arise in each of the six problem classes, $\hat{P}_\varepsilon^+$, $\hat{P}_\varepsilon^{l+}$, $\hat{P}_\varepsilon^0$, $\hat{P}_\varepsilon^{l0}$, $\hat{P}_\varepsilon^{l-}$ and $\hat{P}_\varepsilon^-$, encountered in this section.

3 Numerical Methods

We now construct appropriate numerical methods for generating approximate solutions to problems from each class. Note however that any problem from $\hat{P}_\varepsilon^-$ can be transformed into an equivalent problem in $\hat{P}_\varepsilon^+$, using the change of variables $\tilde{x} = 1 - \hat{x}$. Therefore we need only be concerned with the numerical solution of problems from classes $\hat{P}_\varepsilon^+$, $\hat{P}_\varepsilon^0$, $\hat{P}_\varepsilon^{l0}$ and $\hat{P}_\varepsilon^{l-}$.

Before we introduce the numerical methods we need some criteria to decide whether a given method is adequate for the problem in question. We would ideally like globally-defined, pointwise-accurate, ε-uniform monotone numerical methods. For a discussion of these concepts see Farrell et al. [1].

To generate numerical solutions for problems from all the above classes, we construct a numerical method consisting of a standard finite difference operator and a piecewise uniform fitted mesh. The only exception to this is in the case of class $\hat{P}_\varepsilon^{l-}$ where we use a uniform mesh.

First of all we consider class $\hat{P}_\varepsilon^+$ (all considerations are similar for $\hat{P}_\varepsilon^{l+}$). We use the following piecewise uniform mesh in the x-direction. Divide $\overline{\Omega}$ into two subintervals
$$\overline{\Omega} = \overline{\Omega}_l \cup \overline{\Omega}_r$$

where $\Omega_l = (0, \sigma)$, $\Omega_r = (\sigma, 1)$ and the fitting factor σ is chosen to be

$$\sigma = \min\left\{\frac{1}{2}, \frac{\varepsilon}{\hat{\alpha}}\ln N\right\}$$

where N is the number of mesh elements in the x-direction and $\hat{\alpha}$ is the lower bound on $\hat{a}$. We construct our piecewise uniform mesh Ω_σ^N on Ω by placing a uniform mesh in the subintervals Ω_l, Ω_r using $N/2$ mesh elements in each subinterval. A uniform mesh Ω_u^M with M mesh elements is used on $(0, T)$. We then define the fitted piecewise uniform mesh $\hat{D}_\sigma^N$ to be

$$\hat{D}_\sigma^{N,M} = \Omega_\sigma^N \times \Omega_u^M.$$

The resulting numerical method is thus

$$(\hat{P}_\varepsilon^{+, N}) \quad \hat{L}_\varepsilon^N U^N \equiv \varepsilon \delta_x^2 U^N + \hat{a} D_x^+ U^N - \hat{b} D_t^- U^N - \hat{d} U^N = \hat{f} \quad \text{on} \quad \hat{D}_\sigma^{N,M},$$

$$U^N = u \quad \text{on} \quad \overline{\hat{D}}_\sigma^{N,M} \setminus \hat{D}_\sigma^{N,M}.$$

Theorem 1. *For problems from class $\hat{P}_\varepsilon^+$, which are sufficiently compatible at the corners, the numerical approximations generated by the numerical method defined by $\hat{P}_\varepsilon^{+, N}$ are ε-uniform and satisfies the following error estimate*

$$\sup_{0<\varepsilon\leq 1} \|U - u\|_{\overline{\hat{D}}_\sigma^{N,M}} \leq CN^{-1}(\ln N)^2 + CM^{-1}$$

where C is a constant independent of N, M and ε.

Proof. See Shishkin [3].

To numerically solve problems from the classes $\hat{P}_\varepsilon^0$ and $\hat{P}_\varepsilon^{l0}$, we use the same finite difference operators but the fitted mesh used is different. First of all consider class $\hat{P}_\varepsilon^0$. In this case the interval $\hat{\Omega}$ is divided into three subintervals

$$\overline{\Omega} = \overline{\Omega}_l \cup \overline{\Omega}_c \cup \overline{\Omega}_r$$

where $\Omega_l = (0, \sigma), \Omega_c = (\sigma, 1 - \sigma), \Omega_r = (1 - \sigma, 1)$ and the fitting factor σ is chosen to be

$$\sigma = \min\left\{\frac{1}{4}, 2\sqrt{\varepsilon}\ln N\right\}.$$

The fitted piecewise uniform mesh is then defined as in the previous case. The resulting numerical method is denoted by $\hat{P}_\varepsilon^{0, N}$.

Theorem 2. *For problems from class $\hat{P}_\varepsilon^0$, which are sufficiently compatible at the corners, the numerical approximations generated by the numerical method defined by $\hat{P}_\varepsilon^{0, N}$ is ε-uniform and satisfies the following error estimate*

$$\sup_{0<\varepsilon\leq 1} \|U - u\|_{\overline{\hat{D}}_\sigma^{N,M}} \leq C(N^{-1}\ln N)^2 + CM^{-1}$$

where C is a constant independent of N, M and ε.

Proof. See, for example, Miller et al. [2].

For problems from class $\hat{P}_\varepsilon^{l0}$ we use a similar numerical method as that used for problems from class $\hat{P}_\varepsilon^+$, but with the fitting factor chosen to be

$$\sigma = \min\left\{\frac{1}{2},\ 2\sqrt{\varepsilon}\ \ln N\right\}.$$

We denote the resulting numerical method by $\hat{P}_\varepsilon^{l0,\,N}$.

As noted above for problems from class $\hat{P}_\varepsilon^{l-}$ it suffices to use a uniform mesh, and the standard finite difference operator. This is due to the fact that the layer that arises is a weak interior layer, in the sense that the solution in the layer region does not possess extremely large gradients, as would be the case with the other types of layers considered in this paper. Denote this method by $\hat{P}_\varepsilon^{l-,\,N}$.

In the next section we demonstrate numerically that the methods introduced for the latter two cases are ε-uniform for problems from the appropriate classes.

4 Numerical Results

As a particular example of a problem from class $\hat{P}_\varepsilon^{l0}$, we let ϕ_1 and ϕ_2 be chosen as in §2.2. Take $T = 1$ and $m = 1$ and let the original problem be

$$\varepsilon u_{xx} + u_x - u_t - u = -x - 1, \qquad \text{on } (-t, 1) \times (0, 1], \qquad (14\text{a})$$
$$u(x, 0) = 1 - x^2, \qquad x \in (0, 1), \qquad (14\text{b})$$
$$u(-t, t) = 1, \quad u(1, t) = 0, \qquad t > 0. \qquad (14\text{c})$$

It is clear that we have $\alpha = m\beta$ and thus the transformed problem will be in class $\hat{P}_\varepsilon^{l0}$. Here we have a parabolic boundary layer at $x = 0$.

As a particular example of a problem from class $\hat{P}_\varepsilon^{l-}$, we again let ϕ_1 and ϕ_2 be chosen as in §2.2. Take $T = 1$ and $m = 2$ and let the original problem be

$$\varepsilon u_{xx} + u_x - u_t - u = -x - 1, \qquad \text{on } (-2t, 1) \times (0, 1], \qquad (15\text{a})$$
$$u(x, 0) = 1 - x^2, \qquad x \in (0, 1), \qquad (15\text{b})$$
$$u(-2t, t) = 1, \quad u(1, t) = 0, \qquad t > 0. \qquad (15\text{c})$$

Here $\alpha < m\beta$ and thus the transformed problem will be in the class $\hat{P}_\varepsilon^{l-}$. In this case we have no layer in the main term of an asymptotic expansion (only a weak layer arises due to the compatibility condition being not of a sufficiently high order).

We take $N = M$ and tabulate the computed errors E_ε^N, and the computed ε-uniform errors E^N, for a variety of values of ε and N, for both problems using the methods described in §3 (see Tables 1 and 2). In both cases we use the numerical

Table 1. Table of computed errors E_ε^N using method $\hat{P}_\varepsilon^{l0,N}$ for problem (14)

	\multicolumn{6}{c}{Number of Intervals N}					
ε	8	16	32	64	128	256
1.0	2.09e-02	1.22e-02	7.12e-03	3.69e-03	1.79e-03	7.85e-04
2^{-1}	2.98e-02	1.55e-02	7.81e-03	3.82e-03	1.79e-03	7.69e-04
2^{-2}	4.68e-02	2.47e-02	1.25e-02	6.13e-03	2.88e-03	1.24e-03
2^{-3}	6.19e-02	3.30e-02	1.68e-02	8.31e-03	3.91e-03	1.68e-03
2^{-4}	7.32e-02	3.96e-02	2.04e-02	1.01e-02	4.76e-03	2.05e-03
2^{-5}	8.09e-02	4.42e-02	2.29e-02	1.14e-02	5.39e-03	2.33e-03
2^{-6}	8.54e-02	4.69e-02	2.45e-02	1.22e-02	5.76e-03	2.49e-03
2^{-7}	1.00e-01	4.92e-02	2.53e-02	1.26e-02	5.94e-03	2.56e-03
2^{-8}	1.06e-01	5.88e-02	2.82e-02	1.28e-02	6.03e-03	2.60e-03
2^{-9}	1.14e-01	6.63e-02	3.31e-02	1.55e-02	6.81e-03	2.67e-03
2^{-10}	1.21e-01	7.04e-02	3.65e-02	1.76e-02	8.05e-03	3.35e-03
2^{-11}	1.25e-01	7.25e-02	3.86e-02	1.90e-02	8.82e-03	3.72e-03
2^{-12}	1.29e-01	7.38e-02	4.03e-02	2.00e-02	9.36e-03	3.98e-03
2^{-13}	1.31e-01	7.57e-02	4.13e-02	2.07e-02	9.74e-03	4.17e-03
2^{-14}	1.32e-01	7.70e-02	4.20e-02	2.12e-02	1.00e-02	4.30e-03
2^{-15}	1.33e-01	7.79e-02	4.25e-02	2.15e-02	1.02e-02	4.39e-03
2^{-16}	1.34e-01	7.85e-02	4.29e-02	2.17e-02	1.03e-02	4.46e-03
2^{-17}	1.34e-01	7.89e-02	4.31e-02	2.19e-02	1.04e-02	4.50e-03
$\vdots$	$\vdots$	$\vdots$	$\vdots$	$\vdots$	$\vdots$	$\vdots$
2^{-32}	1.35e-01	8.00e-02	4.36e-02	2.23e-02	1.06e-02	4.61e-03
E^N	1.35e-01	8.00e-02	4.36e-02	2.23e-02	1.06e-02	4.61e-03

Table 2. Table of computed errors E_ε^N using method $\hat{P}_\varepsilon^{l-,N}$ for problem (15)

	\multicolumn{6}{c}{Number of Intervals N}					
ε	8	16	32	64	128	256
2^{-0}	4.78e-02	2.51e-02	1.27e-02	6.19e-03	2.91e-03	1.25e-03
2^{-1}	6.67e-02	3.65e-02	1.84e-02	8.99e-03	4.21e-03	1.81e-03
2^{-2}	8.93e-02	5.06e-02	2.57e-02	1.27e-02	5.95e-03	2.56e-03
2^{-3}	1.17e-01	6.57e-02	3.43e-02	1.70e-02	8.02e-03	3.46e-03
2^{-4}	1.40e-01	7.94e-02	4.24e-02	2.11e-02	1.00e-02	4.34e-03
2^{-5}	1.55e-01	8.98e-02	4.84e-02	2.43e-02	1.16e-02	5.00e-03
2^{-6}	1.64e-01	9.64e-02	5.19e-02	2.61e-02	1.24e-02	5.38e-03
2^{-7}	1.69e-01	1.00e-01	5.37e-02	2.70e-02	1.29e-02	5.57e-03
2^{-8}	1.71e-01	1.02e-01	5.46e-02	2.75e-02	1.31e-02	5.66e-03
2^{-9}	1.73e-01	1.03e-01	5.50e-02	2.77e-02	1.32e-02	5.70e-03
2^{-10}	1.73e-01	1.03e-01	5.53e-02	2.78e-02	1.32e-02	5.73e-03
2^{-11}	1.74e-01	1.03e-01	5.54e-02	2.79e-02	1.33e-02	5.74e-03
2^{-12}	1.74e-01	1.03e-01	5.54e-02	2.79e-02	1.33e-02	5.75e-03
2^{-13}	1.74e-01	1.04e-01	5.55e-02	2.79e-02	1.33e-02	5.75e-03
$\vdots$	$\vdots$	$\vdots$	$\vdots$	$\vdots$	$\vdots$	$\vdots$
2^{-32}	1.74e-01	1.04e-01	5.55e-02	2.79e-02	1.33e-02	5.75e-03
E^N	1.74e-01	1.04e-01	5.55e-02	2.79e-02	1.33e-02	5.75e-03

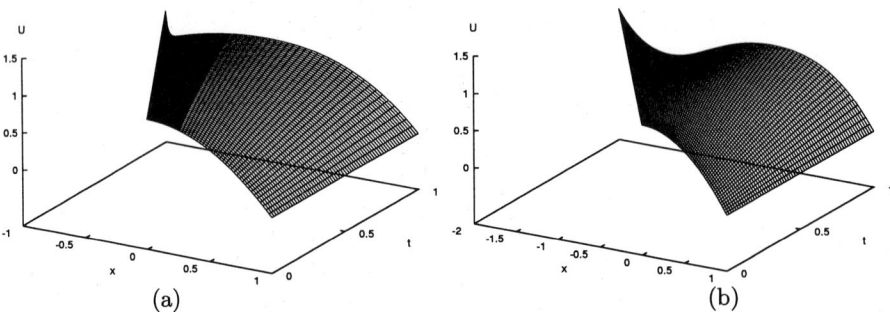

Fig. 1. Numerical solutions generated by (a) $\hat{P}_\varepsilon^{l0,N}$ and (b) $\hat{P}_\varepsilon^{l-,N}$ with $N=64$, $\varepsilon=2^{-10}$ for problems (14) and (15)

solution on the finest mesh available, namely $N=1024$, as the approximation to the exact solution. The computed pointwise errors, E_ε^N and E^N, are defined as

$$E_\varepsilon^N = \max_{0 \leq i,j \leq N} |U_\varepsilon^N(x_i, t_j) - \overline{U}_\varepsilon^{1024}(x_i, t_j)|,$$

$$E^N = \max_\varepsilon E_\varepsilon^N .$$

In both of these tables we see the maximum errors decrease as N increases for each value of ε and that the ε-uniform errors, E^N, also decrease with increasing N. This demonstrates numerically that these methods are ε-uniform for the problem classes in question. In Figure 1 we plot the numerical solution of these problems for particular values of ε and N.

References

1. Farrell, P. A., Hegarty, A. F., Miller, J. J. H., O'Riordan, E., and Shishkin, G. I.: Robust Computational Techniques for Boundary Layers. Chapman and Hall/CRC Press, Boca Raton, U. S. A. (2000)
2. Miller, J. J. H., O'Riordan, E., Shishkin, G. I., and Shishkina, L. P.: Fitted mesh methods for problems with parabolic boundary layers. Mathematical Proceedings of the Royal Irish Academy **98A** (2) (1998) 173–190
3. Shishkin, G. I.: Discrete approximation of singularly perturbed elliptic and parabolic equations. Russian Academy of Sciences, Ural Section, Ekaterinburg (1992)

Special Types of Badly Conditioned Operator Problems in Energy Spaces and Numerical Methods for Them

Eugene G. D'yakonov

Department of Computer Mathematics and Cybernetics
Moscow State University, Moscow, 119899, Russia
dknv@cmc.msk.su

Abstract. Badly conditioned operator problems in Hilbert spaces are characterized by very large condition numbers. For special types of such problems, their reduction to ones with strongly saddle operators leads to remarkable improvement of correctness and to justification of the famous Bakhvalov—Kolmogorov principle about asymptotically optimal algorithms.
The first goal of the present paper is to present a short review of recently obtained results for stationary problems in classical Sobolev and more general energy spaces. The second goal is a study of the approach indicated above to the case of nonstationary problems; special attention is paid to parabolic problems with large jumps in coefficients; the study is based on relatively new extension theorems and special energy methods.

1 Introduction

1.1 Normally Invertible Operators

Only real Hilbert space s and bounded operators are considered in this paper; the normed linear space of linear bounded operators mapping a space U into a space F is denoted by $\mathcal{L}(U; F)$; $\mathcal{L}(H) \equiv \mathcal{L}(H; H)$; Ker $A \equiv \{v : Av = 0\} \equiv$ the kernel (null-space) of the operator A; Im $A \equiv$ the image (range) of the operator A; $I \equiv$ the identity operator; $H^* \equiv$ the linear space of bounded linear functionals l mapping H into $\mathbf{R}$; $A^* \equiv$ the adjoint operator to $A \in \mathcal{L}(H_1; H_2)$; for $A \in \mathcal{L}(H)$, $[A] \equiv 2^{-1}(A + A^*))$; $\mathcal{L}^+(H)$ denotes the set of linear, symmetric, and positive definite operators in $\mathcal{L}(H)$; $H(B) \equiv$ the Hilbert space differing from H only by inner product defined by $B \in \mathcal{L}^+(H)$, namely $(u, v)_{H(B)} \equiv (u, v)_B \equiv (Bu, v)_H \equiv (Bu, v)$.

Operators $A_{2,1} \in \mathcal{L}(H_1; H_2)$ with Im $A_{2,1} = H_2$ are called normally invertible; they correspond to a particular case of normally solvable operators which are defined as operators such that Im $A_{2,1}$ is a subspace in H_2 (operators with closed images); if $A_{2,1}$ is a normally solvable operator, then H_1 is an orthogonal sum of Ker $A_{2,1}$ and Im $A_{2,1}^*$, i.e.,

$$H_1 = \text{Ker } A_{2,1} \oplus \text{Im } A_{1,2}; A_{1,2} \equiv A_{2,1}^*. \tag{1.1}$$

Note that the indicated operators are fundamental in theory of Fredholm's equations (see [1-3] and references therein).

A normally invertible operator $A_{2,1}$ yields a one-to-one mapping of the Hilbert space Im $A_{1,2}$ (orthogonal complement in H_1 to Ker $A_{2,1}$) onto H_2 and, by the Banach theorem, this mapping is invertible and the corresponding inverse (the right inverse) $A_{2,1}^{(-1)} \equiv A_{2,1}^{\dagger}$ is such that

$$\|A_{2,1}{}^{(-1)}\| \equiv \sigma^{-1} < \infty. \tag{1.2}$$

We note that the well-known inf-sup condition

$$\inf_{u_2 \in H_2} \sup_{u_1 \in H_1} \frac{(A_{2,1}u_1, u_2)_{H_2}}{\|u_1\|_{H_1}\|u_2\|_{H_2}} \geq \sigma > 0 \tag{1.3}$$

is often used instead of (1.1), (1.2); (1.3) can be written in the form

$$\|A_{2,1}^* u_2\|_{H_1} \geq \sigma \|u_2\|_{H_2}, \quad \forall u_2 \in H_2 \tag{1.4}$$

(see [3], [4] and references therein) and implies that $A_{2,1} A_{2,1}^* \in \mathcal{L}^+(H_2)$.

1.2 Strongly Saddle Operators and Their Generalizations

In the Hilbert space $H = H_1 \times H_2$, we consider $A_\alpha \in \mathcal{L}(H)$ of the form

$$A_\alpha \equiv \begin{bmatrix} A_{1,1} & A_{1,2} \\ A_{2,1} & -\alpha A_{2,2} \end{bmatrix}, \tag{1.5}$$

where $A_{i,j} \in \mathcal{L}(H_j; H_i)$, $\alpha \geq 0$, $A_{2,1}$ is a normally invertible operator;

$$[A_{1,1}] \in \mathcal{L}^+(H_1), \quad A_{1,2} = A_{2,1}^*; \quad [A_{2,2}] \geq 0.$$

Under above conditions, A_α is called a generalized strongly saddle operator. For such operators, it was proved (see [3], Theorem 7.1.3) that A_α is invertible and

$$\|A_\alpha^{-1}\| \leq K, \tag{1.6}$$

where the constant K can be chosen uniformly for all $\alpha \geq 0$ and all $A_{2,2}$ with $[A_{2,2}] \geq 0$. This implies that problem

$$A_\alpha u = f, \tag{1.7}$$

is correctly posed. Moreover, (1.6) implies that the condition number $\kappa_H(A_\alpha) \equiv \|A_\alpha\|\|A_\alpha^{-1}\| \asymp 1$ if $\alpha \in [0, \alpha_0]$ and that optimal perturbation estimates (see [3,5]) hold; they follow from (1.6) and the known inequality $\|A_\alpha^{-1} - \tilde{A}^{-1}\| \leq \|A_\alpha^{-1}\| \times \|A_\alpha - \tilde{A}\| \times \|\tilde{A}^{-1}\|$; first such results were obtained in [5] (see also [3]) for $A_{2,1}$ associated with the divergence operator.

Note that A_α in (1.5) is a strongly saddle operator if $[A_{i,i}] = A_{i,i}$, $i \in [1,2]$; if additionally $A_{2,2} \in \mathcal{L}^+(H_2)$, $\alpha > 0$ and $f_2 = 0$ then the first component of the solution of (1.7) coincides with the solution of problem

$$\Lambda_\alpha u_1 \equiv A_{1,1} u_1 + \frac{1}{\alpha} A_{1,2} A_{2,2}^{-1} A_{2,1} u_1 = f_1, \tag{1.8}$$

which might serve as a typical example of variational problems involving a large parameter $1/\alpha$; the condition number $\kappa_{H_1}(\Lambda_\alpha)$ is very large ($\kappa_{H_1}(\Lambda_\alpha) \asymp 1/\alpha$); hence it complicates construction of good numerical methods and algorithms very strongly.

1.3 Regularization of Certain Badly Conditioned Operator Problems

Operator problems of type (1.8) with large parameters $1/\alpha \gg 1$ ($\alpha \to +0$) in Sobolev and more general energy spaces H_1 can be found in many important branches of mathematical physics; the corresponding variational problem

$$u_1 = \arg\min_{v_1 \in H_1}[I_2(v_1) - 2l(v_1)], \quad I_2(v_1) \equiv (A_{1,1}v_1, v_1) + \frac{1}{\alpha}\|A_{2,1}v_1\|^2_{A_{2,2}^{-1}} \quad (1.9)$$

can be connected with an application of the standard penalty method for a problem with the linear constraint $A_{2,1}v_1 = 0$; we stress that the classical Lagrange approach (the Lagrange multiplier method) to the variational problem with this constraint yields good conditioned problem (1.7) with $\alpha = 0$, $f_2 = 0$ (the additional function u_2 plays the role of the Lagrangian multiplier).

Thanks to an understanding of the role of (1.7) and its grid analogs in the theory of projective-grid (finite element) methods and iterative processes, it now seems reasonable to regard problems (1.7) in the Hilbert space $H \equiv H_1 \times H_2$ as basic. If

$$H_2 \equiv \prod_{k=1}^{k^*} H_{2,k}; \quad A_{2,1}v_1 \equiv [A_{2,1,1}v_1, \ldots, A_{2,1,k^*}v_1] \quad (1.10)$$

($A_{2,1,k^*}v_1 \in H_{2,k}$) then it is possible even to deal with variational problems with several large $1/\alpha_k$, $k \in [1, k^*]$; for example, from (1.9), (1.10) with

$$I_2(v_1) \equiv (A_{1,1}v_1, v_1) + \sum_{k=1}^{k^*} \frac{1}{\alpha_k}\|A_{2,1,k}v_1\|^2_{H_{2,k}} \quad (1.11)$$

we can pass to problems (1.7) with the block $\alpha A_{2,2}$ in (1.5) replaced by the block-diagonal operator

$$\alpha A_{2,2} \equiv \operatorname{diag}(\alpha_1 I_{2,1}, \ldots, \alpha_{k^*} I_{2,k^*}), \quad (1.12)$$

where $\alpha_k > 0$, $I_{2,k}$—the identity operator in $H_{2,k}$, $k \in [1, k^*]$.

The indicated remarkable improvement of correctness leads sometimes even to the construction of asymptotically optimal numerical methods and algorithms under natural conditions on the smoothness of the solution (to justification of the famous Bakhvalov—Kolmogorov principle about asymptotically optimal algorithms; see [3]). We recall that projective methods for problems with saddle operators make use of a special sequence of finite-dimensional subspaces

$\hat{H}_h \equiv \hat{H}_{1,h} \times \hat{H}_{2,h} \in H$ approximating the original Hilbert space H (H_r is approximated by the sequence $\hat{H}_{r,h} \equiv \hat{H}_r$, $r = 1, 2$); it is required that

$$\inf_{\hat{u}_2 \in \hat{H}_2} \sup_{\hat{u}_1 \in \hat{H}_1} \frac{(A_{2,1}\hat{u}_1, \hat{u}_2)_{H_2}}{\|\hat{u}_1\|_{H_1} \|\hat{u}_2\|_{H_2}} \geq \sigma_0 > 0, \tag{1.13}$$

where σ_0 is independent of h; (1.13) implies that $\|\hat{A}_{2,1}^{(-1)}\| \leq \sigma_0^{-1} < \infty$ ($\hat{A}_{2,1}$ is an approximation of $A_{2,1}$).

1.4 Improved Correctness of Problems with Strongly Saddle Operators and Their Generalizations

Problems (1.8)–(1.12) can be sometimes reduced to those of type (1.7) in a better Hilbert space $G \equiv G_1 \times G_2 \subset H$; such a reduction was indicated in [7] and is based on the following lemma (see [2,7]):

Lemma 1.1. Let $A_{2,1} \in \mathcal{L}(H_1; H_2)$ be normally invertible and the embedding operator of a Hilbert space G_2 into H_2 be bounded. Let G_1 be a subset of H_1 such that $\dim G_1 = \infty$ and $\|v\|_{G_1}^2 \equiv \|v\|_{H_1}^2 + \|A_{2,1}v\|_{G_2}^2 < \infty$. Then G_1 is a Hilbert space and the restriction $A_{2,1,G_1} \in \mathcal{L}(G_1; G_2)$ of A to G_1 is normally invertible.

We recall that a pre-Hilbert space H is called Hilbert space if it is complete and separable and $\dim H = \infty$; if $\dim H < \infty$, the term Euclidean space is usually preferred.

2 Examples of Normally Invertible Operators and Regularized Problems

2.1 The Divergence Operator; Elasticity and Hydrodynamics Problems

In what follows, we assume, for simplicity, that Ω is a bounded domain in the Euclidean space $\mathbf{R}^d$, $d = 2, 3$, with Lipschitz piecewise smooth boundary $\Gamma \equiv \partial \Omega$ and $\bar{\Omega} \equiv \Omega \cup \Gamma$. We write

$$(u,v)_{0,\Omega} \equiv (u,v)_{L_2(\Omega)}, \quad |u|_{0,\Omega} \equiv (u,u)_{0,\Omega}^{1/2}, \quad |u|_{1,\Omega} \equiv (u,u)_{1,\Omega}^{1/2} \equiv (|\nabla u|^2, 1)_{0,\Omega}^{1/2}$$

and make use of the Sobolev space $W_2^1(\Omega) \equiv H^1(\Omega)$ (see [3,4,6]) with the norm

$$\|u\|_{H^1(\Omega)} \equiv \|u\|_{1,\Omega} \equiv [|u|_{1,\Omega}^2 + |u|_{0,\Omega}^2]^{1/2}. \tag{2.1}$$

For vector fields, the norms are defined in the same manner. Examples of problems from hydrodynamics and elasticity associated with the divergence operator $A_{2,1}$ for the corresponding vector fields can be found in [3,4,7] for various choices of boundary conditions; not only the Stokes system but many its generalizations were considered from the point of view of optimization of numerical methods and algorithms; special attention was paid to estimates of accuracy and computational work independent of parameters like α (see [3,7]).

2.2 The Trace Operator; Problems in Strengthened Sobolev Spaces; New Penalty Methods for the Dirichlet Conditions

The strengthened Sobolev spaces are naturally connected, e.g., with such important (two or three-dimensional) problems of mathematical physics as those in theory of plates and shells with stiffeners or in the hydrodynamics involving the surface tension (see [3,7,8,9]). If $d = 2$, the model strengthened Sobolev space $G_{1,m} \equiv G(\Omega; S) \equiv G$, $m = [m] \geq 1$, is defined as a subset of functions in $H^1(\Omega)$ (see (1.1)) such that their traces on each S_r belong to $H^m(S_r) \equiv W_2^m(S_r)$, so we can define the norm in G by

$$\|v\|_G^2 \equiv \|v\|_{H^1(\Omega)}^2 + \sum_{r=1}^{r*} \|v\|_{H^m(S_r)}^2; \tag{2.2}$$

$S \subset \bar{\Omega}$ consists of straight line segments (stiffeners) $S_1, \ldots, S_{r*}$ (smooth arcs are also allowed). It was shown (see [9]) that $G_{1,1}$ is a completion of the space of smooth functions in the sense of norm (2.2).

These nonstandard Hilbert spaces allow to set correct variational and operator problems. Among possible spectral (eigenvalue) problems, we mention those that are reduced to the problems $Mu = \lambda Lu$ with $L \in \mathcal{L}^+(G)$ and symmetric and compact operators M; for such problems in our Hilbert space s G, the classical Hilbert-Schmidt theorem holds (see [1,3]). Examples of problems on more involved composed manifolds of different dimensionality can be found in [8,9]. Special attention was paid to numerical methods based on the use of projective-grid methods and effective iterative methods such as multigrid and cutting methods; in case where the original problems were badly conditioned, the indicated above reduction to problems (1.7) in the Hilbert space $G \equiv G_1 \times G_2$ turned out to be very efficient (see [3,8,9]).

As is known, the homogeneous Dirichlet conditions can be understood in terms of the penalty method as a limit of natural boundary conditions of the type $\frac{\partial u_\alpha}{\partial n} + (1 + 1/\alpha)u_\alpha = 0$, where n is the unit vector of the outer normal to the boundary Γ, $\alpha \to +0$; the latter ones are connected with the additional term (penalty term) $F(u) \equiv (1+1/\alpha)|u|_{0,\Gamma}^2$ in the minimized energy functional. From the mathematical point of view, this penalty term might be considered as rather weak because of additional requirements on smoothness of the solutions in order to obtain optimal perturbation estimates $\|u_\alpha - u\|_{1,\Omega} + |u_\alpha - u|_{0,\Gamma} = O(\alpha)$. It was shown recently (see [9]) that such and even stronger estimates hold under correctness of the original problem if apply the stronger penalty term $\bar{F}(u) \equiv (1 + 1/\alpha)\|u\|_{H^\nu(\Gamma)}^2$ with $\nu \geq 1/2$ and treat the arising problem in the corresponding strengthened Sobolev space. Moreover, this approach leads to important a posteriori estimates with no additional assumptions on the solution in contrast to estimates based on respective estimates of residuals; similar results hold for spectral problems (see [9,10]). The results obtained yield also understanding of mechanism of splitting of the problem under consideration (with large penalty parameters on S) into separate ones in subdomains with the homogeneous Dirichlet conditions. The case with $\nu = 1$ is especially important since efficient numerical methods were indicated.

2.3 The Jump Operator; Problems in Weakened Sobolev Spaces

In what follows, me make use of a partition

$$\bar{\Omega} = \cup_{i=1}^{i^*} \bar{\Omega}_i \tag{2.3}$$

into blocks with Lipschitz and piecewise smooth boundaries $\partial \Omega_i \equiv \Gamma_i$. The factored Sobolev space (associated with (2.3)) is

$$H_f^1 \equiv \prod_{i=1}^{i^*} H^1(\Omega_i).$$

The boundary Γ_i consists of blocks $\Gamma_{i,j} \equiv \Gamma_i \cap \Gamma_j$; they constitute $R \equiv \cup_{i<j} \Gamma_{i,j}$. For $w \in H_f^1$, we define the local jump $J_{i,j} w$ of traces on $\Gamma_{i,j}$ as

$$J_{i,j} w \equiv \text{Tr}_{H^1(\Omega_j) \mapsto L_2(\Gamma_{i,j})} w - \text{Tr}_{H^1(\Omega_i) \mapsto L_2(\Gamma_{i,j})} w.$$

If we assume, for simplicity of presentation, that different $\Gamma_{i,j} \equiv R_r$ are separated, then the weakened Sobolev space $\mathcal{A}_{1,1} \equiv \mathcal{A}_{1,1}(\cup_{i=1}^{i^*} \Omega_i; R)$ (see [9–11]) consists of $w \in H_f^1$ such that the jumps of the traces of w on each $\Gamma_{i,j}$ belong to $H^1(\Gamma_{i,j})$;

$$\|w\|_{\mathcal{A}_{1,1}}^2 \equiv \sum_{i=1}^{i^*} \|w\|_{H^1(\Omega_i)}^2 + \sum_{i<j} \|J_{i,j}\, w\|_{H^1(\Gamma_{i,j})}^2; \tag{2.4}$$

$\mathcal{A}_{1,1}$ is a strengthened H_f^1; $H^1(\Omega)$ is a subspace in $\mathcal{A}_{1,1}$ which explains the term used above. It was proved (see [9–11]) that $\mathcal{A}_{1,1}$ is a completion in norm (2.4) of the space of discontinuous functions such that their restrictions to each Ω_i are continuous and smooth functions; this is the case for more general spaces.

Weakened Sobolev spaces and related H_f^1 are well suited for mathematical modelling of problems in composite structures where discontinuous solutions are allowed (problems with interfaces). Spaces of the relevant type are used in domain decomposition methods, especially in the case of nonmatching grids. Our attention to the above spaces was motivated by the fact that problems in weakened Sobolev spaces have good perspectives from the point of view of obtaining a posteriori error estimates of solutions to classical elliptic boundary and spectral problems under no additional requirements of the solution smoothness. Effective numerical methods for solving elliptic problems in weakened Sobolev spaces were considered in [9–11].

2.4 The Restriction Operator; Elliptic Problems with Large Jumps in Coefficients

Iterative methods of various nature for the discretized elliptic problems, mentioned above, have been considered in many papers (see [12-14] and references therein); probably, the first effective iterations were indicated in [15]. We concentrate here on estimates of accuracy of projective methods and construction of asymptotically optimal algorithms with estimates independent of the jumps in coefficients (such methods and algorithms are referred sometimes as robust).

Below, we use a sufficiently simple part $\Gamma_0 \subset \Gamma$ with $|\Gamma_0|_{(d-1)} > 0$, where $|\ |_{(d-1)}$ denotes the $(d-1)$-dimensional measure (the case $\Gamma_0 = \Gamma$ is allowed); we define $\tilde{H}_1 \equiv H^1(\Omega; \Gamma_0)$ as a standard subspace in $H^1(\Omega)$ (it consists of functions v with zero traces on Γ_0). We also use an open set $C \subset \Omega$ such that

$$\bar{C} = \cup_{k=1}^{k^*} \bar{C}_k, \tag{2.5}$$

where each C_k is a domain with Lipschitz and piecewise smooth boundary ∂C_k; $\bar{C}_k$ consists of blocks $\bar{\Omega}_i$ ($\bar{C}_k = \cup_{i \in \pi(k)} \bar{\Omega}_i$); the distance between different $\bar{C}_k$ is greater then $2\rho' > 0$. We take $\rho \in (0, \rho')$) and define $C_{k,\rho}$ as a set of points whose distances to C_k are smaller then ρ; we define $\tilde{H}_{1,k,\rho}$ as a subspace of $H^1(\Omega; \Gamma_0)$ consisting of functions with supports in $\bar{C}_{k,\rho}$ (the functions vanish at the points whose distances to C_k are greater then ρ); C in (2.5) will be associated with certain large coefficients $1/\alpha_k > 0$, $k \in [1, k^*]$ and the Hilbert space

$$\tag{2.6}$$

where each V_k is ... we introduce $W_k \equiv H^1(C_k; \ldots)$ ... k and we write $k \in \pi_0$, $(\partial C_k \cap \Gamma_0) \equiv \Gamma_{k,0}$, $(\partial C_k \cap \Gamma) \setminus \Gamma_{k,0} \equiv \Gamma_{k,1}$. If $|\partial C_k \cap \Gamma_0|_{(d-1)} = 0$ then $k \in \pi_1$ and $W_k = H^1(C_k)$; V_k is now defined as a set of functions in W_k such that their traces on a piece $\gamma_k \subset \bar{C}_k$ of a smooth $(d-1)$-dimensional surface are orthogonal to 1 in the sense of $L_2(\gamma_k)$ (the orthogonality condition is written in the form $\varphi_k(v) = 0$); here $|\gamma_k|_{(d-1)} > 0$ and the case $\gamma_k \subset \partial C_k$ is allowed. The norm in each V_k is chosen as (see (2.1), (2.6)) $\|v_{2,k}\|_{V_k} \equiv |v_{2,k}|_{1,C_k} \asymp \|v_{2,k}\|_{1,C_k}$; elements of H_2 are written as $v_2 \equiv [v_{2,1}, \ldots, v_{2,k^*}]$. The key Hilbert space H_1 is a subspace of $\tilde{H}_1$ characterized by the conditions

$$\varphi_k(v) = 0, \ \forall k \in \pi_1; \tag{2.7}$$

the restriction operators of elements $v_1 \in H_1$ onto C_k and C are denoted by R_k and R respectively.

Theorem 2.1. *For all $k \in \pi_1$, suppose that the distance between Γ_0 and each $\bar{C}_k$ with $k \in \pi_1$ is greater then $2\rho > 0$ ($\rho < \rho'$). For all $k \in \pi_0$, suppose that the distance between $\Gamma_{k,1}$ and $\Gamma_0 \setminus \Gamma_{k,0}$ is greater then 2ρ and that the extension* theorem of V_k to $\tilde{H}_{1,k,\rho}$ holds (see [14]). Then there exists a constant $K^* > 0$ such that, for each $v_2 \in H_2$, it is possible to indicate a function $u_1 \in H_1$ with properties:

$$Ru_1 = v_2, \quad \|u_1\|_{H_1} \leq K^* [\sum_{k=1}^{k^*} \|v_{2,k}\|_{V_k}^2]^{1/2}. \tag{2.8}$$

Proof. Under the above assumptions, it suffices to construct an extension $u_{1,k} \in H_{1,k,\rho}$ of each $v_{2,k} \in V_k$ to $H_{1,k,\rho}$ separately (the desired extension u_1 can be taken as $u_{1,1} + \cdots + u_{1,k^*}$). For each $k \in \pi_1$, the classical extension theorems (see [6,3]) yield a function $w_{2,k} \in H^1(\mathbf{R}^d)$; its product with a smooth function $g_k(x)$ (it vanishes on $\Omega \setminus \bar{C}_{k,\rho}$ and equals to 1 on C_k) gives $u_{1,k}$. For $k \in \pi_0$, instead of the classical extension theorems, more involved ones should be used (see [14]); they apply harmonic equation in $C_{k,\rho} \setminus \bar{C}_k$ with specially chosen Dirichlet conditions.

Note that Theorem 2.1 implies that the restriction operators R_k and R are normally invertible. It is important that grid extension theorems can also be obtained; they deal, e.g., with piecewise linear functions $\hat{v}_{2,k}$ and $\hat{u}_1$ defined on triangulations $T_h(\bar{C}_k)$ and $T_h(\bar{\Omega})$ of $\bar{C}_k$ and $\bar{\Omega}$ respectively; K^* (see (2.8)) in these grid theorems does not depend on h (see [3,14] and (1.13)); domains with non-Lipschitz boundaries are allowed (see [3,14]).

As an example of elliptic problems with large jumps in coefficients we consider the problem of finding $u \in H_1$ such that

$$b(u; u') \equiv \sum_{i=1}^{i^*} c_i^{(0)} (u/\Omega_i; u'/\Omega_i)_{1,\Omega_i} +$$

$$+ \sum_{k=1}^{k^*} \frac{1}{\alpha_k} \sum_{i \in \pi_{(k)}} c_i^{(1)} (u/\Omega_i; u'/\Omega_i)_{1,\Omega_i} = l(u'), \quad \forall u' \in H_1, \tag{2.9}$$

where $l \in H_1^*$, all $c_i^{(0)}$, $c_i^{(1)}$ and α_k are positive constants, but α_k are relatively small. Correctness of problem (2.9) is obvious (see (1.11)). We can reduce it to problem (1.7), (1.12) if, instead of (2.6), we define H_2 and $H_{2,k}$ as

$$H_2 \equiv \prod_{k=1}^{k^*} H_{2,k}; \tag{2.10}$$

$$\|v_{2,k}\|_{H_{2,k}}^2 \equiv \sum_{i \in \pi_{(k)}} c_i^{(1)} |v_{2,k}/\Omega_i|_{1,\Omega_i}^2. \tag{2.11}$$

It is important that the norms in $H_{2,k}$ and V_k are equivalent; the same holds for the old and new norms in H_2. This enables us to consider the restriction operator $R \in \mathcal{L}(H_1; H_2)$ as normally invertible and deal with basic problem (1.7), (1.12), (2.10), (2.11). It seems natural to assume that

$$\|u\|_{W_2^{1+\gamma}(\Omega_i)} \leq K_{1,i}^* \quad i \in [1, i^*], \tag{2.12}$$

where u—the solution of (2.9) and $\gamma \in (0,1]$. In the same manner as for $\tilde{H}_1$ (see [3]) it can be verified that inequalities (2.12) determine a compact set in H_1; for the $N(\varepsilon)$-width in the sense of Kolmogorov for this set, we have $N(\varepsilon) \asymp \varepsilon^{-d/\gamma}$, where $\varepsilon > 0$ is a prescribed tolerance and $N(\varepsilon)$ corresponds to the dimension of the used approximating subspace of H_1.

Under conditions (2.12), asymptotically optimal projective-grid methods for (1.7), (1.12) can be constructed on the base of quasiuniform triangulation s of blocks $\bar{\Omega}_i$ and spline subspaces $\hat{H}_{r,h}$ of dimensions $N_r = O(N(\varepsilon))$, $r = 1, 2$ (see [3,14]); we assume that $T_h(\bar{\Omega})$ are consistent with geometry of blocks $\bar{\Omega}_i$ and Γ_0 so the subspaces $\tilde{H}_{1,h} \subset \tilde{H}_1$ and $\hat{H}_{1,h} \subset H_1$ yield subspaces $\hat{W}_{k,h} \subset W_k$, $\hat{H}_{2,k,h} \subset H_{2,k}$, $\hat{H}_{2,h} \subset H_2$. Estimates of accuracy $\|\hat{u} - u\|_H \leq K h^\gamma$ for such methods are independent of all $\alpha_k \in [0, c_0]$.

Our projective-grid method yields grid systems

$$A\mathbf{u} \equiv \begin{bmatrix} A_{1,1} & A_{1,2} \\ A_{2,1} & -A_{2,2} \end{bmatrix} \begin{bmatrix} \mathbf{u}_1 \\ \mathbf{u}_2 \end{bmatrix} = \begin{bmatrix} \mathbf{f}_1 \\ 0 \end{bmatrix} \equiv \mathbf{f}, \qquad (2.13)$$

in the standard Euclidean space s $\mathbf{H} \equiv \mathbf{H}_1 \times \mathbf{H}_2$, $\dim \mathbf{H}_1 = \dim \hat{H}_1$, $\dim \mathbf{H}_2 = \dim \hat{H}_2$. In (2.13), $A_{2,2} \equiv \text{diag}(\alpha_1 \Lambda_{2,1}, \ldots, \alpha_{k^*} \Lambda_{2,k^*})$, $\Lambda_{2,k}$ is a corresponding analog of the identity operator in $\hat{H}_{2,k}$ ($\Lambda_{2,k}$ is a Gram matrix), $k \in [1, k^*]$. If $T_h(\bar{\Omega}) \equiv T^{(p)}(\bar{\Omega})$ is obtained as a result of a refinement procedure that is applied recursively p times for an initial coarse triangulation $T^{(0)}(\bar{\Omega})$ with $p \asymp |\ln h|$, then, for $A_{1,1}$ in (2.12), there exists an asymptotically optimal model operator $B_1 \asymp A_{1,1}$ such that the constants of spectral equivalence and the estimates of the required computational work in solving systems with B_1 are independent of all numbers α_k: it is constructed in accordance with theory of model cooperative operators (see [3,9]) based on proper multigrid splittings of the spline space $\hat{H}_{1,h}$ (hierarchical basis for it is used). The same applies to model operators B_2 for the block-diagonal operator $\Lambda_2 \equiv \text{diag}(\Lambda_{2,1}, \ldots, \Lambda_{2,k^*})$; there is only a relatively new problem connected with the use of the basis in $\hat{H}_{2,k}$ if $k \in \pi_1$ (see (2.7)). But it can be reduced (see [3], Section 8.3) to a similar standard problem with the natural basis in $\hat{W}_k$ and the corresponding Gram matrix (it is nonnegative). Hence, we can construct an asymptotically optimal model operator $B \in \mathcal{L}^+(\mathbf{H})$ such that $B\mathbf{u} = [B_1\mathbf{u}_1, B_2\mathbf{u}_2]$, $\forall \mathbf{u} \in \mathbf{H}$, and apply effective iterations

$$B\mathbf{u}^{n+1} = B\mathbf{u}^n - \tau_n A^* B^{-1}(A\mathbf{u}^n - \mathbf{f}). \qquad (2.14)$$

A combination of (2.14) with the multigrid continuation procedure leads to justification of the Bakhvalov—Kolmogorov principle with estimates of computational work independent of all $\alpha_k \in [0, c_0]$ (see [3,14]). Instead of (2.14), the modified conjugate gradient iterations can also be used.

It should be noted that problem (2.9) in H_1 is rather unusual if the set π_1 is nonempty and $m \geq 1$ conditions (2.7) are necessary (we take π_1 as a set of first m indexes). But $\tilde{H}_1 = H_1 + \text{lin}[e_1, \ldots, e_m]$, where basic functions $e_1, \ldots, e_m$ can be easily constructed; moreover, we can assume that $\varphi_k(e_j) = \delta_{k,j}$. This enables us to reduce the original operator problem in $\tilde{H}_1$ to $m+2$ problems in H_1 in the same manner as it is done in the well-known block elimination procedure (see [3], Section 1.5).

3 Parabolic Problems with Large Jumps in Coefficients

3.1 Discretized Parabolic Problems with Large Jumps in Coefficients

For nonstationary problems in $Q_T \equiv \bar{\Omega} \times [0,T]$, we take $\tau \equiv T/n^*$ and write $t_n \equiv n\tau$, $u_r^n \equiv u_r(t_n)$ $(n = 0, \ldots, n^*)$, $\bar{\partial}_0 u_r^n \equiv [u_r^n - u_r^{n-1}]/\tau$ $(n = 1, \ldots, k \leq n^*)$.

Hereafter, H_1, H_2, $H \equiv H_1 \times H_2$ are Hilbert space s or Euclidean space s. On the basis of (2.9) and as a typical example of parabolic problems and their discretizions, we consider a sequence of stationary problems in H_1

$$(\bar{\partial}_0 u_1^n; v_1^n)_{0,\Omega} + b(u^n; v_1^n) = (F_1^n; v_1^n)_{0,\Omega} \quad (3.1)$$

where v_1^n refers to arbitrary elements of H_1, $F_1^n \in L_2(\Omega)$, $n \geq 1$, $u_1^0 = 0$ (this can be assumed without loss of generality). We stress that (3.1) is a problem involving very large parameters $1/\alpha_k$ which makes standard accuracy estimates of numerical methodss (see [16]) rather unsatisfactory. To make them independent of α_k, we apply the same regularization as for (2.9):

$$M\bar{\partial}_0 u_1^n + A_{1,1} u_1^n + A_{1,2} u_2^n = Mg_1^n, \quad (3.2)$$

$$A_{2,1} u_1^n - A_{2,2} u_2^n = f_2^n, \quad (3.3)$$

where $M \in \mathcal{L}(H_1)$, $M = M^* > 0$, $(Mu_1, v_1)_{H_1} \equiv (u_1, v_1)_{0,\Omega}$, $\|u_1\|_M \equiv |u_1|_{0,\Omega}$, $\forall u_1 \in H_1$, $\forall v_1 \in H_1$; $\|Mu_1\|_{H_1} \leq \|M\| |u_1|_{0,\Omega}$; $f_2^0 = 0$.

Theorem 3.1. *There exist constants τ_0 and K, independent of all $\alpha_j \in [0, c_0]$ and such that, for the solution of (3.2), (3.3), with $\tau \leq \tau_0$ and $n = 1, \cdots, k \leq n^*$, the a priori estimate*

$$\|u_1^k\|_{H_1}^2 + \tau \sum_{n=1}^{k} [\|\bar{\partial}_0 u_1^n\|_M^2 + \|u_1^n\|_{H_1}^2 + \|u_2^n\|_{H_2}^2 + \|A_{2,1} \bar{\partial}_0 u_1^n\|_{H_2}^2] \leq K F_k, \quad (3.4)$$

holds, where

$$F_k \equiv \tau \sum_{n=1}^{k} \left(\|g_1^n\|_M^2 + \|\bar{\partial}_0 f_2^n\|_{H_2}^2 \right). \quad (3.5)$$

Proof. Restriction (3.3) implies that

$$-(A_{2,1} \bar{\partial}_0 u_1^n + A_{2,2} \bar{\partial}_0 u_2^n, u_2^n)_{H_2} = -(\bar{\partial}_0 f_2^n, u_2^n)_{H_2}. \quad (3.6)$$

The inner product of each part of (3.2) and $\bar{\partial}_0 u_1^n$ in H_1 yields an equality; summing up the obtained equality and (3.6), we obtain

$$X \equiv \tau \sum_{n=1}^{k} \left(\|\bar{\partial}_0 u_1^n\|_M^2 + (A_{1,1} u_1^n, \bar{\partial}_0 u_1^n)_{H_1} + (A_{2,2} u_2^n, \bar{\partial}_0 u_2^n)_{H_2} \right) =$$

$$= \tau \sum_{n=1}^{k} \left((g_1^n, \bar{\partial}_0 u_1^n)_{M_0} - (\bar{\partial}_0 f_2^n, u_2^n)_{H_2} \right) \equiv Y. \tag{3.7}$$

It can be easily verified that

$$X \geq \tau \sum_{n=1}^{k} \|\bar{\partial}_0 u_1^n\|_{M_0}^2 + \frac{1}{2} \|u_1^k\|_{A_{1,1}}^2; \quad (g_1^n, \bar{\partial}_0 u_1^n)_M \leq \frac{1}{4} \|\bar{\partial}_0 u_1^n\|_M^2 + \|g_1^n\|_M^2 \tag{3.8}$$

For the second term on the right-hand side of (3.7), we have

$$Z \equiv -\tau \sum_{n=1}^{k} (\bar{\partial}_0 f_2^n, u_2^n)_{H_2}; \quad |Z| \leq \tau \sum_{n=1}^{k} \|\bar{\partial}_0 f_2^n\|_{H_2} \|u_2^n\|_{H_2}. \tag{3.9}$$

It is important that $\|u_2^n\|_{H_2}$ can be estimated from above as

$$\|u_2^n\|_{H_2} \leq K' \left(\|M f_1^n\|_{H_1} + \|M \bar{\partial}_0 u_1^n\|_{H_1} + \|A_{1,1} u_1^n\|_{H_1} \right) \tag{3.10}$$

(see (3.2) and fundamental inequalities (1.4), (1.13)). Hence,

$$|Z| \leq \kappa \tau \sum_{n=1}^{k} \left(\frac{\nu}{2} \|\bar{\partial}_0 u_1^n\|_M^2 + (1 + 2/\nu) \|\bar{\partial}_0 f_2^n\|_{H_2}^2 + \|u_1^n\|_{H_1}^2 \right) \tag{3.11}$$

with a $\kappa > 0$ and arbitrary $\nu > 0$. Combination of (3.8), (3.9), (3.11) yields the desired estimate for Y in (3.7) and an unequality of standard type, which, for small enough $\nu > 0$ and τ_0, leads to

$$\tau \sum_{n=1}^{k} \|u_1^n\|_{H_1}^2 \leq K_0 F_k, \quad \|u_1^k\|_{H_1}^2 + \tau \sum_{n=1}^{k} \|\bar{\partial}_0 u_1^n\|_M^2 \leq K_0 F_k.$$

These estimates yield (3.4), (3.5) since we can apply (3.10) and $\|A_{2,1} \bar{\partial}_0 u_1^n\|_{H_2} \leq \|A_{2,2} u_2^n + \bar{\partial}_0 f_2^n\|_{H_2}$ (all $\alpha_j \leq c_0$).

References

1. Kantorovich, L. V., Akilov, G. P.: Functional Analysis in Normed Spaces, Pergamon, London, 1964.
2. Krein, S. G.: Linear Differential Equations in Banach Spaces, Nauka, Moscow, 1971 (in Russian).

3. D'yakonov, E. G.: Optimization in Solving Elliptic Problems. CRC Press, Boca Raton, 1996.
4. Girault, V., Raviart, P. A.: Finite Element Methods for Navier—Stokes Equations. Theory and Algorithms, Springer, Berlin, 1986.
5. D'yakonov, E. G.: Estimates of computational work for boundary value problems with the Stokes operators. Soviet Math. (Iz. VUZ). **27** (1983) 57–71.
6. Besov, O. V., Il'in, V. P., Nikol'skii, S. M.: Integral Representation of Functions and Embedding Theorems. **1** Winston and Sons, Washington, 1978; **2** A Halsted Press Book, John Wiley, New York, 1979.
7. D'yakonov, E. G.: Improved correctness of Stokes and Navie—Stokes type problems and their grid approximations. Vestn. Mosk. Gos. Univ., Ser.15: Vychisl. Mat. Kibern. 1998 N.1 3–9.
8. D'yakonov, E. G.: Operator problems in strengthened Sobolev spaces and numerical methods for them. Lecture Notes in Computer Science, 1196 (1997) 161–169.
9. D'yakonov, E. G.: Strengthened and weakened energy spaces and their applications. Journal of Computational, Civil and Structural Engineering. 1 (2000) N. 1 42–63.
10. D'yakonov, E. G.: New types of a posteriori error estimates in the solution of elliptic boundary and spectral problems. Vestn. Mosk. Gos. Univ., Ser.15: Vychisl. Mat. Kibern. 1998 N.4 3–9.
11. D'yakonov, E. G.: Cutting method for multidimensional stationary problems. Vestn. Mosk. Gos. Univ., Ser.15: Vychisl. Mat. Kibern. 1999 N.2 9–16.
12. Bakhvalov, N. S.: Efficient iterative methods for stiff multidimensional multiparametric problems. Comp. Math. and Math. Phys. **39** (1999) 1938–1966.
13. Graham, I. G., Hagger, M. J.: Unstructured additive Schwarz–conjugate gradient method for elliptic problems with highly discontinuous coefficients. SIAM J. Sci. Comput. **20** (1999) 2041–2066.
14. D'yakonov, E. G.: Elliptic problems with large jumps in coefficients and asymptotically optimal algorithms for their approximate solution. Vestn. Mosk. Gos. Univ., Ser.15: Vychisl. Mat. Kibern. 2000 N.1 5–13.
15. D'yakonov, E. G.: On the triangulations in the finite element and efficient iterative methods. Topics in Numerical Analysis, III, Miller, J. J. H., Ed. (1977) Academic Press, London 103–124.
16. Thomee, V.: Galerkin Finite Element Methods for Parabolic Problems. Springer Series in Computational Mathematics, **25** (1997) Springer-Verlag, Berlin.

Proper Weak Regular Splitting for M-Matrices

István Faragó*

ELTE University, Dept. Applied Analysis
H-1053 Budapest, Hungary

Abstract. The iterative solution of the system of linear algebraic equations $\mathbf{Ax} = \mathbf{b}$ with a nonsingular M-matrix $\mathbf{A}$ is considered. A one-step iterative method is constructed which is based on the special weak regular splitting of the matrix $\mathbf{A}$. We prove that the obtained iterative method is not only convergent but it has also some further advantageous properties: the maximal rate of convergence, the efficiency from the point of view of computational costs and the qualitative adequacy. We also examine the relation between this splitting and the regular splittings. Finally we construct two-sided monotone sequences to the solution of the above system. These sequences are produced by the iteration based on the weak regular splitting of $\mathbf{A}$, with different suitable starting vectors. The method of the possible determination of these vectors are also indicated.

1 Introduction

The solution of the system of linear algebraic equations

$$\mathbf{Ax} = \mathbf{b} \qquad (1)$$

with the given regular matrix $\mathbf{A} \in \mathbb{R}^{n \times n}$ and the nonnegative vector $\mathbf{b} \in \mathbb{R}^n$ is a basic problem of the numerical methods. A considerable part of the applications results in such a system where $\mathbf{A}$ is an M-matrix, that is a matrix with nonpositive offdiagonal elements and there exists a positive vector $\mathbf{f}_{pos}$ such that the vector $\mathbf{Af}_{pos}$ is also positive.

In order to solve this problem usually we construct the one step iteration of the form

$$\mathbf{x}^{(j+1)} = \mathbf{Tx}^{(j)} + \mathbf{g}, \quad j = 0, 1, \dots \qquad (2)$$

Here the question is the construction of the iterative matrix $\mathbf{T}$ and the vector $\mathbf{g}$. The usual approach [1,9] is their determination through the splitting of the matrix $\mathbf{A}$

$$\mathbf{A} = \mathbf{M} - \mathbf{N}, \quad \mathbf{M} \text{ is regular} \qquad (3)$$

---

* This research was supported by the Hungarian National Research Funds OTKA under grant no. T 031807

by the formulas

$$\mathbf{T} = \mathbf{M}^{-1}\mathbf{N}, \quad \mathbf{g} = \mathbf{M}^{-1}\mathbf{b}. \tag{4}$$

Clearly, for convergent iterative matrices the iteration (2) is convergent to the solution of the equation (1).

However, in addition to the convergence we have to impose some additional requirements to the iteration. Our aim is to choose a splitting such that the iteration (2) (4) satisfies some further expectations. Namely,

1. *Maximal rate of convergence.* It means that the spectral radius $\rho(\mathbf{T})$ is as small as possible.
2. *Efficiency.* The iterative method (2) (4) is called efficient if $\mathbf{M}^{-1}$ is easily obtainable and the computational cost of the splitting (3) is low.
3. *Qualitative adequacy.* The iterated vectors $\mathbf{x}^{(j)}$ for all fixed j preserve the main qualitative properties of the solution vector $\mathbf{x}$.

Assume that the nonsingular matrix $\mathbf{A}$ and the convergent matrix $\mathbf{T}$ are any fixed matrices. Clearly, the matrices defined by

$$\mathbf{M} = \mathbf{A}(\mathbf{I} - \mathbf{T})^{-1}, \quad \mathbf{N} = \mathbf{M} - \mathbf{A},$$

where $\mathbf{I}$ denotes the unit matrix, form a splitting of $\mathbf{A}$ and $\mathbf{T} = \mathbf{M}^{-1}\mathbf{N}$ [7]. So, if $\varepsilon \in (0,1)$ is arbitrary and $\mathbf{T} = \varepsilon \mathbf{I}$, then the splitting has the form

$$\mathbf{M} = \frac{1}{1-\varepsilon}\mathbf{A}, \quad \mathbf{N} = \frac{\varepsilon}{1-\varepsilon}\mathbf{A}, \tag{5}$$

and the spectral radius of iteration matrix in the splitting based on (5) is equal to ε, that is arbitrarily small. Therefore this splitting satisfies the first requirement.

With respect to the second requirement, a splitting is called efficient if M has a suitable form and its computation is not too difficult, for instance, in the case when M is triangular. In [8] it is proved that provided the LU-decomposition of the matrix $\mathbf{A}$ exists the splitting of the form

$$\mathbf{M} = \mathbf{L}\mathbf{D}^{-1}, \quad \mathbf{N} = \mathbf{M} - \mathbf{A} \tag{6}$$

satisfies both the first and second requirements, too.

Under the assumptions made the solution of the system (1) is nonnegative. Therefore our aim is to preserve this property during the whole iteration, that is, after stopping the iteration after any step, the approximate solution has to preserve the nonnegativity. In order to guarantee the third requirement, we are able to analyse some other basic qualitative properties of the iteration process [3,4,5]. As the results show in case of preservation of the nonnegativity of the initial vector the basic qualitative properties are also preserved during the iterative process. Therefore our aim is to construct a *weak regular splitting* of the matrix $\mathbf{A}$, that is a splitting of the form (3) with a monotone matrix $\mathbf{M}$ and a nonnegative matrix $M^{-1}N$. Obviously, if M is monotone and N is nonnegative, that is (3) defines a *regular splitting*, then it is a weak regular splitting, too.

In the sequel the decomposition (3) is called a *proper splitting* if the corresponding iteration is convergent and satisfies all the above three requirements.

Usually these requirements raise objections in choosing of the splitting because they result in inconsistent conditions. As one can easily see the splitting (5) has a maximal rate of convergence and on the class of the monotone matrices it is a weak regular splitting. However, it is not efficient: the computation of M^{-1} is equivalent to the inversion of the matrix A. Therefore, even on the monotone matrices it is not a proper splitting. On the other hand, the splitting (6) has a fast convergence and it is efficient, but the third requirement is not satisfied for any matrices A.

Therefore the construction of a proper splitting is a complex task. Since for the M-matrices the iterations defined by weak regular splittings are convergent [1,10] therefore in the following we restrict our consideration to such kind of splittings.

The paper is organised as follows. In Section 2 we prove that the splitting (6) is a proper weak regular splitting on the M-matrices. We also show that this splitting is not a regular splitting of the matrix A. In Section 3 we construct two-sided monotone sequences to the solution of (1). These sequences are produced by the iteration (2)(4) based on the weak regular splitting of A, with different suitable starting vectors. We also present a possible method of choosing the starting vectors.

2 An Efficient Weak Regular Splitting of M-Matrices

In the following we show that for any regular M-matrix $A \in \mathbb{R}^{n \times n}$ there exists an efficient weak regular splitting .

First we formulate a statement, the proof of which follows immediately from the proof of the statement E_{18} of Theorem 2.3 of Chapter 6 in [2].

Lemma 1. *If A is an M-matrix then there exists an LU-decomposition*

$$A = LU. \tag{7}$$

with regular triangular M-matrices L and U.

Assume that $\lambda_1, \lambda_2, \ldots \lambda_n$ are any different fixed numbers on the interval $(0,1)$ and we introduce the notations

$$d_i = \frac{1 - \lambda_i}{u_{i,i}} > 0, \quad D = diag(d_1, d_2, \ldots d_n), \quad \Lambda = diag(\lambda_1, \lambda_2, \ldots \lambda_n). \tag{8}$$

Then the following theorem holds.

Theorem 1. *The matrices*

$$M = LD^{-1}, \quad N = M - A \tag{9}$$

define a weak regular splitting of the matrix A .

Proof. Clearly (9) defines a splitting of $\mathbf{A}$. We prove that it is a weak regular splitting. Since $\mathbf{L}$ is an M-matrix therefore $\mathbf{M}^{-1} = \mathbf{D}\mathbf{L}^{-1}$ is the product of two nonnegative matrices, that is $\mathbf{M}$ is monotone. Clearly, $\mathbf{N} = \mathbf{L}(\mathbf{D}^{-1} - \mathbf{U})$ therefore we have

$$\mathbf{M}^{-1}\mathbf{N} = \mathbf{D}\mathbf{L}^{-1}\mathbf{L}(\mathbf{D}^{-1} - \mathbf{U}) = \mathbf{I} - \mathbf{D}\mathbf{U}. \tag{10}$$

The matrix $\mathbf{D}\mathbf{U}$ is an upper triangular M-matrix with the diagonal elements $1 - \lambda_i$, which proves the statement.

In the following we show the possibility of choosing a weak regular splitting with an arbitrarily small spectral radius of the iteration matrix $\mathbf{M}^{-1}\mathbf{N}$.

Theorem 2. *Assume that ε is an arbitrarily small positive number and the fixed different numbers λ_i satisfy the conditions*

$$\lambda_i \leq \varepsilon \text{ for all } i = 1, 2, \ldots n. \tag{11}$$

Then for the weak regular splitting of the form (9) the relation $\rho(\mathbf{M}^{-1}\mathbf{N}) \leq \varepsilon$ holds.

Proof. As we have proved the diagonal elements of the matrix $\mathbf{D}\mathbf{U}$ are $1 - \lambda_i$. Apparently these numbers are the eigenvalues of the matrix and they are different. Therefore the matrix is diagonalizable in the form

$$\mathbf{D}\mathbf{U} = \mathbf{S}(\mathbf{I} - \mathbf{\Lambda})\mathbf{S}^{-1} \tag{12}$$

where $\mathbf{S}$ is a regular matrix [6]. Using the relations (10) and (12) we get

$$\mathbf{M}^{-1}\mathbf{N} = \mathbf{I} - \mathbf{S}(\mathbf{I} - \mathbf{\Lambda})\mathbf{S}^{-1} = \mathbf{S}\mathbf{\Lambda}\mathbf{S}^{-1}, \tag{13}$$

therefore the relation $\rho(\mathbf{M}^{-1}\mathbf{N}) = \rho(\mathbf{\Lambda}) = \max \lambda_i \leq \varepsilon$ holds.

Corollary 1. *Since in the splitting (9) the matrix $\mathbf{M}$ is triangular, therefore, as a consequence of Theorems 2. and 3., this splitting satisfies our requirements, that is it defines a proper splitting.*

In the following we examine the sign-pattern of the matrix $\mathbf{N}$ in the weak regular splitting (9) for the symmetric positive definite M-matrices $\mathbf{A}$. In this case there exists the Cholesky factorization, that is,

$$\mathbf{A} = \mathbf{L}\mathbf{L}^T \tag{14}$$

with a regular lower triangular M-matrix $\mathbf{L}$. Here the elements of $\mathbf{L}$ are defined by the formulas

$$l_{i,i}^2 = a_{i,i} - \sum_{t=1}^{i-1} l_{i,t}^2, \quad l_{i,j} = \frac{a_{i,j} - \sum_{t=1}^{i-1} l_{j,t} l_{i,t}}{l_{i,i}}, \quad i = 1, \ldots n, \quad j = i+1, \ldots n. \tag{15}$$

Obviously $l_{i,i} > 0$. With arbitrary numbers λ_i from $(0,1)$ we define the notations as before:

$$d_i = \frac{1 - \lambda_i}{l_{i,i}} > 0, \quad \mathbf{D} = diag(d_1, d_2, \ldots d_n), \quad \mathbf{\Lambda} = diag(\lambda_1, \lambda_2, \ldots \lambda_n). \tag{16}$$

Theorem 3. *For the symmetric positive definite M-matrix $\mathbf{A}$ the splitting (9) with (15) and (16) defines a weak regular but not a regular splitting.*

Proof. The weak regularity follows from Theorem 2. Therefore it is sufficient to prove that the condition $\mathbf{N} \geq 0$ cannot be satisfied. Using (15) for the offdiagonal elements of the matrix $\mathbf{N}$ we have

$$n_{j,i} = \frac{l_{j,i}}{d_i} - a_{j,i} = \frac{1}{d_i l_{i,i}} \left(-\sum_{t=1}^{i-1} l_{j,t} l_{i,t} + \lambda_i a_{j,i} \right). \tag{17}$$

Since $\mathbf{A}$ and $\mathbf{L}$ are M-matrices therefore the right side of (17) is nonpositive. Moreover, there exists a negative offdiagonal element of $\mathbf{A}$ which proves our statement.

We remark that by analogical computation for the diagonal elements of the matrix $\mathbf{N}$ we obtain

$$n_{i,i} = \frac{l_{i,i}}{d_i} - a_{i,i} = \frac{1}{d_i l_{i,i}} \left(-\sum_{t=1}^{i-1} l_{i,t}^2 + \lambda_i a_{i,i} \right). \tag{18}$$

This relation shows that the sign of the diagonal elements depends on the choice of the numbers λ_i. However, if these numbers are chosen sufficiently small then the diagonal elements of $\mathbf{N}$ are negative.

3 Two-Sided Iterations

In this section we show the possibility of the construction of two-sided iterations to the solution of (1).

As before we assume that $\mathbf{A}$ is an M-matrix and (3) is a weak regular splitting. Then the iteration (2)(4) is convergent to the solution of (1) for any initial vector $\mathbf{x}^{(0)}$. We show that a suitable choice of the initial vector results in monotonically convergent vector sequences to the solution from both directions.

Theorem 4. *Assume that the vectors $\mathbf{v}^{(0)}$ and $\mathbf{w}^{(0)}$ satisfy the conditions $\mathbf{A}\mathbf{v}^{(0)} \leq \mathbf{b}$ and $\mathbf{A}\mathbf{w}^{(0)} \geq \mathbf{b}$, respectively. Then for the vector sequences $(\mathbf{v}^{(k)})$ and $(\mathbf{w}^{(k)})$ the following statements are true:*

1. *They are convergent to the solution of (1), that is to the vector $\mathbf{x}$.*
2. *The vector sequence $(\mathbf{v}^{(k)})$ monotonically increases, that is $\mathbf{v}^{(k+1)} \geq \mathbf{v}^{(k)}$.*
3. *The vector sequence $(\mathbf{w}^{(k)})$ monotonically decreases, that is $\mathbf{w}^{(k+1)} \leq \mathbf{w}^{(k)}$.*
4. *They form a two-sided bound for the solution, that is for all $k \in \mathbb{N}$ the relation*

$$\mathbf{v}^{(k)} \leq \mathbf{x} \leq \mathbf{w}^{(k)} \tag{19}$$

holds.

Proof. The first statement is already proved. On the base of the iteration the relation $\mathbf{Mv}^{(1)} = \mathbf{Nv}^{(0)} + \mathbf{b}$ holds. On the other hand, due to the assumption we have $\mathbf{Mv}^{(0)} \leq \mathbf{Nv}^{(0)} + \mathbf{b}$. Using the monotonicity of the matrix $\mathbf{M}$ these relations imply the relation $\mathbf{v}^{(1)} \geq \mathbf{v}^{(0)}$. For any k the proof is based on the induction: if $\mathbf{v}^{(k)} \geq \mathbf{v}^{(k-1)}$ then, using the form of the iteration (2) and the nonnegativity of the matrix $\mathbf{T}$, the relation $\mathbf{v}^{(k+1)} - \mathbf{v}^{(k)} = \mathbf{T}\left(\mathbf{v}^{(k)} - \mathbf{v}^{(k-1)}\right) \geq 0$ holds. This proves the second statement. The third statement is proved in a similar manner. The last statement is an obvious consequence of the first three statements.

An important question is the possibility of choosing the suitable vectors $\mathbf{v}^{(0)}$ and $\mathbf{w}^{(0)}$. Using the fact that $\mathbf{A}$ is an M-matrix we can give a method to their determination. Clearly, the diagonal matrix $diag\mathbf{A}$ is a nonsingular, nonnegative matrix. Let us denote by $\mathbf{v}$ the nonnegative solution of the easily solvable equation

$$diag\mathbf{A}\mathbf{v} = \mathbf{b}. \tag{20}$$

Since the offdiagonal elements of $\mathbf{A}$ are nonpositive therefore the relation $0 \leq (diag\mathbf{A} - \mathbf{A})\mathbf{v} = \mathbf{b} - \mathbf{A}\mathbf{v}$ holds, that is $\mathbf{b} \geq \mathbf{A}\mathbf{v}$. So, the choice $\mathbf{v}^{(0)} = \mathbf{v}$ is suitable. Using the notations

$$\mathbf{1}_n = [1,1,\ldots,1]^T \in \mathbb{R}^n, \quad a_{max} = \max_i a_{i,i} > 0, \quad b_{min} = \min_i b_i, \quad \gamma_1 = \frac{b_{min}}{a_{max}}$$

we can observe that the vector $\tilde{\mathbf{v}} = \gamma_1 \mathbf{1}_n$ also satisfies the condition $\mathbf{b} \geq \mathbf{A}\tilde{\mathbf{v}}$, that is we can choose it as the initial vector $\mathbf{v}^{(0)}$.

In order to choose the suitable vector $\mathbf{w}^{(0)}$ we use the positive vector $\mathbf{f}_{pos}$ defined in the definition of the M-matrices. If we introduce the notations

$$b_{max} = \max_i b_i, \quad f_{min} = \min_i (\mathbf{A}\mathbf{f}_{pos})_i, \quad \gamma_2 = \frac{b_{max}}{f_{min}}$$

then for the vector $\mathbf{w} = \gamma_2 \mathbf{f}_{pos}$ the relation

$$(\mathbf{A}\mathbf{w})_i = \gamma_2 (\mathbf{A}\mathbf{f}_{pos})_i \geq b_{max} \geq b_i$$

holds, that is $\mathbf{A}\mathbf{w} \geq \mathbf{b}$. Therefore the choice $\mathbf{w}^{(0)} = \mathbf{w}$ is suitable. We remark that for the diagonally dominant M-matrices the vector $\mathbf{f}_{pos}$ can be chosen by $\mathbf{f}_{pos} = \mathbf{1}_n$. Therefore, in this case the choice $\mathbf{w}^{(0)} = \gamma_2 \mathbf{1}_n$ is suitable with $\gamma_2 = b_{max}/s_{min}$, where $s_{min} > 0$ denotes the minimum of the row-summs of the matrix $\mathbf{A}$.

Finally we remark that the two-sided bound (19) can be successfully applied to construct a stopping criterion of the iteration.

References

1. Axelsson, O.: Iterative Solution Method, Univ.Press, Cambridge (1994)
2. Berman, A., Plemmons, R. J.: Nonnegative matrices in the mathematical sciences. Academic Press, New York (1979)
3. Faragó, I.: Qualitative properties of the numerical solution of linear parabolic problems with nonhomogeneous boundary conditions. Comp. Math. Appl. **26** (1996) 143-150
4. Faragó, I., Tarvainen, P.: Qualitative analysis of one step algebraic models with tridiagonal Toeplitz matrices. Periodica Hung. **31** (1997) 177-192
5. Faragó, I., Tarvainen, P.: Qualitative analysis of matrix splitting methods. Comp. Math. Appl. (to appear)
6. Horn, R. A., Johnson, C. R.: Matrix analysis. Cambridge Univ. Press, London (1986)
7. Lanzkron, P. J., Rose, D. J., Szyld, D. B.: Convergence of nested classical iterative methods for linear systems. Numer. Math. **58** (1991) 685-702
8. Yuan, J. Y.: Iterative refinement using splitting method. Lin. Alg. Appl. **273** (1998) 199-214
9. Varga, R.: Matrix Iterative Analysis. Prentice-Hall, New Jersey (1962)
10. Woznicki, Z.: Nonnegative splitting theory, Japan J. Indust. Appl. Math. **11** (1994) 289-342

Parameter-Uniform Numerical Methods for a Class of Singularly Perturbed Problems with a Neumann Boundary Condition*

P. A. Farrell[1], A. F. Hegarty[2], J. J. H. Miller[3], E. O'Riordan[4], and G. I. Shishkin[5]

[1] Department of Mathematics and Computer Science, Kent State University
Kent, Ohio 44242, USA
[2] Department of Mathematics and Statistics, University of Limerick, Ireland
[3] Department of Mathematics, Trinity College, Dublin, Ireland
[4] School of Mathematical Sciences, Dublin City University, Ireland
[5] Institute for Mathematics and Mechanics, Russian Academy of Sciences
Ekaterinburg, Russia

Abstract. The error generated by the classical upwind finite difference method on a uniform mesh, when applied to a class of singularly perturbed model ordinary differential equations with a singularly perturbed Neumann boundary condition, tends to infinity as the singular perturbation parameter tends to zero. Note that the exact solution is uniformly bounded with respect to the perturbation parameter. For the same classical finite difference operator on an appropriate piecewise–uniform mesh, it is shown that the numerical solutions converge, uniformly with respect to the perturbation parameter, to the exact solution of any problem from this class.

1 Introduction

Consider the following class of linear one dimensional convection–diffusion problems

$$L_\varepsilon u_\varepsilon \equiv \varepsilon u_\varepsilon'' + a(x) u_\varepsilon' = f(x), \quad x \in \Omega = (0, 1), \tag{1a}$$

$$\varepsilon u_\varepsilon'(0) = A, \quad u_\varepsilon(1) = B, \tag{1b}$$

$$a, f \in C^2(\Omega), \quad a(x) \geq \alpha > 0, \quad x \in \overline{\Omega}. \tag{1c}$$

Note that a Neumann boundary condition has been specified at $x = 0$. We recall the comparison principle for this problem (see [2], for example).

Theorem 1. *Assume that* $v \in C^2(\overline{\Omega})$. *Then, if* $v'(0) \leq 0$, $v(1) \geq 0$ *and* $L_\varepsilon v(x) \leq 0$ *for all* $x \in \Omega$, *it follows that* $v(x) \geq 0$ *for all* $x \in \overline{\Omega}$.

---

* This research was supported in part by the Russian Foundation for Basic Research under grant No. 98-01-00362, by the National Science Foundation grant DMS-9627244 and by the Enterprise Ireland grant SC-98-612.

From this we can easily establish the following stability bound on the solution

$$|u_\varepsilon(x)| \leq |u_\varepsilon(1)| + \frac{\varepsilon}{\alpha}|u'_\varepsilon(0)|e^{-\alpha x/\varepsilon} + \frac{1}{\alpha}\|f\|(1-x).$$

Lemma 1. *[2] The derivatives $u_\varepsilon^{(k)}$ of the solution of (1) satisfy the bounds*

$$\|u_\varepsilon^{(k)}\| \leq C\varepsilon^{-k} \max\{\|f\|, \|u_\varepsilon\|\}, \quad k = 1, 2$$
$$\|u_\varepsilon^{(3)}\| \leq C\varepsilon^{-3} \max\{\|f\|, \|f'\|, \|u_\varepsilon\|\}$$

where C depends only on $\|a\|$ and $\|a'\|$.

Consider the following decomposition of the solution u_ε

$$u_\varepsilon = v_\varepsilon + w_\varepsilon, \quad v_\varepsilon = v_0 + \varepsilon v_1 + \varepsilon^2 v_2; \quad w_\varepsilon = w_0 + \varepsilon w_1 \tag{2a}$$

where the components v_0, v_1 and v_2 are the solutions of the problems

$$av'_0 = f, \quad v_0(1) = u_\varepsilon(1), \tag{2b}$$
$$av'_1 = -v''_0, \quad v_1(1) = 0, \tag{2c}$$
$$L_\varepsilon v_2 = -v''_1, \quad \varepsilon v'_2(0) = 0, \quad v_2(1) = 0 \tag{2d}$$

and the components w_0, w_1 are the solutions of

$$L_\varepsilon w_0 = 0, \quad \varepsilon w'_0(0) = \varepsilon u'_\varepsilon(0), \quad w_0(1) = 0. \tag{2e}$$
$$L_\varepsilon w_1 = 0, \quad \varepsilon w'_1(0) = -v'_\varepsilon(0), \quad w_1(1) = 0. \tag{2f}$$

Thus the components v_ε and w_ε are the solutions of the problems

$$L_\varepsilon v_\varepsilon = f, \quad v'_\varepsilon(0) = v'_0(0) + \varepsilon v'_1(0), \quad v_\varepsilon(1) = u_\varepsilon(1)$$
$$L_\varepsilon w_\varepsilon = 0, \quad w'_\varepsilon(0) = u'_\varepsilon(0) - v'_\varepsilon(0), \quad w_\varepsilon(1) = 0.$$

Also, they satisfy the bounds given in the following lemma.

Lemma 2. *The components $v_\varepsilon, w_\varepsilon$ and their derivatives satisfy*

$$\|v_\varepsilon^{(k)}\| \leq C(1 + \varepsilon^{2-k}), \quad k = 0, 1, 2, 3$$
$$|w_\varepsilon^{(k)}(x)| \leq C(\varepsilon|u'_\varepsilon(0)| + \varepsilon)\varepsilon^{-k}e^{-\alpha x/\varepsilon}, \quad k = 0, 1, 2, 3.$$

Proof. Use Lemma 1 and the fact that

$$\varepsilon\psi_\varepsilon(x) = \int_x^1 e^{-A(t)/\varepsilon}dt, \quad \text{where} \quad A(t) = \int_0^t a(s)ds$$

is the exact solution of

$$L_\varepsilon \psi_\varepsilon = 0, \quad \varepsilon\psi'_\varepsilon(0) = -1, \quad \psi_\varepsilon(1) = 0.$$

2 Upwinding on a Uniform Mesh

In this section we examine the convergence behaviour of standard upwinding on a uniform mesh. The Neumann boundary condition is discretized by the scaled discrete derivative $\varepsilon D^+ U_\varepsilon(0)$.

$$L_\varepsilon^N U_\varepsilon \equiv \varepsilon \delta^2 U_\varepsilon + a(x_i) D^+ U_\varepsilon = f(x_i), \quad x_i \in \Omega^N, \tag{3a}$$

$$\varepsilon D^+ U_\varepsilon(0) = \varepsilon u'_\varepsilon(0), \quad U_\varepsilon(1) = u_\varepsilon(1), \tag{3b}$$

where $\overline{\Omega}^N$ is an arbitrary mesh. We now state a discrete comparison principle.

Theorem 2. [2] Let L_ε^N be the upwind finite difference operator defined in (3) and let Ω^N be an arbitrary mesh of $N+1$ mesh points. If V is any mesh function defined on this mesh such that

$$D^+ V(x_0) \leq 0, \quad V(x_N) \geq 0 \quad \text{and} \quad L_\varepsilon^N V \leq 0 \text{ in } \Omega^N,$$

then $V(x_i) \geq 0$ for all $x_i \in \overline{\Omega}^N$.

Hence, on an arbitrary mesh, the discrete solution U_ε satisfies the bound

$$|U_\varepsilon(x_i)| \leq |u_\varepsilon(1)| + \varepsilon |u'_\varepsilon(0)| \Phi_i + \frac{1}{\alpha} \|f\|(1 - x_i).$$

where Φ_i is the solution of the constant coefficient problem

$$\varepsilon \delta^2 \Phi_i + \alpha D^+ \Phi_i = 0, \quad \varepsilon D^+ \Phi_0 = -1, \quad \Phi_N = 0.$$

Theorem 3. Let u_ε be the continuous solution of (1) and let U_ε be the numerical solution generated from the upwind finite difference scheme (3) on a uniform mesh Ω_u^N. Then,
(a) if $|u'_\varepsilon(0)| \leq C$, we have

$$\|\overline{U}_\varepsilon - u_\varepsilon\|_{\overline{\Omega}} \leq C N^{-1}$$

where $\overline{U}_\varepsilon$ is the linear interpolant of U_ε and C is a constant independent of N and ε. Also,
(b) if $\varepsilon |u'_\varepsilon(0)| = C \neq 0$, then for any fixed N,

$$\|U_\varepsilon\| \to \infty \quad \text{as} \quad \varepsilon \to 0$$

Proof. (a) Consider first the case of $|u'_\varepsilon(0)| \leq C$. The discrete solution U_ε can be decomposed into the sum

$$U_\varepsilon = V_\varepsilon + W_\varepsilon$$

where V_ε and W_ε are respectively the solutions of the problems

$$L_\varepsilon^N V_\varepsilon = f(x_i), \quad x_i \in \Omega_u^N, \quad \varepsilon D^+ V_\varepsilon(0) = \varepsilon v'_\varepsilon(0), \quad V_\varepsilon(1) = v_\varepsilon(1)$$
$$L_\varepsilon^N W_\varepsilon = 0, \quad x_i \in \Omega_u^N, \quad \varepsilon D^+ W_\varepsilon(0) = \varepsilon w'_\varepsilon(0), \quad W_\varepsilon(1) = 0.$$

We estimate the errors $V_\varepsilon - v_\varepsilon$ and $W_\varepsilon - w_\varepsilon$ separately. By standard local truncation error estimates, we obtain

$$|L_\varepsilon^N(V_\varepsilon - v_\varepsilon)(x_i)| \leq \frac{\varepsilon}{3}(x_{i+1} - x_{i-1})\|v_\varepsilon^{(3)}\| + \frac{a(x_i)}{2}(x_{i+1} - x_i)\|v_\varepsilon^{(2)}\| \leq CN^{-1}.$$

Note also that

$$|D^+(V_\varepsilon - v_\varepsilon)(0)| = |v_\varepsilon'(0) - D^+v_\varepsilon(0)| = \frac{1}{h}|\int_0^h (s-h)v_\varepsilon''(s)\,ds| \leq CN^{-1}.$$

With the two functions $\psi^\pm(x_i) = CN^{-1}(1-x_i) \pm (V_\varepsilon - v_\varepsilon)(x_i)$, and the discrete minimum principle for L_ε^N we easily derive

$$|(V_\varepsilon - v_\varepsilon)(x_i)| \leq CN^{-1}.$$

Note that if $|u_\varepsilon'(0)| \leq C$ then

$$|w_\varepsilon^{(k)}(x)| \leq C\varepsilon^{1-k}e^{-\alpha x/\varepsilon}, \quad k = 0, 1, 2, 3.$$

The local truncation error for the layer component is given by

$$|L_\varepsilon^N(W_\varepsilon - w_\varepsilon)(x_i)| \leq C\varepsilon^{-1}(x_{i+1} - x_{i-1})e^{-\alpha x_{i-1}/\varepsilon} \leq C\varepsilon^{-1}N^{-1}e^{-\alpha x_{i-1}/\varepsilon}$$

and

$$|\varepsilon D^+(W_\varepsilon - w_\varepsilon)(0)| = \frac{\varepsilon}{h}|\int_0^h (s-h)w_\varepsilon''(s)\,ds| \leq CN^{-1}.$$

Introduce the two mesh functions

$$\Psi_i^\pm = \frac{C\lambda^2}{\gamma(\alpha - \gamma)} N^{-1}Y_i \pm (W_\varepsilon - w_\varepsilon)(x_i)$$

where γ is any constant satisfying $0 < \gamma < \alpha$ and

$$Y_i = \frac{\lambda^{N-i} - 1}{\lambda^N - 1}, \quad \lambda = 1 + \frac{\gamma h}{\varepsilon}, \quad h = 1/N.$$

Note that $Y_i \leq \lambda^{-i}$. It is easy to see that

$$\lambda^2 D^+ Y_i \leq -\frac{\gamma}{\varepsilon}e^{-\gamma x_{i-1}/\varepsilon}$$

and so Y_i decreases monotonically with $0 \leq Y_i \leq 1$. We then have $\varepsilon D^+\Psi_0^\pm \leq 0$, $\Psi_N^\pm = 0$ and using $(\varepsilon\delta^2 + \gamma D^+)Y_i = 0$, we obtain

$$L_\varepsilon^N \Psi_i^\pm = \frac{C\lambda^2}{\gamma(\alpha-\gamma)} N^{-1}(a(x_i) - \gamma)D^+Y_i \pm L_\varepsilon^N(W_\varepsilon - w_\varepsilon)(x_i)$$

$$\leq -C\varepsilon^{-1}N^{-1}\left(\frac{a(x_i) - \gamma}{\alpha - \gamma}e^{-\gamma x_{i-1}/\varepsilon} - e^{-\alpha x_{i-1}/\varepsilon}\right) < 0.$$

By the discrete minimum principle we conclude that $\Psi_i^\pm \geq 0$ and so for all $x_i \in \overline{\Omega}_u^N$

$$|(W_\varepsilon - w_\varepsilon)(x_i)| \leq \frac{C\lambda^2}{\gamma(\alpha - \gamma)} N^{-1} Y_i \leq C\lambda^2 N^{-1}.$$

Thus, we have that

$$|(W_\varepsilon - w_\varepsilon)(x_i)| \leq CN^{-1}, \quad \text{when} \quad h \leq \varepsilon.$$

From an integral representation of the truncation error, we have

$$|L_\varepsilon^N(W_\varepsilon - w_\varepsilon)(x_i)| \leq C \int_{x_{i-1}}^{x_{i+1}} \frac{1}{\varepsilon} e^{-\alpha t/\varepsilon} dt \leq C e^{-\alpha x_{i-1}/\varepsilon}.$$

As before we can establish

$$|(W_\varepsilon - w_\varepsilon)(x_i)| \leq C\varepsilon\lambda^{2-i}.$$

Hence, for $1 \leq i \leq N$ and $\varepsilon \leq h$,

$$|(W_\varepsilon - w_\varepsilon)(x_i)| \leq C\varepsilon\lambda \leq Ch.$$

Note that

$$|D^+(W_\varepsilon - w_\varepsilon)(0)| = \frac{1}{h} |\int_0^h (s-h) w_\varepsilon''(s)\, ds| \leq C$$

which implies that

$$|(W_\varepsilon - w_\varepsilon)(0)| \leq |(W_\varepsilon - w_\varepsilon)(x_1)| + Ch \leq Ch.$$

On the interval $[x_i, x_{i+1}]$ we have

$$|(w_\varepsilon - \bar{w}_\varepsilon)(x)| = |\int_{x_i}^x w_\varepsilon'(t) dt - \frac{x - x_i}{x_{i+1} - x_i} \int_{x_i}^{x_{i+1}} w_\varepsilon'(t) dt| \leq C N^{-1} \|w_\varepsilon'\|$$

Combining this with the argument in [2] completes part (a).

(b) If we discretize the Neumann boundary condition $\varepsilon u_\varepsilon'(0) = C$ by the standard discrete derivative $\varepsilon D^+ U_\varepsilon(0) = C$ on a **uniform** mesh then

$$U_\varepsilon(h) = U_\varepsilon(0) + C\frac{h}{\varepsilon}.$$

For a fixed distance h,

$$\lim_{\varepsilon \to 0} |U_\varepsilon(h) - U_\varepsilon(0)| \to \infty.$$

On a uniform mesh, the discrete solution is not bounded independently of ε.

Remarks. (i) We define a a weak boundary layer by $\|u'_\varepsilon(x)\| \leq C$, that is, the derivative is uniformly bounded with respect to ε. In this case, Theorem 3 states that the solution u_ε can be approximated ε-uniformly on a uniform mesh. However, the first derivative $u'_\varepsilon(x)$ is not approximated ε-uniformly by the discrete derivative $D^+U_\varepsilon(x_i)$ on a uniform mesh. This can be checked by solving a non-trivial constant coefficient continuous problem and its corresponding discrete problem directly, setting $\varepsilon N = 1$ and then taking the limit as $N \to \infty$.
(ii) In the case of the constant coefficient problem (1), we observe that

$$\lim_{\varepsilon \to 0} |\varepsilon D^+ U_\varepsilon(x_i) - \varepsilon u'_\varepsilon(x_i)| = 0.$$

Thus, although the discrete solutions are unbounded as $\varepsilon \to 0$, the scaled discrete derivatives are at least ε-uniformly bounded and, moreover, converge as $N \to \infty$ to $\varepsilon u'(x_i)$ for each fixed ε. However, the scaled discrete derivatives $\varepsilon D^+ U_\varepsilon(0)$ are not ε-uniformly convergent to $\varepsilon u'_\varepsilon(0)$. In contrast, for the problem

$$L_\varepsilon u_\varepsilon \equiv \varepsilon u''_\varepsilon + a(x) u'_\varepsilon = f(x), \quad x \in \Omega, \tag{4a}$$
$$u_\varepsilon(0) = A, \quad u_\varepsilon(1) = B, \tag{4b}$$

with Dirichelet boundary conditions, we have that

$$\lim_{\varepsilon \to 0} |U_\varepsilon(x_i) - u_\varepsilon(x_i)| = 0,$$

and that

$$\lim_{\varepsilon \to 0} |\varepsilon D^+ U_\varepsilon(0) - \varepsilon u'_\varepsilon(0)| = O(1).$$

This can be seen easily from the explicit solutions to the constant coefficient continuous problem.

$$u_\varepsilon(x) = u_\varepsilon(0) + \frac{f}{\alpha} x - (u_\varepsilon(0) - u_\varepsilon(1) + \frac{f}{\alpha}) \left(\frac{1 - e^{-\alpha x/\varepsilon}}{1 - e^{-\alpha/\varepsilon}} \right)$$

and the discrete problem

$$U_\varepsilon(x_i) = u_\varepsilon(0) + \frac{f}{\alpha} x_i - (u_\varepsilon(0) - u_\varepsilon(1) + \frac{f}{\alpha}) \left(\frac{1 - \lambda^{-i}}{1 - \lambda^{-N}} \right), \lambda = 1 + \alpha h/\varepsilon.$$

3 Upwinding on a Piecewise–Uniform Mesh

Consider the same upwind finite difference scheme (3) on the piecewise–uniform mesh

$$\overline{\Omega}_\varepsilon^N = \{x_i | x_i = 2i\sigma/N, \ i \leq N/2; \ x_i = x_{i-1} + 2(1-\sigma)/N, \ N/2 < i\} \tag{5a}$$

where the transition parameter σ is fitted to the boundary layer by taking

$$\sigma = \min\{\frac{1}{2}, \frac{1}{\alpha} \varepsilon \ln N\}. \tag{5b}$$

The next result shows that upwinding on this mesh produces an ε-uniform numerical method.

Theorem 4. Let u_ε be the continuous solution of (1) and let U_ε be the numerical solution generated from an upwind finite difference scheme (3) on the piecewise-uniform mesh (5). Then, for all $N \geq 4$, we have

$$\|U_\varepsilon - u_\varepsilon\|_{\overline{\Omega}_\varepsilon^N} \leq CN^{-1} \ln N$$

where C is a constant independent of N and ε.

Proof. As for the uniform mesh we derive

$$|(V_\varepsilon - v_\varepsilon)(x_i)| \leq CN^{-1}.$$

When $\sigma = 1/2$, the mesh is uniform and applying the argument of the previous theorem, we get

$$|(W_\varepsilon - w_\varepsilon)(x_i)| \leq CN^{-1}\varepsilon^{-1} \leq CN^{-1} \ln N.$$

When $\sigma < 1/2$, the argument is divided between the coarse mesh and fine mesh regions. Consider first the coarse mesh region $[\sigma, 1]$, where

$$|w_\varepsilon(x)| \leq Ce^{-\alpha\sigma/\varepsilon} \leq CN^{-1}.$$

Using the discrete comparison principle, we get

$$W_\varepsilon(x_i) \leq \varepsilon |w'_\varepsilon(0)| \Phi_i$$

where Φ_i is the solution of the constant coefficient problem

$$\varepsilon\delta^2\Phi_i + \alpha D^+\Phi_i = 0, \quad \varepsilon D^+\Phi_0 = -1, \quad \Phi_N = 0.$$

From an explicit representation of Φ_i one can show that

$$|\Phi_{N/2}| \leq CN^{-1}$$

Hence, for $x_i \geq \sigma$

$$|W_\varepsilon(x_i) - w_\varepsilon(x_i)| \leq |W_\varepsilon(x_i)| + |w_\varepsilon(x_i)| \leq CN^{-1}$$

Consider now the fine mesh region, using the same argument as in the previous theorem we get

$$|(W_\varepsilon - w_\varepsilon)(x_i)| \leq C\lambda^2 N^{-1} \ln N \leq CN^{-1} \ln N.$$

This completes the proof.

In [1], an essentially second order scheme is constructed on a piecewise-uniform mesh, using a more complicated finite difference operator. As in [2], the nodal error estimate for the simpler scheme presented here can easily be extended to a global error estimate by simple linear interpolation. That is, we have

$$\|\bar{U}_\varepsilon - u_\varepsilon\|_{\overline{\Omega}} \leq CN^{-1} \ln N$$

where $\bar{U}_\varepsilon$ is the linear interpolant of U_ε. Also, using the techniques in [2], one can deduce that

$$\varepsilon\|D^+U_\varepsilon - u'_\varepsilon\|_{\overline{\Omega}\setminus\{1\}} \leq CN^{-1} \ln N.$$

4 Parabolic Boundary Layers

In this section, we introduce a new class of problems. Let $\Omega = (0,1)$, $D = \Omega \times (0,T]$ and $\Gamma = \Gamma_l \cup \Gamma_b \cup \Gamma_r$ where Γ_l and Γ_r are the left and right sides of the box D and Γ_b is its base. Consider the following linear parabolic partial differential equation in D with Dirichlet-Neumann boundary conditions on Γ

$$L_\varepsilon u_\varepsilon(x,t) \equiv -\varepsilon \frac{\partial^2 u_\varepsilon}{\partial x^2} + b(x,t) u_\varepsilon + d(x,t) \frac{\partial u_\varepsilon}{\partial t} = f(x,t),\ (x,t) \in D, \quad (6a)$$

$$u_\varepsilon = \varphi_b \text{ on } \Gamma_b,\ \sqrt{\varepsilon} \frac{\partial u_\varepsilon}{\partial x} = \varphi_l \text{ on } \Gamma_l,\ u_\varepsilon = \varphi_r \text{ on } \Gamma_r \quad (6b)$$

$$d(x,t) > \delta > 0 \quad \text{and } b(x,t) \geq \beta > 0, (x,t) \in \overline{D} \quad (6c)$$

$$\varphi_l(0) = \sqrt{\varepsilon} \varphi_b'(0), \quad \varphi_b(1) = \varphi_r(0). \quad (6d)$$

We have the comparison principle

Lemma 3. *Assume $b, d \in C^0(\overline{D})$ and $\psi \in C^2(D) \cap C^1(\overline{D})$. Suppose that $\psi \geq 0$ on $\Gamma_b \cup \Gamma_r$ and $\frac{\partial \psi}{\partial x} \leq 0$ on Γ_r. Then $L_\varepsilon \psi \geq 0$ in D implies that $\psi \geq 0$ in $\overline{D}$.*

and the following stability bound

Theorem 5. *Let v be any function in the domain of the differential operator L_ε. Then*

$$\|v\| \leq (1 + \alpha T) \max\{\|L_\varepsilon v\|, \|v\|_{\Gamma_b \cup \Gamma_r}\} + \sqrt{\frac{\varepsilon}{\beta}} \|v_x\|_{\Gamma_l} e^{-\sqrt{\beta} x / \sqrt{\varepsilon}}$$

where $\alpha = \max_{\overline{D}}\{0, (1-b)/d\} \leq 1/\delta$.

Assume that the data b, d, f, φ satisfy sufficient regularity and compatibility conditions so that the problem has a unique solution u_ε and $u_\varepsilon \in C_\lambda^4(\overline{D})$ and, furthermore, such that the derivatives of the solution u_ε satisfy, for all non-negative integers $i, j,\ 0 \leq i + 2j \leq 4$

$$\left\| \frac{\partial^{i+j} u_\varepsilon}{\partial x^i \partial t^j} \right\|_{\overline{D}} \leq C \varepsilon^{-i/2}$$

where the constant C is independent of ε. We write the solution as the sum

$$u_\varepsilon = v_\varepsilon + w_\varepsilon$$

where $v_\varepsilon, w_\varepsilon$ are smooth and singular components of u_ε defined in the following way. The smooth component is further decomposed into the sum

$$v_\varepsilon = v_0 + \varepsilon v_1$$

where v_0, v_1 are defined by

$$b v_0 + d \frac{\partial v_0}{\partial t} = f \text{ in } D,\quad v_0 = u_\varepsilon \text{ on } \Gamma_b \quad (7a)$$

$$L_\varepsilon v_1 = \frac{\partial^2 v_0}{\partial x^2} \text{ in } D,\ v_1 = 0 \text{ on } \Gamma \setminus \Gamma_l,\ \frac{\partial v_1}{\partial x} = 0 \text{ on } \Gamma_l. \quad (7b)$$

The singular component is decomposed into the sum

$$w_\varepsilon = w_l + w_r$$

where w_l and w_r are defined by

$$L_\varepsilon w_r = 0 \text{ in } D, w_r = u_\varepsilon - v_0 \text{ on } \Gamma_r, w_r = 0 \text{ on } \Gamma_b \cup \Gamma_l \quad (8a)$$

$$L_\varepsilon w_l = 0 \text{ in } D, \quad (8b)$$

$$\frac{\partial w_l}{\partial x} = \frac{\partial u_\varepsilon}{\partial x} - \frac{\partial v}{\partial x} - \frac{\partial w_r}{\partial x} \text{ on } \Gamma_l, w_l = 0 \text{ on } \Gamma_r \cup \Gamma_b. \quad (8c)$$

It is clear that w_l, w_r correspond respectively to the boundary layer functions on Γ_l and Γ_r. Assume that the data satisfy sufficient regularity and compatibility conditions so that $v_\varepsilon, w_\varepsilon \in C_\lambda^4(\overline{D})$.

Theorem 6. *[3] For all non-negative integers i, j, such that $0 \leq i + 2j \leq 4$*

$$\left\| \frac{\partial^{i+j} v_\varepsilon}{\partial x^i \partial t^j} \right\|_{\overline{D}} \leq C(1 + \varepsilon^{1-i/2})$$

and for all $(x,t) \in D$,

$$\left| \frac{\partial^{i+j} w_l(x,t)}{\partial x^i \partial t^j} \right| \leq C\varepsilon^{-i/2} e^{-x/\sqrt{\varepsilon}}, \quad \left| \frac{\partial^{i+j} w_r(x,t)}{\partial x^i \partial t^j} \right| \leq C\varepsilon^{-i/2} e^{-(1-x)/\sqrt{\varepsilon}}$$

where C is a constant independent of ε.

Problem (6) is discretized using a standard numerical method composed of a standard finite difference operator on a fitted piecewise uniform mesh.

$$L_\varepsilon^N U_\varepsilon = -\varepsilon \delta_x^2 U_\varepsilon + b U_\varepsilon + d D_t^- U_\varepsilon = f, (x,t) \in D_\sigma^N \quad (9a)$$

$$U_\varepsilon = u_\varepsilon \text{ on } \Gamma_{b,\sigma}^N \cup \Gamma_{r,\sigma}^N, \quad D_x^+ U_\varepsilon = \frac{\partial u_\varepsilon}{\partial x} \text{ on } \Gamma_{l,\sigma}^N \quad (9b)$$

where

$$D_\sigma^N = \Omega_\sigma^{N_x} \times \Omega^{N_t}, \text{ and } \Gamma_\sigma^N = \overline{D}_\sigma^N \cap \Gamma. \quad (9c)$$

A uniform mesh Ω^{N_t} with N_t mesh elements is used on $(0,T)$. A piecewise uniform mesh $\Omega_\sigma^{N_x}$ on Ω with N_x mesh elements is obtained by putting a uniform mesh with $N_x/4$ mesh elements on both $(0, \sigma)$ and $(1-\sigma, 1)$ and one with $N_x/2$ mesh elements on $(\sigma, 1-\sigma)$, with the transition parameter

$$\sigma = \min\left\{ \frac{1}{4}, 2\sqrt{\varepsilon} \ln N_x \right\}. \quad (9d)$$

We have the following discrete comparison principle

Lemma 4. *Assume that the mesh function Ψ satisfies $\Psi \geq 0$ on $\Gamma_{b,\sigma}^N \cup \Gamma_{r,\sigma}^N$ and $D_x^+ \Psi \leq 0$ on $\Gamma_{l,\sigma}^N$. Then $L_\varepsilon^N \Psi \geq 0$ on D_σ^N implies that $\Psi \geq 0$ on $\overline{D}_\sigma^N$.*

The ε–uniform error estimate is contained in

Theorem 7. *Let u_ε be the continuous solution of (6) and let U_ε be the numerical solution generated from (9). Assume that $v_\varepsilon, w_\varepsilon \in C_\lambda^4(\overline{D})$. Then, for all $N \geq 4$, we have*

$$\sup_{0 < \varepsilon \leq 1} \|U_\varepsilon - u_\varepsilon\|_{\overline{D}_\sigma^N} \leq CN_x^{-1} \ln N_x + CN_t^{-1}$$

where C is a constant independent of N_x, N_t and ε.

Proof. The argument follows [3]. The discrete solution U_ε is the sum $U_\varepsilon = V_\varepsilon + W_\varepsilon$ where V_ε and W_ε are the obvious discrete counterparts to v_ε and w_ε. The classical truncation error estimate yields

$$|L_\varepsilon^N(V_\varepsilon - v_\varepsilon)| \leq C\sqrt{\varepsilon}N_x^{-1} + CN_t^{-1} \quad \text{and} \quad |D_x^+(V_\varepsilon - v_\varepsilon)(0,t)| \leq CN_x^{-1}.$$

It follows that $|V_\varepsilon - v_\varepsilon| \leq CN_x^{-1} + CN_t^{-1}$. Note also that

$$|L_\varepsilon^N(W_l - w_l)| \leq CN_x^{-1} \ln N_x + CN_t^{-1}$$

and

$$\sqrt{\varepsilon}|D_x^+(W_l - w_l)(0,t)| \leq CN_x^{-1} \ln N_x.$$

The proof is completed as in [3].

5 Numerical Results

In this section we present numerical results for the following specific elliptic problem

$$\varepsilon \Delta u_\varepsilon + \frac{\partial u_\varepsilon}{\partial x} = 16x(1-x)(1-y)y, \quad (x,y) \in (0,1)^2 \quad (10a)$$

$$u_\varepsilon = 1, \quad (x,y) \in \Gamma_R \cup \Gamma_T \quad (10b)$$

$$\varepsilon\frac{\partial u_\varepsilon}{\partial x} = 0, \ (x,y) \in \Gamma_L, \quad \sqrt{\varepsilon}\frac{\partial u_\varepsilon}{\partial y} = -16x^2(1-x)^2, \ (x,y) \in \Gamma_B \quad (10c)$$

whose solution has a parabolic boundary layer near Γ_B. The nature of the boundary layer function associated with this layer is related to the solutions of the parabolic problems examined in the previous section. In Figure 1 we present the numerical solution generated by applying standard upwinding on a uniform mesh. The numerical solutions are not bounded uniformly with respect to ε as $\varepsilon \to 0$. This should be compared with the accurate approximation given in Figure 2, which was generated by applying standard upwinding on the piecewise–uniform mesh

$$\Omega_{\sigma_2}^{N,N} = \Omega_u^N \times \Omega_\tau^N,$$

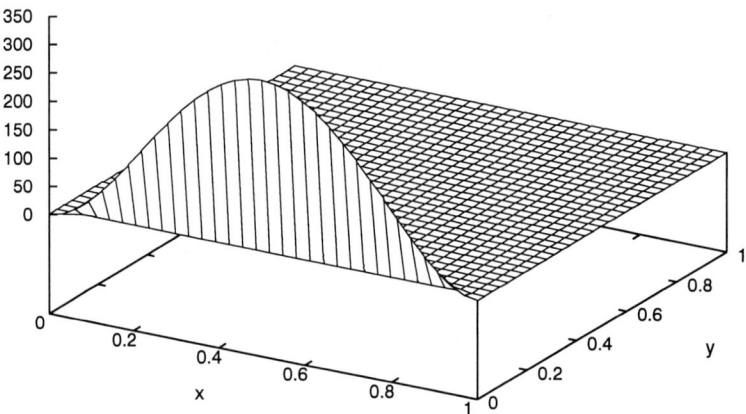

Fig. 1. Numerical solution generated by upwinding on a uniform mesh with $N=32$, $\varepsilon=10^{-8}$ for problem (10)

where

$$\tau = \min\left\{\frac{1}{2}, \sqrt{\varepsilon}\ln N\right\}.$$

Note the significant difference in the vertical scale in these two figures. In Table 1 we present the computed orders of convergence (see [2]) generated by applying standard upwinding on this piecewise–uniform mesh. These indicate that the method is ε–uniformly convergent for problem (10).

References

1. Andreyev V. B. and Savin I. A. (1996). The computation of boundary flow with uniform accuracy with respect to a small parameter. *Comput. Maths. Math. Phys.*, **36** (12) 1687–1692.
2. Farrell, P. A., Hegarty A.F, Miller, J. J. H., O'Riordan, E. ,Shishkin G. I., *Robust Computational techniques for boundary layers*, Chapman and Hall/CRC Press, 2000.
3. Miller, J. J. H.,O'Riordan, E.,Shishkin G. I.and Shishkina,L. P. (1998) *Fitted mesh methods for problems with parabolic boundary layers*, Math. Proc. Royal Irish Academy, **98A** (2) 173–190.

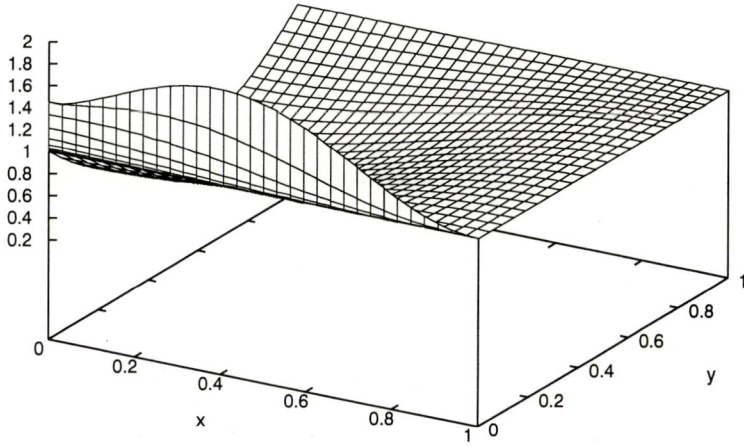

Fig. 2. Numerical solution generated by upwinding on a piecewise–uniform mesh with $N=32$, $\varepsilon=10^{-8}$ for problem (10)

Table 1. Computed orders of convergence generated by upwinding on a piecewise–uniform mesh applied to problem (10)

	\multicolumn{5}{c}{Number of intervals N}				
ε	8	16	32	64	128
1	1.20	1.11	1.06	1.03	1.02
2^{-2}	1.23	1.14	1.07	1.04	1.02
2^{-4}	1.18	1.11	1.06	1.03	1.01
2^{-6}	1.20	1.17	1.09	1.04	1.02
2^{-8}	0.54	0.69	0.70	1.09	1.04
2^{-10}	0.54	0.77	0.81	0.82	0.83
2^{-12}	0.55	0.77	0.81	0.82	0.83
2^{-14}	0.55	0.77	0.81	0.82	0.82
p^N	0.76	0.77	0.74	0.90	0.82

Reynolds–Uniform Numerical Method for Prandtl's Problem with Suction–Blowing Based on Blasius' Approach*

B. Gahan[1], J. J. H. Miller[1], and G. I. Shishkin[2]

[1] Department of Mathematics, Trinity College
Dublin, Ireland
bazzag@maths.tcd.ie
jmiller@tcd.ie

[2] Institute for Mathematics and Mechanics, Russian Academy of Sciences
Ekaterinburg, Russia
shishkin@maths.tcd.ie

Abstract. We construct a new numerical method for computing *reference* numerical solutions to the self–similar solution to the problem of incompressible laminar flow past a thin flat plate with suction–blowing. The method generates global numerical approximations to the velocity components and their scaled derivatives for arbitrary values of the Reynolds number in the range $[1, \infty)$ on a domain including the boundary layer but excluding a neighbourhood of the leading edge. The method is based on Blasius' approach. Using an experimental error estimate technique it is shown that these numerical approximations are pointwise accurate and that they satisfy pointwise error estimates which are independent of the Reynolds number for the flow. The Reynolds–uniform orders of convergence of the reference numerical solutions, with respect to the number of mesh subintervals used in the solution of Blasius' problem, is at least 0.86 and the error constant is not more than 80. The number of iterations required to solve the nonlinear Blasius problem is independent of the Reynolds number. Therefore the method generates reference numerical solutions with ε–uniform errors of any prescribed accuracy.

1 Introduction

The numerical solution of singularly perturbed boundary value problems, for which the solutions exhibit boundary layers, gives rise to significant difficulties. The errors in the numerical solutions of such problems generated by classical numerical methods depend on the value of the singular perturbation parameter ε, and can be large for small values of ε [2]. For representative classes of singular perturbation problems special methods have been constructed and shown

---

* This research was supported in part by the National Science Foundation grant DMS-9627244, by the Enterprise Ireland grant SC-98-612 and by the Russian Foundation for Basic Research grant No. 98-01-00362.

theoretically to generate numerical approximations that converge ε–uniformly. Also, numerical experiments have confirmed the efficacy of such methods in practice [2]. Singularly perturbed boundary value problems, for which the solutions exhibit boundary layers, frequently arise in flow problems with large Reynolds number Re. In such problems the small parameter $\varepsilon = Re^{-1}$. The discretization of such problems gives rise to nonlinear finite difference methods for which there is no known ε–uniform error analysis in the maximum norm. For this reason an experimental method for justifying ε–uniform convergence is the only remaining possibility. To make use of such a technique, especially for large Re, it is essential to have a known ε–uniform reference solution which approximates the exact solution to any prescribed accuracy. For flow problems with boundary layers there is usually no known analytic solution that can be used as a reference solution, and the same is true even for problems with a self–similar solution. Thus the task of constructing a reference *numerical* solution with ε–uniform errors of any prescribed accuracy arises from a wide class of flow problems.

An example of such a problem is flow past a flat plate with suction–blowing, for all Reynolds numbers for which the flow remains laminar and no separation occurs. For this problem it is important to construct a numerical method for which the pointwise errors in the scaled numerical solutions and their scaled derivatives are independent of the Reynolds number. In the present paper we consider the associated Prandtl problem of flow past a flat plate with suction–blowing. For large values of the Reynolds number the solution of this problem exhibits parabolic boundary layers in the neighbourhood of the plate, outside a neighbourhood of the leading edge. At the leading edge new singularities appear due to the incompatibilities of the problem data at the leading edge. Therefore,in the present paper we construct a numerical method which generates Reynolds–uniform reference numerical approximations to the scaled velocity components and their scaled derivatives for arbitrary values of the Reynolds number in a finite rectangular domain including the boundary layer but excluding a neighbourhood of the leading edge. This numerical method is based on the numerical solution of the related Blasius problem on the positive semi–axis. The accuracy of the numerical approximations depends on only the number of mesh subintervals N used for the solution of the Blasius problem. Our method is a development of that described in [2] for flow past a flat plate without suction–blowing.

2 Formulation of the Problem

We are required to find the solution, and its derivatives, of Prandtl's problem for incompressible flow past a semi–infinite flat plate $P = \{(x,0) \in \Re^2 : x \geq 0\}$ with suction–blowing in a bounded domain $\overline{D}$, which adjoins the plate and contains the boundary layer.

Prandtl's problem on the cut plane $\Omega = \Re^2 \setminus P$ is described as follows

(P_P) $\begin{cases} \text{Find } \mathbf{u_P} = (u_P, v_P) \text{ such that for all } (x,y) \in \Omega \\ \mathbf{u_P} \text{ satisfies the differential equations} \\[4pt] \frac{-1}{Re} \frac{\partial^2 u_P(x,y)}{\partial^2 y} + \mathbf{u_P} \cdot \nabla u_P(x,y) = 0 \\[4pt] \nabla \cdot \mathbf{u_P}(x,y) = 0 \\[4pt] \text{with the boundary conditions} \\[4pt] u_P(x,0) = 0, v_P = v_0(x) \text{ for all } x \geq 0 \\[4pt] \lim_{|y| \to \infty} \mathbf{u_P}(x,y) = \lim_{x \to -\infty} \mathbf{u_P}(x,y) = (1,0), \text{ for all } x \in \Re \end{cases}$

where $v_0(x)$ is the vertical component of the suction–blowing velocity. This is a nonlinear system of equations for the unknown components u_P, v_P of the velocity $\mathbf{u_P}$. The solution at all points in the open half plane to the left of the leading edge is $\mathbf{u_P} = (1,0)$. For special choices of the function v_0 the solution of (P_P) is self–similar, see (3) below.

Note that in Prandtl's problem, even without suction–blowing, the vertical component of the velocity tends to infinity as we approach the leading edge. To avoid this singularity, we choose the computational domain $D = (a, A) \times (0, B)$ where a, A and B are fixed positive numbers independent of Re. Our aim is to construct a method for finding reference numerical approximations to the self–similar solution and its derivatives of problem (P_P) for arbitrary $Re \in [1, \infty)$ with error independent of Re.

We now describe conditions under which the solution of (P_P) is self–similar. Using the approach of Blasius, see [1], for example, a solution $\mathbf{u_P} = (u_P, v_P)$ of (P_P) can be written in the form

$$u_P(x,y) \equiv u_B(x,y) = f'(\eta) \tag{1}$$

$$v_P(x,y) \equiv v_B(x,y) = \sqrt{\frac{1}{2xRe}}(\eta f'(\eta) - f(\eta)) \tag{2}$$

where

$$\eta = y\sqrt{Re/2x}$$

and the function f is the solution of the problem

(P_B) $\begin{cases} \text{Find a function } f \in C^3([0,\infty)) \text{ such that for all } \eta \in (0,\infty) \\[4pt] f'''(\eta) + f(\eta)f''(\eta) = 0 \\[4pt] \text{with the boundary conditions} \\[4pt] f(0) = f_0, \ f'(0) = 0, \ \lim_{\eta \to \infty} f'(\eta) = 1. \end{cases}$

(P_B) is known as Blasius' problem and $\mathbf{u}_B = (u_B, v_B)$ is known as the Blasius solution of (P_P). The existence and uniqueness of a solution to this third order nonlinear ordinary differential equation is discussed in [1]. Positive values of f_0 correspond to suction, while negative values of f_0 represent blowing, and f_0 is related to v_0 in (P_P) by the formula (see for example [3])

$$v_0(x) = -f_0 \sqrt{1/2xRe}. \tag{3}$$

The first order derivatives of the velocity components u_P and v_P are given by

$$\frac{\partial u_P}{\partial y}(x, y) = \frac{\partial u_B}{\partial y}(x, y) = \frac{\eta}{y} f''(\eta) \tag{4}$$

$$\frac{\partial v_P}{\partial y}(x, y) = \frac{\partial v_B}{\partial y}(x, y) = \frac{\eta}{2x} f''(\eta) \tag{5}$$

$$\frac{\partial u_P}{\partial x}(x, y) = -\frac{\partial v_P}{\partial y}(x, y) \tag{6}$$

$$\frac{\partial v_P}{\partial x}(x, y) = \frac{\partial v_B}{\partial x}(x, y) = -\frac{1}{2x}[v_B + \sqrt{\frac{1}{2xRe}} \eta^2 f''(\eta)] \tag{7}$$

From (1), (2), (4), (5), (6) and (7) we see that to find the velocity components u_P and v_P, and their first order derivatives, it is necessary to know $f'(\eta)$, $\eta f'(\eta) - f(\eta)$, $\eta f''(\eta)$ and $\eta^2 f''(\eta)$ for all $\eta \in [0, \infty)$. We also observe from these relations that, when Re is large, v_P and $\frac{\partial v_P}{\partial x}$ are small and $\frac{\partial u_P}{\partial y}$ is large. Therefore, in order to have values of order unity, we use the following scaled components: $\sqrt{Re} v_P$, $\sqrt{Re} \frac{\partial v_P}{\partial x}$, and $\frac{1}{\sqrt{Re}} \frac{\partial u_P}{\partial y}$.

In the next section numerical approximations to the solution of (P_B), and its first order derivatives, are constructed on the semi-infinite domain $[0, \infty)$.

3 Numerical Solution of Blasius' Problem

To find u_P and v_P and their first order derivatives we have to solve (P_B) for f and its derivatives on the semi-infinite domain $[0, \infty)$. This is not a trivial matter, since numerical solutions can be obtained at only a finite number of mesh points. For this reason, for each value of the parameter $L \in [1, \infty)$, we introduce the following problem on the finite interval $(0, L)$

$$(P_{B,L}) \begin{cases} \text{Find a function } f_L \in C^3(0, L) \text{ such that for all } \eta \in (0, L) \\ f_L'''(\eta) + f_L(\eta) f_L''(\eta) = 0 \\ \text{with the boundary conditions} \\ f_L(0) = f_0, \quad f_L'(0) = 0, \quad f_L'(L) = 1. \end{cases}$$

The collection of all such problems forms a one-parameter family of problems related to (P_B), where the interval length L is the parameter of the family. Because the values of f_L, f_L' and f_L'' are needed at all points of $[0, \infty)$, we introduce the following extrapolations

$$f_L''(\eta) = 0, \text{ for all } \eta \geq L \tag{8}$$
$$f_L'(\eta) = 1, \text{ for all } \eta \geq L \tag{9}$$
$$f_L(\eta) = (\eta - L) + f_L(L), \text{ for all } \eta \geq L. \tag{10}$$

To solve (P_B), we first obtain a numerical solution F_L of $(P_{B,L})$ on the finite interval $(0, L)$ for an increasing sequence of values of L. Then, we extrapolate F_L to the semi-infinite domain $[0, \infty)$. The sequence of values of L is defined as follows. For each even number $N \geq 4$ define $L_N = \ln N$ (see [2] for motivation for this choice of L_N) and consider the corresponding finite interval $[0, L_N]$. On $[0, L_N]$ a uniform mesh $\overline{I}_u^N = \{\eta_i : \eta_i = iN^{-1}\ln N, 0 \leq i \leq N\}_0^N$ with N mesh subintervals is constructed. Then numerical approximations F_L, D^+F_L, $D^+D^+F_L$ to f_L, f_L', f_L'' respectively, are determined at the mesh points in $\overline{I}_u^N$ using the following non-linear finite difference method

$$(P_{B,L}^N) \begin{cases} \text{Find F on } \overline{I}_u^N \text{ such that, for all } \eta_i \in I_u^N, 2 \leq i \leq N-1, \\ \delta^2(D^-F)(\eta_i) + F(\eta_i)D^+(D^-F)(\eta_i) = 0 \\ F(0) = f_0 \quad D^+F(0) = 0, \text{ and } \quad D^0F(\eta_{N-1}) = 1. \end{cases}$$

We note that, in order to simplify the notation, we have dropped explicit use of the indices L and N. Thus, we denote the solution of $P_{B,L}^N$ by F instead of F_L^N.

Since $(P_{B,L}^N)$ is non-linear, we use the following iterative solver to compute its solution

$$(A_B^N) \begin{cases} \text{For each integer } m, 1 \leq m \leq M, \text{ find } F^m \text{ on } I_u^N \\ \text{such that, for all } \eta_i \in I_u^N \\ \delta^2(D^-F^m)(\eta_i) + F^{m-1}(\eta_i)D^+(D^-F^m)(\eta_i) - D^-(F^m - F^{m-1})(\eta_i) = 0 \\ F^m(0) = f_0, \quad D^+F^m(0) = 0, \text{ and } D^0F^m(\eta_{N-1}) = 1 \\ \text{with the starting values for all mesh points } \eta_i \in \overline{I}_u^N \\ F^0(\eta_i) = \eta_i. \end{cases}$$

Algorithm (A_B^N) involves the solution of a sequence of linear problems, with one linear problem for each value of the iteration index m. The total number of iterations M is taken to be $M = 8\ln N$ The motivation for this choice of M is described in [2]. It is important to note the crucial property that M is independent of the Reynolds number Re. The final output of algorithm (A_B^N) is denoted by F, where again we simplify the notation by omitting explicit mention of the total number of iterations M. We follow the same criterion as in [2] to determine F on the finest required mesh as the "exact" solution. The corresponding value of N is denoted by N_0.

To ensure that F, D^+F and D^+D^+F are defined at all points of each mesh $\overline{I}_u^N$ the following values are assigned: $D^+F(\eta_N) = 1$, $D^+D^+F(\eta_{N-1}) = 0$, $D^+D^+F(\eta_N) = 0$. We then define F, D^+F and D^+D^+F at each point of $[0, L_N]$ using piecewise linear interpolation of the values at the mesh points of $\overline{I}_u^N$. The resulting interpolants are denoted by $\overline{F}$, $\overline{D^+F}$ and $\overline{D^+D^+F}$ respectively.

In order to define $\overline{F}$, $\overline{D^+F}$ and $\overline{D^+D^+F}$ at each point $\eta \in [0, \infty)$ the following extrapolations, analogous to (8), (9) and (10), are introduced

$$\overline{D^+D^+F}(\eta) = 0, \text{ for all } \eta \in [L_N, \infty) \tag{11}$$
$$\overline{D^+F}(\eta) = 1, \text{ for all } \eta \in [L_N, \infty) \tag{12}$$
$$\overline{F}(\eta) = \overline{F}(L_N) + (\eta - L_N), \text{ for all } \eta \in [L_N, \infty). \tag{13}$$

The values of $\overline{F}$, $\overline{D^+F}$ and $\overline{D^+D^+F}$, respectively, are the required numerical approximations to f, f', f'' of the Blasius solution and its derivatives at each point of $[0, \infty)$.

4 Numerical Experiments for Blasius' Problem

In [3] a limiting value for suction is found at $f_0 = 7.07$ and for blowing at $f_0 = -0.875745$. In numerical experiments to illustrate the proposed technique, we take the representative values $f_0 = 3$ and $f_0 = 6$ for suction; $f_0 = -0.25$ and $f_0 = -0.5$ for blowing.

We want to determine error estimates for the approximations $\overline{F}$, $\overline{D^+F}$ and $\overline{D^+D^+F}$ to f, f' and f'', respectively, for all $N \geq 2048$. Consequently, we take $I_u^{N_0}$, where $N_0 = 65536$, to be the finest mesh on which we solve Blasius' problem. Using the experimental numerical technique described in [2] we determine the following computed error estimates

$f_0 = 3$

$$\|\overline{F} - f\|_{[0,\infty)} \leq 2.505 N^{-0.86}$$
$$\|\overline{D^+F} - f'\|_{[0,\infty)} \leq 1.452 N^{-0.86}$$
$$\|\overline{D^+D^+F} - f''\|_{[0,\infty)} \leq 20.427 N^{-0.84}$$

$f_0 = 6$

$$\|\overline{F} - f\|_{[0,\infty)} \leq 2.635 N^{-0.86}$$
$$\|\overline{D^+F} - f'\|_{[0,\infty)} \leq 2.925 N^{-0.86}$$
$$\|\overline{D^+D^+F} - f''\|_{[0,\infty)} \leq 65.927 N^{-0.81}$$

$f_0 = -0.25$

$$\|\overline{F} - f\|_{[0,\infty)} \le 1.066 N^{-0.86}$$
$$\|\overline{D^+F} - f'\|_{[0,\infty)} \le 0.202^{-0.86}$$
$$\|\overline{D^+D^+F} - f''\|_{[0,\infty)} \le 0.453 N^{-0.86}$$

$f_0 = -0.5$

$$\|\overline{F} - f\|_{[0,\infty)} \le 0.603 N^{-0.85}$$
$$\|\overline{D^+F} - f'\|_{[0,\infty)} \le 0.345 N^{-0.87}$$
$$\|\overline{D^+D^+F} - f''\|_{[0,\infty)} \le 0.488 N^{-0.86}.$$

Similarly, the computed error estimates for the approximations $\eta \overline{D^+F}(\eta) - \overline{F}(\eta)$, $\eta \overline{D^+D^+F}(\eta)$ and $\eta^2 \overline{D^+D^+F}(\eta)$ to $(\eta f' - f)(\eta)$, $\eta f''(\eta)$ and $\eta^2 f''(\eta)$, respectively, for all $N \ge 2048$, are

$f_0 = 3$

$$\|(\eta \overline{D^+F} - \overline{F}) - (\eta f' - f)\|_{[0,\infty)} \le 2.505 N^{-0.86}$$
$$\|\eta(\overline{D^+D^+F} - f'')\|_{[0,\infty)} \le 1.8 N^{-0.85}$$
$$\|\eta^2(\overline{D^+D^+F} - f'')\|_{[0,\infty)} \le 0.7 N^{-0.86}$$

$f_0 = 6$

$$\|(\eta \overline{D^+F} - \overline{F}) - (\eta f' - f)\|_{[0,\infty)} \le 2.635 N^{-0.86}$$
$$\|\eta(\overline{D^+D^+F} - f'')\|_{[0,\infty)} \le 3.297 N^{-0.85}$$
$$\|\eta^2(\overline{D^+D^+F} - f'')\|_{[0,\infty)} \le 0.745 N^{-0.86}$$

$f_0 = -0.25$

$$\|(\eta \overline{D^+F} - \overline{F}) - (\eta f' - f)\|_{[0,\infty)} \le 1.066 N^{-0.86}$$
$$\|\eta(\overline{D^+D^+F} - f'')\|_{[0,\infty)} \le 1.178 N^{-0.86}$$
$$\|\eta^2(\overline{D^+D^+F} - f'')\|_{[0,\infty)} \le 3.275 N^{-0.86}$$

$f_0 = -0.5$

$$\|(\eta \overline{D^+F} - \overline{F}) - (\eta f' - f)\|_{[0,\infty)} \le 1.228 N^{-0.86}$$
$$\|\eta(\overline{D^+D^+F} - f'')\|_{[0,\infty)} \le 1.670 N^{-0.86}$$
$$\|\eta^2(\overline{D^+D^+F} - f'')\|_{[0,\infty)} \le 5.952 N^{-0.86}.$$

We see from the above computed error estimates that, in all cases and at each point of $[0,\infty)$, the orders of convergence with respect to N, the number of mesh intervals used to solve Blasius' problem, are not less than 0.81. Similarly, in all cases, the error constants are at most 65.927. The worst cases occur for $f_0 = 6$.

5 Numerical Experiments for Prandtl's Problem

In this section we find reference numerical solutions of Prandtl's problem and computed error estimates for the scaled numerical solutions and their derivatives. In all of the numerical computations we use the specific values $a = 0.1$, $A = 1.1, B = 1.0$.

We construct the approximations $\mathbf{U}_B = (U_B, V_B)$ of the velocity components $\mathbf{u}_B$ of the self–similar solution of Prandtl's problem (P_P) by substituting the approximate expressions $\overline{F}$ and $\overline{D^+F}$ for f and f' respectively, into (1) and (2). Thus, for each (x, y) in the open quarter plane $\{(x,y) : x > 0, y > 0)\}$ we have

$$U_B(x,y) = \overline{D^+F}(\eta) \tag{14}$$

$$V_B(x,y) = \sqrt{\frac{1}{2xRe}}(\eta\overline{D^+F}(\eta) - \overline{F}(\eta)) \tag{15}$$

We call $\mathbf{U}_B = (U_B, V_B)$ the reference numerical solutions of the self–similar solution of Prandtl's problem (P_P).

We now assume that error estimates, for the scaled approximations $(U_B, \sqrt{Re}V_B)$ to $(u_P, \sqrt{Re}v_P)$, of the form

$$\|U_B - u_P\|_{\overline{\Omega}} \leq C_1 N^{-p_1}$$

$$\sqrt{Re}\|V_B - v_P\|_{\overline{\Omega}} \leq C_2 N^{-p_2}$$

are valid for all $N > N_0$ where $p_1 > 0$, $p_2 > 0$, and the constants N_0, p_1, p_2, C_1, C_2 are independent of the total number of iterations M and the number of mesh intervals N used in the numerical solution of Blasius' problem.

The errors in the x-component U_B and the scaled y-component $\sqrt{Re}V_B$ of the velocity corresponding to $M \geq 8 \ln N$ satisfy

$$\|U_B - u_P\|_{\overline{\Omega}} = \|\overline{D^+F} - f'\|_{[0,\infty)}$$

$$\sqrt{Re}\|V_B - v_P\|_{\overline{\Omega}} = \sqrt{Re}\sqrt{\frac{1}{2xRe}}\|[(\eta\overline{D^+F}(\eta) - \overline{F}(\eta)) - (\eta f' - f)]\|_{[0,\infty)}$$

$$\leq \sqrt{5}\|(\eta\overline{D^+F}(\eta) - \overline{F}(\eta)) - (\eta f' - f)\|_{[0,\infty)}.$$

Then, using the experimental numerical technique described in [2] and the computed error estimates for the numerical solutions of Blasius' problem in the previous section, we obtain for all $N \geq 2048$ the following computed error estimates for the reference numerical solutions of Prandtl's problem

$$f_0 = 3$$

$$\|U_B - u_P\|_{\overline{\Omega}} \leq 1.452 N^{-0.86}$$
$$\sqrt{Re}\|V_B - v_P\|_{\overline{\Omega}} \leq 5.601 N^{-0.86}$$

$f_0 = 6$

$$\|U_B - u_P\|_{\overline{\Omega}} \leq 2.925 N^{-0.86}$$
$$\sqrt{Re}\|V_B - v_P\|_{\overline{\Omega}} \leq 5.89 N^{-0.86}$$

$f_0 = -0.25$

$$\|U_B - u_P\|_{\overline{\Omega}} \leq 0.202 N^{-0.86}$$
$$\sqrt{Re}\|V_B - v_P\|_{\overline{\Omega}} \leq 2.38 N^{-0.86}$$

$f_0 = -0.5$

$$\|U_B - u_P\|_{\overline{\Omega}} \leq 0.345 N^{-0.87}$$
$$\sqrt{Re}\|V_B - v_P\|_{\overline{\Omega}} \leq 1.35 N^{-0.86}.$$

We see from these computed error estimates that, in all cases, the orders of convergence with respect to N, the number of mesh intervals used to solve Blasius' problem, are at least 0.86. Similarly, in all cases, the error constants are at most 5.89. The worst case occurs for $f_0 = 6$.

Substituting the appropriate expressions into (4), (5), (6) and (7) we obtain the approximations $D_x U_B, D_y U_B, D_x V_B, D_y V_B$ to the first order derivatives of the velocity components of the self–similar solution of Prandtl's problem (P_P), where

$$D_y U_B(\eta(x,y)) = \frac{\eta}{y} \overline{D_\eta^+ D_\eta^+ F}(\eta)$$

$$D_y V_B(\eta(x,y)) = \frac{\eta}{2x} \overline{D_\eta^+ D_\eta^+ F}(\eta)$$

$$D_x U_B(\eta(x,y)) = -D_y V_B(\eta(x,y))$$

$$D_x V_B(\eta(x,y)) = -\frac{1}{2x}(V_B + \sqrt{\frac{1}{2xRe}} \eta^2 \overline{D_\eta^+ D_\eta^+ F}(\eta)).$$

From the computed error estimates for the numerical solutions of Blasius' problem, in the previous section, we obtain for all $N \geq 2048$ the following computed error estimates for the reference scaled discrete derivatives of the velocity components

$$\frac{1}{\sqrt{Re}}\|D_y U_B - \frac{\partial u_P}{\partial y}\|_{\overline{\Omega}} = \sqrt{\frac{1}{2x}}\|\overline{D_\eta^+ D_\eta^+ F}(\eta) - f''(\eta)\|_{[0,\infty)}$$

$$\leq \sqrt{5}\|\overline{D_\eta^+ D_\eta^+ F}(\eta) - f''(\eta)\|_{[0,\infty)}$$

$$\|D_y V_B - \frac{\partial v_P}{\partial y}\|_{\overline{\Omega}} = \|D_x U_B - \frac{\partial u_P}{\partial x}\|$$

$$= \frac{\eta}{2x} \|\overline{D_\eta^+ D_\eta^+ F(\eta)} - f''(\eta)\|$$

$$\sqrt{Re}\|D_x V_B - \frac{\partial v_P}{\partial x}\|_{\overline{\Omega}} = \frac{\sqrt{Re}}{2x}(\|V_B - v_B\| + \sqrt{\frac{1}{2xRe}}\eta^2 \|\overline{D_\eta^+ D_\eta^+ F(\eta)} - f''(\eta)\|)$$

$$\leq \frac{1}{2x}(\sqrt{Re}\|V_B - v_B\| + \sqrt{\frac{1}{2x}}\eta^2 \|\overline{D_\eta^+ D_\eta^+ F(\eta)} - f''(\eta)\|).$$

Then, for all $N \geq 2048$ we obtain the following estimates

$f_0 = 3$

$$\frac{1}{\sqrt{Re}}\|D_y U_B - \frac{\partial u_P}{\partial y}\|_{\overline{\Omega}} \leq 45.676 N^{-0.86}$$
$$\|D_y V_B - \frac{\partial v_P}{\partial y}\|_{\overline{\Omega}} \leq 9 N^{-0.85}$$
$$\sqrt{Re}\|D_x V_B - \frac{\partial v_P}{\partial x}\|_{\overline{\Omega}} \leq 35.831 N^{-0.86}$$

$f_0 = 6$

$$\frac{1}{\sqrt{Re}}\|D_y U_B - \frac{\partial u_P}{\partial y}\|_{\overline{\Omega}} \leq 147.42 N^{-0.86}$$
$$\|D_y V_B - \frac{\partial v_P}{\partial y}\|_{\overline{\Omega}} \leq 16.49 N^{-0.85}$$
$$\sqrt{Re}\|D_x V_B - \frac{\partial v_P}{\partial x}\|_{\overline{\Omega}} \leq 37.78 N^{-0.86}$$

$f_0 = -0.25$

$$\frac{1}{\sqrt{Re}}\|D_y U_B - \frac{\partial u_P}{\partial y}\|_{\overline{\Omega}} \leq 1.01 N^{-0.86}$$
$$\|D_y V_B - \frac{\partial v_P}{\partial y}\|_{\overline{\Omega}} \leq 5.89 N^{-0.86}$$
$$\sqrt{Re}\|D_x V_B - \frac{\partial v_P}{\partial x}\|_{\overline{\Omega}} \leq 48.52 N^{-0.86}$$

$f_0 = -0.5$

$$\frac{1}{\sqrt{Re}}\|D_y U_B - \frac{\partial u_P}{\partial y}\|_{\overline{\Omega}} \leq 1.09 N^{-0.86}$$
$$\|D_y V_B - \frac{\partial v_P}{\partial y}\|_{\overline{\Omega}} \leq 8.35 N^{-0.86}$$
$$\sqrt{Re}\|D_x V_B - \frac{\partial v_P}{\partial x}\|_{\overline{\Omega}} \leq 73.3 N^{-0.86}.$$

We see from these computed error estimates that, in all cases, the orders of convergence with respect to N, the number of mesh intervals used to solve Blasius' problem, are at least 0.85. Similarly, in all cases, the error constants are at most 73.3. The worst order of convergence occurs for $f_0 = 6$ and the worst error constant for $f_0 = -0.5$.

Remark on Navier-Stokes' Problem It is well known that incompressible flow past a plate $P = \{(x,0) \in \Re^2 : x \geq 0\}$ with suction–blowing in the domain $D = \Re^2 \backslash P$ is governed by the Navier-Stokes equations

$$(P_{NS}) \begin{cases} \text{Find } \mathbf{u}_{NS} = (u_{NS}, v_{NS}),\ p_{NS} \text{ such that for all } (x,y) \in D \\ \mathbf{u}_{NS} \text{ satisfies the differential equations} \\ -\frac{1}{Re}\Delta \mathbf{u}_{NS} + \mathbf{u}_{NS} \cdot \nabla \mathbf{u}_{NS} = -\frac{1}{\rho}\nabla p_{NS} \\ \nabla \cdot \mathbf{u}_{NS} = 0 \\ \text{with the boundary conditions} \\ u_{NS}(x,0) = 0,\ v_{NS} = v_0(x) \text{ for all } x \geq 0 \\ \lim_{|y| \to \infty} \mathbf{u}_{NS}(x,y) = \lim_{x \to -\infty} \mathbf{u}_{NS}(x,y) = (1,0),\ \text{for all } x \in \Re \end{cases}$$

where $\mathbf{u}_{NS}$ is the velocity of the fluid, Re is the Reynolds number, ρ is the density of the fluid and p is the pressure. This is a nonlinear system of equations for the unknowns $\mathbf{u}_{NS}, p_{NS}$. It is known that the solution of (P_P) is a good approximation to the solution of (P_{NS}) in a subdomain excluding the leading edge region, provided that the flow remains laminar and no separation occurs. Moreover, as Re increases the difference between the solutions of problems (P_P) and (P_{NS}) decreases. This means that the reference solution of Prandtl's problem is the leading term in the solution of the above Navier–Stokes' problem.

6 Conclusion

For the problem of incompressible laminar flow past a thin flat plate with suction–blowing we construct a new numerical method for computing *reference* numerical solutions to the self–similar solution of the related Prandtl problem. The method generates global numerical approximations to the velocity components and their scaled derivatives for arbitrary values of the Reynolds number in the range $[1,\infty)$ on a domain including the boundary layer but excluding a neighbourhood of the leading edge. The method is based on Blasius' approach. Using an experimental error estimate technique it is shown that these numerical approximations are pointwise accurate and that they satisfy pointwise error estimates which are independent of the Reynolds number for the flow. The Reynolds–uniform orders of convergence of the reference numerical solutions, with respect to the number of mesh subintervals used in the solution of Blasius' problem, is at least 0.86 and the error constant is not more than 80. The number of iterations required to solve the nonlinear Blasius problem is independent of the Reynolds number. Therefore the method generates reference numerical solutions with ε–uniform errors of any prescribed accuracy.

References

1. H. Schlichting, *Boundary Layer Theory*, 7th Edition, McGraw Hill, 1951.
2. P. Farrell, A. Hegarty, J. J. H. Miller, E. O' Riordan, G. I. Shishkin, *Robust Computational Techniques for Boundary Layers* , Series in Applied Mathematics and Mathematical Computation (Eds. R J Knops and K W Morton), Chapman and Hall/CRC Press, 2000.
3. D. F. Rogers, *Laminar Flow Analysis* , Cambridge University Press, 1992.
4. B. Gahan, J. J. H. Miller, G. I. Shishkin: Accurate numerical method for Blasius' problem for flow past a flat plate with mass transfer. TCD Maths Department Preprint no. 00-xx.
5. D. J. Acheson, *Elementary Fluid Dynamics* , Oxford: Clarendon, 1990.

Multigrid Methods and Finite Difference Schemes for 2D Singularly Perturbed Problems*

F. Gaspar, F. Lisbona, and C. Clavero

Departamento de Matemática Aplicada. Universidad de Zaragoza
Zaragoza, Spain
{fjgaspar,lisbona,clavero}@posta.unizar.es

Abstract. Solving the algebraic linear systems proceeding from the discretization on some condensed meshes of 2D singularly perturbed problems, is a difficult task. In this work we present numerical experiments obtained with the multigrid method for this class of linear systems. On Shishkin meshes, the classical multigrid algorithm is not convergent. We see that modifying only the restriction operator in an appropriate form, the algorithm is convergent, the CPU time increases linearly with the discretization parameter and the number of cycles is independent of the mesh sizes.

1 Introduction

In this paper we are interested in the application of multigrid techniques to solve the algebraic linear systems arising from the discretization of singularly perturbed problems on Shishkin meshes. We consider problems of type

$$L_\varepsilon u \equiv -\varepsilon \Delta u + \mathbf{b} \cdot \nabla u + c u = f, \quad \text{in } \Omega = (0,1)^2,$$
$$u = 0, \quad \text{on } \Gamma_D, \quad \frac{\partial u}{\partial n} = 0, \quad \text{on } \Gamma_N, \quad (1)$$

where $\Gamma = \partial \Omega = \Gamma_D \cup \Gamma_N$ and $0 < \varepsilon \leq 1$. We assume that $\mathbf{b}$, c and f are sufficiently smooth functions satisfying enough compatibility conditions with $c \geq 0$. Thus, depending on the value of the convection term, it is known, [8], that the exact solution of (1) can present regular and/or parabolic layers. In all cases, classical schemes on uniform meshes give a numerical solution reliable only if a very large number of mesh points is taken [8]. To solve efficiently this type of problems, ε-uniformly convergent schemes are needed.

In recent years, schemes based on *a priori* fitted meshes (see [8] and references therein) are commonly used for the numerical approximation of the solution of problems of type (1). Between the different possibilities, Shishkin meshes, [10,11], seem the most adequate because they can be easily constructed. On these meshes, classical numerical schemes are in many cases uniformly convergent [8]. Nevertheless, since the ratio between the mesh sizes is very large for ε sufficiently

---
* This research was supported by the projects DGES-PB97-1013 and P226-68

small, the resolution of the associated linear systems is difficult [9]. The BI-CGSTAB algorithm, [12], is generally an efficient method when a not very large number of mesh points is taken. In general, for large linear systems, the multigrid technique is a good alternative, [1,5,13], but efficient multigrid has not yet been achieved for singular perturbation problems on Shishkin meshes. In [4] we showed that standard multigrid is not convergent when the numerical schemes are constructed on Shishkin meshes. Also, we saw that modifying adequately the restriction operator, the deduced algorithm is very efficient. In this paper we present the modified restriction operator, adapted for use on Shishkin grids. We apply this multigrid algorithm for a hybrid scheme solving a convection-diffusion problem with regular layers and a high order scheme for a problem with regular and parabolic layers. Another approach that will lead to efficient multigrid methods for singular perturbation problems, is presented in [6] and [14]. In this approach the smoother is changed to an incomplete line LU relaxation method (ILLU), which makes classical multigrid more robust. Finally, we would like to mention that the algebraic multigrid methods may also lead to robust solvers for the problem considered here.

2 The Multigrid Algorithm

All components of the multigrid that we consider, except the restriction operator, are standard components, [13], i.e., the smoother is a line Gauss-Seidel of alternating symmetric type, the prolongation operator is the bilinear interpolation and the coefficient matrix of the linear systems, constructed in each level of the algorithm, are obtained by discretization, with the finite difference scheme, of the differential equation on the corresponding associated grid. Let Ω_l, Ω_{l-1} be the spaces of the grid functions respectively defined on the meshes of level l and $l-1$ of the multigrid algorithm. The restriction operator R_l^{l-1} is a linear mapping

$$R_l^{l-1} : \Omega_l \longrightarrow \Omega_{l-1},$$
$$r^l \longmapsto R_l^{l-1} r^l = r^{l-1}$$

which maps fine-grid functions onto coarse-grid functions. They can be represented by the stencil

$$R_l^{l-1} = \begin{bmatrix} \varrho_{-1,1} & \varrho_{0,1} & \varrho_{1,1} \\ \varrho_{-1,0} & \varrho_{0,0} & \varrho_{1,0} \\ \varrho_{-1,-1} & \varrho_{0,-1} & \varrho_{1,-1} \end{bmatrix},$$

which describes the formula

$$r^{l-1}(x_i, y_j) = \sum_{m,n=-1}^{1} \varrho_{m,n} r^l(x_i + m\bar{h}_m^x, y_j + n\bar{h}_n^y), \quad (x_i, y_j) \in \Omega^{l-1}$$

where $\bar{h}_{-1}^x = x_i - x_{i-1}$, $\bar{h}_1^x = x_{i+1} - x_i$, $\bar{h}_{-1}^y = y_j - y_{j-1}$, $\bar{h}_1^y = y_{j+1} - y_j$, $\bar{h}_0^x = \bar{h}_0^y = 0$. To define a general restriction operator we proceed as follows. Let V_P^l be a molecule centered in the point $P = (x_i, y_j)$ on the fine grid. The residual associated to this molecule is calculated by

$$Q^l_{V^l_P}(r^l) = \sum_{P_i \in V^l_P} a_{P_i} r^l_{P_i}, \qquad (2)$$

where a_{P_i} are the weights of a quadrature formula. The restriction operator on the coarse grid at P, r^{l-1}_P, is given by the following discrete conservation equality

$$(\text{area} V^l_P) r^{l-1}_P = Q^l_{V^l_P}(r^l). \qquad (3)$$

Using the composite trapezoidal rule on uniform meshes, we obtain the most commonly used restriction operator, the *full weighting* operator, defined by the stencil

$$R^{l-1}_l = \frac{1}{16} \begin{bmatrix} 1 & 2 & 1 \\ 2 & 4 & 2 \\ 1 & 2 & 1 \end{bmatrix} \qquad (4)$$

To define a different operator (see [4] for details), we only modify the quadrature formula used to calculate the residual. For the x direction (similarly for the y direction), we use the composite trapezoidal rule when the step sizes associated to the point P are equal, i.e., $\bar{h}^x_{-1} = \bar{h}^x_1$. Otherwise, when $\bar{h}^x_{-1} \neq \bar{h}^x_1$, the formula is

$$r^l_{i,j} \bar{h}^x_{-1} + \frac{r^l_{i,j} + r^l_{i+1,j}}{2} \bar{h}^x_1. \qquad (5)$$

Thus, the 2D quadrature formula is the product of the corresponding 1D formulas.

3 A Hybrid Difference Scheme

In this section we want to approximate the solution of problem (1) supposing that $\mathbf{b} = (b_1, b_2) \geq (\beta_1, \beta_2) > (0, 0)$. In this case, since there are regular layers in $x = 1$ and $y = 1$, the Shishkin mesh, $\bar{\Omega}_N$, is constructed as follows. Let $N \geq 4$ be an even number. We define the transition parameters

$$\sigma_x = \min\{1/2, \sigma_{0,x} \varepsilon \log N\}, \quad \sigma_y = \min\{1/2, \sigma_{0,y} \varepsilon \log N\}, \qquad (6)$$

where $\sigma_{0,x} \geq 1/\beta_1$, $\sigma_{0,y} \geq 1/\beta_2$ are constants to be chosen later. Taking $N/2+1$ uniformly distributed points in the intervals $[0, 1-\sigma_x]$ and $[0, 1-\sigma_y]$, and also $N/2 + 1$ equally spaced points in $[1 - \sigma_x, 1]$ and $[1 - \sigma_y, 1]$ we obtain the grid as tensor product of the corresponding one-dimensional meshes. We see that if ε is large enough, the mesh is uniform; otherwise, the points concentrate in the regular layer region, having only two different step sizes in each direction, given by $H_x = 2(1 - \sigma_x)/N$, $h_x = 2\sigma_x/N$, $H_y = 2(1 - \sigma_y)/N$, $h_y = 2\sigma_y/N$. We note that, for each N, only the finest grid is of Shishkin type; the grid associated to level $l-1$ has step sizes, in each direction, which are double of the corresponding step sizes in previous level l.

In the sequel, we denote $h_i^x = x_i - x_{i-1}$, $h_j^y = y_j - y_{j-1}$, D_x^-, D_x^+, D_x^0 the backward, forward and central difference discretizations of the first derivative respectively, and $D_x^- D_x^+$, the second order central difference discretization, and similarly for the variable y. We define the following hybrid difference operator to approximate the first-order derivative:

$$D_x^h U_{i,j} = \begin{cases} D_x^- U_{i,j} & \text{for } 0 < i \leq N/2, \\ D_x^0 U_{i,j} & \text{for } N/2 < i < N, \end{cases}$$

and analogously we can define D_y^h. On $\bar{\Omega}_N$ we consider the scheme

$$L_\varepsilon^N U_{i,j} \equiv -\varepsilon(D_x^- D_x^+ + D_y^- D_y^+)U_{i,j} + \mathbf{b}_{i,j} \cdot (D_x^h U_{i,j}, D_y^h U_{i,j}) + c_{i,j} U_{i,j} = f_{i,j},$$
$$i,j = 1, \ldots, N-1, \quad (7)$$
$$U_{i,j} = 0, \quad \text{on } \Gamma^N = \Gamma \cap \Omega^N. \quad (8)$$

This scheme is uniformly convergent with order 1 (see [7]). Considering the restriction operator given in section 2, we obtain four different expressions depending on where the point is. In the algorithm, these operators must be calculated one time at the beginning. This fact supposes a great simplification in the code in contrast with general non uniform grids. Defining the following sets of points,

$$\Omega^{1-\sigma} = \{(1-\sigma_x, 1-\sigma_y)\},$$
$$\Omega^{x,1-\sigma} = \{(x_i, 1-\sigma_y), i=0,\ldots,N\} \setminus \Omega^{1-\sigma},$$
$$\Omega^{1-\sigma,y} = \{(1-\sigma_x, y_j), j=0,\ldots,N\} \setminus \Omega^{1-\sigma},$$
$$\Omega^r = \Omega \setminus \{\Omega^{x,1-\sigma} \cup \Omega^{1-\sigma,y} \cup \Omega^{1-\sigma}\},$$

the operators are given by

$$R_\sigma = \frac{1}{4} \begin{bmatrix} 0 & (2-\sigma_x)\sigma_y & \sigma_x \sigma_y \\ 0 & (2-\sigma_x)(2-\sigma_y) & \sigma_x(2-\sigma_y) \\ 0 & 0 & 0 \end{bmatrix}, \quad \text{if } (x_i, y_j) \in \Omega^{1-\sigma}, \quad (9)$$

$$R_\sigma = \frac{1}{8} \begin{bmatrix} \sigma_y & 2\sigma_y & \sigma_y \\ 2-\sigma_y & 2(2-\sigma_y) & 2-\sigma_y \\ 0 & 0 & 0 \end{bmatrix}, \quad \text{if } (x_i, y_j) \in \Omega^{x,1-\sigma}, \quad (10)$$

$$R_\sigma = \frac{1}{8} \begin{bmatrix} 0 & 2-\sigma_x & \sigma_x \\ 0 & 2(2-\sigma_x) & 2\sigma_x \\ 0 & 2-\sigma_x & \sigma_x \end{bmatrix}, \quad \text{if } (x_i, y_j) \in \Omega^{1-\sigma,y}, \quad (11)$$

and by (4) if $(x_i, y_j) \in \Omega^r$. To see the good properties of the new multigrid method, we solve the problem

$$-\varepsilon \Delta u + u_x + u_y = f, \quad \text{in } \Omega, \quad (12)$$
$$u = 0, \quad \text{on } \Gamma,$$

where f is such that the exact solution is given by $u(x,y) = xy(e^{(x-1)/\varepsilon} - 1)(e^{(y-1)/\varepsilon} - 1)$. We show the results on 32^2, 64^2, 128^2 and 256^2 Shishkin meshes for some values of ε sufficiently small. In [4] we saw that for large values of diffusion parameter ε the new restriction operator is less efficient that the *full weighting* operator. In Table 1, the spectral radius ρ, the number of iterations needed to obtain a residual of 10^{-5} and the number within brackets corresponding to wall-clock time, are shown. We also show the discrete maximum norm of the global discretization error, i.e., $\| e \|_\infty = \max_{ij} | u(x_i, y_j) - U_{ij} |$, $i, j = 0, 1, \ldots, N$. This table illustrates both the first-order convergence of the discretization given by the hybrid scheme, a linear increment of CPU time and also the independence of the spectral radius with respect to the size of the mesh. Thus, we conclude that the method has all expected good properties of the multigrid technique.

Table 1. Error, spectral radius, number of cycles and CPU time

Grid		$\varepsilon = 10^{-4}$	$\varepsilon = 10^{-5}$	$\varepsilon = 10^{-6}$	$\varepsilon = 10^{-7}$	$\varepsilon = 10^{-8}$
	$\|e\|_\infty$	5.97D-2	5.99D-2	5.99D-2	5.99D-2	5.99D-2
32×32	ρ	0.03	0.03	0.03	0.03	0.03
	CPU	6(0.69)	7(0.82)	8 (0.93)	8 (0.93)	9 (1.05)
	$\|e\|_\infty$	3.10D-2	3.12D-2	3.13D-2	3.13D-2	3.13D-2
64×64	ρ	0.04	0.04	0.04	0.04	0.04
	CPU	7 (3.17)	8 (3.61)	8 (3.61)	9 (4.05)	10(4.49)
	$\|e\|_\infty$	1.56D-2	1.58D-2	1.58D-2	1.58D-2	1.58D-2
128×128	ρ	0.04	0.05	0.05	0.05	0.05
	CPU	7 (12.78)	8 (14.57)	9 (16.35)	10 (18.13)	11 (19.90)
	$\|e\|_\infty$	7.69D-3	7.88D-3	7.90D-3	7.90D-3	7.90D-3
256×256	ρ	0.12	0.06	0.06	0.06	0.06
	CPU	11 (80.70)	8 (59.15)	9 (66.40)	11(80.78)	12(88.20)

4 A High Order Scheme on a Shishkin Mesh

Now we consider the problem

$$-\varepsilon \Delta u + u_x = \sin(\pi x) \sin(\pi y), \quad \text{in } \Omega = (0,1)^2,$$
$$u(0,y) = 0, \quad u(1,y) = 1, \, y \in [0,1], \quad \frac{\partial u(x,0)}{\partial n} = \frac{\partial u(x,1)}{\partial n} = 0, \, x \in [0,1] \quad (13)$$

In this case a regular layer in $x = 1$ and two parabolic layers in $y = 0$ and $y = 1$ appear in the solution. Thus, to construct the Shishkin mesh $\bar\Omega_N$, we take (see [8]) the transition parameters

$$\sigma_x = \min\{1/2, \sigma_{0,x}\varepsilon \log N\}, \quad \sigma_y = \min\{1/4, \sigma_{0,y}\sqrt\varepsilon \log N\}, \quad (14)$$

and we define a piecewise uniform mesh with $N/2 + 1$ points in $[0, 1 - \sigma_x]$ and $[1-\sigma_x, 1]$, $N/4+1$ points in $[0, \sigma_y]$ and $[1-\sigma_y, 1]$ and $N/2+1$ points in $[\sigma_y, 1-\sigma_y]$. Again, for ε large the mesh is uniform and otherwise we have two different step sizes for each space direction, given by $H_x = 2(1-\sigma_x)/N$, $h_x = 2\sigma_x/N$, $H_y = 2(1-2\sigma_y)/N$, $h_y = 4\sigma_y/N$. Considering the following sets of points

$$\begin{aligned}\Omega_{N,1} &= \{(x_i, y_j) \in \Omega : 0 \le i \le N/2\},\\ \Omega_{N,2} &= \{(x_i, y_j) \in \Omega : N/2 < i \le N\},\end{aligned} \quad (15)$$

the scheme that we use (see [2] for details of the construction) is given by

$$L_\varepsilon^N U_{i,j} \equiv r_{i,j}^1 U_{i-1,j} + r_{i,j}^2 U_{i+1,j} + r_{i,j}^3 U_{i,j-1} + r_{i,j}^4 U_{i,j+1} + r_{i,j}^5 U_{i,j} = Q_N(f_{i,j}),\\ (x_i, y_j) \in \Omega_N, \quad (16)$$

where $r_{i,j}^3$ and $r_{i,j}^4$ are

$$r_{i,j}^3 = \frac{-2\varepsilon}{(h_j^y + h_{j+1}^y)\, h_j^y}, \quad r_{i,j}^4 = \frac{-2\varepsilon}{(h_j^y + h_{j+1}^y)\, h_{j+1}^y}, \quad (x_i, y_j) \in \Omega_N$$

and the remaining coefficients are defined, depending on where the point is, as:

$$\begin{aligned}r_{i,j}^1 &= \frac{-2\varepsilon}{(h_i^x + h_{i+1}^x)\, h_i^x} - \frac{Q_N^1(a_{i,j} - b_{i,j} h_i^x/2)}{h_i^x}, \quad r_{i,j}^2 = \frac{-2\varepsilon}{(h_i^x + h_{i+1}^x)\, h_{i+1}^x},\\ r_{i,j}^5 &= -r_{i,j}^1 - r_{i,j}^2 - r_{i,j}^3 - r_{i,j}^4 + Q_N^1(b_{i,j}),\ Q_N(f_{i,j}) = Q_N^1(f_{i,j}),\ (x_i, y_j) \in \Omega_{N,1},\\ r_{i,j}^1 &= \frac{-2\varepsilon}{(h_i^x + h_{i+1}^x)\, h_i^x} - \frac{a_{i,j}}{h_i^x + h_{i+1}^x}, \quad r_{i,j}^2 = \frac{-2\varepsilon}{(h_i^x + h_{i+1}^x)\, h_{i+1}^x} + \frac{a_{i,j}}{h_i^x + h_{i+1}^x},\\ r_{i,j}^5 &= -r_{i,j}^1 - r_{i,j}^2 - r_{i,j}^3 - r_{i,j}^4 + b_{i,j},\ Q_N(f_{i,j}) = f_{i,j},\ (x_i, y_j) \in \Omega_{N,2}.\end{aligned}$$

This method is uniformly convergent with order $3/2$ for ε sufficiently small (see [2]). Now, we have six different restriction operators depending on where the point is localized in the mesh. Distinguishing the following sets of points

$$\begin{aligned}\Omega^{1-\sigma} &= \{(1-\sigma_x, 1-\sigma_y)\},\\ \Omega^\sigma &= \{(1-\sigma_x, \sigma_y)\},\\ \Omega^{x,1-\sigma} &= \{(x_i, 1-\sigma_y), i = 0, \ldots, N\} \setminus \Omega^{1-\sigma},\\ \Omega^{x,\sigma} &= \{(x_i, \sigma_y), i = 0, \ldots, N\} \setminus \Omega^\sigma,\\ \Omega^{1-\sigma, y} &= \{(1-\sigma_x, y_j), j = 0, \ldots, N\} \setminus \{\Omega^{1-\sigma} \cup \Omega^\sigma\},\\ \Omega^r &= \Omega \setminus \{\Omega^{x,1-\sigma} \cup \Omega^{x,\sigma} \cup \Omega^{1-\sigma,y} \cup \Omega^{1-\sigma} \cup \Omega^\sigma\},\end{aligned}$$

the operators are given by

$$R_\sigma = \frac{1}{2}\begin{bmatrix} 0 & (2-\sigma_x)\sigma_y & \sigma_x\sigma_y \\ 0 & (2-\sigma_x)(1-\sigma_y) & \sigma_x(1-\sigma_y) \\ 0 & 0 & 0 \end{bmatrix}, \quad \text{if } (x_i, y_j) \in \Omega^{1-\sigma}, \quad (17)$$

$$R_\sigma = \frac{1}{2}\begin{bmatrix} 0 & 0 & 0 \\ 0 & (2-\sigma_x)(1-\sigma_y) & \sigma_x(1-\sigma_y) \\ 0 & (2-\sigma_x)\sigma_y & \sigma_x\sigma_y \end{bmatrix}, \quad \text{if } (x_i, y_j) \in \Omega^\sigma, \quad (18)$$

$$R_\sigma = \frac{1}{4}\begin{bmatrix} \sigma_y & 2\sigma_y & \sigma_y \\ 1-\sigma_y & 2(1-\sigma_y) & 1-\sigma_y \\ 0 & 0 & 0 \end{bmatrix}, \quad \text{if } (x_i, y_j) \in \Omega^{x,1-\sigma}, \quad (19)$$

$$R_\sigma = \frac{1}{4} \begin{bmatrix} 0 & 0 & 0 \\ 1-\sigma_y & 2(1-\sigma_y) & 1-\sigma_y \\ \sigma_y & 2\sigma_y & \sigma_y \end{bmatrix}, \quad \text{if } (x_i, y_j) \in \Omega^{x,\sigma}, \tag{20}$$

by (11) if $(x_i, y_j) \in \Omega^{1-\sigma,y}$ and by (4) if $(x_i, y_j) \in \Omega^r$. Since we do not know the exact solution, we estimate the errors by $e_{i,j}^N = \max_{i,j} | u_{i,j}^N - u_{2i,2j}^{2N} |$, $i,j = 0, 1, \ldots N$, where U^{2N} is the approximation on the mesh $\bar{\Omega}^{2N} = \{(x_i, y_j), i, j = 0, 1, \ldots, 2N\}$ defined as

$$(x_{2i}, y_{2j}) = (x_i, y_j) \in \bar{\Omega}^N, \quad i, j = 0, 1, \ldots, N$$

$$(x_{2i+1}, y_{2j+1}) = (\frac{x_i + x_{i+1}}{2}, \frac{y_j + y_{j+1}}{2}), \quad i, j = 0, 1, \ldots, N-1,$$

and the numerical order of convergence, calculated using the double mesh principle (see [3]), is given by $p = \log(e_{i,j}^N / e_{i,j}^{2N}) / \log 2$. In Tables 2 and 3 we show the

Table 2. Error and convergence rates outside of layer regions

N	$\varepsilon = 10^{-4}$	$\varepsilon = 10^{-5}$	$\varepsilon = 10^{-6}$	$\varepsilon = 10^{-7}$	$\varepsilon = 10^{-8}$
32	3.052D-3 2.009	3.074D-3 2.000	3.076D-3 1.999	3.076D-3 1.999	3.076D-3 1.999
64	7.583D-4 2.020	7.683D-4 2.003	7.693D-4 2.002	7.694D-4 2.001	7.694D-4 2.001
128	1.869D-4 2.039	1.917D-4 2.005	1.921D-4 2.001	1.922D-4 2.001	1.922D-4 2.001
256	4.547D-5	4.775D-5	4.798D-5	4.800D-5	4.800D-5

maximum point errors and the corresponding rates of convergence of the finite difference scheme in two subdomains: $\Omega_{rl} = [0, 1-\sigma_x) \times (\sigma_y, 1-\sigma_y)$ (outside of layers regions) and $\Omega_{cl} = [1-\sigma_x, 1] \times [0, 1-\sigma_y]$ (in a corner layer). From these

Table 3. Error and convergence rates in a corner layer

N	$\varepsilon = 10^{-4}$	$\varepsilon = 10^{-5}$	$\varepsilon = 10^{-6}$	$\varepsilon = 10^{-7}$	$\varepsilon = 10^{-8}$
32	6.025D-3 1.498	5.548D-3 1.505	5.398D-3 1.507	5.350D-3 1.508	5.335D-3 1.509
64	2.133D-3 1.589	1.955D-3 1.561	1.899D-3 1.552	1.881D-3 1.548	1.875D-3 1.547
128	7.090D-4 1.646	6.625D-4 1.628	6.477D-4 1.621	6.431D-4 1.619	6.416D-4 1.619
256	2.266D-4	2.144D-4	2.105D-4	2.093D-4	2.089D-4

two tables, we deduce that the discretization scheme has order 2 in the subdomain Ω_{rl} while in the corner layer the order is approximately 1.5, according to the theoretical results (see [2]).

To see the efficiency of the new multigrid method, we compare the results with these ones obtained using the BI-CGSTAB method for the value $\varepsilon = 10^{-6}$. In Table 4 we show the iterations number and the CPU time of each one of these methods. ¿From these results, we see that the methods are comparable for meshes with few points, but when the number of points increase, the multigrid method does not increase the number of iterations. Also the CPU time increases linearly for multigrid and more rapidly for the BI-CGSTAB method.

Table 4. Number of iterations and CPU time for $\varepsilon = 10^{-6}$

N	32×32	64×64	128×128	256×256	512×512
BI-CGSTAB	9(0.19)	12(0.80)	21(4.85)	39(37.36)	123(403.64)
MULTIGRID	3(0.4)	3(1.44)	4 (7.49)	4 (29.88)	5 (148.26)

References

1. A. Brandt, Multi-level adaptive solutions to boundary-value problems, *Math. Comput.* **31**, 333-390 (1977).
2. C. Clavero, J. L. Gracia, F. Lisbona, G. I. Shishkin, A method of high order for convection-diffusion problems in domains with characteristic boundaries, *submitted* (2000).
3. P. A. Farrell, A. F. Hegarty, On the determination of the order of uniform convergence, in R. Wichnevetsky and J. J. H. Miller (Eds): Proceeding of the 13th IMACS World Congress for Computation and Applied Mathematics, IMACS, 501-502, (1991).
4. F. J. Gaspar, C. Clavero, F. Lisbona, Some numerical experiments with multigrid methods on Shishkin meshes, *submitted* (2000).
5. W. Hackbusch, Multi-grid methods and applications, Berlin:Springer-Verlag 1985.
6. P. W. Hemker, Multigrid methods for problems with a small parameter in the highest derivative, *Lecture Notes in Mathematics* **1066**,106–121 (1983).
7. T. Linß, M. Stynes, A hybrid difference scheme on a Shishkin mesh for linear convection-diffusion problems, *Appl. Numer. Math.* **31**, 255-270 (1999).
8. J. J. Miller, E. O'Riordan, G. I. Shishkin, Fitted numerical methods for singular perturbation problems. Error estimates in the maximun error for linear problems in one and two dimensions, Singapore: World Scientific, 1996.
9. H. G. Roos, A note on the conditioning of upwind schemes on Shishkin meshes. *IMA J. Numer. Anal.* **16**, 529-538 (1996).
10. G. I. Shishkin, Grid approximation of singularly perturbed boundary value problems with convective terms, *Sov. J. Numer. Anal. Math. Modelling* **5**, 173-187 (1990).
11. G. I. Shishkin, Grid approximation of singularly perturbed elliptic equations in domains with characteristics faces, *Sov. J. Numer. Anal. Math. Modelling* **5**, 327-343 (1990).

12. H. A. Van Der Vorst, BI-CGSTAB: A fast and smoolhly converging variant of BI-CG for the solution of nonsymetric linear systems, *SIAM J. Sci. Stat. Comput.* **3**,v. 2, 631-644 (1992).
13. P. Wesseling, An introduction to multigrid methods, Chichester: Wiley 1992.
14. P. M. De Zeeuw, Matrix-dependent prolongations and restrictions in a black-box multigrid solver, *Journal of Computational and Applied Mathematics* **3**, 1-27 (1990).

Recursive Version of LU Decomposition

K. Georgiev[1] and J. Waśniewski[2]

[1] Central Laboratory for Parallel Processing, Bulgarian Academy of Sciences
Acad. G. Bonchev, Bl. 25-A, 1113 Sofia, Bulgaria
georgiev@parallel.bas.bg
[2] Danish Computing Center for Research and Education,
Technical University of Denmark
Building 304, DK-2800 Lyngby, Denmark
jerzy.wasniewski@uni-c.dk

Abstract. The effective use of the cache memories of the processors is a key component of obtaining high performance algorithms and codes, including here algorithms and codes for parallel computers with shared and distributed memories. The recursive algorithms seem to be a tool for such an action. Unfortunately, worldwide used programming language FORTRAN 77 does not allow explicit recursion.
The paper presents a recursive version of LU factorization algorithm for general matrices using FORTRAN 90. FORTRAN 90 allows writing recursive procedures and the recursion is automatic as it is a duty of the compiler. Usually, recursion speeds up the algorithms. The recursive versions reported in the paper are some modification of the LAPACK algorithms and they transform some basic linear algebra operations from BLAS level 2 to BLAS level 3.

Keywords: numerical linear algebra, recursive algorithms, FORTRAN 90, LU factorization

AMS Subject Classifications: 65F05, 65Y10

1 Introduction

The data flow from the memory to the computational units is the most critical part in the problem of constructing high-speed algorithms. The functional units have to work very close to their peak capacity. The *registers* (very high-speed memory) communicate directly with a small, very fast **cache memory**. This memory is a form of storage that is automatically filled and emptied according to a fixed scheme defined by the hardware system. The cache memory is a buffer between the processor and the main memory. It is many times faster than the main memory. Therefore, the effective use of the cache memory is a key component in designing high-performance numerical algorithms [3]. One way for solving this problem is to use **recursive algorithms**. Unfortunately, the worldwide used programming language FORTRAN 77 does not allow explicit recursion and writing recursive algorithms using this language is a very difficult task. FORTRAN

90/95 support recursion as a language feature [6]. Recursion leads to automatic variable blocking for linear algebra problems with dense coefficient matrices [5]. The algorithms reported in this paper are some modifications of well known LAPACK algorithms [1] where BLAS level 2 version subroutines are transformed into level 3. The rest of the paper is organized as follows. Section 2 describes the recursive version of the LU factorization algorithm. In section 3 and Section4 the recursive versions of the subroutines for matrix-matrix multiplication and solving systems of linear equations with triangular coefficient matrices which are needed inside the LU recursive algorithm are presented.

2   Recursively Partitioned LU Factorization

The algorithm factors an $m \times n$ matrix A into an $m \times n$ lower trapezoidal matrix L (upper triangle part is all zeros) with $1's$ on the main diagonal and an $n \times n$ upper triangular matrix U in the case $m \geq n$ (Fig.1), and into an $m \times m$ lower triangular matrix L with entries $1's$ on the main diagonal and an $m \times n$ upper trapezoidal matrix U (lower triangular part is all zeros) in the case $m \leq n$ (Fig. 2).

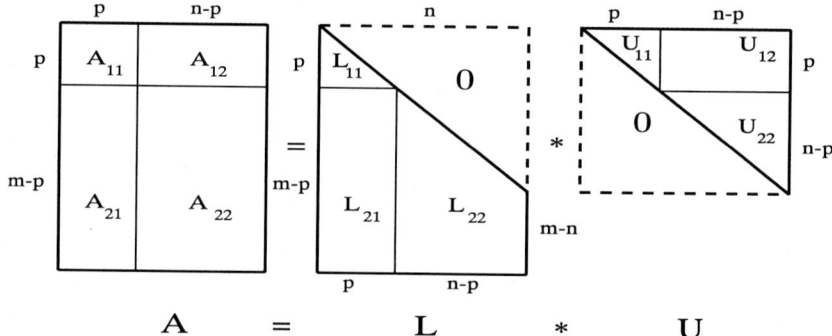

Fig. 1. Partitioning of the matrices in the case $m \geq n$

Let the matrix A be divided into four blocks (1) (see also Fig.1 and Fig.2) and $p = [\min(m, n)/2]$

$$\begin{pmatrix} A_{11} & A_{12} \\ A_{21} & A_{22} \end{pmatrix} = \begin{pmatrix} L_{11} & 0 \\ L_{21} & L_{22} \end{pmatrix} \begin{pmatrix} U_{11} & U_{12} \\ 0 & U_{22} \end{pmatrix} = \begin{pmatrix} L_{11}U_{11} & L_{11}U_{12} \\ L_{21}U_{11} & L_{21}U_{12} + L_{22}U_{22} \end{pmatrix} \quad (1)$$

In order to obtain the entries of the matrices L and U the following four subproblems have to be solved:

$$L_{11}U_{11} = A_{11} \qquad (2)$$

$$L_{11}U_{12} = A_{12} \qquad (3)$$

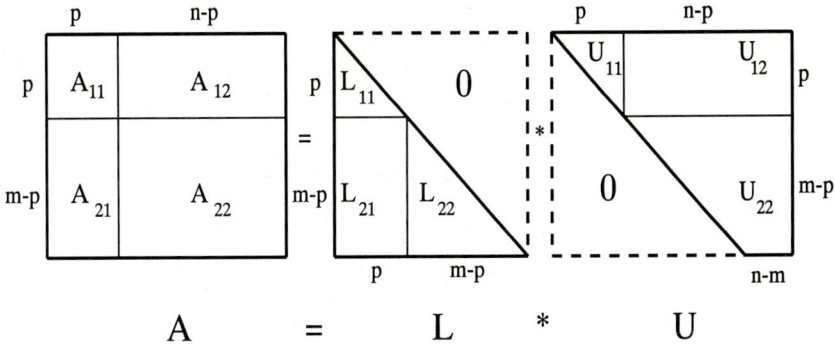

Fig. 2. Partitioning of the matrices in the case $m \leq n$

$$L_{21}U_{11} = A_{21} \implies U_{11}^T L_{21}^T = A_{21}^T \quad (4)$$

$$L_{21}U_{12} + L_{22}U_{22} = A_{22} \implies L_{22}U_{22} = A_{22} - L_{21}U_{12} \quad (5)$$

The sizes of the submatrices in the case $m \geq n$ are as follows: A_{11}, L_{11} and U_{11} are $p \times p$ matrices, A_{21} and L_{21} are $(m-p) \times p$ matrices, A_{12} and U_{12} are $p \times (n-p)$ matrices, A_{22} and L_{22} are $(m-p) \times (n-p)$ matrices and U_{22} is an $(n-p) \times (n-p)$ matrix. In the other case, $m \leq n$ the sizes of the submatrices are as follows: for A_{11}, L_{11} and U_{11} are $p \times p$ matrices, A_{21} and L_{21} are $(m-p) \times p$ matrices, A_{12} and U_{12} are $p \times (n-p)$ matrices, A_{22} and U_{22} are $(m-p) \times (n-p)$ matrices and L_{22} is an $(m-p) \times (m-p)$ matrix. There are standard LAPACK (GETRF) [1,4] and BLAS (TRSM, GEMM) [2] subroutines for solving these problems. Following the main idea, i.e. to go to the cache memory, recursive versions of them will be used here. The recursive algorithms for *matrix-matrix multiplications* and *solving systems of linear equations with triangular (lower or upper) coefficient matrices* will be described in the next sections. The corresponding recursive algorithms and subroutines are RGETRF, RTRSM and RGEMM, respectively. RGETRF is used for solving (2) and (5). RTRSM is used to problems (3) and (4) while RGEMM is used to obtain the right-hand side of (5). One can find bellow the high-level description of the recursive LU-factorization algorithm.

```
RECURSIVE SUBROUTINE RGETRF( A, IPIV, INFO )
! Use Statements:
   USE LA_PRECISION, ONLY: WP => DP
   USE LA_AUXMOD,    ONLY: ERINFO, LSAME
   USE F90_RCF,      ONLY: RLUGETRF => RGETRF, RTRSM, RGEMM
   USE F77_LAPACK,   ONLY: GETRF_F77 => LA_GETRF
! Purpose:
! RGETRF computes an LU factorization of a general M-by-N matrix
! A  using partial pivoting with row interchanges.
! The factorization has the form
!
!        A = P * L * U
!
! where P is a permutation matrix, L is lower triangular with unit
```

```
! diagonal elements (lower trapezoidal if M > N), and U is upper
! triangular (upper trapezoidal if M < N).
! This is the right-looking Level 3 BLAS version of the algorithm.
! Other subroutines used:  RGEMM, RTRSM, DLASWP
! Remark: The parameter N_CASH shows how many double precision
!         real numbers can be put in the "cache memory"

  M      = SIZE(A,1)
  N      = SIZE(A,2)
  MN_MIN = MIN(M,N)
  MEMORY = M*N
  IF( MEMORY <= N_CASH .OR. MN_MIN == 1) THEN
!    Call the standard Fortran'90 routine LA_DGETRF
  ELSE
    P = MN_MIN/2
    CALL RLUGETRF( A(1:P,1:P), IPIV=LIPIV, INFO=LINFO )
    MN_LOC = MIN(N-P,P)
    CALL DLASWP(N-P, A(1:P,P+1:N), P, 1, MN_LOC, LIPIV, 1)
    CALL RTRSM(A(1:P,1:P),A(1:P,P+1:N),UPLO='L',SIDE='L',DIAG='U')
    CALL RTRSM(A(1:P,1:P), A(P+1:M,1:P), UPLO='U', SIDE='R' )
    CALL RGEMM(A(P+1:M,1:P), A(1:P,P+1:N), A(P+1:M,P+1:N), &
               ALPHA=-ONE, CASH=N_CASH )
    CALL RLUGETRF(A(P+1:M,P+1:N), IPIV=LIPIV, INFO=LINFO )
    MN_LOC = MIN(P,M-P)
    CALL DLASWP(P, A(P+1:M,1:P), M-P, 1, MN_LOC, LIPIV, 1)
  ENDIF
END SUBROUTINE RGETRF
```

3 RGEMM: A Recursive Algorithm for Matrix-Matrix Multiplication

RGEMM is a recursive version of the BLAS routine GEMM. RGEMM performs in recursive way one of the following operations:

$$C := \alpha * op(A) * op(B) + \beta * C, \qquad (6)$$

where $op(X) = X$ or $op(X) = X^T$. Here, $op(A)$ is an $M \times K$ matrix, $op(B)$ is a $K \times N$ matrix and C is an $M \times N$ matrix, α and β are scalars. Since we can perform the following three types of actions:

$$C = \beta C + \alpha AB \qquad (7)$$
$$C = \beta C + \alpha AB^T \qquad (8)$$
$$C = \beta C + \alpha A^T B \qquad (9)$$

If the dimensions of the arrays is large enough to be put in the cache memory of the processor then we divide the matrices A, B and C into four by four blocks as follows. If $n_{min} = \min(m, k, n)$ then $p = [n_{min}/2]$ and in the case (7) of the above mentioned types of actions the dimensions of the blocks are:

$A_{11}(p \times p), A_{12}(p \times k - p), A_{21}(m - p \times p), \quad A_{22}(m - p \times k - p)$
$B_{11}(p \times p), B_{12}(p \times n - p), B_{21}(k - p \times p), \quad B_{22}(k - p \times n - p)$
$C_{11}(p \times p), C_{12}(p \times n - p), C_{21}(m - p \times n - p), C_{22}(m - p \times n - p).$

In the case (8) the dimensions are:

$A_{11}(p \times p), A_{12}(p \times k - p), A_{21}(m - p \times p), \quad A_{22}(m - p \times k - p)$
$B_{11}(p \times p), B_{12}(p \times k - p), B_{21}(n - p \times p), \quad B_{22}(n - p \times k - p)$
$C_{11}(p \times p), C_{12}(p \times n - p), C_{21}(m - p \times n - p), C_{22}(m - p \times n - p).$

And in the case (9) the dimensions are:

$A_{11}(p \times p), A_{12}(p \times m - p), A_{21}(k - p \times p), \quad A_{22}(k - p \times m - p)$
$B_{11}(p \times p), B_{12}(p \times n - p), B_{21}(k - p \times p), \quad B_{22}(k - p \times n - p)$
$C_{11}(p \times p), C_{12}(p \times n - p), C_{21}(m - p \times n - p), C_{22}(m - p \times n - p).$

It is well seen that these formulaes lead to eight new problems of the same type but with matrices with smaller dimensions. In the case (7) they are:

$C_{11} = \beta C_{11} + \alpha(A_{11}B_{11} + A_{12}B_{21}) \quad C_{12} = \beta C_{12} + \alpha(A_{11}B_{12} + A_{12}B_{22})$
$C_{21} = \beta C_{21} + \alpha(A_{21}B_{11} + A_{22}B_{21}) \quad C_{22} = \beta C_{22} + \alpha(A_{21}B_{12} + A_{22}B_{22})$

In the other two cases the formulaes are similar. Therefore, we have eight recursive calls to the same algorithm. When the size of the blocks becomes small enough then the standard Fortran 90 subroutine GEMM is used to solve the problem with matrices have being stored in the cache memory of the processor. One can find bellow the high-level description of the RGEMM.

```
RECURSIVE SUBROUTINE RGEMM( A, B, C, ALPHA, BETA, TRA, TRB, CASH)
!   Use Statements:
        USE F90_BLAS,      ONLY: LA_GEMM
        USE F90_RCF,       ONLY: RCFGEMM => RGEMM
!   Other parameters:
!       TRA and TRB - specify the operation to be performed
!       TRA = 'N' =>  op( A ) = A, TRA = 'T' =>  op( A ) = A'
!       TRB = 'N' =>  op( B ) = B, TRB = 'T' =>  op( B ) = B'
!   Other subroutines used:  DSYRK_90, RGEMM, ERINFO
        IF( LSAME(LTRA,'N') ) THEN
            M = SIZE(A,1);   K = SIZE(A,2)
        ELSE
            M = SIZE(A,2);   K = SIZE(A,1)
        ENDIF
        IF( LSAME(LTRB,'N') ) THEN
            N = SIZE(B,2)
        ELSE
            N = SIZE(B,1)
        ENDIF
```

```
      MEMORY = M*K + K*N + M*N;        N_MIN = MIN(M,N,K)
      IF( MEMORY <= CASH .OR. N_MIN == 1) THEN
!        Call the standard Fortran'90 routine LA_GEMM
         call LA_GEMM( A, B, C, TRA=LTRA,TRB=LTRB, ALPHA=LAL )
      ELSE
         P = N_MIN/2
         IF( LSAME(LTRA,'N') .AND. LSAME(LTRB,'N') ) THEN
call RCFGEMM(A(1:P,1:P),B(1:P,1:P),C(1:P,1:P), ...)
call RCFGEMM(A(1:P,P+1:K),B(P+1:K,1:P),C(1:P,1:P), ...)
call RCFGEMM(A(1:P,1:P),B(1:P,P+1:N),C(1:P,P+1:N), ...)
call RCFGEMM(A(1:P,P+1:K),B(P+1:K,P+1:N),C(1:P,P+1:N), ...)
call RCFGEMM(A(P+1:M,1:P),B(1:P,1:P),C(P+1:M,1:P), ...)
call RCFGEMM(A(P+1:M,P+1:K),B(P+1:K,1:P),C(P+1:M,1:P), ...)
call RCFGEMM(A(P+1:M,1:P),B(1:P,P+1:N),C(P+1:M,P+1:N), ...)
call RCFGEMM(A(P+1:M,P+1:K),B(P+1:K,P+1:N),C(P+1:M,P+1:N), ...)
         ENDIF
         IF( LSAME(LTRA,'N') .AND. LSAME(LTRB,'T')) THEN
call RCFGEMM(A(1:P,1:P),B(1:P,1:P),C(1:P,1:P), ...)
call RCFGEMM(A(1:P,P+1:K),B(1:P,P+1:K),C(1:P,1:P), ...)
call RCFGEMM(A(1:P,1:P),B(P+1:N,1:P),C(1:P,P+1:N), ...)
call RCFGEMM(A(1:P,P+1:K),B(P+1:N,P+1:K),C(1:P,P+1:N), ...)
call RCFGEMM(A(P+1:M,1:P),B(1:P,1:P),C(P+1:M,1:P), ...)
call RCFGEMM(A(P+1:M,P+1:K),B(1:P,P+1:K),C(P+1:M,1:P), ...)
call RCFGEMM(A(P+1:M,1:P),B(P+1:N,1:P),C(P+1:M,P+1:N), ...)
call RCFGEMM(A(P+1:M,P+1:K),B(P+1:N,P+1:K),C(P+1:M,P+1:N), ...)
         ENDIF
         IF( LSAME(LTRA,'T') .AND. LSAME(LTRB,'N')) THEN
call RCFGEMM(A(1:P,1:P),B(1:P,1:P),C(1:P,1:P), ...)
call RCFGEMM(A(P+1:K,1:P),B(P+1:K,1:P),C(1:P,1:P), ...)
call RCFGEMM(A(1:P,1:P),B(1:P,P+1:N),C(1:P,P+1:N), ...)
call RCFGEMM(A(P+1:K,1:P),B(P+1:K,P+1:N),C(1:P,P+1:N), ...)
call RCFGEMM(A(1:P,P+1:M),B(1:P,1:P),C(P+1:M,1:P), ...)
call RCFGEMM(A(P+1:K,P+1:M),B(P+1:K,1:P),C(P+1:M,1:P), ...)
call RCFGEMM(A(1:P,P+1:M),B(1:P,P+1:N),C(P+1:M,P+1:N), ...)
call RCFGEMM(A(P+1:K,P+1:M),B(P+1:K,P+1:N),C(P+1:M,P+1:N), ...)
         ENDIF
      ENDIF
   ENDIF
END SUBROUTINE RGEMM
```

4 RTRSM: A Recursive Algorithm for Solving Systems of Linear Equations with Triangular Coefficient Matrices

RTRSM is a recursive version of the BLAS routine TRSM. RTRSM solves systems of linear equations with triangular (lower or upper) coefficient matrices, i.e one of the following operations

$$op(A) X = \alpha B \quad \text{or} \quad X \, op(A) = \alpha B \qquad (10)$$

in recursive way, where $op(A)$ is an $m \times m$ triangular matrix ($op(A) = A$ or $op(A) = A^T$), α is a scalar, X and B are $m \times n$ matrices.

Let $p = \lceil m/2 \rceil$ and for simplicity to look only at the first possible operation in (10), i.e. $AX = \alpha B$. Then we divide the matrix A into four blocks: $A_{11}(1 : p, 1 : p), A_{12}(1 : p, m - p : m), A_{21}(p + 1 : m, 1 : p), A_{22}(p + 1 : m, p + 1 : m)$, the matrices X and B into two blocks: $X_1(1 : p, 1 : n), X_2(p + 1 : m, 1 : n)$ and $B_1(1 : p, 1 : n), B_2(p+1 : m, 1 : n)$. If A is a lower triangular matrix then A_{11} and A_{22} are lower triangular matrices too and $A_{12} = 0$. If A is an upper triangular matrix then A_{11} and A_{22} are upper triangular matrices too and $A_{21} = 0$. In both cases the block algorithm leads to two times using the same algorithm for solving systems with triangular coefficient matrices and ones using procedure for a matrix-matrix multiplication, i.e. using the recursive algorithm RGEMM (see Section 3.). If A is a lower triangular matrix then the block algorithm is:

$$A_{11}X_1 = B_1 \qquad \text{(RTRSM)}$$
$$B_2 - A_{21}X_1 \qquad \text{(RGEMM)}$$
$$A_{22}X_2 = B_2 - A_{21}X_1 \qquad \text{(RTRSM)},$$

and if A is an upper triangular matrix then:

$$A_{22}X_2 = B_2 \qquad \text{(RTRSM)}$$
$$B_1 - A_{12}X_2 \qquad \text{(RGEMM)}$$
$$A_{21}X_1 = B_1 - A_{12}X_2 \qquad \text{(RTRSM)}.$$

One can find bellow the high-level description of the RTRSM.

```
RECURSIVE SUBROUTINE RTRSM( A, B, ALPHA, UPLO, SIDE, TRANSA, DIAG)
!  Use Statements:
   USE F90_BLAS, ONLY: LA_GEMM, LA_TRSM
   USE F90_RCF, ONLY: RCFTRSM => RTRSM, RGEMM

   IF( LSAME(LUP,'U').AND.LSAME(LTRA,'N') .OR. &
       LSAME(LUP,'L').AND.LSAME(LTRA,'T') )THEN
         R1=P+1; R2=L; S1 = 1; S2 = P
   ELSE
         S1=P+1; S2=L; R1 = 1; R2 = P;
   ENDIF
   MEMORY = M*M + 2*M*N
```

```
  N_MIN = MIN(M,N)
  IF( MEMORY <= CASH .OR. N_MIN == 1) THEN
!    Call the standard Fortran'90 routine LA_TRSM
     CALL LA_TRSM( A, B, LAL, LUP, LSIDE, LTRA, LDIAG)
  ELSEIF( LSAME(LSIDE,'L') ) THEN
     CALL RCFTRSM(A(R1:R2,R1:R2),B(R1:R2,1:N),LAL,LUP,LSIDE,LTRA,
                  LDIAG)
     IF( LSAME(LTRA,'N') )THEN
CALL RGEMM(A(S1:S2,R1:R2),B(R1:R2,1:N),B(S1:S2,1:N),AL=-1.0, ...)
     ELSE
CALL RGEMM(A(R1:R2,S1:S2),B(R1:R2,1:N),B(S1:S2,1:N),AL=-1.0, ...)
     ENDIF
CALL RCFTRSM(A(S1:S2,S1:S2),B(S1:S2,1:N),LAL,LUP,LSIDE,LTRA,LDIAG)
  ELSE
CALL RCFTRSM(A(S1:S2,S1:S2),B(1:m,S1:S2),LAL,LUP,LSIDE,LTRA,LDIAG)
        IF( LSAME(LTRA,'N') )THEN
CALL RGEMM(B(1:M,S1:S2),A(S1:S2,R1:R2),B(1:M,R1:R2),AL=-1.0, ...)
        ELSE
CALL RGEMM(B(1:M,S1:S2),A(R1:R2,S1:S2),B(1:M,R1:R2),AL=-1.0, ...)
        ENDIF
CALL RCFTRSM(A(R1:R2,R1:R2),B(1:M,R1:R2),LAL,LUP,LSIDE,LTRA,LDIAG)
  ENDIF
END SUBROUTINE RTRSM
```

Acknowledgments

This research was partly supported by the UNI•C collaboration with the IBM T.J. Watson Research Center at Yorktown Heights and the Ministry of Science and Education of Bulgaria (Grant I-901/99).

References

1. Anderson, E., Bai, Z., et al: LAPACK Users' Guide Release 2.0. SIAM, Philadelphia (1995)
2. Dongara, J., DuCroz, J., Duff, I., Hammarling, S.: A Set of Level 3 Basic Linear Algebra Subprograms. ACm Trans. Math. Softw. **16** (1990) 1-17
3. Dongara, J., Duff, I., Sorensen, D., van der Vorst, H.: Numerical Linear Algebra for High-Performance Computers. SIAM, Philadelphia (1998)
4. Dongara, J., Waśnievski J.: High-Performance Linear Algebra Package - LAPACK90. Report UNIC-98-01, UNI•C, Lyngby, Denmark (1998)
5. Gustavson, F. G.: Recursion Leads to Automatic Variable Blocking for Dense Linear Algebra Algorithms. IBM J. of Research and Development, **41, No. 6** (1997) 737-755
6. Nyhoff, L., Leestma, S.: Introduction to Fortran 90 for Engineers and Scientists. Prentice Hall, New Jersey (1997)

Inversion of Symmetric Matrices in a New Block Packed Storage*

Gabriela Georgieva[1], Fred Gustavson[2], and Plamen Yalamov[1]

[1] Center of Applied Mathematics and Informatics, University of Rousse
7017 Rousse, Bulgaria
{ami94gmg,yalamov}@ami.ru.acad.bg
[2] IBM Watson Research Center
P.O. Box 218, Yorktown Heights, NY 10598, USA
gustav@watson.ibm.com

Abstract. In LAPACK we have two types of subroutines for solving problems with symmetric matrices: with full and packed storage. The performance of the full storage scheme is much better because it allows the usage of BLAS Level 2 and 3, while the memory requirements for the packed scheme are about twice less. Recently a new storage scheme was proposed which combines the advantages of both schemes: it has a performance similar to that of full storage, and the memory requirements are a little bit higher than for packed storage. In this paper we apply the scheme for inversion of symmetric indefinite matrices.

1 Introduction

Nowadays performance of numerical algorithms depends significantly on the computer architecture. Modern processors have a hierarchical memory which, if utilized appropriately, can bring to several times better performance.

One of the ways to use effectively the different levels of memory in the algorithms of numerical linear algebra is to introduce blocking in the algorithm. In this way effectively designed BLAS (Basic Linear Algebra Subroutines) [1995, p. 140] can be used, and improve the performance essentially. This is the approach accepted in LAPACK (Linear Algebra PACKage) [1995]. In many algorithms of LAPACK BLAS Level 3 (matrix-matrix operations) and Level 2 (matrix-vector operations) are used.

In this work we consider the inversion of matrix $A \in \mathcal{R}^{n \times n}$, where A is symmetric indefinite. The most popular algorithm for this problem uses the LDL^T decomposition of matrix A with Bunch-Kaufman pivoting [1996, §4.4], [1996, §10.4.2]. There are two types of subroutines in LAPACK implementing

---

* This research is supported by the UNI•C collaboration with the IBM T.J. Watson Research Center at Yorktown Heights. The last author was partially supported by Grant I-702/97 and Grant MM-707/97 from the Bulgarian Ministry of Education and Science.

this method. In the first one the matrix is stored in a two-dimensional array, and this is called full storage. For example, a 8 × 8 matrix is stored as follows:

```
1  *  *  *  *  *  *  *
2 12  *  *  *  *  *  *
3 13 23  *  *  *  *  *
4 14 24 34  *  *  *  *
5 15 25 35 45  *  *  *
6 16 26 36 46 56  *  *
7 17 27 37 47 57 67  *
8 18 28 38 48 58 68 78
*  *  *  *  *  *  *  *
*  *  *  *  *  *  *  *
```

where the entries denoted by a star are not referenced by the algorithm. The upper triangle of the matrix is not kept because it is symmetric. The last two unreferenced rows are added to illustrate that one can choose a leading dimension of the two-dimensional array in order to achieve a good level 1 cache utilization (in the above example the leading dimesion is 10, and the order of the matrix is 8). For simplicity, the entries are given integer values (the numbers of the places where the used elements are stored if the two-dimensional array is mapped to a one-dimensional array). This is enough to illustrate the two types of storage.

Practical problems can be very large, and in this case memory is an important issue. Clearly in full storage we use about twice more memory than necessary. Therefore, a second type of storage has been designed which is called packed storage. With this type of storage we keep only the essential part of the matrix needed for the computations in a one-dimensional array as follows:

```
1  *  *  *  *  *  *  *
2  9  *  *  *  *  *  *
3 10 16  *  *  *  *  *
4 11 17 22  *  *  *  *
5 12 18 23 27  *  *  *
6 13 19 24 28 31  *  *
7 14 20 25 29 32 34  *
8 15 21 26 30 33 35 36
```

Clearly, packed storage needs about twice less memory than full storage.

The disadvantage of full storage is that it uses more memory but its advantage is that it allows the usage of BLAS Level 3 and Level 2 calls which speeds up the computation essentially. For comparison, with packed storage we can use only BLAS Level 1. To illustrate this we present in Fig. 1 on the left performance results with random matrices for both types of storage. DSYTRI denotes the full storage code, and DSPTRI denotes the packed storage one. It is seen that the performance of DSYTRI is much better. Let us note that these experiments include the time for the LDL^T factorization of matrix A.

We also measured the pure time for inversion only (not including the factorization part). The results are given in Fig. 1 on the right. It can be seen that

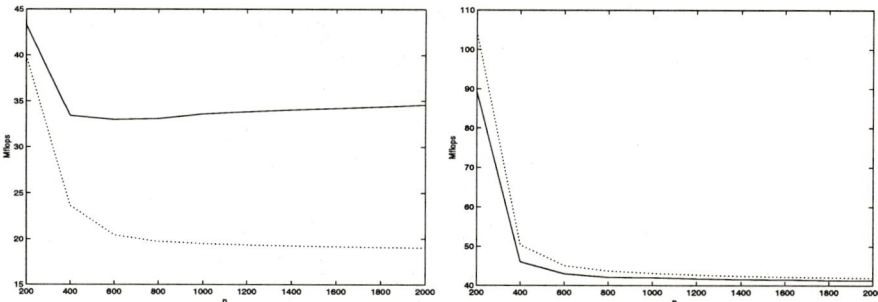

Fig. 1. Performance results for full (DSYTRI, solid line) and packed (DSPTRI, dotted line) storage from factorization plus inversion (left) and inversion only (right) for different sizes n of matrix A

the performance of the inversion part only is even better for the packed storage format. The explanation is that both inversion codes use BLAS Levels 1 and 2 only. So, the better performance of the whole algorithm comes from the usage of BLAS Level 3 in the factorization code _SYTRF.

In the present work we use the proposed in [1999] new type of packed storage. The columns of the matrix are divided into blocks. The blocks are kept in packed storage, i. e. the blocks are stored successively in the memory. Then several successive columns of the matrix are kept inside each block as if they were in full storage. The result of this storage is that it allows the usage of BLAS Level 3. Of course, we need slightly more memory than for the _SPTRI storage scheme but this memory is about 3-5% more on average for problems of practical interest. Thus the new storage scheme combines the two advantages of the storage formats in LAPACK, the smaller size of the memory in _SPTRI, and the better performance of _SYTRI.

Let us note that our storage scheme allows the usage of BLAS Level 3 in the inversion part of the algorithm. In _SPTRI and _SYTRI BLAS Levels 1 and 2 are used. This also improves the performance of the whole algorithm essentially.

The paper is organized as follows. In Sections 2 we present the so called overlapping scheme developed in [1999]. From [1999] one can see that this is the best scheme for problems involving symmetric indefinite matrices. In Section 3 the block inversion algorithm is presented. Finally, in Section 4 we illustrate our results by numerical tests.

2 The Block Rectangular Overlapping (BRO) Storage Scheme

We assume that the LDL^T factorization with the Bunch-Kaufman pivoting is used in the factorization part of the algorithm. Thus we have

$$PAP^T = LDL^T = WL^T, \text{ with } W = LD,$$

where P is a permutation matrix, L is unit lower triangular, and D is block diagonal with 1×1, or 2×2 blocks. More detailed descriptions can be found in [1996, §4.4], [1996, §10.4.2]. The idea in [1999] is similar to the LAPACK _SYTRF algorithm. The columns of matrix A are split into blocks, each block having n_b columns. For simplicity, we assume that n is a multiple of n_b. The results for the opposite case are the same.

We will illustrate the BRO scheme by an example. With the BRO scheme we would have

$$\begin{array}{|ccc|ccc|cc|}
\hline
1 & * & * & * & & & & \\
2 & 12 & * & * & & & & \\
3 & 13 & 23 & * & & & & \\
\hline
4 & 14 & 24 & 34 & * & * & * & \\
5 & 15 & 25 & 35 & 42 & * & * & \\
6 & 16 & 26 & 36 & 43 & 50 & * & \\
\hline
7 & 17 & 27 & 37 & 44 & 51 & 58 & * \\
8 & 18 & 28 & 38 & 45 & 52 & 59 & 61 \\
* & * & * & * & * & * & * & * \\
* & * & * & * & * & * & * & * \\
\hline
\end{array} \qquad (1)$$

The elements denoted by a star are not referenced by the algorithm and inserted to allow blocking and a choice of a good leading dimension. For simplicity the values of the entries show the order of the elements in the one-dimensional array, where we store the matrix. This blocking scheme leads to BLAS Level 3 only. When reaching the boundary between two blocks we can have two situations: 1×1, or 2×2 pivot. To better understand this let us consider the situation when reaching column 3 of the matrix we have a 2×2 pivot. We can see from (1) that $A(5:8, 1:4)$ is stored in such a way that it can be accessed by BLAS Level 3 without any problems. At the same time the block $A(6:8, 3:5)$ can be also accessed by BLAS Level 3 in case we have a 1×1 pivot at column 3. Thus, both situations are handled in a nice way.

The total memory we need (without taking into account the leading dimesnnion) is estimated in [1999]:

$$n(n+1)/2 + nn_b + n - n_b, \qquad (2)$$

which in practice is slightly larger than the memory necessary for the LAPACK packed storage.

3 Inversion of Symmetric Indefinite Matrices in BRO Storage

The inversion code is based on the following fact. Let us first ignore the permutations in the matrix, and assume that $A = LDL^T$ and that L and D are blocked as follows:

$$L = \begin{pmatrix} M & 0 \\ N & Q \end{pmatrix}, \quad D = \begin{pmatrix} D_1 & 0 \\ 0 & D_2 \end{pmatrix}.$$

where M and Q are unit lower triangular (not of equal size in general). Then for the inverse A^{-1} we have

$$A^{-1} = \begin{pmatrix} A_1^{-1} + W^T Y & Y^T \\ Y & A_2^{-1} \end{pmatrix},$$

where
$$A_1^{-1} = M^{-T} D_1^{-1} M^{-1}, \ A_2^{-1} = Q^{-T} D_2^{-1} Q^{-1},$$
$$W = N M^{-1}, \ Y = -A_2^{-1} W.$$

Finally, we apply the permutations stored in matrix P to rows and columns of matrix A^{-1}. For simlicity let us denote the permuted matrix by A^{-1} again.

Now assume that A_2^{-1} is already computed and stored in Q. Then the algorithm for computing A^{-1} is given below. In order to show how much memory we need we use the entries M, N, Q to store the results. We give also in brackets the corresponding BLAS or LAPACK routine. A working array W is introduced because it is necessary for the computation.

Step 1. Copy N to W (_COPY).
Step 2. Compute $W = W M^{-1}$ (_TRSM).
Step 3. Compute $N = Q W$ (_SYMM).
Step 4. Compute $M = M^{-1}$ (_SYTRI).
Step 5. Compute $M = M + W^T N$ (_GEMM).
Step 6. Apply permutations in P to A^{-1}.

This scheme is simple to implement. We have a few BLAS calls and one LAPACK call. Let us point out some of the advantages of this scheme:

– Mostly BLAS Level 3 calls are used (the only exception is the call to _SYTRI inside which Level 1 and 2 calls are used);
– We do not need additional memory for the working array W. The same storage is necessary for the factorization part, and we have already allocated this storage;
– In the present implementation _GEMM is used at Step 5, and the algorithm can be several times faster than _SYTRI, and _SPTRI (see the following section). But matrix M at Step 5 is symmetric, and a special routine can be written for this operation which takes into account symmetry. Thus the flop count for this operation only can be reduced about twice. Such a routine is not present in BLAS now, and we use _GEMM in the numerical experiments.
– During the factorization part instead of the diagonal matrix D we keep its inverse D^{-1}. The reason is that when solving a system of linear equations or inverting a matrix we need D^{-1} only. This improves slightly the overall performace as well.

We presented only one block step of the whole algorithm. Repeating this step recursively we get the whole algorithm. The advantages given above lead to a better performance which is illustrated in the next section.

4 Numerical Tests

The tests are done in Fortran 77 on an IBM 4-way SMP node with PowerPC 604e 332 MHz CPUs. The matrices are generated randomly with a generator which produces uniformly distributed in [0,1] numbers.

We compare the performance (in Mflops) of three algorithms: DSYTRI and DSPTRI from LAPACK, and DBSTRI (the algorithm with the packed BRO storage). In Figs. 2–3 we present tests on the SMP node with 1 CPU, and the SMP node with 4 CPUs, respectively. In Fig. 4 we show also the the speedup of all three algorithms on 4 processors. The results show that with the new packed storage scheme

- the performance is several times better than the performance of the LAPACK packed storage routine DSPTRI while using slightly more amount of memory, and the same number of flops;
- moreover, the performance is up to 2-3 times better than the performance of the LAPACK full storage routine DSYTRI which uses about twice more memory;
- the speedup of the studied in this paper algorithm is larger than the speedups of the two LAPACK routines. This means that our algorithm is better suited for parallel architectures.

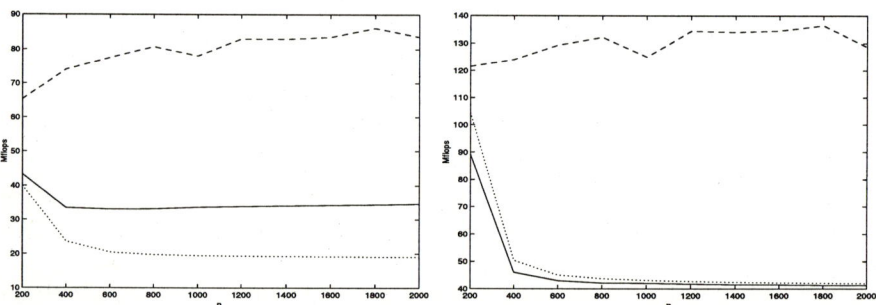

Fig. 2. Performance results for full (DSYSV, solid line), packed (DSPSV, dotted line), and BRO storage (dashed line) from factorization plus inversion (left) and inversion only (right) for different sizes n of matrix A on an IBM SMP node with 1 CPU

Finally, in Fig. 5 we show the memory requirements for different values of n. For this purpose we use the expressions

$$n^2 + n + nn_b, \quad n^2 + n$$

for full (_SYSV) and packed (_SPSV) storage, respectively, and (2) for the BRO storage. In _SYTRF there is also a buffer with n_b columns. Therefore, the term

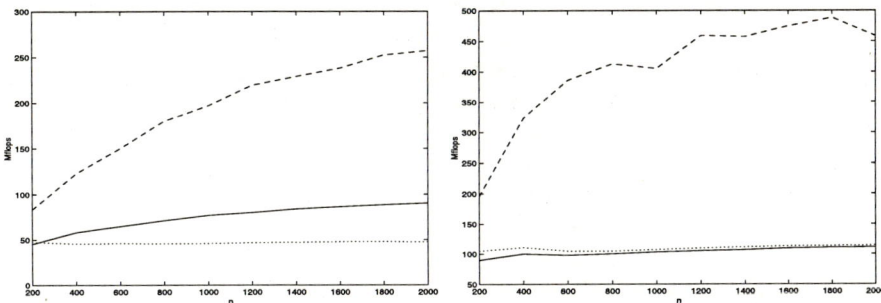

Fig. 3. Performance results for full (DSYSV, solid line), packed (DSPSV, dotted line), and BRO storage (dashed line) from factorization plus inversion (left) and inversion only (right) for different sizes n of matrix A on an IBM SMP node with 4 CPUs

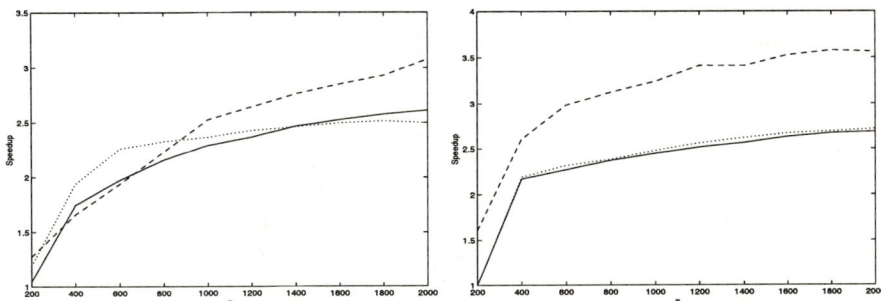

Fig. 4. Speedup of full (DSYSV, solid line), packed (DSPSV, dotted line), and BRO storage (dashed line) from factorization plus inversion (left) and inversion only (right) for different sizes n of matrix A on an IBM SMP node with 4 CPUs

nn_b appears. The value of n_b is chosen from practical experience, so that we have almost best performance. We see that the memory requirements for the BRO scheme are much closer to the _SPSV packed storage than to the _SYSV full storage.

References

1995. Anderson, E., Bai, Z., Bischof, C., Demmel, J., Dongarra, J., Du Croz, J., Greenbaum, A., Hammarling, S., McKenney, A., Ostrouchov, S., Sorensen, D.: LAPACK Users' Guide Release 2.0, SIAM, Philadelphia, 1995
1996. Golub, G. H., Van Loan, C. F.: Matrix Computations, 3rd edition, The John Hopkins University Press, Baltimore, 1996

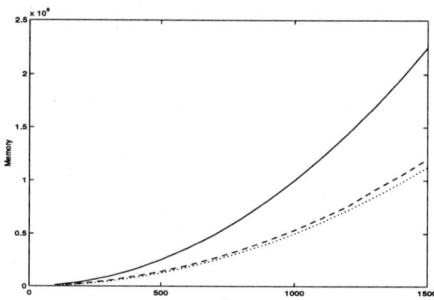

Fig. 5. Memory requirements for full (DSYSV, solid line), packed (DSPSV, dotted line), and BRO storage (dashed line)

1999. Gustavson, F., Karaivanov, A., Wasniewski, J., Yalamov, P.: The efficiency of a new packed storage for symmetric indefinite matrices, 1999 (manuscript)
1996. Higham, N. J.: Accuracy and Stability of Numerical Algorithms, SIAM, Philadelphia, 1996

The Stability Boundary of Certain Two-Layer and Three-Layer Difference Schemes

Alexei V. Goolin

M. V. Lomonosov Moscow State University, Russia

Abstract. The stability with respect to initial data of difference schemes with operator weights is investigated in the frameworks of the general stability theory of operator-difference schemes. The stability is defined as the existence of a selfadjoint positive operator which determines the time-nonincreasing norm of the difference solution. The norm-independent stability criterions are obtained in the form of operator inequalities. The notation of stability boundary in the plane of two grid parameters is introduced for multi-parameter difference schemes which approximate two-dimensional parabolic and hyperbolic differential equations. The noniterative numerical algorithm is suggested for the construction of the stability boundaries of difference schemes with variable weighting factors. The approach is based on finding the smallest eigenvalue of an auxiliary selfadjoint eigenvalue problem.

1 Introduction

The stability theory for time-dependent operator-difference schemes in Hilbert spaces has been suggested originally by A. A. Samarskii [1,2] and was developed in numerous papers (see, e.g., [3]—[6] and references therein).

The stability is referred as the existence of a selfadjoint positive operator that determines the norm in the grid space nonincreasing on the solution of difference problem. This norm is connected as a rule, with the difference schemes under question. Such obtained stability conditions are criterions for stability in this prescribed norm, but it turns out being only rough sufficient conditions for stability in other norms. The set of the so-called symmetrizable difference schemes was singled out in [7,8] for which the theorems about norm-independent necessary and sufficient stability conditions were obtained.

In the present paper the two-layer and symmetrical three-layer difference schemes with operator weight multipliers are studied. We will consider operator equations

$$\frac{y_{n+1} - y_n}{\tau} + \sigma A y_{n+1} + (I - \sigma) A y_n = 0, \quad n = 0, 1, \ldots, \quad y_0 \text{ specified}, \quad (1)$$

$$\frac{y_{n+1} - 2y_n + y_{n-1}}{\tau^2} + \sigma A y_{n+1} + (I - 2\sigma) A y_n + \sigma A y_{n-1} = 0, \quad (2)$$

$n = 1, 2, \ldots,$ y_0 and y_1 specified, where $y_n = y(t_n) \in H$ is an unknown element of the Euclidean space H, $t_n = n\tau$, $\tau > 0$, A and σ are linear operators in H, I is the identity operator. We suppose that operators A and σ are n-independent. Let (y,v) be an inner product in H, $\|y\| = \sqrt{(y,y)}$ is the norm in H and $D: H \to H$ is a selfadjoint positive operator. Denote by H_D the space H with norm $\|y\|_D = \sqrt{(Dy,y)}$.

The difference scheme (1) is called *stable in the space* H_D (or in the norm D), if inequalities $\|y_{n+1}\|_D \le \|y_n\|_D$ hold for the solution of the problem (1) for arbitrary $y_0 \in H$ and $n = 0, 1, \ldots$

The stability definition for the three-layer difference scheme (2) is similar.

2   Stability Theorems

The theory of difference schemes with operator weight multipliers has its own features that are expounded in detail in the book [5].

We quote here the theorem on necessary and sufficient stability conditions of scheme (1) obtained in [7]. Preliminarily we represent the scheme (1) in the canonical form

$$B\frac{y_{n+1} - y_n}{\tau} + Ay_n = 0, \quad n = 0, 1, \ldots, \quad y_0 \text{ specified,} \qquad (3)$$

where $B = I + \tau\sigma A$.

Theorem 1. *Suppose that $A^* = A$, $\sigma^* = \sigma$ and the operators A^{-1} and $B^{-1} = (I + \tau\sigma A)^{-1}$ exist. If the scheme (1) is stable in a space H_D, then the operator inequality*

$$A + \tau A\mu A \ge 0 \qquad (4)$$

holds, where $\mu = \sigma - 0{,}5I$. Conversely, if (4) is fulfilled, then the difference scheme (1) is stable in H_{A^2}.

Proof. Multiplying (3) by the operator A we have the equivalent equation

$$\tilde{B}\frac{y_{n+1} - y_n}{\tau} + \tilde{A}y_n = 0, \quad n = 0, 1, \ldots, \quad y_0 \text{ specified,} \qquad (5)$$

where $\tilde{A} = A^2$ and $\tilde{B} = I + \tau A\sigma A$ are selfadjoint operators, $\tilde{A} > 0$. We have the following theorem (see [4]).

Suppose that the operators $\tilde{A}$ and $\tilde{B}$ are selfadjoint and do not depend on n, and the operator $\tilde{A}$ is positive. If scheme (5) is stable in a space H_D, then these operators satisfy the inequality

$$\tilde{B} \ge 0.5\tau\tilde{A}. \qquad (6)$$

Conversely, if (6) is satisfied, then scheme (5) is stable in H_{A^2}.

In our case the inequality (6) has the form

$$I + \tau A\sigma A \ge 0.5\tau A^2,$$

and coincides with (4).

A similar result is valid for the symmetrical three-layer scheme (2).

Theorem 2. *Let $A^* = A$, $\sigma^* = \sigma$ and suppose that A^{-1} and $(I + \tau^2 \sigma A)^{-1}$ exist. If the scheme (2) is stable in a space H_D, then the operator inequality*

$$\tau^2 A + (\tau^2 A)\mu(\tau^2 A) \geq 0, \tag{7}$$

holds, where $\mu = \sigma - 0,25I$. Conversely, if the inequality

$$\tau^2 A + (\tau^2 A)\mu(\tau^2 A) > 0, \tag{8}$$

is fulfilled, then the difference scheme (2) is stable in the norm

$$\|y_n\|_D = \left(\left\| \frac{y_n + y_{n-1}}{2} \right\|_{A^2}^2 + \left\| \frac{y_n - y_{n-1}}{\tau} \right\|_{A+\tau^2 A\mu A}^2 \right)^{1/2}.$$

Proof. By (2) we have the equivalent equation

$$A \frac{y_{n+1} - 2y_n + y_{n-1}}{\tau^2} + A\sigma A y_{n+1} + A(I - 2\sigma) A y_n + A\sigma A y_{n-1} = 0,$$

which can be represented in the canonical form

$$\tau^2 \tilde{R} y_{\bar{t}t,n} + \tilde{A} y_n = 0, \tag{9}$$

where $y_{\bar{t}t,n} = (y_{n+1} - 2y_n + y_{n-1})/\tau^2$, $\tilde{A} = A^2$ and $\tilde{R} = \tau^{-2} A + A\sigma A$. By a general stability theorem (A. Samarskii [2]), the condition (8) is sufficient for the stability of the scheme (2) in the norm above mentioned. Let us prove the necessity of condition (7). Let us represent the three-layer scheme (9) as an equivalent two-layer scheme $Y_{n+1} = SY_n$, where $Y_n = (y_{n-1} \ y_n)^T$, and $S = (S_{\alpha\beta})$ is the matrix with elements $S_{11} = 0$, $S_{12} = I$, $S_{21} = -I$, $S_{22} = 2I - \tilde{R}^{-1}\tilde{A}$. Let s be an arbitrary eigenvalue of the matrix S and $y = (y^{(1)} \ y^{(2)})^T$ is the corresponding eigenvector. The eigenvalue problem $Sy = sy$ can be reduced to the quadratic problem $((1-s)^2 \tilde{R} + s\tilde{A}) y^{(1)} = 0$ or, more in detail,

$$((1-s)^2 (\tau^{-2} A + A\sigma A) + sA^2) y^{(1)} = 0. \tag{10}$$

After transforming the equation (10) in the form

$$A((1-s)^2 (\tau^{-2} A^{-1} + \sigma) + sI) A y^{(1)} = 0$$

and denoting $x^{(1)} = Ay^{(1)}$, we have the eigenvalue problem

$$(\tau^{-2} A^{-1} + \sigma) x^{(1)} = \lambda x^{(1)},$$

where $\lambda = -s/(1-s)^2$. Note, that $s = 1$ is not the eigenvalue of (10), by the assumption that A^{-1} exists. Further, λ is a real number because $\tau^{-2} A^{-1} + \sigma$ is the selfadjoint operator. If the scheme (9) is stable in a space, then $|s| \leq 1$ and, consequently, $\lambda \geq 0,25$. Fulfilment of inequalities $\lambda_k \geq 0,25$ for all eigenvalues λ_k of the selfadjoint operator $\tau^{-2} A^{-1} + \sigma$ is equivalent to the operator inequality

$$\tau^{-2} A^{-1} + \sigma \geq 0,25 I,$$

which is equivalent to (7).

3 Stability Boundaries of Two-Dimensional Difference Schemes

Before the general discussion we shall treat the following example. Let us consider the first boundary-value problem for the heat conduction equation

$$\frac{\partial u}{\partial t} = \frac{\partial^2 u}{\partial x_1^2} + \frac{\partial^2 u}{\partial x_2^2} \qquad (11)$$

in the rectangle $0 < x_1 < l_1$, $0 < x_2 < l_2$ with zero boundary conditions and arbitrary initial condition.

Let us introduce the timestep $\tau > 0$ and the spacesteps h_1 and h_2 in the directions x_1 and x_2 respectively. Let us denote

$$x_1^{(i)} = ih_1, \quad i = 0, 1, \ldots, N_1, \quad h_1 N_1 = l_1,$$

$$x_2^{(j)} = jh_2, \quad j = 0, 1, \ldots, N_2, \quad h_2 N_2 = l_2,$$

$$t_n = n\tau, \quad n = 0, 1, \ldots, \quad \tau > 0, \quad y_{ij}^n = y(x_1^{(i)}, x_2^{(j)}, t_n),$$

$$y_{\bar{x}_1 x_1, ij}^n = \frac{y_{i-1j}^n - 2y_{ij}^n + y_{i+1j}^n}{h_1^2}, \quad y_{\bar{x}_2 x_2, ij}^n = \frac{y_{ij-1}^n - 2y_{ij}^n + y_{ij+1}^n}{h_2^2},$$

$$\Delta_h y_{ij}^n = y_{\bar{x}_1 x_1, ij}^n + y_{\bar{x}_2 x_2, ij}^n,$$

$$y_{t,ij}^n = \frac{y_{ij}^{n+1} - y_{ij}^n}{\tau}, \quad y_{\bar{t},ij}^n = \frac{y_{ij}^n - y_{ij}^{n-1}}{\tau}, \quad y_{\bar{t}t,ij}^n = \frac{y_{ij}^{n+1} - 2y_{ij}^n + y_{ij}^{n-1}}{\tau^2}.$$

Let σ be a real number and suppose that $l_1 = l_2 = 1$. Let us approximate the original differential problem by the difference one,

$$y_{t,ij}^n = \sigma \Delta_h y_{ij}^{n+1} + (1-\sigma)\Delta_h y_{ij}^n, \qquad (12)$$

$$i = 1, 2, \ldots, N_1 - 1, \quad j = 1, 2, \ldots, N_2 - 1, \quad n = 0, 1, \ldots,$$

$$y_{ij}^n|_{\Gamma_h} = 0, \quad y_{ij}^0 = u_0(x_1^{(i)}, x_2^{(j)}),$$

where Γ_h is the grid boundary and $u_0(x_1^{(i)}, x_2^{(j)})$ is the specified initial value.

It is well known and it follows from Theorem 1 that the inequality

$$\sigma \geq \frac{1}{2} - \frac{1}{\tau \lambda_{\max}}, \qquad (13)$$

is the necessary and sufficient condition for the stability of the difference scheme (12) with respect to initial data. Here

$$\lambda_{\max} = \frac{4}{h_1^2} \cos^2 \frac{\pi}{2N_1} + \frac{4}{h_2^2} \cos^2 \frac{\pi}{2N_2}$$

is the largest eigenvalue of the five-point difference Laplace operator Δ_h. The stability condition (13) can be rewritten as follows:

$$4\mu(\gamma_1 a_1 + \gamma_2 a_2) + 1 \geq 0,$$

where $\mu = \sigma - 0,5$ and

$$\gamma_1 = \frac{\tau}{h_1^2}, \quad \gamma_2 = \frac{\tau}{h_2^2}, \quad a_1 = \cos^2 \frac{\pi}{2N_1}, \quad a_2 = \cos^2 \frac{\pi}{2N_2}.$$

If $\mu \geq 0$, then the scheme (12) is stable for all γ_1 and γ_2. In case $\mu < 0$, the stability condition assume the form

$$\frac{\gamma_1}{\gamma_{1,0}} + \frac{\gamma_2}{\gamma_{2,0}} \leq 1, \tag{14}$$

where

$$\gamma_{1,0} = \frac{1}{4(0,5-\sigma)a_1}, \quad \gamma_{2,0} = \frac{1}{4(0,5-\sigma)a_2}. \tag{15}$$

Thus, in the first quadrant of the $\gamma_1 O \gamma_2$-plane the stability boundary exists, which constitutes the segment of straight line

$$\frac{\gamma_1}{\gamma_{1,0}} + \frac{\gamma_2}{\gamma_{2,0}} = 1. \tag{16}$$

The scheme (12) is unstable in every point (γ_1, γ_2) of the first quadrant, which is situated above this straight line, and it is stable in each underlying point. One can see from (15) that the stability domain enlarges when σ increases, and it coincides with the entire first quadrant for $\sigma = 0,5$.

It follows from expressions for a_1 and a_2 that the stability boundary weakly depends on N_1 and N_2.

4 Numerical Construction of Stability Boundaries for Difference Schemes with Variable Weights

The theorems above mentioned enable us to carried out the numerical stability investigation of difference schemes which approximate the mathematical physics problems and to construct the stability boundaries for such schemes.

Let us consider the problem of numerical construction of stability boundary for the multiparameter family of difference schemes

$$\frac{y_{ij}^{n+1} - y_{ij}^n}{\tau} = \sigma_{ij} \Lambda y_{ij}^{n+1} + (1 - \sigma_{ij}) \Lambda y_{ij}^n, \tag{17}$$

$$i = 1, 2, \ldots, N_1 - 1, \quad j = 1, 2, \ldots, N_2 - 1, \quad n = 0, 1, \ldots,$$

$$y_{ij}^n|_{\Gamma_h} = 0, \quad y_{ij}^0 = u_0(x_1^{(i)}, x_2^{(j)}).$$

Here σ_{ij} are given real numbers (weights) and Λ is a five-point difference operator, namely $\Lambda = \Lambda_1 + \Lambda_2$, where

$$(\Lambda_1 y)^n_{ij} = \frac{1}{h_1}\left(a_{i+1j}\frac{y^n_{i+1j} - y^n_{ij}}{h_1} - a_{ij}\frac{y^n_{ij} - y^n_{i-1j}}{h_1}\right),$$

$$(\Lambda_2 y)^n_{ij} = \frac{1}{h_2}\left(b_{ij+1}\frac{y^n_{ij+1} - y^n_{ij}}{h_2} - b_{ij}\frac{y^n_{ij} - y^n_{ij-1}}{h_2}\right).$$

We suppose that $a_{ij} \geq c_1 > 0$, $b_{ij} \geq c_2 > 0$ for all i, j. The equation (17) depends on two grid parameter, $\gamma_1 = \tau/h_1^2$ and $\gamma_2 = \tau/h_2^2$, which determines along with $\sigma = (\sigma_{ij})$ whether the difference scheme is stable or not.

Let the set $\sigma = (\sigma_{ij})$ be settled. We assume further that weighting factors σ_{ij} are independent of γ_1 and γ_2. We shall accept the following terminology. The difference scheme (17) is said to be *stable in the point* (γ_1, γ_2) of $\gamma_1 O \gamma_2$-plane, if it is stable for these values γ_1 and γ_2. We assume further that $\gamma_1 \geq 0$, $\gamma_2 \geq 0$. Let us call by *stability domain* of the difference scheme (17) the set of all points in the first quadrant where the scheme is stable. Similarly *instability domain* consists of all the other points of the first quadrant.

The curve in the first quadrant $\gamma_1 \geq 0$, $\gamma_2 \geq 0$, which separates stability and instability domains is said to be *the stability boundary* of the difference scheme (17). The scheme is referred to as absolutely stable if it stable for all points $\gamma_1 \geq 0$, $\gamma_2 \geq 0$. Thus, the stability boundary does not exist for absolutely stable schemes.

For the three-layer difference scheme

$$\frac{y^{n+1}_{ij} - 2y^n_{ij} + y^{n-1}_{ij}}{\tau^2} = \sigma_{ij}\Lambda y^{n+1}_{ij} + (1 - 2\sigma_{ij})\Lambda y^n_{ij} + \sigma_{ij}\Lambda y^{n-1}_{ij} \qquad (18)$$

the notation of stability boundary is introduced the same way, but here $\gamma_1 = \tau^2/h_1^2$ and $\gamma_2 = \tau^2/h_2^2$.

On the basis of the method of numerical construction of stability boundaries we will put the transferring from Cartesian coordinate (γ_1, γ_2) to the polar one (r, φ) and the search of the point of stability boundary on the ray $\varphi = \mathrm{const}$. We first consider the two-layer difference scheme (17). It can be represented in the form (1), where $A = -\Lambda$. It is well known that τA is a symmetrical and positively defined matrix. By theorem 1 the scheme (17) is stable if and only if all the eigenvalues of *stability matrix* $P = (\tau A) + (\tau A)\mu(\tau A)$ are nonnegative. The sought parameters γ_1 and γ_2 are included only in matrix τA and are not contained in matrix μ. The matrix τA has the form

$$\tau A = \gamma_1 A_1 + \gamma_2 A_2, \qquad (19)$$

where $\gamma_1 = \tau/h_1^2$, $\gamma_2 = \tau/h_2^2$, $A_1 = -h_1^2\Lambda_1$, $A_2 = -h_2^2\Lambda_2$. It is important to note that A_1 and A_2 are independent of the grid parameters γ_1, γ_2.

Setting $\gamma_1 = r\cos\varphi$, $\gamma_2 = r\sin\varphi$, we have from (19), that $\tau A = rA_\varphi$, where

$$A_\varphi = A_1 \cos\varphi + A_2 \sin\varphi. \qquad (20)$$

Thus, the sought parameter r is included in the matrix τA as a numerical factor, and the matrix A_φ is independent of r. Note that matrix A_φ differed from τA only by a positive factor and therefore A_φ is a symmetrical positively definite matrix. The stability matrix $P = (\tau A) + (\tau A)\mu(\tau A)$ can be written in the form $P = rP_\varphi$, where

$$P_\varphi = A_\varphi + rA_\varphi \mu A_\varphi. \tag{21}$$

The search of stability boundary is based on the following characteristics of the matrix P_φ.

Lemma 1. *The difference scheme (17) is stable in the point $\gamma_1 = r\cos\varphi$, $\gamma_2 = r\sin\varphi$ if and only if all the eigenvalues of matrix P_φ are nonnegative.*

Let us consider the generalized eigenvalue problem

$$A_\varphi \mu A_\varphi x = \lambda A_\varphi x. \tag{22}$$

It turned out, that the property of having fixed sign for the spectrum of the problem (22) is independent of φ.

Lemma 2. *The smallest eigenvalue $\lambda_{\min}$ of (22) is nonnegative if and only if $\mu_{ij} \geq 0$ for all i and j.*

Corollary 1. *If all $\sigma_{ij} \geq 0,5$, then the scheme (17) is absolutely stable.*

In the next statement the coordinates of the point of stability boundary is indicated, which is situated on the ray $\gamma_2 = \gamma_1 \tan\varphi$.

Theorem 3. *Let at least one weighting factor $\sigma_{ij} < 0,5$. Then for all φ the smallest eigenvalue $\lambda_{\min}(\varphi)$ of the problem (22) is negative. The point $\gamma_1 = r\cos\varphi$, $\gamma_2 = r\sin\varphi$ is situated on the stability boundary of difference scheme (17) if and only if $r = -1/\lambda_{\min}(\varphi)$.*

Corollary 2. *If at least one weighting factor $\sigma_{ij} < 0,5$, then the stability boundary $r = r(\varphi)$ exists, which is one-valued function of φ.*

Thus, for constructing a stability boundary of scheme (17) in the case, when at least one weighting factor is less than 0,5, it is sufficient for all $\varphi \in [0, \pi/2]$ to find the smallest eigenvalue $\lambda_{\min} = \lambda_{\min}(\varphi)$ of the problem (22) and to set

$$r = -1/\lambda_{\min}(\varphi), \quad \gamma_1 = r\cos\varphi, \quad \gamma_2 = r\sin\varphi.$$

In the case of three-layer scheme (18) all statements formulated above change slightly. It is necessary only to replace the condition $\sigma_{ij} \geq 0,5$ by $\sigma_{ij} \geq 0,25$ and the condition $\sigma_{ij} < 0,5$ by $\sigma_{ij} < 0,25$. Besides, in the case of three-layer scheme we have $\gamma_1 = \tau^2/h_1^2$ and $\gamma_2 = \tau^2/h_2^2$. The numerical algorithm for stability boundary constructing does not change.

5 Support Points and the Basic Straight Line

The computations, which have been performed in [9], demonstrate, that as a rule, the stability boundary is slightly differed from the segment of a straight line, determinated by parameters σ_{ij}. This straight line crosses the stability boundary in the points $(\gamma_{10}, 0)$ and $(0, \gamma_{20})$. Such a line we call the *basic straight line*, and the points $(\gamma_{10}, 0)$ and $(0, \gamma_{20})$ the *support points*.

It is possible to construct support points according to the 2-dimensional algorithm above described. But in this case the algorithm can be essentially simplified and reduced to find the stability boundary of one-dimensional problems. Let us consider for example how to find the point $(\gamma_{10}, 0)$. In this case the polar angle $\varphi = 0$ and matrices $A_\varphi = A_1$ and $A_\varphi \mu A_\varphi$, including in the main equation (22), assume a block diagonal form, namely $A_1 \mu A_1 = \mathrm{diag}[M_1, M_2, \ldots, M_{N_2-1}]$, where M_j are symmetrical matrices of order $N_1 - 1$. Therefore the spectrum of the problem (22) is the union of the spectra of $N_2 - 1$ matrices of order $N_1 - 1$. Respectively, the j's diagonal block P_j of stability matrix (21) contains only elements of the j's line

$$\sigma^{(j)} = (\sigma_{1j}, \sigma_{2j}, \ldots, \sigma_{N_1-1\,j})$$

of matrix σ.

Thus, the smallest eigenvalue of matrix P_j determines the stability boundary $\gamma_1(j)$ of an one-dimensional problem with weights distribution $\sigma^{(j)}$. The stability boundary of the two-dimensional problem can be found as follows:

$$\gamma_{10} = \min_{1 \leq j \leq N_2 - 1} \gamma_1(j).$$

So, the minimal number from all $\gamma_1(j)$ corresponds to the worst one-dimensional variant along lines (notations of the optimal and the worst variants were introduced in [10]). It follows that the support point $(\gamma_{10}, 0)$ is independent of lines permutation in the matrix σ. In exactly the same way, the support point $(0, \gamma_{20})$ is determined by the worst variant along columns of the matrix σ and is independent of columns permutation.

Supported by the Russian Foundation for Basic Research (grant No 99-01-00958).

References

1. Samarskii, A. A.: Regularisation of difference schemes. USSR Comput.Math.and Math.Phys.**7**(1967) 79-120 (in Russian)
2. Samarskii, A. A.: Classes of stable schemes. USSR Comput.Math.and Math.Phys.**7**(1967) 171-223 (in Russian)
3. Samarskii, A. A., Goolin A. V.: Stability of difference schemes. Moscow, Nauka. 1973 (in Russian)
4. Gulin A. V., Samarskii A. A.: On some results and problems of the stability theory of difference schemes. Math.USSR Sbornic **28**(1976) 263–290

5. Samarskii A. A., Vabishchevich P. N., Matus P. P.: Difference schemes with operator's factors. Minsk. 1998 (in Russian)
6. Samarskii A. A., Vabishchevich P. N., Matus P. P.: Stability of operator-difference schemes. Differentsial'nye Uravnenia **35**(1999) 152–187 (in Russian)
7. Gulin A. V., Samarskii A. A.: On the stability of a class of difference schemes. Differentsial'nye Uravnenia **29**(1993) 1163–1173 (in Russian)
8. Gulin A. V.: On the stability theory of symmetrizable difference schemes. Matem. modelirovanye **6**(1993),no. 6, 9–13 (in Russian)
9. Gulin A. V., Yukhno L. F.: Numerical investigations of the stability of two-layer difference schemes for the two-dimensional heat-conduction equation. Comp. Maths. Math. Phys. **36** (1996) 1079–1085 (in Russian)
10. Gulin A. V., Gulin V. A.: Stability boundaries for difference schemes with variable weights. Izvestia vusov. Matem. (1994), no. 9(388), 28–39 (in Russian)

High Order ε-Uniform Methods for Singularly Perturbed Reaction-Diffusion Problems*

J. L. Gracia, F. Lisbona, and C. Clavero

Departamento de Matemática Aplicada. Universidad de Zaragoza, Zaragoza, Spain
{jlgracia,lisbona,clavero}@posta.unizar.es

Abstract. The central difference scheme for reaction-diffusion problems, when fitted Shishkin type meshes are used, gives uniformly convergent methods of almost second order. In this work, we construct HOC (High Order Compact) compact monotone finite difference schemes, defined on *a priori* Shishkin meshes, uniformly convergent with respect the diffusion parameter ε, which have order three and four except for a logarithmic factor. We show some numerical experiments which support the theoretical results.

1 Introduction

In this paper we are interested in the construction of high order compact schemes to solve problems of type

$$L_\varepsilon[u] \equiv -\varepsilon u'' + b(x)u = f(x), \quad 0 < x < 1, \quad u(0) = u_0, \quad u(1) = u_1, \qquad (1)$$

where $0 < \varepsilon \leq 1$, u_0 and u_1 are given constants and we suppose that b and f are sufficiently smooth functions with $b(x) \geq \beta > 0$, $x \in [0,1]$. The exact solution of (1) satisfies $|u^{(k)}(x)| \leq C\varepsilon^{-k/2}\left(1 + e(x,x,\beta,\varepsilon)\right)$, where $e(x,y,\beta,\varepsilon) = \exp\left(-\sqrt{\beta/\varepsilon}\,x\right) + \exp\left(-\sqrt{\beta/\varepsilon}\,(1-y)\right)$. For sufficiently small ε, classical methods on uniform meshes only work for very large number of mesh points. Nevertheless, if these methods are defined on special fitted meshes, the convergence to the exact solution is uniform in ε [3,4]. Shishkin meshes, [6,7], are simple piecewise uniform meshes of this kind, frequently used for singularly perturbed problems. For the reaction-diffusion problems considered, the corresponding Shishkin mesh, $I_N = \{0 = x_0, \ldots, x_N = 1\}$, is defined as follows. Let $\sigma = \min\left\{\frac{1}{4}, \sigma_0\sqrt{\varepsilon}\log N\right\}$, where σ_0 is a constant to be chosen later. Dividing $[0,1]$ into three intervals $[0, \sigma]$, $[1-\sigma, 1]$ and $[\sigma, 1-\sigma]$, we take a piecewise uniform mesh with $N/4$, $N/4$ and $N/2$ subintervals respectively. We denote by $h_j = x_j - x_{j-1}$, $j = 1, \ldots, N$ and by $h^{(1)} = 2(1-2\sigma)/N$, $h^{(2)} = 4\sigma/N$. In the sequel we will suppose that $\sigma = \sigma_0\sqrt{\varepsilon}\log N$ (otherwise classical analysis proves the convergence of the methods).

The finite difference schemes that we present, are modifications of the central difference scheme, by incorporating some compact difference approximations of

---

* This research was supported by the project DGES-PB97-1013

some terms of its local truncation error. In [5,8] this procedure was used to construct high order schemes for non singularly perturbed problems.

Henceforth, we denote by c, C any positive constant independent of ε and the mesh.

2 A Uniform Convergent Scheme with Order 3

Central difference scheme is given by

$$L^1_{\varepsilon,N}[Z_j] \equiv -\varepsilon D^+ D^- Z_j + b_j Z_j = f_j, \ 1 \leq j \leq N-1, \ Z_0 = u_0, \ Z_N = u_1, \quad (2)$$

where $D^+ D^- Z_j$ is the second order central difference discretization on non uniform meshes. Its local truncation error satisfies

$$T^1_{j,u} \equiv L^1_{\varepsilon,N}[u(x_j,\varepsilon) - Z_j] = \frac{-\varepsilon(h_{j+1} - h_j)}{3} u'''_j - \frac{2\varepsilon(h^3_{j+1} + h^3_j)}{4!(h_j + h_{j+1})} u^{(4)}_j + O(N^{-3}). \quad (3)$$

Therefore, to construct a third order accurate (possibly non uniform) compact scheme, we must to find compact approximations of the first two terms of (3) with order $O(N^{-3})$. Deriving equation (1), we have

$$-\varepsilon u''' = f' - b'u - bu', \qquad -\varepsilon u^{(4)} = f'' - b''u - 2b'u' - bu'', \quad (4)$$

and therefore

$$-\varepsilon u^{(4)} = f'' - b''u - 2b'u' + \frac{bf - b^2 u}{\varepsilon}. \quad (5)$$

Thus, to obtain third order approximations of both u''' and $u^{(4)}$, approximations of the first derivative with order $O(N^{-2})$ are required. For the first term of (3), clearly we must only analyze the transition points σ and $1 - \sigma$. We consider

$$u'_j \approx \delta_{j,N/4} \left(D^- u_j + \frac{h_j}{2} u''_j \right) + \delta_{j,3N/4} \left(D^+ u_j - \frac{h_{j+1}}{2} u''_j \right), \ j = N/4, 3N/4,$$

where $\delta_{jl} = 1$ if $j = l$, $\delta_{jl} = 0$ if $j \neq l$. Using the differential equation, we deduce

$$u'_j \approx \delta_{j,N/4} \left(D^- u_j + \frac{h_j(b_j u_j - f_j)}{2\varepsilon} \right) + \delta_{j,3N/4} \left(D^+ u_j - \frac{h_{j+1}(b_j u_j - f_j)}{2\varepsilon} \right). \quad (6)$$

For the second term of (3), we approximate the first derivative depending of the sign of b'. Thus, we take

$$u'_j \approx \text{sgn } b'_j \left(D^- u_j + \frac{h_j}{2} u''_j \right) + (1 - \text{sgn } b'_j) \left(D^+ u_j - \frac{h_{j+1}}{2} u''_j \right),$$

where sgn $z_j = 1$, if $z_j \geq 0$ and sgn $z_j = 0$, if $z_j < 0$. Using again the differential equation, we use the following approximation

$$u'_j \approx \text{sgn } b'_j \left(D^- u_j + \frac{h_j(b_j u_j - f_j)}{2\varepsilon} \right) + (1 - \text{sgn } b'_j) \left(D^+ u_j - \frac{h_{j+1}(b_j u_j - f_j)}{2\varepsilon} \right). \quad (7)$$

Therefore, we have obtained suitable approximations of the first two terms of (3) depending on U_{j-1}, U_j, U_{j+1} and linear combinations of the data. The incorporation of these approximations to central difference scheme, gives the following scheme:

$$L^2_{\varepsilon,N}[U_j] = Q^2_N(f_j), \quad 1 \le j \le N-1, \quad U_0 = u_0, \quad U_N = u_1, \tag{8}$$

where

$$\begin{aligned} L^2_{\varepsilon,N}[U_j] &\equiv -\varepsilon D^+D^-U_j + \frac{h_{j+1}-h_j}{3}b_j(\delta_{j,N/4}D^-U_j + \delta_{j,3N/4}D^+U_j) + \\ &+ \frac{h_{j+1}^3+h_j^3}{6(h_j+h_{j+1})}b'_j(\operatorname{sgn} b'_j D^-U_j + (1-\operatorname{sgn} b'_j)D^+U_j) + Q^2_N(b_j)U_j, \end{aligned} \tag{9}$$

and

$$\begin{aligned} Q^2_N(z_j) &\equiv z_j + \frac{h_{j+1}-h_j}{3}\left(z'_j + \frac{b_j z_j}{2\varepsilon}(\delta_{j,N/4}h_j - \delta_{j,3N/4}h_{j+1})\right) + \\ &+ \frac{h_{j+1}^3+h_j^3}{12(h_j+h_{j+1})}\left(z''_j + \frac{b_j z_j}{\varepsilon} + \frac{b'_j z_j}{\varepsilon}(h_j \operatorname{sgn} b'_j - (1-\operatorname{sgn} b'_j)h_{j+1})\right). \end{aligned}$$

The local truncation error for this scheme is given by

$$\begin{aligned} \tau^2_{j,u} &= -\varepsilon\left(\frac{2(h_{j+1}^4-h_j^4)}{5!(h_j+h_{j+1})}u_j^{(5)} + \frac{2}{h_j+h_{j+1}}\left(\frac{R_5(x_j,x_{j+1},u)}{h_{j+1}} + \frac{R_5(x_j,x_{j-1},u)}{h_j}\right)\right) + \\ &+ \frac{h_{j+1}-h_j}{3}b_j\left(\delta_{j,N/4}\frac{R_2(x_j,x_{j-1},u)}{h_j} + \delta_{j,3N/4}\frac{R_2(x_j,x_{j+1},u)}{h_{j+1}}\right) + \\ &+ \frac{h_{j+1}^3+h_j^3}{12(h_j+h_{j+1})}b'_j\left(\operatorname{sgn} b'_j\frac{R_2(x_j,x_{j-1},u)}{h_j} + (1-\operatorname{sgn} b'_j)\frac{R_2(x_j,x_{j+1},u)}{h_{j+1}}\right), \end{aligned} \tag{10}$$

where R_n is the remainder of Taylor formula.

Proposition 1. *We suppose that $b'_j \ge 0$ (similar bounds can be obtained for $b'_j \le 0$). Let $d_j = (h_j^3 + h_{j+1}^3)/((h_j + h_{j+1})\varepsilon)$. Then,*

$$|\tau^2_{j,u}| \le \begin{cases} CN^{-4}\sigma_0^4 \log^4 N, & \text{for } 1 \le j < N/4 \text{ and } 3N/4 < j < N, \\ C\left(N^{-4} + N^{-\sqrt{\beta\sigma_0}}\right), & \text{for } N/4 < j < 3N/4, \\ C\left(N^{-3}\sigma_0^2\varepsilon \log^2 N + d_{N/4}N^{-\sqrt{\beta\sigma_0}}\right), & \text{for } j = N/4, \\ C\left(N^{-3}\sigma_0^2\varepsilon \log^2 N + N^{-4}|b'_{3N/4}| + d_{3N/4}N^{-\sqrt{\beta\sigma_0}}\right), & \text{for } j = 3N/4. \end{cases}$$

Proof. We distinguish several cases depending on the localization of the mesh point. For $x_j \in (0,\sigma) \cup (1-\sigma,1)$, from (10) we deduce that

$$|\tau^2_{j,u}| \le C\left[h^{(2)^4} + h^{(2)^4}\varepsilon^{-2}(e(x_j, x_{j+1}, \beta, \varepsilon) + e(x_{j-1}, x_j, \beta, \varepsilon))\right].$$

Using that $h^{(2)} = 4N^{-1}\sigma_0\sqrt{\varepsilon}\log N$ and bounding the exponential functions by constants, we prove that $|\tau^2_{j,u}| \le CN^{-4}\sigma_0^4 \log^4 N$, $1 \le j < N/4$ and $3N/4 < j \le N-1$.

For $x_j \in (\sigma, 1-\sigma)$ we distinguish two cases. First, if ${h^{(1)}}^2 < \varepsilon$, we easily obtain that

$$|\tau^2_{j,u}| \leq C\left[{h^{(1)}}^4 + {h^{(1)}}^4 \varepsilon^{-2}(e(x_j, x_{j+1}, \beta, \varepsilon) + e(x_{j-1}, x_j, \beta, \varepsilon))\right]$$
$$\leq N^{-4} + N^{-\sqrt{\beta}\sigma_0}.$$

Secondly, if ${h^{(1)}}^2 \geq \varepsilon$, we can prove that

$$|\tau^2_{j,u}| \leq C\left({h^{(1)}}^4 + \int_{x_j}^{x_{j+1}} (x_{j+1} - \xi)^3 \varepsilon^{-2} e(\xi, \xi, \beta, \varepsilon) d\xi + \int_{x_{j-1}}^{x_j} (\xi - x_{j-1})^3 \varepsilon^{-2} e(\xi, \xi, \beta, \varepsilon) d\xi\right).$$

Integrating by parts, we deduce

$$\int_{x_j}^{x_{j+1}} (x_{j+1} - \xi)^3 \varepsilon^{-2} e(\xi, \xi, \beta, \varepsilon) d\xi \leq C\left((h^{(1)} \varepsilon^{-1/2})^3 e(x_j, x_j, \beta, \varepsilon) + \int_{x_j}^{x_{j+1}} \varepsilon^{-1/2} e(\xi, \xi, \beta, \varepsilon) d\xi\right) \leq C\left(e(x_j, x_j, \beta, \varepsilon) + e(x_j, x_{j+1}, \beta, \varepsilon)\right)$$
$$\leq CN^{-\sqrt{\beta}\sigma_0}.$$

A bound for the other integral follows in the same form. Using that $h^{(1)} < 2N^{-1}$, we deduce that $|\tau^2_{j,u}| \leq C(N^{-4} + N^{-\sqrt{\beta}\sigma_0})$. Finally, we analyze the error for the transition point $x_{N/4} = \sigma$ (similarly for $x_{3N/4} = 1 - \sigma$). If ${h^{(1)}}^2/(2\varepsilon) < ({h^{(1)}}^3 + {h^{(2)}}^3)/((h^{(1)} + h^{(2)})\varepsilon) < 1$, then

$$|\tau^2_{N/4,u}| \leq C(h^{(1)} {h^{(2)}}^2 + {h^{(1)}}^3 \varepsilon^{-3/2} e(x_{N/4}, x_{N/4+1}, \beta, \varepsilon) +$$
$$+ h^{(1)} {h^{(2)}}^2 \varepsilon^{-3/2} e(x_{N/4-1}, x_{N/4}, \beta, \varepsilon)) \leq C(N^{-3} \varepsilon \sigma_0^2 \log^2 N + N^{-\sqrt{\beta}\sigma_0}).$$

On the other hand, if $1 \leq ({h^{(2)}}^3 + {h^{(1)}}^3)/((h^{(2)} + h^{(1)})\varepsilon) < 2{h^{(1)}}^2/\varepsilon$, we have

$$|\tau^2_{N/4,u}| \leq C\left(h^{(1)} {h^{(2)}}^2 + d_{N/4}\left(\int_{x_{N/4}}^{x_{N/4+1}} \varepsilon^{-1/2} e(\xi, \xi, \beta, \varepsilon) d\xi +\right.\right.$$
$$+ \int_{x_{N/4-1}}^{x_{N/4}} \varepsilon^{-1/2} e(\xi, \xi, \beta, \varepsilon) d\xi\bigg)\bigg) \leq C\left(h^{(1)} {h^{(2)}}^2 + d_{N/4}\left(e(x_{N/4}, x_{N/4+1}, \beta, \varepsilon) + e(x_{N/4-1}, x_{N/4}, \beta, \varepsilon)\right)\right) \leq C(N^{-3} \sigma_0^2 \varepsilon \log^2 N + d_{N/4} N^{-\sqrt{\beta}\sigma_0}).$$

Proposition 2. *Let* $L^2_{\varepsilon,N}[U_j] \equiv r_j^- U_{j-1} + r_j^c U_j + r_j^+ U_{j+1}$. *For N sufficiently large, there exists a constant c such that*

$$0 < c \max\{1, d_j\} \leq r_j^- + r_j^c + r_j^+, \quad r_j^- < 0, \quad r_j^+ < 0. \tag{11}$$

Proof. Expression (9) can be written in the form of the statement with

$$r_j^- = \frac{-2\varepsilon}{(h_j + h_{j+1})h_j} - \delta_{j,N/4}\frac{(h_{j+1} - h_j)b_j}{3h_j} - \frac{(h_{j+1}^3 + h_j^3)b_j'}{6(h_j + h_{j+1})h_j}\operatorname{sgn} b_j', \qquad (12)$$

$$r_j^+ = \frac{-2\varepsilon}{(h_j + h_{j+1})h_{j+1}} + \delta_{j,3N/4}\frac{(h_{j+1} - h_j)b_j}{3h_{j+1}} + \frac{(h_{j+1}^3 + h_j^3)b_j'}{6(h_j + h_{j+1})h_j}(1 - \operatorname{sgn} b_j'), \qquad (13)$$

$$r_j^c = -r_j^- - r_j^+ + Q_N^2(b_j). \qquad (14)$$

Using (12) and (13) it follows readily that $r_j^- < 0$ and $r_j^+ < 0$. By (14), $0 < c\max\{1,d_j\} \leq r_j^- + r_j^c + r_j^+$ is equivalent to $0 < c\max\{1,d_j\} \leq Q_N^2(b_j)$. Since

$$Q_N^2(b_j) > b_j + \frac{h_{j+1} - h_j}{3}b_j' + d_j\frac{b_j''\varepsilon + b_j}{12} > c\max\{1,d_j\},$$

for N sufficiently large, the result holds.

Corollary 1. *The operator (8) is of positive type and therefore it satisfies the maximum discrete principle.*

Theorem 1. *Let $u(x,\varepsilon)$ be the solution of (1) and $\{U_j;\ 0 \leq j \leq N\}$ the solution of the scheme (8). Then*

$$|u(x_j,\varepsilon) - U_j| \leq C\left(N^{-4}\sigma_0^4 \log^4 N + N^{-3}\sigma_0^2\varepsilon \log^2 N + N^{-\sqrt{\beta\sigma_0}}\right), \quad 0 \leq j \leq N.$$

Proof. Defining $\phi_j = C(N^{-4}\sigma_0^4 \log^4 N + N^{-3}\sigma_0^2\varepsilon \log^2 N + N^{-\sqrt{\beta\sigma_0}})$ with C sufficiently large, using that $r_j^- + r_j^c + r_j^+ \geq c\max\{1,d_j\}$ and the maximum principle, the result follows.

3 A Uniform Convergent Scheme with Order 4

From Proposition 1, if $\sqrt{\beta\sigma_0} \geq 4$ the scheme (8) is accurate of order 4, except for the transition points when $1 > (h^{(1)^3} + h^{(2)^3})/(\varepsilon(h^{(1)} + h^{(2)}))$. Then, we must only modify the scheme in this case. To do that, we write the local truncation error of the central difference scheme in the form

$$\tau_{j,u}^1 = -\varepsilon\left[\frac{h_{j+1} - h_j}{3}u_j''' + \frac{2(h_{j+1}^3 + h_j^3)}{4!(h_j + h_{j+1})}u_j^{(4)} + \frac{2(h_{j+1}^4 - h_j^4)}{5!(h_j + h_{j+1})}u^{(5)}\right] + O(N^{-4}).$$

Thus, we must find approximations with order $O(N^{-4})$ of the terms into brackets. For the first term, an approximation of the first derivative is needed. We consider

$$u_j' \approx \delta_{j,N/4}\left(D^- u_j + \frac{h_j(b_ju_j - f_j)}{2\varepsilon} - \frac{h_j^2}{3!}u_j'''\right) + $$
$$+ \delta_{j,3N/4}\left(D^+ u_j - \frac{h_{j+1}(b_ju_j - f_j)}{2\varepsilon} - \frac{h_{j+1}^2}{3!}u_j'''\right). \qquad (15)$$

Combining (15) with (4), we obtain

$$
\begin{aligned}
u'_j \approx \delta_{j,N/4} &\left(D^- u_j + \frac{h_j(b_j u_j - f_j)}{2\varepsilon} - \frac{h_j^2}{3!}\frac{b_j D^+ u_j + b'_j u_j - f'_j}{\varepsilon} \right) + \\
+ \delta_{j,3N/4} &\left(D^+ u_j - \frac{h_{j+1}(b_j u_j - f_j)}{2\varepsilon} - \frac{h_{j+1}^2}{3!}\frac{b_j D^- u_j + b'_j u_j - f'_j}{\varepsilon} \right).
\end{aligned}
\tag{16}
$$

For the second term, we consider the same approximations that for the scheme $L^2_{\varepsilon,N}$. For the last term, deriving (5) and using (1), we can deduce

$$
-\varepsilon u^{(5)} = \frac{\varepsilon(f''' - b'''u - 3b''u') - 4bb'u - b^2 u' + 3b'f + bf'}{\varepsilon}.
\tag{17}
$$

Therefore, it is sufficient to find $O(N^{-1})$ approximations of $b''_j u'_j$ and $b^2_j u'_j/\varepsilon$. We propose the following ones:

$$
\begin{aligned}
b''_j u'_j &\approx b''_j \left(\delta_{j,N/4}(\operatorname{sgn} b''_j\ D^- u_j + (1 - \operatorname{sgn} b''_j)D^+ u_j) + \delta_{j,3N/4}(\operatorname{sgn} b''_j\ D^+ u_j + \right. \\
&\quad \left. + (1 - \operatorname{sgn} b''_j)D^- u_j) \right), \\
\frac{b_j^2 u'_j}{\varepsilon} &\approx \frac{b_j^2}{\varepsilon} \left(\delta_{j,N/4}\left(D^- u_j + \frac{h_j(b_j u_j - f_j)}{2\varepsilon} \right) + \delta_{j,3N/4}\left(D^+ u_j - \frac{h_{j+1}(b_j u_j - f_j)}{2\varepsilon} \right) \right).
\end{aligned}
$$

Therefore, in the transition points the scheme is $L^3_{\varepsilon,N}[U_j] \equiv r^-_j U_{j-1} + r^c_j U_j + r^+_j U_{j+1} = Q^3_N(f_j)$, $j = N/4, 3N/4$, where

$$
\begin{aligned}
Q^3_N(z_j) &\equiv z_j + \frac{h^3_{j+1} + h^3_j}{12(h_j + h_{j+1})}\left[z''_j + \frac{b_j z_j}{\varepsilon} + \frac{b'_j z_j}{\varepsilon}(h_j \operatorname{sgn} b'_j - (1 - \operatorname{sgn} b'_j)h_{j+1}) \right] + \\
&+ \frac{h_{j+1} - h_j}{3}\left[z'_j + \frac{b_j z_j}{2\varepsilon}(h_j \delta_{j,N/4} - h_{j+1}\delta_{j,3N/4}) - \frac{b_j z'_j}{3!\varepsilon}(\delta_{j,N/4}h^2_j + \delta_{j,3N/4}h^2_{j+1}) \right] + \\
&+ \frac{2(h^4_{j+1} - h^4_j)}{5!(h_j + h_{j+1})}\left[z'''_j + \frac{3b'_j z_j}{\varepsilon} + \frac{b_j z'_j}{\varepsilon} + \frac{b^2_j z_j}{2\varepsilon^2}(\delta_{j,N/4}h_j - h_{j+1}\delta_{3N/4,j}) \right],
\end{aligned}
$$

$$
\begin{aligned}
r^-_j &= \frac{-2\varepsilon}{h_j(h_j + h_{j+1})} + \frac{h_{j+1} - h_j}{3}\left(-\frac{\delta_{j,N/4}b_j}{h_j} + \frac{\delta_{j,3N/4}h^2_{j+1}b^2_j}{3!\varepsilon h_j} \right) - \\
&\quad - \frac{\operatorname{sgn} b'_j (h^3_j + h^3_{j+1})b'_j}{6h_j(h_j + h_{j+1})} - \\
&\quad - \frac{2(h^4_{j+1} - h^4_j)}{5!(h_j + h_{j+1})}\left(\frac{3b''_j \operatorname{sgn} b''_j\ \delta_{j,N/4}}{h_j} + \frac{3b''_j(1 - \operatorname{sgn} b''_j)\delta_{j,3N/4}}{h_j} + \frac{\delta_{j,N/4}b^2_j}{\varepsilon h_j} \right),
\end{aligned}
\tag{18}
$$

$$
\begin{aligned}
r^+_j &= \frac{-2\varepsilon}{h_{j+1}(h_j + h_{j+1})} + \frac{h_{j+1} - h_j}{3}\left(\frac{\delta_{j,3N/4}b_j}{h_{j+1}} - \frac{\delta_{j,N/4}h^2_j b^2_j}{3!\varepsilon h_{j+1}} \right) + \\
&\quad + \frac{(1 - \operatorname{sgn} b'_j)(h^3_j + h^3_{j+1})b'_j}{6h_{j+1}(h_j + h_{j+1})} + \\
&\quad + \frac{2(h^4_{j+1} - h^4_j)}{5!(h_j + h_{j+1})}\left(\frac{3b''_j(1 - \operatorname{sgn} b''_j)\delta_{j,N/4}}{h_{j+1}} + \frac{3b''_j \operatorname{sgn} b''_j\ \delta_{j,3N/4}}{h_{j+1}} + \frac{\delta_{j,3N/4}b^2_j}{\varepsilon h_{j+1}} \right),
\end{aligned}
\tag{19}
$$

$$
r^c_j = -r^-_j - r^+_j + Q^3_N(b_j).
\tag{20}
$$

Proposition 3. *For N sufficiently large, there exists c such that*

$$0 < c\max\{1, d_j\} \le r_j^- + r_j^c + r_j^+, \qquad r_j^- < 0, \qquad r_j^+ < 0. \tag{21}$$

Proof. We only must study the cases where the new scheme is different of (8). From (18) and (19) it is obvious that $r_j^- < 0$ and $r_j^+ < 0$. Using that

$$Q_N^3(b_j) > b_j + \frac{h_{j+1} - h_j}{3}\left(b_j' - \frac{b_j b_j'}{3!\varepsilon}(\delta_{j,N/4}h_j^2 + \delta_{j,3N/4}h_{j+1}^2)\right) +$$
$$+ \frac{h_j^3 + h_{j+1}^3}{12(h_j + h_{j+1})}b_j'' + \frac{2(h_{j+1}^4 - h_j^4)}{5!(h_j + h_{j+1})}\left(b_j''' + \frac{4b_j b_j'}{\varepsilon}\right) > c > 0,$$

the result follows from (20).

Corollary 2. *The new operator is of positive type and therefore it satisfies the maximum discrete principle.*

We finally study the local truncation error in the transition points σ and $1 - \sigma$. We show the case $x_j = \sigma$ and we could proceed similarly for $x_j = 1 - \sigma$. The local truncation error in this point is given by

$$\tau_{j,u}^3 = \frac{2\varepsilon}{h_j + h_{j+1}}\left[\frac{R_5(x_j, x_{j+1}, u)}{h_{j+1}} + \frac{R_5(x_j, x_{j-1}, u)}{h_j}\right] + \frac{b_j^2}{\varepsilon}\frac{R_2(x_j, x_{j-1}, u)}{h_j} +$$
$$+ \frac{h_{j+1} - h_j}{3}b_j\left(\frac{R_3(x_j, x_{j-1}, u)}{h_j} + \frac{h_j^2}{3!}\frac{R_1(x_j, x_{j+1}, u)}{h_{j+1}\varepsilon}\right) +$$
$$+ \frac{h_j^3 + h_{j+1}^3}{12(h_j + h_{h_j+1})}b_j'\left[\operatorname{sgn} b_j'\frac{R_2(x_j, x_{j+1}, u)}{h_j} + (1 - \operatorname{sgn} b_j')\frac{R_2(x_j, x_{j+1}, u)}{h_{j+1}}\right] +$$
$$+ \frac{2(h_{j+1}^4 - h_j^4)}{5!(h_j + h_{j+1})}\left[3b_j''\left(\left(\operatorname{sgn} b_j''\frac{R_1(x_j, x_{j-1}, u)}{h_j} + (1 - \operatorname{sgn} b_j'')\frac{R_1(x_j, x_{j+1}, u)}{h_{j+1}}\right)\right)\right]. \tag{22}$$

Proposition 4. *The local truncation error in the transition points satisfies*

$$|\tau_{j,u}^3| \le C\left[N^{-4}\sigma_0^2 \log^2 N + d_j N^{-\sqrt{\beta\sigma_0}}\right], \quad \text{for } j = N/4 \text{ or } j = 3N/4.$$

Proof. In same way as in Proposition 1, using that $h^{(1)}/\sqrt{\varepsilon}$ is bounded, $h^{(1)} < 2N^{-1}$ and $h^{(2)} = 4N^{-1}\sqrt{\varepsilon}\log N$, it is straightforward (see [1] for details) to prove that

$$|\tau_{j,u}^3| \le C\left(\frac{h^{(1)2}h^{(2)2}}{\varepsilon} + e(x_j, x_{j+1}, \beta, \varepsilon) + e(x_{j-1}, x_j, \beta, \varepsilon)\right) \le$$
$$\le C(N^{-4}\sigma_0^2 \log^2 N + N^{-\sqrt{\beta\sigma_0}}).$$

Using the uniform stability of the operator and the bounds for the local truncation error, we obtain the following convergence result.

Theorem 2. *Let $u(x, \varepsilon)$ be the solution of (1) and $\{U_j; 0 \le j \le N\}$ the solution of the new finite difference scheme. Then,*

$$|u(x_j, \varepsilon) - U_j| \le C\left(N^{-4}\sigma_0^4 \log^4 N + N^{-\sqrt{\beta\sigma_0}}\right), \qquad 0 \le j \le N.$$

4 Numerical Experiments

To confirm the theoretical result, we consider the problem

$$-\varepsilon u'' + (1 + x^2 + \cos x)u = x^{4.5} + \sin x, \quad 0 < x < 1, \quad u(0) = 1, \quad u(1) = 1,$$

whose exact solution is not known. Pointwise errors are estimated by $e^{\varepsilon,N} = \|U_i^N - U_{2i}^{2N}\|_\infty$ where U^{2N} is the approximate solution on the mesh $x_{2i} = x_i \in I_N$, $i = 0, 1 \ldots, N$, $x_{2i+1} = (x_i + x_{i+1})/2$, $i = 0, 1 \ldots, N-1$. We denote $E_{\varepsilon,N} = \max_\varepsilon e^{\varepsilon,N}$. The numerical ε-uniform rate of convergence, calculated using the double-mesh principle (see [2]), is given by $p = \log(E_{\varepsilon,N}/E_{\varepsilon,2N})/\log 2$. In Table 1 we give the results obtained with the scheme (8) taking $\sigma_0 = 4$, which are agree with the order given by Theorems 1 and 2.

Table 1. Pointwise errors and numerical order of convergence

ε	N=16	N=32	N=64	N=128	N=256	N=512	N=1024
2^0	1.166E-7	7.285E-9	4.554E-10	2.845E-11	1.778E-12	1.111E-13	6.944E-15
	4.000	4.000	4.000	4.000	4.000	4.000	
2^{-6}	1.106E-4	7.703E-6	4.874E-7	3.056E-8	1.912E-9	1.195E-10	7.469E-12
	3.844	3.982	3.995	3.998	4.000	4.000	
2^{-12}	6.211E-3	4.813E-3	1.347E-3	1.052E-4	7.333E-6	4.650E-7	2.921E-8
	0.368	1.837	3.678	3.842	3.979	3.993	
2^{-18}	6.174E-3	4.801E-3	1.513E-3	2.293E-4	2.663E-5	2.721E-6	2.620E-7
	0.363	1.666	2.722	3.106	3.291	3.377	
2^{-24}	6.174E-3	4.801E-3	1.513E-3	2.293E-4	2.663E-5	2.721E-6	2.620E-7
	0.363	1.666	2.722	3.106	3.291	3.377	
$E_{\varepsilon,N}$	6.211E-3	4.813E-3	1.513E-3	2.293E-4	2.663E-5	2.721E-6	2.620E-7
	0.368	1.669	2.722	3.106	3.291	3.377	

References

1. J. L. Gracia, Esquemas de orden superior para problemas de convección-difusión y reacción-difusión lineales, Ph. Thesis, University of Zaragoza, (1999) (in spanish).
2. A. F. Hegarty, J. J. H. Miller, E. O'Riordan, G. I. Shishkin, Special numerical methods for convection-dominated laminar flows at arbitrary Reynolds number, *East-West J. Numer. Math.* **2** (1994) 65–74.
3. J. J. H. Miller, E. O'Riordan, G. I. Shishkin, *Fitted numerical methods for singular perturbation problems. Error estimates in the maximum norm for linear problems in one and two dimensions*, World Scientific, (1996).
4. H. G. Roos, M. Stynes, L. Tobiska, *Numerical methods for singularly perturbed differential equations*, Springer-Verlag, (1996).
5. A. A. Samarski, V. B. Andréiev, *Métodos en diferencias para las ecuaciones elípticas*, Mir, Moscú, (1979).

6. G. I. Shishkin, *Discrete approximation of singularly perturbed elliptic and parabolic equations*, Russian Academic of Sciences. Ural Section. Ekaterinburg, (1992).
7. G. I. Shishkin, Approximation of solutions of singularly perturbed boundary value problems with a parabolic boundary layer, *USSR Comput. Maths. Maths. Phys.* **29** (1989) 1–10.
8. W. F. Spotz, *High-order compact finite difference schemes for computational mechanics*, Ph. D. Thesis, University of Texas at Austin, (1995).

A Grid Free Monte Carlo Algorithm for Solving Elliptic Boundary Value Problems*

T. Gurov[1,2], P. Whitlock[2], and I. Dimov[1]

[1] Central Laboratory for Parallel Processing, Bulgarian Academy of Sciences
Acad. G. Bonchev St.,bl. 25 A, 1113 Sofia, Bulgaria
{dimov,gurov}@copern.bas.bg

[2] Dep. of Comp. and Inf. Sci., Brooklyn College – CUNY
2900 Bedford Avenue Brooklyn, NY 11210-2889
{whitlock,gurov}@sci.brooklyn.cuny.edu

Abstract. In this work a grid free Monte Carlo algorithm for solving elliptic boundary value problems is investigated. The proposed Monte Carlo approach leads to a random process called a ball process.
In order to generate random variables with the desired distribution, rejection techniques on two levels are used.
Varied numerical tests on a Sun Ultra Enterprise 4000 with 14 Ultra-SPARC processors were performed. The code which implemented the new algorithm was written in JAVA.
The numerical results show that the derived theoretical estimates can be used to predict the behavior of a wide class of elliptic boundary value problems.

1 Introduction

Consider the following three-dimensional elliptic boundary value problem:

$$Mu = -\phi(x), \ x \in \Omega, \ \Omega \subset \mathbb{R}^3 \ \text{and} \ u = \psi(x), \ x \in \partial\Omega, \qquad (1)$$

where the differential operator M is equal to

$$M = \sum_{i=1}^{3} \left(\frac{\partial^2}{\partial x_i^2} + b_i(x)\frac{\partial}{\partial x_i} \right) + c(x).$$

We assume that the regularity conditions for the closed domain $\overline{\Omega}$ and the given functions $\mathbf{b}(x)$, $c(x) \leq 0$, $\phi(x)$ and $\psi(x)$ are satisfied. These conditions guarantee the existence and uniqueness of the solution $u(x)$ in $\mathbf{C}^2(\Omega) \cap \mathbf{C}(\overline{\Omega})$ of problem (1), (see [1,5]), as well as the possibility of its local integral representation (when $div\mathbf{b}(x) = \sum_{i=1}^{3} \frac{\partial b_i(x)}{\partial x_i^2} = 0$) by making use of the Green's function approach for standard domains lying inside the domain Ω (for example - a ball or an ellipsoid).

---

* Supported by ONR Grant N00014-96-1-1-1057 and by the National Science Fund of Bulgaria under Grant # I 811/1998.

Denote by $B(x)$ the ball: $B(x) = B_R(x) = \{y : r = |y - x| \leq R(x)\}$, where $R(x)$ is the radius of the ball. Levy's function for the problem (1) is

$$L_p(y, x) = \mu_p(R) \int_r^R (1/r - 1/\rho) p(\rho) d\rho, \quad r \leq R, \tag{2}$$

where the following notations are used: $p(\rho)$ is a density function;

$$r = |x - y| = \left(\sum_{i=1}^{3} (x_i - y_i)^2 \right)^{1/2}, \quad \mu_p(R) = [4\pi q_p(R)]^{-1}, \quad q_p(R) = \int_0^R p(\rho) d\rho.$$

It is readily seen that Levy's function $L_p(y, x)$, and the parameters $q_p(R)$ and $\mu_p(R)$ depend on the choice of the density function $p(\rho)$. In fact, the Eq.(2) defines a family of functions.

For the Levy's function the following representation holds (see [4]):

$$u(x) = \int_{B(x)} \left(u(y) M_y^* L_p(y, x) + L_p(y, x) \phi(y) \right) dy \tag{3}$$

$$+ \int_{\partial B(x)} \sum_{i=1}^{3} n_i \left[\left(\frac{L_p(y, x) \partial u(y)}{\partial y_i} - \frac{u(y) \partial L_p(y, x)}{\partial y_i} \right) - b_i(y) u(y) L_p(y, x) \right] d_y S,$$

where $\mathbf{n} \equiv (n_1, n_2, n_3)$ is the exterior normal to the boundary $\partial B(x)$ and

$$M^* = \sum_{i=1}^{3} \left(\frac{\partial^2}{\partial x_i^2} - b_i(x) \frac{\partial}{\partial x_i} \right) + c(x).$$

is the adjoint operator to M.

It is proved (see [3]) that the conditions $M_y^* L_p(y, x) \geq 0$ (for any $y \in B(x)$) and $L_p(y, x) = \partial L_p(y, x)/\partial y_i = 0$, $i = 1, 2, 3$ (for any $y \in \partial B(x)$) are satisfied for $p(r) = e^{-kr}$, where

$$k \geq b^* + Rc^*, \quad b^* = \max_{x \in \Omega} |\mathbf{b}(x)|, \quad c^* = \max_{x \in \Omega} |c(x)|$$

and R is the radius of the maximal ball $B(x) \subset \overline{\Omega}$.

This statement shows that it is possible to construct the Levy's function choosing the density $p(\rho)$ such that kernel $M_y^* L_p(y, x)$ is non-negative in $B(x)$ and such that $L_p(y, x)$ and its derivatives vanish on $\partial B(x)$.

It follows that the representation (3) can be written in the form:

$$u(x) = \int_{B(x)} M_y^* L_p(y, x) u(y) dy + \int_{B(x)} L_p(y, x) \phi(y) dy, \tag{4}$$

where

$$M_y^* L_p(y, x) = \mu_p(R) \frac{p(r)}{r^2} - \mu_p(R) c(y) \int_r^R \frac{p(\rho)}{\rho} d\rho$$

$$+ \frac{\mu_p(R)}{r^2} \left[c(y) r + \sum_{i=1}^{3} b_i(y) \frac{y_i - x_i}{r} \right] \int_r^R p(\rho) d\rho.$$

The representation of $u(x)$ in (4) is the basis for the proposed Monte Carlo method. Using it, a biased estimator for the solution can be obtained.

2 Monte Carlo Method

The Monte Carlo procedure for solving Eq.(4) can be defined as a "ball process" or "walk on small spheres". Consider a transition density function

$$p(x,y) \geq 0 \quad \text{and} \quad \int_{B(x)} p(x,y)dy = 1. \tag{5}$$

and define a Markov chain $\xi_0, \xi_1, \ldots$, such that every point $\xi_j, j = 1, 2, \ldots$, is chosen in the maximal ball $B(\xi_{j-1})$ lying in Ω in accordance with the density (5).

Generally, the "walk on small spheres" process can be written as following (see [8]):

$$\xi_j = \xi_{j-1} + \mathbf{w}^j \alpha R(\xi_{j-1}), \quad j = 1, 2, \ldots, \quad \alpha \in (0, 1],$$

where $\mathbf{w}^j$ are independent unit isotropic vectors in $\mathbb{R}^3$. In particular, when $\alpha = 1$ the process is called "walk on spheres" (see [6,8]).

To ensure the convergence of the process under consideration we introduce the ε-strip of the boundary, i.e.

$$\partial \Omega_\varepsilon = \{ y \in \Omega : \exists x \in \partial \Omega \text{ for which } |y - x| \leq \varepsilon \}.$$

Thus the Markov chain terminates when it reaches $\partial \Omega_\varepsilon$ and the final point is $\xi_{l_\varepsilon} \in \partial \Omega_\varepsilon$.

Consider the biased estimate Θ_{l_ε} for the solution of Eq.(4) at the point ξ_0 (see [2]):

$$\Theta_{l_\varepsilon}(\xi_0) = \sum_{j=0}^{l_\varepsilon-1} W_j \int_{B(\xi_j)} L_p(y, \xi_j) \phi(y) dy + W_{l_\varepsilon} \psi(\xi_{l_\varepsilon}), \tag{6}$$

where

$$W_0 = 1, \quad W_j = W_{j-1} \frac{M_y^* L_p(\xi_j, \xi_{j-1})}{p(\xi_{j-1}, \xi_j)}, \quad j = 1, 2, \ldots, l_\varepsilon.$$

If the first derivatives of the solution are bounded in Ω then the following inequality holds (see [6]):

$$|E\Theta_{l_\varepsilon}(\xi_0) - u(\xi_0)|^2 \leq c_1 \varepsilon^2. \tag{7}$$

Using N independent samples we construct a random estimate of the form

$$\overline{\Theta}_{l_\varepsilon}(\xi_0) = \frac{1}{N} \sum_{i=0}^{N} \Theta_{l_\varepsilon}^{(i)}(\xi_0) \approx u(\xi_0).$$

The root mean square deviation is defined by the relation

$$E(\Theta_{l_\varepsilon}(\xi_0) - u(\xi_0))^2 = Var(\Theta_{l_\varepsilon}(\xi_0)) + (u(\xi_0) - E\Theta_{l_\varepsilon}(\xi_0))^2.$$

Hence

$$E(\overline{\Theta}_{l_\varepsilon}(\xi_0) - u(\xi_0))^2 = \frac{Var(\Theta_{l_\varepsilon}(\xi_0))}{N} + (u(\xi_0) - E\Theta_{l_\varepsilon}(\xi_0))^2 \leq \frac{d_0}{N} + c_1 \varepsilon^2 = \mu^2, \tag{8}$$

where μ is the desired error, d_0 is upper boundary of the variance and c_1 is the constant from Eq. (7).

3 A Grid Free Monte Carlo Algorithm

Using spherical coordinates [2] we can express the kernel $k(x,y) = M_y^* L_p(y,x)$ as follows:

$$k(r,\mathbf{w}) = \frac{p(r)\sin\theta}{q_p(R)4\pi} \times$$

$$\times \left[1 + \frac{\sum_{i=1}^{3} b_i(x+r\mathbf{w})w_i + c(x+r\mathbf{w})r}{p(r)} \int_r^R p(\rho)d\rho - \frac{c(x+r\mathbf{w})r^2}{p(r)} \int_r^R \frac{p(\rho)}{\rho}d\rho\right].$$

Here $\mathbf{w} \equiv (w_1, w_2, w_3)$ is an unit isotropic vector in $\mathbb{R}^3$, where $w_1 = \sin\theta\cos\varphi$, $w_2 = \sin\theta\sin\varphi$ and $w_3 = \cos\theta$ ($\theta \in [0, \pi)$ and $\varphi \in [0, 2\pi)$).

Let us consider the following two non-negative functions

$$p_0(r,\mathbf{w}) = \frac{p(r)\sin\theta}{q_p(R)4\pi} \left[1 + \frac{\sum_{i=1}^{3} b_i(x+r\mathbf{w})w_i}{p(r)}\right] \int_r^R p(\rho)d\rho,$$

when $c(x+r\mathbf{w}) \equiv 0$ and $p(r,\mathbf{w}) = k(r,\mathbf{w})$, when $c(x+r\mathbf{w}) \leq 0$.

The following inequalities hold:

$$p(r,\mathbf{w}) \leq p_0(r,\mathbf{w}) \leq \frac{p(r)\sin\theta}{q_p(R)4\pi} \left[1 + \frac{b^*}{p(r)} \int_r^R p(\rho)d\rho\right]. \tag{9}$$

We note that function $p_0(r,\mathbf{w})$ satisfies the condition (5) (see [2]).

Denote by $\bar{p}(r,\mathbf{w})$ the following function:

$$\bar{p}(r,\mathbf{w}) = \frac{p(r,\mathbf{w})}{V}, \text{ where } \int_0^\pi \int_0^{2\pi} \int_0^R p(r,\mathbf{w})drd\theta d\varphi = V < 1. \tag{10}$$

Introduce the functions:

$$\bar{p}(\mathbf{w}/r) = 1 + \frac{\sum_{i=1}^{3} b_i(x+r\mathbf{w})w_i + c(x+r\mathbf{w})r}{p(r)} \int_r^R p(\rho)d\rho - \frac{c(x+r\mathbf{w})r^2}{p(r)} \int_r^R \frac{p(\rho)}{\rho}d\rho;$$

$$\bar{p}_0(\mathbf{w}/r) = 1 + \frac{\sum_{i=1}^{3} b_i(x+r\mathbf{w})w_i}{p(r)} \int_r^R p(\rho)d\rho.$$

Using inequalities (9) we obtain:

$$\bar{p}(\mathbf{w}/r) \leq \bar{p}_0(\mathbf{w}/r) \leq 1 + \frac{b^*}{p(r)} \int_r^R p(\rho)d\rho. \tag{11}$$

Now we can describe the grid free algorithm for simulating the Markov chain with transition density function (10). The Markov chain is started at the fixed point ξ_0. The inequalities in (11) are used to sample the next point ξ_1 by applying a two level acceptance-rejection sampling (ARS) rule.

The ARS rule or the Neumann rule can be used if another density function $v_2(x)$ exists such that $c_2 v_2(x)$ is everywhere a maximum of the density function $v_1(x)$, that is, $c_2 v_2(x) \geq v_1(x)$ for all values x (see for details [2]). The efficiency of this rule depends upon $c_2 v_2(x)$ and how closely it envelopes $v_1(x)$. A two level ARS rule is preferable when $v_1(x)$ is a complex function. In this case a second majorant function must be found which envelopes very closely our density function.

Algorithm 3.1

1. **Compute** the radius $R(\xi_0)$ of the maximal ball lying inside Ω and having center ξ_0.
2. **Generate** a random value r of the random variable τ with the density

$$\frac{p(r)}{q_p(R)} = \frac{ke^{-kr}}{1 - e^{-kR}}. \qquad (12)$$

3. **Calculate** the function

$$h(r) = 1 + \frac{b^*}{p(r)} \int_r^R p(\rho)d\rho = 1 + \frac{b^*}{k}(1 - e^{-k(R-r)}).$$

4. **Generate** the independent random values $\mathbf{w}$ of a unit isotropic vector in $\mathbb{R}^3$.
5. **Generate** the independent random value γ of an uniformly distributed random variable in the interval $[0, 1]$.
6. **Go to** the step 8 **if** the inequality holds: $\gamma h(r) \leq \bar{p}_0(\mathbf{w}/r)\}$.
7. **Go to** the step 4 **otherwise**.
8. **Generate** the independent random value γ of a uniformly distributed random variable in the interval $[0, 1]$.
9. **Go to** the step 11 **if** the inequality holds: $\gamma \bar{p}_0(\mathbf{w}/r) \leq \bar{p}(\mathbf{w}/r)$.
10. **Go to** the step 4 **otherwise**.
11. **Compute** the random point ξ_1, with a density $\bar{p}(\mathbf{w}/r)$ using the following formula: $\xi_1 = \xi_0 + r\mathbf{w}$.
The value $r = |\xi_1 - \xi_0|$ is the radius of the sphere lying inside Ω and having center at ξ.
12. **Repeat** Algorithm 3.1 for new point ξ_1 if $\xi_1 \bar{\in} \partial\Omega_\varepsilon$.
13. **Stop** Algorithm 3.1 if $\xi_1 \in \partial\Omega_\varepsilon$.

The random variable $\Theta_{l_\varepsilon}(\xi_0)$ is calculated using formula (6).

The computational cost of the algorithm under consideration is measured by quantity

$$S = N t_0 E l_\varepsilon,$$

where N is the number of the trajectories performed; El_ε is the average number of balls on a single trajectory; t_0 is the time of modeling a point into the maximal ball lying inside Ω and of computing the weight W which corresponds to this point.

We note that for a wide class of boundaries Ω, (see [8,6]), the following estimate has been obtained on the basis of the restoration theory, $El_\varepsilon = c_2 |\ln \varepsilon|$. If the radius $r = r_0$ is fixed and $r_0/R = \alpha \in (0, 1]$ then the following estimate holds (see [8]):

$$El_\varepsilon = \frac{4R^2 |\ln \varepsilon|}{r_0^2} + O(r_0^4),$$

where R is the radius of the maximal ball lying inside Ω.

It is clear that the algorithmic efficiency of the Algorithm 3.1 depends on the position of the points in the Markov chain. They must be located "not far from the boundary of the ball". Thus, the location of every point depends on the random variable τ with a density Eq.(12).

The following assertion holds (see [2]):

Lemma 1. *Let $\alpha_0 \in (0, 0.5)$. Then $E\tau \in (\alpha_0 R, 0.5R)$, if and only if the radius R of the maximal ball and the parameters b^* and c^* satisfy the inequality*

$$R(b^* + Rc^*) \leq \beta_0, \tag{13}$$

where β_0 is the solution of the equation $g_1(z) = \frac{1}{z} + \frac{1}{1-e^z} = \alpha_0$.

Therefore, after substitution $r_0 = \alpha_0 R$, where α_0 is the parameter from Lemma 1, the average number of balls get

$$El_\varepsilon \asymp \frac{4}{\alpha_0^2} |\ln \varepsilon|. \tag{14}$$

In order to obtain the error of order μ, (see Eq. (8)), the optimal order of the quantities N and ε must be

$$N = O(\mu^{-2}), \quad \varepsilon = O(\mu), \quad S \asymp \frac{4t_0}{\alpha_0^2} \frac{|\ln \mu|}{\mu^2}.$$

Note that this estimate of computational cost is optimal as to the order of magnitude of μ only. It does not take into account the values of the constants in (8).

In order to minimize the computational cost we should solve the conditional minimum problem (see [6]):

$$S = N t_0 E l_\varepsilon \to \min_{N,\varepsilon}, \quad \frac{d_0}{N} + c_1 \varepsilon^2 = \mu^2 \quad \text{or} \quad S = \frac{d_0 t_0}{\mu^2 - c_1 \varepsilon^2} E l_\varepsilon \to \min_\varepsilon, \quad c_1 \varepsilon^2 \leq \mu^2.$$

Having solved this problem we obtain the optimal values of the quantities N, S and ε:

$$N^* = \frac{d_0}{2c_1 \varepsilon_*^2 |\ln \varepsilon_*|}, \quad S^* \asymp \frac{2d_0 t_0}{c_1 \alpha_0^2 \varepsilon_*^2},$$

where ε_* is a solution of the equation

$$c_1 \varepsilon^2 + 2c_1 \varepsilon^2 |\ln \varepsilon| = \mu^2. \tag{15}$$

It is not difficult to estimate the variance $Var(\Theta_{l_\varepsilon}(\xi_0))$ when the function $\phi(x) = 0$. In this case we have $\Theta_{l_\varepsilon}(\xi_0) = W_{l_\varepsilon} \psi(\xi_{l_\varepsilon})$. Thus

$$Var(\Theta_{l_\varepsilon}(\xi_0)) \leq E(\Theta_{l_\varepsilon}^2(\xi_0)) =$$

$$= \int_{B(\xi_0)} \cdots \int_{B(\xi_{l_\varepsilon})} \frac{(M_y^* L_p(\xi_1, \xi_0))^2}{\overline{p}(\xi_0, \xi_1)} \cdots \frac{(M_y^* L_p(\xi_{l_\varepsilon}, \xi_{l_\varepsilon - 1}))^2}{\overline{p}(\xi_{l_\varepsilon - 1}, \xi_{l_\varepsilon})} \psi(\xi_{l_\varepsilon})^2 d\xi_{l_\varepsilon} \cdots d\xi_0.$$

Denote by
$$V = \max_{x \in \Omega} \int_{B(x)} M_y^* L_p(y,x) dy \text{ and } \psi_*^2 = \max_{x \in \Omega} \int_{B(x)} |\psi(y)|^2 dy$$

Now we obtain
$$Var(\Theta_{l_\varepsilon}(\xi_0)) \leq V^{2l_\varepsilon} \psi_*^2 \leq \psi_*^2.$$

Thus in this case, the optimal values of the quantities N, S get:
$$N^* = \frac{\psi_*^2}{2c_1 \varepsilon_*^2 |\ln \varepsilon_*|}, \quad S^* \asymp \frac{2\psi_*^2 t_0}{c_1 \alpha_0^2 \varepsilon_*^2},$$

where the constant c_1 depends on the condition the first derivatives of the solution shall be bounded in Ω and α_0 depends on Eq.(13).

4 Numerical Result

As an example the following boundary value problem was solved in the cube $\Omega = [0,1]^3$:

$$\sum_{i=1}^{3} \left(\frac{\partial^2 u}{\partial x_i^2} + b_i(x) \frac{\partial u}{\partial x_i} \right) + c(x) u = 0,$$

$$u(x_1, x_2, x_3) = e^{a_1(x_1 + x_2 + x_3)}, \quad (x_1, x_2, x_3) \in \partial \Omega_\varepsilon.$$

In our tests $b_1(x) = a_2 x_1 (x_2 - x_3)$, $b_2(x) = a_2 x_2 (x_3 - x_1)$, $b_3(x) = a_2 x_3 (x_1 - x_2)$, and $c(x) = -3a_1^2$, where a_1 and a_2 are parameters.

We note that the condition div $\mathbf{b}(x) = 0$ is satisfied.

The code which implemented the algorithm under consideration was written in JAVA. The multiplicative linear-congruential generator, which was used to obtain a sequence of random numbers distributed uniformly between 0 and 1, is $x_n = 7^5 x_{n-1} \mod(2^{31} - 1)$. It was highly recommended by Park and Miller [7] and they called it the "minimal standard".

Numerical tests on a Sun Ultra Enterprise 4000 with 14 UltraSPARC processors were performed for different values of the parameters a_1 and a_2 (see Tables 1,2). The solution was estimated at the point with coordinates $x = (0.5, 0.5, 0.5)$. In the tables, $u(x)$ is the exact solution, $u_{l_\varepsilon}(x)$ is the estimate of the solution, μ_ε is the estimate of the corresponding mean square error, σ^2 is the estimate $Var(\Theta_{l_\varepsilon}(x))$. The results presented in Table 1 are in good agreement with theoretical one (see Eq's. 14,15). Moreover, the results presented in Table 2 show how important it is to have a good balancing between the stochastic and systematic error. When $N^* = 50533$, the time of estimating solution is: $t_1 = 51m14.50s$ and when $N = 10^6$ the time is: $t_2 = 19h17m44.25s$. Thus, the computational effort in the first case is about twenty times better than second one, while Monte Carlo solutions are approximately equal. On the other hand the numerical tests show that the variance does not depend on the vector-function $b(x)$.

Table 1. $u(x) = 1.4549915$, $a_1 = 0.25$, $c^* = 0.1875$, $R_{max} = 0.5$

$b^* = a_2\sqrt{3}$	$\varepsilon^* = 0.01$, $N^* = 3032$				$\varepsilon^* = 0.001$, $N^* = 202130$			
	$u_{l_\varepsilon}(x)$ μ_ε		σ^2	El_ε	$u_{l_\varepsilon}(x)$ μ_ε		σ^2	El_ε
$8\sqrt{3}$	1.465218 ± 0.029		0.0437	63.73	1.456395 ± 0.0035		0.04455	95.75
$4\sqrt{3}$	1.460257 ± 0.029		0.0427	43.62	1.456704 ± 0.0035		0.04448	74.92
$2\sqrt{3}$	1.465602 ± 0.029		0.0434	38.85	1.457545 ± 0.0035		0.04466	69.46
$\sqrt{3}$	1.456592 ± 0.029		0.0423	36.96	1.456211 ± 0.0035		0.04462	67.95
$\sqrt{3}/4$	1.461289 ± 0.029		0.0432	36.46	1.456149 ± 0.0035		0.04455	67.32
$\sqrt{3}/16$	1.456079 ± 0.029		0.0428	36.72	1.455545 ± 0.0035		0.04450	67.12

Table 2. $u(x) = 2.117$, $a_1 = 0.5$, $c^* = 0.75$, $R_{max} = 0.5$

$b^* = a_2\sqrt{3}$	$\varepsilon^* = 0.001$, $N^* = 50533$				$\varepsilon = 0.001$, $N = 1000000$			
	$u_{l_\varepsilon}(x)$ μ_ε		σ^2	El_ε	$u_{l_\varepsilon}(x)$ μ_ε		σ^2	El_ε
$8\sqrt{3}$	2.12829 ± 0.015		0.374	97.41	2.12749 ± 0.015		0.3770	97.00
$4\sqrt{3}$	2.13151 ± 0.015		0.3782	75.93	2.12972 ± 0.015		0.3787	75.68
$2\sqrt{3}$	2.12878 ± 0.015		0.3775	70.17	2.12832 ± 0.015		0.3790	70.05
$\sqrt{3}$	2.12898 ± 0.015		0.3781	68.39	2.12547 ± 0.015		0.3774	67.95
$\sqrt{3}/4$	2.12227 ± 0.015		0.3771	67.71	2.12125 ± 0.015		0.3760	67.63
$\sqrt{3}/16$	2.12020 ± 0.015		0.3753	67.61	2.11869 ± 0.015		0.3750	67.53

5 Conclusion

In this work it is shown that a grid free Monte Carlo algorithm under consideration can be successfully applied for solving elliptic boundary value problems. An estimate for minimization of computational cost is obtained. The balancing of errors (both systematic and stochastic) either reduces the computational complexity when the desired error is fixed or increases the accuracy of the solution when the desired computational complexity is fixed.

The studied algorithm is easily programmable and parallelizable and can be efficiently implemented on MIMD-machines.

References

1. Bitzadze, A. V.: Equations of the Mathematical Physics. Nauka, Moscow, (1982).
2. Dimov, I., Gurov, T.: Estimates of the computational complexity of iterative Monte Carlo algorithm based on Green's function approach. Mathematics and Computers in Simulation. **47** (2-5) (1998) 183–199.
3. Ermakov, S. M., Nekrutkin V. V., Sipin, A. S.: Random processes for solving classical equations of the mathematical physics. Nauka, Moscow, (1984).
4. Miranda, C.: Equasioni alle dirivate parziali di tipo ellipttico. Springer-Verlag, Berlin, (1955).
5. Mikhailov, V. P.: Partial differential equations. Nauka, Moscow, (1983).

6. Mikhailov, G. A.: New Monte Carlo Methods with Estimating Derivatives. Utrecht, The Netherlands, (1995).
7. Park, S. K., Miller, K. W.: Random Number Generators: Good Ones Are Hard to Find, Communications of the ACM, **31** (10) (1988) 1192–1201.
8. Sabelfeld, K. K.: Monte Carlo Methods in Boundary Value Problems. Springer Verlag, Berlin - Heidelberg - New York - London, (1991).

Newton's Method under Different Lipschitz Conditions

José M. Gutiérrez and Miguel A. Hernández*

Universidad de La Rioja, Departamento de Matemáticas y Computación,
26004 Logroño, SPAIN

Abstract. The classical Kantorovich theorem for Newton's method assumes that the derivative of the involved operator satisfies a Lipschitz condition
$$\|F'(x_0)^{-1}\left[F'(x) - F'(y)\right]\| \leq L\|x - y\|$$
In this communication, we analyse the different modifications of this condition, with a special emphasis in the center-Lipschitz condition:
$$\|F'(x_0)^{-1}\left[F'(x) - F'(x_0)\right]\| \leq \omega(\|x - x_0\|)$$
being ω a positive increasing real function and x_0 the starting point for Newton's iteration.

In this paper we make a survey of the convergence of Newton's method in Banach spaces. So, let X, Y be two Banach spaces and let $F: X \to Y$ be a Fréchet differentiable operator. Starting from $x_0 \in X$, the well-known Newton's method is defined by the iterates

$$x_{n+1} = x_n - F'(x_n)^{-1} F(x_n), \quad n = 0, 1, 2, \ldots \qquad (1)$$

provided that the inverse of the linear operator $F'(x_n)$ is defined at each step.

Under different conditions on the operator F, the starting point x_0 or even on the solution, it is shown that the sequence (1) converges to a solution x^* of the equation $F(x) = 0$.

In broad outline, three types of convergence results can be given:

- *Local convergence*: The existence of solution is assumed. In addition, it is also required the invetibility of $F'(x^*)$ and a Lipschitz-type condition:

$$\|F'(x^*)^{-1}\left[F'(x) - F'(y)\right]\| \leq \beta\|x - y\|, \quad x, y \in \Omega_0 \subseteq X. \qquad (2)$$

- *Semilocal convergence*: Conditions on the starting point x_0 instead of the solution x^* are assumed. In this way, two types of semilocal results can be distinguish:

---
* Research of both authors has been supported by a grant of the Universidad de La Rioja (ref. API-99/B14) and two grants of the DGES (refs. PB98-0198 and PB96-0120-C03-02).

- Kantorovich conditions: There exists the inverse of $F'(x_0)$ and the following conditions are fulfilled:

$$\|F'(x_0)^{-1}F(x_0)\| \leq a,$$

$$\|F'(x_0)^{-1}[F'(x) - F'(y)]\| \leq b\|x - y\|, \quad x, y \in \Omega_1 \subseteq X, \quad (3)$$

$$ab \leq 1/2.$$

- Smale's α-theory: The Lipschitz condition on a domain (3) is replaced by a punctual condition on F and its derivatives. Let

$$\|F'(x_0)^{-1}F(x_0)\| \leq a, \quad \sup_{k \geq 2} \left\| \frac{1}{k!} F'(x_0)^{-1} F^{(k)}(x_0) \right\|^{1/(k-1)} \leq \gamma.$$

Then $a\gamma \leq 3 - 2\sqrt{2}$ is a sufficient condition for the convergence of Newton's method.

- Global convergence: Monotone convergence of (1) is established, in general, under convexity conditions on the operator F.

In this paper we analyze some modifications of the Kantorovich-type conditions, mainly modifications of (3). First at all, let us say that Kantorovich conditions guarantee the existence and uniqueness of the solution in given balls around x_0 and the quadratic convergence of (1). This kind of results can be proved by finding a majorizing sequence $\{t_n\}$, that is, a real sequence satisfying:

$$\|x_{n+1} - x_n\| \leq t_{n+1} - t_n, \quad n \geq 0.$$

The sequence $\{t_n\}$ is shown to be Newton's method applied to the equation

$$p(t) = 0, \text{ where } p(t) = \frac{b}{2}t^2 - t + a. \quad (4)$$

For more information about these topics, consult the basic reference text [5].

The first modification we comment here is due to Wang Zhenda. In his paper [4], he considers Newton's method under the following Lipschitz condition on the second derivative:

$$\|F'(x_0)^{-1}[F''(x) - F''(y)]\| \leq b_1\|x - y\|, \quad x, y \in \Omega_2 \subseteq X. \quad (5)$$

Then, supposing that $\|F'(x_0)^{-1}F(x_0)\| \leq a$, $\|F'(x_0)^{-1}F''(x_0)\| \leq c$ the convergence of (1) follows from the convergence of the Newton sequence applied to a cubic polynomial.

Almost at the same time, we have proved in [2] the convergence of (1) under the weaker condition:

$$\|F'(x_0)^{-1}[F''(x) - F''(x_0)]\| \leq b_2\|x - x_0\|, \quad x \in \Omega_3 \subseteq X. \quad (6)$$

On the one hand, this condition is more restrictive than (3) because it concerns the second derivative instead of the first one. But on the other hand, the Lipschitz condition is weakened, because one of the points is fixed. This kind of conditions are known as center Lipschitz conditions. In addition, notice that (6) is weaker than (5). In fact, (5) implies (6).

The technique followed in [2] is different from the one of [4]. However in both cases, the convergence holds from the study of a third order polynomial and the existence of a positive root for the corresponding polynomial is assumed.

Let us concrete now, by stating both results: the famous Kantorovich theorem and the theorem given in [2].

Theorem 1 ((Kantorovich)). *Let F be a differentiable operator defined in an open ball $\Omega = B(x_0, R) = \{x \in X; \|x - x_0\| < R\}$. Let us assume that Γ_0, the inverse of $F'(x_0)$ is defined and*

$$\|\Gamma_0 F(x_0)\| \leq a, \quad \|\Gamma_0[F'(x) - F'(y)]\| \leq b\|x - y\|, \quad x, y \in \Omega.$$

Then if $ab \leq 1/2$ and $t^* \leq R$, Newton's method (1) converges to x^*, solution of the equation $F(x) = 0$. In addition, the solution is located in $\overline{B(x_0, t^*)}$ and is unique in $B(x_0, \rho_0)$, where $\rho_0 = \min\{t^{**}, R\}$. Here, we have denoted

$$t^* = \frac{1 - \sqrt{1 - 2ab}}{b}, \quad t^{**} = \frac{1 + \sqrt{1 - 2ab}}{b}.$$

Notice that t^* and t^{**} are the roots of the polynomial $p(t)$ defined in (4). The condition $ab \leq 1/2$ guarantees the existence of such roots.

Theorem 2 (([2])). *Let F be a twice differentiable operator defined in an open ball $\Omega = B(x_0, R)$. Let us assume that Γ_0, the inverse of $F'(x_0)$ is defined and*

$$\|F'(x_0)^{-1} F(x_0)\| \leq a, \quad \|F'(x_0)^{-1} F''(x_0)\| \leq c,$$

$$\|F'(x_0)^{-1}[F''(x) - F''(x_0)]\| \leq b_2\|x - x_0\|, \quad x \in \Omega.$$

Then, if the polynomial

$$q(t) = \frac{b_2}{6} t^3 + \frac{c}{2} t^2 - t + a, \tag{7}$$

has two positive roots r_1, r_2 ($r_1 \leq r_2$) and $r_1 \leq R$, then Newton's method (1) converges to x^*, solution of the equation $F(x) = 0$. In addition, the solution is located in $\overline{B(x_0, r_1)}$ and is unique in $B(x_0, \rho_1)$, where $\rho_1 = \min\{r_2, R\}$.

The following condition

$$6ac^3 + 9a^2 b_2^2 + 18acb_2 \leq 3c^2 + 8b_2,$$

is equivalent to the existence of roots for the polynomial $q(t)$ defined in (7).

Theorems 1 and 2 are not comparable, as we show in the following examples.

Example 1. Let $\Omega = X = [-1,1]$, $Y = \mathbb{R}$, $x_0 = 0$ and $f : X \to Y$ the polynomial

$$f(x) = \frac{1}{6}x^3 + \frac{1}{6}x^2 - \frac{5}{6}x + \frac{1}{3}.$$

In this case, $a = 2/5$ and $b = 8/5$. Then $ab = 16/25 > 1/2$ and Kantorovich condition fails. However, with the above notation, we have $a = 2/5$, $c = 2/5$ and $b_2 = 6/5$. Then,

$$6ac^3 + 9a^2b_2^2 + 18acb_2 = 5.6832 < 3c^2 + 8b_2 = 10.08.$$

So, the corresponding polynomial (7) has two positive roots and Theorem 2 can be applied.

Example 2. Let $\Omega = X = Y = \mathbb{R}$, $x_0 = 0$ and $f : X \to Y$ the function

$$f(x) = \sin x - 5x - 8.$$

Now, $a = 2$, $c = 0$ and $b = b_2 = 1/4$. Then $ab = 1/2$ and the hypothesis of the Kantorovich theorem holds. However, in this case, the polynomial (4) is

$$q(t) = \frac{1}{24}t^3 - t + 2,$$

which has not positive roots and we cannot use Theorem 2.

Sometimes, the convergence of (1) can be established by using Theorems 1 or 2 indistinctly. Then we wonder which result gives us more accurate information on the solutions of $F(x) = 0$.

Let us consider the polynomials p and q defined in (4) and (7) respectively. We have denoted t^*, t^{**} the roots of p and r_1, r_2 the roots of q. Then

$$q(t^*) = \frac{(t^*)^2}{2}\left(\frac{b_2}{3}t^* - (b-c)\right), \quad q(t^{**}) = \frac{(t^{**})^2}{2}\left(\frac{b_2}{3}t^{**} - (b-c)\right).$$

Observe that

$$p(t^*) \leq 0 \iff b_2\left(1 - \sqrt{1-2ab}\right) \leq 3b(b-c),$$

$$p(t^{**}) \leq 0 \iff b_2\left(1 + \sqrt{1-2ab}\right) \leq 3b(b-c).$$

Our goal now is to get the smallest region where the solution is located and the biggest one where this solution is unique. We distinguish three situations:

1. $b_2\left(1 + \sqrt{1-2ab}\right) \leq 3b(b-c)$. Then $r_1 \leq t^*$, $t^{**} \leq r_2$ and, consequently, the solution x^* is located in $\overline{B(x_0, r_1)}$ and is unique in $B(x_0, r_2)$.
2. $b_2\left(1 - \sqrt{1-2ab}\right) \leq 3b(b-c) < b_2\left(1 + \sqrt{1-2ab}\right)$. In this situation $r_1 \leq t^*$, $r_2 \leq t^{**}$, then the solution x^* belongs to $\overline{B(x_0, r_1)}$ and is the only one in $B(x_0, t^{**})$.

3. $3b(b-c) \leq b_2\left(1-\sqrt{1-2ab}\right)$. Now we have $t^* \leq r_1$, $r_2 \leq t^{**}$, thus x^* is located in $\overline{B(x_0, t^*)}$ and is unique in $B(x_0, t^{**})$.

In cases 1 and 3 we get the best information from the Theorems 2 and 1 respectively. But in the second case, the best information is obtained by mixing both results.

The next modification we comment here consists in considering a center Lipschitz condition for the first derivative, that is,

$$\|\Gamma_0[F'(x) - F'(x_0)]\| \leq b_3 \|x - x_0\|, \quad x \in \Omega. \tag{8}$$

First, we notice that this condition is weaker than the classical Lipschitz condition (3). Obviously, (3) implies (8) but the reciprocal is not true. So, there are functions satisfying (8) but not (3). For instance, $f(x) = \sqrt{x}$ defined in $\Omega = [0, \infty)$ is not a Lipschitz function in Ω. Nevertheless, if we take $x_0 = 1$, we obtain

$$|f(x) - f(x_0)| = |\sqrt{x} - 1| = \frac{|x-1|}{1+\sqrt{x}} \leq |x - x_0|, \quad \forall x \in \Omega.$$

Newton's method under condition (8) has been studied in [3], where the following result can be found:

Theorem 3 (([3])). *Let F be a differentiable operator defined in an open ball $\Omega = B(x_0, R)$. Let us assume that Γ_0, the inverse of $F'(x_0)$ is defined and*

$$\|\Gamma_0 F(x_0)\| \leq a, \quad \|\Gamma_0[F'(x) - F'(x_0)]\| \leq L\|x - x_0\|, \quad x \in \Omega.$$

Then, if $aL \leq (14 - 4\sqrt{6})/25 = 0.1680816\ldots$ and $\delta_1 \leq R$, we have that Newton's method (1) converges to a solution x^* of the equation $F(x) = 0$. In addition, the solution is located in $\overline{B(x_0, \delta_1)}$ and is unique in $B(x_0, \rho_2)$, where $\rho_2 = \min\{\delta_2, R\}$. Here, we have denoted

$$\delta_1 = \frac{2 + 5aL - \sqrt{25(aL)^2 - 28aL + 4}}{12L}, \quad \delta_2 = \frac{2}{L} - \delta_1.$$

To prove this result, the authors follow a technique based on the use of recurrence relations instead of the classical majorizing sequences. The idea of the proof has also been used by Rokne in his classical paper [6]. In his general Theorem 1, Rokne assumes a center Lipschitz condition together with a Lipschitz condition. Consequently, the hypothesis of Rokne's result are more restrictive than the ones in the previous theorem.

Finally, the last modification we consider here is a generalization of (8):

$$\|\Gamma_0[F'(x) - F'(x_0)]\| \leq \omega(\|x - x_0\|), \quad \forall x \in B(x_0, R), \tag{9}$$

where ω is a real function such that $\omega'(t) > 0$ for $t \in [0, R]$ and $\omega(0) = 0$.

As particular cases in (9) we have the center Lipschitz case ($\omega(t) = Lt$), the center Hölder case ($\omega(t) = Lt^p$, $0 < p < 1$), combinations of both of them, etc.

To study of Newton's method under this condition we must previously define the non-linear second-order recurrence relations

$$r_0 = 0; \quad r_1 = a; \quad r_{k+1} = r_k + \frac{1}{1 - \omega(r_k)} \int_{2r_{k-1}}^{r_{k-1} + r_k} \tilde{\omega}(s)\, ds, \tag{10}$$

where $\tilde{\omega}(s) = \sup\{\omega(u) + \omega(v); u + v = s\}$. This function $\tilde{\omega}$ has been introduced in [1] for the study of Newton's method.

Theorem 4. *Let F be an operator defined in the open ball $\Omega = B(x_0, R)$. Let us assume that F is differentiable in Ω and $\Gamma_0 = F'(x_0)^{-1}$ is defined, with $\|\Gamma_0 F(x_0)\| \leq a$. Let us suppose that condition (9) holds. If the sequence $\{r_k\}$ defined in (10) is increasing, with a limit $r^* \leq R$ such that $\omega(r^*) \leq 1$, then Newton's iterates $\{x_k\}$ are well defined and*

$$\|x_{k+1} - x_k\| \leq r_{k+1} - r_k. \tag{11}$$

Consequently, $\{x_k\}$ converges to a limit x^*, that is a solution of $F(x) = 0$. This solution is located in $\overline{B(x_0, r^*)}$. In addition, if $\omega(x) > 1$ for some $x > r^*$, the solution is unique in $B(x_0, \tau)$ where τ is the only solution of the equation

$$\frac{1}{x - r^*} \int_{r^*}^{x} \omega(s)\, ds = 1, \quad x \geq r^*. \tag{12}$$

Proof. We proceed inductively. First, (11) is clear for $k = 0$:

$$\|x_1 - x_0\| = \|\Gamma_0 F(x_0)\| \leq a = r_1 - r_0.$$

Now, let us assume that $\|x_{j+1} - x_j\| \leq r_{j+1} - r_j$, for $j = 0, 1, \ldots, k-1$. Then

$$\|x_{k+1} - x_k\| \leq \|\Gamma_k F'(x_0)\| \|\Gamma_0 F(x_k)\|.$$

As $\|x_k - x_0\| \leq \|x_k - x_{k-1}\| + \cdots + \|x_1 - x_0\| \leq r_k < r^* \leq R$,

$$\|I - \Gamma_0 F'(x_k)\| = \|\Gamma_0 [F'(x_k) - F'(x_0)]\| \leq \omega(\|x_k - x_0\|) \leq \omega(r_k) < \omega(r^*) \leq 1.$$

Then, there exists $\Gamma_k F'(x_0)$ and $\|\Gamma_k F'(x_0)\| \leq 1/(1 - \omega(r_k))$.

Next, by (1), we have the following expression for $F(x_k)$:

$$F(x_k) = F(x_k) - F(x_{k-1}) - F'(x_{k-1})(x_k - x_{k-1})$$

$$= \int_0^1 [F'(x_{k-1} + t(x_k - x_{k-1})) - F'(x_{k-1})](x_k - x_{k-1})\, dt.$$

So, we have

$$\|\Gamma_0 F(x_k)\| \leq \left\| \int_0^1 \Gamma_0 [F'(x_{k-1} + t(x_k - x_{k-1})) - F'(x_0)](x_k - x_{k-1})\, dt \right\|$$

$$+ \left\| \int_0^1 \Gamma_0[F'(x_{k-1}) - F'(x_0)](x_k - x_{k-1})\, dt \right\|$$

$$\leq \int_0^1 [\omega(r_{k-1} + t(r_k - r_{k-1})) + \omega(r_{k-1})](r_k - r_{k-1})\, dt$$

$$\leq \int_0^1 \tilde{\omega}(2r_{k-1} + t(r_k - r_{k-1}))(r_k - r_{k-1})\, dt = \int_{2r_{k-1}}^{r_k + r_{k-1}} \tilde{\omega}(s)\, ds.$$

Consequently

$$\|x_{k+1} - x_k\| \leq \frac{1}{1 - \omega(r_k)} \int_{2r_{k-1}}^{r_k + r_{k-1}} \tilde{\omega}(s)\, ds = r_{k+1} - r_k.$$

As $\{r_k\}$ converges, $\{x_k\}$ is also a convergent sequence. In addition, if $\lim x_k = x^*$, then

$$\lim_{k \to \infty} \|\Gamma_0 F(x_k)\| = \|\Gamma_0 F(x^*)\| \leq \lim_{k \to \infty} \int_{2r_{k-1}}^{r_k + r_{k-1}} \tilde{\omega}(s)\, ds = 0,$$

and hence, $F(x^*) = 0$.

Finally, from (11) we have $\|x_{k+m} - x_k\| \leq r_{k+m} - r_k$, $\forall m \geq 0$ and then $\|x^* - x_k\| \leq r^* - r_k$. In particular, $\|x^* - x_0\| \leq r^*$.

To show the unicity, notice that under the hypothesis of the theorem, equation (12) has only one solution: r. So, let us suppose that $y^* \in B(x_0, r)$ is another solution of $F(x) = 0$. Then

$$0 = \Gamma_0[F(x^*) - F(y^*)] = A(x^* - y^*), \quad A = \int_0^1 \Gamma_0[F'(y^* + t(x^* - y^*))]\, dt.$$

As the linear operator A is invertible because $\|I - A\| < 1$, $y^* = x^*$. □

As a particular case, let us see the behaviour of the sequences defined in (10) when $\omega(t) = Lt$, with L a positive constant, that is, the center Lipschitz case. So, we can compare this result with Theorem 3. For $\omega(t) = Lt$ we have

$$\tilde{\omega}(s) = \sup\{\omega(u) + \omega(v); u + v = s\} = \sup\{Lu + Lv; u + v = s\} = Ls.$$

Then (10) is now defined by $r_0 = 0$, $r_1 = a$,

$$r_{k+1} = r_k + \frac{1}{1 - Lr_k} \int_{2r_{k-1}}^{r_k + r_{k-1}} Ls\, ds = r_k + \frac{L(r_k + 3r_{k-1})}{2(1 - Lr_k)}(r_k - r_{k-1}).$$

Let us write $t_k = Lr_k$, for $k \geq 0$. Then, the previous sequence can be expressed in the following way:

$$\begin{cases} t_{k+1} - t_k = \dfrac{t_k + 3t_{k-1}}{2(1 - t_k)}(t_k - t_{k-1}), & k \geq 1, \\ t_0 = 0, \quad t_1 = aL = h. \end{cases}$$

To study analytically the convergence of the sequence $\{t_k\}$, let us assume that $t_k \leq T$ for $k \geq 0$, where T is a bound that we have to settle. So, if $T < 1/3$, we have
$$\frac{t_k + 3t_{k-1}}{2(1 - t_k)} \leq \frac{2T}{1 - T} = M < 1.$$

We calculate $t_2 = (2h - h^2)/(2(1 - h))$. Then we can bound t_k in terms of t_2:
$$t_k \leq t_2 + (t_2 - h)\frac{M}{1 - M}, \quad k \geq 3.$$

The question is now, when is true that $t_2 + (t_2 - h)M/(1 - M) \leq T$? This is a equation in T which has a solution if $h \leq 0.187472\ldots$ The solution is then
$$T = \frac{1 + 2h + t_2 - \sqrt{(1 + 2h + t_2)^2 - 12t_2}}{6}.$$

Consequently, we have proved that if $h \leq 0.187472\ldots$, then the sequence $\{t_k\}$ is increasing and satisfies $t_k \leq \left(1 + 2h + t_2 - \sqrt{(1 + 2h + t_2)^2 - 12t_2}\right)/6$, for all $k \geq 0$. Consequently, $\{t_k\}$ is convergent.

Notice that the value of h obtained above improves the value given in Theorem 3. Besides, this technique shows that the value of $h = aL$ can be improved by working with t_3, t_4, etc. In this way, numerical experiments show that this bound for the product aL can be improved until a value close to 0.213854.

Notes and Comments. In this paper we have analysed the convergence of Newton's method by modifying the Lipschitz condition that appears in the classical Kantorovich conditions. We have also studied which is the influence of these changes in the domains of existence and uniqueness of solution. All the results considered here are semilocal, that is, all of them include only a condition on the starting point for Newton's method.

It would be interesting to analyse the influence of similar changes in the local study. For instance, what happens if condition (2) is changed by a condition on the second derivative or by a center Lipschitz condition? One interesting reference for finding some answers to this question is the paper of Wang Xinghua [7], where local results are given under different Lipschitz conditions.

References

1. Appell, J., de Pascale, E., Lysenko, L. V. and Zabrejko, P. P.: New Results on Newton-Kantorovich Approximations with Applications to Nonlinear Integral Equations. Numer. Funct. Anal. Opt. **18 (1&2)** (1997) 1–17.
2. Gutiérrez, J. M.: A New Semilocal Convergence Theorem for Newton's method. J. Comput. Appl. Math. **79** (1997) 131–145.
3. Gutiérrez, J. M. and Hernández, M. A.: Newton's Method under Weak Kantorovich Conditions. To appear in IMA J. Numer. Anal.
4. Huang, Z.: A Note on the Kantorovich Theorem for Newton Iteration. J. Comput. Appl. Math. **47** (1993) 211–217.

5. Kantorovich, L. V. and Akilov, G. P.: *Functional Analysis*. Pergamon Press (New York), 1982.
6. Rokne, J.: Newton's Method Under Mild Differentiability Conditions with Error Analysis. Numer. Math. **18** (1972) 401–412.
7. Wang, X.: Convergence of Newton's Method and Uniqueness of the Solution of Equations in Banach Space. To appear in IMA J. Numer. Anal.

Positive Definite Solutions of the Equation $X + A^*X^{-n}A = I$

Vejdi Hassanov and Ivan Ivanov

Shoumen University, Laboratory Mathematical Modelling,
Shoumen 9712, Bulgaria
vejdi@dir.bg
i.gantchev@fmi.shu-bg.net

Abstract. The general nonlinear matrix equation $X + A^*X^{-n}A = I$ is discussed (n is a positive integer). Some necessary and sufficient conditions for existence a solution are given. Two methods for iterative computing a positive definite solution are investigated. Numerical experiments to illustrate the performance of the methods are reported.

1 Introduction

We consider the nonlinear matrix equation

$$X + A^*X^{-n}A = I \tag{1}$$

where X is a unknown matrix, I is the identity matrix and n is a positive integer.

The equation $X + A^*X^{-1}A = Q$ has many applications (see bibliography [1,2,8,9]). There are necessary and sufficient conditions for the existence of a positive definite solution [1,2]. Effective iterative procedures for solving the equation $X + A^*X^{-1}A = Q$ have been proposed in [5,8]. The iterative positive definite solutions and the properties of the equation $X + A^*X^{-2}A = I$ have been discussed in [6]. The general nonlinear matrix equation $X + A^*\mathcal{F}(X)A = Q$, where $\mathcal{F}$ maps positive definite matrices either into positive definite matrices or into negative definite matrices, and its iterative positive definite solutions have been investigated in [3]. The notation $Z > Y$ ($Z \geq Y$) indicates that $Z - Y$ is positive definite (semidefinite). The cases when the operator $\mathcal{F}(X)$ is monotone (if $0 < X \leq Y$ then $\mathcal{F}(X) \leq \mathcal{F}(Y)$) or anti-monotone (if $0 < X \leq Y$ then $\mathcal{F}(X) \geq \mathcal{F}(Y)$) are considered. For instance, the operator $\mathcal{F}(X) = X^r$ is a monotone one for $0 < r < 1$ and anti-monotone for $r = -1$ [7].

We derive some necessary and sufficient conditions for solutions of (1). Zhan and Xie [9] have derived necessary and sufficient conditions for the matrix equation $X + A^*X^{-1}A = I$ to have a positive definite solution. In this paper we extend these conditions for the general equation (1). Two iterative processes for computing a positive definite solution of the equation (1) are studied.

The following notations are used the paper. The $\rho(A)$ is the spectral radius of A. We denote by $\|.\|$ the spectral norm. The notation $Y = \sqrt[n]{Z}$ means that Y, Z are positive definite and $Z = Y^n$. We can compute Y in the following way.

Since Z is a positive definite matrix we have $Z = UDU^*$ where U is unitary and D is diagonal with positive entries. Hence, $Y = U\sqrt[n]{D}U^*$. If $Y \geq Z > 0$ we have that $\sqrt[n]{Y} \geq \sqrt[n]{Z} > 0$ where n is a positive integer since $\sqrt[n]{Y}$ is a monotone operator [7].

2 Necessary and Sufficient Conditions

In this section we discuss positive definite solutions of the equation (1) where A is a real matrix ($A^* = A^T$).

Theorem 1. *The equation (1) has a solution X if and only if the matrix A has the decomposition*

$$A = \begin{cases} VW^T Z, & n = 2k+1, \, k = 0, 1, \ldots \\ VZ, & n = 2k, \quad k = 1, 2, \ldots \end{cases} \quad (2)$$

where $V = (W^T W)^k$ and W is a nonsingular square matrix and the columns of $\begin{pmatrix} W \\ Z \end{pmatrix}$ are orthonormal. In this case $X = W^T W$ is a solution and all solutions can be obtained in this way.

Proof. If (1) has a solution $X > 0$ then we can write $X = W^T W$ where W is a nonsingular square matrix. We rewrite the equation (1) as

$$W^T W + A^T (W^T W)^{-n} A = I$$
$$W^T W + Z^T Z = I,$$

where $Z = \begin{cases} W^{-T}(W^{-1}W^{-T})^k A, & n = 2k+1, \, k = 0, 1, \ldots \\ (W^{-1}W^{-T})^k A, & n = 2k, \quad k = 1, 2, \ldots \end{cases}$

Hence

$$A = \begin{cases} VW^T Z, & n = 2k+1, \, k = 0, 1, \ldots \\ VZ, & n = 2k, \quad k = 1, 2, \ldots \end{cases}$$

and columns of $\begin{pmatrix} W \\ Z \end{pmatrix}$ are orthonormal.

Conversely, assume A has the decomposition (2) and $X = W^T W$. Then

$$X + A^T X^{-n} A = \begin{cases} W^T W + Z^T W V^T (W^T W)^{-n} V W^T Z, & n = 2k+1 \\ W^T W + Z^T V^T (W^T W)^{-n} V Z, & n = 2k \end{cases}$$
$$= W^T W + Z^T Z = I,$$

since $V = (W^T W)^k$.

Hence X is a solution. □

Theorem 2. *The equation (1) has a solution if and only if there exist orthogonal matrices P and Q and diagonal matrices $\Theta > 0$ and $\Sigma \geq 0$, such that $\Theta^2 + \Sigma^2 = I$ and $A = P^T \Theta^n Q \Sigma P$. In this case $X = P^T \Theta^2 P$ is a solution.*

Proof. Assume the equation (1) has a solution X. From the Theorem 1 it follows that the matrix A has the factorization:

$$A = \begin{cases} VW^T Z, & n = 2k+1, \; k = 0, 1, \ldots \\ VZ, & n = 2k, \quad k = 1, 2, \ldots \end{cases}$$

where $V = (W^T W)^k$ and W is a nonsingular square matrix and the columns of $\begin{pmatrix} W \\ Z \end{pmatrix}$ are orthonormal. In this case $X = W^T W$.

We extend the matrix $\begin{pmatrix} W \\ Z \end{pmatrix}$ to an orthogonal matrix $\Xi = \begin{pmatrix} W & U \\ Z & H \end{pmatrix}$.

The matrix Ξ can be written as [4]

$$\begin{pmatrix} W & U \\ Z & H \end{pmatrix} = \begin{pmatrix} U_1 & 0 \\ 0 & U_2 \end{pmatrix} \begin{pmatrix} \Theta & -\Sigma \\ \Sigma & \Theta \end{pmatrix} \begin{pmatrix} P & 0 \\ 0 & H_2 \end{pmatrix},$$

where U_1, U_2, P and H_2 are orthogonal matrices, Θ and Σ are positive semidefinite which satisfy $\Theta^2 + \Sigma^2 = I$, $W = U_1 \Theta P$ and $Z = U_2 \Sigma P$. Since W is nonsingular then $\Theta > 0$. Define

$$Q = \begin{cases} U_1^T U_2, & n = 2k+1, \; k = 0, 1, \ldots \\ PU_2, & n = 2k, \quad k = 1, 2, \ldots \end{cases}.$$

Hence $A = P^T \Theta^n Q \Sigma P$.

Conversely, suppose $A = P^T \Theta^n Q \Sigma P$ where P and Q are orthogonal, Θ and Σ are diagonal matrices and $\Theta > 0$, $\Sigma \geq 0$, $\Theta^2 + \Sigma^2 = I$.

For $X = P^T \Theta^2 P$ we obtain

$$P^T \Theta^2 P + P^T \Sigma Q^T \Theta^n P P^T \Theta^{-2n} P P^T \Theta^n Q \Sigma P = P^T \Theta^2 P + P^T \Sigma^2 P = I.$$

Hence $X = P^T \Theta^2 P$ is a solution of the equation (1). □

Theorem 3. *If the equation (1) has a solution then* $\|A\| < 1$.

Proof. If (1) has a solution by Theorem 2 we obtain that there exist orthogonal matrices P and Q and diagonal matrices Θ, Σ such that $\Theta^2 + \Sigma^2 = I$ and $A = P^T \Theta^n Q \Sigma P$. Compute

$$\|A\| = \|P^T \Theta^n Q \Sigma P\| \leq \|P^T\| \|\Theta^n\| \|Q\| \|\Sigma\| \|P\| = \|\Theta^n\| \|\Sigma\|.$$

Since Θ is a diagonal nonsingular matrix and $\Theta^2 + \Sigma^2 = I$, $\Theta > 0$, $\Sigma \geq 0$ we obtain $\|\Theta^n\| = \|\Theta\|^n$, $\|\Sigma\| < 1$ and $\|\Theta\| \leq 1$.

Hence $\|A\| \leq \|\Theta^n\| \|\Sigma\| = \|\Theta\|^n \|\Sigma\| < 1$. □

Theorem 4. *If the equation (1) has a solution X then*

(i) $I \geq X > \sqrt[n]{AA^T}$,
(ii) $I - A^T A - \sqrt[n]{AA^T} > 0$,
(iii) $\rho(A) \leq \sqrt{\frac{n^n}{(n+1)^{n+1}}}$,
(iv) $\rho(A + A^T) \leq 1$,
(v) $\rho(A - A^T) \leq 1$.

Proof. Further on, we use $\lambda(A)$ to denote the set of eigenvalues of A.

((i):) Since we discuss positive definite solutions of the (1) then $X > 0$ and $A^T X^{-n} A > 0$. Hence $X \leq I$ and $A^T X^{-n} A < I$, respectively. From Theorem 1 it follows that the solution has the type $X = W^T W$ and

$$A = \begin{cases} (W^T W)^k W^T Z, & n = 2k+1, \ k = 0, 1, \ldots \\ (W^T W)^k Z, & n = 2k, \ k = 1, 2, \ldots \end{cases}.$$

Then

$$X - \sqrt[n]{AA^T} = \begin{cases} W^T W - \sqrt[n]{(W^T W)^k W^T Z Z^T W (W^T W)^k}, & n = 2k+1 \\ W^T W - \sqrt[n]{(W^T W)^k Z Z^T (W^T W)^k}, & n = 2k \end{cases}$$

Since $\lambda(ZZ^T) = \lambda(Z^T Z)$ and $I - Z^T Z = W^T W > 0$ we obtain $ZZ^T < I$.

$$\sqrt[2k+1]{(W^T W)^k W^T Z Z^T W (W^T W)^k} < W^T W,$$

$$\sqrt[2k]{(W^T W)^k Z Z^T (W^T W)^k} < W^T W.$$

Hence $X - \sqrt[n]{AA^T} > 0$.

((ii):) Using ((i)) we have $X < I$, $X^{-1} > I$ and $X > \sqrt[n]{AA^T}$. Thus

$$0 = X + A^T X^{-n} A - I > \sqrt[n]{AA^T} + A^T X^{-n} A - I > \sqrt[n]{AA^T} + A^T A - I.$$

((iii):) For the eigenvalues of A we have

$$\lambda(A) = \lambda(P^T \Theta^n Q \Sigma P) = \lambda(\Theta^n Q \Sigma) = \lambda(Q \Sigma \Theta^n).$$

Moreover

$$\rho(A) = \max |\lambda(Q \Sigma \Theta^n)| \leq \|Q \Sigma \Theta^n\| = \|\Sigma \Theta^n\|.$$

Assume $\Sigma = diag\{\sigma_i\}$, $\Theta = diag\{\theta_i\}$. Then $\sigma_i \geq 0$, $\theta_i > 0$ and $\sigma_i^2 + \theta_i^2 = 1$. We obtain

$$\rho(A) \leq \|\Sigma \Theta^n\| = \max_i |\sigma_i \theta_i^n| = \max_i \sigma_i (1 - \sigma_i^2)^{\frac{n}{2}}$$

$$\leq \max_{x \in [0,1)} x(1 - x^2)^{\frac{n}{2}} = \sqrt{\frac{n^n}{(n+1)^{n+1}}}.$$

((iv):) Consider $I \pm (A^T + A)$ in cases $n = 2k+1$ and $n = 2k$.

$$\begin{aligned}I \pm (A^T + A) &= Z^T Z + W^T W \pm Z^T W(W^T W)^k \pm (W^T W)^k W^T Z \\ &\geq Z^T (WW^T)^{2k} Z + W^T W \pm Z^T W(W^T W)^k \pm (W^T W)^k W^T Z \\ &= (W \pm W(W^T W)^{k-1} W^T Z)^T (W \pm W(W^T W)^{k-1} W^T Z) \geq 0. \\ I \pm (A^T + A) &= Z^T Z + W^T W \pm Z^T (W^T W)^k \pm (W^T W)^k Z \\ &\geq Z^T (W^T W)^{2k-1} Z + W^T W \pm Z^T (W^T W)^k \pm (W^T W)^k Z \\ &= (W \pm W(W^T W)^{k-1} Z)^T (W \pm W(W^T W)^{k-1} Z) \geq 0.\end{aligned}$$

Since $\lambda(WW^T) = \lambda(W^T W)$ and $I - W^T W = Z^T Z \geq 0$ it follows $WW^T \leq I$. Hence $\rho(A^T + A) \leq 1$.

((v):) Consider $\rho(A - A^T)$ in cases $n = 2k+1$ and $n = 2k$.

$$\begin{aligned}\rho(A - A^T) &= \rho((W^T W)^k W^T Z - Z^T W(W^T W)^k) \\ &\leq \rho((W^T W)^{2k+1} + Z^T Z) \leq \rho(W^T W + Z^T Z) = 1, \\ \rho(A - A^T) &= \rho((W^T W)^k Z - Z^T (W^T W)^k) \\ &\leq \rho((W^T W)^{2k} + Z^T Z) \leq \rho(W^T W + Z^T Z) = 1.\end{aligned}$$

Hence $\rho(A - A^T) \leq 1$. □

3 Iteration Methods for Solving the Equation

We consider iterative processes for solving the equation (1). Conditions for convergence of the iterative algorithms are given.

Consider the iterative method

$$X_0 = \gamma I \quad X_{s+1} = I - A^* X_s^{-n} A, \tag{3}$$

where $\gamma \in (\frac{n}{n+1}, 1]$.

Theorem 5. *If there exist numbers α and β such that*

(i) $\frac{n}{n+1} < \alpha \leq \beta \leq 1$;
(ii) $\beta^n (1 - \beta) I \leq A^* A \leq \alpha^n (1 - \alpha) I$.

Then the iterative process (3), with $\alpha \leq \gamma \leq \beta$, converges to a positive definite solution X of (1) with linear convergence rate and $\frac{n}{n+1} I \leq X \leq I$.

Proof. We shall show the matrix sequence $\{X_s\}$ is a Cauchy sequence and for each X_s we have $\alpha I \leq X_s \leq \beta I$.

Suppose $X_0 = \gamma I$, $(\alpha \leq \gamma \leq \beta)$. Obviously $\alpha I \leq X_0 \leq \beta I$. We get

$$\alpha I \leq I - \frac{\alpha^n}{\gamma^n}(1 - \alpha) I \leq X_1 = I - \frac{A^* A}{\gamma^n} \leq I - \frac{\beta^n}{\gamma^n}(1 - \beta) I \leq \beta I.$$

Hence $\alpha I \leq X_1 \leq \beta I$. Assume $\alpha I \leq X_s \leq \beta I$. We obtain

$$\frac{1}{\beta^n} I \leq X_s^{-n} \leq \frac{1}{\alpha^n} I$$

$$\frac{\beta^n}{\beta^n}(1-\beta)I \leq \frac{A^*A}{\beta^n} \leq A^* X_s^{-n} A \leq \frac{A^*A}{\alpha^n} \leq \frac{\alpha^n}{\alpha^n}(1-\alpha)I$$

$$I - \beta I \leq A^* X_s^{-n} A \leq I - \alpha I$$

$$\alpha I \leq I - A^* X_s^{-n} A \leq \beta I.$$

Thus $\alpha I \leq X_s \leq \beta I$, $s = 0, 1, 2, \ldots$.

$$X_{s+1} - X_s = A^*(X_{s-1}^{-n} - X_s^{-n})A = A^* X_s^{-n}(X_s^n - X_{s-1}^n)X_{s-1}^{-n} A$$
$$= A^* X_s^{-n}[X_s^{n-1}(X_s - X_{s-1}) + X_s^{n-2}(X_s - X_{s-1})X_{s-1} + \ldots$$
$$+ X_s(X_s - X_{s-1})X_{s-1}^{n-2} + (X_s - X_{s-1})X_{s-1}^{n-1}]X_{s-1}^{-n} A$$
$$= A^* X_s^{-1}(X_s - X_{s-1})X_{s-1}^{-n} A + \ldots + A^* X_s^{-n}(X_s - X_{s-1})X_{s-1}^{-1} A$$

$$\|X_{s+1} - X_s\| \leq \|A\|^2 \left[\|X_s^{-1}\|\|X_{s-1}^{-n}\| + \ldots + \|X_s^{-n}\|\|X_{s-1}^{-1}\|\right] \|X_s - X_{s-1}\|$$
$$\leq \frac{n\|A\|^2}{\alpha^{n+1}} \|X_s - X_{s-1}\| \leq \frac{n\alpha^n(1-\alpha)}{\alpha^{n+1}} \|X_s - X_{s-1}\|.$$

Since $\alpha > \frac{n}{n+1}$ then $q = \frac{n(1-\alpha)}{\alpha} < 1$. Hence $\|X_{s+1} - X_s\| \leq q^s \|X_1 - X_0\|$. Moreover

$$\|X_{s+p} - X_s\| \leq q^s \|X_p - X_0\| \leq q^s \left(\|X_p - X_{p-1}\| + \ldots + \|X_1 - X_0\|\right)$$
$$\leq q^s(q^{p-1} + q^{p-2} + \ldots + 1)\|X_1 - X_0\| < \frac{q^s}{1-q} \|X_1 - X_0\|.$$

Since $q < 1$ then $\lim_{s\to\infty} \frac{q^s}{1-q} = 0$.

Consequently the $\{X_s\}$ is a Cauchy matrix sequence. Since $\mathcal{C}^{n\times n}$ is a Banach space then $\{X_s\}$ converges to a positive definite solution of the (1). □

We consider the second iterative process

$$Y_0 = \xi I, \quad Y_{s+1} = \sqrt[n]{A(I - Y_s)^{-1} A^*}, \tag{4}$$

where $\xi \in [0, \frac{n}{n+1}]$.

Theorem 6. *If there exist numbers η and γ such that*

(i) $0 \leq \eta \leq \gamma \leq \frac{n}{n+1}$;
(ii) $\eta^n(1-\eta)I \leq AA^* \leq \gamma^n(1-\gamma)I$.

Then the iterative process (4), with $\xi = \eta$ and $\xi = \gamma$, converges to a positive definite solution Y of (1) and $Y \leq \frac{n}{n+1} I$.

Proof. Consider the case $Y_0 = \xi I$ ($\xi = \eta$). We shall show that the matrix sequence $\{Y_s\}$ is a monotonically increasing sequence and for each Y_s we have $\eta I \leq Y_s \leq \gamma I$.

According to (ii) we have

$$\eta I = \sqrt[n]{\frac{\eta^n(1-\eta)}{1-\eta}} I \leq \sqrt[n]{\frac{AA^*}{1-\eta}} = Y_1 \leq \sqrt[n]{\frac{AA^*}{1-\gamma}} \leq \sqrt[n]{\frac{\gamma^n(1-\gamma)}{1-\gamma}} I = \gamma I.$$

Hence $Y_0 \leq Y_1 \leq \gamma I$. Assume that $Y_{s-1} \leq Y_s \leq \gamma I$.
Thus

$$Y_{s+1} = \sqrt[n]{A(I-Y_s)^{-1}A^*} \geq \sqrt[n]{A(I-Y_{s-1})^{-1}A^*} = Y_s.$$

Since $Y_s \leq \gamma I$ it follows $A(I-Y_s)^{-1}A^* \leq \frac{AA^*}{1-\gamma} \leq \frac{\gamma^n(1-\gamma)}{1-\gamma}I = \gamma^n I$. Hence $Y_s \leq Y_{s+1} \leq \gamma I$ for $s = 0, 1, \ldots$. □

Remark 1. If $Y_0 = \gamma I$ it can be proved that Y_s is a monotonically decreasing sequence.

Theorem 7. *If the equation (1) where A is real has a positive definite solution Y, then the iterative process (4) where $Y_0 = \sqrt[n]{AA^T}$ converges to the smallest positive definite solution Y_{min}.*

Proof. We shall show that the sequence $\{Y_s\}$ is a monotonically increasing one and bounded above from any positive definite solution Y. We have
$0 < I - \sqrt[n]{AA^T} \leq I$ since $I > \sqrt[n]{AA^T}$ by Theorem 4. Then $(I - \sqrt[n]{AA^T})^{-1} \geq I$ and $A(I - \sqrt[n]{AA^T})^{-1}A^T \geq AA^T$. We can write

$$Y_1 = \sqrt[n]{A(I-Y_0)^{-1}A^T} \geq Y_0 = \sqrt[n]{AA^T}$$

We assume that $Y_s \geq Y_{s-1}$. It easy to show that $Y_{s+1} \geq Y_s$.
Hence the sequence $\{Y_s\}$ is monotonically increasing.
Let Y be a any positive definite solution of the equation (1). From Theorem 4 (i) we obtain $Y > \sqrt[n]{AA^T} = Y_0$.
We suppose that $Y_s \leq Y$. We shall prove $Y_{s+1} \leq Y$. We have

$$(I-Y_s)^{-1} \leq (I-Y)^{-1}$$
$$Y_{s+1} = \sqrt[n]{A(I-Y_s)^{-1}A^T} \leq \sqrt[n]{A(I-Y)^{-1}A^T} = Y.$$

Hence $Y_{s+1} \leq Y$ for all s and each positive definite solution Y of the (1).
Thus the $\{Y_s\}$ converges to the smallest positive definite solution Y_{min} of the equation (1). □

4 Numerical Experiments

We carry out numerical experiments for computing the positive definite solutions of the equation (1) where $n = 3$ in MATLAB on a PENTIUM computer. We use considered methods (3) and (4). As a practical stopping criterion we use $\varepsilon = \|Z + A^T Z^{-3} A - I\|_\infty \leq tol$ and $tol = 10^{-8}$.

Example 1. Consider the matrix

$$A = \frac{1}{100} \begin{pmatrix} 16 & -9 & -8 \\ 11 & 16 & 5 \\ 4 & -8 & 18 \end{pmatrix}.$$

We compute the solution X using the method (3) with different values of γ and the method (4) with different values of ξ. The method (3) ($X_0 = \gamma I$) with $\gamma = 1$ it needs 8 iterations and for $\gamma = 0,955$ it needs 7 iterations. In case $\gamma = 0,951$ it needs 7 iterations and for $\gamma = 0,75$ it needs 10 iterations. The method (4) ($Y_0 = \xi I$) with $\xi = 0$ it needs 13 iterations and for $\xi = 0,403$ it needs 13 iterations. In case $\xi = 0,414$ it needs 13 iterations and for $\xi = 0,75$ it needs 14 iterations.

5 Conclusion

In this paper we introduced the general nonlinear matrix equation (1). We have studied some properties and two recurrence algorithms. The recurrence equation (3) defines the monotonically matrix sequence $\{X_s\}$ ($\gamma = 1$) for the equation $X + A^*X^{-1}A = I$ [2] which has a limit. It is proven that this limit is the largest positive definite solution. We expect that the matrix sequence $\{X_s\}$ (3) converges to the largest positive definite solution of this general equation. The matrix sequence $\{Y_s\}$ converges to the smallest positive definite solution of (1) for the special initial point.

References

1. Engwerda, J., Ran, A., Rijkeboer, A.: Necessary and Sufficient Conditions for the Existence of a Positive Definite Solution of the Matrix Equation $X + A^*X^{-1}A = Q$. Linear Algebra Appl. **186** (1993) 255–275
2. Engwerda, J. : On the Existence of a Positive Definite Solution of the Matrix Equation $X + A^TX^{-1}A = I$. Linear Algebra Appl. **194** (1993) 91–108
3. El-Sayed, S., Ran, A.: On an Method for Solving a Class of Nonlinear Matrix Equations. SIAM J. on Matrix Analysis /to appear/
4. Golub, G., van Loan C. : Matrix Computations, John Hopkins, Baltimore (1989)
5. Guo, C., Lancaster, P. : Iterative Solution of Two Matrix Equations. Mathematics of Computation **68** (1999) 1589–1603
6. Ivanov, I., El-Sayed, S. : Properties of Positive Definite Solution of the Equation $X + A^*X^{-2}A = I$. Linear Algebra And Appl. **279** (1998) 303–316
7. Kwong, M. : Some Results on Matrix Monotone Functions. Linear Algebra And Appl. **118** (1989) 129–153
8. Zhan, X. : Computing the Extremal positive definite solution of a Matrix Equation. SIAM J. Sci. Comput. **247** (1996) 337–345
9. Zhan, X., Xie, J. : On the Matrix Equation $X + A^TX^{-1}A = I$. Linear Algebra And Appl. **247** (1996) 337–345

Fast and Superfast Algorithms for Hankel-Like Matrices Related to Orthogonal Polynomials

Georg Heinig

Kuwait University, Dept.of Math.& Comp.Sci.
P.O.Box 5969, Safat 13060, Kuwait
georg@mcs.sci.kuniv.edu.kw

Abstract. Matrices are investigated that are Hankel matrices in bases of orthogonal polynomials. With the help of 3 equivalent definitions of this class fast LU-factorization algorithms and superfast solvers are constructed.

Keywords: Hankel matrix, orthogonal polynomials, fast algorithm

AMS(MOS) Subject Classification: 47B35, 15A09, 15A23

1 Introduction

Let a_k, b_k ($k = 0, 1, \ldots, 2n$) be given real numbers, where $b_0 = 0$ and $b_k \neq 0$ for $k > 0$, and $e_{-1} = 0$, $e_0(t) = 1$. Then the three-term recursion

$$b_{j+1} e_{j+1}(t) = (t - a_j) e_j(t) - b_j e_{j-1}(t) \tag{1}$$

defines a system of polynomials $e_j(t) = \sum_{i=0}^{j} e_{ij} t^i$ ($j = 0, 1, \ldots, 2n$), where $e_{jj} \neq 0$. We introduce the matrices $E_n = [e_{ij}]_{i,j=0}^{n}$ with $e_{ij} = 0$ for $i > j$.

In this paper we consider $(n+1) \times (n+1)$ matrices of the form $R_n = E_n^T H_n E_n$ where $H = [h_{j+k}]_{j,k=0}^{n}$ is a Hankel matrix. We call matrices of this form *OP-Hankel matrices*, where "OP" stand for "orthogonal polynomials". This name should point out that orthogonal polynomials satisfy a three-term recursion (1).

Let us mention some instances where OP-Hankel matrices appear. The most familiar one seems to be modified moment problems. In fact, for orthogonal polynomials $\{e_j(t)\}$ on the real line, the modified moment matrices with entries

$$r_{ij} = \int_{\mathbb{R}} e_i(t) e_j(t) d\sigma, \tag{2}$$

where σ is some (not necessarily positive) measure on the real line, are OP-Hankel matrices. Some general references for this are [5], [4], [3]. Then OP-Hankel matrices appear in least square problems for OP expansions. In this connection OP-Hankel matrices were introduced and studied in [8]. In [2] OP-Hankel matrices were used for preconditioning of ill-conditioned Hankel systems. This is based on the remarkable fact that positve definite OP-Hankel matrices can be well conditioned, whereas positive definite Hankel matrices are always ill conditioned. Finally, in [6] it is shown that symmetric Toeplitz matrices and, more general, centrosymmetric Toeplitz-plus-Hankel matrices are unitarily equivalent to a direct sum of two special Chebyshev OP-Hankel matrices.

An inversion algorithm for $n \times n$ OP-Hankel matrices with complexity $O(n^2)$ was, as far as we know, first presented in [8]. More algorithms with this complexity are contained in [3]. The algorithms in [8] and [3] are Levinson-type algorithms. The disadvantage of Levinson-type algorithm compared with Schur-type algorithms is that they cannot fully parallelized and speeded up to superfast algorithms.

A Schur-type algorithm and the corresponding superfast $O(n \log^2 n)$ complexity solver for the special case of Chebyshev-Hankel matrices is presented in [6]. This also leads to a superfast Toeplitz solver based in real arithmetics.

In this paper the approach from [6] is generalized to arbitrary OP-Hankel matrices. The basic fact of our approach is that OP-Hankel matrices can be described in 3 different ways: Firstly, they are matrices of the form $R_n = E_n^T H_n E_n$, secondly they are matrices R_n for which the "displacement" $R_n T_n - T_n R_n$ has rank 2 and a special structure, where T_n is the tridiagonal matrix defined in Section 2, and finally they are the matrices of restricted multiplication operators with respect to the basis $\{e_k(t)\}$.

The last interpretation enables us to derive immediately Levinson- and Schur-type algorithms for LU- factorization of strongly nonsingular OP-Hankel matrices and their inverses. The combination of the Levinson and the Schur-type algorithms can be used to speed up the algorithm for the solution of OP-Hankel systems to complexity $O(n \log^3 n)$.

2 Displacement Structure

Throughout the paper, let T_m denote the $(m+1) \times (m+1)$ tridiagonal matrix

$$T_m = \begin{bmatrix} a_0 & b_1 & & \\ b_1 & a_1 & \ddots & \\ & \ddots & \ddots & b_m \\ & & b_m & a_m \end{bmatrix}.$$

We consider the commutator (or displacement) transformation $\nabla R_n = R_n T_n - T_n R_n$. Since all eigenvalues of T_n are simple, the kernel of ∇ has dimension $n+1$. Furthermore, R_n can be reproduced from ∇R_n and the first or the last column of R_n. In fact, the following is easily checked.

Proposition 1. Let r_k denote the $(k+1)$ th column of R_n and t_k the $(k+1)$ th column of ∇R_n $(k = 0, \ldots, n+1)$, then

$$r_{k+1} = \frac{1}{b_{k+1}}((T_n - a_k I_n)r_k - b_k r_{k-1} + t_k), \quad r_{k-1} = \frac{1}{b_k}((T_n - a_k)r_k - b_{k+1}r_{k+1} + t_k),$$

with $r_{-1} = r_{n+1} = 0$.

Let $\mathcal{H}_n$ denote the space of all matrices R_n for which ∇R_n has the form

$$\nabla R_n = g e_n^T - e_n g^T \tag{3}$$

for some $g \in \mathbb{R}^{n+1}$. Obviously, we may assume that the last component of g is zero. We shall show that $\mathcal{H}_n$ is just the set of all OP-Hankel matrices corresponding to the data a_k and b_k. For this we mention first that from the fact that the kernel of ∇ has dimension $n+1$ and g has n degrees of freedom it follows that

$$\dim \mathcal{H}_n \leq 2n + 1. \tag{4}$$

Next we observe the following.

Proposition 2. *An OP-Hankel matrix $R_n = E_n^T H_n E_n$ belongs to $\mathcal{H}_n$ and (3) holds with*

$$g = b_n r_{n-1} - (T_n - a_n I_{n+1}) r_n.$$

Proof. We may extend H_n to an infinite Hankel matrix H_∞. For the corresponding OP-Hankel matrix R_∞ we have $R_\infty T_\infty = T_\infty R_\infty$. Taking the first $n+1$ rows and columns of this relation we obtain the assertion. $\square$

Since the dimension of the space of all $(n+1) \times (n+1)$ Hankel matrices equals $2n+1$, the space of all $(n+1) \times (n+1)$ OP-Hankel matrices also equals $2n+1$, the following is true.

Corollary 1. *Any matrix $R_n \in \mathcal{H}_n$ admits a representation $R_n = E_n^T H_n E_n$, where H_n is a $(n+1) \times (n+1)$ Hankel matrix.*

We give now a third characterization of OP-Hankel matrices. Let $\mathbb{R}_n[t]$ denote the space of all polynomials of degree less than or equal to n with real coefficients and $\mathcal{P}_n$ the projection defined for polynomials $x(t) = \sum_{k=0}^{N} x_k e_k(t)$ by $\mathcal{P}_n x(t) = \sum_{k=0}^{n} x_k e_k(t)$. Note that $\mathcal{P}_n t^k = 0$ if $k > 2n$. Furthermore, for a given polynomial $x(t)$, let $[x(t)]_k$ denote its coefficient in its expansion by $\{e_k(t)\}$, i.e. if $x(t) = \sum_{k=0}^{N} x_k e_k(t)$, then $[x(t)]_k = x_k$.

For a given polynomial $p(t)$ of degree less than or equal to $2n$, let $\mathcal{R}_n(p)$ denote the operator in $\mathbb{R}_n[t]$ defined by $\mathcal{R}_n(p) x(t) = \mathcal{P}_n p(t) x(t)$. For $p(t) = t$ we set $\mathcal{S}_n := \mathcal{R}_n(p)$.

Proposition 3.

$$(\mathcal{R}_n(p) \mathcal{S}_n - \mathcal{S}_n \mathcal{R}_n(p)) x(t) = [p(t) x(t)]_{n+1} e_n(t) - b_{n+1} (g(t)[x(t)]_n),$$

where $g(t) = \mathcal{P} p(t) e_{n+1}(t)$.

The proof is a straightforward verification.

Let $R_n(p)$ denote the matrix of the operator $\mathcal{R}_n(p)$ with respect to the basis $\{e_k(t)\}$. In particular we have $R_n(1) = I_n$ and $R_n(t) = T_n$. Furthermore,

$$R_n(f) = [\, I_n \; 0 \,] p(T_N) \begin{bmatrix} I_n \\ 0 \end{bmatrix}$$

for any $N > 2n$, and the relation in Proposition 2.2 can be written in the form

$$R_n(p)T_n - T_n R_n(p) = b_{n+1}(g_n e_n^T - e_n g_n^T), \tag{5}$$

where g is the coefficient vector of $g(t)$ with respect to expansion of $g(t)$ by $\{e_k(t)\}$.

That means the matrices $\mathcal{R}_n(p)$ belong to the class $\mathcal{H}_n$ and are, therefore, OP-Hankel matrices. Since the mapping $p(t) \longrightarrow R_n(p)$ is one-to-one for $p(t) \in \mathbb{R}_{2n}[t]$, the dimension of the space of matrices $R_n(p)$ equals $2n+1$. This leads to the main result of this section.

Theorem 1. *For an $(n+1) \times (n+1)$ matrix R, the following are equivalent:*

1. *The matrix R_n is of the form $R_n = E_n^T H_n E_n$ for some Hankel matrix H.*
2. *The commutator ∇R_n satisfies (3) for some $g \in \mathbb{R}^{n+1}$.*
3. *For some polynomial $p(t) \in \mathbb{R}_{2n}[t]$, $R_n = R_n(p)$.*

If R_n is given in the form $R_n = E_n^T H E_n$ with $H = [h_{i+j}]_{i,j=0}^n$, then the coefficient vector p of $p(t)$ with respect to the basis $\{e_k(t)\}$ is given by $p = E_{2n}^T s$, where $h = [h_k]_{k=0}^{2n}$. If R_n is given by (2) then the coefficients of $p(t)$ are the numbers r_{i0} ($i = 0, \ldots, 2n$).

3 Algorithms for LU-Factorization

In this and the next sections we consider only strongly nonsingular OP-Hankel matrices $R_n = R_n(p) = [r_{ij}]_{i,j=0}^n$. That means we assume that the principal subsections $[r_{ij}]_{i,j=0}^k$ are nonsingular for $k = 0, \ldots, n$. This covers, in particular, the case when R_n is positive definite.

We seek fast algorithms for the LU-factorization of R_n and its inverse. More precisely, we are looking for an upper triangular matrix $U_n = [u_{ij}]_{i,j=0}^n$ and a lower triangular matrix $L_n = [l_{ij}]_{i,j=0}^n$ satisfying

$$R_n U_n = L_n \quad \text{and} \quad u_{ii} = 1 \quad (i = 0, \ldots, n). \tag{1}$$

In polynomial language this can be written in the form

$$p(t)u_k(t) = l_k(t), \tag{2}$$

where

$$u_k(t) = \sum_{i=0}^{k} u_{ik} e_i(t), \quad l_k(t) = \sum_{i=k}^{2n} l_{ik} e_i(t).$$

Theorem 2. *The columns of U_n and L_n in (1) can be computed via the recursion*

$$b_{k+1} u_{k+1}(t) = (t - \alpha_k) u_k(t) - \beta_k u_{k-1}(t)$$
$$b_{k+1} l_{k+1}(t) = (t - \alpha_k) l_k(t) - \beta_k l_{k-1}(t),$$

where $k = 0, \ldots, n-1$,

$$\beta_k = \frac{b_k l_{kk}}{l_{k-1,k-1}}, \quad \alpha_k = \frac{b_k l_{kk} l_{k-1,k} - b_{k+1} l_{k,k+1} l_{k-1,k-1}}{l_{kk} l_{k-1,k-1}}.$$

This theorem can be proved by straightforward verification.

The initial polynomials $u_0(t)$, $l_0(t)$ are given by $u_0(t) = 1$ and $l_0(t) = p(t)$.

The recursions can easily be translated into vector language using the fact that matrix of the operator of multiplication by t with respect to the basis $\{e_k(t)\}$ is equal to T_{2n}.

The algorithm emerging from the theorem is a hybrid Levinson-Schur type algorithm. It is in particular convenient for parallel computation and has $O(n)$ complexity if n processors are available.

It is possible to calculate only the columns of the upper factor U_n, and the quantities l_{ij} for $0 \leq j - i \leq 1$ as some inner products of rows of R_n and the u_k. This leads to a Levinson type algorithm. It is also possible to calculate only the lower factor L, which is results in a pure Schur type algorithm. In this case the solution of a system $R_n x = b$ will be obtained by backward substitution.

4 OP-Bezoutians

Apparently it makes no sense to ask for LU-factorization algorithms that require less than $O(n^2)$ operations, but it makes sense to ask for such algorithms to solve systems of equations. In the case of Hankel matrices one can make use of the fact that inverses of Hankel matrices are Bezoutians, which are matrices $[b_{jk}]$ such that the "generating function" $B(t, s) = \sum_{j,k} b_{jk} t^k s^k$ equals $(u(t)v(s) - v(t)u(s))/(t - s)$, where $u(t)$, $v(t)$ are certain polynomials. Onces u and v are given, a Hankel system can be solved by matrix-vector multiplication which can be carries out with $O(n \log n)$ complexity if FFT is used.

This leads us to the definition of OP-Bezoutians. For our given system of polynomials $E = \{e_j(t)\}$ and a given matrix $B = [b_{jk}]$ we define the "E-generating function"

$$B_E(t, s) = \sum_{j,k} b_{jk} e_j(t) e_k(s).$$

A matrix B is called an E-Bezoutian if $B_E(t, s) = (u(t)v(s) - v(t)u(s))/(t - s)$. Since OP-Hankel matrices admit a representation $R_n = E_n^T H_n E_n$, we conclude the following.

Proposition 4. *Inverses of OP-Hankel matrices are OP-Bezoutians.*

Let us mention that the polynomials $u(t)$ and $v(t)$ are, up to a constant factor, equal to $u_n(t)$ and $u_{n+1}(t)$ introduced in the previous section. That means, in order to solve systems of equations with the coefficient matric R_n it is sufficient to store these two polynomials.

5 Fast Polynomial Multiplication

In order to obtain algorithms for the solution of systems with a OP-Hankel coefficient matrix we need an algorithm for fast multiplication of polynomials in OP-expansions. For this we can use the approximative algorithms from [1], but we can also use the exact algorithms described in [13]. In the latter paper $(2N+1) \times (N+1)$ matrices of the form

$$V_N = [e_k(c_j)]_{j=0,\ k=0}^{2N,\ N}$$

with $c_j = \cos \frac{j\pi}{N}$ are considered and algorithms are presented that multiply a vector by V_N or by V_N^T with complexity $O(N \log^2 N)$ and resonable accuracy.

We need the following property, which is mentioned in [13].

Proposition 5. *Let w be the first column of the inverse of the matrix $\widetilde{V}_N = [e_k(c_j)]_{j,k=0}^{2N}$, and let D_w be the diagonal matrix $D_w = \text{diag}\,[w_j]_{j=0}^{2N}$. Then*

$$V_N^T D_w V_N = I_{N+1}.$$

Once the weight vector w is precomputed, it is clear how to multiply two polynomials. First we chose $N > 2(m+n)$ and multiply the matrix V_N by coefficient vectors of $x(t)$ and $y(t)$, which means that the values of $x(t)$ and $y(t)$ at c_j are computed. Then the computed values are multiplied by each other and by w_j, and the V_N^T is applied to obtain the coefficient vector of the product in the expansion by $\{e_k(t)\}$.

Let us note that the algorithms briefly sketched in this section can also be used for fast matrix-vector multiplication by OP-Bezoutians. That means if the data in the OP-Bezoutians are given, then a $n \times n$ system with an OP-Hankel coefficient matrix can be solved with complexity $O(n \log^2 n)$. This complexity reduces to $O(n \log n)$ in the case of Chebyshev polynomials.

6 Superfast Algorithm

We show now how an algorithm with complexity $O(n \log^3 n)$ to find $u_n(t)$ and $u_{n+1}(t)$, which are required for the solution of OP-Hankel systems, can be designed. We introduce 2×2 matrix polynomials

$$U_k(t) = \begin{bmatrix} u_k(t) & u_{k-1}(t) \\ l_k(t) & l_{k-1}(t) \end{bmatrix}, \quad \Theta_k(t) = \frac{1}{b_{k+1}} \begin{bmatrix} t - \alpha_k & b_{k+1} \\ -\beta_k & 0 \end{bmatrix}$$

Then the relation in Theorem 3.1 can be written in the form

$$U_{k+1}(t) = U_k(t)\Theta_k(t). \tag{3}$$

We define, for $j > k$

$$\Theta_{kj}(t) = \Theta_k(t)\Theta_{k+1}(t)\ldots\Theta_{j-1}(t).$$

Then, for $j > i > k$,

$$\Theta_{kj}(t) = \Theta_{ki}(t)\Theta_{ij}(t), \quad U_j(t) = U_k(t)\Theta_{kj}(t). \tag{4}$$

In order to achieve complexity $O(n \log^3 n)$ it is important to carry out the calculations not with the complete polynomials $l_k(t)$ but only with the relevant part of them. We define

$$l_k^j(t) = \sum_{i=k}^{j} l_{ki} e_i(t).$$

It is easily checked that $\Theta_{k,k+1}(t)$ can be computed from $l_{k-1}^k(t)$ and $l_k^{k+1}(t)$, $\Theta_{k,k+2}(t)$ from $l_{k-1}^{k+2}(t)$ and $l_k^{k+3}(t)$ and, in general, $\Theta_{kj}(t)$ from $l_{k-1}^{2j-k-2}(t)$ and $l_k^{2j-k-1}(t)$. Furthermore, the following is true for $k < i < j$:

$$\left[l_{i-1}^{2j-i-2}(t) \; l_i^j(t) \right] = [\mathcal{P}_{2j-i-2} h_{i-1}(t) \quad \mathcal{P}_{2j-i-1} h_i(t)], \tag{5}$$

where

$$[h_{i-1}(t) \quad h_i(t)] = \left[l_{k-1}^j(t) \; l_k^j(t) \right] \Theta_{ki}(t). \tag{6}$$

This leads to the following recursive procedure.

Input: $[l_{k-1}^{2j-k-2}(t) \; l_k^{2j-k-1}(t)]$, **Output:** $\Theta_{kj}(t)$

1. If $j = k+1$ then apply Theorem 3.1.
2. Otherwise choose i with $k < i < j$ and carry out the following steps:
   (a) Apply the Procedure for $[l_k^{2i-k-2}(t) \; l_{k-1}^{2i-k-1}(t)]$. The output is $\Theta_{ki}(t)$.
   (b) Compute $[l_{i-1}^{2j-i-2}(t) \; l_i^{2j-i-1}(t)]$ by (5) and (6).
   (c) Apply the Procedure for $[l_i^j(t) \; l_{i-1}^j(t)]$. The output is $\Theta_{ij}(t)$.
   (d) Compute $\Theta_{kj}(t) = \Theta_{ki}(t)\Theta_{ij}(t)$ using a fast algorithm (as described in Section 4).

It is convenient to choose i close to the average of j and k. Proceeding in this way the problem to compute $\Theta_{kj}(t)$ is reduced to two subproblems of about half the size plus $O((j-k)\log^2(j-k))$ operations for polynomial multiplication. This ends up with complexity $O((j-k)\log^3(j-k))$. In particular, $U_n(t)$ can be computed with $O(n \log^3 n)$ operations.

7 Other Approaches

Let us briefly mention some other approaches to solve linear systems with a OP-Hankel coefficient matrix. The first one is described in [9]. It is based on displacement structure and Schur complements and applicable to matrices R for which the rank of $T_1 R - RT_2$, where T_1 and T_2 are tridiagonal matrices, is small compared with the size of the matrix R. This approach, however, does not fully use the specifics of OP-Hankel matrices.

The second approach is based on transformation into Cauchy-like matrices (see [7]) or into a tangential interpolation problem (as in [6] for Chebyshev-Hankel matrices) and the application of the algorithm described in [12]. For this the eigenvalues and eigenvectors of the matrix T have to be precomputed.

Finally, a convenient basis change transforms a general OP-Hankel matrix into a Chebyshev-Hankel matrix. The basis change can be carried out with the help of the algorithms described in [13] with $O(n \log^2 n)$ complexity. For the resulting Chebyshev-Hankel system one could use the $O(n \log^2 n)$ complexity algorithm described in [6]. This leads to a $O(n \log^2 n)$ complexity algorithm for general OP-Hankel systems. However, it is possible that the change of the basis increases the condition number of the matrix essentially so that the numerical application of this approach might be restricted.

References

1. Dutt, A., Rokhlin, V.: Fast Fourier transforms for nonequispaced data. SIAM J. Sci. Comp. **14** (1993) 1368–1393
2. Fasino, D.: Preconditioning finite moment problems. J. Comp. Appl. Math. **65** (1995) 145–155
3. Gemignani, L.: A fast algorithm for generalized Hankel matrices arising in finite-moment problems. Linear Algebra Appl. **267** (1997) 41–52
4. Golub, G., Gutknecht, M.: Modified moments for indefinite weight functions. Numer. Math. **67** (1994) 71–92
5. Gustafson, S. A.: On computational applications of the theory of moment problems. Rocky Mountain J. Math. **2** (1974) 227–240
6. Heinig, G.: Chebyshev-Hankel matrices and the splitting approach for centrosymmetric Toeplitz-plus-Hankel matrices. Linear Algebra Appl. (to appear)
7. Heinig, G., Bojanczyk, A.: Transformation techniques for Toeplitz and Toeplitz-plus-Hankel matrices, I. Transformations: Linear Algebra Appl. **254** (1997) 193–226, II. Algorithms: Linear Algebra Appl. 278 (1998), 11–36
8. Heinig, G., Hoppe, W., Rost, K.: Structured matrices in interpolation and approximation problems, Wissensch. Zeitschr. d. TU Karl-Marx-Stadt **31** 2 (1989) 196–202
9. Heinig, G., Olshevsky, V.: The Schur algorithm for matrices with Hessenberg displacement structure. (in preparation)
10. Heinig, G., Rost, K.: Algebraic Methods for Toeplitz-like matrices and operators. Akademie-Verlag Berlin and Birkhäuser Basel, Boston, Stuttgart, 1984
11. Kailath, T., Sayed, A.: Displacement structure: Theory and applications. SIAM Revue **37** (1995) 297–386
12. Olshevsky, V., Pan, V.: A unified superfast algorithm for boundary rational tangential interpolation problems and for inversion and factorization of dense structured matrices. Proc. of 39th Annual IEEE Symposium on Foundation of Computer Science 1998, 192–201
13. Potts, D., Steidl, G., Tasche, M.: Fast algorithms for discrete polynomial transforms. Math.Comp. **67** 224 (1998) 1577–1599

Acceleration by Parallel Computations of Solving High-Order Time-Accurate Difference Schemes for Singularly Perturbed Convection-Diffusion Problems*

Pieter W. Hemker[1], Grigorii I. Shishkin[2], and Lidia P. Shishkina[2]

[1] CWI
Amsterdam, The Netherlands
P.W.Hemker@cwi.nl

[2] Institute of Mathematics and Mechanics, Ural Branch of RAS,
Ekaterinburg 620219, Russia
{Grigorii,Lida}@shishkin.ural.ru

Abstract. For singularly perturbed convection-diffusion problems with the perturbation parameter ε multiplying the highest derivatives, we construct a scheme based on the defect correction method and its parallel variant that converge ε-uniformly with second-order accuracy in the time variable. We also give the conditions under which the parallel computation accelerates the solution process with preserving the higher-order accuracy of the original schemes.

1 Introduction

For several singularly perturbed boundary value problems, ε-uniformly convergent finite difference schemes have been constructed and analyzed (see, e.g., [1]–[5]). The time-accuracy of such schemes for nonstationary problems usually do not exceed first order. The use of a defect correction technique allows us to construct ε-uniform numerical methods with a higher order of accuracy in time (see e.g., [6,7]). Parallelization of the numerical method based on decomposition of the problem makes it possible to solve the discrete problem on a computer with several processors that may accelerate the computational process. However, this parallel process introduces additional errors in the numerical solutions. If the numerical method is accurate in time with order more than one, then the errors introduced by the domain decomposition (DD) can essentially exceed the discretization errors. Therefore, it is necessary to construct the parallel method such that the computation time is essentially less, and the accuracy is not lower than those for the corresponding nonparallel method.

---

* This research was supported in part by the Netherlands Organization for Scientific Research NWO, dossiernr. 047.008.007, and by the Russian Foundation for Basic Research under grant N 98-01-00362.

In the case of singularly perturbed problems ε-uniform parallel schemes based on the defect correction principle were studied in [8]. Parallel methods that allowed us to accelerate the numerical solution of the boundary value problems for parabolic reaction-diffusion equations on an interval were developed in [9,8].

In the present paper we consider the Dirichlet problem for a singularly perturbed convection-diffusion equation on a rectangle in that case when characteristics of the reduced equation are parallel to the sides of the rectangle. In this case regular and parabolic layers appear for $\varepsilon \to 0$. To solve the problem, we construct an ε-uniform scheme based on the defect correction method and its parallel variant convergent (ε-uniformly) with second-order accuracy in time. We also write out the conditions under which the parallel computation accelerates the solution process without losing the accuracy of the original schemes. The technique for analysis of difference schemes is similar to that given in [8].

2 Problem Formulation

On the domain $G = D \times (0,T]$, $D = (0,1) \times (0,1)$, with boundary $S = \overline{G} \setminus G$, we consider the Dirichlet problem for the singularly perturbed parabolic equation

$$Lu(x,t) \equiv \left\{ \varepsilon^2 \sum_{s=1,2} a_s(x,t) \frac{\partial^2}{\partial x_s^2} + b_1(x,t) \frac{\partial}{\partial x_1} - c(x,t) - p(x,t) \frac{\partial}{\partial t} \right\} u(x,t) =$$

$$= f(x,t), \quad (x,t) \in G, \quad \text{(1a)}$$

$$u(x,t) = \varphi(x,t), \quad (x,t) \in S. \quad \text{(1b)}$$

Here $a_s(x,t)$, $b_1(x,t)$, $c(x,t)$, $p(x,t)$, $f(x,t)$, $(x,t) \in \overline{G}$, and $\varphi(x,t)$, $(x,t) \in S$ are sufficiently smooth and bounded functions, moreover, $a_s(x,t) \geq a_0 > 0$, $b_1(x,t) \geq b_0 > 0$, $p(x,t) \geq p_0 > 0$, $c(x,t) \geq 0$, $(x,t) \in \overline{G}$; $\varepsilon \in (0,1]$.

Let $S = S^L \cup S_0$, $S_0 = \overline{S}_0$. We distinguish four faces in the lateral boundary S^L: $S^L = \cup_{j=1}^{4} S_j$, $S_j = \Gamma_j \times (0,T]$, where Γ_1, Γ_2, Γ_3 and Γ_4 denote the left, bottom, right and top sides of the rectangle D respectively.

When the perturbation parameter ε tends to zero, regular and parabolic layers appear respectively in the neighborhood of the boundaries S_1 and S_2, S_3.

3 Special Finite Difference Scheme

On $\overline{G}$ we construct the piecewise uniform grid (see, e,g, [10,3])

$$\overline{G}_h = \overline{D}_h \times \overline{\omega}_0, \quad \overline{D}_h = \overline{\omega}_1 \times \overline{\omega}_2. \quad (1)$$

Here $\overline{\omega}_0$ is a uniform mesh on $[0,T]$ with step-size $\tau = T/N_0$, $\overline{\omega}_s = \overline{\omega}_s(\sigma_s)$, $s=1,2$ is a piecewise uniform mesh with N_s intervals on the x_s-axis. To construct the mesh $\overline{\omega}_2(\sigma_2)$, we divide $[0,1]$ in three parts $[0,\sigma_2]$, $[\sigma_2, 1-\sigma_2]$, $[1-\sigma_2, 1]$; we take $\sigma_2 = \min[1/4, m_2 \varepsilon \ln N_2]$. In each part we place a uniform mesh with $N_2/2$

Throughout this paper we denote by M, $M^{(i)}$ (or m_i, $m^{(i)}$) arbitrary, sufficiently large (small) positive constants independent of ε and the discretization parameters.

elements in $[\sigma_2, 1-\sigma_2]$ and with $N_2/4$ elements in each subinterval $[0, \sigma_2]$ and $[1-\sigma_2, 1]$. When constructing $\overline{\omega}_1(\sigma_1)$, we divide $[0,1]$ in two parts with the transition point $\sigma_1 = \min[1/2, m_1^{-1}\varepsilon^2 \ln N_1]$, where $0 < m_1 < m_1^0$, $m_1^0 = \min_{\overline{G}}[a_1^{-1}(x,t)b_1(x,t)]$. We place a uniform mesh in $[0, \sigma_1]$, $[\sigma_1, 1]$ using $N_1/2$ mesh elements in each subinterval.

For problem (1) we use the difference scheme [11]

$$\Lambda_{(2)}z(x,t) = f(x,t), \quad (x,t) \in G_h, \quad z(x,t) = \varphi(x,t), \quad (x,t) \in S_h, \qquad (2)$$

where $\Lambda_{(2)} \equiv \varepsilon^2 \sum_{s=1,2} a_s(x,t)\delta_{\overline{x}s\widehat{x}s} + b_1(x,t)\delta_{x1} - c(x,t) - p(x,t)\delta_{\overline{t}}, \quad \delta_{\overline{t}}z(x,t)$,

$\delta_{x1} z(x,t)$ and $\delta_{\overline{x}s\widehat{x}s} z(x,t)$ are the first and the second differences of $z(x,t)$.

Theorem 1. *The solution of finite difference scheme (2), (1) converges ε-uniformly to the solution of (1) with an error bound given by*

$$|u(x,t) - z(x,t)| \leq M(N_1^{-1} \ln N_1 + N_2^{-2} \ln^2 N_2 + \tau), \quad (x,t) \in \overline{G}_h.$$

Remark 1. Let $u \in C^{\beta, \beta/2}(\overline{G})$, $\beta = K+2+\alpha$, $K \geq 0$, $\alpha > 0$. Then the derivatives $(\partial^{k_0}/\partial t^{k_0})u(x,t)$ and the divided differences $\delta_{l\overline{t}} z(x,t)$ satisfy the estimates

$$\left|\frac{\partial^{k_0}}{\partial t^{k_0}}u(x,t)\right| \leq M_{(3)}^{(k_0)}, \quad (x,t) \in \overline{G}, \quad k_0 \leq K+2; \qquad (3)$$

$$|\delta_{l\overline{t}} z(x,t)| \leq M_{(4)}^{(l)}, \quad (x,t) \in \overline{G}_h, \quad t \geq l\tau, \quad l \leq K+1. \qquad (4)$$

Here $\delta_{l\overline{t}} z(x,t) = (\delta_{l-1\overline{t}} z(x,t) - \delta_{l-1\overline{t}} z(x, t-\tau))/\tau$, $(x,t) \in \overline{G}_h$, $t \geq l\tau$, $l \geq 1$, $\delta_{0\overline{t}} z(x,t) = z(x,t)$, and $\delta_{l\overline{t}}z(x,t)$ denotes the backward difference of order l.

4 Parallelization of Finite Difference Scheme (2), (1)

We derive the difference scheme to be solved on $P \geq 1$ parallel processors [8].
1. First we describe a partitioning of the domain D

$$D = \bigcup_{k=1}^{K} D^k, \quad D^k = (0,1) \times d_2^k, \qquad (1)$$

where d_2^k are open intervals in (0,1) on the x_2-axis. Let $G^k = D^k \times (0,T]$, $k = 1, \ldots, K$. We denote the minimal overlap of the sets D^k and $D^{[k]} = \bigcup_{i=1, i\neq k}^{K} D^i$ by δ^k, and by δ the smallest value of δ^k, i.e.,

$$\min_{k, x^1, x^2} \rho(x^1, x^2) = \delta, \qquad (2)$$

$$x^1 \in \overline{D}^k, \quad x^2 \in \overline{D}^{[k]}, \quad x^1, x^2 \notin \{D^k \cap D^{[k]}\}, \quad k = 1, \ldots, K.$$

In general, the value δ may depend on the parameter ε.

Let each D^k be partitioned into P disjoint (possibly empty) parts

$$D^k = \bigcup_{p=1}^{P} D_p^k, \quad k = 1, \ldots, K, \quad \overline{D}_i^k \cap \overline{D}_j^k = \emptyset, \quad i \neq j; \quad D_p^k = (0,1) \times d_{2p}^k. \qquad (3)$$

We set $G_p^k = D_p^k \times (0, T]$, $p = 1, \ldots, P$, $k = 1, \ldots, K$.

We introduce the rectangular grids on each of the sets $\overline{G}^k$ and $\overline{G}_p^k$:

$$\overline{G}_h^k = \overline{G}^k \cap \overline{G}_{h(1)}, \quad \overline{G}_{ph}^k = \overline{G}_p^k \cap \overline{G}_{h(1)}. \tag{4}$$

We define the prizm $G(t_1)$ with the boundary $S(t_1) = \overline{G}(t_1) \setminus G(t_1)$ by

$$G(t_1) = \{(x,t) : (x,t) \in G, \ t_1 < t \leq t_1 + \tau\}, \quad t_1, \ t_1 + \tau \in \overline{\omega}_0.$$

Let the discrete function $v(x,t;t_1)$ be defined at the boundary mesh points $S_h(t_1) = S(t_1) \cap \overline{G}_h$, $t_1 \in \overline{\omega}_0$. By $\overline{v}(x,t;t_1)$ we denote the extension of this function to the grid set $\overline{G}_h(t_1) = \overline{G}(t_1) \cap \overline{G}_h$. The "prizm" $\overline{G}_h(t_1)$ consists of only two time levels $\overline{G}_h(t_1) = \{\overline{D}_h \times [t = t_1]\} \cup \{\overline{D}_h \times [t = t_1 + \tau]\}$.

2. Before to describe the difference scheme designed for parallel implementation on P processors, we assume that $z(x,t)$ is known for $t \leq t^n$. Then we solve

$$\Lambda_{(2)} z_p^{\frac{k}{K}}(x,t) = f(x,t), \quad (x,t) \in G_{ph}^k(t^n), \tag{5a}$$

$$z_p^{\frac{k}{K}}(x,t) = \begin{cases} \overline{z}(x,t;t^n), & k = 1, \\ z^{\frac{k-1}{K}}(x,t), & k \geq 2 \end{cases}, \quad (x,t) \in S_{ph}^k(t^n), \ p = 1, \ldots, P$$

for $(x,t) \in \overline{G}_{ph}^k(t^n)$, $k = 1, \ldots, K$, $t^n \in \overline{\omega}_0$, $n \leq N_0 - 1$;

$$z^{\frac{k}{K}}(x,t) = \begin{cases} z_p^{\frac{k}{K}}(x,t), & (x,t) \in \overline{G}_{ph}^k(t^n), \ p = 1, \ldots, P, \\ \overline{z}(x,t;t^n), & k = 1, \\ z^{\frac{k-1}{K}}(x,t), & k \geq 2 \end{cases}, \quad (x,t) \in \overline{G}_h(t^n) \setminus \bigcup_{p=1}^{P} \overline{G}_{ph}^k(t^n)$$

for $(x,t) \in \overline{G}_h(t^n)$, $k = 1, \ldots, K$, $t^n \in \overline{\omega}_0$.

We define the function $z_{(5)}(x,t)$ on the prizm $\overline{G}_h(t^n)$ by the relation

$$z_{(5)}(x,t) = z^{\frac{K}{K}}(x,t), \quad (x,t) \in \overline{G}_h(t^n), \ t^n \in \overline{\omega}_0. \tag{5b}$$

The difference scheme (5) can be written in the operator form

$$Q_{(5)}(z_{(5)}(x,t); f(\cdot), \varphi(\cdot), \psi(\cdot)) = 0, \quad (x,t) \in \overline{G}_h. \tag{5c}$$

Here the function $\psi(x,t;t^n)$, $(x,t) \in G(t^n)$ defines the prolonged function

$$\overline{z}(x,t;t^n) = \begin{cases} v(x,t;t^n), & (x,t) \in S(t^n), \\ v(x,t^n;t^n) + \psi(x,t;t^n), & (x,t) \in G(t^n) \end{cases}, \quad (x,t) \in \overline{G}(t^n), \tag{5d}$$

where

$$v(x,t;t^n) = \begin{cases} \varphi(x,t), & (x,t) \in S_h(t^n), & t^n = t^0 = 0, \\ \varphi(x,t), & (x,t) \in S_h(t^n) \cap S_h, \ t \geq t^n, \\ z(x,t), & (x,t) \in S_h(t^n) \setminus S_h, \ t = t^n \end{cases}, \ t^n > 0, \tag{5e}$$

$(x,t) \in S_h(t^n)$, $n = 0, 1, \ldots, N_0 - 1$.

In the specific problem (5) we take $\psi(x,t;t^n) \equiv 0$.

Note that the intermediate problems in the discrete DD method (5), (4) are solved on the subsets $\overline{D}_{ph}^k = \overline{D}_{p(3)}^k \cap \overline{D}_h$ independently of each other ("in parallel") for all $p = 1, \ldots, P$.

Let the following condition be satisfied

$$\delta = \delta_{(2)}(\varepsilon) > 0, \quad \varepsilon \in (0,1], \quad \inf_{\varepsilon \in (0,1]}[\varepsilon^{-1}\delta_{(2)}(\varepsilon)] > 0. \tag{6}$$

A technique similar to the one exposed in [6,7] gives us the error estimate

$$|u(x,t) - z_{(5)}(x,t)| \leq M(N_1^{-1}\ln N_1 + N_2^{-2}\ln^2 N_2 + N_0^{-1}), \quad (x,t) \in \overline{G}_h. \tag{7}$$

Theorem 2. *Under condition (6) and for N, $N_0 \to \infty$, the solution of the difference scheme (5), (4) converges to the solution of (1) ε-uniformly. The estimate (7) holds for the solution of this difference scheme.*

5 Improved Time-Accuracy. Parallel Scheme

1. Constructing the defect-correction difference scheme on $\overline{G}_h$, we rewrite the finite difference scheme (2) as in [7]:

$$\Lambda_{(2)}z^{(1)}(x,t) = f(x,t), \quad (x,t) \in G_h, \quad z^{(1)}(x,t) = \varphi(x,t), \quad (x,t) \in S_h, \tag{1}$$

where $z^{(1)}(x,t)$ is the uncorrected solution. To find the corrected solution $z^{(2)}(x,t)$, we solve the problem

$$\Lambda_{(2)}z^{(2)}(x,t) = f(x,t) + \begin{cases} 2^{-1}p(x,t)\,\tau\,\dfrac{\partial^2}{\partial t^2}u(x,0), & t = \tau, \\ 2^{-1}p(x,t)\,\tau\,\delta_{2\bar{t}}\,z^{(1)}(x,t), & t \geq 2\tau \end{cases}, \quad (x,t) \in G_h,$$
$$z^{(2)}(x,t) = \varphi(x,t), \quad (x,t) \in S_h. \tag{2}$$

Here the derivative $(\partial^2/\partial t^2)u(x,0)$ is obtained from equation (1a).

In the remainder of this section we consider a homogeneous initial condition

$$\varphi(x,0) = 0, \quad x \in \overline{D}. \tag{3}$$

Under this condition, for the solution of problem (2), (1) we have

$$\left|u(x,t) - z^{(2)}(x,t)\right| \leq M\left[N_1^{-1}\ln N_1 + N_2^{-2}\ln^2 N_2 + \tau^2\right], \quad (x,t) \in \overline{G}_h. \tag{4}$$

Proceeding in a similar way, one can construct difference schemes with a higher order of time-accuracy $\mathcal{O}(\tau^l)$, $l > 2$ (see [7,8] for $l = 3$).

2. Let us consider a parallel version for the defect correction scheme. In the operator form the above difference scheme is written as follows

$$\begin{aligned}Q_{(5)}(z^{(1)}(x,t); f^{(1)}(\cdot), \varphi(\cdot), \psi^{(1)}(\cdot)) &= 0, \quad (x,t) \in \overline{G}_h, \\ Q_{(5)}(z^{(2)}(x,t); f^{(2)}(\cdot), \varphi(\cdot), \psi^{(2)}(\cdot)) &= 0, \quad (x,t) \in \overline{G}_h,\end{aligned} \tag{5}$$

where

$$f^{(1)}(x,t) = f(x,t), \quad f^{(2)}(x,t) = f^{(2)}(x,t; z^{(1)}(\cdot)) = f(x,t) +$$
$$+ \begin{cases} 2^{-1} p(x,t)\, \tau\, (\partial^2/\partial t^2) u(x,0), & t = \tau, \\ 2^{-1} p(x,t)\, \tau\, \delta_{2\bar{t}} z^{(1)}(x,t), & t \geq 2\tau \end{cases}, \quad (x,t) \in G_h^k,$$

$$\psi^{(1)}(x,t;t^n) \equiv 0, \quad \psi^{(2)}(x,t;t^n) = \psi^{(2)}(x,t;t^n, z^{(1)}(\cdot)) =$$
$$= z^{(1)}(x,t^{n+1}) - z^{(1)}(x,t^n), \quad (x,t) \in G_h(t^n), \quad t = t^{n+1}.$$

It is easy to see that $z^{(1)}(x,t) \equiv z_{(5;\,4)}(x,t)$.

Following the arguments from [6,7,9] we obtain the main convergence result.

Theorem 3. *Let condition (3) hold. Then, under condition (6), the solution of the difference scheme (5), (4) converges, as N, $N_0 \to \infty$, to the solution of the boundary value problem (1) ε-uniformly. For the discrete solution the estimate (4) holds.*

6 Acceleration of Computations by the Parallel Scheme

To solve the problem (1), we use scheme (2), (1) with improved time-accuracy as the base scheme. One can also use the parallel variant of scheme (5), (4). We say that the use of parallel computations leads to the real acceleration of the solution process if such a scheme with $P > 1$ parallel processors can be found for which the computation time turns out to be smaller and the accuracy of the approximate solution is not lower than those for the base scheme.

We shall consider the difference scheme for P parallel solvers on the meshes

$$\overline{G}_{ph}^{kP} = \overline{G}_p^k \cap \overline{G}_h^P, \quad \overline{G}_h^P = \overline{D}_h \times \overline{\omega}_0^P, \tag{1}$$

where $\overline{D}_h = \overline{D}_{h(1)}$, $\overline{\omega}_0^P$ is a uniform mesh on $[0,T]$ with the number of nodes $N_0^P + 1$ and the mesh step τ^P; generally speaking, $\overline{\omega}_{0(1)}^P \neq \overline{\omega}_{0(1)}$.

1. We now describe the decomposition of the set $\overline{D}$ which can ensure the acceleration of the solution process.

Let the domain D consist of J non-overlapping rectangles

$$D^{<j>}, \quad j = 1, \ldots, J, \tag{2a}$$

where $D^{<i>} \cap D^{<j>} = \emptyset$ for $i \neq j$, $\overline{D} = \bigcup_{j=1}^J \overline{D}^{<j>}$; $J \leq M$. On each of the sets $\overline{G}^{<j>} = \overline{D}^{<j>} \times [0,T]$, the mesh $\overline{G}_h$ with the given distribution of its nodes generates the meshes $\overline{G}_h^{<j>} = \overline{G}^{<j>} \cap \overline{G}_h$, $j = 1, \ldots, J$, $\overline{G}_h = \overline{G}_{h(1)}^P$. For each of the sets $D^{<j>}$ we construct the rectangle D^j containing $D^{<j>}$ together with some neighborhood. This set $\overline{D}^j$ satisfies the three conditions:

(a) $\overline{D}^j$ contains the set of the points distant from $\overline{D}^{<j>}$ on the distance which is not smaller than δ_0, where

$$\delta_0 = m_{(2)}^{(1)} \varepsilon \quad \text{with some fixed } m_{(2)}^{(1)}; \tag{2b}$$

(b) the sides of the set $\overline{G}^j = \overline{D}^j \times [0,T]$ pass through the nodes of the mesh $\overline{G}_h$;

(c) the number of nodes in each of the meshes $\overline{D}_h^j = \overline{D}^j \cap \overline{D}_h$ is the same and it does not depend on the number j.

Let the work time of the processors, assigned for resolving the discrete problem on the level $t = t^1$ of the mesh set $\overline{D}_h^0$ from $\overline{D}_{h(1)}$, be defined by the value $\mu(\overline{D}_h^0)$, that is the number of nodes in the set $\overline{D}_h^0$.

The sets

$$G^j = D^j \times (0,T], \quad j = 1, \ldots, J, \tag{2c}$$

form the preliminary covering of the set G, that is, $G = \bigcup_{j=1}^{J} G^j$. Assume

$$\mu(\overline{D}_h^j) = \mu^0, \quad \mu(\overline{D}_h^j) = (1 + m_{(2)}^{(2)}) \mu(\overline{D}_h^{<j>}), \quad j = 1, \ldots, J. \tag{2d}$$

The sets (2c) are used for the construction of the special DD scheme (5), (1) with P processors. For this, we construct the sets

$$G^{\{k\}}, \quad k = 1, \ldots, K \tag{3a}$$

which cover the set G, where the value $K = K(P)$ is chosen from the condition $KP = J$. The each of the sets $G^{\{k\}}$ is multiply connected (for $P > 1$) and formed by the union of the P non-overlapping domains from (2c). Thus, for the subsets G_p^k which form the sets from (3a), the following condition holds:

$$G_p^k \subset \{G^j, \ j = 1, \ldots, J\}_{(2c)}, \quad k = 1, \ldots, K, \ p = 1, \ldots, P, \tag{3b}$$

where $\mu(\overline{D}_{ph}^k) = \mu^0$, $G^{\{k\}} = \bigcup_{p=1}^{P} G_p^k$. With such decomposition the processors are loaded more effectively.

2. By definition, we denote the work time, which is required to solve problems (2), (1) and (5), (4) respectively, by

$$\vartheta = \vartheta(N_0) \equiv N_0 \mu(\overline{D}_h), \qquad \vartheta^P = \vartheta^P(N_0^P, P) \equiv N_0^P \sum_{k=1}^{K} \max_p \mu(\overline{D}_{ph}^k).$$

Then the rate of acceleraton for our computations is defined by

$$C = C(N_0, N_0^P, P) = \vartheta(\vartheta^P)^{-1} \equiv N_0 (N_0^P)^{-1} \mu(\overline{D}_h) \left\{ \sum_{k=1}^{K(P)} \max_p \mu(\overline{D}_{ph}^k) \right\}^{-1}.$$

3. We now give the conditions ensuring the acceleration of the solution process based on parallelization of scheme (2), (1). Here we assume that the derivative

$(\partial^3/\partial t^3)u(x,t)$ on the set $\overline{G}$ is not too small. Precisely, let the following condition hold

$$\left|\frac{\partial^3}{\partial t^3}u(x,t)\right| \geq m^{(3)}, \quad (x,t) \in \overline{G}^* \tag{4}$$

on some set $\overline{G}^* = \{(x,t): x_s^{*1} \leq x_s \leq x_s^{*2}, \ s = 1, 2, \ t^{*1} \leq t \leq t^{*2}\}$, $\overline{G}^* \subseteq \overline{G}$.

In the case when the number P of processors is sufficiently large, i.e.,

$$P > M\left(1 + m_{(2)}^{(2)}\right)\left(m_{(4)}^{(3)}\right)^{-1}\left[M_{(3)}^{(3)} + M_{(3)}^{(4)} + M_{(3)}^{(5)}\right] \equiv P^*, \tag{5}$$

the acceleration can be really attained for the numerical solution of the boundary value problem. In fact, the acceleration is achieved under the condition

$$N_0^P = \left(1 + m_{(2)}^{(2)}\right)^{-1} N_0 \, P^*. \tag{6}$$

The value of C, which characterizes the attained rate of acceleration, is defined by

$$C = P(P^*)^{-1}, \quad P^* = P_{(5)}^*. \tag{7}$$

Theorem 4. *Let conditions (3), (4), (4) hold for the solutions of the boundary value problem (1) and scheme (2), (1). Then, in the class of difference schemes (5), (1) for P parallel processors, ε-uniform acceleration of solving problem (1), as compared to the base scheme (2), (1), can be achieved in general; in particular, for the decomposition (3), (1) the acceleration is achievable under condition (5). Moreover, for scheme (5), (3), (1) the acceleration is attained under conditions (5), (6), and the rate C of acceleration is defined by (7).*

References

1. Farrell, P. A., Hemker, P. W., Shishkin, G. I.: Discrete approximations for singularly perturbed boundary value problems with parabolic layers. I, J. Comput. Math. **14** (1) (1996) 71–97; II, J. Comput. Math. **14** (2) (1996) 183–194; III, J. Comput. Math. **14** (3) (1996) 273–290
2. Farrell, P. A., Miller, J. J. H., O'Riordan, E., Shishkin, G. I.: A uniformly convergent finite difference scheme for a singularly perturbed semilinear equation. SIAM J. Numer. Anal. **33** (1996) 1135–1149
3. Shishkin, G. I.: Grid Approximations of Singularly Perturbed Elliptic and Parabolic Equations (in Russian). Ural Branch of Russian Acad. Sci., Ekaterinburg (1992)
4. Miller, J. J. H., O'Riordan, E., Shishkin G. I.: Fitted Numerical Methods for Singular Perturbation Problems. World Scientific, Singapore (1996)
5. Roos, H.-G., Stynes, M., Tobiska, L.: Numerical Methods for Singularly Perturbed Differential Equations. Springer-Verlag, Berlin (1996)
6. Hemker, P. W., Shishkin, G. I., Shishkina, L. P.: The use of defect correction for the solution of parabolic singular perturbation problems. ZAMM **77** (1997) 59–74

7. Hemker, P. W., Shishkin, G. I., Shishkina, L. P.: $\mathcal{E}$-uniform schemes with high-order time-accuracy for parabolic singular perturbation problems. IMA J. Numer. Anal. **20** (2000) 99–121
8. Hemker, P. W., Shishkin, G. I., Shishkina, L. P.: Distributing the numerical solution of parabolic singularly perturbed problems with defect-correction over independent processes. Siberian J. Numer. Mathematics, Novosibirsk (to appear)
9. Shishkin, G. I.: Acceleration of the process of the numerical solution to singularly perturbed boundary value problems for parabolic equations on the basis of parallel computations. Russ. J. Numer. Anal. Math. Modelling **12** (1997) 271–291
10. Shishkin, G. I.: Grid approximation of singularly perturbed elliptic equation in domain with characteristic bounds. Sov. J. Numer. Anal. Math. Modelling **5** (1990) 327–343
11. Samarskii, A. A.: Theory of Difference Schemes (in Russian). Nauka, Moscow (1989)

Experience with the Solution of a Finite Difference Discretization on Sparse Grids

P. W. Hemker[1] and F. Sprengel[2]

[1] Centrum voor Wiskunde en Informatica,
P.O.Box 94079, NL–1090 GB Amsterdam, The Netherlands
[2] SCAI Institute for Algorithms and Scientific Computing
GMD German National Research Center for Information Technology
Schloss Birlinghoven, D–53754 Sankt Augustin, Germany

Abstract. In a recent paper [10], we described and analyzed a finite difference discretization on adaptive sparse grids in three space dimensions. In this paper, we show how the discrete equations can be efficiently solved in an iterative process. Several alternatives have been studied before in Sprengel [16], where multigrid algorithms were used. Here, we report on our experience with BiCGStab iteration. It appears that, applied to the hierarchical representation and combined with Nested Iteration in a cascadic algorithm, BiCGStab shows fast convergence, although the convergence rate is not truly independent of the meshsize.

1 Introduction

Recently, the use of sparse grids has drawn considerable attention [4,6,7,10,11,16] because of its prospects for a very efficient treatment of higher dimensional problems. Most attention is directed towards the solution of three-dimensional partial differential equations, because of their importance for scientific and technical problems. The contrast of sparse grids with the classical grids is the fact that on usual regular three-dimensional grids the number of gridpoints grows with $\mathcal{O}(h^{-3})$ with decreasing mesh-width h, whereas the number of mesh-points grows with only $\mathcal{O}(h^{-1}|\log h|^2)$ for sparse grids. For a solution, u, with sufficient smoothness, the loss off accuracy (e.g. with piecewise trilinear approximation) is remarkably small. Viz., with bounded mixed derivatives $D^{2,2,2}u$ (at least in the weak sense) the usual accuracy of $\mathcal{O}(h^2)$ reduces to only $\mathcal{O}(h^2|\log h|^2)$.

Here we should notice that the smoothness requirement is essential, and that, with sufficient smoothness, classical higher order methods may yield even more efficiency. As higher order methods can also be used in combination with sparse grids [4], both regular and sparse grids may have their own areas of application. However, it is clear that proper grid-alignment plays a more important role for sparse grids. Therefore, it is useful to see what grids should be used in practice under what circumstances.

Considering the smoothness conditions required for the different approximations, we see that the usual, regular approximations require $u \in C^k(\Omega)$, i.e., all derivatives up to some constant k should be bounded, whereas the error for sparse grids is bounded mainly by the mixed derivatives. This implies that the error estimates in the former case are essentially direction-independent, whereas the error for the sparse grid case is dependent on the grid orientation. This may show the area of application of sparse grids: the cases where significant features of the solution can be captured by grid positioning.

We do not want to go into detailed arguments on grid selection. However, we want to say that the study of sparse grids has led to new insights in the proper application of semi-refinement, hierarchical representation of functions, and the use of partially ordered sets of spaces for mesh-adaptive approximation.

This paper concerns the solution of linear systems as they arise in the finite difference approximation of PDEs in 3D. The FD approach to the solution of PDEs on sparse grids was initiated by Griebel in [7] and worked out in more detail in [13]. More results are found in [10], where we described how the finite difference discretization is constructed and how the discrete functions can be represented on a nodal and on a hierarchical basis. Other relevant papers on the solution of 3D discrete systems on sparse grids are [6,11].

The emphasis of this note is on the experience with several solution algorithms for the finite difference discretization on sparse grids. The algorithms are based on a basic iterative solver (BiCGStab [1]) and Nested Iteration. The work is inspired by [12], where hierarchical basis preconditioners in three dimensions are described in a finite element context. The difference is that in [12] a classical sequence of meshes is used, constructed from tetrahedral elements and quasi-uniform refinement. It has been shown that, in that case, the condition of the matrix based on the hierarchical representation, preconditioned by a coarse grid operator is $\mathcal{O}(h^{-1}|\log h|)$, where h is the mesh size. By diagonal scaling by levels, the condition number could be reduced to $\mathcal{O}(h^{-1})$. Similarly, in the present paper, we observe also that the hierarchical representation gives a better convergence rate than the usual nodal representation.

2 Adaptive Function Approximation

For an arbitrary $\mathbf{k} = (k_1, k_2, k_3) \in \mathbb{N}_0^3$, we define a dyadic grid $\Omega_\mathbf{k}^+$ over $\Omega \subset \mathbb{R}^3$ by
$$\Omega_\mathbf{k}^+ = \{\mathbf{x}_{\mathbf{k},\mathbf{j}} \mid \mathbf{x}_{\mathbf{k},\mathbf{j}} = \mathbf{j} \cdot \mathbf{h}_\mathbf{k} = (j_1 2^{-k_1}, j_2 2^{-k_2}, j_3 2^{-k_3})\} \cap \overline{\Omega},$$
and we consider tensor-type basis functions $\varphi_{\mathbf{k},\mathbf{j}}(\mathbf{x}) = \prod_{i=1}^3 \varphi(x_i/h_{k_i} - j_i)$, where $\varphi(x) = \max(0, 1 - |x|)$ is the usual hat function. Given a continuous function $u \in C(\Omega)$, we can approximate it by $u_\mathbf{n} \in V_\mathbf{n} = \text{Span}\{\varphi_{\mathbf{n},\mathbf{j}}\}$ by interpolation on $\Omega_\mathbf{n}^+$. Obviously, the function $u_\mathbf{n}$ on $\Omega_\mathbf{n}$ is given by

$$u_\mathbf{n} = \sum_\mathbf{j} a_{\mathbf{n},\mathbf{j}} \varphi_{\mathbf{n},\mathbf{j}}. \tag{1}$$

We can make an approximation (1) for all grids $\Omega_\mathbf{n}^+$ with $\mathbf{n} \geq \mathbf{0}$. For large enough $\mathbf{n}$, the approximation can be arbitrarily accurate, but the number of degrees of freedom increases geometrically with $|\mathbf{n}| = n_1 + n_2 + n_3$. Therefore, in practice we select a 'smallest' **n** such that an accuracy criterion is satisfied. Notice that keeping the representations in *all* coarser $V_\mathbf{k}$ (all $V_\mathbf{k}$, $0 \leq \mathbf{k} \leq \mathbf{n}$) does not take essentially more coefficients than the representation on the finest grid (i.e., in $V_\mathbf{n}$) alone.

In order to obtain an efficient approximation, we can distinguish different areas in the domain Ω, in each of which we make the finest approximation of u in different $V_\mathbf{n}$. We make full and efficient use of the system $\{V_\mathbf{n} \mid \mathbf{n} \in \mathbb{N}_0^3\}$, by *in principle* approximating a given function $u \in C(\Omega)$ in *all* $\{V_\mathbf{n} \mid \mathbf{n} \in \mathbb{N}_0^3\}$, but using *in practice* only those coefficients that contribute to a sufficiently accurate representation. This implies that in practice the function u is represented in a particular $V_\mathbf{n}$ only on part of the domain Ω. To introduce a (minimal) structure in the family of approximating basis functions $\{\varphi_{\mathbf{n},\mathbf{j}}\}$, we introduce the following condition **H**. (*The H condition:*) If a basis function $\varphi_{\mathbf{n},\mathbf{j}}(\mathbf{x})$ is used in the representation (1), then all corresponding coarser basis functions (i.e., functions $\varphi_{\mathbf{k},\mathbf{i}}$ for which $\mathrm{supp}(\varphi_{\mathbf{k},\mathbf{i}}) \supset \mathrm{supp}(\varphi_{\mathbf{n},\mathbf{j}})$) are also used for the representation.

E- and H-Representation. We call the representation of the approximation of a function $u \in C(\Omega)$ by a collection of such (partial) approximations (1) in the family of spaces $\{V_\mathbf{n}\}$, the *nodal representation*, or the *E-representation* of the approximation. This E-representation requires the coefficients $a_{\mathbf{n},\mathbf{j}} = u(\mathbf{x}_{\mathbf{n},\mathbf{j}})$ corresponding with grid-points $\mathbf{x}_{\mathbf{n},\mathbf{j}}$, to be equal on the different grids $\Omega_\mathbf{n}^+$ at coinciding grid-points $\mathbf{x}_{\mathbf{n},\mathbf{j}}$. Thus, because points from coarser grids coincide with those from finer ones, a certain consistency is required (and a redundancy exists) in the E-representation of an approximation.

Another way of representing approximations on the family of grids $\{\Omega_\mathbf{n}^+\}$ is by partitioning the approximation over the different grids. Then, instead of (1) the approximation reads

$$u_h = \sum_\mathbf{n} \sum_\mathbf{j} a_{\mathbf{n},\mathbf{j}} \varphi_{\mathbf{n},\mathbf{j}}.$$

In this case, of course, the set of coefficients $\{a_{\mathbf{n},\mathbf{j}}\}$ always determines a unique function u_h. However, for a given function u_h, now the coefficients $\{a_{\mathbf{n},\mathbf{j}}\}$ are not uniquely determined because the $\{\varphi_{\mathbf{n},\mathbf{j}}\}$ are linearly dependent. One way to select a special unique representation is by choosing the coefficients $a_{\mathbf{n},\mathbf{j}}$ such that $a_{\mathbf{n},\mathbf{j}} \neq 0$ only for those $(\mathbf{n},\mathbf{j})$ for which $\|\mathbf{j}\| = j_1 \cdot j_2 \cdot j_3$ is odd[1]. This implies that $a_{\mathbf{n},\mathbf{j}} = 0$ except for a pair $(\mathbf{n},\mathbf{j})$ for which $\Omega_\mathbf{n}^+$ is the coarsest grid which contains the nodal point $\mathbf{x}_{\mathbf{n},\mathbf{j}}$. This representation

$$u_h = \sum_{(\mathbf{n},\mathbf{j}), \|\mathbf{j}\| \text{ odd}} a_{\mathbf{n},\mathbf{j}} \varphi_{\mathbf{n},\mathbf{j}} \qquad (2)$$

we call the *H-representation* because it represents the approximation in the hierarchical basis

$$\{\varphi_{\mathbf{n},\mathbf{j}} \mid \mathbf{n} \in \mathbb{N}_0^3, \mathbf{j} \in \mathbb{Z}^3, \|\mathbf{j}\| \text{ odd}, \mathbf{x}_{\mathbf{n},\mathbf{j}} \in \Omega_\mathbf{n}^+\}, \qquad (3)$$

---
[1] More precisely, with "$\|\mathbf{j}\|$ is odd" we mean: for all $i = 1,2,3$, either j_i is an odd integer, or $k_i = 0$ (i.e., j_i lives on the coarsest grid in the i-direction).

and the part of u_h in
$$W_{\mathbf{n}} = \operatorname{Span}\{\varphi_{\mathbf{n},\mathbf{j}} \mid \mathbf{j} \in \mathbb{Z}^3, \|\mathbf{j}\| \text{ odd}, \mathbf{x}_{\mathbf{n},\mathbf{j}} \in \Omega_{\mathbf{n}}^+\}$$
is the *hierarchical contribution* from the grid $\Omega_{\mathbf{n}}^+$ to the approximation. We notice that
$$V_{\mathbf{n}} = W_{\mathbf{n}} + \sum_{j=1}^{3} V_{\mathbf{n}-\mathbf{e}_j} = \sum_{0 \le \mathbf{m} \le \mathbf{n}} V_{\mathbf{m}},$$
and the sparse grid space is defined by
$$V_L = \sum_{0 \le |\mathbf{m}| \le L} V_{\mathbf{m}},$$
corresponding to a sparse grid $\Omega_L^+ = \bigcup_{0 \le |\mathbf{m}| \le L} \Omega_{\mathbf{m}}^+$. Interpolating the function u at the nodal points $\mathbf{x}_{\mathbf{n},\mathbf{j}}$, the hierarchical coefficients $a_{\mathbf{n},\mathbf{j}}$ in
$$u(\mathbf{x}_{\mathbf{n},\mathbf{j}}) = \sum_{(\mathbf{n},\mathbf{j}), \|\mathbf{j}\| \text{ odd}} a_{\mathbf{n},\mathbf{j}} \, \varphi_{\mathbf{n},\mathbf{j}}(\mathbf{x}_{\mathbf{n},\mathbf{j}})$$
are determined by (cf. [9])
$$a_{\mathbf{n},\mathbf{j}} = \prod_{i=1}^{3} \left[-\frac{1}{2}, 1, -\frac{1}{2}\right]_{h_{n_i} \mathbf{e}_i} u(\mathbf{j} \mathbf{h}_{\mathbf{n}}),$$
where $\left[-\frac{1}{2}, 1, -\frac{1}{2}\right]_{h_{n_i} \mathbf{e}_i}$ denotes the difference stencil for the mesh-size h_{n_i} in the i-th coordinate direction. Notice that this expression is well-defined for each odd $\mathbf{j}$ because Condition **H** requires that all h_i-neighbors are nodal points in the approximation.

For piecewise multilinear functions, it is often described [5,6,7] how a pyramid algorithm can be used to convert an E-representation to a H-representation, and vice versa. Such a conversion can be executed in $\mathcal{O}(N)$ operations, where N is the total number of coefficients.

The Data Structure. The data structure to implement all the above possibilities of an adaptive (sparse) grid representation can be efficient and relatively simple. For the d-dimensional case ($d = 1, 2, 3$), we use the data structure BASIS3 [8] that takes the 'patch' $P_{\mathbf{n},\mathbf{j}}$ as an elementary entity. This $P_{\mathbf{n},\mathbf{j}}$ takes all information related to a right-open left-closed cell
$$\prod_{k=1}^{3} \left[j_k 2^{-n_k}, (j_k + 1) 2^{-n_k}\right).$$
This implies that there exist as many patches in the data structure as there are points used in the description of the approximation. The patches are related to each other by means of pointers in an intertwined tree structure, where each patch has at most 15 pointers to related patches (3 fathers, 6 neighbors and 6 kids). The data structure is symmetric with respect to any of the coordinate directions.

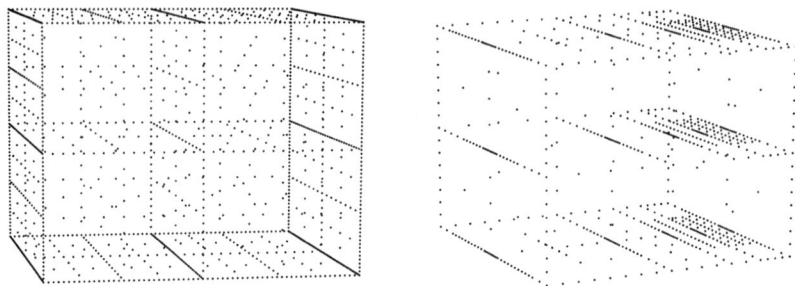

Fig. 1. Regular sparse grid Ω_6^+ for $\Omega = (0,1)^3$ (left) and an adaptive sparse grid (ASG) (right)

3 Difference Operators for ASG Functions

Although finite element discretization of a PDE on a sparse grid is feasible for a constant coefficient problem in two dimensions, finite elements for more-dimensional problems and variable coefficients give problems. The difficulty arises because — with the hierarchical basis (3) for test and trial space — the computational complexity of the evaluation of the discrete operator becomes too large. This is caused by the fact that the intersection of the supports of an arbitrary trial and test function is much smaller than the supports of these functions themselves. This has as a consequence that the advantage of sparse grids is lost if the FEM discrete operator is evaluated.

The alternative, as it was already suggested in [7,13], is the use of a finite difference discretization. Therefore, in order to solve PDEs on sparse grids, we should be able to apply (approximate) differentiation to discrete representations of approximations as described in [10]. The application of linear difference operators of the form

$$L_h u_h \equiv \sum_{i,j} \frac{\partial}{\partial x_i} \left(A_{ij}(\mathbf{x}) \frac{\partial}{\partial x_j} \right) u_h(\mathbf{x}) + \sum_i B_i(\mathbf{x}) \frac{\partial}{\partial x_j} u_h(\mathbf{x}) + C(\mathbf{x}) u_h(\mathbf{x}) \quad (4)$$

comes down to the construction of linear combinations, the pointwise multiplication, and the differentiation of functions (2). In both representations the construction of a linear combination over the real numbers is directly computed by application of the linear combination to the coefficients. Pointwise multiplication is only possible in the E-representation, in which the function values at grid-points are directly available. For a description of the evaluation of first and second order derivatives we again refer to [10].

First and Second Order Interpolation. Because we use piecewise tri-linear basis functions $\varphi_{\mathbf{nj}}(\mathbf{x})$ on the grid $\Omega_{\mathbf{n}}^+$, truncating at a particular level corresponds with tri-linear interpolation between the nodal points included. In this way, piecewise tri-linear interpolation is natural in the finite hierarchical representation.

For $C^{2,2,2}(\Omega)$-functions, the behavior of the coefficients $a_{\mathbf{n}\mathbf{j}}$ is rather predictable for higher levels of approximation because Lemma [9, Lemma 3.2]. gives a precise relation with the second order cross derivatives, or in lower dimensional manifolds (at the coarsest level, at the boundaries, or in mixed H-E-representations over the different coordinate directions) with the second order derivatives. This allows for an efficient quadratic interpolation procedure when a finite hierarchical representation of a discrete function is available. To interpolate the function

$$u_h^\ell(\mathbf{x}) = \sum_{|\mathbf{n}|\leq\ell} \sum_{\mathbf{j},\|\mathbf{j}\| \text{ odd}} a_{\mathbf{n},\mathbf{j}}\varphi_{\mathbf{n},\mathbf{j}}(\mathbf{x}). \tag{5}$$

with second order accuracy to a function $u_h^{\ell+1}(\mathbf{x})$, the coefficients $\{a_{\mathbf{n},\mathbf{j}} \mid |\mathbf{n}| = \ell + 1\}$ can be derived from the coefficients $\{a_{\mathbf{n},\mathbf{j}} \mid |\mathbf{n}| = \ell\}$ by taking the new coefficients $a_{\mathbf{m},\mathbf{k}} = a_{\mathbf{n},\mathbf{j}}/4$, where $|\mathbf{m}| = |\mathbf{n}|+1$ and $\mathbf{m}$ and $\mathbf{j}$ satisfy $|\mathbf{x}_{\mathbf{m},\mathbf{k}}-\mathbf{x}_{\mathbf{n},\mathbf{j}}| \leq 2^{-\ell}$. This corresponds with the extrapolation assumption that the second order derivative is slowly varying (constant) over the smallest covering cell $\Omega_{\mathbf{n},\mathbf{j}}$. In order to maintain symmetry over the coordinate directions, in the case of a non-unique smallest covering cell one may take the mean value of the coefficients of all (at most $d-1$) smallest covering cells. In this way, we introduce the second order interpolation operator $P_{\ell+1,\ell}$, defined by

$$u_h^{\ell+1} = P_{\ell+1,\ell}\, u_h^\ell, \tag{6}$$

where both u_h^ℓ and $u_h^{\ell+1}$ are described by (5). First order interpolation is simply achieved by setting $a_{\mathbf{m},\mathbf{k}} = 0$ for $|\mathbf{m}| = |\mathbf{n}| + 1$.

4 Solution of the Finite Difference Discretization for the Laplacian

In the remaining part of this paper, as an example of (4), we solve the discretized operator equation as it was described in detail in [10]. For simplicity, we restrict ourselves to the model problem of Poisson's equation with homogeneous Dirichlet boundary conditions,

$$-\Delta u = f \quad \text{in } \Omega, \tag{7}$$
$$u|_{\delta\Omega} = 0,$$

on the cube $\Omega = (0,1)^3$ and a regular sparse grid.

Iteration Based on a Galerkin Relation. In [10], an analysis of the discretization was made and multilevel-type algorithms, based on the Galerkin structure of the equations were proposed. The coarse grid operators involved were no longer finite difference operators. In an obvious way, the Galerkin relations lead to iterative (defect correction) solution algorithms that are applied in a multilevel setting. However, no spectral equivalence could be established, and the convergence of the iterative schemes appears to depend on the maximum discretization level used, so that the convergence rate slows down on finer grids. The algorithm is briefly characterized in Figure 2 (for details see [10]). Applied to the 3D-problem (7) with the right-hand side $f(\mathbf{x}) = -3\pi^2 (\prod_{i=1}^{3} \sin \pi x_i + 8 \prod_{i=1}^{3} \sin 8\pi x_i)$ and starting from the zero function $u_L^{(0)} \equiv 0$, we obtain the convergence behavior shown in Figure 3. We see that we get better convergence if we include also lower levels (right). In both cases, however, the speed of convergence slows down with growing levels. Approximately, the reduction factor gets worse with L^2, the square of the highest level. The slow convergence motivates us to see if better convergence could be obtained by cascadic iteration.

for ℓ from L_0 to L
do for $i = 1$ to ν
    do for all $|\mathbf{n}| = \ell$
        do $u_h := u_h + P_{L,\mathbf{n}} \ L_{\mathbf{n}}^{-1} \tilde{R}_{\mathbf{n},L} \ (f_h^L - L_h^L u_h)$ enddo   (G)
    enddo
enddo

Fig. 2. The Galerkin algorithm (G)

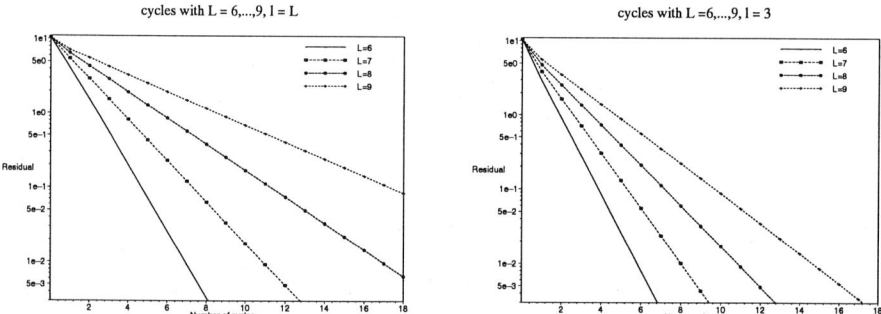

Fig. 3. Left: Convergence of Algorithm (G) for the levels $L = L_0 = 6, \ldots, 9$. Right: Convergence of Algorithm (G) for the levels $L = 6, \ldots, 9$, $L_0 = 3$ with $\nu = 1$

Cascadic Iteration. By construction, the sparse grids and the sparse grid spaces are provided with a multilevel structure, i.e., $\Omega_\ell^+ \subset \Omega_{\ell+1}^+$ and $V_\ell \subset V_{\ell+1}$. Moreover, in [10], we could prove a Galerkin relation

$$L_h^\ell = R_{\ell,\ell+1} \ L_h^{\ell+1} \ \tilde{P}_{\ell+1,\ell}$$

for the discrete Laplace operator L_h^ℓ in hierarchical representation. Here, $R_{\ell,\ell+1}$ denotes the natural hierarchical restriction and $\tilde{P}_{\ell+1,\ell}$ is the first order interpolation. This will be used in a cascadic iteration.

In [2,3], Bornemann and Deuflhard proposed the cascadic multigrid method. In this method, a solution is computed by nested iteration on a sequence of refining grids, without coarse grid corrections applied on the finer grids. In cascadic MG, more basic iterations are used on the coarser than on the finer levels. It has been proved [3,14] that cascadic MG applied to a FEM discretization using P1-conforming elements for the second order 3D problem is accurate with an optimal computational complexity for all conventional iterative methods, like Jacobi or Gauss–Seidel iteration, as well as for the conjugate gradient method as a smoother. However, in the 2D case the cascadic MGM gives accurate solution with optimal complexity for the CG method, but only nearly optimal complexity for the other conventional smoothers.

In [15], it is shown that that this is also true for other conforming or nonconforming elements, provided that $m_l \geq \beta^{L-l} m_L$, with m_l the number of iterations on level l and some constant β depending on the relaxation method.

Fig. 4. Cascadic iteration: the problem is approximately solved on a coarser (lower) grid before interpolation to a finer (higher) grid is made. The cycle over all levels is repeated in an outer defect correction (iterative refinement) process. The levels used are the union of the grids $\Omega_\mathbf{n}$, with $|\mathbf{n}| = k$, $k = 1, 2, \ldots, 10$. The number of points at each level is given in Table 1

For iteration, we use a cascadic application of the BiCGStab algorithm [1] for the solution of $L_h^L u_h = f_h^L$. The algorithm is shown in Figure 5. In the algorithm $R_{\ell,L}$ denotes the natural hierarchical restriction and $P_{\ell+1,\ell}$ is the first order prolongation $\tilde{P}_{\ell+1,\ell}$ or the second order prolongation (6). Computations are made with this algorithm on meshes up to 10 levels. The corresponding number of gridpoints is given in Table 1.

Table 1. The number of points on the different levels

levels k:	0	1	2	3	4	5	6	7	8	9	10
points #:	8	44	158	473	1286	3302	8170	19699	46594	108568	249910

The working horse of the solution algorithm is BiCGStab iteration. Because of the non-sparse structure of the matrix representation of the sparse grid dis-

```
until a convergence criterion is satisfied
do   $f_h^* := f_h^L - L_h^L u_h$
     $c_h := 0$
     for $\ell$ from $L_0$ to $L$
     do   int $i=0$, $i_c=0$;
          until $i > i_{\max}$ do
               real $n,\rho,\beta,\omega$, $\alpha = 0.0$, $\rho_0 = 1.0$, $\omega_0 = 1.0$;
               $r_h := R_{\ell,L} f_h^* - L_h^\ell c_h$
               $n = (r_h, r_h)$
               if j=0 then $n_0 = n$ endif
               if $n < \varepsilon$ then break endif
               $v_h = 0$
               $p_h = v_h$
               $\tilde{r}_h = r_h$
               until $i > i_{\max}$ do
                    $\rho = (\tilde{r}_h, r_h)$
                    if $|\rho| < \varepsilon$ then break endif
                    $\beta = (\rho/\rho_0)(\alpha/\omega_0)$
                    $p_h = r_h + \beta(p_h - \omega v_h)$
                    $v_h = L_h^\ell p_h$
                    $d = (\tilde{r}_h, v_h)$
                    if $|d| < \varepsilon$ then $d = 1.0$ endif
                    $\alpha = \rho/d$
                    $r_h = r_h - \alpha v_h$
                    $t_h = L_h^\ell r_h$
                    $d = (t_h, t_h)$
                    if $|d| < \varepsilon$ then break endif
                    $\omega = (t_h, r_h)/d$
                    $c_h = c_h + \alpha p_h + \omega r_h$
                    $r_h = r_h - \omega t_h$
                    $\rho_0 = \rho$
                    $\omega_0 = \omega$
                    $n = (r_h, r_h)$
                    if $|\omega| < \varepsilon$ then break endif
                    if $i_c > i_{c,\max}$ then $i_c=0$; break; endif
                    $i=i+1$
               enddo
          enddo
          $c_h := P_{\ell+1,\ell} c_h$
     enddo
     $u_h := u_h + c_h$
enddo
```

Fig. 5. The cascadic iteration algorithm with BiCGStab

crete operators, we are only interested in matrix-free methods. This restricts the choice of the applicable preconditioning methods. In fact, for preconditioning we restrict ourselves to diagonal scaling and transformation between E- and H- representation. We exploit the available hierarchical structure of the approximate solution by the computation of a good initial approximation on a given level by interpolation of a sufficiently accurate solution that is computed on a coarser level. Thus, starting from a coarsest grid, we obtain the cascadic algorithm.

First, the algorithm was applied both to the E-representation and to the H-representation of the solution, and it appeared that the solution of the H-representation is much faster. This is in agreement with the findings of Ong [12] for the solution of a FEM discretization with the tetrahedral element and quasi-uniform refinement, as discussed in the introduction. As a consequence we further only considered iteration with the H-representation.

By itself the BiCGStab is not a very efficient solver, but combined with cascadic switching between the levels we obtain an algorithm that solves the equation up to truncation error accuracy in only a few (outer) cycles. This is shown in the Figures 6 and 7. In the Fig. 6, we see the difference between using a large number of (inner) BiCGStab iterations vs using a small number. In Fig. 7, on level 10, we see the difference between the use of the first order prolongation $\tilde{P}_{\ell+1,\ell}$ (left) or the second order formula (6) (right). We clearly see that second order interpolation gives a much better convergence, so that truncation error accuracy is obtained in a small number of (4) outer iteration cycles.

Legend to Figures 6 and 7. Top figures: the logarithm of the two-norm of the measured residual at different levels and in the inner loop, against the number of inner iterations. Bottom figures: logarithm of the residual and the global discretization error of the solution of the target equation against the number of elementary operations (flops). The constant lines indicate the approximation error and the local truncation error.

5 Conclusion

Because the evaluation of finite element stiffness matrices for variable coefficient equations on sparse grids in three dimensions still yields difficulties, finite differences are an interesting alternative instead. In this paper, we show how a cascadic multigrid application of BiCGStab yields an efficient solution method for the resulting discrete equations.

The method applies the BiCGStab-iteration to the H-representation of the discrete solution, it uses second order interpolation between the different levels of discretization and it applies global defect correction (iterative refinement) as an outer iteration cycle. Results for this solution method are presented which show that 3 or 4 iteration cycles may be sufficient to solve the discrete equations up to local truncation error accuracy.

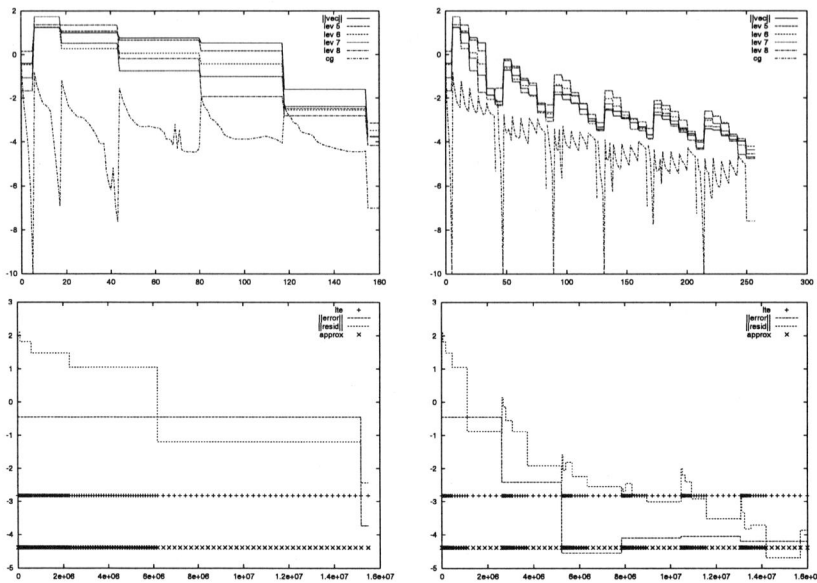

Fig. 6. The advantage of spreading inner iterations over more outer iterations. Left: a single outer iteration with 36 inner iterations at each level. Right: 6 outer iterations with 6 inner iterations each

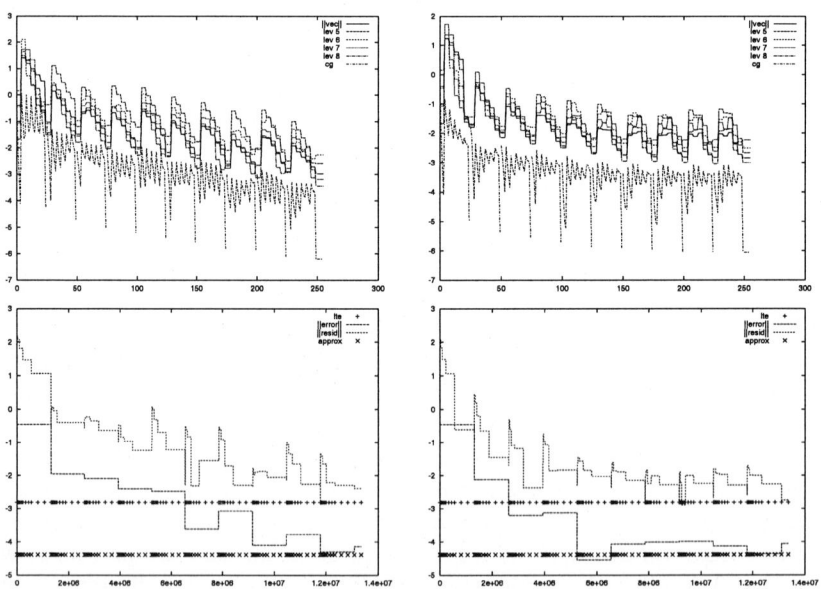

Fig. 7. Convergence at level $k = 10$. Left: first order interpolation between the levels. Right: second order interpolation

References

1. R. Barrett, M. Berry, T. F. Chan, J. Demmel, J. Donato, J. J. Dongarra, V. Eijkhout, R. Pozo, C. Romine, and H. van der Vorst. *Templates for the solution of linear systems: building blocks for iterative methods*. SIAM Books, Philadelphia, 1994.
2. F. Bornemann and P. Deuflhard. The cascadic multigrid method for elliptic problems. *Numerische Mathematik*, 75:135–152, 1996.
3. F. Bornemann and P. Deuflhard. The cascadic multigrid method. In R. Glowinsk, J. PÄéiaux, Z. Shi, and O. Widlund, eds., *The 8th International Conference on Domain Decomposition Methods for Partial Differential Equations*. John Wiley & Sons Ltd., 1997.
4. H.-J. Bungartz and T. Dornseifer. Sparse grids: Recent developments for elliptic partial differential equations. In W. Hackbusch and G. Wittum, eds., *Multigrid Methods V*, Lecture Notes in Computational Science and Engineering 3. Springer, Berlin, 1998.
5. M. Griebel. A parallelizable and vectorizable multi-level algorithm on sparse grids. In *Parallel algorithms for partial differential equations*, Notes Numer. Fluid Mech. 31, 94–100. Vieweg, Wiesbaden, 1991.
6. M. Griebel. *Multilevelmethoden als Iterationsverfahren über Erzeugendensystemen*. Teubner Skripten zur Numerik. Teubner, Stuttgart, 1994.
7. M. Griebel. Adaptive sparse grid multilevel methods for elliptic PDEs based on finite differences. *Computing*, 2000. to appear.
8. P. W. Hemker and P. M. de Zeeuw. BASIS3, a data structure for 3-dimensional sparse grids. In H. Deconinck, ed., *Euler and Navier-Stokes Solvers Using Multidimensional Upwinds Schemes and Multigrid Acceleration*, Notes Numer. Fluid Mech. 57, 443–484. Vieweg, Wiesbaden, 1997.
9. P. W. Hemker and C. Pflaum. Approximation on partially ordered sets of regular grids. *Appl. Numer. Math.*, 25:55–87, 1997.
10. P. W. Hemker and F. Sprengel. On the representation of functions and finite difference operators on adaptive sparse grids. submitted (CWI Report MAS-R9933), 1999.
11. J. Noordmans and P. W. Hemker. Convergence results for 3D sparse grid approaches. In *PRISM'97, Proceedings of the Conference on Preconditioned Iterative Solution Methods for Large Scale Problems in Scientific Computations*, 11–22, Nijmegen, The Netherlands, 1997. Nijmegen University.
12. M. E. Ong. Hierarchical basis preconditioners in three dimensions. *SIAM Journal on Scientific Computing*, 18:479–498, 1997.
13. T. Schiekofer. *Die Methode der Finiten Differenzen auf dünnen Gittern zur Lösung elliptischer und parabolischer partieller Differentialgleichungen*. PhD thesis, Universität Bonn, 1998.
14. V. Shaidurov. Some estimates of the rate of convergence for the cascadic conjugate gradient method. *Comp. Math. Appls*, 31:161–171, 1996.
15. Z.-C. Shi and X. Xu. Cascadic multigrid method for elliptic problems. *East-West J. Numer. Math*, 7:199–209, 1999.
16. F. Sprengel. Some remarks on multilevel algorithms for finite difference discretizations on sparse grids. CWI report MAS-R9924, 1999.

Topology Optimization of Conductive Media Described by Maxwell's Equations*

Ronald H. W. Hoppe[1], Svetozara I. Petrova[1], and Volker H. Schulz[2]

[1] Institut für Mathematik, Universität Augsburg
Universitätsstraße 14, D-86135 Augsburg, Germany
[2] Weierstraß-Institut für Angewandte Analysis und Stochastik (WIAS)
Mohrenstraße 39, D-10117 Berlin, Germany

Abstract. The problem of an energy dissipation optimization in a conductive electromagnetic media is considered. The domain is known a priori and is fixed throughout the optimization process. We apply a perturbed and damped interior–point Newton method for the primal–dual formulation of the nonlinear programming problem. Nonnegative slack variables are added to the inequality constraints in the optimization problem. Computational results concerning a two–dimensional isotropic system are included.

Keywords: Maxwell's equations, topology optimization, nonlinear programming, primal–dual approach, interior–point method

AMS subject classifications: 65K05, 90C05, 90C30.

1 Introduction

Computation of electromagnetic fields in various settings, analysis and different approaches for the spatial discretization of the Maxwell equations have been a subject of intense research in the last decade, see, e.g., [2,7,10]. In this paper we consider problems concerning topology optimization in electromagnetic media. For a general overview on the field of structural optimization and topology design, we refer to [3]. We are looking for an optimal distribution of conductivity in a fixed geometrical configuration.

Let $\Omega \subset R^3$ be a domain occupied by a conductor with a conductivity $\sigma > 0$. The rest of the space is vacuum. To simplify the presentation, we consider the stationary case, i.e., constant currents are available in the conductor (div $\mathbf{J} = 0$). In this case the Maxwell equations read:

$$\mathbf{curl\,E} = -\partial_t \mathbf{B}, \quad \mathbf{curl\,H} = \mathbf{J}, \quad \mathrm{div\,}\mathbf{D} = \rho, \quad \mathrm{div\,}\mathbf{B} = 0, \qquad (1)$$

---

* This work was supported in part by the Alexander von Humboldt Foundation. The second author has also been supported by the Bulgarian Ministry for Education, Science, and Technology under Grant MM–98#801.

supplemented by the following material laws:

$$\mathbf{D} = \varepsilon \mathbf{E}, \quad \mathbf{B} = \mu \mathbf{H}, \quad \mathbf{J} = \sigma \mathbf{E}. \tag{2}$$

Here, the fundamental electromagnetic quantities are the *electric field* $\mathbf{E}$, the *magnetic induction* $\mathbf{B}$, the *magnetic field* $\mathbf{H}$, the *electric induction* $\mathbf{D}$, the *electric current density* $\mathbf{J}$, and the *space charge density* ρ. We consider only linear and isotropic materials, so that the *electric permeability* ε, the *magnetic permeability* μ, and the *electric conductivity* σ are supposed bounded scalar functions of the spatial variable $\mathbf{x}$ with $\varepsilon \geq \varepsilon_0 > 0$, $\mu \geq \mu_0 > 0$, and $\sigma > 0$. Steep jumps of these coefficients may occur at material interfaces. One can introduce a scalar electric potential φ and a magnetic vector potential $\mathbf{A}$, so that

$$\mathbf{E} = -\mathrm{grad}\,\varphi - \partial_t \mathbf{A} \quad \text{and} \quad \mathbf{B} = \mathrm{curl}\,\mathbf{A}. \tag{3}$$

To specify $\mathbf{A}$, which is not uniquely defined, we use the Coulomb *gauge*, namely, $\mathrm{div}\,\mathbf{A} = 0$. From (2) and (3) one gets $\mathbf{J} = \sigma \mathbf{E} = -\sigma \mathrm{grad}\,\varphi - \sigma \partial_t \mathbf{A}$, which yields

$$\mathrm{div}\,\mathbf{J} = \mathrm{div}\,(\mathrm{\mathbf{curl}\,H}) = 0 = -\mathrm{div}\,(\sigma\,\mathrm{\mathbf{grad}}\,\varphi) - \mathrm{div}\,(\sigma\,\partial_t \mathbf{A}). \tag{4}$$

Suppose now that σ is piecewise constant, i.e., independent of the spatial variable $\mathbf{x}$. Then $\mathrm{div}\,\mathbf{A} = 0$ results in $\mathrm{div}\,(\sigma\,\partial_t \mathbf{A}) = 0$. From (4) we get the following coupled system of equations for φ and $\mathbf{A}$:

$$\mathrm{div}(\sigma\,\mathbf{grad}\,\varphi) = 0 \text{ in } \Omega, \quad \mathbf{n} \cdot \sigma\,\mathbf{grad}\,\varphi = \begin{cases} I_\nu \text{ on } \Gamma_\nu \subset \partial\Omega \\ 0 \quad \text{otherwise} \end{cases} \tag{5}$$

$$\sigma\,\partial_t \mathbf{A} + \mathrm{\mathbf{curl}}\,(\mu^{-1}\,\mathrm{\mathbf{curl}A}) = \begin{cases} -\sigma \mathbf{grad}\,\varphi & \text{in } \Omega \\ 0 & \text{in } R^3 \backslash \bar{\Omega} \end{cases}. \tag{6}$$

Here, the unit normal vector is denoted by $\mathbf{n}$. For the given electric current densities $\{I_\nu\}$ on the boundary $\Gamma_\nu \subset \partial\Omega$ we impose the compatibility condition $\sum_\nu I_\nu = 0$. The energy dissipation given by the Joule–Lenz law reads as follows:

$$f(\varphi,\sigma) := \int_\Omega \mathbf{J}\,\mathbf{E}\,dx = -\int_\Omega \mathbf{J} \cdot \mathbf{grad}\,\varphi\,dx = -\int_\Omega \mathrm{div}(\varphi \mathbf{J})\,dx. \tag{7}$$

Using the Gauss–Ostrogradski formula and the Neumann boundary conditions from (5) we get the following expression:

$$f(\varphi,\sigma) = -\int_{\partial\Omega} \mathbf{n} \cdot \mathbf{J}\,\varphi\,ds = \sum_\nu \int_{\Gamma_\nu} I_\nu\,\varphi\,ds. \tag{8}$$

The remainder of this paper is organized as follows. In Section 2 we introduce the primal–dual formulation of our nonlinear nonconvex programming problem. Slack variables are added directly to the optimization problem. In Section 3 we discuss the steplength strategy and give the interior–point algorithm. In the last section, we include some numerical experiments concerning the conductivity distribution for a two–dimensional isotropic system.

2 Primal–Dual Approach

In this section, we formulate the nonlinear nonconvex optimization problem for a minimization of the energy dissipation given by (8).

$$\min_{\varphi,\sigma} f(\varphi,\sigma) = \min_{\varphi,\sigma} \sum_\nu \int_{\Gamma_\nu} I_\nu \varphi \, ds, \tag{9}$$

subject to the following constraints:

φ satisfies (5),
$\int_\Omega \sigma \, dx = C$ (mass constraint), (10)
$\sigma_{\min} \leq \sigma \leq \sigma_{\max}$ (conductivity box constraint).

Here, $\sigma_{\min}$ and $\sigma_{\max}$ are a priori given positive limits for the conductivity and C is a fixed given value. Note that we formulate a constrained optimization problem, where the differential equation for φ (5) is part of the constraints. This is in contrast to many standard optimization approaches, which would consider φ as a function of the independent variable σ via the differential equation. However, this simultaneous optimization approach reduces the overall computational complexity of the resulting optimization algorithm.

We apply the primal–dual interior–point method, originally proposed for linear programs by [8]. This method has been recently extended to nonlinear programming in [1] and started to prove its impressive computational performance for nonlinear programming, see, e.g., [5,6,12]. We deal with the corresponding inequality constraints introducing nonnegative slack variables. This variant of the primal–dual approach has been used, e.g., in [1,9]. After a finite element discretization of the domain we get the following finite dimensional nonlinear programming problem:

$$\min_{\varphi,\sigma} f(\varphi,\sigma), \tag{11}$$

subject to
$$A(\sigma)\varphi - \mathbf{b} = 0, \quad \sigma_{\min}\mathbf{e} - \sigma + \mathbf{s} = 0, \quad \mathbf{s} \geq 0,$$
$$g(\sigma) - C = 0, \quad \sigma - \sigma_{\max}\mathbf{e} + \mathbf{t} = 0, \quad \mathbf{t} \geq 0, \tag{12}$$

where $A(\sigma)$ is the finite element stiffness matrix, $\mathbf{b}$ is the discrete load vector and $g(\sigma)$ is a discrete approximation of $\int_\Omega \sigma \, dx$. Here, $\mathbf{e} \in \mathcal{R}^N$, $\mathbf{e} = (e_1, \ldots, e_N)^T$, $e_i = 1$, $1 \leq i \leq N$, and $\sigma, \mathbf{s}, \mathbf{t} \in \mathcal{R}^N$, where N is the number of finite elements. Note that the lower bound $\sigma_{\min}$ plays a crucial role keeping the ellipticity of the discrete problem.

The *Lagrangian function* associated with problem (11)-(12) is:

$$\mathcal{L}(\varphi,\sigma,\lambda,\eta,\mathbf{z},\mathbf{w},\mathbf{s},\mathbf{t},\alpha,\beta) := f(\varphi,\sigma) + \lambda^T(A(\sigma)\varphi - \mathbf{b}) + \eta(g(\sigma) - C)$$
$$+ \mathbf{z}^T(\sigma_{\min}\mathbf{e} - \sigma + \mathbf{s}) + \mathbf{w}^T(\sigma - \sigma_{\max}\mathbf{e} + \mathbf{t})$$
$$- \alpha^T\mathbf{s} - \beta^T\mathbf{t}. \tag{13}$$

Here, λ, η, $\mathbf{z} \geq 0$, $\mathbf{w} \geq 0$ and $\alpha \geq 0$, $\beta \geq 0$ are the Lagrange multipliers for the equality and inequality constraints in (12), respectively. Our purpose is to find an isolated (locally unique) local minimum of the problem (11)-(12) under the assumption that at least one such point exists. We suppose that the standard conditions for the application of Newton's method, see, e.g., [4], are satisfied. Denote by $\Phi := (\varphi, \sigma, \lambda, \eta, \mathbf{z}, \mathbf{w}, \mathbf{s}, \mathbf{t})$ the vector of the unknown variables. The complementarity conditions $Z\mathbf{s} = 0$ and $W\mathbf{t} = 0$ are replaced by the *perturbed complementarity conditions* $Z\mathbf{s} = p\mathbf{e}$ and $W\mathbf{t} = p\mathbf{e}$. At each iteration, the positive parameter p is decreased by a certain amount.

The necessary first–order Karush–Kuhn–Tucker (KKT) optimality conditions lead to the following nonlinear equation:

$$F_p(\Phi) := \begin{pmatrix} \nabla_\varphi \mathcal{L} \\ \nabla_\sigma \mathcal{L} \\ \nabla_\lambda \mathcal{L} \\ \nabla_\eta \mathcal{L} \\ \nabla_\mathbf{z} \mathcal{L} \\ \nabla_\mathbf{w} \mathcal{L} \\ \nabla_\mathbf{s} \mathcal{L} \\ \nabla_\mathbf{t} \mathcal{L} \end{pmatrix} = \begin{pmatrix} \nabla_\varphi f + A(\sigma)^T \lambda \\ \partial_\sigma (\lambda^T A(\sigma)\varphi) + \eta \nabla g(\sigma) - \mathbf{z} + \mathbf{w} \\ A(\sigma)\varphi - \mathbf{b} \\ g(\sigma) - C \\ \sigma_{min} \mathbf{e} - \sigma + \mathbf{s} \\ \sigma - \sigma_{max} \mathbf{e} + \mathbf{t} \\ Z\mathbf{s} - p\mathbf{e} \\ W\mathbf{t} - p\mathbf{e} \end{pmatrix} = 0, \qquad (14)$$

where $\nabla_\mathbf{s} \mathcal{L} = Z\mathbf{s} - p\mathbf{e}$, $\nabla_\mathbf{t} \mathcal{L} = W\mathbf{t} - p\mathbf{e}$. The *search direction* is given by $\Delta\Phi := (\Delta\varphi, \Delta\sigma, \Delta\lambda, \Delta\eta, \Delta\mathbf{z}, \Delta\mathbf{w}, \Delta\mathbf{s}, \Delta\mathbf{t})$. The update $\Phi \leftarrow \Phi + \Delta\Phi$ is determined by the increment $\Delta\Phi$ computed by using the Newton method for the following p–dependent system of equations.

$$F'_p(\Phi) \Delta\Phi = -F_p(\Phi), \qquad (15)$$

where (15) is often referred to as the *primal–dual* system and solved at each iteration with a decreasing parameter p. More precisely, (15) is equivalent to:

$$\begin{pmatrix} 0 & \mathcal{L}_{\varphi\sigma} & \mathcal{L}_{\varphi\lambda} & 0 & 0 & 0 & 0 & 0 \\ \mathcal{L}_{\sigma\varphi} & \mathcal{L}_{\sigma\sigma} & \mathcal{L}_{\sigma\lambda} & \mathcal{L}_{\sigma\eta} & -I & I & 0 & 0 \\ \mathcal{L}_{\lambda\varphi} & \mathcal{L}_{\lambda\sigma} & 0 & 0 & 0 & 0 & 0 & 0 \\ 0 & \mathcal{L}_{\eta\sigma} & 0 & 0 & 0 & 0 & 0 & 0 \\ 0 & -I & 0 & 0 & 0 & 0 & I & 0 \\ 0 & I & 0 & 0 & 0 & 0 & 0 & I \\ 0 & 0 & 0 & 0 & S & 0 & Z & 0 \\ 0 & 0 & 0 & 0 & 0 & T & 0 & W \end{pmatrix} \begin{pmatrix} \Delta\varphi \\ \Delta\sigma \\ \Delta\lambda \\ \Delta\eta \\ \Delta\mathbf{z} \\ \Delta\mathbf{w} \\ \Delta\mathbf{s} \\ \Delta\mathbf{t} \end{pmatrix} = - \begin{pmatrix} \nabla_\varphi \mathcal{L} \\ \nabla_\sigma \mathcal{L} \\ \nabla_\lambda \mathcal{L} \\ \nabla_\eta \mathcal{L} \\ \nabla_\mathbf{z} \mathcal{L} \\ \nabla_\mathbf{w} \mathcal{L} \\ \nabla_\mathbf{s} \mathcal{L} \\ \nabla_\mathbf{t} \mathcal{L} \end{pmatrix}, \qquad (16)$$

where I stands for the identity matrix, $S = \text{diag}(s_i)$, $Z = \text{diag}(z_i)$, $T = \text{diag}(t_i)$, and $W = \text{diag}(w_i)$ are diagonal matrices. Note that $\mathcal{L}_{\lambda\varphi} = A(\sigma)$ is the stiffness matrix of the electric potential equation, $\mathcal{L}_{\sigma\sigma}$ is a diagonal matrix, and $\mathcal{L}_{\eta\sigma} = \nabla^T g(\sigma)$ is just one row vector.

The primal–dual matrix $F'_p(\Phi)$ in (15) is sparse, nonsymmetric, indefinite, and usually well–conditioned. Our approach is to transform $F'_p(\Phi)$ to a smaller (so called *condensed*) matrix, which is inherently ill–conditioned, but the ill–conditioning should not necessarily be avoided and has no negative consequences. For detailed discussion, see, e.g., [12]. We eliminate the increments for $\mathbf{s}$ and $\mathbf{t}$

from the 5th and 6th rows of (16), namely, $\triangle \mathbf{s} = \triangle\boldsymbol{\sigma} - \nabla_z \mathcal{L}$, $\triangle \mathbf{t} = -\triangle\boldsymbol{\sigma} - \nabla_w \mathcal{L}$. From the last two rows of (16) we obtain the increments for $\mathbf{z}$ and $\mathbf{w}$:

$$\triangle \mathbf{z} = S^{-1}(-\nabla_s \mathcal{L} - Z(\triangle\boldsymbol{\sigma} - \nabla_z \mathcal{L})) \qquad (17)$$
$$\triangle \mathbf{w} = T^{-1}(-\nabla_t \mathcal{L} - W(-\triangle\boldsymbol{\sigma} - \nabla_w \mathcal{L})).$$

Substituting (17) in the second row of (16), we get the following linear system:

$$\begin{pmatrix} 0 & \mathcal{L}_{\varphi\sigma} & \mathcal{L}_{\varphi\lambda} & 0 \\ \mathcal{L}_{\sigma\varphi} & \tilde{\mathcal{L}}_{\sigma\sigma} & \mathcal{L}_{\sigma\lambda} & \mathcal{L}_{\sigma\eta} \\ \mathcal{L}_{\lambda\varphi} & \mathcal{L}_{\lambda\sigma} & 0 & 0 \\ 0 & \mathcal{L}_{\eta\sigma} & 0 & 0 \end{pmatrix} \begin{pmatrix} \triangle\varphi \\ \triangle\sigma \\ \triangle\lambda \\ \triangle\eta \end{pmatrix} = - \begin{pmatrix} \nabla_\varphi \mathcal{L} \\ \tilde{\nabla}_\sigma \mathcal{L} \\ \nabla_\lambda \mathcal{L} \\ \nabla_\eta \mathcal{L} \end{pmatrix}, \qquad (18)$$

where $\tilde{\mathcal{L}}_{\sigma\sigma} = \mathcal{L}_{\sigma\sigma} + S^{-1}Z + T^{-1}W$ and the modified entry for the right-hand side is

$$\tilde{\nabla}_\sigma \mathcal{L} = \nabla_\sigma \mathcal{L} + S^{-1}(\nabla_s \mathcal{L} - Z\nabla_z \mathcal{L}) - T^{-1}(\nabla_t \mathcal{L} - W\nabla_w \mathcal{L}).$$

Transforming iterations, proposed in [11], for the null space decomposition of the condensed matrix, are applied to compute the search direction, see, [9].

3 Interior-Point Method

We apply the *line-search* version of the Newton method. After computation of the search direction $\triangle\boldsymbol{\Phi}$, a common steplength α ($\alpha > 0$) is employed to update the solution $\boldsymbol{\Phi} \leftarrow \boldsymbol{\Phi} + \alpha\triangle\boldsymbol{\Phi}$. In all Newton-type methods, $\alpha = 1$ is almost always the "ideal" value. The method for choosing α at each iteration becomes more complex, as it is well known that for general nonlinear problems with a poor initial estimate, Newton's method may diverge. Complete convergence analysis of the Newton interior-point method for nonlinear programming is given by [1] provided the Jacobian $F'_p(\boldsymbol{\Phi})$ of the system (14) remains nonsingular.

A standard approach for choosing the steplength α is to define a suitable merit function, that measures the progress towards the solution. The squared l_2-norm of the residual as a merit function was introduced in [1] as

$$M(\boldsymbol{\Phi}) = \|F(\boldsymbol{\Phi})\|_2^2, \qquad (19)$$

where $F(\boldsymbol{\Phi}) := F_p(\boldsymbol{\Phi}) + p\hat{\mathbf{e}}$, see (14), and $\hat{\mathbf{e}} = (0,\ldots,0,1,\ldots,1)$ is a vector with $2N$ ones. We accept the following notations: $M_k = M_k(0) = M(\boldsymbol{\Phi}_k)$ and $M_k(\alpha) = M(\boldsymbol{\Phi}_k + \alpha\triangle\boldsymbol{\Phi}_k)$, where $\boldsymbol{\Phi}_k$ is the computed solution at a given iteration.

To specify the selection of α, we apply the algorithm proposed by [1]. Let $\boldsymbol{\Phi}_0 = (\varphi_0, \sigma_0, \lambda_0, \eta_0, \mathbf{z}_0, \mathbf{w}_0, \mathbf{s}_0, \mathbf{t}_0)$ be a given starting point with $(\mathbf{z}_0, \mathbf{w}_0, \mathbf{s}_0, \mathbf{t}_0) > 0$. Let

$$\tau = \min(Z_0\,\mathbf{s}_0, W_0\,\mathbf{t}_0)/[(\mathbf{z}_0^T\mathbf{s}_0 + \mathbf{w}_0^T\mathbf{t}_0)/(2\,N)].$$

We denote by
$$\Phi(\alpha) := (\varphi(\alpha), \sigma(\alpha), \lambda(\alpha), \eta(\alpha), z(\alpha), w(\alpha), s(\alpha), t(\alpha)) = \Phi + \alpha \triangle \Phi.$$

For a given iteration k, we define
$$q_k(\alpha) = \min(Z(\alpha)s(\alpha), W(\alpha)t(\alpha)) - \gamma_k \tau \left(z(\alpha)^T s(\alpha) + w(\alpha)^T t(\alpha)\right)/(2N),$$
where $\gamma_k \in (0,1)$ is a constant. The steplength α_k is determined as
$$\alpha_k = \max_{\alpha \in (0,1]} \{\alpha : q_k(\alpha') \geq 0, \text{ for all } \alpha' \leq \alpha\}. \tag{20}$$

Note that the function $q_k(\alpha)$ is piecewise quadratic and, hence, α_k is either one or the smallest positive root of $q_k(\alpha)$ in $(0,1]$.

We describe now the primal–dual Newton interior–point algorithm.

Interior–point algorithm:

1. Choose $\Phi_0 = (\varphi_0, \sigma_0, \lambda_0, \eta_0, z_0, w_0, s_0, t_0)$ such that $(z_0, w_0, s_0, t_0) > 0$ and $\beta \in (0, 1/2]$. Set $k = 0$, $\gamma_{k-1} = 1$, and compute $M_0 = M(\Phi_0)$. For $k = 0, 1, 2, \ldots$, do the following steps:
2. Test for convergence: if $M_k \leq \epsilon_{\text{exit}}$, stop.
3. Choose $\xi_k \in (0, 1)$; for $\Phi = \Phi_k$, compute the perturbed Newton direction $\triangle \Phi_k$ from (15) with a perturbation parameter
$$p_k = \xi_k (z_k^T s_k + w_k^T t_k)/(2N). \tag{21}$$
4. Steplength selection.
   (4a) Choose $1/2 \leq \gamma_k \leq \gamma_{k-1}$; compute α_k from (20).
   (4b) Let $\alpha_k = \alpha_k/(2^n)$, where $n > 0$ is the smallest integer such that
$$M_k(\alpha_k) \leq M_k(0) + \alpha_k \beta \nabla M_k^T \triangle \Phi_k.$$
5. Let $\Phi_{k+1} = \Phi_k + \alpha_k \triangle \Phi_k$ and $k \leftarrow k+1$. Go to 2.

It was shown in [1] that for the proposed choice of p_k in (21), the search direction $\triangle \Phi_k$, generated by the interior–point algorithm, gives descent for the merit function $M(\Phi_k)$, i.e., $\nabla M_k^T \triangle \Phi_k < 0$, where ∇M_k is the derivative of $M_k(\alpha)$ at $\alpha = 0$.

4 Numerical Experiments

In this section, we give some details concerning our computations. We solve the optimization problem (11)-(12) with an objective function defined in (8). The first equality constraint is related to solving elliptic differential equation for the electric potential φ, see (5). We allow here some modification in the conductivity, namely, we consider
$$\operatorname{div}(h(\sigma) \operatorname{\mathbf{grad}} \varphi) = 0 \text{ in } \Omega, \quad \mathbf{n} \cdot h(\sigma) \operatorname{\mathbf{grad}} \varphi = \begin{cases} I_\nu & \text{on } \Gamma_\nu \subset \partial \Omega \\ 0 & \text{otherwise} \end{cases}, \tag{22}$$

where

$$h(\sigma) = \left(\frac{\sigma - \sigma_{\min} + 0.01}{\sigma_{\max} - \sigma_{\min}}\right)^2 \qquad (23)$$

is treated as a conductivity. Neumann boundary conditions were imposed, assuming that the compatibility condition from Section 1 is satisfied. The computations have been carried through a rectangular domain Ω decomposed into N uniform quadrilateral finite elements. We suppose that the domain is an isotropic conductor. The conductivity is computed at the center points of the finite elements and the electric potential is approximated at the midpoints of the edges. Due to the definition (23), the diagonal matrix $\mathcal{L}\sigma\sigma$ does not vanish.

Our primal–dual code was written in C++ using double precision binary arithmetic. All numerical tests were run on Alpha PC164LX machine. We choose lower and upper limits for the conductivity $\sigma_{\min} = 0.01$ and $\sigma_{\max} = 1$, respectively. In all runs, an initial homogeneous distribution was proposed with $\sigma = 0.45$. The constant C in (11) is computed in accordance with this initialization. The following parameters for the interior–point algorithm in Section 3 are used: $\xi_k = \min(0.2, 100\,(\mathbf{z}_k^T \mathbf{s}_k + \mathbf{w}_k^T \mathbf{t}_k))$, $\beta = 0.0001$, and $\epsilon_{exit} = 10^{-6}$.

The most expensive (in terms of CPU–time) part of the algorithm during a given iteration is to solve the condensed primal–dual system finding the increments. Two transforming iterations have been used with a zero initial guess. The preconditioned conjugate gradient (PCG) method is applied with the symmetric successive overrelaxation (SSOR) iteration as a preconditioner for the stiffness matrix. We choose a relaxation parameter $\omega = 1.5$ and a stopping criterion for both iterative procedures $\mathbf{r}^T A(h(\boldsymbol{\sigma}))\mathbf{r} < 10^{-10}$, where $\mathbf{r}$ is the current residual.

The results from our numerical experiments are reported in Table 1 for various number of contacts NC and various number of finite elements N. The dimension of the stiffness matrix is denoted by NP. We report as well the global number of iterations in the main optimization loop denoted by iter, the perturbation parameter p and the merit function $M(\boldsymbol{\Phi})$ at the last iteration.

Table 1. Results from applications of the interior–point algorithm

NC	N	NP	iter	p	$M(\boldsymbol{\Phi})$
2	30	71	20	1.13e-4	5.11e-7
2	40	93	14	2.17e-5	3.42e-8
2	80	178	18	7.03e-5	5.76e-8
2	80	178	22	5.08e-4	2.05e-7
2	120	262	34	3.93e-5	1.27e-8
3	30	71	25	6.03e-4	4.18e-7
3	64	144	41	5.17e-5	8.03e-8
4	96	212	45	3.12e-4	4.32e-7
5	180	388	42	1.18e-4	2.84e-7

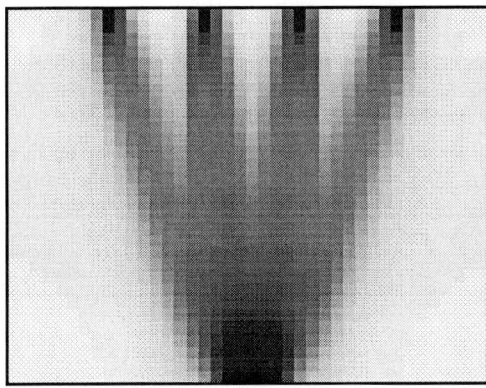

Fig. 1. Conductivity distribution for a mesh 30 × 40 with 5 contacts

Figure 1 shows the conductivity distribution for a mesh 30 × 40 with five contacts. The black color indicates elements where the conductivity is very close to σ_{max} and the white color indicates those elements with a conductivity close to σ_{min}.

References

1. El–Bakry, A. S., Tapia, R. A., Tsuchiya, T., Zhang, Y.: On the formulation and theory of the Newton interior–point method for nonlinear programming. Journal of Optimization Theory and Applications, **89** (1996) 507-541.
2. Beck, R., Deuflhard, P., Hiptmair, R., Hoppe, R. H. W., Wohlmuth, B.: Adaptive miltilevel methods for edge element discretizations of Maxwell's equations. Surveys of Math. in Industry, **8** (1999) 271-312.
3. Bendsøe, M. P.: Optimization of Structural Topology, Shape, and Material. Springer, (1995).
4. Fiacco, A. V., McCormick, G. P.: Nonlinear Programming: Sequential Unconstrained Minimization Techniques. John Wiley and Sons, New York, (1968). Republished by SIAM, Philadelphia, (1990).
5. Forsgen, A., Gill, P.: Primal–dual interior methods for nonconvex nonlinear programming. SIAM Journal on Optimization, **8** (1998) 1132-1152.
6. Gay, D. M., Overton, M. L., Wright, M. H.: A primal–dual interior method for nonconvex nonlinear programming. Technical Report 97-4-08, Computing Science Research Center, Bell Laboratories, Murray Hill, New Jersey, (1997).
7. Hoppe, R. H. W.: Mortar edge elements in R^3. East–West J. Numer. Anal., **7** (1999) 159-173.
8. Kojima, M., Mizuno, S., Yoshise, A.: A primal–dual method for linear programming. Progress in Mathematical Programming Interior Point and Related Methods, Edited by N. Megiddo, Springer Verlag, New York, (1989).
9. Maar, B., Schulz, V.: Interior point multigrid methods for topology optimization. Preprint 98-57, IWR, University of Heidelberg, Germany, (1998), Structural Optimization (to appear).

10. Monk, P.: Analysis of finite element method for Maxwell's equations. SIAM J. Numer. Anal., **29** (1992) 714-729.
11. Wittum, G.: On the convergence of multigrid methods with transforming smoothers. Theory with applications to the Navier–Stokes equations. Numer. Math., **57** (1989) 15-38.
12. Wright, M. H.: Ill–conditioning and computational error in interior methods for nonlinear programming. Technical Report 97-4-04, Computing Science Research Center, Bell Laboratories, Murray Hill, New Jersey, (1997).

Finite Element Simulation of Residual Stresses in Thermo-coupled Wire Drawing Process

R. Iankov[1], A. Van Bael[2], and P. Van Houtte[2]

[1] Institute of Mechanics, Bulgarian Academy of Sciences, Sofia
[2] Catholic Univ. of Leuven, Department of Metallurgy and Materials Eng., Belgium

Abstract. The objective of this paper is to calculate residual stress in drawn wire taking into account induced temperature due to plastic dissipation energy. Finite element analysis (FEA) for the simulation of wire drawing is applied. The general purpose FEA code MARC, is used to analyse thermo-coupled wire drawing processes. The necessary condition for determination of range of steady state flow was proposed.

1 Introduction

Wire drawing forming involves complicated deformation process with material, geometrical and contact nonlinearities. One of the vital characteristic of drawn wire is the distribution of the residual stress in it. For obtaining the optimised design of such forming processes, investigations into details of deformation, namely, residual stress and deformation state, microlevel changes and cracks are extremely important.

The numerical simulation of wire drawing by means of finite element has been dealt by many authors Davas W. and Fischer F.D.; Boris S.; Doege E. and Kroeff A. [9,8,10,1].

In [8] a study-state cold wire drawing model based on Lagrangian incremental elastic-plastic formulation is considered. The general purpose finite element program *ABAQUS* has been used to solve 2D wire drawing finite element model. Two different optimisation problems associated with optimal die design were considered. Minimisation of the total energy in the process and maximisation of the reduction area.

There are metal forming processes in which thermo-mechanical coupling investigations are necessary. For example the deformation and friction during aluminium extrusion cause considerable temperature increases (up to more $100^0 C$) [2].

During the wire drawing process, large nonhomogeneities in deformation and consequently in heat generation, usually occur. Moreover, especially if the dies are at a considerably lower temperature than the workpiece, the heat losses by conduction to the dies and by radiation and convection to the environment contribute to the existence of severe temperature gradients. The friction forces between workpiece and dies are heat source, which is very important when a mild materials are considered. Thus including temperature effects in the analysis of wire drawing problems is very important. Furthermore, at elevated temperatures plastic deformation can induce phase transformations and alterations in grain structures which in turn will modify the flow resistance of the material as well as other mechanical properties. It is necessary to include these metallurgical effects in thermo-mechanical coupling models.

The influence of friction in wire drawing is very important. A new upsetting sliding test is used in [4] for the determination of the friction coefficient by simulating the wire drawing contact conditions. The test is performed on the real workpiece directly from the drawing plant. this result is directly usable for the finite element simulation of the wire drawing process.

The design, control and optimisation of wire-drawing metal forming processes by means of classical trial—errors procedures become increasingly heavy in terms of time and cost in a competitive environment. Simultaneously, the improvement of the final product requires the microstructure, constitutive behaviour and deformability to be known a priori regarding a targeted application. During the last years, numerical simulations have become a very efficient tool to reach these goals. It is well known that the residual stresses are induced by fabrication processes and that those stresses will superimpose on to the service stresses especially in surface layers where, in most cases, fatigue or stress corrosion cracks initiate.

The aim of finite element simulation of wire-drawing process is prediction of local value of strain rates, strains, stresses and temperature during deformation with a view to obtaining some insight into the effect of the process on the final mechanical properties: texture, anisotropy, residual stress and die wear. However, reliable predictions from numerical simulations require reliable input data, including constitutive laws and friction conditions.

2 Residual Stresses in Wire-Drawing Process

Residual stresses are effective static stresses, which are in a state equilibrium, without the action of external forces and/or moments. They always occur whenever a macroscopic cross-sectional area of a component or a microscopic area of a multi-phase material is partially and plastically deformed by external and internal forces. These forces may either be due to thermal loading, processes of diffusion or phase transformation in such a way that, incompatibilities of deformations may caused. [3]

Due to the plastic deformation in wire-drawing, most of the mechanical energy expended in the deformation process is converted into heat and the remainder is stored in the material. The stored energy is associated with residual stress generated in the wire after plastic deformation and unloading as well as with the creation of lattice imperfections. It means that the stress in free-force state are residual stress in wire. The nature of this residual stress is plastic deformation as well as changes on micro-level.

Metal forming processes (especially wire-drawing and extrusion) commonly generate non-homogeneous plastic deformation in the workpiece so that the final product is left in a state of residual stress. In cold working the yield stress of work piece is higher than for warm or hot working so that the residual stresses produced in cold working are in general higher.

The residual stresses can have a deleterious or beneficial effect on fatigue strength. Hence, investigation of the generation of residual stresses in metal forming can be important from the standpoint of either avoiding defects by reducing residual stresses or tailoring the die-geometry to produce high beneficial stresses.

The causes of residual stresses can be classified under the following three main groups: material, manufacturing and loading and service conditions.

Manufacture-induced residual stresses can be determined both by calculation, as well as by experimental. An experimental determination of residual stress in cementite and ferrite phases of high carbon steel was provide by Houtte in [6].

Most of previous analyses of metal-forming processes were on the basis of rigid- plastic theory and models unfortunately they can not provide information concerning residual stresses in wire drawing. [7]

The residual stresses generated by metal-forming processes occur because of variations in the plastic strain distribution which are of the order of elastic strains. Small changes in the forming tool configuration can have a dominant influence on the residual stresses. This suggests that the die design could be chosen to produce beneficial stresses. A basic aspect of the problem, which might become significant in view of the sensitivity of stress distribution to small changes are die geometry and boundary conditions like a friction in boundary value problems. This type of sensitivity analysis could be included in the finite element calculations. It would be useful to develop the models and FE codes to incorporate more general material properties such as anisotropic hardening and influence on texture on material properties.

In present investigation a FE solution is used to obtained fields of stresses, strains, plastic strains and residual stresses.

3   Finite Element Method Application

Finite element approach based on displacement method was applied. The governing matrix equation for the thermo-mechanical couple problem in the case without dynamic effects are as follows:

$$[K(T,u)]\{u\} = \{F\}. \tag{1}$$

$$[C(T)]\dot{T} + [H(T,u)]T = Q + Q^p + Q^f \quad , \tag{2}$$

where T is temperature, $\{u\}$ is displacement vector, $[K(T,u)]$ is the stiffness matrix, $[C(T)]$ is the heat-capacity matrix and $[H(T,u)]$ is the thermal-conductivity matrix are all dependent on temperature and in the case of update-Lagrangian analisys $[K(T,u)]$ and $[H(T,u)]$ are dependent upon prior displacement. Q^p is the internal heat generation due to inelastic deformation, Q^f is the heat generation due to the friction between workpiece and die. The coupling between heat transfer problem and the mechanical problem is due to the temperature-dependent mechanical properties and the internal heat generated.

4 Steady State Flow Area

One of the problem is haw to determine a minimum length of wire piece in which after numerical simulation can be reach steady state flow area. The steady state area is defined as a set of cross sections of wire, where the local error between axial residual stresses satisfied following energy condition:

$$\|g_i - g_j\|_{L_2} = \left[\int_0^R |g_i(\xi) - g_j(\xi)|^2 \, d\xi\right]^{\frac{1}{2}} \leq \varepsilon \qquad (3)$$

or

$$\|g_i - g_j\|_{L_2} = \left[\sum_{k=1}^{N_k} |g_i(x_k) - g_j(x_k)|^2\right]^{\frac{1}{2}} \leq \varepsilon \qquad (4)$$

where g_i and g_j are axial residual stress functions in cross sections i and j in free-force state, N_k is a number of gauss integration points(x_k) or nodal points(x_k) in a cross section, ε is a small constants and $i, j = 1, 2, 3, \cdots N, i \neq j$, N- number of cross sections.

Cross section can be defined as a sequence of finite elements or as a sequence of nodes. In proposed definition L_2 norm is used, which is same as in a finite element approximation theory. It is assume that the deformed wire body is passed through the die and a force-free state have been reached in wire.

5 Numerical Example

A thermo–couple wire–drawing problem is considered. A large deformation problem incorporating thermo-mechanical coupling is performed. The kinematics of deformation is described by update Langrangian approach which is useful in the cases in which rotations are large so that nonlinear terms in the curvature expressions may not longer be neglected and for calculations which the plastic deformations cannot be assumed infinitesimal. The update Lagrange formulation takes the reference configuration at $t = n+1$, Cauchy stress and true strain, are used in the constitutive relationship.

Two examples of thermo-coupled wire drawing problem are considered. The difference is only in die model. In first case die is model as rigid body and in second case the die is model as deformable one. A four-node bilinear axisymmetric finite element is used. Temperature dependent material data and geometrical data a given in table 1.

Finite element simulation of wire drawing was provide and following assumption were made:

- the deformation of the work piece was axisymmetric;

- the material exhibited elastic-plastic behaviour in both the loading process, during which the wire moved through the die, and the unloading process, after which the wire was emerged from the die;
- constant friction coefficient in Coulomb friction law was assumed at the die and workpiece interface. – $k_f = 0.1$.
- material is homogeneous and isotropic with a non-linear hardening;
- as the temperature changes, thermal stresses are developed due to nonzero coefficient of thermal expansion;
- as temperature changes, the mechanical properties changes (softening), it happens because of the temperature-dependent flow stress was assumed;
- as the geometry changes, the heat transfer boundary value problem changes, this includes changes in the contacting interface;
- as plastic work is performed, internal heat is generated ;
  $(Q_p = \int_V \frac{Q_t}{\rho c_v} \bar{\sigma} \dot{\bar{\varepsilon}}^p [N] dV)$
- as the bodies slide, friction generates heat $(Q_f = \int_{S_f} | F_f || v_r | [N] dS_f)$

Only one wire-drawing pass is simulated with 24% reduction in area that is why FE-rezoning procedure was not applied. Fig. 1 and fig. 2 show initial FE

Table 1. Material and geometrical data for the FE model and material data for thermo coupled wire drawing problem for Aluminium 1100 AT

workpiece – data	die – data	FE model data
initial lenght $L = 82mm$	outlet angle $\gamma = 38.66°$	wire: 430 FE
$E = 1000N/mm^2$; $\nu = 0.33$	inlet angle $\alpha = 9.65°$	4 node axis. FE
masss density $1.0g/mm^3$	approach angle $\beta = 9.65°$	1033 nodes
$\sigma_y = 3.4N/mm^2$ at 200°	reduction in area $R_a = 24\%$	Die : 24 FE
coeff. of friction $k_f = 0.1$	inlet radius $D_0 = 15mm$	4 nodes FE
workhardening data	outlet radius $D_f = 13mm$	rigid die
plastic strain – flow stress	approach zone $l_a = 30mm$	
0.00 — 3.400	bearing zone $l_b = 5mm$	
0.15 — 5.100	outlet zone $l_o = 5mm$	
0.70 — 5.780	$E = 1.10^6 N/mm^2$; $\nu = 0.33$	
5.00 — 6.000	mass density $1.0g/mm^3$	
initial temp. - 427°C , Al.	initial temp. - 20°C	
$k_w = 242N/s°K$	$k_d = 19N/s°K$	
$c_w = 2.4255242N/mm^{2°}K$	$c_d = 3.77N/mm^{2°}K$	
$h = 0.007N/smm°K$	$h_{lubr} = 35N/smm°K$	
Teylor-Quiny coeff. $Q_t = 0.9$		
σ_y decrease at rate		
$0.007N/mm^2$ when the temp. increase		

mesh and geometry of die and workpiece in both cases. Fig. 3 and fig. 4 show temperature field distribution in die and workpiece in both cases. Fig. 5 and fig. 6 show residual stresses distribution in both cases.

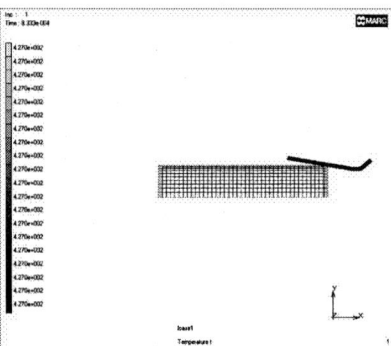

Fig. 1. Initial finite element mesh and geometry of work piece and die in the case with a rigid die

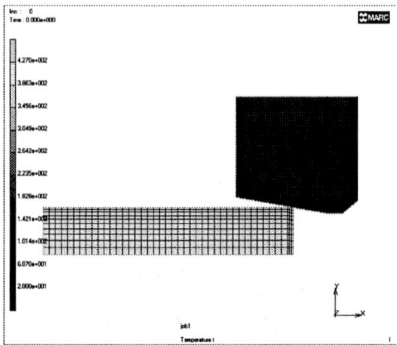

Fig. 2. Initial geometry and FE mesh of workpiece and deformable die

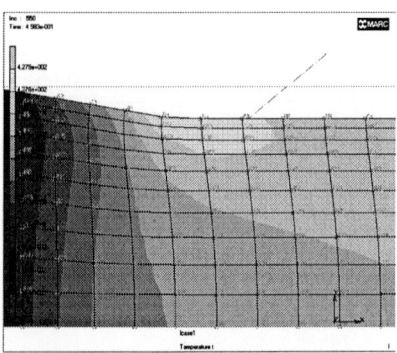

Fig. 3. Temperature field distribution in approach zone and bearing zone for inc. 550

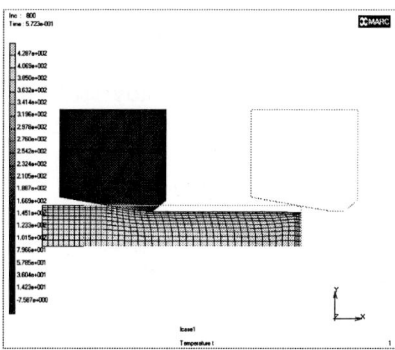

Fig. 4. Temperature field distribution in wire and die for inc. 800

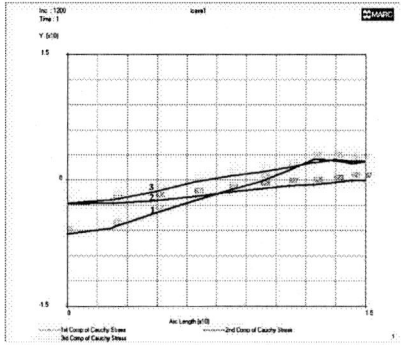

Fig. 5. Axial – 1, radial – 2 and circumference – 3 residual stresses in cross section 26–57 in the case with a rigid die

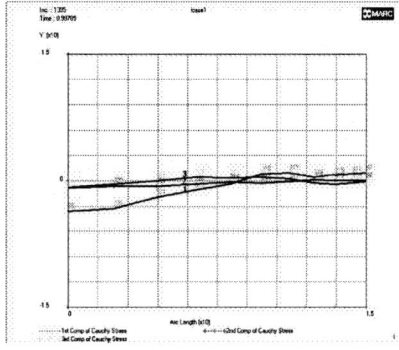

Fig. 6. Distribution of the residual stresses: 1–axial, 2–radial and 3–circumference in cross section 26—57 in the case with a deformable die

6 Conclusions

Proposed numerical condition eq.3 successfully can be used to determine the minimum length of workpiece in which steady state flow will be reached. FE simulation of thermo coupled wire drawing process can be used to predict more homogeneity in a final product, to optimise process parameters – die geometry and load parameters and to increase die wear.

Acknowledgements

This work has been made possible through the financial support trough the Belgian NATO Science Program. The author would like to thank the Belgian Delegate to the NATO Science Committee, for financial support. Partly the research was supported by grant MM-903/99, National Science Fund(Bulgaria).

References

1. Doege E., Kroeff A., and Massai A. Stress and strain analysis of automated miltistage fem- simulation of wiredrawing considering the backward force. *Wire Journal International*, pages 516–522, 1998.
2. Mooi H. G. *Finite Element Simulation of Aluminium Extrusion*. FEBO druk BV, The Netherlands, 1996.
3. Kloos K. H. and Kaiser B. Residual stresses induced by manufacturing. In V. Hauk, H. Hougardy, and E. Macherauch, editors, *Residual Stresses: Measument, Calculation, Evaluation*, pages 205–226. DGM Informationsgesellschaft, 1991.
4. Lazzarotto L., L. Dubar, A. Dubois, P. Ravassard, and J. Oudin. Identification of coulombs friction coefficient in real contact conditions applied to a wire drawing process. *Wear*, 211:54–63, 1997.
5. Rebelo N. and Kobayashi S. A coupled analysis of viscoplastic deformation and heat transfer - i. *Int. J. Mech. Sci.*, 22:707–718, 1980.
6. Van Houtte P., Van Acker K., and Root J. Residual stress determination in cementite and ferrite phases of high carbon steel. *Texture and Microstructures*, 33:187–206, 1999.
7. McMeeking R. M. and Lee E. H. The generation of residual stresses in metal-forming processes. In E. Kula and V. Weiss, editors, *Residual Stresses and Stress Relaxation*, pages 315–329. Plenum Press, 1982.
8. Boris S. and Mihelič A. Optimal design of the die shape using nonlinear finite element analysis. In Shen and Dawson, editors, *Simulation of Materials Processing: Theory, Methods and Applications. NUMIFORM95*, pages 625–630, France, 1995. A. A. Balkema.
9. Daves W. and Fischer F. D. Drawing of a curved wire. In Shen and Dawson, editors, *Simulation of Materials Processing: Theory, Methods and Applications. NUMIFORM95*, pages 693–698, France, 1995. A. A. Balkema.
10. Daves W. and Fischer F. D. Numerical investigation of the influence of alternating bending on the mechanical properties of cold drawn wire. In Shen and Dawson, editors, *Simulation of Materials Processing: Theory, Methods and Applications. NUMIFORM98*, pages 597–601, France, 1998. A. A. Balkema.

Construction and Convergence of Difference Schemes for a Modell Elliptic Equation with Dirac-delta Function Coefficient

B. S. Jovanović[1], J. D. Kandilarov[2], and L. G. Vulkov[2]

[1] University of Belgrade, Faculty of Mathematics
Studentski trg 16, 11000 Belgrade, Yugoslavia
bosko@matf.bg.ac.yu
[2] University of Rousse, Department of Applied Mathematics and Informatics,
Studentska str. 8, 7017 Rousse, Bulgaria
{juri,vulkov}@ami.ru.acad.bg

Abstract. We first discuss the difficulties that arise at the construction of difference schemes on uniform meshes for a specific elliptic interface problem. Estimates for the rate of convergence in discrete energetic Sobolev's norms compatible with the smoothness of the solution are also presented.

1 Introduction

Interface problems occur in many physical applications. We present a model case below to show the characteristic of such type interface problems. Namely, in the region $D = (0, 1)^2$ we consider the Dirichlet problem

$$-\Delta u + c(x)\,\delta_S(x)\,u = f(x), \quad x \in \Omega; \tag{1}$$

$$u = 0, \quad x \in \Gamma = \partial\Omega, \tag{2}$$

where S is a continuous curve (for example closed curve), $S \subset \Omega$ and $\delta_S(x)$ is Dirac–delta function concentrated on S. We suppose that

$$c(x) \in L_\infty(S), \quad 0 < C_0 \le c(x) \le C_1 \tag{3}$$

almost everywhere on S.

We assume for simplicity that the curve S separates Ω into two regions: $\bar{\Omega} = \bar{\Omega}_1 \cup \bar{\Omega}_2$, $\Omega_1 \cap \Omega_2 = \emptyset$. Then, at some assumptions for smoothness, the equation (1) can be rewritten as follows:

$$-\Delta u = f(x), \quad x \in \Omega_1 \cup \Omega_2; \quad [u]_S = 0, \quad \left[\frac{\partial u}{\partial \nu}\right]_S = c(x)\,u, \tag{4}$$

where $\partial u/\partial \nu$ – is the normal derivative.

A classification of interface problems is given in [1]. The most noticeable characteristic of the present interface problem is the **singular coefficient**. This brings up several substantial difficulties in the numerical analysis process some of which are discussed here. The first difficulty arises at the **discretization**. Many current techniques such as harmonic averagion or coefficient smoothing fail to give high accuracy in two or higher dimensions. **The immersed interface method (IIM)** developed in the recent years for many other interface problems [1], [2], [3] is not easy to be applied to (1), (2), see [4], [5] for one-dimensional problems. In order to achieve second-order accuracy one must use 4-points stencil in the one-dimensional case and 12-points for two-dimensional, but with line interface. See also the discussion in Section 2 of the present paper.

Because of the discontinuity and non-smoothness in the solution and the complexity of the interface, it is difficult to perform **error and convergence** analysis in the conventional way. Due to the presence of the interface and the discontinuity or localized nonlinearities, the system of discrete equations lose many nice properties, such as symmetric positive definiteness, and diagonal dominance etc. In [5] algorithms are proposed for decoupling of the linear and nonlinear equations of the discrete systems. For two-dimensional problems with curvelinear interface such decoupling is not clear.

The structure of the article is as follows. In Sect. 2 we derive and compare numerically three difference schemes in the one- and two-dimensional case. In Sect. 3 we formulate some convergence results for linear problems of type (1), (2).

2 Construction of the Difference Scheme

In the present Section three difference schemes are discussed and compared numerically.

2.1 One-Dimensional Case

We will analyze three schemes on the computational example

$$u'' - wu = 1 + K\delta(x - \zeta)u, \quad K > 0, \quad w = \text{const} \geq 0, \quad u(0) = u(1) = 0. \quad (5)$$

Let specify an uniform fixed grid $x_i = ih$, $i = 0, \ldots, N$, $hN = 1$. We wish to solve the equation (5) only on uniform mesh and the point ζ will not lie on a grid point x_I, $x_I < \zeta < x_{I+1}$, $1 \leq I < N$, so that the delta function must be replaced by appropriately discrete approximation $d_h(x)$. For example, by the "hat function" with support $(-h, h)$

$$d_h^{(1)} = \begin{cases} (h - |x|)/h^2, & |x| \leq h, \\ 0, & \text{otherwise.} \end{cases}$$

We shall make use of the following lemma [3].

Lemma 1. *Suppose* $d_h(x)$ *satisfies* $d_h(x) = 0$ *for* $|x| > Mh$ *and also the discrete moment condition* $h \sum_j (x_j - \zeta)^m d_h(x_j - \zeta) = \delta_{m0}$ *for* $m = 0, 1, \ldots,$

$p-1$. If $f \in C^{(p-1)}([\zeta - Mh, \zeta + Mh])$ and $f^{(p-1)}$ is Lipshitz continuous on the interval, then

$$f(\zeta) - h \sum_j f(x_j) d_h(x_j - \zeta) = O(h^p) \quad \text{as} \quad h \to 0.$$

First we discuss an integrointerpolation scheme for the equation (5). For $i \neq I, I+1$ we set

$$y_{\bar{x}x,i} - wy_i = 1,$$

where y_x and $y_{\bar{x}}$ are standard upwind and backward finite differences (see [6]). To get the difference equation in $x_i = x_I$ we integrate (5) from $x_{I-0.5}$ to $x_{I+0.5}$:

$$u'(x_{I+0.5}) - u'(x_{I-0.5}) - w \int_{x_{I-0.5}}^{x_{I+0.5}} u\, dx = h + Ku(\zeta).$$

One can use the formula $u'(x_{I-0.5}) = (u(x_I) - u(x_{I-1}))/h + O(h^2)$, but similar approximation of $u'(x_{I+0.5})$ leads to local truncation error $O(h^{-1})$. In spite of all this integrointerpolation scheme (II) in the following form

$$y_{\bar{x}x,i} - wy_i = 1 + \frac{K}{h}\delta_{I,i}y_i, \quad i = 1, \ldots, N-1, \quad \delta_{I,i} = \begin{cases} 0, & i \neq I, \\ 1, & i = I \end{cases}$$

can be found in the literature.

More accurate scheme one obtains after applying the averaging operator T^2 (defined below) to (5):

$$u_{\bar{x}x,i} - \frac{w}{h}\int_{x_{i-1}}^{x_{i+1}} \left(1 - \frac{|x - x_i|}{h}\right) u(x)\, dx = 1, \quad i \neq I, I+1,$$

$$u_{\bar{x}x,I} - \frac{w}{h^2}\left(\int_{x_{I-1}}^{x_I}(x-x_{I-1})u(x)dx + \int_{x_I}^{x_{I+1}}(x_{I+1}-x)u(x)dx\right) - \frac{K}{h^2}(x_{I+1}-\zeta)u(\zeta) = 1,$$

$$u_{\bar{x}x,I+1} - \frac{w}{h^2}\left(\int_{x_I}^{x_{I+1}}(x-x_I)u(x)dx + \int_{x_{I+1}}^{x_{I+2}}(x_{I+2}-x)u(x)dx\right) - \frac{K}{h^2}(\zeta - x_I)u(\zeta) = 1.$$

After approximations of the integrals and $u(\zeta)$ (by Lemma 1) we get the difference equations:

$$\left(\frac{1}{h^2} - Kda\right)y_{I-1} - \left(\frac{2}{h^2} + w + Kdb\right)y_I + \left(\frac{1}{h^2} - Kdc\right)y_{I+1} - Kday_{I+1} = 1,$$

$$-Keay_{I-1} + \left(\frac{1}{h^2} - Keb\right)y_I - \left(\frac{2}{h^2} + w + Kec\right)y_{I+1} + \left(\frac{1}{h^2} - Kea\right)y_{I+2} = 1,$$

where

$$a = -\rho_I \rho_{I+1}, \quad b = 1 - \rho_I^2, \quad c = 1 - \rho_{I+1}^2,$$

$$d = \frac{\rho_{I+1}}{h}, \quad e = \frac{\rho_I}{h}, \quad \rho_I = \frac{\zeta - x_I}{h}, \quad \rho_{I+1} = \frac{x_{I+1} - \zeta}{h}.$$

Now the truncation error is $O(h)$ and comes only from approximation of the integrals and $u(\zeta)$. This scheme on Table 1 is denoted by AO.

The main idea of the IIM consists in appropriate modification of the difference scheme at the irregular grid points which are near the interface by using the jump condition. For the example (5), we have

$$u''(x_I) = u_{\bar{x}x,I} - \frac{(x_{I-1} - \zeta)}{h^2}[u']_\zeta - \frac{(x_{I+1} - \zeta)^2}{2h^2}[u'']_\zeta -$$

$$- \frac{(x_{I+1} - \zeta)^3}{6h^2}[u''']_\zeta + O(h^2),$$

$$u''(x_{I+1}) = u_{\bar{x}x,I+1} + \frac{(x_I - \zeta)}{h^2}[u']_\zeta + \frac{(x_I - \zeta)^2}{2h^2}[u'']_\zeta +$$

$$+ \frac{(x_{I+1} - \zeta)^3}{6h^2}[u''']_\zeta + O(h^2),$$

$$[u']_\zeta = Ku(\zeta), \quad [u'']_\zeta = 0, \quad [u''']_\zeta = wKu(\zeta).$$

The value $u(\zeta)$ is approximated by Lemma 1. If d_h satisfies Lemma 1 with $p = 2$ then the resulting difference scheme is 4-points near the interface and the truncation error is $O(h)$, see [4], [5]. If $p = 3$, the stencil used is 6-points and the truncation error is $O(h^2)$. Here we present the IIM scheme with 4-points stencil, but on Table 1 the results for the scheme with 6-points stencil are also presented:

$$\frac{y_{i-1}}{h^2} - \left(\frac{2}{h^2} + w\right) y_i + \frac{y_{i+1}}{h^2} = 1, \quad i = 1, \ldots, N-1, \quad i \neq I, I+1; \quad y_0 = y_N = 0,$$

$$\left(\frac{1}{h^2} - k_1 a\right) y_{I-1} - \left(\frac{2}{h^2} + w + k_1 b\right) y_I + \left(\frac{1}{h^2} - k_1 c\right) y_{I+1} - k_1 a y_{I+1} = 1,$$

$$-k_2 a y_{I-1} + \left(\frac{1}{h^2} - k_2 b\right) y_I - \left(\frac{2}{h^2} + w + k_2 c\right) y_{I+1} + \left(\frac{1}{h^2} - k_2 a\right) y_{I+2} = 1,$$

$$k_1 = K(d + fw), \quad k_2 = K(e + gw), \quad f = \frac{h\rho_{I+1}^3}{6}, \quad g = \frac{h\rho_I^3}{6},$$

and a, b, c, d, e are the same as above.

As the results in Table 1 show the schemes AO and IM-4 points have the same order of accuracy. But AO can be easily generalized for the two-dimensional case.

Table 1. Truncation error with using difference schemes

N	II	AO	IIM - 4 points	IIM - 6 points
19	2,8733	$3,6060 \cdot 10^{-3}$	$5,6231 \cdot 10^{-3}$	$8,1478 \cdot 10^{-4}$
39	6,5063	$1,8192 \cdot 10^{-3}$	$2,8544 \cdot 10^{-3}$	$2,0122 \cdot 10^{-4}$
79	13,7544	$9,1281 \cdot 10^{-4}$	$1,4416 \cdot 10^{-3}$	$5,0011 \cdot 10^{-5}$
159	28,2417	$4,5713 \cdot 10^{-4}$	$7,2442 \cdot 10^{-4}$	$1,2467 \cdot 10^{-5}$
319	57,2119	$2,2875 \cdot 10^{-4}$	$3,6221 \cdot 10^{-4}$	$3,1123 \cdot 10^{-6}$

2.2 Two-Dimensional Case

First we present an IIM scheme for the problem (1), (2) when S is a **segment parallel** to one of the coordinate axes, for example: $S = \{(x_1, x_2) : x_2 = \zeta, 0 \leq x_1 \leq 1\}$. We use the uniform mesh $\overline{\omega} = \{(x_{1i}, x_{2j}) : x_{1i} = ih, x_{2j} = jh, i, j = 0, 1, \ldots, N\}$. Let ω be the set of the internal nodes and γ – the set of boundary nodes. We assume that $x_{2J} \leq \zeta < x_{2J+1}$, $1 \leq J \leq N-1$. Now $[u]_{x_2=\zeta} = 0$, $[u_{x_2}]_{x_2=\zeta} = c(x_1, \zeta) u(x_1, \zeta)$ and $[u_{x_2 x_2}]_{x_2=\zeta} = 0$. After some algebra we have

$$-u_{\bar{x}_1 x_1} - u_{\bar{x}_2 x_2} + d_h^{(1)}(x_{2j} - \zeta) c(x_{1i}, \zeta) u(x_{1i}, \zeta) = f(x_{1i}, x_{2j}) + O(h).$$

For the approximation of $u(x_{1i}, \zeta)$ we use Lemma 1. If $y_{ij} \cong u(x_{1i}, x_{2j})$, $\varphi_{ij} \cong f(x_{1i}, x_{2j})$ then the difference scheme is as follows:

$$-\Delta_h y_{ij} = -y_{\bar{x}_1 x_1, ij} - y_{\bar{x}_2 x_2, ij} = \varphi_{ij}, \quad i, j = 1, 2, \ldots, N-1, \quad j \neq J, J+1,$$

$$-y_{i-1,J} - y_{i+1,J} - (1 - adh^2 c_i) y_{i,J-1} + (4 + bdh^2 c_i) y_{i,J} -$$
$$- (1 - cdh^2 c_i) y_{i,J+1} + adh^2 c_i y_{i,J+2} = \varphi_{iJ} h^2,$$

$$ach^2 c_i y_{i,J-1} - (1 - beh^2 c_i) y_{i,J} + (4 + ceh^2 c_i) y_{i,J+1} + aeh^2 y_{i,J+2} -$$
$$- (y_{i-1,J+1} + y_{i+1,J+1}) = \varphi_{i,J+1} h^2.$$

In the case when S is an **arbitrary closed curve** in Ω we will consider difference scheme with averaged right hand side and coefficient. We define the Steklov averaging operators as follows [7]:

$$T_1 f(x_1, x_2) = T_1^- f(x_1 + h/2, x_2) = T_1^+ f(x_1 - h/2, x_2) = \frac{1}{h} \int_{x_1 - h/2}^{x_1 + h/2} f(x_1', x_2) \, dx_1',$$

$$T_2 f(x_1, x_2) = T_2^- f(x_1, x_2 + h/2) = T_2^+ f(x_1, x_2 - h/2) = \frac{1}{h} \int_{x_2 - h/2}^{x_2 + h/2} f(x_1, x_2') \, dx_2'.$$

Notice that these operators commute and map the derivatives of sufficiently smooth function u into finite differences, for example

$$T_i^+ \frac{\partial u}{\partial x_i} = u_{x_i}, \qquad T_i^2 \frac{\partial^2 u}{\partial x_i^2} = u_{\bar{x}_i x_i}.$$

We approximate the boundary value problem (1), (2) on the mesh $\bar{\omega}$ with

$$-\Delta_h y + \alpha y = \varphi \quad \text{in} \quad \omega; \qquad y = 0 \quad \text{on} \quad \gamma, \tag{6}$$

where $\varphi = T_1^2 T_2^2 f$ and $\alpha = T_1^2 T_2^2 (c\delta_S)$. The coefficient α in (6) can be written as follows

$$\alpha(x) = \begin{cases} h^{-2} \int_{S(x)} \kappa(x, x') c(x') dS_{x'}, & x \in S_h, \\ 0, & x \in \omega \setminus S_h, \end{cases}$$

where $\kappa(x, x') = \left(1 - \frac{|x_1' - x_1|}{h}\right)\left(1 - \frac{|x_2' - x_2|}{h}\right)$, $S(x) = S \cap e(x)$, $e(x) = (x_1 - h, x_1 + h) \times (x_2 - h, x_2 + h)$ is the cell attached to the internal node $x \in \omega$, and $S_h = \{x \in \omega : S(x) \neq \emptyset\}$.

3 Convergence of the Difference Schemes

Let assume $f \in W_2^{-1}(\Omega)$. Then the problem (1), (2) can be formulated in the weak form:

$$a(u, v) = \langle f, v \rangle, \qquad \forall v \in \overset{\circ}{W}{}_2^1(\Omega),$$

where

$$a(u, v) = \iint_\Omega \left(\frac{\partial u}{\partial x_1}\frac{\partial v}{\partial x_1} + \frac{\partial u}{\partial x_2}\frac{\partial v}{\partial x_2}\right) dx_1\, dx_2 + \int_S c\, u\, v\, dS$$

and $\langle f, v \rangle$ is duality on $W_2^{-1}(\Omega) \times \overset{\circ}{W}{}_2^1(\Omega)$.
The following assertions hold.

Lemma 2. *For each $f \in W_2^{-1}(\Omega)$, the problem (1), (2) has unique solution $u \in \overset{\circ}{W}{}_2^1(\Omega)$.*

Lemma 3. *If $f \in W_2^{\theta-1}(\Omega)$, $-1/2 < \theta < 1/2$, then the problem (1), (2) has unique weak solution $u \in W_2^{1+\theta}(\Omega)$.*

3.1 Global Estimate

The error $z = u - y$ satisfies the equation

$$-\Delta_h z + \alpha z = -\psi_{1,\bar{x}_1 x_1} - \psi_{2,\bar{x}_2 x_2} + \chi \quad \text{in} \quad \omega; \qquad z = 0 \quad \text{on} \quad \gamma \tag{7}$$

where

$$\psi_i = u - T_{3-i}^2 u, \quad i = 1, 2,$$
$$\chi = \alpha u - h^{-2} \int_{S(x)} \kappa(x, x') c(x') u(x') dS_{x'}, \quad x \in S_h$$
$$\chi = 0, \quad x \in \omega \setminus S_h.$$

Let H_h be the set of mesh functions defined on the mesh $\bar\omega$ and equal to zero on γ. We define the scalar products

$$(y, v)_h = h^2 \sum_{x \in \omega} y(x)\, v(x), \quad [y, v)_{h,i} = h^2 \sum_{x \in \omega^i} y(x)\, v(x)$$

and the corresponding norms $\|v\|_h$ and $|[v\|_{h,i}$, where $\omega^i = \{x \in \bar\omega : 0 \le x_i < 1,\ 0 < x_{3-i} < 1\}$, $i = 1, 2$. Let us introduce also the W_2^1 mesh norm:

$$\|v\|^2_{W^1_{2,h}} = |v|^2_{W^1_{2,h}} + \|v\|^2_h, \qquad |v|^2_{W^1_{2,h}} = |[v_{x_1}\|^2_{h,1} + |[v_{x_2}\|^2_{h,2}.$$

Taking the scalar product of (7) and summing by parts, we get

$$|[z_{x_1}\|^2_{h,1} + |[z_{x_2}\|^2_{h,2} + h^2 \sum_{x \in S_h} \alpha z^2 = [\psi_1,\, z_{x_1})_{h,1} + [\psi_2,\, z_{x_2})_{h,2} + h^2 \sum_{x \in S_h} \chi z.$$

Using the difference analog of the Friedrichs inequality [7], we get the a priori estimate

$$\|z\|_{W^1_{2,h}} \le C \left\{ |[\psi_1,\, x_1\|_{h,1} + |[\psi_2,\, x_2\|_{h,2} + \left(h^2 \sum_{x \in S_h} \frac{\chi^2}{\alpha} \right)^{1/2} \right\}. \tag{8}$$

Estimating the terms in the right-hand side of (8) using methodology proposed in [7] and [8], we obtain the following result.

Theorem 1. *The solution of the scheme (6) converges to the solution of the differential problem (1), (2) and the following convergence rate estimate is valid*

$$\|u - y\|_{W^1_{2,h}} \le C h^\theta \|u\|_{W^{1+\theta}_2(\Omega)}, \quad 0 < \theta < 1/2.$$

3.2 An Improved Estimate for the Rate of Convergence

Let suppose that solution of the problem (1), (2) has raised smoothness in the regions Ω_1 and Ω_2. Now, the following improved estimate for the rate of convergence of the difference scheme can be proved:

$$\|u-y\|_{W^1_{2,h}} \le C h^\theta \left(\|u\|_{W^{1+\theta}_2(\Omega_1)} + \|u\|_{W^{1+\theta}_2(\Omega_2)} + \|u\|_{W^{1/2+\theta}_2(\Omega)} \right), \quad 1/2 \le \theta < 1.$$

3.3 Line Interface

Here we consider the case when S is a segment parallel to one of the coordinate axes. Let, for example, S is given by the equation $x_2 = \zeta$. Contrary to the previous cases, in this section we assume that $\zeta = Jh$. Notice, that then the inequality $C_0/h \le \alpha \le C_1/h$ is fulfilled.

Taking the scalar product of (7) with z and summing by parts, we get a priori estimate

$$\|z\|_{W^1_{2,h}} \le C \left\{ |[\psi_2,\, x_2\|_{h,2} + \left(h^2 \sum_{x \in \omega^1 \cap \Omega_1} \psi^2_{1,\, x_1} \right)^{1/2} + \left(h^2 \sum_{x \in \omega^1 \cap \Omega_2} \psi^2_{1,\, x_1} \right)^{1/2} \right.$$

$$+\left(h^2 \sum_{x \in S_h} \tilde{\psi}_{1,x_1}^2\right)^{1/2} + \left(h^2 \sum_{x \in S_h} \frac{\overline{\psi}_{1,\bar{x}_1 x_1}^2}{\alpha}\right)^{1/2} + \left(h^2 \sum_{x \in S_h} \frac{\chi^2}{\alpha}\right)^{1/2}\Bigg\},$$

where $\overline{\psi}_1(x_1,\zeta) = -\dfrac{h}{6}\left[\dfrac{\partial u}{\partial x_2}\right]_{(x_1,\zeta)} = -\dfrac{h}{6}c(x_1)\,u(x_1,\zeta)$ and $\tilde{\psi}_1(x_1,\zeta) = \psi_1(x_1,\zeta) - \overline{\psi}_1(x_1,\zeta).$

Now, the following estimate for the rate of convergence of the scheme (6) can be established:

$$\|u-y\|_{W_{2,h}^1} \leq C h^2 \left(\left\|\frac{\partial^3 u}{\partial x_1^2 \partial x_2}\right\|_{L_2(\Omega)} + \left\|\frac{\partial^3 u}{\partial x_1 \partial x_2^2}\right\|_{L_2(\Omega_1)} + \left\|\frac{\partial^3 u}{\partial x_1 \partial x_2^2}\right\|_{L_2(\Omega_2)} \right.$$

$$\left. + \|c\|_{W_2^2(0,1)} \|u\|_{W_2^2(S)}\right).$$

Acknowledgment

The research of the first author was supported by Science Fund of Serbia, grant number 04M03/C.

References

1. Li Z.: The Immersed Interface Method – A Numerical Approach for Partial Differential Equations with Interfaces. PhD thesis, University of Washington, 1994.
2. Wiegmann A., Bube K.: The immersed interface method for nonlinear differential equations with discontinuous coefficients and singular sources. SIAM J. Numer. Anal. **35** (1998), 177–200.
3. Beyer R. P., Leveque R. J.: Analysis of one–dimensional model for the immersed boundary method, SIAM J. Numer. Anal. **29** (1992), 332–364.
4. Kandilarov J.: A second–order difference method for solution of diffusion problems with localized chemical reactions. in Finite–Difference methods: Theory and Applications (CFDM 98), Vol. 2, 63–67, Ed. by A. A. Samarskiĭ, P. P. Matus, P. N. Vabishchevich, Inst. of Math., Nat. Acad. of Sci. of Belarus, Minsk 1998.
5. Vulkov L., Kandilarov J.: Construction and implementation of finite–difference schemes for systems of diffusion equations with localized chemical reactions, Comp. Math. and Math. Phys., 40, N 5, (2000), 705–717.
6. Samarskiĭ A. A.: Theory of difference schemes, Nauka, Moscow 1987 (in Russian).
7. Samarskiĭ A. A., Lazarov R. D., Makarov V. L.: Difference schemes for differential equations with generalized solutions, Vyshaya Shkola, Moscow 1989 (in Russian).
8. Jovanović B. S.: Finite difference method for boundary value problems with weak solutions, Mat. Institut, Belgrade 1993.

Operator's Approach to the Problems with Concentrated Factors

Boško S. Jovanović[1] and Lubin G. Vulkov[2]

[1] University of Belgrade, Faculty of Mathematics
Studentski trg 16, 11000 Belgrade, Yugoslavia
[2] University of Rousse, Department of Applied Mathematics and Informatics
Studentska str. 8, 7017 Rousse, Bulgaria

Abstract. In this paper finite–difference schemes approximating the one–dimensional initial–boundary value problems for the heat equation with concentrated capacity are derived. An abstract operator's method is developed for studying such problems. Convergence rate estimates consistent with the smoothness of the data are obtained.

1 Introduction

One interesting class of parabolic initial–boundary value problems (IBVPs) models processes in heat–conduction media with concentrated capacity. In this case the Dirac's Delta distribution is involved in the heat capacity coefficient and, consequently, the jump of the heat flow in the singular point is proportional to the time derivative of the temperature. Dynamical boundary conditions cause similar effect [4], [8]. These problems are non–standard and classical analysis is difficult to apply for convergence analysis.

In the present paper finite–difference schemes (FDSs) approximating the one–dimensional IBVPs for the heat equation with concentrated capacity or dynamical boundary conditions are derived. An abstract operator's method is developed for studying such problems. Sobolev's norms with weight operator, corresponding to norms L_2, $W_2^{1,1/2}$ and $W_2^{2,1}$ are constructed. In these norms convergence rate estimates compatible with the smoothness of the IBVP data are obtained. Analogous results for equation with constant coefficients are obtained in [6]. Convergence of FDSs for the problems with smooth solutions were investigated in [1], [2], [3] and [15].

2 Preliminary Results

Let H be a real separable Hilbert space endowed with inner product $(\cdot,\cdot)$ and norm $\|\cdot\|$ and S – unbounded selfadjoint positive definite linear operator, with domain $D(S)$ dense in H. The product $(u,v)_S = (Su,v)$ $(u, v \in D(S))$ satisfies the inner product axioms. Reinforceing $D(S)$ in the norm $\|u\|_S = (u,u)_S^{1/2}$ we obtain a Hilbert space $H_S \subset H$. The inner product (u,v) continuously extends to $H_S^* \times H_S$, where H_S^* is the adjoint space for H_S. Operator S extends to mapping $S: H_S \to H_S^*$. There exists unbounded selfadjoint positive definite linear operator $S^{1/2}$ [10], [7], such that $D(S^{1/2}) = H_S$ and $(u,v)_S = (Su,v) = (S^{1/2}u, S^{1/2}v)$. We also define the Sobolev spaces $W_2^s(a,b; H)$, $W_2^0(a,b; H) = L_2(a,b; H)$, of the functions $u = u(t)$ mapping interval $(a,b) \subset R$ into H [7].

Let A and B are unbounded selfadjoint positive definite linear operators, not depending on t, in Hilbert space H, with $D(A)$ – dense in H and $H_A \subset H_B$. In general, A and B are noncomutative. We consider an abstract Cauchy problem (comp. [16], [9])

$$B\frac{du}{dt} + Au = f(t), \quad 0 < t < T; \quad u(0) = u_0, \qquad (1)$$

where u_0 is a given element in H_B, $f(t) \in L_2(0,T; H_{A^{-1}})$ – given function and $u(t)$ – unknown function from $(0,T)$ into H_A.

The following proposition holds.

Lemma 1. *The solution u of the problem (1) satisfies a priori estimates:*

$$\int_0^T \left(\|Au(t)\|_{B^{-1}}^2 + \left\|\frac{du(t)}{dt}\right\|_B^2 \right) dt \leq C\left(\|u_0\|_A^2 + \int_0^T \|f(t)\|_{B^{-1}}^2 dt \right),$$

if $u_0 \in H_A$ and $f \in L_2(0,T; H_{B^{-1}})$;

$$\int_0^T \|u(t)\|_A^2 dt + \int_0^T \int_0^T \frac{\|u(t) - u(t')\|_B^2}{|t-t'|^2} dt\, dt' \leq C\left(\|u_0\|_B^2 + \int_0^T \|f(t)\|_{A^{-1}}^2 dt \right),$$

if $u_0 \in H_B$ and $f \in L_2(0,T; H_{A^{-1}})$; and

$$\int_0^T \|u(t)\|_B^2 dt \leq C\left(\|Bu_0\|_{A^{-1}}^2 + \int_0^T \|A^{-1}f(t)\|_B^2 dt \right),$$

if $Bu_0 \in H_{A^{-1}}$ and $A^{-1}f \in L_2(0,T; H_B)$.

Setting in (1) $f(t) = dg(t)/dt$ we get the Cauchy problem

$$B\frac{du}{dt} + Au = \frac{dg}{dt}, \quad 0 < t < T; \quad u(0) = u_0. \qquad (2)$$

The following assertion is valid.

Lemma 2. *The solution u of the problem (2) satisfies a priori estimates:*

$$\int_0^T \|u(t)\|_A^2 dt + \int_0^T \int_0^T \frac{\|u(t) - u(t')\|_B^2}{|t-t'|^2} dt\, dt' \leq C\Bigg[\|u_0\|_B^2 +$$

$$+ \int_0^T \int_0^T \frac{\|g(t) - g(t')\|_{B^{-1}}^2}{|t-t'|^2} dt\, dt' + \int_0^T \left(\frac{1}{t} + \frac{1}{T-t}\right) \|g(t)\|_{B^{-1}}^2 dt \Bigg],$$

if $u_0 \in H_B$ and $g \in W_2^{1/2}(0,T; H_{B^{-1}})$; and

$$\int_0^T \|u(t)\|_B^2 \, dt \leq C\left(\|Bu_0 - g(0)\|_{A^{-1}}^2 + \int_0^T \|g(t)\|_{B^{-1}}^2 \, dt\right),$$

if $Bu_0 - g(0) \in H_{A^{-1}}$ and $g \in L_2(0,T; H_{B^{-1}})$.

Analogous results hold for operator–difference schemes. Let H_h be finite-dimensional real Hilbert space with inner product $(\cdot, \cdot)_h$ and norm $\|\cdot\|_h$. Let A_h and B_h be constant selfadjoint positive linear operators in H_h, in general case noncomutative. By H_{S_h}, where $S_h = S_h^* > 0$, we denote the space $H_{S_h} = H_h$ with inner product $(v,w)_{S_h} = (S_h v, w)_h$ and norm $\|v\|_{S_h} = (S_h v, v)_h^{1/2}$.

Let ω_τ be an uniform mesh on $(0,T)$ with the step size $\tau = T/m$, $\omega_\tau^- = \omega_\tau \cup \{0\}$, $\omega_\tau^+ = \omega_\tau \cup \{T\}$ and $\bar{\omega}_\tau = \omega_\tau \cup \{0,T\}$. Further we shall use standard denotation of the theory of difference schemes [11].

We consider the simplest two-level operator–difference scheme

$$B_h v_{\bar{t}} + A_h v = \varphi(t), \quad t \in \omega_\tau^+; \quad v(0) = v_0, \tag{3}$$

where v_0 is a given element in H_h, $\varphi(t)$ is also given and $v(t)$ – unknown function with values in H_h. Let us also consider the scheme

$$B_h v_{\bar{t}} + A_h v = \psi_{\bar{t}}, \quad t \in \omega_\tau^+; \quad v(0) = v_0, \tag{4}$$

where $\psi(t)$ is a given function with values in H_h.

The following analogues of Lemmas 1 and 2 hold true.

Lemma 3. *The solution v of the problem (3) satisfies a priori estimates:*

$$\tau \sum_{t \in \omega_\tau^+} \|A_h v(t)\|_{B_h^{-1}}^2 + \tau \sum_{t \in \omega_\tau^+} \|v_{\bar{t}}(t)\|_{B_h}^2 \leq C\left(\|v_0\|_{A_h}^2 + \tau \sum_{t \in \omega_\tau^+} \|\varphi(t)\|_{B_h^{-1}}^2\right),$$

$$\tau \sum_{t \in \bar{\omega}_\tau} \|v(t)\|_{A_h}^2 + \tau^2 \sum_{t \in \bar{\omega}_\tau} \sum_{t' \in \bar{\omega}_\tau,\, t' \neq t} \frac{\|v(t) - v(t')\|_{B_h}^2}{|t - t'|^2} \leq$$

$$\leq C\left(\|v_0\|_{B_h}^2 + \tau \|v_0\|_{A_h}^2 + \tau \sum_{t \in \omega_\tau^+} \|\varphi(t)\|_{A_h^{-1}}^2\right),$$

$$\tau \sum_{t \in \omega_\tau^+} \|v(t)\|_{B_h}^2 \leq C\left(\|B_h v_0\|_{A_h^{-1}}^2 + \tau \sum_{t \in \omega_\tau^+} \|A_h^{-1} \varphi(t)\|_{B_h}^2\right).$$

Lemma 4. *The solution v of the problem (4) satisfies a priori estimates:*

$$\tau \sum_{t \in \bar{\omega}_\tau} \|v(t)\|_{A_h}^2 + \tau^2 \sum_{t \in \bar{\omega}_\tau} \sum_{t' \in \bar{\omega}_\tau,\, t' \neq t} \frac{\|v(t) - v(t')\|_{B_h}^2}{|t - t'|^2} \leq C\left[\|v_0\|_{B_h}^2 + \tau \|v_0\|_{A_h}^2 + \right.$$

$$+\tau^2 \sum_{t \in \omega_\tau} \sum_{t' \in \bar\omega_\tau,\ t' \neq t} \frac{\|\psi(t) - \psi(t')\|^2_{B_h^{-1}}}{|t - t'|^2} + \tau \sum_{t \in \omega_\tau} \left(\frac{1}{t} + \frac{1}{T-t}\right) \|\psi(t)\|^2_{B_h^{-1}}\Bigg],$$

$$\tau \sum_{t \in \omega_\tau^+} \|v(t)\|^2_{B_h} \leq C\left(\|B_h v_0 - \psi(0)\|^2_{A_h^{-1}} + \tau \sum_{t \in \omega_\tau^+} \|\psi(t)\|^2_{B_h^{-1}}\right).$$

3 Heat Equation with Concentrated Capacity

Let us consider the IBVP for the heat equation with the presence of concentrated capacity at interior point $x = \xi$ [8]:

$$[c(x) + K\,\delta(x - \xi)]\frac{\partial u}{\partial t} - \frac{\partial}{\partial x}\left(a(x)\frac{\partial u}{\partial x}\right) = f(x,t), \quad (x,t) \in Q, \tag{5}$$

$$u(0,t) = 0, \quad u(1,t) = 0, \quad 0 < t < T \tag{6}$$

$$u(x,0) = u_0(x), \quad x \in (0,1), \tag{7}$$

where $Q = (0,1) \times (0,T)$, $K > 0$, $0 < c_1 \leq a(x) \leq c_2$, $0 < c_3 \leq c(x) \leq c_4$ and $\delta(x)$ is the Dirac's distribution [14]. From (5) follows that the solution of the problem satisfies the equation

$$c(x)\frac{\partial u}{\partial t} - \frac{\partial}{\partial x}\left(a(x)\frac{\partial u}{\partial x}\right) = f(x,t),$$

for $(x,t) \in Q_1 = (0,\xi) \times (0,T)$ and $(x,t) \in Q_2 = (\xi,1) \times (0,T)$, while for $x = \xi$ the conjugation conditions

$$[u]_{x=\xi} \equiv u(\xi+0,t) - u(\xi-0,t) = 0, \quad \left[a\frac{\partial u}{\partial x}\right]_{x=\xi} = K\frac{\partial u(\xi,t)}{\partial t}$$

are fulfilled.

It is easy to see that the IBVP (5)–(7) can be written in the form (1), where $H = L_2(0,1)$, $H_A = \overset{\circ}{W}_2^1(0,1)$, $Au = -\frac{\partial}{\partial x}\left(a(x)\frac{\partial u}{\partial x}\right)$ and $Bu = [c(x) + K\,\delta(x-\xi)]\,u(x,t)$. Further,

$$\|w\|_A^2 = \int_0^1 a(x)\,[w'(x)]^2\,dx \asymp \|w\|^2_{W_2^1(0,1)}, \quad w \in \overset{\circ}{W}_2^1(0,1),$$

$$\|w\|_B^2 = \int_0^1 c(x)\,w^2(x)\,dx + K\,w^2(\xi) \asymp \|w\|^2_{L_2(0,1)} + w^2(\xi), \quad w \in \overset{\circ}{W}_2^1(0,1).$$

Let $\omega_h = \{x_1, x_2, \ldots, x_{n-1}\}$ be a nonuniform mesh in $(0,1)$, containing the node ξ. Denote $\omega_h^- = \omega_h \cup \{x_0\}$, $\omega_h^+ = \omega_h \cup \{x_n\}$, $\bar\omega_h = \omega_h \cup \{x_0, x_n\}$, $x_0 = 0$, $x_n = 1$ and $h_i = x_i - x_{i-1}$. Also denote $v_x = (v_+ - v)/h_+$, $v_{\bar x} = (v - v_-)/h$, $v_{\hat x} = (v_+ - v)/\hbar$, $v = v(x)$, $v_\pm = v(x_\pm)$, $x = x_i$, $x_\pm = x_{i\pm 1}$, $\hbar = (h + h_+)/2$. We assume that $1/c_0 \leq h_+/h \leq c_0$, $c_0 = \text{const} \geq 1$.

We approximate the IBVP (5)–(7) on the mesh $\bar{\omega}_h \times \bar{\omega}_\tau$ by the implicit FDS with averaged right hand side

$$(c + K\,\delta_h)\,v_{\bar{t}} - (\tilde{a}\,v_{\bar{x}})_{\hat{x}} = T_x^2 T_t^- f, \quad (x,t) \in \omega_h \times \omega_\tau^+, \tag{8}$$

$$v(0,t) = 0, \quad v(1,t) = 0, \quad t \in \omega_\tau^+, \tag{9}$$

$$v(x,0) = u_0(x), \quad x \in \bar{\omega}_h, \tag{10}$$

where $\tilde{a}(x) = [a(x) + a(x-h)]/2$, $\delta_h = \delta_h(x - \xi) = \begin{cases} 0, & x \in \omega_h \setminus \{\xi\} \\ 1/\hbar, & x = \xi \end{cases}$ is the mesh Dirac's function, and T_x^2, T_t^- are Steklov averaging operators [12]:

$$T_t^- f(x,t) = T_t^+ f(x, t-\tau) = \frac{1}{\tau} \int_{t-\tau}^{t} f(x,t')\,dt',$$

$$T_x^- f(x,t) = \frac{1}{h_-} \int_{x_-}^{x} f(x',t)\,dx', \quad T_x^+ f(x,t) = \frac{1}{h_+} \int_{x}^{x_+} f(x',t)\,dx',$$

$$T_x^2 f(x,t) = \frac{1}{\hbar} \int_{x_-}^{x_+} \kappa(x,x')\,f(x',t)\,dx', \quad \kappa(x,x') = \begin{cases} \frac{1+(x'-x)}{h_-}, & x_- < x' < x \\ \frac{1-(x'-x)}{h_+}, & x < x' < x_+ \end{cases}.$$

Notice that these operators comute and map partial derivatives into finite differences, for example $T_x^2 \frac{\partial^2 u}{\partial x^2} = u_{\bar{x}\hat{x}}$, $T_t^- \frac{\partial u}{\partial t} = u_{\bar{t}}$.

Let H_h be the set of functions defined on the mesh $\bar{\omega}_h$ and equal to zero at $x = 0$ and $x = 1$. We define the inner products $(v,w)_h = \sum_{x \in \omega_h} v(x)\,w(x)\,\hbar$, $(v,w)_{h*} = \sum_{x \in \omega_h^+} v(x)\,w(x)\,h$ and corresponding norms $\|w\|_h = \|w\|_{L_{2,h}} = (w,w)_h^{1/2}$, $\|w\|_{h*} = (w,w)_{h*}^{1/2}$.

The FDS (8)–(10) can be reduced to the form (3) by setting $A_h v = -(\tilde{a}\,v_{\bar{x}})_{\hat{x}}$ and $B_h v = (c + K\,\delta_h)\,v$. For $w \in H_h$ we have

$$\|w\|_{A_h}^2 = (A_h w, w)_h = \sum_{x \in \omega_h^+} \tilde{a}(x)\,w_{\bar{x}}^2(x)\,h \asymp \|w_{\bar{x}}\|_{h*}^2,$$

$$\|w\|_{B_h}^2 = (B_h w, w)_h = \sum_{x \in \omega_h} c(x)\,w^2(x)\,\hbar + K\,w^2(\xi) \asymp \|w\|_{B_{0h}}^2,$$

$$\|w\|_{B_h^{-1}}^2 = (B_h^{-1} w, w)_h = \sum_{x \in \omega_h \setminus \{\xi\}} \frac{w^2(x)}{c(x)}\,\hbar + \frac{w^2(\xi)}{K + \hbar\,c(\xi)}\,h^2 \asymp \|w\|_{B_{0h}^{-1}}^2,$$

where $B_{0h} w = (1 + \delta_h)\,w$.

Let us introduce the mesh Sobolev norms with weight operator B_{0h}:

$$\|w\|_{L_{2,h}}^2 = \|w\|_{B_{0h}}^2 = \|w\|_{L_{2,h}}^2 + w^2(\xi), \quad \|w\|_{\widetilde{W}_{2,h}^1}^2 = \|w_{\bar{x}}\|_{h*}^2 + \|w\|_{B_{0h}}^2,$$

$$\|w\|_{\widetilde{W}_{2,h}^2}^2 = \|w_{\bar{x}\hat{x}}\|_{B_{0h}^{-1}}^2 + \|w_{\bar{x}}\|_{h*}^2 + \|w\|_{B_{0h}}^2; \quad \|w\|_{L_{2,h\tau}}^2 = \tau \sum_{t \in \bar{\omega}_\tau} \|w(\cdot,t)\|_{L_{2,h}}^2,$$

$$\|w\|^2_{\widetilde{W}^{1,1/2}_{2,h\tau}} = \tau \sum_{t\in\bar\omega_\tau} \|w(\cdot,t)\|^2_{\widetilde{W}^1_{2,h}} + \tau^2 \sum_{t\in\bar\omega_\tau}\sum_{t'\in\bar\omega_\tau,\, t'\neq t} \frac{\|w(\cdot,t)-w(\cdot,t')\|^2_{B^{-1}_{0h}}}{|t-t'|^2},$$

$$\|w\|^2_{\widetilde{W}^{2,1}_{2,h\tau}} = \tau \sum_{t\in\bar\omega_\tau} \|w(\cdot,t)\|^2_{\widetilde{W}^2_{2,h}} + \tau \sum_{t\in\omega^+_\tau} \|w_{\bar t}(\cdot,t)\|^2_{B_{0h}}.$$

Lemma 5. *Let $A_h v = -(\tilde a\, v_{\bar x})_{\hat x}$ and $B_h v = (c + K\,\delta_h)\, v$, where a and c are continuous functions. Then the norm $\|v\|_{A_h}$ is equivalent to the mesh norm $\widetilde{W}^1_{2,h}$. If the function $a(x)$ is continuously differentiable then the norm $\|A_h v\|_{B^{-1}_h}$ is equivalent to the mesh norm $\widetilde{W}^2_{2,h}$.*

3.1 Convergence in $\widetilde{W}^{1,1/2}_{2,h\tau}$

Let u be the solution of the IBVP (5)–(7) and v – the solution of FDS (8)–(10). The error $z = u - v$ satisfies

$$(c + K\,\delta_h)\, z_{\bar t} - (\tilde a\, z_{\bar x})_{\hat x} = \psi_{\bar t} - \chi_{\hat x}, \quad (x,t) \in \omega_h \times \omega^+_\tau, \tag{11}$$

$$z(0,t) = 0, \quad z(1,t) = 0, \quad t \in \omega^+_\tau, \tag{12}$$

$$z(x,0) = 0, \quad x \in \bar\omega_h, \tag{13}$$

where $\psi = c\,u - T^2_x(c\,u) + \left(\frac{h^2}{6}(c\,u)_{\bar x}\right)_{\hat x}$ and $\chi = \tilde a\, u_{\bar x} - T^-_x T^-_t \left(a\,\frac{\partial u}{\partial x}\right) + \frac{h^2}{6}(c\,u)_{\bar x\bar t}$.
From Lemmas 3–5, using inequality

$$\|\chi_{\hat x}\|_{A^{-1}_h} = \max_{w\in H_h} \frac{|(\chi_{\hat x}, w)_h|}{\|w\|_{A_h}} = \max_{w\in H_h} \frac{|-(\chi, w_{\bar x}]_{h\star}|}{\|w\|_{A_h}} \leq \frac{1}{c_1}\|\chi\|_{h\star},$$

one immediately obtains the following a priori estimate for the problem (11)–(13)

$$\|z\|_{\widetilde{W}^{1,1/2}_{2,h\tau}} \leq C \Bigg[\tau^2 \sum_{t\in\bar\omega_\tau}\sum_{t'\in\bar\omega_\tau,\, t'\neq t} \frac{\|\psi(\cdot,t)-\psi(\cdot,t')\|^2_{B^{-1}_{0h}}}{|t-t'|^2} + \tag{14}$$
$$+ \tau \sum_{t\in\omega_\tau} \left(\frac{1}{t} + \frac{1}{T-t}\right) \|\psi(\cdot,t)\|^2_{B^{-1}_{0h}} + \tau \sum_{t\in\omega^+_\tau} \|\chi(\cdot,t)\|^2_{h\star} \Bigg]^{1/2}.$$

Therefore, in order to estimate the convergence rate of FDS (8)–(10) in $\widetilde{W}^{1,1/2}_{2,h\tau}$, it is sufficiently to estimate the right hand side terms in (14). Using integral representations of ψ and χ and the form of corresponding norms, similarly as in [5], [6], we obtain the following convergence rate estimate

$$\|z\|_{\widetilde{W}^{1,1/2}_{2,h\tau}} \leq C\,(h^2_{max} + \tau) \left(\|a\|_{W^1_2(0,1)} + \|c\|_{W^1_2(0,1)}\right) \left\|\frac{\partial^2 u}{\partial x \partial t}\right\|_{L_2(Q)} +$$

$$+ C\, h^2_{max}\, \sqrt{\ln 1/\tau}\, \left(\|a\|_{W^2_2(0,1)} + \|c\|_{W^2_2(0,1)}\right) \left(\|u\|_{W^{3,3/2}_2(Q_1)} + \|u\|_{W^{3,3/2}_2(Q_2)}\right).$$

3.2 Convergence in $\widetilde{L}_{2,h\tau}$

Let us consider the following approximation of the initial condition (7)

$$v(x,0) = \begin{cases} \dfrac{T_x^2(c\,u_0)(x)}{c(x)}, & x \in \omega_h \setminus \{\xi\} \\ \dfrac{K\,u_0(\xi) + \hbar\,T_x^2(c\,u_0)(\xi)}{K + \hbar\,c(\xi)}, & x = \xi. \end{cases} \quad (15)$$

Let u be the solution of IBVP (5)–(7) and v – the solution of FDS (8), (9), (15). The error $z = u - v$ satisfies the conditions (11), (12) and

$$(c(x) + K\,\delta_h(x-\xi))\,z(x,0) = \psi(x,0) - \beta_{\hat{x}}(x,0), \quad x \in \omega_h, \quad (16)$$

where $\beta = \frac{h^2}{6}(c\,u)_{\bar{x}}$. The term $\chi_{\hat{x}}$ can be represented in the form $\chi_{\hat{x}} = (\tilde{a}\mu_{\bar{x}})_{\hat{x}} + \alpha_{\hat{x}} + \beta_{\hat{x}\bar{t}}$, where $\mu = u - T_t^- u$ and $\alpha = \tilde{a}\,T_x^- T_t^- \left(\frac{\partial u}{\partial x}\right) - T_x^- T_t^- \left(a\,\frac{\partial u}{\partial x}\right)$.

Lemma 6. *For the solution of FDS*

$$(c + K\,\delta_h)\,z_{\bar{t}} - (\tilde{a}\,z_{\bar{x}})_{\hat{x}} = -\beta_{\hat{x}\bar{t}}, \quad (x,t) \in \omega_h \times \omega_\tau^+,$$

with homogeneous initial and Dirichlet boundary conditions the following a priori estimate holds:

$$\|z\|'_{\widetilde{L}_{2,h\tau}} = \left\{ \tau \sum_{t \in \omega_\tau^+} \left\| \frac{z(\cdot,t) + z(\cdot, t-\tau)}{2} \right\|^2_{\widetilde{L}_{2,h}} \right\}^{1/2} \le$$

$$\le C \left\{ \tau^2 \sum_{t \in \bar{\omega}_\tau} \sum_{t' \in \bar{\omega}_\tau,\ t' \ne t} \frac{\|\beta(\cdot,t) - \beta(\cdot,t')\|^2_{h\star}}{|t - t'|^2} + \tau \sum_{t \in \omega_\tau} \left(\frac{1}{t} + \frac{1}{T-t}\right) \|\beta(\cdot,t)\|^2_{h\star} \right\}^{1/2}.$$

Using Lemmas 3–6 for FDS (11), (12), (16) we obtain a priori estimate in the form

$$\|z\|'_{\widetilde{L}_{2,h\tau}} \le C \Bigg[\tau \sum_{t \in \omega_\tau^+} \|\psi(\cdot,t)\|^2_{B_{0h}^{-1}} + \tau \sum_{t \in \omega_\tau^+} \|\mu(\cdot,t)\|^2_{B_{0h}} + \tau \sum_{t \in \omega_\tau^+} \|\alpha(\cdot,t)\|^2_{h\star} +$$

$$+ \tau^2 \sum_{t \in \bar{\omega}_\tau} \sum_{t' \in \bar{\omega}_\tau,\ t' \ne t} \frac{\|\beta(\cdot,t) - \beta(\cdot,t')\|^2_{h\star}}{|t-t'|^2} + \tau \sum_{t \in \omega_\tau} \left(\frac{1}{t} + \frac{1}{T-t}\right) \|\beta(\cdot,t)\|^2_{h\star} \Bigg]^{1/2}. \quad (17)$$

From (17), using integral representations of ψ, μ, α and β, we obtain the following convergence rate estimate for FDS (8), (9), (15)

$$\|z\|'_{\widetilde{L}_{2,h\tau}} \le C\left(h^2_{max}\sqrt{\ln 1/\tau} + \tau\right)\left(\|c\|_{W_2^2(0,1)} + \|a\|_{W_2^2(0,1)} + 1\right) \times$$

$$\times \left(\|u\|_{W_2^{2,1}(Q_1)} + \|u\|_{W_2^{2,1}(Q_2)} + \left\| \frac{\partial u(\xi, \cdot)}{\partial t} \right\|_{L_2(0,T)} \right). \quad (18)$$

Remark 1. In estimate (18) requirements on the smoothness of coefficients a and c are overstated. Analogous estimate when $a, c \in W_2^1(0,1)$ can be obtained using so called "exact FDS" [5] for approximation of $\frac{\partial}{\partial x}\left(a(x)\frac{\partial u}{\partial x}\right)$.

3.3 Approximation and Convergence in $\widetilde{W}_{2,h\tau}^{2,1}$

Following [13] we approximate the equation (5) as follows

$$(c+K\,\delta_h)\,v_{\bar{t}}+\frac{h_+ - h}{3}(c\,v)_{x\bar{t}}-(\tilde{a}\,v_{\bar{x}})_{\hat{x}}-\frac{h_+ - h}{6}\left(a_x\,v_{\bar{x}\hat{x}}-a_{\bar{x}\hat{x}}\,v_{\bar{x}}\right)=T_x^2 T_t^- f. \quad (19)$$

Boundary and initial conditions we approximate with (9) and (10). We also assume that $c_0 \leq 2$ and $h_+ = h$ for $x = \xi$.

The error $z = u - v$ satisfies FDS

$$(c + K\,\delta_h)\,z_{\bar{t}} + \frac{h_+ - h}{3}(c\,z)_{x\bar{t}} - (\tilde{a}\,z_{\bar{x}})_{\hat{x}} - \frac{h_+ - h}{6}\left(a_x\,z_{\bar{x}\hat{x}} - a_{\bar{x}\hat{x}}\,z_{\bar{x}}\right) = \varphi,$$

with homogeneous boundary and initial conditions (12) and (13). Here

$$\varphi = \varphi_1 - \varphi_2 = T_t^-\left[c\frac{\partial u}{\partial t} - T_x^2\left(c\frac{\partial u}{\partial t}\right) + \frac{h_+ - h}{3}\left(c\frac{\partial u}{\partial t}\right)_x\right] -$$

$$-\left[(\tilde{a}\,u_{\bar{x}})_{\hat{x}} + \frac{h_+ - h}{6}\left(a_x\,u_{\bar{x}\hat{x}} - a_{\bar{x}\hat{x}}\,u_{\bar{x}}\right) - T_x^2 T_t^- \frac{\partial}{\partial x}\left(a\frac{\partial u}{\partial x}\right)\right].$$

We also denote $A_{1h}z = -\frac{h_+ - h}{6}\left(a_x\,z_{\bar{x}\hat{x}} - a_{\bar{x}\hat{x}}\,z_{\bar{x}}\right)$ and $B_{1h}z = \frac{h_+ - h}{3}(c\,z)_x$. The following assertions hold true.

Lemma 7. *If $c \in C^1[0,1]$ and the maximal step size of the mesh $\bar{\omega}_h$ is sufficiently small ($h_{max} \leq (1/6 - \varepsilon)/\|c\|_{C^1[0,1]}$, $0 < \varepsilon < 1/6$) then the following inequality holds*

$$|(B_{1h}z, z)_h| \leq (1 - \varepsilon)\|z\|_{B_h}^2, \quad z \in H_h.$$

Lemma 8. *If $a \in C^2[0,1]$ then*

$$\|A_{1h}z\|_{B_{0h}^{-1}} \leq \frac{h_{max}}{6}\|a\|_{C^2[0,1]}\|z\|_{\widetilde{W}_{2,h}^2}, \quad z \in H_h.$$

From Lemmas 1, 7 and 8 one obtains the following a priori estimate for the solution of FDS (19), (9), (10), assuming h_{max} is sufficiently small

$$\|z\|_{\widetilde{W}_{2,h\tau}^{2,1}} \leq C\left\{\tau \sum_{t \in \omega_\tau^+} \|\varphi(\cdot, t)\|_{B_{0h}^{-1}}^2\right\}^{1/2}. \quad (20)$$

From (20), using integral representation of φ, one obtains the following convergence rate estimate for FDS (19), (9), (10)

$$\|z\|_{\widetilde{W}_{2,h\tau}^{2,1}} \leq C\,(h_{max}^2 + \tau)\left(\|a\|_{W_2^3(0,1)} + \|c\|_{W_2^2(0,1)}\right)\left\{\|u\|_{W_2^{3,0}(Q_1)} + \|u\|_{W_2^{3,0}(Q_2)} + \right.$$

$$\left. + \left\|\frac{\partial u}{\partial t}\right\|_{W_2^{2,0}(Q_1)} + \left\|\frac{\partial u}{\partial t}\right\|_{W_2^{2,0}(Q_2)} + \left\|\left[\frac{\partial u}{\partial x}\right]_{(\xi,\cdot)}\right\|_{L_2(0,T)} + \left\|\left[\frac{\partial^2 u}{\partial x \partial t}\right]_{(\xi,\cdot)}\right\|_{L_2(0,T)}\right\}.$$

4 Problem with Dynamical Boundary Condition

Let us consider the IBVP for the heat equation with dynamical boundary condition for $x = 0$ (see [15]):

$$c(x)\frac{\partial u}{\partial t} - \frac{\partial}{\partial x}\left(a(x)\frac{\partial u}{\partial x}\right) = f(x,t), \quad x \in (0,1), \quad 0 < t < T, \qquad (21)$$

$$K\frac{\partial u(0,t)}{\partial t} = a(0)\frac{\partial u(0,t)}{\partial x}, \quad u(1,t) = 0, \quad 0 < t < T, \qquad (22)$$

and initial condition (7), where $K > 0$, $0 < c_1 \leq a(x) \leq c_2$ and $0 < c_3 \leq c(x) \leq c_4$.

IBVP (21), (22), (7) can be treated as the limit case of IBVP in $(-\varepsilon, 1) \times (0,T)$, with concentrated capacity in $x = 0$, when $\varepsilon \to +0$:

$$[c(x) + K\,\delta(x)]\frac{\partial u}{\partial t} - \frac{\partial}{\partial x}\left(a(x)\frac{\partial u}{\partial x}\right) = f(x,t), \quad x \in (-\varepsilon, 1), \quad 0 < t < T,$$

$$\frac{\partial u(-\varepsilon,t)}{\partial x} = 0, \quad u(1,t) = 0, \quad 0 < t < T; \quad u(x,0) = u_0(x), \quad x \in (-\varepsilon, 1).$$

IBVP (21), (22), (7) can be written in the form (1) where $H = L_2(0,1)$, $H_A = \widehat{W}_2^1(0,1) = \{w \in W_2^1(0,1) : w(1) = 0\}$, $Au = -\frac{\partial}{\partial x}\left(a(x)\frac{\partial u}{\partial x}\right)$ and $Bu = [c(x) + K\,\delta(x)]\,u(x,t)$. We have

$$\|w\|_A^2 = \int_0^1 a(x)\,[w'(x)]^2\,dx \asymp \|w\|_{W_2^1(0,1)}^2, \quad w \in \widehat{W}_2^1(0,1),$$

$$\|w\|_B^2 = \int_0^1 c(x)\,w^2(x)\,dx + K\,w^2(0) \asymp \|w\|_{L_2(0,1)}^2 + w^2(0), \quad w \in \widehat{W}_2^1(0,1).$$

As in the Sect. 3, we introduce a nonuniform mesh $\bar{\omega}_h$ on $[0,1]$. Let $\widehat{H}_h$ be the space of mesh functions vanishing for $x = 1$. We define the inner product $[v,w)_h = \frac{h_1}{2} v(0)\,w(0) + \sum_{x \in \omega_h} v(x)\,w(x)\,\hbar$ and corresponding norm $\|[w]\|_h = \|[w]\|_{L_{2,h}} = [w,w)_h^{1/2}$.

We extend definition of $v_{\hat{x}}$, $\delta_h(x)$ and $T_x^2 f(x,t)$ to include the point $x = 0$:

$$v_{\hat{x}} = \begin{cases} \frac{v_+ - v}{\hbar}, & x \in \omega_h \\ \frac{2v_+}{h_1}, & x = 0, \end{cases} \qquad \delta_h = \delta_h(x) = \begin{cases} 0, & x \in \omega_h^+ \\ 2/h_1, & x = 0, \end{cases}$$

$$T_x^2 f(0,t) = \frac{2}{h_1}\int_0^{x_1}\left(1 - \frac{x'}{h_1}\right) f(x',t)\,dx'.$$

We approximate the IBVP (21), (22), (7) with the following FDS

$$(c + K\,\delta_h)\,v_{\bar{t}} - (\tilde{a}\,v_{\bar{x}})_{\hat{x}} = T_x^2 T_t^- f, \quad (x,t) \in \omega_h^- \times \omega_\tau^+, \qquad (23)$$

$$v(1,t) = 0, \quad t \in \omega_\tau^+, \qquad (24)$$

with the initial condition (10).

The FDS (23), (24), (10) can be rewritten in the form (3), where difference operators $A_h v = -(\tilde{a} v_{\bar{x}})_{\hat{x}}$ and $B_h v = [c(x) + K \delta_h(x)] v$ are defined also for $x = 0$. For $w \in \widehat{H}_h$ we define energetic norms $|[w]|_{A_h}$, $|[w]|_{B_h}$, $|[w]|_{B_h^{-1}}$ and weighted Sobolev norms $|[w]|_{\widetilde{L}_{2,h\tau}}$, $|[w]|_{\widetilde{W}_{2,h\tau}^{1,1/2}}$, $|[w]|_{\widetilde{W}_{2,h\tau}^{2,1}}$ analogously as in Sect. 3, substituting norm $\|\cdot\|_h$ with $|[\cdot]|_h$.

Analogous results as for the equation with concentrated capacity hold true. The FDS (23), (24), (10) satisfies the convergence rate estimate

$$\|z\|_{\widetilde{W}_{2,h\tau}^{1,1/2}} \leq C \left(h_{max}^2 \sqrt{\ln 1/\tau} + \tau \right) \left(\|a\|_{W_2^2(0,1)} + \|c\|_{W_2^2(0,1)} \right) \|u\|_{W_2^{3,3/2}(Q)}.$$

FDS (23), (24) with the initial condition

$$v(x,0) = \begin{cases} \dfrac{T_x^2 (c u_0)(x)}{c(x)}, & x \in \omega_h^+ \\ \dfrac{2 K u_0(0) + h_1 T_x^2 (c u_0)(0)}{2 K + h_1 c(0)}, & x = 0 \end{cases}$$

satisfies the convergence rate estimate

$$\|z\|'_{\widetilde{L}_{2,h\tau}} \leq C \left(h_{max}^2 \sqrt{\ln 1/\tau} + \tau \right) \Big(\|c\|_{W_2^2(0,1)} +$$

$$+ \|a\|_{W_2^2(0,1)} + 1 \Big) \left(\|u\|_{W_2^{2,1}(Q)} + \left\| \frac{\partial u(0,\cdot)}{\partial t} \right\|_{L_2(0,T)} \right).$$

FDS (19), (24), (10) with the boundary condition

$$(c + K\delta_h) v_{\bar{t}} - (\tilde{a} v_{\bar{x}})_{\hat{x}} = T_x^2 T_t^- f, \qquad x = 0$$

satisfies the convergence rate estimate

$$\|z\|_{\widetilde{W}_{2,h\tau}^{2,1}} \leq C (h_{max}^2 + \tau) \left(\|a\|_{W_2^3(0,1)} + \|c\|_{W_2^2(0,1)} \right) \left(\left\| \frac{\partial u}{\partial t} \right\|_{W_2^{2,0}(Q)} + \|u\|_{W_2^{3,0}(Q)} \right).$$

5 Weakly–Parabolic Equation

Let us consider the following IBVP:

$$K \delta(x-\xi) \frac{\partial u}{\partial t} - \frac{\partial}{\partial x}\left(a(x) \frac{\partial u}{\partial x} \right) = f(x,t), \quad x \in (0,1), \quad 0 < t < T, \qquad (25)$$
$$u(0,t) = 0, \quad u(1,t) = 0, \quad 0 < t < T; \quad u(\xi, 0) = u_0 = \text{const},$$

where $K > 0$, $0 < c_1 \leq a(x) \leq c_2$ and $\delta(x)$ is the Dirac's distribution.

The IBVP (25) can be also rewritten in the form (1), where $Au = -\frac{\partial}{\partial x}\left(a(x) \frac{\partial u}{\partial x} \right)$ and $Bu = K\delta(x-\xi) u(x)$. In this case, A is positive definite operator in the space $H_A = \overset{\circ}{W}_2^1(0,1)$, while B is nonnegative operator in H_A

and $\|u\|_B^2 = K\, u^2(\xi)$. It is easy to see that a priori estimates from Lemmas 1 and 2, not involving inverse operator B^{-1}, still hold.

Using denotations from Sect. 3, we approximate IBVP (25) in the following way

$$K\,\delta_h\, v_{\bar t} - (\tilde a\, v_{\bar x})_{\hat x} = T_x^2 T_t^- f, \quad (x,t) \in \omega_h \times \omega_\tau^+, \tag{26}$$
$$v(0,t)=0, \quad v(1,t)=0, \quad t \in \omega_\tau^+; \qquad v(\xi,0)=u_0.$$

The error $z = u - v$ satisfies the following FDS

$$K\,\delta_h\, z_{\bar t} - (\tilde a\, z_{\bar x})_{\hat x} = -\chi_{1,\hat x}, \quad (x,t) \in \omega_h \times \omega_\tau^+, \tag{27}$$
$$z(0,t)=0, \quad z(1,t)=0, \quad t \in \omega_\tau^+; \qquad z(\xi,0)=0,$$

where $\chi_1 = \tilde a\, u_{\bar x} - T_x^- T_t^-\left(a\, \dfrac{\partial u}{\partial x}\right)$.

FDS (27) is of the form (3), where $A_h v = -(\tilde a\, v_{\bar x})_{\hat x}$ is positive definite linear operator in H_h and $B_h v = K\, \delta_h v$ is nonnegative linear operator in H_h. We have

$$\|v\|_{A_h} = \left\{\sum_{x \in \omega_h^+} \tilde a\, v_{\bar x}^2\, \hbar\right\}^{1/2} \asymp \|v_{\bar x}\|_{h\star}, \qquad \|v\|_{B_h} = \sqrt{K}\,|v(\xi)|.$$

We also define the norm

$$\|v\|_{\widetilde W_{2,h\tau}^{1,1/2}}^2 = \tau \sum_{t \in \omega_\tau^+} \left(\|v(\cdot,t)\|_h^2 + \|v_{\bar x}(\cdot,t)\|_{h\star}^2 + |v(\xi,t)|^2\right) +$$
$$+\tau^2 \sum_{t \in \bar\omega_\tau}\sum_{t' \in \bar\omega_\tau,\, t' \ne t} \dfrac{|v(\xi,t) - v(\xi,t')|^2}{|t - t'|^2}.$$

From Lemma 3, using discrete Friedrichs inequality and imbeding theorem $\max_{x \in \bar\omega_h}|v(x)| \le 0.5\,\|v_{\bar x}\|_{h\star}$ [11], we obtain a priori estimate

$$\|z\|_{\widetilde W_{2,h\tau}^{1,1/2}} \le C\left\{\tau \sum_{t \in \omega_\tau^+}\|\chi_1(\cdot,t)\|_{h\star}^2\right\}^{1/2},$$

giving the following convergence rate estimate of FDS (26)

$$\|z\|_{\widetilde W_{2,h\tau}^{1,1/2}} \le C\,(h_{max}^2 + \tau)\,\|a\|_{W_2^2(0,1)}\left(\left\|\dfrac{\partial^2 u}{\partial x\partial t}\right\|_{L_2(Q)} + \|u\|_{W_2^{2,0}(Q_1)} + \|u\|_{W_2^{2,0}(Q_2)}\right).$$

From Lemma 3 we also obtain the estimate in the "weak" norm

$$\tau \sum_{t \in \omega_\tau^+} |z(\xi,t)|^2 \equiv \dfrac{\tau}{K}\sum_{t \in \omega_\tau^+}\|z(\cdot,t)\|_{B_h}^2 \le C\tau \sum_{t \in \omega_\tau^+}\|A_h^{-1}\chi_{1,\hat x}(\cdot,t)\|_{B_h}^2,$$

and convergence rate estimate

$$\left\{\tau \sum_{t \in \omega_\tau^+}|z(\xi,t)|^2\right\}^{1/2} \le C\tau\left\|\dfrac{\partial u(\xi,\cdot)}{\partial t}\right\|_{L_2(0,T)} +$$
$$+C\,h_{max}^2\,\|a\|_{W_2^2(0,1)}\left(\|u\|_{W_2^{2,0}(Q_1)} + \|u\|_{W_2^{2,0}(Q_2)}\right).$$

Acknowledgement

The research of the first author was supported by MST of Serbia, grant number 04M03 / C.

References

1. Braianov, I.: Convergence of a Crank–Nicolson difference scheme for heat equation with interface in the heat flow and concentrated heat capacity. Lect. Notes Comput. Sci. **1196** (1997), 58–65.
2. Braianov, I.; Vulkov, L.: Finite Difference Schemes with Variable Weights for Parabolic Equations with Concentrated Capacity. Notes on Numerical Fluid Dynamics, Vieweg, **62** (1998), 208-216.
3. Braianov, I. A., Vulkov L. G.: Homogeneous difference schemes for the heat equation with concentrated capacity. Zh. vychisl. mat. mat. fiz. **39** (1999), 254–261 (Russian).
4. Eschet, J.: Quasilinear parabolic systems with dynamical boundary conditions. Communs. Partial Differential Equations **19** (1993), 1309–1364.
5. Jovanović, B. S., Matus, P. P., Shcheglik, V. S.: Difference schemes on nonuniform meshes for the heat equation with variable coefficients and generalizad solutions. Doklady NAN Belarusi **42**, No 6 (1998), 38–44 (Russian).
6. Jovanović, B., Vulkov, L.: On the convergence of finite difference schemes for the heat equation with concentrated capacity. Numerishe Mathematik, in press.
7. Lions, J. L., Magenes, E.: Non homogeneous boundary value problems and applications. Springer–Verlag, Berlin and New York, 1972.
8. Lykov A. V.: Heat–masstransfer. Energiya, Moscow 1978 (Russian).
9. Renardy, M., Rogers, R. C.: An introduction to partial differential equations. Springer-Verlag, Berlin and New York, 1993.
10. Riesz, F., Sz.-Nagy, B.: Leçons d'analyse fonctionelle. Akadémiai Kiadó, Budapest 1972.
11. Samarskiĭ A. A.: Theory of difference schemes. Nauka, Moscow 1989 (Russian).
12. Samarskiĭ A. A., Lazarov R. D., Makarov V. L.: Difference schemes for differential equations with generalized solutions. Vysshaya shkola, Moscow 1987 (Russian).
13. Samarskiĭ, A. A., Mazhukin, V. I., Malafeĭ, D. A., Matus, P. P.: Difference schemes of high order of approximation on nonuniform in space meshes. Doklady RAN **367**, No 3 (1999), 1–4 (Russian).
14. Vladimirov, V. S.: Equations of mathematical physics. Nauka, Moscow 1988 (Russian).
15. Vulkov, L.: Application of Steklov–type eigenvalues problems to convergence of difference schemes for parabolic and hyperbolic equation with dynamical boundary conditions. Lect. Notes Comput. Sci. **1196** (1997), 557–564.
16. Wloka, J.: Partial differential equations. Cambridge Univ. Press, Cambridge 1987.

A Method of Lines Approach to the Numerical Solution of Singularly Perturbed Elliptic Problems

J. D. Kandilarov[1], L. G. Vulkov[1], and A. I.Zadorin[2]

[1] University of Rousse, Bulgaria
{juri,vulkov}@ami.ru.acad.bg
[2] Institute of Mathematics, Russian Academy of Sciences, Siberian branch,
Omsk 644099, Russia
Zadorin@iitam.omsk.net.ru

Abstract. Two elliptic equations with power and interior boundary layers, respectively, in a square are considered. The elliptic problems are reduced to systems of ordinary differential equations by the method of lines. For construction of difference schemes fitted operator technique is used. Uniform convergence for the scheme of the first problem is proved.

1 Introduction and Statement of the Problems

The numerical solution of multi-dimensional singularly perturbed differential equations is highy complicated and problem dependent process. The solution may contain interactions between different layers. A method developed for particular problem may, or not work for another with stronger (or weak) type layers. Methods, which work well in one space dimension may or may not be easily extended to two or three dimensions.

The analysis of this talk is based on solving ODE singularly perturbed problems that generalize to higher spatial dimensions. This is the philosophy of the method of lines (MOL). In MOL applied to two-dimensional elliptic problems the discretizations with respect to the first (x) and second (y) independent variables are deconpled and analyzed independently. We discuss the MOL for the following two problems.

1.1 Formulation of Problem 1 (P1)

Some physical processes, as pollution transfer, lead to elliptic problems including variable coefficient of turbulent diffusion. When diffusion coefficient is linear function of a coordinate, we have a problem with a power layer:

$$Lu = \varepsilon\frac{\partial^2 u}{\partial x^2} + (\varepsilon+y)\frac{\partial^2 u}{\partial y^2} - a(x,y)\frac{\partial u}{\partial x} + w(y)\frac{\partial u}{\partial y} - c(x,y)u = f(x,y), \quad (x,y) \in G,$$

$$G = \{0 < x < 1, 0 < y < 1\}, \; u\mid_{\Gamma_D} = 0, \; u'_x(1,y) + \delta u(1,y) = 0, \; \Gamma = \Gamma_D \cup \Gamma_{Mix}, \tag{1}$$

where functions a, w, c, f are enough smooth,

$$\varepsilon > 0, \ a(x,y) \geq \alpha > 0, \ w(y) \geq \beta > 0, \ c(x,y) \geq 0, \ c(x,y) + 4a'_x(x,y) \geq 0, \ \delta \geq 0. \tag{2}$$

1.2 Formulation of Problem 2 (P2)

The second problem is a singularly perturbed reaction-diffusion equation with known singular sources: f is a smooth at least in $\Omega \backslash S$,

$$\varepsilon^2 \frac{\partial^2 u}{\partial x^2} + \frac{\partial^2 u}{\partial y^2} = f(x,y) + \int_S Q(t) \delta\left(\vec{x} - \vec{X}(t)\right) dt \tag{3}$$

$$u = 0 \quad \text{on} \quad \partial \Omega = \Gamma. \tag{4}$$

The solution to (3) satisfies

$$[u]_S = 0 \quad \text{and} \quad \left[\frac{\partial u}{\partial n}\right]_S = Q, \tag{5}$$

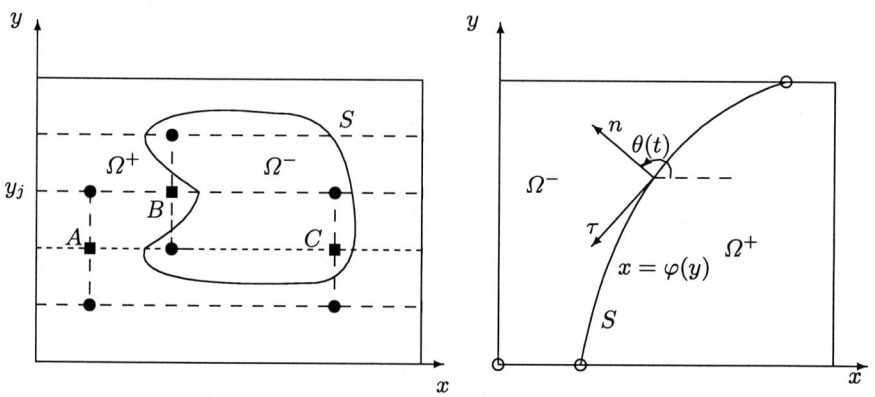

Fig. 1. Semidiscretization: regular A and irregular points B, C

Fig. 2. Local coordinate system

Here $[\cdot]_S$ is the jump of the corresponding quantity across S, $\delta(\cdot)$ is the Dirac-delta function and the interface curve $\Gamma = \vec{X}(t) = (x(t), y(t))$ is parametrized by arclength t. The normalized tangent direction to Γ at t is $\vec{\eta}(t) = (x'(t), y'(t)) = (-\sin\theta(t), \cos\theta(t))$, where $\theta(t)$ is the angle from the vector $(1,0)$ to the outward normal $\vec{\xi}(t) = (\cos\theta(t), \sin\theta(t),)$ to Γ at t. For simplicity we shall explain our construction on the case of interface curve presented on Fig.2.

It is well known that in the case of absent of interface boundary layers at $x = 0, 1$ and corner layers at the four corners of the square appear when ε is small. Now, in additional, interior layers near to the interface and interface corner layers at the intersection points of the interface with the boundary Γ occur.

The main goal of the paper is to construct difference schemes for (P1) and (P2) with the property of **uniform convergence** with respect to the small parameter on any uniform mesh.

In the equation (1) the derivatives with respect to x are discretized. For construction of difference scheme the fitted operator method is applied [1]. The uniform convergence in the strong norm of the difference scheme solution to the differential problem solution is proved.

It is convienient for the problem (P2) the semi-discretization to be done with respect to y. The immersed interface method (IIM) is applied on an uniform Cartesian grid. The main idea of the IIM, [2], consists in modification of the difference scheme on the irregular points, Fig.1, near the interface by using jump conditions. Such obtained ODEs is solved by the fitted operator method again.

The proof of all statements in this paper and numerical experiments are included in forthcoming papers of the authors.

2 Problem (P1)

The following proposition describes the layer which arises in equation (1) as $\varepsilon \to 0$.

Lemma 1. *For solution of (P1) and it's derivatives the following estimates hold:*

$$||u|| = \max_{x,y} |u(x,y)| \le \alpha^{-1}||f||$$

$$\left|\frac{\partial^k u}{\partial x^k}\right| \le C\left[1 + \varepsilon^{1-k}\exp(\alpha\varepsilon^{-1}(x-1))\right], \quad k \le 4, \tag{6}$$

$$\left|\frac{\partial u}{\partial y}\right| \le \frac{C}{\varepsilon + y}. \tag{7}$$

If in addition $\beta > 1$, then

$$\left|\frac{\partial u}{\partial y}\right| \le C + C\left(\frac{\varepsilon}{\varepsilon + y}\right)^{\beta-1} \frac{1}{\varepsilon + y}. \tag{8}$$

Here and everywhere later, C will denote a generic positive constant that is independent of the perturbation parameter ε.

In order to approximate the boundary value problem (1), we introduce the following notation:

$$w_1 = \{x_i = ih_1, \quad i = 0, \ldots, N\}, \quad w_2 = \{y_j = ih_2, \quad j = 0, \ldots, M\},$$

$$w = w_1 \times w_2, \quad v_i(y) = v(x_i, y).$$

The semidiscrete scheme takes the form

$$L_i v = (\varepsilon + y)v_i'' + w(y)v_i' + \varepsilon \frac{v_{i+1} - 2v_i + v_{i-1}}{h_1^2} - a_i(y)\frac{v_i - v_{i-1}}{h_1} - c_i(y)v_i(y) = f_i(y), \tag{9}$$

$$v_i(0) = v_i(1) = 0, \ 0 < i < N, \ v_0(y) = 0, \ D^h v = \frac{v_N - v_{N-1}}{h_1} + \delta v_N = 0, \tag{10}$$

where $a_i = a(x_i, y), c_i = c(x_i, y), f_i = f(x_i, y), \ i = 1, ..., N$.

We begin the investigation of (9), (10) with the following maximum principle

Lemma 2. Let $\varepsilon > 0, a_i \geq 0$. Let there exists a vector-function $\phi(y)$, such that

$$\phi(y) > 0, \ L_i \phi < 0, \ 0 < i < N, \ D^h \phi \geq 0. \tag{11}$$

Let $\Psi(y)$ is sufficiently smooth vector-function and

$$L_i \Psi \leq 0, \ \Psi_i(0) \geq 0, \ \Psi_i(1) \geq 0, 0 < i < N, \ \Psi_0(y) \geq 0, \ D^h \Psi \geq 0. \tag{12}$$

Then for all $i, y \in (0, 1) \ \Psi_i(y) \geq 0$.

The next lemma establishes convergence of the solution v of the semidiscrete problem (9), (10) to the solution u of the differential problem (1).

Lemma 3. The error $z_i(y) = v_i(y) - u(x_i, y)$ satisfies the inequality

$$\max_y |z_i(y)| \leq Ch_1, \quad i = 0, 1, ..., N. \tag{13}$$

The problem (9), (10) can be rewritten as follows

$$(\varepsilon + y)\frac{d^2 v_i}{dy^2} + w(y)\frac{dv_i}{dy} = F_i(y),$$

$$v_i(0) = v_i(1) = 0, \ 0 < i < N, \ v_0(y) = 0, \ D^h v_i(y) = 0, \tag{14}$$

where

$$F_i = f_i + c_i v_i - \varepsilon \frac{v_{i+1} - 2v_i + v_{i-1}}{h_1^2} + a_i \frac{v_i - v_{i-1}}{h_1}.$$

We can use special nonuniform mesh on y (fitted meshes [1], [3], [4]) to get approximations with the property of uniform convergence. Here however, following [5], we derive the fitted operator scheme

$$\left[\frac{p_{i,j}^h - p_{i,j-1}^h}{h_2} - \frac{F_{i,j}}{w_j}\right] D_j(\varepsilon+y_j)^{-w_j} - \left[\frac{p_{i,j+1}^h - p_{i,j}^h}{h_2} - \frac{F_{i,j+1}}{w_{j+1}}\right] D_{j+1}(\varepsilon+y_j)^{-w_{j+1}}$$

$$= \frac{F_{i,j+1}}{w_{j+1}} - \frac{F_{i,j}}{w_j}, \quad p_{i,0}^h = 0, \quad p_{i,M}^h = 0, \quad i = 0, 1, ..., N,$$

$$p_{0,j}^h = 0, \quad \frac{p_{N,j}^h - p_{N-1,j}^h}{h_1} + \delta p_{N,j}^h = 0, \quad j = 0, 1, ..., M, \tag{15}$$

where

$$D_j = (1-w_j)h_2 \left[(\varepsilon+y_j)^{1-w_j} - (\varepsilon+y_{j-1})^{1-w_j}\right]^{-1},$$

$$F_{i,j} = -\varepsilon \frac{p_{i+1,j}^h - 2p_{i,j}^h + p_{i-1,j}^h}{h_1^2} + a(x_i, y_j)\frac{p_{i,j}^h - p_{i-1,j}^h}{h_1} + c(x_i, y_j)p_{i,j}^h + f(x_i, y_j).$$

Finally, the uniform convergence of the fully discrete solution is proved in the next theorem.

Theorem 1. *Let u is solution of problem (1), p^h is solution of the scheme (15). Then*

$$||p^h - [u]_\Omega|| \leq C[h_1 + |\ln h_2|h_2]. \tag{16}$$

In the case $\beta > 1$

$$||p^h - [u]_\Omega|| \leq C(h_1 + h_2). \tag{17}$$

3 Problem (P2)

We introduce the uniform mesh $w_y = \{y_j : y_j = jk, \ j = 0, \ldots, J, \ Jk = 1\}$ and approximate the second derivative of u with respect to y as follows: in regular points A

$$\frac{\partial^2 u}{\partial y^2} = \frac{u(x, y_{j-1}) - 2u(x, y_j) + u(x, y_{j+1})}{k^2} + O(k^2),$$

$$x \in (0,1) \setminus (\varphi(y_{j-1}), \varphi(y_{j+1}));$$

in irregular points B, C

$$\frac{\partial^2 u}{\partial y^2} = \frac{u(x, y_{j-1}) - 2u(x, y_j) + u(x, y_{j+1})}{k^2} +$$

$$\left(\frac{[u]}{k^2} + \frac{y_{j-1} - y(x)}{k^2}\left[\frac{\partial u}{\partial y}\right] + \frac{(y_{j-1} - y(x))^2}{2k^2}\left[\frac{\partial^2 u}{\partial y^2}\right]\right)_{(x,y(x))} + O(k),$$

$$\varphi(y_{j-1}) \leq x \leq \varphi(y_j);$$

$$\frac{\partial^2 u}{\partial y^2} = \frac{u(x, y_{j-1}) - 2u(x, y_j) + u(x, y_{j+1})}{k^2} -$$

$$\left(\frac{[u]}{k^2} + \frac{y_{j+1} - y(x)}{k^2}\left[\frac{\partial u}{\partial y}\right] + \frac{(y_{j+1} - y(x))^2}{k^2}\left[\frac{\partial^2 u}{\partial y^2}\right]\right)_{(x,y(x))} + O(k),$$

$$\varphi(y_j) \leq x \leq \varphi(y_{j+1}).$$

Here we shall describe our construction on the case of Fig 2. For the function $\varphi(y)$, $y \in [0,1]$ we assume that there exists their inverse one $y = \varphi^{-1}(x)$ (or $y = y(x)$), $x \in [0,1]$. Also,

$$\tilde{t}(y) = \int_0^y \sqrt{1 + \varphi'^2} d\rho, \quad y \in [0,1], \quad y(t) = \tilde{t}^{-1}(t), \quad x = \varphi(y) = \varphi(y(t)),$$

$$s = \sin\theta(t), \quad c = \cos\theta(t), 0 \leq t \leq \tilde{t}(1).$$

The jumps are calculated from (3), (5). As a result the following ODEs with zero boundary conditions arises:

$$\varepsilon^2 v'' + Av = F(x) + \tilde{Q}(x)\tilde{\delta}(x - \zeta), \tag{18}$$

$$v = \begin{pmatrix} v_1 \\ \vdots \\ v_{J-1} \end{pmatrix}, \quad A = threediag\left(\frac{1}{k^2}; -\frac{2}{k^2}; \frac{1}{k^2}\right)$$

$$F(x) = \begin{pmatrix} F_1(x) \\ \vdots \\ F_{J-1}(x) \end{pmatrix}, \quad F_j(x) = \begin{cases} f(x, y_j) + 0 & ,0 \leq x \leq \varphi(y_{j-1}), \\ f(x, y_j) + \phi_{j1} & ,\varphi(y_{j-1}) \leq x \leq \varphi(y_j) \\ f(x, y_j) + \phi_{j2} & ,\varphi(y_j) \leq x \leq \varphi(y_{j+1}) \\ f(x, y_j) + 0 & ,\varphi(y_{j+1}) \leq x \leq 1, \end{cases}$$

$$\phi_{j1} = \frac{Q(x, y(x))s}{\varepsilon^2 c^2 + s^2} \cdot \frac{y_{j-1} - y(x)}{h^2} + R,$$

$$\phi_{j2} = \frac{Q(x, y(x)) s}{\varepsilon^2 c^2 + s^2} \cdot \frac{y_{j+1} - y(x)}{h^2} - R,$$

$$R = \left(\frac{s^2 [f]}{\varepsilon^2 c^2 + s^2} + 2cs \left(\frac{\widetilde{Q}}{\varepsilon^2 c^2 + s^2} \right)'_t + \varepsilon^2 \frac{c^2 - s^2}{\varepsilon^2 c^2 + s^2} \frac{Q}{\varepsilon^2 c^2 + s^2} \right) \frac{(y_{j-1} - y(x))^2}{2h^2}$$

$$Q(x) = diag\left(\left. \frac{cQ(x, y(x))}{\varepsilon^2 c^2 + s^2} \right|_{y_1}, \ldots, \left. \frac{cQ(x, y(x))}{\varepsilon^2 c^2 + s^2} \right|_{y_{J-1}} \right),$$

$$\delta(x - \zeta) = \begin{pmatrix} \delta(x - \zeta_j) \\ \vdots \\ \delta(x - \zeta_{J-1}) \end{pmatrix}, \quad \zeta_j = x(y_j), \quad j = 1, \ldots, J-1$$

The eigenvalues and their corresponding eigenvectors of the matrix A are as follows

$$\lambda_j = \frac{2}{\varepsilon k} \sin \frac{j\pi}{2J}, \quad l_j = \begin{pmatrix} l_{j1} \\ \vdots \\ l_{jJ-1} \end{pmatrix}, \quad l_{jm} = \sqrt{\frac{2}{J}} \sin \frac{\pi m j}{J}, \quad m, j = 1, \ldots, J-1.$$

Let $\overline{v}(x)$ is the solution vector function of the ODEs (18) in which $F(x) \simeq \overline{F}(x) = F(x_i), \quad x \in (x_{i-1}, x_i), \quad i = 1, \ldots, I, \quad hI = 1$. Then for it's components we have the representation

$$\overline{v}^i(x) = \sum_{j=1}^{J-1} \left(p^i_{1j}(x) v(x_{i-1}) + p^i_{2j}(x) v(x_i) - \left(P^i_{1j}(x) + P^i_{2j}(x) \right) B^i \right.$$

$$\left. - \left(p^i_{1j}(x) D(x_{i-1}) + p^i_{2j}(x) D(x_i) \right) l_j \right) l_j + B^i + D(x),$$

where

$$p^i_{1j}(x) = \frac{sh \, \lambda_j (x_i - x)}{sh \, \lambda_1 h}, \quad p^i_{2j}(x) = \frac{sh \, \lambda_j (x - x_{i-1})}{sh \, \lambda_1 h},$$

$$B^i = A^{-1} \overline{F}^i, \quad D(x) = \begin{pmatrix} D_1(x) \\ \vdots \\ D_{J-1}(x) \end{pmatrix}, \quad D_j(x) = \begin{cases} \frac{x}{\zeta_j}, & 0 \leq x \leq \zeta_j, \\ \frac{1-x}{1-\zeta_j}, & \zeta_j \leq x < 1, \end{cases}$$

$$d_j = -Q(x, y(x)) \left. \frac{c}{\varepsilon^2 c^2 + s^2} \right|_{(x(y_j), y_j)}.$$

Now our difference scheme can be written in a vector form as follows:

$$\overline{V}^0 = V(0) = 0,$$

$$-\overline{A}_i \overline{V}^i + \overline{C}_i \overline{V}^i - \overline{B}_i \overline{V}^i = \overline{\Psi}_i, \quad i = 1, \ldots, I-1, \tag{19}$$

$$\overline{V}^I = V(1) = 0,$$

where

$$\overline{A}_i = \|a_{nm}^i\|, \quad a_{nm}^i = \sum_{j=1}^{J-1} \frac{\lambda_j}{sh\,\lambda_j h} l_{jn} l_{jm},$$

$$\overline{B}_i = \overline{A}_i, \quad \overline{C}_i = 2 \sum_{j=1}^{J-1} \lambda_j \, cth\, \lambda_j h l_{jn} l_{jm}, \quad n, m = 1, \ldots, J-1$$

$$\overline{\Psi}_i = \left(\overline{A}_i - 0.5\overline{C}_i\right)\left(B^i + B^{i+1}\right) - \overline{A}_i D\left(x_{i-1}\right) + \overline{C}_i D\left(x_i\right) - \overline{B} D\left(x_{i+1}\right) +$$

$$D^1\left(x_i + 0\right) - D^1\left(x_1 - 0\right),$$

The scheme (19) has $O(h+k)$ local approximation. The accuracy with respect to y can be improved if at the approximation of $\partial^2 u/\partial y^2$ on the irregular nodes one adds the jump $[\partial^3 u/\partial y^3]$ and with respect to x piecewise linear approximation of the coefficients is used.

The full proof for uniform convergence of the scheme is still an open question.

References

1. Miller J. J. H., O'Riordan E., Shishkin G. I. Fitted numerical methods for singularly perturbed problems. World scientific, Singapure 1996.
2. Leveque R., Li Zhilin, The immersed interface method for elliptic equations with discontinuous coefficients and singular sources, SIAM J. Numer. Anal. (1994). 1019-1044.
3. Bagaev B. M., Solusenko N. P. Numerical solution for problems with a power boundary layer. Modelling in mechanics, 3, N 2, Novosibirsk, 1989.
4. Liseikin V. D. About numerical solution of problems with power boundary layer. Journal of Comput. Math. and Mathem. Physics, 26, N 12, (1986), 1813-1821.
5. Zadorin A. I. Numerical solution of a boundary value problem for a system of equations with a small parameter Comput. Math. and Math. Physics, 38, 1998, N8, 1201-1211.

Sobolev Space Preconditioning of Strongly Nonlinear 4th Order Elliptic Problems

János Karátson*

ELTE University, Dept. Applied Analysis
H-1053 Budapest, Hungary

Abstract. Infinite-dimensional gradient method is constructed for nonlinear fourth order elliptic BVPs. Earlier results on uniformly elliptic equations are extended to strong nonlinearity when the growth conditions are only limited by the Hilbert space well-posedness. The obtained method is opposite to the usual way of first discretizing the problem. Namely, the theoretical iteration is executed for the BVP itself on the continuous level in the corresponding Sobolev space, reducing the nonlinear BVP to auxiliary linear problems. Thus we obtain a class of numerical methods, in which numerical realization is determined by the method chosen for the auxiliary problems. The biharmonic operator acts as a Sobolev space preconditioner, yielding a fixed ratio of linear convergence of the iteration (i.e. one determined by the original coefficients only, independent of the way of solution of the auxiliary problems), and at the same time reducing computational questions to those for linear problems. A numerical example is given for illustration.

1 Introduction

This paper is devoted to an approach of numerical solution to strongly nonlinear fourth order elliptic boundary value problems. The usual way of the numerical solution of elliptic equations is to discretize the problem and use an iterative method for the solution of the arising nonlinear system of algebraic equations (see e.g. [13]). For the latter suitable preconditioning technique has to be used [3]. For instance, an efficient way of this is the Sobolev gradient technique [15], which relies on using the trace of the Sobolev inner product in the discrete spaces. This technique is a link towards an approach opposite to the above: namely, the iteration can be executed on the continuous level directly in the corresponding Sobolev space, reducing the nonlinear problem to auxiliary linear BVPs. Then discretization may be used for these auxiliary problems. The theoretical background of this approach is the generalization of the gradient method to Hilbert spaces (see e.g. [4,9,10,17], and for a class of non-uniformly monotone operators [12]). Application to uniformly elliptic BVPs is summarized in [11].

The aim of this paper is to construct a class of numerical methods for strongly nonlinear fourth order problems, based on the Sobolev space gradient method.

---
* This research was supported by the Hungarian National Research Funds AMFK under Magyary Zoltán Scholarship and OTKA under grant no. F022228

This result extends the scope of [11] as wide as possible within the Hilbert space well-posedness of our problem. The actual numerical realization is established by the choice of a suitable numerical method for the solution of the auxiliary problems. *This approach can be regarded as infinite-dimensional preconditioning by the biharmonic operator, yielding two main advantages. Firstly, a favourable ratio of convergence is achieved for the iteration. Secondly, computational problems are reduced to those arising for the auxiliary linear problems for the biharmonic operator, since the nonlinearity is entirely handled by the outer simple GM iteration.* The numerical solution of the former is much developed (see e.g. [5,7,16]). This paper focuses on constructing the Sobolev space GM and proving its linear convergence in the corresponding energy norm. A simple numerical example is given to illustrate the growth conditions involved in strong nonlinearity of the lower order terms.

2 Formulation of the Dirichlet Problem

The following notations will be used throughout the paper. For $u \in H^2(\Omega)$ the Hessian is denoted as usual by $D^2 u$. For any $H, V : \Omega \to \mathbf{R}^{N \times N}$ let

$$H \cdot V := \sum_{i,k=1}^{N} H_{ik} V_{ik}, \qquad \mathrm{div}^2 H := \sum_{i,k=1}^{N} \partial_{ik} H_{ik}.$$

We consider the boundary value problem

$$\begin{cases} T(u) \equiv \mathrm{div}^2 A(x, D^2 u) - \mathrm{div}\, f(x, \nabla u) + q(x, u) = g(x) \\ u_{|\partial\Omega} = \partial_\nu u_{|\partial\Omega} = 0 \end{cases} \quad (1)$$

with the following conditions:

(C1) $N = 2$ or 3, $\Omega \subset \mathbf{R}^N$ is a bounded domain, $\partial\Omega \in C^4$ (cf. also the third remark).

(C2) $A \in C^1(\overline{\Omega} \times \mathbf{R}^{N \times N}, \mathbf{R}^{N \times N})$, $f \in C^1(\overline{\Omega} \times \mathbf{R}^N, \mathbf{R}^N)$, $q \in C^1(\overline{\Omega} \times \mathbf{R})$ and $g \in L^2(\Omega)$.

(C3) There exist constants $m' \geq m > 0$ such that for any $(x, \Theta) \in \overline{\Omega} \times \mathbf{R}^{N \times N}$ the Jacobian array

$$A'_\Theta(x, \Theta) = \{\partial_{\Theta_{ik}} A_{rs}(x, \Theta)\}_{i,k,r,s=1}^{N}$$

(in $\mathbf{R}^{N^2 \times N^2}$) is symmetric and its eigenvalues Λ satisfy

$$m \leq \Lambda \leq m'.$$

(C4) There exist constants $\kappa, \beta \geq 0$, further, $2 \leq p$ (if $N = 2$) and $2 \leq p \leq 6$ (if $N = 3$) such that for any $(x, \eta) \in \overline{\Omega} \times \mathbf{R}^N$ the Jacobian matrix

$$f'_\eta(x, \eta) = \{\partial_{\eta_k} f_i(x, \eta)\}_{i,k=1}^{N}$$

(in $\mathbf{R}^{N \times N}$) is symmetric and its eigenvalues μ satisfy

$$0 \leq \mu \leq \kappa + \beta |\eta|^{p-2}.$$

(C5) For any $(x, \xi) \in \overline{\Omega} \times \mathbf{R}$ there holds $0 \leq \partial_\xi q(x, \xi)$.

3 Sobolev Space Background

We introduce the real Hilbert space

$$H_0^2(\Omega) := \{u \in H^2(\Omega) : u_{|\partial\Omega} = \partial_\nu u_{|\partial\Omega} = 0\}. \tag{2}$$

It is well-known that

$$\langle u, v \rangle_{H_0^2} \equiv \int_\Omega D^2 u \cdot D^2 v \tag{3}$$

defines an inner product on $H_0^2(\Omega)$, equivalent to the H^2 one.

Remark 1. (See [1].) Assumptions (C1) and (C4) imply the following Sobolev embeddings. There exist constants $K_\infty > 0$ and $K_p > 0$ such that

$$H_0^2(\Omega) \subset C(\overline{\Omega}), \quad \|u\|_\infty \leq K_\infty \|u\|_{H_0^2} \quad (u \in H_0^2(\Omega)); \tag{4}$$

$$H_0^2(\Omega) \subset W_0^{1,p}(\Omega), \quad \|u\|_{W_0^{1,p}} \leq K_p \|u\|_{H_0^2} \quad (u \in H_0^2(\Omega)), \tag{5}$$

where $\|u\|_{W_0^{1,p}} := \left(\int_\Omega |\nabla u|^p\right)^{1/p}$. Further,

$$\|u\|_{L^2(\Omega)} \leq \lambda^{-1/2} \|u\|_{H_0^2} \quad (u \in H_0^2(\Omega)) \tag{6}$$

where λ denotes the smallest eigenvalue of Δ^2 on $H^4(\Omega) \cap H_0^2(\Omega)$.

4 The Gradient Method

4.1 Gradient Method for the Dirichlet Problem

Proposition 1. *The following equation defines an operator* $F : H_0^2(\Omega) \to H_0^2(\Omega)$:

$$\langle F(u), v \rangle_{H_0^2} \equiv \int_\Omega [A(x, D^2 u) \cdot D^2 v + f(x, \nabla u) \cdot \nabla v + q(x, u)v] \quad (u, v \in H_0^2(\Omega)).$$

Proof. We use conditions (C3)-(C4) for the functions A and f. Lagrange's inequality yields the following estimate for the right side integral (with suitable constants $m_0, \tilde{m}_0, \kappa', \tilde{\kappa}', \beta', \gamma > 0$):

$$\int_\Omega [(m_0 + m'|D^2 u|)|D^2 v| + (\kappa' + \beta'|\nabla u|^{p-1})|\nabla v| + |q(x, u)v|] \leq$$

$$(\tilde{m}_0 + m'\|u\|_{H_0^2})\|v\|_{H_0^2} + \left(\tilde{\kappa}' + \beta'\|\nabla u\|_{L^p}^{p-1}\right)\|\nabla v\|_{L^p} + \gamma \max_{\substack{x \in \overline{\Omega} \\ |u| \leq \|u\|_\infty}} |q(x, u)| \, \|v\|_\infty.$$

The Sobolev embeddings (4)-(5) yield that the norms in this estimate are finite, and for fixed $u \in H_0^2(\Omega)$ the discussed integral defines a bounded linear functional on $H_0^2(\Omega)$. Hence the Riesz theorem provides the existence of $F(u) \in H_0^2(\Omega)$. □

Remark 2. Since the embeddings (4)-(5) are sharp, the growth conditions (C3)-(C5) are the strongest that allow Proposition 1 to hold in $H_0^2(\Omega)$.

A *weak solution* $u^* \in H_0^2(\Omega)$ of problem (1) is defined in the usual way by

$$\langle F(u^*), v \rangle_{H_0^2} = \int_\Omega gv \qquad (v \in H_0^2(\Omega)). \tag{7}$$

Now we formulate and prove our main result on the Sobolev space gradient method for problem (1). For this we introduce the operator Δ^2 in the space $L^2(\Omega)$ with domain $D(\Delta^2) := H^4(\Omega) \cap H_0^2(\Omega)$. Then Green's formula yields

$$\int_\Omega (\Delta^2 u)v = \int_\Omega D^2 u \cdot D^2 v = \langle u, v \rangle_{H_0^2} \qquad (u, v \in H^4(\Omega) \cap H_0^2(\Omega)), \tag{8}$$

hence the energy space H_{Δ^2} of Δ^2 is $H_0^2(\Omega)$. Further, we will use the following notations:

$$\tilde{q}(u) := \max\{q'_\xi(x, \xi) : x \in \overline{\Omega}, |\xi| \leq u\} \qquad (u > 0); \tag{9}$$

$$M(r) = m' + \kappa K_2^2 + \beta K_p^p r^{p-2} + \lambda^{-1} \tilde{q}(K_\infty r) \qquad (r > 0), \tag{10}$$

where K_2, K_p are from (5), and λ denotes the smallest eigenvalue (or lower bound) of Δ^2 on $H^4(\Omega) \cap H_0^2(\Omega)$.

Theorem 1. *Problem (1) has a unique weak solution $u^* \in H_0^2(\Omega)$. Further, let $u_0 \in H^4(\Omega) \cap H_0^2(\Omega)$, and*

$$M_0 := M\left(\|u_0\|_{H_0^2} + \frac{1}{m\sqrt{\lambda}} \|T(u_0) - g\|_{L^2}\right) \tag{11}$$

with $M(r)$ defined in (10). For $n \in \mathbf{N}$ let

$$u_{n+1} = u_n - \frac{2}{M_0 + m} z_n, \tag{12}$$

where $z_n \in H^4(\Omega) \cap H_0^2(\Omega)$ is the solution of the auxiliary problem

$$\begin{cases} \Delta^2 z_n = T(u_n) - g \\ z_n|_{\partial\Omega} = \partial_\nu z_n|_{\partial\Omega} = 0. \end{cases} \tag{13}$$

Then the sequence (u_n) converges linearly to u^*, namely,

$$\|u_n - u^*\|_{H_0^2} \leq \frac{1}{m\sqrt{\lambda}} \|T(u_0) - g\|_{L^2} \left(\frac{M_0 - m}{M_0 + m}\right)^n \qquad (n \in \mathbf{N}). \tag{14}$$

Proof. A Hilbert space result, given in [12], will be applied in the real Hilbert space $H := L^2(\Omega)$. For this purpose first some properties of the operator T in H, defined on $D(T) := H^4(\Omega) \cap H_0^2(\Omega)$, and of the generalized differential operator F, are proved.

(a) There holds $R(\Delta^2) \supset R(T)$ since the operator Δ^2 in $L^2(\Omega)$ is onto by regularity. Namely, condition (C1) implies that for any $g \in L^2(\Omega)$ the weak solution of $\Delta^2 z = g$ with $z_{|\partial\Omega} = \partial_\nu z_{|\partial\Omega} = 0$ is in $D(\Delta^2) = H^4(\Omega) \cap H_0^2(\Omega)$ [2].

(b) Green's formula and (8) yield that for any $u, v \in H^4(\Omega) \cap H_0^2(\Omega)$

$$\langle F(u), v \rangle_{H_0^2} = \int_\Omega T(u)v = \langle \Delta^{-2} T(u), v \rangle_{H_0^2}. \tag{15}$$

Hence $F_{|H^4(\Omega) \cap H_0^2(\Omega)} = \Delta^{-2} T$.

(c) F has a bihemicontinuous Gâteaux derivative F' such that for any $u \in H_0^2(\Omega)$, the operator $F'(u)$ is self-adjoint and satisfies

$$m\|h\|_{H_0^2}^2 \leq \langle F'(u)h, h \rangle_{H_0^2} \leq M(\|u\|_{H_0^2})\|h\|_{H_0^2}^2 \quad (h \in H_0^2(\Omega)) \tag{16}$$

with the increasing function M defined in (10). These properties can be checked by suitably modifying the proof of the corresponding result for uniformly elliptic problems [11] (quoted in the introduction), now using Sobolev embedding estimates. This works in the same way for the existence and bihemicontinuity of F' as for verifying (16), hence for brevity the former is left to the reader. The operators $F'(u)$ are given by the formula

$$\langle F'(u)h, v \rangle_{H_0^2} = \int_\Omega [A'_\Theta(x, D^2 u) D^2 h \cdot D^2 v + f'_\eta(x, \nabla u) \nabla h \cdot \nabla v + q'_\xi(x, u) hv] \tag{17}$$

(for any $u, v, h \in H_0^2(\Omega)$). Now we use conditions (C3)-(C5). The symmetry assumptions on A'_Θ and f'_η imply that $F'(u)$ is self-adjoint. Further, there holds

$$m \int_\Omega |D^2 h|^2 \leq \langle F'(u)h, h \rangle_{H_0^2} \leq \int_\Omega \left[m'|D^2 h|^2 + \left(\kappa + \beta|\nabla u|^{p-2}\right) |\nabla h|^2 + \tilde{q}(u)h^2 \right]$$

$$\leq m'\|h\|_{H_0^2}^2 + \kappa\|h\|_{H_0^1}^2 + \beta\|u\|_{W_0^{1,p}}^{p-2}\|h\|_{W_0^{1,p}}^2 + \tilde{q}(\|u\|_\infty)\|h\|_{L^2}^2$$

$$\leq \left(m' + \kappa K_2^2 + \beta K_p^p \|u\|_{H_0^2}^{p-2} + \lambda^{-1}\tilde{q}(K_\infty \|u\|_{H_0^2}) \right) \|h\|_{H_0^2}^2,$$

using Remark 1. Thus (16) is verified.

The obtained properties (a)-(c) of T, F and the auxiliary operator $B := \Delta^2$ yield that the conditions of Theorem 3 and Corollary 2 in [12] are satisfied in the space $H = L^2(\Omega)$. Hence equation $T(u) = g$ has a unique weak solution $u^* \in H_B = H_0^2(\Omega)$, and for any $u_0 \in D(B)$ the sequence $u_{n+1} = u_n - \frac{2}{M_0 + m} B^{-1}(T(u_n) - g)$ converges to u^* according to the estimate (14).

Remark 3. (13) can be written in the weak form

$$\int_\Omega D^2 z_n \cdot D^2 v = \int_\Omega (T(u_n) - g)\, v \qquad (v \in H_0^2(\Omega)). \tag{18}$$

This is also valid when Ω violates condition $\partial\Omega \in C^4$ in (C1) (i.e. H^4-regularity is not guaranteed for the solutions of (13)). Moreover, numerical realization essentially relies on (18), and the aim of the strong form (13) in the theoretical iteration is rather to indicate clearly the preconditioning role of Δ^2.

4.2 Generalizations

(a) We may set a weight function $w \in L^\infty(\Omega)$ in (18), which means preconditioning formally by the operator $Bz = \operatorname{div}^2(w\, D^2 z)$. For instance, a piecewise constant w may give more accurate approximation of the bounds of T.

(b) The GM in Theorem 2 works similarly, involving the weak formulation, for mixed boundary conditions

$$u_{|\partial\Omega} = \bigl(\alpha(x)\, A(x, D^2 u)\nu \cdot \nu + \gamma(x)\partial_\nu u\bigr)_{|\partial\Omega} = 0, \tag{19}$$

where $\alpha, \gamma \in C(\partial\Omega)$, $\alpha, \gamma \geq 0$, $\alpha^2 + \gamma^2 > 0$ a.e. on $\partial\Omega$. Defining $B := \Delta^2$ with the domain $D(B) := \{u \in H^4(\Omega) : u_{|\partial\Omega} = (\alpha(x)\partial_\nu u^2 + \gamma(x)\partial_\nu u)_{|\partial\Omega} = 0\}$, and letting $\Gamma_\alpha := \{x \in \partial\Omega : \alpha(x) > 0\}$, the energy norm is now

$$\|u\|_B^2 \equiv \int_\Omega |D^2 u|^2 + \int_{\Gamma_\alpha} \frac{\gamma}{\alpha}(\partial_\nu u)^2 d\sigma \qquad (u \in H^2(\Omega),\ u_{|\partial\Omega} = 0,\ \partial_\nu u_{|\partial\Omega \setminus \Gamma_\alpha} = 0).$$

5 Numerical Example

As referred to in the introduction, the GM in Theorem 2 presents a class of methods wherein actual numerical realization is established by the choice of a suitable numerical method for the solution of the linear auxiliary problems. For instance, the latter method may be a FDM or FEM discretization. (The FEM for 4th order linear equations is highly developed [5,7,16], making its coupling to the GM as promising as has already been achieved in this way for 2nd order uniformly elliptic problems [8,9].) An important special case might be the use of one fixed grid for each linear problem, providing a suitably preconditioned nonlinear FEM iteration (cf. the Sobolev gradient technique [15]).

Here we consider the simplest case of realization to illustrate the theoretical convergence result: the auxiliary problems are solved exactly. (Besides simplicity, this actually realizes infinite-dimensional preconditioning.) The main purpose of the model problem is to give an example of the growth conditions involved in strong nonlinearity of the lower order terms.

In the sequel we will use notations

$$v^k := (v_1^k, v_2^k) \quad \text{and} \quad [v^k] := v_1^k + v_2^k \quad \text{for} \quad v = (v_1, v_2) \in \mathbf{R}^2,\ k \in \mathbf{N}^+.$$

We consider the following semilinear model problem:
$$\begin{cases} T(u) \equiv \Delta^2 u - \operatorname{div}(\nabla u)^3 + ue^{u^2} = g(x_1, x_2) & \text{in } \Omega = [0, \pi]^2 \subset \mathbf{R}^2 \\ u_{|\partial\Omega} = \partial_\nu^2 u_{|\partial\Omega} = 0 \end{cases} \quad (20)$$

with
$$g(x_1, x_2) = \frac{2 \sin x_1 \sin x_2}{3(2 - 0.249 \cos 2x_1)(2 - 0.249 \cos 2x_2)}.$$

Using the notations in (C1)-(C5) after (1), we now have $A(x, \Theta) = \Theta$, $f(x, \eta) = \eta^3$, $q(x, \xi) = \xi e^{\xi^2}$, hence $m = m' = 1$, $\kappa = 0$, $\beta = 3$, $p = 4$. The boundary condition (19) holds with $\alpha \equiv 1$, $\gamma \equiv 0$, hence $\Gamma_\alpha = \partial\Omega$. Therefore, defining $B = \Delta^2$ with this BC, the energy space is

$$H_B = \{u \in H^2(\Omega) : u_{|\partial\Omega} = 0\} \qquad \text{with} \qquad \|u\|_B^2 \equiv \int_\Omega |D^2 u|^2.$$

Then (20) is the formal Euler equation of the potential $J : H_B \to \mathbf{R}$,

$$J(u) := \int_\Omega \left(\frac{1}{2} |D^2 u|^2 + \frac{1}{4} [(\nabla u)^4] + \frac{1}{2} e^{u^2} - gu \right).$$

In order to apply Theorem 2, we need the values of the constants in (10). These can be estimated by suitable integration and Schwarz inequality, following [6,14] (for brevity the calculations are omitted), thus we obtain $K_2 = 2^{-1/2}$, $K_4 = 6^{1/2}$, $\lambda = 4$ and $K_\infty = 1.2$. The calculations are made up to accuracy 10^{-4}. We define the Fourier partial sum

$$\tilde{g}(x_1, x_2) = \sum_{\substack{k,l \text{ are odd} \\ k+l \leq 6}} a_{kl} \sin kx_1 \sin lx_2, \qquad a_{kl} = 2.3803 \cdot 4^{-(k+l)}$$

which fulfils $\|g - \tilde{g}\|_{L^2(\Omega)} \leq 0.0001$. The solution of $T(u) = \tilde{g}$ is denoted by $\tilde{u}$. Let $u_0 \equiv 0$. Then (10) and (11) yield

$$M_0 = 3.6903, \qquad \frac{2}{M_0 + m} = 0.4262, \qquad \frac{M_0 - m}{M_0 + m} = 0.5735.$$

Now we are in the position to apply the GM iteration (12)-(13). The main idea of realization is the following. In each step $u_n e^{u_n^2}$ is approximated by $T_{k_n}(u_n)$, where $T_{k_n}(\xi)$ is the k_n-th Taylor polynomial of ξe^{ξ^2}, chosen up to accuracy 10^{-4} for $|\xi| \leq \|u\|_\infty$. Hence, by induction, the sequences (z_n) and (u_n) consist of sine-polynomials (preserving this from $\tilde{g}$ and u_0), and the auxiliary equations (13) are elementary to solve. Namely, if $h(x_1, x_2) = \sum_{k,l=1}^s a_{kl} \sin kx_1 \sin lx_2$, then the solution of $\Delta^2 z = h$ with $z_{|\partial\Omega} = \partial_\nu^2 z_{|\partial\Omega} = 0$ is given by

$$z(x_1, x_2) = \sum_{k,l=1}^s \frac{a_{kl}}{(k^2 + l^2)^2} \sin kx_1 \sin lx_2.$$

The algorithm has been performed in MATLAB.
(The high-index almost zero coefficients were dropped within accuracy 10^{-4}, and the error was calculated from the residual.) The following table contains the error $e_n = \|u_n - \tilde{u}\|_{H_0^2}$ versus the number of steps n.

step n	1	2	3	4	5	6	7
error e_n	0.1173	0.0556	0.0275	0.0158	0.0118	0.0065	0.0037

8	9	10	11	12	13	14	15
0.0021	0.0014	0.0010	0.0008	0.0005	0.0003	0.0002	0.0001

References

1. Adams, R. A.: Sobolev spaces, Academic Press, New York-London, 1975.
2. Agmon, S.: Lectures on elliptic boundary value problems, D. van Nostrand Co., 1965.
3. Axelsson, O.: Iterative solution methods, Cambridge Univ. Press, 1994.
4. Axelsson, O., Chronopoulos, A. T.: On nonlinear generalized conjugate gradient method s, Numer. Math. 69 (1994), No. 1, 1-15.
5. Brezzi, F., Raviart, P. A.: Mixed Finite Element Method for 4th Order Elliptic Equations, in: Topics in Numerical Analysis III (ed.: J.Miller), Academic Press, 1998.
6. Burenkov, V. I., Gusakov, V. A.: On exact constants in Sobolev embeddings III., Proc. Stekl. Inst. Math. 204 (1993), No. 3., 57-67.
7. Ewing, R. E., Margenov, S. D., Vassilevski, P. S.: Preconditioning the biharmonic equation by multilevel iterations, Math. Balkanica (N. S.) 10 (1996), no. 1, 121–132.
8. Faragó, I., Karátson, J.: The gradient–finite element method for elliptic problems, to appear in Comp. Math. Appl.
9. Gajewski, H., Gröger, K., Zacharias, K.: Nichtlineare Operatorgleichungen und Operatordifferentialgleichungen, Akademie-Verlag, Berlin, 1974
10. Kantorovich, L. V., Akilov, G. P.: Functional Analysis, Pergamon Press, 1982.
11. Karátson, J.: The gradient method for non-differentiable operators in product Hilbert spaces and applications to elliptic systems of quasilinear differential equations, J. Appl. Anal., 3 (1997) No.2., 205-217.
12. Karátson, J.: Gradient method for non-uniformly convex functionals in Hilbert space, to appear in Pure Math. Appl.
13. Kelley, C. T.: Iterative methods for linear and nonlinear equations, Frontiers in Appl. Math., SIAM, Philadelphia, 1995.
14. Lions, J. L.: Quelques méthodes de résolution des problèmes aux limites non linéaires, Dunod, Gauthier-Villars, Paris, 1969.
15. Neuberger, J. W.: Sobolev gradients and differential equations, Lecture Notes in Math., No. 1670, Springer, 1997.
16. Temam, R.: Survey of the status of finite element methods for partial differential equations. Finite elements (Hampton, VA, 1986), 1–33, ICASE/NASA LaRC Ser., Springer, New York-Berlin, 1988.
17. Vainberg, M.: Variational Method and the Method of Monotone Operators in the Theory of Nonlinear Equations, J.Wiley, New York, 1973.

Numerical Techniques for the Recovery of an Unknown Dirichlet Data Function in Semilinear Parabolic Problems with Nonstandard Boundary Conditions

Roger Van Keer and Marián Slodička[*]

Department of Mathematical Analysis, Ghent University
Galglaan 2, B-9000 Gent, Belgium
{rvk,ms}@cage.rug.ac.be

Abstract. We study a semilinear parabolic partial differential equation of second order in a bounded domain $\Omega \subset \mathbb{R}^N$, with nonstandard boundary conditions (BCs) on a part Γ_{non} of the boundary $\partial \Omega$. Here, neither the solution nor the flux are prescribed pointwise. Instead, the total flux through Γ_{non} is given and the solution along Γ_{non} has to follow a prescribed shape function, apart from an additive (unknown) space-constant $\alpha(t)$.
Using the semidiscretization in time (so called Rothe's method) we provide a numerical scheme for the recovery of the unknown boundary data.

Keywords: nonlocal boundary condition, parameter identification, semilinear parabolic BVP

2000 MSC: 35K20, 35B30, 65N40

1 Introduction

Let $N \in \mathbb{N}$, $N \geq 2$. We consider a bounded open domain $\Omega \subset \mathbb{R}^N$ with a Lipschitz continuous boundary $\partial \Omega = \overline{\Gamma_{Dir}} \cup \overline{\Gamma_{Neu}} \cup \overline{\Gamma_{non}}$. The index "Dir" stands for the Dirichlet part, "Neu" for the Neumann part of $\partial \Omega$, while "non" is the part of the boundary with a nonlocal type BC. The three parts $\Gamma_{Dir}, \Gamma_{Neu}$ and Γ_{non} are supposed to be mutually disjoint. Moreover, we assume that Γ_{non} is non negligible and that Γ_{non} and Γ_{Dir} are not adjacent, i.e.,

$$|\Gamma_{non}| > 0, \qquad \overline{\Gamma_{non}} \cap \overline{\Gamma_{Dir}} = \emptyset. \tag{1}$$

We study the following semilinear parabolic partial differential equation of second order

$$p(t, \mathbf{x}) \frac{\partial u(t, \mathbf{x})}{\partial t} - \nabla \cdot (K(t, \mathbf{x}) \nabla u(t, \mathbf{x})) = f(t, \mathbf{x}, u(t, \mathbf{x})) \quad \text{in } (0, T) \times \Omega. \tag{2}$$

---
[*] This work was supported by the VEO-project no. 011 VO 697.

We consider nonstandard boundary conditions on Γ_{non} of the type

$$u(t,\mathbf{x}) = g_{non}(t,\mathbf{x}) + \alpha(t) \quad \text{in } (0,T) \times \Gamma_{non}$$
$$-\int_{\Gamma_{non}} K(t,\mathbf{x})\nabla u(t,\mathbf{x}) \cdot \boldsymbol{\nu} \, d\gamma = s(t), \quad \text{in } (0,T). \quad (3)$$

Here, the time dependent enforced total flux $s(t)$ through Γ_{non} is given and the solution along Γ_{non} has to preserve the prescribed shape g_{non}, apart from an additive (*unknown*) time-depending degree of freedom $\alpha(t)$, which has to be determined as a part of the problem.

There are standard pointwise boundary conditions on Γ_{Dir} and Γ_{Neu}:

$$u(t,\mathbf{x}) = g_{Dir}(t,\mathbf{x}) \quad \text{in } (0,T) \times \Gamma_{Dir}$$
$$-K(t,\mathbf{x})\nabla u(t,\mathbf{x}) \cdot \boldsymbol{\nu} - g_{Rob}(t,\mathbf{x})u(t,\mathbf{x}) = g_{Neu}(t,\mathbf{x}) \quad \text{in } (0,T) \times \Gamma_{Neu}. \quad (4)$$

The initial condition is given as

$$u(0,\mathbf{x}) = u_0(\mathbf{x}) \in H^1(\Omega) \qquad \text{in } \Omega. \quad (5)$$

We suppose that for the functions g_{Dir} on $(0,T) \times \Gamma_{Dir}$ and g_{non} on $(0,T) \times \Gamma_{Neu}$ there exists a prolongation $\tilde{g}$ of these functions to the whole domain Ω such that

$$\tilde{g} \in L_2\left((0,T)H^1(\Omega)\right). \quad (6)$$

With respect to this assumption one can easily see that in (1) Γ_{Dir} and Γ_{non} had to be required to be non-adjacent.

The right-hand side f is supposed to be globally Lipschitz continuous in all variables and the data functions g_{Neu}, g_{Rob}, p, K obey

$$\begin{aligned} 0 \leq g_{Rob} \leq C, & \quad \text{a.e. in } (0,T) \times \Gamma_{Neu} \\ g_{Neu} \in L_2\left((0,T), L_2(\Gamma_{Neu})\right) & \\ 0 < C_0 < p(t,\mathbf{x}), K(t,\mathbf{x}) < C & \quad \text{a.e. in } (0,T) \times \Omega. \end{aligned} \quad (7)$$

This type of initial boundary value problems (IBVPs) arises in the determination of the magnetic properties of materials used in electric machinery. In practice, the original problems are highly nonlinear in that memory properties (hysteresis behaviour) of the material must be taken into account. The nonlocal BCs (3) considered in the IBVP (2)-(5) correspond to the situation when the average flux in the lamination is enforced, from which the magnetic field strength at the surface of the lamination must be derived. Such models have been studied e.g. by Van Keer, Dupré & Melkebeek in [6]. In that paper the authors suggested a modified finite element-finite difference scheme for the numerical approximation of the unknown u and α. The existence and uniqueness of the exact solution has not been discussed there. To deal with the nonlocal BC in a variational setting in [6] a space of trial and test functions has been considered with constant traces on Γ_{non}. Therefore, the standard FE packages could not be used for the numerical computations.

In this paper we prove the uniqueness of the solution to the IBVP (2)-(5) and provide a numerical method for the recovery of the unknown boundary data

α. First, we give the variational formulation of the problem (2)-(5). We apply Rothe's method for the time discretization, see Kačur [2] or Rektorys [3]. We have to solve a recurrent system of elliptic BVPs at each successive time point t_i of a suitable time partitioning. We apply the ideas from Slodička & Van Keer [5] to obtain a weak solution $u_i \approx u(t_i)$ at each time step t_i and to determine the unknown value $\alpha_i \approx \alpha(t_i)$.

2 Variational Formulation, Uniqueness

We denote by $(w,z)_M$ the standard L_2-scalar product of real or vector-valued functions w and z on a set M, i.e., $(w,z)_M = \int_M wz$. The corresponding norm is denoted by $\|w\|_{0,M} = \sqrt{(w,w)_M}$. The first-order Sobolev space $H^1(\Omega)$ is equipped with the usual norm $\|\cdot\|_{1,\Omega}$,

$$\|w\|_{1,\Omega}^2 = (w,w)_\Omega + (\nabla w, \nabla w)_\Omega = \|w\|_{0,\Omega}^2 + |w|_{1,\Omega}^2.$$

We define the following space V of test functions (in fact a Hilbert space)

$$V = \{\varphi \in H^1(\Omega); \varphi = 0 \text{ on } \Gamma_{Dir}, \varphi = const \text{ on } \Gamma_{non}\},$$

which is endowed with the induced norm $\|\cdot\|_{1,\Omega}$ from $H^1(\Omega)$.

The variational formulation of the IBVP (2)-(5) reads as follows:

Problem 1. Find a couple (u, α) obeying

1. $u \in C\left([0,T], L_2(\Omega)\right) \cap L_2\left((0,T)H^1(\Omega)\right)$,
2. $\frac{\partial u}{\partial t} \in L_2\left((0,T), L_2(\Omega)\right)$,
3. $u = g_{Dir}$ on $(0,T) \times \Gamma_{Dir}$,
4. $u - g_{non} = \alpha \in L_2\left((0,T)\right)$ on Γ_{non},
5. $u(0) = u_0$ in Ω,

such that for all $\varphi \in V$ and for almost all $t \in [0,T]$ holds

$$\left(p(t)\frac{\partial u(t)}{\partial t}, \varphi\right)_\Omega + (K(t)\nabla u(t), \nabla \varphi)_\Omega + (g_{Rob}(t)u(t), \varphi)_{\Gamma_{Neu}} \quad (8)$$
$$= (f(t, u(t)), \varphi)_\Omega - (g_{Neu}(t), \varphi)_{\Gamma_{Neu}} - s(t)\varphi|_{\Gamma_{non}}.$$

Now, we prove the uniqueness of the solution to the IBVP 1.

Theorem 1. *Let (1), (5), (6) and (7) be satisfied and let f be globally Lipschitz continuous in all variables. Then the IBVP 1 admits at most one weak solution.*

Proof. Suppose that there exist two solutions (u_α, α) and (u_β, β) to the IBVP 1. Subtract the identity (8) for u_α from the corresponding identity for u_β, take $\varphi = (u_\alpha - u_\beta)(t) \in V$ and get

$$\left(p(t)\frac{\partial(u_\alpha - u_\beta)(t)}{\partial t}, (u_\alpha - u_\beta)(t)\right)_\Omega + (K(t)\nabla(u_\alpha - u_\beta)(t), \nabla(u_\alpha - u_\beta)(t))_\Omega$$
$$+ (g_{Rob}(t)(u_\alpha - u_\beta)(t), (u_\alpha - u_\beta)(t))_{\Gamma_{Neu}}$$
$$= (f(t, u_\alpha(t)) - f(t, u_\beta(t)), (u_\alpha - u_\beta)(t))_\Omega.$$

We denote by C a generic positive constant. Integrating this equality over $t \in (0,s)$, for any $s \in (0,T)$, and taking into account the assumption (7) and the Lipschitz continuity of the right-hand side f, we arrive at

$$\|(u_\alpha - u_\beta)(s)\|_{0,\Omega}^2 + \int_0^s |(u_\alpha - u_\beta)(t)|_{1,\Omega}^2 \, dt \leq C \int_0^s \|(u_\alpha - u_\beta)(t)\|_{0,\Omega}^2 \, dt.$$

From Gronwall's lemma we conclude that $u_\alpha = u_\beta$ in the space $C([0,T], L_2(\Omega)) \cap L_2((0,T)H^1(\Omega))$. For α and β we successively deduce that

$$\begin{aligned}
|\alpha(t) - \beta(t)| &= \frac{1}{|\Gamma_{non}|} \int_{\Gamma_{non}} |\alpha(t) - \beta(t)| = \frac{1}{|\Gamma_{non}|} \int_{\Gamma_{non}} |u_\alpha(t) - u_\beta(t)| \\
&\leq C \int_{\partial\Omega} |u_\alpha(t) - u_\beta(t)| \leq C \|u_\alpha(t) - u_\beta(t)\|_{0,\partial\Omega} \\
&\leq C \|u_\alpha(t) - u_\beta(t)\|_{1,\Omega},
\end{aligned}$$

where in the last but one step we used the Cauchy-Schwarz inequality in $L_2(\partial\Omega)$ and in the last step we invoked the trace inequality. Integrating the inequality with respect to the time variable we get

$$\int_0^T |\alpha(t) - \beta(t)|^2 \, dt \leq C \int_0^T \|u_\alpha(t) - u_\beta(t)\|_{1,\Omega}^2 \, dt.$$

Recalling that $u_\alpha = u_\beta$ in the space $L_2((0,T)H^1(\Omega))$, we conclude that $\alpha(t) = \beta(t)$ a.e. in $(0,T)$. □

3 Time Discretization

We divide the time interval $[0,T]$ into $n \in \mathbb{N}$ equal subintervals (t_{i-1}, t_i), where $t_i = i\tau$, with the time step $\tau = \frac{T}{n}$. We introduce the following notation for any abstract function z on $[0,T]$:

$$z_i = z(t_i), \qquad \delta z_i = \frac{z_i - z_{i-1}}{\tau}.$$

The application of the usual Rothe method to the IBVP 1 complicated by the nonlocal BC on Γ_{non}. Here, we apply the ideas from Slodička & Van Keer [5] for elliptic problems. We consider the following *linear* elliptic BVP with nonlocal BCs at each time point t_i, $i = 1, 2, \ldots$.

Problem 2. Find a couple $(u_i, \alpha_i) \in H^1(\Omega) \times \mathbb{R}$ obeying

1. $u_i = g_{Dir_i}$ on Γ_{Dir}
2. $u_i - g_{non_i} = \alpha_i$ on Γ_{non}

such that

$$\begin{aligned}
(p_i \delta u_i, \varphi)_\Omega + (K_i \nabla u_i, \nabla \varphi)_\Omega + (g_{Rob_i} u_i, \varphi)_{\Gamma_{Neu}} \\
= (f(t_i, u_{i-1}), \varphi)_\Omega - (g_{Neu_i}, \varphi)_{\Gamma_{Neu}} - s_i \varphi|_{\Gamma_{non}},
\end{aligned} \qquad \varphi \in V. \qquad (9)$$

The existence of a weak solution $(u_i, \alpha_i) \approx (u(t_i, \boldsymbol{x}), \alpha(t_i))$ at each t_i is shown below by invoking some arguments from [5].

Theorem 2. *Suppose that (1), (6), (7) hold and assume that $u_0 \in L_2(\Omega)$. Then, there exists a unique solution (u_i, α_i) to the BVP 2 for any $i = 1, \ldots, n$.*

Proof. We introduce a subspace V_0 of V by

$$V_0 = \{\varphi \in H^1(\Omega); \; \varphi = 0 \text{ on } \Gamma_{Dir} \cup \Gamma_{non}\}.$$

We define the bilinear form $a : V \times V \to \mathbb{R}$ by means of

$$a(w, \varphi) = \left(\frac{p_i w}{\tau}, \varphi\right)_\Omega + (K_i \nabla w, \nabla \varphi)_\Omega + (g_{Rob_i} w, \varphi)_{\Gamma_{Neu}}$$

and the linear functional F on V by

$$\langle F, \varphi \rangle = (f(t_i, u_{i-1}), \varphi)_\Omega - (g_{Neu_i}, \varphi)_{\Gamma_{Neu}} + \left(\frac{p_i u_{i-1}}{\tau}, \varphi\right)_\Omega.$$

We consider two auxiliary problems at each time step t_i. The first one takes into account the source term and the nonhomogeneous BCs, i.e.:
Find $v_i \in H^1(\Omega)$ obeying

$$\begin{array}{ll} v_i = g_{Dir_i} \text{ on } \Gamma_{Dir}, & v_i = g_{non_i} \text{ on } \Gamma_{non}, \\ a(v_i, \varphi) = \langle F, \varphi \rangle, & \forall \varphi \in V_0. \end{array} \quad (10)$$

In the second problem the right-hand side and the Dirichlet data on Γ_{Dir} are taken to be zero, while the trace of the solution has to take the constant value one throughout Γ_{non}:
Find $z_i \in H^1(\Omega)$ obeying

$$\begin{array}{ll} z_i = 0 \text{ on } \Gamma_{Dir}, & z_i = 1 \text{ on } \Gamma_{non}, \\ a(z_i, \varphi) = 0, & \forall \varphi \in V_0. \end{array} \quad (11)$$

The Lax-Milgram lemma implies the existence and uniqueness of weak solutions v_i and z_i, for all $i = 1, \ldots, n$. Applying the principle of superposition, the function $u_{\alpha_i} \equiv v_i + \alpha_i z_i$, with $\alpha_i \in \mathbb{R}$, satisfies the BVP

$$\begin{array}{ll} u_{\alpha_i} = g_{Dir_i} \text{ on } \Gamma_{Dir}, & u_{\alpha_i} = g_{non_i} + \alpha_i \text{ on } \Gamma_{non}, \\ a(u_{\alpha_i}, \varphi) = \langle F, \varphi \rangle, & \forall \varphi \in V_0. \end{array} \quad (12)$$

Now, we have to find such an α_i for which the total flux of u_{α_i} through Γ_{non} is just s_i. To this end, similarly as in [5], we introduce so called "total flux functionals" on V

$$\begin{aligned} \langle \tilde{G}(z_i), \varphi \rangle &= -a(z_i, \varphi), \\ \langle \tilde{G}(v_i), \varphi \rangle &= -a(v_i, \varphi) + \langle F, \varphi \rangle, \\ \langle \tilde{G}(u_{\alpha_i}), \varphi \rangle &= -a(u_{\alpha_i}, \varphi) + \langle F, \varphi \rangle, \end{aligned} \quad (13)$$

representing the total flux of v_i, z_i and u_{α_i} through Γ_{non}. It follows that the constant α_i must fulfill

$$\langle \tilde{G}(u_{\alpha_i}), \tilde{1} \rangle \equiv \langle \tilde{G}(v_i), \tilde{1} \rangle + \alpha_i \langle \tilde{G}(z_i), \tilde{1} \rangle = s_i,$$

where $\tilde{1}$ is any smooth function satisfying

$$\tilde{1} = \begin{cases} 1 & \text{on } \Gamma_{non} \\ 0 & \text{on } \Gamma_{Dir}. \end{cases} \quad (14)$$

Therefore,

$$\alpha_i = \frac{s_i - \langle \tilde{G}(v_i), \tilde{1} \rangle}{\langle \tilde{G}(z_i), \tilde{1} \rangle}. \quad (15)$$

Here, $\langle \tilde{G}(z_i), \tilde{1} \rangle \neq 0$. Otherwise we would get a contradiction with the trace of z_i on Γ_{non} and the uniqueness of the solution to the BVP (see [4])

$$\begin{aligned} \nabla \cdot (-K_i \nabla w) + \frac{p_i w}{\tau} &= 0 && \text{in } \Omega \\ w &= 0 && \text{on } \Gamma_{Dir} \\ -K_i \nabla w \cdot \boldsymbol{\nu} - g_{Rob_i} w &= 0 && \text{on } \Gamma_{Neu} \\ w &= const && \text{on } \Gamma_{non} \\ \int_{\Gamma_{non}} -K_i \nabla w \cdot \boldsymbol{\nu} &= 0. \end{aligned}$$

□

It was shown that the couple $(u_i, \alpha_i) \approx (u(t_i, \boldsymbol{x}), \alpha(t_i))$ can be constructed from the solution of two auxiliary BVPs with standard (local) BCs. In practice, the auxiliary elliptic problems must be solved numerically.

4 Numerical Experiments

In the previous section we described the numerical scheme for the time discretization. For the space discretization of the two elliptic auxiliary problems at each time step, we use a mixed nonconforming finite element method, where the usual nonconforming basis on a triangle has been enriched by a bubble (polynomial of third order vanishing at the boundary). For details see Arnold and Brezzi [1].

4.1 Example 1

The first example is a linear parabolic problem. Let Ω be the rectangular domain $\Omega = (0, 0.5) \times (0, 0.02)$. Its boundary is splitted into three parts: Γ_{Dir} (right), Γ_{Neu} (top and bottom) and Γ_{non} (left part of $\partial \Omega$). We consider the following IBVP:

$$\begin{aligned} 10^{-3} \frac{\partial u}{\partial t} - \Delta u &= 0 && \text{in } (0,1) \times \Omega \\ u(t) &= 10^3 \sin(2\pi t) && \text{in } (0,1) \times \Gamma_{Dir} \\ -\nabla u \cdot \boldsymbol{\nu} &= 0 && \text{in } (0,1) \times \Gamma_{Neu} \\ u(t) &= \alpha(t) && \text{in } (0,1) \times \Gamma_{non} \\ -\int_{\Gamma_{non}} \nabla u(t) \cdot \boldsymbol{\nu} \, d\gamma &= 2\pi \cos(2\pi t) && \text{in } (0,1) \times \Gamma_{non} \\ u(0) &= 0 && \text{in } \Omega. \end{aligned}$$

We have used the time step $\tau = 0.005$ and a fixed uniform triangular mesh consisting of 200 triangles. In Figure 1, the function $\alpha(t)$ (i.e., the space-constant unknown value on Γ_{non}) is plotted versus the function $s(t)$. The loop in Figure 1

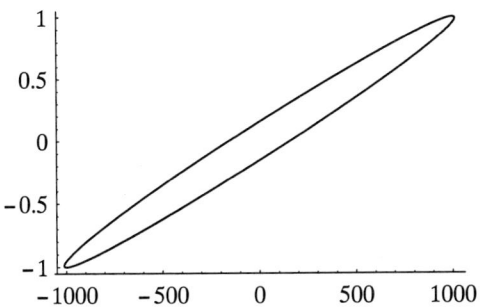

Fig. 1. Example 1: The behaviour of $\alpha(t)$ (at the x-axes) versus $s(t) = 2\pi \int_0^t \cos(2\pi s) \, ds$ (on the y-axes) for $t \in [0,1]$

is a consequence of the periodicity of the boundary conditions in the problem setting. Such curves can be obtained in the computation of the electromagnetic losses in a lamination of an electric machine, based upon the Maxwell equations. The domain Ω can be seen as the cross section of the yoke. The surface enclosed by an (α, s)-loop is a measure of the electromagnetic losses over one time period. In fact, the practical problem setting is nonlinear and should include also the hysteresis behaviour of the material (cf. [6]).

4.2 Example 2

The second example is a semilinear parabolic problem with a nonlinear right-hand side. We consider the unit circle as the domain Ω. The boundary $\partial \Omega$ is splitted into two halves by the x-axis. The top part is Γ_{non} and the bottom part is Γ_{Neu}. We consider the following evolution IBVP for $(u(t,x,y), \alpha(t))$:

$$\frac{\partial u}{\partial t} - \Delta u = u^2 - v^2 + \frac{\partial v}{\partial t} - \Delta v \quad \text{in } (0,1) \times \Omega$$
$$u(t) = v(t) - t^{0.75} + \alpha(t) \quad \text{in } (0,1) \times \Gamma_{non}$$
$$-\int_{\Gamma_{non}} \nabla u \cdot \boldsymbol{\nu} \, d\gamma = -\int_{\Gamma_{non}} \nabla v \cdot \boldsymbol{\nu} \, d\gamma \quad \text{in } (0,1) \times \Gamma_{non}$$
$$-\nabla u(t,\mathbf{x}) \cdot \boldsymbol{\nu} - u(t,\mathbf{x}) = -\nabla v(t,\mathbf{x}) \cdot \boldsymbol{\nu} - v(t,\mathbf{x}) \quad \text{in } (0,1) \times \Gamma_{Neu}.$$
$$u(0) = v(0) \quad \text{in } \Omega,$$

where

$$v(t,x,y) = tx \cos(\pi y) + y \sin(\pi t x) + t^2 x^2.$$

The exact solution is
$$u(t,x,y) = v(t,x,y)$$
$$\alpha(t) = t^{0.75}.$$

We have chosen the time step $\tau = 0.005$ and an unstructured mesh consisting of 11872 triangles. The evolution of the absolute errors for u_i and α_i at subsequent time points in the time interval $[0,1]$ is depicted in Figure 2.

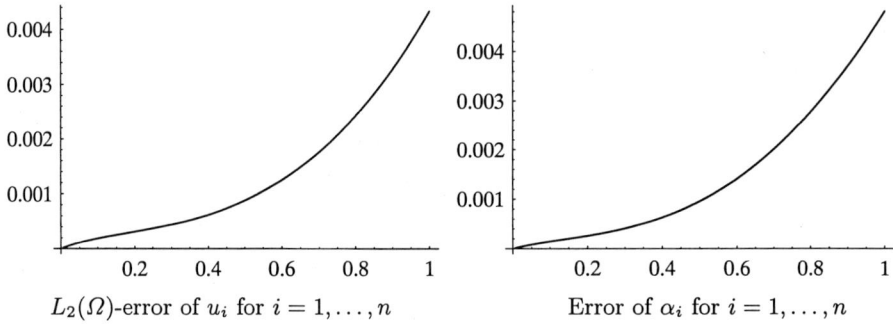

Fig. 2. Example 2: Absolute errors

References

1. D. N. Arnold and F. Brezzi. Mixed and nonconforming finite element methods: implementation, postprocessing and error estimates. *M²AN Math. Modelling and Numer. Anal.*, 19(1):7–32, 1985.
2. J. Kačur. *Method of Rothe in evolution equations*, volume 80 of *Teubner Texte zur Mathematik*. Teubner, Leipzig, 1985.
3. K. Rektorys. *The method of discretization in time and partial differential equations*. Reidel Publishing Company, Dordrecht-Boston-London, 1982.
4. M. Slodička and H. De Schepper. Modelling of pressure from discharges at active wells by soil venting facilities. In I. Troch and F. Breitenecker, editors, *IMACS Symposium on MATHEMATICAL MODELLING*, pages 103–110, Vienna University of Technology, Austria, 2000. ARGESIM Report No. 15. ISBN 3-901608-15-X.
5. M. Slodička and Roger Van Keer. A nonlinear elliptic equation with a nonlocal boundary condition solved by linearization. Preprint 2000/no. 1, 2000. Group of Numerical Functional Analysis and Mathematical Modelling, Department of Mathematical Analysis, Faculty of Engineering, Ghent University, Ghent, Belgium.
6. R. Van Keer, L. Dupré, and J. Melkebeek. Computational methods for the evaluation of the electromagnetic losses in electrical machinery. *Archives of Computational Mathods in Engineering*, 5(4):385–443, 1999.

A Generalized GMRES Iterative Method

David R. Kincaid[1], Jen-Yuan Chen[2], and David M. Young[1]

[1] University of Texas at Austin
Austin, TX 78712, USA
{kincaid,young}@cs.utexas.edu
http://www.cs.utexas.edu/users/{kincaid,young}
[2] I-Shou University
Ta-Hsu, Kaohsiung 840, Taiwan
jchen@isu.edu.tw

Abstract. We describe a generalization of the GMRES iterative method in which the residual vector is no longer minimized in the 2-norm but in a C-norm, where C is a symmetric positive definite matrix. The resulting iterative method call GGMRES is derived in detail and the minimizing property is proven.

1 Introduction

We are interested in iterative methods for solving systems of linear equations of the form $Au = b$, where A is a large sparse nonsingular matrix. When A is symmetric positive definite, conjugate-gradient-type methods are often used and are fairly well understood. On the other hand, when A is nonsymmetric, the choice of iterative method is much more difficult.

We consider a method similar to the Generalized Minimum Residual (GMRES) method that was introduced by Saad and Schultz [9] for solving a linear system where the coefficient matrix A is nonsymmetric. We describe a generalization of the GMRES method called "GGMRES." In this procedure, the residual vector is minimized in a C-norm rather than the 2-norm for some symmetric positive definite (SPD) matrix C. When $C = I$, the GGMRES method reduces to the GMRES method. If one is able to find a matrix C so that CA is symmetric and nonsingular, then the GGMRES method simplifies (an upper Hessenberg matrix reduces to a tridiagonal matrix) and short recurrence relations can be used in the GGMRES algorithm. Additional details can be found in [2] and [3].

First, one chooses an SPD matrix C and an initial approximation $u^{(0)}$ to the true solution, $\bar{u} = A^{-1}b$. Then starting with the initial residual vector $r^{(0)} = b - Au^{(0)}$, one generates a sequence of vectors $w^{(0)}, w^{(1)}, \ldots, w^{(n-1)}$, which are mutually C-orthogonal. The iterates $u^{(1)}, u^{(2)}, \ldots, u^{(n)}$ are chosen so that for each n, the error vector $u^{(n)} - u^{(0)}$ is a linear combination of the vectors $w^{(0)}, w^{(1)}, \ldots, w^{(n-1)}$ and so that the C-norm of the n-th residual vector $r^{(n)} = b - Au^{(n)}$ is minimized. This can be done in a stable manner by the use of Givens rotations applied to a related linear system, which involves an upper Hessenberg matrix. If $C = I$, the procedure reduces to the standard GMRES method. If CA is symmetric, then the vectors $\{w^{(i)}\}_{i=0}^{n-1}$ can be determined by short recurrence relations.

2 Generalized GMRES (GGMRES) Method

First, we choose an SPD matrix C and we generate a C-orthonormal basis $\{w^{(i)}\}_{i=0}^{n-1}$ for the Krylov space $\mathcal{K}_n(r^{(0)}, A)$ by the simplified Arnoldi procedure as following with $u^{(0)}$ arbitrary.

$$\widetilde{w}^{(0)} = r^{(0)} = b - Au^{(0)}$$
$$\sigma_0 = \|\widetilde{w}^{(0)}\|_C$$
$$w^{(0)} = \widetilde{w}^{(0)}/\sigma_0$$
for $i = 1, 2, \ldots, n-1$
  for $j = 0, 1, \ldots, i-1$
    $$b_{i,j} = \langle w^{(i-1)}, w^{(j)}\rangle_{CA}/\|w^{(j)}\|_C^2$$
  end for
  $$\widetilde{w}^{(i)} = Aw^{(i-1)} - \sum_{j=0}^{i-1} b_{i,j} w^{(j)}$$
  $$\sigma_i = \|\widetilde{w}^{(i)}\|_C$$
  $$w^{(i)} = \widetilde{w}^{(i)}/\sigma_i$$
end for

Here the C-norm of a vector x is given by $\|x\|_C^2 = \langle x, x\rangle_C = x^T C x$ using the C-inner product $\langle x, y\rangle_C = \langle x, Cy\rangle = x^T Cy$.

The above procedure is called Phase I and the following two basic relations are obtained

$$AW_{n-1} = W_n H_n \tag{1}$$
$$W_n^T C W_n = I, \tag{2}$$

where

$$W_i = [w^{(0)} \quad w^{(1)} \quad w^{(2)} \quad \cdots \quad w^{(i)}] \tag{3}$$

$$H_n = \begin{bmatrix} b_{10} & b_{20} & b_{30} & \cdots & & b_{n0} \\ \sigma_1 & b_{21} & b_{31} & \ddots & & \vdots \\ 0 & \ddots & \ddots & \ddots & & \vdots \\ & \ddots & \ddots & \ddots & & b_{n,n-2} \\ 0 & & & & \sigma_{n-1} & b_{n,n-1} \\ 0 & \cdots & 0 & 0 & & \sigma_n \end{bmatrix}_{(n+1)\times n}. \tag{4}$$

Next in Phase II, we obtain

$$c^{(n)} = [c_0^{(n)} \quad c_1^{(n)} \quad c_2^{(n)} \quad \cdots \quad c_{n-1}^{(n)}]^T$$
$$u^{(n)} = u^{(0)} + W_{n-1} c^{(n)}.$$

We now show how $c^{(n)}$ is determined. Evidently, we have

$$\begin{aligned} r^{(n)} &= b - Au^{(n)} \\ &= r^{(0)} - AW_{n-1}c^{(n)} \\ &= r^{(0)} - W_n H_n c^{(n)} \\ &= W_n(q - H_n c^{(n)}), \end{aligned} \tag{5}$$

using Equation (1) and letting

$$r^{(0)} = W_n q$$
$$q = \sigma_0 e^{(n+1,1)} = \sigma_0 [1 \ 0 \ 0 \ \cdots \ 0]_{n+1}^T \in \mathbb{R}^{n+1}$$
$$\sigma_0 = \|r^{(0)}\|_C.$$

Then we assume that we can applied $n-1$ Givens rotations $Q_1, Q_2, \ldots, Q_{n-1}$ to the Hessenberg matrix H_n

$$Q_{n-1}\cdots Q_2 Q_1 H_n = \begin{bmatrix} \alpha_1 & \times & \times & \times & \cdots & \times \\ 0 & \alpha_2 & \ddots & \ddots & \ddots & \vdots \\ \vdots & \ddots & \ddots & \ddots & \ddots & \times \\ \vdots & \ddots & \ddots & \ddots & \ddots & \times \\ \vdots & \cdots & \ddots & \ddots & \ddots & \times \\ 0 & \cdots & \cdots & \cdots & 0 & \widetilde{b}_{n,n-1} \\ 0 & \cdots & \cdots & \cdots & \cdots & \sigma_n \end{bmatrix}_{(n+1)\times n}.$$

Here $\widetilde{b}_{n,n-1}$ is the modified entry in this position of the matrix as a result of applying the Givens rotation matrices $Q_1, Q_2, \ldots, Q_{n-1}$. Next we use the n-th Givens rotation matrix

$$Q_n = \begin{bmatrix} 1 & & & & & \\ & 1 & & & & \\ & & \ddots & & & \\ & & & 1 & & \\ & & & & c_n & -s_n \\ & & & & s_n & c_n \end{bmatrix}_{(n+1)\times(n+1)} \tag{6}$$

where $c_n = \widetilde{b}_{n,n-1}/\alpha_n$, $s_n = -\sigma_n/\alpha_n$, $\sigma_n = \|\widetilde{w}^{(n)}\|_C$, and $\alpha_n = [\widetilde{b}_{n,n-1}^2 + \sigma_n^2]^{\frac{1}{2}}$. Since $\sigma_n = \|\widetilde{w}^{(n)}\|_C \neq 0$, then $\alpha_n \neq 0$.

Using Equation (5), we consider

$$H_n c^{(n)} = q.$$

If $Q = Q_n Q_{n-1} \cdots Q_2 Q_1$ is the product of n Givens rotations, we let

$$R_n = QH_n,$$

where R_n is an $(n+1) \times n$ upper triangular matrix with a zero last row

$$R_n = \begin{bmatrix} r_{11} & r_{12} & r_{13} & \cdots & r_{1n} \\ 0 & r_{22} & r_{23} & \ddots & \vdots \\ 0 & \ddots & \ddots & \ddots & \vdots \\ 0 & \ddots & \ddots & \ddots & r_{n-1,n} \\ 0 & \cdots & 0 & 0 & r_{n,n} \\ 0 & \cdots & 0 & 0 & 0 \end{bmatrix}_{(n+1) \times n}.$$

Letting
$$Qq = z = [z_0 \quad z_1 \quad \cdots \quad z_{n-1} \quad z_n]^T,$$
we then solve the first n equations of
$$R_n c^{(n)} = z,$$
for $c^{(n)}$. Moreover, it follows that if z_n is the last element of the vector Qq, then $z_n = \sigma_0 s_1 s_2 \ldots s_n$. Thus, we obtain $\|r^{(n)}\|_C$ for each iteration without having to compute the inner product directly. Finally, we compute $u^{(n)}$ by
$$u^{(n)} = u^{(0)} + W_{n-1} c^{(n)}.$$

We now proof the following theorem.

Theorem 1. *Using the GGMRES method, the approximate solution $u^{(n)}$ for the exact solution $\bar{u}$ of the linear systems $Au = b$ minimizes the C-norm of the nth residual vector $r^{(n)} = b - Au^{(n)}$; namely,*
$$\min_{u^{(n)} - u^{(0)} \in \mathcal{K}_n(r^{(0)}, A)} \|r^{(n)}\|_C^2.$$

Proof. Using Equations (5) and (2), we have
$$\begin{aligned} \|r^{(n)}\|_C^2 &= \langle Cr^{(n)}, r^{(n)} \rangle = \langle CW_n(q - H_n c^{(n)}), W_n(q - H_n c^{(n)}) \rangle \\ &= (q - H_n c^{(n)})^T W_n^T CW_n(q - H_n c^{(n)}) \\ &= \|q - H_n c^{(n)}\|_2^2, \end{aligned}$$

Thus, we have
$$\min_{u^{(n)} - u^{(0)} \in \mathcal{K}_n(r^{(0)}, A)} \|r^{(n)}\|_C^2 = \min_{c^{(n)} \in \mathbb{R}^n} \|q - H_n c^{(n)}\|_2^2.$$

Since H_n can be factored as $Q^T R_n$ where the unitary matrix Q is a product of several Givens rotations and R_n is a $(n+1) \times n$ upper triangular matrix, then we have
$$\begin{aligned} \|q - H_n c^{(n)}\|_2^2 &= \|Q^T(Qq - R_n c^{(n)})\|_2^2 \\ &= (Qq - R_n c^{(n)})^T QQ^T (Qq - R_n c^{(n)}) \\ &= \|Qq - R_n c^{(n)}\|_2^2, \end{aligned}$$

using $Q^T Q = I$. Since $Qq = z = [z_0 \ z_1 \ \cdots \ z_{n-1} \ z_n]^T$, we have

$$\min_{u^{(n)} - u^{(0)} \in \mathbb{R}^n} \|r^{(n)}\|_C^2 = \min_{c^{(n)} \in \mathbb{R}^n} \|R_n c^{(n)} - z\|_2^2.$$

Since the last row of the matrix R_n is zero, it can be written in the form

$$R_n = \begin{bmatrix} \tilde{R}_n \\ 0 \ \cdots \ 0 \end{bmatrix}.$$

We can write $\|R_n c^{(n)} - z\|_C^2$ as

$$\|\tilde{R}_n c^{(n)} - [z_0 \ z_1 \ \cdots \ z_{n-1}]^T\|_2^2 + |z_n|^2.$$

If we choose $c^{(n)}$ such that $\tilde{R}_n c^{(n)} = [z_0 \ z_1 \ \cdots \ z_{n-1}]^T$, then $\|r^{(n)}\|_C^2$ will be minimized and $\|r^{(n)}\|_C = |z_n|$. Thus, $u^{(n)} = u^{(0)} + W_{n-1} c^{(n)}$ is the approximate solution to the exact solution $\bar{u}$ of $Au = b$, which minimizes $\|r^{(n)}\|_C^2$.

Example 1 (Case $n = 3$). We now illustrate the algorithm for a small system. Using Equations (1), (3), and (4), we have

$$AW_2 = W_3 H_3$$
$$A[w^{(0)} \ w^{(1)} \ w^{(2)}] = [w^{(0)} \ w^{(1)} \ w^{(2)} \ w^{(3)}] H_3,$$

where

$$H_3 = \begin{bmatrix} b_{10} & b_{20} & b_{30} \\ \sigma_1 & b_{21} & b_{31} \\ 0 & \sigma_2 & b_{32} \\ 0 & 0 & \sigma_3 \end{bmatrix}_{4 \times 3}.$$

To get the least squares solution of

$$H_3 c^{(3)} = q$$

$$\begin{bmatrix} b_{10} & b_{20} & b_{30} \\ \sigma_1 & b_{21} & b_{31} \\ 0 & \sigma_2 & b_{32} \\ 0 & 0 & \sigma_3 \end{bmatrix} \begin{bmatrix} c_0^{(3)} \\ c_1^{(3)} \\ c_2^{(3)} \end{bmatrix} = \begin{bmatrix} \sigma_0 \\ 0 \\ 0 \\ 0 \end{bmatrix},$$

we apply two Givens rotations to both sides and obtain

$$R_3 c^{(3)} = z$$

$$\begin{bmatrix} r_{11} & \times & \times \\ 0 & r_{22} & \times \\ 0 & 0 & r_{33} \\ 0 & 0 & 0 \end{bmatrix} \begin{bmatrix} c_0^{(3)} \\ c_1^{(3)} \\ c_2^{(3)} \end{bmatrix} = \begin{bmatrix} z_0 \\ z_1 \\ z_2 \\ z_3 \end{bmatrix}.$$

We then solve the first three equations for $c_0^{(3)}, c_1^{(3)}$, and $c_2^{(3)}$. Finally, we have

$$u^{(3)} = u^{(0)} + c_0^{(3)} w^{(0)} + c_1^{(3)} w^{(1)} + c_2^{(3)} w^{(2)}.$$

Example 2 (Case $n = 3$). We next show that $z_3 = \sigma_0 s_1 s_2 s_3$. Clearly from Equation (6), we have

$$Q = Q_3 Q_2 Q_1 = \begin{bmatrix} 1 & & & \\ & 1 & & \\ & & c_3 & -s_3 \\ & & s_3 & c_3 \end{bmatrix} \begin{bmatrix} 1 & & & \\ & c_2 & -s_2 & \\ & s_2 & c_2 & \\ & & & 1 \end{bmatrix} \begin{bmatrix} c_1 & -s_1 & & \\ s_1 & c_1 & & \\ & & 1 & \\ & & & 1 \end{bmatrix}.$$

It follows that the $(3,1)$-element of Q is $s_3 s_2 s_1$ and we have $z = \sigma_0 s_1 s_2 s_3$ since $z = Qq = Q\sigma_0 e^{(4,1)}$.

Notes and Comments. If we let $C = I$, we have the standard GMRES method (Saad and Schultz [9]). Moreover, if CA is symmetric and nonsingular, then the C-orthogonal procedure truncates and the upper Hessenberg matrix H_n reduces to a tridiagonal matrix.

Additional details on the GGMRES method can be found in the dissertation of Chen [2] and the paper by Chen-Kincaid-Young [3]. Chen presents a wide range of numerical examples that illustrate the numerical behavior of various GMRES-type iterative methods and compares their rates of convergence to several other well-known iterative methods.

Acknowledgments

This work was supported, in part, by grants TARP–003658–0197–1997 and ATP–003658–0526–1999 at The University of Texas at Austin from the State of Texas Higher Education Coordinating Board through the Texas Advanced Research/Advanced Technology Program and by grant NSC–89–2115–M–214–005 at I-Shou University from the National Science Council, Taiwan.

We wish to thank the referee for some most useful comments and suggestions.

References

1. Axelsson, O.: Iterative Solution Methods. Cambridge University Press, New York, NY (1994).
2. Chen, J.-Y.: Iterative solution of large nonsymmetric linear systems. Report CNA–285, Center for Numerical Analysis, University of Texas at Austin (1997)
3. Chen, J.-Y., Kincaid, D. R., Young, D. M.: Generalizations and modifications of the GMRES iterative method. *Numerical Algorithms* **21** (1999) 119–146
4. Freund, R. W., Nachtigal, N. M.: QMR: A quasi-minimal residual method for non-hermitian linear systems. *Numerische Mathematik* **60** 315–339 (1991)
5. Hestenes, M. R., Stiefel,E.: Methods of conjugate gradients for solving linear systems. *Journal of Research of the National Bureau of Standards* **49** (6) 409–436 (1952)
6. Jea, K. C.: Generalized conjugate gradient acceleration of iterative methods. Report CNA–176, Center for Numerical Analysis, University of Texas at Austin (1982)

7. Jea, K. C., Young, D. M.: On the simplification of generalized conjugate gradient methods for nonsymmetrizable linear systems. *Linear Algebra Appl.* **52/53** (1983) 399–417
8. Saad, Y.: *Iterative Methods for Sparse Linear Systems*. PWS Publisher, Boston, MA (1996)
9. Saad, Y., Schultz, M. H.: GMRES: A generalized minimal residual algorithm for solving nonsymmetric linear systems. *SIAM J. Sci. Statist. Comput.*, **7** (3) (1986) 856–869
10. Sonneveld, P.: CGS: A fast Lanczos-type solver for nonsymmetric linear systems. *SIAM J. Sci. Stat. Comput.* **10** (1) 36–52 (1989)
11. Van Der Vorst, H. A.: BI-CGSTAB: A fast and smoothly converging variant of BI-CG for the solution of nonsymmetric linear systems. *SIAM J. Sci. Stat. Comput.* **13** (2) 631–644 (1992)
12. Young, D. M., Hayes, L. J., Jea, K. C.: Generalized conjugate gradient acceleration of iterative methods. Part I: The nonsymmetrizable case, Report CNA–162, Center for Numerical Analysis, University of Texas at Austin (1981)
13. Young, D. M., Jea, K. C.: Generalized conjugate gradient acceleration of iterative methods. *Linear Algebra Appl.* **34** (1980) 159–194
14. Young, D. M., Jea, K. C.: Generalized conjugate gradient acceleration of iterative methods. Part II: The nonsymmetrizable case, Report CNA–163, Center for Numerical Analysis, University of Texas at Austin (1981)

AMLI Preconditioning of Pure Displacement Non-conforming Elasticity FEM Systems

Tzanio Kolev[1] and Svetozar Margenov[2]

[1] Department of Mathematics, Texas A&M University,
College Station, TX 77843, USA
[2] Central Laboratory for Parallel Processing, Bulgarian Academy of Sciences
Acad. G. Bontchev Str., Bl. 25A, 1113 Sofia, Bulgaria

Abstract. This paper is concerned with the pure displacement problem of planar linear elasticity. Our interest is focussed to a *locking-free* FEM approximation of the problem in the case when the material is *almost incompressible*. The approximation space is constructed using the Crouzeix-Raviart linear finite elements. Choosing a proper hierarchical basis of this space we define an optimal order algebraic multilevel (AMLI) preconditioner for the related stiffness matrix. Local spectral analysis is applied to find the scaling parameter of the preconditioner as well as to estimate the related constants in the strengthened C.B.S. inequality. A set of numerical tests which illustrate the accuracy of the FEM solution, and the convergence rate of the AMLI PCG method is presented.

Keywords: PCG, multilevel preconditioners, non-conforming FEM

AMS Subject Classifications: 65F10, 65N30

1 Introduction

In this paper we consider the parameter dependent planar linear elasticity problem for *almost incompressible* material. It is known [9,5,7], that when the Poisson ratio ν tends to 0.5, the so called *locking phenomenon* appears, if low order conforming finite elements were used in the construction of the approximation space. Following [9,8], we use the Crouzeix-Raviart linear finite elements to get a *locking-free* FEM solution of the problem. Note that the straightforward FEM discretization works well for the pure displacement problem only [7]. The next important step is the construction of a *locking-free* solution method for the obtained linear algebraic system. Let us note, that the condition number of the FEM stiffness matrix tends to infinity when $\nu \to 0.5$. This means, that if, e.g., the preconditioned conjugate gradient (PCG) method is used as an iterative solver for the algebraic problem, then the relative condition number of the candidates for *good* preconditioners should be uniformly bounded with respect to the Poisson ratio.

An optimal order full multigrid algorithm for the pure displacement problem is presented in [6]. More recently, robust multigrid preconditioning for the problem in primal variables, obtained by the selective and reduced integration (SRI)

method, was proposed and studied in [12]. The SRI method is a particular case of a more general mixed formulation which produces a *locking-free* FEM discretization (see [9], and also [3] for some more recent results about the efficient solution of the related saddle point problems).

Here we study an efficient application of the general framework of the algebraic multilevel iteration (AMLI) method as introduced by Axelsson and Vassilevski (see, e.g. [4,13]). The presented new results complete the last years investigations of the authors devoted to the development of robust preconditioners for the algebraic problem under consideration. A detailed study of the related two-level method was recently published in [11]. The corresponding hierarchical basis multilevel algorithm was first announced in [10].

The remainder of the paper is organized as follows. A short introduction to the Lamé model of elasticity and the *locking effect* is given in the next section. The construction of the two-level and the AMLI preconditioners is described in §3. A model analysis of these preconditioners (obtained for a uniform mesh of half squares) is presented in §4. The last section contains numerical tests, which illustrate both the *locking-free* approximation properties of the Crouzeix-Raviart linear finite elements for the pure displacement problem, and the optimal convergence rate of the AMLI algorithm. Brief concluding remarks are also given at the end of the paper.

2 Lamé Model of Elasticity and the Locking Effect

Several problems in computational mechanics can be written in the form

$$\left(A + \frac{1}{\epsilon}B\right)u = f, \qquad (1)$$

where A is a given well conditioned operator, B is an operator with a non-trivial null-space, and $\epsilon > 0$ is a small parameter. If we study the behavior of the solution when $\epsilon \to 0$, we call (1) a *parameter dependent problem* [1].

Let us consider an elastic body occupying a bounded domain $\Omega \subset \mathcal{R}^2$. We are interested in finding the vector field of the displacements $u : \Omega \mapsto \mathcal{R}^2$, when the field of volume forces $f : \Omega \mapsto \mathcal{R}^2$ is given. Suppose that $f \in [L_2(\Omega)]^2$, and let $\mathcal{V} = [H_0^1(\Omega)]^2$. Then, the weak formulation of the pure displacement problem with homogeneous boundary conditions $u|_{\partial\Omega} = 0$ reads (see [7] for more details)

$$\text{find } u \in \mathcal{V}: \quad a(u,v) = F(v) \quad \forall v \in \mathcal{V}, \qquad (2)$$

where

$$a(u,v) = \int_\Omega \left[\lambda (\nabla u)(\nabla v) + 2\mu\, T_\varepsilon(u) : T_\varepsilon(v)\right] \quad \text{and} \quad F(v) = \int_\Omega fv.$$

Here $T_\varepsilon(u) = (\nabla u + (\nabla u)^T)/2$, and $\lambda > 0$ and $\mu > 0$ stand for the Lamé coefficients, which can be expressed by the elasticity modulus $E > 0$ and the Poisson ratio $\nu \in [0, 1/2)$ as follows: $\lambda = (E\nu)/[(1+\nu)(1-2\nu)]$, $\mu = E/[2(1+\nu)]$.

The case $\nu \to 0.5$ it called *almost incompressible*, and so (2) belongs to the class of parameter dependent problems with $\epsilon = (1 - 2\nu)$. Now suppose that Ω is a polygon, $\mathcal{T}_h$ is a regular family of triangulations of Ω, and FEM is used to get a numerical solution of the considered elasticity problem. As it was mentioned before, the *locking effect* appears for low order conforming FEM discretization of (2). This means that the relative error of the FEM solution is unbounded when $\nu \to 1/2$ for any fixed mesh parameter $h \to 0$ (see [8] for more details). Fortunately, it turns out that *locking* can be overcome if the non-conforming Crouzeix-Raviart finite elements are used. Let $\mathcal{N}(\mathcal{T}_h)$ be the midpoints of the sides of the triangles $T \in \mathcal{T}_h$. First we define the scalar FEM space:

$$\mathcal{V}^0_{cr,h} = \{v : v|_T \text{ is linear}; v \text{ is continuous in } \mathcal{N}(\mathcal{T}_h), v = 0 \text{ in } \mathcal{N}(\mathcal{T}_h) \cap \partial\Omega\}.$$

Now we introduce $\mathcal{V}_h = [\mathcal{V}^0_{cr,h}]^2$. Note that $\mathcal{V}_h \not\subset [C(\Omega)]^2 \Rightarrow \mathcal{V}_h \not\subset [H^1(\Omega)]^2$, that is this finite element space is non-conforming.

Since a pure displacement problem is under consideration, the bilinear form $a(\cdot,\cdot)$ is equivalent to the following one:

$$a^s(u,v) = \int_\Omega [(\lambda + \mu)(\nabla u)(\nabla v) + \mu \nabla u : v\nabla]. \tag{3}$$

Note that the modification of the variational formulation based on the bilinear form (3) is of principal importance. The discrete version of $a^s(\cdot,\cdot)$ in the non-conforming case is defined by element wise splitting of the integral, i.e.

$$a^s_h(u,v) = \sum_{T \in \mathcal{T}_h} \int_T [(\lambda + \mu)(\nabla u|_T)(\nabla v|_T) + \mu \nabla u|_T : v|_T \nabla].$$

If u_h is the solution of the discrete problem

$$\text{find } u_h \in \mathcal{V}_h : \quad a^s_h(u_h, v_h) = F(v_h) \quad \forall v_h \in \mathcal{V}_h,$$

then the following *locking-free* error estimate holds (see [8]):

Theorem 1. *There exists a constant $C_{\Omega,\theta}$ (independent of h, λ, μ) such that*

$$\|u - u_h\|_h \leq C_{\Omega,\theta} \, h \, \|f\|_{[L_2(\Omega)]^2},$$

where $\|\cdot\|_h := \sqrt{a^s_h(\cdot,\cdot)}$, and θ is the smallest angle in the triangulation.

The standard computational procedure leads to a linear system of equations $\mathcal{A}_h \mathbf{u}_h = \mathbf{f}_h$ where $\mathcal{A}_h$ is the corresponding stiffness matrix. At this point we run into a discrete *locking* phenomenon since the condition number $\kappa(\mathcal{A}_h) \to \infty$ as $\nu \to 1/2$. Our next step is the construction of a *locking-free* preconditioner $\mathcal{M}$ for $\mathcal{A}_h$, such that $\kappa(\mathcal{M}^{-1}\mathcal{A}_h) = \mathcal{O}(1)$ uniformly on ν.

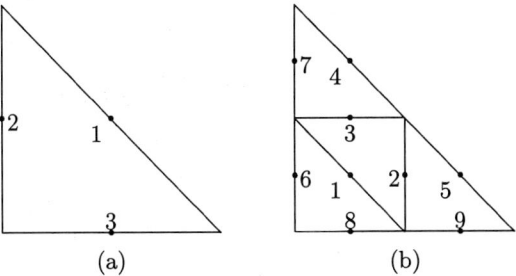

Fig. 1. C.-R. FE: (a) triangle $e \in \mathcal{T}_H$; (b) refined macro-element $E \in \mathcal{T}_h$

3 From Two-Level to AMLI Preconditioner

This section begins with a short presentation of the construction of the two-level algorithm as it was introduced and studied in [11]. Let $\mathcal{T}_h$ be a refinement for a coarser triangulation $\mathcal{T}_H$. Associated with $\mathcal{T}_h$ are the FEM space $\mathcal{V}_h$ and the corresponding nodal basis element stiffness matrix $\mathcal{A}_h$. Observe that $\mathcal{V}_H$ and $\mathcal{V}_h$ are not nested as in the conforming case.

Let $\mathcal{A}_e$ be the element stiffness matrix corresponding to the triangle $e \in \mathcal{T}_H$, and $\mathcal{A}_E$ be the macro-element stiffness matrix where the macro-element $E \in \mathcal{T}_h$ is obtained by a regular bisection refinement of $e \in \mathcal{T}_H$ (see Fig. 1). Following the FEM assembling procedure we have $\mathcal{A}_H = assembl\{\mathcal{A}_e\}_{e \in \mathcal{T}_H}$, $\mathcal{A}_h = assembl\{\mathcal{A}_E\}_{E \in \mathcal{T}_h}$. Let us denote by $\phi_E^t = \{\varphi(x,y)_i\}_{i=1}^{9}$ the macro-element vector of the nodal basis functions. In all local matrices the numbering of the nodes corresponds to Fig. 1. Now, we are ready to define locally the hierarchical two-level basis $\widetilde{\phi}_E$.

$$\widetilde{\phi}_E = J_E \phi_E, \quad J_E = \begin{pmatrix} I & 0 \\ 0 & T \end{pmatrix}, \quad T = \frac{1}{2}\begin{pmatrix} 1 & -1 & & & \\ & & 1 & -1 & \\ 1 & 1 & & & 1 & -1 \\ & & 1 & 1 & \\ & & & & 1 & 1 \end{pmatrix}. \quad (4)$$

Here, and in what follows: I stands for the identity matrix of the appropriate size; all matrices and vectors related to the two-level basis are marked by tilde. The global two-level stiffness matrix reads: $\widetilde{\mathcal{A}}_h = assembl\{\widetilde{\mathcal{A}}_E\}_{E \in \mathcal{T}_k}$, where $\widetilde{\mathcal{A}}_E = J_E \mathcal{A}_E J_E^T$. The global transformation matrix J is also assembled by the local matrices J_E. We now split and factorize $\widetilde{\mathcal{A}}_h$ into 2×2 block form

$$\widetilde{\mathcal{A}}_h = \begin{pmatrix} \widetilde{\mathcal{A}}_{11;h} & \widetilde{\mathcal{A}}_{12;h} \\ \widetilde{\mathcal{A}}_{21;h} & \widetilde{\mathcal{A}}_{22;h} \end{pmatrix} = \begin{pmatrix} \widetilde{\mathcal{A}}_{11;h} & 0 \\ \widetilde{\mathcal{A}}_{21;h} & \widetilde{\mathcal{B}}_h \end{pmatrix} \begin{pmatrix} I & \widetilde{\mathcal{A}}_{11;h}^{-1} \widetilde{\mathcal{A}}_{12;h} \\ 0 & I \end{pmatrix}. \quad (5)$$

Here: the block $\widetilde{\mathcal{A}}_{11}$ corresponds to the interior nodal unknowns with respect to the macro-elements $E \in \mathcal{T}_h$; and $\widetilde{\mathcal{B}}_h$ stands for the Schur complement of this elimination step. Note that $\widetilde{\mathcal{A}}_{11;h}$ is a block-diagonal matrix with blocks which are 6×6 matrices, i.e. this elimination can be performed macro-element by macro-element. The next step of the construction is to approximate the matrix $\widetilde{\mathcal{B}}_h$ written again in 2×2 block form

$$\widetilde{\mathcal{B}}_h = \begin{pmatrix} \widetilde{\mathcal{B}}_{11;h} & \widetilde{\mathcal{B}}_{12;h} \\ \widetilde{\mathcal{B}}_{21;h} & \widetilde{\mathcal{B}}_{22;h} \end{pmatrix} = \begin{pmatrix} \widetilde{\mathcal{B}}_{11;h} & 0 \\ \widetilde{\mathcal{B}}_{21;h} & \widetilde{\mathcal{S}}_h \end{pmatrix} \begin{pmatrix} I & \widetilde{\mathcal{B}}_{11;h}^{-1} \widetilde{\mathcal{B}}_{12;h} \\ 0 & I \end{pmatrix}, \qquad (6)$$

where the first pivot block $\widetilde{\mathcal{B}}_{11;h}$ corresponds to the two-level basis functions which are defined as half–differences of nodal basis functions (see (4)), and $\widetilde{\mathcal{S}}_h$ is the current Schur complement. It is important to note that $wt\mathcal{B}_{22;h}$ has the same sparsity pattern as the true discretization matrix $\mathcal{A}_H$ corresponding to the coarse triangulation $\mathcal{T}_H$.

Definition 1 *The two-level preconditioner is defined as* $\mathcal{M}_{2L} = J^{-1} \widetilde{\mathcal{M}}_{2L} J^{-T}$, *where*

$$\widetilde{\mathcal{M}}_{2L} = \begin{pmatrix} \widetilde{\mathcal{A}}_{11;h} & 0 \\ \widetilde{\mathcal{A}}_{21;h} & \mathcal{M}_{\widetilde{\mathcal{B}}_h} \end{pmatrix} \begin{pmatrix} I & \widetilde{\mathcal{A}}_{11;h}^{-1} \widetilde{\mathcal{A}}_{12;h} \\ 0 & I \end{pmatrix},$$

$$\mathcal{M}_{\widetilde{\mathcal{B}}_h} = \begin{pmatrix} \widetilde{\mathcal{D}}_{11;h} & 0 \\ \widetilde{\mathcal{B}}_{21;h} & \widetilde{\mathcal{B}}_{22;h} \end{pmatrix} \begin{pmatrix} I & \widetilde{\mathcal{D}}_{11;h}^{-1} \widetilde{\mathcal{B}}_{12;h} \\ 0 & I \end{pmatrix},$$

and where: $\widetilde{\mathcal{D}}_{11;h} = \omega \widetilde{\mathcal{B}}_{11;h}^{(d)}$; $\widetilde{\mathcal{B}}_{11;h}^{(d)}$ *stands for the diagonal part of* $\widetilde{\mathcal{B}}_{11;h}$; $\omega > 0$ *is a parameter.*

Let ω be chosen so that $\mathbf{v}^T \widetilde{\mathcal{B}}_{11;h} \mathbf{v} \leq \mathbf{v}^T \widetilde{\mathcal{D}}_{11;h} \mathbf{v} \leq \delta \, \mathbf{v}^T \widetilde{\mathcal{B}}_{11;h} \mathbf{v}$, then the following estimate is a straightforward conclusion from the general result for the convergence of the two-level algorithms [2,13,11]:

$$\kappa(\mathcal{M}_{2L}^{-1} \mathcal{A}_h) \leq \frac{\delta}{1 - \gamma^2}. \qquad (7)$$

Here γ is the constant in the strengthened C.B.S. inequality corresponding to the 2×2 block-presentation (6) of $\widetilde{\mathcal{B}}_h$.

Now, let us assume that the same uniform refinement procedure is used to construct a sequence of nested triangulations $\mathcal{T}_1 \subset \mathcal{T}_2 \subset \ldots \subset \mathcal{T}_\ell$. The final goal of this paper is to construct AMLI preconditioner $\mathcal{M}_{AMLI}$ for the stiffness matrix $\mathcal{A}^{(\ell)}$ corresponding to $\mathcal{T}_\ell$, and to study its convergence behavior.

Definition 2 *The AMLI preconditioner is determined recursively as follows:*

$$\mathcal{M}_{AMLI}^{(1)} = \mathcal{A}^{(1)}$$

for $k = 2, 3, \ldots, \ell$

$$\mathcal{M}_{AMLI}^{(k)} = J^{-1} \widetilde{\mathcal{M}}_{AMLI}^{(k)} J^{-T},$$

where

$$\widetilde{\mathcal{M}}_{AMLI}^{(k)} = \begin{pmatrix} \widetilde{\mathcal{A}}_{11}^{(k)} & 0 \\ \widetilde{\mathcal{A}}_{21}^{(k)} & \mathcal{M}_{\mathcal{B}}^{(k)} \end{pmatrix} \begin{pmatrix} I & \widetilde{\mathcal{A}}_{11}^{(k)^{-1}} \widetilde{\mathcal{A}}_{12}^{(k)} \\ 0 & I \end{pmatrix}, \qquad (8)$$

$$M_{\mathcal{B}}^{(k)} = \begin{pmatrix} \widetilde{\mathcal{D}}_{11}^{(k)} & 0 \\ \widetilde{\mathcal{B}}_{21}^{(k)} & \hat{\mathcal{A}}^{(k)} \end{pmatrix} \begin{pmatrix} I & \widetilde{\mathcal{D}}_{11}^{(k)^{-1}} \widetilde{\mathcal{B}}_{12}^{(k)} \\ 0 & I \end{pmatrix},$$

where

$$\hat{\mathcal{A}}^{(k)^{-1}} = [I - p_\beta(M_{AMLI}^{(k-1)^{-1}} \mathcal{A}^{(k-1)})]\mathcal{A}^{(k-1)^{-1}},$$

and where: $\mathcal{A}^{(k-1)} = \widetilde{\mathcal{B}}_{22}^{(k)}$; $\widetilde{\mathcal{D}}_{11}^{(k)} = \omega \widetilde{\mathcal{B}}_{11}^{(k)(d)}$, $\widetilde{\mathcal{B}}_{11}^{(k)(d)}$ is the diagonal part of $\widetilde{\mathcal{B}}_{11}^{(k)}$; p_β is properly scaled polynomial of degree β.

Following the general scheme of the convergence analysis from [4] one can prove that:

Lemma 1. *The AMLI method (8) is of optimal computational complexity if*

$$(1 - \gamma^2)^{-1/2} < \beta < 4.$$

4 Model Convergence Analysis

The model problem is defined on a convex polygon $\Omega = \bigcup \{T : T \in \mathcal{T}_H\}$ under the additional assumption, that $\mathcal{T}_1$ is obtained by diagonal bisection of square cells of a given uniform rectangle mesh, with mesh lines which are parallel to the coordinate axes. The considerations in this section are aimed to a quantitative analysis of the behavior of the constant in the strengthened C.B.S. inequality. The next two-level estimate was recently published in [11].

Theorem 2. *The two-level constant in the strengthened C.B.S. inequality (for the model problem under consideration) satisfies the estimate*

$$\gamma \leq \gamma_E \leq \frac{\sqrt{8 + \sqrt{8}}}{4} = 0.822\ldots \quad \forall \nu \in [0, 1/2). \tag{9}$$

The general approach for such estimates is based on a local analysis on a macro-element level. For the AMLI algorithm, this estimate is valid only at the first factorization step. Unlike the case when AMLI is applied to conforming linear finite elements, here the *coarse grid* element stiffness matrices are changed at each factorization step. The behavior of γ when k and ν are varied is presented

Table 1. Numerical analysis of the C.B.S. constant γ

k	$\nu = 0.3000$	$\nu = 0.4000$	$\nu = 0.4900$	$\nu = 0.4990$	$\nu = 0.4999$
1	0.786065	0.801174	0.820122	0.822405	0.822638
2	0.631551	0.725086	0.937080	0.993267	0.999322
3	0.277714	0.343513	0.655475	0.940351	0.994058
4	0.224511	0.239849	0.444969	0.632548	0.661470
5	0.165863	0.187030	0.291319	0.318613	0.338222
6	0.094187	0.092353	0.140757	0.193588	0.146164
7	0.085456	0.078540	0.083019	0.102814	0.126900

Table 2. Relative error stability for $\nu \to 1/2$

ν	$\|u-u_h\|_{[L_2]^2}/\|f\|_{[L_2]^2}$	ν	$\|u-u_h\|_{[L_2]^2}/\|f\|_{[L_2]^2}$
0.4	.3108249106503572	0.4999	.3771889077038727
0.49	.3695943747405575	0.49999	.3772591195613628
0.499	.3764879643773666	0.499999	.3772661419401481

Table 3. AMLI preconditioning of non-conforming FEM system ($\beta = 2$)

ℓ	N	$\nu=0.3$	$\nu=0.4$	$\nu=0.49$	$\nu=0.499$	$\nu=0.4999$	$\nu=0.49999$	$\nu=0.499999$
4	1472	13	13	12	13	13	13	13
5	6016	12	12	12	14	13	13	13
6	24320	12	12	12	12	13	13	13
7	97792	11	11	11	12	13	13	13
8	196096	11	11	11	12	12	13	13

in next Table. The conclusion from the test data presented in the table is, that: a) γ strongly decreases with k for moderate values of ν; b) there is an oscillation of γ when ν is near the *incompressible limit* followed again by a stable decreasing.

5 Numerical Tests

The numerical tests presented in this section are to illustrate the behavior of the FEM error as well as the optimal convergence rate of the AMLI algorithm when the size of the discrete problem is varied and $\nu \in [0, 1/2)$ tends to the incompressible limit. The simplest test problem in the unit square $\Omega = (0,1)^2$ with $E = 1$ is considered. The right hand side corresponds to a given exact solution $u(x,y) = (\sin(\pi x)\sin(\pi y), y(y-1)x(x-1))$. The relative stopping criterion $(\mathcal{M}^{-1}r^{N_{it}}, r^{N_{it}})/(\mathcal{M}^{-1}r^0, r^0) < \varepsilon^2$ is used for the PCG algorithm, where r^i stands for the residual at the i-th iteration step.

The relative FEM errors, given in Table 2, well illustrate the *locking-free* approximation. Here $\ell = 4$, $N = 1472$, and $\varepsilon = 10^{-9}$. This Table is presented here for completeness. It was first published in [11].

In Table 3, the number of iterations are presented as a measure of the robustness of the proposed two-level preconditioner. Here $\varepsilon = 10^{-3}$. The optimal order *locking-free* convergence rate of the AMLI algorithm is well expressed.

In the next table a modification of the AMLI algorithm (8) is used where two PCG inner iteration have been applied to stabilize the multilevel algorithm instead of the acceleration polynomial $[I - p_\beta(\cdot)]$. This approach was shown to be even a better candidate in the case under consideration when the constant in the strengthened C.B.S. inequality is varying during the AMLI factorization procedure.

Remark 1. It is important to note once again that the application of the algorithm presented in this article is strictly restricted to the case of pure displacement. The more general case where Neumann boundary conditions are also

Table 4. Modified AMLI preconditioning (fixed PCG iterations)

ℓ	N	$\nu = 0.3$	$\nu = 0.4$	$\nu = 0.49$	$\nu = 0.499$	$\nu = 0.4999$	$\nu = 0.49999$	$\nu = 0.499999$
4	1472	10	10	11	10	10	10	10
5	6016	10	11	11	11	11	11	11
6	24320	10	11	11	11	11	11	11
7	97792	11	11	11	11	11	11	12
8	196096	10	11	11	11	11	11	12

assumed require a modification of the variational formulation. Otherwise the second Korn's inequality does not hold in the case of low-order non-conforming FEM discretization (see [9] for more details for the 2D case).

Acknowledgments

A part of this study has been supported by the Bulgarian NSF Grants MM-801 and MU-901.

References

1. Arnold, D. N.: Discretization by finite elements of a parameter dependent problem. Numer. Math. **37** (1981) 404-421
2. Axelsson, O.: Iterative Solution Methods. Cambridge, University press (1996)
3. Axelsson O., Padiy, A.: On a Robust and Scalable Linear Elasticity Solver Based on a Saddle Point Formulation. Int. J. Numer. Meth. Enging. **44** (1999) 801-818
4. Axelsson O., Vassilevski, P. S.: Algebraic Multilevel Preconditioning Methods, Part II. SIAM J. Numer. Anal. **27** (1990) 1569-1590
5. Babuška, I., Suri, M.: Locking Effects in the Finite Approximation of Elasticity Problems. Numer. Math. **62** (1992) 439-463
6. Brenner, S.: A Non-Conforming Mixed Multigrid Method for the Pure Displacement Problem in Planar Linear Elasticity. SIAM J. Numer. Anal. **30** (1993) 116-135
7. Brenner, S., Scott, L.: The Mathematical Theory of Finite Element Methods. Texts in Applied Mathematics, Vol. 15, Springer-Verlag (1994)
8. Brenner S., Sung, S.: Linear Finite Element Methods for Planar Linear Elasticity. Math. Comp. **59** (1992) 321-338
9. Falk, R. S.: Non-conforming Finite Element Methods for the Equations of Linear Elasticity. Mathematics of Computation **57** (1991) 529-550
10. Kolev, Tz., Margenov, S.: Multilevel HB Preconditioning of Non-Conforming FEM Systems. In: Iliev, O., Kaschiev, M., Margenov, S., Sendov, Bl., Vassilevski, P. S. (eds.): Recent Advances in Numerical Methods and Applications II, World Scientific (1999) 603-610
11. Kolev, Tz., Margenov, S.: Two-Level Preconditioning of Pure Displacement Non-Conforming FEM Systems. Numer. Linear Algebra Appl. **6** (1999) 533-555
12. Schöberl, J.: Robust Multigrid Preconditioner for Parameter-Dependent Problems I: The Stokes-Type Case. Technical Report No 97-2, Johannes Kepler University, Linz, Austria (1997)
13. Vassilevski, P. S.: On Two Ways of Stabilizing the Hierarchical Basis Multilevel Methods. SIAM Review **39** (1997) 18-53

Computationally Efficient Methods for Solving SURE Models*

Erricos J. Kontoghiorghes and Paolo Foschi

Institut d'informatique, Université de Neuchâtel
Rue Emile Argand 11, CH-2007 Neuchâtel, Switzerland
{erricos.kontoghiorghes,paolo.foschi}@info.unine.ch

Abstract. Computationally efficient and numerically stable methods for solving Seemingly Unrelated Regression Equations (SURE) models are proposed. The iterative feasible generalized least squares estimator of SURE models where the regression equations have common exogenous variables is derived. At each iteration an estimator of the SURE model is obtained from the solution of a generalized linear least squares problem. The proposed methods, which have as a basic tool the generalized QR decomposition, are also found to be efficient in the general case where the number of linear independent regressors is smaller than the number of observations.

1 Introduction

The basic computational formulae for deriving the estimators of Seemingly Unrelated Regression Equations (SURE) models involve Kronecker products and direct sums of matrices that make the solution of the models computationally expensive even for modest sized models. Therefore the derivation of numerically stable and computationally efficient methods is of great importance [3,6,17,18].

The SURE model is given by

$$y_i = X_i \beta_i + u_i, \qquad i = 1, 2, \ldots, G, \qquad (1)$$

where $y_i \in \Re^T$ is the endogenous vector, $X_i \in \Re^{T \times k_i}$ is the exogenous matrix with full column rank, $\beta_i \in \Re^{k_i}$ are the coefficients and $u_i \in \Re^T$ is the disturbance vector, having zero mean and variance–covariance matrix $\sigma_{ii} I_T$. Furthermore, the covariance matrix of u_i and u_j is given by $\sigma_{ij} I_T$, i.e. contemporaneous disturbances are correlated.

In the compact form the SURE model can be written as

$$\begin{pmatrix} y_1 \\ y_2 \\ \vdots \\ y_G \end{pmatrix} = \begin{pmatrix} X_1 & & & \\ & X_2 & & \\ & & \ddots & \\ & & & X_G \end{pmatrix} \begin{pmatrix} \beta_1 \\ \beta_2 \\ \vdots \\ \beta_G \end{pmatrix} + \begin{pmatrix} u_1 \\ u_2 \\ \vdots \\ u_G \end{pmatrix} \qquad (2)$$

---

* This work is in part supported by the Swiss National Foundation Grants 21-54109.98 and 1214-056900.99/1.

or
$$\text{vec}(Y) = \left(\oplus_{i=1}^{G} X_i\right) \text{vec}(\{\beta_i\}_G) + \text{vec}(U), \tag{3}$$

where $Y = (y_1, \ldots, y_G)$, $U = (u_1, \ldots, u_G)$, the direct sum of matrices $\oplus_{i=1}^{G} X_i$ is equivalent to the block diagonal matrix $\text{diag}(X_1, \ldots, X_G)$, $\{\beta_i\}_G$ denotes the set of vectors $\beta_1, \ldots, \beta_G$ and $\text{vec}(\cdot)$ is the vector operator which stacks one column under the other of its matrix or set of vectors argument. The disturbance term $\text{vec}(U)$ has zero mean and dispersion matrix $\Sigma \otimes I_T$, where, $\Sigma = [\sigma_{ij}]$ is symmetric and positive semidefinite and $\otimes$ denotes the Kronecker product operator [2,4,15]. That is,

$$\Sigma \otimes I_T = \begin{pmatrix} \sigma_{11} I_T & \sigma_{12} I_T & \cdots & \sigma_{1G} I_T \\ \sigma_{21} I_T & \sigma_{22} I_T & \cdots & \sigma_{2G} I_T \\ \vdots & \vdots & & \vdots \\ \sigma_{G1} I_T & \sigma_{G2} I_T & \cdots & \sigma_{GG} I_T \end{pmatrix}.$$

For notational convenience the subscript G in the set operator $\{\cdot\}$ is dropped and $\oplus_{i=1}^{G}$ is abbreviated to $\oplus_i$.

The Best Linear Unbiased Estimator (BLUE) of $\text{vec}(\{\beta_i\})$ is obtained from the solution of the General Least Squares (GLS) problem

$$\underset{\beta_1,\ldots,\beta_G}{\text{argmin}} \left\| \text{vec}(Y) - \text{vec}(\{X_i \beta_i\}) \right\|_{\Sigma^{-1} \otimes I_T} \tag{4}$$

which is given by

$$\text{vec}(\{\hat{\beta}_i\}) = \left((\oplus_i X_i^T)(\Sigma^{-1} \otimes I_T)(\oplus_i X_i)\right)^{-1} (\oplus_i X_i^T) \text{vec}(Y \Sigma^{-1}). \tag{5}$$

Often Σ is unknown and an iterative procedure is used to obtain the Feasible GLS (FGLS) estimator. Initially, the regression equations of the SURE model are assumed to be unrelated, that is, the correlation among contemporaneous disturbances of the model is ignored and $\Sigma = I_G$. This is equivalent to computing the Ordinary Least Squares (OLS) estimator of $\{\beta_i\}$. Then, from the residuals a new estimator for Σ is derived which is used in (5) to provide another estimator for the coefficients $\{\beta_i\}$. This process is repeated until convergence is achieved [16]. Generally, at the ith iteration the estimator of Σ is computed by

$$\Sigma_{(i+1)} = U_{(i)}^T U_{(i)} / T \tag{6}$$

where $U_{(i)} = \left(u_1^{(i)} \ldots u_G^{(i)}\right)$ and $u_j^{(i)} = y_j - X_j \hat{\beta}_j$ $(j = 1, \ldots, G)$, are the residuals of the jth regression equation.

The regression equations in a SURE model frequently have common exogenous variables (or common regressors). The purpose of this work is to propose computational efficient methods which exploit this possibility.

2 Numerical Solution of SURE Models

The BLUE of the SURE model comes from the solution of the Generalized Linear Least Squares Problem (GLLSP)

$$\underset{V,\{\beta_i\}}{\operatorname{argmin}} \|V\|_F \text{ subject to } \operatorname{vec}(Y) = (\oplus_i X_i)\operatorname{vec}(\{\beta_i\}) + \operatorname{vec}(VC^T), \quad (7)$$

where $\|\cdot\|_F$ denotes the *Frobenius* norm, $\Sigma = CC^T$, the upper triangular $C \in \Re^{G \times G}$ has full rank and the random matrix V is defined as $(C \otimes I_T)\operatorname{vec}(V) = \operatorname{vec}(U)$; that is, $VC^T = U$, which implies that $\operatorname{vec}(V)$ has zero mean and variance-covariance matrix I_{TG} [8,11,12,13]. Without loss of generality it has been assumed that Σ is non-singular.

Consider the GQRD:

$$Q^T(\oplus_i X_i) = \begin{pmatrix} \oplus_i R_i \\ 0 \end{pmatrix} \quad (8a)$$

and

$$Q^T(C \otimes I_T) P = \begin{pmatrix} W_{11} & W_{12} \\ 0 & W_{22} \end{pmatrix} \begin{matrix} K \\ GT - K \end{matrix}, \quad (8b)$$

where $K = \sum_{i=1}^{G} k_i$, $R_i \in \Re^{k_i \times k_i}$ and W_{22} are upper triangular, and $Q, P \in \Re^{GT \times GT}$ are orthogonal [1,14]. Using (8) the GLLSP (7) can be written as

$$\underset{\{\tilde{v}_i\},\{\hat{v}_i\},\{\beta_i\}}{\operatorname{argmin}} \sum_{i=1}^{G} (\|\tilde{v}_i\|_2 + \|\hat{v}_i\|_2) \text{ subject to}$$

$$\begin{pmatrix} \operatorname{vec}(\{\tilde{y}_i\}) \\ \operatorname{vec}(\{\hat{y}_i\}) \end{pmatrix} = \begin{pmatrix} \oplus_i R_i \\ 0 \end{pmatrix} \operatorname{vec}(\{\beta_i\}) + \begin{pmatrix} W_{11} & W_{12} \\ 0 & W_{22} \end{pmatrix} \begin{pmatrix} \operatorname{vec}(\{\tilde{v}_i\}) \\ \operatorname{vec}(\{\hat{v}_i\}) \end{pmatrix}, \quad (9)$$

where

$$Q^T \operatorname{vec}(Y) = \begin{pmatrix} \operatorname{vec}(\{\tilde{y}_i\}) \\ \operatorname{vec}(\{\hat{y}_i\}) \end{pmatrix} \begin{matrix} K \\ GT - K \end{matrix} \text{ and } P^T \operatorname{vec}(V) = \begin{pmatrix} \operatorname{vec}(\{\tilde{v}_i\}) \\ \operatorname{vec}(\{\hat{v}_i\}) \end{pmatrix} \begin{matrix} K \\ GT - K \end{matrix}.$$

From (9) it follows that $\operatorname{vec}(\{\hat{v}_i\}) = W_{22}^{-1}\operatorname{vec}(\{\hat{y}_i\})$ and $\tilde{v}_i = 0$. Thus, the solution of the SURE model comes from solving the triangular system

$$\begin{pmatrix} \operatorname{vec}(\{\tilde{y}_i\}) \\ \operatorname{vec}(\{\hat{y}_i\}) \end{pmatrix} = \begin{pmatrix} \oplus_i R_i & W_{12} \\ 0 & W_{22} \end{pmatrix} \begin{pmatrix} \operatorname{vec}(\{\beta_i\}) \\ \operatorname{vec}(\{\hat{v}_i\}) \end{pmatrix}. \quad (10)$$

Notice that W_{11} is not used. Furthermore, for deriving the iterative FGLS, the RQD of $Q^T(C \otimes I_T)$ in (8b) is the most costly operation as this needs to be recomputed for different C at each iteration.

The matrix Q in (8) is defined as

$$Q = \left(\oplus_i \tilde{Q}_i \; \oplus_i \hat{Q}_i \right) \equiv \begin{pmatrix} \tilde{Q}_1 & & \hat{Q}_1 & \\ & \ddots & & \ddots \\ & & \tilde{Q}_G & & \hat{Q}_G \end{pmatrix},$$

where

$$Q_i^T X_i = \begin{pmatrix} R_i \\ 0 \end{pmatrix}, \quad \text{with} \quad Q_i = \begin{pmatrix} \tilde{Q}_i & \hat{Q}_i \\ {}^{k_i} & {}^{T-k_i} \end{pmatrix},$$

is the QRD of X_i ($i = 1, \ldots, G$),

$$Q_i^T y_i = \begin{pmatrix} \tilde{y}_i \\ \hat{y}_i \end{pmatrix}, {}^{k_i}_{T-k_i} \quad \text{and} \quad Q_i^T v_i = \begin{pmatrix} \tilde{v}_i \\ \hat{v}_i \end{pmatrix}, {}^{k_i}_{T-k_i}.$$

The computation of (8b) derives in two stages. The first stage computes

$$Q^T (C \otimes I) Q = \begin{pmatrix} \widetilde{W}_{11} & \widetilde{W}_{12} \\ \widetilde{W}_{21} & \widetilde{W}_{22} \end{pmatrix} {}^{K}_{GT-K},$$

where $\tilde{W}_{ij}$ ($i, j = 1, 2$) is block upper triangular. Furthermore the main block-diagonals of $\tilde{W}_{12}$ and $\tilde{W}_{21}$ are zero, and the ith ($i = 1, \ldots, G$) block of the main diagonal of $\tilde{W}_{11}$ and $\tilde{W}_{22}$ are given by $C_{ii} I_{k_i}$ and $C_{ii} I_{T-k_i}$, respectively. The second stage computes the RQD

$$\left(\widetilde{W}_{21} \; \widetilde{W}_{22} \right) \tilde{P} = \left(0 \; W_{22} \right) \tag{11a}$$

and

$$\left(\widetilde{W}_{11} \; \widetilde{W}_{12} \right) \tilde{P} = \left(W_{11} \; W_{12} \right). \tag{11b}$$

Thus, in (8b) $P = Q\tilde{P}$. Sequential and parallel strategies for computing the RQD (11) have been described in [5,6]. Figure 1 illustrates the *diagonally-based* strategy, where $G = 3$. At each step of this method a block-diagonal of $\widetilde{W}_{21}$ is annihilated by a series of simultaneous factorizations. Each factorization (denoted by an arc) annihilates a block and affects two block-columns of $\left(\widetilde{W}_{11}^T \; \widetilde{W}_{21}^T \right)^T$ and $\left(\widetilde{W}_{12}^T \; \widetilde{W}_{22}^T \right)^T$. The block triangular structure of $\widetilde{W}_{11}$ and $\widetilde{W}_{22}$ is preserved throughout the annihilation process while $\widetilde{W}_{12}$ becomes full except from its last diagonal block which remains zero.

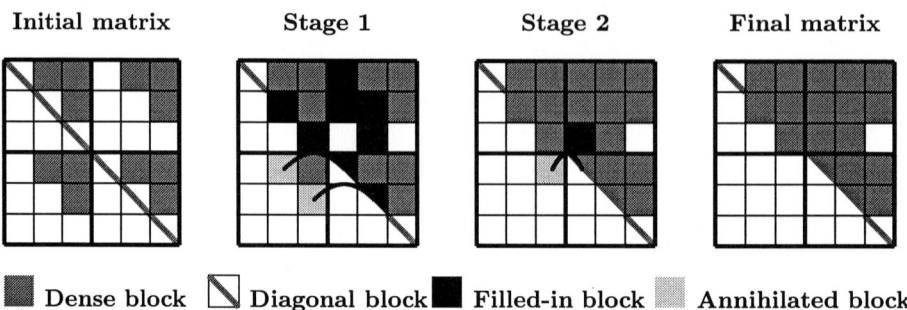

Fig. 1. Computing (11) using the *diagonally-based* method, where $G = 3$

3 SURE Model with Common Regressors

Consider now the SURE model with common regressors

$$\text{vec}(Y) = \left(\oplus_{i=1}^{G} X^d S_i\right) \text{vec}(\{\beta_i\}_G) + \text{vec}(U), \tag{12}$$

where X^d denotes the matrix consisting of the K^d distinct regressors, $K^d \leq K$, $S_i \in \Re^{K^d \times k_i}$ is a selection matrix that comprises relevant columns of the $K^d \times K^d$ identity matrix and the exogenous matrix X_i ($i = 1, \ldots, G$) is defined as $X_i = X^d S_i$ [7,9]. Let the QR decomposition of X^d be given by

$$Q_d^T X^d = \begin{pmatrix} R^d \\ 0 \end{pmatrix} \begin{matrix} r \\ T-r \end{matrix} \quad \text{with} \quad Q_d = \begin{pmatrix} \widetilde{Q}_d & \widehat{Q}_d \end{pmatrix} \begin{matrix} r & T-r \end{matrix}, \tag{13}$$

where $\text{rank}(X^d) = r < T$. Premultiplying (12) from the left by the orthogonal matrix $Q_D = \left(I_G \otimes \widetilde{Q}_d \ \ I_G \otimes \widehat{Q}_d\right)$ gives

$$\begin{pmatrix} \text{vec}(\widetilde{Y}) \\ \text{vec}(\widehat{Y}) \end{pmatrix} = \begin{pmatrix} \oplus_i R^d S_i \\ 0 \end{pmatrix} \text{vec}(\{\beta_i\}) + \begin{pmatrix} \text{vec}(\widetilde{U}) \\ \text{vec}(\widehat{U}) \end{pmatrix}, \tag{14}$$

where

$$Q_d^T Y = \begin{pmatrix} \widetilde{Y} \\ \widehat{Y} \end{pmatrix} \begin{matrix} r \\ T-r \end{matrix} \quad \text{and} \quad Q_d^T U = \begin{pmatrix} \widetilde{U} \\ \widehat{U} \end{pmatrix} \begin{matrix} r \\ T-r \end{matrix}.$$

The covariance matrix of the disturbance term $\text{vec}((\widetilde{U} \ \widehat{U}))$ is given by

$$\begin{pmatrix} \Sigma \otimes I_r & 0 \\ 0 & \Sigma \otimes I_{T-r} \end{pmatrix}. \tag{15}$$

Thus, the SURE model estimators $\{\hat{\beta}_i\}$ arise from the solution of the reduced sized model

$$\operatorname{vec}(\widetilde{Y}) = (\oplus_i R^d S_i) \operatorname{vec}(\{\beta_i\}) + \operatorname{vec}(\widetilde{U}), \tag{16}$$

where the variance-covariance matrix of $\operatorname{vec}(\widetilde{U})$ is given by $\Sigma \otimes I_r$.

From (14) and (16) it follows that the estimator $\Sigma_{(i+1)}$ in (6) is equivalent to

$$\Sigma_{(i+1)} = (\widetilde{U}_{(i)}^T \widetilde{U}_{(i)} + \widehat{Y}^T \widehat{Y})/T, \tag{17}$$

where $\widetilde{U}_{(i)}$ is the residual matrix of (16) at the ith iteration. Thus, the (upper triangular) Cholesky factor of $\Sigma_{(i+1)}$, denoted by $C_{(i+1)}$, can be computed from the QLD

$$Q_C^T \begin{pmatrix} \widetilde{U}_{(i)} \\ \widehat{Y} \end{pmatrix} = \begin{pmatrix} 0 \\ C_{(i+1)}^T \end{pmatrix} \begin{matrix} T-G \\ G \end{matrix}, \tag{18}$$

where $\Sigma_{(i+1)} = C_{(i+1)} C_{(i+1)}^T$ and $Q_C \in \Re^{T \times T}$ is orthogonal. However, if the QLD of $\widehat{Y}$ is given by

$$Q_Y^T \widehat{Y} = \begin{pmatrix} 0 \\ L_Y \end{pmatrix} \begin{matrix} T-r-K^d \\ K^d \end{matrix}, \tag{19}$$

then $C_{(i+1)}$ in (18) can be derived from the updated QLD

$$\widehat{Q}_C^T \begin{pmatrix} \widetilde{U}_{(i)} \\ L_Y \end{pmatrix} = \begin{pmatrix} 0 \\ C_{(i+1)}^T \end{pmatrix}. \tag{20}$$

Notice that if $K^d > T-r$, then $L_Y \in \Re^{(T-r) \times K^d}$ in (19) is lower trapezoidal. Algorithm 1 summarizes the iterative procedure for computing the FGLS estimator of SURE models with common regressors.

Consider now the case where there are no common regressors and $T \gg K$. That is, $X^d = (X_1 \ldots X_G) \in \Re^{T \times K^d}$, $K^d = K$,

$$Q_d^T X^d = \begin{pmatrix} R_d \\ 0 \end{pmatrix} \begin{matrix} K \\ T-K \end{matrix}, \tag{21a}$$

$$R_d = \begin{pmatrix} R_d^{(1)} & R_d^{(2)} & \cdots & R_d^{(G)} \end{pmatrix} \begin{matrix} k_1 & k_2 & \cdots & k_G \end{matrix} \tag{21b}$$

and

$$R_d^{(i)} = \begin{pmatrix} \widetilde{R}_d^{(i)} \\ \widehat{R}_d^{(i)} \\ 0 \end{pmatrix} \begin{matrix} K^{(i-1)} \\ k_i \\ K - K^{(i)} \end{matrix}$$

where $K^{(i)} = \sum_{j=1}^{i} k_j$, $\widehat{R}_d^{(i)}$ is upper triangular and in (16) $RS_i \equiv R_d^{(i)}$. As in the case of SURE models with common regressors, the computational burden of deriving the iterative FGLS estimator can be reduced significantly if the original model is transformed to the smaller in size SURE model (16).

Algorithm 1 Iterative estimation of the SURE model with common regressors

1: Compute the QRD $Q_d^T X^d = \begin{pmatrix} R^d \\ 0 \end{pmatrix}$ and $Q_d^T Y = \begin{pmatrix} \widetilde{Y} \\ \widehat{Y} \end{pmatrix}$

2: Compute the QLD $Q_Y^T \widehat{Y} = \begin{pmatrix} 0 \\ L_Y \end{pmatrix}$

3: Compute the QRDs $Q_i^T R^d S_i = \begin{pmatrix} R_i \\ 0 \end{pmatrix}$ and $Q_i^T \tilde{y}_i = \begin{pmatrix} \tilde{y}_i^* \\ \widehat{y}_i^* \end{pmatrix}$ $(i = 1, \ldots, G)$

4: Let $C_{(0)} = I_G$, $\text{vec}(\{\beta_j^{(0)}\}) = 0$ and $\text{vec}(\{v_j^*\}) = 0$

5: **for** $i = 1, 2, \ldots$ **do**

6:    **if** $i > 1$ **then**

7:       Compute $\begin{pmatrix} \oplus_j \widetilde{Q}_j^T \\ \oplus_j \widehat{Q}_j^T \end{pmatrix} (C_{(i-1)} \otimes I_r) \left(\oplus_j \widetilde{Q}_j \; \oplus_j \widehat{Q}_j \right) = \begin{pmatrix} \widetilde{W}_{11} & \widetilde{W}_{12} \\ \widetilde{W}_{21} & \widetilde{W}_{22} \end{pmatrix}$

8:       Compute the RQD $\begin{pmatrix} \widetilde{W}_{21} & \widetilde{W}_{22} \end{pmatrix} \widetilde{P} = \begin{pmatrix} 0 & W_{22} \end{pmatrix}$

9:       Compute $\begin{pmatrix} \widetilde{W}_{11} & \widetilde{W}_{12} \end{pmatrix} \widetilde{P} = \begin{pmatrix} W_{11} & W_{12} \end{pmatrix}$

10:      Solve the triangular system $W_{22} \text{vec}(\{v_j\}) = \text{vec}(\{\widehat{y}_j^*\})$

11:      Compute $\text{vec}(\{v_j^*\}) = W_{12} \text{vec}(\{v_j\})$

12:    **end if**

13:    Solve the triangular systems $R_j \beta_j^{(i)} = (\tilde{y}_j^* - v_j^*)$ $(j = 1, \ldots, G)$

14:    Compute the residuals $u_j^{(i)} = \tilde{y}_j - X\beta_j^{(i)}$ $(j = 1, \ldots, G)$

15:    Compute the updated QLD $\widehat{Q}_C^T \begin{pmatrix} \widetilde{U}_{(i)} \\ L_Y \end{pmatrix} = \begin{pmatrix} 0 \\ C_{(i)}^T \end{pmatrix}$, where $\widetilde{U}_{(i)} = \begin{pmatrix} u_1^{(i)} \ldots u_G^{(i)} \end{pmatrix}$

16: **end for until** $C_{(i)} = C_{(i-1)}$ **and** $\{\beta_j^{(i)}\} = \{\beta_j^{(i-1)}\}$.

4 Conclusions

A numerical and computational efficient method has been proposed to solve the SURE model with common regressors. The method is based on the GLLSP approach which does not require any matrix inversion and can derive the BLUE of the SURE model when Σ is singular [8,10]. The computation of the iterative FGLS estimator requires the solution of SURE models where the covariance matrix is re-estimated at each step. Thus, at each iteration step the QRD in (8b) for fixed Q and different C is computed. It has been shown how to transform the model to a smaller-in-size one. With this transformation both the computational cost and memory requirements for computing the QRD in (8b) are reduced significantly. Furthermore, this approach is found to be efficient also in the case where there are no common regressors and $T \gg K$.

Currently the complexity analysis of the algorithm, parallel strategies for solving the GQRD (8) and the adaptation of these numerical methods to solve other linear econometric models are investigated.

References

1. E. Anderson, Z. Bai, and J. J. Dongarra. Generalized QR factorization and its applications. *Linear Algebra and its Applications*, 162:243–271, 1992.
2. H. C. Andrews and J. Kane. Kronecker matrices, computer implementation, and generalized spectra. *Journal of the ACM*, 17(2):260–268, 1970.
3. P. J. Dhrymes. *Topics in Advanced Econometrics*, volume Vol.2: Linear and Nonlinear Simultaneous Equations. Springer–Verlag, New York, 1994.
4. Alexander Graham. *Kronecker products and matrix calculus: with applications*. Ellis Horwood Series in Mathematics and its Applications. Chichester: Ellis Horwood Limited, Publishers; New York etc.: Halsted Press: a division of John Wiley & Sons., 1986.
5. E. J. Kontoghiorghes. Parallel strategies for computing the orthogonal factorizations used in the estimation of econometric models. *Algorithmica*, 25:58–74, 1999.
6. E. J. Kontoghiorghes. *Parallel Algorithms for Linear Models: Numerical Methods and Estimation Problems*, volume 15 of *Advances in Computational Economics*. Kluwer Academic Publishers, 2000.
7. E. J. Kontoghiorghes. Parallel strategies for solving SURE models with variance inequalities and positivity of correlations constraints. *Computational Economics*, 15(1+2):89–106, 2000.
8. E. J. Kontoghiorghes and M. R. B. Clarke. An alternative approach for the numerical solution of seemingly unrelated regression equations models. *Computational Statistics & Data Analysis*, 19(4):369–377, 1995.
9. E. J. Kontoghiorghes and E. Dinenis. Solving triangular seemingly unrelated regression equations models on massively parallel systems. In M. Gilli, editor, *Computational Economic Systems: Models, Methods & Econometrics*, volume 5 of *Advances in Computational Economics*, pages 191–201. Kluwer Academic Publishers, 1996.
10. E. J. Kontoghiorghes and E. Dinenis. Computing 3SLS solutions of simultaneous equation models with a possible singular variance–covariance matrix. *Computational Economics*, 10:231–250, 1997.
11. S. Kourouklis and C. C. Paige. A constrained least squares approach to the general Gauss–Markov linear model. *Journal of the American Statistical Association*, 76(375):620–625, 1981.
12. C. C. Paige. Numerically stable computations for general univariate linear models. *Communications on Statistical and Simulation Computation*, 7(5):437–453, 1978.
13. C. C. Paige. Fast numerically stable computations for generalized linear least squares problems. *SIAM Journal on Numerical Analysis*, 16(1):165–171, 1979.
14. C. C. Paige. Some aspects of generalized QR factorizations. In M. G. Cox and S. J. Hammarling, editors, *Reliable Numerical Computation*, pages 71–91. Clarendon Press, Oxford, UK, 1990.
15. P. A. Regalia and S. K. Mitra. Kronecker products, unitary matrices and signal processing applications. *SIAM Review*, 31(4):586–613, 1989.
16. V. K. Srivastava and T. D. Dwivedi. Estimation of seemingly unrelated regression equations Models: a brief survey. *Journal of Econometrics*, 10:15–32, 1979.
17. V. K. Srivastava and D. E. A. Giles. *Seemingly Unrelated Regression Equations Models: Estimation and Inference (Statistics: Textbooks and Monographs)*, volume 80. Marcel Dekker, Inc., 1987.
18. A. Zellner. An efficient method of estimating seemingly unrelated regression equations and tests for aggregation bias. *Journal of the American Statistical Association*, 57:348–368, 1962.

Application of Boundary Collocation Method in Fluid Mechanics to Stokes Flow Problems

Anna Kucaba-Pietal

Department of Fluid Mechanics and Aerodynamics, Rzeszow University of Technology
Rzeszow, W.Pola 2, Poland

Abstract. The aim of this work is to present a short review of application of the boundary collocation technique to some problems in fluid mechanics. The steps used to find interaction between a wall and a sphere moving axisymetrically towards the flat wall in micropolar fluid are outlining to illustrate workability of the method.This problem occurs in modeling flow problems in microdevices as well as in human joins.

1 Introduction

Boundary methods are usually understood as numerical procedures which require the use of trial functions satisfying the differential equation and which reduce the boundary conditions to an approximate form. There are two main possibilities to formulate boundary methods; one is based on the use of the boundary integral equations and the second one, to use a system of trial functions. Methods based on trial functions are typically known as instances of the method of weigh residuals (MWR). Here, the trial functions are used as the basic functions for the truncated series expansion of the solution. The choice of the trial function is one of the features, which distinguish MWR from finite element and finite difference methods. The boundary collocation method (BCM) belongs to the class of MWR and is the most primitive version of this method. Its main disadvantage is that it is applicable only to linear problems. A review of application of BCM in mechanics of continuous media to date can be found in review article by Kolodziej [10].

The Stokes equations in fluid mechanics are linear and often used to describe creeping flows that appear in microhydrodynamics and biomechanics. These flows occur in colloids, suspension rheology, aerosols, microfabricated fluid systems (i.e. pumps, valves, microchannels, computer chips) and bio-flows. In the past few years several important advances have been made in the numerical treatment of some Stokes flow problems by application of collocation techniques. For instance, solution of Stokes equations was presented by Skalak and co-workers [20] for several different flow problems involving an infinite array of identical particles. This method was popularized by the work of Gluckman *et all* [6] which solved flow past finite assemblages of particles of an arbitrary shape. They examined the flows past finite arrays of axisymmetric bodies such as spheres and spheroids, which conform to special natural coordinates systems.

Over the subsequent decade BCM has been used to solve a wide range of problems. in Stokes flow. A few illustrative examples include an arbitrary convex body of revolution in [7], multiple spheres in a cylinder [17] and two spheroids in a uniform stream [18]. Ganatos and coworkers made major modification to the theory and extended it to handle variety of non-axisymmetric creeping flow problems with planar symmetry where the boundaries conform to more than a single orthogonal co-ordinate system. In [3] they studied quasi-steady steady time-dependent motion of three or more spheres settling under gravity in vertical planar configurations. Next the method has also been extended to bounded flow problems. In [4,5] motion of a sphere and spheroids was examined and the effect of the walls on hydrodynamic quantities on sphere motion for various flow geometry studied. Solutions for axisymmetric and three-dimensional motions of a torus in the presence of the walls were obtained in [11,12].

The conjunction of BCM with the boundary integral method permits to solve such problems as: a sphere in a circular orifice [1], hydrodynamic interaction of a three-dimensional finite cluster at arbitrarily sized spherical particles [9] - to cite a few. A fairly complete overview of the range of problems that has been successfully tackled by this method during the period 1978-1990 is available in the review article by Weinbaum et al. [21]. Conjunction of the collocation method with the perturbation method permitted to calculate resistance coefficient of a spherical particle moving in the presence of a deformed wall [15]. Algorithm proposed there can be applied to a wide class of bodies, shape of, which can be described in separable coordinates (ellipsoid, torus, spheroid, and sphere). All results cited above concern Newtonian Fluid. The first solution for Stokes flow past a sphere in bounded flow of non-Newtonian fluid, micropolar fluid, was derived by Kucaba-Pietal [14] (1999).

In general, the BCM is very efficient tool for a class of Stokes flow involving interactions between particles of simple shape. A cardinal rule for the application the collocation technique in solving Stokes flow problem is that the velocity disturbance produced by each co-ordinate boundary may be represented by an ordered sequence of fundamental solutions appropriate to the constant orthogonal surface to be described. These fundamental solutions for the velocity field are known for rectangular, cylindrical, spherical [16] and spheroidal [8] co-ordinates. Fundamental solution for Stokes equation in toroidal coordinates was found by Kucaba-Pietal (1985)[12]. The coefficients that appear in the fundamental solutions have to be calculated from the boundary conditions. The series, which represent solution, can be truncated and the boundary conditions are not applied exactly to the whole body but only at some carefully chosen points – collocation points. For more complicated regions (for example for bounded flows past a sphere) using the boundary conditions imposed on velocities along both confining walls we are able to invert analytically the Fourier-Bessel transform of fundamental solution which represent disturbance produced by the walls. In this manner, the original mixed co-ordinate, infinite domain boundary value problem is reduced to much simpler finite domain problem in which only the two infinite arrays of unknown coefficients, which appeared in fundamental solution described the moving body disturbance, need to be determined so as to satisfy the appropriate boundary conditions on the surface of the body.

The difficulty in construction of a collocation technique is not in formulation, which is conceptually simple, but in the detailed development of the truncation. As was demonstrated in appendix to Gluckman *et al.*, the numerical solution can oscillate and become unstable as the number of collocation points is increased if an inappropriate set of fundamental solution is used.

The aim of this work is to illustrate the power of the boundary collocation technique by outlining the steps used to find interaction between a wall and a sphere moving axisymetrically towards the flat wall in micropolar fluid. This problem occurs in microdevices as well as in human joins.

2 Formulation of the Problem

Let us consider a quasi-steady flow field of an incompressible micropolar fluid [2] due to a translational axisymmetrical motion of a sphere S_a of a radius a towards the wall. Figure 1 shows the separation between the sphere and the wall is denoted d.

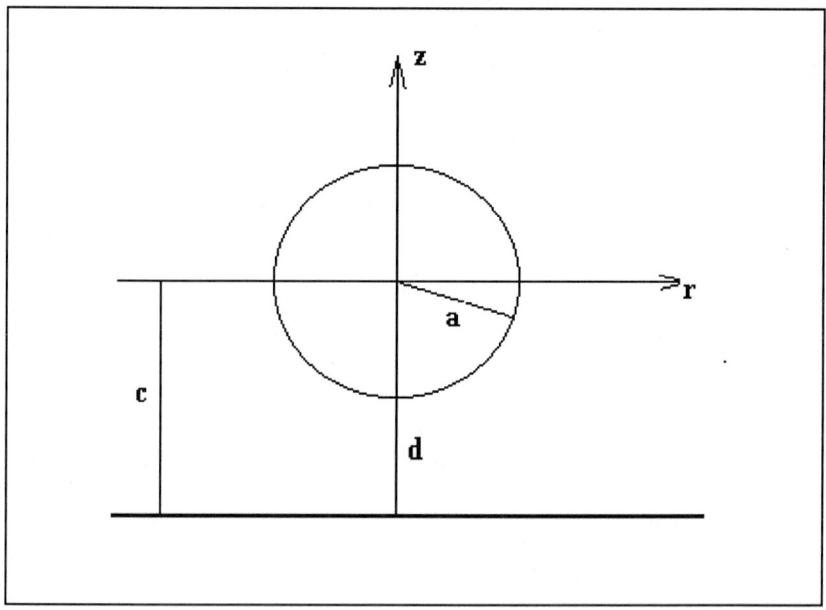

Fig. 1. Geometry of the flow

In the polar coordinate system (r, θ, z) with the origin in the center of the moving sphere the surface of a wall is described as $z = -c$; $c = d+a$. The translational velocity of the sphere S_a is $(0, 0, U)$. The fluid at infinity is at rest. The flow is at low Reynolds number. Because of axisymmetric geometry of the flow, the stream function $\Psi(r, z)$ can be used.

The equations of motion describing this flow are Stokes equations [14], and in terms of the stream function read:

$$-(\mu + \kappa)L_1^2\Psi + \kappa L_1(r\omega) = 0, \tag{1}$$

$$-\gamma L_1(r\omega) + \kappa L_1 \Psi - 2\kappa r \omega = 0 \tag{2}$$

In these equations ω is the microrotation vector. Positive constants μ, k, γ characterize isotropic properties of the micropolar fluid. L1 is the generalized axisymmetric Stokesian operator:

$$\left(\frac{\delta^2}{\delta R^2} - \frac{1}{R}\frac{\delta}{\delta R} + \frac{\delta^2}{\delta Z^2}\right) = L_1 \tag{3}$$

After elimination of the microrotation vector ω from equations (1), (2) we arrive at:

$$L_1^2(L_1 - \lambda^2)\Psi = 0, \tag{4}$$

with the microrotation given by:

$$\omega = \frac{1}{2}r\left(\frac{L_1\Psi + \gamma(\mu + \kappa)}{\kappa^2 L_1^2 \Psi}\right), \tag{5}$$

and constant λ_2 defined as:

$$\lambda^2 = \frac{\kappa(2\mu + \kappa)}{\gamma(\mu + \kappa)}. \tag{6}$$

The boundary conditions for Ψ and ω are on the sphere S_a:

$$\Psi = \frac{1}{2}r^2 U, \quad \frac{\delta \Psi}{\delta z} = 0, \tag{7}$$

$$\omega = \alpha_1 \frac{1}{2} \text{rot } v, \tag{8}$$

and on the wall: $z = -c$

$$\Psi = 0, \tag{9}$$

$$\omega = \alpha_2 \frac{1}{2} \text{rot } v \tag{10}$$

where constants $\alpha_2, \alpha_1 > 0$.

3 Algorithm for Receiving the Flow Field

The technique described below is based on the use of the fundamental solution of the Stokes equation and the application of the reciprocal Hankel transformation. As a consequence the solution can be expressed using a series in which the unknown constants appear. The algorithm for determining flow field can be summarized as follows:

1. *First*, the stream function Ψ is decomposed into two parts

$$\Psi = \Psi_1 + \Psi_2 \qquad (11)$$

a) Ψ_1 is fundamental solution represents an infinite series containing all of the simply separable solutions of Eqs. (1-2) in the spherical coordinates. These solutions are regular in the flow field and given by the formula [14]:

$$\Psi_1 = \sum (B_n \rho^{-n+1} + D_n \rho^{-n+3} + A_n I_{n-\frac{1}{2}}(\rho\lambda)) I_n(\zeta) \qquad (12)$$

where
- $\zeta = cos(\theta)$
- $I_n(\zeta)$ is the Gegenbauer function of the first kind of order n and degree - $\frac{1}{2}$,
- $I_{n-\frac{1}{2}}$ Bessel functions
- (ρ, θ, ζ) are sperical coordinates measured from the center of the sphere.

B_n, A_n, and D_n are unknow constants which will be determined from equation resulting from satisfying the non-slip boundary conditions on the surface of the sphere in the presence of the confining wall.

b) Ψ_2 is fundamental solution of Eqs. (1-2) in terms of cylindrical coordinates and represents an integral of all of the separable solutions which produce the finite velocities everywhere in the flow field and is given by the Fourier-Bessel integral [14]:

$$\Psi_2 = \int [B(\alpha)e^{-\alpha z} + D(\alpha)e^{-\alpha z}\alpha z + G(\alpha)e^{-\delta z}] J_1(\alpha, r) r /, d\alpha \qquad (13)$$

where
- $\delta = \sqrt{\lambda^2 + \alpha^2}$,
- $B(\alpha), \Delta(\alpha), \Gamma(\alpha)$ are unknown functions of the separation variable α
- J_1 denotes the Bessel function of the first kind of order unity.

The disturbances produced by the sphere along the wall can be completely reduced due to the solution of (13) obtained by the proper choice of functions $B(\alpha), \Delta(\alpha), \Gamma(\alpha)$.

2. Second, we apply the boundary condition equations on the wall (9-10) after replacing $\Psi 1$ and $\Psi 2$ by their series (12) and integral (13) representations respectively. As result we get equations which can be easily inverted and integration can be achieved by applying Hankel transforms. We are able now to express unknown functions $B(\alpha)$, $D(\alpha)$, $G(\alpha)$ by the series (12) and the original problem is reduced to the infinity domain of the flow. Thus the axial v_r and radial v_z velocities of the fluid flow and the microrotation ω can be rewritten in terms of the unknown functions: B_n, A_n, and D_n.
3. Third, we truncate the infinite series, which appears in the formulas defining the velocity and the microrotation. In order to obtain a unique solution, the boundary conditions on the sphere (7-8) are applied at a finite number of discrete points on the sphere. Then we solve a derived linear set of equations by numerical method to find B_n, A_n, and D_n. At this stage the solution is known.

4 Force

Very useful for the considered problem is the expression for the force acting on the sphere moving axisymmetrically in micropolar fluid, in terms of the stream function derived by Ramkisson, Majumdar [19]. It reads:

$$Fz = \int [\frac{2\mu + \kappa}{2} r^3 \frac{\delta}{\delta n}(L_1 \frac{\Psi}{r^2}) + \kappa \frac{\delta}{\delta n}(L_1 \Psi) + \frac{2\mu + \kappa}{2\kappa} r \frac{\delta}{\delta n}(L_1^2 \Psi)] ds \qquad (14)$$

5 Numerical Results and Conclusions

The algorithm was implemented in Fortran and run on PC with a 160 MHz Pentium processor. The scheme for spacing the collocation points on the surface of the sphere was based on the paper by Ganatos [5]. A unique feature of the approach was that the convergence and the accuracy of the solution could have been controlled simply by selecting the proper trial set of points on the surface of the sphere. To study this algorithm a series of calculations of the force (14) for various rheological parameters of the fluid and the non-dimensional distance *dis* between a sphere and the wall was performed. The parameter *dis* was defined $dis = d/a$. To investigate the influence of the wall we investigated wall corrector factor WCF as function of *dis*. The WCF was defined as the ratio of vale of calculated force to force acting on the sphere in unbounded flow. Some results are summarized in the Table 1.

Results show *similarity of behavior of force f* acting on a translating sphere in micropolar and classical fluid, but for given values *dis* and *a* the force *f* increases with increase of the ratio $K = \kappa/\mu$. So influence of rheological properties of the fluid on the force can be clearly observed. Summarizing, the following conclusions can be drawn:

Table 1. Drag correction factor WCF for the sphere moving towards a wall in micropolar fluid

WCF Dis	$\kappa/\mu = 2$	$\kappa/\mu = 1.5$	$\kappa/\mu = 0.5$	$\kappa/\mu = 0$ (Newtonian)
1	20.0	17.3	13.34	8.71
1.5	17.81	12.72	5.95	2.32
2	9.33	6.21	3.73	1.82
3	5.42	4.32	2.65	1.43
4	4.85	3.63	1.82	1.21
5	3.83	2.72	1.64	1.01
6	3.51	2.4	1.43	1.01
7	3.36	2.23	1.35	1.001
8	2.75	1.89	1.33	1.001
9	2.51	1.85	1.32	1.001
10	2.49	1.83	1.32	1.001

- Results show that the force acting on the moving body depends on the rheological properties of micropolar fluid and distance of the body from the wall.

- The area of an active interaction between a body and a wall is the most important factor, which increases the drag.

References

1. Dagan, Z. S., Weinbaum, R., Pfeffer, J.: Fluid Mech. **117** (1982) 143
2. Eringen, A. C.: J. Math. Mech. **16**, 1 (1966) 1-16
3. Ganatos, P., Pfeffer R., Weinbaum S. J.: Fluid Mech. **84** 1 (1978) 79-111
4. Ganatos, P., Pfeffer R., Weinbaum, S. J.: Fluid Mech. **99** (1980) 755-793
5. Ganatos, R., Weinbaum S. P., Pfeffer, J.: Fluid Mech. **99** (1980) 739-753
6. Gluckman, M. J., Pfeffer, R., Weinbaum, S. J.,: Fluid Mech. **50** (1971) 739-753
7. Gluckman, M. J., Weinbaum, S., Pfeffer, R. J.: Fluid Mech. **55** (1972) 677-703
8. Happel, H., Brenner, H.: Low Reynolds Number Hydrodynamics. Prentice Hall (1965)
9. Hasspnje, Q., Ganatos, P., Pfeffer, R. J.: Fluid Mech. **197** (1988) 1-37
10. Kolodziej, J.: SM Archives. **12/4** (1987) 187-231
11. Kucaba-Pietal, A.: Arch. Mech. **38** (1986) 647-663
12. Kucaba-Pietal, A.: Bull. of Polish Academy of Sciences. Tech. Sci. **36** (1988) 2-9, 501-511
13. Kucaba-Pietal, A.: Engn. Trans. **31**, 1, (1983) 151-161
14. Kucaba-Pietal, A. J.: Theor. Appl. Mech. **3**, 37, (1999) 593-606
15. Kucaba-Pietal, A. J.: Theor. Appl. Mech. **4**, 35, (1997) 813-827
16. Lamb, H.: Hydrodynamics. (1945)
17. Leichtberg, S., Weinbaum, S., Pfeffer, R., Gluckman, M. J.: Phil Trans R. Soc., London, A282 (1976) 585-613
18. Liao, W. H., Kruger, D. J.: Fluid Mech. **96** (1980) 223-247
19. Ramkisson, H., Majumdar R.: Physics of Fluids. **19**,1 (1976) 16-21
20. Skalak, R., Chen, T. C.: Appl. Sci. Res. **22** (1970) 403-425
21. Weinbaum, S., Ganatos, P., Yan, Z.: Ann. Rev. Fluid Mech.,**22**, (1990) 275-316

Strang-Type Preconditioners for Differential-Algebraic Equations

Siu-Long Lei and Xiao-Qing Jin*

Faculty of Science and Technology, University of Macau,
Macau, P.R. China
{fstlsl,fstxqj}@umac.mo

Abstract. We consider linear constant coefficient differential-algebraic equations (DAEs) $A\boldsymbol{x}'(t) + B\boldsymbol{x}(t) = \boldsymbol{f}(t)$ where A, B are square matrices and A is singular. If $\det(\lambda A + B)$ with $\lambda \in \mathbb{C}$ is not identically zero, the system of DAEs is solvable and can be separated into two uncoupled subsystems. One of them can be solved analytically and the other one is a system of ordinary differential equations (ODEs). We discretize the ODEs by boundary value methods (BVMs) and solve the linear system by using the generalized minimal residual (GMRES) method with Strang-type block-circulant preconditioners. It was shown that the preconditioners are nonsingular when the BVM is $A_{\nu,\mu-\nu}$-stable, and the eigenvalues of preconditioned matrices are clustered. Therefore, the number of iterations for solving the preconditioned systems by the GMRES method is bounded by a constant that is independent of the discretization mesh. Numerical results are also given.

Keywords: GMRES method, Strang-type block-circulant preconditioner, convergence rate, clustered spectrum, DAEs, ODEs, BVMs

AMS(MOS) Subject Classifications: 65F10, 65N22, 65L05, 65F15, 15A18

1 Introduction to DAE Solver

Consider the linear DAEs

$$\begin{cases} A\boldsymbol{x}'(t) + B\boldsymbol{x}(t) = \boldsymbol{f}(t) , & t \in (t_0, T] , \\ \boldsymbol{x}(t_0) = \boldsymbol{z} , \end{cases} \quad (1)$$

where A, B are $n \times n$ matrices and A is singular. This kind of problems arises in a wide variety of applications in electrical engineering and control theory, see [4].

A matrix pencil is defined by $\lambda A + B$ with $\lambda \in \mathbb{C}$. A pencil is said to be regular if $\det(\lambda A + B)$ is not identically zero. When $\lambda A + B$ is regular, then the

---
* Authors are supported by the research grant No. RG010/99-00S/JXQ/FST from the University of Macau.

equation (1) is solvable and there exists two nonsingular matrices P and Q such that
$$PAQ = \begin{bmatrix} I & 0 \\ 0 & N \end{bmatrix}, \qquad PBQ = \begin{bmatrix} G & 0 \\ 0 & I \end{bmatrix}.$$
Here the sum of the matrix sizes of N and G is n and N is a nilpotent matrix, i.e., there exists a positive integer ν such that $N^\nu = 0$ and $N^{\nu-1} \neq 0$, see [2]. To compute the matrix P and Q, we have the following constructive approach given in [7]:

(i) Let $B_1 = cA + B$ be nonsingular for some $c \in \mathbb{C}$. Then
$$B_1^{-1}(\lambda A + B) = B_1^{-1}(B_1 + (\lambda - c)A) = I + (\lambda - c)B_1^{-1}A$$
for all $\lambda \in \mathbb{C}$.

(ii) Let R be an invertible matrix such that $R^{-1}B_1^{-1}AR$ is in Jordan form. Here, R can be found by using the "jordan" command in Maple after doing with "linalg". By interchanging the columns of R, we can assume that $R^{-1}B_1^{-1}AR = \text{diag}\{J_1, J_0\}$. Here J_1 and J_0 are Jordan matrices where all the main diagonal entries of J_1 are nonzero and all the main diagonal entries of J_0 are zeros. Therefore,
$$R^{-1}(I + (\lambda - c)B_1^{-1}A)R = \begin{bmatrix} I + (\lambda - c)J_1 & 0 \\ 0 & (I - cJ_0) + \lambda J_0 \end{bmatrix}.$$

(iii) Then compute
$$\begin{bmatrix} I & 0 \\ 0 & (I - cJ_0)^{-1} \end{bmatrix} \begin{bmatrix} I + (\lambda - c)J_1 & 0 \\ 0 & (I - cJ_0) + \lambda J_0 \end{bmatrix}$$
$$= \begin{bmatrix} I + (\lambda - c)J_1 & 0 \\ 0 & I + \lambda(I - cJ_0)^{-1}J_0 \end{bmatrix}.$$

(iv) Since J_0 is nilpotent and $(I - cJ_0)^{-1}$ commutes with J_0, the matrix $(I - cJ_0)^{-1}J_0$ is also nilpotent. Let E be an invertible matrix such that $E^{-1}(I - cJ_0)^{-1}J_0 E = N$ is in Jordan form. Then we have
$$\begin{bmatrix} J_1^{-1} & 0 \\ 0 & E^{-1} \end{bmatrix} \begin{bmatrix} I + (\lambda - c)J_1 & 0 \\ 0 & I + \lambda(I - cJ_0)^{-1}J_0 \end{bmatrix} \begin{bmatrix} I & 0 \\ 0 & E \end{bmatrix}$$
$$= \begin{bmatrix} J_1^{-1} + (\lambda - c)I & 0 \\ 0 & I + \lambda N \end{bmatrix} = \begin{bmatrix} G & 0 \\ 0 & I \end{bmatrix} + \lambda \begin{bmatrix} I & 0 \\ 0 & N \end{bmatrix},$$
where $G = J_1^{-1} - cI$.

(v) Let P be the product of all the matrices used to multiply the matrix pencil $\lambda A + B$ on the left in steps (i)–(iv) and let Q be the product of all the matrices used to multiply the matrix pencil on the right in steps (i)–(iv). We have
$$P(\lambda A + B)Q = \lambda \begin{bmatrix} I & 0 \\ 0 & N \end{bmatrix} + \begin{bmatrix} G & 0 \\ 0 & I \end{bmatrix}.$$

The P and Q are our desired matrices.

Remark 1. Using this method to construct the matrices P and Q is only efficient when the system size is small.

Applying the coordinate changes P and Q to the DAEs in (1), we have

$$\begin{aligned} y_1' + Gy_1 &= g_1(t) , \\ Ny_2' + y_2 &= g_2(t) , \end{aligned} \quad (2)$$

where $Q^{-1}x = [y_1^T, y_2^T]^T$ and $Pf = [g_1^T, g_2^T]^T$. The first equation in (2) is a system of ODEs and a solution exists for any initial value of y_1. The second equation has only one solution

$$y_2(t) = \sum_{i=0}^{\nu-1}(-1)^i N^i g_2^{(i)}(t)$$

where $g_2^{(i)}(t)$ denotes the i-th order derivative of $g_2(t)$ with respect to t.

In the remainder of this paper, we concentrate on the first equation in (2) with a given initial condition.

2 The Matrix Forms of BVMs

Now, we consider the following general initial value problem (IVP),

$$\begin{cases} y'(t) = J_m y(t) + g(t) , & t \in (t_0, T] , \\ y(t_0) = z , \end{cases} \quad (3)$$

where $y(t), g(t) : \mathbb{R} \to \mathbb{R}^m$, and $J_m \in \mathbb{R}^{m \times m}$.

BVMs are methods based on linear multistep formulae (LMF) for solving ODEs, see [3]. For given IVP in (3), a BVM approximates its solution by means of a discrete boundary value problem. By using a μ-step LMF over a uniform mesh $t_j = t_0 + jh$, $j = 0, \cdots, s$, with $h = (T - t_0)/s$, we have

$$\sum_{i=-\nu}^{\mu-\nu} \alpha_{i+\nu} y_{n+i} = h \sum_{i=-\nu}^{\mu-\nu} \beta_{i+\nu} f_{n+i} , \quad n = \nu, \cdots, s - \mu + \nu . \quad (4)$$

Here, y_n is the discrete approximation to $y(t_n)$, $f_n = J_m y_n + g_n$ and $g_n = g(t_n)$.

The BVM in (4) must be used with ν initial conditions and $\mu - \nu$ final conditions. The initial condition in (3) only provides us with one value, we have to provide additional $(\mu - 1)$ equations:

$$\sum_{i=0}^{\mu} \alpha_i^{(j)} y_i = h \sum_{i=0}^{\mu} \beta_i^{(j)} f_i , \quad j = 1, \cdots, \nu - 1 , \quad (5)$$

and

$$\sum_{i=0}^{\mu} \alpha_{\mu-i}^{(j)} y_{s-i} = h \sum_{i=0}^{\mu} \beta_{\mu-i}^{(j)} f_{s-i} , \quad j = s - \mu + \nu + 1, \cdots, s . \quad (6)$$

By combining (4), (5) and (6), we obtain a linear system $My = b$ where
$$M = G \otimes I_m - hH \otimes J_m , \qquad (7)$$
$y = [y_0^T, \cdots, y_s^T]^T \in \mathbb{R}^{(s+1)m}$, and $b = e_1 \otimes z + h(H \otimes J_m)g$ with $e_1 = [1, 0, \cdots, 0]^T \in \mathbb{R}^{(s+1)}$ and $g = [g_0^T, \cdots, g_s^T]^T \in \mathbb{R}^{(s+1)m}$. The matrix $G \in \mathbb{R}^{(s+1) \times (s+1)}$ in (7) is defined by:

$$G = \begin{bmatrix} 1 & \cdots & 0 & & & & & & \\ \alpha_0^{(1)} & \cdots & \alpha_\mu^{(1)} & & & & & & \\ \vdots & & \vdots & & & & & & \\ \alpha_0^{(\nu-1)} & \cdots & \alpha_\mu^{(\nu-1)} & & & & 0 & & \\ \alpha_0 & \cdots & \alpha_\mu & & & & & & \\ & \alpha_0 & \cdots & & \alpha_\mu & & & & \\ & & \ddots & \ddots & & \ddots & & & \\ & & & \ddots & \ddots & & \ddots & & \\ & & & & \alpha_0 & \cdots & & \alpha_\mu & \\ & 0 & & & \alpha_0^{(s-\mu+\nu+1)} & \cdots & & \alpha_\mu^{(s-\mu+\nu+1)} & \\ & & & & \vdots & & & \vdots & \\ & & & & \alpha_0^{(s)} & \cdots & & \alpha_\mu^{(s)} & \end{bmatrix}$$

and $H \in \mathbb{R}^{(s+1) \times (s+1)}$ in (7) is defined similarly by using $\{\beta_i^{(j)}\}$ instead of $\{\alpha_i^{(j)}\}$ in G for all $i = 1, 2, \cdots, \mu$ and $j = 1, 2, \cdots, s$, and the first row of H is zero. The advantage in using BVMs is that they have much better stability properties than traditional initial value methods, see [3].

3 Construction of Preconditioner

The following preconditioner for (7) is proposed:
$$S \equiv \tilde{s}(G) \otimes I_m - h\tilde{s}(H) \otimes J_m , \qquad (8)$$
where $\tilde{s}(G) \in \mathbb{R}^{(s+1) \times (s+1)}$ is defined by

$$\tilde{s}(G) = \begin{bmatrix} \alpha_\nu & \cdots & \alpha_\mu & & & \alpha_0 & \cdots & \alpha_{\nu-1} \\ \vdots & \ddots & & \ddots & & & \ddots & \vdots \\ \alpha_0 & & \ddots & & \ddots & & & \alpha_0 \\ & \ddots & & \ddots & & \ddots & 0 & \\ & & \ddots & & \ddots & & \ddots & \\ & 0 & & \ddots & & \ddots & & \\ \alpha_\mu & & \ddots & & \ddots & & & \alpha_\mu \\ \vdots & \ddots & & \ddots & & \ddots & & \vdots \\ \alpha_{\nu+1} & \cdots & \alpha_\mu & & & \alpha_0 & \cdots & \alpha_\nu \end{bmatrix}$$

and $\tilde{s}(H) \in \mathbb{R}^{(s+1)\times(s+1)}$ is defined similarly by using $\{\beta_i\}$ instead of $\{\alpha_i\}$ in $\tilde{s}(G)$. The sequences $\{\alpha_i\}_{i=0}^{\mu}$ and $\{\beta_i\}_{i=0}^{\mu}$ are the coefficients in (4). We note that S is the Strang-type block-circulant preconditioner proposed in [5].

The invertibility of S depends on the stability of the BVM that we used to discretize (3). The stability of a BVM is closely related to two characteristic polynomials defined as follows:

$$\rho(z) \equiv z^\nu \sum_{j=-\nu}^{\mu-\nu} \alpha_{j+\nu} z^j \quad \text{and} \quad \sigma(z) \equiv z^\nu \sum_{j=-\nu}^{\mu-\nu} \beta_{j+\nu} z^j . \qquad (9)$$

Definition 1 ([3]). *Consider a BVM with the characteristic polynomials $\rho(z)$ and $\sigma(z)$ defined by (9). The region*

$$\mathcal{D}_{\nu,\mu-\nu} \equiv \{q \in \mathbb{C} : \rho(z) - q\sigma(z) \text{ has } \nu \text{ zeros inside } |z| = 1$$
$$\text{and } \mu - \nu \text{ zeros outside } |z| = 1\}$$

is called the region of $A_{\nu,\mu-\nu}$-stability of the given BVM. Moreover, the BVM is said to be $A_{\nu,\mu-\nu}$-stable if $\mathbb{C}^- \equiv \{q \in \mathbb{C} : Re(q) < 0\} \subseteq \mathcal{D}_{\nu,\mu-\nu}$.

Theorem 1 ([5]). *If the BVM for (3) is $A_{\nu,\mu-\nu}$-stable and $h\lambda_k \in \mathcal{D}_{\nu,\mu-\nu}$ where λ_k ($k = 1, \cdots, m$) are the eigenvalues of J_m, then the preconditioner $S = \tilde{s}(G) \otimes I_m - h\tilde{s}(H) \otimes J_m$ is nonsingular. In particular, S is nonsingular if $\lambda_k \in \mathbb{C}^-$.*

It is well known that if the spectrum of the preconditioned system is clustered, then the GMRES method applied for solving the preconditioned system will converge very fast.

Theorem 2 ([5]). *All the eigenvalues of the preconditioned matrix $S^{-1}M$ are 1 except for at most $2m\mu$ outliers. The GMRES method, when applied for solving the preconditioned system $S^{-1}M\mathbf{y} = S^{-1}\mathbf{b}$, will converge in at most $2m\mu + 1$ iterations in exact arithmetic.*

Regarding the operation cost of the method, we refer to [5].

4 Numerical Example

In this section, we compare the Strang-type block-circulant preconditioner with other preconditioners by solving the subsystem of ODEs extracted from a system of DAEs. All the experiments are preformed in MATLAB with machine precision 10^{-16}. The GMRES method [6] is employed to solve linear systems. We use the MATLAB-provided M-file "gmres" (see MATLAB on-line documentation) in our implementation. In our example, the zero vector is the initial guess and the stopping criterion is $||\mathbf{r}_q||_2/||\mathbf{r}_0||_2 < 10^{-6}$ where $\mathbf{r}_q$ is the residual after q iterations.

Example 1. Consider

$$\begin{cases} A\boldsymbol{x}'(t) + B\boldsymbol{x}(t) = \boldsymbol{0}, & t \in (0,1], \\ \boldsymbol{x}(0) = [1,1,1,1,1,1,1]^T, \end{cases}$$

where

$$A = \begin{bmatrix} 50 & 114 & 95 & 140 & 129 & 91 & 43 \\ 101 & 198 & 149 & 155 & 223 & 183 & 138 \\ 97 & 206 & 156 & 197 & 187 & 156 & 87 \\ 82 & 185 & 148 & 164 & 156 & 129 & 81 \\ 82 & 202 & 167 & 186 & 201 & 180 & 114 \\ 111 & 226 & 193 & 197 & 229 & 198 & 138 \\ 32 & 122 & 107 & 100 & 115 & 100 & 74 \end{bmatrix}, \quad B = \begin{bmatrix} 79 & 156 & 158 & 209 & 188 & 69 & 47 \\ 87 & 256 & 161 & 162 & 241 & 203 & 162 \\ 168 & 264 & 203 & 272 & 223 & 78 & 52 \\ 180 & 260 & 189 & 229 & 255 & 142 & 111 \\ 135 & 295 & 243 & 250 & 282 & 200 & 158 \\ 188 & 357 & 268 & 298 & 337 & 261 & 185 \\ 53 & 167 & 141 & 88 & 196 & 174 & 166 \end{bmatrix}.$$

Then there are two invertible matrices

$$P^{-1} = \begin{bmatrix} 9 & 0 & 1 & 3 & 2 & 4 & 6 \\ 2 & 8 & 4 & 8 & 1 & 8 & 3 \\ 6 & 4 & 9 & 0 & 0 & 5 & 8 \\ 4 & 6 & 9 & 1 & 7 & 2 & 5 \\ 8 & 7 & 4 & 2 & 4 & 6 & 7 \\ 7 & 9 & 8 & 1 & 9 & 8 & 4 \\ 4 & 7 & 0 & 6 & 4 & 0 & 3 \end{bmatrix}, \quad Q^{-1} = \begin{bmatrix} 1 & 3 & 6 & 7 & 4 & 1 & 0 \\ 1 & 8 & 8 & 3 & 6 & 9 & 8 \\ 6 & 8 & 6 & 8 & 6 & 2 & 1 \\ 3 & 5 & 3 & 5 & 7 & 2 & 2 \\ 5 & 4 & 2 & 3 & 9 & 8 & 6 \\ 1 & 8 & 3 & 7 & 5 & 7 & 2 \\ 6 & 8 & 5 & 5 & 8 & 1 & 4 \end{bmatrix}$$

such that

$$PAQ = \begin{bmatrix} I_4 & 0 \\ 0 & N \end{bmatrix}, \quad PBQ = \begin{bmatrix} C & 0 \\ 0 & I_3 \end{bmatrix}$$

where

$$N = \begin{bmatrix} 0 & 0 & 0 \\ 1 & 0 & 0 \\ 0 & 1 & 0 \end{bmatrix} \quad \text{and} \quad C = \begin{bmatrix} 2 & 0 & 0 & 0 \\ -1 & 2 & 0 & 0 \\ 0 & -1 & 2 & 0 \\ 0 & 0 & -1 & 2 \end{bmatrix}.$$

Now, we show the efficiency of solving the following IVP,

$$\begin{cases} \boldsymbol{y}'(t) = -C\boldsymbol{y}(t), & t \in (0,1], \\ \boldsymbol{y}(0) = [22, 43, 37, 27]^T. \end{cases}$$

The third order generalized backward differentiation formula is used to solve this system of ODEs. The formulae and the additional initial and final equations can be found in [3].

Table 1 lists the number of iterations required for convergence of the GMRES method with different preconditioners. In the table, I means no preconditioner is used and S denotes the Strang-type block-circulant preconditioner defined as in (8). For a comparison, we introduce T. Chan's preconditioner and Bertaccini's preconditioner, see [1]. T. Chan's block-circulant preconditioner T is defined as

$$T \equiv c(G) \otimes I_m - hc(H) \otimes J_m$$

where the diagonals $\hat{\alpha}_j$ of $c(G)$ are given by

$$\hat{\alpha}_j = \left(1 - \frac{j}{s+1}\right)\alpha_{j+\nu} + \frac{j}{s+1}\alpha_{j+\nu-(s+1)}, \quad j = 0, \cdots, s, \tag{10}$$

and the diagonals $\hat{\beta}_j$ of $c(H)$ are defined similarly by replacing $\alpha_{j+\nu}$ by $\beta_{j+\nu}$ and $\alpha_{j+\nu-(s+1)}$ by $\beta_{j+\nu-(s+1)}$ in (10). And Bertaccini's block-circulant preconditioner P is defined as

$$P \equiv \tilde{G} \otimes I_m - h\tilde{H} \otimes J_m$$

where the diagonals $\tilde{\alpha}_j$ of $\tilde{G}$ are given by

$$\tilde{\alpha}_j = \left(1 + \frac{j}{s+1}\right)\alpha_{j+\nu} + \frac{j}{s+1}\alpha_{j+\nu-(s+1)}, \quad j = 0, \cdots, s, \tag{11}$$

and the diagonals $\tilde{\beta}_j$ of $\tilde{H}$ are defined similarly by replacing $\alpha_{j+\nu}$ by $\beta_{j+\nu}$ and $\alpha_{j+\nu-(s+1)}$ by $\beta_{j+\nu-(s+1)}$ in (11).

We see from Table 1 that when s increases, the numbers of iterations required for convergence stay almost the same when preconditioners are used. However, the number of iterations increases when no preconditioner is used. It is also clear that the Strang-type block-circulant preconditioner performs better than T. Chan's and Bertaccini's preconditioners. To further illustrate the clustering property in Theorem 2, we give in Figures 1,2 the spectra of the preconditioned matrices with the preconditioner S and the spectra with no preconditioner, for $s = 12, 48$.

Table 1. Number of iterations required for convergence

s	GMRES method			
	I	S	T	P
6	22	11	13	15
12	31	11	13	15
24	44	11	14	16
48	73	11	14	16
96	132	11	14	16

References

1. Bertaccini, D.: A Circulant Preconditioner for the Systems of LMF-Based ODE Codes. SIAM J. Sci. Comput. (to appear)
2. Brenan, K., Campbell, S., Petzold, L.: Numerical Solution of Initial-Value Problems in Differential-Algebraic Equations. SIAM Press, Philadelphia, 1996
3. Brugnano, L., Trigiante, D.: Solving Differential Problems by Multistep Initial and Boundary Value Methods. Gordan and Berach Science Publishers, Amsterdam, 1998

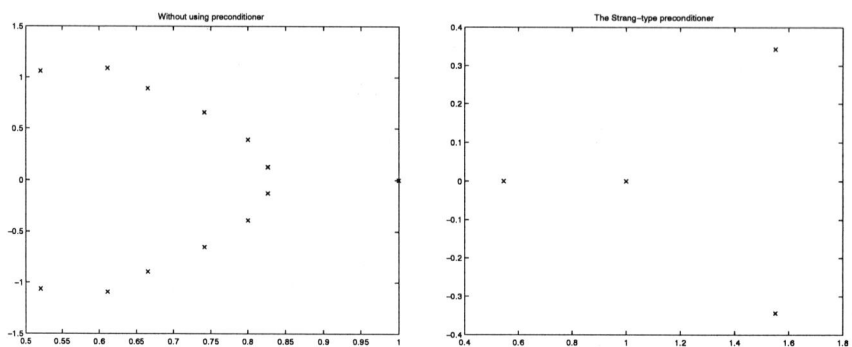

Fig. 1. Spectra of matrices M(left) and $S^{-1}M$(right) when $s = 12$

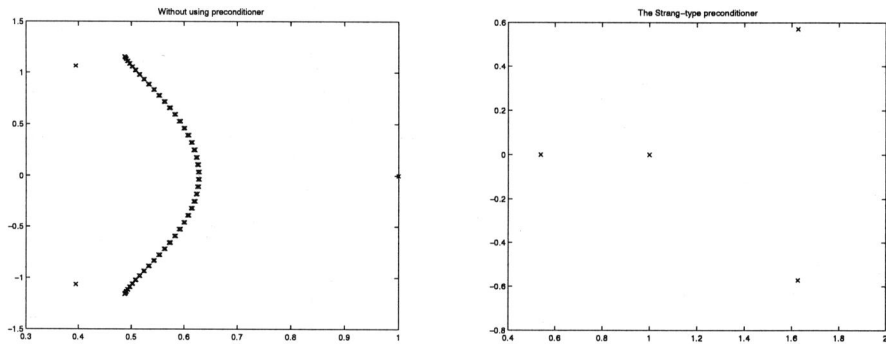

Fig. 2. Spectra of matrices M(left) and $S^{-1}M$(right) when $s = 48$

4. Campbell, S.: Singular Systems of Differential Equations. Pitman, 1980
5. Chan, R., Ng, M., Jin, X.: Strang-type Preconditioners for Systems of LMF-Based ODE Codes. IMA J. Numer. Anal. (to appear)
6. Saad, Y., Schultz, M.: GMRES: a Generalized Minimal Residual Algorithm for Solving Non-Symmetric Linear Systems. SIAM J. Sci. Stat. Comput. **7** (1986) 856–869
7. Shirvani, M., So, J.: Solutions of Linear Differential Algebraic Equations. SIAM Review **40** (1998) 344–346

Solvability of Runge-Kutta and Block-BVMs Systems Applied to Scalar ODEs*

G. Di Lena and F. Iavernaro

Dipartimento di Matematica, Università di Bari,
Via E. Orabona 4, I-70125 Bari, Italy

Abstract. A characterization of P_0 matrices is reported and used to derive simple necessary and sufficient conditions for the unique solvability of a class of nonlinear systems of equations depending on a parameter. An application to the problem of existence and uniqueness of the solutions of one-step implicit schemes applied to scalar ODEs is also presented.

1 Introduction

A square, real matrix $\mathcal{A} = (\alpha_{ij}) \in \mathbb{R}^{n \times n}$ is said to be a P matrix (P_0 matrix) if all its principal minors are positive (nonnegative). A well known characterization of P matrices is the following (see [8]): $\mathcal{A}$ is a P matrix if and only if for each real vector $\mathbf{x} \in \mathbb{R}^n$, $\mathbf{x} \neq 0$, there exists an index $i \in \{1, \ldots, n\}$ such that $x_i \sum_{j=1}^{n} \alpha_{ij} x_j > 0$.

The analogous characterization of P_0 matrices is reported in the following theorem.

Theorem 1. *The following statements are equivalent:*

(i) $\mathcal{A}$ *is a* P_0 *matrix;*
(ii) for each real vector $\mathbf{x} \in \mathbb{R}^n$, $\mathbf{x} \neq 0$, *there exists an index* $i \in \{1, \ldots, n\}$ *such that*

$$x_i \neq 0 \quad \text{and} \quad x_i \sum_{j=1}^{n} \alpha_{ij} x_j \geq 0.$$

Proof. It is sufficient to observe that $\mathcal{A}$ is a P_0 matrix if and only if $\mathcal{A} + \epsilon I$ is a P matrix for any $\epsilon > 0$ (I stands for the identity matrix of dimension n). The assertion is then a direct consequence of the above mentioned characterization of P matrices. □

Stated differently, a P_0 matrix $\mathcal{A}$ is a P matrix or a vector $\mathbf{x}$ may be found with $x_i \neq 0$ and $\sum_{j=1}^{n} \alpha_{ij} x_j = 0$. In the next section we consider an application of this result to the problem of unique solvability of the nonlinear system of equations

$$A\mathbf{y} - hBF(\mathbf{y}) = \mathbf{b}, \tag{1}$$

---
* Work supported by MURST and GNIM.

where A and B are real square matrices of dimension n with $det(A) \neq 0$, $\mathbf{y} = [y_1, \ldots, y_n]^T$ is a vector of unknowns, $\mathbf{b} = [b_1, \ldots, b_n]^T$ is a known term, h is a nonnegative parameter and $F(\mathbf{y}) = [f_1(y_1), f_2(y_2), \ldots, f_n(y_n)]^T$, where $f_i : \mathbb{R} \longrightarrow \mathbb{R}$ are continuous functions satisfying the monotonicity condition

$$\frac{f_i(x) - f_i(y)}{x - y} \leq \mu \qquad \forall x, y \in \mathbb{R}, \; x \neq y. \tag{2}$$

We denote by $\mathcal{F}_\mu$ the set of all functions F whose components f_i satisfy (2). Systems of the form (1) arise, for example, when an implicit one-step method (such as a Runge-Kutta (RK) or a Boundary Value Method (BVM)) is used to solve the scalar Initial Value Problem (IVP)

$$\begin{cases} y'(t) = f(t, y) & t \in [t_0, t_f] \\ y(t_0) = y_0 \end{cases}, \tag{3}$$

where f satisfies a one-sided Lipschitz condition with one-sided Lipschitz constant μ. In this case the parameter h stands for the stepsize of integration. The more general problem of determining equivalent conditions to the existence and uniqueness of solutions of the internal stages of a RK-method when f is a vector function, has been fully studied (see for example [2,3,4,6,7,9,10] and references therein). These conditions essentially lead to certain restrictions on the choice of the stepsize h which are independent of the dimension of the IVP. In Section 3 we show that such restrictions become weaker when the scalar case is considered. For a reference on Boundary Value Methods, their definition, properties and implementation techniques see for example [1,11,12].

For reasons that will be clear in the sequel it is more convenient to recast Theorem 1 into an equivalent form. We recall that the Hadamard (or Schur) product $\mathbf{x} \circ \mathbf{y}$ of two vectors $\mathbf{x}, \mathbf{y} \in \mathbb{R}^n$, is a vector of length n whose components are $(\mathbf{x} \circ \mathbf{y})_i = x_i y_i$ (see [8]). The standard notation $\mathbf{x} > 0$ ($\mathbf{x} \geq 0$, $\mathbf{x} < 0$, $\mathbf{x} \leq 0$) is used to consider vectors with positive (nonnegative, negative, nonpositive) components. Associated to a real matrix $\mathcal{A}$ of dimension n we consider the following set of vectors:

$$\chi(\mathcal{A}) = \{ \mathbf{x} \in \mathbb{R}^n | \mathbf{x} \circ \mathcal{A}\mathbf{x} \leq 0 \; \text{ and } \; ((\mathcal{A}\mathbf{x})_i = 0 \Rightarrow x_i = 0) \}.$$

Observe that $\chi(\mathcal{A}) \neq \emptyset$ since $\mathbf{0} \in \chi(\mathcal{A})$. It follows that

$$\mathbf{x} \in \chi(\mathcal{A}) \Leftrightarrow \begin{cases} x_i \sum_{j=1}^n \alpha_{ij} x_j \leq 0, & \forall i = 1, \ldots, n, \\ \sum_{j=1}^n \alpha_{ij} x_j = 0 \Rightarrow x_i = 0, \end{cases}$$

and Theorem 1 may be rewritten in a more compact notation as follows:

$$\mathcal{A} \text{ is a } P_0 \text{ matrix} \quad \Leftrightarrow \quad \chi(\mathcal{A}) = \{\mathbf{0}\}. \tag{4}$$

We finally observe that $\chi(0) = \{\mathbf{0}\}$ and for each $\beta \in \mathbb{R}$ and $\mathbf{x} \in \chi(\mathcal{A})$, $\beta \mathbf{x} \in \chi(\mathcal{A})$, that is $\chi(\mathcal{A})$ is star shaped with respect to the origin.

2 P_0 Matrices and the Solution of the System (1)

Hereafter our interest is in the characterization of the unique solvability of the system (1) under the assumption (2) in terms of conditions related to the matrices A and B and the scalars h and μ. It is well known that this problem poses an upper bound on the product $h\mu$. If γ is a given real number, the system (1) is said to be uniquely solvable on the interval $]-\infty, \gamma[$ if it admits a unique solution whenever $h\mu \in]-\infty, \gamma[$. Our goal is to find the optimal value of γ, say $\bar{\gamma}$, defined as

$$\bar{\gamma} = \sup\{\gamma|\text{ the system (1) is uniquely solvable on the interval }]-\infty, \gamma[\}. \tag{5}$$

The results reported in the present section will be later compared to those known in literature and derived in a more general context.

Since the existence of the solution of (1) on the open interval $]-\infty, \gamma[$ is a consequence of its uniqueness (see [5,7,10]), we confine our study to this latter question.

Suppose $\mathbf{b}_1, \mathbf{b}_2 \in \mathbb{R}^n$ are given and consider the systems $A\mathbf{y}^{(1)} - hBF(\mathbf{y}^{(1)}) = \mathbf{b}_1$ and $A\mathbf{y}^{(2)} - hBF(\mathbf{y}^{(2)}) = \mathbf{b}_2$. Subtracting yields

$$A\Delta\mathbf{y} - hB\Delta F = \Delta\mathbf{b}, \tag{6}$$

with $\Delta\mathbf{y} = \mathbf{y}^{(2)} - \mathbf{y}^{(1)}$, $\Delta F = F(\mathbf{y}^{(2)}) - F(\mathbf{y}^{(1)})$ and $\Delta\mathbf{b} = \mathbf{b}_2 - \mathbf{b}_1$. Uniqueness of the solution of (1) is then equivalent to requiring $\Delta\mathbf{b} = \mathbf{0} \Rightarrow \Delta\mathbf{y} = \mathbf{0}$. Observe that choosing $F(\mathbf{y}) = \mu\mathbf{y}$ (the linear case), a necessary condition for the uniqueness is founded to be $det(A - h\mu B) \neq 0$. Under this assumption the equation (6) with $\Delta\mathbf{b} = \mathbf{0}$ is equivalent to

$$\Delta\mathbf{y} = h(A - h\mu B)^{-1} B(\Delta F - \mu\Delta\mathbf{y}). \tag{7}$$

The theorem below establishes the link between the present problem and P_0 matrices.

Theorem 2. *Assume that $(A - h\mu B)$ is nonsingular and define $\mathcal{A} = h(A - h\mu B)^{-1}B$. The following statements are equivalent:*

(a) the solution of (1) is unique for any $F \in \mathcal{F}_\mu$;
(b) $\chi(\mathcal{A}) = \{\mathbf{0}\}$.

Proof. (a) $\Rightarrow$ (b). Assume there exists $\mathbf{x} \neq \mathbf{0}$ such that $\mathbf{x} \in \chi(\mathcal{A})$ and set $\Delta\mathbf{y} = \mathcal{A}\mathbf{x}$. For $i = 1, \ldots, n$ consider the linear functions $f_i(z) = \alpha_i z$, where

$$\alpha_i = \begin{cases} \mu + \dfrac{x_i}{\Delta y_i} & \text{if } x_i \neq 0, \\ \mu & \text{if } x_i = 0. \end{cases}$$

This definition is well posed since $x_i \neq 0 \Rightarrow \Delta y_i \neq 0$. Furthermore, observing that for each $i = 1, \ldots, n$, $x_i \Delta y_i \leq 0$, it is easily seen that $F(\mathbf{y}) =$

$[f_1(y_1), \ldots, f_n(y_n)]^T \in \mathcal{F}_\mu$ and the vector $\Delta \mathbf{y} \neq \mathbf{0}$ satisfies (7) which compared to (a) gives a contradiction.

$(b) \Rightarrow (a)$. Let $\Delta \mathbf{y}$ be solution of (7), with $F \in \mathcal{F}_\mu$. Defining $\mathbf{x} = \Delta F - \mu \Delta \mathbf{y}$, equation (7) assumes the form

$$\Delta \mathbf{y} = \mathcal{A} \mathbf{x}. \tag{8}$$

From the expression of $\mathbf{x}$, one can verify that $\Delta y_i x_i \leq 0$, and $\Delta y_i = 0 \Rightarrow x_i = 0$, that together with formula (8) imply that $\mathbf{x} \in \chi(\mathcal{A})$. From (b) we deduce $\mathbf{x} = \mathbf{0}$ and consequently $\Delta \mathbf{y} = \mathbf{0}$. □

Now we consider the family of matrices $\mathcal{A}_\gamma = (A - \gamma B)^{-1} B$ depending on the parameter $\gamma \in \Omega \subset \mathbb{R}$. The set Ω consists of all values of γ such that the corresponding elements of the family are well defined, that is $A - \gamma B$ is nonsingular. However it should be observed that if for a given $\gamma_1 \in \mathbb{R}$, $\mathcal{A}_{\gamma_1}$ exists and is a P_0 matrix, then for each $\gamma \leq \gamma_1$, $A - \gamma B$ is nonsingular and $\mathcal{A}_\gamma$ is still a P_0 matrix. This property is a direct consequence of Theorem 2 and the fact that uniqueness of the solution of (1) in the class $\mathcal{F}_{\mu_1}$ also implies uniqueness in the class $\mathcal{F}_\mu$, for each $\mu \leq \mu_1$. It follows that the values of the parameter γ which make $\mathcal{A}_\gamma$ a P_0 matrix (if any exists), form an interval of the form $]-\infty, \delta]$ if $\delta \in \Omega$ or $]-\infty, \delta[$ if $\delta \notin \Omega$: this provides the basis for characterizing the number $\bar{\gamma}$ as defined in (5). In the sequel we adopt the convention $\sup \emptyset = -\infty$.

Theorem 3. *The following expressions hold true for the scalar $\bar{\gamma}$:*

$$\bar{\gamma} = \sup\{\gamma \in \Omega \mid \mathcal{A}_\gamma \text{ is a } P_0 \text{ matrix }\}; \tag{9}$$

if B is nonsingular,

$$\bar{\gamma} = \min\{\lambda \mid \lambda \text{ is a real eigenvalue of any principal submatrix of } B^{-1}A\}. \tag{10}$$

Proof. Formula (9) is a natural consequence of Theorem 2 and the previous discussion (we observe that if $\det(A - \bar{\gamma}B) \neq 0$ then $\bar{\gamma}$ is a maximum). When B is nonsingular, after a simple manipulation one can check that $\mathcal{A}_\gamma = (B^{-1}A - \gamma I)^{-1}$. From (9) and considering that the inverse of a P_0 matrix is a P_0 matrix, we are conducted to seek the values of γ that make $(B^{-1}A - \gamma I)$ a P_0 matrix. Using the fact that P matrices are characterized by having positive all the real eigenvalues of their principal submatrices we arrive at formula (10). □

In the case when B is nonsingular, the number $\bar{\gamma}$ may be explicitly determined by formula (10). Alternatively, working on the principal minors of the matrix $\mathcal{A}_\gamma$, one can locate $\bar{\gamma}$ by means of an iterative procedure (such as the bisection method) that produces a sequence $\{\gamma_i\}$ convergent to $\bar{\gamma}$. We consider further applications in the next section.

3 Unique Solvability of RK and BVM Systems

As mentioned in the introduction the study of conditions for the existence and uniqueness of systems of the form (1) has been conducted by a number of authors when the vectors $\mathbf{y}$, $F(\mathbf{y})$ and $\mathbf{b}$ have components in $\mathbb{R}^m$ with m a given positive integer. In such a general case formula (1) is replaced by $<\mathbf{f}_i(\mathbf{x})-\mathbf{f}_i(\mathbf{y}),\mathbf{x}-\mathbf{y}>\leq \mu||\mathbf{x}-\mathbf{y}||^2$, $\mathbf{x},\mathbf{y}\in\mathbb{R}^m$, where $<\cdot,\cdot>$ is an inner product on $\mathbb{R}^m$ and the matrices A and B are viewed as linear operators $(\mathbb{R}^m)^n \longrightarrow (\mathbb{R}^m)^n$, that is for example, $A\mathbf{y}$ stands for $(A\otimes I)\mathbf{y}$ with I the m-dimensional identity matrix. The application of a RK or a BVM formula to the solution of m-dimensional ODEs leads to such kind of systems. In particular for RK methods the solution of (1) represents the vector of internal stages so that the matrix A is the identity matrix while the coefficient of B are defined by the Bucher array. In the case of BVMs the system (1) itself represents the discrete counterpart of the continuous problem and the entries of A and B define the way this discretization is carried out. In all concrete cases the matrix A is nonsingular and therefore all the results obtained in the RK context are also available for BVMs (via a left multiplication of the system by A^{-1}). Kraaijevanger and Schneid [10] proved that when $G=A^{-1}B$ is irreducible and nonsingular, the unique solvability on the interval $]-\infty,\gamma[$ corresponds to requiring that the matrix $G^{-1}-\gamma I$ is Lyapunov diagonally semi-stable, namely that a positive diagonal matrix D exists such that the matrix $(G^{-1}-\gamma I)^T D + D(G^{-1}-\gamma I)$ is positive semidefinite. More precisely they introduced the definition of suitability of the system (1) on intervals of the form $]-\infty,\gamma[$ which means that it is uniquely solvable on that interval whatever the choice of the dimension m. How one should expect (see for example lemma 2.3 in [10], page 135), the conditions for the unique solvability for a given dimension m also imply unique solvability for lower dimensions, while the converse is true only for dimensions $m\geq n$. Their result is in fact independent of the dimension m and so is the value of $\bar\gamma$. To avoid confusion about the number $\bar\gamma$ in the scalar and vector case, we introduce the notation $\bar\gamma_v$ when referring to this latter. The above mentioned authors showed that $\bar\gamma_v = \sup_{D>0}\gamma_D(G)$, with $\gamma_D(G)=\sup\{\gamma|(G^{-1}-\gamma I)^T D+D(G^{-1}-\gamma I)$ is positive semi-definite $\}$. In our discussion m has been fixed to one (scalar problems) and, since semi-stable matrices form a proper subset of the wider class of P_0 matrices, we realize that weaker restrictions on the product $h\mu$ occur in this case, namely methods could in principle occur for which $\bar\gamma_v < \bar\gamma$. While it is known that $\gamma_D(G)$ is the smallest eigenvalue of $(D^{1/2}G^{-1}D^{-1/2}+(D^{1/2}G^{-1}D^{-1/2})^T)/2$, the computation of $\sup_{D>0}\gamma_D(G)$ is difficult because the optimal D is not known a priori. However for a wide class of Runge-Kutta schemes the number $\bar\gamma_v$ has been successfully determined showing that the upper bound $\bar\gamma_v \leq \min_i (G^{-1})_{ii}$, is indeed attained if $DG^{-1}+(DG^{-1})^T$ is a diagonal matrix. This is the case of Gauss, Radau IA, Radau IIA, Lobatto IIIC methods (see for example [3,7,4]). Now, since $\bar\gamma\leq \min_i(G^{-1})_{ii}$ as well, it also follows that for these methods $\bar\gamma=\bar\gamma_v$. The same holds true for DIRK-methods for which the relation $\bar\gamma_v = \min_i(1/G_{ii})$ has been proved. No concrete weakness emerges when passing from the vector to the scalar problem for these methods; the following example shows that this does not represent the general rule.

Example 1. Here we consider the Runge-Kutta formula used by Hairer and Wanner [7] to state that a B-stable method on a contractive problem ($\mu = 0$) does not always admit a solution. The matrix G is defined as

$$G = \frac{1}{48} \begin{pmatrix} 3 & 0 & 3 & -6 \\ 6 & 9 & 0 & 1 \\ 5 & 18 & 9 & 0 \\ 12 & 15 & 18 & 3 \end{pmatrix}$$

while the abscissae and the weights are respectively $\mathbf{c} = (0, 1/3, 2/3, 1)$ and $\mathbf{b} = (1/8, 3/8, 3/8, 1/8)$. It has been shown that a problem of dimension two exists in the class $\mathcal{F}_0$ such that the system (1) does not have a solution: this also shows that $\bar{\gamma}_v \leq 0$. It is easy to verify that G is a P-matrix and hence from (9) it follows that $\bar{\gamma} > 0$. A direct computation based on formula (10) gives $\bar{\gamma} = 3/4$.

Now we turn our attention to a class of block-Boundary Value Methods, namely the Generalized Adams Methods (GAMs) which have been inserted into a code (the code GAM [13]) that implements a variable-step variable-order strategy to determine numerical solutions of Initial Value Problems. Firstly we give a brief account on how such methods are defined (for simplicity they will be considered as applied to the scalar problem (3)).

Starting from a known estimation of the true solution $\hat{y}(t)$ at a given time (without loss of generality they may be assumed to be y_0 and t_0 respectively), an order p block-BVM computes, through system (1), a vector $\mathbf{y}$ of approximations of order $p+1$ to $\hat{y}(t)$ on the uniform mesh $t_j = t_0 + jh$, $j = 1, \ldots, n$, that is $y_j = \hat{y}(t_j) + O(h^{p+1})$, $j = 1, \ldots, n$. In particular, denoting by $\{\beta_{ij}\}_{i,j=1,\ldots,n}$ and $\{\beta_{i0}\}_{i=1,\ldots,n}$ the entries of B and $\mathbf{b}$ respectively, for a GAM of odd order p and dimension n, the i-th component of the system (1) is a liner multistep formula of the form

$$y_i - y_{i-1} = h \sum_{j=-k_1^{(i)}}^{k_2^{(i)}} \beta_{ij} f_{i+j}, \qquad i = 1, \ldots, n, \tag{11}$$

with $k_1^{(i)}$ and $k_2^{(i)}$ nonnegative integers such that $k_1^{(i)} + k_2^{(i)} = p - 1$ and

$$k_1^{(i)} = \begin{cases} i & \text{for } i = 1, \ldots, (p-3)/2, \\ (p-1)/2 & \text{for } i = (p-1)/2, \ldots, n - (p-1)/2, \\ i - n + p - 1 & \text{for } i = n - (p-3)/2, \ldots, n; \end{cases}$$

From the above definition it is deduced that A is bidiagonal and Toeplitz with 1 and -1 as diagonal and lower diagonal entries. We also observe that for GAMs (and in general for block-BVMs) the only link between the order p and the dimension n of (1) is that the latter must be sufficiently large in order that the matrices A and B may contain the coefficient of each formula (11). The

Table 1. Estimated values of $\bar{\gamma}$ for GAMs

order	$\bar{\gamma}^{(u)}$	$\bar{\gamma}^{(c)}$	$\bar{\gamma}^{(o)}$
3	0.9011	0.2943	0.9502
5	0.2646	0.2374	0.6674
7	-0.2228	0.1032	0.4591
9	-0.5532	0.1728	0.3577

Table 2. Optimal symmetric distributions of the stepsizes

order	diagonal entries of $\widetilde{H}$
3	1.0449 0.9551 0.9551 1.0449
5	0.8247 1.1138 1.0615 1.0615 1.1138 0.8247
7	0.6622 1.0070 1.1335 1.1972 1.1972 1.1335 1.0070 0.6622
9	0.6660 0.8630 1.1092 1.2336 1.2563 1.2336 1.1092 0.8630 0.6660

code GAM is based on the Generalized Adams Methods of orders 3, 5, 7 and 9 and dimensions 4,6,8 and 9 respectively. For an explicit list of the coefficients β_{ij} see [11]; more about definitions, properties and implementation techniques of GAMs may be found in [1,11,12].

In the first column of Table 1 we report the values of $\bar{\gamma}$ for the GAMs used in the code. Unfortunately, negative values of $\bar{\gamma}$ occur for the GAMs of order 7 and 9. A negative value of $\bar{\gamma}$ means that given a stepsize $h > 0$ and a constant $\mu \geq 0$, a scalar function $f(t, y)$ can be found in the class $\mathcal{F}_\mu$ for which the system (1) does not admit a unique solution. This circumstance could produce irregularity during the execution of the code, due to the unpleasant situation in which the predicted stepsize may be rejected because the scheme that provides the solution of (1) does not attain convergence.

A way to overcome this problem is to allow different stepsizes inside each formula (11). A variable-step block-GAM takes the form

$$A\mathbf{y} - H\widetilde{B}F(\mathbf{y}) = \mathbf{b}, \qquad (12)$$

where $H = diag(h_1, h_2, \ldots, h_n)$. Considering that a single step of a block-BVM covers a time interval of length nh, the system (12) is recast in the form (1) by setting $h = 1/n\sum_{i=1}^{n} h_i$, $\widetilde{H} = 1/hH$, and $B = \widetilde{H}\widetilde{B}$. The diagonal elements $\tilde{h}_i$ in the matrix $\widetilde{H}$ define the mesh $_0 = 0$, $_i = _{i-1} + \tilde{h}_i$, $i = 1, \ldots, n$ over the interval $[0, n]$ and the question is how to choose the values of the abscissae $_i$ in order that the corresponding system (12) admits a positive value of $\bar{\gamma}$. A first attempt is to consider smaller stepsizes at the beginning and at the end of the time interval, a technique that has been successfully used to improve convergence and stability properties of block-BVMs. A Chebyshev distribution of the abscissae $_i$ satisfies this requirement and the corresponding values of $\bar{\gamma}$ are reported in the column of Table 1 labeled by $\bar{\gamma}^{(c)}$. We see that positiveness of $\bar{\gamma}$ is achieved for the orders 7 and 9 although $\bar{\gamma}^{(c)}$ is worse than $\bar{\gamma}^{(u)}$ (uniform mesh) for the orders 3 and 5. The values $\bar{\gamma}^{(o)}$ solve the problem $\bar{\gamma}^{(o)} = \max_{\widetilde{H}} \left\{ \bar{\gamma} | \tilde{h}_i = \tilde{h}_{n-i}, \ i = 1, \ldots, n/2 \right\}$, and have been determined using the Matlab optimization toolbox. The constraint of a symmetric distribution avoids an undesirable growth of the condition number of the matrix $A^{-1}B$, a prerequisite that guarantees well-conditioning of some problems that emerge when handling block-BVMs (see for example [1,12]). The values $\tilde{h}_i$ that correspond to the optimal $\bar{\gamma}$ are reported in Table 2.

References

1. L. Brugnano, D. Trigiante, *Solving ODEs by Linear Multistep Initial and Boundary Value Methods*, Gordon & Breach, Amsterdam, (1998).
2. J. C. Butcher, *The numerical analysis of Ordinary Differential Equations , Runge-Kutta and General Linear Methods*, John Wiley & Sons, Chichester, (1987).
3. K. Dekker, *Error bounds for the solution to the algebraic equations in Runge-Kutta methods*, BIT, **24**, (1984), 347–356.
4. K. Dekker, J. G. Verwer, *Stability of Runge-Kutta methods for stiff nonlinear differential equations*, CWI Monographs, North-Holland, Amsterdam, (1994).
5. G. Di Lena, F. Iavernaro, F. Mazzia, *On the unique solvability of the systems arising from Boundary Value Methods*, Nonlinear Studies **4 (1)** (1997), 1–12.
6. G. Di Lena, R. Peluso, *On conditions for the existence and uniqueness of solutions to the algebraic equations in Runge-Kutta methods*, BIT **25** (1985) 223–232.
7. E. Hairer, G. Wanner, *Solving Ordinary Differential Equations II, Stiff and Differential-Algebraic Problems*, Springer–Verlag, Berlin Heidelberg, (1991).
8. R. A. Horn, C. A. Johnson, *Topics in Matrix Analysis*, Cambridge University Press, New York, (1991).
9. J. F. B. M. Kraaijevanger, *A characterization of Lyapunov diagonal stability using Hadamard products*, Linear Algebra Appl., **151**, (1991) 245–254.
10. J. F. B. M. Kraaijevanger, J. Schneid, *On the unique solvability of the Runge-Kutta equations*, Numer. Math. **59** (1991) 129–157.
11. F. Iavernaro, F. Mazzia, *Solving ordinary differential equations by block Boundary Value Methods: properties and implementation techniques*, Appl. Num. Math., Vol. **28**, Issue 2-4,(1998) 107-126.
12. F. Iavernaro, F. Mazzia, *Block-Boundary Value Methods for the solution of Ordinary Differential Equations*, Siam J. Sci. Comput. **21** (1999), 323–339.
13. F. Iavernaro, F. Mazzia, *GAM* August 1997. Available via WWW at URL http://www.dm.uniba.it/~mazzia/ode/readme.html.

On the Local Sensitivity of the Lyapunov Equations

Suzanne Lesecq[1], Alain Barraud[1], and Nicolai Christov[2]

[1] Laboratoire d'Automatique de Grenoble
BP46, 38402 Saint Martin d'Hères, France
[2] Department of Automatics, Technical University of Sofia
1756 Sofia, Bulgaria

Abstract. This paper presents a new local perturbation bound for the continuous-time Lyapunov matrix equations, which is not formulated in terms of condition numbers. The new bound is a nonlinear, first order homogeneous function of the absolute perturbations in the data and is sharper than the linear local bounds based on condition numbers.

1 Introduction

The Lyapunov matrix equations (LME) are fundamental in the theory of linear systems. That is why, the problem of their reliable solution, including derivation of perturbation bounds, is of great practical interest. The conditioning of LME is well studied and different types of condition numbers are derived [1] - [6]. Unfortunately, perturbation bounds, based on condition numbers, may eventually produce pessimistic results.

In this paper a new local perturbation bound for the continuous-time LME is presented. It is a non-linear, first order homogeneous and tighter than the local bounds based on condition numbers. A comparative study of the new and existing local perturbation bounds is performed.

The following notations are used later on: $\mathcal{R}^{m \times n}$ – the space of real $m \times n$ matrices; I_n – the unit $n \times n$ matrix; $A^\top = [a_{ji}]$ – the transpose of the matrix $A = [a_{ij}]$; $\mathrm{vec}(A) \in \mathcal{R}^{mn}$ – the column-wise vector representation of the matrix $A \in \mathcal{R}^{m \times n}$; $\Pi_{n^2} \in \mathcal{R}^{n^2 \times n^2}$ – the vec-permutation matrix, such that $\mathrm{vec}(X^\top) = \Pi \mathrm{vec}(X)$ for all $X \in \mathcal{R}^{n \times n}$; $A \otimes B = [a_{ij}B]$ – the Kronecker product of the matrices A and B; $\|\cdot\|_2$ – the spectral (or 2-) norm in $\mathcal{R}^{m \times n}$; $\|\cdot\|_F$ – the Frobenius (or F-) norm in $\mathcal{R}^{m \times n}$. The notation ':=' stands for 'equal by definition'.

2 Problem Statement

Consider the LME
$$F(X, P) := A^\top X + XA + Q = 0 \tag{1}$$
where $X \in \mathcal{R}^{n \times n}$ is the unknown matrix, $A \in \mathcal{R}^{n \times n}$ and $Q = Q^\top \in \mathcal{R}^{n \times n}$ are given matrices and $P := (A, Q)$.

We suppose that $0 \notin \{\lambda_i(A) + \lambda_k(A) : i \in \overline{1,n},\ k \in \overline{1,n}\}$, where $\lambda_i(A)$ are the eigenvalues of the matrix A. Under this assumption the partial Fréchet derivative F_X of F in X is invertible and (1) has a unique solution $X = X^\top$.

Let the matrices A and Q be perturbed as

$$A \mapsto A + \Delta A,\quad Q \mapsto Q + \Delta Q$$

and denote by $P + \Delta P$ the perturbed matrix pair P in which A and Q are replaced by $A + \Delta A$ and $Q + \Delta Q$. Then the perturbed equation is

$$F(Y, P + \Delta P) = 0. \tag{2}$$

Since the operator F_X is invertible, the perturbed equation (2) has an unique solution $Y = X + \Delta X$, $Y = Y^\top$, in the neighborhood of X if the perturbation ΔP is sufficiently small.

Denote by

$$\Delta := [\Delta_A, \Delta_Q]^\top \in \mathcal{R}_+^2$$

the vector of absolute norm perturbations $\Delta_A := \|\Delta A\|_F$ and $\Delta_Q := \|\Delta Q\|_F$ in the data matrices A and Q.

In this paper we consider local bounds for the perturbation $\Delta_X := \|\Delta X\|_F$ in the solution of (1). These are bounds of the type

$$\Delta_X \leq f(\Delta) + O(\|\Delta\|^2),\quad \Delta \to 0 \tag{3}$$

where f is a continuous function, non-decreasing in each of its arguments and satisfying $f(0) = 0$. Particular cases of (3) are the well known linear perturbation bounds [2] - [6]

$$\Delta_X \leq K_A \Delta_A + K_Q \Delta_Q + O(\|\Delta\|^2) \tag{4}$$

and

$$\Delta_X \leq \sqrt{2} K_\mathcal{L} \Delta_{\max} + O(\|\Delta\|^2) \tag{5}$$

where K_A and K_Q are the individual absolute condition numbers of (1) relative to the perturbations in A and Q, $K_\mathcal{L}$ is the overall absolute condition number of (1) and $\Delta_{\max} = \max\{\Delta_A, \Delta_Q\}$.

In what follows the local linear bounds (4) and (5) are first derived using the approach developed in [2,3]. Then a new perturbation bound of the type (3) is given, where f is not a linear but a first order homogeneous function of the vector of absolute perturbations Δ.

3 Condition Numbers

Consider the conditioning of the LME (1). Since $F(X, P) = 0$, the perturbed equation (2) may be written as

$$F(X + \Delta X, P + \Delta P) :=$$
$$F_X(\Delta X) + F_A(\Delta A) + F_Q(\Delta Q) + G(\Delta X, \Delta P) = 0$$

where
$$F_X(Z) = A^\top Z + ZA, \quad F_A(Z) = ZX + XZ, \quad F_Q(Z) = Z$$
are the partial Fréchet derivatives of F in the corresponding matrix arguments, computed at the point (X, P), and $G(\Delta X, \Delta P)$ contains second and higher order terms in ΔX, ΔP.

Since the operator $F_X(.)$ is invertible we get
$$\Delta X = \Phi(\Delta X, \Delta P) := -F_X^{-1} \circ F_A(\Delta A) - F_X^{-1} \circ F_Q(\Delta Q) - F_X^{-1}(G(\Delta X, \Delta P)). \tag{6}$$
The relation (6) gives
$$\Delta_X \leq K_A \Delta_A + K_Q \Delta_Q + O(\|\Delta\|^2) \tag{7}$$
where the quantities
$$K_A = \left\| F_X^{-1} \circ F_A \right\|_2, \quad K_Q = \left\| F_X^{-1} \circ F_Q \right\|_2$$
are the absolute condition numbers of (1) relative to the perturbations in A and Q, respectively. Here $\|\mathcal{F}\|$ is the norm of the operator $\mathcal{F}$, induced by the F-norm.

The calculation of the condition numbers K_A and K_Q is straightforward. Denote by M_X, M_A and M_Q
the matrix representations of the operators $F_X(\cdot)$, $F_A(\cdot)$ and $F_Q(\cdot)$:
$$M_X = A^\top \otimes I_n + I_n \otimes A^\top, \quad M_A = (I_{n^2} + \Pi_{n^2})(I_n \otimes X), \quad M_Q = I_{n^2}.$$
Then
$$K_A = \left\| M_X^{-1} M_A \right\|_2, \quad K_Q = \left\| M_X^{-1} \right\|_2. \tag{8}$$
Relation (6) also gives
$$\Delta_X \leq \sqrt{2}\, K_{\mathcal{L}}\, \Delta_{\max} + O(\|\Delta\|^2), \quad \Delta \to 0 \tag{9}$$
where
$$K_{\mathcal{L}} = \left\| M_X^{-1}[M_A, I_{n^2}] \right\|_2 \tag{10}$$
is the overall absolute condition number of LME (1).

4 First Order Homogeneous Perturbation Bound

The local linear bounds (4) and (5) may eventually produce pessimistic results. At the same time it is possible to derive a local,
first order homogeneous bound which is sharper in general.
The operator equation (6) may be written in a vector form as
$$\operatorname{vec}(\Delta X) = N_1 \operatorname{vec}(\Delta A) + N_2 \operatorname{vec}(\Delta Q) - M_X^{-1} \operatorname{vec}(G(\Delta X, \Delta P)) \tag{11}$$
where
$$N_1 := -M_X^{-1} M_A, \quad N_2 := -M_X^{-1}.$$

The local linear bound (7), (8) is a corollary of (11):

$$\begin{aligned}
\Delta_X &= \|\Delta X\|_F = \|\text{vec}(\Delta X)\|_2 \\
&\leq \text{est}_1(\Delta, N) + O(\|\Delta\|^2) \\
&:= \|N_1\|_2 \Delta_A + \|N_2\|_2 \Delta_Q + O(\|\Delta\|^2) \\
&= K_A \Delta_A + K_Q \Delta_Q + O(\|\Delta\|^2), \quad \Delta \to 0
\end{aligned}$$

where $N := [N_1, N_2]$.

Relation (11) also gives

$$\Delta_X \leq \text{est}_2(\Delta, N) + O(\|\Delta\|^2) := \|N\|_2 \|\Delta\|_2 + O(\|\Delta\|^2), \quad \Delta \to 0.$$

Since $\|\Delta\|_2 \leq \sqrt{2}\,\Delta_{\max}$, the bound $\text{est}_2(\Delta, N)$ is less than or equal to the local linear bound (9), (10).

The bounds $\text{est}_1(\Delta, N)$ and $\text{est}_2(\Delta, N)$ are alternative, i.e. which one is less depends on the particular value of Δ.

There is also a third bound, which is always less than or equal to $\text{est}_1(\Delta, N)$. We have

$$\Delta_X \leq \text{est}_3(\Delta, N) := \sqrt{\Delta^T S(N) \Delta} + O(\|\Delta\|^2), \quad \Delta \to 0$$

where $S(N)$ is the 2×2 matrix with elements $s_{ij}(N) = \|N_i^T N_j\|_2$. Since

$$\|N_i^T N_j\|_2 \leq \|N_i\|_2 \|N_j\|_2$$

we get $\text{est}_3(\Delta, N) \leq \text{est}_1(\Delta, N)$. Hence we have the overall estimate

$$\Delta_X \leq \text{est}(\Delta, N) + O(\|\Delta\|^2), \quad \Delta \to 0 \tag{12}$$

where

$$\text{est}(\Delta, N) := \min\{\text{est}_2(\Delta, N), \text{est}_3(\Delta, N)\}. \tag{13}$$

The local bound $\text{est}(\Delta, N)$ in (12), (13) is a non-linear, first order homogeneous and piece-wise real analytic function in Δ.

5 Numerical Exemples

Among many numerical experiments, the following one is reported. According to the Matlab syntax, some block matrices are defined by : M=invhilb(n); Z=zeros(n,n); J=ones(n,n). Now the A Lyapunov equation parameter is set to :

$$A^T = \begin{bmatrix} M & Z \\ J & M \end{bmatrix};$$

Then the matrix Q is computed from A and X with :

$$X = scale * \begin{bmatrix} 1 & 1 & \ldots & 1 \\ 1 & 1 & \ldots & 1 \\ \ldots & \ldots & \ldots & \ldots \\ 1 & 1 & 1 & 1 \end{bmatrix}$$

where scale is a integer scaling factor. Lastly, the perturbations ΔA and ΔQ are generated from A and Q, by randomly perturbating their least significant bits. Setting here $n = 4$, a moderately ill conditionned 8×8 Lyapunov equation is obtained. Numerical results are summarised bellow :

scale	elin	est2	est1
1	$4.1091e-010$	$2.9161e-010$	$2.9142e-010$
10	$4.2985e-009$	$4.0133e-009$	$3.0711e-009$
100	$2.6129e-007$	$1.8672e-007$	$2.7182e-008$
1e4	$2.9207e-003$	$2.0652e-003$	$1.6813e-006$
1e6	$3.2627e+001$	$2.3070e+001$	$1.8975e-004$
1e10	$3.3388e+009$	$2.3609e+009$	$1.9326e+000$

scale	est3	Δ_X
1	$2.9142e-010$	$6.6102e-011$
10	$3.0711e-009$	$3.7805e-010$
100	$2.7182e-008$	$7.9174e-009$
1e4	$1.6813e-006$	$6.8222e-007$
1e6	$1.8975e-004$	$7.5801e-005$
1e10	$1.9326e+000$	$9.4309e-001$

Clearly, the new local bounds can give much sharper results than the standard linear perturbation bound "elin"(5). For comparison purpose, Δ_X has been computed, from its definition, through the numerical solution of the original Lypunov equation and its perturbed version. If relative errors are invoked, the new bounds remain, for this numerical test, 300% to 900% better than the linear one.

6 Conclusion

New local perturbations bounds has been presented. These non linear bounds can be much sharper than their linear counterpart, depending on the problem data. This conclusion remains true if relative errors $\Delta_X / \|X\|_F$ are considered. In such a developpement K_A, K_Q, K_L must be respectively replaced by $K_A / \|X\|_F$, $K_Q / \|X\|_F$, $K_L / \|X\|_F$, and M_A, M_Q, by $\|A\|_F M_A$, $\|Q\|_F M_Q$. Extensions of these results to non local analysis is under investigation.

References

1. Byers, R.: A Linpack-style condition estimator for the equation $AX - XB^\top = C$. IEEE Trans. Automat. Contr. **AC-29** (1984) 926-928
2. Konstantinov, M. M., Petkov, P.Hr., Christov, N. D.: Perturbation analysis of the continuous and discrete matrix Riccati equations. Proc. 1986 ACC, Seattle, vol. 1, 636-639

3. M. M. Konstantinov, N. D. Christov and P.Hr. Petkov. Perturbation analysis of linear control problems. Prepr. 10th IFAC World Congress, Munich, 1987, vol. 9, 16-21
4. Hewer, G., Kenney, C.: The sensitivity of the stable Lyapunov equation. SIAM J. Contr. Optim. **26** (1988) 321-343
5. B. Kagstrom. A perturbation analysis of the generalized Sylvester equation $(AR - LB, DR - LE) = (C, F)$. SIAM J. Matrix Anal. Appl. **15** (1994) 1045-1060
6. Ghavimi, A., Laub, A.: Backward error, sensitivity and refinement of computed solutions of algebraic Riccati equations. Numer. Lin. Alg. Appl. **2** (1995) 29-49

A Level Set-Boundary Element Method for Simulation of Dynamic Powder Consolidation of Metals

Zhilin Li[1] and Wei Cai[2]

[1] Center for Research in Scientific Computation & Department of Mathematics
North Carolina State University, Raleigh, NC 27695, USA
zhilin@math.ncsu.edu http://ww4.ncsu.edu/~zhilin
[2] Department of Mathematics, University of North Carolina at Charlotte
Charlotte, NC 28223, USA
wcai@uncc.edu

Abstract. In this paper, the level set method is coupled with the boundary element method to simulate dynamic powder consolidation of metals based on linear elastostatics theory. We focus on the case of two particles that are in contact. The boundaries of the two particles are expressed as the zero level curves of two level set functions. The boundary integral equations are discretized using the piecewise linear elements at some projections of irregular grid points on the boundaries of the two particles. Numerical examples are also provided.

1 Introduction

The application of large amplitude stress waves for materials processing and powder compaction has been of increasing interest in recent years [3,7]. The technique is also used for materials synthesis where the stress wave can promote metallurgical reactions between two or more pure powders to produce alloy phases. When powder consolidation is of interest, it is important to understand the interaction, deformation, and bonding of particles in response to the stress wave. But the understanding of the dynamic process is far more complete. There are few papers on numerical simulations in the literature. A finite element method [1] gives some microscope analysis of a few particles. However, it seems that one can not afford to generate a body fitting grid for thousands particles at every time steps.

In this paper, we develop a boundary element–level set method to simulate the solidification process of metal particles. The choice of the boundary element method is based on the fact that we are only interested in the motion (deformation) of the boundaries of the particles, and the boundary integral equations are available and well understood. The use of the level set method [6] is to eliminate the cost of the grid generation and to simplify simulations for three dimensional problems. The projections of irregular grid points serve as a bridge between the boundaries of the particles and the underline Cartesian grid.

2 The Boundary Integral Equations

We consider a simplified model that involves only two interacting particles, see Fig. 1 for an illustration. We assume that a small but fixed traction/pressure is applied to a portion of the boundary of a left particle while a portion of a second particle is fixed against a wall. We expect the particle will deform and want to know the position of the particles and the traction along the boundaries. For a problem with many particles, we can decompose the particle as groups of two particles using domain decomposition techniques.

We assume that the deformation is small, then from the linear elastostatics theory, see for example, [2], the traction **p** and the deformation **u** in the vector form

$$\mathbf{u} = \begin{bmatrix} u_1 \\ u_2 \end{bmatrix}, \quad \mathbf{p} = \begin{bmatrix} p_1 \\ p_2 \end{bmatrix} \tag{1}$$

are coupled by the following boundary integral equations:

$$c\,\mathbf{u}(\xi) + \int_\Gamma \mathbf{p}^*(\xi,\mathbf{x})\,\mathbf{u}(\mathbf{x})\,d\Gamma(\mathbf{x}) = \int_\Gamma \mathbf{u}^*(\xi,\mathbf{x})\,\mathbf{p}(\mathbf{x})\,d\Gamma(\mathbf{x}) \\ + \int_\Omega \mathbf{u}^*(\xi,\mathbf{x})\,\mathbf{b}(\mathbf{x})\,d\Omega(\mathbf{x}), \tag{2}$$

where **b** is the body force, $c^i_{lk} = \frac{1}{2}\delta_{lk}$, where δ_{lk} is the Kronecker delta, $\mathbf{p}^*$ and $\mathbf{u}^*$ are the Green's functions

$$\mathbf{p}^* = \begin{bmatrix} p^*_{11} & p^*_{12} \\ p^*_{21} & p^*_{22} \end{bmatrix}, \quad \mathbf{u}^* = \begin{bmatrix} u^*_{11} & u^*_{12} \\ u^*_{21} & u^*_{22} \end{bmatrix}, \tag{3}$$

with

$$u^*_{lk} = \frac{1}{8\pi\mu(1-\nu)}\left((3-4\nu)\log\frac{1}{r}\,\delta_{lk} + r_{,l}\,r_{,k}\right),$$

$$p^*_{lk} = -\frac{1}{4\pi(1-\nu)\,r}\left(\frac{\partial r}{\partial n}\left[(1-2\nu)\,\delta_{lk} + 2r_{,l}\,r_{,k}\right] + (1-2\nu)\,(n_l\,r_{,k} - n_k\,r_{,l})\right),$$

where

$$r = |\xi - \mathbf{x}|, \quad r_{,k} = \frac{\partial r}{\partial x_k}.$$

The Lamé's constant can be expressed in terms of the more familiar shear modulus G, modulus of elasticity E and Poisson's ratio ν by the following formulae,

$$\mu = G = \frac{E}{2(1+\nu)}; \quad \lambda = \frac{\nu E}{(1+\nu)(1-2\nu)}. \tag{4}$$

In our simulation, the body force is zero. $\Gamma = \Gamma_k$, $k=1$ or $k=2$, is one of the boundaries of the two particles. We want to evaluate the displacement of **u** along the boundaries so that we can determine the location of the boundaries.

3 Numerical Method

We use a rectangular domain $\Omega = [a, b] \times [c, d]$ to enclose the two particles and generate a Cartesian grid

$$x_i = a + ih_x, \quad i = 0, 1, \cdots m, \tag{5}$$
$$y_j = c + jh_y, \quad j = 0, 1, \cdots n. \tag{6}$$

We use the zero level sets of two functions $\varphi_1(x, y)$ and $\varphi_2(x, y)$ to express the two particles respectively

$$\varphi_k(x, y) \begin{cases} \leq 0 & \text{inside the } k\text{-th particle,} \\ = 0 & \text{on the boundary of the } k\text{-th particle,} \\ \geq 0 & \text{outside the } k\text{-th particle,} \end{cases} \tag{7}$$

where $k = 1$ or $k = 2$. Since the two articles are immiscible, they share part of common boundary.

To use the boundary element method, we need to have some discrete points on the boundaries. We choose some of the *projections* of irregular grid points on the boundaries. An irregular grid point (x_i, y_j) is a grid point at which one of the level set functions $\varphi_k(x_i, y_j)$ changes signs in the standard five point stencil centered at (x_i, y_j).

The projection of an irregular grid point $\mathbf{x} = (x_i, y_j)$ on the boundary is determined by

$$\mathbf{x}^* = \mathbf{x} + \alpha\mathbf{q}, \quad \text{where} \quad \mathbf{q} = \frac{\nabla\varphi_k}{||\nabla\varphi_k||_2}, \tag{8}$$

and α is determined from the following quadratic equation:

$$\varphi_k(\mathbf{x}) + ||\nabla\varphi_k||_2\, \alpha + \frac{1}{2}\left(\mathbf{q}^T He(\varphi_k)\mathbf{q}\right)\alpha^2 = 0, \tag{9}$$

where $He(\varphi_k)$ is the Hessian matrix of φ_k evaluated at $\mathbf{x}$. All the first and second order derivatives at the irregular grid point are evaluated using the second order central finite difference schemes.

3.1 Contact of Two Particles

We use two level set functions to represent two immiscible particles and update their positions. On the part of the contact, both the level set functions should be zero.

Given two level set functions whose zero level curves intersect with each other, that is, $\varphi_{1,ij} \leq 0$ and $\varphi_{2,ij} \leq 0$, we modify the level set functions in the following way

$$\varphi_{k,ij} \longleftarrow \varphi_{k,ij} + \delta \tag{10}$$

where

$$\delta = \frac{|\varphi_{1,ij} + \varphi_{2,ij}|}{2}. \tag{11}$$

After such adjustment, the two level set functions can only have contact but not overlap, see Fig. 1 for an illustration. Note that it may be necessary to re-initialize the level set functions after such an adjustment.

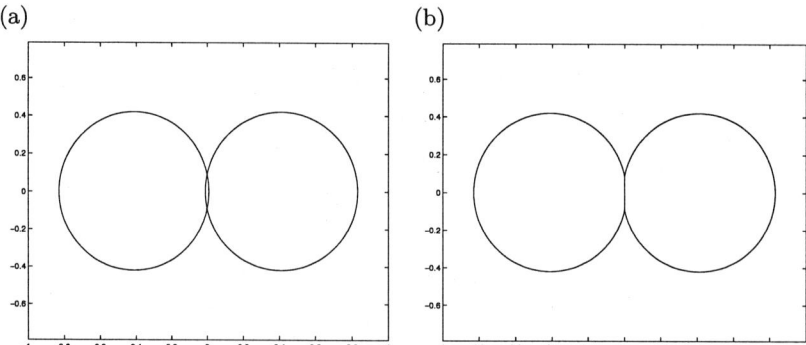

Fig. 1. Contact adjustment of the boundaries of two particles. (a): Two zero level set functions that overlap with each other. (b): The zero level curves (boundaries) of two particles after the adjustment

3.2 Set-up the Linear System of Equations

Since the projections on the zero level sets are not equally spaced, we use the piecewise linear basis functions to approximate **u** and **p**, and discretize the boundary integral equation to get second order accurate method. For example, the displacement **u** between two nodal points can be interpolated using

$$\mathbf{u}(\xi) = \phi_1 \mathbf{u}^1 + \phi_2 \mathbf{u}^2 = [\phi_1, \ \phi_2] \begin{bmatrix} \mathbf{u}^1 \\ \mathbf{u}^2 \end{bmatrix}. \tag{12}$$

In the expression above, ξ is the dimensionless coordinate varying from -1 to $+1$ and the two interpolation functions are

$$\phi_1 = \frac{1}{2}(1-\xi); \quad \phi_2 = \frac{1}{2}(1+\xi). \tag{13}$$

Given a projection (x_l^*, y_l^*) on one of the boundaries $\varphi_k = 0$, the next point (x_{l+1}^*, y_{l+1}^*) is determined as the closest projection in a small neighborhood of (x_l^*, y_l^*) that satisfies

$$\nabla \varphi_k(x_l^*, y_l^*) \cdot \nabla \varphi_k(x_{l+1}^*, y_{l+1}^*) \geq 0,$$

where $k = 1$ or $k = 2$. The gradient at the projection is computed using the bi-linear interpolation from those at the neighboring grid points evaluated with the standard central finite difference scheme. We refer the reader to [4,5] for detailed information on the bi-linear interpolation between projections and grid points.

The matrix-vector form of the linear system of equations at a particular node i can be written as

$$\mathbf{c}^i \mathbf{u}^i + \sum_{j=1}^{N} \hat{\mathbf{H}}^{ij} \mathbf{u}^j = \sum_{j=1}^{N} \mathbf{G}^{ij} \mathbf{p}^j, \qquad (14)$$

where N is the total number of nodes or the projections that are used to set up the equations, $\hat{\mathbf{H}}^{ij}$ and $\mathbf{G}^{ij}$ (both are 2×2 matrices), are influence matrices. To avoid an ill-conditioned system and reduce the size of the linear system of equations, we use **ONLY** the projections of irregular grid points from one particular side. For example, we use the projections of irregular grid points where $\varphi_1(x_i, y_j) \leq 0$ and those projections of irregular grid points where $\varphi_2(x_i, y_j) \geq 0$. In this way, along the contact of the two particles, we can use just one projection at irregular grid points.

We use Gaussian quadrature of order four, which is an open formula, to evaluate the integrals $\mathbf{G}^{ij}$ and $\hat{\mathbf{H}}^{ij}$ for $i \neq j$. If $i = j$, the integral is a singular Cauchy principal integral and we use the *rigid body condition*

$$\hat{\mathbf{H}}^{ii} = - \sum_{j=1, j \neq i} \hat{\mathbf{H}}^{ij} - \mathbf{c}^i, \qquad (15)$$

to evaluate the diagonal entries. The details about the boundary element method can be found in Chapter 3 of Brebbia and Dominguez's book [2].

3.3 Velocity Evaluation

In order to use the level set function, we need to evaluate the velocity at the grid points in a computational tube. At those grid points where the projections are used to set up the linear system of equations, we directly shift the velocity to the grid points. At those grid points where the projections are not used, for example, $\varphi_1(x_i, y_j) > 0$, we use the velocity at the closest projection from the other side where $\varphi_1(x_i, y_j) \leq 0$.

After we have evaluated the velocity $\mathbf{u}_k = \begin{bmatrix} u_{k,1} & u_{k,2} \end{bmatrix}^T$ at irregular grid points, we need to extend the normal velocity

$$V_k = u_{k,1} n_x + u_{k,2} n_y, \qquad k = 1 \text{ or } k = 2, \qquad (16)$$

where $(n_x, n_y) = \nabla \varphi_k / \|\nabla \varphi_k\|_2$ is the unit normal direction, to all grid points inside a computational tube $|\varphi_k| \leq \delta$ surrounding the boundary of the particle, where $\delta = Ch$ is the width of the computational domain. This is done through an upwind scheme

$$\frac{\partial V_k}{\partial t} \pm \nabla V_k \cdot \frac{\nabla \varphi_k}{\|\nabla \varphi_k\|_2} = 0, \qquad (17)$$

$k = 1$ or $k = 2$, which propagates V_k along the normal direction away from the interface. The sign is determined from the normal direction of the level set function.

3.4 Update the Level Set Functions

Once we have obtained the normal velocity in the computational domain, we can update the level set functions by solving one step of the Hamilton Jacobi equation

$$\frac{\partial \varphi_k}{\partial t} + V_k ||\nabla \varphi_k||_2 = 0, \qquad k = 1 \quad \text{or} \quad k = 2. \tag{18}$$

The zero level sets $\varphi_k = 0$ then gives the new location of the boundaries of the two particles.

We summarize our algorithm below:

- Set up the problem that includes input of the material parameters, initialization of the two level functions that represent the two particles.
- Adjust the level set functions at grid points where the two level set functions are both non-negative to treat the contact part.
- Find the projections of irregular grid points inside the first particle and outside the second particle.
- Find the next point for each projection on the boundaries to form the line segment needed in the boundary element method.
- Set-up the system of equations using the Gaussian quadrature of order four at all selected projections for each level set function. If **p** is known then **u** is unknown and vise versa. At contact, both **p** and **u** are unknowns. Use the rigid body condition to compute the diagonal entries.
- Shift the velocity to irregular grid points.
- Extend the normal velocity to a computational tube with a pre-selected width δ.
- Update the two level set functions by solving the Hamilton-Jacobi equation.
- Repeat the process if necessary.

4 Numerical Examples

We have done a number of numerical experiments. The results are reasonable good and are within the regime of the linear elasticity.

Example 1. The material parameters for the first and second particles are

$$G_1 = \mu_1 = 26, \quad \nu_1 = 0.33, \quad G_2 = \mu_2 = 83, \quad \nu_2 = 0.27.$$

The boundaries of the initial two particles are the circles

$$(x - 0.31)^2 + y^2 = 0.32^2; \qquad (x - 0.21)^2 + y^2 = 0.22^2,$$

before the adjustment. From the left, we apply a constant p

$$p(x,y) = C\cos(\pi x/2\epsilon), \quad \text{for} \quad |x| \leq \epsilon,$$

on the part of the boundary of the left particle, where we take $C = 5$ and $\epsilon = 0.1$. On the right, we fix the displacement $\mathbf{u} = \mathbf{0}$ if $|x - x_{max}| \leq 0.05$ along the part of the boundary of the right particle, where x_{max} is the largest x coordinates of the projections of irregular grid points. Fig. 2 is the computational result using our method.

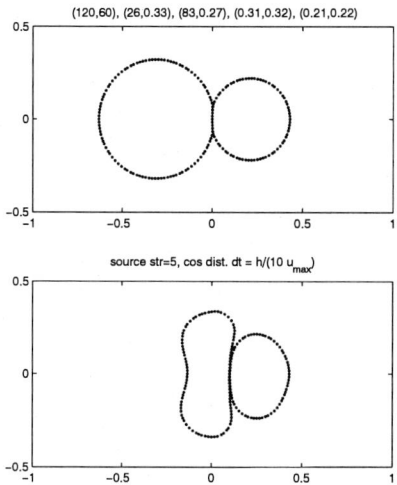

Fig. 2. Numerical result of Example 1 using a 120 by 60 grid. The upper half picture is the original particles; the lower half is the computed result

Example 2. The material parameters for the first and second particles are

$$G_1 = \mu_1 = 83, \quad \nu_1 = 0.33, \quad G_2 = \mu_2 = 26, \quad \nu_2 = 0.27.$$

The boundaries of the initial two particles are the circles

$$(x+0.21)^2 + y^2 = 0.22^2; \quad (x-0.31)^2 + y^2 = 0.32^2,$$

before the adjustment. The rest of set-up is the same as Example 1. Fig. 3 is the computational result using our method.

5   Conclusion and Acknowledgment

A new numerical method that couples the boundary element method with the level set method is proposed in this paper to simulate multi-particles of liner elasticity. The new method can handle the contact of two particles easily.

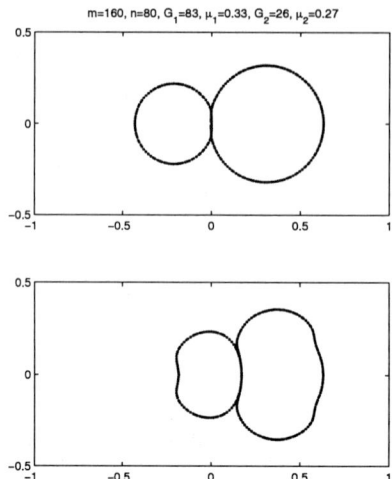

Fig. 3. Numerical result of Example 2 using a 160 by 80 grid. The upper half picture is the original particles; the lower half is the computed result

The first author was partially supported by a USA ARO grant 39676-MA and an NSF grant, DMS-96-26703. The second author was partially supported by a USA NSF grant CCR-9972251.

References

1. D. J. Benson, W. Tong, and G. Ravichandran. Particle-level modeling of dynamic consolidation of Ti-SiC powders. *Model. Simul. Mater. Sci. Eng.*, 3:771–796, 1995.
2. C. A. Brebbia and J. Dominguez. *Boundary elements, An introductory course.* MxGraw-Hill Book Company, 1992.
3. J. E. Flinn, R. L. Williamson, R. A. Berry, and R. N. Wright. Dynamics consolidation of type 304 stainless-steel powders in gas gun experienments. *J. Appl. Phys.*, 64(3):1446–1456, 1988.
4. T. Hou, Z. Li, S. Osher, and H. Zhao. A hybrid method for moving interface problems with application to the Hele-Shaw flow. *J. Comput. Phys.*, 134:236–252, 1997.
5. Z. Li, H. Zhao, and H. Gao. A numerical study of electro-migration voiding by evolving level set functions on a fixed cartesian grid. *J. Comput. Phys.*, 152:281–304, 1999.
6. S. Osher and J. A. Sethian. Fronts propagating with curvature-dependent speed: Algorithms based on Hamilton-Jacobi formulations. *J. Comput. Phys.*, 79:12–49, 1988.
7. R. L. Williamson, J. R. Knibloe, and R. N. Wright. Particle-level investigation of densification during uniaxial hot pressing: Continuum modeling and experiments. *J. Eng. Mat. Tech.*, 114:105–110, 1992.

Parallel Performance of a 3D Elliptic Solver

Ivan Lirkov[1], Svetozar Margenov[1], and Marcin Paprzycki[2]

[1] Central Laboratory for Parallel Processing, Bulgarian Academy of Sciences
Acad.G.Bonchev, block 25A, 1113 Sofia, Bulgaria
{ivan,margenov}@cantor.bas.bg

[2] Department of Computer Science and Statistics, University of Southern Mississippi,
Hattiesburg, Mississippi, 39406-5106, USA
marcin@orca.st.usm.edu

Abstract. It was recently shown that block-circulant preconditioners applied to a conjugate gradient method used to solve structured sparse linear systems arising from 2D or 3D elliptic problems have good numerical properties and a potential for high parallel efficiency. In this note parallel performance of a circulant block-factorization based preconditioner applied to a 3D model problem is investigated. The aim of the presentation is to report on the experimental data obtained on SUN Enterprise 3000, SGI/Cray Origin 2000, Cray J-9x, Cray T3E computers and on two PC clusters.

1 Introduction

Let us consider numerical solution of a self-adjoint second order 3D linear boundary value problem of elliptic type. After discretization, such a problem results in a linear system $A\mathbf{x} = \mathbf{b}$, where A is a sparse symmetric positive definite matrix. In the computational practice, large-scale problems of this class are most often solved by Krylov subspace iterative (e.g. conjugate gradient) methods. Each step of such a method requires only a single matrix-vector product and allows exploitation of sparsity of A. The rate of convergence of these methods depends on the condition number κ of the matrix A (smaller $\kappa(A)$ results in faster convergence). Unfortunately, for second order 3D elliptic problems, usually $\kappa(A) = \mathcal{O}(N^{2/3})$, where N is the size of the discrete problem, and hence it grows rapidly with N. To alleviate this problem, iterative methods are almost always used with a preconditioner M. The preconditioner is chosen with two criteria in mind: to minimize $\kappa(M^{-1}A)$ and to allow efficient computation of the product $M^{-1}\mathbf{v}$ for any given vector $\mathbf{v}$. These two goals are often in conflict and a lot of research has been done devising preconditioners that strike a balance between them. Recently, a third aspect has been added to the above two, namely, the parallel efficiency of the iterative method (and thus the preconditioner).

One of the most popular and the most successful preconditioners are the incomplete LU (ILU) factorizations. Unfortunately, standard ILU preconditioners have limited degree of parallelism. Some attempts to modify them and introduce more parallelism often result in a deterioration of the convergence rate. R. Chan and T. F. Chan [2] proposed another class of preconditioners based on averaging coefficients of A to form a block-circulant approximation. The block-circulant preconditioners are highly parallelizable but they are very sensitive to a possible high variation of the coefficients of the elliptic operator. To reduce this sensitivity a new class of circulant block-factorization (CBF) preconditioners [5] was introduced by Lirkov, Margenov and Vassilevski. Recently a new CBF preconditioner for 3D problems was introduced in [3,4].

The main goal of this note is to report on the parallel performance of the PCG method with a circulant block-factorization preconditioner applied to a model 3D linear PDE of elliptic type. Results of experiments performed on Sun Ultra-Enterprise, Crays J-9x and T3E, SGI/Cray Origin 2000 high performance computers and on two PC clusters are presented and analyzed.

We proceed as follows. In Section 2 we sketch the algorithm of the parallel preconditioner (for more details see [3,4]). Section 3 contains the theoretical estimate of its arithmetical complexity. Finally, in Section 4 we report the results of our experiments.

2 Circulant Block-Factorization

Let us recall that a circulant matrix C has the form $(C_{k,j}) = (c_{(j-k) \bmod m})$, where m is the dimension of C. Let us also denote by $C = (c_0, c_1, \ldots, c_{m-1})$ the circulant matrix with the first row $(c_0, c_1, \ldots, c_{m-1})$. Any circulant matrix can be factorized as $C = F \Lambda F^*$ where Λ is a diagonal matrix containing the eigenvalues of C, and F is the Fourier matrix of the form

$$F_{jk} = \frac{1}{\sqrt{m}} e^{2\pi \frac{jk}{m} \mathbf{i}}, \qquad (1)$$

where $F^* = \overline{F}^T$ denotes the adjoint matrix of F.

The CBF preconditioning technique incorporates the circulant approximations into the framework of LU block-factorization. Let us consider a 3D elliptic problem (see also [3]) on the unit cube with Dirichlet boundary conditions. If the domain is discretized on a uniform grid with n_1, n_2 and n_3 grid points along the coordinate directions, and if a standard (for such a problem) seven-point FDM (FEM) approximation is used, then the stiffness matrix A admits a block-tridiagonal structure. The matrix A can be written in the form

$$A = tridiag(-A_{i,i-1}, A_{i,i}, -A_{i,i+1}) \qquad i = 1, 2, \ldots, n_1,$$

where $A_{i,i}$ are block-tridiagonal matrices which correspond to the x_1-plane and the off-diagonal blocks are diagonal matrices. In this case the general CBF preconditioning approach is applied to construct the preconditioner M_{CBF} in the form

$$M_{CBF} = tridiag(-C_{i,i-1}, C_{i,i}, -C_{i,i+1}) \qquad i = 1, 2, \ldots n_1, \qquad (2)$$

where $C_{i,j} = $ Block-Circulant$(A_{i,j})$ is a block-circulant approximation of the corresponding block $A_{i,j}$. The stiffness matrix A and the preconditioner M_{CBF}

are $N \times N$ matrices where $N = n_1 n_2 n_3$. The relative condition number of the CBF preconditioner for the model (Laplace) 3D problem for $n_1 = n_2 = n_3 = n$ is (for derivation see [3]):
$$\kappa(M_0^{-1} A_0) \leq 4n. \qquad (3)$$

2.1 Parallel Circulant Block-Factorization Preconditioner

The basic advantage of circulant preconditioners is their inherent parallelism. Let us now describe how to implement in parallel an application of the inverse of the preconditioner to a given vector. Using the standard LU factorization procedure, we can first split $M = D - L - U$ into its block-diagonal and strictly block-triangular parts respectively. Then the *exact* block-factorization of M can be written in the form

$$M = (X - L)(I - X^{-1}U),$$

where $X = diag(X_1, X_2, \ldots, X_n)$ and the blocks X_i are determined by the recursion

$$X_1 = C_{1,1}, \quad \text{and} \quad X_i = C_{i,i} - C_{i,i-1} X_{i-1}^{-1} C_{i-1,i}, \quad i = 2, \ldots, n_1. \qquad (4)$$

It is easy to observe here that X_i are also block-circulant matrices.

In order to compute $M^{-1}\mathbf{v}$ we rewrite the block-circulant blocks of the preconditioner as

$$C_{i,j} = (F \otimes F) \Lambda_{i,j} (F^* \otimes F^*).$$

Here $\otimes$ denotes the Kronecker product. It can be observed that for X_i we have

$$X_i = (F \otimes F) D_i^{-1} (F^* \otimes F^*)$$

and the latter yields

$$D_1^{-1} = \Lambda_{1,1},$$
$$D_i^{-1} = \Lambda_{i,i} - \Lambda_{i,i-1} D_{i-1} \Lambda_{i-1,i}.$$

Let $\Lambda = tridiag(\Lambda_{i,i-1}, \Lambda_{i,i}, \Lambda_{i,i+1})$. Then the following relation holds

$$M\mathbf{u} = \mathbf{v} \quad \Longleftrightarrow \quad (I \otimes F \otimes F) \Lambda (I \otimes F^* \otimes F^*) \mathbf{u} = \mathbf{v}.$$

The above system can be rewritten as

$$\begin{pmatrix} \mathcal{F} & & & \\ & \mathcal{F} & & \\ & & \mathcal{F} & \\ & & & \ddots \\ & & & & \mathcal{F} \end{pmatrix} \begin{pmatrix} \Lambda_{11} \Lambda_{12} & & & \\ \Lambda_{21} \Lambda_{22} \Lambda_{23} & & \\ & \Lambda_{32} \Lambda_{33} & \\ & & \ddots \\ & & & \Lambda_{nn} \end{pmatrix} \begin{pmatrix} \mathcal{F}^* & & & \\ & \mathcal{F}^* & & \\ & & \mathcal{F}^* & \\ & & & \ddots \\ & & & & \mathcal{F}^* \end{pmatrix} \begin{pmatrix} \mathbf{u}_1 \\ \mathbf{u}_2 \\ \mathbf{u}_3 \\ \vdots \\ \mathbf{u}_n \end{pmatrix} = \begin{pmatrix} \mathbf{v}_1 \\ \mathbf{v}_2 \\ \mathbf{v}_3 \\ \vdots \\ \mathbf{v}_n \end{pmatrix}$$

where $\mathcal{F} = F \otimes F$.

We can distinguish three stages in computing $\mathbf{u} = M^{-1}\mathbf{v}$:

$$1) \quad \hat{\mathbf{v}} = (I \otimes F^* \otimes F^*)\mathbf{v}$$
$$2) \quad \Lambda \hat{\mathbf{u}} = \hat{\mathbf{v}} \quad (5)$$
$$3) \quad \mathbf{u} = (I \otimes F \otimes F)\hat{\mathbf{u}}.$$

Due to the special form of F (see (1) above), we can use a fast Fourier transform to perform the first and third stages of the algorithm. Namely, we use a standard two-dimensional block-FFT which is easily parallelizable (see [6]). The second stage consist of solving two recurrence equations

$$\begin{vmatrix} \hat{\mathbf{w}}_1 = D_1 \hat{\mathbf{v}}_1 \\ \hat{\mathbf{w}}_i = D_i(\hat{\mathbf{v}}_i - \Lambda_{i,i-1}\hat{\mathbf{w}}_{i-1}) \\ i = 2,3,\ldots n_1 \end{vmatrix} \quad \begin{vmatrix} \hat{\mathbf{u}}_n = \hat{\mathbf{w}}_n \\ \hat{\mathbf{u}}_i = \hat{\mathbf{w}}_i - D_i \Lambda_{i,i+1}\hat{\mathbf{u}}_{i+1} \\ i = n_1-1, n_1-2, \ldots 1 \end{vmatrix} \quad (6)$$

Since blocks D_i and $\Lambda_{i,j}$ in the recurrences (6) are diagonal the solution of $n_2 n_3$ independent linear systems can be calculated in parallel.

3 Parallel Complexity

Let us present the theoretical estimate of the total execution time T_{PCG} for one PCG iteration for the proposed circulant block-factorization preconditioner on a parallel system with p processors (detailed analysis of parallel complexity can be found in [4]). Each iteration consists of one matrix vector multiplication involving matrix A, one multiplication involving the inverse of the preconditioner M_{CBF} (solving a system of equations with matrix M), two inner products and three linked triads (a vector updated by a vector multiplied by a scalar). Consequently

$$T_{PCG}(p) = T_{mult} + T_{prec} + 2T_{inn\_prod} + 3T_{triads}.$$

For simplicity we assume that the mesh dimensions are equal and they are equal to an exact power of two, i.e., $n_1 = n_2 = n_3 = n = 2^l$. We also assume that the time to execute K arithmetic operations on one processor is $T_a = K * t_a$, where t_a is an average time of one arithmetic operation. In addition, the communication time of a transfer of K words between two neighbor processors is $T_{local} = t_s + K * t_c$, where t_s is the start-up time and t_c is the time for each word to be sent/received. Finally, let us assume that a 2-radix algorithm is used to calculate the FFT's and thus the cost per processor is $T_{FFT}(n) = 5n \log n t_a$. Then the formula for computational complexity has the form

$$T_{PCG}(p) = 5(7 + 4\log n)\frac{n^3}{p}t_a + 4\left(t_s + n^2 t_c\right) + 2g(\frac{n^3}{p}, p) + 2g(p,p) + 2b(p),$$

where $b(p)$ denotes time to broadcast a single value from one processor to all other processors and $g(K,p)$ denotes time to gather $\frac{K}{p}$ words from all processors into one processor. It can be shown that, for instance, when only the leading terms are taken into consideration, for the shared memory parallel computer the above function simplifies to

$$T_{PCG}(p) \approx 2pt_s + 2(1 - \frac{1}{p})\frac{n^3}{p}t_c + 5(7 + 4\log n)\frac{n^3}{p}t_a. \qquad (7)$$

Next we analyze the relative speedup S_p and the relative efficiency E_p, where $S_p = \frac{T(1)}{T(p)} \leq p$ and $E_p = \frac{S_p}{p} \leq 1$. Thus the formula for the speedup becomes

$$S_p \approx \frac{5(7 + 4\log n)}{2\frac{p^2}{n^3}\frac{t_s}{t_a} + 2(1 - \frac{1}{p})\frac{t_c}{t_a} + 5(7 + 4\log n)}p. \qquad (8)$$

Obviously, $\lim_{n\to\infty} S_p = p$ and $\lim_{n\to\infty} E_p = 1$, i.e., the algorithm is asymptotically optimal. More precisely, if $\log n \gg \frac{p^2}{n^3}\frac{t_s}{t_a} + \frac{t_c}{t_a}$, then E_p approaches 1. Unfortunately, the start-up time t_s is usually much larger than t_a, and for relatively small n the first term of the denominators in (8) is significant, in this case the efficiency is much smaller than 1.

4 Experimental Results

In this section we report the results of the experiments executed on Sun Ultra-Enterprise 3000, Cray J-9x and T3E, SGI Origin 2000 computers and on two PC clusters. The code has been implemented in C and the parallelization has been facilitated using the MPI [7] library. In all cases the manufacturer provided MPI kernels have been used. No machine-dependent optimization has been applied to the code itself. Instead, in all cases, the most aggressive optimization options of the compiler have been turned on. Times have been collected using the MPI provided timer and, as verification, the *clock* Unix timer. Results reported by both timers were very close to each other. In all cases we report the best results from multiple runs in interactive and batch modes. In Table 1 we report results obtained on the Sun, the vector-Cray and the SGI computers for $n_1 = n_2 = n_3 = 64, 96, 128, 144, 160$ and for the number of processors p that exactly divides the dimensions of the problem (a temporary limitation of the experimental code). The Sun has 8 processors. On the Cray J-9x and the SGI Origin we could effectively use only up to 16 processors. On the Cray, for larger problems, due to the memory limitation, we could not even use these 16 processors. We report time $T(p)$, speedup S_p (calculated as time on one processor divided by the time on p processors), and efficiency E_p.

A number of observations can be made. First, the proposed implementation, which in a natural way follows the algorithm description, is clearly not appropriate for the vector computer. To be able to achieve a respectable performance on the Cray a vector-oriented implementation would be necessary. Second, for small problems, the proposed approach parallelizes rather well on both shared memory (Sun) and dynamic shared (SMP) machines (SGI). However, as the problem size increases parallel efficiency of the Sun decreases. It can be assumed that this

Table 1. Parallel performance on the SUN Enterprise 3000 superserver, the Cray J-9x vector-computer and the SGI Origin 2000 dynamic shared memory parallel computer

n	p	SUN $T(p)$	SUN S_p	SUN E_p	Cray J-9x $T(p)$	Cray J-9x S_p	Cray J-9x E_p	SGI $T(p)$	SGI S_p	SGI E_p
64	1	2.39			14.07			0.92		
	2	1.16	2.06	1.03	7.32	1.92	0.96	0.46	2.00	1.00
	4	0.60	3.99	1.00	3.87	3.63	0.91	0.23	4.00	1.00
	8	0.31	7.66	0.96	2.21	6.36	0.80	0.12	7.66	0.96
	16				1.86	7.56	0.47	0.09	9.38	0.64
96	1	18.38			44.81			5.56		
	2	9.02	2.04	1.02	23.10	1.93	0.97	2.75	2.02	1.01
	3	6.08	3.02	1.01	16.14	2.77	0.93	1.96	2.83	0.95
	4	4.68	3.93	0.98	12.04	3.72	0.93	1.38	4.02	1.01
	6	3.19	5.76	0.96	8.76	5.11	0.85	0.96	5.67	0.97
	8	2.90	6.34	0.79	6.79	6.59	0.82	0.74	7.51	0.94
	12				5.38	8.32	0.69	0.54	10.29	0.86
	16				5.61	7.98	0.50	0.47	11.82	0.74
128	1	27.67			130.75			10.64		
	2	12.85	2.15	1.08	69.12	1.89	0.95	5.41	1.96	0.98
	4	9.33	2.97	0.74	35.36	3.69	0.92	3.11	3.42	0.86
	8	6.17	4.49	0.56	20.09	6.50	0.81	1.33	8.00	1.00
	16				12.85	10.17	0.64	0.78	13.64	0.85
144	1	70.19			167.55			20.92		
	2	35.21	1.99	1.00	96.23	1.74	0.87	10.64	1.96	0.98
	3	23.79	2.95	0.98	58.32	2.87	0.96	7.05	2.96	0.99
	4	21.52	3.26	0.82	47.37	3.53	0.88	5.57	3.75	0.94
	6	21.45	3.27	0.55	36.55	4.58	0.76	3.55	5.89	0.98
	8	15.39	4.56	0.57	34.76	4.82	0.60	2.67	7.83	0.98
	12				31.82	5.26	0.44	1.84	11.36	0.95
	16							1.46	14.32	0.90
160	1	112.66			223.03			31.85		
	2	46.63	2.42	1.21	116.43	1.87	0.96	14.74	2.16	1.08
	4	24.39	4.62	1.15	61.60	3.77	0.91	7.34	4.33	1.08
	5	28.06	4.01	0.80	50.96	4.65	0.88	6.01	5.29	1.06
	8	21.36	5.27	0.66	36.48	6.83	0.76	3.84	8.29	1.04
	10				32.51	8.43	0.69	2.99	10.65	1.07
	16							2.01	15.84	0.99

is due to the communication overhead which saturates the memory-processor pathways. In addition, the single processor performance follows the same pattern. While for $n = 64$ it takes the Sun twice as long to solve the problem, this ratio increases to almost four times longer for $n = 160$. This observation should also be related to the appearance of super-linear speedup. This effect is visible not only on the Sun, but also, for the largest problem, on the SGI. This effect has a relatively simple explanation. It has been observed many times that, on the RISC based hierarchical memory computers, as the problem size increases their efficiency rapidly decreases (see for instance [1]).

In Table 2 we present the results of our experiments on the Cray T3E and the two PC clusters: the Beowulf cluster of 16 233 MHz PII processors and the Scali cluster of 16 450 MHz PIII processors. The reason for this combination

Table 2. Parallel performance on the Cray T3E and the PC clusters

n	p	Cray T3E $T(p)$	S_p	E_p	Beowulf cluster $T(p)$	S_p	E_p	Scali cluster $T(p)$	S_p	E_p
64	1	1.39			2.81			0.90		
	2	0.68	2.04	1.02	1.84	1.52	0.76	0.48	1.88	0.94
	4	0.35	3.97	0.99	1.01	2.78	0.70	0.25	3.60	0.90
	8	0.20	6.95	0.87	0.70	4.01	0.50	0.12	7.50	0.94
	16	0.11	12.63	0.79	0.49	5.73	0.36	0.06	15.00	0.94
96	1	7.46			17.06			5.34		
	2	3.74	1.99	1.00	10.14	1.68	0.84	2.75	1.94	0.96
	3	2.54	2.94	0.98	6.99	2.44	0.81	1.90	2.83	0.94
	4	1.90	3.92	0.98	5.31	3.21	0.80	1.42	3.76	0.93
	6	1.31	5.69	0.95	4.06	4.20	0.70	0.97	5.57	0.93
	8	0.98	7.61	0.95	3.26	5.23	0.65	0.73	7.31	0.91
	12	0.67	11.13	0.93	2.35	7.25	0.60	0.49	10.96	0.91
	16	0.52	14.34	0.90	1.98	8.61	0.54	0.37	14.43	0.89

is that the Cray in the NERSC center has only 256 Mbytes of memory per processor (which is exactly the same amount of memory as we had per node in both clusters) and thus we were able to run on them only the smaller problems. In addition, all three machines represent pure message passing environments.

The results are rather surprising. The Cray is only 3-4 times faster that the 233 MHz PII cluster and slower than the 450 MHz PIII cluster. It should be also added here, that the code on the Beowulf was compiled using the GNU compiler, while the code on the Scali cluster was compiled using the Portland Group compiler and thus the Beowulf results could have been somewhat better if the better quality compiler was used. Observe also that for $n = 96$ the Beowulf cluster has a performance comparable to the Sun (see Table 1). Interestingly, the Scali cluster slightly outperforms the SGI supercomputer. It is a pity that the distributed memory machines did not have more memory per node as it would be very interesting to find out if this relationship holds also for larger problems.

5 Concluding Remarks

In this note we have reported on the parallel performance of a new preconditioner applied to the conjugate gradient method used to solve a sparse linear system arising from a 3D elliptic model problem. We have shown that the code parallelizes well on a number of machines representing shared memory, dynamic shared memory (SMP) and message passing environments. In the near future

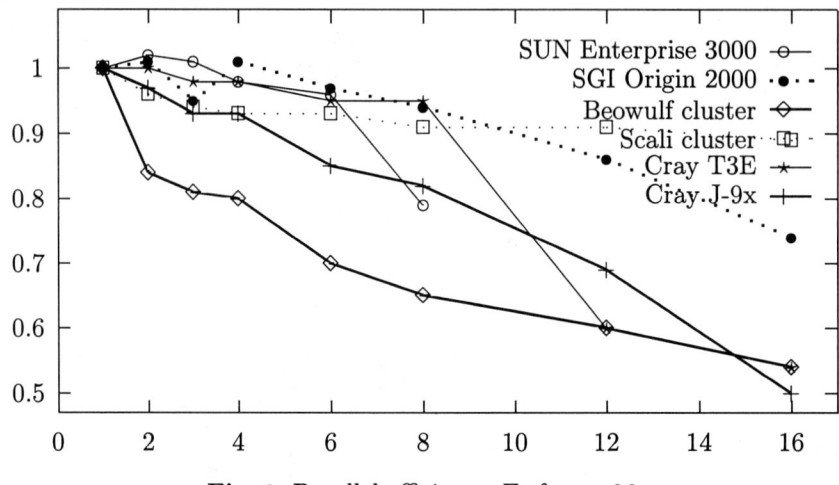

Fig. 1. Parallel efficiency E_p for n=96

we plan, first, to complete the performance studies by running our code on a number of additional machines (e.g. IBM SP2, HP SPP 1000, Compaq Alpha Cluster etc.). Second, we will extend our work to non-uniformly shaped domains, non-uniform discretizations as well as situations when the proposed approach is embedded in a solver for non-linear problems.

Acknowledgments

This research has been supported by Bulgarian NSF Grant I-701/97. Computer time grants from NERSC and MCSR are kindly acknowledged. We would also like to thank Prof. Horst Simon for helping us obtaining accounts on the NERSC machines and to Mr. Anders Liverud from Scali for running the experiments on their cluster.

References

1. I. Bar-On, M. Paprzycki, High Performance Solution of Complex Symmetric Eigenproblem, *Numerical Algorithms*, **18** (1998) 195–208.
2. R. H. Chan, T. F. Chan, Circulant preconditioners for elliptic problems, *J. Num. Lin. Alg. Appl.*, **1** (1992) 77–101.
3. I. Lirkov, S. Margenov, Conjugate gradient method with circulant block-factorization preconditioners for 3D elliptic problems, In *Proceedings of the Workshop # 3 Copernicus 94-0820 "HiPerGeoS" project meeting*, Rožnov pod Radhoštěm, 1996.
4. I. Lirkov, S. Margenov, Parallel complexity of conjugate gradient method with circulant block-factorization preconditioners for 3D elliptic problems. In *Recent Advances in Numerical Methods and Applications*, O. P. Iliev, M. S. Kaschiev, Bl. Sendov, P. V. Vassilevski, eds., World Scientific, Singapore, (1999) 455–463.

5. I. Lirkov, S. Margenov, P. S. Vassilevski, Circulant block-factorization preconditioners for elliptic problems, *Computing*, **53** 1 (1994) 59–74.
6. C. Van Loan, *Computational frameworks for the fast Fourier transform*, SIAM, Philadelphia, 1992.
7. M. Snir, St. Otto, St. Huss-Lederman, D. Walker, J. Dongara, *MPI: The Complete Reference*, Scientific and engineering computation series, The MIT Press, Cambridge, Massachusetts, 1997, Second printing.

Schwarz Methods for Convection-Diffusion Problems*

H. MacMullen[1], E. O'Riordan[1], and G. I. Shishkin[2]

[1] School of Mathematical Sciences, Dublin City University
Dublin, Ireland
[2] Institute for Mathematics & Mechanics, Russian Academy of Sciences
Ekaterinburg, Russia

Abstract. Various variants of Schwarz methods for a singularly perturbed two dimensional stationary convection-diffusion problem are constructed and analysed. The iteration counts, the errors in the discrete solutions and the convergence behaviour of the numerical solutions are analysed in terms of their dependence on the singular perturbation parameter of the Schwarz methods. Conditions for the methods to converge parameter uniformly and for the number of iterations to be independent of the perturbation parameter are discussed.

1 Introduction

Consider the following two dimensional convection-diffusion problem

$$-\varepsilon \Delta u_\varepsilon + \boldsymbol{a} \cdot \nabla u_\varepsilon = f \quad \text{on} \quad \Omega = (0,1) \times (0,1), \tag{1a}$$

$$u = g \quad \text{on} \quad \partial\Omega, \tag{1b}$$

$$\boldsymbol{a} = (a_1, a_2) > (0,0) \quad \text{on} \quad \bar{\Omega}, \tag{1c}$$

where $\boldsymbol{a}$, f and g are sufficiently smooth and f and g are sufficiently compatible at the four corners.

We wish to examine the suitability or otherwise of various Schwarz domain decomposition methods for this problem. It is well known that if one uses a monotone finite difference operator on an appropriately fitted piecewise-uniform mesh [2], the piecewise bilinear interpolant of the discrete solution satisfies the error bound $\|u_\varepsilon - \bar{U}_\varepsilon^N\| \leq CN^{-1} \ln N$, where C is a constant independent of ε. When an iterative numerical method is employed then both the disretization error and the iteration counts should be examined as functions of the small parameter ε. The number of mesh intervals in any coordinate direction is denoted by N and k denotes the iteration counter. Our goal is to design an iterative numerical method for (1) that satisfies an estimate of the form: $\|u_\varepsilon - \bar{U}_\varepsilon^{N,[k]}\| \leq CN^{-p} + Cq^k$, $p > 0$, $q < 1$, where C, p and q are independent of ε and N.

---

* This research was supported in part by the DCU Research Grant RC98-SPRJ-12EOR, by the Enterprise Ireland grant SC-98-612 and by the Russian Foundation for Basic Research under grant No. 98-01-00362.

2 One Dimensional Convection-Diffusion

We begin by examining Schwarz methods for the one dimensional convection-diffusion problem

$$-\varepsilon u_\varepsilon'' + a(x)u_\varepsilon' = f, \quad x \in \Omega = (0,1), \tag{2a}$$

$$u_\varepsilon(0) = A, \quad u_\varepsilon(1) = B. \tag{2b}$$

where $a(x) > \alpha > 0$. The solution $u_\varepsilon^{[k]}$, of an overlapping continuous Schwarz method, described in [9], where solution domain, Ω, is partitioned into two overlapping subdomains,

$$\Omega_0 = (0, \xi^+) \quad \text{and} \quad \Omega_1 = (\xi^-, 1),$$

satisfies the following error estimate. This is a well known result (see, for example [1,8,7,3]).

Lemma 1. *[9] Let u_ε be the solution of (2) and let $\{u_\varepsilon^{[k]}\}_{k=1}^\infty$ be the sequence of Schwarz iterates. Then, for all $k \geq 1$,*

$$\|u_\varepsilon^{[k]} - u_\varepsilon\|_{\bar\Omega} \leq C q_1^k$$

where C is independent of k and ε and

$$q_1 = e^{-\alpha(\xi^+ - \xi^-)/\varepsilon} < 1.$$

When the constants $\xi^- = 1 - \tau$ and $\xi^+ = 1 - 2\tau$ are chosen using the Shishkin transition point τ,

$$\tau = \min\{1/3, \frac{\varepsilon}{\alpha} \ln N\}, \tag{3}$$

the solution of this method converges to the exact solution u_ε, independently of both ε and N. However, in [6] numerical computations are presented which demonstrate that the discrete analogue of this method, in which uniform meshes discretise both subdomains, does not produce ε-uniform convergent approximations [5]. An alternative overlapping Schwarz method using a special piecewise uniform mesh in the layer subdomain was proposed in [6], and theoretical analysis showed this method to be ε-uniform. However, the width of the overlapping region was $O(N^{-1})$ and so the iteration counts became large as the number of grid nodes increased. Also the analysis of this method was cumbersome. Therefore, we examined less complicated non-overlapping methods using uniform grids, with the intention of developing and analysing a two dimensional Schwarz approach. We now describe two such methods and present theoretical convergence results.

2.1 A Non-overlapping Schwarz Method with Dirichlet Boundary Conditions

The solution domain is partitioned into two non-overlapping subdomains;

$$\Omega_0 = (0, \xi^+), \quad \Omega_1 = (\xi^+, 1) \quad \text{where} \quad \xi^+ = 1 - \tau,$$

and τ is given by (3). We also define $\xi^- = \frac{N-1}{N}(1-\tau)$ to be the $(N-1)^{\text{th}}$ grid node on $\bar{\Omega}_0$. The method is now formally described as follows.

Method 1. *The exact solution u_ε is approximated by the limit $\bar{U}_\varepsilon$ of a sequence of discrete Schwarz iterates $\{\bar{U}_\varepsilon^{[k]}\}_{k=0}^\infty$, which are defined as follows. For each $k \geq 1$,*

$$\bar{U}_\varepsilon^{[k]}(x) = \begin{cases} \bar{U}_0^{[k]}(x), & x \in \bar{\Omega}_0, \\ \bar{U}_1^{[k]}(x), & x \in \bar{\Omega}_1 \end{cases}$$

where $\bar{U}_i^{[k]}$ is the linear interpolant of $U_i^{[k]}$. Let $\bar{\Omega}_0^N = \{x_i\}_0^N$ be a uniform mesh on Ω_0 with $x_i = i\xi^+/N$ and $\bar{\Omega}_1^N = \{x_i\}_0^N$ be a uniform mesh on Ω_1 with $x_i = \xi^+ + i(1-\xi^+)/N$. Then for $k = 1$

$$L_\varepsilon^N U_0^{[1]} = f \quad \text{in } \Omega_0^N, \quad U_0^{[1]}(0) = u_\varepsilon(0), \quad U_0^{[1]}(\xi^+) = 0,$$

$$L_\varepsilon^N U_1^{[1]} = f \quad \text{in } \Omega_1^N, \quad U_1^{[1]}(\xi^+) = U_0^{[1]}(\xi^-), \quad U_1^{[1]}(1) = u_\varepsilon(1),$$

and for $k > 1$

$$L_\varepsilon^N U_0^{[k]} = f \quad \text{in } \Omega_0^N, \quad U_0^{[k]}(0) = u_\varepsilon(0), \quad U_0^{[k]}(\xi^+) = U_1^{[k-1]}(\xi^+),$$

$$L_\varepsilon^N U_1^{[k]} = f \quad \text{in } \Omega_1^N, \quad U_1^{[k]}(\xi^+) = U_0^{[k]}(\xi^-), \quad U_1^{[k]}(1) = u_\varepsilon(1).$$

The simple Dirichlet interface conditions on $\bar{\Omega}_1$, $U_1^{[k]}(\xi^+) = U_0^{[k]}(\xi^-)$, mean that the error reduction attained at $x = \xi^-$, when the method is applied in Ω_0, is transferred to $\bar{\Omega}_1$ but since no values are passed from $\bar{\Omega}_1$ to $\bar{\Omega}_0$, the problem of an accumulating error, see [5], in the iteration process is avoided. The following theorem gives error estimates for this method.

Theorem 1. *[5] Assume $\tau = \frac{\varepsilon}{\alpha} \ln N < 1/3$. For all $k \geq 1$,*

$$\|\bar{U}_\varepsilon^{[k]} - u_\varepsilon\| \leq CN^{-1}(\ln N)^2 + CN^{-1}\sum_{j=1}^{k-1}\lambda^{-j} + C\lambda^{-k}$$

$$\leq CN^{-1}(\ln N)^2 + C\varepsilon + C\lambda^{-k}$$

where C is a constant independent of k, N and ε and $\lambda = 1 + \frac{\alpha(1-\tau)}{\varepsilon N}$.

Consequently, this method is first order convergent for $\varepsilon \leq N^{-1}$. However, at each iteration the error reduction, λ^{-1}, is proportional to $1 - N^{-1}$ and so, the iterations become large for large N.

2.2 A Non-overlapping, Non-iterative Schwarz Method

In this section we discuss a Schwarz method which uses the same domain structure as described in the previous section and applies the Neumann interface condition, $D^-U_0(\xi^+) = 0$ on $\bar{\Omega}_0^N$. Therefore, no solution values are passed between subdomains and the method does not iterate. This approach uses a minimum amount of information and yet, for a singularly perturbed problem, produces accurate approximations for small ε. The method is defined as follows.

Method 2. *The exact solution u_ε is approximated by $\bar{U}_\varepsilon$ which is defined as follows*

$$\bar{U}_\varepsilon(x) = \begin{cases} \bar{U}_0(x), & x \in \bar{\Omega}_0, \\ \bar{U}_1(x), & x \in \bar{\Omega}_1 \end{cases}$$

where $\bar{U}_i$ is the linear interpolant of U_i. Let $\bar{\Omega}_0^N = \{x_i\}_0^N$ be a uniform mesh on Ω_0 with $x_i = i\xi^+/N$ and $\bar{\Omega}_1^N = \{x_i\}_0^N$ be a piecewise uniform mesh on Ω_1 with $x_i = \xi^+ + i(1-\xi^+)/N$. Then

$$L_\varepsilon^N U_0 = f \quad \text{in } \Omega_0^N, \quad U_0(0) = u_\varepsilon(0), \quad D^-U_0(\xi^+) = 0,$$

$$L_\varepsilon^N U_1 = f \quad \text{in } \Omega_1^N, \quad U_1(\xi^+) = U_0(\xi^+), \quad U_0(1) = u_\varepsilon(1),$$

The following theorem states the convergence behaviour for this approach.

Theorem 2. *[5] Assume $\tau < 1/3$. Then*

$$\|\bar{U}_\varepsilon - u_\varepsilon\| \leq CN^{-1}(\ln N)^2 + C\varepsilon$$

where C is a constant independent of N and ε.

We remark that numerical experiments have been carried out in [5] which demonstrate that the numerical approximations produced by this method are equivalent to those from Method 1.

Finally, Methods 1 and 2, although efficient for small values of ε, fail to be convergent for $\varepsilon > N^{-1}$. This is due to the non-matching of the interface conditions to the true solution for large values of ε. Numerical experiments are presented in [5] for a non-overlapping Schwarz method which uses the interface condition,

$$D^+U_0^{[k]}(\xi^-) = D^-U_0^{[k-1]}(\xi^-), \quad U_1^{[k]}(\xi^+) = U_0^{[k]}(\xi^+).$$

These computations show the numerical solutions converge for larger values of ε than solutions of Methods 1 and 2. However, an important observation is that the iterations required by this method are proportional to N and so the computational cost grows with the number of grid nodes required.

Remark 1. Recently, there has been a considerable interest in examining various different types of interface conditions (e.g., Robin–Robin or Dirichlet–Robin) for singularly perturbed convection–diffusion problems (see, for example, [10,4,11] and the references therein). However, in many cases, the analysis is restricted to the continuous Schwarz algorithm and/or the errors are examined in an L_2–norm, as opposed to a pointwise norm. Note that the L_2–norm is an inappropriate norm for singularly perturbed problems [2]. In this paper, we determine the explicit dependence of the pointwise errors and the iteration counts on both ε and N for two discrete Schwarz algorithms. These methods are not optimal, but are of theoretical interest. We expose the explicit error bounds that highlight the difference between the discrete and the continuous Schwarz algorithms and display the intricate nature of how the pointwise discretization error depends on the three variables ε, k and N.

3 Two Dimensional Convection-Diffusion

We extend Method 1 to the two dimensional problem (1). The solution domain $\Omega = (0,1)^2$ is partitioned into four non-overlapping subdomains $\Omega_a, \Omega_b, \Omega_c$ and Ω_d defined by

$$\Omega_a = (0, 1-\tau_1) \times (0, 1-\tau_2), \quad \Omega_b = (1-\tau_1, 1) \times (0, 1-\tau_2),$$
$$\Omega_c = (0, 1-\tau_1) \times (1-\tau_2, 1), \quad \Omega_d = (1-\tau_1, 1) \times (1-\tau_2, 1).$$

where the transition parameters τ_1, τ_2 are chosen to satisfy for $i=1$ and $i=2$,

$$\tau_i = \min\{\frac{1}{2}, \frac{\varepsilon}{\alpha_i} \ln N\}.$$

We use the notation, $\xi_1^+ = 1 - \tau_1$, $\xi_2^+ = 1 - \tau_2$, $\xi_1^- = \frac{N-1}{N}(1-\tau_1)$ and $\xi_2^- = \frac{N-1}{N}(1-\tau_2)$.

Method 3. *For each $k \geq 1$, $\bar{U}_\varepsilon^{[k]}(x,y) = \bar{U}_i^{[k]}(x,y)$, $(x,y) \in \bar{\Omega}_i$, $i = a,b,c,d$ where $\bar{U}_i^{[k]}$ is the bilinear interpolant of $U_i^{[k]}$. Let $\Omega_i^N = \{x_i, y_j\}$ be a uniform mesh on Ω_i. On $\Omega_i \setminus \Gamma_i$, $U_i^{[k]} = g \ \forall \, k \geq 1$. Then for $k = 1$*

$L_\varepsilon^N U_a^{[1]} = f$ *in* Ω_a^N, $U_a^{[1]}(x_i, y_j) = \Psi$ *on* Γ_a

$L_\varepsilon^N U_b^{[1]} = f$ *in* Ω_b^N, $U_b^{[1]}(x_i, \xi_2^+) = \Psi$, $U_b^{[1]}(\xi_1^+, y_j) = U_a^{[1]}(\xi_1^-, y_j)$

$L_\varepsilon^N U_c^{[1]} = f$ *in* Ω_c^N, $U_c^{[1]}(\xi_1^+, y_j) = \Psi$, $U_c^{[1]}(x_i, \xi_2^+) = U_a^{[1]}(x_i, \xi_2^-)$

$L_\varepsilon^N U_d^{[1]} = f$ *in* Ω_d^N, $U_d^{[1]}(\xi_1^+, y_j) = U_c^{[1]}(\xi_1^-, y_j)$, $U_d^{[1]}(x_i, \xi_2^+) = U_b^{[1]}(x_i, \xi_2^-)$.

Then for $k > 1$,

$L_\varepsilon^N U_a^{[k]} = f$ *in* Ω_a^N, $U_a^{[k]}(\xi_1^+, y_j) = U_b^{[k-1]}(\xi_1^+, y_j)$, $U_a^{[k]}(x_i, \xi_2^+) = U_c^{[k-1]}(x_i, \xi_2^+)$

$L_\varepsilon^N U_b^{[k]} = f$ *in* Ω_b^N, $U_b^{[k]}(x_i, \xi_2^+) = U_d^{[k-1]}(x_i, \xi_2^+)$, $U_b^{[k]}(\xi_1^+, y_j) = U_a^{[k]}(\xi_1^-, y_j)$

$L_\varepsilon^N U_c^{[k]} = f$ *in* Ω_c^N, $U_c^{[k]}(\xi_1^+, y_j) = U_d^{[k-1]}(\xi_1^+, y_j)$, $U_c^{[k]}(x_i, \xi_2^+) = U_a^{[k]}(x_i, \xi_2^-)$

$L_\varepsilon^N U_d^{[k]} = f$ *in* Ω_d^N, $U_d^{[k]}(\xi_1^+, y_j) = U_c^{[k]}(\xi_1^-, y_j)$, $U_d^{[k]}(x_i, \xi_2^+) = U_b^{[k]}(x_i, \xi_2^-)$

where Ψ is some arbitrary function with sufficient smoothness.

The following convergence result for the numerical solution generated is derived in [5], using comparison principle arguments and appropriate estimates on the solution components.

Theorem 3. *Assume $\tau_i < 1/3$. For $k \geq 1$,*
$$\|\bar{U}_\varepsilon^{[k]} - u_\varepsilon\| \leq CN^{-1}(\ln N)^2 + C\lambda_1^{-k} + C\lambda_2^{-k} + C\varepsilon$$
where $\lambda_i = 1 + \frac{\alpha_i(1-\tau_i)}{\varepsilon N}$ $i = 1,2$ and C is a constant independent of k, N and ε.

This theorem reveals a natural extension of Method 1 to two dimensions.

Remark 2. A parameter uniform overlapping method can be designed in which the overlapping regions would be fixed independently of ε and N, and a Shishkin fitted mesh placed in the subdomains containing layer regions. For the problem class 1, this Schwarz method would have no advantages over the fitted non-iterative Shishkin mesh. However, a problem involving a complex domain structure in higher dimensions, in which a fitted mesh may not be viable, may require this type of Schwarz approach.

3.1 Numerical Results

Numerical computations are carried out on the following problem for a sequences of meshes Ω_i^N, $i = a, b, c, d$ corresponding to $N = 4, 8, 16, 32, 64, 128$:

$$\varepsilon \Delta u_\varepsilon + (2 + x^2 y)u_x + (1 + xy)u_y = x^2 + y^3 + \cos(x + 2y), \quad (4a)$$

with boundary conditions

$$u(x,0) = 0, \quad u(x,1) = \begin{cases} 4x(1-x), & x < 1/2, \\ 1, & x \geq 1/2, \end{cases} \quad (4b)$$

$$u(0,y) = 0, \quad u(1,y) = \begin{cases} 8(y - 2y^2), & x < 1/4, \\ 1, & x \geq 1/4. \end{cases} \quad (4c)$$

In Figure 1 the numerical solution U_ε^{16}, with $N = 16$ intervals in each subdomain and $\varepsilon = 0.001$, is shown. The orders of convergence presented in Table 1 are computed using the double mesh principle (see [2]),

$$p_\varepsilon^N = \log_2\left(\frac{D_\varepsilon^N}{D_\varepsilon^{2N}}\right) \quad \text{where} \quad D_\varepsilon^N = \max_{x_i \in \bar{\Omega}_i^N} |U_\varepsilon^N(x_i) - \bar{U}_\varepsilon^{2N}(x_i)|.$$

In Table 2, the required iteration counts are given for a tolerance level of

$$\max_{x_i \in \bar{\Omega}_\varepsilon^N} |U_\varepsilon^{[k]}(x_i) - U_\varepsilon^{[k-1]}(x_i)| \leq 10^{-8}.$$

These results show experimentally that, for small values of ε, this method produces accurate numerical approximations and is computationally efficient.

Table 1. Computed orders of convergence p_ε^N for Method 3 applied to Problem (4) for various values of ε and N, where N is the number of intervals in each subdomain

ε	N			
	4	8	16	32
2^{-6}	0.08	0.24	0.45	0.57
2^{-7}	0.09	0.28	0.52	0.63
2^{-8}	0.08	0.31	0.59	0.73
2^{-9}	0.06	0.32	0.63	0.80
2^{-10}	0.05	0.31	0.65	0.84
2^{-11}	0.05	0.31	0.65	0.85
2^{-12}	0.04	0.31	0.65	0.85
$\vdots$	$\vdots$	$\vdots$	$\vdots$	$\vdots$
2^{-19}	0.04	0.31	0.65	0.85

Table 2. Iteration count for Method 3 applied to Problem (4) for various values of ε and N, where N is the number of intervals in each subdomain

ε	N					
	4	8	16	32	64	128
2^{-6}	7	9	12	18	28	46
2^{-7}	6	8	10	13	18	28
2^{-8}	6	7	8	10	13	18
2^{-9}	5	6	7	8	10	13
2^{-10}	5	5	6	7	8	10
2^{-11}	4	5	5	6	7	8
2^{-12}	4	4	5	5	6	7
2^{-13}	4	4	4	5	5	6
2^{-14}	4	4	4	4	5	5
2^{-15}	4	5	4	4	4	5
2^{-16}	3	4	6	4	4	4
2^{-17}	3	3	4	6	11	22
2^{-18}	3	3	3	4	6	10
2^{-19}	3	3	3	3	4	5

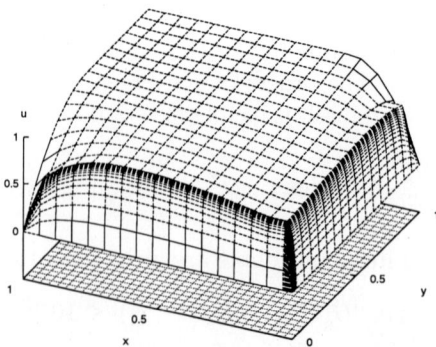

Fig. 1. Numerical solution of problem 4 with Ω_i^{16} and $\varepsilon = 0.001$

References

1. Farrell, P. A., Boglaev, I. G., and Sirotkin, V. V.: Parallel domain decomposition methods for semi-linear singularly perturbed differential equations. Comput. Fluid Dynamic Journal **2** (4) (1994) 423–433
2. Farrell, P. A., Hegarty, A. F., Miller, J. J. H., O'Riordan, E., and Shishkin, G. I.: Robust Computational Techniques for Boundary Layers. Chapman and Hall/CRC Press, Boca Raton, U. S. A. (2000)
3. Garbey, M.: A Schwarz alternating procedure for singular perturtation problems. J. Sci. Comput. **17** (5) (1996) 1175–1201
4. Lube, G., Otto, F. C. and Muller, H. : A non-overlapping domain decomposition method for parabolic initial-boundary value problems. Appli. Numer. Math. **28** (1998) 359–369
5. MacMullen, H.: Schwarz domain decomposition methods for singularly perturbed differential equations. PhD thesis, Mathematical Sciences, D. C.U Dublin (2000)
6. MacMullen, H., Miller, J. J. H., O'Riordan E., and Shishkin, G. I.: Schwarz iterative methods for convection-diffusion problems with boundary layers. In: J. J. H. Miller and G.I Shishkin and L.G Vulkov eds. *Analytical and Numerical Methods for Convection-Dominated and Singularly Perturbed Problems*. Nova, New York (to appear)
7. MacMullen, H., Miller, J. J. H., O'Riordan E., and Shishkin, G. I.: A second order parameter-uniform overlapping Schwarz method for reaction-diffusion problems with boundary layers. J. Comput. Appl. Math. (to appear)
8. Mathew, T. P.: Uniform convergence of the Schwarz alternating method for solving singularly perturbed advection-diffusion equations. SIAM J. Numer. Anal. **35** (4) (1998) 1663–1683
9. Miller, J. J. H., O'Riordan, E., and Shishkin, G. I.: Fitted Numerical Methods for Singular Perturbation Problems. World Scientific Publishing Co., Singapore (1996)
10. Nataf, F. and Rogier, F. : Factorization of the convection-diffusion operator and the Schwarz algorithm. Math. Models Meth. Appli. Sci. **67** (1), (1995) 67–93
11. Quarteroni, A. and A. Valli, A. : Domain Decomposition Methods for Partial Differential Equations. Oxford University Press, (1999)

Matrix Computations Using Quasirandom Sequences

Michael Mascagni[1] and Aneta Karaivanova[1,2]

[1] Department of Computer Science, Florida State University
203 Love Building, Tallahassee, FL 32306-4530, USA
mascagni@cs.fsu.edu
http://www.cs.fsu.edu/~mascagni

[2] Central Laboratory for Parallel Processing, Bulgarian Academy of Sciences
Acad. G. Bonchev St.,bl. 25 A, 1113, Sofia, Bulgaria
aneta@scri.fsu.edu

Abstract. The convergence of Monte Carlo method for numerical integration can often be improved by replacing pseudorandom numbers (PRNs) with more uniformly distributed numbers known as quasirandom numbers(QRNs). Standard Monte Carlo methods use pseudorandom sequences and provide a convergence rate of $O(N^{-1/2})$ using N samples. Quasi-Monte Carlo methods use quasirandom sequences with the resulting convergence rate for numerical integration as good as $O((log N)^k) N^{-1})$.
In this paper we study the possibility of using QRNs for computing matrix-vector products, solving systems of linear algebraic equations and calculating the extreme eigenvalues of matrices. Several algorithms using the same Markov chains with different random variables are described. We have shown, theoretically and through numerical tests, that the use of quasirandom sequences improves both the magnitude of the error and the convergence rate of the corresponding Monte Carlo methods. Numerical tests are performed on sparse matrices using PRNs and Soboĺ, Halton, and Faure QRNs.

1 Introduction

Monte Carlo methods (MCMs) are based on the simulation of stochastic processes whose expected values are equal to computationally interesting quantities. Despite the universality of MCMs, a serious drawback is their slow convergence, which is based on the $O(N^{-1/2})$ behavior of the size of statistical sampling errors. This represents a great opportunity for researchers in computational science. Even modest improvements in the MCM can have substantial impact on the efficiency and range of applicability for MCM. Much of the effort in the development of Monte Carlo methods has been in construction of variance reduction methods which speed up the computation by reducing the constant in front of the $O(N^{-1/2})$. An alternative approach to acceleration is to change the choice of sequence and hence improve the behavior with N. Quasi-Monte Carlo methods (QMCMs) use quasirandom (also known as low-discrepancy) sequences instead of pseudorandom sequences.

QRNs are constructed to minimize a measure of their deviation from uniformity called discrepancy. There are many different discrepancies, but let us consider the most common, the star discrepancy. Let us define the star discrepancy of a one-dimensional point set, $\{x_n\}_{n=1}^N$, by

$$D^\star = D^\star(x_1,\ldots,x_N) = \sup_{0 \leq u \leq 1} |\frac{1}{N} \sum_{n=1}^N \chi_{[0,u)}(x_n) - u| \quad (1)$$

where $\chi_{[0,u)}$ is the characteristic function of the half open interval $[0,u)$. The mathematical motivation for quasirandom numbers can be found in the classic Monte Carlo application of numerical integration. We detail this for the trivial example of one-dimensional integration for illustrative simplicity.

Theorem (Koksma-Hlawka, [6]): if $f(x)$ has bounded variation, $V(f)$, on $[0,1)$, and $x_1,\ldots,x_N \in [0,1]$ have star discrepancy $D^\star$, then:

$$|\frac{1}{N} \sum_{n=1}^N Nf(x_n) - \int_0^1 f(x)\,dx| \leq V(f)D^\star, \quad (2)$$

The star discrepancy of a point set of N truly random numbers in one dimension is $O(N^{-1/2}(\log\log N)^{1/2})$, while the discrepancy of N quasirandom numbers can be as low as N^{-1}.[1] In $s > 3$ dimensions it is rigorously known that the discrepancy of a point set with N elements can be no smaller than a constant depending only on s times $N^{-1}(\log N)^{(s-1)/2}$. This remarkable result of Roth, [10], has motivated mathematicians to seek point sets and sequences with discrepancies as close to this lower bound as possible. Since Roth's remarkable results, there have been many constructions of low discrepancy point sets that have achieved star discrepancies as small as $O(N^{-1}(\log N)^{s-1})$. Most notably there are the constructions of Hammersley, Halton, [5], Sobol, [11], Faure, [3], and Niederreiter, [9].

While QRNs do improve the convergence of applications like numerical integration, it is by no means trivial to enhance the convergence of all MCMs. In fact, even with numerical integration, enhanced convergence is by no means assured in all situations with the näive use of quasirandom numbers, [1,8].

In this paper we study the applicability of quasirandom sequences for solving some linear algebra problems. We have already produced encouraging theoretical and empirical results with QMCMs for linear algebra problems and we believe that this initial work can be improved.

Solving Systems of Linear Algebraic Equations via Neumann Series

Assume that a system of linear algebraic equations (SLAE) can be transformed into the following form: $x = Ax + \varphi$, where A is a real square, $n \times n$, matrix,

---

[1] Of course, the N optimal quasirandom points in $[0,1)$ are the obvious: $\frac{1}{(N+1)}, \frac{2}{(N+1)}, \ldots, \frac{N}{(N+1)}$.

$x = (x_1, x_2, ..., x_n)^t$ is the $1 \times n$ solution vector and $\varphi = (\varphi_1, \varphi_2, ..., \varphi_n)^t$ is the given right-hand side vector.[2] In addition, assume that A satisfies either the condition $\max_{1 \leq i \leq n} \sum_{j=1}^n |a_{ij}| < 1$, or, that all the eigenvalues of A lie within the unit circle.

Now consider the sequence $x^{(1)}, x^{(2)}, \ldots$ defined by the following recursion:

$$x^{(k)} = Ax^{(k-1)} + \varphi, \ k = 1, 2, \ldots.$$

Given initial vector $x^{(0)}$, the approximate solution to the system $x = Ax + \varphi$ can be developed via a truncated Neumann series:

$$x^{(k)} = \varphi + A\varphi + A^2\varphi + \ldots + A^{(k-1)}\varphi + A^k x^{(0)}, \ k > 0 \tag{3}$$

with a truncation error of $x^{(k)} - x = A^k(x^{(0)} - x)$.

This iterative process (3) of applying the matrix A repeatedly is the basis for deriving a Monte Carlo approach for this problem.

The Monte Carlo Method

Consider the problem of evaluating the inner product of a given vector, g, with the vector solution of the considered system

$$(g, x) = \sum_{\alpha=1}^n g_\alpha x_\alpha. \tag{4}$$

To solve this problem via a MCM (see, for example, [12]) one has to construct a random process with mean equal to the solution of the desired problem. This requires the construction of a finite-state Markov chain. Consider the following Markov chain:

$$k_0 \to k_1 \to \ldots \to k_i, \tag{5}$$

where $k_j = 1, 2, \ldots, n$ for $j = 1, \ldots, i$ are natural numbers. The rules for constructing the chain (5) are: $P(k_0 = \alpha) = p_\alpha, P(k_j = \beta | k_{j-1} = \alpha) = p_{\alpha\beta}$ where p_α is the probability that the chain starts in state α and $p_{\alpha\beta}$ is the transition probability from state α to state β. Probabilities $p_{\alpha\beta}$ define a transition matrix P. We require that $\sum_{\alpha=1}^n p_\alpha = 1$, $\sum_{\beta=1}^n p_{\alpha\beta} = 1$ for any $\alpha = 1, 2, ..., n$, and that the distribution $(p_1, ..., p_n)^t$ is permissible to the vector g and similarly the distribution $p_{\alpha\beta}$ is permissible to A [12]. Common constructions are to choose $p_{\alpha\beta} = \frac{|a_{\alpha\beta}|}{\sum_\beta |a_{\alpha\beta}|}$ for $\alpha, \beta = 1, 2, ..., n$, which corresponds to an *importance sampling* MCM (MCM with a reduced variance), or to choose $p_{\alpha\beta} = 1/n$ for $\alpha, \beta = 1, 2, ..., n$ which corresponds to standard MCM.

Now define the random variables $\theta[g]$:

$$\theta[g] = \frac{g_{k_0}}{p_{k_0}} \sum_{j=0}^\infty W_j \varphi_{k_j} \tag{6}$$

where $W_0 = 1$, $W_j = W_{j-1} \frac{a_{k_{j-1}k_j}}{p_{k_{j-1}k_j}}$,

---

[2] If we consider a given system $Lx = b$, then it is possible to find a non-singular matrix, M, such that $ML = I - A$ and $Mb = \varphi$. Thus without loss of generality the system $Lx = b$ can always be recast as $x = Ax + \varphi$.

It is known [12] that the mathematical expectation $E[\theta[g]]$ of the random variable $\theta[g]$ is:
$$E[\theta[g]] = (g, x).$$
The partial sum corresponding to (6) is defined as $\theta_i[g] = \frac{g_{k_0}}{p_{k_0}} \sum_{j=0}^{i} W_j \varphi_{k_j}$. Thus the Monte Carlo estimate for (g, x) is $(g, x) \approx \frac{1}{N} \sum_{s=1}^{N} \theta_i[g]_s$, where N is the number of chains and $\theta_i[g]_s$ is the value of $\theta_i[g]$ taken over the s-th chain, and a statistical error of size $O(Var(\theta_i)^{1/2} N^{-1/2})$.

Computing the Extremal Eigenvalues

Let A be an $n \times n$ large, sparse, matrix. Consider the problem of computing one or more eigenvalues of A, i.e., the values of λ for which $Au = \lambda u$ holds. Suppose the eigenvalues are ordered $|\lambda_1| > |\lambda_2| \geq \ldots \geq |\lambda_{n-1}| > |\lambda_n|$. There are two deterministic numerical methods that can efficiently compute only the extremal eigenvalues - the *power method* and *Lanczos-type methods*. (Note that, the Lanczos method is applicable to only *symmetric* eigenproblems, [4].)
Computational Complexity: If k iterations are required for convergence, the number of arithmethic operations is $O(kn^2)$ for the power method and $O(n^3 + kn^2)$ for both the inverse and inverse shifted power method.

The Monte Carlo Method

Consider MCMs based on the power method. When computing eigenvalues, we work with the matrix A and its *resolvent* matrix $R_q = [I - qA]^{-1} \in \mathbb{R}^{n \times n}$. If $|q\lambda| < 1$, R_q may be expanded as a series via the binomial theorem:

$$[I - qA]^{-m} = \sum_{i=1}^{\infty} q^i C_{m+i-1}^i, \quad |q\lambda| < 1. \quad (7)$$

The eigenvalues of the matrices R_q and A are connected by the equality $\mu = \frac{1}{1-q\lambda}$, and the eigenvectors of the two matrices coincide[3]. Let $f \in \mathbb{R}^n, h \in \mathbb{R}^n$. Applying the power method, ([2]), leads to the following iterative processes:

$$\lambda^{(m)} = \frac{(h, A^i f)}{(h, A^{i-1} f)} \xrightarrow[m \to \infty]{} \lambda_{max} \quad (8)$$

$$\mu^{(m)} = \frac{([I - qA]^{-m} f, h)}{([I - qA]^{-(m-1)} f, h)} \xrightarrow[m \to \infty]{} \mu_{max} = \frac{1}{1 - q\lambda}. \quad (9)$$

Construct the same Markov chain as before with the initial density vector, $p = \{p_\alpha\}_{\alpha=1}^n$, and the transition density matrix, $P = \{p_{\alpha\beta}\}_{\alpha\beta=1}^n$. Define the following

---
[3] If $q > 0$, the largest eigenvalue μ_{max} of the resolvent matrix corresponds to the largest eigenvalue, λ_{max}, of the matrix A, but if $q < 0$, then μ_{max}, corresponds to the smallest eigenvalue, λ_{min}, of the matrix A.

random variable: $W_0 = \frac{h_{k_0}}{p_{k_0}}$, $W_j = W_{j-1}\frac{a_{k_{j-1}k_j}}{p_{k_{j-1}k_j}}$, $j = 1,\ldots,i$. This has the desired expected values ([2]):

$$E[W_i f_{k_i}] = (h, A^i f), \quad i = 1, 2, \ldots,$$

$$E[\sum_{i=0}^{\infty} q^i C_{i+m-1}^i W_i f(x_i)] = (h, [I - qA]^{-m} f), \quad m = 1, 2, \ldots,$$

and allows us to estimate the desired eigenvalues as:

$$\lambda_{max} \approx \frac{E[W_i f_{k_i}]}{E[W_{i-1} f_{k_{i-1}}]}. \tag{10}$$

and

$$\lambda \approx \frac{1}{q}\left(1 - \frac{1}{\mu^{(m)}}\right) = \frac{E[\sum_{i=1}^{\infty} q^{i-1} C_{i+m-2}^{i-1} W_i f(x_i)]}{E[\sum_{i=0}^{\infty} q^i C_{i+m-1}^i W_i f(x_i)]}. \tag{11}$$

We remark that in (10) the length of the Markov chain, l, is equal to the number of iterations, i, in the power method. However in (11) the length of the Markov chain is equal to the number of terms in truncated series for the resolvent matrix. In this second case the parameter m corresponds to the number of iterations.

Table 1. Monte Carlo estimations using PRNs and QRN sequences for the dominant eigenvalue of matrices of size 128 and 2000

	PRN	QRN(Faur)	QRN(Sobol)	QRN(Halton)
Estimated $\lambda 128_{max}$	61.2851	63.0789	63.5916	65.1777
Relative Error	0.0424	0.0143	0.0063	0.0184
Estimated $\lambda 2000_{max}$	58.8838	62.7721	65.2831	65.377
Relative Error	0.0799	0.01918	0.0200	0.0215

Quasi-Monte Carlo Methods for Matrix Computations

Recall that power method iterations are based on computing $h^T A^i f$ (see (8) and (9)). Even if we are interested in evaluating the inner product (4), substituting x with $x^{(k)}$ from (3) will give $(g, x) \approx g^T \varphi + g^T A \varphi + g^T A^2 \varphi + \ldots + g^T A^{(k-1)} \varphi + g^T A^k x^{(0)}$, $k > 0$. Define the sets $G = [0, n)$ and $G_i = [i-1, i)$, $i = 1, \ldots, n$, and likewize define the piecewise continous functions $f(x) = f_i$, $x \in G_i, i = 1, \ldots, n$, $a(x, y) = a_{ij}, x \in G_i, y \in G_j, i, j = 1, \ldots, n$ and $h(x) = h_i$, $x \in G_i, i = 1, \ldots, n$.

Computing $h^T A^i f$ is equivalent to computing an $(i+1)$-dimensional integral. Thus we may analyze using QRNs in this case with bounds from numerical integration. We do not know A^i explicitly, but we do know A and can use a random walk on the elements of the matrix to compute approximately $h^T A^i f$.

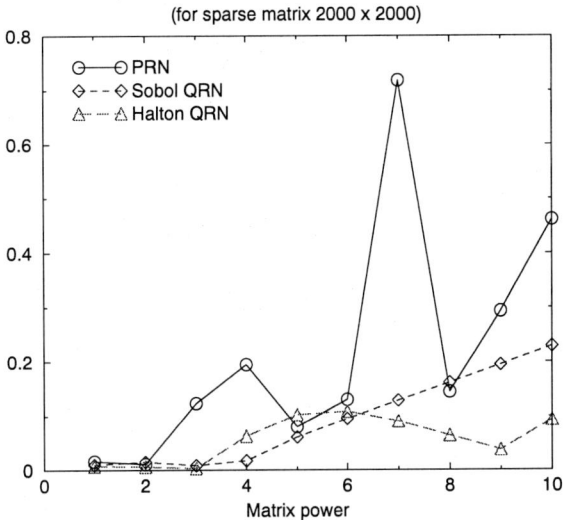

Fig. 1. Relative errors in computing $h^T A^k f$ for $k = 1, 2, \ldots, 10$ for a sparse matrix 2000×2000. The corresponding Markov chains are realized using PRN, Soboĺ and Halton sequences

Consider $h^T A^i f$ and an $(i+1)$-dimensional QRN sequence. Normalizing A with $\frac{1}{n}$, and h and f with $\frac{1}{\sqrt{n}}$, we have the following error bound (for proof see [7]):

$$|h_N^T A_N^l f_N - \frac{1}{N} \sum_{s=1}^{N} h(x_s) a(x_s, y_s) \ldots a(z_s, w_s) f(w_s)| \leq |h|^T |A|^l |f| D_N^*.$$

If A is a general sparse matrix with d nonzero elements per row, and $d \ll n$, then *importance sampling method* can be used; the normalizing factors in the error bound (3) are then $1/d$ for the matrix and $\frac{1}{\sqrt{(n)}}$ for the vectors.

2 Numerical Results

Why are we interested in quasi-MCMs for the eigenvalue problem? Because the computational complexity of QMCMs is bounded by $O(lN)$ where N is the number of chains, and l is the mathematical expectation of the length of the Markov chains, both of which are independent of matrix size n. This makes QMCMs very efficient for large, sparse, problems, for which deterministic methods are not computationally efficient.

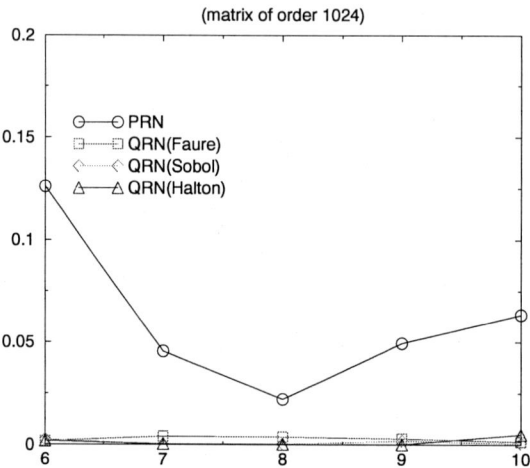

Fig. 2. Relative errors in computing λ_{max} using different length of Markov chains for a sparse matrix 1024×1024. The random walks are realized using PRN, Faure, Soboĺ and Halton sequences

Numerical tests were performed on general sparse matrices using PRNs and Soboĺ, Halton and Faure QRNs. The relative errors in computing $h^T A^k f$ with A a sparse matrix of order 2000 and $h = f = (1, 1, \ldots, 1)$, are presented in *Figure 1*. The results confirm that the QRNs produce higher precision results than PRNs. The more important fact is the smoothness of the quasirandom "iterations" with k. This is important because these eigenvalue algorithms compute a Raleigh quotient which requires the division of values from consecutive iterations.

The estimated λ_{max} and the corresponding relative errors using MCM and QMCM are presented in *Table 1*. The exact value of λ_{max} for all test matrices is 64.0000153. The results show improvement of the accuracy. Numerical experiments using resolvent MCM and resolvent QMCM have been also performed - the relative errors in computing λ_{max} using Markov chains with different lengths are presented in *Figures 2 and 3*.

Acknowledgements

This paper is based upon work supported by the North Atlantic Treaty Organization under a Grant awarded in 1999.

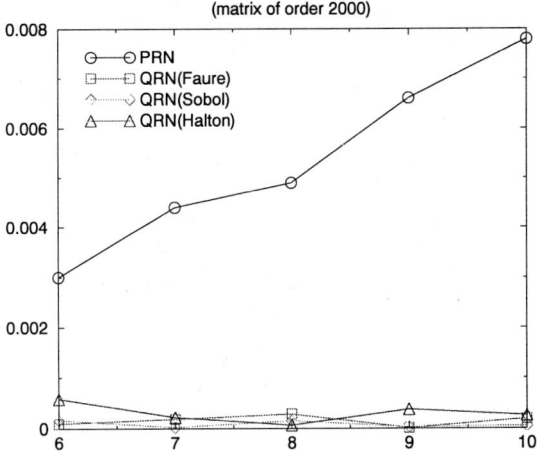

Fig. 3. Relative errors in computing λ_{max} using different length of Markov chains for a sparse matrix 2000 × 2000. The random walks are realized using PRN, Faure, Soboĺ and Halton sequences

References

1. R. E. CAFLISCH, "Monte Carlo and quasi-Monte Carlo methods," *Acta Numerica*, **7**: 1–49, 1998.
2. I. DIMOV, A. KARAIVANOVA, "Parallel computations of eigenvalues based on a Monte Carlo approach," *J. of MC Methods and Appl.*, **4**, Num.1: 33–52, 1998.
3. H. FAURE, "Discrépance de suites associées à un système de numération (en dimension s)," *Acta Arithmetica*, **XLI**: 337–351, 1992.
4. G. H. GOLUB, C. F. VAN LOAN, "Matrix computations", *The Johns Hopkins Univ. Press*, Baltimore, 1996.
5. J. H. HALTON, "On the efficiency of certain quasi-random sequences of points in evaluating multi-dimensional integrals," *Numer. Math.*, **2**: 84–90, 1960.
6. J. F. KOKSMA, "Een algemeene stelling uit de theorie der gelijkmatige verdeeling modulo 1," *Mathematica B (Zutphen)*, **11**: 7–11, 1942/43.
7. M. MASCAGNI, A. KARAIVANOVA, "Are Quasirandom Numbers Good for Anything Besides Integration?" to appear in *Proc. of Advances in Reactor Physics and Mathematics and Computation into the Next Millennium (PHYSOR2000)*, 2000.
8. B. MOSKOWITZ, R. E. CAFLISCH, "Smoothness and dimension reduction in quasi-Monte Carlo methods", *J. Math. Comput. Modeling*, **23**: 37–54, 1996.
9. H. NIEDERREITER, *Random number generation and quasi-Monte Carlo methods*, SIAM: Philadelphia, 1992.
10. K. F. ROTH, "On irregularities of distribution," *Mathematika*, **1**: 73–79, 1954.
11. I. M. SOBOĹ, "The distribution of points in a cube and approximate evaluation of integrals," *Zh. Vychisl. Mat. Mat. Fiz.*, **7**: 784–802, 1967.
12. I. M. SOBOĹ *Monte Carlo numerical methods*, Nauka, Moscow, 1973.

On the Stability of the Generalized Schur Algorithm*

Nicola Mastronardi[1,2], Paul Van Dooren[3], and Sabine Van Huffel[1]

[1] Department of Electrical Engineering, ESAT-SISTA/COSIC
Katholieke Universiteit Leuven, Kardinaal Mercierlaan 94, 3001 Leuven, Belgium
[2] Dipartimento di Matematica, Università della Basilicata
via N. Sauro 85, 85100 Potenza, Italy
[3] Department of Mathematical Engineering, Université Catholique de Louvain
Avenue Georges Lemaitre 4, B-1348 Louvain-la-Neuve, Belgium

Abstract. The generalized Schur algorithm (GSA) is a fast method to compute the Cholesky factorization of a wide variety of structured matrices. The stability property of the GSA depends on the way it is implemented. In [15] GSA was shown to be as stable as the Schur algorithm, provided one hyperbolic rotation in factored form [3] is performed at each iteration. Fast and efficient algorithms for solving Structured Total Least Squares problems [14,15] are based on a particular implementation of GSA requiring two hyperbolic transformations at each iteration. In this paper the authors prove the stability property of such implementation provided the hyperbolic transformations are performed in factored form [3].

1 Introduction

The generalized Schur algorithm (GSA) is a fast method to compute the Cholesky decomposition of a wide variety of symmetric positive definite structured matrices, i.e., block–Toeplitz and Toeplitz–block matrices, matrices of the form $T^T T$, where T is a rectangular Toeplitz matrix [9,7] and to compute the LDL^T factorization of *strongly regular* [1] structured matrices, where L is a triangular matrix and $D = \text{diag}(\pm 1, \ldots, \pm 1)$. The stability property of the GSA depends

---

* S. Van Huffel is a Senior Research Associate with the F.W.O. (Fund for Scientific Research - Flanders). This paper presents research results of the Belgian Programme on Interuniversity Poles of Attraction (IUAP P4-02 and P4-24), initiated by the Belgian State, Prime Minister's Office - Federal Office for Scientific, Technical and Cultural Affairs, of the European Community TMR Programme, Networks, project CHRX-CT97-0160, of the Brite Euram Programme, Thematic Network BRRT-CT97-5040 'Niconet', of the Concerted Research Action (GOA) projects of the Flemish Government MEFISTO-666 (Mathematical Engineering for Information and Communication Systems Technology), of the IDO/99/03 project (K.U.Leuven) "Predictive computer models for medical classification problems using patient data and expert knowledge", of the FWO "Krediet aan navorsers" G.0326.98 and the FWO project G.0200.00.

on the way it is implemented [11,5]. In [15] the GSA was shown to be stable, provided one hyperbolic rotation in factored form [3] is performed at each iteration. Similar results were obtained in [5], using the *OD procedure* or the *H procedure* instead of using the hyperbolic rotations in factored form. The computational complexity of GSA is $\mathcal{O}(\alpha N^2)$, where N is the order of the involved matrix and α is its displacement rank (see §2).

The Structured Total Least Squares problem, as described in [14], solves overdetermined Toeplitz systems of equations with unstructured right-hand side and can be formulated as follows:

$$\min_{\Delta A, \Delta b, x} \|[\Delta A \ \Delta b]\|_F^2$$

such that $(A + \Delta A)x = b + \Delta b$, $A, \Delta A \in \mathbb{R}^{m \times n}, m \gg n$,

with $A, \Delta A$ Toeplitz matrices. The kernel of the algorithm proposed in [14], is the solution of a least squares problem, where the coefficient matrix is a rectangular Toeplitz-block matrix, with dimensions $(2m+n-1) \times (m+2n-1)$. The sparsity structure of the corresponding generators is such that when using at each iteration two hyperbolic rotations rather than one, the complexity of the GSA can be reduced from $\mathcal{O}(\alpha(m+n)^2)$ to $\mathcal{O}(\alpha mn)$. It is therefore worth studying the stability of such a modification of the GSA, which is what we do in this paper. Although the sparsity of the generators is exploited in this modified GSA, we need not keep track of this zero pattern in order to perform the error analysis. The paper is organized as follows. In §2 we describe the implementation of the GSA when using two hyperbolic rotations; we analyze its stability property in §3 and we terminate with some conclusions in §4.

2 The Generalized Schur Algorithm

In this section we introduce the GSA to compute the $R^T R$ factorization of a symmetric positive definite matrix A, where R is an upper triangular matrix, using two hyperbolic rotations at each iteration.

Given an $n \times n$ symmetric positive definite matrix A, define $D_A = A - ZAZ^T$. We say that the displacement rank of A with respect to Z is α if rank$(D_A) = \alpha$, where Z is the lower triangular (block) shift matrix of order n (for a more general choice of the matrix Z, see [9,6]). Clearly D_A will have a decomposition of the form $D_A = G^T J_A G$, where

$$G = \begin{bmatrix} u_{1,1}^{(1)} & u_{1,2}^{(1)}{}^T \\ u_{2,1}^{(1)} & U_{2,2}^{(1)} \\ v_{1,1}^{(1)} & v_{1,2}^{(1)}{}^T \\ z_{1,1}^{(1)} & z_{1,2}^{(1)}{}^T \\ v_{2,1}^{(1)} & V_{2,2}^{(1)} \end{bmatrix}, \quad J_A = I_p \oplus -I_q, \ q = \alpha - p,$$

where $u_{1,1}^{(1)}, v_{1,1}^{(1)}, z_{1,1}^{(1)} \in \mathbb{R}$, $u_{1,2}^{(1)}, v_{1,2}^{(1)}, z_{1,2}^{(1)} \in \mathbb{R}^{n-1}$, $u_{2,1}^{(1)} \in \mathbb{R}^{p-1}$, $v_{2,1}^{(1)} \in \mathbb{R}^{q-2}$, $U_{2,2}^{(1)} \in \mathbb{R}^{(p-1) \times (n-1)}$, $V_{2,2}^{(1)} \in \mathbb{R}^{(q-2) \times (n-1)}$, and I_k is the identity matrix of order k. The pair $(G, J_A), G \in \mathbb{R}^{\alpha \times n}$ is said to be a generator pair for A [12]. A matrix Θ is said J_A–orthogonal if $\Theta^T J_A \Theta = J_A$.

The GSA requires n iterations to compute the factor R. Let $G_{0,Z} = G$. At the ith iteration, $i = 1, \ldots, n$, a J_A–orthogonal matrix Θ_i is chosen such that the ith column of $G_i = \Theta_i G_{i-1,Z}$ has all the elements equal to zero with the exception of a single pivot element in the first row (the first $i-1$ columns of G_i are zero). The generator matrix G_i is said to be in a proper form. Then the first row of G_i becomes the ith row of R. The generator matrix $G_{i,Z}$ at the next iteration is given by

$$G_{i,Z}(1,:) = G_i(1,:)Z^T, \quad G_{i,Z}([2:\alpha],:) = G_i([2:\alpha],:).$$

Without loss of generality, the matrices Θ_i, $i = 1, \ldots, n$, can be factored as the product of two hyperbolic rotations and an orthogonal one, i.e.,

$$\Theta_i = H_{i,1} H_{i,2} Q_i, \quad \text{where} \quad Q_i = \begin{bmatrix} Q_{i,1} & \\ & Q_{i,2} \end{bmatrix},$$

with $Q_{i,1}$ and $Q_{i,2}$ orthogonal matrices of order p and q, such that

$$Q_i G_{i-1,Z} = \begin{bmatrix} 0_{i-1}^T & u_{1,1}^{(i,1)} & u_{1,2}^{(i,1)}{}^T \\ 0_{p-1,i-1} & 0_{p-1} & U_{2,2}^{(i,1)} \\ 0_{i-1}^T & v_{1,1}^{(i,1)} & v_{1,2}^{(i,1)}{}^T \\ 0_{i-1}^T & z_{1,1}^{(i,1)} & z_{1,2}^{(i,1)}{}^T \\ 0_{q-2,i-1} & 0_{q-2} & V_{2,2}^{(i,1)} \end{bmatrix} \tag{1}$$

respectively, and

$$H_{i,1} = \frac{1}{\sqrt{1-\rho_{i,1}^2}} \begin{bmatrix} 1 & 0_{p-1}^T & \rho_{i,1} & 0_{q-1}^T \\ 0_{p-1} & I_{p-1} & 0 & 0_{p-1,q-1} \\ \rho_{i,1} & 0_{p-1}^T & 1 & 0_{q-1}^T \\ 0_{q-1} & 0_{q-1,p-1} & 0_{q-1} & I_{q-1} \end{bmatrix},$$

$$H_{i,2} = \frac{1}{\sqrt{1-\rho_{i,2}^2}} \begin{bmatrix} 1 & 0_p^T & \rho_{i,2} & 0_{q-2}^T \\ 0_p & I_p & 0 & 0_{p,q-2} \\ \rho_{i,2} & 0_p^T & 1 & 0_{q-2}^T \\ 0_{q-2} & 0_{q-2,p} & 0_{q-2} & I_{q-2} \end{bmatrix},$$

$|\rho_{i,1}|, |\rho_{i,2}| < 1$, such that

$$G_i = H_{i,1} H_{i,2} Q_i G_{i-1,Z} = \begin{bmatrix} 0_{i-1}^T & u_{1,1}^{(i,2)} & u_{1,2}^{(i,2)}{}^T \\ 0_{p-1,i-1} & 0_{p-1} & U_{2,2}^{(i,2)} \\ 0_{i-1}^T & 0 & v_{1,2}^{(i,2)}{}^T \\ 0_{i-1}^T & 0 & z_{1,2}^{(i,2)}{}^T \\ 0_{q-2,i-1} & 0_{q-2} & V_{2,2}^{(i,2)} \end{bmatrix},$$

where $0_{r,s}$ denotes the rectangular null matrix with r rows and s columns. As mentioned in §1, the computation of the hyperbolic rotation in a stable way is crucial for the stability of the algorithm. For implementation details of hyperbolic rotations in factored form see [3,15]. In the next section we will show that this variant of the GSA is stable, provided in each iteration the J_A-orthogonal matrix is computed as previously described, and the hyperbolic rotations are implemented in factored form. Similar stability results hold considering either the H-procedure or the OD-procedure to implement the hyperbolic rotations [5,12].

3 Stability Analysis

A stability analysis of the GSA with a single hyperbolic rotation in factored form per iteration is presented in [15]. The stability analysis for the algorithm described in the previous section can be done in a similar way. It is split up into two parts : one which shows how local errors propagate through the algorithm and one which bounds the local errors. We consider the same notation as introduced in §2 but denote by the superscript the corresponding quantities as stored in the computer. Hence $\tilde{G}_i = \begin{bmatrix} \tilde{u}_i & \tilde{U}_i^T & \tilde{v}_i & \tilde{z}_i^T & \tilde{V}_i \end{bmatrix}^T$. The local errors, generated by computing $\tilde{G}_{i+1}$ by means of orthogonal and hyperbolic transformations, are given by

$$\epsilon F_i = \tilde{G}_{i+1}^T J_A \tilde{G}_{i+1} - \tilde{G}_{i,Z}^T J_A \tilde{G}_{i,Z} + O(\epsilon^2), \quad i=1,\ldots,n, \qquad (2)$$

where ϵ is the machine precision. In [15] is proved that

$$A - \tilde{R}^T \tilde{R} = \sum_{j=0}^{n-1} Z_j (G^T J_A G - \tilde{G}^T J_A \tilde{G}) Z_j^T - \epsilon \sum_{j=0}^{n-1} \sum_{k=1}^{n-j-1} Z_j F_k Z_j^T + O(\epsilon^2), \qquad (3)$$

where $Z_j = Z^j$ and $\tilde{R}$ is the computed Cholesky factor. This means that if the errors in the computation of the initial generator matrix and the local errors are bounded, the algorithm is stable. The error in the initial generator matrix is not a problem, since often it is explicitly known or can be computed in a backward stable way [8]. Below, we assume that the initial generator matrix is computed exactly and restrict ourselves to the effects of local errors due to the orthogonal and hyperbolic transformations.

Because any bounds on the errors produced by the transformations will depend on the norm of the generators, it is essential to bound the generators.

Theorem 1. *When the generators are computed by applying a block diagonal orthogonal matrix and two hyperbolic transformations, they satisfy*

$$\|G_i\|_F \leq 2\sqrt{i-1}\|A\|_F + \|G\|_F \qquad (4)$$

Proof. Let $\hat{u}_i, \hat{v}_i$ and $\hat{z}_i$ be the generator vectors in (1) that will be modified by the two hyperbolic rotations $H_{i,2}$ and $H_{i,1}$,

$$\begin{bmatrix} u_i^T \\ v_i^T \\ z_i^T \end{bmatrix} = \frac{1}{\delta_{i,1}\delta_{i,2}} \begin{bmatrix} 1 & 0 & \rho_{i,2} \\ 0 & \delta_{i,2} & 0 \\ \rho_{i,2} & 0 & 1 \end{bmatrix} \begin{bmatrix} 1 & \rho_{i,1} & 0 \\ \rho_{i,1} & 1 & 0 \\ 0 & 0 & \delta_{i,1} \end{bmatrix} \begin{bmatrix} \hat{u}_i^T \\ \hat{v}_i^T \\ \hat{z}_i^T \end{bmatrix}$$

$$= \frac{1}{\delta_{i,1}\delta_{i,2}} \begin{bmatrix} 1 & \rho_{i,1} & \rho_{i,2}\delta_{i,1} \\ \rho_{i,1}\delta_{i,2} & \delta_{i,2} & 0 \\ \rho_{i,2} & \rho_{i,1}\rho_{i,2} & \delta_{i,1} \end{bmatrix} \begin{bmatrix} \hat{u}_i^T \\ \hat{v}_i^T \\ \hat{z}_i^T \end{bmatrix},$$

where $\delta_{i,k} = \sqrt{1 - \rho_{i,k}^2}$, $k = 1, 2$. Then we have

$$\begin{bmatrix} u_i^T \\ v_i^T \\ z_i^T \end{bmatrix} = \frac{1}{\delta_{i,1}\delta_{i,2}} \begin{bmatrix} 1 & \rho_{i,1} & \rho_{i,2}\delta_{i,1} \\ \rho_{i,1}\delta_{i,2} & \rho_{i,1}^2\delta_{i,2} & \rho_{i,1}\rho_{i,2}\delta_{i,1}\delta_{i,2} \\ \rho_{i,2} & \rho_{i,1}\rho_{i,2} & \rho_{i,2}^2\delta_{i,1} \end{bmatrix} \begin{bmatrix} \hat{u}_i^T \\ \hat{v}_i^T \\ \hat{z}_i^T \end{bmatrix}$$

$$+ \frac{1}{\delta_{i,1}\delta_{i,2}} \begin{bmatrix} 0 & 0 & 0 \\ 0 & \delta_{i,1}^2\delta_{i,2} & -\rho_{i,1}\rho_{i,2}\delta_{i,1}\delta_{i,2} \\ 0 & 0 & \delta_{i,1}\delta_{i,2}^2 \end{bmatrix} \begin{bmatrix} \hat{u}_i^T \\ \hat{v}_i^T \\ \hat{z}_i^T \end{bmatrix}$$

$$= \begin{bmatrix} u_i^T \\ \rho_{i,1}\delta_{i,2}u_i^T \\ \rho_{i,2}u_i^T \end{bmatrix} + \begin{bmatrix} 0 & 0 & 0 \\ 0 & \delta_{i,1} & -\rho_{i,1}\rho_{i,2} \\ 0 & 0 & \delta_{i,2} \end{bmatrix} \begin{bmatrix} \hat{u}_i^T \\ \hat{v}_i^T \\ \hat{z}_i^T \end{bmatrix}.$$

Consider the Givens rotations

$$U = \begin{bmatrix} 1 & 0 & 0 \\ 0 & c_l & -s_l \\ 0 & s_l & c_l \end{bmatrix}, \quad \begin{cases} c_l = \rho_{i,2}/\sqrt{1 - \delta_{i,1}^2\delta_{i,2}^2} \\ s_l = \rho_{i,1}\delta_{i,2}/\sqrt{1 - \delta_{i,1}^2\delta_{i,2}^2} \end{cases}$$

$$V = \begin{bmatrix} 1 & 0 & 0 \\ 0 & c_r & -s_r \\ 0 & s_r & c_r \end{bmatrix}, \quad \begin{cases} c_r = \rho_{i,2}\delta_{i,1}/\sqrt{1 - \delta_{i,1}^2\delta_{i,2}^2} \\ s_r = \rho_{i,1}/\sqrt{1 - \delta_{i,1}^2\delta_{i,2}^2} \end{cases}.$$

Then

$$U \begin{bmatrix} u_i^T \\ v_i^T \\ z_i^T \end{bmatrix} = U \begin{bmatrix} u_i^T \\ \rho_{i,1}\delta_{i,2}u_i^T \\ \rho_{i,2}u_i^T \end{bmatrix} + U \begin{bmatrix} 0 & 0 & 0 \\ 0 & \delta_{i,1} & -\rho_{i,1}\rho_{i,2} \\ 0 & 0 & \delta_{i,2} \end{bmatrix} V^T V \begin{bmatrix} \hat{u}_i^T \\ \hat{v}_i^T \\ \hat{z}_i^T \end{bmatrix}$$

$$= \begin{bmatrix} u_i^T \\ 0 \\ \sqrt{1 - \delta_{i,1}^2\delta_{i,2}^2}u_i^T \end{bmatrix} + \begin{bmatrix} 0 & 0 & 0 \\ 0 & 1 & 0 \\ 0 & 0 & \delta_{i,1}\delta_{i,2} \end{bmatrix} \begin{bmatrix} \tilde{u}_i^T \\ \tilde{v}_i^T \\ \tilde{z}_i^T \end{bmatrix}$$

$$= \begin{bmatrix} u_i^T \\ \tilde{v}^T \\ \sqrt{1 - \delta_{i,1}^2\delta_{i,2}^2}u_i^T + \delta_{i,1}\delta_{i,2}\tilde{z}_i^T \end{bmatrix},$$

where

$$\begin{bmatrix} \tilde{u}_i^T \\ \tilde{v}_i^T \\ \tilde{z}_i^T \end{bmatrix} = V \begin{bmatrix} \hat{u}_i^T \\ \hat{v}_i^T \\ \hat{z}_i^T \end{bmatrix}.$$

Then

$$\left\| U \begin{bmatrix} u_i^T \\ v_i^T \\ z_i^T \end{bmatrix} \right\|_F^2 = \left\| \begin{bmatrix} u_i^T \\ v_i^T \\ z_i^T \end{bmatrix} \right\|_F^2 = \left\| \begin{bmatrix} u_i^T \\ \tilde{v}_i^T \\ \sqrt{1-\delta_{i,1}^2 \delta_{i,2}^2} u_i^T + \delta_{i,1}\delta_{i,2}\tilde{z}_i^T \end{bmatrix} \right\|_F^2,$$

Applying the inequality

$$\left\| [\sqrt{1-\alpha^2}, \alpha] \begin{bmatrix} u_i^T \\ z_i^T \end{bmatrix} \right\|_2^2 \le \left\| \begin{bmatrix} u_i^T \\ z_i^T \end{bmatrix} \right\|_2^2$$

with $\alpha = \delta_{i,1}\delta_{i,2}$, $|\alpha| \le 1$, we finally obtain

$$\left\| \begin{bmatrix} u_i^T \\ v_i^T \\ z_i^T \end{bmatrix} \right\|_F^2 \le 2\|u_i\|_2^2 + \left\| \begin{bmatrix} \tilde{v}_i^T \\ \tilde{z}_i^T \end{bmatrix} \right\|_F^2 = 2\|u_i\|_2^2 + \left\| \begin{bmatrix} \hat{v}_i^T \\ \hat{z}_i^T \end{bmatrix} \right\|_F^2 \le 2\|u_i\|_2^2 + \left\| \begin{bmatrix} \hat{u}_i^T \\ \hat{v}_i^T \\ \hat{z}_i^T \end{bmatrix} \right\|_F^2.$$

Since the orthogonal transformations do not affect the norm of the generators and $\|Z\|_2 = 1$, then $\|G_i\|_F^2 \le 2\|u_i\|^2 + \|G_{i-1}\|_F^2$, and recursively we have

$$\|G_i\|_F^2 \le 2\sum_{j=1}^{i} \|u_j\|_2^2 + \|G\|_F^2 = 2\|R(1:k,:)\|_F^2 + \|G\|_F^2.$$

Then (4) follows since, for an arbitrary positive semi–definite, rank $i-1$ matrix with a factorization $A = R^T R$, (see [15]), $\|R\|_F^2 \le \sqrt{i}\|A\|_F^2$. □

To complete the stability analysis we need to show that the orthogonal and hyperbolic transformations, applied in factored form, produce a local error, ϵF_i, which is proportional to the norm of the generator matrix. An error analysis of hyperbolic transformations applied in factored form is given in [3]. Denoted by

$$H_{i,j} = \frac{1}{\sqrt{1-\rho_{i,j}^2}} \begin{bmatrix} 1 & 0 \\ \rho_{i,j} & \sqrt{1-\rho_{i,j}^2} \end{bmatrix} \begin{bmatrix} \frac{1}{\sqrt{1-\rho_{i,j}^2}} & 0 \\ 0 & 1 \end{bmatrix} \begin{bmatrix} 1 & \rho_{i,j} \\ 0 & 1 \end{bmatrix}, \quad j=1,2,$$

the hyperbolic transformations applied in factored form, then

$$\begin{bmatrix} \tilde{u}_{1,i+1}^T + \widehat{\Delta u}^T \\ \tilde{v}_{1,i+1}^T \\ \tilde{z}_{1,i+1}^T \end{bmatrix} = H_{i,2}H_{i,1} \begin{bmatrix} \hat{u}_{1,i}^T \\ \hat{v}_{1,i}^T + \widehat{\Delta v}^T \\ \hat{z}_{1,i}^T + \widehat{\Delta z}^T \end{bmatrix}. \quad (5)$$

The mixed error vectors $\widehat{\Delta u}$, $\widehat{\Delta v}$ and $\widehat{\Delta z}$ satisfy

$$\left\| \begin{bmatrix} \widehat{\Delta u}^T \\ \widehat{\Delta v}^T \\ \widehat{\Delta z}^T \end{bmatrix} \right\|_F \le 12.5\epsilon \left\| \begin{bmatrix} \tilde{u}_{1,i+1}^T \\ \hat{v}_{1,i}^T \\ \hat{z}_{1,i}^T \end{bmatrix} \right\|_F, \quad (6)$$

where ϵ is the roundoff unit. Furthermore, concerning the application of the orthogonal transformations, it can be proved [15,16] that there exist orthogonal matrices $\hat{Q}_{i,1}$ and $\hat{Q}_{i,2}$ such that

$$\left[\begin{array}{c|c}\hat{Q}_{i,1} & \\ \hline & \hat{Q}_{i,2}\end{array}\right]\begin{bmatrix}\tilde{u}_{1,i}^T + \Delta u_1^T \\ \tilde{U}_{2,i} + \Delta U_2 \\ \tilde{v}_{1,i}^T + \Delta v_1 \\ \tilde{z}_{1,i}^T + \Delta z_1 \\ \tilde{V}_{2,i} + \Delta V_2\end{bmatrix} = \begin{bmatrix}\hat{u}_{1,i}^T \\ \hat{U}_{2,i} \\ \hat{v}_{1,i}^T \\ \hat{z}_{1,i}^T \\ \hat{V}_{2,i}\end{bmatrix} = \hat{G}_i, \qquad (7)$$

where, for $m = \max\{p, q-1\}$,

$$\left\|\begin{bmatrix}\Delta u_1^T \\ \Delta U_2\end{bmatrix}\right\|_F \le 6m\epsilon \left\|\begin{bmatrix}\tilde{u}_{1,i}^T Z^T \\ \tilde{U}_{2,i}\end{bmatrix}\right\|_F \quad \text{and} \quad \left\|\begin{bmatrix}\Delta v_1^T \\ \Delta z_1^T \\ \Delta V_2\end{bmatrix}\right\|_F \le 6m\epsilon \left\|\begin{bmatrix}\tilde{v}_{1,i}^T Z^T \\ \tilde{z}_{1,i}^T Z^T \\ \tilde{V}_{2,i}\end{bmatrix}\right\|_F.$$

Letting $\Delta G_i = [\Delta u_1 \ \Delta U_2^T \ \Delta v_1 \ \Delta z_1 \ \Delta V_2^T]^T$, then $\|\Delta G_i\|_F \le 6m\epsilon\|G_{i,Z}\|_F \le 6m\epsilon\|G_i\|_F$. Analogously, letting $\widehat{\Delta G_i} = [\widehat{\Delta u} \ \widehat{\Delta v} \ \widehat{\Delta z}]^T$, then the error bounds (5) and (7) can be used to show that

$$\hat{G}_i^T J_A \hat{G}_i = (\tilde{G}_{i,Z} + \Delta G_i)^T J_A (\tilde{G}_{i,Z} + \Delta G_i),$$
$$(\tilde{G}_{i+1} + e_1 \widehat{\Delta u}^T)^T J_A (\tilde{G}_{i+1} + e_1 \widehat{\Delta u}^T) = (\hat{G}_i^T + e_{p+1}\widehat{\Delta v}^T + e_{p+2}\widehat{\Delta z}^T)^T J_A$$
$$\times (\hat{G}_i^T + e_{p+1}\widehat{\Delta v}^T + e_{p+2}\widehat{\Delta z}^T),$$

where e_1, e_{p+1} and e_{p+2} are standard basis vectors. Then

$$\epsilon F_i = \tilde{G}_{i,Z}^T J_A \Delta G_i + \Delta G_i^T J_A \tilde{G}_{i,Z} - [\tilde{u}_{1,i+1} \ \hat{v}_{1,i} \ \hat{z}_{1,i}]\widehat{\Delta G_i} - \widehat{\Delta G_i}^T \begin{bmatrix}\tilde{u}_{1,i+1}^T \\ \hat{v}_{1,i}^T \\ \hat{z}_{1,i}^T\end{bmatrix},$$

corresponding to the bound

$$\|\epsilon F_i\|_F \le 2\|\tilde{G}_{i,Z}\|_F\|\Delta G_i\|_F + 2\left(\|\hat{G}_i\|_F + \|\tilde{G}_{i+1}\|_F\right)\|\widehat{\Delta G_i}\|_F$$
$$\le 12m\epsilon\|\tilde{G}_{i,Z}\|_F^2 + 25\epsilon\left(\|\hat{G}_i\|_F + \|\tilde{G}_{i+1}\|_F\right)^2$$
$$\le 12m\epsilon\|G_i\|_F^2 + 25\epsilon(\|G_i\|_F + \|G_{i+1}\|_F)^2 + O(\epsilon^2).$$

By Theorem 1, $\|G_i\|_F^2, \|G_{i+1}\|_F^2 \le 2\sqrt{i}\|A\|_F + \|G\|_F^2$, the following bound holds

$$\|\epsilon F_i\|_F \le (12m + 100)\epsilon \left(2\sqrt{i}\|A\|_F + \|G\|_F^2\right).$$

From (3), we have $\|A - \tilde{R}^T \tilde{R}\|_F \le (6m+50)(n-1)n\epsilon \left(2\sqrt{n}\|A\|_F + \|G\|_F^2\right).$

4 Conclusion

Fast and efficient algorithms for solving Structured Total Least Squares problems [14,15] are based on a particular implementation of the GSA requiring two hyperbolic transformations at each iteration.

In this paper the stability of such implementation is discussed. It is proved that if the hyperbolic transformations are performed in factored form, the considered implementation is as stable as the implementation studied in [15] that requires only one hyperbolic transformation at each iteration.

References

1. Bojanczyk, A. W., Brent, R. P., De Hoog, F. R., Sweet, D. R.: On the stability of the Bareiss and related Toeplitz factorization algorithms. SIAM J. Matrix Anal. Appl. **16** (1995) 40–57.
2. Bojanczyk, A. W., Brent, R. P., De Hoog, F. R.: Stability analysis of a general Toeplitz systems solver. Numerical Algorithms **10** (1995) 225–244.
3. Bojanczyk, A. W., Brent, R. P., Van Dooren, P., De Hoog,F. R.: A note on downdating the Cholesky factorization. SIAM J. Sci. Stat. Comput. **1** (1980) 210–220.
4. Bunch, J.: Stability of methods for solving Toeplitz systems of Equations. SIAM J. Sci. Stat. Comput. **6** (1985) 349–364.
5. Chandrasekaran, S., Sayed, A. H.:, Stabilizing the generalized Schur algorithm. SIAM J. Matrix Anal. Appl. **17** no. 4, (1996) 950–983.
6. Chandrasekaran, S., Sayed, A. H.:, A fast stable solver for nonsymmetric Toeplitz and quasi-Toeplitz systems of linear equations. SIAM J. Mat. Anal. and Appl. **19** (1998) 107–139.
7. Chun, J., Kailath, T., Lev-ari, H.:, Fast parallel algorithms for QR and triangular factorization. SIAM J. Sci. and Stat. Comp. **8** (1987) 899–913.
8. Golub, G. H., Van Loan, C. F.:, Matrix Computations. Third ed., The John Hopkins University Press, Baltimore, MD, 1996.
9. Kailath, T., Chun, J.: Generalized displacement structure for block–Toeplitz, Toeplitz–block and Toeplitz–derived matrices. SIAM J. Matrix Anal. Appl. **15** (1994), 114–128.
10. Kailath, T., Kung, S., Morf., M.: Displacement ranks of matrices and linear equations. J. Math. Anal. Appl. **68** (1979) 395–407.
11. Kailath, T., Sayed, A. H.: Displacement structure: Theory and applications. SIAM Review **37** (1995) 297–386.
12. Kailath, T.: Displacement structure and array algorithms, in Fast Reliable Algorithms for Matrices with Structure, T. Kailath and A. H. Sayed, Ed., SIAM, Philadelpia, 1999.
13. Lemmerling, P., Mastronardi, N., Van Huffel, S.: Fast algorithm for solving the Hankel/Toeplitz structured total least squares problem. Numerical Algorithms (to appear).
14. Mastronardi, N., Lemmerling, P., Van Huffel,S.: Fast structured total least squares algorithm for solving the basic deconvolution problem. SIAM J. Matrix Anal. Appl.(to appear).
15. Stewart, M., Van Dooren, P.: Stability Issues in the Factorization of Structured Matrices. SIAM J. Matrix Anal. Appl. **18** (1997) 104–118.
16. Wilkinson, J. H.: The Algebraic Eigenvalue Problem, Oxford University Press, (1965) London.

Stability of Finite Difference Schemes on Non-uniform Spatial-Time-Grids

Piotr P. Matus[1], Vladimir I. Mazhukin[2], and Igor E. Mozolevsky[3]

[1] Institute of Mathematics of NAS of Belarus
Surganov 12 St., 220072 Minsk, Belarus
matus@im.bas-net.by
[2] Institute of Mathematical Modelling of Russian AS
[3] Federal University of Santa Catarina, Brazil

Abstract. The three level operator finite difference schemes on non-uniform on time grids in Hilbert spaces of finite dimension are considered. A priori estimates for uniform stability on the initial conditions are received under natural assumptions on operators and non uniform time grids. The results obtained here are applied to study stability of the three levels weighted schemes of second order approximation $O\left(h^2 + \tau_n\right)$ for some hyperbolic and parabolic equations of the second order. It is essential to note that the schemes of raised order of approximation are constructed here on standard stencils which are used in finite difference approximation techniques.

1 Introduction

Contemporaneous computational methods of mathematical physics alongside with the traditional requirements, such as stability and conservativity, have to satisfy also the adaptivity requirement. Application of adaptive grids first of all means that one have to use non uniform grid instead of uniform one which is adapted to behaviour of the singularities of the solution. It is known , that at use of non-uniform grids the order of local approximation becomes lower. One can increase the order of approximation by simple use of more extended stencils or by considering more restricted classes of solutions of differential problem. Let us to call attention to an another opportunity to increase accuracy expanding the approximation of initial differential equations from the points of a computational grid to some intermediate points of computational domain [1]. At present computational methods on non-uniform spatial grids have been widely studied for wide class of equations of mathematical physics with preservation of the second order local approximation with respect to the spatial variable [1] —[6]. Nevertheless the theoretical aspects of the three-level schemes on non-uniform time grids are less investigated [7,8].

This communication is devoted to investigation of the three level operator finite difference schemes on non-uniform on time grids with the operators acting in Hilbert spaces of finite dimension. The stability on initial conditions is proved and also a priori estimations in grid energy norms are obtained. Examples of the three level finite difference schemes of the second order of local approximations on time and spatial variables for parabolic and hyperbolic equation of the second order are presented. Especially we emphasize, that increase of the order of local approximation on non-uniform grids is achieved without increases of a standard stencil of the finite difference scheme.

2 Three Level Operator Finite Difference Schemes

Let us consider real Hilbert space H of finite dimension of real valued functions defined on non-uniform time grid

$$\hat{\omega}_\tau = \{t_n = t_{n-1} + \tau_n, \ n \in 1,2,\ldots, N_0, \ t_0 = 0, \ t_{N_0} = T\} = \hat{\omega}_\tau \bigcup \{0,T\} \ .$$

We designate as $D(t)$, $B(t)$, $A : H \to H$ linear operators in H. Let us consider a Cauchy problem for homogeneous finite difference operator equation

$$D y_{\bar{t}t} + B y_t + A y = 0, \quad y_0 = u_0, \quad y_1 = u_1 \ , \tag{1}$$

where $y = y_n = y(t_n) \in H$ is the unknown function, and $u_0, u_1 \in H$ are given functions. Here and in the following index-economic notations are used:

$$y_{\bar{t}t} = (y_t - y_{\bar{t}})/\tau^*, \quad y_t = (y_{n+1} - y_n)/\tau_{n+1}, \quad y_{\bar{t}} = (y_n - y_{n-1})/\tau_n,$$

$$\hat{y} = y_{n+1}, \quad \check{y} = y_{n-1}, \quad \tau^* = 0{,}5\,(\tau_{n+1} + \tau_n), \quad y_{\stackrel{\circ}{t}} = \frac{y_{n+1} - y_{n-1}}{\tau_n + \tau_{n+1}} \ .$$

Let us designate as H_{R_k}, where $R_k^* = R_k \geq 0$ a space with inner product $(y,v)_{R_k}$, $y,v \in H$, and with semi-norms $\|y\|_{R_k}^2 = (R_k y, y)$. Let us suppose that the operators entering in the scheme (1) satisfy the following conditions :

$$D(t) = D^*(t) \geq 0, \quad B(t) \geq 0, \quad t \in \hat{\omega}_\tau, \quad A = A^* > 0 \ , \tag{2}$$

$$D(t+\tau) \leq D(t), \quad \frac{\tau_{n+1}}{\tau_{n+2}} \leq \frac{\tau_n}{\tau_{n+1}}, \quad B(t_n) \geq 0{,}5 \tau_{n+1} A \ , \tag{3}$$

where $A(t) = A$ is a constant operator. Concerning conditions (3) we shall make some observations.

Remark 1. Usually in the theory of stability of the three level finite difference schemes [1] with the variable operator $D(t)$ its Lipschitz-continuity on variable t is required. However, if one studies the stability , for example, of the weighted three level scheme [2]

$$y_{\bar{t}t} + A y^{(\sigma_1, \sigma_2)} = 0, \quad y_0 = u_0, \quad y_1 = u_1 \ , \tag{4}$$

$$y^{(\sigma_1, \sigma_2)} = \sigma_1 \hat{y} + (1 - \sigma_1 - \sigma_2) y + \sigma_2 \check{y} \ , \tag{5}$$

than this requirement implies undesirable requirement of quasi-uniformity of the time grid

$$|\tau_{n+1} - \tau_n| \leq c_0 \tau_n^2, \quad n = 1, 2, \ldots, N_0 - 1 \ . \tag{6}$$

Remark 2. The second restriction from (3) $\tau_{n+1}/\tau_{n+2} \leq \tau_n/\tau_{n+1}$ is not rigid. Really, let the steps of a grid are chosen satisfying the geometrical progression law $\tau_{n+1} = q\tau_n$. Then the given inequality is valid for any $q = \text{const} > 0$.

Before we formulate results, we shall give definition of stability of the finite difference scheme (1) in case of the linear operators D, B, A.

Definition 1. *The operator finite difference scheme (1) is called unconditionally stable on initial conditions if there exist positive constants $M_1 > 0$, $M_2 > 0$, independent of τ_n and $u_0 \in H$, $u_1 \in H$, such that for all sufficiently small $\tau_n^* < \tau_0$, $n = 1, 2, \ldots, N_0$, the solution of the Cauchy problem (1) satisfies the estimate*

$$\|y_{\bar{t},n}\|_{R_{1n}}^2 + \|y_n\|_{R_{2n}}^2 \leq M_1 \|y_{\bar{t},1}\|_{R_{11}}^2 + M_2 \|y_1\|_{R_{21}}^2 . \tag{7}$$

If the inequality (7) is valid for every τ_n, then the scheme is called absolutely stable and when $M_1 = 1$, $M_2 = 1$ — uniformly stable.

Let us prove the following affirmation.

Theorem 1. *Let us suppose that the conditions (2), (3) are valid. Then the finite difference scheme (1) is uniformly stable with respect to initial conditions and the following estimate is valid*

$$\|y_{\bar{t},n+1}\|_{D_{n+1}}^2 + \|y_{n+1}\|_{R_{n+1}}^2 \leq \|y_{\bar{t},1}\|_{D_1}^2 + \|y_1\|_{R_1}^2 , \tag{8}$$

where $R_n = 0{,}5(1 + \tau_n/\tau_{n+1})A$.

Proof. Considering inner product of both parts of the equation (1) with $2\tau^* y_t$ and using the first condition from (4), one have

$$2\tau^* (Dy_{\bar{t}\bar{t}}, y_t) = 2\tau^* (Dy_{\bar{t}\bar{t}}, 0{,}5(y_t + y_{\bar{t}}) + 0{,}5\tau^* y_{\bar{t}\bar{t}}) =$$
$$= \|y_t\|_D^2 - \|y_{\bar{t}}\|_D^2 + \tau^{*2}\|y_{\bar{t}\bar{t}}\|_D^2 \geq \|y_{\bar{t},n+1}\|_{D_{n+1}}^2 - \|y_{\bar{t},n}\|_{D_n}^2 ,$$
$$2\tau^* (By_t, y_t) = 2\tau_n^* \|y_{t,n}\|_{B_n}^2 .$$

If the second condition from (3) is satisfied then one has inequality $\tau_n^*/\tau_{n+1} \geq \tau_{n+1}^*/\tau_{n+2}$. Therefore using the last estimate one obtain

$$2\tau^* (Ay, y_t) = \frac{\tau_n^*}{\tau_{n+1}} \left(\|y_{n+1}\|_A^2 - \|y_n\|_A^2\right) - \tau_{n+1}\tau_n^* \|y_{t,n}\|_A^2 \geq$$
$$\geq \|y_{n+1}\|_{R_{n+1}}^2 - \|y_n\|_{R_n}^2 - 2\tau_n^* \|y_{t,n}\|_{0{,}5\tau_{n+1}A}^2 .$$

Summing these estimates and using the third condition from (3), one has the following relation

$$\|y_{\bar{t},n+1}\|_{D_{n+1}}^2 + \|y_{n+1}\|_{R_{n+1}}^2 \leq \|y_{\bar{t},n}\|_{D_n}^2 + \|y_n\|_{R_n}^2 ,$$

which is valid for every $n = 1, \ldots, N_0 - 1$. This immediately implicates the desired estimate (8). □

Example 1. Let us consider weighted three level operator finite difference scheme (4). Using the identities

$$y^{(\sigma_1,\sigma_2)} = y_n + (\sigma_1\tau_{n+1} - \sigma_2\tau_n) y_t + \sigma_2\tau_n\tau_n^* y_{\bar{t}\hat{t}} \tag{9}$$

this scheme can be reduced to its canonical form (1) with

$$D_n = E + \tau_n\tau_n^*\sigma_2 A, \quad B_n = (\sigma_1\tau_{n+1} - \sigma_2\tau_n) A.$$

One can note that the conditions of the Theorem 1

$$D_{n+1} - D_n = \sigma_2 \left(\tau_{n+1}\tau_{n+1}^* - \tau_n\tau_n^*\right) A \le 0,$$

$$B_n - 0{,}5\tau_{n+1}A = (\tau_{n+1}(\sigma_1 - 0{,}5) - \tau_n\sigma_2) A \ge 0$$

are satisfied if

$$\sigma_2\tau_{n+1}\tau_{n+1}^* \le \sigma_2\tau_n\tau_n^*, \quad \sigma_1 \ge \frac{1}{2} + \frac{\tau_n}{\tau_{n+1}}\sigma_2 \; .$$

On the harmonic grid $\tau_{n+1} = q\tau_n$ the first of above inequalities for $\sigma_2 > 0$ is satisfied on condensing grid with $q \le 1$, and for $\sigma_2 < 0$ this inequality is satisfied on dilating grid. If $\sigma_2 = 0$, $\sigma_1 = \sigma$, then the scheme (4) could be transformed to the following form (with constant operator $D_n = E$)

$$y_{\bar{t}t} + Ay^{(\sigma)} = 0, \quad y_0 = u_0, \quad y_1 = u_1 \; . \tag{10}$$

Here $y^{(\sigma)} = \sigma y_{n+1} + (1-\sigma)y_n$. As the Theorem 1 affirms, its solution satisfies the a priori estimate

$$\|y_{\bar{t},n}\|^2 + \|y_n\|_{R_n}^2 \le \|y_{\bar{t},1}\|^2 + \|y_1\|_{R_1}^2, \quad n = 1, 2, \ldots, N_0 \;, \tag{11}$$

(here still $R_n = 0{,}5(1 + \tau_n/\tau_{n+1})A$, and $A^* = A > 0$ — is constant operator) if the conditions

$$\sigma \ge \frac{1}{2} + \frac{\tau_n}{\tau_{n+1}}\sigma_2, \quad \frac{\tau_n}{\tau_{n+1}} \ge \frac{\tau_{n+1}}{\tau_{n+2}} \; . \tag{12}$$

are satisfied.

Example 2. The second order of local approximation scheme on non-uniform time grid. In rectangle $\overline{Q}_T = \overline{\Omega} \times [0,T]$, $\overline{\Omega} = \{x : 0 \le x \le l\}$, $0 \le t \le T$ let us consider the first initial boundary value problem for one dimensional parabolic equation

$$\frac{\partial u}{\partial t} = \frac{\partial}{\partial x}\left(k(x)\frac{\partial u}{\partial x}\right), \quad 0 < x < l, \quad 0 < t \le T \;, \tag{13}$$

$$u(0,t) = u(l,t) = 0, \quad t > 0, \quad u(x,0) = u_0(x), \quad 0 \le x \le l \;, \tag{14}$$

where $0 < c_1 \le k(x) \le c_2$, $c_1, c_2 = $ const. On uniform in space and time variable grid

$$\overline{\omega} = \overline{\omega}_h \times \hat{\omega}_\tau, \quad \overline{\omega}_h = \{x_i = ih, \; i = 0, 1, \ldots, N, \; hN = l\} \;,$$

let us approximate the differential problem (13), (14) by the following finite difference sheme

$$y_t + 0{,}5\tau_+ y_{\bar{t}t} = (a\hat{y}_{\bar{x}})_x, \quad \tau_+ = \tau_{n+1}, \tag{15}$$

$$\hat{y}_0 = \hat{y}_N = 0, \quad y(x,0) = u_0(x), \quad x \in \bar{\omega}_h. \tag{16}$$

Here

$$a = a_i = 0{,}5(k_{i-1} + k_i), \quad k_i = k(x_i), \quad y = y_i^n = y(x_i, t_n),$$
$$(ay_{\bar{x}})_x = (a_{i+1} y_{\bar{x},i+1}^n - a_i y_{\bar{x},i}^n)/h, \quad y_{\bar{x},i} = (y_i^n - y_{i-1}^n)/h.$$

It is easy to verify that at the node (x_i, t_{n+1}) the three level scheme (13), (16) approximates differential problem with the second order, that is

$$\psi_i^{n+1} = -u_{t,i} - 0{,}5\tau_{n+1} u_{\bar{t}t,i} + (au_{\bar{x}}^{n+1})_{x,i} = O(h^2 + \tau_{n+1}^2).$$

The scheme (13) is one generalization of well known asymptotically stable scheme [3, p.309] at non-uniform time grid

$$\frac{3}{2} y_t - \frac{1}{2} y_{\bar{t}} = (ay_{\bar{x}})_x.$$

The scheme (13), (16) could be transformed to operator finite difference scheme (1), by putting $y = y_n = (y_1^n, y_2^n, \ldots, y_{N-1}^n)$, $(Ay)_i = -(ay_{\bar{x}})_{x,i}$, $i = 1, \ldots, N-1$, $y_0 = y_N = 0$, $D_n = 0{,}5\tau_{n+1}E$, $B_n = E + \tau_{n+1}A$. In this example the space $H = H_h$ consists in grid functions which are defined on the grid $\bar{\omega}_h$ and which are equal to zero on the boundary. Scalar product and norm are defined by expressions:

$$(y,v) = \sum_{i=1}^{N-1} h y_i v_i, \quad \|y\| = \sqrt{(y,y)}.$$

The properties of the operator A are well investigated [3]. In particular, $A^* = A > \delta E$, $\delta = 8c_1/l^2$. Let's check up the conditions of Theorem 1. It is obvious, that $D_n \leq D_{n-1}$ pri $\tau_{n+1} \leq \tau_n$, $B_n - 0{,}5\tau_{n+1} = E + 0{,}5\tau_{n+1}A > 0$ for every $\tau_{n+1} > 0$. Hence, the scheme (13), (16) is uniformly stable on initial data if

$$\frac{\tau_{n+1}}{\tau_{n+2}} \leq \frac{\tau_n}{\tau_{n+1}}, \quad \tau_{n+1} \leq \tau_n. \tag{17}$$

Let us note, that the conditions (17) are satisfied on harmonic grid $\tau_{n+1} = q\tau_n$ with arbitrary $0 < q \leq 1$.

3 Finite Difference Schemes of Raised Order of Approximation on Non-uniform on Time and Space Grids

Suppose that in the domain $\bar{Q}_T$ it is required to find continuous function $u(x,t)$, satisfying following initial boundary value problem

$$\frac{\partial^2 u}{\partial t^2} = \frac{\partial^2 u}{\partial x^2}, \quad 0 < x < l, \quad 0 < t \leq T,$$

$$u(0,t) = u(l,t) = 0, \quad u(x,0) = u_0(x), \quad \frac{\partial u}{\partial t}(x,0) = \bar{u}_0(x) .$$

Let us consider next non-uniform spatial - time grid $\bar{\omega} = \hat{\bar{\omega}}_h \times \hat{\bar{\omega}}_\tau$:

$$\hat{\bar{\omega}}_h = \{x_i = x_{i-1} + h_i, \quad i = 1, 2, \ldots, N, \quad x_0 = 0, \; x_N = l\} =$$
$$= \hat{\omega}_h \bigcup \{x_0 = 0, \; x_N = l\} ,$$
$$\hat{\bar{\omega}}_\tau = \{t_n = t_{n-1} + \tau_n, \quad n = 1, 2, \ldots, N_0, \quad t_0 = 0, \; t_{N_0} = T\} =$$
$$= \hat{\omega}_\tau \bigcup \{t_0 = 0, \; t_{N_0} = T\} .$$

We approximate on this grid the differential problem by the finite difference one

$$y_{\bar{t}\hat{t}} + \frac{h_+ - h}{3} y_{\bar{t}\hat{t}\bar{x}} = y_{\bar{x}\hat{x}}^{(\sigma_1,\sigma_2)} , \tag{18}$$

$$y_0^{n+1} = y_N^{n+1} = 0, \quad y_i^0 = u_i^0, \quad y_{t,i}^0 = \tilde{u}_{0i}, \quad i = 0, 1, \ldots, N . \tag{19}$$

Let us note, that $\tilde{u}_0(x)$, $x \in \hat{\omega}_h$, is chosen in such a way that the error of approximation of the second initial condition has order $O(\tau_1^2)$:

$$\tilde{u}_0(x) = \bar{u}_0(x) + 0{,}5\tau_1 u_0''(x) .$$

Here usual designations are used [1]:

$$h_+ = h_{i+1}, \quad h = h_i, \quad y = y_i^n = y(x_i, t_n), \quad y_{\bar{x}} = (y_i^n - y_{i-1}^n)/h_i ,$$

$$y_{\bar{x}\hat{x}} = (y_x - y_{\bar{x}})/\hbar, \quad y_x = (y_+ - y)/h_+, \quad y_\pm = y(x_{i\pm 1}, t_n), \quad \hbar = 0{,}5(h_+ + h) .$$

Let us show, that in supplementary node $(\bar{x}_i, \bar{t}_n)$:

$$\bar{x}_i = \frac{1}{3}(x_{i-1} + x_i + x_{i+1}) = x_i + \frac{h_{i+1} - h_i}{3} ,$$
$$\bar{t}_n = \frac{1}{3}(t_{n-1} + t_n + t_{n+1}) = t_n + \frac{\tau_{n+1} - \tau_n}{3} , \tag{20}$$

with

$$\sigma_1 \tau_{n+1} - \sigma_2 \tau_n = \frac{\tau_{n+1} - \tau_n}{3} \tag{21}$$

the finite difference scheme (18), (19) approximates the differential problem with the second order $O(\hbar^2 + \tau_n^{*2})$. For this purpose we shall rewrite the residual ψ as

$$\psi = u_{\bar{x}\hat{x}}^{(\sigma_1,\sigma_2)} - \left(u_{\bar{t}\hat{t}} + \frac{h_+ - h}{3} u_{\bar{t}\hat{t}\bar{x}}\right) = \psi_1 + \psi_2 ,$$

$$\psi_1 = u_{\bar{x}\hat{x}}^{(\sigma_1,\sigma_2)} - \frac{\partial^2 \bar{u}}{\partial x^2}, \quad \psi_2 = \frac{\partial^2 \bar{u}}{\partial t^2} - \left(u_{\bar{t}\hat{t}} + \frac{h_+ - h}{3} u_{\bar{t}\hat{t}\bar{x}}\right) .$$

Here $\bar{u} = u(\bar{x}, \bar{t})$, $\bar{x} = x + (h_+ - h)/3$, $\bar{t} = t + (\tau_+ - \tau)/3$. Let us note that an advantage of the scheme (18)(19), (21) is that fact, that for uniform grids ω_h, ω_τ

(t.e. $\tau_+ = \tau$, $h_+ = h$) this scheme reduces to the classical scheme of the order $O(h^2 + \tau^2)$ on a uniform grid [3]. We proceed with the analysis of ψ_1, ψ_2. Using the identity (9) and the weight conditions (21), , we conclude, that for any grid function $v(x_i, t_n)$ the following relation is valid

$$v^{(\sigma_1,\sigma_2)} = v + \frac{\tau_+ - \tau}{3} v_t + \sigma_2 \tau \tau^* v_{t\bar{t}} = v(x,\bar{t}) + O(\tau^{*2}) \ . \tag{22}$$

Hence,

$$\psi_1 = \psi_3 + O(\tau^{*2}), \quad \psi_3 = u_{\bar{x}\hat{x}}(x_i, \bar{t}) - \frac{\partial^2 \bar{u}}{\partial x^2} \ . \tag{23}$$

Using the Taylor series decomposition it is easy to show, that

$$\psi_3 = u_{\bar{x}\hat{x}}(x_i, \bar{t}_n) - \frac{\partial^2 u}{\partial x^2}(\bar{x}_i, \bar{t}_n) = \frac{\partial^2 u}{\partial x^2}(x_i, \bar{t}_n) + \frac{h_{i+1} - h_i}{3} \frac{\partial^3 u}{\partial x^3}(x_i, \bar{t}_n) -$$

$$- \frac{\partial^2 u}{\partial x^2}(\bar{x}_i, \bar{t}_n) + O(\hbar_i^2) = O(\hbar_i^2) \ . \tag{24}$$

By virtue of the next relations (which one can easy obtain with the help of the Taylor's formula)

$$v(x_i, t_n) + \frac{h_+ - h}{3} v_{\bar{x},i} = v(\bar{x}_i, t_n) + O(\hbar_i^2) \ ,$$

$$\frac{\partial^2 \bar{u}}{\partial t^2} - v_{\bar{t}\hat{t}}(\bar{x}_i, t_n) = O(\tau^{*2}) \ ,$$

one conclude, that the grid function ψ_2 is an infinitesimal of the second order, that is.

$$\psi_2 = O(\hbar_i^2 + \tau^{*2}) \ . \tag{25}$$

On the basis of the formulas (22) — (25) one conclude that finite difference scheme (18), (19), (21) approximates the initial boundary value problem for the wave equation on the standard 9-points stencil (see Fig. 1) with the second order (for sufficiently smooth function $u(x,t)$):

$$\psi = O(\hbar_i^2 + \tau_n^{*2}) \ .$$

For further investigation of the finite difference scheme (18), (19) some known formulas and identities are required:

$$y = \frac{\hat{y} + y}{4} + \frac{y + \check{y}}{4} - \frac{\tau_+ - \tau}{4} y_{\overset{\circ}{t}} - \frac{\tau \tau_+}{4} y_{\bar{t}\hat{t}} \ , \tag{26}$$

$$y^{(\sigma_1,\sigma_2)} = y + (\sigma_1 \tau_+ - \sigma_2 \tau) y_{\overset{\circ}{t}} + \frac{\sigma_1 + \sigma_2}{2} \tau \tau_+ y_{\bar{t}\hat{t}} \ , \tag{27}$$

$$y_{\overset{\circ}{t}} = \frac{y_t + y_{\bar{t}}}{2} + \frac{\tau_+ - \tau}{4} y_{\bar{t}\hat{t}} \ . \tag{28}$$

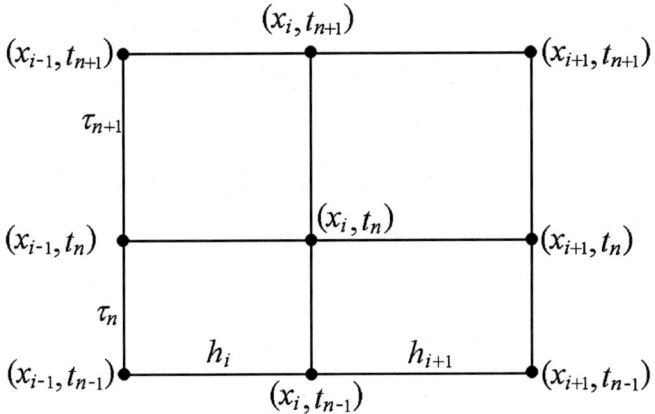

Fig. 1.

Let us introduce scalar products and norms of functions defined over a non-uniform spatial grid:

$$(y,v)_* = \sum_{i=1}^{N-1} \hbar_i y_i v_i, \quad \|y\|^2 = (y,y)_*, \quad (y,v] = \sum_{i=1}^{N} h_i y_i v_i, \quad \|y\| = (y,y] \ .$$

Lemma 1 (First finite difference Green's formula). *For any grid function $y(x)$, which is defined on non-uniform grid $\hat{\bar{\omega}}_h$ and vanish at $x = 0$ and at $x = l$ the next formula is valid*

$$(y, v_{\bar{x}\hat{x}})_* = (y_{\bar{x}}, v_{\bar{x}}] \ . \tag{29}$$

One has the following theorem.

Theorem 2. *Let us suppose that*

$$\|\hbar/h\|_C \leq c, \ \|\cdot\|_C = \max_{x \in \hat{\omega}_h} |\cdot|, \ \tau_{n+1} - \tau_n \geq \sqrt{\frac{2c}{3}} \|h_+ - h\|_C, \ n = 1, \ldots, N_0 - 1 \ , \tag{30}$$

and

$$\sigma_1^n = \frac{2\tau_{n+1} + \tau_n}{6(\tau_{n+1} + \tau_n)}, \quad \sigma_2^n = \frac{\tau_{n+1} + 2\tau_n}{6(\tau_{n+1} + \tau_n)} \ . \tag{31}$$

Then the finite difference scheme (18), (19) of the second order of local approximation $O(\hbar_i^2 + \tau^{*2})$ is uniformly stable and one has the estimation

$$\|y_{t\bar{x}}^n]\|^2 + \left\|y_{\bar{x}\hat{x}}^{(0,5)}(t_n)\right\|^2 \leq \|y_{t\bar{x}}^0]\|^2 + \left\|y_{\bar{x}\hat{x}}^{(0,5)}(0)\right\|^2 \ , \tag{32}$$

where $v^{(0,5)}(t_n) = 0, 5(v^{n+1} + v^n)$, $v_t^n = (v^{n+1} - v^n)/\tau_{n+1}$.

Proof. Let us note that σ_1^n, σ_2^n are defined with the formula (31) and satisfy the relation (21), which is necessary for increase the approximation order on a non-uniform grid. Let us multiply now the finite difference equation (18) by $-2\tau^* \hbar_i y_{\overset{\circ}{t\bar{x}\hat{x}},i}$ and sum at inner nodes of non-uniform space grid $\hat{\omega}_h$. After application the formula (29) we obtain the energy identity:

$$2\tau^* \left(y_{\bar{t}t\bar{x}}, y_{\overset{\circ}{t\bar{x}}} \right] - 2\tau^* \left(\frac{h_+ - h}{3} y_{\bar{t}t\bar{x}}, y_{\overset{\circ}{t\bar{x}\hat{x}}} \right)_* + 2\tau^* \left(y_{\bar{x}\hat{x}}^{(\sigma_1, \sigma_2)}, y_{\overset{\circ}{t\bar{x}\hat{x}}} \right)_* = 0 \ . \quad (33)$$

Applying identity (28), one finds the equality

$$2\tau^* \left(y_{\bar{t}t\bar{x}}, y_{\overset{\circ}{t\bar{x}}} \right] = ||y_{t\bar{x}}^n||^2 - ||y_{t\bar{x}}^{n-1}||^2 + 0.5\tau_n^* (\tau_{n+1} - \tau_n) ||y_{\bar{t}t\bar{x}}^n||^2 \ , \quad (34)$$

Using now formulas (26) (, 27) and condition of the second order approximations (21), we obtain for $y^{(\sigma_1, \sigma_2)}$ the following representation

$$y^{(\sigma_1, \sigma_2)} = \frac{1}{2} \left(y^{(0,5)} + \check{y}^{(0,5)} \right) + \frac{\tau_+ - \tau}{12} y_{\overset{\circ}{t}} \ . \quad (35)$$

In deriving the formula (35) we used the property

$$\sigma_1^n + \sigma_2^n = \frac{1}{2} \ . \quad (36)$$

Let us note, that if the variable weight multipliers do not satisfy to equality (36), then it is possible prove stability of the finite difference scheme (18),(19) only on quasi-uniform in time grid (6). In this case the estimation of stability will not carry uniform character, that is the constant $M_1 = \exp c_0 T$, appearing in definition 1 (see.(7)) will be much more than unit. Taking into account, that $y_{\overset{\circ}{t}} = (y^{(0,5)} - \check{y}^{(0,5)})/\tau^*$ and using (35) for third term in (34) one can find the following equality:

$$2\tau^* \left(y_{\bar{x}\hat{x}}^{(\sigma_1, \sigma_2)}, y_{\overset{\circ}{t\bar{x}\hat{x}}} \right)_* = \left\| y_{\bar{x}\hat{x}}^{(0,5)}(t_n) \right\|^2 - \left\| y_{\bar{x}\hat{x}}^{(0,5)}(t_{n-1}) \right\|^2 + \tau_n^* \frac{\tau_{n+1} - \tau_n}{6} \left\| y_{\overset{\circ}{t\bar{x}\hat{x}}} \right\|^2 \ . \quad (37)$$

Using the algebraic inequality $2ab \geq -a^2 - b^2$, we shall estimate the last remaining scalar product in (33):

$$-2\tau^* \left(\frac{h_+ - h}{3} y_{\bar{t}t\bar{x}}, y_{\overset{\circ}{t\bar{x}\hat{x}}} \right)_* \geq -\tau^* \frac{\tau_{n+1} - \tau_n}{2} ||y_{\bar{t}t\bar{x}}||^2 - \frac{2}{9} \left(\frac{(h_+ - h)^2}{\tau_{n+1} - \tau_n} \frac{\hbar_i}{h_i}, y_{\overset{\circ}{t\bar{x}\hat{x}}}^2 \right)_* \ . \quad (38)$$

Substituting obtained estimations (34) (,37) (,38) in energy identity (33), we come to recurrent relation

$$||y_{t\bar{x}}^n||^2 + \left\| y_{\bar{x}\hat{x}}^{(0,5)}(t_n) \right\|^2 \leq ||y_{t\bar{x}}^{n-1}||^2 + \left\| y_{\bar{x}\hat{x}}^{(0,5)}(t_{n_1}) \right\|^2 \ ,$$

rrom which immediately follows the estimation (32). □

This work was supported by Byelorussian Republican Fund of Fundamental Researches (project F99R-153).

References

1. Samarskii A. A., Vabishchevich P. N., Matus P. P.: Difference schemes with operator factors. Minsk. TSOTZH (1998) (in russian)
2. Samarskii A. A., Makarevich E. L., Matus P. P., Vabishchevich P. N.: Stability of three-level difference schemes on non-uniform in time grids. Dokl. RAN (to appear) (in russian)
3. Samarskii A. A.: Theory of difference schemes. Moscow. Nauka (1977) (in russian)
4. Samarskii A. A., Vabishchevich P. N., Zyl A. N., Matus P. P.: The difference scheme of raised order approximation for Dirichlet problem in arbitrary domain. Dokl. NAN Belarusi. **42(1)** (1998) 13—17 (in russian)
5. Samarskii A. A., Vabishchevich P. N., Matus P. P.: Difference schemes of raised order accuracy on non-uniform grids. Differents. uravnenia. **32(2)** (1996) 265—274 (in russian)
6. Samarskii A. A., Vabishchevich P. N., Matus P. P.: Difference schemes of second order accuracy on non-uniform grids. Zhurn. Vychislit. Matematiki i Matem. Fiziki. **38(3)** (1998) 413—424 (in russian)
7. Bokov A. G. Godishn. Vissh. Uchebni Zabed. Prilozh. mat. **12(2)** (1976) 87—96
8. Diakonov E. G., Bokov A. G. Dokl. Bolg. AN **28(2)** (1975) 157—160

Matrix Equations and Structures: Efficient Solution of Special Discrete Algebraic Riccati Equations

Beatrice Meini

Dipartimento di Matematica, Università di Pisa,
via Buonarroti 2, 56127 Pisa, Italy
meini@dm.unipi.it
www.dm.unipi.it/~meini

Abstract. We propose a new quadratically convergent algorithm, having a low computational cost per step and good numerical stability properties, that allows the computation of the maximal solutions of the matrix equations $X + C^*X^{-1}C = Q$, $X - C^*X^{-1}C = Q$, $X + C^*(R + B^*XB)^{-1}C = Q$. The algorithm is based on the cyclic reduction method.

1 Introduction

We consider the problem of the computation of the maximal Hermitian solutions of the matrix equations

$$X + C^*X^{-1}C = Q, \tag{1}$$
$$X - C^*X^{-1}C = Q, \tag{2}$$

where Q is an $m \times m$ Hermitian positive definite matrix, C is an $m \times m$ matrix, and C^* denotes the conjugate transposed of C.

For Hermitian matrices X, Y, we write $X > Y$ ($X \geq Y$) if $X - Y$ is positive definite (semidefinite); we say that X_+ is the maximal Hermitian solution of an equation if $X_+ \geq X$ for any Hermitian solution X.

Equations (1) and (2) arise in a wide variety of research areas, that include control theory, ladder networks, dynamic programming, stochastic filtering and statistics (see [1,18,8] for a list of references).

Equation (1) is a special discrete algebraic Riccati equation

$$-X + A^*XA + Q - (C + B^*XA)^*(R + B^*XB)^{-1}(C + B^*XA) = 0, \tag{3}$$

where A, B, C, Q, R are $m \times m$ matrices, Q and R are Hermitian, obtained by setting $A = R = 0$ and $B = I$.

The available numerical methods for the solution of (1) and (2) are based on fixed point iterations, or on applications of Newton's algorithm [10,8,18,19,6,7,1]. In particular, in [10] the authors adapt the algorithm proposed in [9] for the computation of the maximal solution of (3), based on Newton's iteration. Newton's iteration for solving matrix equations generates a sequence that quadratically converges to the seeked solution, but generally has a large computational cost per iteration; fixed point iterations have a lower computational cost at each step, but have linear convergence.

Here we derive a new algorithm for computing the maximal solution of (1) and (2), that has a double exponential convergence, like Newton's method, and a low computational cost per step, like fixed point iterations.

The idea consists in rewriting equations (1) and (2) in terms of infinite block tridiagonal block Toeplitz systems and in applying a modification of the cyclic reduction algorithm to the above systems [12]. The computational cost per iteration is $O(m^3)$ arithmetic operations (ops), like fixed point iterations. Moreover, the algorithm shows good numerical stability properties.

Finally we analyze the problem of the solution of the Riccati equation (3) in the particular case where $A = 0$, i.e.,

$$-X + Q - C^*(R + B^*XB)^{-1}C = 0. \tag{4}$$

We show, that, if B is nonsingular, the problem of the computation of the maximal solution of (4) can be reduced to the problem of the computation of the maximal solution of a matrix equation of the form (1). Thus, the efficient solution of (1) allows the efficient solution of the equation (3) in the case $A = 0$ and $\det B \neq 0$.

A possible extension of these results for the solution of a general Riccati equation (3) is under study.

The paper is organized as follows. In Section 2 we recall conditions for the existence of the maximal solution of (1), (2) and (3), and some spectral properties of the solution, that will be used in the subsequent sections to show the convergence of our algorithm. In Sections 3 and 4 we present the algorithm for the solution of (1) and (2), respectively. In Section 5 we analyze the problem of the solution of (4).

2 Existence and Properties of the Maximal Solution

In this section we recall conditions about the existence of the maximal solution X_+ of (1), (2) and (3), and some spectral properties of matrices related to X_+, that will be used in the subsequent sections to show the convergence of our algorithm.

Necessary and sufficient conditions for the existence of a positive definite solution of (1) are provided in [7]. More specifically, let us introduce the rational matrix function

$$\psi(\lambda) = \lambda C + Q + \lambda^{-1}C^*, \tag{5}$$

defined on the unit circle S of the complex plane, that is Hermitian for any $\lambda \in S$. This function is said regular if there exists at least a $\lambda \in S$ such that $\det \psi(\lambda) \neq 0$.

The following fundamental results hold [7,18]:

Theorem 1. *Equation (1) has a positive definite solution X if and only if $\psi(\lambda)$ is regular and $\psi(\lambda) \geq 0$ for all $\lambda \in S$. In that case (1) has a largest and unique solution X_+ such that $X + \lambda C$ is nonsingular for $|\lambda| < 1$, and $\rho(X_+^{-1}C) \leq 1$, where the symbol $\rho(\cdot)$ denotes the spectral radius.*

In [10,7] the authors characterize the eigenvalues of $X_+^{-1}C$ and show that $X_+^{-1}C$ has spectral radius strictly less than one if and only if $\psi(\lambda)$ is positive definite on the unit circle:

Theorem 2. *It holds $\rho(X_+^{-1}C) < 1$ if and only if $\psi(\lambda) > 0$ for all $\lambda \in S$.*

Concerning equation (2), in [8] the authors prove the following results:

Theorem 3. *The set of solutions of (2) is non-empty, and admits a maximal element X_+, and X_+ is the unique positive definite solution. Moreover, the spectral radius of $X_+^{-1}C$ is strictly less than one.*

Consider now the discrete algebraic Riccati equation (3). Let us denote by $\mathcal{R}(X)$ the linear application defined by the left-hand size of (3). For $A \in \mathbf{C}^{m \times k}$, $B \in \mathbf{C}^{k \times m}$, the pair (A, B) is said d-stabilizable if there exists a $K \in \mathbf{C}^{m \times k}$ such that $A - KB$ is d-stable, i.e., all its eigenvalues are in the open unit disk.

The following result holds [9,11]:

Theorem 4. *Let (A, B) a be d-stabilizable pair and suppose that there is a Hermitian solution $\widetilde{X}$ of the inequality $\mathcal{R}(\widetilde{X}) \geq 0$ for which $R + B^*\widetilde{X}B > 0$. Then there exists a maximal Hermitian solution X_+ of (3). Moreover, $R + B^*XB > 0$ and all the eigenvalues of $A - B(R + B^*X_+B)^{-1}(C + B^*X_+A)$ lie in S.*

3 Computation of the Maximal Solution of $X + C^*X^{-1}C = Q$

In this section we describe the new algorithm, based on cyclic reduction, that has a computational cost roughly larger of a factor two with respect to fixed point iterations [8,18,19,6,7,1], and a double exponential convergence, like Newton's method [10]. For more details on the new algorithm we refer the reader to [12].

Throughout this section we suppose $\psi(\lambda)$ regular and positive semidefinite for any $\lambda \in S$, where $\psi(\lambda)$ is given in (5), so that the conditions of Theorem 1 are satisfied.

Let X be a solution of (1). Then, by multiplying on the right both sides of (1) by X^{-1}, we find that

$$-I + QX^{-1} - C^*X^{-1}CX^{-1} = 0. \tag{6}$$

Thus, the matrix $G = X^{-1}C$ solves the quadratic matrix equation

$$-C + QG - C^*G^2 = 0. \tag{7}$$

In particular, if X_+ is the maximal solution of (1), since $\rho(X_+^{-1}C) \leq 1$, then the matrix equation (7) has a solution $G_+ = X_+^{-1}C$, with spectral radius at most 1.

The nice relation between the matrix equation (1) and the quadratic matrix equation (7) together with the spectral properties of the solutions of the latter equations, allow us to derive a fast algorithm for the computation of X_+.

The matrix X_+ can be efficiently computed by rewriting the matrix equations (6), (7), in terms of linear systems, and by applying the cyclic reduction algorithm, according to the ideas developed in [2,3,4]. In fact, we observe that the following system of equations is verified:

$$\begin{bmatrix} Q & -C^* & & 0 \\ -C & Q & -C^* & \\ & -C & Q & \ddots \\ 0 & & \ddots & \ddots \end{bmatrix} \begin{bmatrix} I \\ G_+ \\ G_+^2 \\ \vdots \end{bmatrix} X_+^{-1} = \begin{bmatrix} I \\ 0 \\ 0 \\ \vdots \end{bmatrix}. \quad (8)$$

By following the strategy successfully devised in [2,3,4] for solving nonlinear matrix equations arising in Markov chains, we apply the cyclic reduction algorithm to the above systems. This consists in performing an even-odd permutation of the block rows and columns, followed by one step of Gaussian elimination, thus generating the sequence of systems:

$$\begin{bmatrix} X_n & -C_n^* & & 0 \\ -C_n & Q_n & -C_n^* & \\ & -C_n & Q_n & \ddots \\ 0 & & \ddots & \ddots \end{bmatrix} \begin{bmatrix} I \\ G_+^{2^n} \\ G_+^{2 \cdot 2^n} \\ \vdots \end{bmatrix} X_+^{-1} = \begin{bmatrix} I \\ 0 \\ 0 \\ \vdots \end{bmatrix}. \quad (9)$$

The block entries of each system are defined by the following recursions:

$$\begin{aligned} & C_0 = C, \; Q_0 = X_0 = Q, \\ & C_{n+1} = C_n Q_n^{-1} C_n, \\ & Q_{n+1} = Q_n - C_n Q_n^{-1} C_n^* - C_n^* Q_n^{-1} C_n, \\ & X_{n+1} = X_n - C_n^* Q_n^{-1} C_n, \; n \geq 0. \end{aligned} \quad (10)$$

Observe that the matrices Q_n, X_n are Hermitian, and thus the matrices in (9) are Hermitian.

The spectral theory of Hermitian block Toeplitz matrices [5,15,16,14,13,17] guarantees the positive definitiveness, and thus the nonsingularity, of the blocks Q_n. Indeed, let us define the function $f : (-\pi, \pi) \to \boldsymbol{H}^m$, $f(\theta) = -e^{i\theta}C + Q - e^{-i\theta}C^*$, where i is the imaginary unit, and $\boldsymbol{H}^m$ is the set of $m \times m$ Hermitian matrices, and denote by

$$\mu_1 = \inf_{\theta \in (-\pi,\pi)} \lambda_{\min}(f(\theta)), \quad \mu_2 = \sup_{\theta \in (-\pi,\pi)} \lambda_{\max}(f(\theta)),$$

where $\lambda_{\min}(f(\theta))$ ($\lambda_{\max}(f(\theta))$) is the minimum (maximum) eigenvalue of $f(\theta)$.

Since $\psi(\lambda) \geq 0$ for any $\lambda \in S$, it holds $f(\theta) \geq 0$ and $\mu_1 \geq 0$. Moreover, since $\psi(\lambda)$ is regular, the set where $f(\theta)$ is positive definite is given by $(-\pi, \pi)$ except at most a finite number of points.

From these properties, the following result, that guarantees the applicability of the cyclic reduction algorithm and the boundness in norm of Q_n and X_n, can be proved [12]:

Theorem 5. *The matrices Q_n, X_n, $n \geq 0$, are positive definite, and their eigenvalues belong to the interval $[\mu_1, \mu_2]$. Moreover, it holds $0 < Q_{n+1} \leq Q_n$, $0 < X_{n+1} \leq X_n$, for $n \geq 0$.*

If $\psi(\lambda) > 0$ for any $\lambda \in S$ then $\mu_1 > 0$, thus also Q_n^{-1} is bounded in norm, and the condition number of Q_n is bounded. In this case the sequence $\{X_n\}_n$ quadratically converges to X_+ (see [12]):

Theorem 6. *If $\psi(\lambda) > 0$ for any $\lambda \in S$, then for any operator norm $\|\cdot\|$ and for any σ, $\rho(X_+^{-1}C) < \sigma < 1$, it holds $\|I - X_n X_+^{-1}\| = O\left(\sigma^{2\cdot 2^n}\right)$ and $\|C_n\| = O\left(\sigma^{2^n}\right)$.*

These nice properties allow us to design a quadratically convergent algorithm for the computation of the maximal solution X_+. Each step consists in generating the blocks defined in formula (10); hence it requires the solution of two Hermitian linear systems, i.e., the computation of $C_n Q_n^{-1}$ and $C_n^* Q_n^{-1}$, where the matrices Q_n have bounded condition number, and the computation of three matrix products.

If the hypothesis of Theorem 6 are not satisfied, then $\rho(X_+^{-1}C) = 1$. In this case X_n and Q_n are still Hermitian positive definite and bounded in norm. In general, the sequence $\{Q_n^{-1}\}_n$ may be not bounded. However, if the sequence $\{C_n\}_n$ converges to zero, and the sequence $\{(X_+^{-1}C)^{2^n}\}_n$ is bounded, then the sequence $\{X_n\}_n$ still converges to X_+.

4 Computation of the Maximal Solution of $X - C^*X^{-1}C = Q$

For the computation of the maximal solution X_+ of $X - A^*X^{-1}A = Q$ we can apply a technique similar to the one used in the previous section. Specifically we observe that
$$-I + QX_+^{-1} + C^*X_+^{-1}CX_+^{-1} = 0.$$
Thus, by setting $G_+ = X_+^{-1}C$, the following linear system is verified:

$$\begin{bmatrix} Q & C^* & 0 & \\ -C & Q & C^* & \\ & -C & Q & \ddots \\ 0 & & \ddots & \ddots \end{bmatrix} \begin{bmatrix} I \\ G_+ \\ G_+^2 \\ \vdots \end{bmatrix} X_+^{-1} = \begin{bmatrix} I \\ 0 \\ 0 \\ \vdots \end{bmatrix}. \qquad (11)$$

The infinite matrices in the above systems are block Toeplitz, but are not Hermitian. However, if we apply one step of cyclic reduction, we obtain the following system:

$$\begin{bmatrix} X_1 & -C_1^* & 0 & \\ -C_1 & Q_1 & -C_1^* & \\ & -C_1 & Q_1 & \ddots \\ 0 & & \ddots & \ddots \end{bmatrix} \begin{bmatrix} I \\ G_+^2 \\ G_+^4 \\ \vdots \end{bmatrix} X_+^{-1} = \begin{bmatrix} I \\ 0 \\ 0 \\ \vdots \end{bmatrix}, \qquad (12)$$

where

$$\begin{aligned} C_1 &= CQ^{-1}C, \\ Q_1 &= Q + CQ^{-1}C^* + C^*Q^{-1}C, \\ X_1 &= Q + C^*Q^{-1}C. \end{aligned} \qquad (13)$$

Thus, after one step of cyclic reduction, we obtain a Hermitian system, with the structure of (9), where the diagonal blocks are positive definite matrices. Moreover, observe that the function

$$\psi_1(\lambda) = \lambda C_1 + Q_1 + \lambda^{-1} C_1^*,$$

is such that $\psi_1(\lambda) > 0$ for any $\lambda \in S$, since $\psi_1(\lambda) = Q + (C - \bar{\lambda}C^*)Q^{-1}(C^* - \lambda C)$. If we apply cyclic reduction to system (12), we generate the sequence (9), for $n \geq 1$, where

$$\begin{aligned} C_{n+1} &= C_n Q_n^{-1} C_n, \\ Q_{n+1} &= Q_n - C_n Q_n^{-1} C_n^* - C_n^* Q_n^{-1} C_n, \\ X_{n+1} &= X_n - C_n^* Q_n^{-1} C_n, \quad n \geq 1, \end{aligned} \qquad (14)$$

and C_1, Q_1 and X_1 are defined in (13).

Without any assumption, the following convergence result can be proved [12]:

Theorem 7. *For the matrices X_n, Q_n, $n \geq 1$, defined in (14), it holds:*

1. $0 < X_{n+1} \leq X_n$, $0 < Q_{n+1} \leq Q_n$, $n = 1, 2, \ldots$;
2. Q_n and Q_n^{-1} are bounded in norm.

Moreover, for any operator norm $\|\cdot\|$ and for any σ, $\rho(X_+^{-1}C) < \sigma < 1$, it holds

$$\|I - X_n X_+^{-1}\| = O\left(\sigma^{2 \cdot 2^n}\right), \quad \|C_n\| = O\left(\sigma^{2^n}\right).$$

From the above theorem, the quadratic convergence is always guaranteed, and the condition number of Q_n is always bounded. The resulting algorithm has the same nice features, in terms of computational cost and convergence properties, of the algorithm for the solution of (1).

5 Computation of the Maximal Solution of $-X + Q - C^*(R + B^*XB)^{-1}C = 0$

Let us assume that the hypothesis of Theorem 4 are satisfied, and that the matrix B is nonsingular. Hence X is a solution of (4) if and only if X solves

$$-B^*XB + B^*QB - B^*C^*(R + B^*XB)^{-1}CB = 0.$$

Thus, if we define $Y = R + B^*XB$, then the matrix Y solves the matrix equation

$$-Y + R + B^*QB - B^*C^*Y^{-1}CB = 0. \tag{15}$$

The latter equation is of the form (1). Since for Theorem 4 $Y_+ = R + B^*X_+B > 0$, equation (15) has a positive definite solution. Thus, from Theorem 1 it follows that the function $\psi(\lambda)$ of (5), associated with equation (15) is regular and $\psi(\lambda) \geq 0$ for all $\lambda \in S$. Hence, if $R + B^*QB$ is positive definite, we can apply the algorithm described in Section 3 for the computation of the maximal solution Y_+ of (15), and we can recover X_+ by solving the linear equation $Y_+ = R + B^*X_+B$.

References

1. W. N. Anderson Jr., T. D. Morley, and G. E. Trapp. Positive solutions to $X = A - BX^{-1}B^*$. *Linear Algebra Appl.*, 134:53–62, 1990.
2. D. A. Bini and B. Meini. On the solution of a nonlinear matrix equation arising in queueing problems. *SIAM J. Matrix Anal. Appl.*, 17:906–926, 1996.
3. D. A. Bini and B. Meini. Improved cyclic reduction for solving queueing problems. *Numerical Algorithms*, 15:57–74, 1997.
4. D. A. Bini and B. Meini. Effective methods for solving banded Toeplitz systems. *SIAM J. Matrix Anal. Appl.*, 20:700–719, 1999.
5. H. Dym. Hermitian block toeplitz matrices, orthogonal polynomials, reproducing kernel pontryagin spaces, interpolation and extension. *Oper. Theory, Adv. Appl.*, 34:79–135, 1998. Orthogonal matrix-valued polynomials and applications, Pap. Semin. Oper. Theory, Tel Aviv/Isr.
6. J. C. Engwerda. On the existence of a positive definite solution of the matrix equation $X + A^TX^{-1}A = I$. *Linear Algebra Appl.*, 194:91–108, 1993.
7. J. C. Engwerda, A. C. M. Ran, and A. L. Rijkeboer. Necessary and sufficient conditions for the existence of a positive definite solution of the matrix equation $X + A^*X^{-1}A = Q$. *Linear Algebra Appl.*, 186:255–275, 1993.
8. A. Ferrante and B. C. Levy. Hermitian solutions of the equation $X = Q + NX^{-1}N^*$. *Linear Algebra Appl.*, 247:359–373, 1996.
9. C.-H. Guo. Newtons's method for discrete algebraic Riccati equations when the closed-loop matrix has eigenvalues on the unit circle. *SIAM J. Matrix Anal. Appl.*, 20:279–294, 1999.
10. C.-H. Guo and P. Lancaster. Iterative solution of two matrix equations. *Math. Comp.*, 68:1589–1603, 1999.
11. P. Lancaster and L. Rodman. *Algebraic Riccati equations*. Clarendon Press, Oxford, 1995.
12. B. Meini. Efficient computation of the extreme solutions of $X + A^*X^{-1}A = Q$, $X - A^*X^{-1}A = Q$. Submitted for publication, 1999.
13. M. Miranda and P. Tilli. Block Toeplitz matrices and preconditioning. *Calcolo*, 33(1-2):79–86 (1998), 1996. Toeplitz matrices: structures, algorithms and applications (Cortona, 1996).
14. M. Miranda and P. Tilli. Asymptotic spectra of Hermitian block Toeplitz matrices and preconditioning results. *SIAM J. Matrix Anal. Appl.*, 21(3):867–881 (electronic), 2000.
15. S. Serra. Asymptotic results on the spectra of block Toeplitz preconditioned matrices. *SIAM J. Matrix Anal. Appl.*, 20(1):31–44 (electronic), 1999.

16. S. Serra. Spectral and computational analysis of block Toeplitz matrices having nonnegative definite matrix-valued generating functions. *BIT*, 39(1):152–175, 1999.
17. P. Tilli. Asymptotic spectral distribution of Toeplitz-related matrices. In T. Kailath and A. H. Sayed, editors, *Fast reliable algorithms for matrices with structure*, chapter 6, pages 153–187. SIAM, Philadelphia, 1999.
18. X. Zhan. Computing the extremal positive definite solutions of a matrix equation. *SIAM J. Sci. Comput.*, 17:1167–1174, 1996.
19. X. Zhan and J. Xie. On the matrix equation $X + A^T X^{-1} A = I$. *Linear Algebra Appl.*, 247:337–345, 1996.

A Numerical Comparison between Multi-revolution Algorithms for First-Order and Second-Order ODE Systems

M. Begoña Melendo

Dpt. Matemática Aplicada, CPS, University of Zaragoza
María de Luna 3, E–50015 Zaragoza, Spain
bmelendo@posta.unizar.es

Abstract. We test some envelope-following methods for first-order differential systems against their counterparts for second-order systems. While this latter kind of methods are more efficient than the usual second-order designed solvers, they only are more efficient than the envelope-following methods for first order systems in problems without dissipation of orbital dinamics.

1 Introduction

The numerical solution of highly oscillatory systems is today one of the most challenging problems in the numerical solution of ordinary differential equations. Hamiltonian systems of the classical mechanics close enough to integrables share this type of solution. A complete review of the methods and problems that arise when dealing with these 'high oscillations' can be seen in [10].

The basic fact, roughly speaking, is that some of the components of the solution –the angular variable in orbital problems– is a high-frecuency oscillation while the others are not, forcing the high-one to advance the solution very slowly. Furthermore, in these latter problems, the solutions are quasi-periodics and the variations of the rest of the components are small, even when the period of the angular variable is not small.

Between them, envelope-following methods for first-order ODE systems, originally known as multi-revolution or generalizazed RK and multistep methods, have shown to be very efficient in long-term orbit calculations, as well as some generalizations of linear multistep of the Störmer-Cowell type [3,9,7,8].

In previous papers we have seen that methods for second-order systems of envelope-following type are faster than the usual second-order solvers but no comparison between both types of methods was done. This paper is aimed at comparing their performances.

Description of the Methods

Let's suppose that the IPV

$$\begin{cases} y'(t) = f(t, y(t)), & t \in [t_0, L], \\ y(t_0) = y_0 \in \mathbb{R}^s. \end{cases} \quad (1)$$

has a highly-oscillatory quasi-periodic type solution.

In the envelope-following methods for first order systems, a discretization of (1) is made by substituting the difference equation for the IVP

$$y_{n+1} - y_n = g_n, \quad n \in \mathbb{N}, \tag{2}$$

where y_n stands for the approximation to the solution $y(t_n)$ at each of the nodes of a grid $\{t_j = t_0 + jT \mid j = 0, 1 \ldots\}$, T is an approximation to the quasi-period of the solution and g_n to the definite integral $g(t_n, y(t_n)) = \int_{t_n}^{t_n+T} f(\tau, y(\tau))\, d\tau$ where $y(\tau)$ is the solution to the initial differential equation with $(t_n, y(t_n))$ as an initial condition.

A numerical solution to (2) is obtained in a uniform grid of stepwise M, positive integer, $\{\bar{t}_j = t_0 + jH \mid j \geq 0,\ H = MT\}$ by using a linear relationship which can be written in the form

$$\rho(\bar{E}, M)\bar{y}_n = M\sigma(\bar{E}, M)\bar{g}_n, \quad n \geq 0, \tag{3}$$

with $\rho(\zeta, M) = \sum_{j=0}^{\bar{k}} \bar{\alpha}_j(1/M)\zeta^j$, $\sigma(\zeta, M) = \sum_{j=0}^{\bar{k}} \bar{\beta}_j(1/M)\zeta^j$, the first and second characteristic polinomials of the method and $\bar{E} = E^M$, where E stands for the shift operator of length T.

For the special IVP

$$\begin{cases} y''(t) = f(t, y(t)), & t \in [t_0, L], \\ y(t_0) = y_0 \in \mathbb{R}^s, & y'(t_0) = y'_0 \in \mathbb{R}^s, \end{cases} \tag{4}$$

the envelope-following methods that we proposed in [8] follow the same path, i.e., the discretization of (4) is made by substituting the difference equation for the IVP

$$y_{n+2} - 2y_{n+1} + y_n = g_n, \quad n \in \mathbb{N}, \tag{5}$$

whose solution is found in the points of a grid of stepsize M as the previous one through the relation

$$\rho(\bar{E}, M)\bar{y}_n = M^2 \sigma(\bar{E}, M)\bar{g}_n.$$

if we introduce the characteristic polinomials, as before, and where g_n stands now for an approximation to the definite integral $g(t_n, y(t_n))$ that now admits the integral expression

$$T^2 \int_0^1 (1-s)\big[f(t_n + (1+s)T, y(t_n + (1+s)T)) + \\ f(t_n + (1-s)T, y(t_n + (1-s)T))\big]\, ds,$$

where $y(t_n + (1\pm s)T)$, $s \in (0, 1)$, is the solution to the initial differential equation with $(t_n, y(t_n))$ as an initial condition.

Finally, given the IVP

$$\begin{cases} y''(t) = f(t, y(t), y'(t)), & t \in [t_0, L], \\ y(t_0) = y_0 \in \mathbb{R}^s, & y'(t_0) = y'_0 \in \mathbb{R}^s, \end{cases} \quad (6)$$

the following discretization can be done

$$y_{n+2}^{(2)} - 2y_{n+1}^{(2)} + y_n^{(2)} = g_n^{(2)}, \quad n \in \mathbb{N},$$

$$y_{n+1}^{(1)} - y_n^{(1)} = g_n^{(1)}, \quad n \in \mathbb{N},$$

where $y_n^{(2)} \approx y(t_n), y_n^{(1)} \approx y'(t_n)$ and, in order to obtain its numerical solution, consider a couple of multi-revolution algorithms or simply apply a formula for second-order systems like those derived in [8]; that means considering $g_n^{(1)} = y_{n+2}^{(1)} - 2y_{n+1}^{(1)} + y_n^{(1)}$.

Consequently, we will generally have one or at most two sequences of difference equations of the form

$$\rho_i(\bar{E}, M)\bar{y}_n = M^i \sigma_i(\bar{E}, M)\bar{g}_n, \quad n \geq 0, \quad i = 1, 2 \quad (7)$$

where $\rho_i(\bar{E}, M), \sigma_i(\bar{E}, M), i = 1, 2$ stands for the characteristic polinomials of an envelope-following method for a first-order or a second-order ODE system, respectively.

2 Numerical Properties

The accuracy of the numerical solution provided by those methods is measured by the asymtotic expansion in powers of $\bar{h} = M\bar{\varepsilon}$ of the operator

$$L[\bar{y}_n^{(i)}, \bar{h}] = \rho_i(\bar{E}, M)y_{nM}^{(i)} - M^i \sigma_i(\bar{E}, M)g_{nM}^{(i)}, \quad i = 1, 2. \quad (8)$$

where $\bar{\varepsilon}$ is a measure of the variation of the true solution along a quasi-period, which can generally be done through the value of the 'period', T, in a strictly 'highly oscillatory' problem or numerically.

In [8] the following result is proved

Theorem 1. *The envelope-following method (7) is of order p if and only if one of the following conditions is satisfied:*

i) *The linear forms*

$$\bar{C}_r^{(i)} = M_r(\rho_i) - 2^{i-1} \sum_{j=i}^{r} \binom{r}{j} (2^{j-1} - 1)^{i-1} \frac{M_{r-j}(\sigma_i)}{M^{j-i}}$$

satisfy

$$\bar{C}_r^{(i)} = 0, \quad 0 \leq r \leq p+i-1, \quad i = 1, 2.$$

$$\bar{C}_{p+i}^{(i)} \neq 0, \quad i = 1 \text{ or } 2.$$

ii) $\rho_i(e^{\bar{h}}) - M^i(e^{\bar{h}/M} - 1)^i \sigma_i(e^{\bar{h}}) = O(\bar{h}^{p+i-1})$ $(\bar{h} \longrightarrow 0)$, $i = 1, 2$.

iii) $\dfrac{\rho_i(\zeta)}{M^i[\zeta^{1/M} - 1]^i} - \sigma_i(\zeta)$

have in $\zeta = 1$ a cero of order p.

where
$$M_r(\rho_i) = \sum_{j=0}^{\bar{k}} \bar{\alpha}_j^{(i)} j^r, \quad M_r(\sigma_i) = \sum_{j=0}^{\bar{k}} \bar{\beta}_j^{(i)} j^r.$$

are the momenta of the polinomials ρ_i and σ_i.

The stability of the algorithms can be characterized in terms of the roots condition of Dahlquist, i. e., the roots of $\rho_i(\zeta)$ are inside the unit circle and the roots of module one have at most an order of multiplicity i.

Equally, it can be seen that the methods are convergent if their coefficients are related through the relation stated in the previous theorem and the stability condition.

Taking $\rho_i(\zeta) = (\zeta - 1)^i \zeta^{k-i}$ the coefficients of the Adams-Bashforth and Adams-Moulton-like methods are obtained when $i = 1$ and of the Störmer and Cowell-like when $i = 2$.

3 Implementation of the Methods

The methods have been implemented in the predictor-corrector form $P(EC)^m E$, where E means an evaluation of both f_n and g_n at each step. This latter requires an integration over a period in the algorithm for first order and along two periods in the second order algorithms with the approximation provided by the explicit method used as a predictor and follows the same pattern as described in [2,6].

The inner integration is performed by the codes DOP853 [5] and DGEAR of the IMSL library, in the first-order implemetation and by the Nyström of order 10 due to Hairer [4] and DGEAR in the second order.

Predictors are the $\bar{k} = 11$ steps of the Adams-Bashforth and Störmer-like families and correctors the $\bar{k} = 10$ steps methods of the Adams-Moulton and Cowell-like families, respectively.

4 Numerical Tests

We consider as a first test problem the Kepler's planar problem in cartesian coordinates with initial conditions $p_1 = 0, p_2 = \sqrt{\dfrac{1+e}{1-e}}, q_1 = 1 - e, q_2 = 0$, which is a standard in orbital dinamic problems.

The following precision-work diagrams show the performance of the integrators. The abscisa is the global error in uniform norm at the end point of integration and the ordinate is the number of function evaluations in a double logarithmic scale.

Figure 1 show the performance of the multi-revolution algorithms for first order systems, denoted MRABM10, and the DOP853 solver. We have implemented with variable stepsizes and with constant stepsizes the inner integration. Graphs on the top, and on the bottom rigth corner of the figure correspond to eccentricity $e = 0.1$ and the graph to the bottom left corner to $e = 0.5$.

Simbols in figures are ○ for the MRABM10 method with $M = 4$, + for the DOPRI8 and DOP853, ◇ for the MRABM10 method with $M = 8$, * for the MRABM10 method with $M = 16$, ▽ for the MRABM10 method with $M = 32$, × for the MRABM10 method with $M = 64$ and ▷ for the MRABM10 method with $M = 128$.

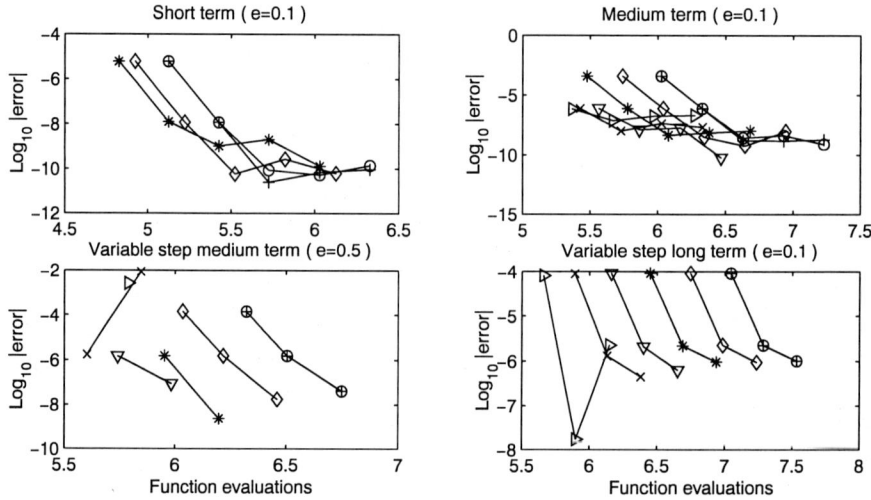

Fig. 1. Performance of multirevolution algorithms for first order systems

The short term integration is taken for 320 periods and stepsizes of external integration of $4, 8$ and 16 periods. The medium and long term integration for 2560 and 20000 periods and stepsizes of $4, 8, 16, 32, 64$ and 128. The $P(EC)^m E$ mode considered take $m = 2$. The inner integration is performed by the DOPRI8 solver with a fixed number of steps of $32, 64, 128, 256$ and 512 by period in the short and medium term integration and with the variable step solver DOP853 in the long term case with tolerances of the same order of magnitude of the local error in the step-fixed implementation. As a conclusion we can say that the multi-revolution algorithm matches the global error propagation of the inner integration but with greater efficiency.

When the eccentricity grows, a greater precision is needed in the inner integration in order to achieve the same global error and smaller stepsizes must be taken in the outer integration. The same behaviour is shown when a longer time of integration is considered.

Figure 2 shows the performance of the algorithms for second-order systems. The graph on the left corresponds to the implementation with a couple of multi-revolution algorithms, denoted MRABMSC10, and the graph on the rigth corresponds to the implementation with only an Störmer-Cowell-like formula, denoted MRSC10. Simbols are as before, joined by a dash-dot line in the implementation with a couple of multi-revolution algorithms and with a dashed line in the direct implementation, with $\triangle$ for the Nyström method. In view of this greater stability shown by the SC-like formula it was the only one used to carry out the following numerical experiments. One posible reason of this behaviour can be that exposed in [1], i. e., the eigenvalues of the numerical operator, when a pair of difference equations is used, are not directly related to the eigenvalues of the initial differential operator.

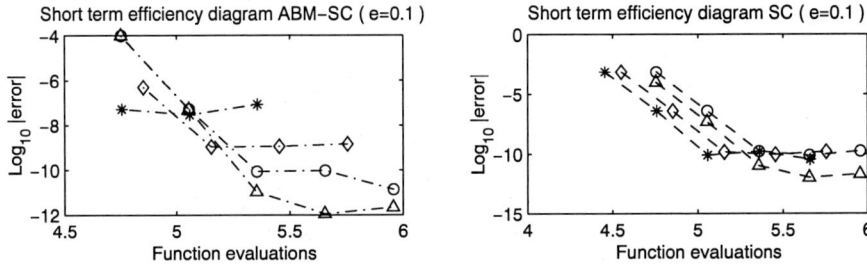

Fig. 2. Performance of multirevolution algorithms for second order systems

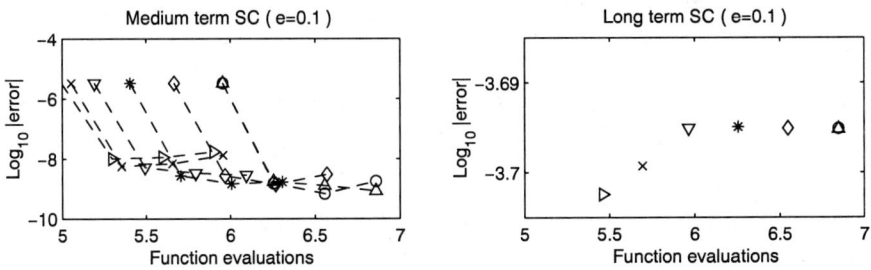

Fig. 3. Performance of multirevolution algorithms for second order systems

The performance of the envelope-following methods for second-order systems reproduces the obtained with their counterparts for first order systems as is shown in Figure 3. The inner integrations is performed a fixed stepsize of $32, 64, 128$ and 256 by period by the Nyström code in the medium term integration and with 32 steps in a period in the long-term run. Though two numerical

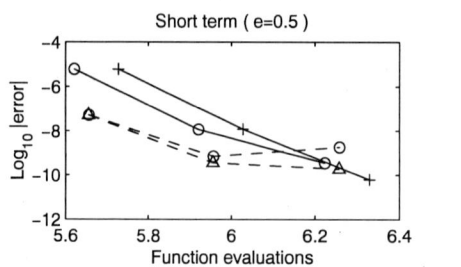 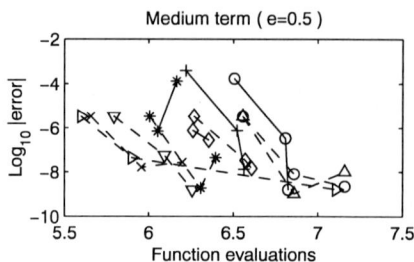

Fig. 4. Multirevolution algoritms for first-order systems versus the algorithms for second-order systems

integrations are needed at each step of the algorithms for second-order equations, a greater precision is got with fewer function evaluations due to the higher order of the inner solver and its smaller number of stages by step.

Finally in Figure 4 a comparison of the efficiency shown for the envelope-following methods for first-order systems versus the algorithms for second-order systems is shown. The mode considered now in the $P(EC)^m E$ implementation has been $m = 1$ for both.

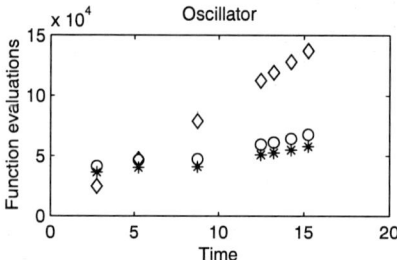 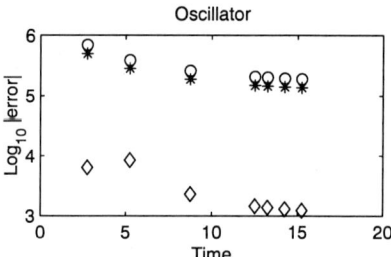

Fig. 5. Envelope-following methods for first-order systems versus the algorithms for second-order systems

Figure 5 corresponds to the scalar problem of pure resonance of harmonic motion described by IVP:

$$y'' + \omega^2 y = 100 \sin \omega t, \quad y(0) = 1, \quad y'(0) = -0.05$$

with $\omega = 10^3$. The inner integrations have been carried out with tolerances of 10^{-8} and the run last about 2500 periods. The inner solver considered has been DGEAR. Simbols are ◇ for the DGEAR code, ○ for the MRSC10 code and ∗ for the MRABM10 code. Even though the number of function evaluations in this problem grows faster for second order algorithms, a better precision is maintained.

5 Conclusions

Both envelope-following methods for first-order and second-order ODE systems match the global error propagation of the inner integration but with greater efficiency. In problems without dissipation a greater precision is obtained with the algorithms specifically designed for second-order ODE systems due to the possibility of also using inner solvers specifically designed for second-order systems as they reach higher orders with a smaller number of stages. When the same kind of solvers is used in the inner integration, the first-order algorithms require a smaller number of function evaluations in the run although less precision is also obtained but, in any case, a deeper experimentation is needed.

Acknowledgements

I'm very grateful to Dr. Roberto Barrio for his encouragement and advice during the development of this paper.

References

1. Gear, C. W.: The stability of numerical methods for second order ordinary differential equations, SIAM J. Numer. Anal., **15**, (1978), 188-197.
2. Graf, O. F.: Application of the multirevolution orbit prediction method, in: Analytical and Computational Mathematics, ACM Schänis AG, Switzerland, Prepared under ESOC-Contract No. 753775/T for European Space Agency (1976).
3. Graf, O. F. and Bettis, D. G.: Modified multirevolution integration methods for satellite orbit computation, Celestial Mech. Dynam. Astronom., **11**, (1975), 433-448.
4. Hairer, E.: A one-step method of order 10 for $y'' = f(x, y)$, IMA J. Numer. Anal., **2**, (1982), 83–94.
5. Hairer, E., Norsett, S., P and Wanner, G.: Solving Ordinary Differential Equations I, Nonstiff Problems (Springer, Berlin, 2nd ed., 1993).
6. Melendo, B. and Palacios, M.: A new approach to the construction of multirevolution methods and their implementation, Appl. Numer. Math., **23**, (1997), 259-274.
7. Melendo, B. and Palacios, M.: Multirevolution algorithms for the long-term integration of quasiperiodic problems. (Spanish) XV Congress on Differential Equations and Applications/V Congress on Applied Mathematics, Vol. I, II (Spanish) (Vigo, 1997), 847–852.
8. Melendo, B.: Envelope-following methods for second-order ODE systems, submitted for publication.
9. Petzold, L. R.: An efficient numerical method for highly oscillatory ordinary differential equations, SIAM J. Numer. Anal., **18**, (1981), 455-479.
10. Petzold, L. R., Jay, L. O. and Yen, J.: Numerical solution of highly oscillatory ordinary differential equations, Acta Numerica (1997), 437-483.

A Robust Layer-Resolving Numerical Method for Plane Stagnation Point Flow

John J. H. Miller[1], Alison P. Musgrave[1], and Grigorii I. Shishkin[2]

[1] Department of Mathematics, University of Dublin, Trinity College
Dublin 2, Ireland
{jmiller,musgrava}@tcd.ie
http://www.maths.tcd.ie/
[2] Institute of Mathematics and Mechanics, Russian Academy of Sciences
Ural Branch, Ekaterinburg 620219, Russia
Grigorii@shishkin.ural.ru
http://dcs.imm.uran.ru/~u1504/Shishkin.html

Abstract. Plane stagnation point flow is one of a small class of problems for which a self-similar solution of the incompressible Navier-Stokes equations exists. The self-similar solution and its derivatives can be expressed in terms of the solution of a transformed problem comprising a partially coupled system of quasilinear ordinary differential equations defined on a semi-infinite interval. In this paper a novel iterative numerical method for the solution of the transformed problem is described and used to compute numerical approximations to the self-similar solution and derivatives. The numerical method is layer-resolving which means that for each of the components, error bounds of the form $C_p N^{-p}$ can be calculated where C_p and p are independent of the Reynolds number, showing that these numerical approximations are of controllable accuracy.

1 Introduction

Plane stagnation point flow arises when fluid flowing in the direction of the negative y-axis impacts on an infinite flat plate $P = \{(x,0) \in \mathbb{R}^2\}$. The fluid separates into two streams which flow in opposite directions along the plate away from the stagnation point at $x = y = 0$. For large values of the Reynolds number Re, the vorticity is limited to a thin parabolic boundary layer on the plate whose thickness is independent of x [1].

This flow is one of a small class of problems for which an exact solution of the steady, incompressible Navier-Stokes equations exists. The exact solution is self-similar and therefore can be written in terms of the solution of a transformed problem, consisting of a system of ordinary differential equations defined on a semi-infinite domain $\eta \in (0, \infty)$, where $\eta = y\sqrt{Re}$. Standard numerical methods approximate the solution of this transformed problem on a finite domain, whose length is determined through a numerical process [2]. This means that the similarity solution and its derivatives are available for only a limited range of Re and so Re-uniform bounds cannot be obtained.

In this paper, a numerical method that is robust and layer-resolving in the sense of [3] is described for the construction of approximate solutions of the transformed problem on the entire semi-infinite interval. The ordinary differential equations are solved on a finite interval whose length is dependent on the numerical width of the boundary layer, and the solution values are extended to the infinite domain using carefully chosen extrapolation formulae. The similarity solution and derivatives are then formed in terms of the transformed solution for values of Re in the range $[1, \infty)$. Experimental Re-uniform error bounds, which are calculated in the maximum norm, are presented for the similarity solution and derivatives. These show that numerical approximations of any desired accuracy can be computed with this method. It should be noted that the use of the maximum norm is essential because other norms, for example the energy norm, do not detect parabolic boundary layers [3].

2 Problem Formulation

Incompressible plane stagnation point flow in the domain $D = \{(x, y) \in \mathbb{R}^2 : y > 0\}$ is governed by the Navier-Stokes equations, which can be written in the dimensionless form

$$(P_{NS}) \begin{cases} \text{Find } u, v \text{ and } p \text{ such that for all } (x,y) \in D \\ \dfrac{\partial u}{\partial x} + \dfrac{\partial v}{\partial y} = 0 \\ u\dfrac{\partial u}{\partial x} + v\dfrac{\partial u}{\partial y} = -\dfrac{\partial p}{\partial x} + \dfrac{1}{Re}\left(\dfrac{\partial^2 u}{\partial x^2} + \dfrac{\partial^2 u}{\partial y^2}\right) \\ u\dfrac{\partial v}{\partial x} + v\dfrac{\partial v}{\partial y} = -\dfrac{\partial p}{\partial y} + \dfrac{1}{Re}\left(\dfrac{\partial^2 v}{\partial x^2} + \dfrac{\partial^2 v}{\partial y^2}\right) \\ x, y = 0: \quad \Delta p = 0 \\ y = 0: \quad u = v = 0 \\ y \to \infty: \quad u \to x, \quad v \to -y, \quad \Delta p \to \tfrac{1}{2}(x^2 + y^2). \end{cases} \quad (1)$$

Here $\Delta p = p_0 - p$ and p_0 is the pressure at the stagnation point $x = y = 0$. The boundary conditions for u, v and p far above the plate are given by the solution of the corresponding irrotational flow problem, and the no-slip condition is satisfied on the surface of the plate [4].

2.1 Transformed Problem

The partial differential equations in (P_{NS}) are reduced to two ordinary differential equations by performing a separation of variables and introducing a simple transformation of variables to rid the resulting equations and boundary conditions of constants. The self-similar solution of (P_{NS}) has the form

$$u \equiv x\, f'(\eta) \qquad (2)$$

$$v \equiv -f(\eta)/\sqrt{Re} \tag{3}$$

$$\Delta p \equiv \frac{1}{2}\left(x^2 + \frac{2}{Re}g(\eta)\right), \tag{4}$$

where

$$\eta = y\sqrt{Re} \tag{5}$$

and $f(\eta)$, $g(\eta)$ satisfy the transformed problem

$$(P_T) \begin{cases} \text{Find } f \text{ and } g \text{ such that for all } \eta \in (0, \infty) \\ f''' + ff'' - (f')^2 + 1 = 0 \\ g' = f'' + ff' \\ \eta = 0: \quad f(0) = f'(0) = g(0) = 0 \\ \eta \to \infty: \quad f'(\eta) \to 1. \end{cases} \tag{6}$$

Here the prime denotes differentiation with respect to η. An expression for g is obtained by integrating the g' equation and using the boundary condition $g(0) = 0$,

$$g = f' + \frac{f^2}{2}. \tag{7}$$

In practice, Re-uniform numerical approximations to the derivatives of the self-similar solution are required. Expressions for these derivatives which are of order one as $Re \to \infty$ are given in equations (8) – (11) below. It should be noted that scaling $\partial u/\partial y$ by $\sqrt{Re}$ is necessary to ensure finite values for large values of Re.

$$\frac{\partial u}{\partial x} = -\frac{\partial v}{\partial y} \equiv f'(\eta) \tag{8}$$

$$\frac{1}{\sqrt{Re}}\frac{\partial u}{\partial y} \equiv x f''(\eta) \tag{9}$$

$$\frac{\partial \Delta p}{\partial x} \equiv x \tag{10}$$

$$\frac{\partial \Delta p}{\partial y} \equiv g'(\eta)/\sqrt{Re}. \tag{11}$$

Note that $\partial v/\partial x \equiv 0$.

In the next section, a numerical method for the solution of the third order quasilinear equation for f in (P_T) is described. The numerical approximations to f are then used to calculate approximations to g' and g.

3 Robust Layer-Resolving Numerical Method

To obtain similarity solutions for all values $Re \in [1, \infty)$, solutions of the transformed problem (P_T) must be found for all values $\eta \in (0, \infty)$. In Farrell et al. [3], a problem that is similar to (P_T) is solved numerically over a finite interval $\eta \in (0, L)$ and extrapolation formulae are used for $\eta \geq L$. The appropriate choice of L is related to the numerical width of the boundary layer and is found by studying the singularly perturbed nature of the f equation. In the present case it is given by $L = L(N) = \ln N$ where N is the discretisation parameter of the transformed problem. The following extrapolation formulae for f, g and their derivatives are derived using the asymptotic nature of the solution of (P_T)

$$f(\eta) = \eta - L + f(L) \qquad \text{for all } \eta \geq L \qquad (12)$$

$$f'(\eta) = 1 \qquad \text{for all } \eta \geq L \qquad (13)$$

$$f''(\eta) = 0 \qquad \text{for all } \eta \geq L \qquad (14)$$

$$g(\eta) = 1 + \frac{1}{2}(\eta - L + f(L))^2 \qquad \text{for all } \eta \geq L \qquad (15)$$

$$g'(\eta) = \eta - L + f(L) \qquad \text{for all } \eta \geq L. \qquad (16)$$

The transformed problem (P_T) is discretised by replacing derivatives in the f and g equations by finite difference operators, which are defined on a uniform mesh in $(0, L)$. The mesh is given by

$$\overline{I}_u^N = \{\eta_i : \eta_i = i N^{-1} \ln N, \ 0 \leq i \leq N\}. \qquad (17)$$

An iterative method is required for the solution of the quasilinear f equation. Here the following continuation algorithm, analogous to that described in [3], is used

$$(A_T^N) \begin{cases} \text{For each integer } m, \ 1 \leq m \leq M, \text{ find } F^m \text{ on } I_u^N \\ \text{such that for all } \eta_i \in I_u^N, \ 2 \leq i \leq N - 1 \\ \delta^2(D^-F^m) + F^{m-1}D^+(D^-F^m) - (D^-F^{m-1})(D^-F^m) = -1 \quad (18) \\ F^m(0) = D^+F^m(0) = 0 \text{ and } D^0F^m(\eta_{N-1}) = 1 \\ \text{with starting values for all mesh points } \eta_i \in \overline{I}_u^N \text{ of } F^0(\eta_i) = \eta_i, \end{cases}$$

where $F(\eta_i) = F^M(\eta_i)$, and for any mesh function $\Phi_i = \Phi(\eta_i)$

$$\delta^2 \Phi_i = \frac{(D^+ - D^-)\Phi_i}{(\eta_{i+1} - \eta_{i-1})/2}, \qquad D^+\Phi_i = \frac{\Phi_{i+1} - \Phi_i}{\eta_{i+1} - \eta_i}, \qquad D^-\Phi_i = \frac{\Phi_i - \Phi_{i-1}}{\eta_i - \eta_{i-1}}.$$

Once F is obtained the values of G and DG are calculated using the formulae

$$G(\eta_i) = D^+F(\eta_i) + \frac{1}{2}F^2(\eta_i) \qquad (19)$$

$$DG(\eta_i) = D^+D^+F(\eta_i) + F(\eta_i)D^+F(\eta_i) \qquad (20)$$

and numerical approximations to the self-similar solutions are found using equations (2) – (4) and (8) – (11).

4 Results and Discussion

In this section the errors in the numerical approximations to the self-similar solution and its derivatives are determined on the rectangular domain $\Omega = (0, L_x) \times (0, L_y)$ in the $x-y$ plane, where L_x and L_y are independent of Re. The global error $\overline{E}$ in $\overline{\Phi}$, the numerical approximation to a function ϕ, on the closed domain $\overline{\Omega}$ is defined by

$$\overline{E}(\overline{\Phi}) = \|\overline{\Phi} - \phi\|_{\overline{\Omega}} = \max_{(x,y)\in\overline{\Omega}} |\overline{\Phi}(x,y) - \phi(x,y)|, \qquad (21)$$

where $\overline{\Phi}$ denotes the piecewise linear interpolant of Φ to each point of $\overline{\Omega}$.

Consider first the global error in $\overline{U}^N$, the numerical approximation to u calculated on the mesh $\overline{I}_u^N$ with N subintervals. Recalling that $\eta = y\sqrt{Re}$, by (2) for each value of Re and N the error can be written as

$$\overline{E}_{Re}^N(\overline{U}^N) = \|\overline{U}^N - u\|_{\overline{\Omega}} = \|x\left(\overline{D^+F}^N - f'\right)\|_{\overline{\Omega}}$$

$$\leq L_x \|\overline{D^+F}^N - f'\|_{\eta \leq L_y \sqrt{Re}}. \qquad (22)$$

This means that the global error in $\overline{U}^N$ can be found directly from the global error in $\overline{D^+F}^N$ evaluated for $\eta \leq L_y\sqrt{Re}$. Analogous formulae for the global errors in the remaining numerical approximations are

$$\|\overline{V}^N - v\|_{\overline{\Omega}} = \frac{1}{\sqrt{Re}} \|\overline{F}^N - f\|_{\eta \leq L_y \sqrt{Re}} \qquad (23)$$

$$\|\overline{\Delta P}^N - \Delta p\|_{\overline{\Omega}} = \frac{1}{Re} \|\overline{G}^N - g\|_{\eta \leq L_y \sqrt{Re}} \qquad (24)$$

$$\|\overline{\partial_x U}^N - \frac{\partial u}{\partial x}\|_{\overline{\Omega}} = \|\overline{D^+F}^N - f'\|_{\eta \leq L_y \sqrt{Re}} \qquad (25)$$

$$\frac{1}{\sqrt{Re}} \|\overline{\partial_y U}^N - \frac{\partial u}{\partial y}\|_{\overline{\Omega}} \leq L_x \|\overline{D^+D^+F}^N - f''\|_{\eta \leq L_y \sqrt{Re}} \qquad (26)$$

$$\|\overline{\partial_y V}^N - \frac{\partial v}{\partial y}\|_{\overline{\Omega}} = \|\overline{D^+F}^N - f'\|_{\eta \leq L_y \sqrt{Re}} \qquad (27)$$

$$\|\overline{\partial_y \Delta P}^N - \frac{\partial \Delta p}{\partial y}\|_{\overline{\Omega}} = \frac{1}{\sqrt{Re}} \|\overline{DG}^N - g'\|_{\eta \leq L_y \sqrt{Re}}, \qquad (28)$$

where ∂ is the forward difference D_η^+ transformed to the $x-y$ plane. By definition

$$\|\overline{\partial_x V}^N - \frac{\partial v}{\partial x}\|_{\overline{\Omega}} = 0 \qquad (29)$$

$$\|\overline{\partial_x \Delta P}^N - \frac{\partial \Delta p}{\partial x}\|_{\overline{\Omega}} = 0 \tag{30}$$

as $\partial v/\partial x = \overline{\partial_x V}^N = 0$ and $\partial \Delta p/\partial x = \overline{\partial_x \Delta P}^N = x$. In the results that follow, L_x and L_y are equal to one and the numerical approximations have been constructed for a range of values of $Re \in R_{Re} = \{2^j\}_{j=0...20}$ and $N \in R_N = \{2^j\}_{j=9...19}$.

Global errors in $\overline{U}^N$ are presented in Table 1 for various values of $Re \in R_{Re}$ and $N \in R_N$. As the exact solution component u in closed form is unknown for this problem, we replace it in the formula for $\overline{E}_{Re}^N$ by the numerical approximation calculated on the finest available mesh, $\overline{U}^{N_{\max}}$, where $N_{\max} = 524288$. For $N = 512, 2048, 8192, 32768$, the Re-uniform global error

$$\overline{E}^N = \max_{Re \in R_{Re}} \overline{E}_{Re}^N \tag{31}$$

is shown in the last row of the table. $\overline{E}^N$ is the maximum global error in $\overline{U}^N$ for a particular N and all available values of Re. It is seen that its values decrease rapidly as N increases which implies that an error bound of the form $C_p N^{-p}$ can be found, where C_p and p are independent of Re. This demonstrates computationally that $\overline{U}^N$ converges Re-uniformly to u. Analogous results are obtained for each of the other components.

Table 1. $\overline{E}_{Re}^N(\overline{U}^N)$ and $\overline{E}^N(\overline{U}^N)$ for various values of $Re \in R_{Re}$ and $N \in R_N$

$Re \setminus N$	512	2048	8192	32768
2^0	1.676×10^{-04}	5.162×10^{-05}	1.506×10^{-05}	4.099×10^{-06}
2^1	1.676×10^{-04}	5.162×10^{-05}	1.506×10^{-05}	4.099×10^{-06}
2^2	1.786×10^{-04}	5.337×10^{-05}	1.543×10^{-05}	4.187×10^{-06}
.	.	.	.	.
.	.	.	.	.
2^{20}	1.786×10^{-04}	5.337×10^{-05}	1.543×10^{-05}	4.187×10^{-06}
$\overline{E}^N$	1.786×10^{-04}	5.337×10^{-05}	1.543×10^{-05}	4.187×10^{-06}

Realistic estimates of the Re-uniform global error parameters C_p and p are found experimentally using the double mesh technique, a complete description of which is contained, for example, in [3]. First, for any mesh function $\overline{\Phi}^N$ defined on $\overline{\Omega}$, the global two-mesh differences

$$\overline{D}_{Re}^N(\overline{\Phi}^N) = \max_{(x,y) \in \overline{\Omega}} \left| \overline{\Phi}^N - \overline{\Phi}^{2N} \right| \tag{32}$$

are calculated for each N satisfying $N, 2N \in R_N$ and $Re \in R_{Re}$, and the Re-uniform global two-mesh differences

$$\overline{D}^N = \max_{Re \in R_{Re}} \overline{D}^N_{Re} \qquad (33)$$

are determined. The Re-uniform order of convergence is then taken to be the minimum value $\check{p}$ of $\overline{p}^N$, where for each N satisfying $N, 2N, 4N \in R_N$

$$\overline{p}^N = \log_2\left(\frac{\overline{D}^N}{\overline{D}^{2N}}\right). \qquad (34)$$

The Re-uniform error constant $\check{C}_{\check{p}}$ is given by

$$\check{C}_{\check{p}} = \max_{N \in R_N} C^N_{\check{p}} = \max_{N \in R_N} \frac{\overline{D}^N N^{\check{p}}}{1 - 2^{-\check{p}}}. \qquad (35)$$

Values of $\overline{D}^N_{Re}$, $\overline{D}^N$ and $\overline{p}^N$ are presented in Table 2 for $\overline{U}$. The corresponding values of $\check{p}$ and $\check{C}_{\check{p}}$ are 8.728×10^{-1} and 4.211×10^{-2} for all $N \geq 2048$. Analogous results are obtained for each of the other components.

Table 2. $\overline{D}^N_{Re}(\overline{U}^N)$, $\overline{D}^N(\overline{U}^N)$ and $\overline{p}^N(\overline{U}^N)$ for various values of $Re \in R_{Re}$ and $N \in R_N$

$Re \setminus N$	512	2048	8192	32768
2^0	7.856×10^{-05}	2.400×10^{-05}	7.145×10^{-06}	2.080×10^{-06}
2^1	7.856×10^{-05}	2.400×10^{-05}	7.145×10^{-06}	2.080×10^{-06}
2^2	8.174×10^{-05}	2.462×10^{-05}	7.303×10^{-06}	2.123×10^{-06}
.	.	.	.	.
2^{20}	8.174×10^{-05}	2.462×10^{-05}	7.303×10^{-06}	2.123×10^{-06}
$\overline{D}^N$	8.174×10^{-05}	2.462×10^{-05}	7.303×10^{-06}	2.123×10^{-06}
$\overline{p}^N$	8.641×10^{-01}	8.728×10^{-01}	8.876×10^{-01}	9.014×10^{-01}

The following computed error bounds for the self-similar solution and its derivatives, of the form $\check{C}_{\check{p}} N^{-\check{p}}$, are valid for all $N \geq 2048$,

$$\|\overline{U}^N - u\|_{\overline{\Omega}} \leq 0.042\, N^{-0.87} \qquad (36)$$

$$\|\overline{V}^N - v\|_{\overline{\Omega}} \leq 1.03\, N^{-0.86} \qquad (37)$$

$$\|\overline{\Delta P}^N - \Delta p\|_{\overline{\Omega}} \leq 0.487\, N^{-0.86} \qquad (38)$$

$$\|\overline{\partial_x U}^N - \frac{\partial u}{\partial x}\|_{\overline{\Omega}} \leq 0.042\, N^{-0.87} \tag{39}$$

$$\frac{1}{\sqrt{Re}}\|\overline{\partial_y U}^N - \frac{\partial u}{\partial y}\|_{\overline{\Omega}} \leq 1.22\, N^{-0.86} \tag{40}$$

$$\|\overline{\partial_y V}^N - \frac{\partial v}{\partial y}\|_{\overline{\Omega}} \leq 0.042\, N^{-0.87} \tag{41}$$

$$\|\overline{\partial_y \Delta P}^N - \frac{\partial \Delta p}{\partial y}\|_{\overline{\Omega}} \leq 1.67\, N^{-0.86}. \tag{42}$$

These error bounds are Re-uniform and allow numerical approximations to the self-similar solutions to be calculated with controllable accuracy. For example, for $\overline{U}^{2048}$ the upper bound on the error is 5.5×10^{-5} which implies that at least 4 digits in $\overline{U}^{2048}$ are accurate provided that N is chosen so that $N \geq 2048$. Accuracy can be increased by increasing N. Error bounds valid for lower values of N can also be obtained.

5 Conclusions

In the case of plane stagnation point flow, Re-uniform numerical approximations to the self-similar solution and its derivatives have been generated. Error bounds for these components show that the numerical method is robust and layer-resolving, allowing numerical approximations of controllable accuracy to be computed independently of the value of Re.

Acknowledgements

This research was supported in part by the Russian Foundation for Fundamental Investigations Grant No. 98-01-00362, the National Science Foundation Grant No. DMS-9627244 and Forbairt Grants SC/97/630 and SC/98/612.

References

1. Rosenhead, L. (ed.): Laminar Boundary Layers. Oxford University Press (1963)
2. Rogers, D. F.: Laminar Flow Analysis. Cambridge University Press (1992)
3. Farrell, P. A., Hegarty, A. F., Miller, J. J. H., O'Riordan, E., Shishkin, G. I.: Robust Computational Techniques for Boundary Layers. CRC Press (2000)
4. Schlichting, H.: Boundary-Layer Theory. McGraw-Hill (1979)

On the Complete Pivoting Conjecture for Hadamard Matrices of Order 16

M. Mitrouli

Department of Mathematics, University of Athens, Panepistemiopolis,
15784 Athens, Greece
mmitroul@cc.uoa.gr

Abstract. In this paper we study explicitly the pivot structure of the Hadamard matrices of order 16. We examine for each representative of the five equivalent classes the appearing pivot structures and we give tables summarising the attained 34 different structures. We give ten examples of 16×16 Hadamard matrices all coming from Class I, for which when Gaussian Elimination with complete pivoting is applied on them, the fourth last pivot is $\frac{16}{2}$.

Keywords: Gaussian elimination, pivot size, complete pivoting, Hadamard matrices

AMS Subject Classification: 65F05, 65G05, 05B20.

1 Introduction

Let A be an $n \times n$ real matrix, let $A^{(1)} = A$, and let $A^{(k+1)}$, $k = 1, \ldots, n-1$, be the $(n-k) \times (n-k)$ matrix derived from A by the Gaussian Elimination (GE). If we partition $A^{(k)}$ as

$$A^{(k)} = \begin{bmatrix} a^{(k)} & a_c^{(k)T} \\ a_r^{(k)} & A_B^{(k)} \end{bmatrix},$$

where the scalar $a^{(k)}$ is known as the *pivot* at the k-th stage of the elimination, then

$$A^{(k+1)} = A_B^{(k)} - a_r^{(k)}[a^{(k)}]^{-1}a_c^{(k)T}.$$

We say that a matrix A is completely pivoted (CP) or feasible if the rows and columns have been permuted so that Gaussian elimination with no pivoting satisfies the requirements for complete pivoting. Let $g(n,A) = \max_{i,j,k} |a_{ij}^{(k)}|/|a_{11}^{(0)}|$ denote the growth factor of the Gaussian elimination on a CP $n \times n$ matrix A and $g(n) = \sup\{g(n,A)\}$. The problem of determining $g(n)$ for various values of n is called the *growth problem*.

The determination of $g(n)$ remains a challenging problem. Wilkinson in [9,10] noted that there were no known examples of matrices for which $g(n) > n$.

In [1] Cryer conjectured that "$g(n, A) \leq n$, with equality if and only if A is a Hadamard matrix". In [6] a matrix of order 13 is given having growth larger than 13. Interesting results on the size of pivots appear when GE is applied to CP skew-Hadamard and weighing matrices of order n and weight $n-1$. In these matrices the growth is also large and experimentally it is believed that equals $n-1$ [7].

An Hadamard matrix H of order n is an $n \times n$ matrix with elements ± 1 and $HH^T = nI$. For more details and construction methods of Hadamard matrices we refer the reader to the book [5].

Since Wilkinson's initial conjecture seems to be connected with Hadamard matrices it is important to study the growth problem for these matrices (see [1,2,8]). In the present paper we study the pivot structures that arises when we apply GE operations on CP Hadamard matrices of order 16. After testing at least 200000 Hadamard matrices, the following conjecture was posed:

Conjecture (The growth conjecture for Hadamard matrices of order 16)

Let A be an 16×16 CP Hadamard matrix. Reduce A by GE. Then

1. $g(16, A) = 16$.
2. The four last pivots are equal to $\frac{16}{2}$ or $\frac{16}{4}, \frac{16}{2}, \frac{16}{2}, 16$.
3. The fifth last pivot can take the values $\frac{16}{3}$ or $\frac{16}{2}$.
4. The sixth last pivot can take the values $\frac{16}{4}, \frac{16}{10/3}$, or $\frac{16}{8/3}$.
5. The seventh last pivot can take the values $\frac{16}{4}, \frac{16}{18/5}$, or $\frac{16}{16/5}$.
6. The eighth last pivot can take the values $2, \frac{9}{2}, \frac{16}{4}, \frac{8}{3}, \frac{16}{3}$, or $\frac{16}{5}$.
7. The first six pivots are equal to $1, 2, 2, 4, 2$ or $3, \frac{10}{3}$ or $\frac{8}{3}$ or 4.
8. The seventh pivot can take the values $2, 4, \frac{16}{5}, \frac{8}{10/3}$, or $\frac{18}{5}$.
9. The eighth pivot can take the values $4, \frac{9}{2}, \frac{16}{2}, \frac{16}{16/5}$, or $\frac{16}{8/3}$.

The equality in 1. above has been proved for a certain class of 16×16 Hadamard matrices [2]. Cryer [1] has shown 2. for the three last pivots. Day and Peterson [2] have shown that the values $\frac{n}{2}$ or $\frac{n}{4}$ appear in the fourth pivot when Gaussian Elimination (not necessarily with complete pivoting) is applied to a Hadamard matrix of order n. They posed the conjecture that when Gaussian elimination with complete pivoting is done on a Hadamard matrix the value of $\frac{n}{2}$ is impossible for the fourth last pivot. In [3] a Hadamard matrix of order 16 is given which has fourth last pivot $\frac{16}{2}$. We found 10 matrices of order 16 having as fourth last pivot $\frac{16}{2}$.

The values in 7 are proved in [2] for the first five values, $1, 2, 2, 4, 2$ or 3, and experimental evidence in [8] and this paper strongly supports the next values and also the values in 5., 6., 8. and 9.

Notation 1. We use $-$ for -1 in matrices in this paper.

2 Pivot Structures for Hadamard Matrices of Order 16

A Hadamard matrix H of order n is an $n \times n$ matrix of $+1$'s and -1's such that

$$H \cdot H^T = nI$$

This equation is equivalent to the assertion that any two rows of H are orthogonal. Clearly, permuting rows or columns of H or multiplying rows or columns of H by -1 leaves this property unchanged, and we consider such matrices equivalent. If H_1 and H_2 are equivalent Hadamard matrices, then

$$H_2 = P \cdot H_1 \cdot Q$$

where P, Q are monomial permutation matrices of $+1$'s and -1's. By this we mean that P and Q have exactly one nonzero entry in every row and in every column, and this nonzero entry is $+1$ or -1. P gives the permutation and change of sign of rows; Q of columns. Given a Hadamard matrix, we can always find one equivalent to it whose first row and first column consist entirely of $+1$'s. Such a Hadamard matrix is called "normalized". Permuting rows except the first, or columns except the first, leaves a normalized matrix normalized, but in general there may be equivalent normalized matrices that are not equivalent by merely permuting rows and columns.

When GECP is applied to equivalent matrices different pivot structures are attained. For Hadamard matrices of order 16 it is proved in [4] that there are 5 equivalent classes and examples of each are given.

In the sequel for each representative of each class we applied GECP to it and we took 40000 equivalent matrices. For class I we found 9 different pivot patterns. For class II we found 18 different pivot patterns, for class III we found 21 different pivot patterns whereas classes IV and V gave 12 different pivot patterns which were the same for both classes since classes IV and V are transpose to each other and thus are identical for the purpose of GECP [2] The following tables summarizes the different pivot structures attained for each class that are also different among all classes.

3 The Fourth Last Pivot

The following matrices are CP Hadamard matrices. When Gaussian Elimination is applied on them they give the following pivot structure

$$(1, 2, 2, 4, 3, \frac{8}{3}, 2, 4, 4, 4, 4, 8, 8, 8, 8, 16).$$

Thus they have their fourth last pivot equal to $\frac{16}{2}$. All of them belong to Class I. The matrix in [3] which also gives as fourth last pivot 8 also belongs to Class I.

Table 1.

	growth	Class I- Pivot Pattern
1	16	$(1, 2, 2, 4, 2, 4, 4, \ 8, \ 2, 4, \ \ 4, \ \ 8, 4, 8, 8, 16)$
2	16	$(1, 2, 2, 4, 2, 4, 4, \frac{16}{8/3}, \frac{8}{3}, 4, \frac{16}{8/3}, \frac{16}{3}, 4, 8, 8, 16)$
3	16	$(1, 2, 2, 4, 2, 4, 4, \frac{16}{8/3}, \frac{8}{3}, 4, \ \ 4, \ \ 8, 4, 8, 8, 16)$
4	16	$(1, 2, 2, 4, 3, \frac{8}{3}, 4, \frac{16}{8/3}, \frac{8}{3}, 4, \ \ 4, \ \ 8, 4, 8, 8, 16)$
5	16	$(1, 2, 2, 4, 3, \frac{8}{3}, 4, \frac{16}{8/3}, \frac{8}{3}, 4, \frac{16}{8/3}, \frac{16}{3}, 4, 8, 8, 16)$
6	16	$(1, 2, 2, 4, 3, \frac{8}{3}, 2, \ \ 4, \ \ 4, 8, \frac{16}{8/3}, \frac{16}{3}, 4, 8, 8, 16)$
7	16	$(1, 2, 2, 4, 3, \frac{8}{3}, 2, \ \ 4, \ \ 4, 8, \ \ 4, \ \ 8, 4, 8, 8, 16)$
8	16	$(1, 2, 2, 4, 3, \frac{8}{3}, 2, \ \ 4, \ \ 4, 4, \ \ 8, \ \ 8, 4, 8, 8, 16)$
9	16	$(1, 2, 2, 4, 3, \frac{8}{3}, 2, \ \ 4, \ \ 4, 4, \ \ 4, \ \ 8, 8, 8, 8, 16)$

Table 2.

	growth	Class II- Pivot Pattern
1	16	$(1, 2, 2, 4, 2, 4, 4, \frac{16}{16/5}, \frac{16}{5}, \frac{16}{16/5}, \frac{16}{10/3}, \frac{16}{3}, 4, 8, 8, 16)$
2	16	$(1, 2, 2, 4, 2, 4, 4, \frac{16}{16/5}, \frac{16}{5}, 4, \frac{16}{8/3}, \frac{16}{3}, 4, 8, 8, 16)$
3	16	$(1, 2, 2, 4, 2, 4, 4, \ \ 4, \ \ 4, \frac{16}{16/5}, \frac{16}{10/3}, \frac{16}{3}, 4, 8, 8, 16)$
4	16	$(1, 2, 2, 4, 2, 4, 4, \ \ 4, \ \ 4, \ \ 4, \frac{16}{8/3}, \frac{16}{3}, 4, 8, 8, 16)$
5	16	$(1, 2, 2, 4, 2, 4, 4, \ \ 4, \ \ 4, \ \ 4, \ \ 8, \ \ 4, 8, 8, 16)$
6	16	$(1, 2, 2, 4, 2, 4, 4, \frac{16}{16/5}, \frac{16}{5}, 4, \ \ 4, \ \ 8, 4, 8, 8, 16)$
7	16	$(1, 2, 2, 4, 3, \frac{10}{3}, \frac{8}{10/3}, 4, \frac{16}{3}, \frac{16}{16/5}, \frac{16}{10/3}, \frac{16}{3}, 4, 8, 8, 16)$
8	16	$(1, 2, 2, 4, 3, \frac{10}{3}, \frac{16}{5}, 4, 4, 4, \frac{16}{8/3}, \frac{16}{3}, 4, 8, 8, 16)$
9	16	$(1, 2, 2, 4, 3, \frac{10}{3}, \frac{8}{10/3}, 4, \frac{16}{3}, 4, 4, 8, 4, 8, 8, 16)$
10	16	$(1, 2, 2, 4, 3, \frac{10}{3}, \frac{16}{5}, 4, 4, \frac{16}{16/5}, \frac{16}{10/3}, \frac{16}{3}, 4, 8, 8, 16)$
11	16	$(1, 2, 2, 4, 3, \frac{10}{3}, \frac{16}{5}, 4, 4, 4, 4, 8, 4, 8, 8, 16)$
12	16	$(1, 2, 2, 4, 3, \frac{10}{3}, \frac{16}{5}, \frac{16}{16/5}, \frac{16}{5}, 4, \frac{16}{8/3}, \frac{16}{3}, 4, 8, 8, 16)$
13	16	$(1, 2, 2, 4, 3, \frac{10}{3}, \frac{16}{5}, \frac{16}{16/5}, \frac{16}{5}, \frac{16}{16/5}, \frac{16}{10/3}, \frac{16}{3}, 4, 8, 8, 16)$
14	16	$(1, 2, 2, 4, 3, \frac{10}{3}, \frac{16}{5}, \frac{16}{16/5}, \frac{16}{5}, \ \ 4, \ \ 4, \ \ 8, 4, 8, 8, 16)$
15	16	$(1, 2, 2, 4, 3, \frac{10}{3}, \frac{8}{10/3}, 4, \frac{16}{3}, 4, \frac{16}{8/3}, \frac{16}{3}, 4, 8, 8, 16)$

Table 3.

	growth	Class III- Pivot Pattern
1	16	$(1, 2, 2, 4, 2, 4, \ \ 4, \ \ 4, \frac{9}{2}, \frac{16}{18/5}, \frac{16}{10/3}, \frac{16}{3}, 4, 8, 8, 16)$
2	16	$(1, 2, 2, 4, 2, 4, \ \ 4, \frac{9}{2}, 4, \frac{16}{18/5}, \frac{16}{10/3}, \frac{16}{3}, 4, 8, 8, 16)$
3	16	$(1, 2, 2, 4, 3, \frac{10}{3}, \frac{18}{5}, 4, 4, \frac{16}{18/5}, \frac{16}{10/3}, \frac{16}{3}, 4, 8, 8, 16)$
4	16	$(1, 2, 2, 4, 3, \frac{8}{3}, 4, 4, 4, \frac{16}{16/5}, \frac{16}{10/3}, \frac{16}{3}, 4, 8, 8, 16)$
5	16	$(1, 2, 2, 4, 3, \frac{8}{3}, 4, 4, 4, 4, \frac{16}{8/3}, \frac{16}{3}, 4, 8, 8, 16)$
6	16	$(1, 2, 2, 4, 3, \frac{8}{3}, 4, 4, 4, 4, 4, 8, 4, 8, 8, 16)$
7	16	$(1, 2, 2, 4, 3, \frac{8}{3}, 4, \frac{16}{16/5}, \frac{16}{5}, 4, \frac{16}{8/3}, \frac{16}{3}, 4, 8, 8, 16)$
8	16	$(1, 2, 2, 4, 3, \frac{8}{3}, 4, \frac{16}{16/5}, \frac{16}{5}, \frac{16}{16/5}, \frac{16}{10/3}, \frac{16}{3}, 4, 8, 8, 16)$
9	16	$(1, 2, 2, 4, 3, \frac{8}{3}, 4, \frac{16}{16/5}, \frac{16}{5}, 4, 4, 8, 4, 8, 8, 16)$
10	16	$(1, 2, 2, 4, 3, \frac{8}{3}, 4, 4, \frac{9}{2}, \frac{16}{18/5}, \frac{16}{10/3}, \frac{16}{3}, 4, 8, 8, 16)$

$$\begin{bmatrix}-&1&1&1&1&1&1&-&1&1&1&1&-&-&-&-\\-&-&1&-&1&1&1&-&1&1&-&1&-&-&1&1\\-&-&-&1&1&1&-&1&1&1&1&-&-&1&-&1\\-&1&-&-&-&-&-&1&1&1&1&1&-&1&1&1\\1&1&-&-&1&-&1&1&-&1&-&1&-&1&-&1\\-&-&1&1&1&-&-&1&-&-&-&1&1&1&1&1\\1&1&-&1&1&1&-&-&1&-&-&1&1&-&-&1\\1&-&-&1&1&-&1&-&1&1&1&1&1&1&1&-\\1&1&1&1&-&1&-&1&1&1&-&1&-&1&1&-\\1&-&1&1&-&-&1&1&1&-&1&1&-&-&-&-\\-&1&-&1&-&1&1&-&-&1&1&-&1&1&1\\-&1&-&1&1&-&1&1&1&-&-&-&-&-&1&-\\-&1&1&-&1&-&-&-&1&-&1&1&-&1&-&-\\-&-&-&1&-&-&-&-&-&-&1&-&1&-&-&-\\-&1&1&1&-&-&1&-&1&1&-&-&1&1&-&1\\-&-&-&-&-&1&1&1&1&-&-&1&1&1&-&-\end{bmatrix}, \begin{bmatrix}1&-&1&-&1&1&1&1&1&1&-&-&1&-&-\\1&1&1&1&-&-&1&1&1&1&-&1&-&1&1&1\\-&-&1&1&-&-&-&1&1&-&1&1&1&1&-&-\\-&1&1&-&-&-&1&-&1&-&1&-&-&-&-&1\\1&-&-&-&-&1&1&-&-&1&1&1&1&-&1\\1&-&-&-&-&1&-&-&1&1&-&1&1&-&-&1&1\\-&1&-&-&1&-&1&1&1&1&-&1&1&-&-&-\\1&1&1&-&-&1&1&1&-&-&1&1&1&-&1&-\\-&1&1&-&1&1&-&1&1&-&-&-&1&1&1&1\\-&-&1&1&1&1&1&-&-&1&-&1&-&1&-\\1&-&1&-&-&-&-&1&1&-&-&1&-&1&-\\1&1&1&1&1&1&-&-&1&1&1&1&-&-&1\\-&-&1&-&-&1&-&1&-&1&-&1&-&-&-&1\\-&-&1&-&1&-&1&-&-&1&1&1&1&1&1\\1&-&1&1&1&-&1&1&-&-&-&-&1&-&-&1\\1&1&1&-&1&-&-&-&-&-&-&-&1&-&1&-&-\end{bmatrix},$$

4 Conclusions

Finally there were found at least 34 different pivot patterns. From the above results we see that the magnitudes of all the intermediate pivot elements are less than 16 and this gives strong evidence that the growth for the Hadamard matrix of order 16 is 16.

It is interesting to study the pivot structures for each class. Class I gave always as sixth pivot 4 or $\frac{8}{3}$ and the fourth last pivot equal to 8 arised only from matrices coming from the first class. Class II gave always as sixth pivot 4 or $\frac{10}{3}$ whereas Class III gave always as sixth pivot 4, $\frac{8}{3}$, or $\frac{10}{3}$. A thorough classification of the appearing pivot structures for each class still remains an open issue.

References

1. Cryer, C. W.: Pivot size in Gaussian elimination. Numer. Math. **12** (1968) 335–345.
2. Day, J. and Peterson, B.: Growth in Gaussian elimination. Amer. Math. Monthly **95** (1988) 489–513.
3. Edelman, A. and Friedman, D.: A counterexample to a Hadamard matrix pivot conjecture. Linear and Multilinear Algebra, **44** (1998) 53–56.
4. Hall, M.: Hadamard matrices of order 16. Jet Propulsion Lab., Res. Summ., 36-10 **Vol. 1**, 21–26, Pasadena, CA, 1961.
5. Geramita, A. V. and Seberry, J.: *Orthogonal Designs: Quadratic forms and Hadamard matrices*. Marcel Dekker, New York-Basel, 1979.
6. Gould, N.: On growth in Gaussian elimination with pivoting. SIAM J. Matrix Anal. Appl. **12** (1991) 354–361.
7. Koukouvinos, C. Mitrouli, M. and Seberry, J.: Growth in Gaussian elimination for weighing matrices $W(n, n-1)$. Lin. Alg. and its Appl. **306** (2000) 189–202.
8. Mitrouli, M. and Koukouvinos, C.: On the growth problem for D-optimal designs. Proceedings of the First Workshop on Numerical Analysis and Applications, Lecture Notes in Computer Science, **Vol. 1196**, Springer Verlag, Heidelberg, (1996), 341–348.
9. Wilkinson, J. H.: Rounding Errors in Algebraic Processes. Her Majesty's Stationery Office, London, 1963.
10. Wilkinson, J. H.: The Algebraic Eigenvalue Problem. Oxford University Press, London, 1988.

Regularization Method by Rank Revealing QR Factorization and Its Optimization

Susumu Nakata[1], Takashi Kitagawa[1], and Yohsuke Hosoda[2]

[1] Institute of Information Sciences and Electronics, University of Tsukuba
Ibaraki, 305-8573, Japan
[2] Faculty of Engineering, Fukui University
Fukui, 910-8507, Japan

Abstract. Tikhonov regularization using SVD (Singular Value Decomposition) is an effective method for discrete ill-posed linear operator equations. We propose a new regularization method using Rank Revealing QR Factorization which requires far less computational cost than that of SVD. It is important to choose regularization parameter to obtain a good approximate solution for the equation. For the choice of the regularization parameter, Generalized cross-validation (GCV) and the L-curve method are often used. We apply these two methods to the regularization using rank revealing QR factorization to produce a reasonable solution.

1 Introduction

We consider the approximate solution of linear discrete ill-posed problem

$$A\boldsymbol{x} = \boldsymbol{b}, \qquad A \in \mathbb{R}^{m \times n}, \quad m \geq n, \qquad (1)$$

where A is an ill-conditioned matrix. The equation (1) arises as discretizations of the Fredholm integral equations of the first kind:

$$\int_a^b K(s,t)f(t)dt = g(s), \qquad s \in [s_{\min}, s_{\max}],$$

where $K(s,t)$ and $g(s)$ are known L_2 functions and $f(t)$ is the unknown function in $L_2[a,b]$. Tikhonov regularization [2,5] is one of the practical methods for this problem. This method uses the singular value decomposition (SVD) of the coefficient matrix A:

$$A = \bar{U}\bar{\Sigma}\bar{V}^T = \sum_{i=1}^n \sigma_i \bar{\boldsymbol{u}}_i \bar{\boldsymbol{v}}_i^T, \qquad \sigma_1 \geq \cdots \geq \sigma_n \geq 0, \qquad (2)$$

where σ_i are the singular values of A and $\bar{\boldsymbol{u}}_i$ and $\bar{\boldsymbol{v}}$ are the ith left and right singular vector, respectively. Using this decomposition, the regularized solution $\bar{\boldsymbol{x}}_\lambda$ is given as follows:

$$\bar{\boldsymbol{x}}_\lambda = \sum_{i=1}^n \bar{f}_{\lambda,i} \frac{\bar{\boldsymbol{u}}_i^T \boldsymbol{b}}{\sigma_i} \bar{\boldsymbol{v}}_i, \qquad \bar{f}_{\lambda,i} \stackrel{\mathrm{d}}{\equiv} \frac{\sigma_i^2}{\sigma_i^2 + \lambda^2}, \qquad (3)$$

where $\bar{f}_{\lambda,i}$ are called filter factors [5] which depend on the regularization parameter λ.

In this paper, we propose a regularization method by a decomposition using rank revealing QR factorization (RRQR) [1,3], which requires less computational cost than that of SVD.

The regularized solution depends on a regularization parameter which controls the influence of the noise in right-hand side b. Hence the choice of the regularization parameter is important to obtain a good approximate solution. Generalized cross-validation (GCV) [4] and the L-curve method [5] are methods to estimate the optimal regularization parameter for the Tikhonov regularization. We apply these two methods to the regularization using rank revealing QR factorization to determine a reasonable parameter.

2 Regularization Method Using Rank Revealing QR Factorization

In this section, we show the regularization method using a rank revealing QR factorization defined using machine precision $\mu > 0$ as follows :

Definition 1. [3] *Assume that a matrix $A \in \mathbb{R}^{m \times n}$ ($m \geq n$) has numerical rank r. If there exists a permutation Π such that*

$$A\Pi = QR, \qquad R = \begin{bmatrix} R_{11} & R_{12} \\ 0 & R_{22} \end{bmatrix}, \qquad R_{11} \in \mathbb{R}^{r \times r},$$

and

$$\sigma_{\min}(R_{11}) \gg \|R_{22}\|_2 = O(\mu),$$

then the factorization $A\Pi = QR$ is called a Rank Revealing QR factorization of A.

Here, numerical rank r of a matrix A is defined as follows :

Definition 2. [3] *A matrix $A \in \mathbb{R}^{m \times n}$ ($m \geq n$) has numerical rank r if*

$$\sigma_r \gg \sigma_{r+1} = O(\mu).$$

We consider the decomposition

$$A = UDRV^T, \qquad D = \mathrm{diag}(d_1, \ldots, d_r), \qquad d_1 \geq \cdots \geq d_r > 0 \qquad (4)$$

where

$$U = [u_1, \ldots, u_r] \in \mathbb{R}^{m \times r}, \qquad V = [v_1, \ldots, v_r] \in \mathbb{R}^{n \times r}$$

are orthogonal matrices and $R \in \mathbb{R}^{r \times r}$ is a well-conditioned matrix. This decomposition can be obtained using rank revealing QR factorization as following algorithm :

1. Compute the rank revealing QR factorization of A^T :

$$A^T \Pi = [V, \tilde{V}] \begin{bmatrix} \tilde{R}_{11} & \tilde{R}_{12} \\ 0 & \tilde{R}_{22} \end{bmatrix},$$

and let $\tilde{R} \leftarrow [\tilde{R}_{11}, \tilde{R}_{12}]$.

2. Compute the decomposition

$$\tilde{R}^T = \hat{L}D, \qquad \hat{L} \in \mathbb{R}^{m \times r},$$

where $\hat{L}$ is a lower triangular matrix whose diagonal elements are 1 and D is a diagonal matrix.

3. $L \leftarrow \Pi \hat{L}$.
4. Compute the QR factorization $L = U\hat{R}$ where $\hat{R} \in \mathbb{R}^{r \times r}$ is an upper triangular matrix.
5. $R \leftarrow D^{-1}\hat{R}D$.

Then the least squares solution of minimal norm x_{LS} for (1) can be given as

$$x_{\text{LS}} = \sum_{i=1}^{r} \frac{u_i^T b}{d_i} w_i, \qquad W \stackrel{\text{d}}{=} VR^{-1} = [w_1, \ldots, w_r]. \tag{5}$$

We define the regularized solution using the decomposition (4) as follows :

$$x_\lambda = \sum_{i=1}^{r} f_{\lambda,i} \frac{u_i^T b}{d_i} w_i, \qquad f_{\lambda,i} \stackrel{\text{d}}{=} \frac{d_i^2}{d_i^2 + \lambda^2}, \tag{6}$$

where $f_{\lambda,i}$ are the filter factors given by the regularization parameter λ. The regularized solution (6), obviously, has the following characterization :

$$x_\lambda \longrightarrow x_{\text{LS}} \quad \text{as} \quad \lambda \longrightarrow 0,$$

and this solution satisfies the following theorem.

Theorem 1. *Let x_λ be as in (6). Then x_λ is the unique solution of the minimization problem*

$$\min_{x \in X} \{\|Ax - b\|_2^2 + \lambda^2 \|RV^T x\|_2^2\}, \quad X \stackrel{\text{d}}{=} \text{span}\{v_1, \ldots, v_r\}. \tag{7}$$

3 Choice of the Regularization Parameter

To obtain the regularized solution (6), the regularization parameter λ has to be chosen properly. In this section, we apply the two methods, GCV and the L-curve method, to the regularization using RRQR.

3.1 GCV

GCV [4] is a method to choose a regularization parameter which is the minimizer of the following function :

$$G(\lambda) \stackrel{\mathrm{d}}{\equiv} \frac{\|A\boldsymbol{x}_\lambda - \boldsymbol{b}\|_2^2}{(\mathrm{trace}(I - AB(\lambda)))^2},$$

where $B(\lambda)$ is a matrix which satisfies $\boldsymbol{x}_\lambda = B(\lambda)\boldsymbol{b}$. Using the decomposition (4) and the filter factors (6), this function is given as follows :

$$G(\lambda) = \frac{\sum_{i=1}^{r}\{(1 - f_{\lambda,i})\boldsymbol{u}_i^T\boldsymbol{b}\}^2 + \|\Delta\boldsymbol{b}\|_2^2}{\left(m - \sum_{i=1}^{r} f_{\lambda,i}\right)^2}, \quad \Delta\boldsymbol{b} \stackrel{\mathrm{d}}{\equiv} \boldsymbol{b} - UU^T\boldsymbol{b}. \tag{8}$$

3.2 L-Curve Method

L-curve [5] is given by plotting two values in the functional (7). This is the graph of $(\|A\boldsymbol{x}_\lambda - \boldsymbol{b}\|_2, \|RV^T\boldsymbol{x}_\lambda\|_2)$ for a large range of λ. The L-curve has a corner and the corresponding regularization parameter is a good compromise between the residual norm and the influence of the noise in $\boldsymbol{b}$. Here, we use the parameter which is the point of the L-curve with maximal curvature. Using the decomposition (4), the curvature $\kappa(\lambda)$ of the L-curve is given as follows:

$$\kappa(\lambda) = \frac{|\alpha(\lambda)\beta(\lambda)/\Sigma - \lambda^2(\alpha(\lambda) + \lambda^2\beta(\lambda))|}{(\alpha(\lambda) + \lambda^4\beta(\lambda))^{\frac{3}{2}}}, \tag{9}$$

where

$$\alpha(\lambda) \stackrel{\mathrm{d}}{\equiv} \|A\boldsymbol{x}_\lambda - \boldsymbol{b}\|_2^2 = \sum_{i=1}^{r}\left(\frac{\lambda^2}{d_i^2 + \lambda^2}\boldsymbol{u}_i^T\boldsymbol{b}\right)^2 + \|\Delta\boldsymbol{b}\|_2^2,$$

$$\beta(\lambda) \stackrel{\mathrm{d}}{\equiv} \|RV^T\boldsymbol{x}_\lambda\|_2^2 = \sum_{i=1}^{r}\left(\frac{d_i}{d_i^2 + \lambda^2}\boldsymbol{u}_i^T\boldsymbol{b}\right)^2$$

and

$$\Sigma = \sum_{i=1}^{r} \frac{(d_i\boldsymbol{u}_i^T\boldsymbol{b})^2}{(d_i^2 + \lambda^2)^3}.$$

The costs for the estimation using these two methods are the same order as that of SVD.

4 Numerical Results

In this section, we test the linear operator equation obtained from the discretization of a Fredholm integral equation of the first kind :

$$\int_0^1 e^{st} f(t) dt = \frac{e^{s+1} - 1}{s + 1}, \quad 0 \le s \le 1, \tag{10}$$

which has a solution $f(t) = e^t$. For sample pints of s, we used 200 random numbers distributed uniformly and for the discretization of the integral form, we used a Gauss integral rule with 100 points. Thus, the size of A is $(m, n) = (200, 100)$. In order to test influence of errors in the right-hand side $\boldsymbol{b}$, we added a normal distribution with zero mean and standard deviation 10^{-4} to each element of $\boldsymbol{b}$.

We compare the properties of the regularization by RRQR with that of the Tikhonov's method. The comparison of computational cost for each decomposition of the coefficient matrix is shown in Table. 1.

Table 1. CPU-time for the calculation of each decomposition

	Avarage time (sec)
$A = \bar{U}\bar{\Sigma}\bar{V}^T$	0.63
$A = UDRV^T$	0.06

Here, the numerical rank of A is $r = 9$ where $\mu = 1.0 \times 10^{-16}$. Diagonal elements of each diagonal matrix and the coefficients of $\boldsymbol{b}$ corresponding to the orthogonal basis $\boldsymbol{u}_i$ are shown in Fig. 1.

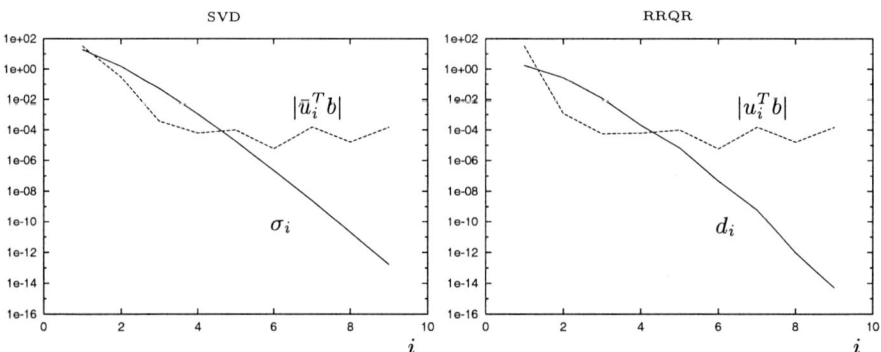

Fig. 1. Diagonal elements and coefficients for the right-hand side

As shown in Fig. 1, the diagonal elements and the corresponding coefficients of both methods have almost the same properties for every point i.

4.1 GCV

Here, we define the error function $e(\lambda)$ for the regularization parameter λ as follows :

$$e(\lambda) \stackrel{\mathrm{d}}{=} \frac{\|\boldsymbol{x}_\lambda - \boldsymbol{x}_0\|_2}{\|\boldsymbol{x}_0\|_2},$$

where x_0 is the true solution of the equation. Fig. 2 shows the errors $e(\lambda)$ and the GCV function (8). The dashed-dotted lines in Fig. 2 are the estimations of the optimal regularization parameter given by GCV method. The regularized solution for each method with the estimated parameter are almost optimal.

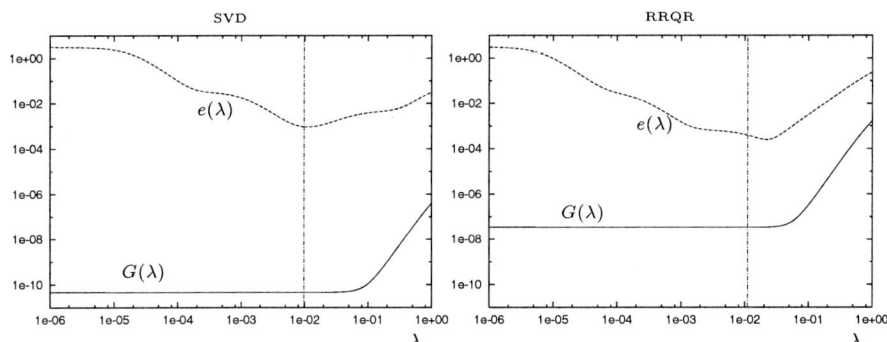

Fig. 2. GCV function

4.2 L-Curve

The L-curves for both methods, $(\|A\bar{x}_\lambda - b\|_2, \|\bar{x}_\lambda\|_2)$ and $(\|Ax_\lambda - b\|_2, \|RV^T x_\lambda\|_2)$ respectively, are shown in Fig. 3. Both of the L-curves have a corner at the point of $\|Ax_\lambda - b\|_2 \approx 1.0 \times 10^{-3}$.

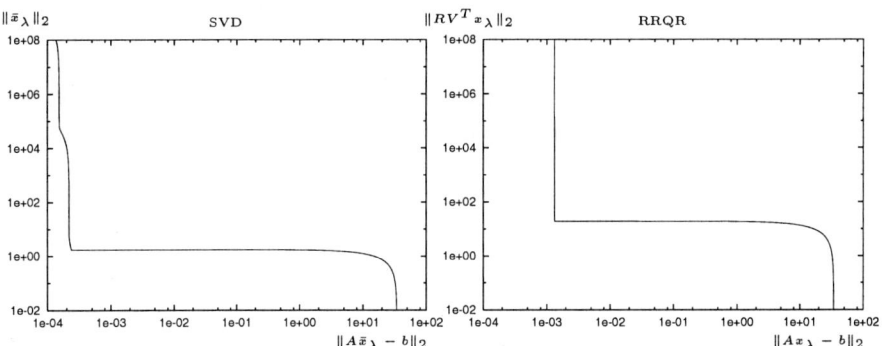

Fig. 3. L-curve

The curvatures of the L-curves are shown in Fig. 4. The dashed-dotted lines in Fig. 4 are the regularization parameters with maximal curvatures. The parameters given by the L-curve method are almost optimal.

Table. 2 shows the comparison of the Optimal parameter, the parameter estimated by GCV and the parameter by L-curve and the corresponding errors for each regularization method.

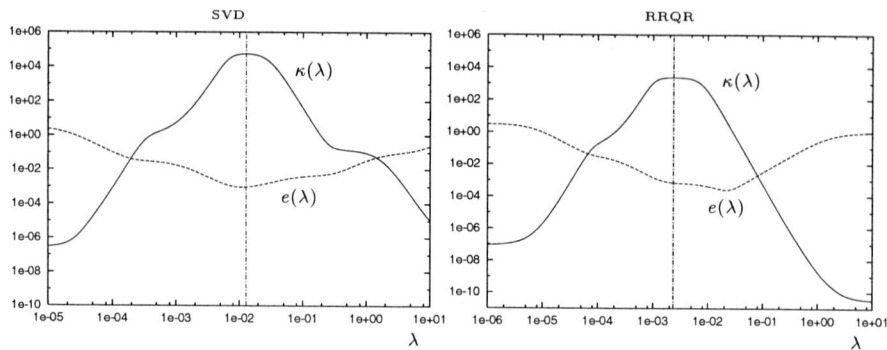

Fig. 4. Curvature of L-curve

Table 2. Regularization parameters and errors

		Optimal	GCV	L-curve
SVD	λ	1.10×10^{-2}	9.77×10^{-3}	1.29×10^{-2}
	$e(\lambda)$	9.27×10^{-4}	9.37×10^{-4}	9.48×10^{-4}
RRQR	λ	2.19×10^{-2}	1.10×10^{-2}	2.34×10^{-3}
	$e(\lambda)$	2.50×10^{-4}	3.90×10^{-4}	6.86×10^{-4}

5 Conclusions

In this paper, we proposed a regularization method for discrete ill-posed linear operator equations using rank revealing QR factorization. The decomposition of the coefficient matrix requires far less computational cost than that of SVD. This method needs a good value of regularization parameter to obtain a good approximate solution. We applied the two methods, generalized cross-validation and the L-curve method, to obtain good estimations of the optimal regularization parameters. The costs for the choice of the parameters are the same order as that of Tikhonov's method. In the numerical example of the Fredholm integral equation of the first kind, we have shown that the errors of the regularized solution are almost the same as that of Tikhonov regularized solution.

References

1. Chan, T. F. : Rank revealing QR factorization, Linear Algebra Appl., **88/89** (1987) 67–82
2. Engl, H. W., Hanke, E., Neubauer, A. : Regularization of inverse problems. Mathematics and its applications, Vol. 375. Kluwer Academic Publishers, Boston (1996)
3. Hong, T. P., Pan, C.-T.: Rank-revealing QR factorizations and the singular value decomposition. Math. Comp., **58** (1992) 213–232
4. Golub, G. H., Heath, M., Wahba, G.,: Generalized cross-validation as a method for choosing a good ridge parameter. Technometrics, **21** (1979) 215–223
5. Hansen, P. C., O'Leary, D. P.: The use of the L-curve in the regularization of discrete ill-posed problems. SIAM J. Sci. Comput. **14** (1993) 1487–1503

A Fast Algorithm for High-Resolution Color Image Reconstruction with Multisensors

Michael K. Ng[1]*, Wilson C. Kwan[1], and Raymond H. Chan[2]**

[1] Department of Mathematics, The University of Hong Kong
Pokfulam Road, Hong Kong
mng@maths.hku.hk
[2] Department of Mathematics, The Chinese University of Hong Kong
Shatin, Hong Kong
rchan@math.cuhk.edu.hk

Abstract. This paper studies the application of preconditioned conjugate gradient methods in high-resolution color image reconstruction problems. The high-resolution color images are reconstructed from multiple undersampled, shifted, degraded color frames with subpixel displacements. The resulting degradation matrices are spatially variant. To capture the changes of reflectivity across color channels, the weighted H_1 regularization functional is used in the Tikhonov regularization. The Neumann boundary condition is also employed to reduce the boundary artifacts. The preconditioners are derived by taking the cosine transform approximation of the degradation matrices. Numerical examples are given to illustrate the fast convergence of the preconditioned conjugate gradient method.

1 Introduction

In this paper, we consider the reconstruction of high-resolution color images from multiple undersampled, shifted, degraded and noisy color images which are obtained by using multiple identical color image sensors shifted from each other by subpixel displacements. We remark that color can be regarded as a set of three images in their primary color components: red, green and blue. The reconstruction of high-resolution color images can be modeled as solving

$$g = Af + \eta, \qquad (1)$$

where A is the reconstruction matrix, η represents unknown Gaussian noise or measurement errors, g is the observed high resolution color image formed from the low resolution color images and f is the desired high resolution color image.

---

* Research supported by Hong Kong Research Grants Council Grant No. HKU 7147/99P and HKU CRCG Grant No. 10202720.
** Research supported in part by Hong Kong Research Grants Council Grant No. CUHK 4207/97P and CUHK DAG Grant No. 2060143.

The observed and original color images can be expressed as

$$g = \begin{pmatrix} g^{(r)} \\ g^{(g)} \\ g^{(b)} \end{pmatrix}, \quad f = \begin{pmatrix} f^{(r)} \\ f^{(g)} \\ f^{(b)} \end{pmatrix},$$

where $g^{(i)}$ and $f^{(i)}$ ($i \in \{r,g,b\}$) are the observed and the original color images from the red, green and blue channels respectively. The multichannel degradation matrix A is given by

$$A = \begin{pmatrix} A^{rr} & A^{rg} & A^{rb} \\ A^{gr} & A^{gg} & A^{gb} \\ A^{br} & A^{bg} & A^{bb} \end{pmatrix}. \qquad (2)$$

Here the matrices A^{ii} and A^{ij} ($i \neq j$) represent the within-channel and the cross-channel degradation matrices respectively. We remark that this formulation of multichannel degradation was considered in [4].

In the case of grey-level high-resolution image reconstruction, where the model was proposed in [2], we have already developed a fast algorithm that is based on the preconditioned conjugate gradient method with cosine transform preconditioners, see [8]. In particular, we have shown that when the L_2 or H_1 norm regularization functional is used, the spectra of the preconditioned normal systems are clustered around 1 and hence the conjugate gradient method converges very quickly. For grey-level images, the use of the Neumann boundary condition can reduce the boundary artifacts and we have shown that solving such systems is much faster than solving those with zero and periodic boundary conditions, see [8]. In the literature, the Neumann boundary condition has also been studied in image restoration [7,1,6].

The main aim of this paper is to extend our results in [8] from grey-level images to color images which are vector-valued grey-level images. We will extend our fast and stable gray-level image processing algorithm with cosine transform preconditioners to the color image reconstruction problem.

The outline of the paper is as follows. In Section 2, we give a mathematical formulation of the problem. In Section 3, we consider the image reconstruction problem when there are no errors in the subpixel displacements. An introduction on the cosine transform preconditioners will be given in Section 4. In Section 5, numerical results are presented to demonstrate the effectiveness of our method.

2 The Mathematical Model

We begin with a brief introduction of the mathematical model in high-resolution image reconstruction. Details can be found in [2,10].

Consider a sensor array with $L_1 \times L_2$ sensors, each sensor has $N_1 \times N_2$ sensing elements (pixels) and the size of each sensing element is $T_1 \times T_2$. Our aim is to reconstruct an image of resolution $M_1 \times M_2$, where $M_1 = L_1 \times N_1$ and $M_2 = L_2 \times N_2$. To maintain the aspect ratio of the reconstructed image, we consider the case where $L_1 = L_2 = L$ only. For simplicity, we assume that L is an even number in the following discussion.

In order to have enough information to resolve the high-resolution image, there are subpixel displacements between the sensors. In the ideal case, the sensors are shifted from each other by a value proportional to $T_1/L \times T_2/L$. However, in practice there can be small perturbations around these ideal subpixel locations due to imperfection of the mechanical imaging system. Thus, for $l_1, l_2 = 0, 1, \cdots, L-1$ with $(l_1, l_2) \neq (0, 0)$, the horizontal and vertical displacements $d^x_{l_1 l_2}$ and $d^y_{l_1 l_2}$ of the $[l_1, l_2]$-th sensor array with respect to the $[0, 0]$-th reference sensor array are given by

$$d^x_{l_1 l_2} = \frac{T_1}{L}(l_1 + \epsilon^x_{l_1 l_2}) \quad \text{and} \quad d^y_{l_1 l_2} = \frac{T_2}{L}(l_2 + \epsilon^y_{l_1 l_2}).$$

Here $\epsilon^x_{l_1 l_2}$ and $\epsilon^y_{l_1 l_2}$ denote respectively the normalized horizontal and vertical displacement errors.

We remark that the parameters $\epsilon^x_{l_1 l_2}$ and $\epsilon^y_{l_1 l_2}$ can be obtained by manufacturers during camera calibration. We assume that

$$|\epsilon^x_{l_1 l_2}| < \frac{1}{2} \quad \text{and} \quad |\epsilon^y_{l_1 l_2}| < \frac{1}{2}, \quad 0 \leq l_1, l_2 \leq L-1. \tag{3}$$

For if not, the low resolution images observed from two different sensor arrays will be overlapped so much that the reconstruction of the high resolution image is rendered impossible.

Let $f^{(r)}$, $f^{(g)}$ and $f^{(b)}$ be the original scene in red, green and blue channels respectively. Then the observed low resolution image in the i-th ($i \in \{r, g, b\}$) channel $g^{(i)}_{l_1 l_2}$ for the (l_1, l_2)-th sensor is modeled by:

$$g^{(i)}_{l_1 l_2}[n_1, n_2] = \sum_{j \in \{r,g,b\}} w_{ij} \left\{ \int_{T_2(n_2 - \frac{1}{2}) + d^y_{l_1 l_2}}^{T_2(n_2 + \frac{1}{2}) + d^y_{l_1 l_2}} \int_{T_1(n_1 - \frac{1}{2}) + d^x_{l_1 l_2}}^{T_1(n_1 + \frac{1}{2}) + d^x_{l_1 l_2}} f^{(j)}(x_1, x_2) dx_1 dx_2 \right\}$$
$$+ \eta^{(i)}_{l_1 l_2}[n_1, n_2], \tag{4}$$

for $n_1 = 1, \ldots, N_1$ and $n_2 = 1, \ldots, N_2$. Here $\eta^{(i)}_{l_1 l_2}$ is the noise corresponding to the (l_1, l_2)-th sensor in the i-th channel, and w_{ii} and w_{ij} ($i \neq j$) are the within-channel and the cross-channel degradation parameters. We note that

$$w_{ij} \geq 0, \quad i, j \in \{r, g, b\} \quad \text{and} \quad \sum_{j=r,g,b} w_{ij} = 1, \quad i \in \{r, g, b\}. \tag{5}$$

Details about these degradation parameters can be found in the multichannel restoration model [4].

To get the matrix representation (1), we intersperse the low resolution images $g^{(i)}_{l_1 l_2}[n_1, n_2]$ to form an $M_1 \times M_2$ image by assigning

$$g^{(i)}[L(n_1 - 1) + l_1, L(n_2 - 1) + l_2] = g^{(i)}_{l_1 l_2}[n_1, n_2], \quad i \in \{r, g, b\}.$$

The image $g^{(i)}$ so formed is called the *observed high-resolution image from the i-th channel*. Similarly, we define $\eta^{(i)}$. Using a column by column ordering for $g^{(i)}$, $f^{(i)}$ and $\eta^{(i)}$, (4) becomes

$$g^{(i)} = \sum_{j \in \{r,g,b\}} w_{ij} H_L(\epsilon) f^{(j)} + \eta^{(i)}.$$

Writing it in matrix form, we get (1) with $A^{(ij)}$ in (2) given by

$$A^{(ij)} = w_{ij} H_L(\epsilon), \quad i, j \in \{r, g, b\}. \tag{6}$$

Under the Neumann boundary condition assumption, the degradation matrix corresponding to the (l_1, l_2)-th sensor is given by

$$H_{l_1 l_2}(\epsilon) = H^x_{l_1 l_2}(\epsilon) \otimes H^y_{l_1 l_2}(\epsilon).$$

Here $H^x_{l_1 l_2}$ and $H^y_{l_1 l_2}$ are banded Toeplitz-plus-Hankel matrices:

$$H^x_{l_1 l_2}(\epsilon) = \frac{1}{L} \begin{pmatrix} 1 & \cdots & 1 & h^{x+}_{l_1 l_2} & & 0 \\ \vdots & \ddots & \ddots & \ddots & \ddots & \\ 1 & \ddots & \ddots & \ddots & \ddots & h^{x+}_{l_1 l_2} \\ h^{x-}_{l_1 l_2} & \ddots & \ddots & \ddots & \ddots & 1 \\ & \ddots & \ddots & \ddots & \ddots & \vdots \\ 0 & & h^{x-}_{l_1 l_2} & 1 & \cdots & 1 \end{pmatrix} + \frac{1}{L} \begin{pmatrix} 1 & \cdots & 1 & h^{x-}_{l_1 l_2} & & 0 \\ \vdots & \ddots & \ddots & & & \\ 1 & \ddots & & & & h^{x+}_{l_1 l_2} \\ h^{x-}_{l_1 l_2} & & & & \ddots & 1 \\ & & & & \ddots & \vdots \\ 0 & & h^{x+}_{l_1 l_2} & 1 & \cdots & 1 \end{pmatrix}, \tag{7}$$

and $H^y_{l_1 l_2}(\epsilon)$ is defined similarly.

The degradation matrix for the whole sensor array is made up of degradation matrices from each sensor:

$$H_L(\epsilon) = \sum_{l_1=0}^{L-1} \sum_{l_2=0}^{L-1} D_{l_1 l_2} H_{l_1 l_2}(\epsilon), \quad i, j \in \{r, b, g\}. \tag{8}$$

Here $D_{l_1 l_2}$ are diagonal matrices with diagonal elements equal to 1 if the corresponding component of the observed low resolution image comes from the (l_1, l_2)-th sensor and zero otherwise, see [2] for more details. From (6), we see that we have the same matrix $H_L(\epsilon)$ within the channels and across the channel. Therefore by (2), the overall degradation matrix is given by

$$A_L(\epsilon) = \begin{pmatrix} w_{rr} & w_{rg} & w_{rb} \\ w_{gr} & w_{gg} & w_{gb} \\ w_{br} & w_{bg} & w_{bb} \end{pmatrix} \otimes H_L(\epsilon) \equiv W \otimes H_L(\epsilon). \tag{9}$$

In the next subsection, we will show that $A_L(\epsilon)$ is ill-conditioned.

2.1 Ill-Conditioning of the Degradation Matrices

When there are no subpixel displacement errors, i.e., when all $\epsilon^x_{l_1,l_2} = \epsilon^y_{l_1,l_2} = 0$, the matrices $H^x_{l_1 l_2}(0)$ and also $H^y_{l_1 l_2}(0)$ are the same for all l_1 and l_2. We will denote them simply by H^x_L and H^y_L.

In this particular case, the eigenvalues of $H_L = H^x_L \otimes H^y_L$ can be computed easily as the matrix can be diagonalized by the 2-dimensional cosine transform $C_{M_1} \otimes C_{M_2}$ [8].

Lemma 1. *[8, Theorem 1] Under the Neumann boundary condition, the eigenvalues of H_L are given by*

$$\lambda_{(i-1)M_2+j}(H_L) = \left(\frac{4}{L}\right)^2 \cos^2\left(\frac{(i-1)\pi}{2M_1}\right) \cdot \cos^2\left(\frac{(j-1)\pi}{2M_2}\right) \cdot p_L\left(\frac{(i-1)\pi}{M_1}\right) \cdot p_L\left(\frac{(j-1)\pi}{M_2}\right) \tag{10}$$

for $1 \leq i \leq M_1, 1 \leq j \leq M_2$. Here

$$p_L\left(\frac{(i-1)\pi}{M_1}\right) = \begin{cases} \sum_{k=1}^{L/4} \cos\left(\frac{(i-1)(2k-1)\pi}{M_1}\right), & \text{when } L = 4l, \\ \frac{1}{2} + \sum_{k=1}^{(L-2)/4} \cos\left(\frac{(i-1)2k\pi}{M_1}\right), & \text{otherwise.} \end{cases}$$

In particular, by choosing $i = M_1$ with $j = M_2$ and $i = j = 1$, we have

$$0 \leq \lambda_{\min}(H_L) \leq O\left(\frac{1}{M_1^2 M_2^2}\right) \quad \text{and} \quad \lambda_{\max}(H_L) = 1. \tag{11}$$

In practical applications, see [4], the within-channel degradation is always stronger than the cross-channel degradation, i.e.,

$$w_{ii} > w_{ij}, \quad \text{for} \quad j \neq i, \quad \text{and} \quad i \in \{r, g, b\}. \tag{12}$$

Under this assumption and using (5), we can prove that W is nonsingular.

Lemma 2. *Let W be a matrix with entries satisfying (5) and (12). Then W is nonsingular. Moreover, we have*

$$0 < \delta = \lambda_{\min}\{W^t W\} \leq \lambda_{\max}\{W^t W\} \leq 2, \tag{13}$$

where δ is a positive constant independent of M_1 and M_2.

Proof. By (5), it is easy to show that 1 is an eigenvalue of W with corresponding eigenvector $[1,1,1]^t$. Since the coefficients of the characteristic polynomial of W are real, the other two eigenvalues of W are in a conjugate pair. Suppose that W is a singular matrix, then W must be a rank one matrix, i.e.,

$$W = [u_1, u_2, u_3]^t [v_1, v_2, v_3],$$

for some u_i, v_i. By (5), we can choose all $u_i, v_i \geq 0$. Also by (5), we have

$$u_1(v_1 + v_2 + v_3) = u_2(v_1 + v_2 + v_3) = u_3(v_1 + v_2 + v_3) = 1$$

or $u_1 = u_2 = u_3 = 1/(v_1 + v_2 + v_3)$. It implies that $w_{ij} = v_j/(v_1 + v_2 + v_3)$. However, this contradicts assumption (12) and hence W is nonsingular. In particular, we have the first inequality in (13). Since all the entries w_{ij} of W are independent of M_1 and M_2, we see that δ is also independent of M_1 and M_2. By (5) and (12), we have $\|W\|_1 \leq 2$ and $\|W\|_\infty = 1$. It follows that $\|W\|_2^2 \leq \|W\|_1 \|W\|_\infty \leq 2$.

Combining Lemmas 1 and 2 and using the tensor product structure (9) of A_L, we get its condition number.

Theorem 1. *Let W be a matrix with entries satisfying (5) and (12). Under the Neumann boundary condition, if H_L is nonsingular, then the condition number $\kappa(A_L)$ of A_L satisfies*

$$\kappa(A_L) \geq O(M_1^2 M_2^2).$$

According to Theorem 1, A_L can be very ill-conditioned or singular. For example, when $L = 4$ and $M_1 = M_2 = 64$, $\lambda_{33}(H_L) = 0$. By continuity arguments, $A_L(\epsilon)$ will still be ill-conditioned if the displacement errors are small. Therefore, a regularization procedure should be imposed to obtain a reasonable estimate of the original image.

2.2 Regularization

In the case of grey-level image reconstruction, the regularization operator only needs to enforce the spatial smoothness of the image. The most usual form of this operator is the discrete version of the 2-dimensional Laplacian. However, in color image reconstruction, in addition to the within-channel spatial smoothness, the cross-channel smoothness must also be enforced. One may incorporate the 3-dimensional discrete Laplacian here. However, color planes are highly correlated and this operator may fail to capture the cross-channel similarities, see [4].

In [4], Galatsanos et al. have proposed the following weighted discrete Laplacian matrix R as the regularization matrix:

$$[Rf]_{r,j,k} = 6[f^{(r)}]_{j,k} - [f^{(r)}]_{j-1,k} - [f^{(r)}]_{j+1,k} - [f^{(r)}]_{j,k-1} - [f^{(r)}]_{j,k+1} - \frac{\|\tilde{f}^{(r)}\|_2}{\|\tilde{f}^{(g)}\|_2}[f^{(g)}]_{j,k} - \frac{\|\tilde{f}^{(r)}\|_2}{\|\tilde{f}^{(b)}\|_2}[f^{(b)}]_{j,k},$$

$$[Rf]_{g,j,k} = 6[f^{(g)}]_{j,k} - [f^{(g)}]_{j-1,k} - [f^{(g)}]_{j+1,k} - [f^{(g)}]_{j,k-1} - [f^{(g)}]_{j,k+1} - \frac{\|\tilde{f}^{(g)}\|_2}{\|\tilde{f}^{(r)}\|_2}[f^{(r)}]_{j,k} - \frac{\|\tilde{f}^{(g)}\|_2}{\|\tilde{f}^{(b)}\|_2}[f^{(b)}]_{j,k},$$

and

$$[Rf]_{b,j,k} = 6[f^{(b)}]_{j,k} - [f^{(b)}]_{j-1,k} - [f^{(b)}]_{j+1,k} - [f^{(b)}]_{j,k-1} - [f^{(b)}]_{j,k+1} - \frac{\|\tilde{f}^{(b)}\|_2}{\|\tilde{f}^{(g)}\|_2}[f^{(g)}]_{j,k} - \frac{\|\tilde{f}^{(b)}\|_2}{\|\tilde{f}^{(r)}\|_2}[f^{(r)}]_{j,k},$$

for $1 \leq j \leq M_1$ and $1 \leq k \leq M_2$. Here $\|\tilde{f}^{(r)}\|_2$, $\|\tilde{f}^{(g)}\|_2$ and $\|\tilde{f}^{(b)}\|_2$ are the estimates of the $\|f^{(r)}\|_2$, $\|f^{(g)}\|_2$ and $\|f^{(b)}\|_2$ respectively and are assumed to be nonzero. The cross-channel weights of this regularization matrix capture the changes of reflectivity across the channels. In practice, we set $\|\tilde{f}^{(i)}\|_2 = \|g^{(i)}\|_2$ for $i \in \{r, g, b\}$, where $g^{(i)}$ is the observed image, see [4].

To sum up, the regularization matrix R is given by

$$R = \begin{pmatrix} 2 & -\frac{\|\tilde{f}^{(r)}\|_2}{\|\tilde{f}^{(g)}\|_2} & -\frac{\|\tilde{f}^{(r)}\|_2}{\|\tilde{f}^{(b)}\|_2} \\ -\frac{\|\tilde{f}^{(g)}\|_2}{\|\tilde{f}^{(r)}\|_2} & 2 & -\frac{\|\tilde{f}^{(g)}\|_2}{\|\tilde{f}^{(b)}\|_2} \\ -\frac{\|\tilde{f}^{(b)}\|_2}{\|\tilde{f}^{(r)}\|_2} & -\frac{\|\tilde{f}^{(b)}\|_2}{\|\tilde{f}^{(g)}\|_2} & 2 \end{pmatrix} \otimes I + I \otimes \Delta \equiv S \otimes I + I \otimes \Delta, \quad (14)$$

where Δ is the 2-dimensional discrete Laplacian matrix with the Neumann boundary condition. We note that Δ can be diagonalized by the 2-dimensional cosine transform matrix $C_{M_1} \otimes C_{M_2}$ [8].

Using Tikhonov regularization, our problem becomes:

$$(A_L(\epsilon)^t \Upsilon A_L(\epsilon) + R^t R)f = A_L(\epsilon)^t \Upsilon g, \quad (15)$$

where $A_L(\epsilon)$ is given in (9),

$$\Upsilon = \begin{pmatrix} \alpha_r I & 0 & 0 \\ 0 & \alpha_g I & 0 \\ 0 & 0 & \alpha_b I \end{pmatrix} = \begin{pmatrix} \alpha_r & 0 & 0 \\ 0 & \alpha_g & 0 \\ 0 & 0 & \alpha_b \end{pmatrix} \otimes I \equiv \Omega \otimes I,$$

and α_r, α_g and α_b are the regularization parameters which are assumed to be positive scalars.

Next we show that the regularized system

$$A_L^t \Upsilon A_L + R^t R = W\Omega W \otimes H_L^t H_L + (S \otimes I + I \otimes \Delta)^t (S \otimes I + I \otimes \Delta) \quad (16)$$

is well-conditioned.

Theorem 2. *Let W be a matrix with entries satisfying (5) and (12). Then there exists a positive scalar γ, independent of M_1 and M_2, such that*

$$\lambda_{\min}\{A_L^t \Upsilon A_L + R^t R\} \geq \gamma > 0. \quad (17)$$

Proof. Under the Neumann boundary condition, the matrices H_L and Δ are symmetric and can be diagonalized by $C_{M_1} \otimes C_{M_2}$. From (16), it therefore suffices to consider the smallest eigenvalue of the matrix

$$W^t \Omega W \otimes \Lambda^2 + (S \otimes I + I \otimes \Sigma)^t (S \otimes I + I \otimes \Sigma) \quad (18)$$

where Λ and Σ are diagonal matrices with diagonal entries given by the eigenvalues of H_L and Δ respectively. More precisely, the diagonal entries Λ_{ij} of Λ are given in (10) and the diagonal entries of Σ are given by

$$\lambda_{(i-1)M_2+j}(\Delta) \equiv \Sigma_{ij} = 4\sin^2\left(\frac{(i-1)\pi}{2M_1}\right) + 4\sin^2\left(\frac{(j-1)\pi}{2M_2}\right), \quad (19)$$

for $1 \leq i \leq M_1$ and $1 \leq j \leq M_2$.

By permutation, we see that the eigenvalues of the matrix in (18) are the same as the eigenvalues of

$$B \equiv \Lambda^2 \otimes W^t \Omega W + (I \otimes S + \Sigma \otimes I)^t (I \otimes S + \Sigma \otimes I), \quad (20)$$

which is a block-diagonal matrix, i.e., all off-diagonal blocks are zero. It therefore suffices to estimate the smallest eigenvalues of the main diagonal blocks of B. For $1 \leq i \leq M_1$, $1 \leq j \leq M_2$, the $((i-1)M_2+j, (i-1)M_2+j)$-th main diagonal block of B is equal to

$$B_{ij} \equiv \Lambda_{ij}^2 \cdot W^t \Omega W + (S + \Sigma_{ij} I)^t (S + \Sigma_{ij} I),$$

where Λ_{ij} is given by the expression in (10) and Σ_{ij} by (19). Since $\lambda_{\min}(X+Y) \geq \lambda_{\min}(X) + \lambda_{\min}(Y)$ for any Hermitian matrices X and Y (see [5, Corollary 8.1.3, p.411]), we have

$$\lambda_{\min}(B_{ij}) \geq \Lambda_{ij}^2 \lambda_{\min}(W^t \Omega W) + \lambda_{\min}\{(S + \Sigma_{ij} I)^t (S + \Sigma_{ij} I)\}.$$

By (13),

$$\lambda_{\min}(W^t \Omega W) \geq \min\{\alpha_r, \alpha_g, \alpha_b\} \cdot \lambda_{\min}(W^t W) = \delta \min\{\alpha_r, \alpha_g, \alpha_b\} \equiv \delta_0 > 0, \quad (21)$$

where δ_0 is a positive constant independent of M_1 and M_2. Hence

$$\lambda_{\min}(B_{ij}) \geq \delta_0 \Lambda_{ij}^2 + \lambda_{\min}\{(S + \Sigma_{ij} I)^t (S + \Sigma_{ij} I)\}. \quad (22)$$

In view of (10) and (19), we define for simplicity

$$\chi(x,y) = \delta_0 \left(\frac{4}{L}\right)^4 \cos^4 x \cos^4 y \cdot p_L^2(2x) p_L^2(2y), \quad (23)$$

$$\phi(x,y) = 4\sin^2 x + 4\sin^2 y, \quad (24)$$

and

$$\psi(x,y) = \lambda_{\min}\left\{(S + \phi(x,y)I)^t (S + \phi(x,y)I)\right\}. \quad (25)$$

With these notations, (22) becomes

$$\lambda_{\min}(B_{ij}) \geq \chi\left(\frac{(i-1)\pi}{2M_1}, \frac{(j-1)\pi}{2M_2}\right) + \psi\left(\frac{(i-1)\pi}{2M_1}, \frac{(j-1)\pi}{2M_2}\right), \quad (26)$$

with $1 \leq i \leq M_1, 1 \leq j \leq M_2$. To complete the proof, we now show that $\chi(x,y) + \psi(x,y) > 0$ for all $(x,y) \in [0, \pi/2]^2$.

From (14), it is easy to check that the eigenvalues of S are 0, 3 and 3 and their corresponding eigenvectors are

$$\left[\frac{||\tilde{f}^{(r)}||_2}{||\tilde{f}^{(b)}||_2}, \frac{||\tilde{f}^{(g)}||_2}{||\tilde{f}^{(b)}||_2}, 1\right]^t, \quad \left[-\frac{||\tilde{f}^{(r)}||_2}{||\tilde{f}^{(g)}||_2}, 1, 0\right]^t \quad \text{and} \quad \left[-\frac{||\tilde{f}^{(r)}||_2}{||\tilde{f}^{(b)}||_2}, 0, 1\right]^t$$

respectively. Therefore, in view of definition (24), for all $(x,y) \in [0,\pi/2]^2$, the matrix $S + \phi(x,y)I$ is nonsingular except when $x = y = 0$. In particular, by definition (25), $\psi(x,y) > 0$ for all $(x,y) \in [0,\pi/2]^2$ except at $x = y = 0$. Moreover, since the entries $||\tilde{f}^{(r)}||_2$, $||\tilde{f}^{(g)}||_2$ and $||\tilde{f}^{(b)}||_2$ of S are constants independent on M_1 and M_2, $\psi(x,y)$ depends only on x, y, $||\tilde{f}^{(r)}||_2$, $||\tilde{f}^{(g)}||_2$ and $||\tilde{f}^{(b)}||_2$ but does not depend on M_1 and M_2. On the other hand, since $\cos^4(x)p_L^2(2x) \geq 1/4$ at $x = 0$ and is nonnegative in $[0,\pi/2]$, by (23), $\chi(x,y) > 0$ at $x = y = 0$ and nonnegative in $[0,\pi/2]^2$. Therefore there exists a positive scalar γ independent of M_1 and M_2 such that $\chi(x,y) + \psi(x,y) \geq \gamma > 0$ for all $(x,y) \in [0,\pi/2]^2$. It follows from (26) that $\lambda_{\min}(B_{ij}) \geq \gamma > 0$ for all $1 \leq i \leq M_1$, $1 \leq j \leq M_2$.

When there are errors in the subpixel displacements, the regularized matrix is given by

$$A_L(\epsilon)^t \Upsilon A_L(\epsilon) + R^t R = W^t \Omega W \otimes H_L(\epsilon)^t H_L(\epsilon) + R^t R.$$

By using arguments similar to that in [8, Theorem 3], we can easily show that this regularized matrix is well-conditioned when the errors are sufficiently small:

Corollary 1. *Let $\epsilon^* = \max_{0 \leq l_1, l_2 \leq L-1} \{|\epsilon^x_{l_1,l_2}|, |\epsilon^y_{l_1,l_2}|\}$ and W be a matrix with entries satisfying (5) and (12). If ϵ^* is sufficiently small, then the smallest eigenvalue of $W^t \Omega W \otimes H_L(\epsilon)^t H_L(\epsilon) + R^t R$ is uniformly bounded away from 0 by a positive constant independent of M_1 and M_2.*

3 Spatially Invariant Case

When there are no subpixel displacement errors, i.e., when all $\epsilon^x_{l_1,l_2} = \epsilon^y_{l_1,l_2} = 0$, we have to solve $(A_L^t \Upsilon A_L + R^t R)f = A_L^t \Upsilon g$ which according to (16) can be simplified to

$$(W^t \Omega W \otimes H_L^t H_L + R^t R)f = (W^t \Omega \otimes H_L^t)g. \tag{27}$$

Recall that if we use the Neumann boundary condition for both H_L and Δ, then both matrices can be diagonalized by discrete cosine transform matrices. From (16) and (18), we see that (27) is equivalent to

$$[W^t \Omega W \otimes \Lambda^2 + (S \otimes I + I \otimes \Sigma)^t(S \otimes I + I \otimes \Sigma)]\hat{f} = (W^t \Omega \otimes \Lambda)\hat{g}, \tag{28}$$

where $\hat{f} = (I \otimes C_{M_1} \otimes C_{M_2})f$ and $\hat{g} = (I \otimes C_{M_1} \otimes C_{M_2})g$. The system in (28) is a block-diagonalized system of $M_1 M_2$ decoupled subsystems. The vector $\hat{f}$ can be computed by solving a set of $M_1 M_2$ decoupled 3-by-3 matrix equations (cf. (20)). The total cost of solving the system is therefore of $O(M_1 M_2 \log M_1 M_2)$ operations.

4 Spatially Variant Case

When there are subpixel displacement errors, the matrix $H_L(\epsilon)$ has the same banded structure as that of H_L, but with some entries slightly perturbed. It is a near block-Toeplitz-Toeplitz-block matrix but it can no longer be diagonalized by the cosine transform matrix. Therefore we solve the linear system in (15) by the preconditioned conjugate gradient method. For an $M_1 \times M_1$ block matrix $H_L(\epsilon)$ with the size of each block equal to $M_2 \times M_2$, the cosine transform preconditioner $c(H_L(\epsilon))$ of $H_L(\epsilon)$ is defined to be the matrix $(C_{M_1} \otimes C_{M_2})\Phi(C_{M_1} \otimes C_{M_2})$ that minimizes

$$\|(C_{M_1} \otimes C_{M_2})\Phi(C_{M_1} \otimes C_{M_2}) - H_L(\epsilon)\|_F$$

over all diagonal matrices Φ, where $\|\cdot\|_F$ is the Frobenius norm, see [3]. Clearly, the cost of computing $c(H_L(\epsilon))^{-1}y$ for any vector y is $O(M_1 M_2 \log M_1 M_2)$ operations. Since $H_L(\epsilon)$ in (8) is a banded matrix with $(L+1)^2$ non-zero diagonals and is of size $M_1 M_2 \times M_1 M_2$, the cost of constructing $c(H_L(\epsilon))$ is of $O(L^2 M_1 M_2)$ operations only, see [3].

We will employ the cosine transform preconditioner $c(H_L(\epsilon))$ of $H_L(\epsilon)$ in our preconditioner. Thus we have to study the convergence rate of the conjugate gradient method for solving the preconditioned system

$$[W^t \Omega W \otimes c(H_L(\epsilon))^t c(H_L(\epsilon)) + R^t R]^{-1}[W^t \Omega W \otimes H_L(\epsilon)^t H_L(\epsilon) + R^t R]f$$
$$= [W^t \Omega \otimes H_L(\epsilon)^t]g. \qquad (29)$$

By using the similar arguments as in [8], we can show that the spectra of the preconditioned normal system are clustered around 1 for sufficiently small subpixel displacement errors. A detail proofs can be found in [9].

Theorem 3. *Let $\epsilon^* = \max_{0 \leq l_1, l_2 \leq L-1}\{|\epsilon^x_{l_1 l_2}|, |\epsilon^y_{l_1 l_2}|\}$ and W be a matrix with entries satisfying (5) and (12). If ϵ^* is sufficiently small, then the spectra of the preconditioned matrices*

$$[W^t \Omega W \otimes c(H_L(\epsilon))^t c(H_L(\epsilon)) + R^t R]^{-1}[W^t \Omega W \otimes H_L(\epsilon)^t H_L(\epsilon) + R^t R]$$

are clustered around 1 and their smallest eigenvalues are uniformly bounded away from 0 by a positive constant independent of M_1 and M_2.

Using standard convergence analysis of the conjugate gradient method, see for instance [5, p.525], we conclude that the conjugate gradient method applied to the preconditioned system (29) will converge superlinearly for sufficiently small displacement errors. Since $H_L(\epsilon)$ has only $(L+1)^2$ non-zero diagonals, the matrix-vector product $A_L(\epsilon)x$ can be done in $O(L^2 M_1 M_2)$. Thus the cost per each PCG iteration is $O(M_1 M_2 \log M_1 M_2 + L^2 M_1 M_2)$ operations, see [5, p.529]. Hence the total cost for finding the high resolution image vector is of $O(M_1 M_2 \log M_1 M_2 + L^2 M_1 M_2)$ operations.

5 Numerical Examples

In this section, we illustrate the effectiveness of using cosine transform preconditioners for solving high resolution color image reconstruction problems. The conjugate gradient method is employed to solving the preconditioned system (29). The cross-channel weights for R (see (14)) are computed from the observed high-resolution image, i.e., $\|\tilde{f}^{(i)}\|_2 = \|g^{(i)}\|_2$, for $i \in \{r, g, b\}$. We tried the following two different degradation matrices to degrade the original color image

$$(i) \begin{pmatrix} 0.8 & 0.1 & 0.1 \\ 0.1 & 0.8 & 0.1 \\ 0.1 & 0.1 & 0.8 \end{pmatrix} \otimes H_L(\epsilon) \quad \text{and} \quad (ii) \begin{pmatrix} 0.5 & 0.3 & 0.2 \\ 0.25 & 0.5 & 0.25 \\ 0.3 & 0.2 & 0.5 \end{pmatrix} \otimes H_L(\epsilon). \quad (30)$$

The interdependency between cross-channels of the first degradation matrix is higher than that of the second degradation matrix. Gaussian white noises with signal-to-noise ratio of 30dB were added to each degraded image plane. We remark that the second degradation matrix W (cf. (9)) has been used to test the least squares restoration of multichannel images [4].

In the tests, we used the same regularization parameter for each channel, i.e., $\alpha_r = \alpha_g = \alpha_b = \alpha$. The initial guess was the zero vector and the stopping criteria was $\|r^{(j)}\|_2/\|r^{(0)}\|_2 < 10^{-6}$, where $r^{(j)}$ is the normal equations residual after j iterations. Tables 1–4 show the numbers of iterations required for convergence for $L = 2$ and 4, i.e., the number of sensor array used is 2×2 and 4×4 respectively. In the tables, "cos", "cir" or "no" signify that the cosine transform preconditioner, the level-2 circulant preconditioner [3] or no preconditioner is used respectively.

We see from the tables that for both degradation matrices, the cosine transform preconditioner converges much faster than the circulant preconditioners for different M, α and $\epsilon^{x,y}_{l_1 l_2}$, where $M(= M_1 = M_2)$ is the size of the reconstructed image and $\epsilon^{x,y}_{l_1 l_2}$ are the subpixel displacement errors. Also the convergence rate is independent of M for fixed α or $\epsilon^{x,y}_{l_1 l_2}$. These results show that our method is very efficient.

Restored color images using our method can be found on-line in [9]. One will see that the details in the image are much better reconstructed under the Neumann boundary condition than that under the zero and periodic boundary conditions. Moreover, the boundary artifacts under the Neumann boundary condition are less prominent too.

References

1. M. Banham and A. Katsaggelos, *Digital Image Restoration*, IEEE Signal Processing Magazine, 14 (1997), pp. 24–41.
2. N. Bose and K. Boo, *High-resolution Image Reconstruction with Multisensors*, International Journal of Imaging Systems and Technology, 9 (1998), pp. 294–304.
3. R. Chan and M. Ng, *Conjugate Gradient Method for Toeplitz Systems*, SIAM Review, 38 (1996), pp. 427–482.

Table 1. Number of iterations for degradation matrix (i) with $L = 2$ and $\epsilon^x_{l_1 l_2} = \epsilon^y_{l_1 l_2} = 0.1$ (left) and $\epsilon^x_{l_1 l_2} = \epsilon^y_{l_1 l_2} = 0.2$ (right)

α	2×10^2			2×10^3			2×10^4			α	2×10^2			2×10^3			2×10^4		
M	cos	cir	no	cos	cir	no	cos	cir	no	M	cos	cir	no	cos	cir	no	cos	cir	no
32	6	16	19	8	38	47	12	95	122	32	9	17	20	13	40	48	24	95	120
64	6	16	19	8	38	46	11	97	121	64	9	16	20	13	39	48	21	99	122
128	6	15	19	7	37	47	12	97	121	128	8	16	19	12	39	47	23	99	121
256	6	15	19	7	38	47	10	98	121	256	8	16	20	11	39	48	21	100	122

Table 2. Number of iterations for degradation matrix (ii) with $L = 2$ and $\epsilon^x_{l_1 l_2} = \epsilon^y_{l_1 l_2} = 0.1$ (left) and $\epsilon^x_{l_1 l_2} = \epsilon^y_{l_1 l_2} = 0.2$ (right)

α	2×10^2			2×10^3			2×10^4			α	2×10^2			2×10^3			2×10^4		
M	cos	cir	no	cos	cir	no	cos	cir	no	M	cos	cir	no	cos	cir	no	cos	cir	no
32	6	16	21	8	39	48	12	95	121	32	9	17	20	13	40	49	24	96	120
64	6	16	21	8	38	47	11	99	122	64	9	17	20	13	40	48	21	100	122
128	6	15	20	7	38	46	12	97	121	128	8	16	20	12	39	47	23	99	121
256	6	15	20	7	38	47	10	98	122	256	8	16	20	11	39	48	21	99	122

Table 3. Number of iterations for degradation matrix (i) with $L = 4$ and $\epsilon^x_{l_1 l_2} = \epsilon^y_{l_1 l_2} = 0.1$ (left) and $\epsilon^x_{l_1 l_2} = \epsilon^y_{l_1 l_2} = 0.2$ (right)

α	2×10^2			2×10^3			2×10^4			α	2×10^2			2×10^3			2×10^4		
M	cos	cir	no	cos	cir	no	cos	cir	no	M	cos	cir	no	cos	cir	no	cos	cir	no
32	5	27	34	7	55	64	11	107	130	32	7	26	34	11	56	66	20	109	134
64	5	28	34	7	64	79	10	147	161	64	7	29	35	11	66	81	20	151	162
128	5	25	33	7	64	81	10	169	181	128	7	25	33	10	66	82	19	166	182
256	5	23	33	6	60	81	10	167	211	256	6	24	33	10	59	82	19	171	209

Table 4. Number of iterations for degradation matrix (ii) with $L = 4$ and $\epsilon^x_{l_1 l_2} = \epsilon^y_{l_1 l_2} = 0.1$ (left) and $\epsilon^x_{l_1 l_2} = \epsilon^y_{l_1 l_2} = 0.2$ (right)

α	2×10^2			2×10^3			2×10^4			α	2×10^2			2×10^3			2×10^4		
M	cos	cir	no	cos	cir	no	cos	cir	no	M	cos	cir	no	cos	cir	no	cos	cir	no
32	5	26	34	7	46	64	11	90	129	32	7	26	35	11	49	64	20	92	131
64	5	28	34	7	63	80	10	131	159	64	7	28	35	11	64	80	20	136	162
128	5	24	33	7	61	82	10	152	183	128	7	24	34	10	63	81	19	156	183
256	5	22	33	6	58	81	10	162	209	256	6	23	33	10	59	82	18	163	209

4. N. Galatsanos, A. Katsaggelosm, R. Chin, and A. Hillery, *Least Squares Restoration of Mulitchannel Images*, IEEE Trans. Signal Processing, 39 (1991), pp. 2222–2236.
5. G. Golub and C. Van Loan, *Matrix Computations*, 2rd ed., The Johns Hopkins University Press, 1993.
6. R. Lagendijk and J. Biemond, *Iterative Identification and Restoration of Images*, Kluwer Academic Publishers, 1991.
7. F. Luk and D. Vandevoorde, *Reducing Boundary Distortion in Image Restoration*, Proc. SPIE 2296, Advanced Signal Processing Algorithms, Architectures and Implementations VI, 1994.

8. M. Ng, R. Chan, T. Chan, and A. Yip, *Cosine Transform Preconditioners for High Resolution Image Reconstruction*, Res. Rept. 99-10, Dept. Math., The Chinese University of Hong Kong, or Linear Algebra Appls., to appear.
9. M. Ng, W. Kwan, and R. Chan, *A Fast Algorithm for High-Resolution Color Image Reconstruction with Multisensors*, Math. Dept. Res. Rep. #99-30, Chinese University of Hong Kong, (ftp://ftp.math.cuhk.edu.hk/report/1999-30.ps.Z).
10. R. Tsai and T. Huang, *Multiframe Image Restoration and Registration*, Advances in Computer Vision and Image Processing, 1 (1984), pp. 317–339.

A Performance Study on a Single Processing Node of the HITACHI SR8000

Seiji Nishimura, Daisuke Takahashi, Takaomi Shigehara,
Hiroshi Mizoguchi, and Taketoshi Mishima

Department of Information and Computer Sciences, Saitama University
Shimo-Okubo 255, Urawa, Saitama 338-8570, Japan.
{seiji,sigehara,hm,mishima}@me.ics.saitama-u.ac.jp
daisuke@ics.saitama-u.ac.jp

Abstract. We carry out a performance study on a single processing node of the HITACHI SR8000. Each processing node of the SR8000 is a shared memory parallel computer which is composed of eight scalar processors with a pseudo-vector processing facility. In this study, we implement highly optimized codes for basic linear operations including matrix-matrix product, matrix-vector product and vector inner-product. As a practical application of matrix-vector product, we examine the performance of two iterative methods for linear systems: the conjugate gradient (CG) method and the conjugate residual (CR) method.

1 Introduction

The significance of a large-scale numerical computation is rapidly growing in various scientific and technological fields such as structural analysis, fluid dynamics and quantum chemistry. In particular, high performance solvers for linear problems are highly desired, since the problem is frequently turned into a linear system after a suitable discretization of space and time. The purpose of this study is to develop highly optimized linear operation codes on the HITACHI SR8000, which is one of up-to-date parallel supercomputers. We restrict ourselves to a single processing node in this paper. A single node of the SR8000 can be considered as a shared memory parallel computer which is composed of eight scalar processors with a pseudo-vector processing facility [1,2]. In this sense, the present work takes a complementary role to ATLAS (Automatically Tuned Linear Algebra Software) in [3], which is intended mainly for RISC processors. After examining the size dependence of the performance of the tuned codes for some basic linear operations, we apply them to the conjugate gradient (CG) and the conjugate residual (CR) methods, which are typical iterative methods for linear systems.

The paper is organized as follows. We summarize experimental environment in Sect.2. In Sect.3, we discuss tuning techniques for basic linear operations on the SR8000 and examine the performance of the tuned codes, which are applied to the CG and CR methods in Sect.4. The current work is summarized in Sect.5.

2 Experimental Environment

We summarize experimental environment of this work and also give specifications of the HITACHI SR8000. Numerical experiments were performed at the Computer Centre Division, Information Technology Center, the University of Tokyo. The SR8000 is composed of 128 processing nodes interconnected through a three-dimensional hyper-crossbar network. The communication bandwidth available to each node is 1GB/sec for a single direction. Each processing node is a shared memory parallel computer with eight scalar processors (Instruction Processor, IP) which is based on RISC architecture. Each IP has two *multiply-add* arithmetic units with machine cycle of 4nsec. As a result, the theoretical peak performance of each IP is 1GFLOPS. The total theoretical peak performance of each processing node is 8GFLOPS. Each IP is designed to achieve a similar performance to a vector processor by adopting a pseudo-vector processing facility, which suppresses a delay caused by cache misses.

We use a single processing node for numerical experiments. The programming language is FORTRAN77. The compile options are "-64 -nolimit -noscope -Oss -procnum=8 -pvfunc=3". These options instruct the compiler to use 64-bit addressing mode ("-64"), to remove limits of memory and time for compilation ("-nolimit"), to forbid dividing a source code into multiple parts when it is compiled ("-noscope"), to set the optimize level to the highest ("-Oss"), to use 8 IP's ("-procnum=8") and to set the pseudo-vectorize level to the highest ("-pvfunc=3"), respectively. We also give the compiler a directive concerning a parallelization among IP's, which is described in the subsequent sections.

3 Basic Linear Operations

In this section, we discuss the basic linear operations including vector operations, matrix-vector product and matrix-matrix product on a single processing node of the SR8000.

We begin with the vector operations. Table 1 is a summary of four basic vector operations which are often used to solve linear systems. In Table 1, **x** and **y** are real n-vectors, while α is a real scalar. The usual inner-product is denoted by $(\cdot, \cdot)$; $(\mathbf{x}, \mathbf{y}) = \sum_{i=1}^{n} x_i y_i$. The first column shows the name of the corresponding subroutines in BLAS (Basic Linear Algebra Subprograms) [4,5,6,7], which is a standard library for basic linear operations. The BLAS routines are classified into three categories; Level 1 (Vector Operations), Level 2 (Matrix-Vector Operations) and Level 3 (Matrix-Matrix Operations). The operations in Table 1 belong to Level 1 BLAS.

In the following, we assume the vector length n to be a multiple of eight, namely the number of IP's in a single node of the SR8000. The vector operations in Table 1 can be written by using a single loop, which is pseudo-vectorized and parallelized with a block distribution in our implementation. For a parallelization of loops on a single node of the SR8000, there are two directives for data distribution to each processor. The directive "*POPTION NOCYCLIC" to the

Table 1. List of basic vector operations

Name	Function
daxpy	$\mathbf{y} := \mathbf{y} + \alpha \mathbf{x}$
ddot	$(\mathbf{x}, \mathbf{y})$
dnrm2	$\|\mathbf{x}\|_2 := \sqrt{(\mathbf{x}, \mathbf{x})}$
dscal	$\mathbf{x} := \alpha \mathbf{x}$

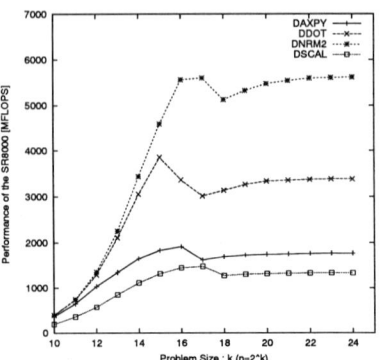

Fig. 1. Performance of basic vector operations

compiler indicates a block distribution; If the loop index i runs from 1 to n, the operations for $i = (k-1)n/8+1, (k-1)n/8+2, \cdots, kn/8$ are performed on k-th processor, $(k = 1, 2, \cdots, 8)$. On the other hand, the directive "*POPTION CYCLIC" indicates a cyclic distribution; The operations for $i = k, k+8, \cdots, n+k-8$ are performed on k-th processor, $(k = 1, 2, \cdots, 8)$. A cyclic distribution is useful in such a case that computational load is not balanced among IP's with a block distribution due to, for example, a branching statement inside the loop. Otherwise, a block distribution is preferable to a cyclic distribution, because the latter requires an overhead to calculate loop-index. For the vector operations in Table 1, a block distribution runs about 10% faster than a cyclic distribution.

We leave the loop-unrolling to the compiler for the operations in Table 1. For a single loop, the compiler can recognize the optimum depth for the loop-unrolling, since it can be determined from the number of floating-point registers on each IP. We have checked through numerical experiment that the compiler indeed gives rise to the optimum loop-unrolling even without any explicit directive to the compiler. (The directive "*SOPTION UNROLL(k)" indicates the loop-unrolling to a depth of k.)

Fig.1 shows the performance of daxpy, ddot, dnrm2 and dscal. The horizontal axis is $k = \log_2 n$ with the problem size n, while the vertical axis shows the performance in units of MFLOPS. One can see that the performance of the operations for a single vector (dnrm2 and dscal) is saturated at $k = 17$. This is because the data cache memory for each processing node is 128KB/IP $\times$ 8IP's in the SR8000. For the operations for two vectors (daxpy and ddot), the saturation occurs around a half of the problem size; $k = 15 \sim 16$. For each operation, the performance is kept at a high level even for a larger problem size, owing to a pseudo-vector facility. For $k = 24$ ($n = 2^{24} = 16,777,216$), the performance of daxpy, ddot, dnrm2 and dscal is 1755.6MFLOPS, 3359.8MFLOPS, 5565.5MFLOPS and 1322.6MFLOPS, respectively. From the viewpoint of the arithmetic operations, the Euclidean norm dnrm2 is the same as the inner-

```
      do 10 i=1,m-1,2                     do 10 i=1,n-1,2
         dtmp1=0.d0                          dtmp1=0.d0
         dtmp2=0.d0                          dtmp2=0.d0
         do 20 j=1,n                         do 20 j=1,m
            dtmp1=dtmp1+a(i  ,j)*v(j)           dtmp1=dtmp1+a(j,i  )*v(j)
            dtmp2=dtmp2+a(i+1,j)*v(j)           dtmp2=dtmp2+a(j,i+1)*v(j)
  20     continue                     20     continue
         u(i  )=beta*u(i  )+alpha*dtmp1      u(i  )=beta*u(i  )+alpha*dtmp1
         u(i+1)=beta*u(i+1)+alpha*dtmp2      u(i+1)=beta*u(i+1)+alpha*dtmp2
  10  continue                         10  continue
```

Fig. 2. Kernel code of dgemv for op$(A) = A$. Additional statements are required if m is odd

Fig. 3. Kernel code of dgemv for op$(A) = A^T$. Additional statements are required if n is odd

product ddot. However, dnrm2 is about 1.6 times faster than ddot. This is due to the fact that the statement $s := s + x_i y_i$ requires two *load* and one *multiply-add* operations. As a result, the performance of ddot is at most 4GFLOPS, namely 50% of the peak performance of a single processing node. On the other hand, dnrm2 requires only one *load* operation for a single *multiply-add* operation. This explains the ratio of the performance between dnrm2 and ddot. One can also observe that ddot is almost twice faster than daxpy. This is because daxpy requires a *store* operation after two *load* and one *multiply-add* operations, which is unnecessary for ddot. Similarly, dscal requires a *store* operation after one *load* and one *multiplication* operations. Thus the ratio of arithmetic operations to data operations is the smallest in dscal. This is the reason why the score of dscal is the poorest in Fig.1.

We proceed to the general matrix-vector product dgemv, $\mathbf{u} := \alpha \, \text{op}(A)\mathbf{v} + \beta \mathbf{u}$ in Level 2 BLAS. Here, α and β are real scalars, A is a real $m \times n$ matrix, $\mathbf{u}$ and $\mathbf{v}$ are real vectors, and op(A) is A or A^T. Clearly, the operations for each component of dgemv can be performed independently. Figs.2 and 3 show kernel double loops for op$(A) = A$ and op$(A) = A^T$, respectively. We parallelize the outer loop with index i with a block distribution and also pseudo-vectorize the inner loop for vector inner-product. In the source codes in Figs.2 and 3, we employ loop-unrolling to a depth of two for the outer loop. For the inner loop, we leave the unrolling to the compiler, as in vector operations. As a result, the inner loop is unrolled to a depth of four. The loop-unrolling to a depth of two for the outer loop makes it possible to reduce the number of *load* operations for the vector $\mathbf{v}$ to the half in the inner loop. As well, the length of the outer loop is reduced by 50%. As a result, the performance is improved by about 10%. We have examined unrolling of the outer loop to greater depth in numerical experiment. The performance for a depth of four is almost the same as for a depth of two, while a depth of eight gives rise to only 10% performance compared to the case of a depth of two. A depth of eight is too large to store relevant elements in the floating-point registers on each IP.

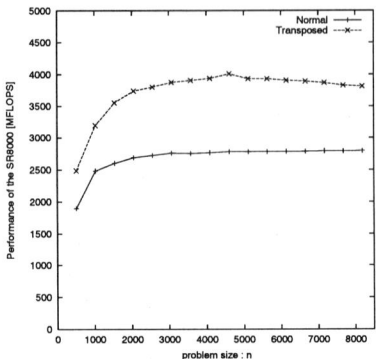

Fig. 4. Performance of `dgemv`. Solid and broken lines show the performance for op(A) = A and op(A) = A^T, respectively

Fig.4 shows the performance of the `dgemv` routine. The matrix is assumed to be square ($m = n$). The size of the matrix is changed as $n = 256 \times i$; $i = 1, 2, \cdots, 32$. For $n = 8192$, the performance is 2797.1 MFLOPS and 3811.5 MFLOPS for op(A) = A and op(A) = A^T, respectively. Since the matrix is stored by columns in FORTRAN, the memory access is continuous for op(A) = A^T. Thus op(A) = A^T is expected to show high performance, comparable to the inner product. Indeed, one can see that the broken line in Fig.4 reproduces the performance of the inner product `ddot` in Fig.1. Recall that since the loop for vector operations in Fig.1 is parallelized among eight IP's, each IP processes only one eighth of the vector elements. Thus, Fig.4 should be compared with the performance of `ddot` for $k = 12 \sim 16$ in Fig.1. In case of op(A) = A, the memory access to the matrix A is not continuous, as seen from Fig.2. This is the reason why op(A) = A is about 75% of op(A) = A^T in performance. The drawback might be settled by changing the order of outer and inner loops. In such case, however, the variable `dtmp1` as well as `dtmp2` should be a vector instead of a scalar. This causes additional *store* operations and a substantial delay is observed; The asymptotic performance is about 2000MFLOPS.

Finally, we consider the matrix-matrix product. The matrix-matrix product $A = BC$ is given by $a_{ij} = \sum_{k=1}^{n} b_{ik} c_{kj}$; $i, j = 1, 2, \cdots, n$. For simplicity, we assume that $A = (a_{ij})$, $B = (b_{ik})$ and $C = (c_{kj})$ are $n \times n$ square matrices. Clearly, the source code of matrix-matrix product is composed of a triply nested loop with indices i, j, k. Numerical experiment shows that the *jki* form together with the options described below gives rise to the best performance. Here the loop indices are ordered as j, k and i from the outer loop to the inner loop in the *jki* form. This corresponds to the linear combination algorithm. We parallelize the outermost loop with index j with a block distribution and also pseudo-vectorize the innermost loop with index i.

Fig.5 shows the performance of matrix-matrix product for problem size $n = 256 \times i$; $i = 1, 2, \cdots, 32$. One can see that the performance is extremely high.

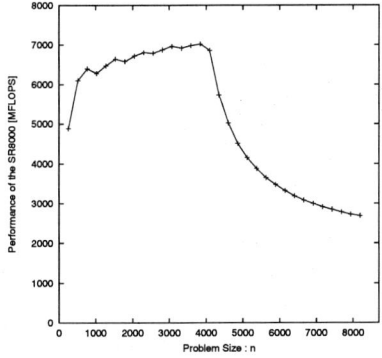
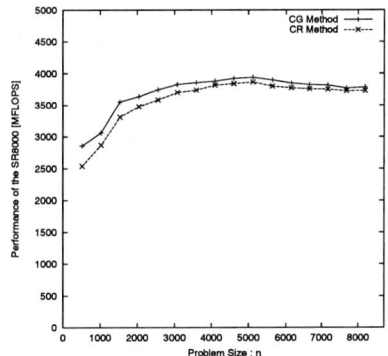

Fig. 5. Performance of matrix-matrix product

Fig. 6. Performance of CG and CR methods

Indeed it is above 75% of the theoretical peak performance of a single processing node of the SR8000 for the problem size $n \simeq 500 \sim 4000$; 6101.6MFLOPS for $n = 512$ and 7021.5MFLOPS for $n = 3840$. However, a further increase of vector length leads to a sudden slow down caused by cache misses; The performance is 2695.5MFLOPS for $n = 8192$ for instance. The so-called block algorithms that utilize submatrices rather than just columns or rows are expected to remedy the situation.

4 CG and CR Methods

As an example of the basic linear codes in Sect.3 at a practical level, we examine the performance of the CG and CR methods. We consider a linear system $A\mathbf{x} = \mathbf{b}$. Here $\mathbf{x}$ is an unknown n-vector which should be determined when an $n \times n$ nonsingular matrix A and an n-vector $\mathbf{b}$ are given. Figs.7 and 8 show the algorithms of the CG and CR methods, respectively. We assume the coefficient matrix A to be dense and we use in the numerical experiment the Frank matrix $A = (a_{ij})$ with $a_{ij} = \min\{i, j\}$; $i, j = 1, 2, \cdots, n$, which is widely used for benchmark.

For implementation of the CG and CR methods on the SR8000, we use the optimized codes in the previous section. We transpose the matrix A and calculate the matrix-vector product $(A\mathbf{x})_k$ by $\sum_{i=1}^{n} (A^T)_{ik} x_i$. This ensures continuous memory access to the matrix A^T. Note that the transposed matrix A^T can be overwritten on the original A, since one uses A only in a form of the matrix-vector product $A\mathbf{x}$. We set the relative residual to $\epsilon = 10^{-12}$ in Figs.7 and 8.

Fig.6 shows the performance of the CG and CR methods for the problem size $n = 256 \times i$; $i = 1, 2, \cdots, 32$. For $n = 8192$, the performance of the CG and CR methods are 3785.8MFLOPS and 3729.2MFLOPS, respectively. They are close to the asymptotic performance of dgemv subroutine, nearly 50% of the peak performance of a single processing node of the SR8000. If we do not transpose

Take an initial guess $\mathbf{x}_0$;
$\mathbf{r}_0 := \mathbf{b} - A\mathbf{x}_0$; $\mathbf{p}_0 := \mathbf{r}_0$;
for $k := 0, 1, 2, \cdots$
until $||\mathbf{r}_k|| < \epsilon ||\mathbf{b}||$ do
begin

$$\mathbf{a} := A\mathbf{p}_k;$$
$$\mu := (\mathbf{p}_k, \mathbf{a});$$
$$\alpha_k := (\mathbf{r}_k, \mathbf{p}_k)/\mu;$$
$$\mathbf{x}_{k+1} := \mathbf{x}_k + \alpha_k \mathbf{p}_k;$$
$$\mathbf{r}_{k+1} := \mathbf{r}_k - \alpha_k \mathbf{a};$$
$$\beta_k := -(\mathbf{r}_{k+1}, \mathbf{a})/\mu;$$
$$\mathbf{p}_{k+1} := \mathbf{r}_{k+1} + \beta_k \mathbf{p}_k;$$

end

Fig. 7. Algorithm of CG method

Take an initial guess $\mathbf{x}_0$;
$\mathbf{r}_0 := \mathbf{b} - A\mathbf{x}_0$; $\mathbf{p}_0 := \mathbf{r}_0$; $\mathbf{q} := A\mathbf{p}_0$;
for $k := 0, 1, 2, \cdots$
until $||\mathbf{r}_k|| < \epsilon ||\mathbf{b}||$ do
begin

$$\mu := (\mathbf{q}, \mathbf{q});$$
$$\alpha_k := (\mathbf{r}_k, \mathbf{q})/\mu;$$
$$\mathbf{x}_{k+1} := \mathbf{x}_k + \alpha_k \mathbf{p}_k;$$
$$\mathbf{r}_{k+1} := \mathbf{r}_k - \alpha_k \mathbf{q};$$
$$\mathbf{a} := A\mathbf{r}_{k+1};$$
$$\beta_k := -(\mathbf{a}, \mathbf{q})/\mu;$$
$$\mathbf{p}_{k+1} := \mathbf{r}_{k+1} + \beta_k \mathbf{p}_k;$$
$$\mathbf{q} := \mathbf{a} + \beta_k \mathbf{q};$$

end

Fig. 8. Algorithm of CR method

the coefficient matrix and use dgemv with op$(A) = A$, the performance is at most 2700MFLOPS, which we have checked in experiment.

We emphasize that it is not always expensive to use iterative methods for dense linear systems. If we apply the diagonal preconditioner (where the preconditioner is just the diagonal of A) to the Frank matrix in CG method, only 100 $\sim$ 600 iterations are required in a wide range of the problem size $512 \sim 8192$, where right-hand side $\mathbf{b}$ and initial guess $\mathbf{x}_0$ are chosen as generic. In such cases, our tuned code for CG method obtains the solution considerably faster than LU decomposition, even if LU decomposition should attain the theoretical peak performance. This implies that highly optimized codes of iterative methods are useful even for dense linear systems, at least for some classes where a suitable preconditioner is known.

5 Summary

In this paper, we implemented the highly optimized codes for basic linear operations on a single processing node of the HITACHI SR8000 and also evaluated their performance. Concerning the tuning techniques, we should select a suitable technique according to loop structure. For vector operations like Level 1 BLAS, which is described by a single loop, an adequate selection of data distribution by a directive to the compiler takes an important role to get good performance in parallel processing. By adopting a block distribution, we attained 25%, 50%, 75% and 17% of the theoretical peak performance for daxpy, ddot, dnrm2 and dscal, respectively. On the other hand, the loop-unrolling of the outer loop is efficient for multiple nested loops such as the matrix-vector product (dgemv) and matrix-matrix product. This is because one can save the *load* and *store* operations in the inner loop. For matrix-vector product, we observed 35% of the peak performance. It is enhanced to nearly 50% by using the transposition of the matrix. This is because the use of the transposed matrix in matrix-vector product ensures continuous memory access to the matrix elements in FORTRAN. Concerning matrix-matrix product, we observed 85% of the peak performance. Although cache misses slow down the present code of matrix-matrix product, the block algorithms are expected to prevent cache misses. As a realistic application of basic linear codes, we implemented the CG and CR methods. They show the same performance as in matrix-vector product with the transposition of the matrix.

Putting every sort of experimental facts together reveals the essence of a single processing node of the SR8000. Unlike on a vector processor such as the HITAC S-3800, the effect of different aspects such as a parallelization by a directive, cache misses and memory access should be take into account for optimization, according to the structure of loops. The cache misses cause a considerable loss of the performance. However, unlike on the usual scalar machines, pseudo-vector processing facility serves to partly suppress the loss by cache misses and it indeed minimizes the effect of cache misses for a certain range of the problem size, as shown by a plateau in Fig.5. In summary, we should remember that a single node of the SR8000 is a *shared memory parallel computer* composed of eight *scalar (RISC) processors* equipped with a *pseudo-vector processing facility*.

In a future work, we shall (1) tune other BLAS routines, (2) implement the block algorithms for matrix-matrix product, and (3) extend the present work to multiple processing nodes.

References

1. K. Nakazawa, H. Nakamura, H. Imori, S. Kawabe: *Pseudo vector processor based on register-windowed superscalar pipeline*, Proceedings of Supercomputing '92, pp. 642-651, 1992.
2. K. Nakazawa, H. Nakamura, T. Boku, I. Nakata, Y. Yamashita: *CP-PACS: A massively parallel processor at the University of Tsukuba*, Parallel Computing, vol. 25, pp. 1635-1661, 1999.
3. R. C. Whaley, J. J. Dongarra: *Automatically Tuned Linear Algebra Software*, http://netlib.org/atlas/atlas.ps.gz.
4. C. Lawson, R. Hanson, D. Kincaid, F. Krogh: *Basic Linear Algebra Subprograms for Fortran Usage*, ACM Trans. on Math. Soft. vol. 5, pp. 308-325, 1979.
5. J. J. Dongarra, J. DuCroz, S. Hammarling, R. Hanson: *An Extended Set of Fortran Basic Linear Algebra Subprograms*, ACM Trans. on Math. Soft. vol. 1, pp. 1-32, 1988.
6. J. J. Dongarra, I. Duff, J. DuCroz, S. Hammarling: *A Set of Level 3 Basic Linear Algebra Subprograms*, ACM Trans. on Math. Soft., 1989.
7. J. J. Dongarra, I. S. Duff, D. C. Sorensen, and H. A. van der Vorst: *Solving Linear Systems on Vector and Shared Memory Computers*, SIAM, Philadelphia, 1991.

Estimation of the Wheat Losses Caused by the Tropospheric Ozone in Bulgaria and Denmark

Tzvetan Ostromsky[1], Ivan Dimov[1], Ilia Tzvetanov[2], and Zahari Zlatev[3]

[1] Central Laboratory for Parallel Processing, Bulgarian Academy of Sciences
Acad. G. Bonchev str., bl. 25-A, 1113 Sofia, Bulgaria
ceco@cantor.bas.bg
dimov@copern.bas.bg
http://copern.bas.bg
[2] Institute of Economics of the Bulgarian Academy of Sciences
3 Aksakov str., 1000 Sofia, Bulgaria
ineco@iki.bas.bg
[3] National Environmental Research Institute, Department of Atmospheric Environment
Frederiksborgvej 399 P. O. Box 358, DK-4000 Roskilde, Denmark
zz@dmu.dk
http://www.dmu.dk/AtmosphericEnvironment

Abstract. Long-term exposures to high ozone concentrations have harmful effect on the crops and reduce the yield. The exposures are measured in terms of AOT40 (Accumulated exposure Over Threshold of 40 ppb). The threshold of 40 ppb has been accepted after several years of experimental research in open top chambers throughout Europe. As a result of these experiments a critical level of 3000 ppb.hours has been established for the crops. The sensitivity of the wheat to exposures above the critical level has been studied in more detail and a linear dependence between the relative yield and the AOT40 has been found. This relationship is used in the paper to estimate the wheat losses in Bulgaria and Denmark by regions in several consequtive years.

The Danish Eulerian Model is used to calculate the AOT40 values on the EMEP grid (a 96×96 square grid with step 50 km., which covers Europe). The results on parts of this grid (covering Bulgaria and Denmark) are only used. In addition regional information about the actual yield and the prices is also needed for the target years. The algorithm for economical evaluation of the losses can be applied with different scenarios for partial reduction of the emissions and some other key parameters. The results can be used as a ground for a cost-benefit analysis of possible ozone reduction measures when effects on vegetation are studied.

1 Introduction

The damaging effects of high ozone concentrations on agricultural crops is known for a long time, but little was known until 1990 on the extent of these damages worldwide. Extended research on this problem has been conducted during the past ten years. Newly designed open top chambers (OTC) allow systematic study, leading to quantifiable estimates for use in policy analysis. To report on early progress, a number of meetings were carried out during the 1990's, e.g. in Switzerland in 1993 ([4]), and in Finland in 1995 ([6]). Among the recommendations from these meetings, a new parameter called AOT40 was introduced ([4] and [6]). This parameter was suggested to be applied to agricultural and economic assessments and subsequent modeling of benefits associated with reduced ozone exposure. The AOT40 parameter is commonly accepted now, also in the discussions of the forthcoming EU Ozone Directive (see [1,3]).

The value 40 ppb is a practically determined threshold, below which the losses of crops due to ozone exposure could be neglected, and above which the losses are assumed to be linear with respect to the exposure. The choice of AOT40 is based on a large number of OTC experiments in Europe and in the United States.

In this work wheat losses in Bulgaria and Denmark for a period of ten years are calculated by using AOT40 values produced by the Danish Eulerian Model (DEM) [10,11]. Our work on these problems started a year ago [2]. Since then the model was improved significantly in its chemical part and in the vertical exchange through the boundary layers in accordance with the knowledge obtained by analyzing new measurement data. In addition a new release of emissions data from the EMEP inventories is used as an input in DEM now. All these developments lead to certain differences in the results, presented in this paper, in comparison with [2]. We believe our new estimates are much more accurate, although many uncertainties still remain.

2 Data Sets Used in the Study

The following data files are used in the calculations:

- The AOT40 values on the EMEP grid for each year in the period 1989 – 1998 (only the values on the parts of the grid covering Bulgaria and Denmark are used). These are produced by DEM and verified by comparisons with measurements and with results, obtained by other models. The maps with the AOT40 values over Bulgaria for 1990 and 1994 are shown in Fig. 1;
- The relationship between the AOT40 and the wheat loss, based on experiments in OTC, presented at meetings in Switzerland and Finland [5,4,7];
- The wheat yield in Bulgaria and Denmark by regions for the years under consideration, taken from the corresponding Statistical Yearbooks [8,9].

3 Calculating the Losses by Regions

The basic assumption behind introduction of the excess ozone concept and AOT40 in particular is that the *relative yield* of wheat linerly depends on the value of AOT40. Provided that y is the actual yield, $y + z$ is the expected yield

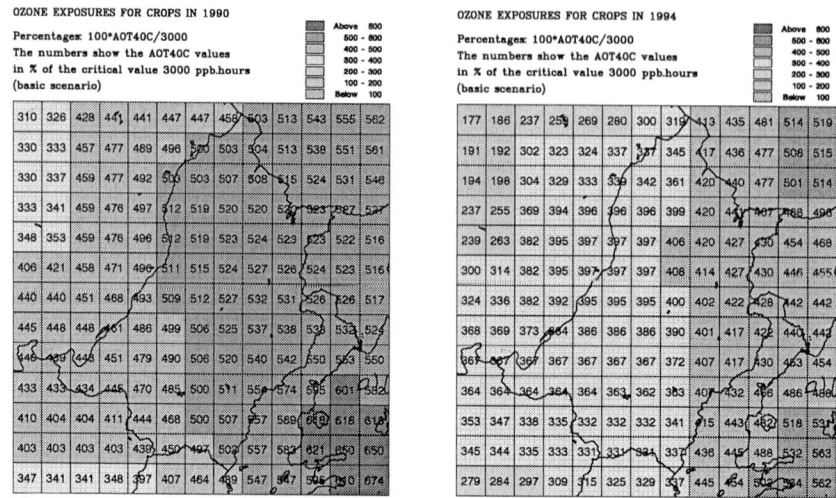

Fig. 1. AOT40 maps of Bulgarian region for 1990 and 1994

without any ozone exposure, ξ is the AOT40 in ppb-hours, this *linear regression* can be expressed as follows:

$$100\, y/(y+z) = \alpha \xi + \beta \quad (\alpha < 0,\ \ \beta \approx 100) \tag{1}$$

where α and β are empirically determined coefficients. The values $\alpha = -0.00151$ and $\beta - 99.5$, derived from OTC experiments performed in the Scandinavian countries, are used for calculating the losses in Denmark. The mean values, obtained by analyzing a wide set of OTC experiments representative for Europe, are $\alpha = -0.00177$ and $\beta = 99.6$. The latter values are used in calculating the losses in Bulgaria. The specific coefficients for South-European countries (including Bulgaria) could be slightly different. Due to the small number of OTC experiments in these countries such specific coefficients cannot be determined yet. The above values are due to Pleijel [7].

Let us consider first a simplified (scalar) version of the task: to find the loss z of given crop yield y from a single region with a constant value ξ of AOT40. The linear regression (1) gives the actual yield y (in our task it is given) as a function of $x = y + z$, where z is the unknown to be calculated.

$$y = \frac{(\alpha \xi + \beta)(y+z)}{100}$$
$$100\, y = (\alpha \xi + \beta)\, y + (\alpha \xi + \beta)\, z$$
$$(100 - \beta - \alpha \xi)\, y = (\alpha \xi + \beta)\, z$$
$$z = \frac{(100 - \beta - \alpha \xi)}{(\alpha \xi + \beta)}\, y = f(\xi)\, y \tag{2}$$

Consider now the real task: to find the losses from each of the m regions of the country, covered by n grid cells, taking into account that each region is covered by several grid cells with different value of AOT40. Denote by

$\Xi = (\xi_j)_{j=1}^n$ – the AOT40 vector ($m \times 1$), per grid cell, calculated by the DEM;
A – the regional division matrix ($m \times n$);
Y – the yield matrix ($m \times k$), yields of k crops per regions;
Z – loss matrix ($m \times k$), per regions – unknown

The calculations can be done in the following way (not unique):

$$\Theta = (f(\xi_j))_{j=1}^n, \quad S = A\Theta, \quad Z = \text{diag}(S)Y$$

Applying f from (2) to Ξ componentwise we find first the relative losses by grid cells Θ (with respect to the actual yields). The matrix-vector product $A\Theta$ gives the relative losses by regions (S), and multiplying with them the rows of Y we obtain the corresponding losses Z.

4 Reduced Traffic Scenario

Traffic is known to be one of the primary sources of ozone pollution in the developed countries. In order to evaluate the contribution of the traffic to the overall ozone pollution and the resulting economical losses, a scenario with 90% reduction of the actual traffic emissions is included in our study. This scenario is called hereafter *traffic scenario*, unlike the *basic scenario*, which denotes the actual situation and the corresponding actual losses. The traffic scenario is applied to the same 10-year period and the corresponding AOT40 values are calculated by using the Danish Eulerian Model [10,11]. The flexibility of the model allows us to calculate these values by proper reduction of the actual emissions and keeping all the other input data unchanged. Simple calculations show that the traffic scenario leads approximately to the following global reductions of the anthropogenic emissions:

- 45% reduction of the anthropogenic NO_x emissions;
- 40% reduction of the anthropogenic VOC emissions;
- 54% reduction of the anthropogenic CO emissions;
- no change in the anthropogenic SO_2 and NH_3 emissions as well as in all biogenic emissions.

The results for the estimated wheat losses in Bulgaria and Denmark, obtained both with the basic and the traffic scenario, are given in the next two sections.

5 Estimated Wheat Losses in Bulgaria

Numerical results for the estimated wheat losses due to high ozone levels in Bulgaria during a ten-year period (1989 - 1998) are presented in this section.

Table 1. Ozone-caused wheat losses in Bulgaria for 1990

Wheat yield in 1990		Losses of wheat (in thousand tons and %)				
Region	Yield	Basic scenario		Traffic scenario		Savings
1 Sofia City	34.5	12.7	26.9%	8.6	19.9%	4.1
2 Burgas	671.1	231.7	25.7%	161.6	19.4%	70.1
3 Varna	1257.2	430.4	25.5%	300.7	19.3%	129.7
4 Lovech	802.4	285.0	26.2%	199.2	19.9%	85.8
5 Montana	667.9	276.9	29.3%	199.6	23.0%	77.3
6 Plovdiv	277.7	105.6	27.5%	75.8	21.4%	29.8
7 Russe	887.7	350.0	28.3%	253.5	22.2%	96.5
8 Sofia	209.8	79.0	27.4%	56.5	21.2%	22.5
9 Haskovo	483.8	173.6	26.4%	125.1	20.5%	48.5
Whole country	5292.	1945.0	26.9%	1380.6	20.7%	564.4

Table 2. Ozone-caused wheat losses in Bulgaria for 1994

Wheat yield in 1994		Losses of wheat (in thousand tons and %)				
Region	Yield	Basic scenario		Traffic scenario		Savings
1 Sofia City	26.9	7.5	21.8%	5.0	15.5%	2.5
2 Burgas	442.3	105.9	19.3%	74.4	14.4%	31.5
3 Varna	956.4	251.5	20.8%	176.5	15.6%	75.0
4 Lovech	485.1	128.5	20.9%	90.7	15.7%	37.8
5 Montana	469.6	133.1	22.1%	99.3	17.5%	33.8
6 Plovdiv	209.6	58.9	21.9%	43.0	17.0%	15.9
7 Russe	664.0	189.2	22.2%	137.0	17.1%	52.2
8 Sofia	143.4	40.4	22.0%	29.4	17.0%	11.0
9 Haskovo	357.0	95.7	21.1%	71.4	16.7%	24.3
Whole country	3754.	1011.0	21.2%	726.7	16.2%	284.3

Table 3. Ozone-caused wheat losses in Bulgaria for 1998

Wheat yield in 1998		Losses of wheat (in thousand tons and %)				
Region	Yield	Basic scenario		Traffic scenario		Savings
1 Sofia City	32.4	10.4	24.3%	5.8	15.2%	4.6
2 Burgas	452.4	122.1	21.3%	68.0	13.1%	54.1
3 Varna	733.7	219.7	23.0%	121.0	14.2%	98.7
4 Lovech	455.4	151.0	24.9%	87.4	16.1%	63.6
5 Montana	326.2	97.8	23.1%	61.3	15.8%	36.5
6 Plovdiv	217.8	71.4	24.7%	40.3	15.6%	31.1
7 Russe	547.7	182.6	25.0%	111.6	16.9%	71.0
8 Sofia	132.0	42.5	24.4%	24.3	15.6%	18.2
9 Haskovo	315.2	96.9	23.5%	57.3	15.4%	39.6
Whole country	3213.	994.0	23.6%	577.0	15.2%	417.0

The results for 1990, 1994, 1998 are presented in more detail in Tables 1, 2 and 3. In Table 4 the mean values for the ten-year period under consideration are given.

Table 4. Average ozone-caused wheat losses in Bulgaria for the period 1989 – 1998

Average yield (1989–98)		Average losses (in thousand tons and %)			
Region	Yield	Basic scenario		Traffic scenario	Savings
1 Sofia City	21.1	7.4	26.0%	5.1 19.6%	2.3
2 Burgas	485.4	131.0	21.3%	96.4 16.6%	34.6
3 Varna	871.4	253.5	22.5%	186.5 17.6%	67.0
4 Lovech	465.6	141.4	23.3%	103.7 18.2%	37.7
5 Montana	391.5	136.2	25.8%	103.7 20.9%	32.5
6 Plovdiv	237.9	70.8	22.9%	53.2 18.3%	17.6
7 Russe	593.1	199.3	25.2%	150.0 20.2%	49.3
8 Sofia	141.1	42.6	23.2%	31.8 18.4%	10.8
9 Haskovo	356.1	107.9	23.3%	81.6 18.7%	26.3
Whole country	3563.	1090.0	23.4%	812.2 18.6%	277.8

The yield of wheat (in thousand tons) in the Bulgarian regions as well as in the whole country is given in column 2 of these tables. The estimated wheat losses (in thousand tons and %) are given in the third column. The virtual losses in case that the traffic emissions in Europe are reduced by 90 % are given in the next column, and the corresponding savings (in thousand tons) are given in the last column of the tables.

6 Estimated Wheat Losses in Denmark

The wheat losses in Denmark (by regions as well as for the whole country) have also been studied. Instead of tables the losses (in %) for 1990, 1994, 1998 and the average values for the period 1989 – 1998 are presented as plots in Fig. 2. The percentages of losses in Denmark are about twice as small as those in Bulgaria, because of the lower AOT40 values for Denmark. The losses in Denmark seem to be more sensitive to reduction of the traffic emissions, as seen from the results for the Traffic scenario (the right-hand side plots in the figure).

7 Concluding Remarks

The results reported in this paper indicate that the current levels of AOT40 are causing rather big losses of the wheat yield, especially in Bulgaria. The study has been carried out over a time interval of ten years (1989-1998). The amount of losses varies considerably from one year to another. The variations are caused both by the fact that the meteorological conditions are changing from one year to another and by the fact that the European anthropogenic emissions were gradually reduced in the studied ten-year period.

The effect of reduction of the traffic emissions is stronger in Denmark, compared to Bulgaria. This can be explained with the higher gradient of the AOT40

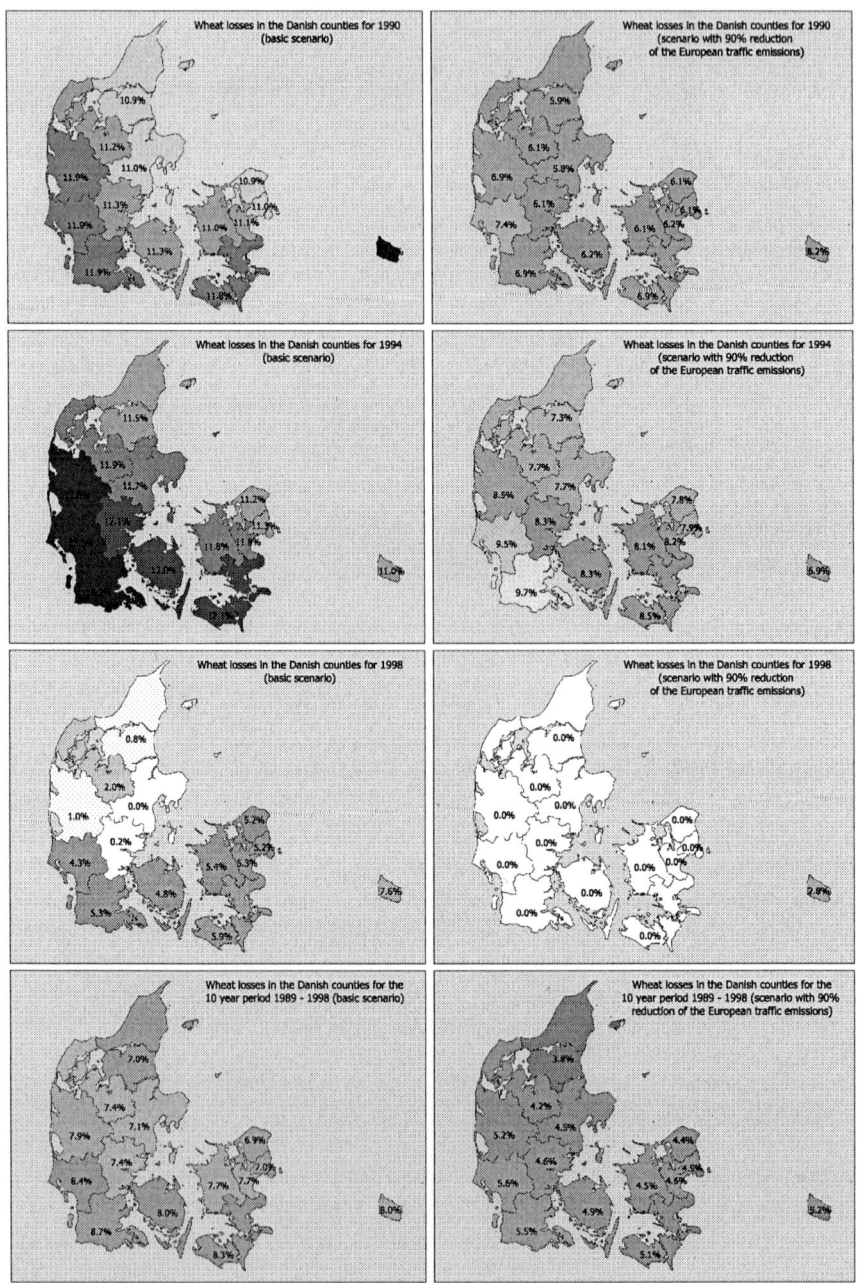

Fig. 2. Wheat losses in the Danish regions under the Basic scenario (left plots) and the Traffic scenario (right plots) for 1990, 1994, 1998, and the average losses for the period 1989–98

values over Denmark due to its location between the zone of high ozone levels around German-Polish border and the zone of clean air over Central Scandinavia.

This work can be generalized in at least two directions. The method and the algorithms described can easily be adjusted to cover other countries or groups of countries, even whole Europe. They can also be used to calculate the total agricultural losses from the ozone exposure, taking into account the various sensitivity of different types of crops. The main obstacles are obtaining the necessary input data as well as the lack of experimental study on the sensitivity of a wider variety of crops to ozone exposure.

Acknowledgments

This research was partially supported by NATO under projects ENVIR.CGR 930449 and OUTS.CGR.960312, by the EU ESPRIT projects WEPTEL (#22727) and EUROAIR (#24618), and by the Ministry of Education and Science of Bulgaria under grants I-811/98 and I-901/99. It is also partly supported by a grant from the Nordic Council of Ministers. A grant from the Danish Natural Sciences Research Council gave us access to all Danish supercomputers.

References

1. M. Amann, I. Bertok, J. Cofala, F. Gyartis, C. Heyes, Z. Kilmont, M. Makowski, W. Schöp and S. Syri. *Cost-effective control of acidification and ground-level ozone.* Seventh Interim Report of IIASA, A-2361 Laxenburg, Austria, 1999.
2. I. Dimov, Tz. Ostromsky, I. Tzvetanov, Z. Zlatev. *Economical Estimation of the Losses of Crops Due to High Ozone Levels.* In Notes on Numerical Fluid Mechanics, Vol.73, 2000, pp. 275–282.
3. *Position paper for ozone.* European Commission, Directorate XI: "Environment, Nuclear Safety and Civil Protection", Brussels, 1999.
4. J. Fuhrer and B. Achermann (eds.). *Critical levels for ozone.* Proc. UN-ECE Workshop on Critical Levels for Ozone, Swiss Federal Research Station for Agricultural Chemistry and Environmental Higyene, Liebefeld-Bern, Switzerland, 1994.
5. J. Fuhrer, L. Skarby and M. R. Ashmore. *Critical levels for ozone effects on vegetation in Europe.* Environmental Pollution, Vol.97, **1-2**, 1997, pp. 91-106.
6. L.Karenlampi and L. Skarby (eds.). *Critical Levels for Ozone in Europe: Testing and Finalizing the Concepts.* Proc. UN-ECE Workshop on Critical Levels for Ozone, University of Kuopio, Finland, 1996.
7. H. Pleijel. *Statistical aspects of critical levels for ozone based on yield reductions in crops.* In "Critical Levels for Ozone in Europe: Testing and Finalizing the Concepts" (L.Karenlampi and L. Skarby, eds.), University of Kuopio, Finland, 1996, pp. 138-150.
8. Statistical Yearbook of Bulgaria, Vol. 90, ...99, Statistical Institute – BAS, Sofia
9. Statistisk Årbog – Danmark, Vol. 90, ...99, Danmarks Statistic, Copenhagen.
10. Z. Zlatev, J. Christensen and Ø. Hov, An Eulerian model for Europe with nonlinear chemistry, J. Atmos. Chem., **15**, 1992, pp. 1-37.
11. Z. Zlatev, I. Dimov and K. Georgiev, Studying long-range transport of air pollutants, Computational Sci. & Eng., **1**, 1994, pp. 45-52.

A Homotopic Residual Correction Process*

V. Y. Pan

Mathematics and Computer Science Department, Lehman College, CUNY
Bronx, NY 10468, USA
vpan@lehman.cuny.edu

Abstract. We present a homotopic residual correction algorithm for the computation of the inverses and generalized inverses of structured matrices. The algorithm simplifies the process proposed in [P92], and so does our analysis of its convergence rate, compared to [P92]. The algorithm promises to be practically useful.

Keywords: residual correction, matrix inverse, Newton' iteration, homotopic algorithms, structured matrices

2000 AMS Math. Subject Classification: 65F10, 65F30

1 Introduction

Residual corection processes for matrix inversion (in particular Newton's iteration) have been known for long time [S33,IK66] but remained unpopular because they involved expensive operations of matrix multiplication in each step. (However, they can be effectively implemented on parallel computers, and they have advantage of converging to the Moore-Penrose generalized inverse where the input matrix is singular.) It was recognized later on [P92,P93,P93a,PZHD97], [PBRZ99,PR00,PRW00,PRW,a,BM,a,P00] that such processes are highly effective in the case of structured input matrices M because multiplication of structured matrices is inexpensive.

It was required, however, to modify the processes in order to preserve the structure, which without special care deteriorates rapidly in the process of the computation. In particular, the displacement rank of a Toeplitz-like matrix can be tripled in each step of Newton's iteration. To counter the problem, it was proposed in [P92,P93,P93a] (cf. also [PBRZ99] and the extensions to non-Toeplitz-like structures in [PZHD97,PR00,PRW00,PRW,a], and [P00]) to recover the structure by periodically zeroing a few smallest singular values of the displacement matrices of the computed approximations to M^{-1}, that is, to rely on the numerical (SVD based) displacement rank.

It was proved in [P92] and [P93] that the truncation of the s smallest singular values of the displacement matrix associated with a computed approximation to M^{-1} increased the residual norm by at most the factor of sn for an $n \times n$

---
* Supported by NSF Grant CCR9732206 and PSC CUNY Award 61393-0030.

matrix M. Such an estimated growth was immaterial where the residual norm was small enough because it was at least squared in each iteration step, but the convergence can be destroyed by the truncation of the singular values unless a sufficiently close initial approximation to M^{-1} is available.

The latter requirement, however, can be partly relaxed based on the homotopy technique proposed in [P92]. The idea is to start with an easily invertible matrix (say, with the identity matrix $M_0 = I$) and to invert recursively the matrices $M_i = t_i M + (1 - t_i) I$ where t_i were monotone increasing from 0 to 1 and where $t_i - t_{i-1}$ were sufficiently small, so that M_{i-1}^{-1} served as a close initial approximation to M_i^{-1}.

In [P92] a variant of this approach was specified and analyzed for the Toeplitz-like input though in a way directed towards asymptotic computational complexity estimates rather than effective practical implementation.

In the present paper, we simplify the latter variant, make it more convenient for the implementation, and generalize the choice of the intial approximation to cover non-Toeplitz-like structures as well.

Our attention is to the design and the study of a homotopic process, where the residual correction algorithm is used as a black box subroutine (based, e.g., on the algorithms of [PBRZ99] in the Toeplitz or Toeplitz-like cases), for which we just supply the required input matrices M_i and M_{i-1}^{-1} and obtain the output matrix M_i^{-1}. We describe this process for a general real symmetric (or Hermitian) positive definite matrix M, though the promising applications are where M is a Toeplitz, Toeplitz-like, or another structured matrix, because in the latter case the residual correction process is most effective [P92,P93,P93a,PBRZ99,PR00,PRW00,PRW,a,BM,a,P00].

The required positive bound θ on the initial residual norm is in our hands – we may choose it as small as we like, but the number of the required homotopic steps is roughly proportional to $1/\theta$ (see section 4).

Our analysis shows the efficacy of the proposed approach.

We organize the presentation as follows. In section 2, we briefly recall the residual correction process. In section 3, we describe our basic homotopic approach. In section 4, we estimate the number of its required homotopic steps and comment on the choice of θ. In section 5, we generalize its initial step and analyze the generalized process.

2 Residual Correction Processes

A crude initial approximation to the inverse of a real symmetric (or Hermitian) positive definite matrix M can be rapidly improved by the method of residual correction:

$$X_{i+1} = X_i \sum_{k=0}^{p-1} R_i^k, \quad i = 0, 1, \ldots, \qquad (1)$$

where we write
$$R_i = R(M, X_i) = I - MX_i. \qquad (2)$$
(1) and (2) imply that
$$R_h = (R_0)^{p^h}, \quad h = 1, 2, \ldots, \qquad (3)$$
which shows the order of p convergence of process (1) to the matrix M^{-1} provided that
$$\rho(R_0) = \|R_0\|_2 \le \theta < 1, \quad R_0 = R(M, X_0), \qquad (4)$$
for a fixed real θ. To optimize both the computational work per step (1) and the number of steps required to ensure the desired upper bound on $\rho(R_h)$, one should choose $p = 3$ [IK66], pages 86-88, but even better results can be reached by using scaled process (1) for $p = 2$, which is actually Newton's scaled process:
$$X_{i+1} = c_{i+1} X_i (I + R_i), \qquad (5)$$
for appropriate scalars c_{i+1} [PS91]. We will apply algorithms (1) or (5) as black box subroutines in a homotopic process, which starts with the trivial inversion of the identity matrix I, ends in computing M^{-1}, and fulfils (4) at every intermediate step. We recall that the residual correction algorithms are strongly stable numerically (even when the input matrix M is singular [PS91]) and our process inherits this property.

3 A Homotopic Residual Correction Process

Let spectrum$(M) = \{\lambda_1, \ldots, \lambda_n\}$, where
$$\lambda_1^+ \ge \lambda_1 \ge \lambda_2 \ge \cdots \ge \lambda_n \ge \lambda_n^- > 0 \qquad (6)$$
and where λ_1^+ and λ_n^- are known values. Let us write $M_0 = M + t_0 I$, $t_0 = \lambda_1^+/\theta$, $0 < \theta < 1$. Then
$$R(M_0, t_0^{-1} I) = I - t_0^{-1} M_0 = t_0^{-1} M, \quad \rho(R(M_0, t_0^{-1} I)) = \|R(M_0, t_0^{-1} I)\|_2 \le \theta,$$
and M_0 is inverted rapidly by processes (1), (5). Let us further write
$$M_{h+1} = t_{h+1} I + M = M_h - \Delta_h I, \quad \Delta_k = t_h - t_{h+1} > 0, \quad h = 0, 1, \ldots. \qquad (7)$$
Then we have
$$R(M_{h+1}, M_h^{-1}) = \Delta_h M_h^{-1},$$
$$r_{h+1} = \|R(M_{h+1}, M_h^{-1})\|_2 \le \Delta_h \|M_h^{-1}\|_2 \le \Delta_h/(t_h + \lambda_n^-).$$
We choose
$$\Delta_h = (t_h + \lambda_n^-)\theta, \quad h = 1, 2, \ldots, H - 1, \qquad (8)$$
which implies that $r_{h+1} \le \theta$ for all h, and we recursively invert the matrices M_{h+1} by applying process (1), (5) as long as t_{h+1} remains positive. As soon as we arrive at $t_H \le 0$, we invert M instead of M_H.

Remark 1. The requirement of nonsingularity of M can be relaxed if we simply replace (6) by the requirement that

$$\lambda_1^+ \geq \lambda_1 \geq \cdots \geq \lambda_r^+ > \lambda_r^- > 0, \quad \lambda_i = 0 \text{ for } i > r = \operatorname{rank} M.$$

The only resulting changes in the homotopic process is that its convergence will be to the Moore-Penrose generalized inverse of M, and λ_r^- will replace λ_n^- in (8) and in our subsequent estimates for the number of homotopic steps.

Remark 2. The approach allows variations. For instance, instead of (7), we may apply the following dual process:

$$M_{h+1} = I + t_{h+1}M = M_h + (t_{h+1} - t_h)M, \quad h = 0, 1, \ldots,$$

followed at the end by a single step (7) or a few steps (7). The resulting computations can be analyzed similarly to (7).

4 Estimating the Number of Homotopic Steps

By (7) and (8) we have, $t_{h+1} = (1-\theta)t_h - \theta\lambda_n^-$, $h = 1, \ldots, H-1$. Therefore,

$$t_1 = (1-\theta)t_0 - \theta\lambda_n^-,$$
$$t_2 = (1-\theta)t_1 - \theta\lambda_n^- = (1-\theta)^2 t_0 - ((1-\theta) + 1)\theta\lambda_n^-,$$
$$t_3 = (1-\theta)t_2 - \theta\lambda_n^- = (1-\theta)^3 t_0 - ((1-\theta)^2 + (1-\theta) + 1)\theta\lambda_n^-,$$

and recursively, we obtain that

$$t_h = (1-\theta)^h t_0 - \sum_{i=0}^{h-1}(1-\theta)^i \theta\lambda_n^- = (1-\theta)^h t_0 - (1-(1-\theta)^h)\lambda_n^-, \quad h = 1, 2, \ldots .$$

We have $t_H \leq 0$ if $(1-\theta)^H t_0 \leq (1-(1-\theta)^H)\lambda_n^-$. Substitute $t_0 = \lambda_1^+/\theta$ and rewrite the latter bound as follows:

$$\lambda_1^+/(\theta\lambda_n^-) \leq \frac{1}{(1-\theta)^H} - 1,$$

$$\frac{1}{(1-\theta)^H} \geq \lambda_1^+/(\theta\lambda_n^-) + 1,$$

$$H \geq -\log(1 + \lambda_1^+/(\theta\lambda_n^-))/\log(1-\theta).$$

We choose the minimum integer H satisfying this bound, that is,

$$H = \lceil \log(1 + \lambda_1^+/(\theta\lambda_n^-))/\log(1/(1-\theta)) \rceil.$$

The scaled Newton's iteration of [PS91] yields the bound of roughly $\log(\lambda_1^+/\lambda_n^-)$ on the overall number of steps (5) for $p = 2$, which is superior to the above bound on H because each homotopic step generally requires a few steps (5). Our homotopic process, however, has an important advantage in application to structured matrices M, as we explained in the introduction. By the latter estimate, we should choose a larger θ to decrease the number of the homotopic steps, H, but in applications to the inversion of structured matrices we should keep θ small enough to preserve the convergence under the truncation of the smallest singular values. In the Toeplitz-like case, recall that such a truncation increases the residual norm by at most the factor of sn (see [P93] and our section 1) and conclude that the choice of the value $\theta = 0.5/(sn)^2$ is clearly sufficient. The bound sn, however, is overly pessimistic according to the extensive numerical experiments reported in [BM,a]. Thus the choice of much larger values of θ should be sufficient, and heuristics seem to be most appropriate here.

5 Unified Initialization Rule for the Homotopic Process

The initial choice of $M_0 = M + t_0 I$ preserves the Toeplitz-like structure of M but may destroy some other matrix structures such as Cauchy-like or Vandermonde-like ones. This choice, however, can be generalized as follows:

Choose a real symmetric (or Hermitian) positive definite and well conditioned matrix M_0. Let it also be readily invertible and let it share its structure with the input matrix M, so that the matrices $tM_0 + M$ are structured for any scalar t. Recursively define the matrices

$$M_{h+1} = t_{h+1} M_0 + M = M_h + (t_{h+1} - t_h) M_0, \quad h = 0, 1, \ldots, H-1,$$

where $t_1 = 1 > t_2 > \cdots > t_{H-1} > t_H = 0$, and write $t_0 = 0$.

Now, let spectrum$(M_0) = \{\mu_1, \ldots, \mu_n\}$, where $\mu_1^+ \geq \mu_2 \geq \cdots \geq \mu_n \geq \mu_n^- > 0$, and μ_1^+ and μ_n^- are available. Let us write $\kappa^+ = \mu_1^+/\mu_n^-$, recall that $\|M_h^{-1}\|_2 \leq 1/(t_h \mu_n^- + \lambda_n^-)$ for all h (cf., e.g., [Par80], p.191), $\|M_0\|_2 = \mu_1 \leq \mu_1^+$, and deduce that

$$\|I - (t_1 M_0)^{-1} M_1\|_2 \leq \|M_0^{-1} M/t_1\|_2 \leq \|M_0^{-1}\|_2 \|M\|_2/t_1 \leq \lambda_1^+/(t_1 \mu_n^-).$$

Now, we choose

$$t_1 = \lambda_1^+/(\mu_n^- \theta), \tag{9}$$

so that $\|I - (t_1 M_0)^{-1} M_1\|_2 \leq \theta$, and we invert M_1 by applying processes (1) or (5) for $X_0 = t_1 M_0$.

Next we deduce that

$$I - M_h^{-1} M_{h+1} = (t_h - t_{h+1}) M_h^{-1} M_0,$$
$$\|I - M_h^{-1} M_{h+1}\|_2 \leq (t_h - t_{h+1}) \|M_h^{-1}\|_2 \|M_0\|_2.$$

Substitute

$$\|M_h^{-1}\|_2 \leq 1/(t_h \mu_n^- + \lambda_n^-),$$
$$\|M_0\|_2 \leq \mu_1^+,$$

and obtain that $||I - M_h^{-1}M_{h+1}||_2 \le \theta$ if $(t_h - t_{h+1})\mu_1^+/(t_h\mu_n^- + \lambda_n^-) \le \theta$ or, equivalently, if $t_{h+1} \ge t_h(1 - \theta/\kappa^+) - \theta\lambda_n^-/\mu_1^+$.

Thus, we choose

$$t_{h+1} = t_h(1 - \theta/\kappa^+) - \theta\lambda_n^-/\mu_1^+. \tag{10}$$

and invert M_{h+1} by applying processes (1) or (5) for $X_0 = M_h$ and for $h = 1, 2, \ldots, H - 2$, until t_{h+1} of (10) becomes nonpositive for $h = H - 1$. By (9) and (10), this must occur for $H \le 1 + \lceil (\log t_1)/\log(1 - \theta/\kappa^+) \rceil$ and t_1 of (9).

References

BM,a.   D. A. Bini, B. Meini, Approximate Displacement Rank and Applications, preprint.

IK66.   E. Issacson, H. B. Keller, *Analysis of Numerical Methods*, Wiley, New York, 1966.

P92.   V. Y. Pan, Parallel Solution of Toeplitz-like Linear Systems, *J. of Complexity*, **8**, 1-21, 1992.

P93.   V. Y. Pan, Decreasing the Displacement Rank of a Matrix, *SIAM J. Matrix Anal. Appl.*, **14, 1**, 118-121, 1993.

P93a.   V. Y. Pan, Concurrent Iterative Algorithm for Toepliz-like Linear Systems, *IEEE Trans. on Parallel and Distributed Systems*, **4, 5**, 592-600, 1993.

P00.   V. Y. Pan, Superfast Computations with Structured Matrices: Unified Study, preprint, 2000.

Par80.   B. N. Parlett, *The Symmetric Eigenvalue Problem*, Prentice-Hall, Englewood Cliffs, NJ, 1980.

PBRZ99.   V. Y. Pan, S. Branham, R. Rosholt, A. Zheng, Newton's Iteration for Structured Matrices and Linear Systems of Equations, *SIAM volume on Fast Reliable Algorithms for Matrices with Structure*, SIAM Publications, Philadelphia, 1999.

PR00.   V. Y. Pan, Y. Rami, Newton's iteration for the Inversion of Structured Matrices, *Structured Matrices: Recent Developments in Theory and Computation*, edited by D.Bini, E. Tyrtyshnikov and P. Yalamov, Nova Science Publishers, USA, 2000.

PRW00.   V. Y. Pan, Y. Rami, X. Wang, Newton's iteration for the Inversion of Structured Matrices, Proc.14th Intern. Symposium on Math. Theory of Network and Systems (MTNS'2000), June 2000.

PRW,a.   V. Y. Pan, Y. Rami, X. Wang, Structured Matrices and Newton's Iteration: Unified Approach, preprint.

PS91.   V. Y. Pan, R. Schreiber, An Improved Newton Iteration for the Generalized Inverse of a Matrix, with Applications, *SIAM J. on Scientific and Statistical Computing*, **12, 5**, 1109-1131, 1991.

PZHD97.   V. Y. Pan, A. Zheng, X. Huang, O. Dias, Newton's Iteration for Inversion of Cauchy-like and Other Structured Matrices, *J. of Complexity*, **13**, 108-124, 1997.

S33.   G. Schultz, Iterative Berechnung der Reciproken Matrix, *Z. Angew. Meth. Mech.*, **13**, 57-59, 1933.

Parallel Monte Carlo Methods for Derivative Security Pricing

Giorgio Pauletto

Department of Econometrics, University of Geneva,
40, boulevard du Pont-d'Arve, CH–1211 Genève-4, Switzerland
Giorgio.Pauletto@metri.unige.ch

Abstract. Monte Carlo (MC) methods have proved to be flexible, robust and very useful techniques in computational finance. Several studies have investigated ways to achieve greater efficiency of such methods for serial computers. In this paper, we concentrate on the parallelization potentials of the MC methods. While MC is generally thought to be "embarrassingly parallel", the results eventually depend on the quality of the underlying parallel pseudo-random number generators. There are several methods for obtaining pseudo-random numbers on a parallel computer and we briefly present some alternatives. Then, we turn to an application of security pricing where we empirically investigate the pros and cons of the different generators. This also allows us to assess the potentials of parallel MC in the computational finance framework.

1 Introduction

The Monte Carlo (MC) method [18,10] is widely applied to large and complex problems to obtain approximate solutions. This method has been successfully applied to problems in physical sciences and, more recently, in finance. Many difficult financial engineering problems such as the valuation of multidimensional options, path-dependent options, stochastic volatility or interest rate options can be tackled thanks to this technique.

An option (also called derivative security) is a security the payoff of which depends on one or several other underlying securities. The prices of these underlying securities are often modeled as continuous-time stochastic processes. Assuming that no arbitrage exists, one can show that the price of such an option is the discounted expected value of the payoffs under the risk neutral measure, see e.g. [5]. In such a framework, pricing an option that can be written as an expectation of a random variable lends itself naturally to a numerical procedure that estimates this expected value through simulation.

Generally, the MC procedure involves generating a large number of realizations of the underlying process and, using the law of large numbers, estimating the expected value as the mean of the sample. In our framework this translates into Algorithm 1.

We note that the standard deviation of the MC estimation $\hat{C}$ decreases at the order $O(1/\sqrt{N})$ and thus that a reduction of a factor 10 requires an increase of the number of simulation runs N of 100 times.

Algorithm 1 Monte Carlo

1: for j=1 to N do
2:   Simulate sample paths of the underlying variables (asset prices, interest rates, etc.) using the risk neutral measure over the time frame of the option. For each simulated path, evaluate the discounted cash flows of the derivative C_j
3: end for
4: Average the discounted cash flows over the sample paths $\hat{C} = \dfrac{1}{N}\sum_{j=1}^{N} C_j$
5: Compute the standard deviation $\hat{\sigma}_{\hat{C}} = \sqrt{\dfrac{1}{(N-1)}\sum_{j=1}^{N}(C_j - \hat{C})^2}$

The major advantage of the MC approach is that it easily accommodates options with complex payoff functions. Asian options and lookback options are two typical examples of path dependent options. The Asian option depends on the average value of the underlying asset price and the lookback option depends on the maximum or minimum of the underlying asset. In such cases, analytic formulas do not always exist and are difficult to construct or to approximate. However, it is straightforward to adapt the MC procedure to price these options by changing the payoff function.

Monte Carlo can also be helpful when considering the valuation of multi-asset options, i.e. options depending on several underlying securities, such as for instance index options, basket options or options on the extremum of several assets. As mentioned earlier, the price of the option can be expressed as an expectation, which is in this case a multidimensional integral. The higher dimension of the problem very quickly becomes a limiting factor with other methods, since the complexity of the computations generally grows exponentially with the dimension. In this case too, MC is essentially the method of choice, since its complexity does not depend on the dimension of the problem.[1]

Since the principal drawback of the MC method is its slow convergence, different strategies have been devised to speed up the process. Variance reduction techniques, such as antithetic variates, control variables, stratified and importance sampling, can be applied. More recently, the use of low discrepancy sequences has also helped in certain cases. Several papers have described and analyzed the use of Monte Carlo techniques in finance [1,11,3]. Improvements in the efficiency using variance reduction techniques is thoroughly discussed in [2].

The MC method relies on the use of a pseudo-random number generator (RNG) to produce its results. The generation of random numbers is known to be difficult since deterministic algorithms are used to obtain "random" quantities. Bad RNGs are detrimental to MC simulations. The sequence of an ideal RNG should:

---

[1] The only assumption is that the function should be square integrable, which usually is not a very stringent condition.

- be uniformly distributed,
- not be correlated,
- pass statistical tests for randomness,
- have a long period,
- be reproducible,
- be fast,
- be portable,
- require limited memory.

It is difficult for an RNG to satisfy all these qualities simultaneously. Two main approaches are used to assess the quality of an RNG: a theoretical study of the properties of the sequence and a statistical test of the sequence. Yet, certain generators that perform well in such studies may prove unreliable in certain MC applications, see [9,20].

The generation of random numbers on parallel computers is usually worse than in the serial case. The streams of numbers produced by different processors could be correlated, which is referred to as *inter-processor correlation*. Such a situation would not appear in the serial case.

In this paper, we will present some of the issues related to the generation of random numbers on parallel machines. In a second part, a parallel application of MC to pricing derivative securities will show some of the problems one may encounter. This complements the more general analysis that can be found on parallel RNGs, since even generators that perform well in the standard tests may prove unreliable in certain applications, particularly in MC simulations.

2 Serial Random Number Generators

Generating random numbers using computers is a difficult topic, but many studies that help better understand the issues involved have been carried out [13,14]. The most commonly used RNG is the *linear congruential generator* (LCG). It is based on the recurrence $y_n = (ay_{n-1} + c) \mod m$ where $m > 0$ is the modulus, $a > 0$ the multiplier and c the additive constant. It is usually denoted $\text{LCG}(m, a, c, y_0)$ where y_0 is the seed. This produces integers in the range $(0, m)$ and to obtain random numbers uniformly distributed in the interval $(0, 1)$, one usually divides these integers by m, i.e. $x_n = y_n/m$.

These numbers also cycle after at most m steps. When the parameter $b = 0$, this RNG is sometimes called a *multiplicative linear congruential generator* and is denoted by MLCG. For appropriately chosen parameters, these RNGs produce a sequence of numbers of maximal period, see [13,19]. There is no unique and undisputed choice of the parameters that guarantees a sequence with maximal period and has good theoretical and statistical properties. For 32-bit machines the choice $\text{LCG}(2^{31} - 1, 16807, 0, 1)$ proposed in [19], also known as MINSTD for *minimal standard*, is a popular one.

The main drawback of LCGs is that the numbers produced have a lattice structure that affects MC simulations [13]. The d-tuples $(x_i, \ldots, x_{i+d-1})$ lie on parallel hyperplanes of the unit hypercube. Since the gaps between the planes are never sampled, the numbers produced can be correlated; furthermore, the higher the dimension d, the worse the problem becomes. This bad behavior can be detected through the *spectral test* and the LCG parameters should be such that the distance between the planes is minimized [15].

Many LCGs use a modulus which is a power of 2, since this allows easier programming and faster execution of binary computers. For instance, the generator employed by the ANSI C language BSD version called by the drand48 function, has a modulus of 2^{48}. This generator is actually exactly described by the parameterization $\text{LCG}(2^{48}, 25214903917, 11, 0)$. Power of 2 moduli have deficiencies since they produce random numbers that have highly correlated low order bits and that can show long-range correlations [6]. LCGs with prime modulus have better randomness properties, but they are more difficult to implement.

Another type of linear generator of interest is the *multiple-recursive generator* (MRG) proposed by L'Ecuyer. It generalizes the MLCG generator by adding k terms in the recurrence $y_n = (a_1 y_{n-1} + a_2 y_{n-2} + \cdots + a_k y_{n-k}) \bmod m$. The coefficients $(a_i)_{i=1}^{k}$ are integers in the range $[-(m-1), (m-1)]$. The period and randomness are generally much improved compared with an MRG at the cost of an increase of computation time.

It is possible to combine such generators to produce sequences that are equivalent to an MRG with very large moduli and therefore very large periods. Details and floating point implementations in C for 32-bit about these generators can be found in [15]. The specific generator MRG32k3a has period length $2^{191} \approx 10^{57}$, whereas MRG32k5a has period length $2^{319} \approx 10^{96}$.

3 Parallel Random Number Generators

As mentioned earlier, a parallel random number generator (PRNG) should have extra qualities. The PRNG should also:
- have the same qualities as serial RNG on one processor,
- show no inter-processor correlation of the streams,
- generate the same stream of numbers for a different number of processors,
- work for any number of processors,
- keep the communication between processors to the minimum.

The generation of random numbers on a parallel computer can be based upon a serial RNG by distributing the numbers produced among the processors. A more modern approach is to parameterize the RNG differently on each processor so that different streams of numbers are generated, see [7] for a survey.

3.1 Leapfrog

The leapfrog method distributes the numbers of a serial RNG in a cyclic fashion to each processor, like a deck of cards dealt to players. If we denote by $(x_i)_{i=0,1,2,\ldots}$ the original sequence and L the lag, then the subsequence processor p gets is

$$\tilde{x}_i = x_{iL+p} \quad \text{with} \quad p = 0, 1, 2, \ldots, P \leq L-1 \ .$$

If the original sequence is

$$x_0, x_1, \ldots, x_{L-1}, x_L, x_{L+1}, \ldots, x_{2L-1}, x_{2L}, x_{2L+1}, \ldots$$

then the subsequence obtained by processor 0 is

$$\boxed{x_0}, x_1, \ldots, x_{L-1}, \boxed{x_L}, x_{L+1}, \ldots, x_{2L-1}, \boxed{x_{2L}}, x_{2L+1}, \ldots \ .$$

A first problem is that long-range correlations embedded in the RNG can become short-range correlations in the new sequence and destroy the quality of the PRNG see [8].

Secondly, such a scheme is not scalable since when the total number of processors P increases, the length of the sequence $(\tilde{x}_i)_{i=0,1,2\ldots}$ decreases.

For this method, we need to easily jump ahead L steps to get the next random number. This can be carried out with an MLCG since we have

$$y_n = ay_{n-1} \bmod m = (a^n \bmod m) y_0 \bmod m$$

$$y_{iL+p} = (a^L \bmod m)^i y_p \bmod m \ .$$

This shows that the sequence used in the processors has now multiplier a^L instead of a and we cannot ensure that this will be a multiplier with good properties for all values of L. Jumping forward in the sequence can also be done for LCGs.

3.2 Sequence Splitting

In this case, the original sequence is split into blocks and distributed to each processor. Let us denote the period of the generator by ρ, the number of processors by P and the block length by $L = \lfloor \rho/P \rfloor$, we have

$$\hat{x}_i = x_{pL+i} \quad p = 0, 1, 2, \ldots, P \ .$$

Then the original sequence

$$x_0, x_1, \ldots, x_{L-1}, x_L, x_{L+1}, \ldots, x_{2L-1}, x_{2L}, x_{2L+1}, \ldots$$

is distributed as follows to processors 0,1,2, ...

$$\boxed{x_0, x_1, \ldots, x_{L-1}}, \boxed{x_L, x_{L+1}, \ldots, x_{2L-1}}, \boxed{x_{2L}, x_{2L+1}, \ldots, x_{3L-1}}, \ldots \ .$$

For this method, long-range correlations can be emphasized and become inter-processor correlations. We know that the sequences produced will not overlap, but cannot be sure that they will not show some correlation. This may again adversely affect the MC simulations see [4,8]. Scalability is an issue once again, as in the previous case.

We need to be able to jump ahead by P steps to get to the new starting point for each processor. This can be done with an MLCG by using a different seed for each processor (see [16])

$$y_n = ay_{n-1} \bmod m$$

$$y_{pL} = (a^{pL} \bmod m) y_0 \quad \bmod m$$

$$y_{pL+i} = (a^i \bmod m) y_{pL} \bmod m \ .$$

3.3 Parameterization

A more recent approach to generate parallel streams of random numbers is based on a parameterization of each stream. This can be done in two ways: in certain generators, the seed value provides a natural way of dividing the sequence of an RNG into independent cycles; the function that outputs the next value in the sequence can be parameterized to give a different stream for a different value.

These ideas are developed in e.g. [17] and implemented in the free package SPRNG available at <http://daniel.scri.fsu.edu/RNG/>. This library of programs contains several RNGs that can be used in parallel and are scalable. The different generators are the following:
- Modified additive lagged Fibonacci generator,
- Multiplicative lagged Fibonacci generator,
- 48-bit linear congruential generator with prime addend,
- 64-bit linear congruential generator with prime addend,
- Combined multiple recursive generator,
- Prime modulus linear congruential generator (requires special multi-precision library).

The authors of SPRNG provide a large number of tests and sound theoretical background for this package.

4 Parallel Monte Carlo Option Pricing

The prices of derivative securities, such as options, are often found analytically by imposing simplifying assumptions. More recently the advent of powerful numerical procedures and computers has made possible the princing of more complex and more realistic derivatives.

As explained in the introduction, the representation of an option price as an expectation naturally provides a way to evaluate the price via MC simulation. The dimension of the integral depends on the number of underlyings and can become large.

The goal of parallel programming is generally to speed up the computation. Two important concepts are the *speed-up* defined as $S_p = T_1/T_p$ where T_1 is the serial execution time and T_p is the parallel execution time using p processors; the efficiency $E_p = S_p/p$ is the proportion of the time devoted to performing useful computational work and ranges from 0 to 1. In the best case, the speed up is linear in the number of processors used and the efficiency stays constant and close to 1. The problem is said to be *scalable* if the efficiency can be kept constant when increasing the problem size together with the number of processors.

In an MC simulation, no communication takes place if the RNG is well designed. Therefore the algorithm scales perfectly and adding a processor will generally decrease the computation time. However, if correlations appear in the computations of the random numbers, the results may be biased, see [8].

4.1 Description of the Option

The problem we investigate is the pricing of multi-asset options, i.e. options depending on several underlying assets. We will in particular consider the pricing of a European call option on the maximum of n risky assets. Even though closed form solutions exist [12], the computations quickly become burdensome when the dimension increases.

The underlying assets have prices $S_1(t), S_2(t), \ldots, S_n(t)$ at time $t = 0, \ldots, T$ and the respective strike prices are $K_1, K_2, \ldots, K_n$. We also assume the usual lognormal diffusion process

$$\frac{dS_i}{S_i} = \mu_i + \sigma_i dZ_i \quad i = 1, 2, \ldots, n ,$$

where μ_i and σ_i denote respectively the expected rate of return and volatility and dZ_i is the Wiener process for asset i. These processes can be correlated and ρ_{ij} denotes the correlation coefficient between dZ_i and dZ_j.

The price of the call at maturity time T is

$$C(T) = \max \{ \max \left(S_1(T) - K_1, S_2(T) - K_2, \ldots, S_n(T) - K_n \right), 0 \}$$

and what we look for is the value of this option at time 0, $C(0)$.

The steps for pricing such options with MC are described in Algorithm 2.

Algorithm 2 Monte Carlo pricing of a multi-asset option

```
1: Decompose the correlation matrix with Cholesky Σ = LL'
2: for j=1 to N do
3:     Generate a n dimensional vector of unit normal random values z
4:     Transform z̃ = Lz
5:     Compute the discounted cash flows of the derivative C_j
6: end for
```
7: Average the discounted cash flows over the sample paths $\hat{C} = \dfrac{1}{N} \sum_{j=1}^{N} C_j$

8: Compute the standard deviation $\hat{\sigma}_{\hat{C}} = \sqrt{\dfrac{1}{(N-1)} \sum_{j=1}^{N} (C_j - \hat{C})^2}$

As one can see the parallel part in this computation is only offered in the main MC loop since the random variables have to be combined into a multivariate normal vector. This is in contrast with other studies such as [8] that distribute the computation along the dimension n.

We nonetheless expect to show that long-range correlations among multiple parallel streams from LCGs produce spurious results when using consecutive blocks. The use of the SPRNG package should resolve the problem since one can generate many non correlated streams on different processors. The package implements the algorithms so that they are scalable which should also remove a second drawback of the splitting or blocking schemes.

References

1. Boyle, P. (1977). "Options: a Monte Carlo approach", *Journal of Financial Economics* 4, 323–338.
2. Boyle, P., M. Broadie and P. Glasserman (1997). "Monte Carlo methods for security pricing", *Journal of Economic Dynamics and Control* 21, 1267–1321.
3. Carverhill, A. and K. Pang (1995). "Efficient and flexible bond option valuation in the Heath, Jarrow and Morton framework", *Journal of Fixed Income* 5, 70–77.
4. Coddington, P. D. (1994). "Analysis of random number generators using Monte Carlo simulation", *International Journal of Modern Physics* C 5, 547.
5. Cox, J. C. and S. A. Ross (1976). "The valuation of options for alternative stochastic processes", *Journal of Financial Economics* 3, 145–166.
6. De Matteis, A., J. Eichenauer-Herrmann and H. Grothe (1992). "Computation of critical distances within multiplicative congruential pseudorandom number sequences", *Journal of Computational and Applied Mathematics* 39, 49–55.
7. Eddy, W. F. (1990). "Random number generators for parallel processors", *Journal of Computational and Applied Mathematics* 31, 63–71.
8. Entacher, K., A. Uhl and S. Wegenkittl (1999). "Parallel random number generation: long-range correlations among multiple processors", in P. Zinterhof, M. Vajteršic and A. Uhl (eds) *ACPC'99, LNCS 1557*, Springer-Verlag, 107–116.
9. Ferrenberg, A. M. and Landau D. P. Landau (1992). "Monte Carlo simulations: hidden errors from 'good' random number generators", *Physical Review Letters* 69 (23), 3382–3384.
10. Hammersley, J. M. and D. C. Handscomb (1964). *Monte Carlo Methods*, Methuen's Monographs and Applied Probability and Statistics, Wiley, New York, NY.
11. Hull, J. and A. White (1987). "The pricing of options on assets with stochastic volatilities", *Journal of Finance* 42, 281–300.
12. Johnson, H. (1987). "Option on the maximum or the minimum of several assets", *Journal of Financial and Quantitative Analysis* 22, 227–283.
13. Knuth, D. E. (1981). *The Art of Computer Programming, Volume 2: Seminumerical Algorithms*, 2nd edition, Addison-Wesley, Reading, MA.
14. L'Ecuyer, P. (1998). "Random number generation", in J. Banks (ed.) *Handbook on Simulation*, Chapter 4, Wiley, New York, NY.
15. L'Ecuyer, P. (1999). "Good Parameter Sets for Combined Multiple Recursive Random Number Generators", *Operations Research*, 47, 1, 159–164.
16. L'Ecuyer, P. and P. Côté (1991). "Implementing a random number package with splitting facilities", *ACM Transactions on Mathematical Software* 17 (1), 98–111.
17. Mascagni M. (1997). "Some Methods of Parallel Pseudorandom Number Generation", in *Proceedings of the IMA Workshop on Algorithms for Parallel Processing*, R. Schreiber, M. Heath and A. Ranade (eds), Springer Verlag, New York.
18. Metropolis, N. and S. Ulam (1949). "The Monte Carlo Method", *Journal of the American Statistical Association* 247 (44), 335–341.
19. Park, S. K. and K. W. Miller (1988). "Random number generators: good ones are hard to find", *Communications of the ACM* 31 (10), 1192–1201.
20. Vattulainen, I., T. Ala-Nissila and K. Kankaala (1995). "Physical models as tests of randomness", *Physical Review E* 52 (3), 3205–3214.

Stability of a Parallel Partitioning Algorithm for Special Classes of Banded Linear Systems*

Velisar Pavlov

Center of Applied Mathematics and Informatics, University of Rousse
7017 Rousse, Bulgaria
velisar@ami.ru.acad.bg

Abstract. The main results of a componentwise error analysis for a parallel partitioning algorithm [7] in the case of banded linear systems are presented. It is shown that for some special classes of matrices, i.e. diagonally dominant (row or column), symmetric positive definite, and M-matrices, the algorithm is numerically stable. In the case when the matrix of the system does not belong to the considered classes is presented a stabilized version of the algorithm.

1 Introduction

A well-known algorithm for solving tridiagonal systems in parallel is the method of Wang [7]. Full roundoff error analysis of this algorithm can be found in [8].

A generalized version of this parallel partitioning algorithm later has been applied from the other authors [2,5]. Backward componentwise error analysis of this generalized version can be found in [9]. In this work are obtained bound on the equivalent perturbations depending on three constants and then are presented bound on the forward error as well depending on two types of condition numbers.

In the present work we consider more precisely the case when matrix of the system belongs to one of the following classes: diagonally dominant, symmetric positive definite, or M-matrices.

First, we present a brief description of the algorithm (for banded systems only). Let the linear system under consideration be denoted by

$$Ax = d, \qquad (1)$$

where $A \in \mathcal{R}^{n \times n}$, which bandwidth is $2j + 1$. For simplicity we assume that $n = ks - j$ for some integer k, if s is the number of the parallel processors we want to use. We partition matrix A and the right hand side d of system (1) as

---

* This work was supported by Grants MM-707/97 and I-702/97 from the National Scientific Research Fund of the Bulgarian Ministry of Education and Science.

follows:

$$\begin{pmatrix} B_1 & \bar{c}_1 & & & & & & \\ a_k & b_k & c_k & & & & & \\ \bar{a}_2 & B_2 & \bar{c}_2 & & & & & \\ & a_{2k} & b_{2k} & c_{2k} & & & & \\ & & \ddots & \ddots & \ddots & & & \\ & & & \bar{a}_{s-1} & B_{s-1} & \bar{c}_{s-1} & & \\ & & & & a_{(s-1)k} & b_{(s-1)k} & c_{(s-1)k} & \\ & & & & & & \bar{a}_s & B_s \end{pmatrix} \begin{pmatrix} X_1 \\ x_k \\ X_2 \\ x_{2k} \\ \vdots \\ X_{s-1} \\ x_{(s-1)k} \\ X_s \end{pmatrix} = \begin{pmatrix} D_1 \\ d_k \\ D_2 \\ d_{2k} \\ \vdots \\ D_{s-1} \\ d_{(s-1)k} \\ D_s \end{pmatrix},$$

where $B_i \in \mathcal{R}^{(k-j) \times (k-j)}$ are band matrices with the same bandwidth as matrix A, $\bar{a}_i, \bar{c}_i \in \mathcal{R}^{(k-j) \times j}$, $a_{ik}, b_{ik}, c_{ik} \in \mathcal{R}^{j \times j}$, $X_i, D_i \in \mathcal{R}^{(k-j) \times 1}$, $x_{ik}, d_{ik} \in \mathcal{R}^{j \times 1}$.

After suitable permutation of the rows and columns of matrix A we obtain the system

$$\mathcal{A}\mathcal{P}x = \mathcal{P}d, \quad \mathcal{A} = \mathcal{P}A\mathcal{P}^T = \begin{pmatrix} A_{11} & A_{12} \\ A_{21} & A_{22} \end{pmatrix},$$

where $\mathcal{P}$ is a permutation matrix, $A_{11} = \text{diag}\{B_1, B_2, \ldots, B_s\} \in \mathcal{R}^{s(k-j) \times s(k-j)}$, $A_{22} = \text{diag}(b_k, b_{2k}, \ldots, b_{(s-1)k}) \in \mathcal{R}^{j(s-1) \times j(s-1)}$, and $A_{12} \in \mathcal{R}^{s(k-j) \times j(s-1)}$, $A_{21} \in \mathcal{R}^{j(s-1) \times s(k-j)}$ are sparse matrices. Evidently, the permutation does not influence the roundoff error analysis.

The algorithm can be presented as follows.

Stage 1. Obtain the block LU-factorization

$$\mathcal{A} = \begin{pmatrix} A_{11} & A_{12} \\ A_{21} & A_{22} \end{pmatrix} = LU = \begin{pmatrix} A_{11} & 0 \\ A_{21} & I_{j(s-1)} \end{pmatrix} \begin{pmatrix} I_{s(k-j)} & R \\ 0 & S \end{pmatrix}$$

by the following steps:

1. Obtain the LU-factorization of $A_{11} = \mathcal{P}_1 L_1 U_1$ with partial pivoting, if necessary. Here $\mathcal{P}_1$ is a permutation matrix, L_1 is unit lower triangular, and U_1 is upper triangular.
2. Solve $A_{11} R = A_{12}$ using the LU-factorization from the previous item, and compute $S = A_{22} - A_{21}R$, which is the Schur complement of A_{11} in $\mathcal{A}$.

Stage 2. Solve $Ly = d$ by using the LU-factorization of A_{11} (Stage 1).
Stage 3. Solve $Ux = y$ by applying Gaussian elimination to the block S.
The block R is quite sparse in the following kind

$$R = \begin{pmatrix} p^{(1)} & & & & \\ q^{(2)} & p^{(2)} & & & \\ & \ddots & \ddots & & \\ & & & p^{(s-1)} & \\ & & & & q^{(s)} \end{pmatrix} \in \mathcal{R}^{s(k-j) \times j(s-1)},$$

where

$$p^{(i)} = (p_{(i-1)k+1}, p_{(i-1)k+2}, \ldots, p_{ik-1})^T \in \mathcal{R}^{(k-j)\times j},$$
$$q^{(i)} = (q_{(i-1)k+1}, q_{(i-1)k+2}, \ldots, q_{ik-1})^T \in \mathcal{R}^{(k-j)\times j}.$$

Let us note that matrix S (the so called reduced matrix) is block tridiagonal, and banded with bandwith $4j - 1$

$$S = \begin{pmatrix} v_1 & w_1 & & & \\ u_2 & v_2 & w_2 & & \\ & \ddots & \ddots & \ddots & \\ & & \ddots & \ddots & w_{s-2} \\ & & & u_{s-1} & v_{s-1} \end{pmatrix} \in \mathcal{R}^{j(s-1)\times j(s-1)},$$

where the entries are computed in the following way

$$u_i = -a_{ik}q_{ik-1}, \quad v_i = b_{ik} - a_{ik}p_{ik-1} - c_{ik}q_{ik+1}, \quad w_i = -c_{ik}p_{ik+1}. \quad (2)$$

2 Main Stability Results

In the following by a hat we denote the computed quantities. By ΔT we denote an equivalent perturbation in matrix T, and by ρ_0 we denote the roundoff unit. The matrix inequalities are understood componentwise.

In other our previous work (see [9]) has been obtained bounds first for the backward error:

$$|\Delta A| \leq |A|h_1(\rho_0) + |A||N|h_2(\rho_0),$$

where

$$h_1(\rho_0) = K_1 f(\rho_0) + K_2 h(\rho_0) + K_1 K_2 f(\rho_0) h(\rho_0)$$
$$+ K_1 f(\rho_0) g(\rho_0) + K_2 h(\rho_0) g(\rho_0) + K_1 K_2 f(\rho_0) h(\rho_0) g(\rho_0),$$
$$h_2(\rho_0) = 3K_1 f(\rho_0) + 2K_2 h(\rho_0) + 2K_1 K_2 f(\rho_0) h(\rho_0)$$
$$+ 3K_1 f(\rho_0) g(\rho_0) + 3K_2 h(\rho_0) g(\rho_0) + 3K_1 K_2 f(\rho_0) h(\rho_0) g(\rho_0)$$
$$+ K_1 f(\rho_0) g^2(\rho_0) + K_2 h(\rho_0) g^2(\rho_0) + K_1 K_2 f(\rho_0) h(\rho_0) g^2(\rho_0),$$

and for the forward error it is true that

$$\frac{\|\delta x\|}{\|\hat{x}\|} = \frac{\|\hat{x}-x\|_\infty}{\|\hat{x}\|_\infty} \leq \text{cond}(A,\hat{x}) h_1(\rho_0) + \text{cond}^*(A,x^*) r h_2(\rho_0). \quad (3)$$

In the above bounds we denote:

$$r = \max\{\|\hat{R}\|_\infty, 1\}, K_1 = \max\{k_1, 1\}, K_2 = \max\{k_2, 1\},$$

where k_1 bounds the growth of elements when we obtain the LU factorization of A_{11} (Stage 1), k_2 bounds the growth of elements of the Gaussian elimination for the reduced system (Stage 3), and

$$f(\rho_0) = \gamma_{j+1} + \gamma_{2j+1}, \quad g(\rho_0) = \gamma_{j+1} + \rho_0, \quad h(\rho_0) = \gamma_{2j-1} + \gamma_{4j-1},$$

where $\gamma_n = n\rho_0/(1 - n\rho_0)$, and $N = \begin{pmatrix} 0 & \hat{R} \\ 0 & I_{j(s-1)} \end{pmatrix}$. The condition number $\text{cond}^*(A, x^*)$ is defined below

$$\text{cond}^*(A, x^*) = \frac{\| \, |A^{-1}| \, |A| \, x^* \, \|_\infty}{\|\hat{x}\|_\infty},$$

where the vector x^* is constructed in the following way

$$x^* = (\|\hat{x}_k\|_\infty e, |\hat{x}_k^T|, \max\{\|\hat{x}_k\|_\infty, \|\hat{x}_{2k}\|_\infty\} e, \ldots,$$
$$|\hat{x}_{(s-1)k}^T|, \max\{\|\hat{x}_{(s-2)k}\|_\infty, \|\hat{x}_{(s-1)k}\|_\infty\} e)^T.$$

Here $e = (1, 1, \ldots, 1) \in \mathcal{R}^{1 \times (k-1)}$. The other condition number is known as the Skeel's conditioning number:

$$\text{cond}(A, \hat{x}) = \frac{\| \, |A^{-1}| \, |A| \, |\hat{x}| \, \|_\infty}{\|\hat{x}\|_\infty}.$$

The condition number $\text{cond}^*(A, x^*)$ is introduced to make the obtained bounds more realistic in some cases. As we shall see in the bounds of the forward error the condition number $\text{cond}^*(A, x^*)$ is multiplied by the factor r (which can be large sometimes) while the condition number $\text{cond}(A, \hat{x})$ is not. So, when $\text{cond}^*(A, x^*)$ is small the influence of r should be negligible.

3 Special Classes of Matrices

In this section we consider more precisely the case when matrix A belongs to one of the following classes: diagonally dominant, symmetric positive definite, or M-matrices.

For the following bounds of $\|\hat{R}\|_\infty$ and k_2 we need to analyze what is the type of the reduced matrix S if matrix A belongs to one of the above mentioned classes. First we analyze the type of S in exact arithmetic because we need this to bound $\|\hat{R}\|_\infty$. Then at the end of this section we consider the roundoff error implementation and comment on the growth of the constant k_2.

First we use well known fact that (see [1, p. 94] and [1, p. 209], respectively) if matrix A is either

- symmetric positive definite, or
- a nonsingular M-matrix,

then the reduced matrix S (the Schur complement) preserves the same property.

It remains to prove that when A is a diagonally dominant matrix then S preserves this property. Let us note that the case when A is a block row diagonally dominant matrix is considered in [4, p. 252]. Here for diagonally dominant matrices $A = \{a_{ij}\}$ we assume row diagonal dominance in the sense that

$$\sum_{i \neq j} |a_{ij}| \leq |a_{ii}|, \quad i, j = 1, 2, \ldots, n,$$

which is wider than the block row diagonally dominant matrices analyzed in [4, p. 252].

Theorem 1. *Let $A \in \mathcal{R}^{n \times n}$ be a nonsingular row diagonally dominant band matrix. Then the reduced matrix S (the Schur complement) preserves the same property.*

Proof. Let us construct the matrix $\mathcal{B}_i^{(1)} = (B_i, \bar{a}_i, \bar{c}_i)$. It is obvious that this matrix possesses the property of row diagonal dominance. Now the question is if $\mathcal{B}_i^{(1)}$ preserves the property of row diagonal dominance when the Gaussian elimination is applied to matrix B_i? A similar problem in the case of an arbitrary dense matrix is studied in [3], where it is shown that the property of diagonal dominance is preserved after forward Gaussian elimination. For the backward Gaussian elimination in analogous way it follows that the same property is preserved. Hence it is true that the matrix $\mathcal{B}_i^{(1)}$ preserves the property of row diagonal dominance when the forward Gaussian elimination and back substitution are applied to matrix B_i. Let us denote the result of this phase as

$$\mathcal{B}_i^{(2)} = (I_{k-j}, q^{(i)}, p^{(i)}). \tag{4}$$

Now we will prove that the reduced matrix S also preserves the property of row diagonal dominance. Let us consider an arbitrary l-th row of S:

$$0, \ldots, 0, u_1^{(l)}, \ldots, u_j^{(l)}, v_1^{(l)}, \ldots, v_j^{(l)}, w_1^{(l)}, \ldots, w_j^{(l)}, 0, \ldots, 0,$$

where without loss of generality it is assumed that the diagonal element is $v_1^{(l)}$. Then from (2) for the entries of S we obtain (for simplicity some of the indexes are omitted):

$$v_i^{(l)} = b_i^{(l)} - a^{(l)} p_{i-}^{(l)} - c^{(l)} q_{i+}^{(l)}, \quad u_i^{(l)} = -a^{(l)} q_{i-}^{(l)}, \quad w_i^{(l)} = -c^{(l)} p_{i+}^{(l)}, \tag{5}$$

where $a^{(l)}, b^{(l)}, c^{(l)} \in \mathcal{R}^{1 \times j}$, $p_{i-}^{(l)}, p_{i+}^{(l)}, q_{i-}^{(l)}, q_{i+}^{(l)} \in \mathcal{R}^{j \times 1}$. From the fact that $\mathcal{B}_i^{(2)}$ is a row diagonally dominant matrix it follows that

$$\sum_{i=1}^{j} |p_{i-}^{(l)}| + \sum_{i=1}^{j} |q_{i-}^{(l)}| \leq 1, \tag{6}$$

$$\sum_{i=1}^{j} |p_{i+}^{(l)}| + \sum_{i=1}^{j} |q_{i+}^{(l)}| \leq 1. \tag{7}$$

Let us introduce the vector $e = (1, 1, \ldots, 1)^T$ of size j. Then from (5), (6) and (7) we obtain

$$|v_1| \geq |b_1^{(l)}| - |a^{(l)}||p_{1-}^{(l)}| - |c^{(l)}||q_{1+}^{(l)}|$$

$$\geq |b_1^{(l)}| - |a^{(l)}|\left(e - \sum_{i=2}^{j}|p_{i-}^{(l)}| - \sum_{i=1}^{j}|q_{i-}^{(l)}|\right) - |c^{(l)}|\left(e - \sum_{i=1}^{j}|p_{i+}^{(l)}| - \sum_{i=2}^{j}|q_{i+}^{(l)}|\right)$$

$$\geq |b_1^{(l)}| - |a^{(l)}|e + |a^{(l)}|\sum_{i=2}^{j}|p_{i-}^{(l)}| + |a^{(l)}|\sum_{i=1}^{j}|q_{i-}^{(l)}|$$

$$- |c^{(l)}|e + |c^{(l)}|\sum_{i=1}^{j}|p_{i+}^{(l)}| + |c^{(l)}|\sum_{i=2}^{j}|q_{i+}^{(l)}|$$

$$\geq |b_1^{(l)}| - \sum_{i=2}^{j}|b_i^{(l)}| - |a^{(l)}|e - |c^{(l)}|e + \sum_{i=2}^{j}|v_i^{(l)}| + \sum_{i=1}^{j}|u_i^{(l)}| + \sum_{i=1}^{j}|w_i^{(l)}|. \quad (8)$$

But A is a row diagonally dominant matrix, i. e.

$$|b_1^{(l)}| - \sum_{i=2}^{j}|b_i^{(l)}| - |a^{(l)}|e - |c^{(l)}|e \geq 0. \quad (9)$$

Then from (8) and (9) we get

$$|v_1^{(l)}| \geq \sum_{i=2}^{j}|v_i^{(l)}| + \sum_{i=1}^{j}|u_i^{(l)}| + \sum_{i=1}^{j}|w_i^{(l)}|.$$

Hence the reduced matrix S is row diagonally dominant.

As we saw from (3) the error bound depends not only on the growth factors K_1 and K_2, but also on the quantity r, which measures the growth in the matrix $\hat{R}$. Clearly, when some of the blocks B_i are ill conditioned (although the whole matrix A is well conditioned) the factor r can be large. This will lead to large errors even for well conditioned matrices. So, we need some bounds for r, or, equivalently $\|\hat{R}\|_\infty$. In the following we show that $\|\hat{R}\|_\infty$ is bounded by not large constants for the above mentioned three classes of matrices.

The proofs of the next four theorems are similar to the proofs of Theorems 5 – 8 in [8].

Theorem 2. *Let $A \in \mathcal{R}^{n \times n}$ be a nonsingular banded M-matrix and $k_1 cond(A) f(\rho_0) < 1$. Then it is true that*

$$\|\hat{R}\|_\infty \leq \frac{cond(A)}{1 - k_1 cond(A_{11}) f(\rho_0)} \leq \frac{cond(A)}{1 - k_1 cond(A) f(\rho_0)}.$$

Theorem 3. *Let $A \in \mathcal{R}^{n \times n}$ be a nonsingular, row diagonally dominant banded matrix, and $k_1 cond(A) f(\rho_0) < 1$. Then we have*

$$\|\hat{R}\|_\infty \leq \frac{1}{1 - k_1 cond(A_{11}) f(\rho_0)} \leq \frac{1}{1 - 2k_1 cond(A) f(\rho_0)}.$$

Theorem 4. *Let $A \in \mathcal{R}^{n \times n}$ be a symmetric positive definite banded matrix and $k_1(k-1)cond_2(A)f(\rho_0) < 1$, where $cond_2(A) = \|A^{-1}\|_2 \|A\|_2$. Then we have*

$$\|\hat{R}\|_\infty \leq \frac{\sqrt{j(s-1)cond_2(A)}}{1 - k_1 cond(A_{11})f(\rho_0)} \leq \frac{\sqrt{j(s-1)cond_2(A)}}{1 - k_1(k-1)cond_2(A)f(\rho_0)}.$$

Theorems 2 - 4 show that $\|\hat{R}\|_\infty$ is bounded by not large constants for the three classes of matrices, if the whole matrix A is well-conditioned. In order to bound k_2 we can use the already obtained bounds for the Gaussian elimination in [4, p. 181, p. 206, p. 198], the already cited (in the begining of this section) properties of matrix S and Theorem 1. However, in practice we obtain the computed matrix $\hat{S}$ instead of the exact one. It is important to know what is the distance between S and $\hat{S}$. This question is answered in Theorem 5.

Theorem 5. *For the error $\Omega S = \hat{S} - S$ in the computed reduced matrix $\hat{S}$ it holds that*

$$\frac{\|\Omega S\|_\infty}{\|S\|_\infty} \leq K_1 cond(A) r f(\rho_0).$$

So, our conclusion of this section is that the algorithm is numerically stable for the considered three classes of matrices.

Unfortunately when the matrix of the system does not belong to the above mention classes, the algorithm can breaks down or behaves poorly. In our paper we present also a stabilization version of the generalized Wang's algorithm for banded linear systems.

4 The Stabilized Algorithm

As was noticed in the previous section the algorithm can break down, or behave poorly, when $u_i^{(1)}$, for $i = 1, \ldots, s(k-j)$ and are zero or small. So, we can perturb them in such a way that it would be away from zero. The stabilization step can be summarized as follows:

$$\begin{aligned}
&\text{if } (|u_i^{(1)}| < \delta) \\
&\quad \text{if } (|u_i^{(1)}| = 0) \\
&\quad\quad u_i^{(1)} = \delta; \\
&\quad \text{else} \\
&\quad\quad u_i^{(1)} = u_i^{(1)} + sign(u_i^{(1)})\delta; \\
&\quad \text{end} \\
&\text{end}
\end{aligned}$$

In this way we shift $u_i^{(1)}$ away from zero. Hence, the algorithm ensures that we do not divide by a small number.

From the other side the obtained solution is perturbed. Then we apply the usual iterative refinement from [3], with some modification:

$$x^{(0)} = \hat{x};$$
for $m = 1, 2, \ldots$
$$r^{(m-1)} = b - Ax^{(m-1)};$$
$$(A + \Delta)y^{(m)} = r^{(m-1)};$$
$$x^{(m)} = x^{(m-1)} + y^{(m)};$$
end

The difference here is that instead of A we solve perturbed systems with the matrix $A + \Delta$, where Δ is a diagonal matrix with all such perturbations, and $\hat{x}$ is the result of the perturbed algorithm before the iterative refinement is applied. We note that, when $\delta = \sqrt{\rho_0} \approx 10^{-8}$ (in double precision), in practice the perturbed solution is very close to the exact one and we need usually only one or two steps of iterative refinement, depending on what accuracy we require. Here by ρ_0 we denote the machine roundoff unit.

Taking into account [6] the condition of convergence of iterative refinement is
$$C \mathrm{cond}(A)\delta < 1,$$
where $\mathrm{cond}(A)$ is a condition number of matrix A and C is a constant of the following kind
$$C = \frac{\max_i(|A||x|)_i}{\min_i(|A||x|)_i}, \quad i = 1, 2, \ldots, n.$$

A number of numerical experiments which confirm theoretical results and the effectiveness of the stabilized algorithm are available from the author.

References

1. Axelsson, O.: Iterative solution methods. Cambridge University Press, New York, 1994.
2. Conroy, J.: Parallel Algorithms for the solution of narrow banded systems. Appl. Numer. Math. **5** (1989) 409–421.
3. Golub, G., Van Loan, C.: Matrix computations. The John Hopkins University Press, Baltimore, 1989.
4. Higham, N.: Accuracy and Stability of Numerical Algorithms. SIAM, Philadelphia, 1996.
5. Meier, U.: A parallel partition method for solving banded linear systems. Parallel Comput. **2** (1985) 33–43.
6. Skeel, R.: Scaling for numerical stability in Gaussian elimination. J. Assoc. Comput. Mach. **26** (1979) 494–526.
7. Wang, H.: A parallel method for tridiagonal linear systems. ACM Transactions on Mathematical Software **7** (1981) 170–183.
8. Yalamov, P., Pavlov, V.: On the Stabilty of a Partitioning Algorithm for Tridiagonal Systems. SIAM J. Matrix Anal. Appl. **20** (1999) 159–181.
9. Yalamov, P., Pavlov V.: Backward Stability of a Parallel Partitioning Algorithm for Banded Linear Systems. Proc. of 4th International Conference on Numerical Methods and Applications, Sofia, August 19–23, 1998, World Scientific Publ. (1999) 655–663.

Numerical Solution of ODEs with Distributed Maple

Dana Petcu

Western University of Timişoara, Computer Science Department
B-dul V.Pârvan 4, RO-1900 Timişoara, Romania,
tel./fax.:++40-56-194002
e-mail: petcu@info.uvt.ro
http://www.info.uvt.ro/~petcu

Abstract. We describe a Maple package named D-NODE (Distributed Numerical solver for ODEs), implementing a number of difference methods for initial value problems. The distribution of the computational effort follows the idea of parallelism across method. We have benchmark the package in a cluster environment. Distributed Maple ensures the inter-processor communications. Numerical experiments show that parallel implicit Runge-Kutta methods can attain speed-ups close to the ideal values when the initial value problem is stiff and has between ten and hundred equations. The stage equations of the implicit methods are solved on different processors using Maple's facilities.

Keywords: parallel numerical methods for ordinary differential equations, distributed computer algebra systems, performance analysis.

1 Introduction

We will concerned with the numerical solution of systems of initial value ordinary differential equations (IVPs for ODEs) of the form

$$y'(t) = f(t, y(t)), \ t \in [t_0, t_0 + T], \ y(t_0) = y_0, \ y_0 \in R^m, \ f : R \times R^m \to R^m. \quad (1)$$

In the numerical solution of ordinary differential equations by implicit timestepping methods a system of linear or nonlinear equations has to be solved each step. The costs of the linear algebra associated with the implementation of the implicit equation solver generally dominate the overall cost of the computation.

The numerical integration of large IVPs is also time consuming. Such large (and stiff) problems often arise in the modeling of mechanical and electrical engineering systems or in the solution of semi-discretization of convection-diffusion problems [7] associated to time-dependent parabolic PDEs. The stiffness of these problems requires that the numerical methods to be used should be unconditionally stable, and therefore implicit. The methods are computationally demanding and require today's fastest high performance computers for practical implementations. However, access to a fast high-speed computer is not sufficient. One must also ensure that the great potential power of the computer is correctly exploited.

The aim of this paper is to investigate in what extent parallel implicit Runge-Kutta methods can be used to solve stiff initial value problems of ten to hundred equations using `Distributed Maple`. Stage systems to be solved are distributed among the processors of a cluster system. Tables and figures illustrate the performance of the implemented methods.

The paper is organized as follows. Section 2 motivates the present work. Section 3 describes the objectives of a `Maple` package named D-NODE (Distributed Numerical solver for ODEs), implementing a number of difference methods designed in the idea of parallelism across method. In Section 4 we report the numerical results obtained using some known parallel implicit Runge-Kutta methods. We have benchmark the package in a cluster environment.

2 On IVP Solving Strategies

One iterative step of many implicit schemes for IVPs of the form (1) requests the solution of a system of algebraic equations of the form

$$Y - h(C \otimes I_m)F - G(Y_0) = 0 \qquad (2)$$

with h the step-size, C a $s \times s$ matrix, I_m the identity matrix, G a known function, $F = (f_1, \ldots, f_s)^T$, $f_i = f(t_i, y_i)$, and $Y = (y_1, \ldots, y_s)^T$, the unknown approximations to the exact solution on $t_1, \ldots, t_s$. It is common practice to use fixed-point iterations or, in the stiff case, some modified Newton iterations. The convergence rate of such methods depends on the method step-size. Implicit Runge-Kutta schemes (IRK) are among the numerical techniques commonly considered as efficient ones in stiff IVP case. The use of a s-stage IRK method for ODEs requires the solution of nonlinear systems of algebraic equations of dimension sm (m defined in 1). Usually, the solution of this system represents the most time-consuming section in the implementation of such method.

A general way of devising parallel ODE solvers is that of considering methods whose work per step can be split over a certain number of processors. The so-called solvers with parallelism across the method are then obtained. Such methods are essentially Runge-Kutta schemes. For a parallel implicit Runge-Kutta methods the system (2) can be split into a number $k \leq s$ independent subsystems. From the computational point of view, the diagonally implicit RK methods (DIRK methods) are the most attractive methods since they have suitable stability properties and the implementation can be carried out with a lower computational cost than fully IRK methods. Block diagonally implicit RK methods (BDIRK) are also used. The so-called PDIRK methods are parallel diagonally iterated RK methods. The computational cost involved in their implementation is similar to DIRK methods. PDIRK methods are able to produce accurate results at a relatively high price. Unfortunately these methods are not the most suitable for solving semi-discretized PDEs in which it is necessary to generate relatively low-accuracy results at low price [2]. The construction of a some PDIRK using `Maple` is presented in [4]. Parallel singly diagonally iterated RK methods (PSDIRK) are particular methods of PDIRK type.

Computer algebra systems (CAS) can be used with success in prototyping sequential algorithms for symbolic or numeric solution of mathematical problems. Maple is such a CAS. Constructing prototypes for parallel algorithms in Maple for numeric solution of ODEs is a challenging problem. Distributed Maple [12] is a portable system for writing parallel programs, in a CAS, which allows to create concurrent tasks and have them executed by Maple kernels running on different machines of a network. The system can be used in any network environment where Maple and Java are available. The user interacts with the system via the text oriented Maple front-end. It also provides facilities for the online visualization of load distribution and for post-execution analysis of a session.

We know that solving systems of algebraic or differential equations of order several hundreds can be an unsolvable problem for an actual CAS. Systems of order several tens equations can be solved with Maple but the long running-time may be a great problem for the user. A correct use of extensions like Distributed Maple can improve the solution computation time. The computer facilities required by such an extension are reasonable for any user since it not supposes access to super-computers.

In general, the system (2) is solved numerically using repeated evaluation of the function F at different values (in the case of a stiff system, they are also required some repeated Jacobian matrix evaluations). In a message-passing computing environment these values must be communicated between different processors participating to a time-step integration of the IVP. Sending to a working processor the algebraic expressions of the part of F for which it is responsible can be a better solution eliminating a significant quantity of values to be communicated between the supervisor-processor storing F and the worker-processors. The interpretation of an algebraic expression requires at a worker processor side at least a small specific expression interpreter (like Maple kernel).

The implicit equation solver can substantially affect the global error of the numerical solution of an IVP. Take for example the fixed-point iterations which usually do not converge in the stiff IVP case to the exact solution of the system (2). Using a fixed-point iterations and ignoring this remark and also the use error control strategies, we can obtain a numerical solution far from the real solution. In practical implementation of implicit time-stepping methods the hardest parts are the implicit equation solver implementation and the error control mechanism combined with variable step-size strategies. Using numerical facilities of CAS systems to do the first job it can simplify the programmer work.

We propose the use of implicit equation solver of Maple for the solution of system (2). In the case of parallel IRK, independent stage-subsystems in Maple algebraic form are to be send to some worker-processors in order to solve them.

3 D-NODE Objectives

The project of a Maple package, D-NODE (Distributed Numerical solver for ODEs) is intended to be an update to the ODEtools Maple package. It implements a number of difference methods designed in the idea of parallelism across method [15].

The package is a part of a bigger project of an expert system for numerical solution of ODEs [10] and it is expected to be finalized at the end of this year.

The facilities of ODEtools from Maple, and the similar tools from other CAS, are far to cover all the user needs (for example, the stiff IVP solving case). Recent reports demonstrate the effort to improve these tools. For example, the paper [13] describes mathematical and software developments for a suite of programs for solving ODEs in Matlab.

D-NODE package has similar facilities with EpODE (ExPert system for ODEs), recently presented [10] and available at http://www.info.uvt.ro/~petcu/epode: a large collection of parallel methods working in a distributed computing environment, automatic detection of method properties including method classification, order, error constant and stability, degree of parallelism, method-interpreter for describing new methods, automatic detection of problem properties (like stiffness), step-size selection mechanism according the method and problem properties, numerical solution computation on a distributed network of workstations (in EpODE based on PVM [8]).

4 Numerical Experiments

This section is devoted to the interpretation of the test results in the integration of large non-linear ODE systems and to the comparisons with the test results of other similar tools (one of them being EpODE, part of the same project [9]).

We consider four methods representative for their class of parallel IRK and which were included in D-NODE. We have benchmark the corresponding package functions in a cluster environment. The cluster comprises 4 dual-processor Silicon Graphics Octanes (2 R10000 at 250 MHz each) linked by three 10 Mbit Ethernet subnetworks connected to a router.

The first scheme is the 4-stage, 2-processor, 4th-order, A-stable DIRK method described in [6]. The second one is a 6-stage, 2-processor, 3th-order, A-stable PDIRK method based on Radau IIA corrector and presented in [14]. The third one is the 4-stage, 2-processor, 4th-order, L-stable Hammer-Hollinworth BDIRK method [5]. The last one is the 9-stages, 3-processor, 4th-order, A-stable PS-DIRK presented in [2]. Details about these methods can be found also in [11].

The degree of parallelism of a method can be detected by applying the direct-graph method proposed in [5]. Figure 1, generated by EpODE, presents the proposed distributions of the computations on processes and parallel stages for the above mentioned methods.

In order to show the performance of the methods on semi-discrete PDEs we include in our tests the linear IVP obtained from the following PDE [13]:

$$\frac{\partial u}{\partial t} = e^t \frac{\partial^2 u}{\partial x^2}, \quad \begin{aligned} x &\in [0,\pi], \\ t &\in [0,10], \end{aligned} \quad \begin{aligned} u(x,0) &= \sin(x), \\ u(0,t) &= u(\pi,t) = 0. \end{aligned} \quad (3)$$

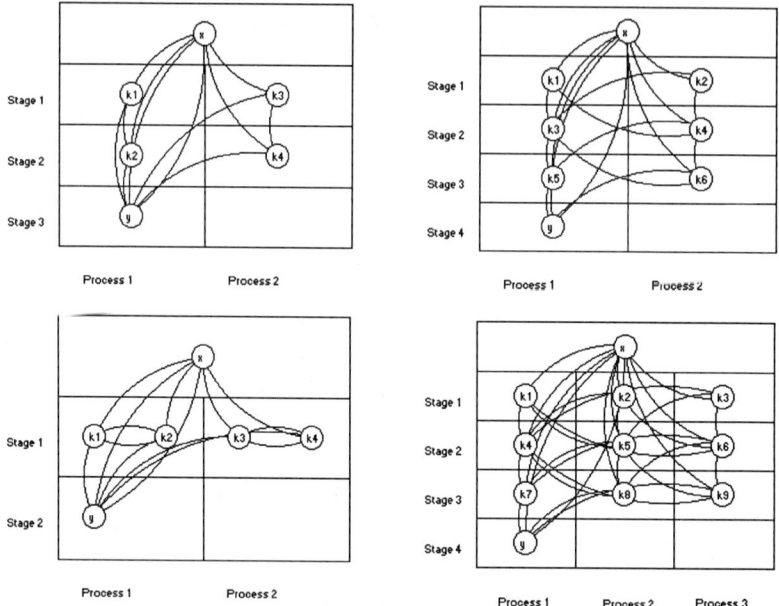

Fig. 1. Data-flow graphs reported by EpODE [10]: from left to right and from top to bottom, the four methods – An arc from the left-part of a circle means a dependency from the top variable to the bottom variable, an arc from the right-part of a circle means a dependency from the bottom to the top variable when starting the next integration step, and an almost horizontal arc indicates an interdependence between the two linked variables. A k-labeled node refers the solving procedure for obtaining the value of the variable k using the previous (above) computed labeled node-values. More than one labeled-node in a computational cell indicates that a system formed with those variables must be solved

As the second test problem we take the nonlinear IVP obtained by the semi-discretization of the following nonlinear convection-diffusion problem [1]:

$$\frac{\partial u}{\partial t} = u\frac{\partial^2 u}{\partial x^2} - x\cos(t)\frac{\partial u}{\partial x} - x^2 \sin(t), \quad \begin{array}{l} x \in [0,1], \ u(0,t) = 0, \ u(x,0) = x^2, \\ t \in [0,1], \ u(1,t) = \cos(t). \end{array} \quad (4)$$

In order to solve both problems, we carry out a semi-discretization on the spatial variable by using second-order symmetric differences on an uniform grid with mesh size $\Delta x = 1/(m+1)$. This method (of lines) leads to IVPs with m ODEs. As the third problem we take a real one. The selected PLEI [3] problem (28 ODEs) is the celestial mechanics problem of seven stars. Similar IVPs have been studied in [7] for the case of a shared-memory parallel computer.

Figure 2 obtained by using the visualize procedures from Distributed Maple shows the ratio between sequential and distributed time measurements corre-

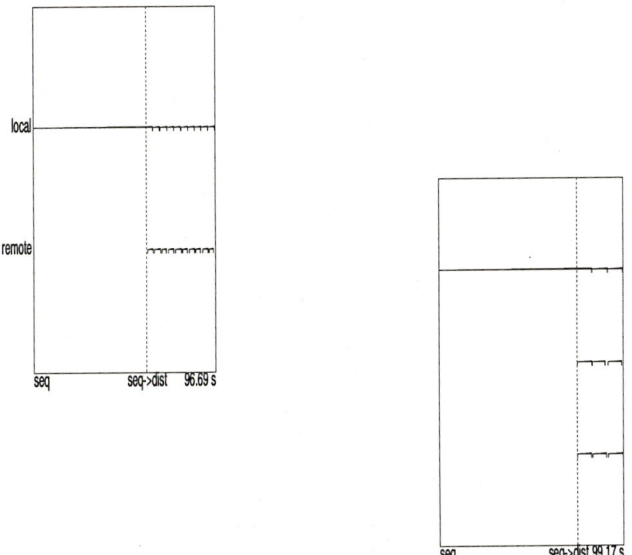

Fig. 2. Time diagrams and processor load per k integration-steps: left figure for the first method and the linear problem with $m = 60$ and $k = 5$ steps, right figure for the fourth method and the nonlinear problem with $m = 20$ and $k = 1$ steps

sponding to one or more arbitrary integration step. First vertical block of each figure corresponds to one sequential-integration, and the second one to the distributed integration. A horizontal line corresponding to a local or remote processor indicates the time when that processor is busy (continuous tasks). The time in seconds reported in the bottom-right corner of each figure represents the total time, the sequential one plus the distributed one. The time difference between a local and a remote task can be explained by the fact that the local processor must compute explicitly the approximate solution y_{n+1} (y in Figure 1) from computed Y vector (x and k_i), must send the tasks to the other processors and the must prepare the algebraic systems to be solved.

Figure 3 also produced by `Distributed Maple` offers more details about the load-balancing between the running processes. Analyzing the top images we see that small linear IVPs (at least for our test problem with $m = 10 \div 20$ equations), cannot be integrated in a distributed computational environment faster than using a sequential computer, since the distributed task are small relative to the overall time spent in one distributed integration step (including the necessary communications). In the case of nonlinear problems of similar dimensions, almost all computation time is spent on computing stage solutions (continuous horizontal lines).

The efficiency measurements of the distributed implementation of the selected method are shortly presented in Table 1. The vertical lines split the ineffi-

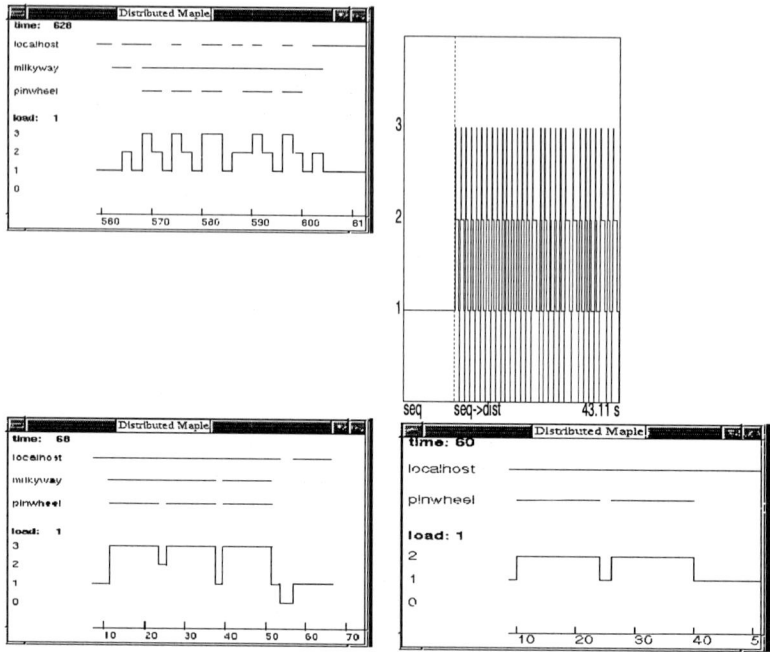

Fig. 3. Load balancing for k-integration steps: PSDIRK for linear problem with $m = 20$ and $k = 10$ steps (top-left), with $m = 10$ and $k = 10$ (top-right), and for nonlinear real problem with $m = 28$ and $k = 1$ (bottom-left), respectively method DIRK method for nonlinear problem with $m = 25$ and $k = 1$ (bottom-right)

Table 1. Efficiency results $E_p = T_s/(pT_p)$ (in percents), where p is the number of processors, T_s, T_p are the mean times necessary to perform one-integration step using one, respectively p processors

Method	$p \backslash m$	Linear problem				Nonlinear problem			Real
		10	20	40	60	5	10	15	20	28
DIRK	2	9.84	25.38	53.85	81.25	24.62	49.97	86.11	94.12	97.00
PDIRK	2	13.96	29.05	66.66	76.72	28.26	51.43	87.38	97.00	94.12
BDIRK	2	34.48	63.08	86.57	88.32	63.08	92.00	99.98	99.99	95.54
PSDIRK	3	10.18	22.67	50.43	67.70	24.31	51.88	81.96	99.10	97.33

cient values (left) from the efficient values (right). We can arrange the analyzed methods in a increasing trust order depending on the order in which they attain the vertical lines: DIRK, PDIRK, BDIRK (we must prefer the BDIRK method). These methods appear in the reverse order if we sort them by the moment when they complete a time-step (DIRK is the faster one). Therefore supplementary parameters (like recommended step-size) must be take into account when we select a distributed methods. We see also that the 3-processor PSDIRK method can be almost so efficient as a two-processor method when we solve a nonlinear problem.

We cannot expect to obtain similar efficiency results when we use explicit Runge-Kutta methods (or explicit multistep methods), since the solution of a stage equation involves only a small number of function evaluations and vector operations.

Comparing the above efficiency results with those reported [9] for similar problems using EpODE written in C, we must remark here a lowest barrier in IVP sizes between efficient and inefficient implementation of distributed solvers. This fact is due to the implicit equation solver implemented in Maple which is more time-consumer than some modified Newton iterations written in programming language like C. On other hand we can have more trust in the Maple solution of implicit equation system. Using the accurate solution of implicit stage-equations produced by Maple we can apply the error control strategies for ODE solvers often reported in literature (usually the great influence of the implicit equation solver on the global error of the numerical ODE solution is neglected).

5 Conclusions

D-NODE, a Maple package using Distributed Maple extends the numerical ODE solving capabilities of Maple to systems of order tens or order hundreds of equations by exploiting the computational power of a local network of workstations. A strategy was adopted in which parts of some nonlinear systems to be solved at each time-step are send in algebraic forms to the workers. The solution accuracy compensates the supplementary time required by this non-classical procedure. Efficiency measurements indicate that the parallel implicit Runge-Kutta methods are fitted with this strategy.

Acknowledgments

This work was supported by Austrian Science Foundation project FWF/ SFB F1303, and by the ERASMUS EU-grant SOCRATES 1999/ 2000. The author would like to express her appreciation to Wolfgang Schreiner, the creator of Distributed Maple and to thank him for the fruitful discussions.

References

1. Cong, N., A Parallel DIRK Method for Stiff Initial-Value Problems. *J. Comp. Appl. Math.* **54** (1994), 121-127.
2. Franco, J. M., Gomez, I., Two Three-Parallel and Three-Processor SDIRK Methods for Stiff Initial-Value Problems, *J. Comp. Appl. Math.* **87** (1997), 119-134.

3. Lioen, W. M., de Swart, J. J. B., van der Veen, W. A., Test Set for IVP Solvers, Report NM-R9615, CWI, August 1996, http://www.cwi.nl/cwi/projects/ IVPtestest.
4. Lioen, W. M., On the diagonal approximation of full matrices, *J. Comp. Appl. Math.* **75** (1996), 35-42.
5. Iserles, A., Nørsett, S. P., On the Theory of Parallel Runge-Kutta Methods. *IMA J. Numer. Anal.* **10** (1990), 463-488.
6. Jackson, K. R., Nørsett, S. P., The Potential for Parallelism in Runge-Kutta Methods, *SIAM J. Numer. Anal.* **32**, No. 1 (1995), 49-82.
7. Kahaner, D. K., Ng, E., Schiesser, W. E., Thompson, S., Experiments with an ODE Solver in the Parallel Solution of MOL Problems on a Shared-Memory Parallel Computer, *J. Comp. Appl. Math.* **38** (1991), 231–253.
8. Petcu, D., Implementations of Some Multiprocessor Algorithms for ODEs Using PVM. In *LNCS* **1332** (1997): Recent Advances in PVM and MPI, eds. M. Bubak, J. Dongarra, J. Waśniewski, Springer Verlag, Berlin, 375-382.
9. Petcu, D., Solving Initial Value Problems with a Multiprocessor Code. In *LNCS* **1662** (1999): Parallel Computing Technologies, ed. Victor Malyshkin, Springer Verlag, Berlin, 452-466.
10. Petcu, D., Drăgan, M., Designing an ODE Solving Environment. In *LNCSE* **10** (2000): Advances in Software Tools for Scientific Computing, eds. H. P. Langtangen, A. M. Bruaset, E. Quak, Springer-Verlag, Berlin, 319-338.
11. Petcu, D., Numerical Solution of ODEs with Distributed Maple, Technical Report 00-09 (2000), Research Institute for Symbolic Computation, Linz, 12 pages.
12. Schreiner, W., Distributed Maple – User and Reference Manual. Technical Report 98-05 (1998), Research Institute for Symbolic Computation, Linz, and http://www.risc.uni-linz.ac.at/software/distmaple.
13. Shampine, L. F., Reichelt, M. W., The Matlab ODE Suite. *SIAM J. Sci. Comput.* **18**, No. 1 (1997), 1-22.
14. Sommeijer, B. P., Parallel Iterated Runge-Kutta Methods for Stiff Ordinary Differential Equations. *J. Comp. Appl. Math.* **45** (1993), 151-168.
15. Van der Houwen, P. J., Parallel Step-by-Step Methods. *Appl. Num. Math.* **11** (1983), 69-81.

The Boundary Layer Problem of Triple Deck Type

Laurent Plantié

CERFACS
42 Avenue Gaspard Coriolis, 31057 Toulouse cedex 01, France
plantie@cerfacs.fr
http://www.cerfacs.fr/algor

Abstract. We give the formulation of the Von Mises problem of the boundary layer of triple deck type. An original non-local condition appears. We prove the existence of a solution by studying a semi-discrete scheme in which we consider the pressure gradient as a parameter. We then obtain a solution in physical variables but the condition $v(x,0) = 0$ is not proved. Besides, the numerical simulations give a surprising non-uniqueness result with given pressure in the case of a break-away.

1 Introduction

The triple deck model was introduced by Stewarston and Williams [7] in 1969 for supersonic flows. Several other models of this type have been introduced later (see references in [4] and [6]). All these models describe the behaviour of a newtonian flow around a perturbation at high Reynolds numbers. In [3] and [6], we introduce a model for a Couette flow in a channel.

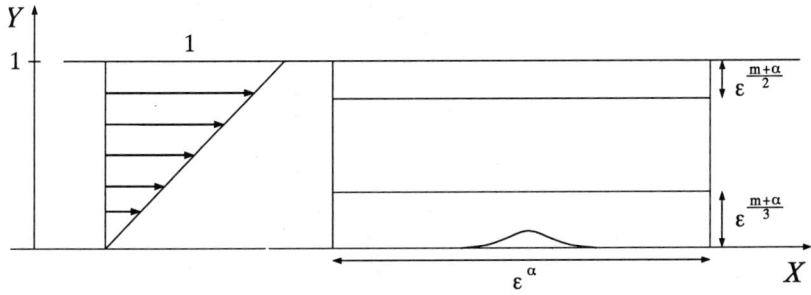

Fig. 1. Boundary layer model for a Couette flow

The lower wall is fixed and has a small perturbation. The upper wall is a flat plate moving with velocity 1 (after adimensionalisation). The entering velocity profile is $U(X,Y) = Y$. The size of the perturbation and of the associated layers are related to the Reynolds number : see Figure 1 where we set $Re = \epsilon^{-m}$. More precisely, the pair (m,α) must verify $-m < \alpha \leq 0$.

The boundary layer of triple deck type is the inner layer or deck (size $\epsilon^{(m+\alpha)/3}$) in Figure 1. We isolate a canonical problem for the boundary layers of this type and we show how to solve it.

2 The Canonical Problem

We now use the inner variables of the boundary layer. Let x and y be the longitudinal and transversal coordinates, u and v the longitudinal and transversal velocities and p the pressure. The canonical problem consists in the Prandtl equations

$$u\frac{\partial u}{\partial x} + v\frac{\partial u}{\partial y} = -\frac{\partial p}{\partial x} + \frac{\partial^2 u}{\partial y^2}, \qquad (1)$$

$$\frac{\partial u}{\partial x} + \frac{\partial v}{\partial y} = 0, \qquad (2)$$

$$\frac{\partial p}{\partial y} = 0, \qquad (3)$$

the initial condition

$$u(0, y) = y \qquad (4)$$

and the boundary conditions

$$\lim_{y \to +\infty} u(x, y) - y = A_d(x) \qquad (5)$$

$$\text{and} \qquad u(x, 0) = v(x, 0) = 0. \qquad (6)$$

The data is the displacement A_d and the unknowns are u, v on $[0, x_0] \times [0, +\infty[$ and p on $[0, x_0]$ (so that (3) is already taken into account).

In (5), the term y represents the non-perturbed velocity profile. The Prandtl transformation

$$\begin{cases} \tilde{x} = x \\ \tilde{y} = y + A_d(x) \end{cases} \quad \begin{cases} \tilde{u}(\tilde{x}, \tilde{y}) = u(x, y) \\ \tilde{v}(\tilde{x}, \tilde{y}) = v(x, y) + A_d{}'(x)\, u(x, y) \end{cases} \quad \tilde{p}(\tilde{x}) = p(x) \qquad (7)$$

does not change (1)-(4) and enables to interpret A_d as a geometrical perturbation. Then, (1)-(6) becomes exactly the problem of the Poiseuille or of the Couette flows. The pressure never appears as a data in the physical problems.

There is no direct relation between A_d and p as it appears in the sequel. The situation is quite different from that of the classical Prandtl problem where p is given and where the condition which replaces (5) is automatically verified [2].

As always done for the Prandtl problem, we use the Von Mises transformation defined by the change of variables and the change of functions

$$\begin{cases} \xi(x, y) = x \\ \psi(x, y) = \int_0^y u(x, t)\, dt \end{cases} \quad \text{and} \quad \sqrt{w(\xi, \psi)} = u(x(\xi, \psi), y(\xi, \psi)).$$

We solve the problem in physical variables by first solving the Von Mises problem and then applying the inverse Von Mises transformation. We therefore look for positive solutions.

The first original result lies precisely in the formulation of the Von Mises problem where a new non-local condition appears [4]. It consists in the Von Mises equation

$$\frac{\partial w}{\partial \xi} - \sqrt{w}\frac{\partial^2 w}{\partial \psi^2} = -2p' \tag{8}$$

and the conditions

$$w(0, \psi) = 2\psi, \tag{9}$$

$$w(\xi, 0) = 0 \tag{10}$$

and $\quad A_d(\xi) = \int_0^{+\infty}\left(\frac{1}{\sqrt{2\psi}} - \frac{1}{\sqrt{w(\xi,\psi)}}\right) d\psi. \tag{11}$

The difficulties of (8)-(9)-(10)-(11) come from the nonlinearity of the equations, their degenerating at $\psi = 0$, the semi-infinite domain and the condition (11) associated to the determination of the additional unknown p. The results of Oleinik, Nickel, Walter, Fife, Serrin (see references in [1], [4]) cannot be used.

An original method based on the study of a semi-discrete scheme is developed. The pressure gradient is considered as a parameter and the problem is solved by induction. We did not find any method for solving directly the continuous problem (8)-(9)-(10)-(11).

2.1 The Problem with Given Pressure

The sequences of pressure gradients $(p'^n)_{n\geq 0}$ are first considered as data and the displacement does not appear. We suppose there exist $M_1 \geq 0$ and $M_2 \geq 0$ such that $-M_1 \leq p'^n \leq M_2$ for all n. The problem considered here consists in finding the sequences $(w^n)_{n\geq 0}$ solution of the equation

$$\frac{w^n - w^{n-1}}{\Delta\xi} - \sqrt{w^n}\frac{\partial^2 w^n}{\partial \psi^2} = -2p'^n \tag{12}$$

which satisfy

$$w^0(\psi) = 2\psi, \qquad w^n(0) = 0 \quad \text{and} \quad |w^n - 2\psi| \text{ bounded}. \tag{13}$$

The study is particularly difficult for positive pressure gradients. We prove [4]

Theorem 1. *a) Let $k_1 \in \,]0,2[$. There exists $\xi_0 = \xi_0(M_2, k_1) > 0$ and there exists a sequence $(w^n)_{n\geq 0}$ in $C^1([0,+\infty[) \cap C^\infty(]0,+\infty[)$ which is solution of (12)-(13) and verifies $w^n \geq k_1\psi$ for all n such that $n\,\Delta\xi \leq \xi_0$.*

Moreover, there exists $k_2 = k_2(M_1, \xi_0) \geq 2$ *such that*

$$-2M_2\,\xi_0 \leq w^n - 2\psi \leq 2M_1\,\xi_0 \quad \text{and} \quad k_1 \leq \frac{\partial w^n}{\partial \psi} \leq k_2. \tag{14}$$

b) If $M_2 = 0$, *any* $\xi_0 > 0$ *is admissible and* $\partial_\psi w^n \geq 2$.

Thus, any break-away is avoided on $[0, \xi_0]$.

We obtain this result by solving a regularized truncated problem. This method was introduced by Oleinik [2]. However, she uses specific techniques for parabolic problems. Here, we use in particular the monotone iteration method.

We also prove the asymptotic behaviours [4]

Theorem 2. *For all* $\gamma > 0$, *there exist* $k_\gamma = k_\gamma(k_2, \gamma) > 0$ *and* $h_\gamma = h_\gamma(k_1, k_2, \gamma) > 0$ *such that*

$$\left|\frac{\partial w^n}{\partial \psi} - 2\right| \leq \max(2 - k_1, k_2 - 2)\, e^{k_\gamma\,\xi_0 + \gamma - \gamma\,(2\psi+1)^{3/4}} \tag{15}$$

and $\quad |w^n - 2\psi + 2p^n| \leq 2(M_1 + M_2)\xi_0\, e^{k_\gamma\,\xi_0 + \gamma - \gamma\,(2\psi+1)^{3/4}} \tag{16}$

for all $\psi \geq 0$ *and all* n *such that* $n\,\Delta\xi \leq \xi_0$ *if* $\Delta\xi \leq h_\gamma$.

We need then very precise bounds of the diffusive term $q^n = \sqrt{w^n}\,\partial^2_\psi w^n$. We obtained them only in the case $-M_1 \leq p'^n \leq 0$. In the sequel, we then focus on this case which corresponds also to a specific case for the displacements. Let us set $p'^0 = 0$. We show [3] [5]

Theorem 3. *Let us suppose* $-M_1 \leq p'^n \leq 0$. *There exists* $h_0 = h_0(M_1, \xi_0) > 0$ *such that*

$$-2M_1 \leq \min_{0 \leq i \leq n} 2p'^i \leq q^n \leq 0 \tag{17}$$

and $\quad |q^n| \leq 2M_1\, e^{\sqrt{M_q} + \xi_0\,\psi^{1/4} - \sqrt{\psi}} \tag{18}$

where $M_q = \max(2M_1, (\xi_0^2 + 1)^2(16 + \xi_0))$ *for all* n *such that* $n\,\Delta\xi \leq \xi_0$ *if* $\Delta\xi \leq h_0$.

The lower bound of q^n in (17) is optimal and essential for solving the inverse problem where the displacement is given. Using the boundedness of q^n, we can then obtain estimates depending on γ as in (15) and (16). In (18), we fixed γ and we obtain

$$q^n \geq -M_1 \tag{19}$$

for $\psi \geq 6\,M_q$ and $\Delta\xi \leq h_0$. Thus, only the data M_1 and ξ_0 appear in (19). This inequality is also important for the inverse problem.

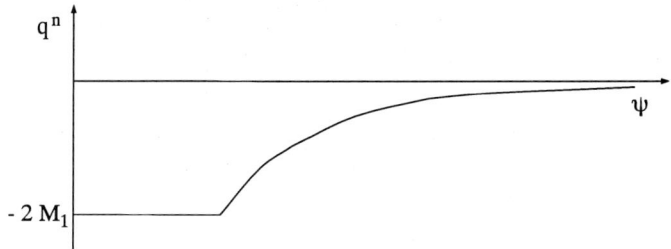

Fig. 2. Bounds of q^n in the case $-M_1 \leq p'^n \leq 0$

Remark 1. It would be possible to solve the inverse problem in the general case corresponding to pressure gradients of any sign if $q^n \leq \max_{0 \leq i \leq n} 2 p'^i$ was true. It is only established in the case of non-decreasing pressure gradients, which is the worst case *a priori* [3]. Using a regularized scheme [5], we obtain a bound greater than $\max_{0 \leq i \leq n} 2 p'^i$ and we cannot apply the method exposed below.

We also need the uniqueness and the continuity of w^n with respect to p'^n. They are obtained in a class of function $\mathcal{S}$ whose elements verify the boundedness and regularity properties of the solution we constructed up to here [5]. The boundedness properties are

$$0 \leq w^n - 2\psi \leq 2M_1\xi_0, \quad 2 \leq \partial_\psi w^n \leq k_2 \quad \text{and} \quad -2M_1 \leq q^n \leq 0. \quad (20)$$

Theorem 4. *a) There exists $h_1 > 0$ such that the solution w^n of (12)-(13) is unique in $\mathcal{S}$ for all n such that $n \, \Delta\xi \leq \xi_0$ if $\Delta\xi \leq h_1$.*

b) Let $n \geq 1$ such that $n \, \Delta\xi \leq \xi_0$ and let $-M_1 \leq p'^n_2 \leq p'^n_1 \leq 0$. There exists $h_2 \in \,]0, h_1]$ such that

$$w^n_1 \leq w^n_2 \leq w^n_1 + 2\,(p'^n_1 - p'^n_2)\,\Delta\xi \quad (21)$$

if $\Delta\xi \leq h_2$ where w^n_1 and w^n_2 are the solutions in $\mathcal{S}$ corresponding to p'^n_1 and p'^n_2 respectively and to the same antecedent w^{n-1}.

2.2 The Problem with Given Displacement

We suppose here that the displacement A_d is lipschitzian non-decreasing, $0 \leq A_d'^n \leq \mathcal{L}$, and verifies $A_d(0) = 0$. We state the inverse problem by introducing a function which links pressure and displacement.

Let w^n be the solution in $\mathcal{S} = \mathcal{S}(M, \xi_0)$ of (12)-(13) which corresponds to a pressure gradient p'^n in $[-M, 0]$ and to a fixed antecedent w^{n-1}. After (11), we define the function A'^n by

$$A'^n(p'^n) = \frac{1}{\Delta\xi} \int_0^{+\infty} \left(\frac{1}{\sqrt{w^{n-1}}} - \frac{1}{\sqrt{w^n}}\right) d\psi = \int_0^{+\infty} (q^n - 2p'^n)\gamma^n \, d\psi \quad (22)$$

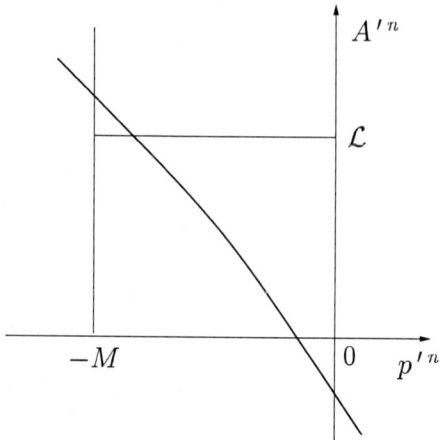

Fig. 3. Function $A'^n(p'^n)$ for a non-decreasing displacement

where $\gamma^n = (\sqrt{w^{n-1}}\sqrt{w^n}(\sqrt{w^{n-1}} + \sqrt{w^n}))^{-1}$. This function is well defined. Indeed, w^n exists for all n and is unique in $\mathcal{S}$ and the first estimate (20) shows that the integral (22) exits.

The problem with given displacement consists in finding sequences $(w^n)_{n \geq 0}$ and $(p'^n)_{n \geq 0}$ which are solution of (12)-(13) and verify $A'^n(p'^n) = A_d'^n$. We show that the frame $-M \leq p'^n \leq 0$ is adapted to this problem.

The function A'^n summarizes the whole problem. We prove a local continuity property and a global coercivity property [3]. The first one is brought by Theorem 4 which implies the continuity of each function A'^n in $[-M, 0]$. The coercivity property is stated: for all $\mathcal{L} \geq 0$ and all $\xi_0 \geq 0$, there exists $M = M(\mathcal{L}, \xi_0) \geq 0$ such that

$$A'^n(0) \leq 0 \quad \text{and} \quad A'^n(-M) \geq \mathcal{L} \qquad (23)$$

if $-M \leq p'^i \leq 0$ for $i \leq n-1$ and if $n \Delta \xi \leq \xi_0$.

The first inequality in (23) follows directly from (22) and $q^n \leq 0$ if $p'^n \leq 0$. For the second, we use $q^n \geq -2M$ if $p'^n \geq -M$. Indeed, (22) already shows $A'^n(-M) \geq 0$. The conclusion finally arises using (19). The lower bound of q^n in (17) is essential.

We can then apply the theorem of the intermediate values and we solve the problem with given displacement by induction. We then obtain sequences $(p'^n)_{n \geq 0}$ in $[-M, 0]$ and $(w^n)_{n \geq 0}$ in $\mathcal{S}$. Then, we can take the limit when $\Delta \xi \to 0$. The Von Mises problem is entirely solved and we obtain [3]

Theorem 5. *Let A_d be a lipschitzian non-decreasing function verifying $A_d(0) = 0$. For all $\xi_0 > 0$, there exist a lipschitzian non-increasing function p and a lipschitzian concave function $w \geq 2\psi$ which is once differentiable with respect to ξ and twice with respect to ψ almost everywhere in the strong sense, such that*

(9)-(10)-(11) are verified strongly and such that (8) is verified almost everywhere in $]0,\xi_0[\,\times\,]0,+\infty[$. For all $\gamma > 0$, there exists $k > 0$ such that

$$|w - 2\psi + 2p|, \quad |\partial_\psi w - 2|, \quad |\sqrt{w}\,\partial_\psi^2 w| \leq k\,e^{-\gamma\psi^{3/4}}. \qquad (24)$$

The inequalities $p' \leq 0$ and $w \geq 2\psi$ were expected. It would be possible to solve the problem in the general case of lipschitzian displacements if the inequality $q^n \leq \max_{0 \leq i \leq n} 2\,p'^{\,i}$ was proved [3].

Then, we look for a solution in physical variables. The inverse Von Mises transformation is well defined since $w^n \geq 2\psi$. We prove its regularity using an original expression of y where the displacement appears [3]

$$y(\xi,\psi) = \sqrt{2\psi} - A_d(\xi) + \int_\psi^{+\infty} \left(\frac{1}{\sqrt{2t}} - \frac{1}{\sqrt{w(\xi,t)}} \right) dt. \qquad (25)$$

Theorem 6. *Let A_d be a lipschitzian non-decreasing function verifying $A_d(0) = 0$. For all $\xi_0 > 0$, there exist a lipschitzian non-increasing function p, a concave function $u \geq y$ which is once differentiable with respect to x and twice with respect to y a. e. in the strong sense and a function v once differentiable with respect to y a. e. in the strong sense, such that the following holds. The equations (1)-(2) are verified almost everywhere in $]0,\xi_0[\,\times\,]0,+\infty[$ and the conditions (4)-(5)-(6) are verified strongly except $v(x,0) = 0$ which is not proved. For all $\gamma > 0$, there exists $k > 0$ such that*

$$\left. \begin{array}{l} |u - y - A_d|, \quad |\partial_y u - 1|, \quad |\partial_y^2 u| \\ |\partial_x u - A_d'|, \quad |v + A_d'\,y + A_d\,A_d' + p'| \end{array} \right\} \leq k\,e^{-\gamma y^{3/2}}.$$

The condition $v(x,0) = 0$ could be proved if the solution was more regular. This result is probable. Indeed, the study of the function $A'^{\,n}(p'^{\,n})$ suggests that the pressure gradients corresponding to lipschitzian displacements are $\frac{1}{3}$-holderian. The available estimates suffice to analyse the behaviours of the solution when $y \to +\infty$ but not when $y \to 0$.

3 Numerical Simulations

We consider the equations in physical variables in order to compute recirculations. *Finite element* schemes and *finite difference* schemes have been written. Small recirculations have been computed and the stability of the scheme is similar to that of the other known schemes when $u < 0$.

Figures 4 and 5 represent the streamlines and the pressure gradient corresponding to a null displacement and to the geometrical perturbation (see (7))

$$A_d(x) = 0.8 \left[1 + \cos\left(\frac{2\pi x}{0.64} + \pi \right) \right]. \qquad (26)$$

A very surprising result of non-uniqueness with given pressure has been observed [3]. Let $(u_1, v_1, p_1{}')$ be the solution corresponding to a displacement A_{d1}.

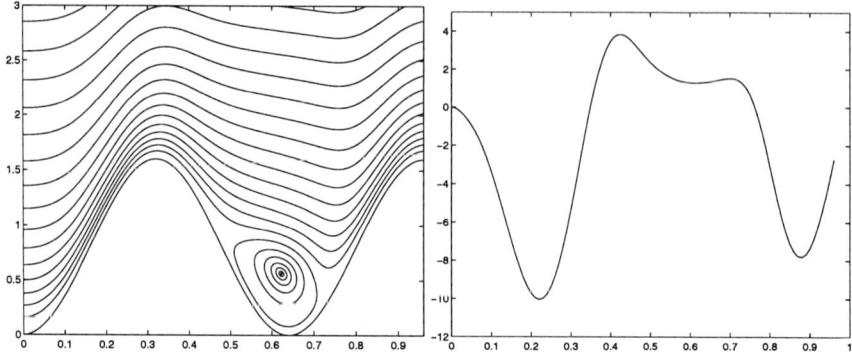

Fig. 4. Streamlines for (26)   **Fig. 5.** Pressure gradient for (26)

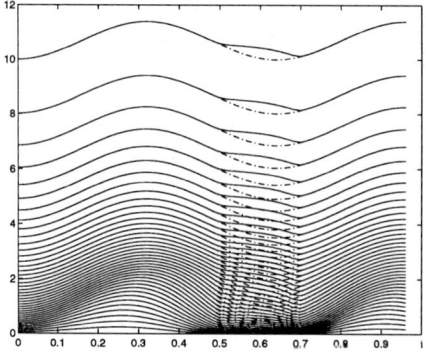

Fig. 6. Curves $x \to u_1(x, y_j)$ and $x \to u_2(x, y_j)$ for $0 \leq y_j \leq 10$ corresponding to the same pressure gradient $p_1{}'$. Dotted lines: solution u_1 with recirculation for A_{d1} given where $A_{d1} = 0.7 \left[1 + \cos\left(\frac{2\pi x}{0.64} + \pi\right)\right]$. Continuous lines: solution u_2 without recirculation for $p_1{}'$ given. Using (5), one can retrieve A_{d1} and A_{d2}

Let us suppose this flow contains a recirculation. Then, we solve the problem with given pressure using $p_1{}'$ so that the Goldstein singularity is avoided and we obtain a displacement A_{d2} and a solution $(u_2, v_2, p_1{}')$ which coincide with A_{d1} and $(u_1, v_1, p_1{}')$ before and after the recirculation but u_2 is always nonnegative. This second solution is then a solution without recirculation. The identity between the solutions after the reattachment point tends to confirm the validity of the solution with recirculation.

This non-uniqueness result is consistent with the absence of direct relation between A_d and p and their difference of regulatity (A_d lipschitzian and p' $\frac{1}{3}$-holderian). This strengthens the difference with the Prandtl problem.

References

1. Nickel, K.: Mathematische Entwicklungen in der Grenzschichttheorie während der letzten 25 Jahre. Z. angew. Math. und Mech. **64** (1984) 18–33.
2. Oleinik, Olga, A.: On a system of equations in boundary layer theory. Zhurn. Vychislit. Mat. Fiz. n° **3** (Engl. transl. in: USSR Comput. Math. Math. Phys. n° **3** (1963) 650–673).
3. Plantié, L.: Le problème de la couche interne des modèles asymptotiques de type triple couche : modèle, analyse et simulations numériques. Ph.D. thesis, Department of Applied Mathematics, Université Paul Sabatier, Toulouse (1997).
4. Plantié, L.: A semi-discrete problem for the boundary layer of triple deck type (part I). CERFACS Report TR/PA/00/45 (2000) `http://www.cerfacs.fr/algor`.
5. Plantié, L.: A semi-discrete problem for the boundary layer of triple deck type (part II). CERFACS Report TR/PA/00/46 (2000) `http://www.cerfacs.fr/algor`.
6. Plantié, L., Mauss, J.: Couches limites interactives pour l'écoulement de Couette dans un canal indenté. C. R. Acad. Sci. Paris, t. **325**, Série II *b* (1997) 693–699.
7. Stewartson, K., Williams, P., G.: Self induced separation. Proc. Roy. Soc. London, A **312** (1969) 181–206.

Cellular Neural Network Model for Nonlinear Waves in Medium with Exponential Memory

Peter Popivanov and Angela Slavova

Institute of Mathematics, Bulgarian Academy of Sciences
Sofia 1113, Bulgaria

1 Nonlinear Waves in Medium with Memory

This paper deals with one dimensional waves in medium with memory. Following [1] we shall denote by x a co-ordinate of a point belonging to a solid body, by t- the time variable, by ε- the deformation, by σ- the tension and b

$$\varepsilon(t) = \int_{-\infty}^{t} \sqrt{1+K^*}\sqrt{a'(\sigma)}\sqrt{1+K^*}\sqrt{a'(\sigma)}\sigma'_t \, dt. \tag{1}$$

In the previous equality K^* is the convolution operator:

$$K^*u(t) = \int_{-\infty}^{t} K(t-\tau)u(\tau)\,d\tau, \tag{2}$$

$\sqrt{1+K^*}$ stands for the development of the operator $1+K^*$ into a power series and the integral operator $\sqrt{1+K^*}$ as well as the multiplication operator $a'(\sigma)$ are acting on the function σ'_t.

It is well known from classical mechanics that the next equation holds:

$$\frac{\partial^2 \varepsilon}{\partial t^2} - \frac{\partial^2 \sigma}{\partial x^2} = 0, \tag{3}$$

supposing ε and σ to be smooth functions of (t,x).

Putting (1) into (3) we conclude that the tension $\sigma(t,x)$ satisfies a rather complicated nonlinear integro-differential equation. According to Theorem 7.1 from [1] the equation (3) with ε given by (1) can be sharply factorized into two first order factors describing the propagation of two waves of tension to the left and to the right-hand side respectively.

Here are the factors:

$$\frac{\partial}{\partial t}\sqrt{1+K^*}\sqrt{a'(\sigma)} \pm \frac{\partial}{\partial x},$$

$$\sqrt{1+K^*}\sqrt{a'(\sigma)}\frac{\partial}{\partial t} \pm \frac{\partial}{\partial x}. \tag{4}$$

We shall concentrate our attention to (4).

Putting $(\sqrt{1+K^*})^{-1} = 1 - \Phi^*$,

$$\Phi^* u = \int_{-\infty}^{t} \Phi(t-\tau) u(\tau)\, d\tau$$

we see that each smooth solution σ of the nonlinear integro-differential equation

$$\sqrt{a'(\sigma)}\frac{\partial \sigma}{\partial t} \pm (1 - \Phi^*)\frac{\partial \sigma}{\partial x} = 0, \ \sigma \in C^2(x \geq 0) \quad (5)$$

will satisfy (3) with ε given by (1).

According to the mechanical terminology the function Φ is called "kernel of heredity". Assume that

$$\Phi(t) = k e^{-kt}, \ k > 0.$$

So we have that a wave of tension, propagating "to the right-hand side" is given by next nonlinear first order equation:

$$\sqrt{a'(\sigma)}\frac{\partial \sigma}{\partial t} + \frac{\partial \sigma}{\partial x} - k \int_{-\infty}^{t} e^{-k(t-\tau)} \frac{\partial \sigma(\tau, x)}{\partial x}\, d\tau = 0. \quad (6)$$

We shall assume, moreover, that

$$a(0) = 0,\ a \in C^2,\ a'(\sigma) > 0 \text{ and } \sigma \in C^2(x \geq 0). \quad (7)$$

$$\sigma = 0 \text{ for } x \geq 0,\ t \leq 0, \sigma(t,0) = \sigma_0(t) \in C^2(R). \quad (8)$$

Obviously, $\sigma_0(t) \equiv 0$ for $t \leq 0$.

We shall construct a classical solution of the mixed problem (6), (8) and we shall prove results for globaly existence in time $t \geq 0$, $x \geq 0$ and for blow up of the corresponding solution. The symbol $\|\sigma_0\|_{C^0(R^1)}$ stands for the uniform norm of function σ_0. We suppose further on that $\|\sigma_0\|_{C^0(R)}, \|\sigma_0'\|_{C^0(R)} < \infty$.

This is our main result.

Theorem 1. *Consider the mixed problem (6), (8). Then*

(i) There exists a constant $C(\|\sigma_0(t)\|_{C^0})$ and such that if $\frac{\|\sigma_0'\|_{C^0}}{k}$
$\times\, C(\|\sigma_0\|_{C^0}) < 1$ *then the problem (6), (8) possesses a unique global classical solution $\sigma \in C^2(x \geq 0, t \geq 0)$.*

The constant $C(\|\sigma_0\|_{C^0})$ can be estimated in the following way:

$$C(\|\sigma_0\|_{C^0}) \leq max_{|\bar{\sigma}| \leq \|\sigma_0\|_{C^0}} \sqrt{a'(\bar{\sigma})} \cdot \frac{1}{min_{|\bar{\sigma}| \leq \|\sigma_0\|_{C^0}} \sqrt{a'(\bar{\sigma})}}.$$

$$.1/2 max_{|\bar{\sigma}| \leq \|\sigma_0\|_{C^0}} \frac{|a''(\bar{\sigma})|}{|a'(\bar{\sigma})|}.$$

(ii) σ_t blows up for a finite $X > 0$ if

a). one can find a point $\beta_0 > 0$ with the property

$$1 + \frac{\sigma_0'(\beta_0)}{kw_0(\beta_0)}(\sqrt{a'(\sigma_0(\beta_0))} - \sqrt{a'(0)}) < 0,$$

where $w_0(\beta_0) = \int_0^{\sigma_0(\beta_0)} \sqrt{a'(\lambda)}\, d\lambda$, $\sigma_0(\beta_0) > 0$.
b). one can find a point $\beta_0 > 0$ such that $\sigma_0(\beta_0) = 0$,

$$-1/2 < \frac{ka'(0)}{\sigma_0'(\beta_0)a''(0)} < 0, \ \sigma_0'(\beta_0) \neq 0, a''(0) \neq 0.$$

The life span X of the corresponding solution in case b). can be estimated in the next way:

$$X \leq \bar{x} = -\frac{1}{\sqrt{a'(0)}} \ln(1 + \frac{2ka'(0)}{\sigma_0'(\beta_0)a''(0)}) > 0.$$

Differentiating (6) in t we have

$$\frac{\partial}{\partial t}(\sqrt{a'(\sigma)}\frac{\partial \sigma}{\partial t}) + \frac{\partial}{\partial t}\frac{\partial \sigma}{\partial x} - k\frac{\partial \sigma}{\partial x} + k^2 \int_{-\infty}^{t} e^{-k(t-\tau)}\frac{\partial \sigma}{\partial x}(\tau, x)d\tau = 0.$$

So

$$\frac{\partial}{\partial t}(\sqrt{a'(\sigma)}\frac{\partial \sigma}{\partial t} + \frac{\partial \sigma}{\partial x} + k\int_0^{\sigma(t,x)} \sqrt{a'(\lambda)}\, d\lambda) = 0,$$

i.e.

$$\sqrt{a'(\sigma)}\frac{\partial \sigma}{\partial t} + \frac{\partial \sigma}{\partial x} + k\int_0^{\sigma(t,x)} \sqrt{a'(\lambda)}\, d\lambda = f(x).$$

According to (8): $f(x) = 0$.
So we reduced the mixed problem (6), (8) to the following nonlinear equation:

$$\frac{\partial}{\partial t}\int_0^{\sigma(t,x)} \sqrt{a'(\lambda)}\, d\lambda + \frac{\partial \sigma}{\partial x} + k\int_0^{\sigma(t,x)} \sqrt{a'(\lambda)}\, d\lambda = 0, \quad (9)$$

$\sigma(t,0) = \sigma_0(t)$, $\sigma_0(t) \equiv 0$, $t \leq 0$, $\sigma = 0$, for $x \geq 0, t \leq 0$, $\sigma_0 \in C^2(R)$.

Let us make the change of the unknown function

$$w = \int_0^{\sigma} \sqrt{a'(\lambda)}\, d\lambda. \quad (10)$$

Obviously, $w' = \sqrt{a'} > 0 \Rightarrow$ there exists $\varphi \in C^2$, such that

$$\sigma = \varphi(w) \quad (11)$$

(i.e. φ is the inverse function, defined by (10)).

Then (9) will be rewritten in the form

$$\frac{\partial w}{\partial t} + \varphi'(w)\frac{\partial w}{\partial x} + kw = 0, \qquad (12)$$

$$w(t,0) = \int_0^{\sigma_0(t)} \sqrt{a'(\lambda)}\, d\lambda \equiv w_0(t),$$

$w_0(t) \equiv 0$ for $t \leq 0$, $w_0 \in C^2(R)$, $w = 0$ for $x \geq 0$, $t \leq 0$.
In fact, $w(t,x) = w(\sigma(t,x)) = \int_0^{\sigma(t,x)} \sqrt{a'(\lambda)}\, d\lambda$ and therefore

$$\sigma(t,x) = \varphi(w(t,x)), \quad \frac{\partial \sigma}{\partial x} = \varphi'(w)\frac{\partial w}{\partial x}, \quad \frac{\partial w}{\partial t} = \frac{\partial \sigma}{\partial t}\sqrt{a'(\sigma)}.$$

Remark. The function $G(\sigma) = \int_0^\sigma \sqrt{a'(\lambda)}\, d\lambda$ is a diffeomorphism: $G : (-\infty, +\infty)$
$\to (A, B)$, where $A = \int_0^{-\infty} \sqrt{a'(\lambda)}\, d\lambda$, $B = \int_0^{\infty} \sqrt{a'(\lambda)}\, d\lambda$, $-\infty \leq A < 0 < B \leq \infty$. Thus, $\sigma = G^{-1}(w) = \varphi(w)$ is well defined and smooth on the open interval (A, B), $w_0(t) = G(\sigma_0(t)) \iff \sigma_0(t) = \varphi(w_0(t))$.

2 Cellular Neural Networks (CNNs)

Cellular Neural Networks (CNNs) are nonlinear, continuous computing array structures well suited for nonlinear signal processing. Since its invention in 1988 [2,3], the investigation of CNNs has envolved to cover a very broad class of problems and frameworks. Many researchers have made significant contributions to the study of CNN phenomena using different mathematical tools.

Definition 1. *The CNN is a*
  i). 2-, 3-, or n- dimensional **array of**
  ii). mainly identical **dynamical systems,** *called cells, which satisfies two properties:*
  iii). most **interactions are local** *within a finite radius r, and*
  iv). all **state variables are continuous valued** *signals.*

Let us consider a two-dimensional grid with 3×3 neighborhood system as it is shown on Fig.1.

The squares are the circuit units - cells, and the links between the cells indicate that there are interactions between linked cells. One of the key features of a CNN is that the individual cells are nonlinear dynamical systems, but that the coupling between them is linear. Roughly speaking, one could say that these arrays are nonlinear but have a linear spatial structure, which makes the use of techniques for their investigation common in engineering or physics attractive.

Definition 2. *An $M \times M$ cellular neural network is defined mathematically by four specifications:*

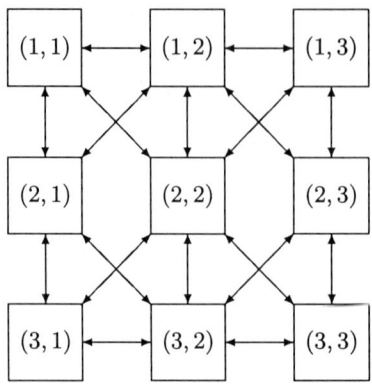

Fig. 1.

1). CNN cell dynamics;
2). CNN synaptic law which represents the interactions (spatial coupling) within the neighbor cells;
3). Boundary conditions;
4). Initial conditions.

Suppose for simplicity that the processing elements of a CNN are arranged on a 2- dimensional (2-D) grid (Fig.1). Then the dynamics of a CNN, in general, can be described by:

$$\dot{x}_{ij}(t) = -x_{ij}(t) + \sum_{C(kl) \in N_r(ij)} \tilde{A}_{ij,kl}(y_{kl}(t), y_{ij}(t)) + \quad (13)$$

$$+ \sum_{C(kl) \in N_r(ij)} \tilde{B}_{ij,kl}(u_{kl}, u_{ij}) + I_{ij},$$

$$y_{ij}(t) = f(x_{ij}), \quad (14)$$

$$1 \leq i \leq M, 1 \leq j \leq M,$$

x_{ij}, y_{ij}, u_{ij} refer to the state, output and input voltage of a cell $C(i,j)$; $C(ij)$ refers to a grid point associated with a cell on the 2-D grid, $C(kl) \in N_r(ij)$ is a grid point (cell) in the neighborhood within a radius r of the cell $C(ij)$, I_{ij} is an independent current sourse. $\tilde{A}$ and $\tilde{B}$ are nonlinear cloning templates, which specify the interactions between each cell and all its neighbor cells in terms of their input, state, and output variables [9,10].

Now in terms of definition 2 we can make a generalization of the above dynamical systems describing CNNs. For a general CNN whose cells are made of time-invariant circuit elements, each cell $C(ij)$ is characterized by its CNN cell dynamics :

$$\dot{x}_{ij} = -g(x_{ij}, u_{ij}, I^s_{ij}), \quad (15)$$

where $x_{ij} \in \mathbf{R}^m$, u_{ij} is usualy a scalar. In most cases, the interactions (spatial coupling) with the neighbor cell $C(i+k, j+l)$ are specified by a CNN synaptic law:

$$I_{ij}^s = A_{ij,kl} x_{i+k,j+l} + \tilde{A}_{ij,kl} * f_{kl}(x_{ij}, x_{i+k,j+l}) + \qquad (16)$$
$$+ \tilde{B}_{ij,kl} * u_{i+k,j+l}(t).$$

The first term $A_{ij,kl} x_{i+k,j+l}$ is simply a linear feedback of the states of the neighborhood nodes. The second term provides an arbitrary nonlinear coupling, and the third term accounts for the contributions from the external inputs of each neighbor cell that is located in the N_r neighborhood.

As it was stated in [4,8], some autonomous CNNs (there are no inputs, i.e. $u_{ij} \equiv 0$) represent an excellent approximation to the nonlinear partial diffrential equations (PDEs). Although the CNN equations describing reaction-diffusion systems are with the large number of cells, they can exhibit new phenomena that can not be obtained from their limiting PDEs. This demonstrates that an autonomous CNN is in some sense more general than its associated nonlinear PDE.

3 CNN Model for Nonlinear Waves in Medium with Memory

Let us consider equation (12) in the following form:

$$\frac{\partial w}{\partial t} = -\varphi'(w) \frac{\partial w}{\partial x} - kw. \qquad (17)$$

For solving such an equation spatial discretization has to be applied. The PDE is transformed into a system of ODEs which is identified as the state equations of an autonomous CNN with appropriate templates. The discretization in space is made in equidistant discrete steps h. We map $w(x,t)$ into a CNN layer such that the state voltage of a CNN cell $x_{ij}(t)$ at a grid point (i,j) is associated with $w(ih, t)$, $h = \Delta x$. Hence, the following CNN model is obtained:

$$\frac{dw_i}{dt} = -\varphi'(w_i) \frac{(w_{i+1} - w_{i-1})}{h} - kw_i. \qquad (18)$$

If we compare the above equation with the state equation of nonlinear CNN we directly find the templates:

$$\hat{A} = [\frac{\varphi'}{h} \quad -k \quad -\frac{\varphi'}{h}].$$

We will consider the following examples for our CNN model (18):

Let $a(\lambda) = \frac{e^{2\lambda}}{2} - \frac{1}{2}$. Then $w = \int_0^\sigma \sqrt{a'} d\lambda = e^\sigma - 1 \Rightarrow \sigma = ln(w+1)$ and $\varphi'(w) = \frac{1}{w+1}$.

a). The initial condition is:

$$w_0 = \begin{cases} 0, & t \leq 0, \\ -\frac{1}{2}\sin t, & t > 0. \end{cases}$$

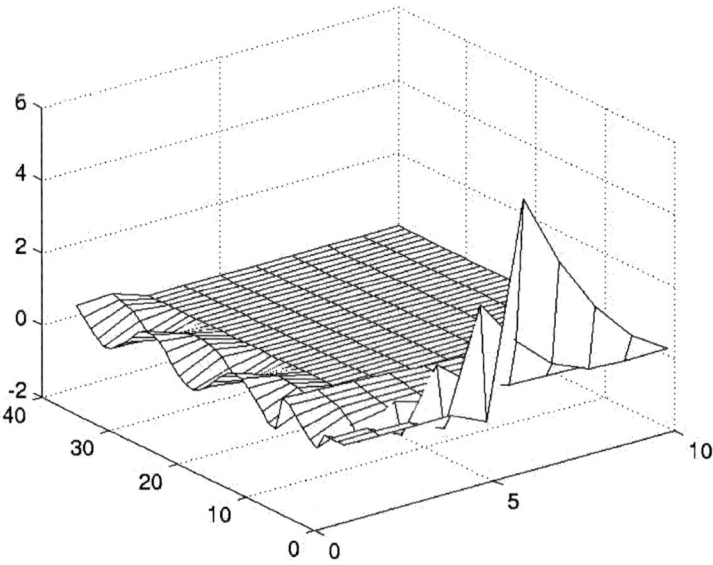

Fig. 2.

b). The initial condition is:

$$w_0 = \begin{cases} 0, & t \leq 0, \\ 1 - \cos t, & t > 0. \end{cases}$$

Acknowledgments

This paper is partially supported by Grant MM-706.

References

1. Alinhac, S.: Blow up for nonlinear hyperbolic equations, Birkhauser, Progress in Nonlinear Diff. Eq. and their Appl., **17** (1995)
2. Chua L. O., Yang L.: Cellular neural networks: Theory, IEEE Trans. Circuit Syst., **35** Oct. (1988) 1257–1271

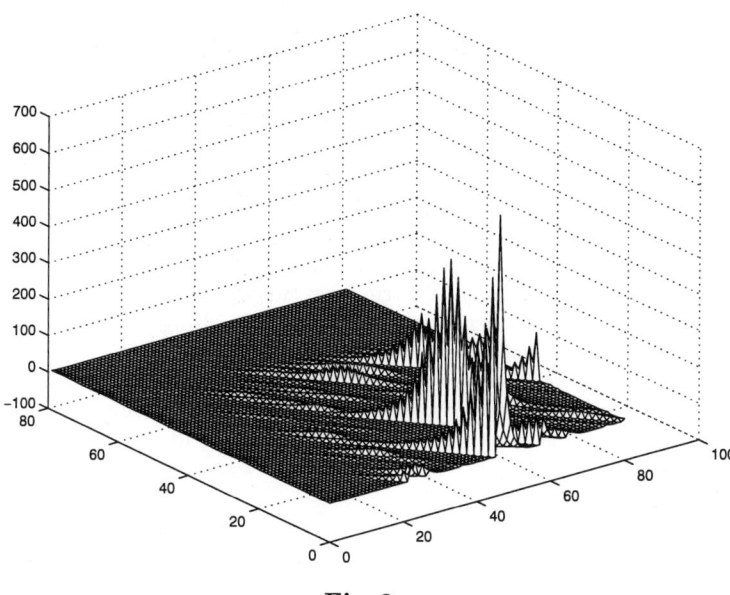

Fig. 3.

3. Chua L. O., Yang L.: Cellular neural networks: Applications, IEEE Trans. Circuit Syst., **35** Oct. (1988) 1272–1290
4. Lokshin, A., Sagomonian, S.: Nonlinear waves in the mechanics of solid bodies, Edition of the Moscow State University, Moscow, 1989 (in Russian)
5. Ta-tsien, Li: Global classical solutions for quasilinear hyperbolic systems, Research in Appl. Math., John Wiley & Sons, Masson, 1994
6. Roska, T., Chua, L., Wolf, D., Kozek, T., Tetzlaff, R., Puffer, F.: Simulating nonlinear waves and PDEs via CNN - Part I: Basic Techniques, Part II: Typical Examples, IEEE Trans. Circuit and Syst. - I, **42** N 10 (1995) 809–820

Numerical Analysis of the Nonlinear Instability of One-Dimensional Compound Capillary Jet

St. Radev[1], M. Kaschiev[2], M. Koleva[2], L. Tadrist[3], and F. Onofri[3]

[1] Institute of Mechanics, Bulgarian Academy of Sciences
Acad. G.Bonchev str. Bl.4, 1113 Sofia, Bulgaria
stradev@sradev1.imbm.bas.bg

[2] Institute of Mathematics and Informatics, Bulgarian Academy of Sciences
Acad. G.Bonchev str. Bl.8, 1113 Sofia, Bulgaria
{kaschievd,mkoleva}@math.bas.bg

[3] Institut Universitaire des Systemes Thermiqus Industriels
Universite de Provence, Technopole de Chateau Gombert
5, rue Enrico Fermi, 13453 Marseille Cedex 13, France
{ltadrist,onofri}@iusti.univ-mrs.fr

Abstract. The nonlinear instability of a compound jet consisting of a liquid core and immiscible coaxial liquid layer is studied. The equations of motion for both liquids (phases) are used in one-dimensional (1-D) approximation similar to that known for one-layer jet. A numerical method is proposed for calculation the radiuses of both interfaces and axial velocities of the core and outer layer. The method is tested for determining the typical forms of compound jet disintegration.

1 Introduction

The compound jet generation principles and a qualitative description of the hydrodynamic of the jet have been given by Hertz and Hermanrud [1]. In their experiments they observed three different types of compound jet instability, namely capillary, sinuous and varicose instability depending on the jet velocity. The present paper is restricted to the analysis of the capillary instability only. The latter manifests itself into disintegration of the jet into drops of different configurations and sizes.

The first models developed to study this kind of compound jet instability are based on the one–dimensional approximation of the Navier-Stokes equations. Based on this approximation in Radev and Shkadov [2] a linear analysis of the jet instability is performed which reveals three different break-up regimes, namely breaking as a single jet, breaking of the core and disintegration by meeting of the interfaces. (Further on for brevity these regimes will be referred as First, Second and Third break-up regimes, respectively). Similar analysis is proposed by Sanz and Meseguer [3].

As it could be expected the above linear models are well suited to the initial evolution of the perturbations along the jet but failed to predict the final break-up configuration, which is strongly controlled by the nonlinear effects. The latter are taken into account in Epikhin et al.[4] and Radev et al. [5] in which the jet flow is assumed of uniform velocity profile and approximated by one–dimensional equations of motion. The disturbances are considered periodical in space of a given wave length, whose amplitude increases in time. The analysis in Epikhin et al. [4] is made by a decomposition of the disturbances in a Fourier series with unknown amplitudes, while in Radev et al. [5] a spline-difference numerical method is proposed. The experimental observations that the jet break-up gives rise of both main and satellite drops are confirmed numerically as well. Moreover it is shown that the satellites for the First disintegration regime are formed from the core liquid only and are entrained by the layer flow. In the Second regime the compound satellite drops appear consisting of a core and concentric layer formed from the jet core and surrounding layer respectively.

For completeness it should be mentioned that 2-D models of the compound jet instability are proposed in Tchavdarov and Radev[6] and Tchavdarov et al. [7]. In the former a linear analysis is performed while the latter is concerned with a direct numerical simulation.

The present paper deals with the nonlinear instability of a one-dimensional compound jet. A numerical method is proposed for calculating the evolution in time of both the interface radiuses and core and layer velocities. It allows accounting for a stepwise profile of the undisturbed velocity. The method is illustrated by the typical disintegration forms of the jet.

2 Statement of the Problem

The compound jet shown in Fig. 1 consists of an axisymmetrical liquid core of (undisturbed) radius H_1 and density ρ_1 and a surrounding coaxial layer of another immiscible liquid of outer radius H_2 and density ρ_2. Both liquids are assumed incompressible and nonviscous. Hereafter the subscript $j = 1$ is set for the core, whereas $j = 2$ is used for the layer.

The jet flow is related to a cylindrical coordinate system (r, z), whose z – axis is directed along the jet axis. By using H_* and U_* as respectively linear and velocity scales the 1-D equations of motion of the jet could be written in the following nondimensional form (for more details see Radev and Shkadov (1985)[2])

$$\frac{\partial u_j}{\partial t} + u_j \frac{\partial u_j}{\partial z} = -\frac{\partial p_j}{\partial z}, \quad j = 1, 2, \tag{1}$$

where the axial velocities $u_j = u_j(t, z)$ and the pressures $p_j = p_j(t, z)$ are unknown functions of the time and axial coordinate.

Partial differential equations for the unknown radiuses $r = h_j(t, z)$ of the inner and outer interfaces are derived from the mass-conservation equation written simultaneously for the core and layer

$$\frac{\partial h_1}{\partial t} + u_1 \frac{\partial h_1}{\partial z} + \frac{1}{2} h_1 \frac{\partial u_1}{\partial z} = 0, \tag{2}$$

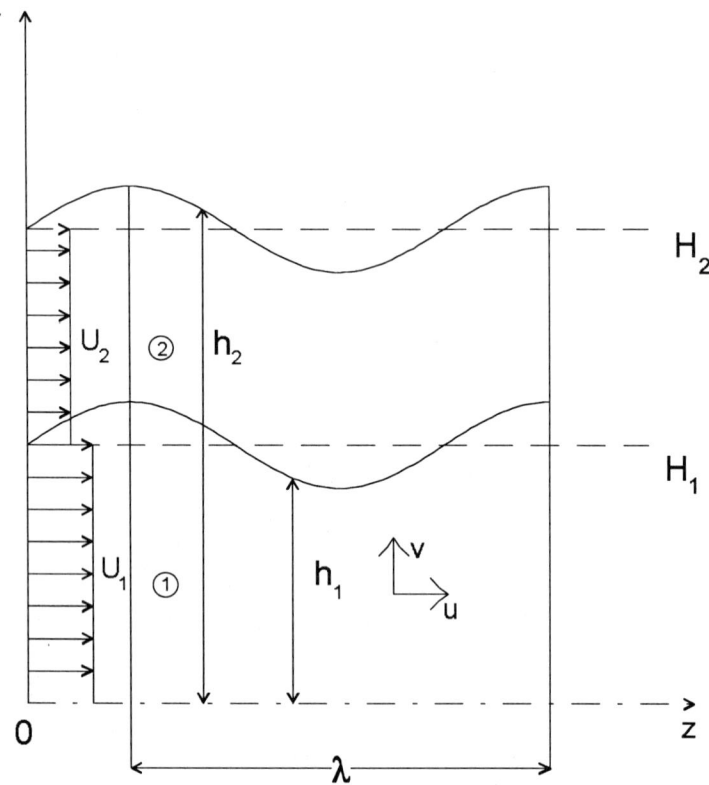

Fig. 1. Compound jet section of length λ related to a cylindrical coordinate system. The undisturbed core and jet are assumed of constant radiuses (H_1 and H_2 respectively) and of uniform axial velocities (U_1 and U_2), the latter allowing for a velocity jump (discontinuity) $\Delta U = U_1 - U_2 > 0$; λ stands for the wave length of the imposed disturbances

$$\frac{\partial h_2}{\partial t} + u_2 \frac{\partial h_2}{\partial z} + \frac{1}{2}\left[1 - \left(\frac{h_1}{h_2}\right)^2\right] h_2 \frac{\partial u_2}{\partial z} + \frac{1}{2}\frac{h_1}{h_2} h_1 \frac{\partial u_1}{\partial z} + (u_1 - u_2)\frac{h_1}{h_2}\frac{\partial h_1}{\partial z} = 0. \quad (3)$$

The pressure terms in eq. (1) are given in the form

$$p_j = \frac{\rho_2}{\rho_j} p_{j+1} + \sigma_j \kappa_j, \quad j = 1, 2, \quad (4)$$

where κ_j are the mean curvature of the interfaces

$$\kappa_j = \left[1 + \left(\frac{\partial h_j}{\partial z}\right)^2\right]^{-1/2} \left\{\frac{1}{h_j} - \left[1 + \left(\frac{\partial h_j}{\partial z}\right)^2\right]^{-1} \frac{\partial^2 h_j}{\partial z^2}\right\}, \quad (5)$$

while $\sigma_j = T_j/(\rho_j H_* U_*^2)$ denote the corresponding inverse Weber numbers related to the inner and outer surface tensions T_j.

In the absence of gravity it is convenient to seek spatially periodical solutions of the above system of partial differential equations, that is

$$h_j(t, z+\lambda) = h_j(t,z), \qquad u_j(t, z+\lambda) = u_j(t,z),$$

$$\frac{\partial h_j}{\partial z}(t, z+\lambda) = \frac{\partial h_j}{\partial z}(t,z), \qquad \frac{\partial u_j}{\partial z}(t, z+\lambda) = \frac{\partial u_j}{\partial z}(t,z), \qquad (6)$$

$$\frac{\partial^2 h_j}{\partial z^2}(t, z+\lambda) = \frac{\partial^2 h_j}{\partial z^2}(t,z),$$

where λ represents the wave length.

3 Linear Instability Analysis of a Compound Jet

In the context of the linear instability analysis the jet flow is decomposed into a steady and nonsteady (disturbed) part. In the steady case the system (1)-(5) allows a simple solution of the form

$$h_j(t,z) = H_j, \qquad u_j(t,z) = U_j, \qquad (7)$$

representing a compound jet of constant radiuses and uniform axial velocities of the core and coaxial layer.

The perturbed flow is given in the form

$$h_j(t,z) = H_j + \tilde{h}_j(t,z), \quad u_j(t,z) = U_j + \tilde{u}_j(t,z), \quad p_j(t,z) = P_j + \tilde{p}_j(t,z) \quad (8)$$

assuming that the nonlinear terms in respect to the disturbances are small enough to be neglected. The solution of the linearized boundary value problem (1)-(8) appears in an analytical form

$$(\tilde{h}_j, \tilde{u}_j, \tilde{p}_j)(t,z) = (\bar{h}_j, \bar{u}_j, \bar{p}_j) \exp[i\alpha(z-ct)], \qquad (9)$$

where $\alpha = 2\pi/\lambda$ is a given wave number while the complex amplitudes $\bar{h}_j, \bar{u}_j, \bar{p}_j$ and complex phase velocity of the perturbations

$$c = \frac{\omega}{\alpha} + i\frac{q}{\alpha} \qquad (10)$$

are unknown. In equation (10) ω denotes the angular frequency, while $c_r = \omega/\alpha$ stands for the phase velocity and $\alpha c_i = q$ - for the growth rate of the disturbances. The complex phase velocity and the wave number are connected in the following (usually called dispersion) equation

$$\begin{aligned}(U_1-c)^4 - 2(U_1-U_2)(U_1-c)^3 + \\ \left[(U_1-U_2)^2 + \tfrac{1}{2}\sigma_2(1-\delta^2)(1-\alpha^2) + A_1\right](U_1-c)^2 - \\ 2A_1(U_1-U_2)(U_1-c) + \\ \left[A_1(U_1-U_2)^2 + \tfrac{1}{4}\sigma_1\sigma_2\delta^{-1}(1-\delta^2)(1-\delta^2\alpha^2)(1-\alpha^2)\right] = 0,\end{aligned} \qquad (11)$$

where
$$A_1 = \frac{1}{2}\left[\sigma_1\delta^{-1}(1-\delta^2\alpha^2) + \sigma_2\delta^2\frac{\rho_2}{\rho_1}(1-\alpha^2)\right]. \quad (12)$$

In principle the initial conditions for the system (1)-(5) should satisfy the equation (6), otherwise they could be chosen arbitrary. However from a physical point of view it will be of interest to have a possibility to study the evolution of initially small disturbances up to the break-up point. Following the linear instability theory in Radev and Shkadov [2] the form of the jet perturbations of sufficiently small amplitudes is derived from the linearized equations (1)-(5). Below on we briefly present some details concerning the linear instability analysis of a compound jet, which will be used in the formulation of initial conditions for the equations (1)-(5) fitted to the linear solution. For our further considerations we will need some details concerning the solutions of the dispersion equation.

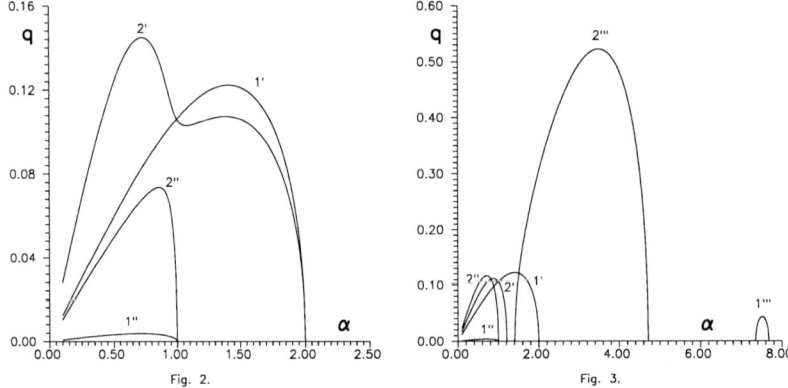

Fig. 2. Amplification rate of the disturbances versus wave number α at a zero undisturbed velocity jump. $\Delta U = 0, \sigma_1 = 0.015, \delta = 0.5, \rho_2/\rho_1 = 1$; Curves $1'$ and $1''$: $\sigma_1/\sigma_2 = 100$, $2'$ and $2''$: $\sigma_1/\sigma_2 = 0.1$. The superscript $'$ and $''$ above denote the first and second linear modes respectively. The maximum growth rate within the curve $1'$ is controlled by the inner surface tension. When the outer surface tension increases this maximum moves into the range of the long waves (curve $2'$)

Fig. 3. The effect of the undisturbed velocity jump on the growth rate. $\Delta U = 0.5, \sigma_1 = 0.015, \delta = 0.5, \rho_2/\rho_1 = 1$. Curves $1'$, $1''$ and $1'''$: $\sigma_1/\sigma_2 = 100$, $2'$, $2''$ and $2'''$: $\sigma_1/\sigma_2 = 0.1$. The superscripts $'$, $''$ and $'''$ above denote the first, second and third linear modes respectively. In the interval of the very short waves a third mode is burned (curve $1'''$). When the outer surface tension increases this mode moves into the range of the long waves with the highest growth rate inside it (curve $2'''$). Simultaneously the second mode (curve $2''$) tends to move above the first mode (curve $2'$) at the begining of long wave interval

This is an algebraic equation of fourth order for calculating the complex phase velocity c as a function of the wave number at given values of the nondimensional parameters $\sigma_j, U_j, \delta = H_1/H_2, \rho_0 = \rho_2/\rho_1$. After determining the complex phase velocity the unknown complex amplitudes $\bar{h}_j, \bar{u}_j$ could be found from the linearized equations (1)-(8) provided that the value of one of these amplitudes is given.

In the particular case when the undisturbed velocity profile is uniform in the both phases ($U_1 = U_2$) eq. (11) is reduced to a biquadratic equation. It is easily seen that in general this equation has two pairs of complex conjugate roots: the first one is defined within the wave number interval $0 < \alpha < \delta^{-1}$, while the second - in $0 < \alpha < 1$. The two branches (further on called modes) with positive imaginary parts c_i define two families of disturbances which grow with amplification rates equal to $q = \alpha c_i$ and propagate with one and the same phase velocity $c_r = U_1$. In Fig. 2 the "$q - \alpha$" curves for both modes are illustrated for two characteristic values of the ratio σ_2/σ_1 of the surface tensions. If as usually we assume that in natural conditions the jet is disintegrated by the disturbances of a higher amplification rate then in Fig. 2 they correspond to the maximum of "$q - \alpha$" curve related to the first mode. However in the case of $\sigma_2/\sigma_1 \gg 1$ this maximum (q_I^*) is attached to the wave number close to the Rayleigh one $\alpha^* \approx \sqrt{2}/2$ and is controlled by the outer surface tension. In the case $\sigma_2/\sigma_1 \ll 1$ the maximum (q_I^*) moves to the range of the shorter waves ($\alpha^{**} \approx \sqrt{2}/2\delta$) being controlled by the inner interface.

The "$q - \alpha$" curves in the case of a stepwise velocity profile are shown in Fig. 3 for a given value of the velocity jump $\Delta U = U_1 - U_2 > 0$. The main difference in respect to the case of a continuous velocity profile manifests itself in the appearance in the range of the short waves of a new unstable mode, resulting in a third family of growing disturbances. The maximum growth rate of the disturbances q_{III}^* and the corresponding wave number α^{***} depends on the value of the velocity jump ΔU: when the latter increases the maximum growth rate increases as well, while the wave number α^{***} moves into the direction of the longer waves. Looking at Fig. 3 it should be mentioned that at sufficiently high values of ΔU the maximum growth rate corresponding to the second mode (q_{II}^*) may become higher than to the first mode (q_I^*)).

Coming back to the nonlinear boundary-value problem (1)-(6), it is quit natural to apply equations (8) and (9) as initial conditions for this problem. It is important to note that in the conditions (8) and (9) one of the complex amplitudes say $\bar{h}_j$ must be considered as an additional input parameter of the nonlinear problem. It will be denoted by $\bar{h}_{j0}$ to point out that this is the initial amplitude of the corresponding interface radius at time $t = 0$. As far as the complex phase velocity is explicitly involved in the linearized form of equations (1)-(5) (not written in the paper) the number of the selected mode will act as a second input parameter in the initial conditions (8) and (9).

4 Numerical Method

In order to eliminate the disturbance translation along the jet axis it is convenient to introduce new independent variables (ξ, τ) and new dependent variables (w_j, Π_j) as follows

$$\xi = \alpha z - \omega t, \quad \tau = \alpha\sqrt{\sigma_*}t, \quad 0 \leq \xi \leq 2\pi. \tag{13}$$

$$u_j = U_j + \sqrt{\sigma_*}w_j, \quad \Pi_j = \sigma_*^{-1}p_j. \tag{14}$$

In these expresstions $\omega = \alpha c_r$ and σ_* stands for σ_2 (or σ_1).

Following [8], for solving the nonlinear boundary value problem (1)-(6), written in new variables, we use the Continuous Analog of Newton Method (CANM). A finite difference method of second order for discretization the obtained CANM problem is applied. All results, shown in figures, are obtained using the Crank-Nikolson difference scheme with steps $h_\xi = \pi/200, h_\tau = 0.01$. The CAMN needs 2-3 iterations to solve the problem in each layer $\tau_k = kh_\tau$.

The jet disintegration time τ_b is determined when one of the following conditions is satisfied

$$\min_\xi h_1(\tau_b, \xi) \leq 10^{-2}, \quad or \quad \min_\xi (h_2(\tau_b, \xi) - h_1(\tau_b, \xi)) \leq 10^{-2}.$$

5 Results and Discussion

Due to the fact that the above described problem appears as multiparametric one, it is rather difficult to illustrate the effect of all entering parameters. For that we will limit our discussion to the case of zero velocity jump ΔU. In these conditions the jet instability is mainly controlled by the ratio σ_1/σ_2 of the surface tensions, whose effect will be analysied below. The values of the remaining nondimensional parameters will be fixed as follows:

$$\sigma_1 = 0.015, \delta = 0.5, \rho_2/\rho_1 = 1, h_{20} = 0.01. \tag{15}$$

Moreover we will concentrate our attention to the cases when the jet is initially excited by the perturbations (8) and (9) related to the first mode of the dispersion equation. In general the calculations will be performed for the wave number of the highest amplification rate. The effect of the second and third mode remains to be studied additionally.

5.1 Compound Jet Disintegration at $\sigma_1/\sigma_2 \ll 1$

In this case the jet instability is controlled by the outer surface tension. The jet disintegration behaves like one-layer jet break-up, as shown in Fig.4, whose parameters correspond to the curve $2'$ in Fig.2. The resulting main and satellite drops are compound as well and consist of a core and concentric layer formed by the inner and outer liquid respectively.

5.2 Compound Jet Disintegration at $\sigma_1/\sigma_2 \gg 1$

When the inner surface tension prevaluates the jet instability appears as a core disintegration resulting into main and satellite drop, which after breaking are entrained by the surrounding liquid. This disintegration regime of the compound jet is demonstrated in Fig.5, whose parameters correspond to curve $1'$ in Fig.2. It should be mentioned that after the core break-up the jet still remains continuous up to the breaking of the outer interface. However this break-up regime is out of the scope of our model.

5.3 Compound Jet Disintegration at $\rho_2 < \rho_1$

A new type of jet disintegration appears if in the range $\sigma_1/\sigma_2 \ll 1$ the density of the outer liquid is decreased below the density of the core. As shown in Fig.6 the minimum distance between the interfaces becomes zero, while the inner interface is still far from the jet axis. This form of a jet disintegration is admissible in the numerical experiments only if $\rho_2 < \rho_1$. However in contrast to the disintegration regimes shown in Fig.4 and Fig.5, this in Fig.6 remains to be demonstrated experimentally.

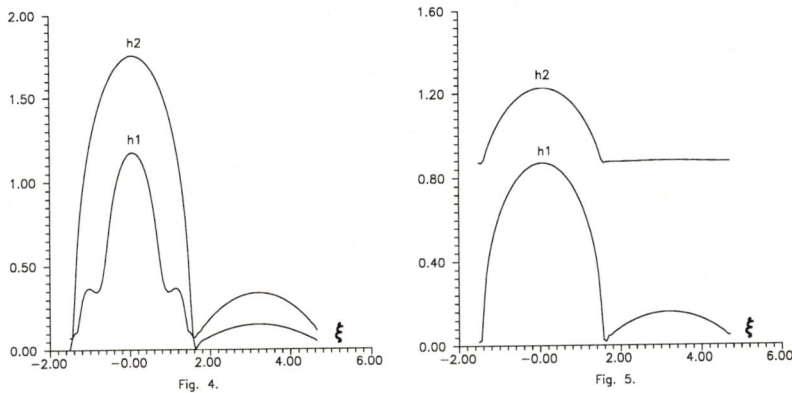

Fig. 4. Compound jet break-up as one-layer jet. $\sigma_1/\sigma_2 = 0.1, \alpha = 0.707, \Delta U = 0, \tau_b = 8.24$. The remaining input parameters are given in (15). The jet is amplified by the corresponding first mode (curve $2'$ in Fig.2). Both interfaces break-up simultaneously at the same points forming one main and one satellite compound drop within one wave length

Fig. 5. Compound jet disintegration due to the core break-up. $\sigma_1/\sigma_2 = 100, \alpha = 1.41, \Delta U = 0, \tau_b = 0.41$. The remaining input parameters are given in (15). The jet is amplified by the corresponding first mode (curve $1'$ in Fig.2). The core breaks-up the first while the layer still exists as a coherent portion. The main and satellite drops detached from the core are entrained by outer flow

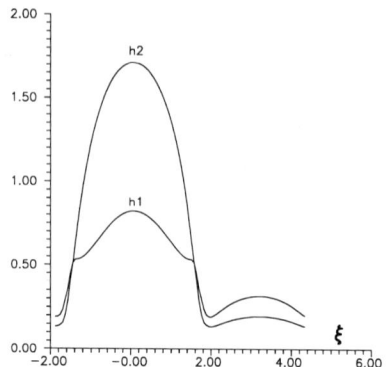

Fig. 6. Compound jet disintegration due to the meeting of the interfaces. $\sigma_1/\sigma_2 = 0.1, \alpha = 0.73, \tau_b = 9.06, \rho_2/\rho_1 = 0.5$ The values of σ_1, δ and h_{20} are given in (15). The jet is amplified by the corresponding first mode. The outer interface approaches the inner one faster than the latter reaches the jet axis

6 Conclusion

The nonlinear instability of a compound jet is studied as a solution of initially small disturbances up to the jet desintegration. It is shown that the nonlinear effects significantly affect the final stages of the jet desintegration. The type of the latter as well as the type of satellite formation is mainly controlled by the ratio of the inner and outer surface tensions. The numerical method developed on the basis of one-dimensional equations of motion accounts for discontinuity (jump) of the velocity in both phases. However the effect of the velocity jump on the jet instability remains to be studied separately.

References

1. Hertz, C. H., Hermanrud, B.: A liquid compound jet, J. Fluid Mech., **131** (1983) 271–287
2. Radev, S. P., Shkadov, V.Ya.: On the stability of two–layer capillary jets, Theor. and Appl. Mech., Bulg. Acad. Sci., **3** (1985) 68–75 (in russian)
3. Sanz, A., Meseguer, J.: One-dimensional linear analysis of the compound jet, J. Fluid Mech., **159** (1985) 55–68
4. Epikhin, V. E., Radev, S. P., Shkadov, V.Ya.: Instability and break–up of two-layer capillary jets, Izv. AN SSSR, Mech. Jidkosti I Gaza, **3** (1987) 29–35 (in russian)
5. Radev, S. P., Boyadjiev, T. L., Puzynin, I. V.: Numerical study of the nonlinear instability of a two–layer capillary jet, JINR Communications P5-86-699, Dubna, 1986 (in russian)
6. Radev, S., Tchavdarov, B.,: Linear capillary instability of compound jets, Int. J. Multiphase Flow, **14** (1988) 67–79

7. Tchavdarov, B., Radev, S., Minev, P.: Numerical analysis of compoud jet disintegration, Comput. Methods Appl. Mech. Engrg., **118** (1994) 121–132
8. St. Radev, M. Koleva, M. Kaschiev, L. Tadrist, Initial Perturbation Effects on the Instability of a Viscous Capillary Jet, Recent Advances in Numerical Methods and Applications, Proc. of 4th Int. Conf. Num. Meth. Appl., 1998, Sofia, Bulgaria, (ed. O. Iliev, M. Kaschiev, S. Margenov, Bl. Sendov, P. Vassilevski), pp.774-882, World Scientific Publ.

Modelling of Equiaxed Microstructure Formation in Solidifying Two–Component Alloys

Norbert Sczygiol

Technical University of Częstochowa,
ul. Dąbrowskiego 73, 42–200 Częstochowa, Poland
norbert.sczygiol@imipkm.pcz.czest.pl

Abstract. The paper deals with a numerical modelling of equiaxed microstructure formation during the solidification of two–component alloys, poured into metal forms. The basic enthalpy formulation was applied to model the solidification. The formulation allows the characteristic dimensions of computed microstructure in thermal calculations to take into account. The so–called indirect model of solidification (solid phase growth), which allows the modelling of all possible solidification courses, from equilibrium to non–equilibrium solidification, was used to model the equiaxed microstructure formation. This model was worked out from an approximate solution of the diffusion equation of solute in a single grain. The equiaxed grain size was dependent on the average velocity of cooling at the moment when the liquid metal reached the temperature of the beginning of solidification. The above simulation was performed using the *NuscaS* computer program, which has been developed at the Technical University of Częstochowa.

1 Introduction

Casting is one of the production methods for machine elements and equipment. Cast products are characterised by the fact that their shapes and properties are formed when liquid metal is passing to the solid state. The casting solidification is a heterogeneous process. This means that solidification proceeds differently in every point of the casting. All possible solidification courses are situated between two extreme cases. The first describes equilibrium and the second non–equilibrium solidification.

The solidification courses, characteristic for the majority of castings, are present between these two extreme cases, both of which are generally difficult to reach in real casting. Significant solute diffusion in the solid phase of growing grains occurs widely in solidification courses. This type of solidification can be called indirect solidification. The solute diffusion has a great influence on the microstructure formed during solidification.

The casting microstructure is mainly composed of three zones of grains: equiaxed chill, columnar and equiaxed. The last one can have a dendritic structure. In many cases the microstructure of whole castings is composed only from equiaxed grains. This often occurs in non–ferrous metal castings.

2 Solidification Model

Solidification is stated by a quasi–linear heat conduction equation containing the heat source term, which describes the rate of latent heat evolution

$$\nabla \cdot (\lambda \nabla T) + \rho_s L \frac{\partial f_s}{\partial t} = c\rho \frac{\partial T}{\partial t}, \qquad (1)$$

where λ is the thermal conductivity coefficient, c is the specific heat, ρ is the density (subscript s refers to the solid phase, l would denoted the liquid phase and f would denoted the pass from the liquid to the solid state), L is the latent heat of solidification and f_s is the solid phase fraction. This equation forms the basis of the thermal description of solidification. Taking into consideration the enthalpy, defined as follows [1,2]

$$H(T) = \int_{T_{\text{ref}}}^{T} c\rho \, dT + \rho_s L \left(1 - f_s(T)\right), \qquad (2)$$

where T_{ref} is the reference temperature, one can pass to the enthalpy formulations of solidification. A few types of enthalpy solidification exist [1,2,3]. The so–called basic enthalpy formulation, which can be presented as [1,2,3,4,5]

$$\nabla \cdot (\lambda \nabla T) = \frac{\partial H}{\partial t}, \qquad (3)$$

is applied in this paper. Eq. (3) is obtained by differentiating the enthalpy given by Eq. (2) with respect to time

$$\frac{\partial H}{\partial t} = c\rho \frac{\partial T}{\partial t} - \rho_s L \frac{\partial f_s}{\partial t}, \qquad (4)$$

and substituting the result into Eq. (1).

The finite element method was used to solve numerically Eq. (3). As a result of semi–discretisation, using the Bubnov–Galerkin method, the following equation was obtained

$$\boldsymbol{M}\dot{\boldsymbol{H}} + \boldsymbol{K}(T)\boldsymbol{T} = \boldsymbol{b}(T), \qquad (5)$$

where $\boldsymbol{M}$ is the mass matrix, $\boldsymbol{K}$ is the conductivity matrix, $\boldsymbol{H}$ is the enthalpy vector, $\boldsymbol{T}$ is the temperature vector and $\boldsymbol{b}$ is the right–hand side vector. This equation must be integrated over time. As the properties of the casting material depend on temperature, it is best to apply a time integration scheme that eliminates the necessity of finding the actual values of the material properties for the calculated temperatures iterativelly. The two–step Dupont II scheme can be applied for this purpose [1]. However, the application of a two–step scheme requires the use of a one–step scheme, i.e. the modified Euler–backward scheme [6], in which the values of material properties are calculated on the basis of a known temperature.

The final form of Eq. (5), after the application of the modified Euler–backward scheme, is as follows [4,5]

$$\left(M + \Delta t K^n \left[\frac{dT}{dH}\right]^n\right) H^{n+1} = \left(M + \Delta t K^n \left[\frac{dT}{dH}\right]^n\right) H^n - \Delta t K^n T^n + \Delta t b^{n+1}, \qquad (6)$$

while the application of the Dupont II scheme gives

$$\left(M + \frac{3}{4}\Delta t K^0 \left[\frac{dT}{dH}\right]^{n+1}\right) H^{n+2} = \left(M + \frac{3}{4}\Delta t K^0 \left[\frac{dT}{dH}\right]^{n+1}\right) H^{n+1} - \frac{3}{4}\Delta t K^0 T^{n+1} - \frac{1}{4}\Delta t K^0 T^n + \frac{3}{4}\Delta t b^{n+2} + \frac{1}{4}\Delta t b^n. \qquad (7)$$

The superscript (^{0}) denotes that the thermal conductivity coefficient is calculated for an extrapolated temperature according to the equation

$$T = \frac{3}{2}T^{n+1} - \frac{1}{2}T^n. \qquad (8)$$

The mass matrix does not contain any of material properties because this properties are placed in the enthalpy. The dT/dH matrix arises from the development of temperature function into a Taylor series, for the time level $n+1$ in Eq. (6) and $n+2$ in Eq. (7). It is a diagonal matrix with coefficients calculated for particular nodes of a finite element. This coefficients are calculated on the basis of equations obtained as a result of differentiating Eq. (2) with respect to temperature in the appropriate temperature intervals. For the interval, in which solidification takes place, one can obtain

$$\frac{dT}{dH} = \frac{1}{c\rho_f - \rho_s L \frac{df_s}{dT}}, \qquad T_S \leq T \leq T_L, \qquad (9)$$

where T_L is the temperature of the begining of solidification (liquidus temperature) and T_S is the temperature of the end of solidification. The application of the above expression in Eqs. (6) and (7) requires a knowledge of the relationship of the solid phase fraction to temperature. Moreover, it is possible to take the forming microstructure directly into account in the above formulation.

From the solution of Eqs. (6) and (7) the enthalpies are obtained. These enthalpies are recalculated into temperatures on the basis of the functions educed from Eq. (2) for particular temperature ranges [5].

3 Solid Phase Growth Model

The behaviour of metal alloys in terms of temperature and chemical constitution is presented with the help of phase diagrams (Fig. 1). The solidus temperature for the equilibrium solidification model is shown as T_S, and the solidus temperature for the indirect solidification model is shown as T_{SE}. The possible solidification runs, between solidus and liquidus lines, are schematically shown for an alloy in which the solute concentration is equal to C_0.

In the case of the non-equilibrium solidification model the eutectic temperature, T_E, is always reached by the solidifying alloy (line 1). This means that a certain last portion of the metal solidifies at a constant temperature. In the case of the equilibrium solidification model (line 2) the temperature of the end of solidification depends on the chemical composition of the alloy. For the indirect solidification model (line 3) the solidification run depends on the diffusion path length of the solute and so on the grain size in the solidifying microstructure.

It is possible to obtain an analytical function which describes the relationship between the solid phase fraction and temperature for two-component metal alloys. This function can be obtained from the solution of the balance equations for the solute mass in a single grain. The balance of the solute mass for the indirect solidification model is as follows [7]

$$m\eta^{m-1}(t)\frac{d\eta(t)}{dt}C_s(\eta(t),t) + m\frac{D_s}{r_g}\eta^{m-1}(t)\frac{\partial C_s(\eta(t),t)}{\partial \xi} + \\ +(1-\eta^m(t))\frac{dC_l}{dt} - m\eta^{m-1}(t)\frac{d\eta(t)}{dt}C_l(t) = 0, \tag{10}$$

where m is a coefficient which equals 1 for plane, 2 for cylindrical and 3 for spherical coordinate systems, C is the solute concentration, η is the current thickness or radius of the solidified part of the grain, r_g is the final thickness or final grain radius, D_s is the solute diffusion coefficient in the solid phase and ξ is the current coordinate.

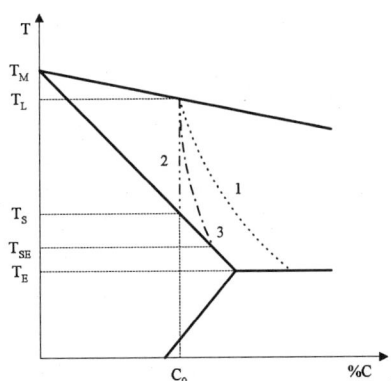

Fig. 1. The solid phase growth models in the two-component alloys (1 – non-equilibrium, 2 – equilibrium, 3 – indirect)

The solution of Eq. (10), after introducing the term of so-called local solidification time t_f and using the relationships received from the phase diagram (connecting the solute concentration with temperature), can be written as [4,5]

$$f_s(T) = \frac{1}{1-nk\alpha}\left(1 - \left(\frac{T_M - T}{T_M - T_L}\right)^{\frac{1-nk\alpha}{k-1}}\right), \tag{11}$$

where n is a coefficient engaging the grain shape ($n = 2$ for plane grain, $n = 4$ for cylindrical (columnar) grain and $n = 6$ for spherical grain) and k is the solute partition coefficient. The α coefficient is defined as

$$\alpha = \frac{D_s}{r_g^2} t_f. \tag{12}$$

The application of Eq. (11) gives physically unrealistic results for α coefficient values above a certain limit value depending on the grain shape. This means that the solid phase fraction is equal to 1 for a temperatures higher than the solidus temperature. One can avoid this inconvenience by introducing an appropriate correction for the α value. In this paper the correction was introduced only for the plane grains, this means for $n = 2$. It equals [8]

$$\Omega(\alpha) = \alpha \left(1 - \exp\left(-\frac{1}{\alpha}\right)\right) - \frac{1}{2}\exp\left(-\frac{1}{2\alpha}\right). \tag{13}$$

The coefficient α can accept any positive value after the application of the above correction, while the coefficient Ω can accept values from 0 to 0.5. The application of correction relies on the replacement of α coefficient with Ω coefficient in Eq. (11).

Substituting $\alpha = 0$ into Eq. (11) one can obtain the relationship of the solid phase fraction for the equilibrium solidification model, while for $\alpha = 1/n$ the relationship of the solid phase fraction for the non–equilibrium solidification model.

4 Equiaxed Microstructure Modelling

The extent of zones with different types of microstructure, as well as the characteristic dimensions of grains in those zones, depend on the degree of undercooling of the melt at the beginning of solidification. The undercooling depends on the velocity of carrying away heat from the casting. Directly taking into account the melt undercooling leads to many numerical difficulties in the solidification model. The assumption, that solidification starts at the liquidus temperature and that the undercooling quantity, represented by the cooling velocity, decides the characteristic dimensions of the created microstructure, is a much better solution.

In the paper it was assumed that only equiaxed microstructure is formed in the casting. Then the final grain radius depends on the cooling velocity, i.e. in the following form [5]

$$r_g = r_b \left(1 - \exp\left(-1/\dot{T}\right)\right), \tag{14}$$

here r_b is the maximal grain radius in the calculated microstructure, while $\dot{T}$ is the average cooling velocity, calculated from the beginning of the cooling process till the liquidus temperature is reached. In Eq. (14) the maximal grain radius depends on the constitution of the casting alloy and should be established experimentally.

5 Example of Computer Simulation

An example computer simulation was carried out for Al – 2% Cu alloy, solidifying in a metal mould. This alloy was chosen because of its wide range of solidification temperatures (40 K). The following values of material properties were used in the calculation: $\rho_s = 2824$ and $\rho_l = 2498$ kg/m^3, $c_s = 1077$ and $c_l = 1275$ J/kg K, $\lambda_s = 262$ and $\lambda = 104$ W/m K, $L = 390000$ J/kg and $k = 0.125$. The linear dependence of the thermal conductivity coefficient with respect to temperature was assumed in the range from liquidus temperature to the temperature of the end of solidification. Temperatures, needed to carry out the numerical simulation, were taken from a phase diagram for the Al – Cu alloys. They are equal to: $T_M = 933$ K, $T_L = 262$ K, $T_S = 886$ K and $T_E = 821$ K.

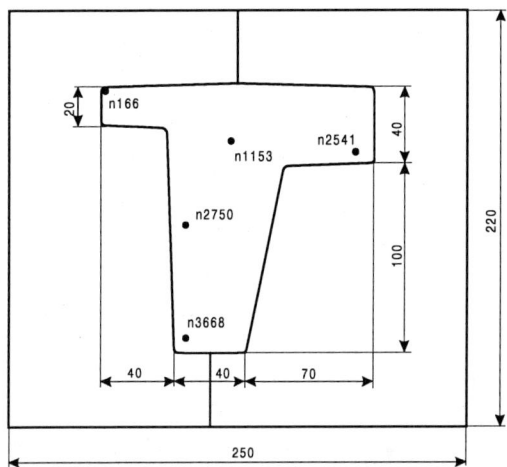

Fig. 2. The analysed casting in the mould

In the calculation it was assumed that the maximal grain radius equals $5 \cdot 10^{-4}$ m. The initial casting temperature was 960 K, while the initial mould temperature was 590 K. The analysed casting together with the mould is shown in Fig. 2. The region was divided into 8609 triangular finite elements, receiving 4659 nodes, with 5815 elements and 3060 nodes in the casting. The continuity conditions were assumed for both the contact between the casting and mould, and two parts of the mould. The heat exchange coefficient through the layer, which separated the casting from the mould, was assumed to be equal to 1000 W/m^2 K, while the heat exchange coefficient between two parts of mould was equal to 800 W/m^2 K. The third type of boundary condition was established on the remaining boundaries. It was assumed that the ambient temperature equals 300 K, while the exchange coefficient with the environment equals 100 W/m^2 K on the top and side–boundaries and 50 W/m^2 K on the bottom boundary. In the calculation of the α coefficient it was assumed that the $D_s t_f$ product is equal to $6 \cdot 10^{-9}$ m^2, while the coefficient engaging the grain shape equals 2. A time step equal to 0.05 s was applied.

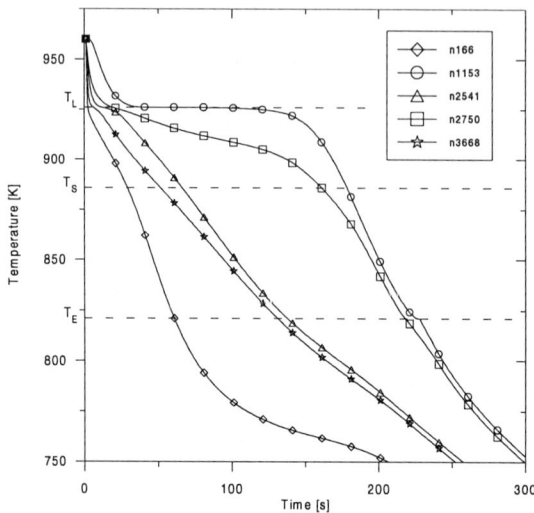

Fig. 3. The cooling curves in the chosen nodes of the casting

The full solidification time equals 235 s. The diagrams showing the cooling curves of the chosen nodes of the finite element mesh displays the solidification course differences in the different casting regions (Fig. 3). The solidification proceeds rapidly in the nodes closest to the mould wall. The biggest grains are formed there (Fig. 4). There is a very wide range of the temperatures for the end of solidification, from 876 K (10 K lower than equilibrium solidus temperature) to eutectic temperature. The values of Ω coefficient also varied widely, from 0.0433 in the central regions to 0.4872 in the layers in contact with the metal mould. In turn, the cooling velocities vary from 0.78 to 29.88 K/s. Because the average radius is the cooling velocity function, there is considerable difference in the grain sizes occurring in the casting. The radii of the smallest grains equals 17, 71 μm, while the radii of the biggest ones equals 368.30 μm.

6 Summary

A new method of numerical modelling of equiaxed microstructure formation in solidifying two–component alloys was presented in this paper. The above mentioned method is based on the so–called indirect solidification (solid phase growth) model. The indirect model, in contrast to commonly used non–equilibrium and equilibrium solidification models, makes it possible to take grain sizes into consideration in the calculation of temperature fields and solidification kinetics. The main advantage of the indirect solidification model is that the temperatures of the end of solidification, determined by this model, can cover the complete range from the equilibrium solidus temperature to eutectic temperature.

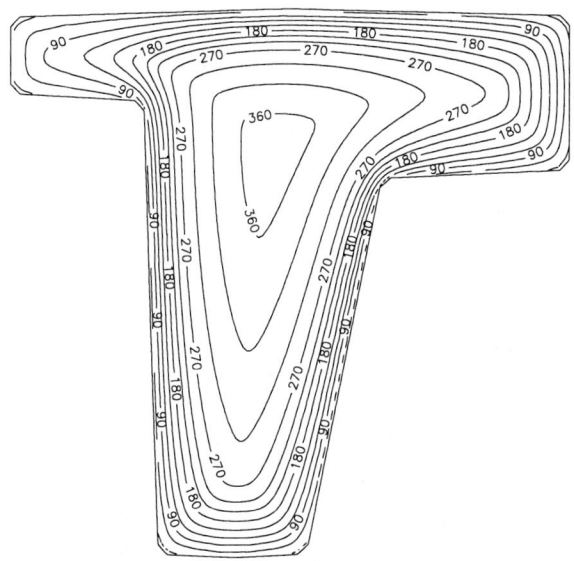

Fig. 4. The distribution of the average radii of grains [μm]

References

1. Dalhuijsen, A. J., Segal, A.: Comparison of finite element techniques for solidification problems. Int. J. Numer. Meth Engng. **23** (1986), 1807–1829
2. Thévoz, Ph., Desbiolles, J. L., Rappaz, M.: Modeling of Equiaxed Microstructure Formation in Casting. Metall. Trans. A **20A** (1989) 311–322
3. Voller, V. R., Swaminathan, C. R., Thomas, B. G.: Fixed grid techniques for phase change problems: a review. Int. J. Numer. Meth Engng. **30** (1990) 875–898
4. Sczygiol, N.: Object-oriented analysis of the numerical modelling of castings solidification. Computer Assisted Mechanics and Engineering Sciences, 2000 (in print)
5. Sczygiol, N., Szwarc, G.: Application of enthalpy formulation for numerical simulation of castings solidification. Computer Assisted Mechanics and Engineering Sciences, 2000 (in print)
6. Wood, W. L.: Practical Time-stepping Schemes. Clarendon Press, Oxford (1990)
7. Sczygiol, N.: Das rechnerische Modellieren der Beimengungsverteilung in erstarrenden Gußstücken aus binären Legierungen. ZAMM **74** (1994) T619–T622
8. Clyne, T. W., Kurz, W.: Solute Redistribution During Solidification with Rapid Solid State Diffusion, Metall. Trans. A **12A** (1981) 965–971

A Posteriori and *a Priori* Techniques of Local Grid Refinement for Parabolic Problems with Boundary and Transition Layers*

Grigorii I. Shishkin

Institute of Mathematics and Mechanics, Russian Academy of Sciences,
Ural Branch, Ekaterinburg 620219, Russia
Grigorii@shishkin.ural.ru

Abstract. A Dirichlet problem for a singularly perturbed parabolic reaction-diffusion equation is considered on a segment and, in particular, in a composite domain. The solution of such a problem exhibits boundary and transition (in the case of the composite domain) parabolic layers. For this problem we study classical difference approximations on sequentially locally refined meshes. The correction of the discrete solutions is performed only on the subdomains subjected to refinement (their boundaries pass through the grid nodes); uniform meshes are used in these adaptation subdomains. For *a posteriori* grid refinement we apply, as indicators, auxiliary functions majorizing the singular component of the solution. As was shown, in this class of the finite difference schemes there exist no schemes which converge independently of the singular perturbation parameter ε (or ε-uniformly). We construct special schemes, which allow us to obtain the approximations that converge "almost ε-uniformly", i.e., with an error weakly depending on ε.

1 Introduction

For a wide variety of singular perturbation problems, special finite difference schemes which converge ε-uniformly have been well developed and analyzed in the last years (see, for example, [1–4]). Usually such numerical methods require *a priori* information about singularities of the solution and are somehow adapted (e.g., by *a priori* refinement of meshes). On the other hand, *a posteriori* technique is often used in computational practice for regular problems in order to improve the accuracy by local grid refinement in those (sufficiently small) subregions where the solution gradients are large (see, e.g., [5,6]). By such a way, no *a priori* knowledge about the solution is required. To compute the improved solution, this method uses uniform meshes that provides the efficiency of calculations. By this argument, it would be of significant interest to develop such techniques for representative classes of singular perturbation problems. The author can mention only [7], in which a similar approach was firstly applied.

---

* This research was supported in part by the Russian Foundation for Basic Research under grant No. 98-01-00362 and by the NWO grant dossiernr. 047.008.007.

In the present paper we consider one approach how to increase the accuracy of numerical solutions for a parabolic singularly perturbed equation of reaction-diffusion type. We use standard finite difference approximations on locally refined grids. Note that, besides boundary layers, for $\varepsilon \to 0$ there appears a transition parabolic layer in the case of a composed domain. We apply two types of local grid refinement in regions of these singularities: either *a priori*, i.e., before all computations, or *a posteriori*, after certain computations and the analysis of intermediate solutions. These *a posteriori* methods, whose errors are weakly depending on the parameter ε (in other words, weakly sensitive methods), are alternative to classical and special ε-uniform *a priori* ones. Contrary to ε-uniform methods, for which the use of meshes abruptly condensing in a parabolic boundary layer is necessary [1,8], weakly sensitive methods comprise only simple uniform meshes. Contrary to classical schemes, which converge only if the mesh width is substantially less than the parameter ε (that is very restrictive for the method), weakly sensitive schemes converge for even not too small values of ε.

To construct *a posteriori* condensing meshes, we use indicator functions which are majorants for the singular component of the solution; these functions obey *parabolic* singularly perturbed equations. In [7], we made use of functions which are solutions of *ordinary* singularly perturbed equations; such indicators are sufficiently rough in order to evaluate exactly the subdomain subject to refinement. It should be noted that boundary value problems in composed domains, i.e., problems with transition layers, were not considered in [7].

2 Problem Formulation

2.1. In the domain $\overline{G}$ with boundary $S = \overline{G} \setminus G$, where

$$G = D \times (0, T], \quad D = \{x : 0 < x < d\}, \tag{1}$$

we consider the following boundary value problem for the parabolic equation

$$Lu(x,t) \equiv \left\{ \varepsilon^2 a(x,t) \frac{\partial^2}{\partial x^2} - c(x,t) - p(x,t) \frac{\partial}{\partial t} \right\} u(x,t) = f(x,t), \quad (x,t) \in G,$$

$$u(x,t) = \varphi(x,t), \quad (x,t) \in S. \tag{2}$$

Here $a(x,t)$, $c(x,t)$, $p(x,t)$, $f(x,t)$, $(x,t) \in \overline{G}$, $\varphi(x,t)$, $(x,t) \in S$ are sufficiently smooth functions, and also $a_0 \leq a(x,t) \leq a^0$, $c(x,t) \geq 0$, $p_0 \leq p(x,t) \leq p^0$, $(x,t) \in \overline{G}$, $a_0, p_0 > 0$; ε is the singular perturbation parameter, $\varepsilon \in (0,1]$. Assume that $f(x,t)$ and $\varphi(x,t)$ satisfy sufficient compatibility conditions on the set $\gamma_0 = \overline{G}_h \times \{t = 0\}$, $\overline{G}_h = \overline{D} \setminus D$, i.e., at the corner points $(0,0), (d,0)$.

We suppose that the boundary S consists of two parts, namely, $S = S_0 \cup S^L$, where S_0 is the lower base of the set $\overline{G}$, $S_0 = \overline{S}_0$, S^L is the lateral boundary. As $\varepsilon \to 0$, parabolic boundary layers appear in a neighbourhood of S^L.

2.2. Let us give a classical finite difference scheme for problem (2), (1) and discuss some difficulties arising in the numerical solution of this problem. On the set $\overline{G}$, we introduce the rectangular grid

$$\overline{G}_h = \overline{\omega}_1 \times \overline{\omega}_0, \qquad (3)$$

where $\overline{\omega}_1$ is a mesh on $[0, d]$ with arbitrary distribution of its nodes satisfying only the condition [1] $h \leq MN^{-1}$, where $h = \max_i h^i$, $h^i = x^{i+1} - x^i$, $x^i, x^{i+1} \in \overline{\omega}_1$; $\overline{\omega}_0$ is a uniform mesh on $[0, T]$ with step-size $h_t = TN_0^{-1}$. Here $N+1$ and N_0+1 are the number of nodes in the meshes $\overline{\omega}_1$ and $\overline{\omega}_0$. An especial attention will be paid to the meshes

$$\overline{G}_h = \overline{G}_{h(3)}, \text{ where } \overline{\omega}_1 \text{ is a piecewise uniform mesh.} \qquad (4)$$

To solve the problem, we use the implicit scheme [9]

$$\Lambda z(x,t) \equiv \{\varepsilon^2 a(x,t)\delta_{\overline{x}\widehat{x}} - c(x,t) - p(x,t)\delta_{\overline{t}}\} z(x,t) = f(x,t), \quad (x,t) \in G_h,$$
$$z(x,t) = \varphi(x,t), \quad (x,t) \in S_h, \qquad (5)$$

where $G_h = G \cap \overline{G}_h$, $S_h = S \cap \overline{G}_h$; $\delta_{\overline{x}\widehat{x}}z(x,t)$, $\delta_{\overline{t}}z(x,t)$ are the second and first difference derivatives, e.g., $\delta_{\overline{x}\widehat{x}}z(x,t) = 2(h^i + h^{i-1})^{-1}[\delta_x - \delta_{\overline{x}}]z(x,t)$, $x = x^i$.

We say that the numerical solution $z(x,t)$ converges *almost ε-uniformly* if for any arbitrarily small number $\nu > 0$ one can find a function $\lambda(\varepsilon^{-\nu}N^{-1}, N_0^{-1})$ such that

$$|u(x,t) - \overline{z}(x,t)| \leq M\lambda(\varepsilon^{-\nu}N^{-1}, N_0^{-1}), \quad (x,t) \in \overline{G},$$

where $\overline{z}(x,t)$, $(x,t) \in \overline{G}$ is the linear (with respect to x and t) interpolant constructed from $z(x,t)$, $(x,t) \in \overline{G}_h$; $\lambda(N^{-1}, N_0^{-1}) \to 0$ for $N, N_0 \to \infty$ ε-uniformly. In other words, the difference scheme converges almost ε-uniformly with defect ν (for $\nu = 0$ the scheme converges ε-uniformly).

For the solution of scheme (5), (3) such an estimate is true:

$$|u(x,t) - z(x,t)| \leq M[(\varepsilon + N^{-1})^{-1}N^{-1} + N_0^{-1}], \quad (x,t) \in \overline{G}_h. \qquad (6)$$

In the case of the difference scheme (5), (4) we have the estimate

$$|u(x,t) - z(x,t)| \leq M[(\varepsilon + N^{-1})^{-2}N^{-2} + N_0^{-1}], \quad (x,t) \in \overline{G}_h. \qquad (7)$$

It follows from estimates (6) and (7) that the schemes under consideration converge if $N^{-1}\lambda(N^{-1}, N_0^{-1})\varepsilon$ or

$$\varepsilon^{-1} = o(N). \qquad (8)$$

If this condition is violated, e.g., for $\varepsilon^{-1} = O(N)$, the solutions of schemes (5), (3) and (5), (4), generally speaking, do not converge to the solution of problem (2). By this argument, there appears such a theoretical problem: to construct

---
[1] Here and below M denote sufficiently large positive constants independent of ε and the discretization parameters. In what follows, the notation $G_{h(i.j)}$ ($\Lambda_{(i.j)}$, $m_{(i.j)}$) indicates that these grids (operators, numbers) are first defined in equation (i.j).

special difference schemes whose errors do not depend on the parameter ε. In particular, it is of interest to develop such schemes that converge under a weaker condition than condition (8).

2.3. In [1,8] the author introduced a special piecewise uniform mesh condensing in the boundary layer, on which the scheme (5) converges ε-uniformly with rate $O\left(N^{-2} ln^2 N + N_0^{-1}\right)$. On the grid (3) where $\overline{\omega}_1$ is a Bakhvalov-type graded mesh from [10], the scheme converges with rate $O\left(N^{-2} + N_0^{-1}\right)$.

In several regular problems having local singularities, locally *a priori* or *a posteriori* refined meshes are used to improve the accuracy of numerical solutions [6]. A *posteriori* refined meshes are also attractive to be applied to singularly perturbed problems, in particular, to problem (2), (1). We consider some algorithms of local grid refinement and study their applicability to the construction of approximate solutions with an error depending weakly on the parameter ε.

3 On ε-Uniformly Convergent Difference Schemes

Let us describe one base algorithm of constructing a locally (in the boundary layer region) refined mesh and show some relevant issues. To construct grids on the subdomains subject to refinement, we use uniform meshes in space and time.

On the set $\overline{G}$ we introduce the uniform rectangular grid

$$\overline{G}_{1h} = \overline{\omega}_1 \times \overline{\omega}_0, \tag{1a}$$

where $\overline{\omega}_1$ is a uniform mesh with step-size $h = dN^{-1}$, $\overline{\omega}_0 = \overline{\omega}_{0(3)}$. For convenience we denote the solution of (5) on the grid (1a) by $z_1(x,t)$, $(x,t) \in \overline{G}_{1h}$.

Let two values d_1^1 and d_1^2 have been found by some way, $d_1^1, d_1^2 \in \overline{\omega}_1$, $d_1^1 \leq d_1^2$ such that for $d_1^1 \leq x \leq d_1^2$ the grid solution $z_1(x,t)$, $(x,t) \in \overline{G}_{1h}$ is a satisfactory numerical approximation to the solution $u(x,t)$ of problem (1), (2). If it appears that $d_1^1 > 0$, $d_1^2 < d$, then we define the subdomains $G_{(2)}^1 = (0, d_1^1) \times (0,T]$, $G_{(2)}^2 = (d_1^2, d) \times (0,T]$.

On the subsets $\overline{G}_{(2)}^i$ we introduce the grids $\overline{G}_{(2)h}^i = \overline{\omega}_{(2)}^i \times \overline{\omega}_0$, $i = 1, 2$, where $\overline{\omega}_{(2)}^1$ and $\overline{\omega}_{(2)}^2$ are uniform meshes each with the number of nodes $N+1$. Let $z_{(2)}^i(x,t)$, $(x,t) \in \overline{G}_{(2)h}^i$ be the solution of the grid problem

$$\Lambda_{(5)} z_{(2)}^i(x,t) = f(x,t), \quad (x,t) \in G_{(2)h}^i,$$

$$z_{(2)}^i(x,t) = \begin{cases} z_1(x,t), & (x,t) \in S_{(2)h}^i \setminus S, \\ \varphi(x,t), & (x,t) \in S_{(2)h}^i \cap S, \end{cases} \quad i = 1, 2,$$

with $G_{(2)h}^i = G_{(2)}^i \cap \overline{G}_{(2)h}^i$, $S_{(2)h}^i = S_{(2)}^i \cap \overline{G}_{(2)h}^i$, $S_{(2)}^i = \overline{G}_{(2)}^i \setminus G_{(2)}^i$. Then we define the grid $\overline{G}_{2h}$ and the function $z_2(x,t)$, $(x,t) \in \overline{G}_{2h}$ by the relations:
$\overline{G}_{2h} = \overline{G}_{(2)h}^1 \cup \overline{G}_{(2)h}^2 \cup \{\overline{G}_{1h} \setminus \{\overline{G}_{(2)}^1 \cup \overline{G}_{(2)}^2\}\}$, and

$$z_2(x,t) = \begin{cases} z_{(2)}^i(x,t), & (x,t) \in \overline{G}_{(2)h}^i, \; i=1,2, \\ z_1(x,t), & (x,t) \in \overline{G}_{1h} \setminus \{\overline{G}_{(2)}^1 \cup \overline{G}_{(2)}^2\}; \end{cases} \quad (x,t) \in \overline{G}_{2h}.$$

Let the grid $\overline{G}_{k-1,h}$ and the function $z_{k-1}(x,t)$ on $\overline{G}_{k-1,h}$ have been already constructed for $k \geq 3$, and assume, similarly to what has been said above, that the grid solution $z_{k-1}(x,t)$, $(x,t) \in \overline{G}_{k-1,h}$ gives a satisfactory approximation to $u(x,t)$ for $d^1_{k-1} \leq x \leq d^2_{k-1}$. If $d^1_{k-1} > 0$, $d^2_{k-1} < d$, we define the domains

$$G^1_{(k)} = (0, d^1_{k-1}) \times (0,T], \quad G^2_{(k)} = (d^2_{k-1}, d) \times (0,T]. \tag{1b}$$

On the sets $\overline{G}^i_{(k)}$ we introduce the grids

$$\overline{G}^i_{(k)h} = \overline{\omega}^i_{(k)} \times \overline{\omega}_0, \quad i = 1, 2, \tag{1c}$$

where $\overline{\omega}^1_{(k)}$ and $\overline{\omega}^2_{(k)}$ are uniform meshes each with the number of nodes $N+1$. Let $z^i_{(k)}(x,t)$, $(x,t) \in \overline{G}^i_{(k)h}$ be the solution of the grid problem

$$\Lambda_{(5)} z^i_{(k)}(x,t) = f(x,t), \quad (x,t) \in G^i_{(k)h},$$

$$z^i_{(k)}(x,t) = \begin{cases} z_{k-1}(x,t), & (x,t) \in S^i_{(k)h} \setminus S, \\ \varphi(x,t), & (x,t) \in S^i_{(k)h} \cap S, \end{cases} \quad i = 1, 2. \tag{1d}$$

Suppose $\overline{G}_{kh} = \overline{G}^1_{(k)h} \cup \overline{G}^2_{(k)h} \cup \{\overline{G}_{k-1,h} \setminus \{\overline{G}^1_{(k)} \cup \overline{G}^2_{(k)}\}\}$,

$$z_k(x,t) = \begin{cases} z^i_{(k)}(x,t), & (x,t) \in \overline{G}^i_{(k)h}, \quad i = 1, 2, \\ z_{k-1}(x,t), & (x,t) \in \overline{G}_{k-1,h} \setminus \{\overline{G}^1_{(k)} \cup \overline{G}^2_{(k)}\}. \end{cases}$$

If for some values $i = j$ and $k = K_0(j)$, $j = 1, 2$ it turned out that $d^1_{K_0(1)} = 0$ (or $d^2_{K_0(2)} = d$), then we suppose $d^1_k = 0$ for $k \geq K_0(1)$ ($d^2_k = d$ for $k \geq K_0(2)$ respectively); let $K_0 = \max[K_0(1), K_0(2)]$. For $k \geq K_0(j) + 1$ the sets $\overline{G}^j_{(k)}$ are assumed to be empty, and we do not compute the functions $z^j_{(k)}(x,t)$. For example, for $k \geq K_0$ we have $z_k(x,t) = z_{K_0}(x,t)$, $\overline{G}_{kh} = \overline{G}_{K_0 h}$.

For $k = K$, where $K \geq 1$ is the given fixed number, we set

$$z^K(x,t) = z_K(x,t) \equiv z(x,t), \quad (x,t) \in \overline{G}_h, \quad \overline{G}^K_h = \overline{G}_{Kh} \equiv \overline{G}_h. \tag{1e}$$

We call the function $z_{(1)}(x,t)$, $(x,t) \in \overline{G}_{h(1)}$ the solution of scheme (5), (1). The given algorithm $A_{(1)}$ allows us to construct meshes condensing in the boundary layers. The number of nodes $N_K + 1$ in the mesh $\overline{\omega}_K$ generating $\overline{G}_{Kh}$ does not exceed $(2K-1)(N+1)$. Thus, the grid $\overline{G}_{Kh}$ belongs to the family $\overline{G}_{h(4)}$. Note that the solution of intermediate problems (1d) requires no interpolation to define the functions $z^i_{(k)}(x,t)$ on the boundary $S^i_{(k)h}$.

The grids $\overline{G}_{kh}$, $k = 1, ..., K$ generated by the algorithm $A_{(1)}$ are defined by the way of choosing the values d^i_k, $i = 1, 2$, $k = 1, ..., K-1$, and also by the values K and N, N_0. Thus, this algorithm $A_{(1)}$ determines the class of finite difference schemes (5), (1). In this class the boundary of the subdomain subject to refinement passes through the nodes of a coarser grid. Note that the smallest step of the mesh $\overline{\omega}_K$ is not less than dN^{-K}. The meshes generated by the algorithm $A_{(1)}$, in which the values d_k^i and K are defined before the start of computations (or in the course of calculations, by relying on intermediate results), belong to *a priori (a posteriori)* refined meshes.

The schemes from the class (5), (1) satisfy the maximum principle [9]. The following theorem states the "negative" result mentioned in the abstract.

Theorem 1. *In the class of difference scheme (5), (1) for the boundary value problem (2), (1) there exist no schemes that converge ε-uniformly.*

Remark 1. The statement of Theorem 1 remains valid when monotone *finite element* or *finite volume* operators are used to approximate the operator $L_{(2)}$.

4 Schemes on *a Priori* Condensing Meshes

In this section we construct finite difference schemes from the class (5), (1) by prescribing principles how to choose the values d_k^i, K.

Let $K \geq 1$. We define the values $d_{k(1)}^i$:

$$d_k^1 = \sigma_k, \quad d_k^2 = d - \sigma_k, \quad \sigma_k = \sigma_k(N) \equiv N^{-k\lambda}, \quad k=1,...,K, \quad i=1,2, \qquad (1)$$

where λ is an arbitrary number from $(0,1)$. Then we get the following estimate for the components $z_k(x,t)$ of the solution of scheme (5), (1), (1):

$$|u(x,t) - z_k(x,t)| \leq Mk\bigl[\varepsilon^{-2} N^{-2(1+(k-1)\lambda)} + N^{-2+2\mu} + N_0^{-1}\bigr], \quad (x,t)\in \overline{G}_{kh},$$
$$k = 1,...,K, \qquad (2)$$

where μ is any number from the interval $(\lambda, 1)$. Note that $z_K(x,t) = z(x,t)$.

On the set $\overline{G}_{kh}$ the k-th component $z_k(x,t)$ converges to the exact solution $u(x,t)$ if such a condition is satisfied:

$$\varepsilon^{-1} = o(N^{1+(k-1)\lambda}), \quad k=1,...,K, \quad \lambda = \lambda_{(1)}. \qquad (3)$$

Thus, for $K \geq 2$ the solution of the scheme and its components $z_k(x,t)$ converge, respectively, on $\overline{G}_h$ and $\overline{G}_{kh}$ (for $k \geq 2$) under the condition weaker than (8).

But if for some k the parameter ε satisfies the condition

$$\varepsilon \in [\varepsilon_k, 1], \quad \varepsilon_k = \varepsilon_k(N) = MN^{-k\beta}, \quad k=1,...,K, \qquad (4)$$

where β is an arbitrary number from the interval $(0, \mu]$, then for the component $z_k(x,t)$, $(x,t) \in \overline{G}_{kh}$ the following estimate is valid:

$$|u(x,t) - z_k(x,t)| \leq Mk\bigl[N^{-2+2\mu} + N_0^{-1}\bigr], \quad (x,t) \in \overline{G}_{kh}, \quad k=1,...,K. \qquad (5)$$

For sufficiently large K satisfying the condition

$$K \geq K_{(6)}(\nu,\lambda), \quad K_{(6)}(\nu,\lambda) = 1 + \lambda^{-1}\nu^{-1}(1-\nu), \quad \lambda = \lambda_{(1)}, \qquad (6)$$

where $\nu > 0$ is an arbitrarily small number, the difference scheme (5), (1), (1) converges almost ε-uniformly with defect ν

$$|u(x,t) - z(x,t)| \leq MK\bigl[(\varepsilon^{-\nu} N^{-1})^{2/\nu} + N^{-2+2\mu} + N_0^{-1}\bigr], \quad (x,t) \in \overline{G}_h. \qquad (7)$$

Theorem 2. Let the solution $u(x,t)$ of the boundary value problem (2), (1) and its regular part $U(x,t)$ satisfy the inclusions $u \in C^{l+\alpha,\,(l+\alpha)/2}(\overline{G})$, $\alpha > 0$, $l \geq 6$ and $U \in C^{l+\alpha,\,(l+\alpha)/2}(\overline{G})$, $\alpha > 0$, $l \geq 4$. Then (i) the solution of scheme (5), (1), (1) and its components, viz. the functions $z(x,t)$, $(x,t) \in \overline{G}_h$ and $z_k(x,t)$, $(x,t) \in \overline{G}_{kh}$, $k=1,...,K$ converge to the solution $u(x,t)$ of problem (2), (1) under condition (3); (ii) scheme (5), (1), (1), (6) converges almost ε-uniformly with defect ν. The discrete solutions satisfy estimates (2), (7) and, besides, (5), if condition (4) is fulfilled.

5 Schemes on *a Posteriori* Condensing Meshes

To construct schemes on *a posteriori* condensing meshes, we apply the algorithm $A_{(1)}$, where we use, as indicators for computing d_k^i, auxiliary grid functions which majorize the singular component of the problem solution.

First we decompose the solution $u(x,t)$ into regular and singular parts: $u(x,t) = U(x,t) + V(x,t)$, $(x,t) \in \overline{G}$. We now estimate the function $V(x,t)$.

Let the function $U_0(x,t)$ be the solution of the two ordinary differential equations (on each of the sides S_i^L, $S^L = S_1^L \cup S_2^L$):

$$\{-p(x,t)\frac{\partial}{\partial t} - c(x,t)\}U_0(x,t) = f(x,t), \quad (x,t) \in S^L, \qquad (1a)$$
$$U_0(x,t) = \varphi(x,t), \quad (x,t) \in \overline{S}^L \cap S_0.$$

Then the boundary-layer function $V(x,t)$ is the solution of the problem

$$L_{(2)}V(x,t) = 0, \quad (x,t) \in G,$$
$$V(x,t) = \varphi^L(x,t), \quad (x,t) \in S^L, \quad V(x,t) = 0, \quad (x,t) \in S_0, \qquad (1b)$$

where $\varphi^L(x,t) = \varphi(x,t) - U_0(x,t)$, $(x,t) \in S^L$.

We represent the function $\varphi^L(x,t)$ as a sum of two functions $\varphi^L(x,t) = \varphi^{L+}(x,t) + \varphi^{L-}(x,t)$, $(x,t) \in S^L$, where $\frac{\partial^{k_0}}{\partial t^{k_0}}\varphi^{L+}(x,t) \geq 0$, $\frac{\partial^{k_0}}{\partial t^{k_0}}\varphi^{L-}(x,t) \leq 0$, $(x,t) \in S^L$, $k_0 = 0, 1, 2$. By $z^\pm(x,t)$ we denote the solution of the problem

$$\Lambda_{(2)}\, z^\pm(x,t) \equiv \{\varepsilon^2 \delta_{\bar{x}\hat{x}} - p\delta_{\bar{t}}\}\,z^\pm(x,t) = 0, \quad (x,t) \in G_h,$$
$$z^\pm(x,t) = \varphi^{L\pm}(x,t), \quad (x,t) \in S_h^L, \quad z^\pm(x,t) = 0, \quad (x,t) \in S_{0h}. \qquad (2)$$

where $p = \min_{\overline{G}}[a^{-1}(x,t)p(x,t)]$. The grid functions $z^+(x,t)$ and $z^-(x,t)$, $(x,t) \in \overline{G}_{1h}(1a)$ majorize the solution of problem (1b):

$$z^-(x,t) - MN_0^{-1} \leq V(x,t) \leq z^+(x,t) + MN_0^{-1}, \quad (x,t) \in \overline{G}_{1h}.$$

We define the values $d_{k(1)}^i$ by

$$d_k^1 = \sigma_k^1, \quad d_k^2 = d - \sigma_k^2, \quad k = 1,...,K. \qquad (3a)$$

Let us determine σ_k^i. Assume $\varepsilon \leq MN^{-k\beta}$, $\beta = \beta_{(4)}$. Let σ_k^{i*}, $k = 1, ..., K$, $i = 1, 2$ be the minimal value of σ^i for which the following inequality holds:

$$z_k^{+i}(x,t) - z_k^{-i}(x,t) \leq Mk[N^{-2+2\mu} + N_0^{-1}], \quad (x,t) \in \overline{G}_{(k)h}^i, \quad i=1,2,$$
$$\sigma^1 \leq x \leq d - \sigma^2. \tag{3b}$$

Here $\overline{G}_{(1)h}^i = \overline{G}_{1h}$, $z_1^{\pm i}(x,t) = z_{(2)}^{\pm}(x,t)$, $(x,t) \in \overline{G}_{1h}$, the functions $z_k^{\pm i}(x,t)$, $(x,t) \in \overline{G}_{(k)h}^i$, $k = 2, ..., K$, $i = 1, 2$ are the solutions of the problems

$$\Lambda_{(2)} z_k^{\pm i}(x,t) = 0, \quad (x,t) \in \overline{G}_{(k)h}^i \setminus \overline{S}_{(k)}^i, \quad z_k^{\pm i}(x,t) = z_{k-1}^{\pm i}(x,t), \quad (x,t) \in \overline{S}_{(k)h}^{iL},$$
$$z_k^{\pm i}(x,t) = 0, \quad (x,t) \in S_{0(k)h}^i, \quad k = 2, ..., K, \quad i = 1, 2.$$

If for some index $i = j$ the inequality (3b) is false for any value σ^j, we suppose $\sigma_k^{j*} = 2^{-1}d$. In (3b) $\mu = \mu_{(2)}$. For $\varepsilon > MN^{-k\beta}$ we take $\sigma_k^{1*} = \sigma_k^{2*} = 0$. Finally, we define σ_k^i: $\sigma_k^i = \min[\sigma_k^{i*}, \sigma_{k(1)}]$.

The solution of difference scheme (5), (1), (3) satisfies the estimate

$$|u(x,t) - z_k(x,t)| \leq Mk[N^{-2+2\mu} + N_0^{-1}], \quad (x,t) \in \overline{G}_{kh},$$
$$\text{for } \sigma_k^1 \leq x \leq d - \sigma_k^2, \quad k = 1, ..., K, \quad \mu = \mu_{(2)}. \tag{4}$$

For sufficiently large N and sufficiently small ε we have the inequalities

$$\sigma_{k(3)}^i, \sigma_{k(3)}^{i*} \leq \sigma_{k(1)} \quad \text{for} \quad \mu = \mu_{(2)}, \quad k = 1, ..., K, \quad i = 1, 2. \tag{5}$$

Taking (5) into account, for the functions $z_k(x,t)$, $(x,t) \in \overline{G}_{kh}$ we obtain

$$|u(x,t) - z_k(x,t)| \leq Mk[\varepsilon^{-2}N^{-2(1+(k-1)\lambda)} + N^{-2+2\mu} + N_0^{-1}], \quad (x,t) \in \overline{G}_{kh},$$
$$k = 1, ..., K, \quad \mu = \mu_{(2)}, \quad \lambda = \lambda_{(1)}. \tag{6}$$

The component $z_k(x,t)$, $(x,t) \in \overline{G}_{kh}$ converges to the solution $u(x,t)$ of the boundary value problem under condition (3). If condition (4) holds, we have

$$|u(x,t) - z_k(x,t)| \leq Mk[N^{-2+2\mu} + N_0^{-1}], \quad (x,t) \in \overline{G}_{kh}, \tag{7}$$
$$\mu = \mu_{(2)}, \quad k = 1, ..., K.$$

It follows from estimates (6) that the solution of scheme (5), (1), (3), (6) converges almost ε-uniformly with defect ν (below $K = K_{(6)}$)

$$|u(x,t) - z(x,t)| \leq MK[(\varepsilon^{-\nu}N^{-1})^{2/\nu} + N^{-2+2\mu} + N_0^{-1}], \quad (x,t) \in \overline{G}_h. \tag{8}$$

Theorem 3. *Let the hypothesis ot Theorem 2 be fulfilled. Then (i) the functions $z(x,t)$, $(x,t) \in \overline{G}_h$ and $z_k(x,t)$, $(x,t) \in \overline{G}_{kh}$, $k = 1, ..., K$, i.e., the solution of scheme (5), (1), (3) and its components, converge to the solution of problem (2), (1) under condition (3); (ii) scheme (5), (1), (3), (6) for $\mu = \mu_{(2)}$ converges almost ε-uiformly with defect ν. The discrete solutions satisfy estimates (4), (6), (8) and, besides, (7), if condition (4) is fulfilled.*

Remark 2. For sufficiently large N and sufficiently small ε the upper bound (5) is fulfilled. Thus, the schemes on *a posteriori* refined meshes defined by (3) are more effective than the schemes on *a priori* refined meshes defined by (1). Note that the use of indicator functions obeying singularly perturbed ODEs (see, e.g., [7]) substantially overstates the values σ_k^i that reduces the efficiency of the numerical method.

6   Boundary Value Problem with a Transition Layer

6.1. In the composed domain $\overline{G}$, where

$$G = G^1 \cup G^2, \quad G^r = D^r \times (0,T], \quad D^1 = (-d, 0), \quad D^2 = (0, d), \tag{1}$$

it is required to find the solution of the problem

$$L\,u(x,t) \equiv \left\{ \varepsilon^2 a(x,t)\frac{\partial^2}{\partial x} - c(x,t) - p(x,t)\frac{\partial}{\partial t} \right\} u(x,t) = f(x,t), \quad (x,t) \in G, \tag{2}$$

$$[\,u(x)\,] = \left[a(x,t)\frac{\partial}{\partial x} u(x,t) \right] = 0, \quad (x,t) \in S^*, \quad u(x,t) = \varphi(x,t), \quad (x,t) \in S.$$

Here $S^* = \{x = 0\} \times (0,T]$, $S = \overline{G} \setminus \{G \cup S^*\}$, $a(x,t) = a_r(x,t), \ldots, f(x,t) = f_r(x,t)$, $(x,t) \in \overline{G}^r$, $r = 1, 2$, and also $0 < a_0 \le a(x,t) \le a^0$, $0 \le c(x,t) \le c^0$, $0 < p_0 \le p(x,t) \le p^0$, $(x,t) \in \overline{G}$. The coefficients and the data of the problem are assumed to be sufficiently smooth. The symbol $[\,v(x,t)\,]$ denotes the jump of the function $v(x,t)$ when crossing S^*: $[\,u(x,t)\,] = \lim_{x_2 \to x} u(x_2, t) - \lim_{x_1 \to x} u(x_1, t)$,

$$\left[a(x,t)\frac{\partial}{\partial x} u(x,t) \right] = \lim_{x_2 \to x} a_2(x_2,t)\frac{\partial}{\partial x} u(x_2,t) - \lim_{x_1 \to x} a_1(x_1,t)\frac{\partial}{\partial x} u(x_1,t),$$

$$x_r \in \overline{D}^r, \quad r = 1, 2, \quad (x,t) \in S^*.$$

We consider that the compatibility conditions are fulfilled on the sets $\gamma_0 = \{(-d, 0) \cup (0, d)\}$ and $\gamma^* = \{(0, 0)\}$ to ensure sufficient smoothness of the solution $u(x,t)$ on the subsets $\overline{G}^1$, $\overline{G}^2$ for each ε.

As $\varepsilon \to 0$, parabolic boundary and transition layers appear in a neighbourhood of the sets S^L and S^* respectively.

6.2. On the set $\overline{G}$, we introduce the grid

$$\overline{G}_h = \overline{\omega}_1 \times \overline{\omega}_0, \tag{3}$$

where $\overline{\omega}_0 = \overline{\omega}_{0(3)}$, $\overline{\omega}_1$ is a mesh on $[-d, d]$ with $N+1$ nodes. We denote the node $x = 0 \in \overline{\omega}_1$ by x^{i_0}. On the grid (3) we construct the difference scheme

$$\Lambda\, z(x,t) = f(x,t), \quad (x,t) \in G_h, \quad z(x,t) = \varphi(x,t), \quad (x,t) \in S_h. \tag{4}$$

Here

$$\Lambda \equiv \varepsilon^2 a(x,t)\delta_{\bar{x}\hat{x}} - c(x,t) - p(x,t)\delta_{\bar{t}}, \quad (x,t) \in G_h \setminus S^*,$$

$$\Lambda \equiv \varepsilon^2 2(h^{io} + h^{io-1})^{-1}\{a_2(x,t)\delta_x - a_1(x,t)\delta_{\bar{x}}\} - \hat{c}(x,t) - \hat{p}(x,t)\delta_{\bar{t}}, \quad (x,t) \in S_h^*;$$

$$f(x,t) = \begin{cases} f_{(2)}(x,t), & (x,t) \in G_h \setminus S^*, \\ \hat{f}(x,t), & (x,t) \in S_h^*, \end{cases}$$

We designate $\hat{v}(x,t) = (h^{io} + h^{io-1})^{-1}(h^{io}v_2(x,t) - h^{io-1}v_1(x,t))$, $x = x^{io} = 0$.
The difference scheme (4), (3) is ε-uniformly monotone.
In the case of the grids

$$\overline{G}_{1h} = \overline{\omega}_1 \times \overline{\omega}_0, \tag{5a}$$

uniform in x (or piecewise uniform with a finite number of intervals where the step-size is constant) we have the estimate

$$|u(x,t) - z(x,t)| \le M\left[(\varepsilon + N^{-1})^{-2}N^{-2} + N_0^{-1}\right], \quad (x,t) \in \overline{G}_{1h},$$

i.e., the scheme converges under condition (8).

6.3. Similarly to scheme (5), (1), we construct the scheme on locally refined (in the boundary and transition layers) meshes replacing problem (5), (1a) by problem (4), (5a), and the domains $\overline{G}^i_{(k)h}$, $\overline{G}_{kh}$ and also the values d^i_k respectively by

$$\overline{G}^{\pm i}_{(k)h}, \quad \overline{G}^{\pm}_{kh} \quad \text{and} \quad d^{\pm i}_k, \tag{5b}$$

where $\overline{G}^{+i}_{(k)h} = \overline{G}^i_{(k)h(1)}$, $\overline{G}^+_{kh} = \overline{G}_{kh(1)}$, $d^{+i}_k = d^i_{k(1)}$; $\overline{G}^{-i}_{(k)h}$, $\overline{G}^-_{kh}$ and d^{-i}_k are constructed by the same way. The grids

$$\overline{G}_{kh} = \overline{G}^-_{kh} \cup \overline{G}^+_{kh}, \quad k = 1, \ldots, K, \tag{5c}$$

obtained by the algorithm $A_{(5)}$, are determined by the choice of the values

$$d^i_k, \quad i = 1, \ldots, 4, \quad k = 1, \ldots, K-1, \tag{5d}$$

where $d^i_k = d^{-i}_k$ for $i = 1, 2$, $d^i_k = d^{+i-2}_k$ for $i = 3, 4$. The solution of the difference scheme, viz. the function $z(x,t)$, $(x,t) \in \overline{G}_h$, where $\overline{G}_h = \overline{G}^K_h = \overline{G}_{Kh}$, is defined by (1e).

In the case of *a priori* condensing meshes we define $d^i_{k(5)}$ by

$$d^1_k = -d + \sigma_k, \quad d^2_k = -\sigma_k, \quad d^3_k = \sigma_k, \quad d^4_k = d - \sigma_k, \quad \sigma_k = \sigma_{k(1)}(N). \tag{6}$$

Theorem 4. *Let the solution $u(x,t)$ of the boundary value problem (2), (1) and its regular part $U(x,t)$ satisfy the inclusions $u \in C^{l+\alpha,(l+\alpha)/2}(\overline{G}^r)$, $r = 1, 2$, $\alpha > 0$, $l \ge 6$ and $U \in C^{l+\alpha,(l+\alpha)/2}(\overline{G}^r)$, $r = 1, 2$, $\alpha > 0$, $l \ge 4$. Then (i) the functions*

$z(x,t)$, $(x,t) \in \overline{G}_h$ and $z_k(x,t)$, $(x,t) \in \overline{G}_{kh}$, $k = 1, \ldots, K$, i.e., the solution of scheme (4), (5), (6) and its components, converge to the solution $u(x,t)$ of problem (2), (1) under condition (3); (ii) scheme (4), (5), (6), (6) converges almost ε-uniformly with defect ν. The discrete solutions satisfy estimates (2), (7) and, besides, (5), if condition (4) is fulfilled.

6.4. Let us consider a scheme on *a posteriori* condensing meshes.

The solution of problem (2), (1) can be represented as a sum: $u(x,t) = U(x,t) + V(x,t)$, $(x,t) \in \overline{G}^r$, $r = 1, 2$. Let us estimate the function $V(x,t)$.

Let the function $U_0(x,t)$, $(x,t) \in S^{Lr} \cup S^{*r}$, $r = 1, 2$ be the solution of the fourth ODEs (on each part S^{Lr} forming S^L, $S^L = S^{L1} \cup S^{L2}$, and on each side S^{*r} of the interface boundary S^*, $S^* = S^{*1} \cup S^{*2}$):

$$\{-p(x,t)\frac{\partial}{\partial t} - c(x,t)\}U_0(x,t) = f(x,t), \quad (x,t) \in S^L \cup S^{*1} \cup S^{*2}, \quad (7\text{a})$$
$$U_0(x,t) = \varphi(x,t), \quad (x,t) \in \{S^L \cup S^{*1} \cup S^{*2}\} \cap S_0.$$

Then $V(x,t)$ is the solution of the problem

$$L_{(2)} V(x,t) = 0, \quad (x,t) \in G^r,$$
$$V(x,t) = \begin{cases} \varphi^{Lr}(x,t), & (x,t) \in S^{Lr}, \\ \varphi^{*r}(x,t), & (x,t) \in S^{*r}, \end{cases} \quad (7\text{b})$$
$$V(x,t) = 0, \quad (x,t) \in S_0^r, \quad r = 1, 2,$$

where $\varphi^{Lr}(x,t) = \varphi(x,t) - U_0(x,t)$, $(x,t) \in S^{Lr}$, $\varphi^{*r}(x,t) = u(x,t) - U_0(x,t)$, $(x,t) \in S^{*r}$. Assume $\varphi^*(x,t) = U_0(x+0,t) - U_0(x-0,t)$, $(x,t) \in S^*$.

Further we decompose the functions $\varphi^{Lr}(x,t)$ and $\varphi^*(x,t)$ as follows:

$$\varphi^{Lr}(x,t) = \varphi^{Lr+}(x,t) + \varphi^{Lr-}(x,t), \quad (x,t) \in S^{Lr}, \quad r = 1, 2,$$
$$\varphi^*(x,t) = \varphi^{*+}(x,t) + \varphi^{*-}(x,t), \quad (x,t) \in S^*,$$

where

$$\frac{\partial^{k_0}}{\partial t^{k_0}} \varphi^{Lr+}(x,t) \geq 0, \quad \frac{\partial^{k_0}}{\partial t^{k_0}} \varphi^{Lr-}(x,t) \leq 0, \quad (x,t) \in S^{Lr}, \quad r = 1, 2;$$

$$\frac{\partial^{k_0}}{\partial t^{k_0}} \varphi^{*+}(x,t) \geq 0, \quad \frac{\partial^{k_0}}{\partial t^{k_0}} \varphi^{*-}(x,t) \leq 0, \quad (x,t) \in S^*; \quad k_0 = 0, 1, 2.$$

Note that the functions considered on the set S^{*r} are limiting on S^* from G^r. By $z^{\pm}(x,t)$ we denote the solution of the problem

$$\Lambda_{(2)} z^{\pm}(x,t) = 0, \quad (x,t) \in G_h^r,$$
$$z^{\pm}(x,t) = \Psi(x,t), \quad (x,t) \in S_h^{Lr} \cup S_h^{*r}, \quad (8)$$
$$z^{\pm}(x,t) = 0, \quad (x,t) \in S_{0h}^r, \quad r = 1, 2.$$

where

$$\Psi(x,t) \equiv \begin{cases} \varphi^{Lr\pm}(x,t), & (x,t) \in S^{Lr}, \\ \varphi^{*\pm}(x,t), & (x,t) \in S^{*1}, \\ -\varphi^{*\pm}(x,t), & (x,t) \in S^{*2}. \end{cases}$$

The functions $z^+(x,t)$, $z^-(x,t)$, $(x,t) \in \overline{G}_h^r$ majorize the solution of problem (7b):

$$z^-(x,t) - M(\varepsilon - N_0^{-1}) \leq V(x,t) \leq z^+(x,t) + M(\varepsilon + N_0^{-1}), \quad (x,t) \in \overline{G}_{1h}^r, \quad r = 1, 2,$$

where $\overline{G}_{1h}^r = \overline{G}^r \cap \overline{G}_{1h(5)}$. We use the functions $z^+(x,t)$, $z^-(x,t)$ as indicators. We now choose the control parameters $d_{k(5)}^i$ by such a way:

$$d_k^1 = -d + \sigma_k^1, \quad d_k^2 = -\sigma_k^2, \quad d_k^3 = \sigma_k^3, \quad d_k^4 = d - \sigma_k^4, \quad k = 1, \ldots, K. \tag{9a}$$

Let us determine σ_k^i. Assume $\varepsilon \leq MN^{-k\beta}$, $\beta = \beta_{(4)}$. Let σ_k^{i*}, $k = 1, \ldots, K$, $i = i(r)$, $i = 1, 2$ for $r = 1$, and $i = 3, 4$ for $r = 2$, be the minimal value of σ^i for which the following inequality is fulfilled:

$$z_k^{+i}(x,t) - z_k^{-i}(x,t) \leq Mk[N^{-1+\mu} + N_0^{-1}], \quad (x,t) \in \overline{G}_{(k)h}^{ri}, \quad i = i(r),$$
$$\text{for } -d + \sigma_k^1 \leq x \leq -\sigma_k^2, \quad \sigma_k^3 \leq x \leq d - \sigma_k^4. \tag{9b}$$

Here $\overline{G}_{(1)h}^{ri} = \overline{G}_{1h}^r$, $z_1^{\pm i}(x,t) = z_{(8)}^{\pm}(x,t)$, $(x,t) \in \overline{G}_{1h}^r$, $r = 1, 2$, the functions $z_k^{\pm i}(x,t)$, $(x,t) \in \overline{G}_{(k)h}^{ri}$, $k = 2, \ldots, K$, $i = i(r)$ are the solutions of the problems

$$\Lambda_{(2)} z(x,t) = 0, \quad (x,t) \in G_{(k)h}^{ri}, \quad z(x,t) = z_{k-1}^{\pm i}(x,t), \quad (x,t) \in \overline{S}_{(k)h}^{iL},$$
$$z(x,t) = 0, \quad (x,t) \in S_{0(k)h}^i, \quad k = 2, \ldots, K, \quad i = i(r), \quad r = 1, 2.$$

If for some index $i = j$ the inequality (9b) is false for any value σ^j, we suppose $\sigma_k^{i*} = 0$, $i = i(r)$. We take $\sigma_k^i = \min[\sigma_k^{i*}, \sigma_{k(1)}]$.

We come to the following estimate for the solution of scheme (4), (5), (9):

$$|u(x,t) - z_k(x,t)| \leq Mk\left[\varepsilon^{-1}N^{-1-(k-1)\lambda} + N^{-1+\mu} + N_0^{-1}\right], \quad (x,t) \in \overline{G}_{kh},$$
$$k = 1, \ldots, K, \quad \mu = \mu_{(2)}, \quad \lambda = \lambda_{(1)}. \tag{10}$$

Thus, the component $z_k(x,t)$, $(x,t) \in \overline{G}_{kh}$ converges to the solution of the boundary value problem under condition (3). Under condition (4) we have

$$|u(x,t) - z_k(x,t)| \leq Mk[N^{-1+\mu} + N_0^{-1}], \quad (x,t) \in \overline{G}_{kh}, \quad k = 1, \ldots, K. \tag{11}$$

For the solution of scheme (4), (5), (9), (6) we obtain the estimate

$$|u(x,t) - z_k(x,t)| \leq MK\left[(\varepsilon^{-\nu}N^{-1})^{1/\nu} + N^{-1+\mu} + N_0^{-1}\right], \quad (x,t) \in \overline{G}_h,$$
$$K = K_{(6)}. \tag{12}$$

Theorem 5. *Let the hypothesis of Theorem 4 be fulfilled. Then (i) the functions $z(x,t)$, $(x,t) \in \overline{G}_h$ and $z_k(x,t)$, $(x,t) \in \overline{G}_{kh}$, $k = 1, \ldots, K$, i.e., the solution of scheme (4), (5), (9) and its components, converge to the solution $u(x,t)$ of the boundary value problem (2), (1) under condition (3); (ii) scheme (4), (5), (9), (6) for $\mu = \mu_{(2)}$ converges almost ε-uniformly with defect ν. The discrete solutions satisfy estimates (10), (12) and, besides, (11), if condition (4) is fulfilled.*

Remark 3. For sufficiently large N and sufficiently small ε the upper bound $\sigma^{i*}_{k(9)} \leq \sigma_{k(1)}$ holds for $i = 1, \ldots, 4$, $k = 1, \ldots, K$.

Remark 4. Let us assume that the function $\varphi^*(x,t)$ satisfies the condition: $\max_{S^*} |\varphi^*(x,t)| \geq m\varepsilon$. In this case we apply scheme (4), (5), (9), where we replace $N^{-1+\mu}$ by $N^{-2+2\mu}$ in the right-hand side of (9b). Then estimates like (4), (6) and (8) are valid for the approximate solutions.

References

1. Shishkin, G. I.: Grid Approximations of Singularly Perturbed Elliptic and Parabolic Equations (in Russian). Ural Branch of Russian Acad. Sci., Ekaterinburg (1992)
2. Miller, J. J. H., O'Riordan, E., Shishkin, G. I.: Fitted Numerical Methods for Singular Perturbation Problems. World Scientific, Singapore (1996)
3. Roos, H.-G., Stynes, M., Tobiska, L.: Numerical Methods for Singularly Perturbed Differential Equations. Convection-Diffusion and Flow Problems. Springer-Verlag, Berlin (1996)
4. Farrell, P. A., Hegarty, A. F., Miller, J. J. H., O'Riordan, E., Shishkin, G. I.: Robust Computational Techniques for Boundary Layers. CRC Press, Boca Raton (2000)
5. Marchuk, G. I., Shaidurov, V. V.: The Increase in Accuracy of the Solutions of Finite Difference Schemes. Nauka, Moscow (1979)
6. Birkhoff, G., Lynch, R. E..: Numerical Solution of Elliptic Problems. SIAM, Philadelphia (1984)
7. Shishkin, G. I.: Grid approximation of singularly perturbed boundary value problems on locally refined meshes. Reaction-diffusion equations. Mat. Model. **11** (12) (1999) 87–104 (in Russian)
8. Shishkin, G. I.: Approximation of solutions of singularly perturbed boundary value problems with a parabolic boundary layer. USSR Comput. Maths. Math. Phys. **29** (4) (1989) 1–10
9. Samarskii, A. A.: Theory of Difference Schemes (in Russian). Nauka, Moscow (1989); German transl.: Leipzig Akademische Verlag, Geest & Portig (1984)
10. Bakhvalov, N. S.: On the optimization of the methods for solving boundary value problems in the presence of a boundary layer. Zh. Vychisl. Mat. Mat. Fiz. **9** (1969) 841–859 (in Russian)

On a Necessary Requirement for *Re*-Uniform Numerical Methods to Solve Boundary Layer Equations for Flow along a Flat Plate

Grigorii I. Shishkin[1], Paul A. Farrell[2], Alan F. Hegarty[3], John J. H. Miller[4], and Eugene O'Riordan[5]

[1] Institute of Mathematics and Mechanics, Russian Academy of Sciences,
Ural Branch, Ekaterinburg 620219, Russia
Grigorii@shishkin.ural.ru
[2] Department of Mathematics and Computer Science, Kent State University
Kent, Ohio 44242, USA
farrell@mcs.kent.edu
[3] Department of Mathematics and Statistics, University of Limerick
Limerick, Ireland
Alan.Hegarty@ul.ie
[4] Department of Mathematics, University of Dublin, Trinity College
Dublin 2, Ireland
jmiller@tcd.ie
[5] School of Mathematical Sciences, Dublin City University
Dublin 9, Ireland
oriordae@ccmail.dcu.ie

Abstract. We consider grid approximations of a boundary value problem for the boundary layer equations modeling flow along a flat plate in a region excluding a neighbourhood of the leading edge. The problem is singularly perturbed with the perturbation parameter $\varepsilon = 1/Re$ multiplying the highest derivative. Here the parameter ε takes any values from the half-interval (0,1], and Re is the Reynolds number. It would be of interest to construct an Re-uniform numerical method using the simplest grids, i.e., uniform rectangular grids, that could provide effective computational methods. To this end, we are free to use any technique even up to fitted operator methods, however, with fitting factors independent of the problem solution. We show that for the Prandtl problem, even in the case when its solution is self-similar, there does not exist a fitted operator method that converges Re-uniformly. Thus, combining a fitted operator and uniform meshes, we do not succeed in achieving Re-uniform convergence. Therefore, the use of the fitted mesh technique, based on meshes condensing in a parabolic boundary layer, is a necessity in constructing Re-uniform numerical methods for the above class of flow problems.

1 Introduction

The boundary layer equations for laminar flow are a suitable model for Navier-Stokes equations with large Reynolds numbers Re. Boundary value problems for these nonlinear equations are singularly perturbed, with the perturbation parameter ε defined by $\varepsilon = Re^{-1}$. The presence of parabolic boundary layers, i.e., layers described by parabolic equations, is typical for such problems [1,2].

The application of numerical methods, developed for regular boundary value problems (see, for example, [3,11]), even to linear singularly perturbed problems yields errors which essentially depend on the perturbation parameter ε. For small values of ε, the errors in such numerical methods may be comparable to, or even much larger than the exact solution. This behaviour of the approximate solutions creates the need to develop numerical methods with errors that are independent of the perturbation parameter ε, i.e. ε-uniform methods.

The presence of a nonlinearity in the differential equations makes it considerably more difficult to construct ε-uniform numerical methods. For example, even in the case of ordinary differential quasilinear equations, ε-uniform fitted operator methods (see, e.g., [5,6]) do not exist. It should be pointed out that even for linear singularly perturbed problems with parabolic boundary layers there are no ε-uniform fitted schemes (see, for example, [7–9]). Thus, the development of special ε-uniform numerical methods for resolving the Navier-Stokes and boundary layer equations has considerable scientific and practical interest.

In this paper, we consider grid approximations of a boundary value problem for boundary layer equations for a flat plate on a bounded domain outside a neighbourhood of its leading edge. The solution of the Prandtl problem is self-similar and exhibits a parabolic boundary layer in the considered domain. We study a wide class of discrete approximations consistent with the differential equations, i.e., the coefficients in the finite difference operators related to the differential coefficients do not depend on the problem solution. It is shown that the use of special meshes condensing in the boundary layer region is necessary. Also, no technique for the construction of the discrete equations leads to an ε-uniform method, unless it uses condensing grids.

2 Problem Formulation

Let us formulate a boundary value problem for Prandtl's boundary layer equations on a bounded domain. Consider a flat semi-infinite plate in the place of the semiaxis $P = \{(x,y) : x \geq 0, \ y = 0\}$. The problem is considered to be symmetric with respect to the plane $y = 0$; we discuss the steady flow of an incompressible fluid on both sides of P, which is laminar and parallel to the plate (no separation occurs on the plate).

As is well known, singularities in such a problem arise for a large Re number. A typical singularity is the appearance of a parabolic boundary layer in a neighbourhood of the flat plate outside some neighbourhood of its leading edge. In a neighbourhood of the leading edge, another type of singularity is generated because the compatability conditions are violated at the leading edge. In order to concentrate on the boundary layer region under consideration, we skip a small neighbourhood of the leading edge.

We consider the solution of the problem on the bounded set

$$\overline{G}, \quad \text{where} \quad G = \{(x,y): x \in (d_1, d_2], \, y \in (0, d_0)\}, \quad d_1 > 0 \tag{1}$$

with the boundary $S = \overline{G} \setminus G$. Let $G^0 = \{(x,y): x \in [d_1, d_2], \, y \in (0, d_0]\}$; $\overline{G}^0 = \overline{G}$; and let $S^0 = \overline{G} \setminus G^0$ be the boundary of the set G^0. Assume $S = \cup S_j$, $j = 0, 1, 2$, where $S_0 = \{(x,y): x \in [d_1, d_2], \, y = 0\}$, $S_1 = \{(x,y): x = d_1, \, y \in (0, d_0]\}$, $S_2 = \{(x,y): x \in (d_1, d_2], \, y = d_0\}$, $\overline{S}_0 = S_0$. Thus, the boundary $S^0 = S_0$ belongs to P. On the set $\overline{G}$, it is required to find a function $U(x,y) = (u(x,y), v(x,y))$ which is the solution of the following Prandtl problem:

$$L^1(U(x,y)) \equiv \varepsilon \frac{\partial^2}{\partial y^2} u(x,y) - u(x,y) \frac{\partial}{\partial x} u(x,y) -$$
$$- v(x,y) \frac{\partial}{\partial y} u(x,y) = 0, \quad (x,y) \in G, \tag{2a}$$

$$L^2 U(x,y) \equiv \frac{\partial}{\partial x} u(x,y) + \frac{\partial}{\partial y} v(x,y) = 0, \quad (x,y) \in G^0, \tag{2b}$$

$$u(x,y) = \varphi(x,y), \quad (x,y) \in S, \tag{2c}$$

$$v(x,y) = \psi(x,y), \quad (x,y) \in S^0. \tag{2d}$$

Here $\varepsilon = Re^{-1}$; the parameter ε takes arbitrary values in the half-interval (0,1]. We now wish to define the functions $\varphi(x,y)$ and $\psi(x,y)$ more precisely.
In the quarter plane

$$\overline{\Omega}, \quad \text{where} \quad \Omega = \{(x,y): x, \, y > 0\} \tag{3}$$

we consider the following Prandtl problem which has a self-similar solution [1]:

$$\begin{array}{ll} L^1(U(x,y)) = 0, & (x,y) \in \Omega, \quad L^2 U(x,y) = 0, \quad (x,y) \in \overline{\Omega} \setminus P, \\ u(x,y) = u_\infty, \quad x = 0, \, y \geq 0, \quad U(x,y) = (0,0), \quad (x,y) \in P. \end{array} \tag{4}$$

The solution of problem (4), (3) can be written in terms of some function $f(\eta)$ and its derivative

$$u(x,y) = u_\infty f'(\eta), \quad v(x,y) = \varepsilon^{1/2} \left(2^{-1} u_\infty x^{-1}\right)^{1/2} (\eta f'(\eta) - f(\eta)) \tag{5}$$

where $\eta = \varepsilon^{-1/2} \left(2^{-1} u_\infty x^{-1}\right)^{1/2} y$. The function $f(\eta)$ is the solution of the Blasius problem

$$\begin{array}{l} L(f(\eta)) \equiv f'''(\eta) + f(\eta) f''(\eta) = 0, \quad \eta \in (0, \infty), \\ f(0) = f'(0) = 0, \quad \lim_{\eta \to \infty} f'(\eta) = 1. \end{array} \tag{6}$$

The functions $\varphi(x,y), \psi(x,y)$ are defined by [1]

$$\varphi(x,y) = u_{(5)}(x,y), \, (x,y) \in S; \quad \psi(x,y) = v_{(5)}(x,y), \, (x,y) \in S^0; \tag{7}$$

note that $\varphi(x,y) = 0, \, \psi(x,y) = 0, \, (x,y) \in S^0$.

---
[1] Here and below the notation $w_{(j.k)}$ indicates that w is first defined in equation (j.k).

In the case of problem (2), (7), (1), as ε tends to zero, a parabolic boundary layer appears in a neighbourhood of the set S^0.

To solve problem (2), (7), (1) numerically, we wish to construct an ε-uniform finite difference scheme.

3 Classical Difference Scheme for the Prandtl Problem

For the boundary value problem (2), (7), (1) we use a classical finite difference scheme. At first we introduce the rectangular grid on the set $\overline{G}$:

$$\overline{G}_h = \overline{\omega}_1 \times \overline{\omega}_2 \tag{1}$$

where $\overline{\omega}_1$ and $\overline{\omega}_2$ are meshes on the segments $[d_1, d_2]$ and $[0, d_0]$, respectively; $\overline{\omega}_1 = \{x^i : i = 0, ..., N_1, x^0 = d_1, x^{N_1} = d_2\}$, $\overline{\omega}_2 = \{y^j : j = 0, ..., N_2, y^0 = 0, y^{N_2} = d_0\}$; $N_1 + 1$ and $N_2 + 1$ are the number of nodes in the meshes $\overline{\omega}_1$ and $\overline{\omega}_2$. Define $h_1^i = x^{i+1} - x^i$, $x^i, x^{i+1} \in \overline{\omega}_1$, $h_2^j = y^{j+1} - y^j$, $y^j, y^{j+1} \in \overline{\omega}_2$, $h_1 = \max_i h_1^i$, $h_2 = \max_j h_2^j$, $h = \max[h_1, h_2]$. We assume that [2] $h \leq MN^{-1}$, where $N = \min[N_1, N_2]$.

We approximate the boundary value problem by the difference scheme

$$\Lambda^1 \left(U^h(x,y)\right) \equiv \varepsilon \delta_{\overline{y}\widehat{y}} u^h(x,y) - u^h(x,y) \delta_{\overline{x}} u^h(x,y) -$$
$$- v^h(x,y) \delta_{\overline{y}} u^h(x,y) = 0, \quad (x,y) \in G_h, \tag{2a}$$

$$\Lambda_1^2 U^h(x,y) \equiv \delta_{\overline{x}} u^h(x,y) + \delta_{\overline{y}} v^h(x,y) = 0, \quad (x,y) \in G_h^0, \; x > d_1, \tag{2b}$$

$$\Lambda_2^2 U^h(x,y) \equiv \delta_x u^h(x,y) + \delta_{\overline{y}} v^h(x,y) = 0, \quad (x,y) \in S_{1h};$$

$$u^h(x,y) = \varphi(x,y), \quad (x,y) \in S_h, \tag{2c}$$

$$v^h(x,y) = \psi(x,y), \quad (x,y) \in S_h^0. \tag{2d}$$

Here $\delta_{\overline{y}\widehat{y}} z(x,y)$ and $\delta_x z(x,y), ..., \delta_{\overline{y}} z(x,y)$ are the second and first (forward and backward) difference derivatives (the bar denotes the backward difference), as follows: $\delta_{\overline{y}\widehat{y}} z(x,y) = 2(h_2^{j-1} + h_2^j)^{-1} (\delta_y z(x,y) - \delta_{\overline{y}} z(x,y))$, $\delta_x z(x,y) = (h_1^i)^{-1} (z(x^{i+1}, y) - z(x,y))$, ..., $\delta_{\overline{y}} z(x,y) = (h_2^{j-1})^{-1} (z(x,y) - z(x, y^{j-1}))$, $(x,y) = (x^i, y^j)$.

The difference scheme (2), (1) approximates problem (2), (1) with the first order of accuracy for a fixed value of the parameter ε.

When the "coefficients" multiplying the differences $\delta_{\overline{x}}$ and $\delta_{\overline{y}}$ in the operator Λ^1 are known (let these be the functions $u_0^h(x,y)$ and $v_0^h(x,y)$), and if they satisfy the condition $u_0^h(x,y), v_0^h(x,y) \geq 0$, $(x,y) \in \overline{G}_h$, we know that the operator Λ^1 is monotone [4]. In such a case we say that the discrete momentum equation (2a) is monotone.

It is known that even for linear singularly perturbed problems, if we consider the heat equation with y and x being the space and time variables, the errors of

---
[2] Throughout this paper M (m) denote sufficiently large (small) positive constants independent of ε and the discretization parameters.

the discrete solution depend on the perturbation parameter and become comparable with the exact solution itself when the value $\varepsilon^{1/2}$ has the same order of magnitude as the mesh step-size h_2 on uniform grids. Therefore, it is not surprising that for the solution of the difference scheme (2), (1) we have the following lower bounds

$$\max_{\overline{G}_h} |u(x,y) - u^h(x,y)| \geq m, \qquad \max_{\overline{G}_h} |v^*(x,y) - v^{*h}(x,y)| \geq m.$$

Here $v^*(x,y) = \varepsilon^{-1/2} v(x,y)$, $v^{*h}(x,y) = \varepsilon^{-1/2} v^h(x,y)$. We call the functions $u(x,y)$, $u^h(x,y)$ and $v^*(x,y)$, $v^{*h}(x,y)$ the normalized components of the solutions of problems (2), (7), (1) and (2), (1). Note that the functions $u(x,y)$, $v^*(x,y)$ are ε-uniformly bounded on $\overline{G}$ and have order of unity, whereas the function $v(x,y)$ tends to zero when $\varepsilon \to 0$.

Thus, our aim is to try to find a numerical method which yields the discrete solutions $u^{0h}(x,y)$, $v^{0*h}(x,y)$ satisfying the error estimates

$$|u(x,y) - u^{0h}(x,y)| \leq M \left[N_1^{-\nu_1} + N_2^{-\nu_2} \right], \tag{3}$$

$$|v^*(x,y) - v^{0*h}(x,y)| \leq M \left[N_1^{-\nu_1} + N_2^{-\nu_2} \right], \quad (x,y) \in \overline{G}_h^0 \tag{4}$$

where $\overline{G}_h^0$ is some grid on $\overline{G}$, and $u^{0h}(x,y)$, $v^{0h}(x,y)$, $(x,y) \in \overline{G}_h^0$ is the solution of some discrete problem on $\overline{G}_h^0$, ν_1 and ν_2 are any positive ε-independent numbers. Throughout this paper M (m) denote sufficiently large (small) positive constants which do not depend on ε and on the discretization parameters.

We say that the method is ε-uniform if the errors in the discrete normalized solutions are independent of the parameter ε. By a robust layer-resolving method we mean a numerical method that generates approximate normalized solutions that are globally defined, pointwise-accurate and parameter-uniformly convergent at each point of the domain, including the boundary layers. Thus, the errors of these normalized solutions in the L_∞-norm are independent of ε, they depend only on N_1 and N_2 and tend to zero when $N_1, N_2 \to \infty$.

Here we try to find a robust layer-resolving numerical method for the particular problem (2), (7), (1). It would be attractive to find a method using uniform grids. Obviously the method (2), (1) is too simple to be a robust layer-resolving method.

4 On Fitted Operator Schemes for the Prandtl Problem

In order to have freedom as much as possibly in the construction of an ε-uniform method, we use a fitted operator method. Such a method admits any technique to be used to construct discrete approximations to the solution of problem (2), (7), (1).

Before we proceed, we make a few remarks.

As was shown in [8,10] (see also [7,9,11]) for a singularly perturbed parabolic equation with parabolic boundary layers, there exist no fitted operator schemes on uniform meshes that are ε-uniform. Note that the coefficients in the terms with first-order derivatives in time and second-order derivatives in the space variables do not vanish in the equations discussed in [8,10]. However, for the Prandtl problem the coefficient multiplying the first derivative with respect to the variable x, which plays the role of the time variable, vanishes on the boundary lying on the x-axis. Unlike the problems studied in [8,10], where the boundary conditions do not obey any restriction, besides the requirement of sufficient smoothness, problem (2), (7), (1) is essentially simpler. Its solution depends only on the one parameter u_∞. In [12] an ε-uniform fitted operator method was constructed for a linear parabolic equation with a discontinuous initial condition in the presence of a parabolic (transient) layer. Such fitted operator schemes have been successfully constructed because all of the singular components of the solution (their main parts) are defined, up to some multiplier, by just one function. Because of the simple (depending on u_∞ only) representation of the solution for the Prandtl problem, it is not obvious that for this problem there are no ε-uniform fitted operator schemes. So it is of interest to establish whether such fitted operator schemes on uniform meshes do exist for the Prandtl problem.

We try to construct a fitted operator scheme starting from equation (2a) under the simplifying assumption that the function $v^h(x, y)$ is known, and also that $v^h(x, y) = v(x, y)$. Let us consider a fitted operator scheme of the form

$$\Lambda^{1*}\left(u^h(x,y)\right) \equiv \varepsilon\gamma_{(2)}\delta_{y\bar{y}}u^h(x,y) - u^h(x,y)\delta_{\bar{x}}u^h(x,y) -$$
$$-\gamma_{(1)}v(x,y)\delta_{\bar{y}}u^h(x,y) = 0, \quad (x,y) \in G_h, \tag{1a}$$

$$u^h(x,y) = \varphi(x,y), \quad (x,y) \in S_h \tag{1b}$$

where

$$\overline{G}_h \text{ is a } uniform \text{ rectangular grid} \tag{2}$$

with steps h_1 and h_2 in x and y respectively; the parameters

$$\gamma_{(i)} = \gamma_{(i)}(x, y; \varepsilon, h_1, h_2), \quad i = 1, 2 \tag{1c}$$

are the fitting coefficients. The discrete equation (1a) is based on the classical monotone discrete momentum equation (2a). We emphasize that $\gamma_{(i)}$ are independent of the unknown solution that is defined by the parameter u_∞. Note that equation (2a) is a particular grid equation from the class of discrete approximations (1a), (1c), namely, when $\gamma_{(i)} = 1$, $i = 1, 2$.

Relying on *a priori* estimates for the solution of problem (2), (7), (1) and its derivatives, we establish, in a similar manner as in [8,13], that there is no ε-uniform fitted operator scheme of the form (1), (2).

Theorem 1. *In the class of finite difference schemes (1), (2) there exists no scheme whose solutions converges ε-uniformly, as $N \to \infty$, to the solution of the boundary value problem (2), (7), (1).*

Remark 1. We conclude that to construct an ε-uniform scheme for the Prandtl problem (2), (1), provided that the coefficients $\gamma_{(i)}$ are independent of the problem solution, it is necessary to use meshes condensing in the neighbourhood of the parabolic boundary layer. No matter, whether *finite elements* or *finite differences* are used.

5 Condensing Mesh Technique

Here we briefly describe the approach to the construction of an ε-uniform method with piecewise uniform condensing meshes that originated in [8].

We introduce a piecewise uniform mesh, which is refined in a neighbourhood of the boundary layer, i.e. of the set S^0. On the set $\overline{G}$, we consider the grid

$$\overline{G}_h^* = \overline{\omega}_1 \times \overline{\omega}_2^* \tag{1}$$

where $\overline{\omega}_1$ is a uniform mesh on $[d_1, d_2]$, $\overline{\omega}_2^* = \overline{\omega}_2^*(\sigma)$ is a special piecewise uniform mesh depending on the parameter σ and the value N_2. The mesh $\overline{\omega}_2^*$ is constructed as follows. We divide the segment $[0, d_0]$ in two parts $[0, \sigma]$ and $[\sigma, d_0]$. The step-size of the mesh $\overline{\omega}_2^*$ is constant on the segments $[0, \sigma]$ and $[\sigma, d_0]$, given by $h_2^{(1)} = 2\sigma N_2^{-1}$ and $h_2^{(2)} = 2(d_0 - \sigma)N_2^{-1}$, respectively. The value of σ is defined by

$$\sigma = \min\left[2^{-1}d_0,\ m\varepsilon^{1/2}\ln N_2\right]$$

where m is an arbitrary positive number.

In the case of the boundary value problem (2), (7), (1), it is required to study whether the solutions of the difference scheme (2), (1) converge to the exact solution.

We mention certain difficulties that arise in the analysis of the convergence.

Note that the difference scheme (2), (1), as well as the boundary value problem (2), (1), is nonlinear. To find an approximate solution of this scheme, we must construct an appropriate iterative numerical method. It is of interest to investigate the influence of the parameter ε upon the number of iterations required for the convergence of this iterative process.

In the case of ε-uniform difference schemes for linear singular perturbation problems, methods are well developed to determine theoretically and numerically the parameters in the error bounds (orders of convergence and error constants) for fixed values of ε and also ε-uniformly (see and compare, for example, [14,15]). In this technique, ε-uniform convergence is ascertained from theoretical investigations. Formally these methods are inapplicable to problem (2), (7), (1) because ε-uniform convergence of scheme (2), (1) was not justified by theory. Nevertheless, the results of such investigations of the error bounds seem to be interesting for practical use.

In [16] we give an experimental technique to study if this method is ε-uniform, and if so, to find the ε-uniform order of convergence. Some relevant ideas are discussed in [13]. This technique is used in [16] to analyze the scheme (2), (1). It was shown that this scheme gives discrete solutions that allow us to approximate the normalized component and its first derivatives in x, y for problem (2), (7), (1). The order of ε-uniform convergence is close to one.

6 Conclusion

We have discussed the problems that arise when the direct method is designed to solve the Prandtl problem for flow along a flat plate. We have shown that, using a uniform mesh, it is impossible to construct a Re-uniform method if the coefficients of the fitted operator are independent of the problem solution. For such operators, in particular, for classical finite difference operators, the use of the grids condensing in the boundary layer region is necessary for the method to be Re-uniform.

Acknowledgements

This research was supported in part by the Russian Foundation for Basic Research under grant No. 98-01-00362, by the National Science Foundation grant DMS-9627244 and by the Enterprise Ireland grant SC-98-612.

References

1. Schlichting, H.: Boundary-Layer Theory. McGraw-Hill, New York (1979)
2. Oleinik, O. A., Samohin, V. N: Mathemathical Methods in the Theory of Boundary Layer. CRC Press, Boca Raton (2000)
3. Marchuk, G. I.: Methods of Numerical Mathematics. Springer, New York (1982)
4. Samarskii, A. A.: Theory of Difference Schemes (in Russian). Nauka, Moscow (1989); German transl. Leipzig Akademische Verlag, Geest & Portig (1984)
5. Shishkin, G. I.: Difference approximation of a singularly perturbed boundary value problem for quasilinear elliptic equations degenerating into a first-order equation. USSR Comput. Maths. Math. Phys. **32** (1992) 467–480
6. Farrell, P. A., Miller, J. J. H., O'Riordan, E., Shishkin, G. I.: On the non-existence of ε-uniform finite difference methods on uniform meshes for semilinear two-point boundary value problems. Math. Comp. **67** (222) (1998) 603–617
7. Shishkin, G. I.: Approximation of solutions of singularly perturbed boundary value problems with a parabolic boundary layer. USSR Comput. Maths. Math. Phys. **29** (1989) 1–10
8. Shishkin, G. I.: Grid Approximations of Singularly Perturbed Elliptic and Parabolic Equations (in Russian). Ural Branch of Russian Acad. Sci., Ekaterinburg (1992)
9. Miller, J. J. H., O'Riordan, E., Shishkin, G. I.: Fitted Numerical Methods for Singular Perturbation Problems. Error Estimates in the Maximum Norm for Linear Problems in One and Two Dimensions. World Scientific, Singapore (1996)
10. Shishkin, G. I.: A difference scheme for a singularly perturbed equation of parabolic type with a discontinuous initial condition. Soviet Math. Dokl. **37** (1988) 792–796
11. Roos, H.-G., Stynes, M., Tobiska, L.: Numerical Methods for Singularly Perturbed Differential Equations. Convection-Diffusion and Flow Problems. Springer-Verlag, Berlin (1996)
12. Hemker, P. W., Shishkin, G. I.: Discrete approximation of singularly perturbed parabolic PDEs with a discontinuous initial condition. Comp. Fluid Dynamics J. **2** (1994) 375–392
13. Farrell, P. A., Hegarty, A. F., Miller, J. J. H., O'Riordan, E., Shishkin, G. I.: Numerical analysis of Re–uniform convergence for boundary layer equations for a flat plate. Computing (to appear).
14. Shishkin, G. I.: Grid approximation of singularly perturbed parabolic equations degenerating on the boundary. USSR Comput. Maths. Math. Phys. **31** (1991) 53–63

15. Hegarty, A. F., Miller, J. J. H., O'Riordan, E., Shishkin, G. I.: Numerical solution of elliptic convection-diffusion problems on fitted meshes. CWI Quarterly **10** (1997) 239–251
16. Farrell, P. A., Hegarty, A. F., Miller, J. J. H., O'Riordan, E., Shishkin, G. I.: Robust Computational Techniques for Boundary Layers. CRC Press, Boca Raton (2000).

A Godunov-Ryabenkii Instability for a Quickest Scheme

Ercília Sousa

Oxford University Computing Laboratory,
Wolfson Building, Parks Road, Oxford OX1 3QD, England
sousa@comlab.ox.ac.uk

Abstract. We consider a finite difference scheme, called Quickest, introduced by Leonard in 1979, for the convection-diffusion equation. Quickest uses an explicit, Leith-type differencing and third-order upwinding on the convective derivatives yielding a four-point scheme. For that reason the method requires careful treatment on the inflow boundary considering the fact that we need to introduce numerical boundary conditions and that they could lead us to instability phenomena. The stability region is found with the help of one of the most powerful methods for local analysis of the influence of boundary conditions – the Godunov-Ryabenkii theory.

1 Introduction

Quickest is a finite difference scheme due to Leonard [8] that deduces this scheme using control volume arguments. Davis and Moore [2] have shown that Quickest can also be derived by considering the Δt^3 in the Taylor expansion of the time derivative and make some subsequent approximations. Morton and Sobey [10] using the exact solution of the convection diffusion equation, derived Quickest based on a cubic local approximation. Quickest scheme uses an explicit, Leith-type differencing and third-order upwinding on the convective derivatives yielding a four-point scheme. In the limit $D \to 0$ is third order accurate in time. The use of third-order upwind differencing for convection greatly reduces the numerical diffusion associated with first-order upwinding [1]. Some of the literature about Quickest used in a flow simulation can be found in [1,2,6,8,9]. The major difficulties associated with the use of Quickest scheme in multidimensions are in the application of boundary conditions, being the major reason to study the influence of a numerical boundary condition on the stability of the numerical scheme.

Fourier analysis is the standard method for analysing the stability of discretisations of an initial value on a regular structured grid. This model problem has Fourier eigenmodes whose stability needs to be analysed. If they are stable at all points in the grid, and the discretisation of the boundary conditions is also stable then for most applications the overall discretisation is stable, in the sense of Lax [12].

The influence of the boundaries can be analysed using the Godunov-Ryabenkii theory. The Godunov-Ryabenkii theory was introduced by Godunov and Ryabenkii [3] and developed by Kreiss [7], Osher [11] and Gustafsson et al [4] (now also called GKS theory). In this paper we find the stability region for the Quickest scheme subject to a numerical boundary condition by applying the Godunov-Ryabenkii theory.

Consider the one-dimensional problem of convection with velocity V in the x-direction and diffusion with coefficient D:

$$\frac{\partial u}{\partial t} + V\frac{\partial u}{\partial x} = D\frac{\partial^2 u}{\partial x^2} \quad 0 \leq x < \infty, \quad t \geq 0 \tag{1}$$

$$u(x,0) = f(x) \tag{2}$$

$$u(0,t) = 0 \tag{3}$$

$$\|u(\cdot,t)\| < \infty \tag{4}$$

If we choose a uniform space step Δx and time step Δt, there are two dimensionless quantities very important in the properties of the scheme:

$$\mu = \frac{D\Delta t}{(\Delta x)^2}, \quad \nu = \frac{V\Delta t}{\Delta x}$$

ν is called the Courant (or CFL) number.

Before we describe the Quickest scheme and its numerical boundary condition, we give in the next section, a brief overview of the Godunov-Ryabenkii theory.

2 Godunov-Ryabenkii Stability Analysis

Two essential aspects of normal mode analysis for the investigation of the influence of boundary conditions on the stability of a scheme are that the initial value problem needs to be stable for the Cauchy problem which is best analysed with the von Neumann method (this means the interior scheme needs to be stable) and that its stability could be destroyed by the boundary conditions, but the converse its not possible.

In this section we give a brief description of the Godunov-Ryabenkii theory. For more detailed information about the theory we suggest [12,13,14] and specially [4]. A particular note is to be made of the work [15,16], establishing a relation between the GKS theory and group velocity.

We can approximate the problem (1)– (4) by the difference scheme

$$QU_j^n = U_j^{n+1}, \quad j = r, r+1, \ldots \tag{5}$$

$$Q = \sum_{j=-r}^{p} a_j E^j, \quad EU_j^n = U_{j+1}^n. \tag{6}$$

where a_j are scalars.

Two important assumptions are made:
a) The scalars a_{-r} and a_p are non-singular;
b) The finite difference scheme (5) is von Neumann stable.

As Q uses r points to the left, the basic approximation can not be used at $x_0, x_1, x_2, \ldots, x_{r-1}$, so there we will have to apply boundary conditions. These can be the conditions that are given for the original problem (in our particular case is associated only with the point x_0), but they can also be difference schemes, which will then be called numerical boundary conditions. The choice of numerical boundary conditions is crucial for the stability.

Let us assume that the boundary conditions can be written as

$$U_\beta^{n+1} = \sum_{j=1}^{q} l_{\beta j} U_j^n \quad \beta = 0, 1, \ldots, r-1 \tag{7}$$

where $l_{\beta j}$ are scalars.

The eigenvalue problem associated with our approximation is:

$$z\phi_j = Q\phi_j \quad j = r, r+1, \ldots \tag{8}$$

$$z\phi_\beta = \sum_{j=1}^{q} l_{\beta j} \phi_j \quad \beta = 0, 1, \ldots, r-1 \tag{9}$$

$$\|\phi\|_h < \infty \tag{10}$$

Lemma 1 Godunov-Ryabenkii Condition The approximation is unstable if the eigenvalue problem (8) – (10) has an eigenvalue z with $|z| > 1$.

Consider the characteristic equation of the interior scheme

$$z - \sum_{j=-r}^{p} a_j k^j = 0. \tag{11}$$

Lemma 2 For z such that $|z| > 1$, there is no solution of equation (11) with $|k| = 1$ and there are exactly r solutions, counted according to their multiplicity, with $|k| < 1$.

A general solution of (8) – (10) is of the form

$$\phi_j = \sum_{|k_a|<1} P_a(j) k_a^j, \quad k_a = k_a(z), \quad |z| > 1 \tag{12}$$

where k_a are solutions of the characteristic equation (11). This solution depends on r free parameters $\sigma = (\sigma_1, \ldots, \sigma_r)$. $P_a(j)$ is a polynomial in j. Its order is at most $m_a - 1$ where m_a is the multiplicity of k_a.

Note that if the solutions are simple, this implies that the solution has the form
$$\phi_j = \sum_{|k_i|<1} \sigma_i k_i^j. \tag{13}$$

This form of the solution is the one that usually arises in practice.

Substituting (12) into the boundary conditions (7) yields a system of equations
$$C(z)\sigma = 0,$$
$\sigma = (\sigma_1, \ldots, \sigma_r)$ and we can rephrase Lemma 2 in the following form:

Lemma 3 *The approximation is unstable if*
$$\text{Det } C(z) = 0 \quad \text{for some } z \in \mathbf{C} \text{ with} \quad |z| > 1.$$

Summarising, this theory is a generalisation of the von Neumann stability analysis taking into account the influence of boundary conditions. It states that the interior scheme needs to be von Neumann stable and when considered in the half-plane $x \geq 0$, a mode k^j with $|k| > 1$ will lead to an unbounded solution in space, that is, k^j will increase without bound when j goes to infinity. Therefore $|k|$ should be lower than one, and the Godunov-Ryabenkii stability condition states that all the modes with $|k| \leq 1$, generated by the boundary conditions, should correspond to $|z| < 1$.

3 Instability of a Quickest Scheme

Consider the interior difference scheme Quickest:
$$U_j^{n+1} = [1 - \nu\Delta_0 + (\frac{1}{2}\nu^2 + \mu)\delta^2 + \nu(\frac{1}{6} - \frac{\nu^2}{6} - \mu)\delta^2\Delta_-]U_j^n, \tag{14}$$

where we use the central, backward and second difference operators: $\Delta_0 U_j := (U_{j+1} - U_{j-1})/2$, $\Delta_- U_j := U_j - U_{j-1}$ and $\delta^2 U_j := U_{j+1} - 2U_j + U_{j-1}$.

We will consider two boundary conditions: the Dirichlet boundary condition associated with the original problem, $U_0^n = 0$ and the numerical boundary condition that we need at the first point of the mesh,
$$U_1^{n+1} = [1 - \nu\Delta_0 + (\frac{1}{2}\nu^2 + \mu)\delta^2 + \nu(\frac{1}{6} - \frac{\nu^2}{6} - \mu)\delta^2\Delta_+]U_1^n, \tag{15}$$

where Δ_+ is the forward operator defined by $\Delta_+ U_j := U_{j+1} - U_j$. This numerical boundary condition is deduced by a similar method used in [10] to obtain the Quickest scheme, using a local cubic interpolation of the points U_{j-2}^n, U_{j-1}^n, U_j^n, U_{j+1}^n. On the first point we can not use this interpolation since we do not have the point U_{-1}. We do instead an interpolation of the points U_0^n, U_1^n, U_2^n, U_3^n and it gives the difference scheme (15). The use of this downwind third difference at $x = \Delta x$ does not affect accuracy because it stills based on a cubic local approximation near $x = \Delta x$ as the interior scheme. However, as we shall show, it does have penalties in terms of stability.

Let us consider the corresponding eigenvalue problem:

$$z\phi_j = [1 - \nu\Delta_0 + (\frac{1}{2}\nu^2 + \mu)\delta^2 + \nu(\frac{1}{6} - \frac{\nu^2}{6} - \mu)\delta^2\Delta_-]\phi_j, \quad j \geq 2$$

$$\phi_0 = 0$$

$$z\phi_1 = [1 - \nu\Delta_0 + (\frac{1}{2}\nu^2 + \mu)\delta^2 + \nu(\frac{1}{6} - \frac{\nu^2}{6} - \mu)\delta^2\Delta_+]\phi_1. \qquad (16)$$

The Godunov-Ryabenkii condition tell us that the system (16) has an eigenvalue z with $|z| > 1$, then the approximation (14) – (15) is not stable. By Lemma 2 we have for this approximation that the characteristic equation for the interior scheme (14) has not $k = e^{i\xi}$, ξ real for $|z| > 1$ and there are exactly two solutions k_i, $i = 1, 2$ with $|k_i| < 1$ for $|z| > 1$.

Consider the characteristic equation for the interior scheme (14)

$$k^3(-c_1 + c_2 + c_3) + k^2(-z + 1 - 2c_2 - 3c_3) + k(c_1 + c_2 + 3c_3) - c_3 = 0. \qquad (17)$$

where $c_1 = \nu/2$, $c_2 = \nu^2/2 + \mu$ and $c_3 = \nu(1 - \nu^2 - 6\mu)/6$.

Assuming that the two solutions of the characteristic equation are distinct, any solution of (16) has the form

$$\phi_j = \sigma_1 k_1^j(z) + \sigma_2 k_2^j(z).$$

We want to find the solutions k_i, $i = 1, 2$ of (17), such that $|k_i(z)| < 1$, $i = 1, 2$ and the linear and homogeneous system

$$\sigma_1 + \sigma_2 = 0$$
$$\sigma_1 g(k_1, z, \mu, \nu) + \sigma_2 g(k_2, z, \mu, \nu) = 0 \qquad (18)$$

has a solution z with $|z| > 1$. The function $g(k, z, \mu, \nu)$ is the polynomial:

$$g(k, z, \mu, \nu) = k^3 c_3 + k^2(-c_1 + c_2 - 3c_3) + k(1 - 2c_2 + 3c_3 - z).$$

Since the first equation gives $\sigma_1 = -\sigma_2$, the linear homogeneous system (18) has a non-trivial solution if

$$g(k_1, z, \mu, \nu) - g(k_2, z, \mu, \nu) = 0.$$

Consider $k_1(z)$ and $k_2(z)$ defined as:

$$k_1(z) = \frac{r_1}{2} + \frac{\sqrt{-3r_1^2 + 4r_2}}{2} \qquad k_2(z) = \frac{r_1}{2} - \frac{\sqrt{-3r_1^2 + 4r_2}}{2}$$

where r_1 and r_2 are:

$$r_1(z, \mu, \nu) = \frac{(1-z)(-c_1 + c_2 + c_3) - 4c_1c_3 + 2c_2(c_1 - c_2)}{(1-z)c_3 - 2c_1c_3 - (c_1 - c_2)^2} \quad (19)$$

$$r_2(z, \mu, \nu) = \frac{(1-z)(z-1+4c_2) - (c_1^2 + 6c_1c_3 + 3c_2^2)}{(1-z)c_3 - 2c_1c_3 - (c_1 - c_2)^2} \quad (20)$$

Let $f(k, z, \mu, \nu)$ denote the characteristic polynomial for the interior scheme (see (17)). After some algebraic manipulations we can prove that for $c_3 \neq 0$, $k_1(z)$ and $k_2(z)$ are solutions of

$$f(k_1, z, \mu, \nu) - f(k_2, z, \mu, \nu) = 0 \quad (21)$$
$$g(k_1, z, \mu, \nu) - g(k_2, z, \mu, \nu) = 0. \quad (22)$$

If additionally to (21) $k_1(z)$ and $k_2(z)$ verify $f(k_1, z, \mu, \nu) + f(k_2, z, \mu, \nu) = 0$ then $k_1(z)$ and $k_2(z)$ are solutions of f. In that way we have two solutions of f that verify (22). Note that the characteristic polynomial f is a third order polynomial, which means we expect three roots, although we only find the analytical solution of two of them. Let $C(z, \mu, \nu) = f(k_1, z, \mu, \nu) + f(k_2, z, \mu, \nu)$. For each (μ, ν) we want to find $z_{\mu\nu}$ such that $C(z_{\mu\nu}, \mu, \nu) = 0$. The requirement for instability is $|z_{\mu\nu}| > 1$. Experimentally we observe that the solution $z(\mu, \nu)$ lies inside $|z| = 1$ for certain values of μ and ν and then crosses it at $z = -1$. We can say $z = -1$ is the value of transition from stable to unstable.

The function $C(z, \mu, \nu)$ as the form

$$C(z, \mu, \nu) = r_1(z, \mu, \nu)(3r_2(z, \mu, \nu) - 2r_1^2(z, \mu, \nu))(-c_1 + c_2 + c_3)$$
$$(2r_2(z, \mu, \nu) - r_1^2(z, \mu, \nu))(-z + 1 - 2c_2 - 3c_3)$$
$$+ r_1(z, \mu, \nu)(c_1 + c_2 + 3c_3) - 2c_3.$$

Let $p(\mu, \nu) = C(-1, \mu, \nu)$. We plot $p(\mu, \nu) = 0$ in Fig. 1. a).

For (μ, ν) such that $p(\mu, \nu) < 0$ there exists an eigenmode $z_{\mu\nu} < -1$ such that $C(z_{\mu\nu}, \mu, \nu) = 0$ (Fig. 1.b)).

This means that for $S_1 = \{(\mu, \nu) : p(\mu, \nu) < 0\}$ there exists $z_{\mu\nu}$ real and less than -1 such that $k_1(z_{\mu\nu}, \mu, \nu)$ and $k_2(z_{\mu\nu}, \mu, \nu)$ are solutions of f and verify (22). To assure that this eigenmode $z_{\mu\nu}$ which absolute value is bigger than one, determine an instable region we still need to verify that for these (μ, ν) we do have $|k_i(z_{\mu\nu}, \mu, \nu)| < 1, i = 1, 2$.

For z fixed let us define the following sets: $A_z = \{(\mu, \nu) : |k_1(z, \mu, \nu)| < 1\}$ and $B_z = \{(\mu, \nu) : |k_2(z, \mu, \nu)| < 1.\}$ For $z < -1$, $B_z \subset A_z$, i. e., if $|k_2(z, \mu, \nu)| < 1$ then $|k_1(z, \mu, \nu)| < 1$. We plot $C(z, \mu, \nu) = 0$ and B_z for $z = -1, -1.5$ in Fig. 2.

From the figure we observe that in the region B_{-1} the root $k_2(-1, \mu, \nu)$, for $(\mu, \nu) : p(-1, \mu, \nu) = 0$, become bigger than one approximately for $\nu < 0.09$. For $z = -1.5$ the same happens but for ν even smaller. Since one of the roots we found become larger than one we can not conclude anything about the instability of the method for $\nu < 0.09$. This is not a big problem since the von Neumann condition give us a stability limit for this region. We will plot the curve $p(\mu, \nu) =$

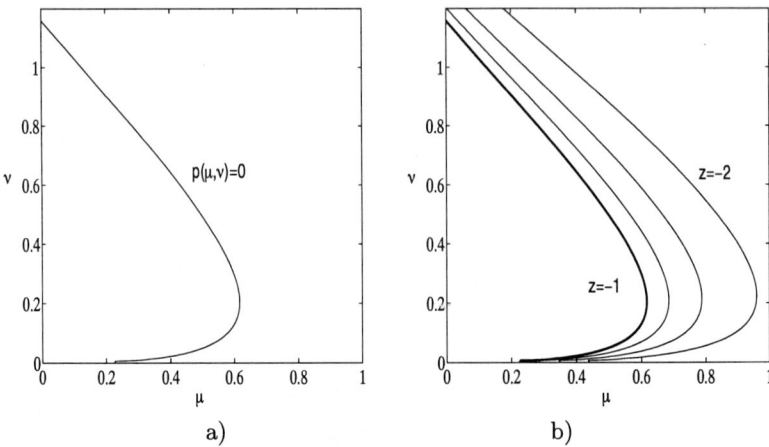

Fig. 1. a) $p(\mu,\nu) = 0$; b) $C(z,\mu,\nu) = 0$ for $z = -1, -1.2, -1.5, -2$

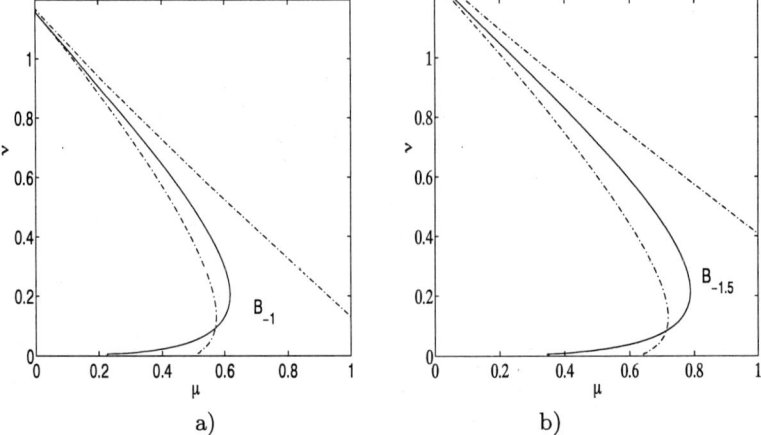

Fig. 2. a) $C(-1,\mu,\nu) = 0$ is the line (–) and B_{-1} is the region between the lines (---); b) $C(-1.5,\mu,\nu) = 0$ is the line (–) and $B_{-1.5}$ is the region between the lines (---)

0 for $\nu > 0.09$ and the von Neumann stability condition. We can see the unstable region plotted in Fig. 3. In fact running experiments numerically the region called stable in Fig. 3 is the exact region of practical stability.

Acknowledgements

Many thanks to Prof. L.N. Trefethen for all the crucial discussions and to Dr. I.J. Sobey for his kind support. The author acknowledges financial support from Sub-Programa Ciência e Tecnologia do 2 Quadro Comunitário de Apoio and Coimbra University, Portugal.

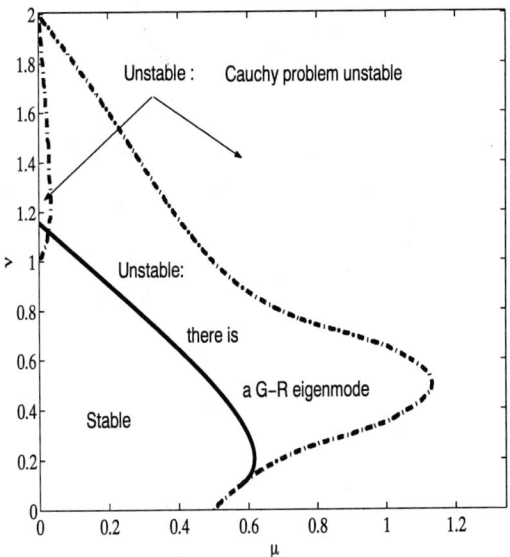

Fig. 3. Stability region: von Neumann condition (-··-) and Godunov-Ryabenkii condition (−)

References

1. Baum, H. R., Ciment, M., Davis, R. W. and Moore, E. F.: (1981), Numerical solutions for a moving shear layer in a swirling axisymmetric flow. Proc. 7th Int. Conf. on Numerical Methods in Fluid Dyn. (ed. W. C. Reynolds & R. W. MacCormack). Lect. Notes in Physics **141** (1981) 74-79.
2. Davis, R. W. and Moore, E. F.: A numerical study of vortex shedding from rectangles. Journal of Fluid Mechanics **116** (1982) 475-506.
3. Godunov, S. K., Ryabenkii: Spectral criteria for the stability of boundary problems for non-self-adjoint difference equations. Uspekhi Mat. Nauk., 18, 3 (1963) (In Russian).
4. Gustafsson, B., Kreiss, H.-O. and Sundstrom, A.: Stability theory of difference approximations for mixed initial boundary value problems, II. Mathematics of Computation **26** (1972) 649-686.
5. Gustafsson, B. , Kreiss, H.-O. and Oliger, J.: Time-dependent problems and difference methods, Wiley-Interscience (1995)
6. Johnson, R. W. and MacKinnon, R. J.: Equivalent versions of the Quick scheme for finite-difference and finite-volume numerical methods. Communications in applied numerical methods **8** (1992) 841-847.
7. Kreiss, H.-O.: Stability theory for difference approximations of mixed initial boundary value problems I. Mathematics of Computation **22** (1968) 703-714.
8. Leonard,B. P.: A stable and accurate convective modelling procedure based on quadratic upstream interpolation. Computer methods in applied mechanics and engineering **19** (1979) 59-98.

9. Leonard, B. P. and Mokhtari, S.: Beyond first-order upwinding the ultra-sharp alternative for non-oscillatory steady-state simulation of convection. International Journal for Numerical Methods in Engineering **30** (1990) 729-766.
10. Morton, K. W. and Sobey, I. J.: Discretisation of a convection-diffusion equation. IMA Journal of Numerical Analysis **13** (1993) 141-160.
11. Osher, S.: Stability of difference approximations of dissipative type for mixed initial-boundary value problems. Mathematics of computation **23** (1969) 335-340.
12. Richtmyer, R. D. and Morton, K. W.: Difference methods for initial-value problems, 2nd edn, Wiley-Interscience, New York (1967)
13. Sod, G. A.: Numerical methods in fluid dynamics: initial and initial boundary-value problems, Cambridge University Press, Cambridge (1988)
14. Strikwerda, J.: Finite difference schemes and partial differential equations, Wadsworth & Brooks, California, (1989)
15. Trefethen, L. N.: Group velocity interpretation of the stability theory of Gustafsson, Kreiss and Sundstrom. Journal of Computational Physics **49** (1983) 199-217.
16. Trefethen, L. N.: Instability of difference models for hyperbolic initial boundary value problems. Comm. Pure and applied Mathematics **37** (1984) 329-367.

Modelling Torsional Properties of Human Bones by Multipoint Padé Approximants

J. J. Telega, S. Tokarzewski, and A. Gałka

Institute of Fundamental Technological Research
Świętokrzyska 21, 00-049 Warsaw, Poland

Abstract. The macroscopic modelling of the macroscopic behaviour of inhomegeneous media requires evaluation of the effective moduli. In the relevant literature, many papers were concerned with estimating of macroscopic moduli $\lambda_e(x)$) for two-phase materials. The main aim of this paper is to find the best estimation of $\lambda_e(x)$) from a given finite number of coefficients of power expansions of $\lambda_e(x)$ at $x = x_i$, $i = 1, 2, ..., N, N+1$, and apply them to model a torsional behaviour of a human cancellous bone filled with marrow. Errors of numerical evaluations are calculated and discussed.

1 Introduction

Macroscopic modelling of microinhomogeneous media requires the evaluation of effective moduli. However their exact values are available only in specific cases; for instance in one-dimensional periodic homogenization. In the relevant literature, many papers were concerned with estimating of the effective coefficients $\lambda_e(x)$, also for bio-materials such as human bones.

The main aim of this contribution is to establish general bounds on the coefficients $\lambda_e(x)$ generated by an arbitrary number of coefficients of power expansions of $\lambda_e(x)$ at $x = x_1, x_2, ..., x_N < \infty$ and $x_{N+1} = \infty$ and apply them to biomechanical problem of torsion of cancellous bone filled with marrow.

2 Preliminaries

Let us consider Stieltjes function represented by

$$f_1(x) = \int_0^{1/(1+x_i)} \frac{d\gamma_{x_i}(u)}{1 + (x - x_i)u}, \quad i = 1, 2, ..., N, \tag{1}$$

and satisfying the inequality $f_1(-1) \leq 1$. Here spectra $\gamma_{x_i}(u)$, $i = 1, 2, ..., N$, $\gamma_\infty(u)$ are real, bounded and non-decreasing functions. Power expansions of $f_1(x)$ at $x = x_i$, $i = 1, 2, ..., N$, and $x_{N+1} = \infty$ are given by

$$f_1(x) = \sum_{k=0}^{\infty} c_{i,1,k+1}(x - x_i)^k, \quad i = 0, 1, ..., N,$$
$$f_1(x) = \frac{1}{x-x_N} \sum_{k=0}^{\infty} d_{\infty,1,k} \left(\frac{1}{x-x_N}\right)^k. \tag{2}$$

Table 1. Discrete values of the elastic torsional modulus $\mu_e/\mu_1 - 1$ for hexagonal array of cylinders, after [6]

x	$\varphi=0.76$	$\varphi=0.80$	$\varphi=0.84$	$\varphi=0.88$
-1	-0.8711	-0.8996	-0.9286	-0.9607
0	0.0000	0.0000	0.0000	0.0000
9	3.3778	3.9489	4.6887	5.7225
49	5.7076	7.2600	9.7931	5.1565
∞	6.7600	8.9586	3.0093	24.4508

The coefficients $c_{i,1,k}$, $i=1,2,...,N$ and $d_{\infty,1,k}$ are assumed to be finite for any fixed k and i. Let us introduce the rational functions $g_C(x; p_{x_1}, p_{x_2}, ..., p_{x_N}, p_\infty)$ and $h_D(x; , p_{x_1}, p_{x_2}, ..., p_{x_N}, p_\infty)$ defined by the relations:

$$g_C(x; p_{x_1}, p_{x_2}, ..., p_{x_N}, p_\infty) = \frac{a'_1 x + a'_2 x^2 + ... + a'_{E[(P+1)/2]} x^{E[(P+\Delta)/2]}}{1 + b'_1 x + b'_2 x^2 + ... + b'_{E(P/2)} x^{E(P/2)}},$$

$$h_D(x; p_{x_1}, p_{x_2}, ..., p_{x_N}, p_\infty) = \frac{a''_1 x + a''_2 x^2 + ... + a''_{E[(P+1+\Delta)/2]} x^{E[(P+1+\Delta)/2]}}{1 + b''_1 x + b''_2 x^2 + ... + b''_{E[(P+1)/2]} x^{E[(P+1)/2]}},$$

$$h_D(-1; p_{x_1}, p_{x_2}, ..., p_{x_N}, p_\infty) = -1,$$

(3)

where

$$C = E[(P+\Delta)/2] + E(P/2); \quad D = E[(P+1+\Delta)/2] + E[(P+1)/2]$$

$$P = \sum_{j=1}^{N} p_j + p_\infty, \quad \Delta = \begin{cases} 1 & \text{if } p_\infty = 0 \\ 0 & \text{if } p_\infty > 0 \end{cases}, \quad E(\zeta) = \max_{U \in \mathbf{N}} \{U < \zeta\}.$$

(4)

The parameters $p_{x_1}, p_{x_2}, ..., p_{x_N}, p_\infty$, appearing in (3) and (4) denote the numbers of coefficients of power expansions of $f_1(x)$ at $x_1, x_2, ..., x_N, x_{N+1} = \infty$. The functions $g_C(x; p_{x_1}, p_{x_2}, ..., p_{x_N}, p_\infty)$ and $h_D(x; p_{x_1}, p_{x_2}, ..., p_{x_N}, p_\infty)$ become diagonal and subdiagonal multipoint Padé approximants to Stieltjes function $xf_1(x)$, if they satisfy:

$$\begin{aligned} xf_1(x) - g_C(x) &= O\left((x-x_i)^{p_1+1}\right), \quad i=1,2,...,N, \\ xf_1(x) - g_C(x) &= O\left(\left(\tfrac{1}{(x-x_N)}\right)^{q_\infty+1}\right), \quad q_\infty = \begin{cases} p_\infty & \text{if } P \text{ is even} \\ p_\infty - 1 & \text{if } P \text{ is odd} \end{cases}, \\ xf_1(x) - h_D(x) &= O\left((x-x_i)^{p_1+1}\right), \quad i=1,2,...,N, \\ xf_1(x) - h_D(x) &= O\left(\left(\tfrac{1}{(x-x_N)}\right)^{r_\infty+1}\right), \quad r_\infty = \begin{cases} p_\infty - 1 & \text{if } P \text{ is even} \\ p_\infty & \text{if } P \text{ is odd} \end{cases}, \end{aligned}$$

(5)

where

$$g_C(x) = g_C(x; p_{x_1}, p_{x_2}, ..., p_{x_N}, p_\infty); \quad h_D(x) = h_D(x; p_{x_1}, p_{x_2}, ..., p_{x_N}, p_\infty). \quad (6)$$

For the sake of simplicity the notations $g_C(x; p_{x_1}, p_{x_2}, ..., p_{x_N}, p_\infty)$, $g_C(x)$ and $h_D(x; p_{x_1}, p_{x_2}, ..., p_{x_N}, p_\infty)$, $h_D(x)$ will equivalently be used.

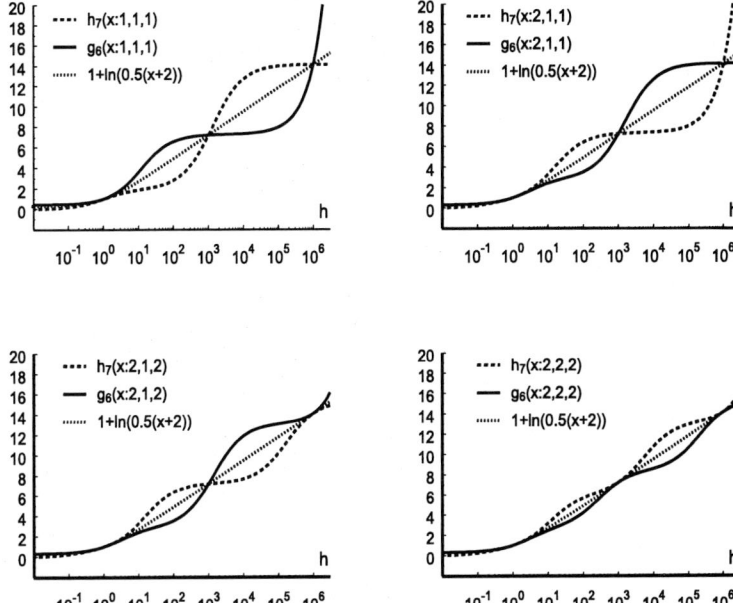

Fig. 1. Multipoint Padé upper and lower bounds on $(\lambda_e(x)/\lambda_1) - 1 = \ln(0.5(x+2))$ predicted by Theorem 1

3 Inequalities for Multipoint Padé Approximants

In our paper [12] it has been proved the following theorem establishing the general inequalities for multipoint Padè approximants to $\lambda_e(x)/\lambda_1 - 1$

Theorem 1. *For any fixed* $x \in (-1, x_1), (x_1, x_2), ..., (x_K, x_{K+1}), ..., (x_N, \infty)$ *the multipointpoint Padé approximants* $g_C(x; p_{x_1}, p_{x_2}, ..., p_{x_N}, p_\infty)$ *and* $h_D(x; p_{x_1}, p_{x_2}, ..., p_{x_N}, p_\infty)$ *to the expansions of* $(\lambda_e(x)/\lambda_1) - 1$ *available at* $x = x_1, x_2, ..., x_K, x_{K+1}, ..., x_N, x_{N+1} = \infty$ *obey the following inequalities:*
(i) *If* $x \in (-1, x_1)$ *then*

$$g_C(x) > \left(\frac{\lambda_e}{\lambda_1} - 1\right) > h_D(x). \tag{7}$$

(ii) *If* $x \in (x_K, x_{K+1}), K = 1, 2, ..., N$ *then*

$$(-1)^{P^K} g_C(x) < (-1)^{P^K} \left(\frac{\lambda_e}{\lambda_1} - 1\right) < (-1)^{P^K} h_D(x), P^K = \sum_{k=1}^{K \leq N} p_{x_K}, \tag{8}$$

where $\lambda_e(x)/\lambda_1$ *stands for the limit as* $P = \sum_{i=1}^{N} p_i + p_\infty$ *goes to infinity of* $1 + g_C(x; p_{x_1}, p_{x_2}, ..., p_{x_N}, p_\infty)$ *or* $1 + h_D(x; p_{x_1}, p_{x_2}, ..., p_{x_N}, p_\infty)$ *in* $x \in (-1, \infty)$.

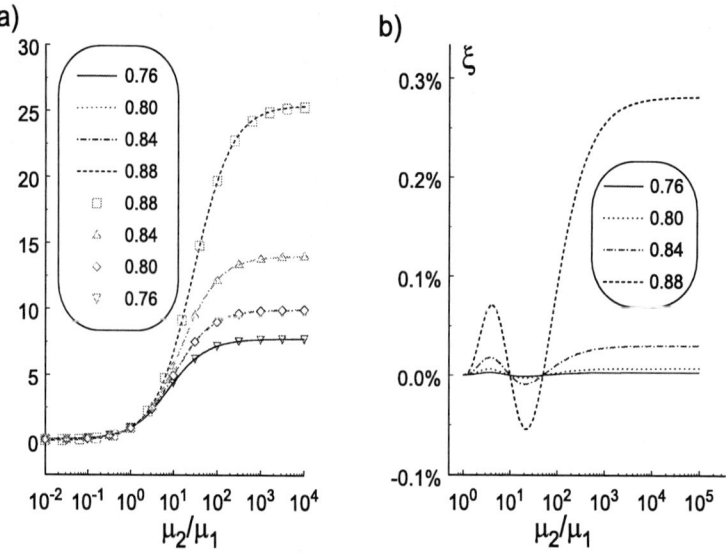

Fig. 2. Hexagonal array of elastic cylinders with volume fraction φ and physical parameter $x = \mu_2/\mu_1 - 1$. (a) Multipoint Padé bounds $(1 + g_4(x; 2, 1, 1, 1)$ – *solid lines*) and $(1 + h_6(x; 2, 1, 1, 1)$– *scattered lines*) on torsional modulus μ_e/μ_1, upper and lower bounds almost coincide. (b) An error $\xi = (g_4(x; 2, 1, 1, 1) - h_6(x; 2, 1, 1, 1)) / (1 + g_4(x; 2, 1, 1, 1))$ for μ_e/μ_1

4 Hexagonal Array of Elastic Cylinders

Let us consider an elastic beam reinforced with elastic fibers arranged in a hexagonal lattice. Assume that β_1 and μ_1 are Lamé constants of the matrix, while β_2 and μ_2 Lamé coefficients of fibers. By φ we denote the volume fraction of inclusions. By using classical homogenization procedure the following equations defining the effective torsion modulus μ_e/μ_1 have been derived, see [11]

$$\frac{\mu_e}{\mu_1} = \frac{1}{|Y|} \int_Y U(y) \frac{\partial T(y)}{\partial y_1} dy,$$

$$\frac{\partial}{\partial y_1}\left(U(y)\frac{\partial T(y)}{\partial y_1}\right) + \frac{\partial}{\partial y_2}\left(U(y)\frac{\partial T(y)}{\partial y_2}\right) = 0, \tag{9}$$

$$\frac{1}{|Y|}\int_Y \frac{\partial T(y)}{\partial y_1} dy = 1, \quad \frac{1}{|Y|}\int_Y \frac{\partial T(y)}{\partial y_2} dy = 0,$$

where $U(y)$ is a characteristic function. The effective modulus μ_e/μ_1 has a Stieltjes integral representation given by(1). For three dimentional materials the power expansion of $\mu_e/\mu_1 - 1$ at $x = 0$ ($x = h - 1$, $h = \mu_2/\mu_1$) takes form

$$\mu_e/\mu_1 - 1 = \varphi x + \frac{1}{3}\varphi(1-\varphi)x^2 + O(x^3) \tag{10}$$

The discrete values of $\mu_e/\mu_1 - 1$ are reported in [6], cf.Table also 1. Multipoint Padè approximants $g_4(x; p_0, p_9, p_{49}, p_\infty) = g_4(x; 2, 1, 1, 1)$ and

$h_6(x; p_0, p_9, p_{49}, p_\infty) = h_6(x; 2, 1, 1, 1)$ evaluated from the input data (10) and Table 1 estimate the effective shear modulus μ_e/μ_1 from above and below, see Fig. 2a. From Fig.2b we conclude that the torsional modulus μ_e/μ_1 differs from the multpoint Padé approximants $1 + g_4(x; 2, 1, 1, 1)$ and $1 + h_6(x; 2, 1, 1, 1)$ less then 0.3%. On account of that we take the rational function

$$\mu_e/\mu_1 - 1 = h_6(x; p_0, p_9, p_{49}, p_\infty) = h_6(x; 2, 1, 1, 1), \quad \varphi \leq 0.88. \quad (11)$$

as a solution of a system of Eqs (9).

5 Modelling of Torsional Behaviour of Cancellous Bone

Let us consider an inhomogeneous beam consisting of viscoelastic cylinders regularly spaced in a viscoelastic phase.

For the investigation of the macroscopic responses of that porous beam the well known elastic-viscoelastic correspondence principle will be used, cf. [3]. That principle reads: the complex torsional modulus μ_e^*/μ_1^* of a viscoelastic system one obtains by replacing in (11) a real variable x by a complex one z. Hence we get

$$\mu_e^*(z)/\mu_1^* = 1 + h_6^*(z; 2, 1, 1, 1), \text{ for } \varphi \leq 0.88, z = x + iy = \mu_2^*/\mu_1^* - 1. \quad (12)$$

Of interest is the composite consisting of fluid cylinders of viscosity μ_2 regularly spaced in an elastic matrix of shear modulus μ_1. Such a composite material models a cancellous human bone filled with a marrow, see Fig 3. By substituting $\mu_1^* = \mu_1$, $\mu_2^* = I\omega\mu_2$ into (12) we obtain complex modulus of a prismatic porous beam filled with viscous fluid

$$\mu_e^*(z)/\mu_1 = 1 + h_6^*\left(\frac{I\omega\mu_2}{\mu_1} - 1; 2, 1, 1, 1\right), \text{ for } \varphi \leq 0.88 \quad (13)$$

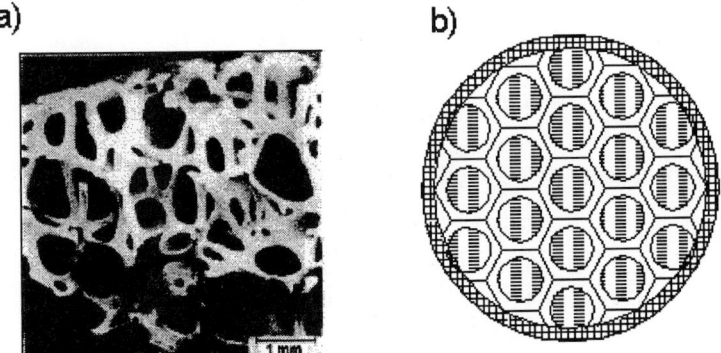

Fig. 3. (a) The scanning electron micrograph showing a prismatic structure of cancellous bone for a sample taken from the femoral head, cf. [4] pp. 318. (b) An idealized structural model of a prism-like cancellous bone

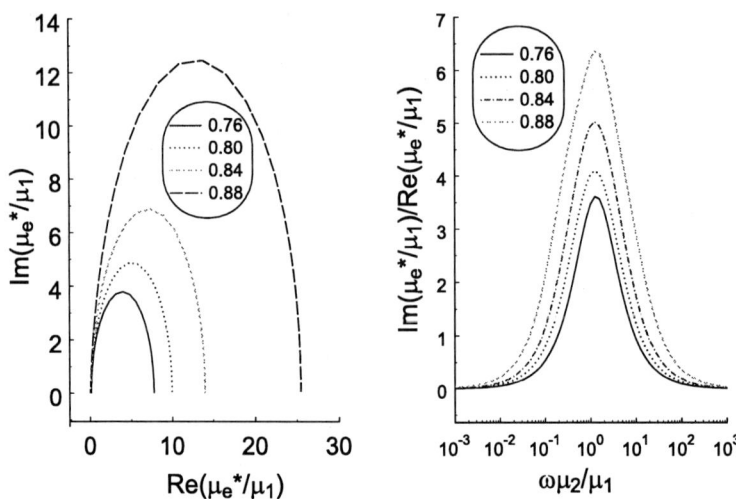

Fig. 4. Complex torsional modulus for the elastic porous beam filled with viscous fluid; $\varphi = 0.76, 0.80, 0.84, 0.88$

Figs 4 and 5 depict complex modulus $\mu_e^*(z)/\mu_1$ and also the real and imaginary parts of it. Note that moduli $\mu_e^*(z)/(\mu_1)$ and compliances $\mu_1/\mu_e^*(z)$, $z = (I\omega/\kappa) - 1$, $\kappa = \mu_1/\mu_2$ divided by $I\omega$ are Fourier transformations of the torsional creep function $\Phi(t)$ and torsional relaxation function $\Psi(t)$, respectively, cf. [3]. Hence we can write

$$\mu_1 \overline{\Phi(I\omega)} = \frac{\mu_1}{I\omega \mu_e^*(z)}, \quad \frac{\overline{\Psi(I\omega)}}{\mu_1} = \frac{\mu_e^*(z)}{I\omega \mu_1}, \quad z = \frac{I\omega \mu_2}{\mu_1} - 1. \quad (14)$$

The inverse of Fourier transformations of $\overline{\Phi(I\omega)}$ and $\overline{\Psi(I\omega)}$ are given by

$$\mu_1 \Phi(t) = d^c + \sum_{n=1}^{3} \frac{b_n^c}{a_n^c} \left(1 - (1 + a_n^c \kappa t) e^{-\kappa t}\right),$$
$$\frac{\Psi(t)}{\mu_1} = d^r - \sum_{n=1}^{3} \frac{b_n^r}{a_n^r} \left(1 - (1 + a_n^r \kappa t) e^{-\kappa t}\right). \quad (15)$$

Here the coefficients $d^c, d^r, b_n^c, b_n^r, a_n^c$ and a_n^r take values listed below

φ	d^c	b_1^c	b_2^c	b_3^c	a_1^c	a_2^c	a_3^c
0.76	0.1289	0.0146	0.0655	0.9348	2.1958	0.8867	0.1238
0.80	0.1004	0.0476	0.0666	0.9226	1.9109	0.6432	0.0948
0.84	0.0714	0.0755	0.0925	0.9011	2.3972	0.5213	0.0656
0.88	0.0393	0.1339	0.1471	0.8608	3.9565	0.4500	0.0344

(16)

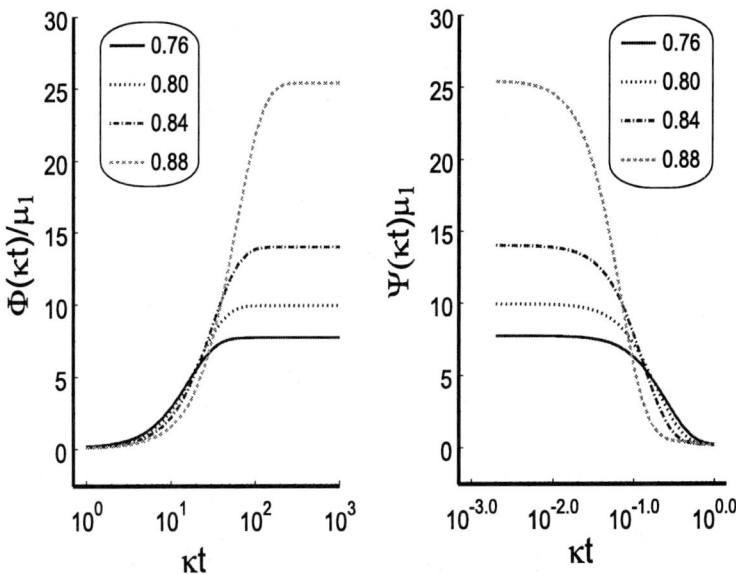

Fig. 5. The torsional creep function $\Phi(t)$ and torsional relaxation function $\Psi(t)$ for a porous beam consisting of hexagonal array of viscous fluid cylinders spaced in a linear elastic matrix

and

φ	d^r	b_1^r	b_2^r	b_3^r	a_1^r	a_2^r	a_3^r
0.76	7.7600	60.980	0.0974	0.0431	8.0939	2.1575	0.8312
0.80	9.9586	102.62	0.1831	0.0218	10.557	1.8143	0.6035
0.84	14.009	209.39	0.4192	0.0192	15.275	2.2101	0.4768
0.88	25.451	737.96	1.6569	0.0223	29.669	3.4456	0.3877

(17)

The torsional creep function $\Phi(t)$ given by (15)-(17) and the relaxation function $\Psi(t)$ determined by (15)-(16) have been depicted in Fig. 5.

6 Summary and Conclusions

By applying the multipoint Padé approximants discussed in [2] the new upper and lower bounds on real-valued transport coefficients of two-phase media have been established (Th.1). The bounds obtained incorporate an arbitrary number of coefficients of power expansions of $\mu_e(x)/\mu_1$ available at finite number of points. Consequently the estimates (7)-(8) generalize the bounds reported earlier in literature.

Multipoint Padé bounds (7)-(8) have been used to study the torsional behaviour an idealized model of cancellous human bone. The torsional rigidities: complex modulus and creep and relaxation functions have been evaluated. By analyzing graphs we observe a hydraulic stiffening of a bone due to the presence of bone marrow.

Multipoint Padé aproximants are particularly suitable for implementation to mechanical problems. Their evaluation from the given coefficients of power expansions of analytical functions leads to fast, accurate numerical algorithms, which are simply recursive and do not involve the solution of large number of equations, see [8]

Acknowledgment

The authors were supported by the State Committee for Scientific Research (KBN Poland) through the grants No7 T07A 021 15 and No 8 T11F 01718.

References

1. I. Andrianov, G. Starushenko, V. Danishevsky, S. Tokarzewski, Homogenization procedure and Padè approximants for effective heat conductivity of composite materials with cylindrical inclusions having square cross-section, *Proc. R. Soc. Lond. A*, **455**, 3401-3413, 1999.
2. G. A. Baker, P. Graves-Morris, Padé Approximants, Second Edition, in:Gian-Carlo Rota, Ed., *Encyclopedia of Mathematics and its Applications*, **59**, Cambridge University Press 1996.
3. Christensen R. M., Mechanics of composite materials, John Wiley&Sons, New York, 1979.
4. Gibson J. L., Ashby M. F., *Cellular Solids: Structure and Properties*, Pergamon Press, 1988.
5. K. Golden and G. Papanicolaou, Bounds for effective parameters of heterogeneous media by analytic continuation, *Comm.Math.Phys.*, **90**, 473-491, 1983.
6. W. T. Perrins, D. R. McKenzie, R. C. Mc Phedran, Transport properties of regular array of cylinders, *Proc. R. Soc. Lond. A* 369, 207-225, 1979.
7. R. C. McPhedran, G. W. Milton, Bounds and exact theories for the transport properties of inhomogeneous media, *Appl. Phys.A* **26**, 207-220, 1981.
8. S. Tokarzewski, A. Gałka, I. Andrianov, Bounds on the effective transport coefficients of two- phase media from discrete theoretical and experimental data, *Comp. Assis. Mech. and Eng. Sc.*, **4**, 229- 241, 1997.
9. S. Tokarzewski, J. J. Telega, Bounds on effective moduli by analytical continuation of the Stieltjes function expanded at zero and infinity, *Z. angew. Math. Phys.* 48, 1-20, 1997.
10. S. Tokarzewski, J. J. Telega, Inequalities for two- point Padè approximants to the expansions of Stieltjes functions in a real domain, *Acta Applicandae Mathematicae*, **48**, 285-297, 1997.
11. S. Tokarzewski, J. J. Telega, A. Gałka, Torsional rigidities of cancellous bone filled with marrow: The application of multipoint Padé approximants, Engineering Trasaction, (in press).
12. S. Tokarzewski, J. J. Telega, A. Gałka, General inequalities for multipoint Padè approximants, (in preparation)

Numerical Algorithm for Studying Hydrodynamics in a Chemical Reactor with a Mixer

I. Zheleva and A. Lecheva

Rousse University, Technology College
7200 Razgrad, POB110, Bulgaria

Abstract. The mixing in stirred vessels is acknowledged to be one of the most important characteristics for many industrial technologies. So that research in this area has concentrated for a long time on visual flow studies and on the measurements of overall properties such as power consumption, the overall gas holdup and overall mass transfer rate. Although, some features of the flow regimes inside the vessels can be obtained in this way. Mathematical modeling of the mixing processes recently becomes a useful and effective method for investigation.
This paper presents a mathematical model for hydrodynamics in a cylindrical reactor with one Rashton mixer. The liquid is supposed to be incompressible and the process 2D steady state. The Navier- Stokes equations are described by means of the stream-function and vorticity.
A numerical methodology for simulating mixing processes in stirred vessels is also presented. The numerical algorithm for solving these equations is based on an alternating direction implicit method for irregular mesh. The proposed algorithm is tested for different model tasks. The initial numerical results for hydrodynamics are presented graphically.

Introduction

Mixing is one of the most important processes, used in many chemical productions, especially in fermentation processes for obtaining antibiotics.

For these technologies usually the biggest part of the power consumption is spent for mixing. Because of this a detailed investigation of mixing processes is still needed. For many real technologies physical measurements and natural experiments are very expensive and at the same time they are not accurate enough. So, recently, mathematical modelling becomes a proper reliable tool for studying complex technological processes.

This paper presents a mathematical model and numerical algorithm for hydrodynamics of a chemical reactor with a mixer.

1 Mathematical Formulation

1.1 Scheme of the Chemical Reactor with a Mixer

The chemical reactor with a mixer is a cylindrical tank with radius R and height Z (Fig.1.). The mixer is situated at H1 height on the cylinder axis, the radius of the mixer spades is R1 and its thickness is L1. We assume firstly that the mixer is a disc with radius R1 and thickness L1.

The chemical reactor is filled with viscous incompressible fluid. The mixer rotates with a given constant angle velocity Ω.

We introduce a cylindrical coordinate system $(\overline{r}, \overline{\varphi}, \overline{z})$, which is also given in Fig.1.

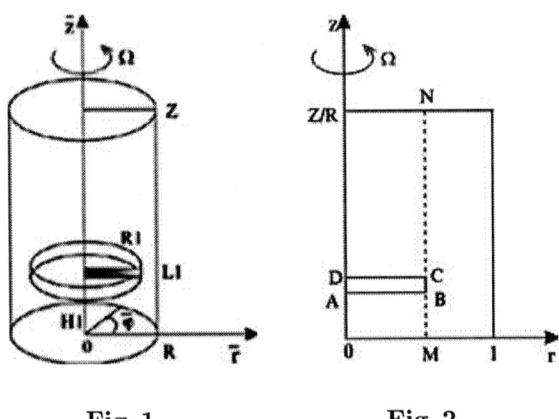

Fig. 1.   Fig. 2.

1.2 Basic Equations

For the formulation of the mathematical model it is assumed that the fluid in the chemical reactor is an incompressible homogeneous Nutonian fluid and its rotating motion is steady-state and axissymmetric.

The mass and the momentum conservation equations can express the dynamic behavior of the fluid. The Navier-Stokes equations are written in the introduced cylindrical coordinate system in the axis symmetrical case:

$$\frac{\partial(\overline{r}\overline{V_r})}{\partial \overline{r}} + \frac{\partial \overline{V_\varphi}}{\partial \overline{\varphi}} + \frac{\partial(\overline{z}\overline{V_z})}{\partial \overline{z}} = 0 \qquad (1)$$

$$\frac{\partial \overline{V_r}}{\partial \overline{t}} + \overline{V_r}\frac{\partial \overline{V_r}}{\partial \overline{r}} + \overline{V_z}\frac{\partial \overline{V_r}}{\partial \overline{z}} - \frac{\overline{V_\varphi}^2}{\overline{r}} = -\frac{1}{\rho}\frac{\partial \overline{p}}{\partial \overline{r}} + \overline{\nu}\left[\nabla^2 \overline{V_r} - \frac{\overline{V_r}}{\overline{r}^2}\right] \qquad (2)$$

$$\frac{\partial \overline{V_\varphi}}{\partial \overline{t}} + \overline{V_r}\frac{\partial \overline{V_\varphi}}{\partial \overline{r}} + \overline{V_z}\frac{\partial \overline{V_\varphi}}{\partial \overline{z}} + \frac{\overline{V_r}\overline{V_\varphi}}{\overline{r}} = \overline{\nu}\left[\nabla^2 \overline{V_\varphi} - \frac{\overline{V_\varphi}}{\overline{r}^2}\right] \qquad (3)$$

$$\frac{\partial \overline{V_z}}{\partial \overline{t}} + \overline{V_r}\frac{\partial \overline{V_z}}{\partial \overline{r}} + \overline{V_z}\frac{\partial \overline{V_z}}{\partial \overline{z}} = -\frac{1}{\rho}\frac{\partial \overline{p}}{\partial \overline{z}} + \overline{\nu}\nabla^2 \overline{V_z} \qquad (4)$$

where $\nabla^2 \equiv \frac{\partial^2}{\partial \overline{r}^2} + \frac{1}{\overline{r}}\frac{\partial}{\partial \overline{r}} + \frac{\partial^2}{\partial \overline{z}^2} = \frac{1}{\overline{r}}\frac{\partial}{\partial \overline{r}}\left(\overline{r}\frac{\partial}{\partial \overline{r}}\right) + \frac{\partial^2}{\partial \overline{z}^2}$, $\overline{\rho}$ is the mass density, $\overline{V}(\overline{V_r}, \overline{V_\varphi}, \overline{V_z})$-the velocity vector which is a function only of $\overline{r}$ and $\overline{z}$, $\overline{p}$- the pressure, $\overline{\nu}$- cinematic viscosity of the fluid.

We will look only for a stationary solution of these equations because we will examine the work of the reactor after its started. Then, obviously, the motion is stationary and axisymmetric also.

1.3 Boundary Conditions

On the solid walls of the reactor the velocity components are equal to zero

$$\overline{V} = 0 \qquad (5)$$

and the mixer is rotating with a given constant velocity.

2 Numerical Algorithm

2.1 Another form of the Equations

For describing the numerical technology of simulating rotating processes in the reactor the equations (1)-(4) are written in another form by introducing the stream line function ψ and vorticity ω in dimensionless form:

$$\frac{\partial}{\partial r}\left(\frac{1}{r}\frac{\partial \psi}{\partial r}\right) + \frac{\partial}{\partial z}\left(\frac{1}{r}\frac{\partial \psi}{\partial z}\right) = -\omega \qquad (6)$$

$$\frac{\partial M}{\partial t} + \frac{1}{r}\left(U\frac{\partial (rM)}{\partial r}\right) + W\frac{\partial M}{\partial z} = \frac{1}{Re}\left[\frac{\partial}{\partial r}\left(\frac{1}{r}\frac{\partial (rM)}{\partial r}\right) + \frac{\partial^2 M}{\partial z^2}\right] \qquad (7)$$

$$\frac{\partial \omega}{\partial t} + U\frac{\partial \omega}{\partial r} + W\frac{\partial \omega}{\partial z} - \frac{1}{r^3}\frac{\partial M^2}{\partial z} = \frac{1}{Re}\left[\nabla^2 \omega - \frac{1}{r^2}\frac{\partial U}{\partial z}\right] \qquad (8)$$

$$U = -\frac{1}{r}\frac{\partial \psi}{\partial z}, W = \frac{1}{r}\frac{\partial \psi}{\partial r} \qquad (9)$$

Here $V(U,V,W)$ is the velocity vector with components in the introduced coordinate system, M is the momentum of the tangential velocity, $r = \frac{\overline{r}}{R}, z = \frac{\overline{z}}{R}, U = \frac{\overline{U}}{\Omega R}, W = \frac{\overline{W}}{\Omega R}, \omega = \frac{\overline{\omega}}{R}, M = \frac{\overline{M}}{\Omega R^2}$ $(M = \overline{V_\varphi}.\overline{r}), t = \overline{t}\Omega$ and Re is the Reynolds number.

The geometrical area for the axisymmetric task described above is shown in Fig.2.

The boundary conditions are:

$$r = 0, \ 0 \leq z \leq \frac{z}{R}, \ \text{then} \ \psi = 0, M = 0, U = 0, W = 0, \omega = 0;$$
$$r = 1, \ 0 \leq z \leq \frac{z}{R}; \ \text{then} \ \psi = 0, M = 0, U = 0, W = 0;$$
$$z = \frac{z}{R}, \ 0 \leq r \leq 1, \ \text{then} \ \psi = 0, M = 0, U = 0, W = 0; \qquad (10)$$
$$z = 0, \ 0 \leq r \leq 1, \ \text{then} \ \psi = 0, M = 0, U = 0, W = 0;$$
$$0 \leq r \leq \frac{R1}{R}, z = \frac{H1}{R}, z = \frac{(H1+L1)}{R}, \ \text{then} \ \psi = 0, M = \frac{r^2}{R^2}, U = W = 0.$$

As we will look for a stationary solution of these equations afterwards for describing the numerical scheme we will think for t as a fictive time parameter.

2.2 Grid

We construct a non-uniform grid. The points of the area, given in Fig.2, namely A, B, C, D, M, N have to be nodes of the grid. For the describing of the grid the geometrical area is divided into three parts in z direction.

The grid can be modified as the number of the points increases in the critical areas - near the mixer, near the walls of the reactor and near the symmetry line Oz.

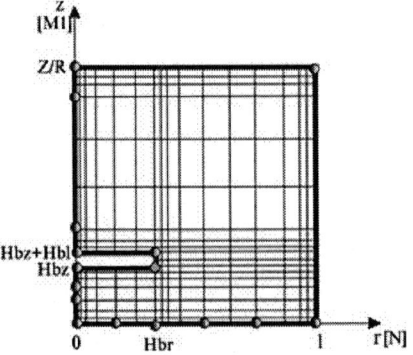

Fig. 3. The grid (N - number of points in r direction, $M1$- number of points in z direction)

2.3 Approximation of the Vorticity Equation

We write the following scheme for the vorticity equation (6)-(9):

$$\frac{\omega_{i,j}^{n+1/2} - \omega_{i,j}^n}{0.5\tau} = \frac{1}{Re}\delta_r^2 \omega_{i,j}^{n+1/2} + \frac{1}{Re}\delta_z^2 \omega_{i,j}^n + C_{i,j}^n \quad (11)$$

$$\frac{\omega_{i,j}^{n+1} - \omega_{i,j}^{n+1/2}}{0.5\tau} = \frac{1}{Re}\delta_z^2 \omega_{i,j}^{n+1} + \frac{1}{Re}\delta_r^2 \omega_{i,j}^{n+1/2} + C_{i,j}^n \quad (12)$$

where $C = -\frac{1}{Re}\frac{1}{r^2}\frac{\partial U}{\partial z} - U\frac{\partial \omega}{\partial r} - W\frac{\partial \omega}{\partial z} + \frac{1}{r^3}\frac{\partial M^2}{\partial z}, \delta_r^2 = \frac{\partial^2}{\partial r^2} + \frac{1}{r}\frac{\partial}{\partial r}, \delta_z^2 = \frac{\partial^2}{\partial z^2}$.

The boundary conditions (10) for the vorticity have to be calculated on the base of the known boundary conditions for the stream-line function. The connection between the vorticity and the stream function can be found from the equation (6) if we assume that this equation is valid on the boundaries. Then we can write the boundary conditions for the vorticity as [1]:

$\omega_{i,N} = -\frac{2\psi_{i,N-1}}{h_r^2} + O(h_r)$ on the right wall of the reactor,

$\omega_{M3,j} = -\frac{2\psi_{M_3-1,j}}{r_j h_z^2} + O(h_z)$ on the top wall of the reactor,

$\omega_{1,j} = -\frac{2\psi_{2,j}}{r_j h_z^2} + O(h_z)$ on the bottom wall of the reactor,

$\omega = 0$ on the axis symmetrical line,

$\omega_{Hbz,j} = -\frac{2\psi_{Hbz-1,j}}{r_j h_z^2} + O(h_z)$ on the bottom side of the spade,

$\omega_{Hbz+Hbl,j} = -\frac{2\psi_{Hbz+Hbl+1,j}}{r_j h_z^2} + O(h_z)$ on the top side of the spade.

2.4 Approximation of the Stream Function Equation

$$\frac{\psi_{i,j}^{n+1/2} - \psi_{i,j}^n}{0.5\tau} = \frac{\partial}{\partial r}\left(\frac{1}{r}\frac{\partial\psi}{\partial r}\right)\Big|_{i,j}^{n+1/2} + \frac{\partial}{\partial z}\left(\frac{1}{r}\frac{\partial\psi}{\partial z}\right)\Big|_{i,j}^{n+1/2} + \omega_{i,j}^n \quad (13)$$

$$\frac{\psi_{i,j}^{n+1} - \psi_{i,j}^{n+1/2}}{0.5\tau} = \frac{\partial}{\partial r}\left(\frac{1}{r}\frac{\partial\psi}{\partial r}\right)\Big|_{i,j}^{n+1} + \frac{\partial}{\partial z}\left(\frac{1}{r}\frac{\partial\psi}{\partial z}\right)\Big|_{i,j}^{n+1} + \omega_{i,j}^n \quad (14)$$

2.5 Approximation of the Momentum Equation

$$\frac{M_{i,j}^{n+1/2} - M_{i,j}^n}{0.5\tau} = \frac{1}{Re}\delta_r^2 M_{i,j}^{n+1/2} + \frac{1}{Re}\delta_z^2 M_{i,j}^n - D_{i,j}^n \quad (15)$$

$$\frac{M_{i,j}^{n+1} - M_{i,j}^{n+1/2}}{0.5\tau} = \frac{1}{Re}\delta_z^2 M_{i,j}^{n+1} + \frac{1}{Re}\delta_r^2 M_{i,j}^{n+1/2} - D_{i,j}^n \quad (16)$$

where $D = -\frac{1}{Re}\frac{1}{r^2}M - \frac{U}{r}M - U\frac{\partial M}{\partial r} - W\frac{\partial M}{\partial z}$, the operators δ_r^2, δ_z^2 are the same like in equation (11).

2.6 Alternating Direction Implicit Method

Many methods for numerical studying of the Navier-Stokes equations are elaborated [1, 4]. We use the Alternating Direction Implicit Method for solving the stream function, the momentum and the vorticity equations (13)-(16). The specialty of this method is that the time of range t is realized in two time layers - n+1/2 and n+1. The priority of this method is that each difference equation in this scheme has only tri-diagonal matrix form. For example the equation (11) connects implicit unknowns $\omega_{i,j+1}^{n+1/2}, \omega_{i,j}^{n+1/2}, \omega_{i,j-1}^{n+1/2}$ and the equation (12) connects implicit unknowns $\omega_{i+1,j}^{n+1}, \omega_{i,j}^{n+1}, \omega_{i-1,j}^{n+1}$. The scheme is effective and stable [1]

2.7 Numerical Algorithm

We use iterative procedure, in which the equations (11), (12) is solved firstly and then ω, ψ, U and W are calculated. The components of the velocity U and W are

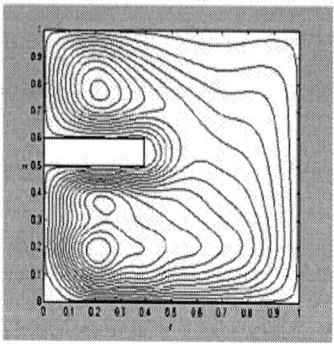
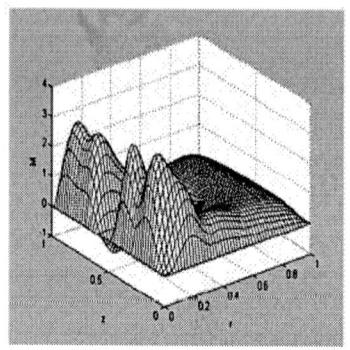

Fig. 4. Stream function isolines in (r, z) plane for Re=10

Fig. 5. Function $M(r, z)$; Re=10

defined on the base of the calculated stream functions ψ. Therefor the vorticity equation and the stream function must be solved together on the time layers n+1/2 and n+1.

The iterative procedure converges to an acceptable accuracy result, if the proper grid and parameters are specified in the calculations. The solution is considered to be converging if

$$\varepsilon = \max \left| \frac{\psi_{i,j}^{n+1} - \psi_{i,j}^n}{\psi_{i,j}^n} \right| \leq \varepsilon_\psi ;$$

$$\varepsilon = \max \left| \frac{\omega_{i,j}^{n+1} - \omega_{i,j}^n}{\omega_{i,j}^n} \right| \leq \varepsilon_\omega ;$$

$$\varepsilon = \max \left| \frac{M_{i,j}^{n+1} - M_{i,j}^n}{M_{i,j}^n} \right| \leq \varepsilon_M .$$

3 Numerical Results

The real problem is solved with the following denominations of physical and geometrical parameters Re = 10, 100; mixer length R1 = 0.4, 0.5; mixer height L1= 0.1; position of the mixer H1= 0.4, 0.5. The accuracy parameter for these calculations is $\varepsilon = \varepsilon_\psi = \varepsilon_\omega = \varepsilon_M$= 0.01, 0.05.

We provide some tests to verify the numerical algorithm which clearly indicate that the developed algorithm works well for the test examples.

The result for the stream function is given in Fig.4.
The results for the momentum is given in Fig. 5.
The result for the vorticity is given in Fig. 6a. and Fig. 6b.

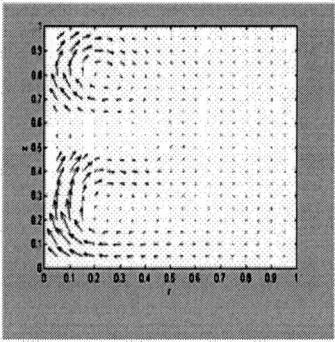 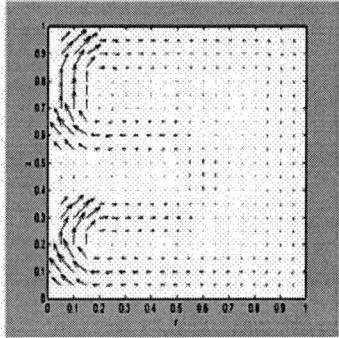

Fig. 6. Vector field in (r,z) plane, Re=10

Fig. 7. Vector field in (r,z) plane, Re=100

The calculated characteristics of examined motion in the chemical reactor with a mixer(shown in Fig.4,5,6) corespond to visual and other examination of the hydrodynamics of the reactor [6].

4 Concluding Remarks

A reliable numerical algorithm for investigation of the hydrodynamics of a chemical reactor with a mixer is developed. The algorithm is based on the alternating direction implicit method. The algorithm is tested and results are acceptable and promising for studying hydrodynamics of mixing vessels.

Acknowledgements

We thank Prof. M. Kaschiev, Dr. V. Kambourova for useful discussions and Mr. M. Nedelchev for technical support of this work.

References

1. Roach P. *Computational Fluid Dynamics*. Hermosa Publishers. Albuquerque, 1976
2. Samarskii A., Theory of Difference Scheme, Moskow, 1989
3. Birukov V., Kentere V., Optimization of periodical processes of microbiological syntheses,Moskow, 1985
4. Christov C. I., Ridha A. *Splitting Scheme for Iterative Solution of Bi-Harmonic Equation. Application to 2D Navier- Stokes Problems. Advances in Numerical Methods and Applications*, Sofia, Bulgaria, World Scientific Publishing Company, 341-352, 1994
5. Christov C. I., Ridha A. *Splitting Scheme for the stream-function formulation of 2D unsteady Navier-Stokes equations*. C. R. Acad. Sci. Paris, t.320, II b, p.441-446, 1995
6. Bekker A., Hydrodynamics of Stirred Gas-Liquid Dispersions, 1992

A Domain Decomposition Finite Difference Method for Singularly Perturbed Elliptic Equations in Composed Domains*

Irina V. Tselishcheva and Grigorii I. Shishkin

Institute of Mathematics and Mechanics, Ural Branch of
Russian Academy of Sciences, Ekaterinburg 620219, Russia
{tsi,Grigorii}@shishkin.ural.ru

Abstract. Numerical modelling of stationary heat and mass transfer processes in composite materials often leads to singularly perturbed problems in composed domains, that is, to elliptic equations with discontinuous coefficients and a small parameter ε multiplying the highest derivatives. The concentrated source acts on the interface boundary. For such problems the application of domain decomposition (DD) methods seems quite reasonable: the original domain is naturally partitioned into several non-overlapping subdomains with smooth coefficients. Due to the presence of transition and boundary layers, standard numerical methods yield large errors for small ε. By this reason, we need for special methods whose errors are independent of the parameter ε. To construct such DD schemes possessing the property of ε-uniform convergence, we use standard finite difference approximations on piecewise uniform grids, which are *a priori* refined in the transition and boundary layers.

1 Introduction

The solutions of boundary value problems in composed domains have singularities generated by the presence of discontinuities in the coefficients and right-hand sides. When the problem in question is singualry perturbed, there appear redundant singularities: the solution of such a problem typically contains transition layers in the neighbourhood of concentrated sources, besides of boundary layers. Classical numerical methods from [1] developed for regular problems are inapplicable for resolving boundary and interior layers because of large errors (up to the exact solution) for small values of the perturbation parameter ε (see, e.g., [2]–[4]). We are interested in parameter-robust numerical methods that converge independently of ε (or ε-uniformly). Here more attractive are methods based on domain decomposition. Clearly, specific attention should be paid to non-overlapping DD techniques where involved subdomains are the subdomains forming the composed domain. Emphasize that we are to develop a DD method whose solutions converge to the solution of the original problem ε-uniformly with

---

* This research was supported by the Russian Foundation for Basic Research (grant No. 98-01-00362) and partially by the NWO grant (dossiernr. 047.008.007).

respect to both the number of mesh points and the number of iterations.

Let us formulate the problem under consideration. On the vertical strip $\overline{D}$,
$$D = \{x: \; -d_0 < x_1 < d^0, \; x_2 \in R\}, \quad d_0, \, d^0 > 0, \tag{1.1}$$
which consists of two subdomains-strips $\overline{D}_1$ and $\overline{D}_2$, where $D_1 = D \cap \{x_1 < 0\}$, $D_2 = D \cap \{x_1 > 0\}$, we consider the Dirichlet problem for the following singularly perturbed elliptic equation of reaction-diffusion type
$$L_k u(x) \equiv \left\{ \varepsilon^2 \sum_{s=1,2} a_{ks}(x) \frac{\partial^2}{\partial x_s^2} - c_k(x) \right\} u(x) = f_k(x), \quad x \in D_k, \tag{1.2a}$$
$$u(x) = \varphi(x), \quad x \in \Gamma, \quad k = 1, 2. \tag{1.2b}$$

The concentrated source acts on the interface boundary $\Gamma^* = \{x_1 = 0\} \times R$
$$[u(x)] = 0, \quad l u(x) \equiv \varepsilon \left[a_1(x) \frac{\partial}{\partial x_1} u(x) \right] = -q(x), \quad x \in \Gamma^*. \tag{1.2c}$$

Here $\Gamma = \overline{D} \setminus D$, the functions $a_{ks}(x)$, $c_k(x)$, $f_k(x)$ are assumed to be sufficiently smooth on $\overline{D}_k$, and the functions $\varphi(x)$ and $q(x)$ on Γ and Γ^*, respectively, moreover [1]
$$\begin{aligned} 0 < a_0 \le a_{ks}(x) \le a^0, \;\; 0 < c_0 \le c_k(x) \le c^0, \;\; |f_k(x)| \le M, \quad x \in \overline{D}_k, \\ |\varphi(x)| \le M, \;\; x \in \Gamma; \quad |q(x)| \le M, \;\; x \in \Gamma^*; \quad k, s = 1, 2; \end{aligned} \tag{1.3}$$

the parameter ε takes arbitrary values from the half-interval $(0,1]$. The symbol $[v(x)]$ denotes the jump of the function $v(x)$ when passing through Γ^* from D_1 to D_2: $[u(x)] = \lim\limits_{\substack{x^+ \to x \\ x^+ \in D_2}} u(x^+) - \lim\limits_{\substack{x^- \to x \\ x^- \in D_1}} u(x^-), \; x \in \Gamma^*$ and
$$\left[a_1(x) \frac{\partial}{\partial x_1} u(x) \right] = \lim_{\substack{x^+ \to x \\ x^+ \in D_2}} a_{21}(x^+) \frac{\partial}{\partial x_1} u(x^+) - \lim_{\substack{x^- \to x \\ x^- \in D_1}} a_{11}(x^-) \frac{\partial}{\partial x_1} u(x^-), \; x \in \Gamma^*.$$

It is convenient also to write (1.2a) in such a form: $L u(x) = f(x), \; x \in D \setminus \Gamma^*$, where $L = L_k$, $f(x) = f_k(x)$ for $x \in D_k$.

As $\varepsilon \to 0$, boundary and transition layers appear in a neighbourhood of the sets Γ and Γ_* respectively.

For problem (1.2), (1.1) we are to construct a domain decomposition scheme by using, as a base scheme, the ε-uniformly convergent scheme from [2].

2 The Base Scheme for Problem (1.2), (1.1)

Let us first give an iteration-free difference scheme. On the set $\overline{D}$ we introduce the rectangular grid
$$\overline{D}_h = \overline{\omega}_1 \times \omega_2, \tag{2.1}$$

---
[1] Here and below we denote by M (m) sufficiently large (small) positive constants which are independent of ε and the discretization parameters. Throughout this paper, the notation $w_{(j.k)}$ indicates that w is first defined in equation (j.k).

where $\overline{\omega}_1 = \{x_1^i : -d_0 = x_1^0 < \ldots < x_1^{N_1} = d^0\}$, $\omega_2 = \{x_2^j : x_2^j < x_2^{j+1}, j = \ldots, -1, 0, 1, 2, \ldots\}$ are arbitrary (possibly) nonuniform meshes on $[-d_0, d^0]$ and on the x_2-axis respectively; the point $x_1 = 0$ belongs to $\overline{\omega}_1$. Assume $h \leq MN^{-1}$, where h is the maximum of the mesh-sizes, $N = \min[N_1, N_2]$, $N_1 + 1$ and $N_2 + 1$ are the number of nodes in the mesh $\overline{\omega}_1$ and the minimal number of nodes in ω_2 on a unit interval. Denote $x_1 = 0$ by $x_1^{i_0}$.

We approximate problem (1.2), (1.1) by the difference scheme [1]

$$\Lambda z(x) = f^h(x), \quad x \in D_h, \quad (2.2a)$$
$$z(x) = \varphi(x), \quad x \in \Gamma_h. \quad (2.2b)$$

Here $D_h = D \cap \overline{D}_h$, $\Gamma_h = \Gamma \cap \overline{D}_h$,

$$\Lambda \equiv \varepsilon^2 \sum_{s=1,2} a_{ks}(x) \delta_{\overline{xs}\,\widehat{xs}} - c_k(x), \quad x \in D_{kh}, \quad k = 1, 2; \quad (2.2c)$$

$$\Lambda \equiv \varepsilon^2 2 \left(h_1^{i_0} + h_1^{i_0-1}\right)^{-1} \{a_{21}(x)\delta_{x1} - a_{11}(x)\delta_{\overline{x1}}\}, \quad x = (x_1^{i_0}, x_2) \in \Gamma_h^*; \quad (2.2d)$$

$$f^h(x) = f_k(x), \quad x \in D_{kh}, \quad k = 1, 2; \quad (2.2e)$$

$$f^h(x) = -2\varepsilon \left(h_1^{i_0} + h_1^{i_0-1}\right)^{-1} q(x), \quad x \in \Gamma_h^*, \quad (2.2f)$$

with $\delta_{\overline{xs}\,\widehat{xs}} z(x)$ being the second (central) difference derivative, e.g., $\delta_{\overline{x1}\,\widehat{x1}} z(x) = 2\left(h_1^{i-1} + h_1^i\right)^{-1} (\delta_{x1} z(x) - \delta_{\overline{x1}} z(x))$, $\delta_{x1} z(x) = \left(h_1^i\right)^{-1} \times (z(x_1^{i+1}, x_2) - z(x))$, $\delta_{\overline{x1}} z(x) = \left(h_1^{i-1}\right)^{-1} (z(x) - z(x_1^{i-1}, x_2))$, $x = (x_1^i, x_2)$ (cf. [1]).

Scheme (2.2), (2.1) is monotone ε-uniformly [1]. By applying the majorizing technique and taking account of a-priori estimates, we find the error bound: $|u(x) - z(x)| \leq M \left[\varepsilon^{-1} N_1^{-1} + N_2^{-1}\right]$, $x \in \overline{D}_h$. Thus, scheme (2.2), (2.1) does not converge ε-uniformly (for all values of ε no matter how small).

We are now in a position to define the piecewise uniform grid from [2] which is condensed in the neighbourhood of the boundary and transition layers:

$$\overline{D}_h^* = \overline{\omega}_1^* \times \omega_2, \quad (2.3)$$

where $\omega_2 = \omega_{2(2.1)}$, $\overline{\omega}_1^* = \overline{\omega}_1^*(\sigma)$ is a piecewise uniform mesh on $[-d_0, d^0]$. To construct $\overline{\omega}_1^*(\sigma)$, we divide $[-d_0, d^0]$ in five parts $[-d_0, -d_0 + \sigma]$, $[-d_0 + \sigma, -\sigma]$, $[-\sigma, \sigma]$, $[\sigma, d^0 - \sigma]$ and $[d^0 - \sigma, d^0]$. In each part we use a uniform mesh, with step-sizes $h_1^{(1)} = 8\sigma N_1^{-1}$ on the subintervals $[-d_0, -d_0 + \sigma]$, $[-\sigma, \sigma]$, $[d^0 - \sigma, d^0]$ and $h_1^{(2)} = 2\left(d_0 + d^0 - 4\sigma\right) N_1^{-1}$ on $[-d_0 + \sigma, -\sigma]$, $[\sigma, d^0 - \sigma]$. We take $\sigma = \min\left[3^{-1} d_0, 3^{-1} d^0, m^{-1} \varepsilon \ln N_1\right]$ as a function of ε and N_1, where $0 < m < m^0$, $m_0 = \min_{k, \overline{D}_k} \left[a_{k1}^{-1}(x) c_k(x)\right]^{1/2}$. See also [3,4] for more details of this fitted mesh technique.

Theorem 1. *Let the data of problem (1.2), (1.1) satisfy condition (1.3), and assume a_{ks}, c_k, $f_k \in C^{K+2+\alpha}(\overline{D}_k)$, $\varphi \in C^{K+2+\alpha}(\Gamma)$, $\psi \in C^{K+2+\alpha}(\Gamma^*)$, $s, k = 1, 2$, $\alpha > 0$ with $K = 4$. Then the solution of scheme (2.2), (2.3) converges ε-uniformly to the solution of (1.2), (1.1) with an error bound given by*

$$|u(x) - z(x)| \leq M \left[N_1^{-1} \ln N_1 + N_2^{-1}\right], \quad x \in \overline{D}_h^*. \quad (2.4)$$

3 Relaxation Scheme

In this section we consider a relaxation scheme assuming that the original domain D is partitioned into the non-overlapping subdomains D_1 and D_2.

1. At first, with problem (1.2) we associate the "nonstationary" problem

$$L_{(1.2)} w(x,t) = f(x), \quad (x,t) \in G \setminus S^*, \tag{3.1a}$$

$$l\, w(x,t) = -q(x), \quad [w(x,t)] = 0, \quad (x,t) \in S^*, \tag{3.1b}$$

$$w(x,t) = \psi(x,t), \quad (x,t) \in S. \tag{3.1c}$$

Here

$$\overline{G} = G \bigcup S, \quad G = G_1 \bigcup G_2 \bigcup S^*, \quad G_k = D_k \times (0,\infty), \quad k=1,2, \tag{3.2}$$

$$S^* = \Gamma^* \times (0,\infty), \quad S = S^L \bigcup S_0, \quad S^L = \Gamma \times (0,\infty), \quad S_0 = \Gamma^* \times \{t=0\},$$

the operator l is defined by $l = l_{(1.2)} - p\dfrac{\partial}{\partial t}$, $p > 0$, the function $\psi(x,t)$ is sufficiently smooth and coincides with $\varphi(x)$ on S^L, besides, it is bounded on S_0.

The maximum principle holds for problem (3.1), (3.2). By estimating $|u(x) - w(x,t)|$ one can verify that the function $w(x,t)$, as $t \to \infty$, converges ε-uniformly to the steady-state solution $u(x)$ of problem (1.2), (1.1).

2. For problem (3.1), (3.2) we apply the method of lines along x with explicit approximation of equation (3.1b) along t.

We discretize the set $\overline{G}$ as follows. On $\overline{D}$ we construct the sets $D_1^0 = (-d_0, x_1^{i_0-1}) \times R$, $D_2^0 = (x_1^{i_0+1}, d^0) \times R$, $\Gamma^{*-} = x_1^{i_0-1} \times R$, $\Gamma^{*+} = x_1^{i_0+1} \times R$, where $x_1^{i_0-1}$, $x_1^{i_0} = 0$, $x_1^{i_0+1}$ are the nodes of the space mesh $\overline{\omega}_1$. Suppose $D^0 = D_1^0 \bigcup D_2^0 \bigcup \Gamma^{*0}$, $\Gamma^{*0} = \Gamma^{*-} \bigcup \Gamma^{*+} \bigcup \Gamma^*$. On the semiaxis t we introduce a uniform mesh $\overline{\omega}_0$ with step τ. Further we construct such a "grid" in time

$$\overline{G}_\tau = G_\tau \bigcup S_\tau, \quad G_\tau = G_{1\tau} \bigcup G_{2\tau} \bigcup S_\tau^{*0}, \quad S_\tau = S_\tau^L \bigcup S_{0\tau}, \tag{3.3}$$

where $G_{k\tau} = D_k^0 \times \omega_0$, $S_\tau^{*0} = \Gamma^{*0} \times \omega_0$, $S_\tau^L = \Gamma \times \omega_0$, $S_{0\tau} = \Gamma^* \times \{t=0,\tau\}$. In this way, the grid $\overline{G}_{\tau(3.3)}$ is controlled by δ_1, δ_2 and τ, where $\delta_1 = x_1^{i_0} - x_1^{i_0-1}$, $\delta_2 = x_1^{i_0+1} - x_1^{i_0}$, $x_1^{i_0} = 0$. Note that the values of δ_1 and δ_2 may depend on the perturbation parameter ε. We use the notation $\widetilde{\delta}_i = \varepsilon^{-1}\delta_i$, $i=1,2$.

By applying the method of lines in x, we construct the semi-discrete scheme

$$\Lambda_\tau\, w^\tau(x,t) = f^\tau(x), \quad (x,t) \in G_\tau, \quad w^\tau(x,t) = \psi^\tau(x,t), \quad (x,t) \in S_\tau. \tag{3.4}$$

Here $\Lambda_\tau = L_{(1.2)}$, $(x,t) \in G_{k\tau}$, $\Lambda_\tau = \Lambda_{(2.2d)} - p\,\delta_t$, $(x,t) \in S_\tau^*$,

$$\Lambda_\tau = \varepsilon^2 a_{21}(x)\Big(\dfrac{\partial}{\partial x_1} - \delta_{\overline{x1}}\Big) + \varepsilon^2 a_{22}(x)\dfrac{\partial^2}{\partial x_2^2} - c_2(x), \quad (x,t) \in S_\tau^{*0}, \ x \in \Gamma^{*+},$$

$$\Lambda_\tau = \varepsilon^2 a_{11}(x)\Big(\delta_{x1} - \dfrac{\partial}{\partial x_1}\Big) + \varepsilon^2 a_{12}(x)\dfrac{\partial^2}{\partial x_2^2} - c_1(x), \quad (x,t) \in S_\tau^{*0}, \ x \in \Gamma^{*-},$$

$$\psi^\tau(x,t) = \psi_{(3.1)}(x,t), \ (x,t) \in S_\tau^L, \quad f^\tau(x) = \begin{cases} f_{(1.2)}(x), & (x,t) \in G_\tau \setminus S_\tau^*, \\ f^h_{(2.2f)}(x), & (x,t) \in S_\tau^*, \end{cases}$$

on the set $S_{0\tau}$ $\psi^\tau(x,t) = \psi_{(3.1)}(x,t)$ for $t = 0$, and for $t = \tau$ $\psi^\tau(x,t)$ is any sufficiently smooth function, moreover, $\psi^\tau(x,t)$ satisfies the Lipschitz condition with respect to t.

3. Let us study scheme (3.4), (3.3). The condition

$$\tau \le p \left(\sup_\varepsilon \max_{x \in \overline{\Gamma}^*} \left\{ \widetilde{\delta}_1^{-1} a_{11}(x) + \widetilde{\delta}_2^{-1} a_{21}(x) \right\} \right)^{-1} \equiv \tau^*_{(3.5)}(\widetilde{\delta}_i) \tag{3.5}$$

is necessary and sufficient for scheme (3.4), (3.3) to be ε-uniformly monotone. The condition

$$\tau \le m_1 p \inf_\varepsilon \min_i \widetilde{\delta}_i \equiv \tau^*_{(3.6)}(\widetilde{\delta}_i), \quad m_1 = 2^{-1}(a^0)^{-1} \tag{3.6}$$

is sufficient and, up to a constant factor, necessary for ε-uniform monotonicity. If $\delta_i = 0$, then $\tau = 0$ and the process loses its stability.

To scheme (3.4), (3.3) we put in correspondence the stationary scheme

$$\Lambda^0 w^0(x) = f^0(x), \quad x \in D^0, \quad w^0(x) = \varphi(x), \quad x \in \Gamma. \tag{3.7}$$

Here D^0 is defined above, $\overline{D}^0 = D^0 \cup \Gamma$,

$$\Lambda^0 = \begin{cases} \Lambda_{\tau(3.4)}, & x \in D^0 \setminus \Gamma^*, \\ \Lambda_{(2.2d)}, & x \in \Gamma^* \end{cases}, \quad f^0(x) = f^\tau(x), \quad x \in D^0.$$

The scheme (3.7) approximating the boundary value problem (1.2), (1.1) is ε-uniformly monotone. The following estimate is valid:

$$|u(x) - w^0(x)| \le M (\widetilde{\delta}_1 + \widetilde{\delta}_2), \quad x \in \overline{D}^0. \tag{3.8}$$

Thus, for $\delta_1, \delta_2 \to 0$ the function $w^0(x)$ converges to $u(x)$ for fixed values of the parameter ε, and it does ε-uniformly under the condition $\widetilde{\delta}_1, \widetilde{\delta}_2 = o(1)$. The last condition is also necessary for ε-uniform convergence of scheme (3.7).

Under condition (3.5) we obtain the estimate

$$|w^0(x) - w^\tau(x,t)| \le M \left[1 + mp^{-1}(1 + \widetilde{\delta}_1 + \widetilde{\delta}_2)^{-1} \tau \right]^{-t/\tau}, \quad x \in \overline{D}^0. \tag{3.9}$$

For fixed values of $\widetilde{\delta}_1, \widetilde{\delta}_2, \tau$ and for $t \to \infty$ the function $w^\tau(x,t)$ converges to the function $w^0(x)$. By (3.9) the condition $t \inf_\varepsilon (1 + \widetilde{\delta}_1 + \widetilde{\delta}_2)^{-1} \to 0$ is necessary and sufficient for ε-uniform proximity of the functions $w^0(x)$ and $w^\tau(x,t)$.

By virtue of estimates (3.8) and (3.9) we have

$$|u(x) - w^\tau(x,t)| \le M \left(\widetilde{\delta}_1 + \widetilde{\delta}_2 + \left[1 + mp^{-1}(1 + \widetilde{\delta}_1 + \widetilde{\delta}_2)^{-1} \tau \right]^{-t/\tau} \right), \quad x \in \overline{D}^0.$$

Consequently, the function $w^\tau(x,t)$ converges ε-uniformly to $u(x)$ under the condition

$$\widetilde{\delta}_1, \widetilde{\delta}_2 \to 0, \quad t \to \infty. \tag{3.10}$$

Recall that this result is valid in the case of condition (3.5) imposed on τ. The condition (3.10) is also necessary for ε-uniform convergence of scheme (3.4), (3.3).

4 Non-overlapping Schwarz Method

The results obtained in Section 3 can be rigorously written in terms of non-overlapping domain decomposition Schwarz-like methods.

1. We begin with consideration of the continuous Schwarz method. As a preliminary, on the set $\overline{D}$ we introduce the sets

$$\overline{D}^1 = \overline{D}_1^1 \bigcup \overline{D}_2^1, \quad \overline{D}_k^1 = D_k^1 \bigcup \Gamma_k^1, \quad k = 1, 2 \quad \text{and} \quad \Gamma^{*1}, \tag{4.1}$$

where $D_1^1 = (-d_0, x_1^{i_0-1}] \times R$, $D_2^1 = [x_1^{i_0+1}, d^0) \times R$, $\Gamma_1^1 = \{-d_0 \times R\} \bigcup \Gamma^*$, $\Gamma_2^1 = \Gamma^* \bigcup \{d^0 \times R\}$, $\Gamma^{*1} = \Gamma^{*-} \bigcup \Gamma^{*+} \bigcup \Gamma^*$. These sets with upper index 1 only slightly differ from the sets with upper index 0 considered in Section 3.

Let us introduce the functions $u_0^n(x)$, $x \in \Gamma^*$, $u_k^n(x)$, $x \in \overline{D}_k^1$, $k = 1, 2$, assuming $u_0^n(x) = w^\tau(x, t^n)$, $x \in \Gamma^*$, $u_k^n(x) = w^\tau(x, t^n)$, $x \in \overline{D}_k^1$, $k = 1, 2$ for $t^n = n\tau$, $n = 1, 2, \ldots$. We find the functions $u_k^n(x)$, $u_0^n(x)$ by solving the problem

$$\Lambda u_k^n(x) = f^1(x), \quad x \in D_k^1,$$
$$u_k^n(x) = \begin{cases} \varphi(x), & x \in \Gamma_k^1 \cap \Gamma, \\ u_0^n(x), & x \in \Gamma_k^1 \setminus \Gamma, \end{cases} \quad k = 1, 2; \tag{4.2}$$
$$u_0^n(x) = F^1\left(f^1(x), u^{*\,n-1}(x)\right), \quad x \in \Gamma^*, \quad n = 1, 2, \ldots.$$

Here $\Lambda = \Lambda_{\tau(3.4)}$, $x \in D_1^1 \bigcup D_2^1$; $f^1(x) = f^\tau_{(3.4)}(x)$, $x \in \overline{D}^1 \setminus \Gamma$,

$$F^1\left(f^1(x), u^{*\,n-1}(x)\right) = u_0^{n-1}(x) + p^{-1}\tau\left\{\Lambda_{(2.2d)} u^{*\,n-1}(x) - f^1(x)\right\}, x \in \Gamma^*;$$

$$u^{*\,n}(x) = \begin{cases} u_0^n(x), & x \in \Gamma^*, \\ u_1^n(x), & x \in \Gamma^{*-}, \\ u_2^n(x), & x \in \Gamma^{*+} \end{cases}, \quad x \in \Gamma^{*1}.$$

Note that the function $u_0^n(x)$, $x \in \Gamma^*$ for $n = 0, 1$ is given according to the problem formulation. We call the function $u^n(x) = \{u_k^n(x), x \in \overline{D}_k^1, k = 1, 2\}$, $x \in \overline{D}^1$, $n = 1, 2, \ldots$ the solution of scheme (4.2), (4.1). The value n defines the the current iteration in the iterative scheme (4.2), (4.1).

All the considerations of Section 3 (estimates and conditions) remains valid with replacing $w^\tau(x, t)$ by $u^n(x)$ and t by $n\tau$.

We say that scheme (4.2), (4.1) for $n = n^*$ is *consistent* with respect to both the limiting accuracy (for $n = \infty$) of the solution and the number of iterations (or, briefly, consistent), if such an estimate is true: $\max_{\overline{D}^1} |u^n(x) - u^\infty(x)| \leq M \max_{\overline{D}^1} |u(x) - u^\infty(x)|$, $n \geq n^*$, where $u^\infty(x) = w^0(x)$.

For consistent monotone scheme (4.2), (4.1) the estimate like (3.8) holds:

$$|u(x) - u^n(x)| \leq M\left(\tilde{\delta}_1 + \tilde{\delta}_2\right), \quad x \in \overline{D}^1, \quad n \geq n^*,$$

where $n^* \leq M \max_i \left(\widetilde{\delta}_i^{-1}\right) \ln \left(\widetilde{\delta}_1 + \widetilde{\delta}_2\right)^{-1}$ for $m \min_i \widetilde{\delta}_i \leq \tau \leq M \min_i \widetilde{\delta}_i$.

2. Let us construct the discrete Schwarz method. For this we define the computational grids on the sets $\overline{D}^1$, $\overline{D}_k^1$ and Γ^{*1}

$$\overline{D}_h^1 = \overline{D}^1 \cap \overline{D}_h, \quad \overline{D}_{kh}^1 = \overline{D}_k^1 \cap \overline{D}_h, \quad \Gamma_h^{*1} = \Gamma^{*1} \cap \overline{D}_h, \qquad (4.3)$$

where either $\overline{D}_h = \overline{D}_{h(2.1)}$ or $\overline{D}_h = \overline{D}_{h(2.3)}^{\star}$. We approximate problem (4.2), (4.1) by the totally discrete scheme

$$\Lambda z_k^n(x) = f^1(x), \quad x \in D_{kh}^1,$$

$$z_k^n(x) = \begin{cases} \varphi(x), & x \in \Gamma_{kh}^1 \cap \Gamma, \\ z_0^n(x), & x \in \Gamma_{kh}^1 \setminus \Gamma, \end{cases} \quad k = 1, 2; \qquad (4.4)$$

$$z_0^n(x) = F^1(f^1(x), z^{*n-1}(x)), \quad x \in \Gamma_h^*, \quad n = 1, 2, \ldots.$$

Here $\Lambda = \Lambda_{(2.2)}$, $x \in D_{1h}^1 \cup D_{2h}^1$; $f^1(x) = f^1_{(4.2)}(x)$, $x \in \overline{D}_h^1 \setminus \Gamma$,

$$F^1(f^1(x), z^{*n-1}(x)) = z_0^{n-1}(x) + p^{-1}\tau\{\Lambda_{(2.2\text{d})} z^{*n-1}(x) - f^1(x)\}, \quad x \in \Gamma_h^*,$$

$$z^{*n}(x) = \begin{cases} z_0^n(x), & x \in \Gamma_h^*, \\ z_1^n(x), & x \in \Gamma_h^{*-}, \\ z_2^n(x), & x \in \Gamma_h^{*+} \end{cases}, \quad x \in \Gamma_h^{*1};$$

the function $z_0^n(x)$, $x \in \Gamma_h^*$ for $n=0,1$ is assumed to be given: $z_0^n(x) = u_{0(4.2)}^n(x)$, $x \in \Gamma_h^*$, $n = 0, 1$. We call the function $z^n(x) = \{z_k^n(x),\ x \in \overline{D}_{kh}^1,\ k = 1, 2\}$, $x \in \overline{D}_h^1$, $n = 1, 2, \ldots$ the solution of difference scheme (4.4), (4.3). It should be noted that the functions $z_1^n(x)$, $x \in \overline{D}_{1h}^1$ and $z_2^n(x)$, $x \in \overline{D}_{2h}^1$ can be computed in parallel on each n-th iteration.

3. We confine the convergence analysis to scheme (4.4), (4.3) in the class of piecewise uniform grids (2.3).

Under condition (3.5) (condition (3.6)) for the solutions of scheme (4.4), (4.3) we obtain the estimate

$$|z(x) - z^n(x)| \leq M \left[1 + mp^{-1}\left(1 + \widetilde{\delta}_1 + \widetilde{\delta}_2\right)^{-1}\tau\right]^{-n}, \quad x \in \overline{D}_h, \qquad (4.5)$$

no matter whether $\overline{D}_{h(2.1)}$ or $\overline{D}_{h(2.3)}^{\star}$ is used.

On the grid $\overline{D}_{h(2.3)}^{\star}$ condition (3.6) takes the form

$$\tau \leq m_2\, p\, N_1^{-1} \ln N_1 \equiv \tau^*_{(4.6)}(N_1). \qquad (4.6)$$

In this way, taking account of estimates (2.4) and (4.5), we find

$$|u(x) - z^n(x)| \leq M\left(N_1^{-1}\ln N_1 + N_2^{-1} + \left[1 + mp^{-1}\tau\right]^{-n}\right), \quad x \in \overline{D}_h^{\star}, \qquad (4.7\text{a})$$

$$|z(x) - z^n(x)| \leq M\left[1 + mp^{-1}\tau\right]^{-n}, \quad x \in \overline{D}_h^{\star}. \qquad (4.7\text{b})$$

Thus, under condition (3.5) (condition (4.6)) and, besides this, provided that

$$n\tau \to \infty, \quad \text{for} \quad N, \, n \to \infty \tag{4.8}$$

the difference scheme is ε-uniformly monotone, and the functions $z^n(x)$ converge to $u(x)$ ε-uniformly. In the case of (4.6) complemented by the condition

$$n, \, N \to \infty \tag{4.9}$$

the scheme possesses the property of ε-uniform monotonicity, however, it does not converge even for fixed values of the parameter ε.

Under the condition

$$\tau = \tau^*_{(4.6)}(N_1) \tag{4.10}$$

the solutions of scheme (4.4), (4.3), (2.3) satisfy the estimates

$$|u(x) - z^n(x)| \leq M \left(N_1^{-1} \ln N_1 + N_2^{-1} + \left[1 + m \, N_1^{-1} \ln N_1 \right]^{-n} \right),$$
$$x \in \overline{D}_h^*; \tag{4.11a}$$

$$|z(x) - z^n(x)| \leq M \left[1 + m \, N_1^{-1} \ln N_1 \right]^{-n}, \quad x \in \overline{D}_h^*. \tag{4.11b}$$

This scheme converges ε-uniformly under the condition $n N_1^{-1} \ln N_1 \to \infty$ for n, $N \to \infty$. If this condition is violated, we have no convergence. Thus, the number of iterations n of the monotone scheme (4.4), (4.3), (2.3), (4.10), required for ε-uniform convergence, is independent of ε and unboundedly grows as $N \to \infty$.

For consistent scheme (4.4), (4.3), (2.3) under condition (4.10) we obtain

$$|u(x) - z^n(x)|, \, |z(x) - z^n(x)| \leq M \left[N_1^{-1} \ln N_1 + N_2^{-1} \right], \quad x \in \overline{D}_h^*, \tag{4.12}$$

where $n \geq n^*$, and also $n^* \leq M N_1$.

Theorem 2. *Conditions (3.5) and (4.8) are necessary and sufficient for ε-uniform monotonicity of scheme (4.4), (4.3), (2.3) (just condition (3.5)) and for ε-uniform convergence of the functions $z^n_{(4.4;\,4.3)}(x)$, $x \in \overline{D}^*_{h(2.3)}$ to the solution $u(x)$ of problem (1.2), (1.1) and to the function $z_{(2.2)}(x)$, $x \in \overline{D}^*_{h(2.3)}$. Conditions (4.6), (4.9) are not sufficient for convergence of $z^n(x)$ to $u(x)$ and $z(x)$ for fixed values of the parameter ε. The functions $z^n(x)$ under condition (3.5) satisfy estimates (4.7b), (4.11b), and also (4.7a), (4.11a), (4.12), if, besides, the hypotheses of Theorem 1 are fulfilled for $K = 3$.*

References

1. Samarskii, A. A.: Theory of Difference Schemes (in Russian). Nauka, Moscow (1989); German transl. Leipzig Akademische Verlag, Geest & Portig (1984)
2. Shishkin, G. I.: Grid Approximations of Singularly Perturbed Elliptic and Parabolic Equations (in Russian). Ural Branch of Russian Acad. Sci., Ekaterinburg (1992)
3. Miller, J. J. H., O'Riordan, E., Shishkin, G. I.: Fitted Numerical Methods for Singular Perturbation Problems. World Scientific, Singapore (1996)
4. Farrell, P. A., Hegarty, A. F., Miller, J. J. H., O'Riordan, E., Shishkin, G. I.: Robust Computational Techniques for Boundary Layers. CRC Press, Boca Raton (2000)

Numerical Analysis of Solid and Shell Models of Human Pelvic Bone*

Antoni John

Silesian University of Technology,
Department for Strength of Material and Computational Mechanics,
Konarskiego 18A, 44–100 Gliwice, Poland
ajohn@rmt4.kmt.polsl.gliwice.pl

Abstract. Numerical modeling of human pelvic bone makes possibilities to determine the stress and strain distribution in bone tissue. The general problems are: complex geometry, material structure and boundary conditions. In the present paper some simplifications in numerical model are performed. Homogeneous elastic properties of bone tissue are assumed. The shell model and solid model of pelvic bone are analyzed. The finite element method is applied. Some numerical results for solid and shell model are presented.

1 Introduction

Pelvic bone is an important supporting element in locomotion system of human. By linking with the spine through sacral bone and the head thighbone in pelvic joint, pelvic bone transfers not only gravity of the upper body parts but static and dynamic loads following stabilized body stance and locomotion as well. So, in physiological conditions in the pelvic bone there is a certain stress distribution that changes under influence of loads and changed anatomical structures.

It is very difficult or impossible to measure the strain and stress "in vivo" because the safety of patient should be taken into account. There are only two possibilities: model testing and numerical calculations. Complex geometry and material structure of bone tissue as well as its state of load or physiological reactions complexity, cause huge variety of acceptable assumption in 3D numerical models [1,4,5,6] and shell models [8,9] which exerts an influence on the calculation outcomes. It is well known that stress distribution depend on boundary conditions. It can be observed during numerical analysis of human pelvic bone ([8,11,12,13]). There is one important question: how to model the boundary conditions in pelvic bone? It causes the next questions: How to model the contact with others elements of bone system? What we know about the stiffness of support? How to model the load? And many others.

When the stresses are analyzed it appears that the stress distribution depend not only on the boundary conditions but on the yield criterion too. The values

---

* The work was done within the confines of research project 8T11F02618

of maximal reduced stresses are changed and the regions of thier application so on.

The present work defines stress and strain distribution in human pelvic bone using earlier studies [10] made in Department for Strength of Material and Computational Mechanics in Silesian University of Technology, adding new elements in numerical model's structure and load caused by muscle tensions [3,4].

2 Numerical Model

Numerical models are performed on the base of data from anatomical specimen. 3D numerical model of pelvic bone have been worked out relying upon programs, PATRAN/NASTRAN. Eight–nodes solid elements illustrating 3D stress distribution were used for modeling. Figure 1 shows the view of solid model. Separate solid elements layers are modeled by cortical and trabecular bone. Figure 2 shows the view in cross–section of solid model of pelvic bone. The shadow elements model the cortical bone and the light trabecular bone respectively. At present homogeneous elastic properties within a certain group of tissue as well as continuum are assumed. Cortical bone is modeled by one layer of elements while trabecular bone one or more, depending on model's bone tissue's thickness. On the basis of data from [1] assumed Young's modulus 15GPa and 100MPa for cortical and trabecular bone respectively.

In shell model assumed Young's modulus 10GPa, Poisson's ratio 0.3, and quotient of compression strength to tensile strength 1.5. The thickness of elements depend on real dimension of bone tissue.

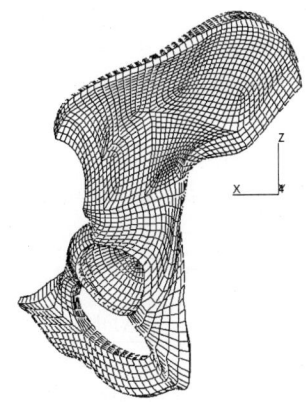

Fig. 1. Solid model of human pelvic bone

Stress and strain distribution of human pelvic bone is a result of external load coming from upper body part's weight and muscles forces (Table 1). Referring

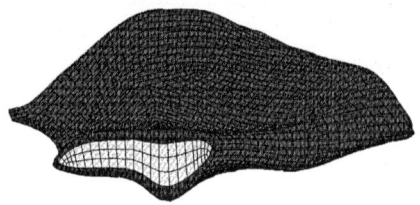

Fig. 2. Cross–section of solid model of pelvic bone

to earlier works [4,5], the model takes up 23 muscle tensions influencing through pelvic bone and tendons on insertions' surfaces.

Muscle forces are depicted in the numerical model as load spread out on nods on insertions' surfaces. The load slants to surface of pelvic bone under angle determined by directive cosines of muscle tensions effect line. Muscle force values are assumed in isomeric conditions [3,4]. Calculations took place with minimum and maximum load: i.e. loaded only by P force and by P force with every muscle force the same time. Muscle tensions load does not take components caused by passive fiber stretch into consideration.

The results show in the present paper are obtained for shell and solid model of human pelvic bone. The results hardly depend on boundary conditions [8,9,12,14]. Here, in acetabulum boundary conditions are given using 20 axial elements (in radial co–ordinate). In contact area with sacral bone boundary conditions are given using axial elements in two co–ordinates, respectively. In pubic symphysis boundary conditions are given in symmetry plane as restraints in selected co–ordinates (selected components in nodes).

Table 1. Maximum values of active muscle forces, muscle tensions interacting on pelvic bone

	RF	S	IP–1	IP–2	GRA
$F_{\max}$ [N]	835	148	1006	1006	165
	GMx–1	GMx–2	ST	SM	BCL
$F_{\max}$ [N]	1559	780	226	1359	745
	ADM–1	ADM–2	ADM–3	ADL	ADB	PC
$F_{\max}$ [N]	354	1063	354	593	452	188
	GMd–1	GMd–2	GMd–3	Gmu–1	Gmu–2	Gmu–3	TFL
$F_{\max}$ [N]	425	425	425	249	249	249	286

In Table 1 the following muscle actons symbols were taken: flexors: RF – rectus femoris, S – sartorius, IP – iliopsoas, IP-2 – psoas maior, GRA – gracilis extensors: GMx – gluteus maximus, ST – semitendinosous, SM – semimembranosous, BCL – biceps femoris caput longum abductors: ADM – adductor magnus, ADL – adductor longus, ADB – adductor brevis, PC – pectineus adductors muscles and stabilising the pelvis: GMd – gluteus medius, GMu – gluteus minimus, TFL – tensor fasciae–latae

3 Numerical Analysis

Numerical results for shell model are obtained for many cases of boundary conditions. The stiffness of axial elements are changed. There is only one load–maximal load. Analysis of results is performed using three yield criteria: maximum shear–stress (Tresca) criterion, shear–strain energy (von Mises) criterion and Burzynski's criterion (modification of shear–strain energy criterion with respect to different value of compression strength and tensile strength).

In table 2 the maximal values of reduced stresses for selected models are depicted. The stress distributions for model 5 are shown in figures 3 and 4 for shear–strain energy (von Mises) criterion and Burzynski's criterion respectively.

Table 2. Maximum values of reduced stresses for selected models

Model	Maximal values of reduced stresse in MPa for given yield criteria		
NR	Tresca	Mises	Burzyński
1	69	62	56
2	87	84	55
3	143	133	93
4	149	144	145
5	157	150	112
6	171	162	157

For all cases figure a) shows the result on outer surface and figure b) shows on inner surface. It can be observed that stress distribution in numerical model of human pelvic bone depend on boundary conditions and yield criteria. Not only the values of maximal reduced stresses are changed but the regions of their application too.

Numerical analysis for solid model is performed only for one case of boundary conditions and for many load cases. Results presented in this paper are obtained for one load case only. Figure 5 shows reduced stresses (von Mises) on outer surface of pelvic bone and fig. 6 shows displacements.

The value of reduced stresses is printed in kPa and value of displacement in meters. The next figure (7) shows distribution of reduced stresses (von Mises) in

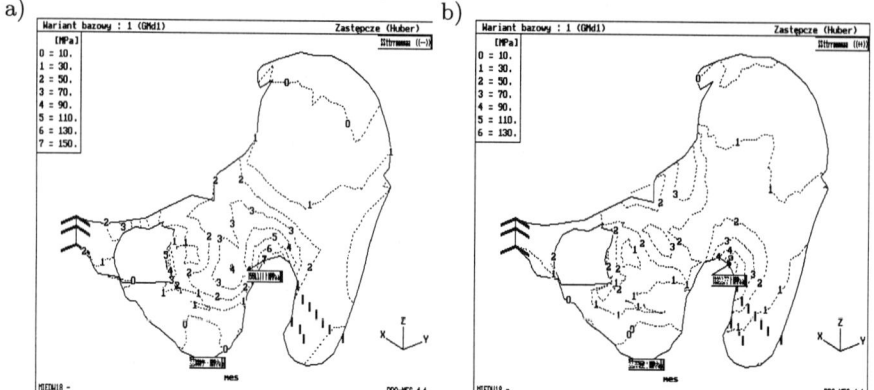

Fig. 3. Von Mises reduced stresses in human pelvic bone:
a) outer surface b) inner surface

cross–section of solid model of pelvic bone. There, we can see the maximal value of stresses on outer surface in cross–section and minimal value in inner area. We can compare the results of solid and shell model. It can be observed that maximal value of reduced stresses and the stress concentration areas are very closer for the same load case and boundary conditions. When the solid model is analyzed not only stresses on the surface of pelvic bone can be taken into account but the inner stresses in cross–section too.

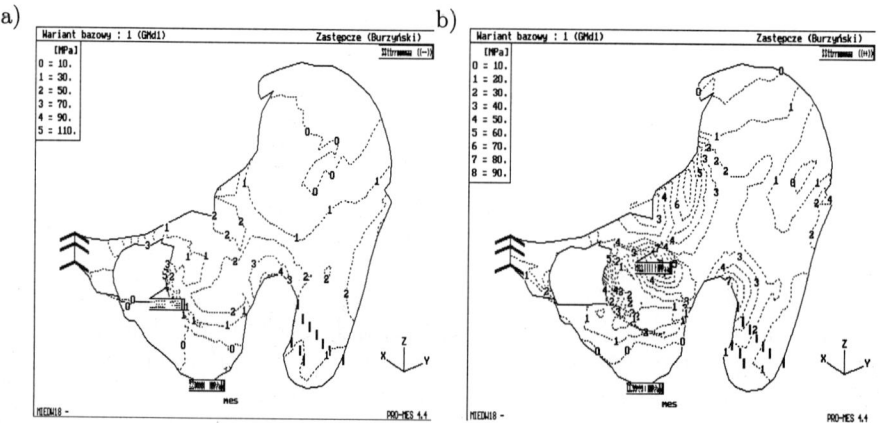

Fig. 4. Burzyński's reduced stresses in human pelvic bone:
a) outer surface b) inner surface

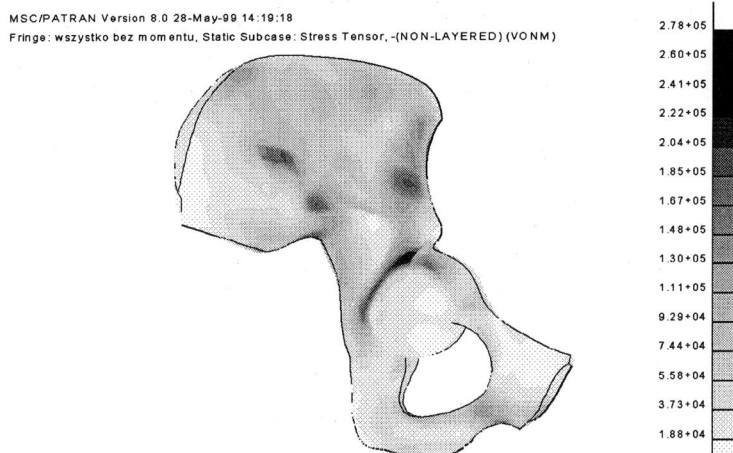

Fig. 5. Distribution of reduced stresses (von Mises) for solid model of pelvic bone

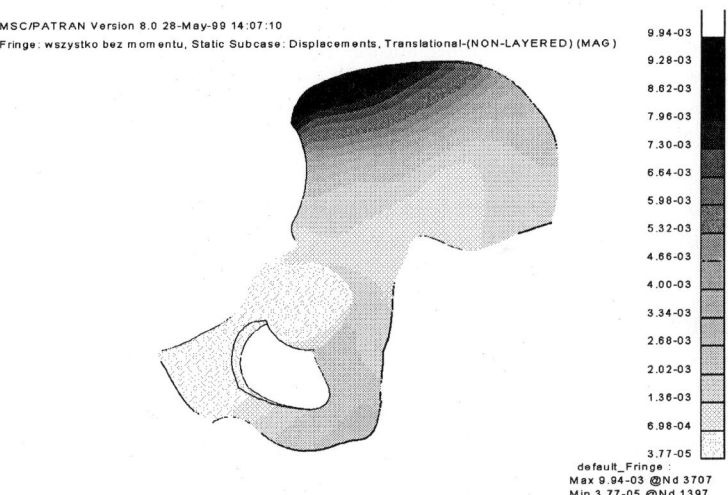

Fig. 6. Displacement diagram for solid model of pelvic bone

4 Conclusions

Presented numerical models of pelvic bone are performed using finite elements and assumed constraints, with a little approximation mapping anatomical shape of the bone and its character of joint in pubic symphysis, on the point of contact with sacral bone and thighbone's head in acetabulum of pelvic joint.

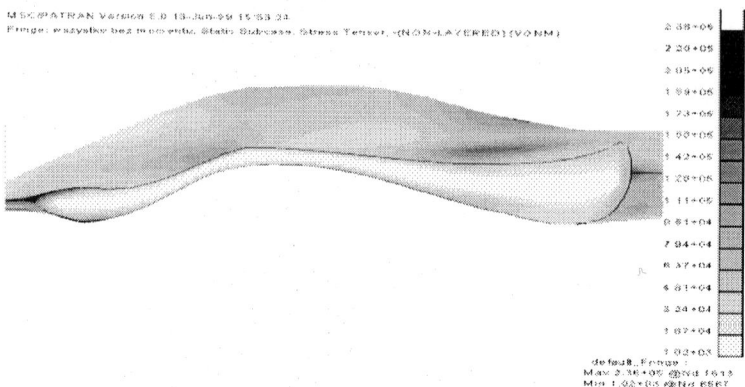

Fig. 7. Distribution of reduced stresses (von Mises) in cross-section of solid model of pelvic bone

Numerical analysis of human pelvic bone shows that stress distribution depend on boundary conditions, e.g. on stiffness of given restraints. There is also problem: how to model the contact with others elements of bone system and what value of material coefficients should be assumed.

The stress distribution and maximal value of reduced stresses depend on yield criteria. In selected model the difference increases over 50MPa, e.g. over 30%. When the Burzynski's criterion is applied the maximal value of reduced stresses decrees. It seems that Burzynski's criterion is closer to real existing conditions.

Shell model of pelvic bone is easy in implementation and the maximal values and distribution of reduced stresses are very closer to solid model.

References

1. Będziński R. (1997) Biomechanika inżynierska. Zagadnienia wybrane. Oficyna Wydawnicza Politechniki Wrocławskiej, Wrocław
2. Dąbrowska-Tkaczyk A. M. (1997) Modelowanie mechanicznych własności tkanek biologicznych, Proc. of Seminar "Biomechanika w Implantologii", Ustroń, Pages 38-45
3. Dąbrowska-Tkaczyk A. M. (1994) Wybrane przykłady modeli mieśnia szkieletowego. Proc. III-th Int. Scient. Conf. "Achievements in the Mechanical and Material Engineering., Gliwice, Pages 85-95
4. Dąbrowska-Tkaczyk A.M, John A. (1997) Numerical modelling of the human pelvis bone. Scientific Papers of Department of Applied Mechanics, vol. 7, Gliwice, pp. 71-78
5. Dalstra M., Huiskes R. Load transfer across the pelvic bone, J. Biomechanics, vol. 28, No 6, 1995, Pages 715-724
6. Dąbrowska-Tkaczyk A.M, John A. (1998) Stan naprężenia i odksztalcenia w przestrzennym modelu numerycznym kości miednicy cz3owieka. Biology of Sport, vol. 15, Supl. 8, Pages 200-205

7. Dąbrowska–Tkaczyk A.M, John A. (1998) Stress and strain distribution in the 3D numerical model of human pelvic bone. Proceedings of the VII–th International Conference Numerical Methods in Continuum Mechanics, Slovak Republic , Pages 422–427
8. Dąbrowska–Tkaczyk A., John A. (1998) Wybrane aspekty numerycznego modelowania kości miednicy człowieka. ,Proceedings Conference on Biomechanics– Modeling, computational methods, experiments and bio– medical applications, Łd, Pages 83–90
9. Dąbrowska–Tkaczyk A. M., Grajek K.,John A. (1998) Stan naprężenia i odksztalcenia w powłokowym modelu numerycznym kości miednicy człowieka. , Proceedings Conference on Biomechanics– Modeling, computational methods, experiments and biomedical applications, Łd, Pages 75–82
10. Jakubowicz A., Rzytka J, Baryluk M. (1990) Wpływ rekonstrukcji stawu biodrowego metodą osteotomii miednicy na warunki biomechaniczne w zespole miednicy i kości udowej. In: Problemy Biocybernetyki i Inżynierii Biomedycznej, Editor: M. Nałęcz, vol. V, Biomechanika, Editors: A. Morecki, W. Ramotowski, WKiŁ,, Pages 224–237
11. John A. (1999) Yield criteria in numerical analysis of human pelvic bone. Acta of Bioengineering and Biomechanics, Vol.1, Supplement 1, pp. 215–218
12. John A. (1999) The boundary conditions in numerical model of human pelvic bone. Acta of Bioengineering and Biomechanics, Vol.1, Supplement 1, pp. 918–222
13. John A. (1999) Yield criteria and stress concentration in numerical analysis of human pelvic bone. Cax Techniques, Proceedings of the 4th International Scientific Colloquium, Fachhochschule Bielefeld, Germany, pp. 289–296
14. Seireg A. (1989) Biomechanical analysis of the musculoskeletal structure for medicine and sport. Hemisphere Publishing Corporation, New York

FEM in Numerical Analysis of Stress and Displacement Distributions in Planetary Wheel of Cycloidal Gear

Manfred Chmurawa [1], Antoni John [2]

[1] Institute of Transport, Silesian Technical University and R & D Centre of Hoisting Machinery „Detrans", Poland
[2] Department for Strengh of Material and Computational Mechanics, Silesian Technical University, Poland

Abstract: Implementation of high speed engines requires application of high ratio mechanical gears. Relatively, the smallest mechanical gear is the cycloidal planetary gear known as Cyclo gear [2, 8- 11]. The complex construction of planet wheels in cycloidal planetary gear (Cyclo) practically makes impossible its optimal design. To calculate distribution of displacements and stresses in planet wheels with co-operating elements FEM has been implemented. There were series of numerical models of planet wheels generated and for example of real model of gear it has been calculated proper values of forces, strains and stresses. In the paper forces and strains calculated with FEM have been used to check the assumptions which have been applied only in analytical so far.

1 Introduction

The Cyclo gear consist of planetary gear fig. 1a and straight-line mechanism fig. 1b in series connection. Because of that kind of connection we get compact gear with stationary central gear (2), which is mating with one or two planet wheels (1, 1'). Planet wheels are driven by the eccentric yoke(3), fig. 1c. In case of immovable stationary wheel (2), a kinematics ratio is given

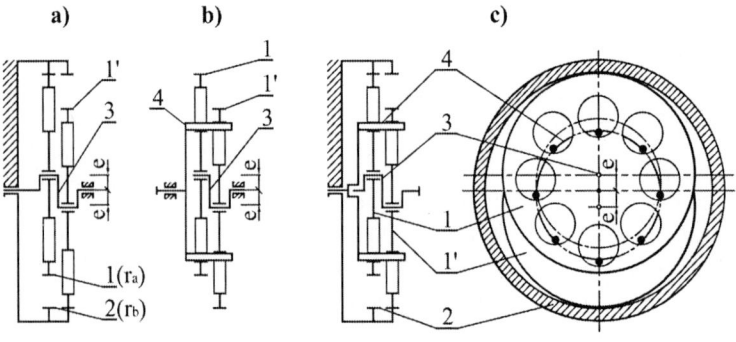

Fig.1 Kinematic scheme of planetary cycloidal gear (Cyclo)

as follows [8, 11]:

$$i = -\frac{z_1}{z_2 - z_1} = -\frac{z_1}{\Delta z} \quad (1.1)$$

where:
$z_1 = z_s$ is a number of teeth of planet wheel 1 or 1',
$z_2 = z_k$ is a number of teeth (rolls) of stationary gear 2.

The main element of the Cyclo gear connecting others elements is the planet wheel 1 and 1'. Outline of planet wheel (meshing) is a shape of an equidistant of shortened epicycloid, (abbreviation ESE) and central gear 2 consists of set of rolls [2, 4, 8-11]. Open-work shape of planet wheel, complex state of load and lack of more precise methods of calculations gives the reasons to apply of FEM for design of Cyclo gear.

In the paper it has been presented a trial of implementation of FEM for calculation of loads in meshing and for distribution of stresses and displacements in high effort points of planetary gear.

2 Distribution of loads and state of equilibrium for planet wheels

Torques acting on three shafts of the Cyclo gear must fulfil condition [10, 11], fig. 1 and 2:

$$M_1 - M_2 + M_h = 0, \quad (2.1)$$

where:
$M_1 = 2M_c$ – torque, arising in planet wheels 1 i 1',
M_2 – torque giving load on interacting central wheel 2,
M_h – input torque (driving) on eccentric shaft (yoke shaft),

Torques M_2, $M_1 = 2M_c$ and M_h occurring in Cyclo gear produce 3 unknown load distributions reacting on planet wheels and other elements:
- load distribution in meshing, distribution of forces P_i between teeth;
- load distribution (Q_j), acting on bolts of straight-line mechanism ;
- load distribution of eccentric R on Q_{ri} loading roller elements (rolls) in bearing hole.

Figure 2 shows how to balance the forces acting on the planet wheel 1 or 1'. Forces between teeth P_i and forces Q_j are function of displacements δ_i and δ_j which arise in points of application of forces. And forces Q_{ri} depending on resolving of force R are the function of geometrical features of roller bearing and mainly depend on radial clearance [4, 5, 12].

To calculate forces between teeth P_i and reaction forces Q_j there is applied analytical method, which has been described by Kudriavcev and Lehmann [8, 9]. Analytical method applies a few simplifying assumptions, fig. 2:
- strains in planet wheel are omitted, wheel is treated as a rigid disk without holes;
- potential strains δ_i in place of acting of meshing forces P_i result from slight angular displacement β of planet wheel as rigid plate and strains δ_j plate

of straight-line mechanism from angle $\Delta\varphi$ coming from bolts of straight line mechanism;
- eccentric reaction force R, burdening the gear is a concentrated force and is not distributed into components Q_{ri} and results from conditions of equilibrium.

3 Modelling of meshing of real planet wheels with strainable rolls and bolts in numerical method

Assumptions done for numerical method, fig. 2:
- planet wheel and co-operating elements are strainable; it has been assumed linear elastic model of material with $E = 2,08\Delta 10^5\ MPa$ and $\nu = 0,3$;
- displacements δ_i and δ_j in place of acting of forces P_i and Q_j result from open-work construction of planet wheel and deflections rolls of stationary wheel and bolts of straight line mechanism,
- active gear loading force is eccentric reaction force R resulting from input torque Mh and conditions of equilibrium [2-4, 8, 9];
- eccentric reaction force R loads gear by setting of pressures Q_{ri} of active rolls of eccentric bearing. Way of calculating forces Q_{ri} in function of radial clearance of central bearing has been presented by Chmurawa [3-5], table 1.

Table 1 Load distribution for n-active rolls in central bearing joint on example of Cyclo gear with ratio $i = 19$, power $N = 6,4\ kW$, $M_1 = 2M_c = 880\ Nm$, $R = 10,3\ kN$ and $d_m = 76,5\ mm$

No	Pressure force [N]	Angle of applying force α_i [°]	Angle of load distribution ψ_ε [°] for (n) - active rolls				
			37,68 (3,4)	48,07 (5)	57,14 (5)	61,44 (5,6)	66,98 (5,6)
1	$Q_{r1}=Q$	$\alpha_1=\alpha_R=42,5°$	5122	4468	3725	3518	3325
2	Q_{r2}	$\alpha_2=18,5°$	2842	3195	2952	2876	2805
3	Q_{r3}	$\alpha_3=-5,5°$	0	6	894	1152	1392
4	Q'_{r3}	$\alpha'_3=90,5°$	0	6	894	1152	1392
5	Q'_{r2}	$\alpha'_2=66,5°$	2842	3195	2952	2876	2805
6	Radial clearance g [mm]		0,19	0,09	0,045	0,033	0,022

Meshing of toothed wheels with elements of Cyclo gear is characterised by coplanar forces and can be modelled in coplanar state of stress [1, 3, 4, 7]. Knowing the profile of outside edge (equidistant) and inside edges (circles) it has been created the geometrical model of planet wheel as the surface which represents real model which has been discretized basing on 8 node surface elements 2D. Model of wheel has been divided into 5590 elements and grid posses 18526 nodes. There was high density of grid applied particularly near edges. It was assured covering of some nodes with characteristic points of

FEM in Numerical Analysis of Stress and Displacement Distributions 775

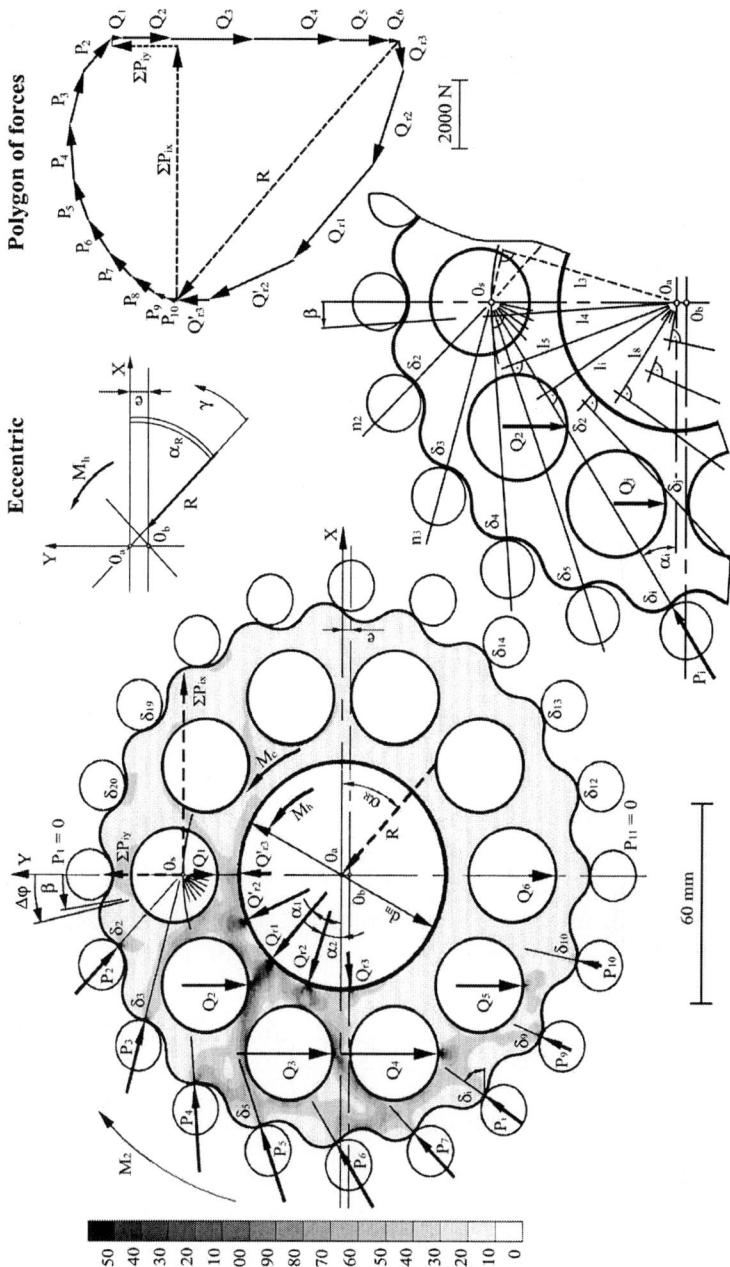

Fig.2 System of forces, distribution of stresses σ_{red} in MPa, displacements, torques and rule of balancing forces acting on planet wheels

meshing of toothed wheels. It concerns contact points of planet wheel teeth with rolls and contact points of bolts and bearing rollers in internal holes.

Meshing of strainable planet wheels 1 and 1' of nominal meshing with rolls of stationary wheel and bolts of straight line mechanism depends on the way of taking over loads. That is why it has been considered 8 different, possible cases (models) of co-operation external 1' and internal 1 planet wheels with remaining elements of gear [3], fig. 2, 4, 5:
- model 1: planet wheel made of 2D surface elements, rolls and bolts modelled as rigid rod elements;
- model 2: planet wheel with rolls and bolts modelled by 2D elements;
- models 3,4,..9,10: planet wheel made of 2D elements, rolls and bolts modelled as strainable rod elements with different substitute stiffness resulting from level of wear coming from long lasting operation. For example model 3 – rolls and bolts take over load without participation of sleeves (at relatively high clearances) and model 9 – rolls and bolts take over load together with sleeves.

4 Results and comparison analyse of results of numerical calculations

FEM calculations were made by MSC Patran/Nastran software for given size of Cyclo gear with parameters as table. Results of calculations (for assumed above numerical models) includes:
- distribution of meshing force P_i and reaction force Q_j,
- distribution of reduces stresses Huber-Mises σ_{red} in planet wheel,
- distribution of place displacement δ_i of meshing.

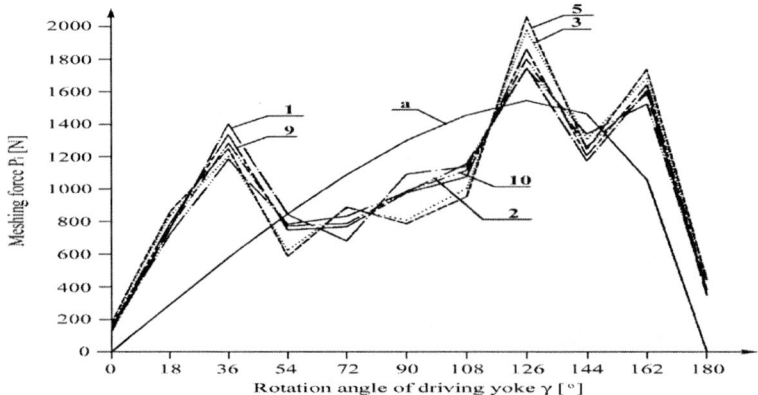

Fig. 3 Changes of meshing force value P_i in Cyclo gear for different numerical models of co-operating planet wheel 1' with elements of gear

Distribution of meshing force P_i illustrates its change during half-cycle of tooth's load and can be shown as the function of rotational angle of driving shaft γ, fig. 3. In the fig. 3 there are set up proper values calculated analytically (symbol a) and numerically (symbols 1 – 10). For half-turn of driving shaft there are 3 (but not 1) cycles of change of meshing force and the highest value P_i = Pmax is $9, 5 - 33, 5$ analytically. Similarly for half-turn of the planet wheel there are 2 (but not 1) cycles of change of force Q_j and the highest value Q_j = Qmax is $5, 5 - 43$ comparing the values it can be noticed high rigidity of rolls and bolts in the gear, diagrams 1 and 9, fig. 3. For example the differences of force values P_i for ideally rigid (model 1) and real rolls and bolts (model3) are small and are only $0, 3 - 1, 8$

Application of FEM enabled also determination and visualisation of distribution of reduced Huber- Mises stresses σ_{red} together with distribution of forces creating them. For example in the fig. 2, 4 there are shown distribution of loads and distribution of stresses σ_{red} which can occur in planet wheel 1' individually for model 9.

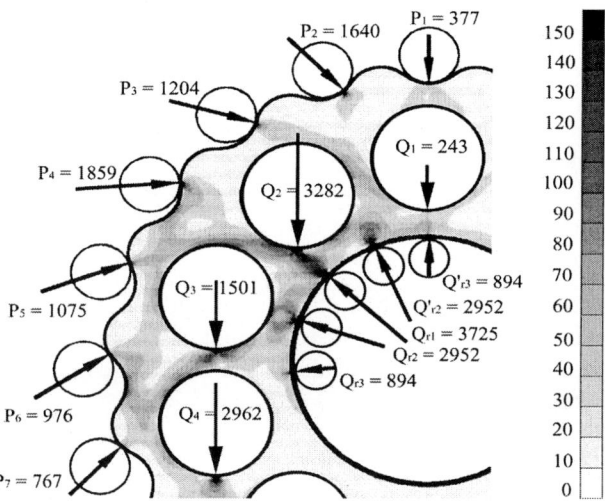

Fig.4 Local distribution of stresses σ_{red} in MPa and forces P_i and Q_j in N in the most effort fragment of planet wheel 1' (for model 9)

To identify regions of occurring maximal stresses σ_{red} there are enlarged fragments of planet wheels prepared. The highest stress level occurs in inside edge of the bearing $\sigma_{red} = 150\ MPa$, in outside edge (equidistant) $\sigma_{red} = 140\ MPa$, between holes $\sigma_{red} = 60 - 90\ MPa$ and also at unloaded side approx. 0 MPa.

Analysis of points of contact between planet wheel teeth and co-operating wheel rolls needs the values of the points knowing displacements δ_i (take a look at fig. 2). Fig. 5 shows for example distribution of displacement δ_i of active and passive points of meshing, for model 3 (rolls and bolts without sleeves) and model 9 (rolls and bolts with sleeves).

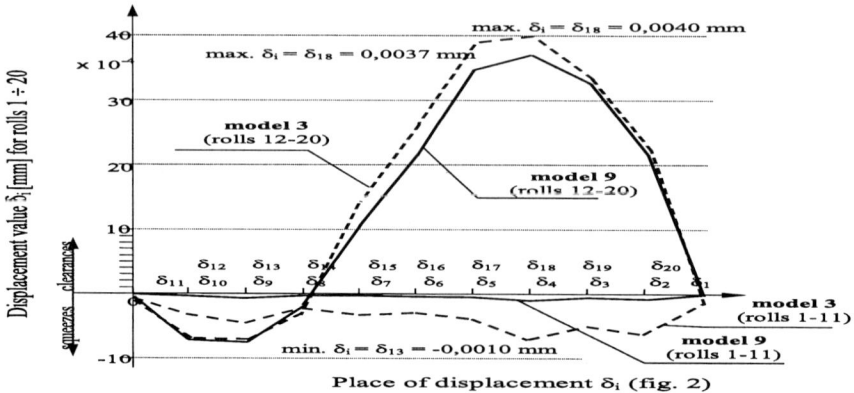

Fig.5 Distribution of displacement δ_i of points of mashing of planet wheel with co-operating wheel

5 Conclusions

- Application of FEM enables calculation and visualisation of unknown stress distribution σ_{red} and displacement distribution δ_i practically of each point on planet wheel. Calculation of stresses analytically is practically impossible.

- Dominant influence on distribution of meshing forces P_i and reaction forces Q_j in cycloidal gear is coming from construction and elastic features of planet wheel. In relatively lower degree elastic features of interacting elements (rolls and bolts) influence differentiating forces P_i and Q_j.

- Distribution of meshing forces P_i, reaction forces Q_j and distribution of displacement δ_i calculated with FEM have different traces comparing it with distributions determined analytically. It results from omitting in analytical method real shape and material features of planet wheels.

- Curvilinear outside edge and regular inside holes of planet wheels in the gear create higher frequency of changes and relatively higher values of meshing forces and reaction forces of the bolts of the straight line mechanism. Presented results concern cycloidal gear with meshing with transmission ratio $\delta_i? = 19$ and power $N = 6,4$ kW.

- Presented numerical analysis of loads, stresses and displacements can be applied in optimisation of distribution of loads using modification of meshing in Cyclo gear.

References

1. Chandrupatla T.R., Belegundu A.D. (1991) Introduction to FEM in engineering. Prentice Hall, London.

2. Chmurawa M., Olejek G. (1994) Zazębienie cykloidalne przekładni planetarnej. Zeszyty Naukowe Pol. Śl., seria Transport Z. 22/94, Gliwice.

3. Chmurawa M., John A., Kokot G. (1999) The influence of numerical model on distribution of loads and stress in cycloidal planetary gear. Proc. 4th International Scientific Colloquium Cax Techniques, Bielefeld, Germany.

4. Chmurawa M. (1999) Distribution of loads in cycloidal planetary gear. Proc. International Conference „Mechanics'99", Kaunas University, Lithuania.

5. Freda–Krzemiński H. (1985) Łożyska toczne. PWN. Warszawa.

6. Hamerak K. (1979) Das Cyclogetriebe–eine geniale Idee und ihre technische Verwirklichung. Technik Heute. Verlag Christiani, nr 6, Bonn.

7. Kleiber M. (Ed.) (1998) Handbook of Computational Solid Mechanics. Springer Verlag, Berlin–Heidelberg.

8. Kudriavcev V.N. (1966) Planetarnyje peredaci. Masinostroenije, Moskva –Leningrad.

9. Lehmann M. (1976) Berechnung und Messung der Krafte in einen Zykloiden–Kurvenscheiben Getriebe. Dissertation. Technische Universität, München.

10. Müller H.W. (1971) Die Umlaufgetriebe. Springer Verlag, Berlin.

11. Müller L. (1983) Przekładnie obiegowe. PWN, Warszawa.

12. Palmgren A. (1964) Grundlagen der Waltzlagertechnik, Francklische Verlagshandlung, Stuttgart.

This Labour has been made in the frames of KBN, project No 7T07C03815

Author Index

Abdallah, H. 1
Amodio, P. 10
Ansari, A. R. 18

Bael, A. Van 423
Barel, M. Van 27
Barraud, A. 35, 521
Barrio, R. 42, 51
Baryamureeba, V. 59
Begoña Melendo, M. 586
Bellavia, S. 68
Bencheva, G. 76
Bergamaschi, L. 84
Bertaccini, D. 93
Blanes, S. 102
Bojović, D. 110
Braianov, I. A. 117
Brainman, I. 125
Bujanda, B. 133

Cai, W. 527
Cano, B. 144
Capizzano, S. S. 152
Cardoso, J. R. 160
Carpentieri, B. 170
Casas, F. 102
Cervantes, L. 179
Chaitin-Chatelin, F. 187
Chan, R. H. 615
Charlier, R. 222
Chen, J.-Y. 475
Chmurawa, M. 772
Christov, N. 35, 521
Clavero, C. 316, 350
Coakley, J. 198
Collar, A. F. 179, 207
Collin, F. 222
Condevaux-Lanloy, Ch. 214

D'yakonov, E. G. 273
Datcheva, M. 222
Dent, D. 230
Di Lena, G. 513
Diderich, C. 238

Dimitriu, G. 246
Dimov, I. 359, 636
Dobrev, V. 253
Dooren, P. Van 560
Duff, I. S. 170
Dunne, R. K. 265

Epelly, O. 214

Faragó, I. 285
Farrell, P. A. 292, 723
Foschi, P. 490
Fragnière, E. 214
Fuertes, A.-M. 198

Gahan, B. 304
Gałka, A. 741
Gaspar, F. 316
Georgiev, K. 325
Georgieva, G. 333
Grebénnikov, A. I. 207
Giraud, L. 170
Goolin, A. V. 341
Gracia, J. L. 350
Gurov, T. 359
Gustavson, F. 333
Gutiérrez, J. M. 368

Hamza, M. 1
Hassanov, V. 377
Hegarty, A. F. 18, 292, 723
Heinig, G. 27, 385
Hemker, P. W. 393, 402
Hernández, M. A. 368
Hoppe, R. H. W. 414
Hosoda, Y. 608
Houtte, P. Van 423
Huffel, S. Van 560

Iankov, R. 423
Iavernaro, F. 513
Ivanov, I. 377

Jin, X.-Q. 505
John, A. 764, 772

Author Index

Jorge, J. C. 133
Jovanović, B. S. 431, 439

Kandilarov, J. D. 431, 451
Karátson, J. 459
Karaivanova, A. 552
Kaschiev, M. 692
Keer, R. Van 467
Kincaid, D. R. 475
Kitagawa, T. 608
Kolev, T. 482
Koleva, M. 692
Kontoghiorghes, E. J. 490
Kravanja, P. 27
Kucaba-Pietal, A. 230, 498
Kwan, W. C. 615

Lecheva, A. 749
Lei, S.-L. 505
Leite, F. S. 160
Lesecq, S. 35, 521
Li, Z. 527
Lirkov, I. 535
Lisbona, F. 316, 350

MacMullen, H. 544
Macconi, M. 68
Margenov, S. 482, 535
Marty, W. 238
Mascagni, M. 552
Mastronardi, N. 560
Matus, P. P. 568
Mazhukin, V. I. 568
Meini, B. 578
Miller, J. J. H. .. 292, 304, 594, 723
Mishima, T. 628
Mitrouli, M. 602
Mizoguchi, H. 628
Morini, B. 68
Mozolevsky, I. E. 568
Musgrave, A. P. 594

Nakata, S. 608
Ng, M. K. 93, 615
Nishimura, S. 628

O'Riordan, E. ... 265, 292, 544, 723
Oliveros, J. J. O. 207

Onofri, F. 692
Ostromsky, T. 636

Pérez, M.-T. 198
Pan, V. Y. 644
Paprzycki, M. 230, 535
Pauletto, G. 650
Pavlov, V. 658
Petcu, D. 666
Petrova, S. I. 414
Plantié, L. 187, 675
Popivanov, P. 684
Possio, C. T. 152

Radev, St. 692
Ros, J. 102

Schulz, V. H. 414
Sczygiol, N. 702
Shigehara, T. 628
Shishkin, G. I. ... 18, 265, 292, 304,
393, 544, 594, 710, 723, 756
Shishkina, L. P. 393
Slavova, A. 684
Slodička, M. 467
Sousa, E. 732
Sprengel, F. 402

Tadrist, L. 692
Takahashi, D. 628
Telega, J. J. 741
Tokarzewski, S. 741
Toledo, S. 125
Traviesas, E. 187
Tselishcheva, I. V. 756
Tzvetanov, I. 636

Vassilevski, P. 253
Vulkov, L. G. 431, 439, 451

Waśniewski, J. 325
Whitlock, P. 359

Yalamov, P. 51, 333
Young, D. M. 475

Zadorin, A. I. 451
Zheleva, I. 749
Zilli, G. 84
Zlatev, Z. 636

Lecture Notes in Computer Science

For information about Vols. 1–1928
please contact your bookseller or Springer-Verlag

Vol. 1929: R. Laurini (Ed.), Advances in Visual Information Systems. Proceedings, 2000. XII, 542 pages. 2000.

Vol. 1931: E. Horlait (Ed.), Mobile Agents for Telecommunication Applications. Proceedings, 2000. IX, 271 pages. 2000.

Vol. 1658: J. Baumann, Mobile Agents: Control Algorithms. XIX, 161 pages. 2000.

Vol. 1756: G. Ruhe, F. Bomarius (Eds.), Learning Software Organization. Proceedings, 1999. VIII, 226 pages. 2000.

Vol. 1766: M. Jazayeri, R.G.K. Loos, D.R. Musser (Eds.), Generic Programming. Proceedings, 1998. X, 269 pages. 2000.

Vol. 1791: D. Fensel, Problem-Solving Methods. XII, 153 pages. 2000. (Subseries LNAI).

Vol. 1799: K. Czarnecki, U.W. Eisenecker, Generative and Component-Based Software Engineering. Proceedings, 1999. VIII, 225 pages. 2000.

Vol. 1812: J. Wyatt, J. Demiris (Eds.), Advances in Robot Learning. Proceedings, 1999. VII, 165 pages. 2000. (Subseries LNAI).

Vol. 1932: Z.W. Raś, S. Ohsuga (Eds.), Foundations of Intelligent Systems. Proceedings, 2000. XII, 646 pages. (Subseries LNAI).

Vol. 1933: R.W. Brause, E. Hanisch (Eds.), Medical Data Analysis. Proceedings, 2000. XI, 316 pages. 2000.

Vol. 1934: J.S. White (Ed.), Envisioning Machine Translation in the Information Future. Proceedings, 2000. XV, 254 pages. 2000. (Subseries LNAI).

Vol. 1935: S.L. Delp, A.M. DiGioia, B. Jaramaz (Eds.), Medical Image Computing and Computer-Assisted Intervention – MICCAI 2000. Proceedings, 2000. XXV, 1250 pages. 2000.

Vol. 1936: P. Robertson, H. Shrobe, R. Laddaga (Eds.), Self-Adaptive Software. Proceedings, 2000. VIII, 249 pages. 2001.

Vol. 1937: R. Dieng, O. Corby (Eds.), Knowledge Engineering and Knowledge Management. Proceedings, 2000. XIII, 457 pages. 2000. (Subseries LNAI).

Vol. 1938: S. Rao, K.I. Sletta (Eds.), Next Generation Networks. Proceedings, 2000. XI, 392 pages. 2000.

Vol. 1939: A. Evans, S. Kent, B. Selic (Eds.), «UML» – The Unified Modeling Language. Proceedings, 2000. XIV, 572 pages. 2000.

Vol. 1940: M. Valero, K. Joe, M. Kitsuregawa, H. Tanaka (Eds.), High Performance Computing. Proceedings, 2000. XV, 595 pages. 2000.

Vol. 1941: A.K. Chhabra, D. Dori (Eds.), Graphics Recognition. Proceedings, 1999. XI, 346 pages. 2000.

Vol. 1942: H. Yasuda (Ed.), Active Networks. Proceedings, 2000. XI, 424 pages. 2000.

Vol. 1943: F. Koornneef, M. van der Meulen (Eds.), Computer Safety, Reliability and Security. Proceedings, 2000. X, 432 pages. 2000.

Vol. 1944: K.R. Dittrich, G. Guerrini, I. Merlo, M. Oliva, M.E. Rodriguez (Eds.), Objects and Databases. Proceedings, 2000. X, 199 pages. 2001.

Vol. 1945: W. Grieskamp, T. Santen, B. Stoddart (Eds.), Integrated Formal Methods. Proceedings, 2000. X, 441 pages. 2000.

Vol. 1946: P. Palanque, F. Paternò (Eds.), Interactive Systems. Proceedings, 2000. X, 251 pages. 2001.

Vol. 1947: T. Sørevik, F. Manne, R. Moe, A.H. Gebremedhin (Eds.), Applied Parallel Computing. Proceedings, 2000. XII, 400 pages. 2001.

Vol. 1948: T. Tan, Y. Shi, W. Gao (Eds.), Advances in Multimodal Interfaces – ICMI 2000. Proceedings, 2000. XVI, 678 pages. 2000.

Vol. 1949: R. Connor, A. Mendelzon (Eds.), Research Issues in Structured and Semistructured Database Programming. Proceedings, 1999. XII, 325 pages. 2000.

Vol. 1950: D. van Melkebeek, Randomness and Completeness in Computational Complexity. XV, 196 pages. 2000.

Vol. 1951: F. van der Linden (Ed.), Software Architectures for Product Families. Proceedings, 2000. VIII, 255 pages. 2000.

Vol. 1952: M.C. Monard, J. Simão Sichman (Eds.), Advances in Artificial Intelligence. Proceedings, 2000. XV, 498 pages. 2000. (Subseries LNAI).

Vol. 1953: G. Borgefors, I. Nyström, G. Sanniti di Baja (Eds.), Discrete Geometry for Computer Imagery. Proceedings, 2000. XI, 544 pages. 2000.

Vol. 1954: W.A. Hunt, Jr., S.D. Johnson (Eds.), Formal Methods in Computer-Aided Design. Proceedings, 2000. XI, 539 pages. 2000.

Vol. 1955: M. Parigot, A. Voronkov (Eds.), Logic for Programming and Automated Reasoning. Proceedings, 2000. XIII, 487 pages. 2000. (Subseries LNAI).

Vol. 1956: T. Coquand, P. Dybjer, B. Nordström, J. Smith (Eds.), Types for Proofs and Programs. Proceedings, 1999. VII, 195 pages. 2000.

Vol. 1957: P. Ciancarini, M. Wooldridge (Eds.), Agent-Oriented Software Engineering. Proceedings, 2000. X, 323 pages. 2001.

Vol. 1960: A. Ambler, S.B. Calo, G. Kar (Eds.), Services Management in Intelligent Networks. Proceedings, 2000. X, 259 pages. 2000.

Vol. 1961: J. He, M. Sato (Eds.), Advances in Computing Science – ASIAN 2000. Proceedings, 2000. X, 299 pages. 2000.

Vol. 1963: V. Hlaváč, K.G. Jeffery, J. Wiedermann (Eds.), SOFSEM 2000: Theory and Practice of Informatics. Proceedings, 2000. XI, 460 pages. 2000.

Vol. 1964: J. Malenfant, S. Moisan, A. Moreira (Eds.), Object-Oriented Technology. Proceedings, 2000. XI, 309 pages. 2000.

Vol. 1965: Ç. K. Koç, C. Paar (Eds.), Cryptographic Hardware and Embedded Systems – CHES 2000. Proceedings, 2000. XI, 355 pages. 2000.

Vol. 1966: S. Bhalla (Ed.), Databases in Networked Information Systems. Proceedings, 2000. VIII, 247 pages. 2000.

Vol. 1967: S. Arikawa, S. Morishita (Eds.), Discovery Science. Proceedings, 2000. XII, 332 pages. 2000. (Subseries LNAI).

Vol. 1968: H. Arimura, S. Jain, A. Sharma (Eds.), Algorithmic Learning Theory. Proceedings, 2000. XI, 335 pages. 2000. (Subseries LNAI).

Vol. 1969: D.T. Lee, S.-H. Teng (Eds.), Algorithms and Computation. Proceedings, 2000. XIV, 578 pages. 2000.

Vol. 1970: M. Valero, V.K. Prasanna, S. Vajapeyam (Eds.), High Performance Computing – HiPC 2000. Proceedings, 2000. XVIII, 568 pages. 2000.

Vol. 1971: R. Buyya, M. Baker (Eds.), Grid Computing – GRID 2000. Proceedings, 2000. XIV, 229 pages. 2000.

Vol. 1972: A. Omicini, R. Tolksdorf, F. Zambonelli (Eds.), Engineering Societies in the Agents World. Proceedings, 2000. IX, 143 pages. 2000. (Subseries LNAI).

Vol. 1973: J. Van den Bussche, V. Vianu (Eds.), Database Theory – ICDT 2001. Proceedings, 2001. X, 451 pages. 2001.

Vol. 1974: S. Kapoor, S. Prasad (Eds.), FST TCS 2000: Foundations of Software Technology and Theoretical Computer Science. Proceedings, 2000. XIII, 532 pages. 2000.

Vol. 1975: J. Pieprzyk, E. Okamoto, J. Seberry (Eds.), Information Security. Proceedings, 2000. X, 323 pages. 2000.

Vol. 1976: T. Okamoto (Ed.), Advances in Cryptology – ASIACRYPT 2000. Proceedings, 2000. XII, 630 pages. 2000.

Vol. 1977: B. Roy, E. Okamoto (Eds.), Progress in Cryptology – INDOCRYPT 2000. Proceedings, 2000. X, 295 pages. 2000.

Vol. 1978: B. Schneier (Ed.), Fast Software Encryption. Proceedings, 2000. VIII, 315 pages. 2001.

Vol. 1979: S. Moss, P. Davidsson (Eds.), Multi-Agent-Based Simulation. Proceedings, 2000. VIII, 267 pages. 2001. (Subseries LNAI).

Vol. 1983: K.S. Leung, L.-W. Chan, H. Meng (Eds.), Intelligent Data Engineering and Automated Learning – IDEAL 2000. Proceedings, 2000. XVI, 573 pages. 2000.

Vol. 1984: J. Marks (Ed.), Graph Drawing. Proceedings, 2001. XII, 419 pages. 2001.

Vol. 1985: J. Davidson, S.L. Min (Eds.), Languages, Compilers, and Tools for Embedded Systems. Proceedings, 2000. VIII, 221 pages. 2001.

Vol. 1987: K.-L. Tan, M.J. Franklin, J. C.-S. Lui (Eds.), Mobile Data Management. Proceedings, 2001. XIII, 289 pages. 2001.

Vol. 1988: L. Vulkov, J. Waśniewski, P. Yalamov (Eds.), Numerical Analysis and Its Applications. Proceedings, 2000. XIII, 782 pages. 2001.

Vol. 1989: M. Ajmone Marsan, A. Bianco (Eds.), Quality of Service in Multiservice IP Networks. Proceedings, 2001. XII, 440 pages. 2001.

Vol. 1990: I.V. Ramakrishnan (Ed.), Practical Aspects of Declarative Languages. Proceedings, 2001. VIII, 353 pages. 2001.

Vol. 1991: F. Dignum, C. Sierra (Eds.), Agent Mediated Electronic Commerce. VIII, 241 pages. 2001. (Subseries LNAI).

Vol. 1992: K. Kim (Ed.), Public Key Cryptography. Proceedings, 2001. XI, 423 pages. 2001.

Vol. 1993: E. Zitzler, K. Deb, L. Thiele, C.A.Coello Coello, D. Corne (Eds.), Evolutionary Multi-Criterion Optimization. Proceedings, 2001. XIII, 712 pages. 2001.

Vol. 1995: M. Sloman, J. Lobo, E.C. Lupu (Eds.), Policies for Distributed Systems and Networks. Proceedings, 2001. X, 263 pages. 2001.

Vol. 1998: R. Klette, S. Peleg, G. Sommer (Eds.), Robot Vision. Proceedings, 2001. IX, 285 pages. 2001.

Vol. 1999: W. Emmerich, S. Tai (Eds.), Engineering Distributed Objects. Proceedings, 2000. VIII, 271 pages. 2001.

Vol. 2000: R. Wilhelm (Ed.), Informatics: 10 Years Back, 10 Years Ahead. IX, 369 pages. 2001.

Vol. 2003: F. Dignum, U. Cortés (Eds.), Agent Mediated Electronic Commerce III. XII, 193 pages. 2001. (Subseries LNAI).

Vol. 2004: A. Gelbukh (Ed.), Computational Linguistics and Intelligent Text Processing. Proceedings, 2001. XII, 528 pages. 2001.

Vol. 2006: R. Dunke, A. Abran (Eds.), New Approaches in Software Measurement. Proceedings, 2000. VIII, 245 pages. 2001.

Vol. 2007: J.F. Roddick, K. Hornsby (Eds.), Temporal, Spatial, and Spatio-Temporal Data Mining. Proceedings, 2000. VII, 165 pages. 2001. (Subseries LNAI).

Vol. 2009: H. Federrath (Ed.), Designing Privacy Enhancing Technologies. Proceedings, 2000. X, 231 pages. 2001.

Vol. 2010: A. Ferreira, H. Reichel (Eds.), STACS 2001. Proceedings, 2001. XV, 576 pages. 2001.

Vol. 2013: S. Singh, N. Murshed, W. Kropatsch (Eds.), Advances in Pattern Recognition – ICAPR 2001. Proceedings, 2001. XIV, 476 pages. 2001.

Vol. 2015: D. Won (Ed.), Information Security and Cryptology – ICISC 2000. Proceedings, 2000. X, 261 pages. 2001.

Vol. 2021: J. N. Oliveira, P. Zave (Eds.), FME 2001: Formal Methods for Increasing Software Productivity. Proceedings, 2001. XIII, 629 pages. 2001.

Vol. 2024: H. Kuchen, K. Ueda (Eds.), Functional and Logic Programming. Proceedings, 2001. X, 391 pages. 2001.

Vol. 2027: R. Wilhelm (Ed.), Compiler Construction. Proceedings, 2001. XIII, 433 pages. 2001.